MODULATION AND CODING TECHNIQUES IN WIRELESS COMMUNICATIONS

MODULATION AND CODING TECHNIQUES IN WIRELESS COMMUNICATIONS

Edited by

Evgenii Krouk

Dean of the Information Systems and Data Protection Faculty, St Petersburg State University of Aerospace Instrumentation, Russia

Sergei Semenov

Specialist, Nokia Corporation, Finland

A John Wiley and Sons, Ltd., Publication

This edition first published 2011

Registered Office
John Wiley & Sons Ltd, The Atrium, Southern Gate, Chichester, West Sussex, PO19 8SQ, United Kingdom

For details of our global editorial offices, for customer services and for information about how to apply for permission to reuse the copyright material in this book please see our website at www.wiley.com.

Library of Congress Cataloging-in-Publication Data

Modulation and coding techniques in wireless communications / edited by Evgenii Krouk, Sergei Semenov.
p. cm.
Includes bibliographical references and index.
ISBN 978-0-470-74505-2 (cloth)
1. Coding theory. 2. Modulation (Electronics). 3. Wireless communication systems. I. Krouk, E. II. Semenov, S.
TK5102.92.M63 2011
621.384–dc22

2010033601

A catalogue record for this book is available from the British Library.

Print ISBN: 9780470745052 [HB]
ePDF ISBN:9780470976760
oBook ISBN: 9780470976777
ePub ISBN: 9780470976715

Typeset in 9/11pt Times by Aptara Inc., New Delhi, India.
Printed and bound in Singapore by Markono Print Media Pte Ltd.

Contents

About the Editors

Evgenii Krouk
Professor E. Krouk has worked in the field of communication theory and techniques for more than 30 years. His areas of interest include coding theory, the mathematical theory of communications and cryptography. He is now the Dean of the Information Systems and Data Protection Faculty of the St Petersburg State University of Aerospace Instrumentation. He is author of three books, more than 100 scientific articles and 30 international and Russian patents.

Sergei Semenov
Sergei Semenov received his PhD degree from the St Petersburg State University for Airspace Instrumentation (SUAI), Russia in 1993. Dr Semenov joined Nokia Corporation in 1999 and is currently a Specialist in Modem Algorithm Design/Wireless Modem. His research interests include coding and communication theory and their application to communication systems.

Contributors

Asbjørn Grøvlen
Nokia, Denmark

Kari Hooli
Nokia Siemens Networks, Finland

Matti Jokimies
Nokia Corporation, Finland

Grigorii Kabatiansky
Institute for Information Transmission Problems, Russian Academy of Sciences, Russia

Tuomas Laine
Nokia Corporation, Finland

Zexian Li
Nokia Corporation, Finland

Andrei Malkov
Nokia Corporation, Finland

Dmitry Osipov
Institute for Information Transmission Problems, Russian Academy of Sciences, Russia

Andrei Ovchinnikov
St Petersburg State University of Aerospace Instrumentation, Russia

Jarkko Paavola
Department of Information Technology, University of Turku, Finland

Kari Pajukoski
Nokia Siemens Networks, Finland

Jussi Henrikki Poikonen
Department of Information Technology, University of Turku, Finland

Esa Tapani Tiirola
Nokia Siemens Networks, Finland

Andrey Trofimov
St Petersburg State University of Aerospace Instrumentation, Russia

Prabodh Varshney
Nokia, USA

Acknowledgements

We would like to thank all the authors who took part in this project, who sacrificed some part of their spare time to make the realization of this book possible.

We also would like to thank the Wiley team who have worked with us.

Introduction

Major achievements in the field of creating digital devices made possible the implementation of algorithms and systems that were considered unfeasible until recent times. Modern communication systems and especially the systems of radiocommunication support this statement. Transmitters and receivers comprising, until recently, bulky and unique devices now can be easily fitted to the body of a small mobile phone and many manufacturers have started to mass produce these devices. This raises the problem of compatibility of devices from different manufacturers.

The solution to this issue is the system of international standards. The modern standards on communications comprise a large number of specifications, and some of them are quite cumbersome. The reason for this is the fact that these specifications are the result of complex and time consuming processes of reconciling comprehensive technical solutions with a large number of contributors.

There is no doubt that the impressive achievements in the development of communication systems are not only the result of development of digital devices but can be explained by significant progress in the field of creation and implementation of the new communication technologies.

These new technologies are based on theoretical results obtained with the help of serious and sometimes non-traditional mathematic apparatus. Understanding the fundamental works on modulation, equalization and coding theory, sophisticated results on multiple access and multiple antenna systems comprising the basis of modern communication standards requires significant efforts and high mathematical culture.

On the other hand, the great number of technical details that must be mentioned in standards specifications sometimes make it difficult to find the correlation between the standard specifications and the theoretical results even for the prepared reader.

Due to this fact, the idea of writing the book uniting both the theoretical results and material of standards on wireless communication was considered as quite fruitful. The goal of this book is to reveal some regular trends in the latest results on communication theory and show how these trends are implemented in contemporary wireless communication standards. It is obvious that to carry out this idea first of all it is necessary to collect in one team, not only the specialists on communication theory, but also people dealing with practical implementation of standards specifications. We are happy that we did manage to solve this tricky problem. The present book is the result of the work carried out by this team of authors.

In line with the above mentioned goal the book consists of two parts. Part 1 is devoted to the review of the basis of communication theory (Chapters 1–9), and Part 2 to the review of modern wireless communication standards.

In Chapter 1 the main definitions in the field of communication theory and typical models of communication channels can be found. In Chapter 2 the main principles of modulation theory are presented and the main modulation methods used in practice are discussed. Chapter 3 is devoted to the coding theory. In this chapter the main constructions of block codes and methods of decoding the block codes are considered. The convolutional and turbo codes are discussed in Chapter 4. In Chapter 5 the materials on equalization theory and channel estimation are collected. In Chapter 6 the main schemes of systems with feedback are considered. The principles and algorithms of coding modulation are presented in Chapter 7. Chapter 8 is devoted to the description of multiple antenna systems. In Chapter 9 the multiple access methods are outlined. Thus, quite thorough review of basis algorithms and technologies of communication theory can be found in Part 1 of the book. These results are to some extent redundant for the description of contemporary standards. However, the presence of these results in the book reflects the authors' confidence that they can be used in industry in the near future.

The usage of layer 1 procedures in the wide range of wireless communication standards is considered in Part 2. In this part authors try to consider the standards which have the most significant impact (in the authors' opinion) to evolution of modern wireless communication. In Chapter 10 the review of communication technologies used in standards IEEE 802.11 and 802.16 can be found. In Chapter 11 the review of 3GPP standards on WCDMA and LTE is presented. Chapter 12 is devoted to layer 1 procedures used in 3GPP2 CDMA2000 standards. Thus, the layer 1 procedures used in the main standards of wireless communication can be inferred from the second part of the book.

We hope that this book will be useful for communication system designers and specialists in communication theory as well. Also it may be used by students of communication systems.

1

Channel Models and Reliable Communication

Evgenii Krouk[1], Andrei Ovchinnikov[1], and Jussi Poikonen[2]
[1]St Petersburg State University of Aerospace Instrumentation, Russia
[2]Department of Information Technology, University of Turku, Finland

1.1 Principles of Reliable Communication

Ideally, design, development and deployment of communication systems aims at maximally efficient utilization of available resources for transferring information reliably between a sender and a recipient. In real systems, typically some amount of unreliability is tolerated in this transfer to achieve a predefined level of consumption of limited resources. In modern communication systems, primary resources are time, space, and power and frequency bandwidth of the electromagnetic radiation used to convey information. Given such resources, systems must be designed to overcome distortions to transmitted information caused mainly by elements within the system itself, possible external communications, and the environment through which the information propagates. To achieve efficient utilization of available resources, knowledge of the mechanisms that cause interference in a given transmission scenario must be available in designing and analyzing a communication system.

In performance evaluation of wireless communication systems, significance of the communication channel is emphasized, since the degradation of a signal propagating from a transmitter to a receiver is strongly dependent on their locations relative to the external environment. Wireless mobile communication, where either the transmitter or the receiver is in motion, presents additional challenges to channel modelling, as it is necessary to account for variation in the signal distortion as a function of time for each transmitter–receiver pair. In developing and analyzing such systems, comprehensively modelling the transmitter–receiver link is a complicated task.

In the following, distortions caused by typical communication channels to transmitted signals are described. A common property of all communication channels is that the received signal contains *noise,* which fundamentally limits the rate of communication. Noise is typically modelled as a Gaussian stochastic process. The *additive white Gaussian noise (AWGN)* channel and its effects on typical digital modulation methods are presented in Section 1.2. Noise is added to transmitted signals at the receiver. Before arriving at the receiver terminal, signals are typically distorted according to various physical

Modulation and Coding Techniques in Wireless Communications Edited by Evgenii Krouk and Sergei Semenov

characteristics of the propagation medium. These distortions attenuate the received signal, and thus increase the detrimental effect of additive noise on the reliability of communication. In Section 1.3 to 1.5 typical cases of distortion in wireless communication channels and models for the effects of such distortion on transmitted signals are presented.

1.2 AWGN

Distortions occurring in typical communication systems can be divided into multiplicative and additive components. In the following, some remarks and relevant results concerning additive distortion – also referred to simply as noise – are presented.

Additive noise is introduced to a wireless communication system both from outside sources – such as atmospheric effects, cosmic radiation and electrical devices – and from internal components of the receiver hardware, which produce thermal and shot noise [9]. Typically, additive distortion in a received signal consists of a sum of a large number of independent random components, and is modelled as additive white Gaussian noise, where the term *white* means that the noise is assumed to have a constant power spectral density. The Gaussian, or normal, distribution of noise is motivated by the central limit theorem (one of the fundamental theorems of probability theory), according to which the distribution of a sum of a large number of random variables approaches a normal distribution, given that these variables fulfill *Lyapunov's condition* (for details, see for example [10]).

In some cases, the received signal is also distorted by a channel-induced superposition of different components of the useful transmission, or by signals from other transmission systems. Such distortions are called *interference*, and differ from additive noise in that typically some source-specific statistical characteristics of interference are known. Thus interference is not in all cases best approximated as an additive white Gaussian process. Interference effects are strongly dependent on the communication systems and transmission scenarios under consideration. Later in this chapter, interference-causing effects of wireless communication channels are considered. In the following, we focus on considering the effects of additive white Gaussian noise on *complex baseband* modulation symbols. Principles of digital modulation methods and the effects of noise on the reception of various types of transmitted signals will be considered in more detail in Chapter 2; the following simple examples are meant to illustrate the concept of additive noise and its effect on digital communication.

1.2.1 Baseband Representation of AWGN

In the following examples, we consider digital data which is mapped to binary phase shift keying (BPSK), quaternary phase shift keying (QPSK/4-QAM), and 16-point quadrature amplitude modulation (16-QAM) symbols. We consider complex baseband signals, that is, for our purposes the transmitted modulation symbols corresponding to a given digital modulation scheme are represented simply as complex numbers. The constellation diagrams for these examples are illustrated in Figure 1.1. The effect of an AWGN channel is to shift these numbers in the complex plane. The receiver has to decide, based on an observed shifted complex number, the most likely transmitted symbol. This decision is performed by finding which, out of the set of known transmitted symbols, is the one with the smallest Euclidian distance to the received noisy symbol. This is a rather abstract representation of digital signals and noise, but sufficient for performing error performance analyses of different modulation schemes. For a more detailed discussion on basic modulation methods and the corresponding signal forms, see Chapter 2.

As outlined above, in complex baseband signal-space representations, the effect of additive white Gaussian noise in the receiver can be described as a complex number added to each transmitted modulation symbol value. The real and imaginary parts of these complex numbers are independent and identically distributed Gaussian random variables with zero mean and variance equal to $\sigma_N^2 = \bar{P}_N/2$, where $\bar{P}_N$ denotes the total average power of the complex noise process (that is, the power of the noise

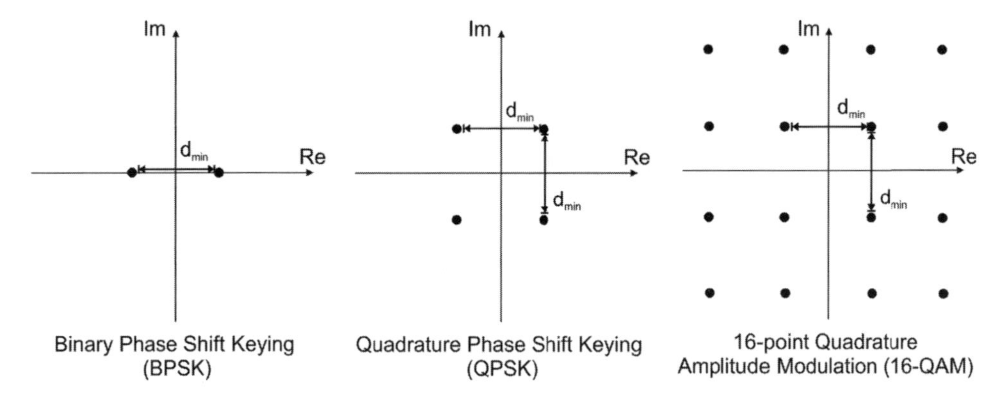

Figure 1.1 Example of signal constellations corresponding to BPSK, QPSK, and 16-QAM modulation schemes

is evenly distributed into the two signalling dimensions). In the following, the orthogonal components of the noise process are denoted by a common notation $X_N \sim N\left(0, \bar{P}_N/2\right)$.

If the absolute value of either the real or the imaginary noise component is larger than half of the Euclidian distance d between adjacent modulation symbols, a transmitted symbol may be erroneously decoded into any symbol within a complex half-plane, as illustrated in the QPSK example of Figure 1.2. The probability of one of the independent and identically distributed noise components having such values can be written as:

$$P\left(X_N > d/2\right) = \int_{d/2}^{\infty} \frac{1}{\sqrt{\pi \bar{P}_N}} \exp\left(-\frac{x^2}{\bar{P}_N}\right)dx = 1 - \int_{-\infty}^{\frac{d}{\sqrt{2\bar{P}_N}}} \frac{1}{\sqrt{2\pi}} \exp\left(-\frac{x^2}{2}\right)dx \tag{1.1}$$

where the final expression is given in terms of the cumulative distribution function of a normalized Gaussian random variable. Error probabilities are usually specified in this form, since the Q-function $Q\left(\alpha\right) = 1 - \int_{-\infty}^{\alpha} \frac{1}{\sqrt{2\pi}} \exp(-\frac{x^2}{2})$ is widely tabulated in mathematical reference books, and easily calculated with programs such as Matlab. The expression (1.1) gives directly the probability of error for BPSK, and can be used to calculate the average probability of error for larger QAM constellations. In Figure 1.3, the principle of calculating the symbol error probability of QPSK using (1.1) is illustrated. The same principle is applied in Figure 1.4 to 16-QAM, where several different error cases have to be considered, and averaged to obtain the total probability of symbol error.

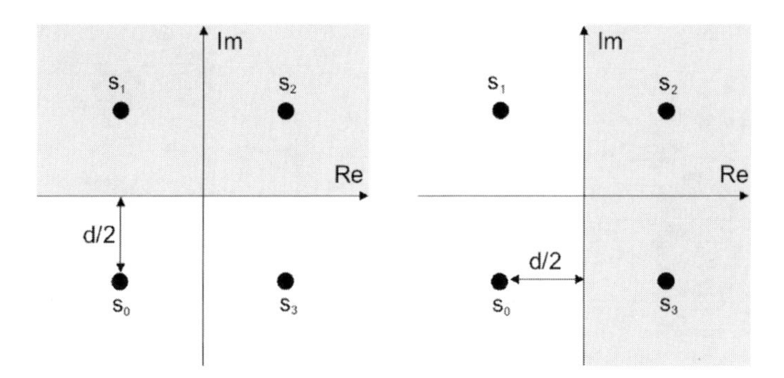

Figure 1.2 The effect of noise on a QPSK signal constellation. Left: imaginary component of noise is larger than d/2 – transmitted symbols s_0 and s_3 will be erroneously decoded either as s_1 or s_2. Right: real component of noise is larger than d/2 – transmitted symbols s_0 and s_1 will be erroneously decoded either as s_2 or s_3

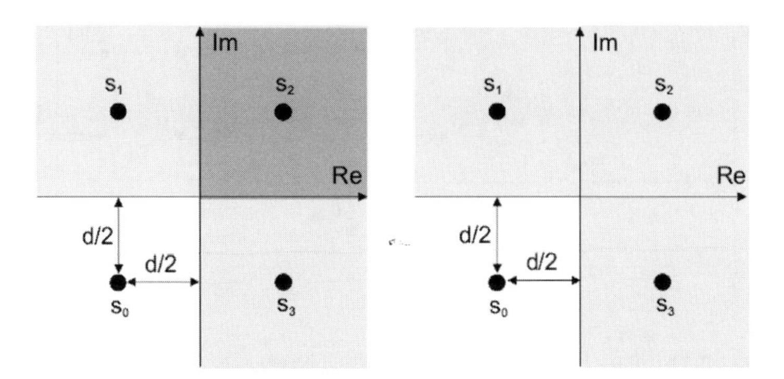

Figure 1.3 Principle of calculating the probability of symbol error for a QPSK signal constellation, assuming s_0 is transmitted. Left: $2P(X_N > d/2)$ includes twice the probability of receiving a value in the diagonally opposite quadrant. Right $2P(X_N > d/2) - P(X_N > d/2)^2$ is the correct probability of symbol error

In the preceeding examples, the error probabilities are calculated in terms of the minimum distance of the constellations and the average noise power. However, it is more convenient to consider error probabilities in terms of the ratio of average signal and noise powers. For any uniform QAM constellation, the distance between any pair of neighbouring symbols (that is, the minimum distance) is easily obtained as a function of the average transmitted signal power $\bar{P}_S$ – which is calculated as the average over the squared absolute values of the complex-valued constellation points – as:

$$d = \begin{cases} 2\sqrt{\bar{P}_S} & \text{(BPSK)} \\ 2\sqrt{\bar{P}_S/2} & \text{(QPSK)} \\ 2\sqrt{\bar{P}_S/10} & (16-\text{QAM}) \end{cases}$$

The average symbol error probability for each of the cases above is now obtained by calculating averages over demodulation error probabilities for the signal sets as a function of the average signal-to-noise ratio, given by $\bar{P}_S/\bar{P}_N \hat{=} \lambda$. Using the equations given above, the average symbol error probabilities are

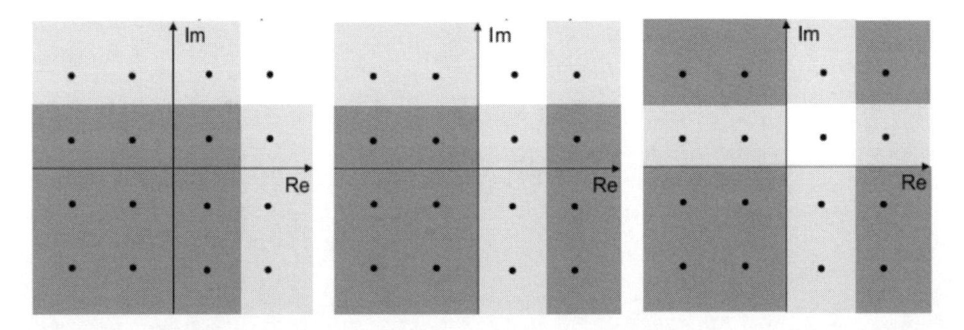

Figure 1.4 Principle of calculating the probability of symbol error for a 16-QAM signal constellation. Left: for the four corner symbols, the probability of symbol error is $2P(X_N > d/2) - P(X_N > d/2)^2$. Center: for the eight outer symbols, $3P(X_N > d/2) - 2P(X_N > d/2)^2$. Right: for the middle symbols, $4P(X_N > d/2) - 4P(X_N > d/2)^2$. The total probability of symbol error is the weighted average of these probabilities

obtained, following the principle outlined in the examples of Figures 1.3 and 1.4, as:

$$p_s(\lambda) = \begin{cases} Q\left(\sqrt{2\lambda}\right) & \text{(BPSK)} \\ 2Q\left(\sqrt{\lambda}\right) - Q\left(\sqrt{\lambda}\right)^2 & \text{(QPSK)} \\ 3Q\left(\sqrt{\lambda/5}\right) - \frac{9}{4}Q\left(\sqrt{\lambda/5}\right)^2 & \text{(16 - QAM)} \end{cases}$$

1.2.2 From Sample SNR to E_b/N_0

Assume the transmitted symbols are mapped to rectangular baseband signal pulses of duration T_{symb}, sampled with frequency f_{sampl}, with complex envelopes corresponding to the constellation points of the signal-space representation used above. These rectangular pulses are then modulated by a given carrier frequency, transmitted through a noisy channel, downconverted in a receiver and passed to a matched filter or correlator for signal detection.

Figure 1.5 shows an example of two BPSK symbols transmitted and received as described above. In this example, the signal-to-noise ratio per sample is defined as $SNR = A^2/\sigma_n^2$, where σ_n^2 is the sample variance of the real-valued noise process. It can be seen that, based on any individual sample of the received signals, the probability of error is quite large. However, calculating the averages (plotted with dashed lines in Figure 1.5) of the signals over their entire durations (0.1 s, containing 100 samples) gives values for the signal envelopes that are very close to the correct values -1 and 1, thus reducing the effect of the added noise considerably. It is clear that in this case, the sample SNR is no longer enough to determine the probability of error at the receiver. The relevant question is how should the sample SNR be scaled to obtain the correct error probability? We study this using BPSK as an example.

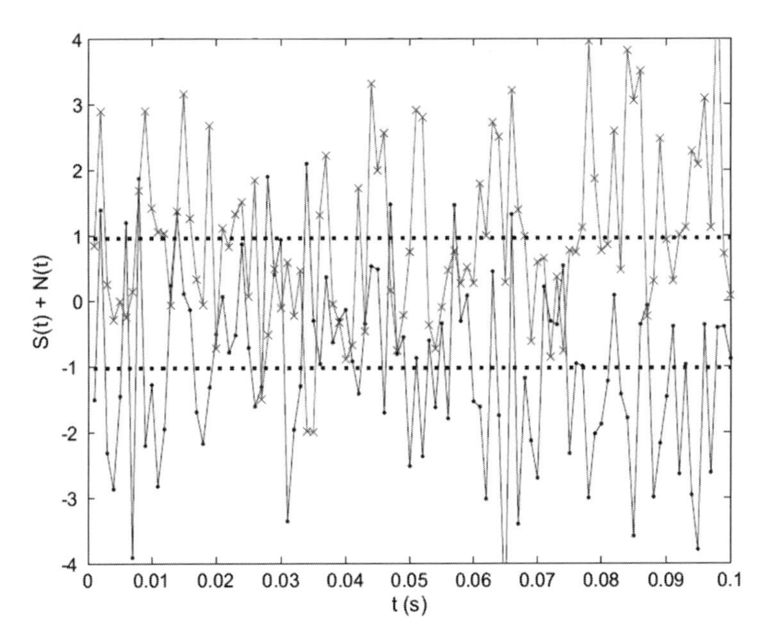

Figure 1.5 Two noisy signal envelopes and their averages. For this example, $T_{symb} = 0.1$ s, $f_{sampl} = 1000$ Hz, $A_1 = -A_0 = 1$, $E_b/N_0 = 15$ dB $\leftrightarrow$ SNR = -5 dB

As above, the probability of symbol (bit) error based on the signal-space representation for BPSK over an AWGN channel is:

$$P_e = P(N < -A_1) = \int_{-\infty}^{-A_1} \frac{1}{\sigma_n\sqrt{2\pi}} \exp\left(-\frac{x^2}{2\sigma_n^2}\right) dx$$

where N is a normally distributed random variable with standard deviation σ_n and zero mean, and it is assumed (without loss of generality) that the signal amplitude $A_1 > 0$ (corresponding to a 1 being sent). This can be thought of as transmitting a single sample of the signal envelope. Sampling a received signal envelope $S(t) + N(t)$ at k points produces a sequence of samples $S(i \cdot T_{sampl}) + N(i \cdot T_{sampl})$, where $T_{sampl} = 1/f_{sampl}$, and $i = 1 \ldots k$. A correlator receiver for BPSK may use the following test statistic to decide whether a 1 was most likely to be transmitted:

$$Z = A_1\left(\sum_{i=1}^{k}\left(S\left(i \cdot T_{sampl}\right) + N\left(i \cdot T_{sampl}\right)\right)\right) = \sum_{i=1}^{k}\left(A_1 S\left(i \cdot T_{sampl}\right)\right) + \sum_{i=1}^{k}\left(A_1 N\left(i \cdot T_{sampl}\right)\right)$$

Assuming that a 1 was indeed sent, a false decision will be made if:

$$A_1 \sum_{i=1}^{k} N\left(i \cdot T_{sampl}\right) < -k \cdot A_1^2 \Leftrightarrow \frac{1}{k} N\left(i \cdot T_{sampl}\right) < -A_1$$

or

$$\bar{N} < -A_1$$

denoting the sample mean of the noise as $\bar{N}$. We note that the expression is the same as for the single sample case, only with the normal random variable replaced by the sample mean of k samples from a normal distribution. Basic results of statistics state that this sample mean is also normally distributed, in this case with mean zero and standard deviation $\sigma_{\bar{N}} = \sigma_n/\sqrt{k}$. We thus find that the error probability in this example is determined by the ratio $k \cdot (\sigma_s^2/\sigma_n^2)$, or k times the sample SNR.

It should be noted that although we used BPSK as an example to simplify the relevant expressions, the above result is not restricted only to BPSK. In fact, the obtained expression $k \cdot (\sigma_s^2/\sigma_n^2)$ is generally used in a form derived as follows:

$$k \cdot \frac{\sigma_s^2}{\sigma_n^2} = \frac{T_{symb}}{T_{sampl}} \cdot \frac{\bar{P}_S}{\bar{P}_N} = \frac{\bar{P}_S \cdot T_{symb}}{\left(1/f_{sampl}\right) \cdot N_0 B_n} = \frac{E_S}{N_0}$$

In the above, N_0 is the noise power spectral density and B_n is the noise bandwidth. Note that the signal energy $E_S = \bar{P}_S \cdot T_{symb}$, and that $B_n = f_{sampl}$ (this is based on the Shannon-Nyquist sampling theorem applied for complex samples). Note also that here it is implicitly assumed that the signal bandwidth corresponds to the Nyquist frequency; if the signal is oversampled, care should be taken in performance analysis to include only the noise bandwidth which overlaps with the spectrum of the signal. Finally, the ratio of energy per bit to noise power spectral density E_b/N_0, very commonly used as a measure for signal quality, is obtained as:

$$\frac{E_b}{N_0} = \frac{1}{n_b}\frac{E_s}{N_0}$$

where n_b is the number of bits per transmitted symbol.

1.3 Fading Processes in Wireless Communication Channels

Additive noise is present in all communication systems. It is a fundamental result of information theory that the ratio of signal and noise powers at the receiver determines the capacity, or maximum

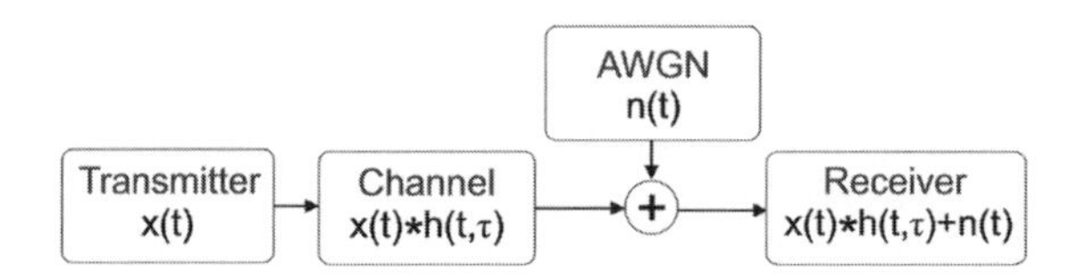

Figure 1.6 System model for transmitting information through a channel with additive white Gaussian noise

achievable rate of error-free transmission of information, of a channel. Generally, multiplicative effects of a communication channel, or *fading*, can be represented as a convolution of the transmitted signal with the *channel impulse response*, as illustrated in Figure 1.6. A general effect of fading is to reduce the signal power arriving at the receiver. Since the noise power at the receiver is independent of the useful signal, and the noise component does not experience fading, a fading channel generally reduces the ratio of the signal power to the noise power at the receiver, thus also reducing the transmission capacity.

The distortion, or noise, caused by a communication channel to the transmitted signal can be divided into multiplicative and additive components; the latter was considered above. Multiplicative noise, or fading, can be defined as the relative difference between the powers contained in corresponding sections of the transmitted and received signals. Factors that typically contribute to the fading in wireless communication systems are the transmitter and receiver antenna and analog front-end characteristics, absorption of the signal power by the propagation media, and reflection, refraction, scattering and diffraction caused by obstacles in the propagation path. The receiver experiences the combined effect of all these physical factors, which vary according to the positions of the receiver and transmitter within the propagation environment. It should be noted that it is generally possible to describe the effects of a communication channel entirely by its impulse response as illustrated in Figure 1.6. However, it is typical that estimation of the average power conveyed by a transmission channel is performed separately from the modelling of the channel's impulse response, which is then power-normalized. We also apply this principle in the following discussion on *fading processes* in wireless channels.

Fading in wireless channels is in literature typically characterized as a concatenation or superposition of several types of fading processes. These processes are often classified using the qualitative terms *path loss*, *shadowing*, and *multipath fading*, which is also often referred to as *fast fading*. However, these fading processes cannot in general be considered fully independent of each other, and indeed in many references (for example in [1],[12]) path loss and shadowing are not considered as separate processes. Justification for this will be subsequently considered in more detail. In the following, fading is primarily classified according to the typical variation from the mean attenuation over a spatial region of given magnitude. The terms large-scale, medium-scale, and small-scale fading are thus used.

Small-scale fading corresponds directly to multipath fading, and involves signal power variations of magnitude up to 40 dB on a spatial scale of a half-wavelength (for example 50 cm at 300 MHz). Averaging the total fading in the receiver over a spatial interval significantly larger than a half-wavelength provides information on the medium-scale fading, or shadowing. Over spatial intervals of magnitude hundreds of meters, medium-scale fading involves signal power variations up to magnitude 20 dB. Again, averaging the total fading over a spatial interval of several hundred meters provides an estimate for the large-scale fading, which may vary up to 150 dB over the considered coverage area. [9] These denominations do not suggest a different origin or effect for the fading types, but rather signify that typically different variation around the mean attenuation is observed at different spatial scales, or observation windows.

1.3.1 Large-Scale Fading (Path Loss)

Large-scale fading, or path loss, is commonly modelled for signals at a given carrier frequency as a deterministic function of the distance between the transmitter and receiver, and is affected by several parameters such as antenna gains and properties of the propagation environment between the transmitter and receiver. Main physical factors that contribute to large-scale fading are free-space loss, or the dispersion of the transmitted signal power into surrounding space, plane earth loss, and absorption of the signal power by the propagation medium.

Free-space loss corresponds to dispersion of transmitted signal power into the space surrounding the transmitter antenna. The most simple free-space loss estimation is obtained by assuming that signals are transmitted omnidirectionally, that is, power is radiated equally to all directions, and there are no obstacles within or around the transmission area, which would affect the propagation of electromagnetic signals. With such assumptions, the *power density* at a distance d meters from the transmitter can be written as:

$$p_R = \frac{P_T}{4\pi d^2} \text{ (watts/m}^2\text{)}$$

where P_T is the total transmitted signal power. This expression is obtained simply by dividing the transmitted power over the surface area of a sphere surrounding the transmitter antenna.

The assumptions specified above are not practical in most communication scenarios. Ignoring for now the likely presence of obstacles around the transmitter and receiver, the free-space loss defined above can be modified into a more realistic expression by taking into account the antenna characteristics of the transmitter and receiver. Specifically, the actual received power depends on the *effective aperture area* of the receiver antenna, which can be written as:

$$A_R = \frac{\lambda^2 G_R}{4\pi} \text{ (m}^2\text{)}$$

where λ is the wavelength of the transmitted signal and G_R is the receiver *antenna gain*, which is affected by the directivity of the antenna – specifically the antenna radiation patterns in the direction of the arriving signal. It should be noted that the above expression means that the received power decreases along with an increase in the carrier frequency. Finally, taking into account the transmitter antenna gain factor G_T, the received power after free-space loss can be written as:

$$P_R = G_T p_R A_R = \frac{P_T G_T G_R \lambda^2}{(4\pi d)^2} \text{ (W)}$$

Note that in the above, the variables are assumed to be given in the linear scale, that is, not in decibels. Figure 1.7 shows examples of the received power as a function of distance from the transmitted antenna for different carrier frequencies, with the antenna gains and transmitted signal power normalized to unity. Formally, an expression for the path loss P_L, that is, attenuation of the transmitted signal, is obtained from the above in decibels as:

$$P_{L,dB} = 10\log_{10}\frac{P_T}{P_R} = 10\log_{10}\frac{(4\pi d)^2}{G_T G_R \lambda^2} = 20\log_{10}(4\pi d) - 10\log_{10}(G_T) - 10\log_{10}(G_R) - 20\log_{10}(\lambda)$$

Real signals do not follow the simple free-space attenuation model partly due to the presence of the ground plane close to the transmitter and receiver. This causes so called plane earth loss, where signal components reflected from the ground plane destructively interfere with the received useful signal. The amount of plane earth loss depends on the distance and heights of the transmitter and receiver antennas. Another significant cause for attenuation is the absorption of signal power by atmospheric gases and hydrometeors (such as clouds, rain, snow etc.).

In addition to these factors, large-scale fading is typically defined to include the average of the shadowing and multipath fading effects. Thus the type of propagation environment must be taken into account in the total power loss. This has been done for example in the widely used Okumura-Hata [13],[14] and COST 231 [15] models by approximating the parameters for the propagation loss for specific environments and transmission setups from sets of field measurements [1]. As an example, the

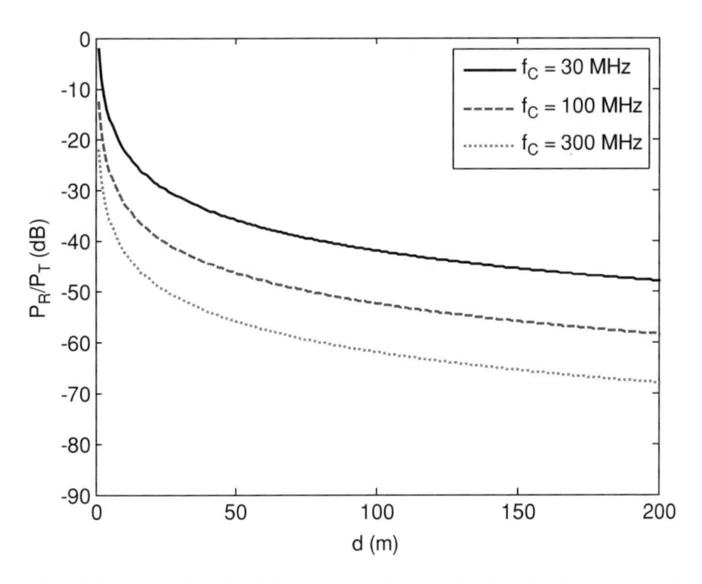

Figure 1.7 Examples of free-space loss for different carrier frequencies f_C, with the antenna gains and transmitted power normalized to 1

Hata model, which is based on the empirical work of Okumura, gives the following expression for path loss in urban areas (in dB):

$$P_{L,dB}(d) = 69.55 + 26.16\log_{10}(f_C) - 13.82\log_{10}(h_T) - a(h_R) + \left[44.9 - 6.55\log_{10}(h_T)\right]\log_{10}(d)$$

where f_C is again the carrier frequency, h_T and h_R are the heights of the transmitter and receiver antennas, respectively, and $a(h_R)$ is a correction factor, which is specified according to the size of the considered reception area. For large urban areas with the carrier frequencies of magnitude $f_C > 300$ MHz, this correction factor is given as:

$$a(h_R) = 3.2(\log_{10}(11.75h_R))^2 - 4.97$$

For other propagation scenarios, such as suburban and various rural areas, correction terms for the path loss expression given above are specified.

The Hata model is considered to be valid roughly in the carrier frequency range 150—1500 MHz and for distances $d > 1$ km, which means that it is not generally valid for example for modelling modern cellular systems with high carrier frequencies and small cell sizes [19]. In the European COST 231 cooperation, an extension to the Hata model was specified, where the carrier frequency is restricted between 1.5 and 2 GHz, the transmitter antenna height between 30 to 200 m, the receiver antenna between 1 and 10 m, and the transmission distance between 1 and 20 km. With these limitations, the path loss for an urban scenario according to the COST 231 extension to the Hata model is obtained from [15] as:

$$P_{L,dB}(d) = 46.3 + 33.9\log_{10}(f_C) - 13.82\log_{10}(h_T) - a(h_R) + \left[44.9 - 6.55\log_{10}(h_T)\right]\log_{10}(d) + C$$

where $a(h_R)$ is as in the Hata model, and C is 0 dB for medium-size urban areas and suburbs, and 3 dB for metropolitan areas.

The *empirical path loss models* outlined above are determined by averaging the results of large sets of measurements performed in propagation environments with specific characteristics. Similar path loss models can be obtained using analytic methods by assuming a statistical terrain description, where obstacles of suitable geometry are distributed randomly in the propagation environment, and by calculating the average propagation loss based on such approximations. For example, [11] contains a detailed description of deriving functions for path loss in various land environments using analytic methods. The physical mechanisms that cause the environment-specific propagation loss are the same for large-scale fading as for medium-scale fading, and are considered in more detail shortly.

Deterministic large-scale fading models – where estimations of the path loss are obtained as functions of the propagation distance – are useful in applications where it is sufficient to have rough estimates on the average attenuation of signal power over a large transmission area, or it is impractical to approximate signal attenuation in more detail. These models are typically used for example in radio resource management and planning of large wireless networks. It should be noted that expressions for large-scale fading can be obtained for generic environments using statistical methods as outlined above or for specific transmission sites by averaging over a site-specific approximation of medium-scale fading. However, this is typically a computationally involved task, as described in the following.

1.3.2 Medium-Scale Fading (Shadowing)

As with large-scale fading, methods for modelling medium-scale fading can typically be categorized as statistical or site-specific. In the statistical approach, the fading is typically assumed – based on empirical data – to follow a lognormal distribution. The mean for this distribution can be obtained for a given carrier frequency and distance from the transmitter using expressions for large-scale fading as outlined in the previous subsection. The standard deviation and autocorrelation of the lognormal distribution are model parameters, which must be selected according to the propagation environment. This standard deviation is known as the location variability, and it determines the range of fluctuation of the signal field strength around the mean value. Its value increases with frequency, and is also dependent on the propagation scenario – for example, the standard deviation is typically larger in suburban areas than in open areas. The standard deviation is typically in the range of 5 to 12 dB. Spatial correlation of shadowing is usually modelled using a first-order exponential model [20]:

$$\rho(d) = e^{-d/dcorr}$$

where d_{corr} is the distance over which the correlation is reduced by e^{-1}. This distance is typically of the same order as the sizes of blocking objects or object clusters within the transmission area.

An intuitive justification for the applicability of a lognormal model for medium-scale, or shadow fading, can be obtained by considering the total attenuation of the signal components arriving at the receiver in an environment with a large number of surrounding obstacles. Typically the signal components arriving at the receiver have passed through a number of obstacles of random dimensions, each attenuating the signal power by some multiplicative factor. The product of these fading factors contributes to the total power attenuation. In the logarithmic scale, the product of several fading components is represented as the sum of their logarithms, and again according to the central limit theorem the distribution of this sum approaches a normal distribution. Figure 1.8 shows examples of log-normal medium-scale fading for standard deviation 10 dB, and correlation distances 20 and 50 meters.

If site-specific data on the terrain profile and obstructions along the propagation path from the transmitter to the receiver are available, an approximation for medium-scale fading can be calculated as summarized in [9]:

1. Locate the positions and heights of the antennas.
2. Construct the great circle – or geodesic – path between the antennas. This represents the shortest distance between the two terminals measured across the Earth's surface.
3. Derive the terrain path profile. These are readily obtained from digital terrain maps, but it is of course also possible to use traditional contour profile maps.

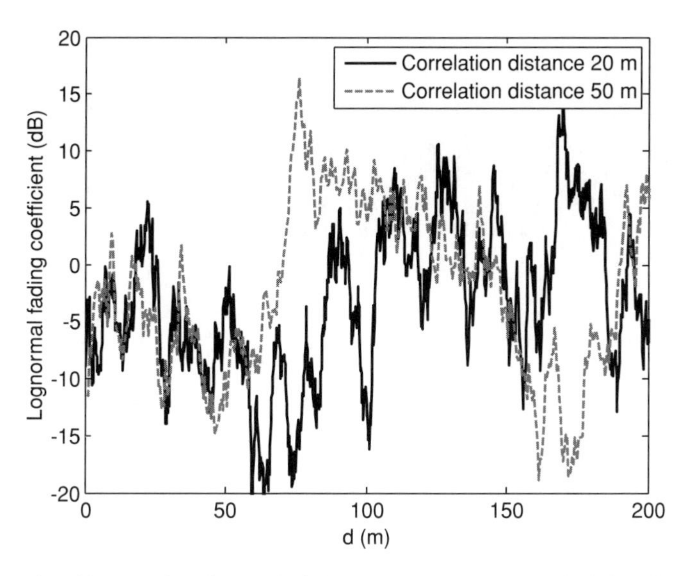

Figure 1.8 Examples of lognormal medium-scale fading processes with standard deviation 10 dB, and correlation distances 20 and 50 meters

4. Uplift the terrain profile by representative heights for any known buildings along the path.
5. Select a value for the effective Earth radius factor appropriate to the percentage of time being designed for; modify the path profile by this value. The effective Earth radius factor is a constant used to increase the effective radius of the Earth as seen by the propagating signal. This is due to tropospheric refraction, which makes the propagation paths curve slightly towards the ground. Since the atmospheric refractivity varies with pressure, temperature and water vapour pressure of the atmosphere, the correct effective Earth radius factor will vary according to location and time.
6. Calculate the free-space loss for the path.
7. If any obstructions exist within 0.6 times the first Fresnel zone of the propagation path, calculate diffraction over these obstructions and account for the excess loss in the fading. The Fresnel zones can be thought of as containing the main propagating energy in the wave; obstructions occupying less than 0.6 times the first Fresnel zone lead to an approximately 0 dB loss of signal power.
8. Compute the path length which passes through trees and add the corresponding extra loss.

Detailed descriptions for each of the steps above are given in [9]. It should be noted that the approach outlined above accounts only for obstructions along the direct propagation path between the transmitter and receiver. Considering propagation paths corresponding to reflections from objects not along the direct path leads to small-scale fading models, considered in the following sections.

1.3.3 Small-Scale Fading (Multipath Propagation)

Small-scale fading is caused by the interference between several reflected, diffracted or scattered signals arriving at the receiver. This effect is commonly called *multipath propagation.* Since the reflected propagation paths may be of different lengths, corresponding to different arrival times for variously faded copies of the transmitted signal at the receiver, the effect of small-scale fading is in the digital domain similar to a finite impulse response (FIR) filter with complex-valued coefficients between the transmitter and receiver. Thus, depending on the path delay profile of the channel, small-scale fading

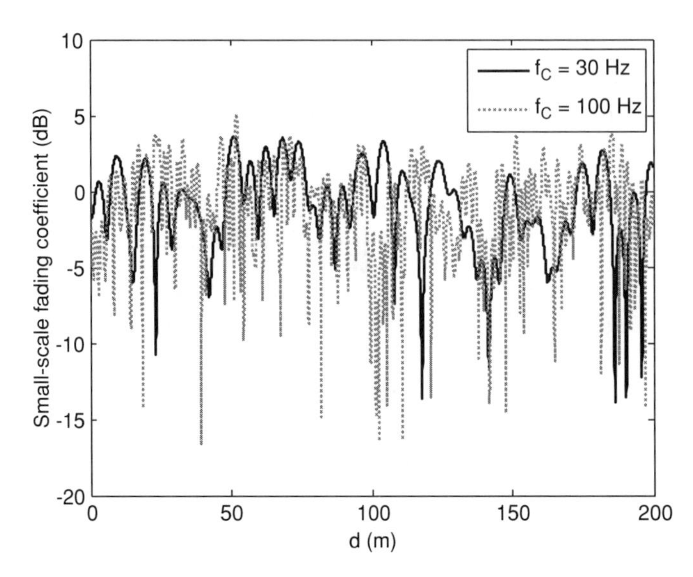

Figure 1.9 Examples of small-scale Rayleigh fading processes with normalized average power, and carrier frequencies of 30 and 100 Hz

may vary rapidly not only in the temporal and spatial domains, but also in the frequency domain. The rate of spatial variation of small-scale power attenuation is generally dependent on the carrier frequency of the transmitted signal. Figure 1.9 shows examples of small-scale fading for carrier frequencies 30 and 100 Hz. Figure 1.10 shows an example of the combined effects of large- medium- and small-scale fading.

Again, small-scale fading models can be divided into statistical and site-specific approaches. Site-specific models typically apply ray-tracing methods, where detailed three-dimensional models of the propagation environment are used to calculate propagation paths between the transmitter and receiver. Such techniques were originally developed for indoor environments, but have also been extended to dense urban outdoor areas [1]. Especially for modelling unconfined outdoor environments, ray-tracing models require large amounts of data and are computationally demanding. In the rest of this chapter we focus on statistical models for multipath propagation.

The causes of multipath propagation may be different in different channels. For example, it may be caused by reflections from buildings, objects or the ground surface in wireless communication channels, the reflection from walls and objects in wireless local area networks, reflections from the ionosphere in high-frequency radio transmission, and so on. Multipath propagation may be schematically described as in Figure 1.11, and mechanisms causing it are listed in Table 1.1 [21],[22]. As can be seen from Table 1.1, fading and propagation delay dispersion may arise even during wired transmission.

Small-scale fading is caused by the interference of multiple signals with random relative phases. Such interference causes random variation of the amplitude of the received signal. This increases the error probability in the system, since it reduces the signal-to-noise ratio. Dispersion of the delays of signal components arriving at a receiver is caused by the difference in the lengths of different propagation paths. If the delay difference is comparable with the symbol period, then the delayed responses from one signal may impose on the next signal, causing *intersymbol interference (ISI)* and frequency-selective fading.

One of the most common models for delay-dispersive wireless propagation channels is the representation of the channel as a *linear filter*. The channel is described by a time-varying impulse response $h(\tau, t)$. Applying the Fourier transform to $h(\tau, t)$ by the variable τ gives the *time-varying frequency response*

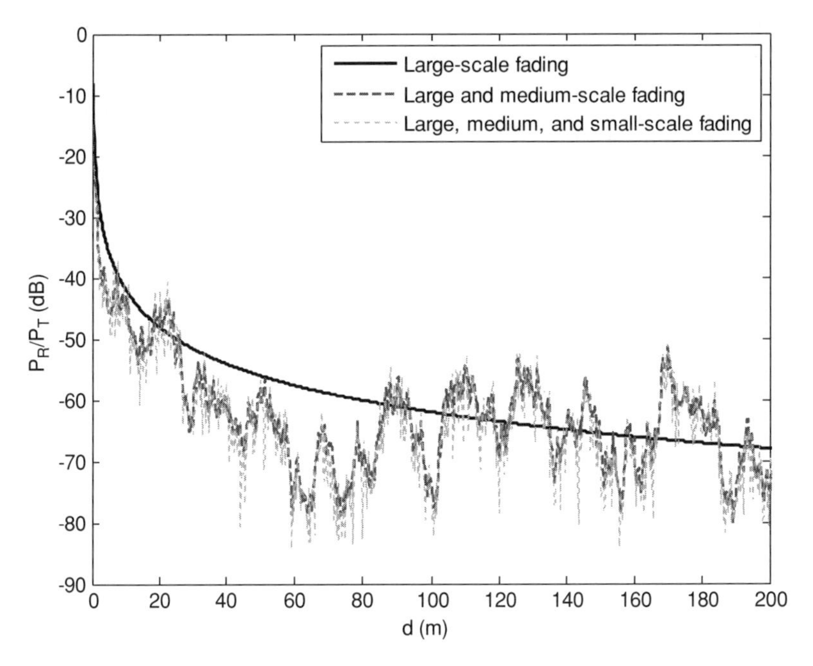

Figure 1.10 Combined effects of the three considered fading processes (large/medium/small-scale). The normalized received power is shown for carrier frequency 300 Hz, transmitted power and antenna gains set to unity, standard deviation of lognormal shadowing 10 dB with correlation distance 20 m, and average power of Rayleigh-distributed small-scale fading set to unity

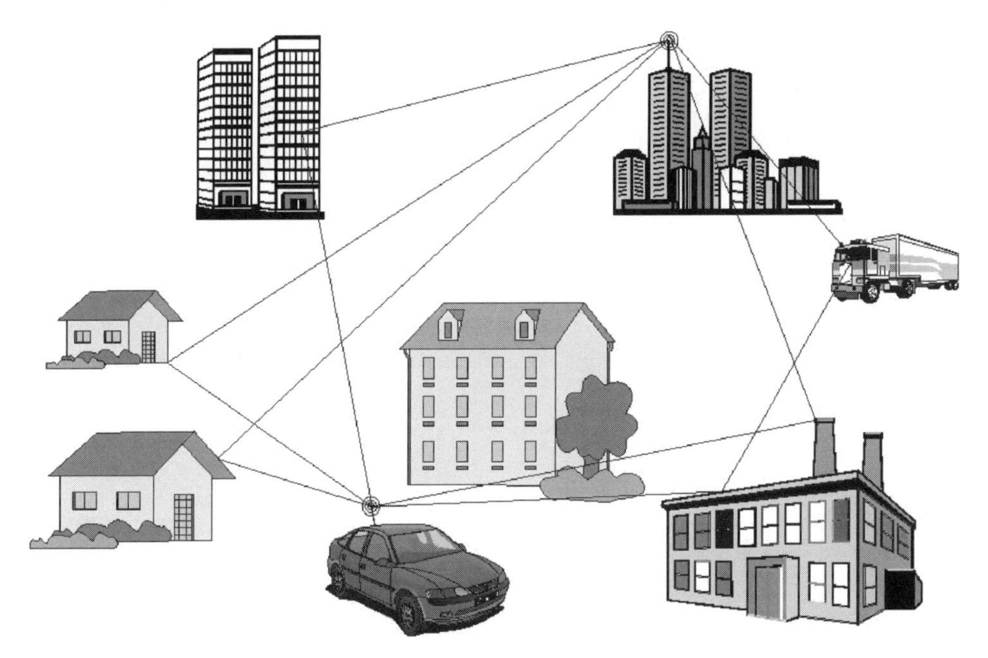

Figure 1.11 Multi-level spreading

Table 1.1 Multipath mechanisms

	System	Multipath mechanism
1	HF radio	Reflection from multiple ionospheric layers
2	Mobile and personal radio	Reflection and scattering from buildings, terrain, etc.
3	Microwave point-to-point links	Atmospheric refraction and reflection
4	Atmospheric refraction and reflection	Ground and building reflection
5	Radio LAN/indoor radio	Reflection from walls and building structure
6	Diffuse infra-red	Reflection from walls
7	Multimode optical fibre	Multimode propagation
8	Telephone/cable network	Reflections from terminations

$H(f, t)$, while the Fourier transform of $h(\tau, t)$ by the variable t gives the *scattering function* $S(\tau, \nu)$, which determines the Doppler spectrum of received signal as a function of the delay. The mean squared amplitudes of the channel impulse response define the *power delay profile* of the channel. An example of such a profile is shown in Figure 1.12. In the following section we consider in more detail this statistical model for multipath propagation outlined above.

1.4 Modelling Frequency-Nonselective Fading

1.4.1 Rayleigh and Rice Distributions

Let μ_1, μ_2 be two normally distributed random variables with zero mean and variance σ_0^2: $\mu_1, \mu_2 \sim N(0, \sigma_0^2)$. A random variable R_1, defined as $R_1 = \sqrt{\mu_1^2 + \mu_2^2}$ has the probability density function:

$$p_{R_1}(r) = \begin{cases} \dfrac{r}{\sigma_0^2} \exp\left(-\dfrac{r^2}{2\sigma_0^2}\right), & r \geq 0 \\ 0 & r < 0 \end{cases}$$

and is said to be *Rayleigh distributed.*

Defining a random variable R_2 as $R_2 = \sqrt{(\mu_1 + a)^2 + \mu_2^2}, a \in \Re$ results in the probability density function:

$$p_{R_1}(r) = \begin{cases} \dfrac{r}{\sigma_0^2} \exp\left(-\dfrac{r^2 + a^2}{2\sigma_0^2}\right) I_0\left(\dfrac{ra}{\sigma_0^2}\right), & r \geq 0 \\ 0 & r < 0 \end{cases}$$

where I_0 is the modified Bessel function of the first kind and zero order. R_2 is said to be *Rice distributed.*

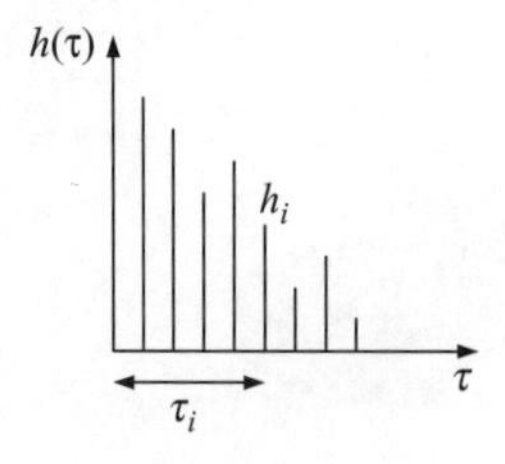

Figure 1.12 Impulse response of a delay-dispersive propagation channel

1.4.2 Maximum Doppler Frequency Shift

In the following, the maximum frequency shift of the received signal experienced by a moving receiver due to the Doppler effect is denoted as f_D, and can be obtained from:

$$f_D = \frac{|\vec{v}|}{c} f_C \tag{1.2}$$

where $\vec{v}$ is the receiver velocity (a stationary transmitter is assumed), c is the speed of light, and f_C is the carrier frequency of the signal. Of course, different frequency components in wideband signals experience different Doppler shifts, which causes some spreading of the signal bandwidth, but this difference is considered small enough to be neglected, as the ratio between the signal bandwidth and carrier frequency is typically small.

The maximum Doppler shift given by (1.2) is an absolute value that corresponds to situations where the receiver is moving radially towards (corresponding to a Doppler shift of $+f_D$) or away from the transmitter (corresponding to $-f_D$). The Doppler shift corresponding to other directions of movement can be obtained as:

$$f_{D,\alpha} = f_D \cos\alpha$$

where α is the angle between directions of the receiver velocity and the arriving signal.

1.4.3 Wide-Sense Stationary Stochastic Processes

A stochastic process is a family of random variables$\{X_t\}_{t\in T}$, where T can be any set. If $T = \aleph$, the stochastic process is simply a sequence of random variables X_n. Such a sequence is called *strict-sense stationary* if and only if for any $n = 0,1, \ldots,$ and any $k = 1,2, \ldots,$ $(X_0, \ldots, X_n)$ and $(X_k, \ldots, X_{k+n})$ have the same distribution. [10]

A stochastic process is said to be *wide-sense stationary (WSS)* if the mean and autocorrelation of the random variables X_n are invariant to a shift of the origin. More specifically, for any $n = 0,1, \ldots, k = 0,1, \ldots,$ $E[X_n]$ has a constant value and:

$$E\left[X_n^* X_k\right] = r_{XX}\left(|n-k|\right)$$

where the asterisk denotes complex conjugation and $r_{XX}(\tau)$ is an autocorrelation function whose value depends only on the time shift τ [16]. Wide-sense stationarity is a weaker condition than strict-sense stationarity, that is, every strict-sense stationary process is wide-sense stationarity, but not vice versa.

For the small-scale fading models described in the following, it is assumed that signals propagate to the receiver antenna along a horizontal plane. Furthermore, it is assumed that the angles of arrival of electromagnetic waves at the receiver antenna are uniformly distributed and that the receiver antenna has a circular-symmetric radiation pattern. As mentioned before, the transmitter antenna is assumed to be stationary, while the receiver moves with velocity $\vec{v}$.

1.4.4 Rayleigh and Rice Models for Frequency-Nonselective Fading

If the propagation delay differences of the reflected and scattered signal components at the receiver are negligible compared to the symbol interval of the transmission, the channel impulse response can in practice be approximated by a single delta function multiplied by a random variable that describes the amplitude fading. Thus there will be no significant interference caused by overlapping successive transmitted symbols at the receiver, or intersymbol interference, ISI. Also, the channel affects all frequency components of the signal similarly, and the term *frequency-nonselective fading* can be used. In this case, the small-scale fading is wholly characterized by the distribution and time-variant behaviour of the channel coefficient random variable.

In non-line-of-sight (NLOS) conditions, where there is no direct, unobstructed propagation path from the transmitter to the receiver, both the in-phase and quadrature parts of the received signal are assumed to consist of sums of large numbers of independently faded scattered components. Thus, by the central limit theorem, the fading of the in-phase and quadrature – or real and imaginary – components of the signal can be approximated as independent normally distributed random variables. As described above, this leads to a Rayleigh distribution for the amplitude of the complex fading coefficient. The phase for the complex fading is uniformly distributed between 0 and 2π.

In line-of-sight (LOS) conditions, the received signal can be characterized as a sum of Rayleigh faded NLOS components as described above, and a coherent LOS component with relatively constant power determined by the medium-scale fading. This can be approximated by adding a constant representing the amplitude of the line-of-sight signal contribution to the real part of the complex fading coefficient, which leads to a Rice distribution for the fading amplitude.

As small-scale fading is a function of the receiver location, it is clear that the rate of variation of the fading in time is dependent on the speed of the receiver. Analytically, the receiver velocity determines the Doppler frequency shift of the received signal as given in Section 1.2.3. Based on the maximum Doppler frequency and the angular probability distribution of the received signal components the probability density function of the Doppler frequencies can be calculated. This probability density function is directly proportional to the Doppler power spectral density of the received in-phase and quadrature signal components, the inverse Fourier transform of which gives the autocorrelation function of the channel fading coefficient [16]. It should also be noted that the primary detrimental effect of the Doppler shift in a wireless channel is due to the random directions of arrival of the reflected signal components arriving at the receiver. This randomness means that the received signals are randomly Doppler shifted between $-f_D$ and f_D, which causes a nontrivial broadening of the signal spectrum, and corresponding interference between signal components adjacent in the frequency domain.

Given the assumptions specified in Section 1.2.3, for the NLOS case the Doppler power spectral distribution is completely determined by the maximum Doppler frequency shift given by (1.2), and follows the so called *Jakes power spectral density*, or *Clarke power spectral density.* The LOS case differs from the above in that the Doppler power spectrum also contains a component corresponding to the power and Doppler shift of the line-of-sight signal component. Figure 1.13 illustrates the probability density function of the Doppler shifts and the corresponding autocorrelation function for the Rayleigh fading process. The autocorrelation function can be written as given in [16]:

$$r_{XX}(\tau) = 2\sigma_0^2 J_0\,(2\pi f_D \tau)$$

where $J_0(.)$ is the zeroth-order Bessel function of the first kind. The *coherence time* T_C of the fading process can be defined as the time interval that fulfills $|r_{XX}(T_C)| = 0.5\, r_{XX}(0)$, that is, the time interval after which the value of the autocorrelation has decreased to half of the value at the origin. For the above, $J_0\,(2\pi f_D \tau) \approx 0.5 \Leftrightarrow 2\pi f_D \tau \approx 1.52$. Thus $T_C \approx 1.52/(2\pi f_D)$. It should however be noted that the coefficient 0.5 assumed above is in no way unique, and also other values for the coherence time could be assigned. However, regardless of the numerical definition, it is important to note that the coherence time is reciprocally proportional to the maximum Doppler frequency f_D.

In practice, the time-variant channel coefficient for small-scale frequency-nonselective fading can be generated by drawing two sequences of normally distributed random numbers – or white Gaussian noise – corresponding to the components of the desired Rayleigh or Rice fading. One way to obtain the correct autocorrelation for the fading is then to low pass filter both of these sequences of random numbers according to the Jakes Doppler spectrum, producing coloured Gaussian noise. Using the filtered sequences as the real and imaginary components of the complex fading coefficient results in approximately the desired probability distribution and autocorrelation described above. Non-ideality

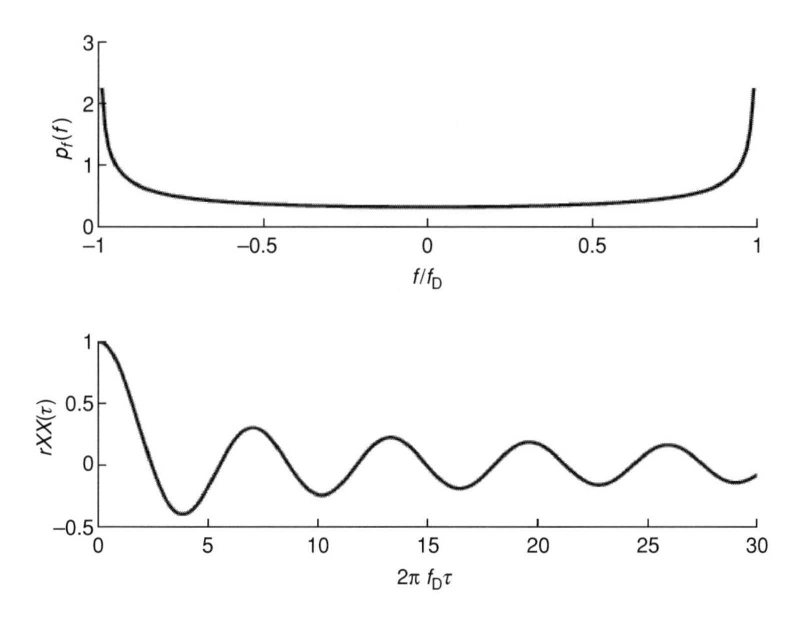

Figure 1.13 Probability density function for Doppler frequency shifts corresponding to the Jakes power spectral density (upper plot), and the corresponding autocorrelation function (lower plot)

arises with this approach mainly from the fact that neither white Gaussian noise nor ideal filters can be realized exactly.

In modelling small-scale fading as described above, the mean and autocorrelation function are typically kept constant, which means that the fading process is wide-sense stationary as defined in Section 1.4.3. It is clear that the mean value of the small-scale fading is dependent on medium-and large-scale fading processes, and thus the assumption of wide-sense stationarity is not generally valid. However, it has been empirically found that small-scale fading can be approximated as a WSS process for short distances (of order tens of wavelengths).

1.4.5 SNR in Rayleigh Fading Channels

In fading channels, the average power of the received signal should be considered a random variable as specified above. Thus, denoting for example the instantaneous signal-to-noise ratio for a Rayleigh channel by random variable Λ_R, the SNR can be written as:

$$\Lambda_R = \frac{\bar{P}_S R^2}{\bar{P}_N}$$

where R is the Rayleigh distributed channel coefficient amplitude. The average SNR, denoted by ρ, is obtained as:

$$\rho = \frac{E\left[R^2\right]\bar{P}_S}{\bar{P}_N} = \frac{2\sigma_0^2\bar{P}_S}{\bar{P}_N}$$

where σ_0 is the variance of the Gaussian components used to define the Rayleigh distribution in Section 1.2.3. In simulations, the average power conveyed by a Rayleigh channel can thus be normalized by selecting $\sigma_0 = 1/\sqrt{2}$. The probability distribution function of Λ_R is obtained as presented for example in reference [9], and can be written as:

$$p_\Lambda(\lambda_R) = \frac{1}{\rho}\exp\left(\frac{-\lambda_R}{\rho}\right), \quad \lambda_R > 0$$

1.5 WSSUS Models for Frequency-Selective Fading

1.5.1 Basic Principles

If the range of propagation delay times from the transmitter to the receiver is not negligible compared to the symbol duration of the transmitted signal, additional distortions of the received signal, such as intersymbol interference and frequency-selective fading, are introduced. In such cases, the frequency-nonselective fading models described above are not generally sufficient to describe the channel.

Adhering to the assumptions given in the previous section, a physical basis for modelling a frequency-selective channel can be found in the ellipses model illustrated for example in [9] and [16]. In this simplified representation of the scattering environment the transmitter and receiver are thought to be at the focal points of elliptical scattering zones, where each ellipse – or set of points with a fixed value for the sum of distances to the transmitter and receiver – defines the geometries of all propagation paths corresponding to a given propagation delay value. This principle is illustrated in Figure 1.14. Thus it is possible to consider the signal components corresponding to each discrete delay value as sums of large numbers of scattered signals with uniform distributions for the angle of arrival at the receiver. This in turn makes it possible to determine the time-variant fading coefficient for each discrete delay value as specified in the previous subsection for frequency-nonselective fading.

Frequency-selective channel models are typically implemented as FIR filters with time-variant complex coefficients by selecting N_t fixed discrete delay values corresponding to the nonzero filter coefficients. The average power for each of the N_t delayed signal components is selected according to a specific *power-delay profile (PDP)*. Given the average powers of each of the nonzero components of the FIR filter, the time-variant complex values for the filter coefficients are generated as WSS fading processes, as described in the previous section. The N_t discrete scattering components of the channel are typically defined as being statistically uncorrelated, which leads to the denomination wide-sense stationary uncorrelated scattering, or WSSUS, models.

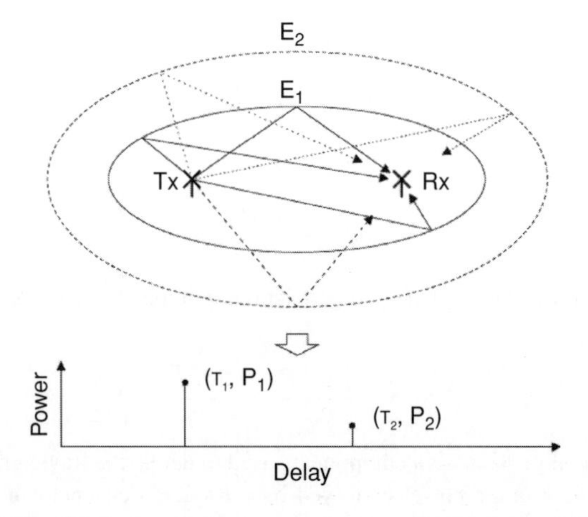

Figure 1.14 Elliptical scattering zones and a corresponding power delay profile

The parameters that characterize a WSSUS channel model according to a given transmission scenario are the PDP, the number and delays of the discrete filter coefficients, or taps, and the types of fading processes and Doppler spectra applied for the individual taps. Typically the continuous-time power delay profile is selected according to an exponential decrease of received signal power as a function of the propagation delay, although sometimes also mixtures of exponential distributions are used. Furthermore, the tap amplitudes are most commonly modelled as Rayleigh fading processes, with possibly a short-delay component defined as line-of-sight, and thus Rice distributed. The Doppler spectra for the independent tap fading processes are typically assumed to have the Jakes distribution, although for long-delay components, or far echoes, Gaussian power spectral densities have been found more accurate in some cases.

1.5.2 Definitions

For a channel with a continuous exponential power delay profile of the form $S(\tau) = (1/\alpha)\exp(-\tau/\alpha)$, $\tau > 0, \alpha > 0$, the mean delay is obtained as $\bar{\tau} = \alpha$ and the *delay spread* S_D as the square root of the second moment $S_D^2 = \alpha^2$. The Fourier transform of $S(\tau)$ is:

$$\psi(f) = \frac{1/\alpha}{1/\alpha + j2\pi f}$$

which gives a measure of the correlation of the fading for a frequency separation of f Hz. The *coherence bandwidth* B_C can be defined – similarly to the coherence time T_C considered previously – as the frequency interval that fulfils $|\psi(B_C)| = 0.5\,\psi(0)$, which is obtained by writing $\sqrt{(1/\alpha)^2 + (2\pi B_C)^2} = 2/\alpha$, and results in:

$$B_C = \frac{\sqrt{3}}{2\pi S_D}$$

Again, regardless of the numerical definition it is clear that the coherence bandwidth is reciprocally proportional to the delay spread of an exponential-PDP WSSUS channel. It is also evident that a continuous exponential distribution is characterized by the single free parameter α, which can be selected according to a given delay spread S_D or coherence bandwidth B_C. After defining the continuous distribution, it still remains to determine a discrete power-delay profile that sufficiently describes the selected distribution.

A discrete power-delay profile consists of sets of propagation delay values τ_i, and average scattered signal powers $P_i, i = 1, \ldots, N_t$. In the discrete case, the delay spread can be written as:

$$S_D = \sqrt{\frac{1}{P_T}\sum_{i=1}^{N_t} P_i \tau_i^2 - \left(\frac{1}{P_T}\sum_{i=1}^{N_t} P_i \tau_i\right)^2}$$

where P_T is the total power conveyed by the channel, given by $P_T = \sum_{i=1}^{N_t} P_i$. Unless more specific information on the propagation scenario to be modelled is available, it is reasonable to assume that the delay times between consecutive nonzero components in the discrete PDP follow an exponential distribution. This means that the number of channel components within a given delay range follows a Poisson distribution, and those components are uniformly distributed within the given delay range.

References

[1] M. C. Jeruchim, P. Balaban, K. S. Shanmugan, Simulation *of Communication Systems*, 2nd edition, Kluwer Academic, New York, 2000.

[2] J. G. Proakis, *Digital Communications*, 3rd edition, McGraw-Hill, 1995.

[3] R. E. Ziemer, R. W. Peterson, *Introduction to Digital Communication*, Prentice Hall, 2001.

[4] S. Haykin, *Communication Systems*, John Wiley & Sons, Ltd., 2001.
[5] T. K. Moon, *Error Correction Coding – Mathematical Methods and Algorithms*, John Wiley & Sons, Ltd., 2005.
[6] Y. Q. Shi, X. M. Zhang, Z.-C. Ni, N. Ansari, "Interleaving for combating bursts of errors," IEEE Circuits and Systems Magazine, vol. 4, First Quarter 2004, 29–42.
[7] C. Oestges, B. Clerckx, *MIMO Wireless Communications: From Real-World Propagation to Space-Time Code Design*, Elsevier, 2007.
[8] E. Lutz, D. Cygan, M. Dippold, F. Dolainsky, W. Papke, "The Land Mobile Satellite Communication Channel – Recording, Statistics and Channel Model," IEEE Trans. Veh. Technol., vol. 40, May 1991, 375–386.
[9] S. R. Saunders, A. Aragón-Zavala, *Antennas and Propagation for Wireless Communication Systems*, 2nd edition, John Wiley & Sons, Ltd., Chichester, 2007.
[10] R. B. Ash, C. A. Doléans-Dade, *Probability & Measure Theory, Second Edition*, Academic Press, San Diego, 2000.
[11] N. Blaunstein, J. B. Andersen, *Multipath Phenomena in Cellular Networks*, Artech House, Boston, 2002.
[12] B. Sklar, "Rayleigh Fading Channels in Mobile Digital Communication Systems Part I: Characterization," IEEE Communications Magazine, September 1997, 136-146.
[13] Y. Okumura, E. Ohmori, K. Fukuda, "Field Strength and its Variability in VHF and UHF Land Mobile Radio Service," Rev. Elec. Commun. Lab., vol. 16, 1968, 825–873.
[14] M. Hata, "Empirical Formulae for Propagation Loss in Land Mobile Radio Services," IEEE Trans. Veh. Technol., vol. VT-29, 1980, 317–325.
[15] COST 231, "Urban Transmission Loss Models for Mobile Radio in the 900 MHz and 1800 MHz Bands (rev. 2)," COST 231 TD(90), 119 Rev. 2, Den Haag, 1991.
[16] M. Pätzold, *Mobile Fading Channels*, John Wiley & Sons, Ltd., Chichester, 2002.
[17] M. R. Spiegel, *Mathematical Handbook of Formulas and Tables*, McGraw-Hill, Inc., New York, 1994.
[18] T. J.Wang, J. G. Proakis, E. Masry, J. R. Zeidler, "Performance Degradation of OFDM Systems Due to Doppler Spreading," IEEE Trans. Wireless Comm., vol. 5, June 2006, 1422–1432.
[19] A. Goldsmith, *Wireless Communications*, Cambridge University Press, New York, 2005.
[20] M. Gudmunson, "Correlation Model for Shadow Fading in Mobile Radio Systems," Electronic Letters, vol. 37, no. 23, pp. 2145–2146, Nov. 1991.
[21] A. Burr. The multipath problem: an overview. In IEE Colloquium on Multipath Countermeasures. London, 23 May 1996, Colloquium Digest 1996/120.
[22] A. Burr. *Modulation and Coding for Wireless Communication*. Prentice Hall, 2001.
[23] P. Bello. Characterization of randomly time-variant linear channels. IEEE Transactions on Communication Systems, CS-11; 36–393, 1963.

2

Modulation

Sergei Semenov
Nokia Corporation, Finland

The aim of modulation is to transfer a source data over a channel in a way most suitable for this channel. That is, the original data should be translated into a form that is compatible with the channel. Since the scope of this book is wireless communication only radio channel is under consideration. In this case the data modulates a radio frequency bearer in the form of a sinusoid which is called a *carrier wave*. This kind of modulation is called the bandpass modulation since it deals with a bandpass channel. It is possible to divide the modulation process into two stages: *baseband modulation* and *bandpass modulation*. In this way the baseband modulation consists of translating the original data (analogue or digital) into some waveforms of low frequency and bandpass modulation consists of modifying the high frequency carrier wave, or simply *carrier*, in accordance with waveforms obtained at the output of the baseband modulation process. Why do we need such a complicated process? Why not transmit the waveforms directly over the radio channel? One of the reasons, and possibly the main one, is the antenna size. The typical antenna size is $\lambda/4$, where λ is the wavelength. Assume that the waveform at the output of baseband modulation is the sinusoid with frequency $f = 1000$ Hz. Then the corresponding wavelength is $\lambda = c/f$, where c is the speed of light. It is easy to verify that the antenna size in this case should be $\lambda/4 \approx \frac{3 \cdot 10^8}{4 \cdot 10^3} = 7.5 \cdot 10^4$ m $= 75$ km. Obviously, this antenna size is unacceptable. However, if the baseband waveform is used for bandpass modulation of 2.5 GHz carrier the needed antenna size is only about 3 cm.

2.1 Basic Principles of Bandpass Modulation

The wave carrier can be represented in the following form:

$$s(t) = A(t) \cdot \cos(\omega_c t + \phi(t)) \tag{2.1}$$

where $A(t)$ is the *amplitude*, $\omega_c = 2\pi f_c$ is the *radian frequency* of the carrier (f_c is the carrier frequency), $\phi(t)$ is the *phase*. The bandpass modulation is based on modifying these parameters. In accordance with whichever parameter is being varied we can distinguish amplitude modulation (AM), frequency modulation (FM) and phase modulation (PM) or a combination of some of these basic modulation types. Actually the carrier frequency in (2.1) is constant and in the case of FM the deviation from the carrier

Modulation and Coding Techniques in Wireless Communications Edited by Evgenii Krouk and Sergei Semenov

frequency is varying. And this deviation is defined by a derivative of the phase. In this sense it is possible to consider FM just as a case of PM.

It is possible to distinguish between two main types of modulation: *analog modulation* and *digital modulation*. The aim of analog modulation is to transfer analog signal, such as speech or TV signal, over bandpass channel and in this case there is an infinite number of possible states of analog signal to modulate some parameter of a carrier. The changing of the carrier parameter in this case is continuous in time in accordance with the changing of original analog signal. The example of analog amplitude modulation is represented in Figure 2.1. In the case of digital modulation a digital bit stream should be transferred over the bandpass channel and there is only a limited number of digital symbols to be represented by the changing of a carrier parameter(s). Each digital symbol has time duration T and the changing of carrier parameter occurred on the boundary of this time interval. The analog modulation is beyond the scope of this book and hereafter we will refer to digital modulation as simply modulation.

The examples of different digital modulation types are depicted in Figure 2.2. Usually, the term *shift keying* stands for modulation in names of different modulation types when we are referring to digital modulation. The amplitude modulation is referred to as *amplitude-shift keying (ASK)*, frequency modulation as *frequency-shift keying (FSK)*, and so on. Shifting here means the changing (modulation) of some parameter and the word "keying" reflects back to the history of the communication: the telegraph.

Speaking about the demodulation process it is possible to recognize two main types: *coherent* and *noncoherent* demodulation. If the demodulator exploits the reference of each possible transmitted signal including not only the set of used waveforms but also the carrier reference to detect the signal the demodulation is coherent. In cases when the demodulator does not require the carrier reference the demodulation is noncoherent. The noncoherent demodulation reduces the implementation complexity of a demodulator, but the coherent demodulation provides better performance than the noncoherent one.

2.1.1 The Complex Representation of a Bandpass Signal

Very often it is convenient to represent the bandpass signal as a complex exponential function. Moreover it is possible to consider the modulated bandpass signal as the product of an exponential function representing carrier with an exponential function representing a baseband signal. In accordance with Euler's formula [1]:

$$\exp(jx) = \cos(x) + j \cdot \sin(x) \tag{2.2}$$

it is possible to write the representation of carrier in (2.1) as follows:

$$\begin{aligned} s(t) &= A(t) \cdot \cos(\omega_c t + \phi(t)) = \mathrm{Re}\{A(t) \cdot \exp(j(\omega_c t + \phi(t)))\} \\ &= \mathrm{Re}\{A(t) \cdot \exp(j\phi(t)) \cdot \exp(j\omega_c t)\} = \mathrm{Re}\{b(t) \cdot \exp(j\omega_c t)\} \end{aligned} \tag{2.3}$$

where function $b(t) = A(t) \cdot \exp(j\phi(t))$ is the baseband signal, $\exp(j\omega_c t)$ is an unmodulated carrier and $\mathrm{Re}\{z\}$ is the real part of the complex number z. The baseband signal $b(t)$ can be a complex function but amplitude $A(t)$ is a real one. Therefore, the modulated bandpass signal can be represented as the product of two phasors: $A(t) \cdot \exp(j\phi(t))$ with amplitude $A(t)$ and $\exp(j\omega_c t)$ with unit amplitude. This notation is more compact and in many cases is more convenient than (2.1). With the help of phasor notation it is easy to visualize the bandpass signal since any phasor $A \cdot \exp(j\theta)$ with a magnitude A and a phase θ can be conveniently represented as an *Argand diagram*, that is, a vector of length A deviated from the abscissa by angle θ in a complex plane. Actually the signal can be described in two ways: in polar form, that is, by its magnitude and phase or by its rectangular projections to the axes. If phase is a function of time $\theta = \omega t$ then we can regard phasor as a vector rotating counterclockwise at the constant rate ω as it is depicted in Figure 2.3. The projection of this vector to Cartesian coordinates is the inphase (real) and the quadrature (imaginary) components of signal that are orthogonal to each other. These components are used in real-world modulators for generation of modulated bandpass signal.

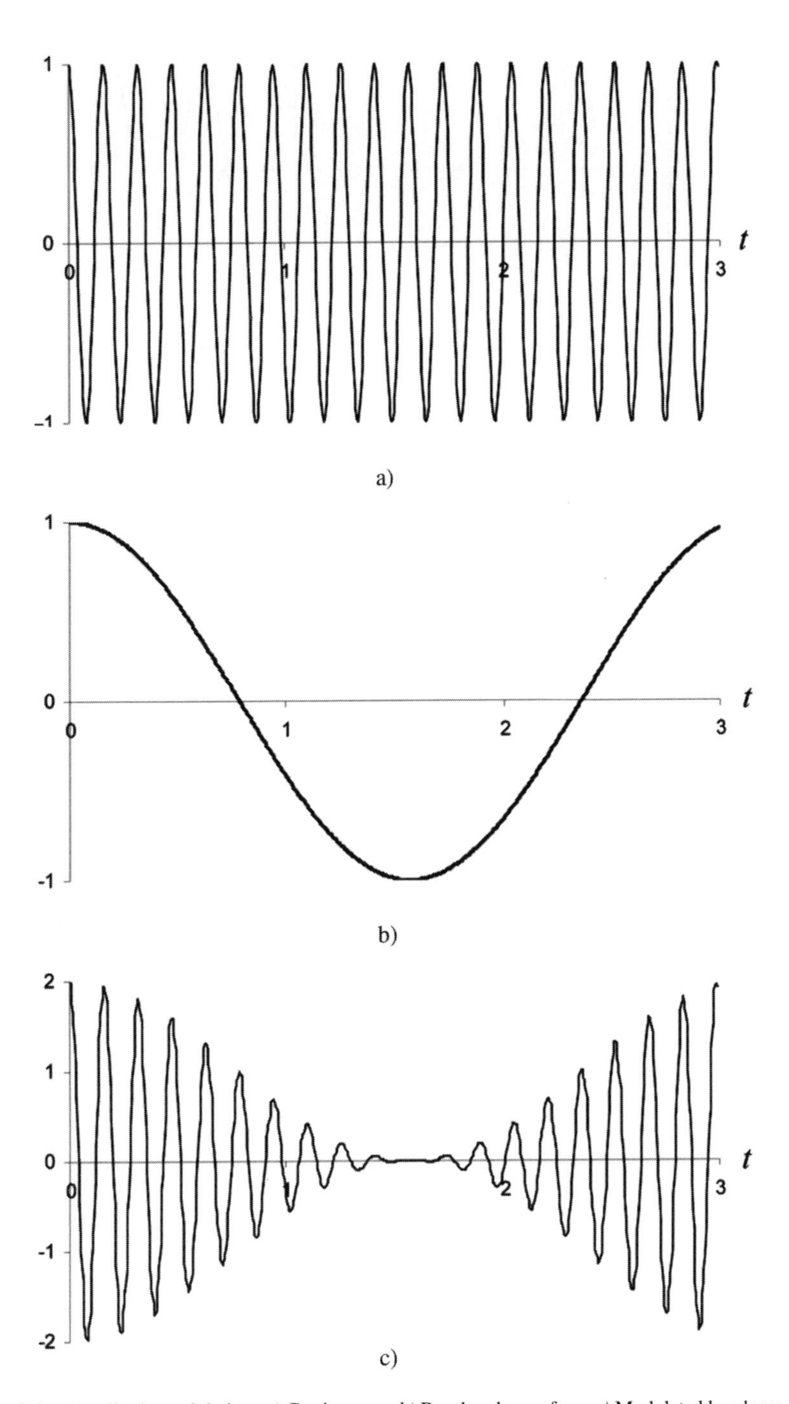

Figure 2.1 Amplitude modulation: *a)* Carrier wave *b)* Baseband waveform *c)* Modulated bandpass signal

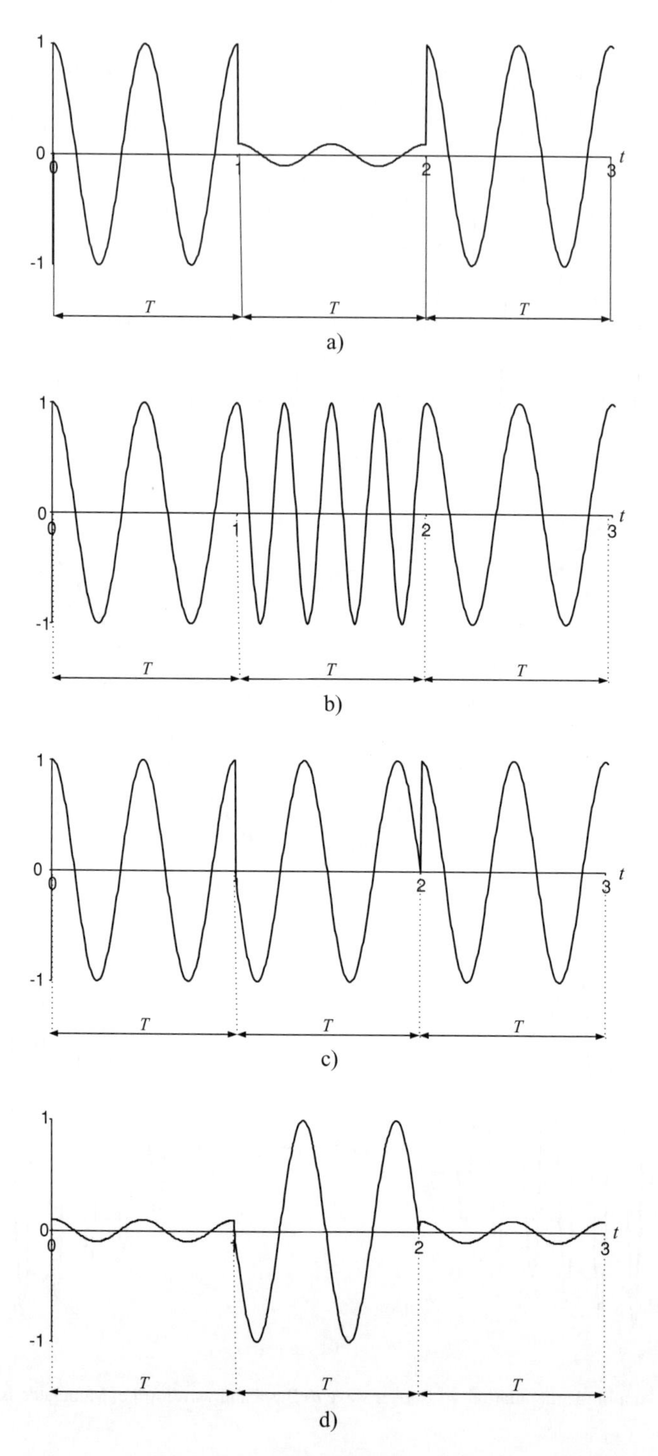

Figure 2.2 Digital modulation types. *a)* ASK *b)* FSK *c)* PSK d) ASK/PSK

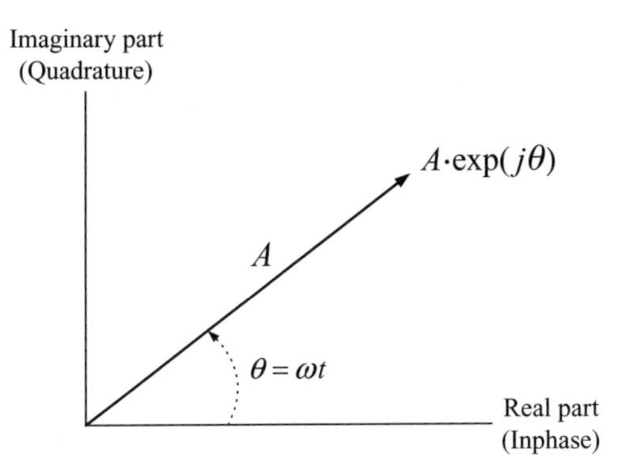

Figure 2.3 Phasor representation of signal

The magnitude of vector representing the signal can be expressed via inphase (I) and quadrature (Q) components as $A = \sqrt{\mathrm{I}^2 + \mathrm{Q}^2}$ and the phase as $\theta = \arctan\left(\frac{\mathrm{Q}}{\mathrm{I}}\right)$. Let us consider some examples showing the usefulness of phasor representation of bandpass signal.

Consider the example of phase modulation. Assume that the baseband signal has the form of sinusoid having unit amplitude and phase $\phi_\Delta x(t)$. In this case it is convenient to represent the modulated carrier in the general form of (2.1) with $\phi(t) = \phi_\Delta x(t)$:

$$s_{PM}(t) = \cos(\omega_c t + \phi_\Delta x(t)) = \mathrm{Re}\{\exp(j\phi_\Delta x(t)) \cdot \exp(j\omega_c t)\} \tag{2.4}$$

where ϕ_Δ is the instantaneous phase shift and $x(t)$ is some periodic function. It is obvious that at any time moment t the phase modulation leads just to phase shift of carrier, i.e. to the additional rotation of the carrier phasor as depicted in (2.4).

Now consider the example of amplitude modulation. Assume that the carrier is modulated with sinusoid waveform. Let the amplitude of this waveform sinusoid be 1 and radian frequency $\omega_m \ll \omega_c$. Then the modulated signal can written as follows:

$$s_{AM}(t) = \cos(\omega_c t) + \cos(\omega_c t) \cdot \cos(\omega_m t) = \cos(\omega_c t) + \frac{\cos((\omega_c + \omega_m)t)}{2} + \frac{\cos((\omega_c - \omega_m)t)}{2} \tag{2.5}$$

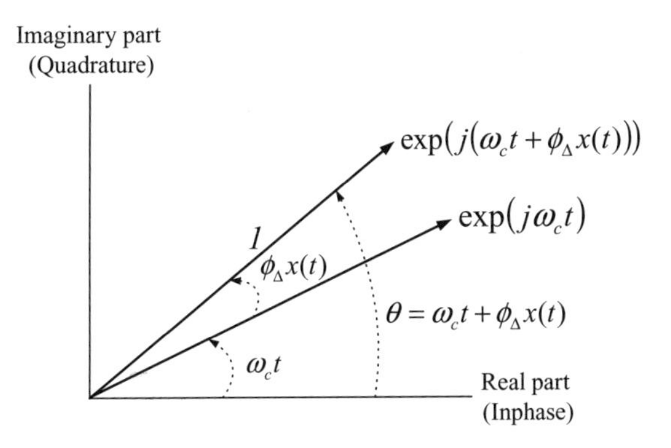

Figure 2.4 Phasor representation of phase modulation

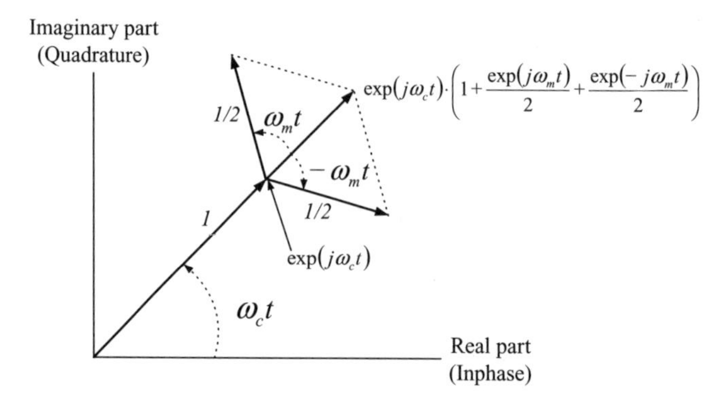

Figure 2.5 Phasor representation of amplitude modulation

Using the phasor representation it is possible to write (2.5) as:

$$s_{AM}(t) = \text{Re}\left\{\exp(j\omega_c t) + \frac{\exp(j(\omega_c + \omega_m)t)}{2} + \frac{\exp(j(\omega_c - \omega_m)t)}{2}\right\} \tag{2.6}$$

The (2.6) in turn can be represented as follows:

$$s_{AM}(t) = \text{Re}\left\{\exp(j\omega_c t) \cdot \left(1 + \frac{\exp(j\omega_m t)}{2} + \frac{\exp(-j\omega_m t)}{2}\right)\right\} \tag{2.7}$$

The representation of bandpass signal in the form of (2.7) helps to visualize the modulation process. Now we can see as depicted in Figure 2.5 that the phasor $\exp(j\omega_c t)$ representing the unmodulated carrier is variated by two sideband terms $\frac{\exp(j\omega_m t)}{2}$ and $\frac{\exp(-j\omega_m t)}{2}$ rotating counterclockwise and clockwise correspondingly. These sideband terms deviate the carrier phasor in both sides by the same angle as can be seen in Figure 2.5. As a result the phase and frequency of the carrier remains the same but the amplitude is changed in accordance with sidebands' radian frequency ω_m.

The next example is narrowband frequency modulation. Assume that the carrier is modulated by a sinusoid waveform having unit amplitude and radian frequency $\omega_m \ll \omega_c$. In this case the modulated signal is:

$$s_{FM}(t) = \cos(\omega_c t) - \beta \sin(\omega_c t) \cdot \sin(\omega_m t) = \cos(\omega_c t) - \beta \frac{\cos((\omega_c - \omega_m)t)}{2} + \beta \frac{\cos((\omega_c + \omega_m)t)}{2} \tag{2.8}$$

where $\beta \ll 1$ is the modulation index. With the help of phasor representation (2.8) can be written as:

$$s_{FM}(t) = \text{Re}\left\{\exp(j\omega_c t) \cdot \left(1 - \frac{\beta}{2}\exp(-j\omega_m t) + \frac{\beta}{2}\exp(j\omega_m t)\right)\right\} \tag{2.9}$$

In this case again as can be seen in Figure 2.6 the carrier phasor is variated by two sideband terms but now the terms have opposite signs which leads to different symmetry of sideband terms in comparison with AM case. Actually the sideband terms are rotated by $\pi/2$ radians in comparison with AM sidebands. Due to this the resulting phasor is rotated relatively to the carrier phasor. This rotation is changing in accordance with sidebands' radian frequency ω_m and modulation index β. The magnitude of the resulting phasor differs from the magnitude of the carrier phasor but since the modulation index $\beta \ll 1$ the difference is almost negligible. So, we can say that the amplitude of the modulated signal in the case of narrowband frequency modulation remains basically the same.

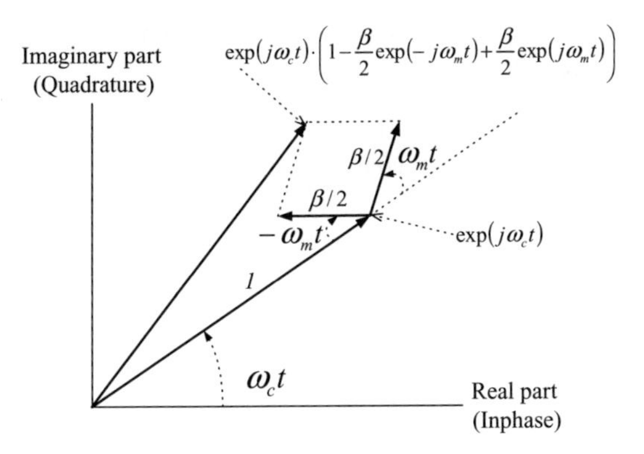

Figure 2.6 Phasor representation of frequency modulation

Note that all the considered examples are the examples of the analog modulation. In the case of digital modulation the phasor movement is restricted to "jumps" between several points on the plane corresponding to the limited number of used waveforms.

2.1.2 *Representation of Signal with Basis Functions*

Consider the power limited signals, that is, signals $s(t)$ for which the following condition holds:

$$\int_0^T |s(t)|^2 dt < \infty \tag{2.10}$$

It was shown in the previous section that any real signal $s(t)$ can be represented as a complex function. Then we can treat the signals as "square-integrable" complex-valued functions on the real interval [0, T]. It is known that these functions can be considered as an example of a *Hilbert space* [7]. The main things that distinguish the Hilbert space from the vector space are the existence of inner product except multiplication and addition and the completeness. Actually, our interest is to find some tool to measure how "far" or how "close" some particular signals (for example, the received and the transmitted signals) are, that is, to find the "distance" between two signals. We can define a *distance* formally as a *metric* on a set of signals. Metric or distance function maps any pair of set elements to some real number. Denote the set by X, then for all $x, y, z \in X$ distance $d(\cdot,\cdot)$ must satisfy the following conditions:

1) Non-negativity

$$d(x, y) \geq 0$$

2) Identity

$$d(x, y) = 0 \text{ iff } x = y$$

3) Symmetry

$$d(x, y) = d(y, x)$$

4) Triangle inequality

$$d(x, z) \leq d(x, y) + d(y, z)$$

However, this is a very general definition which fits any set. Recall that we are using Hilbert space as a model of signals. Then, it is the inner product that helps to measure the distance between signals. Inner product is a generalization of dot product. Let $x(t)$ and $y(t)$ be complex valued functions with $t \in [a, b]$. Then the inner product can be defined as:

$$\langle x(t), y(t) \rangle \equiv \int_a^b x(t) y^*(t) dt \tag{2.11}$$

If the interval $[a, b]$ is not specified then $t \in [-\infty, \infty]$. The inner product has the following properties:

1) Positivity

$$\langle x(t), x(t) \rangle \geq 0 \text{ for all } x(t)$$

2) Definiteness

$$\langle x(t), x(t) \rangle = 0 \text{ iff} x(t) = 0$$

3) Additivity in first slot

$$\langle x(t) + y(t), u(t) \rangle = \langle x(t), u(t) \rangle + \langle y(t), u(t) \rangle$$

4) Homogeneity in first slot

$$\langle ax(t), y(t) \rangle = a \langle x(t), y(t) \rangle$$

5) Conjugate interchange

$$\langle x(t), y(t) \rangle = \langle y(t), x(t) \rangle^*$$

Functions $x(t)$ and $y(t)$ are said to be orthogonal if $\langle x(t), y(t) \rangle = 0$. The distance in Hilbert space is measured by calculating the norm. The norm of function $x(t)$ is defined as:

$$\|x(t)\| = \sqrt{\langle x(t), x(t) \rangle} \tag{2.12}$$

We can consider norm as a "length" of $x(t)$. There are the following properties of norm:

1) $\|ax(t)\| = |a| \, \|x(t)\|$.
2) *Pythagorean Theorem*: If $x(t)$ and $y(t)$ are orthogonal then:

$$\|x(t) + y(t)\|^2 = \|x(t)\|^2 + \|y(t)\|^2$$

3) *Cauchy-Bunyakovsky-Schwarz inequality*:

$$|\langle x(t), y(t) \rangle| \leq \|x(t)\| \, \|y(t)\|$$

4) *Triangle inequality*:

$$\|x(t) + y(t)\| \leq \|x(t)\| + \|y(t)\|$$

5) *Parallelogram equality*:

$$\|x(t) + y(t)\|^2 + \|x(t) - y(t)\|^2 = 2 \|x(t)\|^2 + 2 \|y(t)\|^2$$

For any Hilbert space it is possible to find a set of N orthogonal functions $\{\psi_i(t)\}_{i=1}^N$ called *basis functions*, such that any function $x(t)$ from this space can be expressed as a linear combination of basis functions:

$$x(t) = \sum_{i=1}^{N} a_i \psi_i(t) \tag{2.13}$$

If (2.13) is satisfied, it is said that the set of basis functions $\{\psi_i(t)\}_{i=1}^N$ spans the corresponding Hilbert space. In principle the set of basis functions can comprise an infinite set, that is, $N = \infty$, then the corresponding Hilbert space is infinite-dimensional. If the basis comprises a finite number of basis functions the corresponding Hilbert space is finite-dimensional. If all we know about the signal $s(t)$ is that it is a power limited signal defined on the interval $[0, T]$ then we can consider $s(t)$ to belong to the N-dimensional Hilbert space of complex functions defined on the interval $[0, T]$. This Hilbert space is usually called *signal space*. The properties of inner product and norm of signal space are very useful for calculation of signal energy and the distance between two signals. In accordance with (2.13) any signal $s(t)$ can be represented as a linear combination of basis functions. To form a basis these functions must satisfy the orthogonality condition:

$$\begin{aligned} &\int_0^T \psi_i(t)\psi_j(t)dt = K_i\delta_{ij}, \quad K_i \neq 0, \quad 0 \leq t \leq T, \quad i, j = 1, \ldots, N, \\ &\delta_{ij} = \begin{cases} 1, & i = j, \\ 0, & \mathrm{i} \neq \mathrm{j} \end{cases} \end{aligned} \tag{2.14}$$

If the basis functions are normalized, i.e. $K_i = 1, \quad i = 1, \ldots, N$, then the signal space is called an *orthonormal space* and the corresponding basis is called the *orthonormal basis*. Orthonormal basis is particularly easy to work with, as illustrated by the next orthonormal basis' properties:

1) If $\{\psi_i(t)\}_{i=1}^N$ is an orthonormal basis, then:

$$\left\| \sum_{i=1}^{N} a_i \psi_i(t) \right\|^2 = \sum_{i=1}^{N} |a_i|^2 \tag{2.15}$$

2)

$$\|s(t)\|^2 = \sum_{i=1}^{N} |\langle s(t), \psi_i(t)\rangle|^2 \tag{2.16}$$

The representation of a signal as a linear combination of orthonormal basis functions corresponds to the the representation of signal as a vector in orthonormal signal space. An example of this kind of representation in orthonormal signal space generated by the set of two basis functions $\{\psi_i(t)\}_{i=1}^2$ is depicted in Figure 2.7. The signal $s_1(t) = a_{11}\psi_1(t) + a_{12}\psi_2(t)$ corresponds to vector $\mathbf{s}_1 = (a_{11}, a_{12})$ and the signal $s_2(t) = a_{21}\psi_1(t) + a_{22}\psi_2(t)$ corresponds to vector $\mathbf{s}_2 = (a_{21}, a_{22})$.

Vector components a_{ij} are the projections of the signal $s_i(t)$ to the corresponding basis function $\psi_j(t)$:

$$a_{ij} = \int_0^T s_i(t)\psi_j(t)dt \tag{2.17}$$

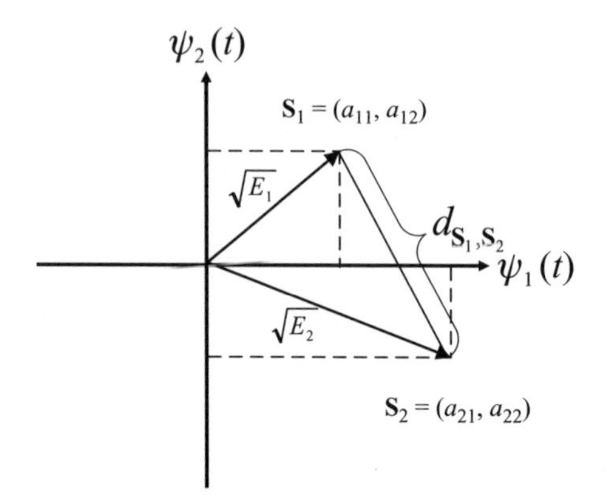

Figure 2.7 Signal representation as a vector in orthonormal signal space

As mentioned above the norm of signal $\|s_i(t)\|$ shows the length of the corresponding vector $\mathbf{s}_i$. Then the signal energy can be calculated in accordance with (2.15) as:

$$E_i = \|s_i(t)\|^2 = \sum_{i=1}^{N} \left|a_{ij}\right|^2 \tag{2.18}$$

The Euclidian distance between two signals $s_1(t)$ and $s_2(t)$ can be calculated in orthonormal signal space as follows:

$$d_{s_1(t),s_2(t)} = d_{\mathbf{s}_1,\mathbf{s}_2} = \|s_1(t) - s_2(t)\| = \sqrt{\sum_{i=1}^{N} (a_{1j} - a_{2j})^2} \tag{2.19}$$

The correlation between two signals $s_1(t)$ and $s_2(t)$, which is the measure of similarity between these two signals, is given by:

$$R_{s_1(t),s_2(t)} = \frac{\langle s_1(t), s_2(t)\rangle}{\|s_1(t)\|\,\|s_2(t)\|} = \frac{\langle s_1(t), s_2(t)\rangle}{\sqrt{E_1 E_2}} \tag{2.20}$$

The more typical case for the communication system is where $s(t)$ is one of M possible signals $s_1(t)$, $s_2(t), \ldots, s_M(t)$. Then signal $s(t)$ from the set $\{s_i(t)\}_{i=1}^{M}$ can be considered to belong to the N-dimensional orthonormal space, where $N \leq M$. The orthonormal basis $\{\psi_i(t)\}_{i=1}^{N}$ of this N-dimensional orthonormal space can be found with the help of the *Gram-Schmidt procedure* [7]:

1) Define $\psi_1(t) = \dfrac{s_1(t)}{\|s_1(t)\|} = \dfrac{s_1(t)}{\sqrt{E_1}}$

2) For $i = 2, \ldots, M$ compute $g_i = s_i(t) - \sum_{j=1}^{i-1} \langle s_i(t), \psi_j(t)\rangle\, \psi_j(t)$, $\begin{cases} \psi_i(t) = \dfrac{g_i}{\|g_i\|}, & \text{if } \|g_i\| \neq 0 \\ \text{stop}, & \text{if } \|g_i\| = 0 \end{cases}$

When the orthonormal basis is known the set of signals can be visualized with the help of a *signal space diagram* which is also often called a *constellation diagram* where the signals are represented in a plot of one basis function against another and so on, as in Figure 2.7. Of course, it helps to visualize the signals only in case the dimension of signal space is no more than 3. However, many signals in today's communication systems can be expressed in only 2-dimensional signal space [5] which makes the signal space diagram quite useful for visualization of signals.

Additive white Gaussian noise (AWGN) can be represented as a linear combination of basis functions in the same way as signals [5]. We will consider the zero-mean AWGN. The noise can be split in two components:

$$n(t) = \hat{n}(t) + \tilde{n}(t) \quad (2.21)$$

where $\hat{n}(t)$ is the noise projected to the signal space and $\tilde{n}(t)$ is the noise outside the signal space. In this case the signal is affected only by $\hat{n}(t)$. This splitting can be considered, for example, as a result of applying low-pass filtering at the receiver input which filters out the part of noise outside the signal bandwidth. Then in accordance with (2.13) $\hat{n}(t)$ can be expressed as:

$$\hat{n}(t) = \sum_{i=1}^{N} n_i \psi_i(t) \quad (2.22)$$

where $n_i = \langle n(t), \psi_i(t) \rangle$ and $\langle \tilde{n}(t), \psi_i(t) \rangle = 0$ for all $i = 1, \ldots, N$.

White noise has a two-sided power spectral density equal to a constant $\frac{N_0}{2}$, for all frequencies from $-\infty$ to ∞. Then the noise variance is given by:

$$\sigma^2 = \int_{-\infty}^{\infty} \frac{N_0}{2} dt = \infty \quad (2.23)$$

However, the variance of filtered noise is finite. For example, if we consider the noise projection to one of orthonormal functions, the variance of this filtered AWGN is as follows [5]:

$$\sigma_i^2 = \text{var}(n_i) = E\left\{\left[\int_0^T n(t)\psi_i(t)dt\right]^2\right\} = \frac{N_0}{2} \quad (2.24)$$

Then $\hat{n}(t)$ can be represented as vector $\hat{n}(t) = \mathbf{n} = (n_1, n_2, \ldots, n_N)$ each component of which n_i is independent zero-mean Gaussian random variable with variance σ_i^2.

2.1.3 Pulse Shaping

Consider the baseband waveforms for the representation of original digital information. As was mentioned in section 2.1.1 each symbol has the same time duration T. Then the easiest and compact in time domain way of representation of digital information in baseband signal is to use the rectangular pulses. Since any digital information can be regarded as sequence of bits we may assume that we deal with bitstream. Then, for example, bit 1 can be represented by positive rectangular pulse and bit 0 by negative rectangular pulse of amplitude A as it is depicted in Figure 2.8.

The use of rectangular pulses as was mentioned above leads to very compact representation of bitstream in time domain. The rectangular pulse is a *strictly time limited* signal, that is, all signal power is concentrated in some limited time interval. Outside this time interval the power of the signal is equal to zero. However, in frequency domain this signal occupies the infinite bandwidth. In other words rectangular pulse is not a *strictly bandlimited* signal, that is, the power of this signal is not concentrated in some limited frequency interval. The representation of signal in frequency domain $B(f)$ can be obtained from the time domain representation $b(t)$ with the help of Fourier transform [2]:

$$B(f) = \int_{-\infty}^{\infty} b(t) \cdot exp(-j2\pi f t)dt \quad (2.25)$$

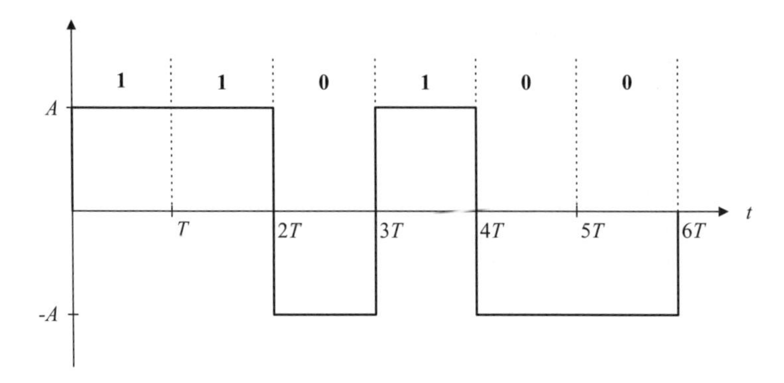

Figure 2.8 Representation of bitstream with rectangular pulses

It is easy to verify that rectangular function and sinc function forms the Fourier transform pair, where $\mathrm{sinc}(x) = \frac{\sin(\pi x)}{\pi x}$. That is, if signal in time domain is represented by the rectangular function (rectangular pulse) then in frequency domain it is represented by the sinc function, conversely if signal is represented by the sinc function in time domain (Nyquist pulse), in frequency domain it is represented by the rectangular function. Actually, any strictly time limited signal cannot be strictly bandlimited and vice versa signals that are strictly bandlimited cannot be strictly time limited [3]. The rectangular pulse in time domain is depicted in Figure 2.9a and in frequency domain in Figure 2.9b. The Nyquist pulse in time domain is depicted in Figure 2.10a and in frequency domain in Figure 2.10b.

The problem is that in real life the radio channel always is limited to some bandwidth due to different regulations. In this case the infinite bandwidth associated with a rectangular pulse is not acceptable. To cope with this problem it is possible to apply a low-pass filtering to the rectangular pulse. This filtering can change the shape of rectangular pulse and make it close to the shape of Nyquist pulse. This kind of filtering is called *pulse shaping*. On the other hand, as was discussed above any strictly bandlimited signal has the infinite time duration. This fact causes the effect called *intersymbol interference* (ISI). That means the pulses can overlap one another which results in distortion of the received symbols. However, it is possible to avoid the ISI with the help of one quite simple method. The idea of this method belongs to Nyquist [4]. It is not possible to get rid of ISI if the *Nyquist pulses* are transmitted since the span of each Nyquist pulse is infinite. However, as can be seen in Figure 2.11 if Nyquist pulses are shifted by integer multiples of symbol period T there are some points where ISI is zero.

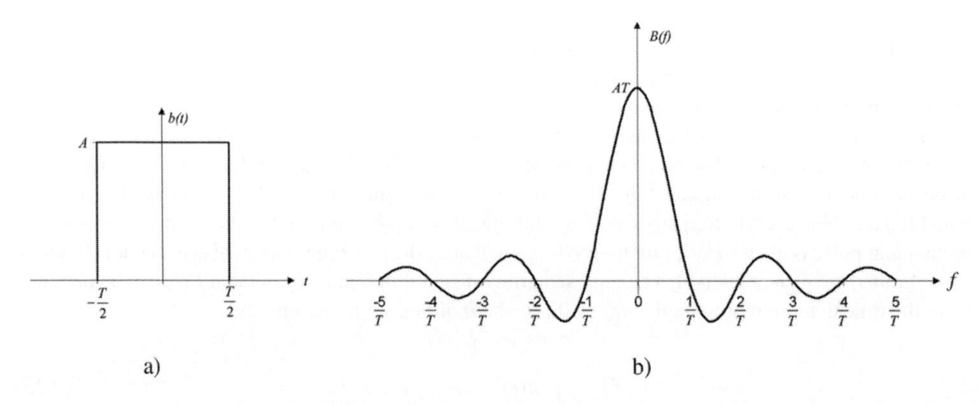

Figure 2.9 Example of strictly time limited signal. Rectangular pulse. a) Time domain. b) Frequency domain

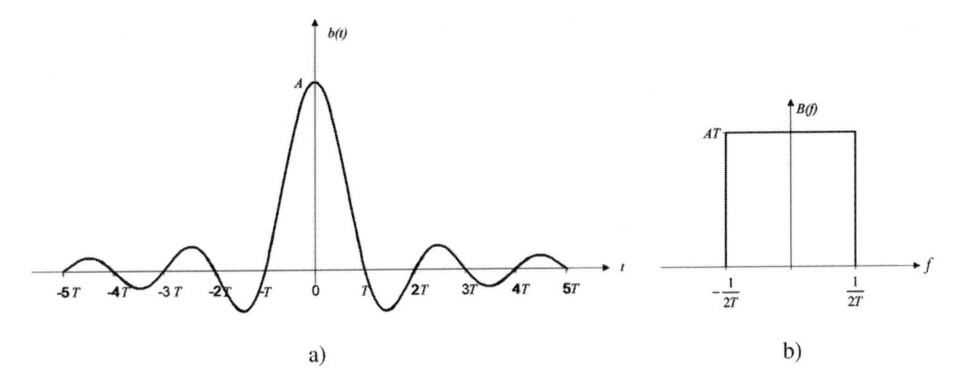

Figure 2.10 Example of strictly bandlimited signal. Nyquist pulse. a) Time domain. b) Frequency domain

Since the Nyquist pulse passes through zero at all points $t = \pm kT$, $k = 0, 1, 2, \ldots$ the ISI also is zero at these points if pulses are shifted by integer multiples of symbol period T. That means if the receiver will sample the received signal exactly in the centre of each symbol period (for example, at time $t = 2T$ for the first pulse and at $t = 3T$ for the second pulse depicted in Figure 2.11) the received symbol will not be affected by other symbols. Then, assuming that the sampling timing is perfect no ISI is introduced by using Nyquist pulses for information transmission. Actually, this effect can be reached not only with the help of Nyquist pulse but with any waveform $b(t)$ obeying the following *Nyquist criterion* in time domain:

$$b(kT) = \begin{cases} A, & k = 0 \\ 0, & k \neq 0 \end{cases} \tag{2.26}$$

The necessary and sufficient condition for $b(t)$ to satisfy (2.26) is that its Fourier transform $B(f)$ satisfy [3]:

$$\sum_{k=-\infty}^{\infty} B\left(f + \frac{k}{T}\right) = AT, |f| \leq \frac{1}{2T} \tag{2.27}$$

The condition in (2.27) is called the Nyquist criterion in frequency domain.

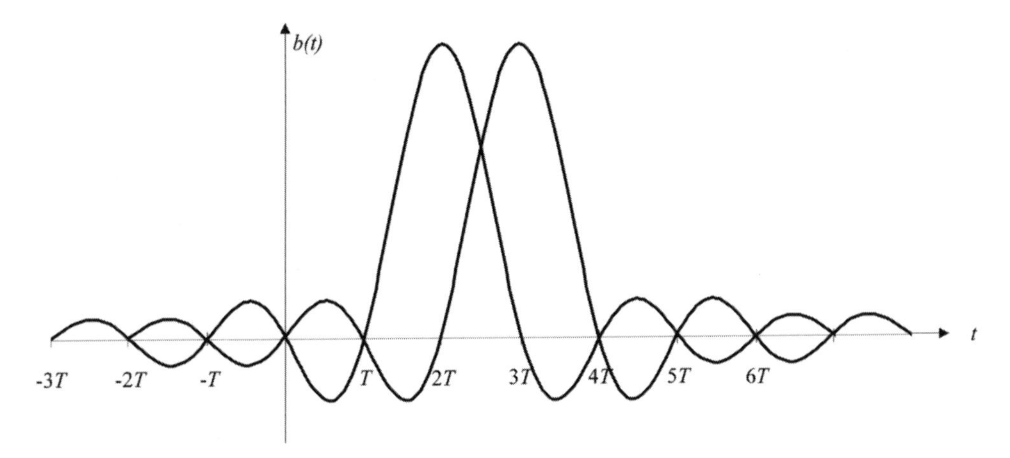

Figure 2.11 Nyquist pulses shifted by T

The bandwidth of Nyquist pulse as can be seen in Figure 2.10b $W = \frac{1}{T} = R_s$, where R_s is the symbol rate. Therefore, it is possible to provide the transmission with the help of Nyquist pulses without ISI by transmitting 1 symbol per second per hertz. Nyquist showed in his work [4] that it is the maximum possible transmitting rate per hertz to transmit signals without ISI. Actually for baseband channel the symbol rate per hertz $\frac{R_s}{W}$ called *symbol-rate packing* is twice that of bandpass channel because for baseband signal only the nonnegative part of signal in frequency domain can be considered [5].

As already mentioned above it is possible to transmit the information with the help of Nyquist pulses (or applying Nyquist filtering to rectangular pulses) without ISI only in the case of perfect synchronization of receiver and transmitter. Even a small error at the receiver sample timing leads to introducing quite a large amount of ISI due to the fact that the tails of Nyquist pulse have large amplitude near the main lobe. So, the communication system using Nyquist pulse provides the optimum bandwidth utilization but it suffers a lot from timing errors. On the other hand, the transmission with the help of rectangular pulses suffers much less from the synchronization errors. To decrease the impact of non-perfect synchronization and still use bandlimited signals it is necessary to find some trade-off between smoothness of signal shape and used bandwidth. This kind of trade-off provides *raised-cosine filter*. The raised-cosine filter belongs to the class of Nyquist filters, which means the impulse response of such a filter passes through zero at the same points as the Nyquist pulse. The impulse response $h(t)$ of raised-cosine filter is given by:

$$h(t) = \frac{\sin\left(\frac{\pi t}{T}\right)}{\frac{\pi t}{T}} \cdot \frac{\cos\left(\frac{\pi t \beta}{T}\right)}{1 - \left(\frac{2t\beta}{T}\right)^2} \tag{2.28}$$

and the frequency function of this filter is:

$$H(f) = \begin{cases} T, & |f| \leq \frac{1-\beta}{2T} \\ \frac{T}{2} \cdot \left[1 + \cos\left(\frac{\pi T}{\beta}\left(|f| - \frac{1-\beta}{2T}\right)\right)\right], & \frac{1-\beta}{2T} < |f| < \frac{1+\beta}{2T} \\ 0, & |f| \geq \frac{1+\beta}{2T} \end{cases} \tag{2.29}$$

where $0 \leq \beta \leq 1$ is the *roll-off factor*. The roll-off factor shows the excess bandwidth of the filter W in comparison with the bandwidth of Nyquist filter W_0 $\beta = \frac{W - W_0}{W_0}$.

Then the total bandwidth of raised-cosine filter is:

$$W = (1 + \beta) \cdot W_0 = \frac{1+\beta}{T} = (1 + \beta) \cdot R_s \tag{2.30}$$

Applying the raised-cosine filter to pulse shaping rather than Nyquist filter allows a trade-off between bandwidth and smoothness of the filter impulse response. The frequency function of raised-cosine filter for different values of roll-off factor is shown in Figure 2.12 and impulse response in Figure 2.13.

When roll-off factor $\beta = 0$ the raised-cosine filter coincides with the Nyquist filter and when $\beta = 1$ the frequency function comprises simply a cosine function raised above the abscissa. When roll-off factor is zero, (2.30) describes the minimum required bandwidth for pulse shaping providing zero ISI. As was mentioned previously in this case the synchronization should be perfect. If we would like to make the receiver more tolerant to synchronization errors and still provide zero ISI we have to sacrifice bandwidth by increasing the roll-off factor and in doing so we decrease the ripples both before and after the pulse interval. Since filtering is a linear operation the pulse shaping filtering can be applied to the output of bandpass modulation rather than to the output of the baseband modulation. Taking into account that usually in real life wireless communication systems the baseband and bandpass modulation are joined in one block called modulator, the pulse shaping is applied to the modulated signal at the output of the modulator.

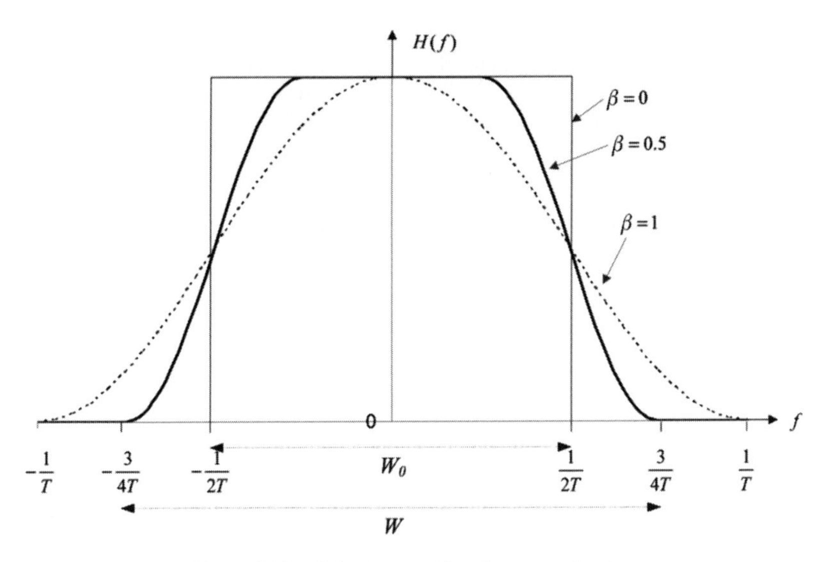

Figure 2.12 Raised-cosine filter frequency function

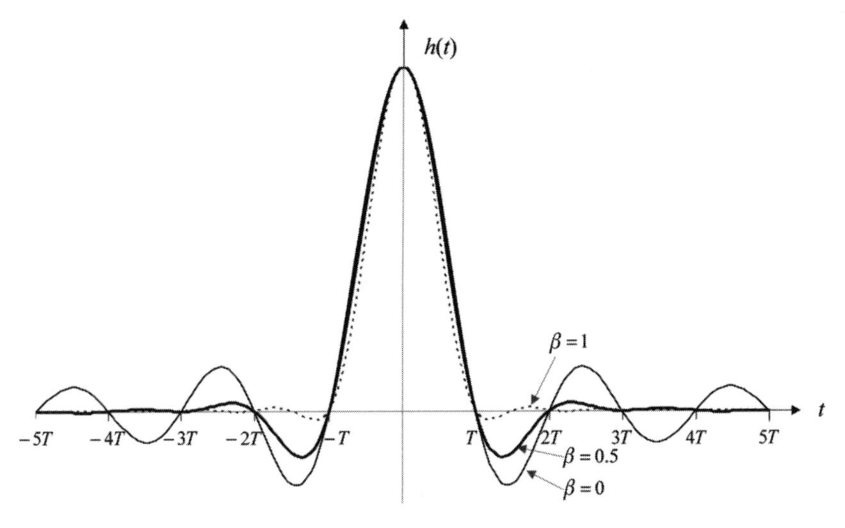

Figure 2.13 Raised-cosine filter impulse response

2.1.4 Matched Filter

The general view of typical wireless communication system is represented in Figure 2.14. As can be seen in Figure 2.14 the output of the modulator is filtered by the transmitter filter with frequency function H_{Tx}, then the signal is transmitted over a radio channel, which can be represented as a filter with frequency function H_C and after that the signal is filtered by the receiver filter with frequency function H_{Rx}. The transmitter filter is a bandlimiting (pulse shaping) filter and the aim of the receiver filter is to minimize the noise. Consider the extreme case when the channel frequency function is constant, the channel is a AWGN channel. The receiver filter is followed by a sampler that provides estimates of signal once per

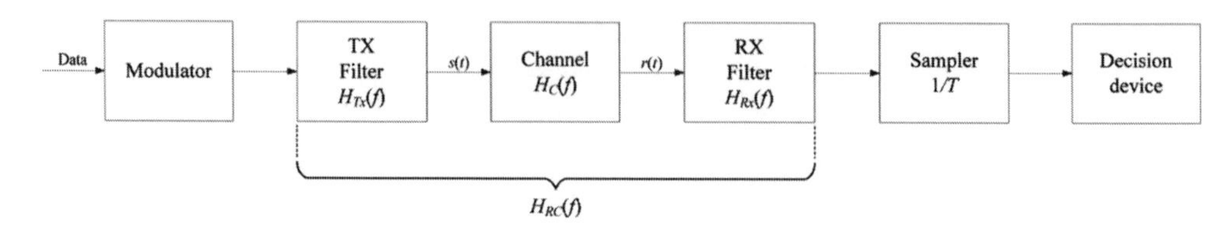

Figure 2.14 Wireless communication system as a cascade of filters

symbol period T. Then the task of the receiver filter is to maximize the signal to noise ratio (SNR) at the end of the symbol period $t = T$. That means the filter should maximize signal instantaneous power at time $t = T$ and minimize noise average power:

$$\mathrm{SNR} = \frac{s^2(T)}{\sigma_0^2} \tag{2.31}$$

where $s(T)$ is the transmitted signal at the output of the receiver filter at time $t = T$ and σ_0 is variance of filtered noise. Then the design of an optimum receiver filter maximizing SNR (2.31) can be derived as follows [5].

The signal at the filter output $s(t)$ can be expressed in terms of frequency function of receiver and transmitter filter:

$$s(t) = \int_{-\infty}^{\infty} H_{Tx}(f) H_{Rx}(f) \exp(j2\pi f t) df \tag{2.32}$$

The power of filtered noise can be written as follows:

$$\sigma_0^2 = \frac{N_0}{2} \int_{-\infty}^{\infty} |H_{Rx}(f)|^2 \, df \tag{2.33}$$

where N_0 is the noise power spectral density and the coefficient $\frac{1}{2}$ indicates that power spectral density is double sided. Then substituting (2.32) and (2.33) into (2.31) obtain:

$$\mathrm{SNR} = \frac{\left| \int_{-\infty}^{\infty} H_{Tx}(f) H_{Rx}(f) \exp(j2\pi f t) df \right|^2}{\frac{N_0}{2} \int_{-\infty}^{\infty} |H_{Rx}(f)|^2 \, df} \tag{2.34}$$

Now applying the Cauchy-Bunyakovsky-Schwarz inequality [6] stating that for any two functions $f_1(x)$ and $f_2(x)$:

$$\left| \int_{-\infty}^{\infty} f_1(x) f_2(x) dx \right|^2 \leq \int_{-\infty}^{\infty} |f_1(x)|^2 \, dx \int_{-\infty}^{\infty} |f_2(x)|^2 \, dx \tag{2.35}$$

to the nominator in (2.34) obtain the following estimate of SNR:

$$\mathrm{SNR} \leq \frac{\int_{-\infty}^{\infty} |H_{Tx}(f) \exp(j2\pi f t)|^2 \, df \int_{-\infty}^{\infty} |H_{Rx}(f)|^2 \, df}{\frac{N_0}{2} \int_{-\infty}^{\infty} |H_{Rx}(f)|^2 \, df} = \frac{2 \int_{-\infty}^{\infty} |H_{Tx}(f) \exp(j2\pi f t)|^2 \, df}{N_0} \tag{2.36}$$

Since the complex exponential $\exp(j2\pi ft)$ has the unit magnitude the integral in the nominator of (2.36) the energy of the transmitted signal $E = \int_{-\infty}^{\infty} |H_{Tx}(f)|^2\, df$. Thus, the SNR at the output of a receiver filter depends only on the transmitted signal energy E and the power spectral density of the noise N_0 and not on the shape of the signal. The equality in (2.35) holds if $f_1(x) = kf_2^*(x)$, where k is a constant. Then the maximum possible value of SNR can be obtained if:

$$H_{Rx}(f) = kH_{Tx}^*(f)\exp(-j2\pi fT) \tag{2.37}$$

that is, $|H_{Rx}(f)| = |H_{Tx}(f)|$ and the receiver filter maximizes the SNR at its output at time $t = T$ if the receiver frequency function matches the transmitter frequency function. Due to this fact the optimum receiver filter is called *matched filter*. In the time domain the matched filter impulse response obtained as inverse Fourier transform of (2.37) can be written as:

$$h_{Rx}(t) = \begin{cases} h_{Tx}(T-t), & 0 \le t \le T \\ 0, & \text{otherwise} \end{cases} \tag{2.38}$$

If the receiver filter is matched to the transmitter filter the equality in (2.36) holds and SNR at the output of matched filter in case of AWGN channel is:

$$\text{SNR} = \frac{2E}{N_0} \tag{2.39}$$

Actually the matched filter can be implemented in a correlator form rather than in a direct form provided by (2.38). Consider the output of the receiver filter:

$$z(t) = r(t)^* h_{Rx}(t) = \int_0^t r(u)\cdot h_{Rx}(t-u)du \tag{2.40}$$

where $r(t) = s(t) + n(t)$ is the received signal, $n(t)$ is the AWGN. Then substituting (2.38) in (2.40) obtain:

$$z(t) = \int_0^t r(u)\cdot h_{Tx}(T-t+u)du \tag{2.41}$$

and at the time $t = T$ (2.41) can be written as:

$$z(T) = \int_0^T r(u)\cdot h_{Tx}(u)du \tag{2.42}$$

That means at the end of the symbol period the output of the matched filter coincides with the output of the corresponding correlator. And in this sense the matched filter can be implemented as a correlator.

Now recall the pulse shaping. It helps to eliminate the ISI at the output of a pulse shaping filter. As was discussed previously, to make it possible to do this the filter frequency function should obey the Nyquist criterion (2.27). It is convenient to choose the raised-cosine filter as a pulse shaping filter. However, our goal is to eliminate the ISI at the output of a receiver filter rather than at the output of the transmitter filter. That means the frequency function of the cascade of all three filters (transmitter filter, channel and receiver filter) must be the frequency function of raised-cosine filter $H_{RC}(f)$ (2.29):

$$H_{RC}(f) = H_{Tx}(f)\cdot H_C(f)\cdot H_{Rx}(f) \tag{2.43}$$

In the special case where the channel is ideal, that is, $H_C(f) = 1, |f| \le W$, we have:

$$H_{RC}(f) = H_{Tx}(f)\cdot H_{Rx}(f) \tag{2.44}$$

On the other hand, the receiver filter should match the transmitter filter to provide maximum SNR:

$$H_{Rx}(f) = H_{Tx}^{*}(f) \tag{2.45}$$

Substituting (2.45) into (2.44) obtain:

$$|H_{Tx}(f)| = |H_{Rx}(f)| = \sqrt{|H_{RC}(f)|} \tag{2.46}$$

Thus, the overall raised-cosine spectral characteristic is split evenly between the transmitter filter and the receiver filter. This kind of filter is called *root raised-cosine filter*.

In case the channel is not ideal the receiver filter must also compensate the distortions inserted into the signal by the channel. These distortions also lead to ISI but this time channel-induced. Usually it is possible to cope with this problem by adding one more filter to the receiver design: an equalizing filter. Then the equalizing filter frequency function $H_E(f)$ should be the reversal to the channel frequency function:

$$H_E(f) = \frac{1}{H_C(f)} \tag{2.47}$$

In this case the overall frequency function of the cascade of channel and equalizer is 1 and (2.46) still holds true.

2.2 PSK

Phase shift keying (PSK) comprises the manipulation of a carrier's phase in accordance with the transmitted bitstream. The general expression for PSK is:

$$s_i(t) = h_{Tx}(t)\cos(\omega_c t + \phi_i),\ i = 1, \ldots, M \tag{2.48}$$

where $h_{Tx}(t)$ is the impulse response of the pulse shaping filter, M is the number of possible values of a phase and ϕ_i usually is chosen as:

$$\phi_i = \frac{2\pi i}{M} \tag{2.49}$$

In some cases it is more convenient to use more general expression of (2.49):

$$\phi_i = \frac{2\pi i}{M} + \phi_0 \tag{2.50}$$

where ϕ_0 is an arbitrary constant phase. An alternative expression of (2.48) is:

$$s_i(t) = \mathrm{Re}\left\{\exp(j\phi_i)h_{Tx}(t)\exp(\omega_c t)\right\},\ i = 1, \ldots, M \tag{2.51}$$

First consider some particular cases of PSK.

2.2.1 BPSK

For *binary phase shift keying (BPSK)* system $M = 2$ and $\phi_1 = 0$, $\phi_2 = \pi$ (or $\phi_1 = \phi_0$, $\phi_2 = \pi + \phi_0$). Then the corresponding signals are:

$$s_i(t) = \begin{cases} h_{Tx}(t)\cos(\omega_c t), & \text{if 1 is emitted} \\ -h_{Tx}(t)\cos(\omega_c t), & \text{if 0 is emitted} \end{cases} \tag{2.52}$$

Assuming that the pulse shaping filter has the impulse response in the form of rectangular pulse (2.52) can be written as:

$$s_i(t) = \begin{cases} A\cos(\omega_c t), & \text{if 1 is emitted} \\ -A\cos(\omega_c t), & \text{if 0 is emitted} \end{cases} \tag{2.53}$$

The average power of sinusoidal signal is $P = \frac{A^2}{2}$, so that $A = \sqrt{2P}$. Then (2.53) can be represented as:

$$s_i(t) = \pm\sqrt{2P}\cos(\omega_c t) = \pm\sqrt{\frac{2E}{T}}\cos(\omega_c t) \tag{2.54}$$

where $E = P \cdot T$ is the energy of signal per symbol period.

The example of BPSK signalling in this case is depicted in Figure 2.15a. Of course in real life this kind of signal is not possible to use since the bandwidth in this case is infinite. The BPSK signal after pulse shaping with raised-cosine filter is depicted in Figure 2.15b. The bandwidth in this case in accordance with (2.30) can be expressed as:

$$W = (1 + \beta) \cdot R_b \tag{2.55}$$

where R_b is bit rate. For BPSK one symbol comprises one bit and therefore bit rate coincides with symbol rate.

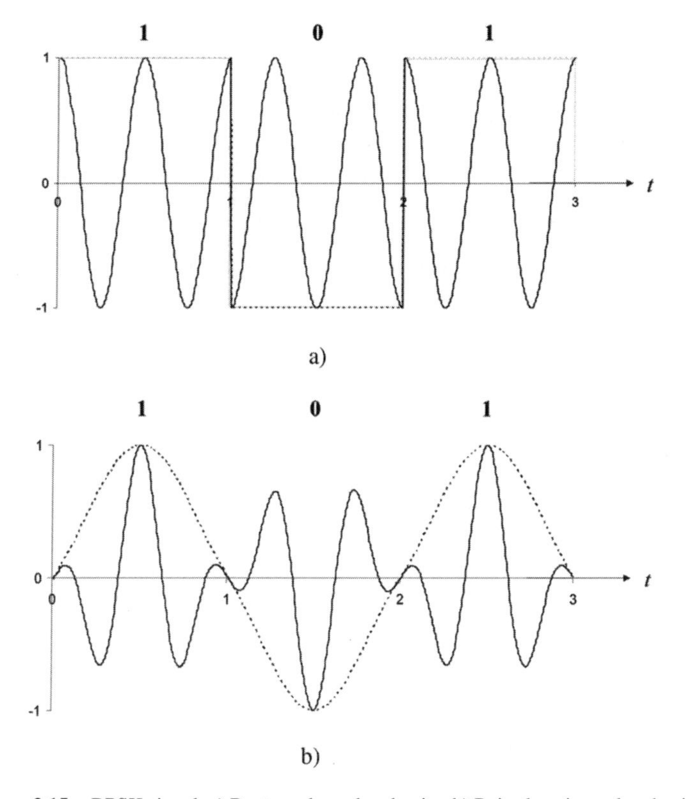

Figure 2.15 BPSK signal. a) Rectangular pulse shaping b) Raised-cosine pulse shaping

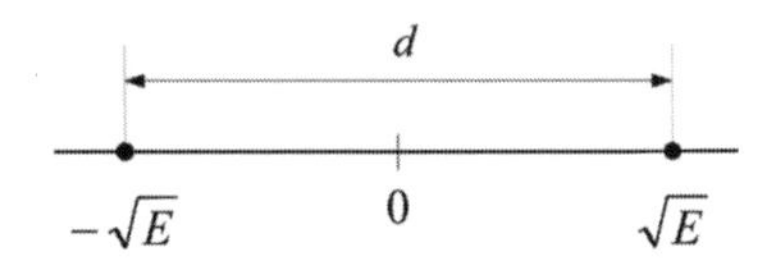

Figure 2.16 Constellation diagram of BPSK signals

Applying the Gram-Schmidt procedure to the set of signals $s_1(t)$, $s_2(t)$ obtains the orthonormal basis consisting of one basis function $\psi_1(t) = \sqrt{\frac{2}{T}} \cos(\omega_c t)$. Then the constellation diagram of the signals comprises just two dots in one line corresponding to function $\psi_1(t)$ as depicted in Figure 2.16.

As can be seen from Figure 2.16 the BPSK signals lie totally in one axis, there is no y-axis projection. The vector flip-flops on the x-axis depending on the value of the transmitted bit.

The BPSK modulator is represented in Figure 2.17. Depending on input data bit value one of two inputs multipliers is chosen $\sqrt{\frac{2E}{T}}$ or $-\sqrt{\frac{2E}{T}}$.

Two possible demodulation schemes for BPSK are depicted in Figure 2.18 and Figure 2.19.

The correlator demodulator using two correlators is depicted in Figure 2.18. In this case the received signal is correlated with two possible transmitted signals $s_1(t)$ and $s_2(t)$. Then the correlator output $z_i(T)$ with maximum value defines the decision about which signal was transmitted. It is possible to decrease the number of correlators to one if the received signal is correlated with basis function $\psi_1(t)$ rather than with the set of possible transmitted signals. This solution is depicted in Figure 2.19. In this case the correlator output $z(T)$ comprises the projection of the transmitted signal $s_i(t)$ to the basis function $\psi_1(t)$ corrupted by the AWGN. Then it is enough to check the sign of $z(T)$ to make the decision. The performance of both solutions is the same. Also the performance will not change if the correlators were replaced by the matched filters, as was shown in section 2.1.4 where a correlator can be regarded simply as a form of a matched filter. Note that in all these cases the coherent demodulation is used since the demodulator utilizes the modulated carrier signals as a reference. Now consider the correlation demodulator with one correlator. As was mentioned above the correlator output can be expressed as:

$$z(T) = a_i(T) + n_1(T) \tag{2.56}$$

where $a_i(T)$ is the projection of the transmitted signal to the function $\psi_1(t)$ and $n_1(T)$ is the projection of the Gaussian noise $n(t)$ to the same function. The filtered noise $n_1(t)$ is also a Gaussian process and $a_i(T)$ is a constant ($\sqrt{E}$ or $-\sqrt{E}$). That means the correlator output $z(T)$ is a random Gaussian variable

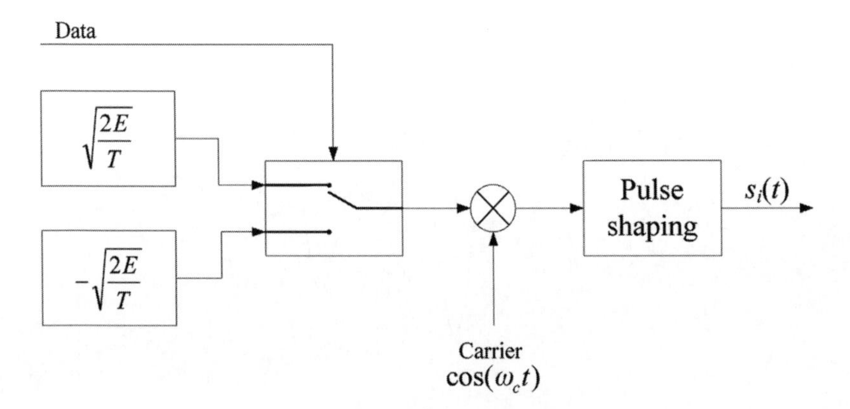

Figure 2.17 BPSK modulator

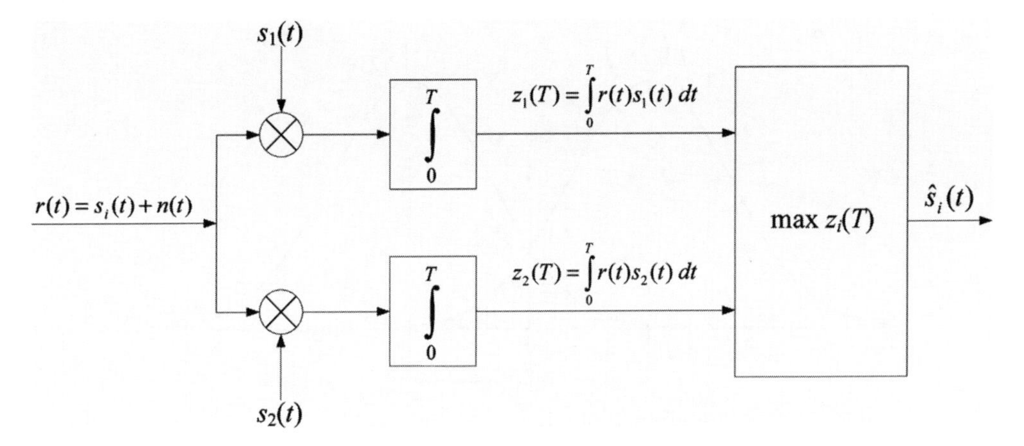

Figure 2.18 BPSK demodulator. Two correlators

with mean $\sqrt{E}$ or $-\sqrt{E}$. The corresponding conditional probability density functions can be written as follows:

$$p(z|s_1) = \frac{1}{\sigma\sqrt{2\pi}} \exp\left(-\frac{1}{2}\left(\frac{z-\sqrt{E}}{\sigma}\right)^2\right) \tag{2.57}$$

$$p(z|s_2) = \frac{1}{\sigma\sqrt{2\pi}} \exp\left(-\frac{1}{2}\left(\frac{z+\sqrt{E}}{\sigma}\right)^2\right) \tag{2.58}$$

where σ^2 is the noise variance. The plots of these pdfs are represented in Figure 2.20.

It can be seen from this plot that the demodulator makes the erroneous decision when the value of the correlator output $z(T)$ is less than zero, given the signal $s_1(t)$ was transmitted or greater than zero, given the signal $s_2(t)$ was transmitted. The probabilities of these events correspond to the shadowed areas in Figure 2.20. These probabilities can be expressed as follows:

$$P(s_1 \to s_2) = \int_{-\infty}^{0} p(z|s_1)\,dz \tag{2.59}$$

$$P(s_2 \to s_1) = \int_{0}^{\infty} p(z|s_2)\,dz \tag{2.60}$$

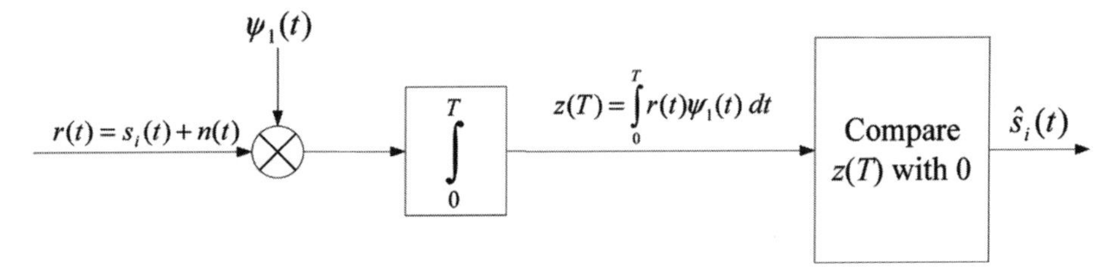

Figure 2.19 BPSK demodulator. 1 correlator

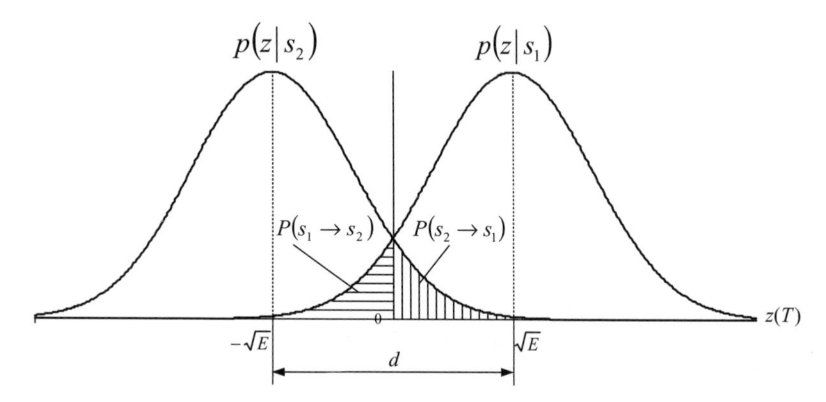

Figure 2.20 Conditional pdfs of the correlator output

Since the pdf $p(z|s_2)$ is just a shifted copy of $p(z|s_1)$, the probabilities of these two events are equal:

$$P(s_2 \to s_1) = P(s_1 \to s_2) \tag{2.61}$$

Then substituting (2.58) into (2.60) obtain:

$$P(s_2 \to s_1) = \int_0^{\infty} \frac{1}{\sigma\sqrt{2\pi}} \exp\left(-\frac{1}{2}\left(\frac{z+\sqrt{E}}{\sigma}\right)^2\right) dz = Q\left(\frac{\sqrt{E}-(-\sqrt{E})}{2\sigma}\right) = Q\left(\frac{d}{2\sigma}\right) \tag{2.62}$$

where d is the Euclidean distance between signals $s_1(t)$ and $s_2(t)$ in signal space and $Q(x)$ is the function defined as:

$$Q(x) = \int_x^{\infty} \frac{1}{\sqrt{2\pi}} \exp\left(-\frac{z^2}{2}\right) dz \tag{2.63}$$

The Q-function is related to the complementary error function:

$$Q(x) = \frac{1}{2}\operatorname{erfc}\left(\frac{x}{\sqrt{2}}\right) \tag{2.64}$$

Then taking in account (2.61) and that the probability of transmitting the signals $s_1(t)$ and $s_2(t)$ is equal to $\frac{1}{2}$ the symbol error probability for BPSK under the conditions of AWGN channel is:

$$P_e = P(s_1)P(s_1 \to s_2) + P(s_2)P(s_2 \to s_1) = \frac{1}{2}Q\left(\frac{d}{2\sigma}\right) + \frac{1}{2}Q\left(\frac{d}{2\sigma}\right) = Q\left(\frac{d}{2\sigma}\right) = Q\left(\sqrt{\frac{2E}{N_0}}\right) \tag{2.65}$$

Since the symbol in BPSK comprises one bit the bit error probability P_b coincides with P_e.

2.2.2 QPSK

The *quadrature phase shift keying (QPSK)* is a simple extension of the BPSK. In this case $M = 4$ and $\phi_i = \frac{\pi i}{2}$, $(i = 1, \ldots, 4)$. Assuming the rectangular pulse shaping the QPSK signal can be written as:

$$s_i(t) = \sqrt{\frac{2E}{T}} \cos\left(\omega_c t + \frac{\pi i}{2}\right) \tag{2.66}$$

Applying to (2.66) the trigonometric identity $\cos(x + y) = \cos x \cos y - \sin x \sin y$ obtain:

$$s_i(t) = \sqrt{\frac{2E}{T}} \left[\cos(\omega_c t) \cos\left(\frac{\pi i}{2}\right) - \sin(\omega_c t) \sin\left(\frac{\pi i}{2}\right)\right] \tag{2.67}$$

It is easy to verify with the help of the Gram-Schmidt procedure that the set of functions:

$$\begin{aligned} \psi_1(t) &= \sqrt{\frac{2}{T}} \cos(\omega_c t), \\ \psi_2(t) &= \sqrt{\frac{2}{T}} \sin(\omega_c t) \end{aligned} \tag{2.68}$$

forms the orthonormal basis for QPSK signals. The constellation diagram of QPSK signals corresponding to (2.67) is depicted in Figure 2.21. With this choice of basis functions the projection of signal to function $\psi_1(t)$ corresponds to the real (Inphase) component of the signal and the projection to function $\psi_2(t)$ corresponds to the imaginary (Quadrature) component.

In accordance with (2.50) the QPSK signals can be written in the following form:

$$s_i(t) = \sqrt{\frac{2E}{T}} \cos\left(\omega_c t + \frac{\pi i}{2} + \frac{\pi}{4}\right) \tag{2.69}$$

The corresponding constellation diagram is represented in Figure 2.22. As can be seen the constellation diagram depicted in Figure 2.22 is simply the rotation of the constellation diagram in Figure 2.21. This rotation does not change any properties of the corresponding system. And both constellations represent the QPSK. However, more often the representation depicted in Figure 2.22 is used.

Two possible methods of QPSK modulator design are represented in Figure 2.23 and Figure 2.24.

In the method corresponding to the polar representation of QPSK modulation first the phase shift of the carrier signal $\phi_i = \frac{\pi i}{2} + \frac{\pi}{4}$, $(i = 1, \ldots, 4)$ is calculated. Since there are four possible phase shifts in

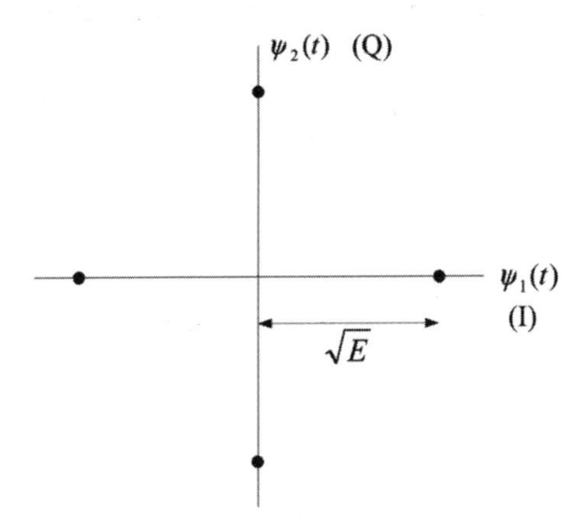

Figure 2.21 QPSK constellation diagram

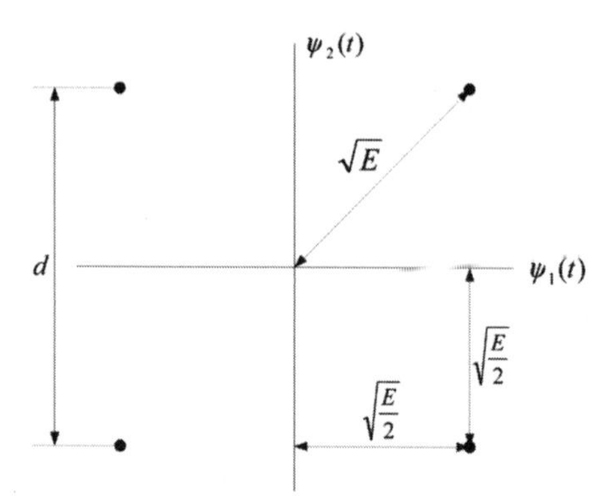

Figure 2.22 Another form of QPSK constellation diagram

QPSK each 2 bits of the input data stream defines the transmitted signal that is the phase shift ϕ_i. Then the phase modulator provides the phase shift of the carrier signal. This is the most straightforward method of the implementation of the QPSK modulation but it requires more operations (since it is necessary to change directly the carrier's phase) than the quadrature method. In the quadrature method the concept of I and Q channels is used. That is the modulated signal is obtained as a sum of its projections to basis functions. It is easy to verify that each channel I or Q comprises simply the BPSK modulation. The example of signals corresponding to the QPSK I and Q channels are depicted in Figure 2.25a and b. The QPSK signal $s_i(t)$ depicted in Figure 2.25c can be obtained as $s_i(t) = \mathrm{I}(t) - \mathrm{Q}(t)$.

It can be seen from Figure 2.22 that the constellation points of QPSK can be represented as two orthogonal sets of BPSK constellation points. The quadrature method is easier to implement and for this reason it is more frequently used in real life QPSK modulators.

The QPSK demodulator can be designed with the help of the same schemes which were discussed above for BPSK demodulation (depicted in Figures 2.18 and 2.19). In this case the demodulator based on the correlation of the set of possible transmitted signals requires four correlators (or matched filters) and the demodulator based on the correlation of the basis functions requires only two correlators.

The error probability per symbol in AWGN channel can be calculated in the same way as for QPSK. As can be seen from Figure 2.22 the minimum Euclidean distance between two signals $d = 2\sqrt{\frac{E}{2}}$. Then the probability that the demodulator makes the incorrect decision due to the fact that noise displaces the

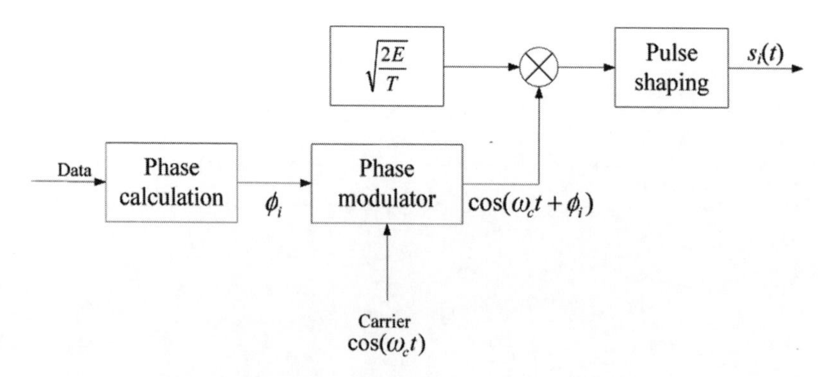

Figure 2.23 Polar form of QPSK modulation

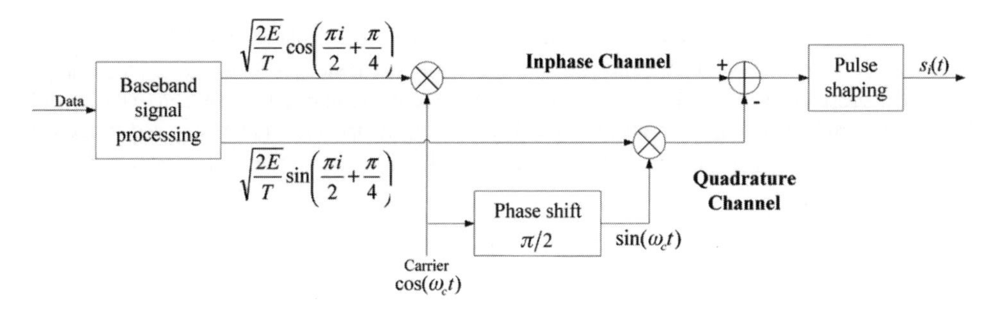

Figure 2.24 Quadrature form of QPSK modulation

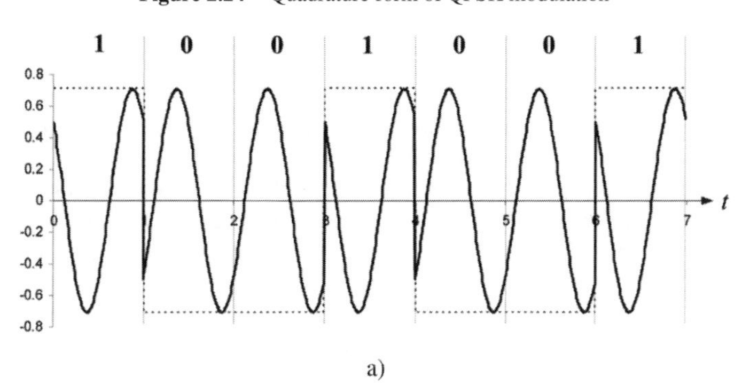

a)

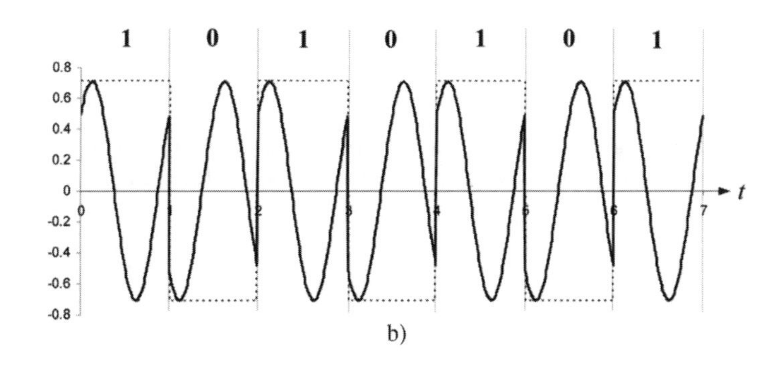

b)

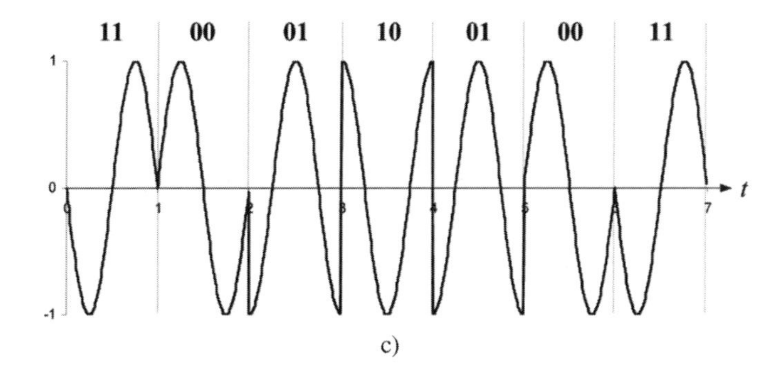

c)

Figure 2.25 Representation of QPSK as a sum of I and Q channels: a) I channel, b) Q channel, c) QPSK signal (I – Q)

received signal to the nearest signal across one of the axis can be calculated as $Q\left(\frac{d}{2\sigma}\right) = Q\left(\sqrt{\frac{E}{N_0}}\right)$. However, the error can occur if noise displaces the received signal across any of two axes rather than just one as in the case of BPSK. Taking into account that the probability of displacement across the axis I is the same as across the axis Q the probability that both I and Q component will be estimated correctly is $\left(1 - Q\left(\frac{d}{2\sigma}\right)\right)^2 = 1 - 2Q\left(\frac{d}{2\sigma}\right) + Q\left(\frac{d}{2\sigma}\right)^2$. Then the overall symbol error probability is as follows:

$$P_e = 2Q\left(\frac{d}{2\sigma}\right) - Q\left(\frac{d}{2\sigma}\right)^2 \tag{2.70}$$

Usually the term $Q\left(\frac{d}{2\sigma}\right)^2$ is almost negligible and more often (2.70) is used in the form of:

$$P_e \approx 2Q\left(\frac{d}{2\sigma}\right) = 2Q\left(\sqrt{\frac{E}{N_0}}\right) \tag{2.71}$$

The bit error probability depends on the code used to put in correspondence the bits and signals, that is, on labelling the constellation points. For example, the labelling depicted in Figure 2.26a is not the best one since in this case if the signal was incorrectly detected as the adjacent one, two bits can be corrupted. In Figure 2.26b the constellation points are labelled in accordance with the *Gray code* providing the Hamming distance of 1 between adjacent code words. This is used to make sure that adjacent symbols differ by only a single digit, and in this case the symbol error (the probability of misinterpreting the symbol which is not directly adjacent is negligible) causes the corruption of only 1 bit.

Of course the labeling, with the help of the Gray code, is preferable since it provides the smallest bit error probability (BER). Assuming that constellation points are labelled in accordance with Gray code the bit error probability is on average half the symbol error probability:

$$P_b = \frac{P_e}{2} \approx Q\left(\sqrt{\frac{E}{N_0}}\right) \tag{2.72}$$

Note that for BPSK the energy per symbol is the same as the energy per bit. For QPSK, the energy per symbol is twice the energy per bit, because there are two bits per symbol $E = 2E_b$. Then the QPSK BER

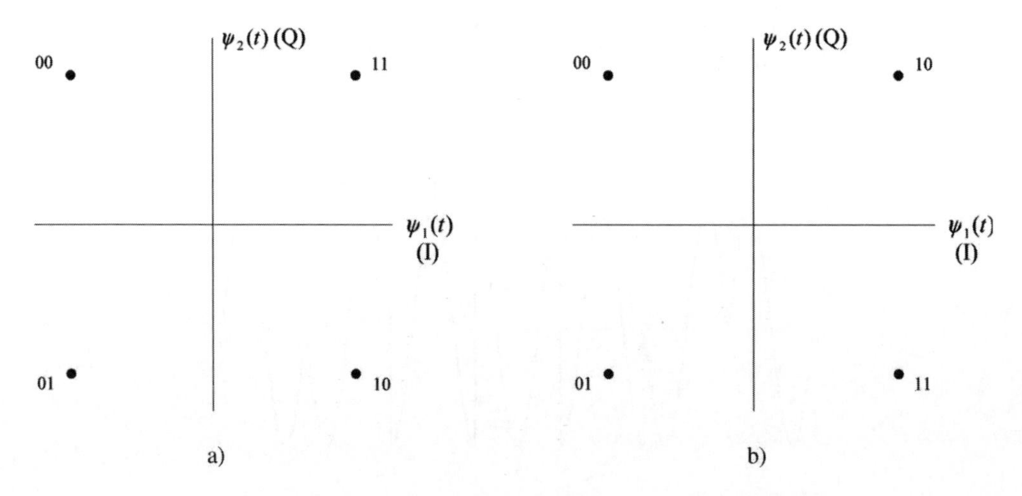

Figure 2.26 Different labeling of QPSK constellation points

can be expressed as follows:

$$P_b = \frac{P_e}{2} \approx Q\left(\sqrt{\frac{2E_b}{N_0}}\right) \tag{2.73}$$

That means QPSK provides the same BER as BPSK. This has not come as a surprise taking in account the fact that the QPSK signals can be regarded as a set of two orthogonal BPSK signals. On the other hand, this fact shows that BPSK utilizes the bandwidth in quite an inefficient way since the QPSK provides the same BER with twice the bit rate.

2.2.3 M-PSK

The general expression for M-PSK signals is given by (2.48)–(2.50). It can be expressed in the same way as for QPSK with the help of a set of the same two basis functions defined in (2.68) or I and Q channels:

$$s_i(t) = \sqrt{E}\left[\psi_1(t)\cos\left(\frac{2\pi i}{M}\right) - \psi_2(t)\sin\left(\frac{2\pi i}{M}\right)\right], i = 1, \ldots, M \tag{2.74}$$

The example of 8-PSK constellation is depicted in Figure 2.27.

Taking into account (2.48)–(2.50) and (2.74) the modulators depicted in Figures 2.23–2.24 can be used (with little changes) for modulation of M-PSK signals. In the same way the demodulators depicted in Figures 2.18–2.19 can be generalized for demodulation of M-PSK signals. In the case of implementing the demodulator depicted in Figure 2.18 the bank of M correlators (matched filters) should be used. In the case of the correlation of basis functions only two correlators are needed, but the complexity of logic block selecting the signal which better corresponds to the output of basis function correlators increases with increasing M.

The symbol error probability under the conditions of AWGN channel can be calculated with the help of the same technique as was used for calculation error probability for QPSK. In the case of M-PSK each constellation point has two adjacent points and it is possible to assume that the probability of misinterpreting the symbol which is not directly adjacent is negligible (however, this assumption is not

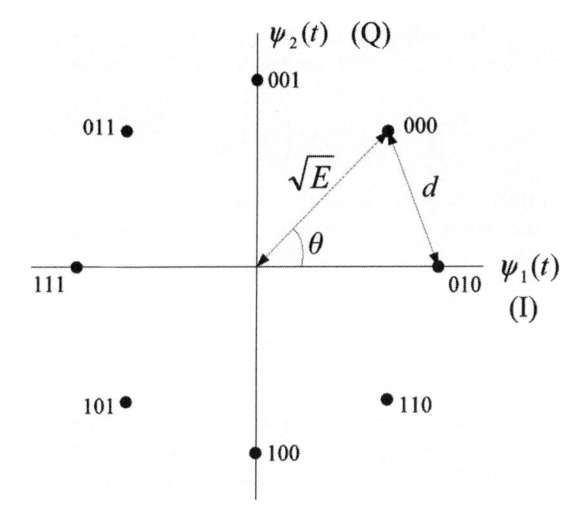

Figure 2.27 8-PSK constellation

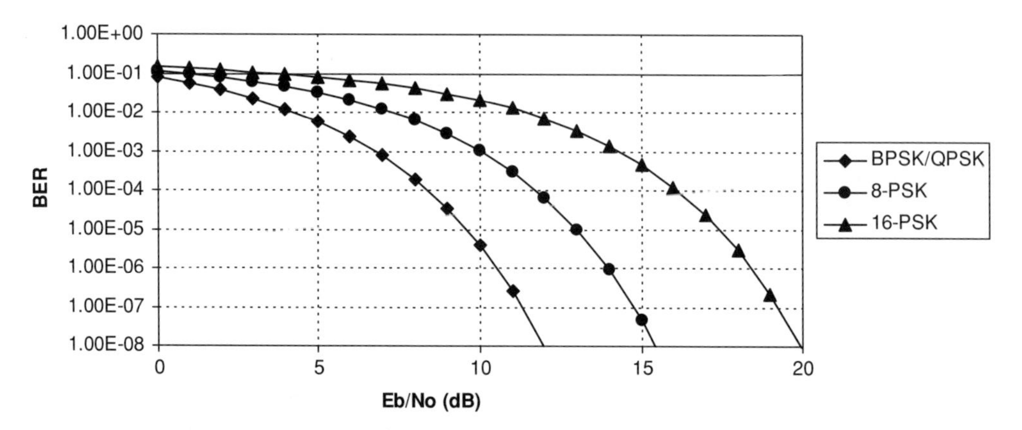

Figure 2.28 BER curves for BPSK, QPSK, 8-PSK and 16-PSK

correct for high values of M). Then the symbol error probability can be calculated in the same way:

$$P_e \approx 2Q\left(\frac{d}{2\sigma}\right) \tag{2.75}$$

As can be seen from Figure 2.27 the Euclidean distance between the adjacent constellation points can be expressed as:

$$d = 2\sqrt{E}\sin\left(\frac{\theta}{2}\right) \tag{2.76}$$

where θ in accordance with (2.49) is $\frac{2\pi}{M}$. Then (2.75) can be written as:

$$P_e \approx 2Q\left(\sin\left(\frac{\pi}{M}\right)\sqrt{\frac{2E}{N_0}}\right) = 2Q\left(\sin\left(\frac{\pi}{M}\right)\sqrt{\frac{2E_b\log_2 M}{N_0}}\right) \tag{2.77}$$

Assuming that the constellation points are labelled with the help of Gray code the symbol error (that is, misinterpreting the correct symbol as the adjacent one) causes the corruption of only 1 bit out of $\log_2 M$ bits comprising the symbol. Therefore, the BER for M-PSK can be estimated as:

$$P_b = \frac{P_e}{\log_2 M} \approx \frac{2}{\log_2 M}Q\left(\sin\left(\frac{\pi}{M}\right)\sqrt{\frac{2E_b\log_2 M}{N_0}}\right) \tag{2.78}$$

In practice the high-order PSK with $M > 8$ is not used since the BER becomes too high and in this case the usage of for example, quadrature amplitude modulation is more beneficiary. The comparison of PSK schemes performance is represented in Figure 2.28.

2.2.4 DPSK

The methods of PSK demodulation considered above represent the coherent demodulation. That means the demodulator exploits the carrier reference to detect the signal. To be more precise the demodulator utilizes the knowledge of the carrier signal phase and then calculates the phase shift between the modulated signal and the carrier. It is possible to decrease the demodulator implementation complexity if using the noncoherent demodulation. In this case the demodulator uses the previous symbol as a phase reference for the current symbol and the knowledge of the carrier phase is not required. To make the

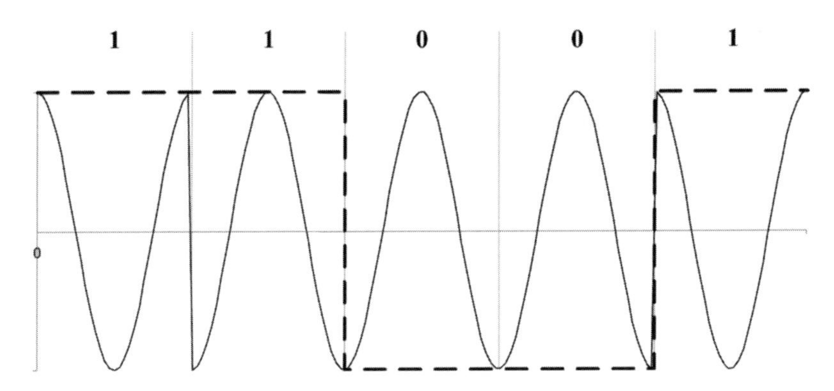

Figure 2.29 DBPSK as an example of differential encoding

signal independent of the carrier phase the transmitted data is *differentially encoded*. The differential encoding consists in changing the phase of the current transmitting signal in accordance with the input bitstream relative to the phase of the previous transmitted signal rather than to the phase of the carrier. For example, in differentially encoded BPSK the transmission of a binary "1" corresponds to the phase shift of the current signal by π relative to the previous signal and in the case of transmitting a binary "0" the phase of the signal remains the same as it was in the previous signal. This kind of modulation is called the *differential phase shift keying (DPSK)*. The example of DBPSK signal is represented in Figure 2.29.

Actually it is possible to use the coherent demodulation for the detection of the DPSK. In this case the same demodulator as for demodulation of PSK can be used to detect the signal and then transmitted data can be derived by comparing the phase of the current signal with the phase of the previous one. The performance of this method is a little bit worse than the performance of the coherent demodulation of the PSK. This small degradation is explained by differential encoding, since single detection error can cause two decision errors. The complexity of the demodulator is slightly increased due to the fact that the additional comparator is required to detect the phase difference between two successive signals. However, this method is quite often used to prevent the error propagation due to carrier recovery failure [8].

More often the noncoherent demodulation is used for the DPSK detection. The noncoherent demodulation of DPBSK is depicted in Figure 2.30.

As can be seen from Figure 2.30 the delayed version of the received signal is used as a reference signal and the correlator output $z(T) = \int_0^T r(t)r^*(t-T)dt$ is positive if the current signal has the same phase as the previous one (i.e., 0 was transmitted) and negative otherwise (i.e., 1 was transmitted). Actually the complex conjugate block is not needed since for BPSK $r^*(t) = r(t)$ but for the higher order modulations

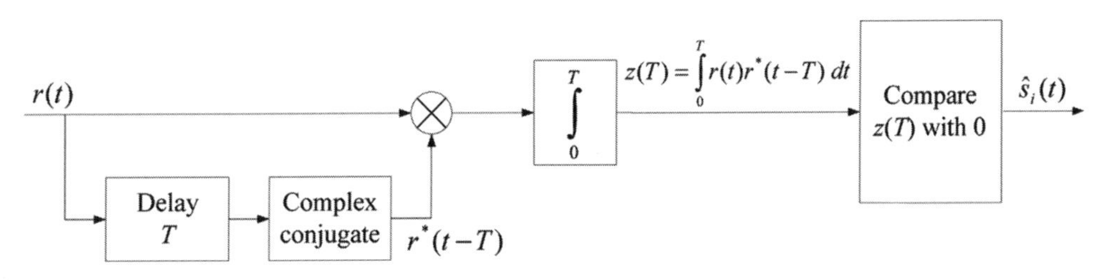

Figure 2.30 DBPSK demodulator

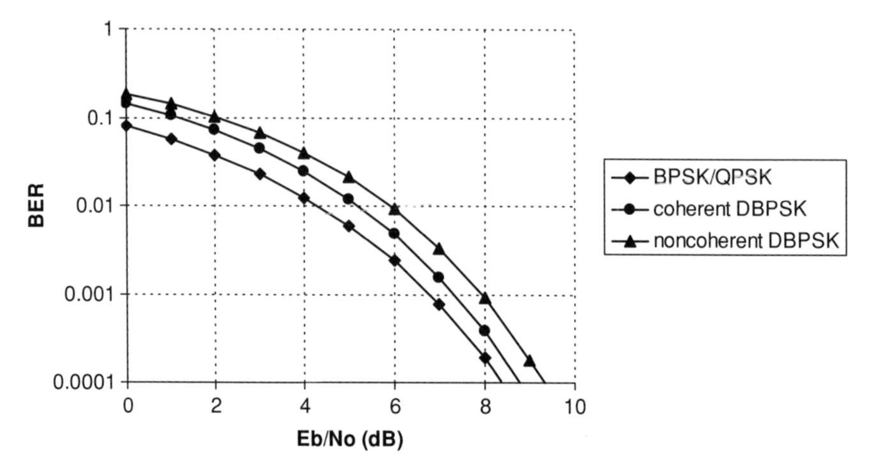

Figure 2.31 BER curves for BPSK/QPSK, coherent DBPSK and noncoherent DBPSK

(DQPSK, D8PSK, etc.) the complex conjugate of the delayed reference signal should be used. In this case the output of the correlator will correspond to the phase difference of received signal and the delayed version of the signal and the comparison should be done with several thresholds. Still the implementation complexity of the noncoherent demodulation remains quite low in comparison with the complexity of the coherent demodulation. The symbol error probability under the conditions of AWGN channel for the coherent detection of differential encoded PSK is given by [3]:

$$P_e = P_b = 2Q\left(\sqrt{\frac{2E_b}{N_0}}\right) \cdot \left(1 - Q\left(\sqrt{\frac{2E_b}{N_0}}\right)\right) \tag{2.79}$$

And the symbol error probability for the noncoherent detection of DBPSK is given by [3]:

$$P_e = P_b = \frac{1}{2}\exp\left(-\frac{E_b}{N_0}\right) \tag{2.80}$$

The BER for coherent and noncoherent detection in AWGN are plotted in Figure 2.31. As can be seen from these plots the performance of noncoherent detection of DBPSK is less than 1 dB worse than coherent detection of differentially encoded BPSK and about 2 dB worse than the performance of coherent detection of BPSK/QPSK at low values of SNR, with high SNR values the difference in performance becomes less visible.

2.2.5 OQPSK

Offset quadrature shift keying (OQPSK) is a minor variation of QPSK. In OQPSK the I and Q channels are shifted by half symbol and thus I and Q channel signals do not transition at the same time. It is done to decrease the signal envelope variations. The reason for this is the nonlinearity of the amplifier. In most cases the amplifier transfer function is linear only in some range of amplitude. That means the amplifier inserts some distortion in signal when crossing the boundary of this range. The more the signal amplitude varies, the more distorted is the signal at the amplifier output. Due to this fact the most beneficial is the usage of the signals with constant envelope. The example of I and Q channel signals of OQPSK is represented in Figure 2.32.

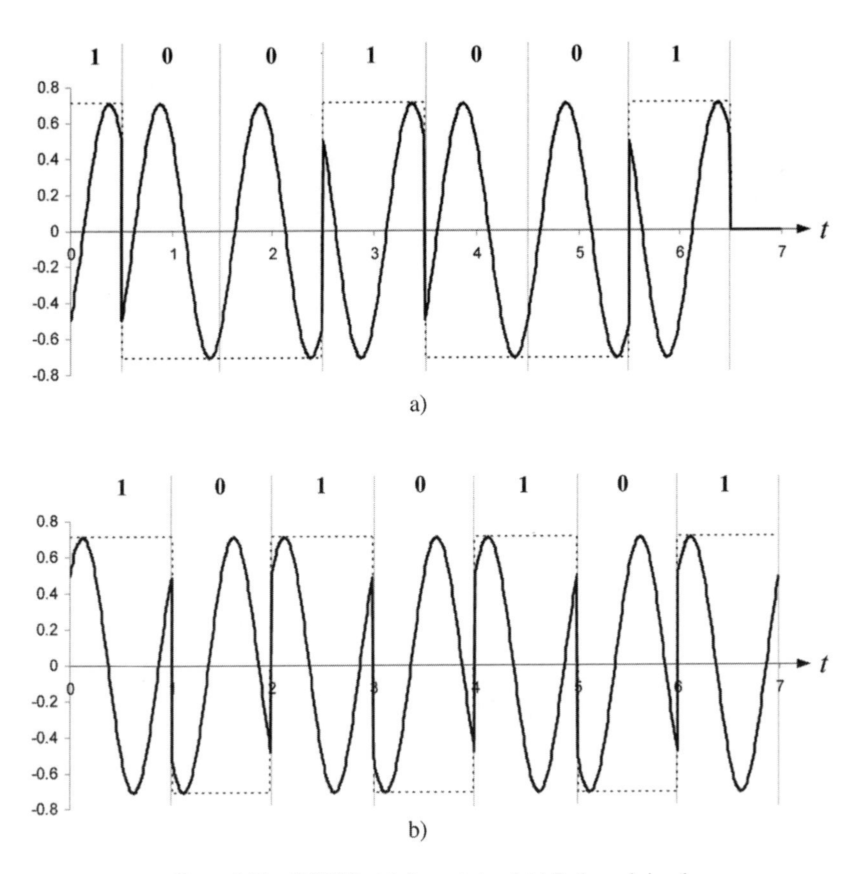

Figure 2.32 OQPSK: a) I channel signal, b) Q channel signal

As can be seen from Figure 2.32 the transitions in I and Q channels do not occur simultaneously. That means the only possible transition in constellation diagram is the transition to the adjacent symbol. This is depicted in Figure 2.33. Comparing the possible transitions in QPSK constellation diagram and OQPSK constellation diagram one can see that in QPSK the transitions between all symbols are possible, whereas in OQPSK the transitions crossing zero point do not exist.

The examples of QPSK and OQPSK signals are shown in Figure 2.34. The phase shifts by π in QPSK signal are highlighted. As can be seen the OQPSK signal is smoother than the QPSK. The maximum possible phase change in OQPSK signal is $\frac{\pi}{2}$. Unlike the QPSK where transitions can occur in every symbol, in OQPSK they can occur more frequently: every half-symbol.

The performance of OQPSK coincides with the performance of the QPSK if the linear amplifier is used in the transmitter. In cases where the nonlinear amplifier is used the OQPSK provides better results. The obvious problem with using OQPSK is that it implies more strict limitations on symbol timing recovery due to staggering of I and Q channels. Sometimes OQPSK is called staggered QPSK (SQPSK).

2.2.6 π/4-QPSK

The $\pi/4$-QPSK comprises two QPSK rotated by the angle $\pi/4$. The $\pi/4$-QPSK solves the same problem as OQPSK: decreasing the phase shifts to make the signal smoother. The constellation diagram of

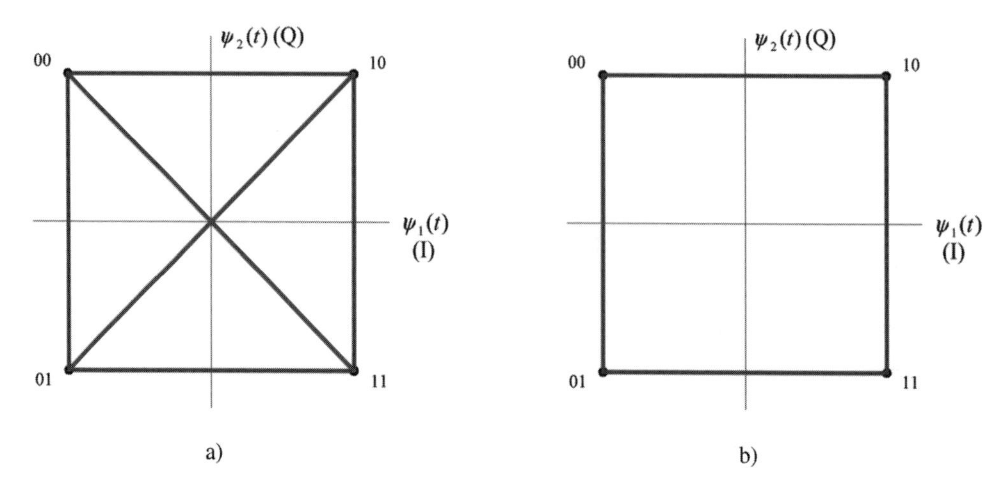

Figure 2.33 Possible transitions between symbols: a) in QPSK, b) in OQPSK

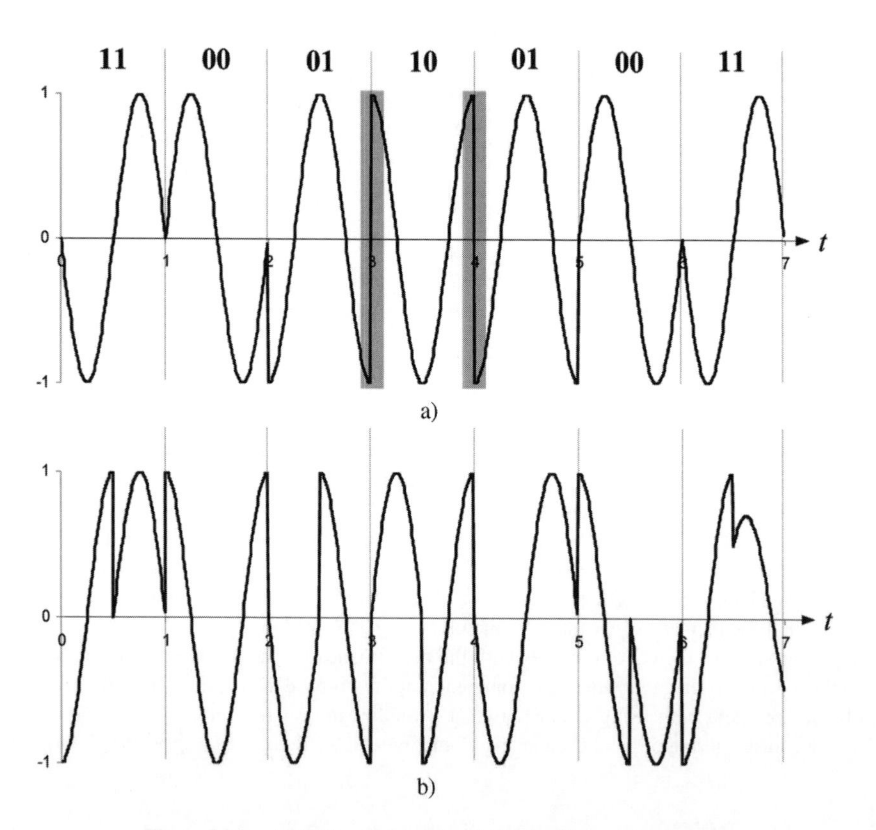

Figure 2.34 a) QPSK modulated signal. b) OQPSK modulated signal

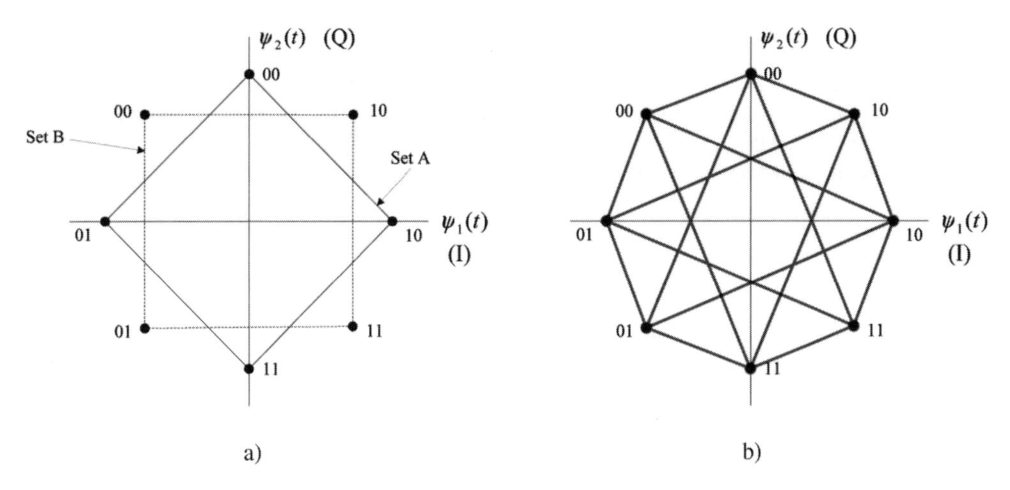

Figure 2.35 $\pi/4$-QPSK. a) Two constellation sets, b) $\pi/4$-QPSK allowed transitions

$\pi/4$-QPSK depicted in Figure 2.35a looks like 8-PSK constellation but in the same way as in OQPSK not all possible transitions existing in 8-PSK are allowed in $\pi/4$-QPSK. Unlike 8-PSK where we have eight possible symbols (or signals) to transmit in $\pi/4$-QPSK we have two sets of four signals and these sets are used in turn. For example, if the sequence 10 00 11 01 ... is transmitted the first symbol 10 is transmitted with the help of set A, the second 00 with the help of set B, third 11 with the help of set A, the fourth with the help of set B and so on. Since two QPSK constellation sets are used in an interleaved way the symbol rate and bit rate and BER for $\pi/4$-QPSK is exactly the same as for QPSK.

The possible transitions in $\pi/4$-QPSK are represented in Figure 2.35b. Since only transitions between set A and set B are allowed the transitions do not cross the centre of the constellation diagram. Due to this fact the phase transitions are less than in QPSK and the signal is smoother. Unlike the OQPSK the $\pi/4$-QPSK does not add restrictions to symbol timing recovery but the carrier recovery may suffer from the increased constellation size.

The example of $\pi/4$-QPSK signal is represented in Figure 2.36.

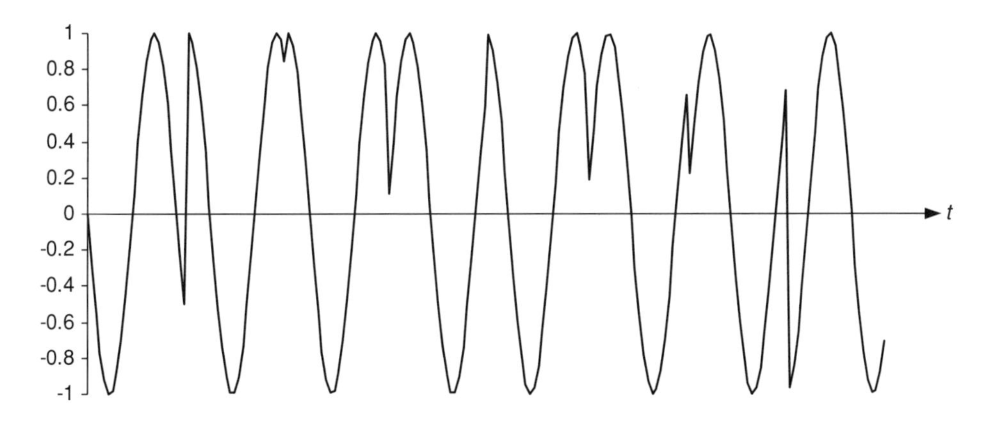

Figure 2.36 $\pi/4$-QPSK modulated signal

2.3 MSK

Minimum shift keying (MSK) can be considered as a modification of the frequency shift keying (FSK) or as a modification of OQPSK due to the dual nature of FSK and PSK. Here we will consider MSK as the evolution of OQPSK. MSK can be considered as OQPSK with one significant change: each channel I and Q is pulse shaped with a half-sinusoid. The impulse response $h(t)$ of the pulse shaping filter is given by:

$$h(t) = A \cdot \text{rect}\left(\frac{t-T}{2T}\right) \cdot \sin\left(\frac{\pi t}{2T}\right) \tag{2.81}$$

where rect () is the rectangular function, $\text{rect}(x) = \begin{cases} 0, & \text{if } |x| \geq 0.5 \\ 1, & \text{if } |x| < 0.5 \end{cases}$. Therefore, the period of pulse shaping is twice the symbol period and the lag between I and Q channel is T. That means the bit rate of MSK is twice as low as the QPSK bit rate, that is, MSK can be regarded as the binary modulation scheme rather than the quadrature one. The MSK signal can be written as follows:

$$s_i(t) = a_{i,\text{I}}(t)\cos\left(\omega_c t + \frac{\pi}{4}\right)\cos\left(\frac{\pi t}{2T}\right) - a_{i,\text{Q}}(t)\sin\left(\omega_c t + \frac{\pi}{4}\right)\sin\left(\frac{\pi t}{2T}\right) \tag{2.82}$$

where $a_{i,\text{I}}(t)$ takes values $\pm\sqrt{\frac{2E}{T}}$ (depending on i value) on the interval $(2k-1)T \leq t < (2k+1)T, (k = 0, 1, 2, \ldots)$, and $a_{i,\text{Q}}(t)$ takes values $\pm\sqrt{\frac{2E}{T}}$ (depending on i value) on the interval $2kT \leq t < (2k+2)T$. Figure 2.37 shows the MSK I and Q channels.

As can be seen from Figure 2.37 MSK can be regarded as shaped OQPSK. The MSK modulation makes the phase change linear and limited to $\pm(\pi/2)$ over a bit interval T. The MSK signal is represented in Figure 2.38.

Fourier transform of (2.81) is given by:

$$H(f) = A \cdot \frac{4T\cos(2\pi fT)}{\pi\left(1-(4fT)^2\right)} \cdot \exp(-j2\pi fT) \tag{2.83}$$

Then, the power spectral density (PSD) of MSK can be calculated as follows [5]:

$$G(f) = \frac{16TA^2}{\pi^2} \cdot \left(\frac{\cos(2\pi fT)}{1-(4fT)^2}\right)^2 \tag{2.84}$$

The comparison of the MSK spectrum with the spectrum of QPSK/OQPSK is represented in Figure 2.39. In this case it is assumed that QPSK is shaped by the rectangular pulse. Obviously, MSK, having a smoother pulse, has lower side-lobes than QPSK and OQPSK.

However, the MSK spectrum has side-lobes extending well above the data rate. And for wireless systems which require more efficient use of RF channel bandwidth, it is necessary to reduce the energy of the upper side-lobes. Moreover, as can be seen from Figure 2.39, the MSK spectrum has a wider main-lobe than QPSK and OQPSK due to the fact that the symbol period of MSK is twice that of QPSK/OQPSK. These drawbacks can be avoided by using a pre-modulation filter. The PSK demodulator depicted in Figure 2.19 can be used for the detection of MSK signal. The only difference is that the matched filters or correlators should match (2.81). In case the linear amplifier is used the performance of MSK is the same as the performance of QPSK.

2.3.1 GMSK

Gaussian minimum shift keying (GMSK) is the development of MSK. The difference from the MSK is that in GMSK the Gaussian pulse is used for pulse shaping instead of half sinusoid. The impulse response

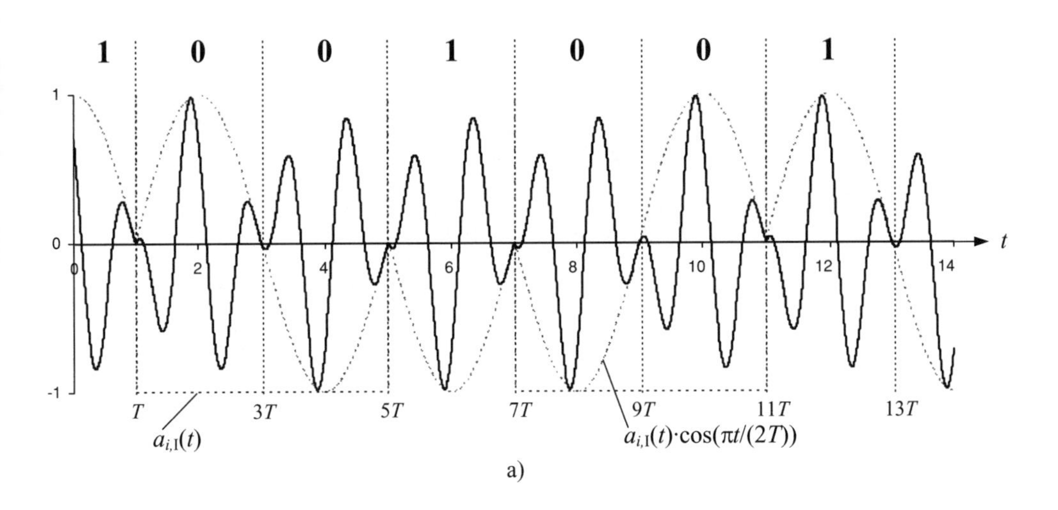

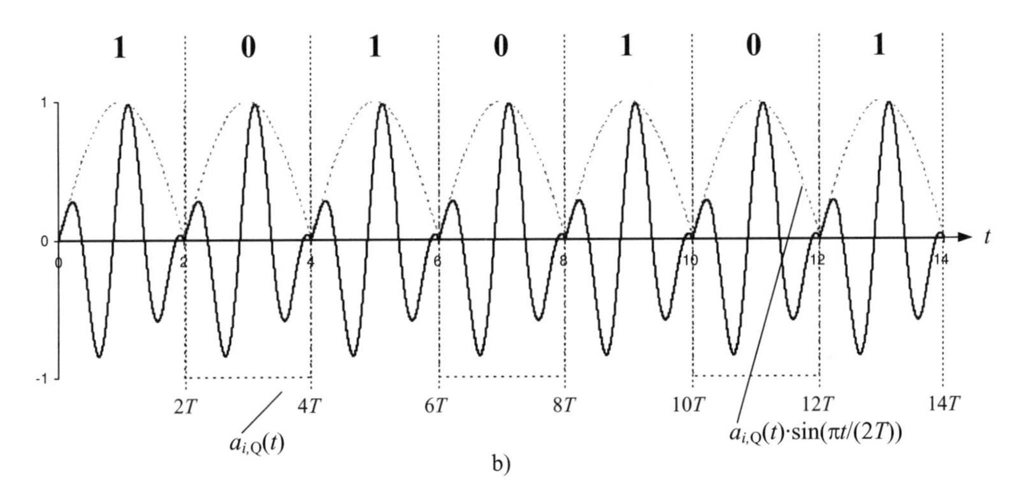

Figure 2.37 MSK: a) I channel, b) Q channel

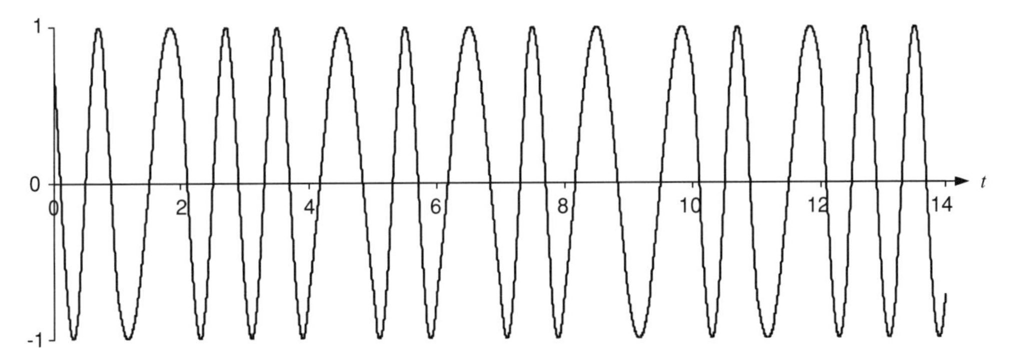

Figure 2.38 MSK modulated signal

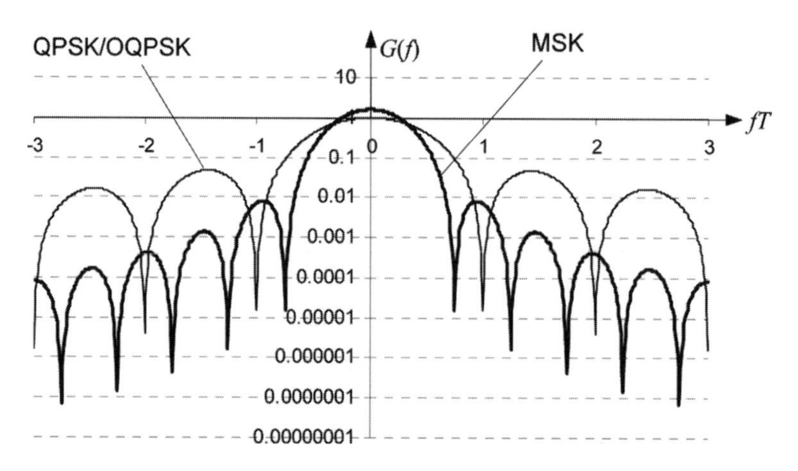

Figure 2.39 Spectrum of MSK and QPSK/OQPSK

of the Gaussian filter is given by:

$$h(t) = \frac{1}{2T}\left(Q\left(2\pi B\frac{t-0.5T}{\sqrt{\ln 2}}\right) - Q\left(2\pi B\frac{t+0.5T}{\sqrt{\ln 2}}\right)\right) \tag{2.85}$$

where $Q(x)$ is the function defined in (2.63), B is the filter bandwidth. The Fourier transform of the Gaussian filter is given by:

$$H(f) = \frac{1}{2}\cdot \text{sinc}(fT)\cdot \exp\left(-\frac{f^2 \ln 2}{2B^2}\right)\cdot \exp(-j\pi fT) \tag{2.86}$$

The spectrum of a Gaussian shaped signal has low side lobes and a narrower main lobe than the spectrum of rectangular pulse. The impulse response of Gaussian filter is depicted in Figure 2.40 and the GMSK spectrum in Figure 2.41. Usually instead of bandwidth of low-pass Gaussian filter it is more convenient to use the *normalized bandwidth*, which is the product of the filter bandwidth B and the symbol period T. This normalized bandwidth or simply *time-bandwidth product BT* plays the same role for the Gaussian filter as the roll-off factor does for the raised cosine filter.

As can be seen from Figure 2.40 the Gaussian filter with $BT = \infty$ corresponds to the rectangular filter. The small values of BT provide improvement in bandwidth efficiency, but on the other hand the power efficiency with very low values of BT degrades significantly. Usually values of BT from 0.3 to 0.5 are used for GMSK. These values provide a good compromise between bandwidth and power efficiency.

The GMSK modulator is represented in Figure 2.42.

Data at the modulator input is represented by the rectangular antipodal pulses as it is depicted in Figure 2.43.

The Gaussian pulses corresponding to input data represented in Figure 2.43 are depicted in Figure 2.44.

At the output of the Gaussian low-pass filter the individual pulses are summed up and forms function $a(t)$ depicted in Figure 2.45.

The output of the Gaussian filter is integrated from time t to infinity. The result of integration $b(t)$ is represented in Figure 2.46.

The inphase and quadrature channels obtained from function $b(t)$ are depicted in Figure 2.47a and 2.47b correspondingly. The modulated GMSK signal is depicted in Figure 2.47c.

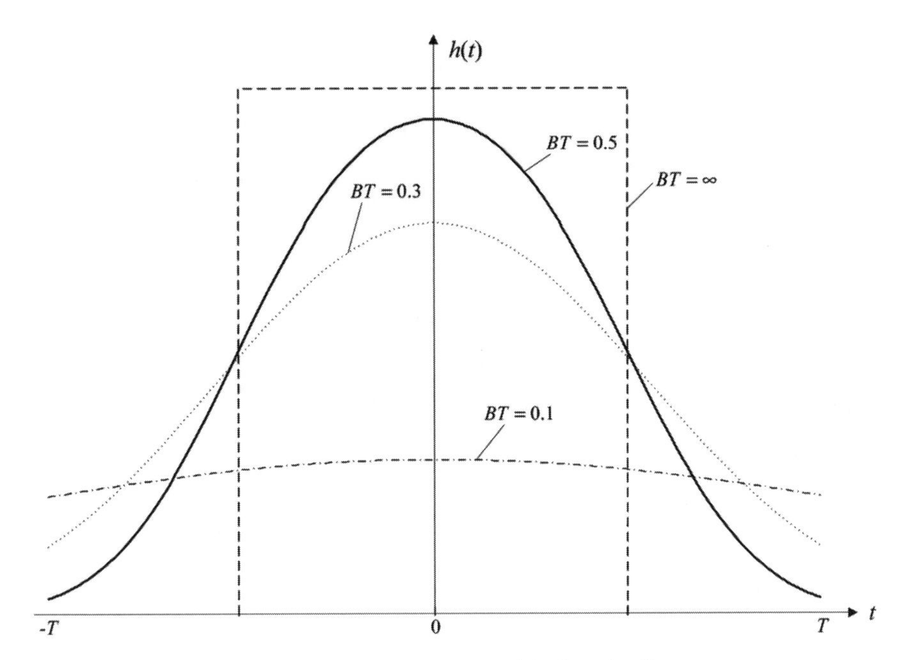

Figure 2.40 Impulse response of the Gaussian filter

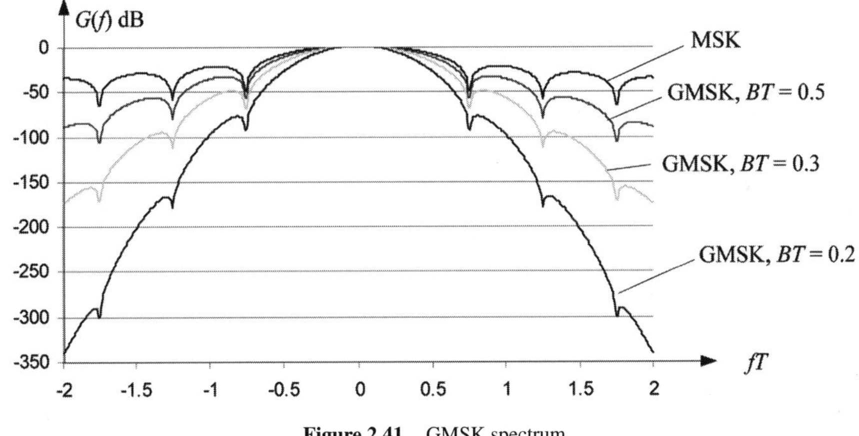

Figure 2.41 GMSK spectrum

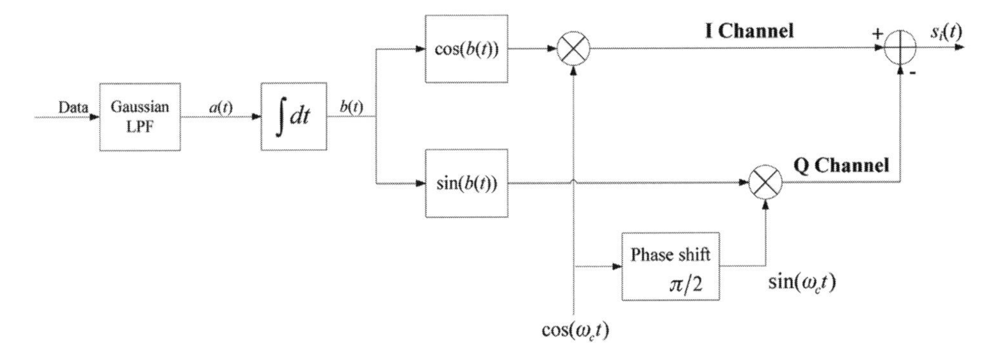

Figure 2.42 GMSK modulator

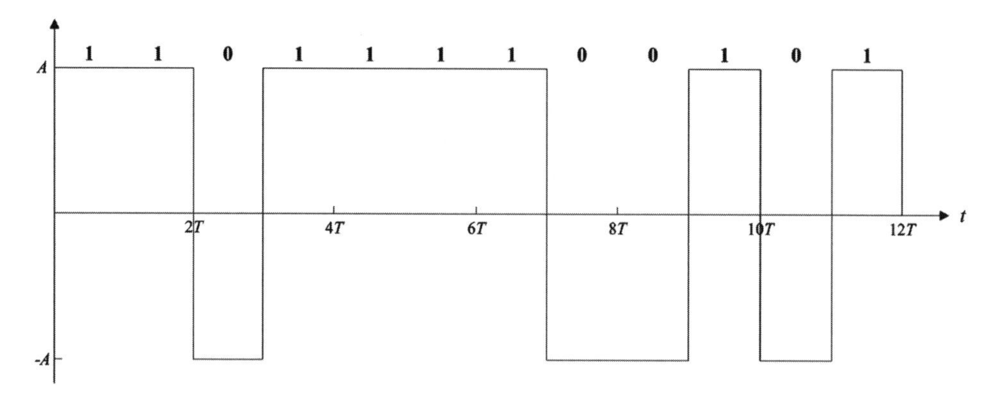

Figure 2.43 Data at the input of the GMSK modulator

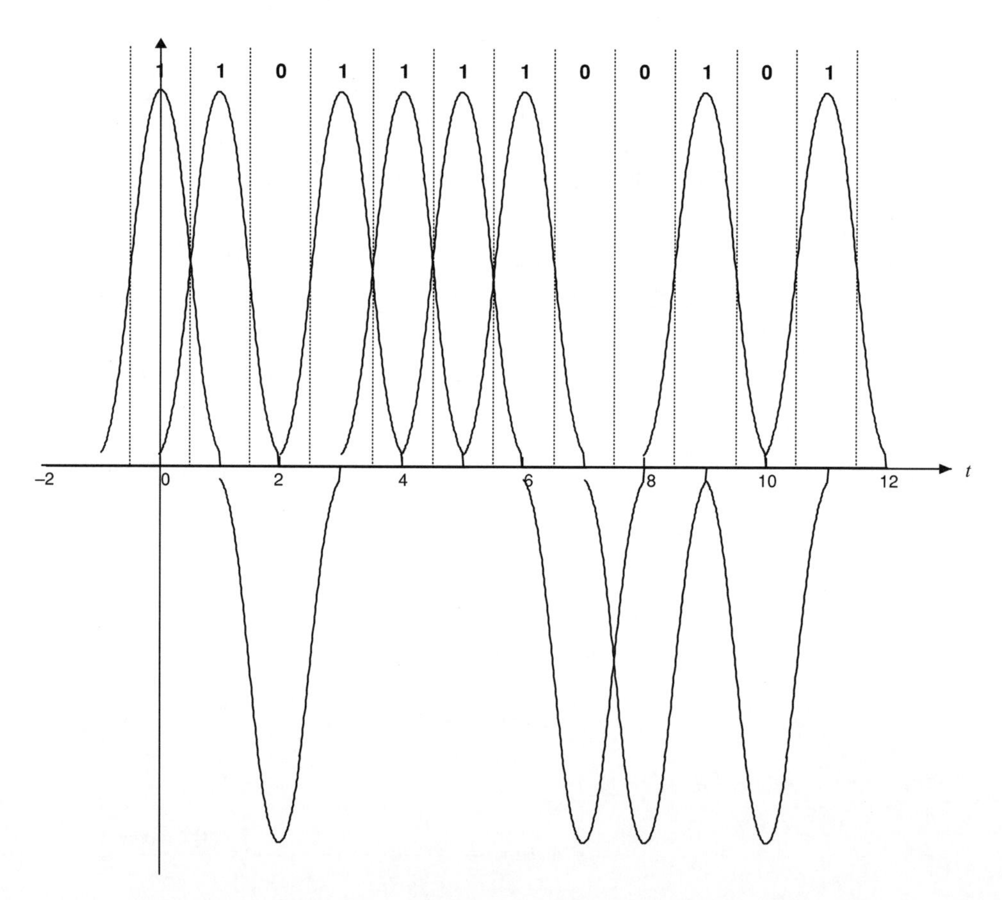

Figure 2.44 Individual Gaussian shaped pulses corresponding to the data in Figure 2.43

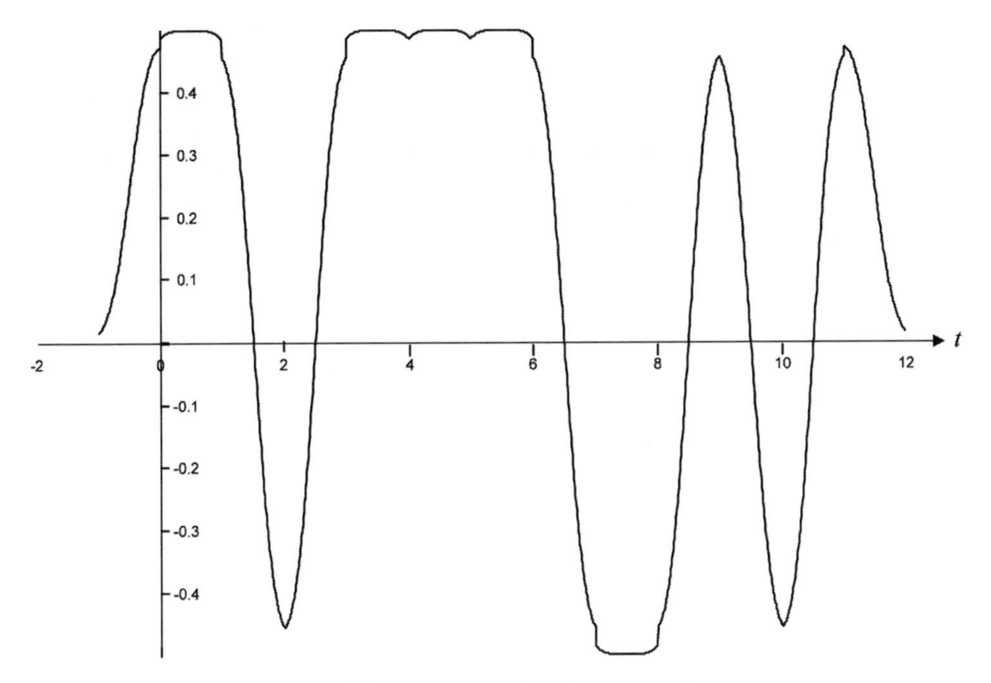

Figure 2.45 Output of the Gaussian low-pass filter

GMSK can be coherently detected with the help of the same type of demodulator as is used for coherent detection of OQPSK or GMSK. In the case of GMSK demodulation the demodulator matched filter of course should match the Gaussian pulse. As was mentioned above the power efficiency of GMSK degrades with decreasing the value of BT product. Due to this fact the performance provided by the GMSK with low values of BT product is slightly worse than the performance of MSK or BPSK/QPSK. The BER curves for MSK and GMSK are represented in Figure 2.48.

As can be seen from Figure 2.47c there are no discontinuous phase changes in GMSK modulated signal. This fact makes GMSK very attractive for using with high power amplifiers. The usage of high

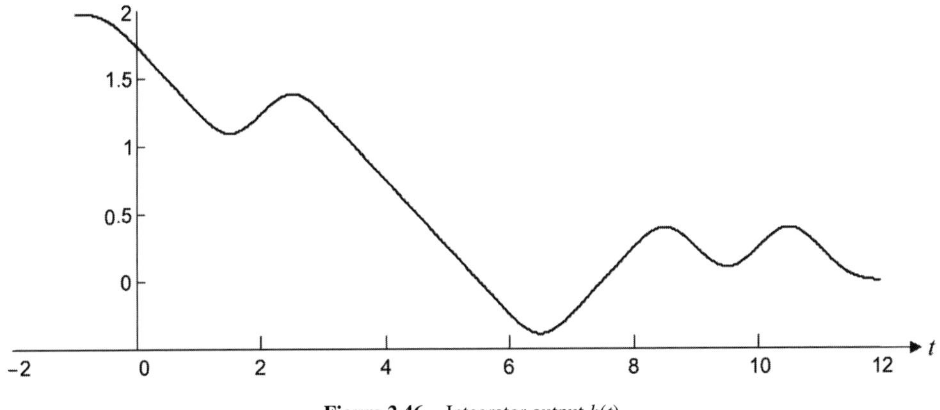

Figure 2.46 Integrator output $b(t)$

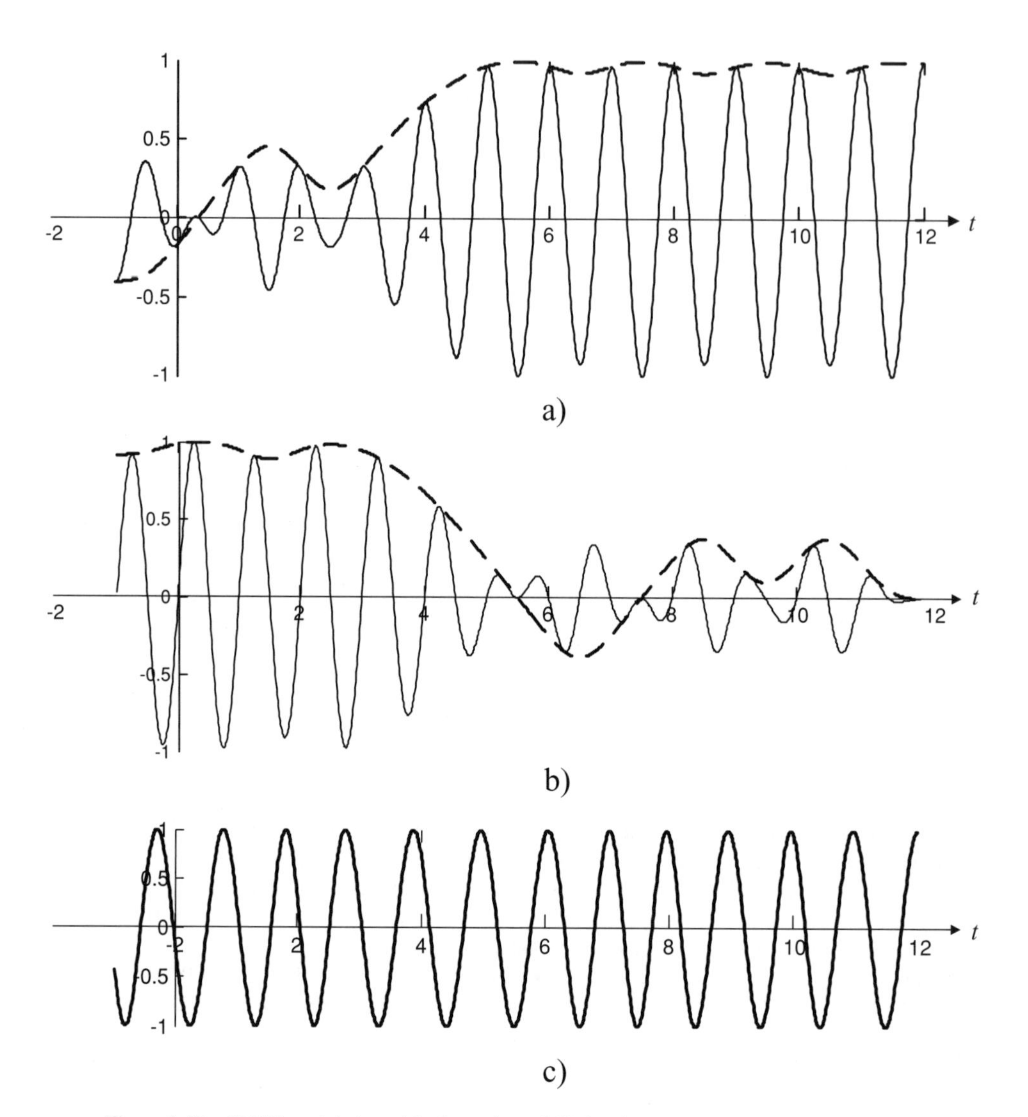

Figure 2.47 GMSK modulation: a) Inphase channel, b) Quadrature channel, c) modulated signal

power amplifiers as well as the low sideband power of GMSK is especially important for wireless communication. GMSK with $BT = 0.3$ is used in GSM and HIPERLAN standards, and with $BT = 0.5$ is used in DECT standard.

2.4 QAM

Quadrature amplitude modulation (QAM) can be considered as an extension of QPSK. In the same way as in QPSK the signal in QAM can be represented as a combination of inphase and quadrature components but the constellation points are distributed over the whole area of the constellation diagram rather than over the circle like in PSK. This is achieved by multilevel amplitude modulation of each I

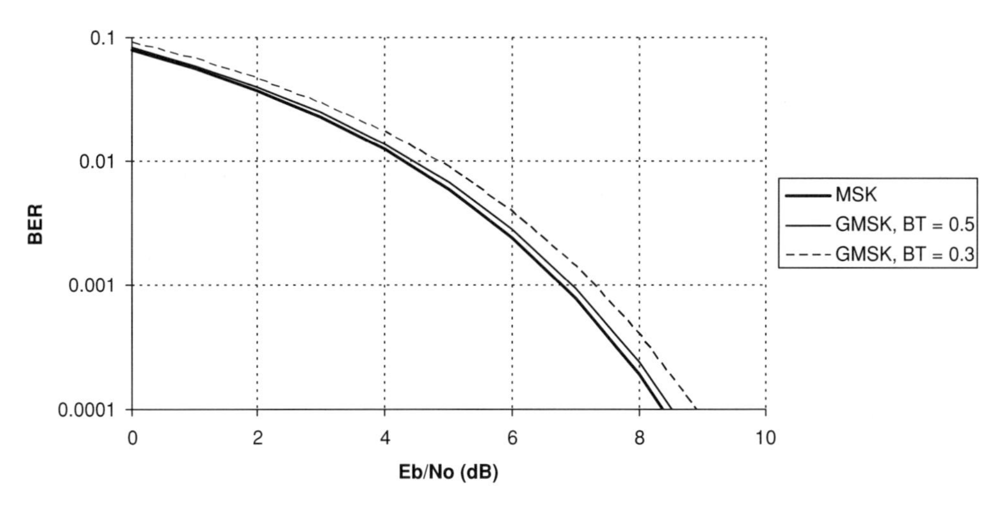

Figure 2.48 BER curves for GMSK and MSK modulation

and Q component. The M-QAM signal can be written as follows:

$$s_i(t) = \underbrace{b_{i,\mathrm{I}}(t) \cdot \cos(\omega_c t)}_{\mathrm{I}_i(t)} - \underbrace{b_{i,\mathrm{Q}}(t) \cdot \sin(\omega_c t)}_{\mathrm{Q}_i(t)}, i = 1, \ldots, M \tag{2.87}$$

where $M = 2^k$ is the modulation cardinality, k is the number of bits per symbol, $b_{i,\mathrm{I}}(t)$ and $b_{i,\mathrm{Q}}(t)$ are $\sqrt{M}$-level amplitude modulated baseband signals corresponding to I and Q channel. The typical QAM modulator is depicted in Figure 2.49.

Each M-QAM symbol corresponds to the block of k bits. The data flow is split in blocks by $k/2$ bits, this process generates two independent signals, which are fed to I and Q channels. The mapper maps the block of $k/2$ bits to the $\sqrt{M}$-level amplitude modulated baseband signal. Usually the pulse shaping is done with a baseband signal but it can be done with a bandpass signal too. Then inphase baseband signal is multiplied by $\cos(\omega_c t)$ and quadrature baseband signal by $\sin(\omega_c t)$ and summed up. The orthonormal basis for QAM coincides with PSK orthonormal basis (2.68). Obviously, the QPSK modulation can be represented as 4-QAM modulation with bipolar baseband signaling.

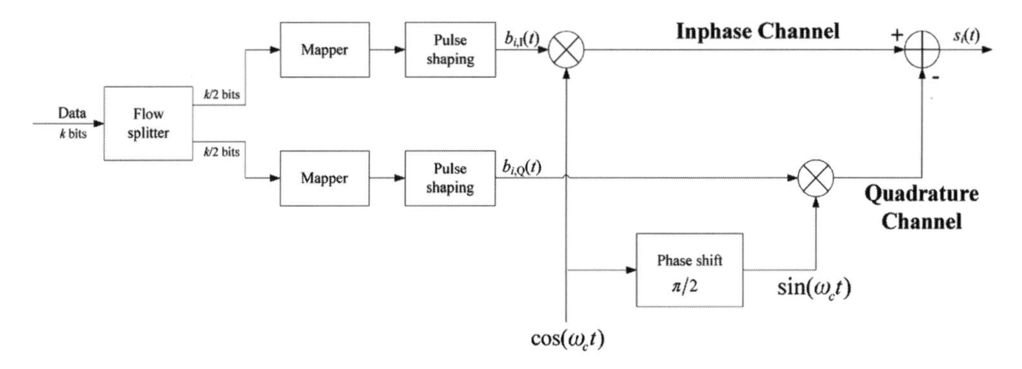

Figure 2.49 M-QAM modulator

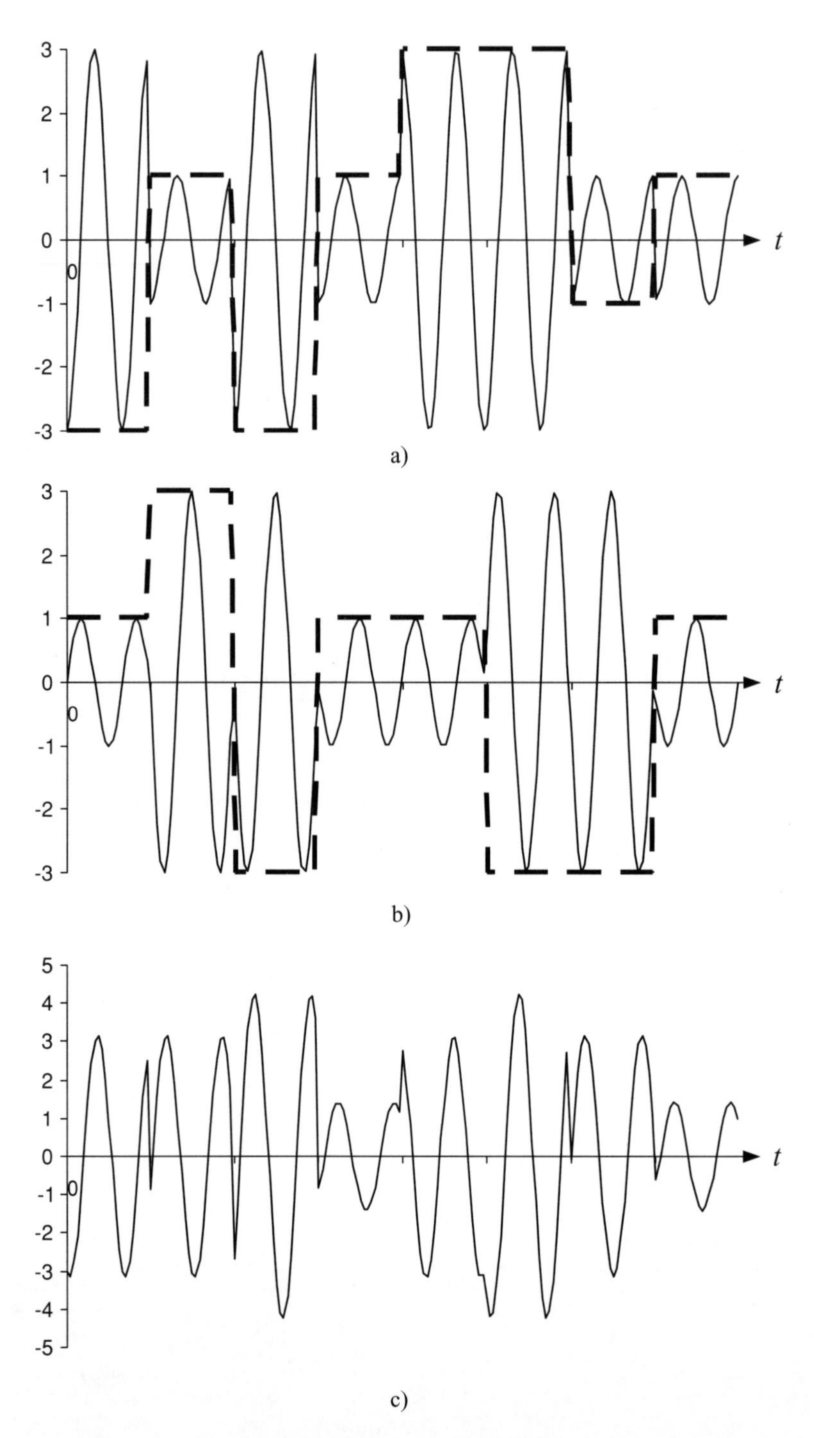

Figure 2.50 16-QAM modulation: a) $b_{i,\mathrm{I}}(t)$ and $\mathrm{I}_i(t)$, b) $b_{i,\mathrm{Q}}(t)$ and $\mathrm{Q}_i(t)$, c) $s_i(t)$

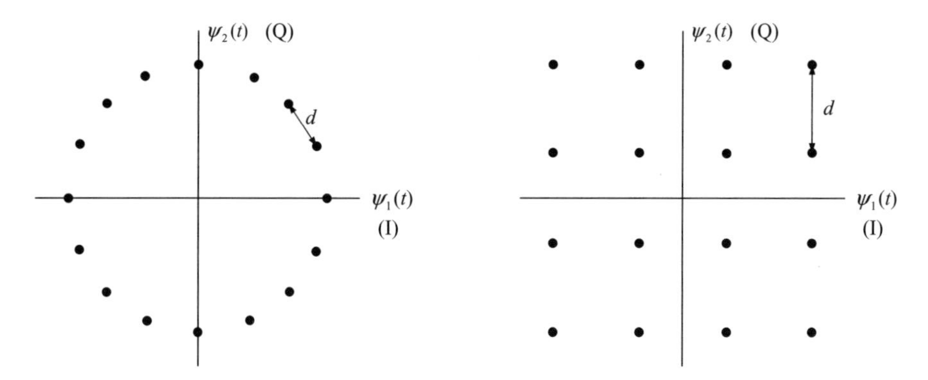

Figure 2.51 16-PSK and 16-QAM constellation

The examples of baseband signals $b_{i,\mathrm{I}}(t)$, $b_{i,\mathrm{Q}}(t)$ and I and Q channels for 16-QAM are depicted in Figure 2.50a and b. The resulting 16-QAM signal $s_i(t) = \mathrm{I}_i(t) - \mathrm{Q}_i(t)$ is depicted in Figure 2.50c.

Due to more effective use of signal space in QAM it is possible to increase the minimum distance between the constellation points in comparison with PSK, which leads to better performance. The comparison of 16-PSK and 16-QAM constellations is represented in Figure 2.51.

Multiplication of the signal $s_i(t)$ by the term $\cos(\omega_c t)$ allows extracting the value of $b_{i,\mathrm{I}}(t)$ from the signal. Using trigonometric identities:

$$\begin{aligned} \cos^2(x) &= \frac{1+\cos(2x)}{2}; \\ \sin^2(x) &= \frac{1-\cos(2x)}{2}; \\ \sin(x)\cdot\cos(x) &= \frac{\sin(2x)}{2} \end{aligned} \tag{2.88}$$

it can be shown that:

$$\begin{aligned} s_i(t)\cdot\cos(\omega_c t) &= b_{i,\mathrm{I}}(t)\cdot\cos^2(\omega_c t) - b_{i,\mathrm{Q}}(t)\cdot\sin(\omega_c t)\cdot\cos(\omega_c t) = \\ &= \frac{1}{2}b_{i,\mathrm{I}}(t) + \frac{b_{i,\mathrm{I}}(t)\cdot\cos(2\omega_c t) - b_{i,\mathrm{Q}}(t)\cdot\sin(2\omega_c t)}{2} \end{aligned} \tag{2.89}$$

Then the low-pass filter can remove the high frequency terms from the signal. Thus the term $b_{i,\mathrm{I}}(t)$ can be recovered from the signal. In the same way, multiplying the signal by the term $-\sin(\omega_c t)$ obtain:

$$\begin{aligned} s_i(t)\cdot(-\sin(\omega_c t)) &= b_{i,\mathrm{Q}}(t)\cdot\sin^2(\omega_c t) - b_{i,\mathrm{I}}(t)\cdot\sin(\omega_c t)\cdot\cos(\omega_c t) = \\ &= \frac{1}{2}b_{i,\mathrm{Q}}(t) - \frac{b_{i,\mathrm{Q}}(t)\cdot\cos(2\omega_c t) + b_{i,\mathrm{I}}(t)\cdot\sin(2\omega_c t)}{2} \end{aligned} \tag{2.90}$$

and the low-pass filtering in this case also solves the problem of extracting the value of $b_{i,\mathrm{Q}}(t)$ from the signal. This property can be used in the demodulation process. The corresponding M-QAM demodulator is depicted in Figure 2.52. It should be mentioned that the receiver matched filter itself is a low-pass filter and in real life the low-pass filtering and matched filtering usually is combined in one block. The obtained estimates of inphase and quadrature baseband signals are compared with thresholds to extract the bit estimates. It is not enough like in QPSK demodulation to check the sign of the inphase or quadrature component to obtain the information about the corresponding transmitted bit. For example, for 16-QAM the threshold corresponding to the value of $2a$ (see Figure 2.53) should be calculated. This is especially important under the conditions of fading channel where the QAM constellation is “pumping” due to the

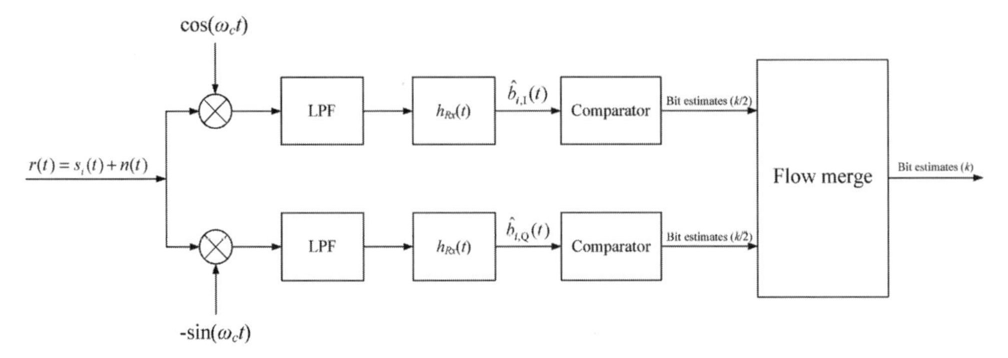

Figure 2.52 M-QAM demodulator, $M = 2^{\text{k}}$

automatic gain control (AGC) procedure, thus the threshold values should be recalculated on fly. After comparison with the thresholds the bit estimates corresponding to the inphase and quadrature component can be obtained. As can be seen the demodulation procedure for QAM is pretty simple and the same is regarded for the implementation complexity. However, the receiver should have exact information about the carrier, that is, the demodulation is coherent.

The calculation of error probability for QAM is not as straightforward as for PSK. As can be seen from Figure 2.53 the QAM signals have different power and different number of neighbouring signals (constellation points). For example, the constellation point with Gray code label 1100 has four neighbours at distance $d = 2a$, while the points labelled 1010 and 1111 have only three and two neighbours at the same distance. So, the size of decision regions for these points differs from each other. Correspondingly the error probability for point 1100 under the conditions of AWGN channel can be estimated as $4Q\left(\frac{d}{2\sigma}\right)$,

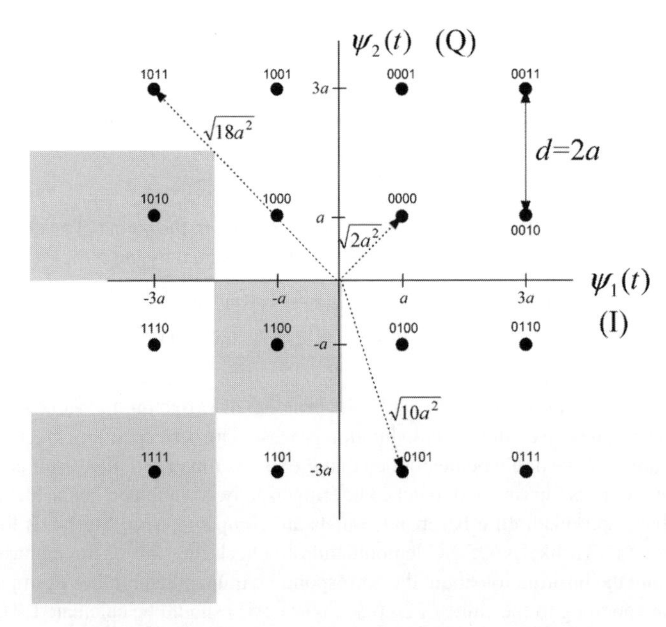

Figure 2.53 Gray code labelling for 16-QAM constellation

for point 1010 as $3Q\left(\frac{d}{2\sigma}\right)$ and for point 1111 as $2Q\left(\frac{d}{2\sigma}\right)$, where $Q(x)$ is the function defined in (2.63). Again in this case we do not take into account the case when the error causes the transition to the symbol, which is not the nearest one.

For 16-QAM there are four inner points with error probability $4Q\left(\frac{d}{2\sigma}\right)$, four corner points with error probability $2Q\left(\frac{d}{2\sigma}\right)$ and eight points with error probability $3Q\left(\frac{d}{2\sigma}\right)$. Assuming all symbols are equally likely, the total symbol error probability can be estimated as:

$$P_e \approx \frac{1}{4}\cdot 2Q\left(\frac{d}{2\sigma}\right)+\frac{1}{2}\cdot 3Q\left(\frac{d}{2\sigma}\right)+\frac{1}{4}\cdot 4Q\left(\frac{d}{2\sigma}\right)=3Q\left(\frac{d}{2\sigma}\right) \tag{2.91}$$

Assuming the constellation points are labelled with the help of Gray code, for example as in Figure 2.53, the symbol error corrupts only one bit out of four. Then the bit error rate under the conditions of AWGN channel can be written as:

$$P_b \approx \frac{3}{4}Q\left(\frac{d}{2\sigma}\right) \tag{2.92}$$

Taking into account that in 16-QAM constellation each of the four inner points has energy $2a^2$, each of the four corner points $18a^2$, and each of the other eight points $10a^2$ the average signal energy for 16-QAM can be calculated as follows:

$$E=\frac{4\cdot 2a^2+4\cdot 18a^2+8\cdot 10a^2}{16}=10a^2=\frac{5d^2}{2} \tag{2.93}$$

Then (2.91) can be written as:

$$P_e \approx 3Q\left(\sqrt{\frac{E}{5N_0}}\right)=3Q\left(\sqrt{\frac{4E_b}{5N_0}}\right) \tag{2.94}$$

where E_b is the average energy per bit. In the same way (2.92) can be written as:

$$P_b \approx \frac{3}{4}Q\left(\sqrt{\frac{4E_b}{5N_0}}\right) \tag{2.95}$$

In general, the constellation points of M-QAM can be represented as:

$$\alpha = a\left(2K-\sqrt{M}+1\right)+ja\left(2L-\sqrt{M}+1\right), \quad K,L=0,\ldots,\sqrt{M}-1 \tag{2.96}$$

Then the average signal energy can be calculated as follows [8]:

$$E=\frac{\sum_{K=0}^{\sqrt{M}-1}\sum_{L=0}^{\sqrt{M}-1} a^2\left(2K-\sqrt{M}+1\right)^2+a^2\left(2L-\sqrt{M}+1\right)^2}{M}=\frac{2(M-1)}{3}a^2 \tag{2.97}$$

The minimum distance between constellation points $d=2a$. The number of corner constellation points with error probability $2Q\left(\frac{d}{2\sigma}\right)$ is 4, the number of inner constellation points with error probability $4Q\left(\frac{d}{2\sigma}\right)$ is $\left(\sqrt{M}-2\right)^2$, and the number of constellation points lying on the outer constellation border except the corners with error probability $3Q\left(\frac{d}{2\sigma}\right)$ is $4\left(\sqrt{M}-2\right)$. Then the total symbol error probability is given by:

$$\begin{aligned}P_e &\approx \frac{4}{M}\cdot 2Q\left(\frac{d}{2\sigma}\right)+\frac{4\left(\sqrt{M}-2\right)}{M}\cdot 3Q\left(\frac{d}{2\sigma}\right)+\frac{\left(\sqrt{M}-2\right)^2}{M}\cdot 4Q\left(\frac{d}{2\sigma}\right)\\ &=4\left(1-\frac{1}{\sqrt{M}}\right)Q\left(\frac{d}{2\sigma}\right)\end{aligned} \tag{2.98}$$

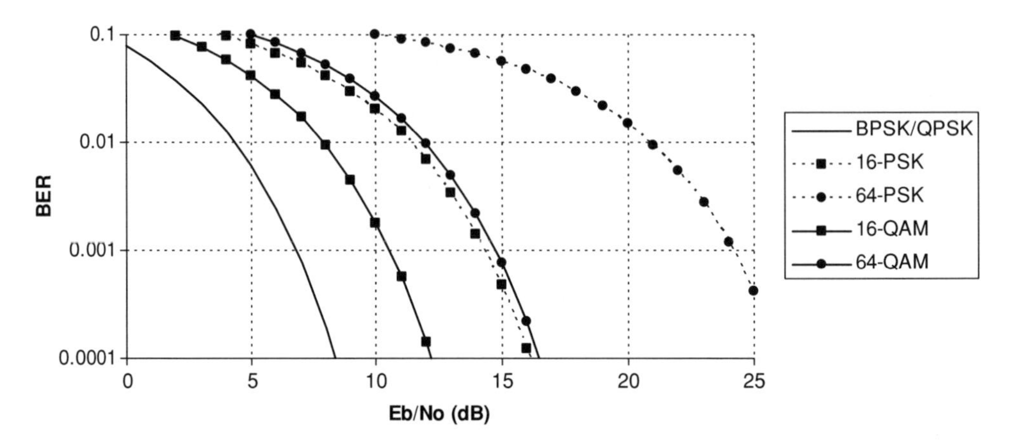

Figure 2.54 BER curves for BPSK/QPSK, 16-PSK, 64-PSK, 16-QAM, 64-QAM. AWGN channel

Taking into account (2.97) (2.98) can be written as:

$$P_e \approx 4\left(1-\frac{1}{\sqrt{M}}\right)Q\left(\sqrt{\frac{3E}{(M-1)N_0}}\right)=4\left(1-\frac{1}{\sqrt{M}}\right)Q\left(\sqrt{\frac{3E_b\log_2 M}{(M-1)N_0}}\right) \tag{2.99}$$

Assuming the constellation points are labelled with the help of Gray code, the BER for M-QAM under the conditions of the AWGN channel is given by:

$$P_b \approx \frac{4}{\log_2 M}\left(1-\frac{1}{\sqrt{M}}\right)Q\left(\sqrt{\frac{3E_b\log_2 M}{(M-1)N_0}}\right) \tag{2.100}$$

See BER curves represented in Figure 2.54 to compare the performance of QAM and PSK, recall that the QPSK can be considered both as PSK and as 4-QAM. Obviously, QAM system outperforms PSK system with the number of levels greater than four.

2.5 OFDM

Orthogonal frequency division multiplexing (*OFDM*) as it suggests from its name is the extension of the *frequency division multiplexing* (*FDM*) technique. The basic idea of the FDM is to divide the available bandwidth into many narrow sub-bands and to use a large number of parallel narrow-band subcarriers rather than a single wide-band carrier to transfer the information as is shown in Figure 2.55 [9].

The main advantage of FDM is the robustness against frequency-selective fading and narrowband interference. If the number of subcarriers is large enough, each subcarrier deals with flat fading rather than with frequency-selective fading as a wideband carrier does. And the narrowband interference will affect only one or two subcarriers of the whole bunch of subcarriers. The other subcarriers will not be corrupted by the interference. Another advantage of the FDM is the much greater symbol length in comparison with wideband carrier. Due to this fact the intersymbol interference in the case of FDM is comparatively short, which helps to cope with this problem. Unfortunately the spectral efficiency of FDM is quite low due to guard intervals (see Figure 2.55b). To eliminate this problem it is possible to use the orthogonal subcarriers as shown in Figure 2.56. Note that at the point corresponding to the peak of each subcarrier's spectrum all other spectra cross the zero, that is, the subcarriers are orthogonal.

The use of orthogonal subcarriers allows the subcarriers' spectra to overlap. Due to the orthogonality it is possible to recover the individual subcarriers' signals despite the overlapping spectra and thus

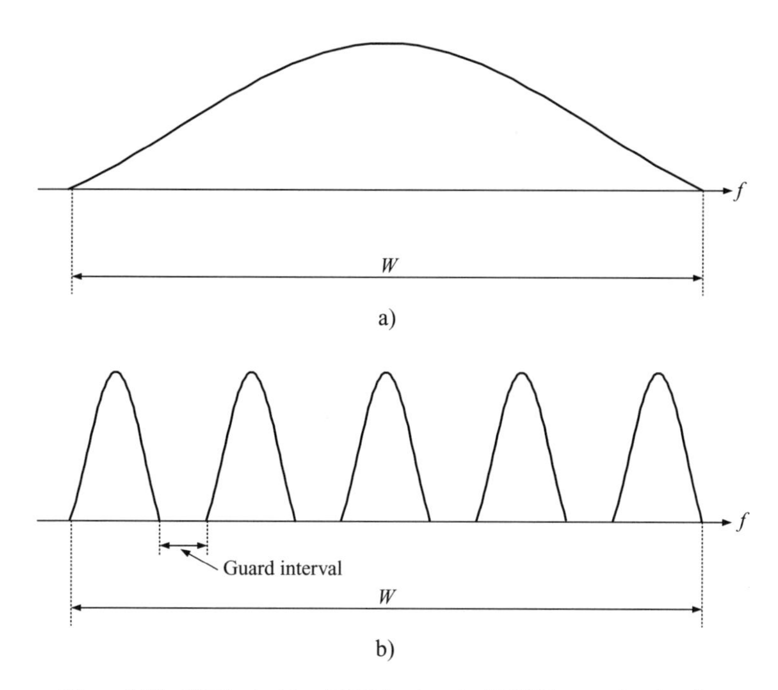

Figure 2.55 FDM principle. a) Wideband carrier b) FDM narrowband carriers

there is no need in guard intervals as in FDM. The usage of the orthogonal subcarriers also helps to decrease the implementation complexity of both transmitter and receiver. In FDM each subcarrier needs a separate pair of matched filters at the transmitter and the receiver to make it possible to eliminate the inter-carrier interference. In OFDM the inter-carrier interference is eliminated due to the orthogonality of the subcarriers and there is no need to use separate filters for each subcarrier. The idea of using orthogonal subcarriers was suggested more than 40 years ago [10], [11], [12] but in practice it was not used for a long time mostly due to its complexity.

Note that the set of OFDM subcarriers forms the set of orthogonal sinusoids. This is quite common with *discrete Fourier transform (DFT)* where the set of orthogonal sinusoids is used for the representation of the signal. The DFT is given by the following formula:

$$X_k = \sum_{n=0}^{N-1} x_n \cdot \exp\left(-\frac{2\pi j}{N}kn\right) = \sum_{n=0}^{N-1} x_n \cdot \cos\left(\frac{2\pi}{N}kn\right) - j\sum_{n=0}^{N-1} x_n \cdot \sin\left(\frac{2\pi}{N}kn\right), \quad k = 0, \ldots, N-1 \tag{2.101}$$

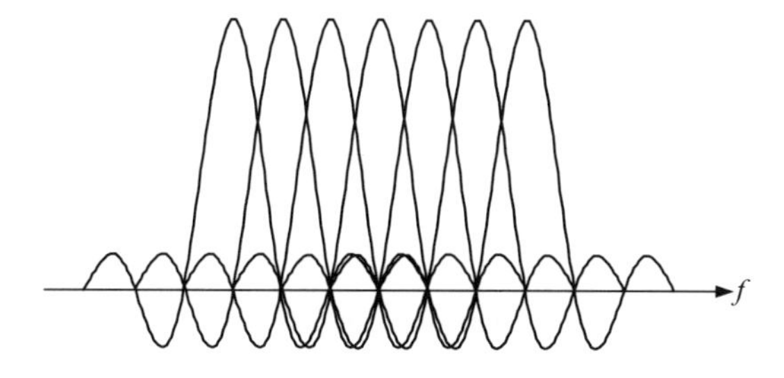

Figure 2.56 OFDM subcarriers

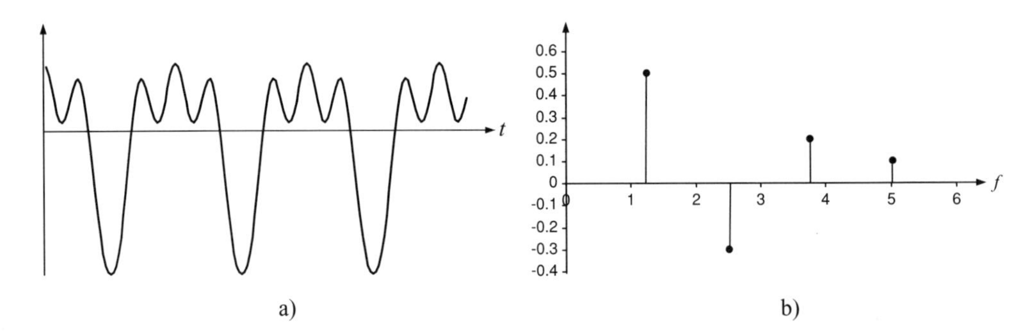

Figure 2.57 The representation of signal: a) Time domain, b) Frequency domain

That is, the DFT correlates the input signal with each of set of orthogonal sinusoids or basis functions. If the input signal has some energy at a certain frequency k, it will be reflected at the correlation of the input signal and this frequency, that is, in the value of spectrum for kth frequency X_k. In other words the DFT converts the time domain representation of signal to the frequency domain representation. The examples of the time domain and frequency domain representation of the same signal are depicted in Figure 2.57.

The *inverse DFT (IDFT)* converts the signal spectrum, that is, the frequency domain signal representation to the time domain. IDFT is given by:

$$x_n = \frac{1}{N}\sum_{k=0}^{N-1} X_k \cdot \exp\left(\frac{2\pi j}{N}kn\right) = \frac{1}{N}\sum_{k=0}^{N-1} X_k \cdot \cos\left(\frac{2\pi}{N}kn\right) + j\frac{1}{N}\sum_{n=0}^{N-1} X_k \cdot \sin\left(\frac{2\pi}{N}kn\right), \quad (2.102)$$
$$n = 0, \ldots, N-1$$

The simple OFDM system without usage of DFT is depicted in Figure 2.58.

The input of this system is QAM modulated signal $X_k = I_k + jQ_k$ with symbol rate R. In fact, it is not necessary to use QAM modulated signal as input, other modulation types can also be used. With the help of a serial-to-parallel convertor the block of N input QAM symbols $X_0, \ldots, X_{N-1}$ simultaneously modulates N orthogonal subcarriers $\exp(j\omega_k t) = \cos(\omega_k t) + j\sin(\omega_k t)$, $k = 0, \ldots, N-1$. Notice that the symbol rate of subcarriers R_s is N times less than the symbol rate of the input signal $R_s = \frac{R}{N}$. In other words the subcarriers' sampling rate is N times less than the input signal sampling rate. The obvious choice of the set of N orthogonal subcarriers is to choose the subcarrier frequencies in accordance with the following rule:

$$\omega_k = k\omega_0 = 2\pi k f_0 \quad (2.103)$$

where $f_0 = R_s$. For the sake of simplicity we can assume that the transmitter filters have the rectangular impulse responses. Then the transmitted signal $x(t)$ can be represented as follows:

$$x(t) = \frac{1}{\sqrt{N}}\sum_{k=0}^{N-1} X_k \cdot \exp(j2\pi k f_0 t) \quad (2.104)$$

The scaling factor $\frac{1}{\sqrt{N}}$ is used for the power normalization. Recall that the subcarrier symbol period T is N times longer than the input symbol period. Now introduce the discrete time $t = n\Delta t$ $(n = 0, \ldots, N-1)$ covering the period T, where $\Delta t = \frac{1}{f_s}$ is the reciprocal of the sampling frequency f_s. Since $T = N\Delta t$ the

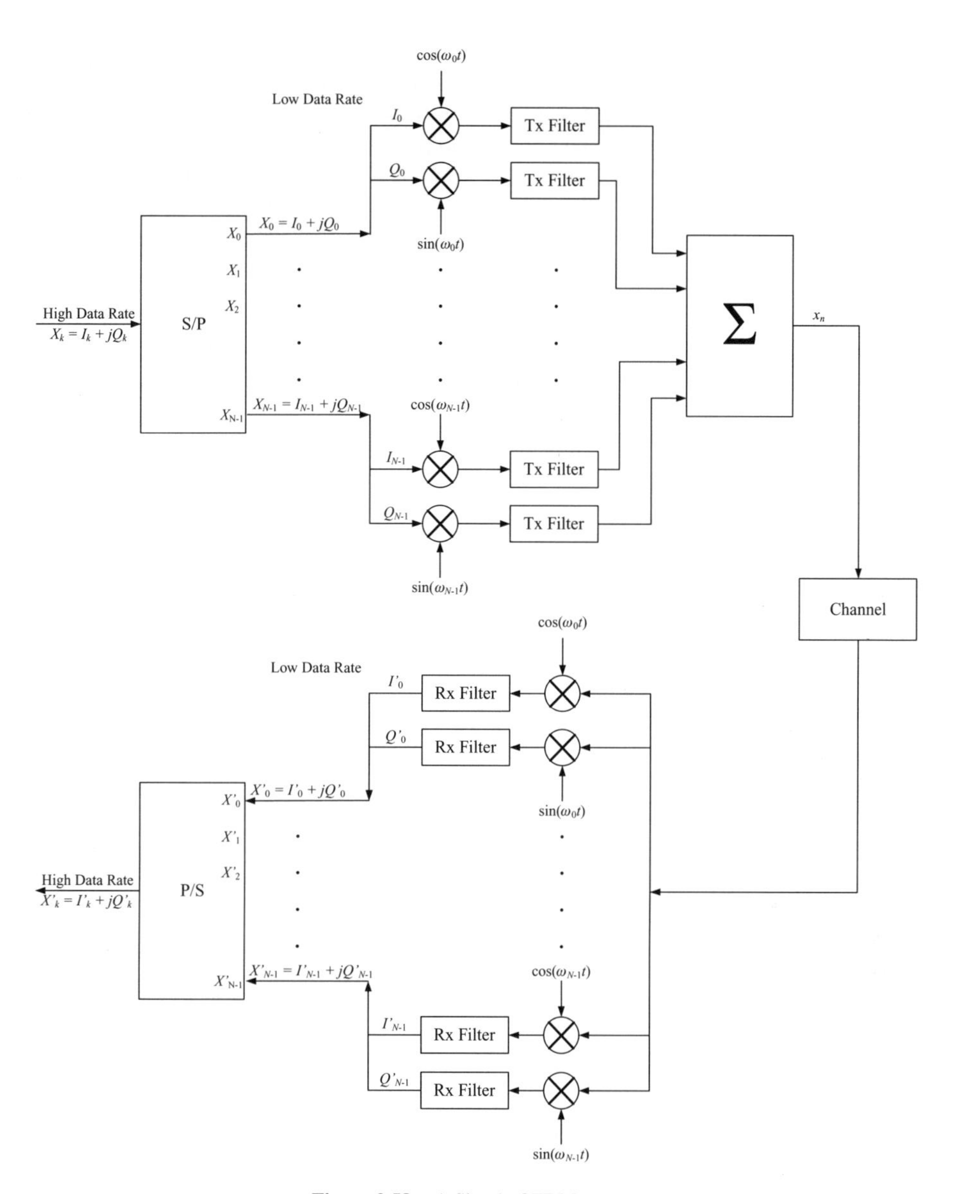

Figure 2.58 A Simple OFDM system

sampling frequency f_s should correspond to the input symbol rate R. Then $f_s = R = N \cdot R_s = N \cdot f_0$ and:

$$\Delta t = \frac{1}{f_s} = \frac{1}{N f_0} \tag{2.105}$$

Substituting (2.105) in (2.104) obtain:

$$x_n = x(n\Delta t) = \frac{1}{\sqrt{N}} \sum_{k=0}^{N-1} X_k \cdot \exp\left(\frac{2\pi j}{N} kn\right), \quad n = 0, \ldots, N-1 \tag{2.106}$$

Comparing (2.106) and (2.102) one can see that OFDM signal can be represented in the form of IDFT. The only difference is the scaling $\frac{1}{\sqrt{N}}$ instead of $\frac{1}{N}$ but it can be taken into account at the receiver. Thus, if there were N OFDM subcarriers, it is possible to treat the modulated symbols of these carriers as frequency domain symbols. These symbols are used as input for the N-dimension IDFT to obtain the time domain OFDM symbol to be transmitted over the channel. At the receiver it is possible to use the DFT function to extract the frequency domain symbols X_k from the received time domain signal [13]. If the number of subcarriers N is big enough the implementation complexity of the transmitter and the receiver using the simple form of OFDM becomes unaffordable. However, if the DFT form is used it is possible to use fast Fourier transform (FFT) instead of discrete Fourier transform and in this case the implementation complexity significantly decreases. The FFT is mathematically equivalent to DFT but it is much more efficient for the implementation. The only restriction implied by the FFT is that the transform length N should be the power of two $N = 2^m$. The OFDM system using FFT and IFFT functions is depicted in Figure 2.59.

The input of the system is the QAM modulated data with high symbol rate R. The symbol period of the input data is Δt. After serial-to-parallel processing N symbols X_k treated as the frequency domain symbols are taken as input for the IFFT block that converts it to time domain block of N symbols x_n. The block of N output symbols x_n forms the OFDM symbol of period $T = N\Delta t$. The stream of OFDM symbols has the symbol rate $R_s = \frac{R}{N}$. Then after the low-pass filtering the OFDM symbol is transmitted over the channel. At the receiver the time domain OFDM symbol distorted by the channel is used as the input for the FFT block, which transforms it to frequency domain and the obtained estimates $\hat{X}_k$ after parallel-to-serial processing forms the data stream with high rate R, which is demodulated by the QAM demodulator. If the channel is noiseless the estimates $\hat{X}_k$ coincide with the original symbols X_k. Actually, as was mentioned above it is not necessary that all subcarriers are modulated in the same way. For example, some subcarriers can be modulated with the help of QPSK and others with the help of 16QAM or higher order modulations.

Most wireless systems are characterized by the presence of a multipath channel. The multipath channel is the origin of the two types of interference in the OFDM system. The first type is the intersymbol interference (ISI). This type of interference occurs when the received OFDM symbol is distorted by the previously transmitted OFDM symbol. It is a common problem for both single-carrier and multicarrier systems but the multicarrier systems are more robust in this case since the symbol period T for the multicarrier system is much longer than the symbol period of the single-carrier system. If the number of subcarriers N is big enough the OFDM symbol period T is longer than the time span of the channel. In this case only part of the received OFDM symbol will be corrupted by the delayed copy of the previously transmitted symbol. Whereas the single-carrier symbol will be corrupted by the several previously transmitted symbols due to the short symbol period, which should be used to keep the same high data rate as in a multicarrier system. The very simple method of coping with ISI in OFDM system is the usage of guard intervals. If the time span of the channel is $T_c < T$ then as was mentioned above only the first part of duration T_c of each OFDM symbol is corrupted by the ISI. In this case it is enough to add the guard interval of length $T_g \geq T_c$ in the beginning of each symbol to get rid of ISI. This can be considered simply as an extension of the OFDM symbol period as shown in Figure 2.60.

The guard interval does not contain any information and should be discarded at the receiver. That means the data rate is decreased $\left(1 + \frac{T_g}{T}\right)$ times. However, if the length of the guard interval is relatively small in comparison with the original OFDM symbol period T this loss is not severe.

Another type of interference caused by the multipath channel is the *inter-carrier interference* (*ICI*). The ICI occurs as a result of the fact that the multipath channel impacts the subcarriers in such a way that they lose the orthogonality. Consider the multipath channel model as a filter of length L with the coefficients $\mathbf{h} = (h_0, \ldots, h_{L-1})$. For simplicity consider the channel model without AWGN. Assume that the OFDM symbol comprising block of N samples x_n, $n = 0, \ldots, N - 1$ is extended with guard interval of L zeros. Now the OFDM symbol transmitted over the multipath channel comprises block of

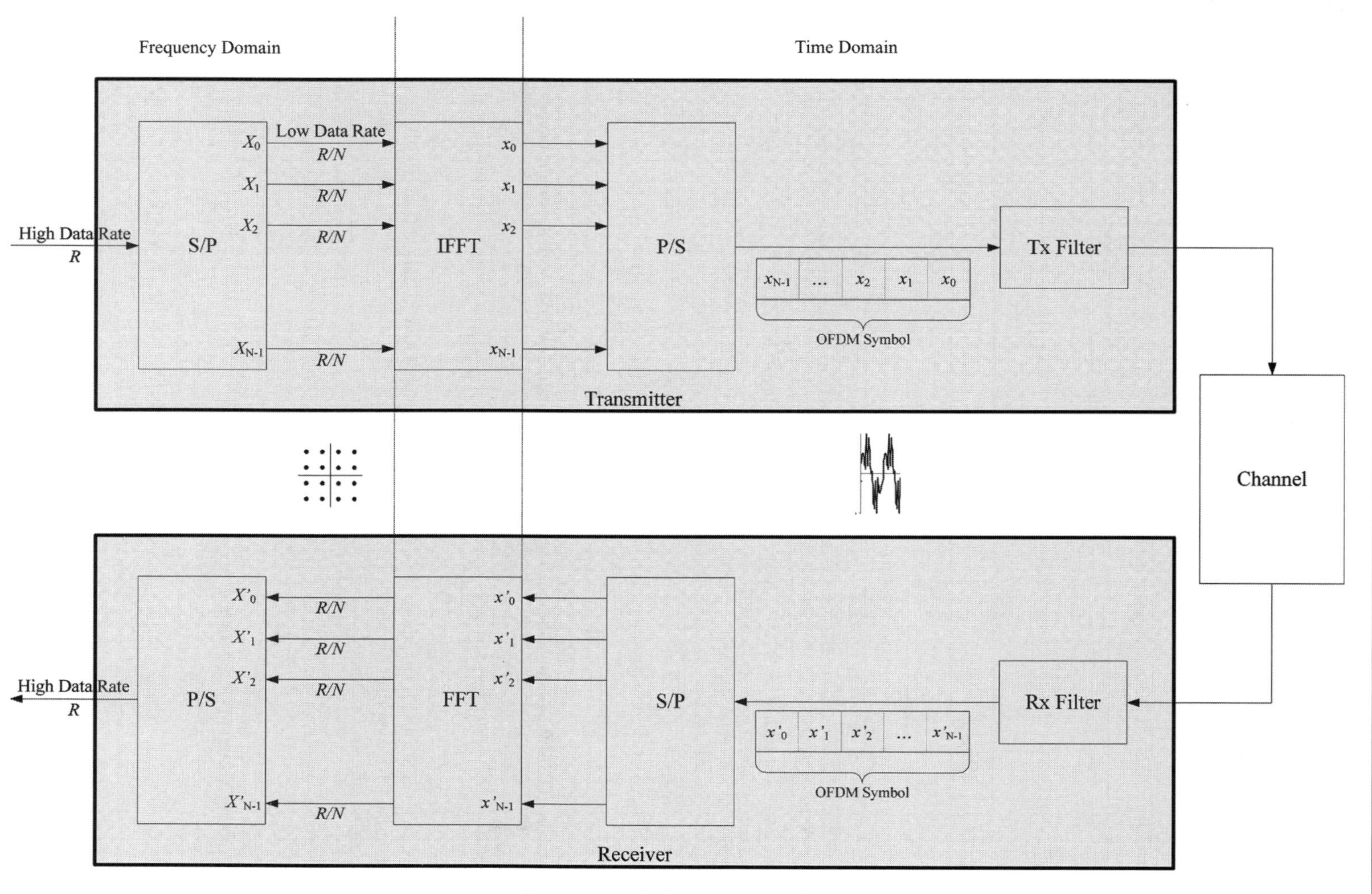

Figure 2.59 OFDM system using FFT

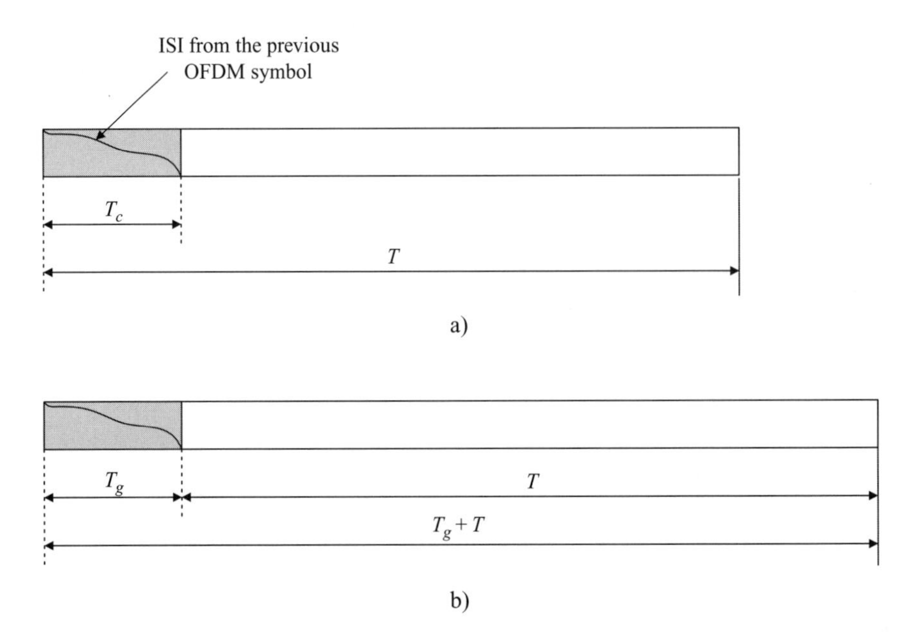

Figure 2.60 Extension of an OFDM symbol to cope with ISI. a) Impact of the ISI to the OFDM symbol of length T, b) Extended OFDM symbol of length $T+T_g$, ISI is eliminated

$N + L$ samples:

$$\begin{cases} z_n = 0, & n = 0, \ldots, L-1 \\ z_n = x_{n-L}, & n = L, \ldots, N+L-1 \end{cases} \tag{2.107}$$

The received OFDM symbol is the convolution of the transmitted OFDM symbol with the channel:

$$r_n = \sum_{l=0}^{L-1} h_l \cdot z_{n-l}, \quad n = 0, \ldots, N+L-1 \tag{2.108}$$

At the receiver the first L samples of the received OFDM symbol r are discarded to avoid ISI. Then the input of the DFT comprises the block of N samples $r_n, n = L, \ldots, N+L-1$. The DFT output in this case can be written as follows:

$$\hat{X}_k = \sum_{n=0}^{N-1} r_{n+L} \cdot \exp\left(-\frac{2\pi j}{N} kn\right), \quad k = 0, \ldots, N-1 \tag{2.109}$$

Substituting (2.108) in (2.109) obtain:

$$\hat{X}_k = \sum_{n=0}^{N-1} \sum_{l=0}^{L-1} h_l \cdot z_{n+L-l} \cdot \exp\left(-\frac{2\pi j}{N} kn\right), \quad k = 0, \ldots, N-1 \tag{2.110}$$

Notice that for $l = 0$ z_{n+L-l} values are taken from x_0 to x_{N-1}, for $l = 1$ z_{n+L-l} takes values $0, x_0, \ldots, x_{N-2}$, and for $l = L$ - 1 z_{n+L-l} takes values $\underbrace{0, \ldots, 0}_{L-1}, x_0, \ldots, x_{N-L+1}$. Then (2.110) can be written

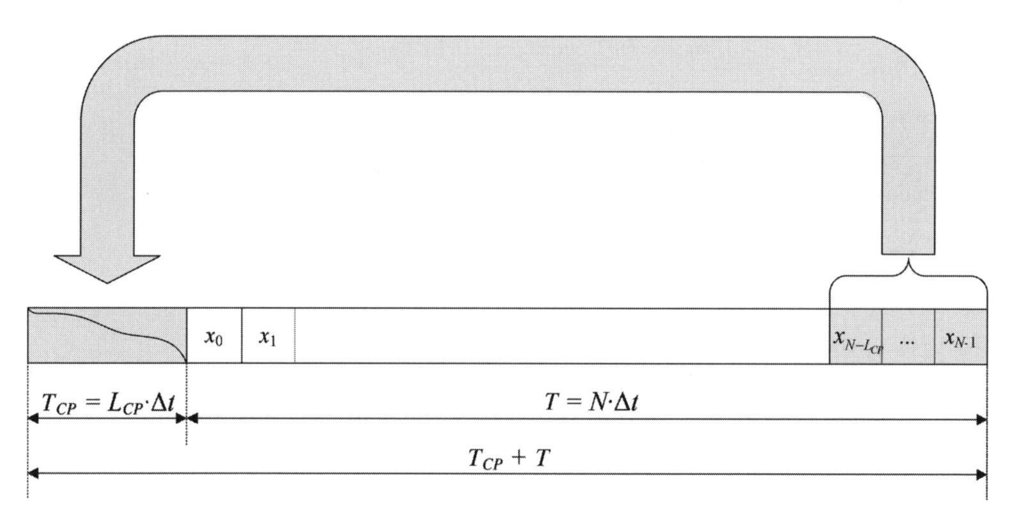

Figure 2.61 Extension of an OFDM symbol with cyclic prefix (CP)

as follows:

$$\hat{X}_k = \sum_{n=0}^{N-1} \sum_{l=0}^{L-1} h_l \cdot x_{(n-l) \bmod N} \cdot \exp\left(-\frac{2\pi j}{N} kn\right) - \underbrace{\sum_{n=0}^{L-2} \sum_{l=n+1}^{L-1} h_l \cdot x_{N-l+n} \cdot \exp\left(-\frac{2\pi j}{N} kn\right)}_{\text{ICI}}, \tag{2.111}$$

$$k = 0, \ldots, N-1$$

The first term in the expression (2.111) is the desired signal from which as will be shown later the transmitted symbol X_k can be easily extracted. The second term in (2.111) is the ICI. Obviously the ICI in this case can be easily eliminated if the zeros in the guard interval, that is, the elements $z_n = 0, n = 0, \ldots, L-1$ in (2.107) will be substituted by the last L samples of the OFDM symbol $x_{N-L}, \ldots, x_{N-1}$ as shown in Figure 2.61. The sequence $x_{N-L}, \ldots, x_{N-1}$ transmitted in the guard period is called *cyclic prefix* [14]. The number of samples in the cyclic prefix L_{CP} should be no less than the channel length $L_{CP} \geq L$.

In this case the obtained OFDM symbol before the transmission is extended with the cyclic prefix of length $L_{CP} \geq L$ and no guard interval is used. Then (2.107) takes the form of:

$$\begin{cases} z_n = x_{N-L_{CP}+n}, & n = 0, \ldots, L_{CP}-1 \\ z_n = x_{n-L_{CP}}, & n = L_{CP}, \ldots, N+L_{CP}-1 \end{cases} \tag{2.112}$$

Then the extended OFDM symbol is convolved with the channel of length L ($L \leq L_{CP}$) as described in (2.108). Now let us consider the multipath channel model including AWGN. Then the samples of the received OFDM symbol can be written as follows:

$$r_n = \sum_{l=0}^{L-1} h_l \cdot z_{n-l} + w_n, \quad n = 0, \ldots, N+L-1 \tag{2.113}$$

where w_n is AWGN. Now at the receiver the first L_{CP} rather than L samples of the received OFDM symbol $\mathbf{r}$ are discarded. Obviously, by this the ISI is eliminated since the length of cyclic prefix is greater than the channel span. Then the input of the DFT comprises the block of N samples r_n,

$n = L_{CP}, \ldots, N + L_{CP} - 1$, and the DFT output is given by:

$$\hat{X}_k = \sum_{n=0}^{N-1} r_{n+L_{CP}} \cdot \exp\left(-\frac{2\pi j}{N}kn\right), \quad k = 0, \ldots, N-1 \tag{2.114}$$

Substituting (2.113) in (2.114) obtain:

$$\hat{X}_k = \sum_{n=0}^{N-1}\sum_{l=0}^{L-1} h_l \cdot z_{n+L_{CP}-l} \cdot \exp\left(-\frac{2\pi j}{N}kn\right) + \sum_{n=0}^{N-1} w_{n+L_{CP}} \cdot \exp\left(-\frac{2\pi j}{N}kn\right), \quad k = 0, \ldots, N-1 \tag{2.115}$$

Taking into account (2.112) the expression (2.115) can be written as:

$$\hat{X}_k = \sum_{n=0}^{N-1}\sum_{l=0}^{L-1} h_l \cdot x_{(n-l) \bmod N} \cdot \exp\left(-\frac{2\pi j}{N}kn\right) + w_k, \quad k = 0, \ldots, N-1 \tag{2.116}$$

where $w_k = \sum_{n=0}^{N-1} w_{n+L_{CP}} \cdot \exp\left(-\frac{2\pi j}{N}kn\right)$, $k = 0, \ldots, N-1$ is the result of the DFT operation on N AWGN samples. The DFT of AWGN results in AWGN. So, w_k can be regarded simply as an AWGN sample. As can be seen from (2.116) the usage of cyclic prefix allows the circular convolution to be faked. Then substituting (2.102) in the expression (2.116) yields the following formula:

$$\hat{X}_k = \sum_{n=0}^{N-1}\left(\sum_{l=0}^{L-1} h_l \cdot \frac{1}{N}\sum_{m=0}^{N-1} X_m \cdot \exp\left(\frac{2\pi j}{N}m(n-l)\right)\right) \cdot \exp\left(-\frac{2\pi j}{N}kn\right) + w_k, \quad k = 0, \ldots, N-1 \tag{2.117}$$

Taking in account that $h_l = 0$ for $l \geq L$, we can allow l to take values from 0 to $N-1$ instead of $L-1$. Then changing the order of summation by l and k the expression (2.117) can be written as follows:

$$\hat{X}_k = \sum_{n=0}^{N-1}\left(\frac{1}{N}\sum_{m=0}^{N-1}\left(\sum_{l=0}^{N-1} h_l \cdot \exp\left(-\frac{2\pi j}{N}ml\right)\right) \cdot X_m \cdot \exp\left(\frac{2\pi j}{N}mn\right)\right) \cdot \exp\left(-\frac{2\pi j}{N}kn\right) + w_k,$$
$$k = 0, \ldots, N-1 \tag{2.118}$$

Notice that $\sum_{l=0}^{N-1} h_l \cdot \exp\left(-\frac{2\pi j}{N}ml\right)$ is simply the DFT of the channel coefficients and the first part of the expression (2.118) contains the IDFT of X_k nested in DFT, which gives X_k. Then (2.118) may be rewritten in the following way:

$$\hat{X}_k = H_k \cdot X_k + w_k, \quad k = 0, \ldots, N-1 \tag{2.119}$$

where $H_k = \sum_{l=0}^{N-1} h_l \cdot \exp\left(-\frac{2\pi j}{N}kl\right)$. Due to the usage of the cyclic prefix it is quite easy now to recover the original symbols X_k. It is enough to calculate the DFT of channel estimates and obtain the estimates of values H_k, which can be used for the simple channel equalization consisting of the division of values $\hat{X}_k$ by these estimates, that is, it is enough to use just the one tap equalizer. The OFDM system using cyclic prefix is depicted in Figure 2.62.

As was mentioned above the usage of cyclic prefix decreases the system efficiency due to some capacity loss, since no new data is transmitted in the cyclic prefix. Except for the loss in data rate there is some loss in SNR due to insertion of the cyclic prefix. This loss is caused by the fact that the number of samples in the extended OFDM symbol increases with the same signal energy. The SNR loss due to cyclic prefix SNR_{CP} can be calculated as follows:

$$SNR_{CP} = \frac{T}{T_{CP} + T} = \frac{N \cdot \Delta t}{L_{CP} \cdot \Delta t + N \cdot \Delta t} = \frac{N}{L_{CP} + N} \tag{2.120}$$

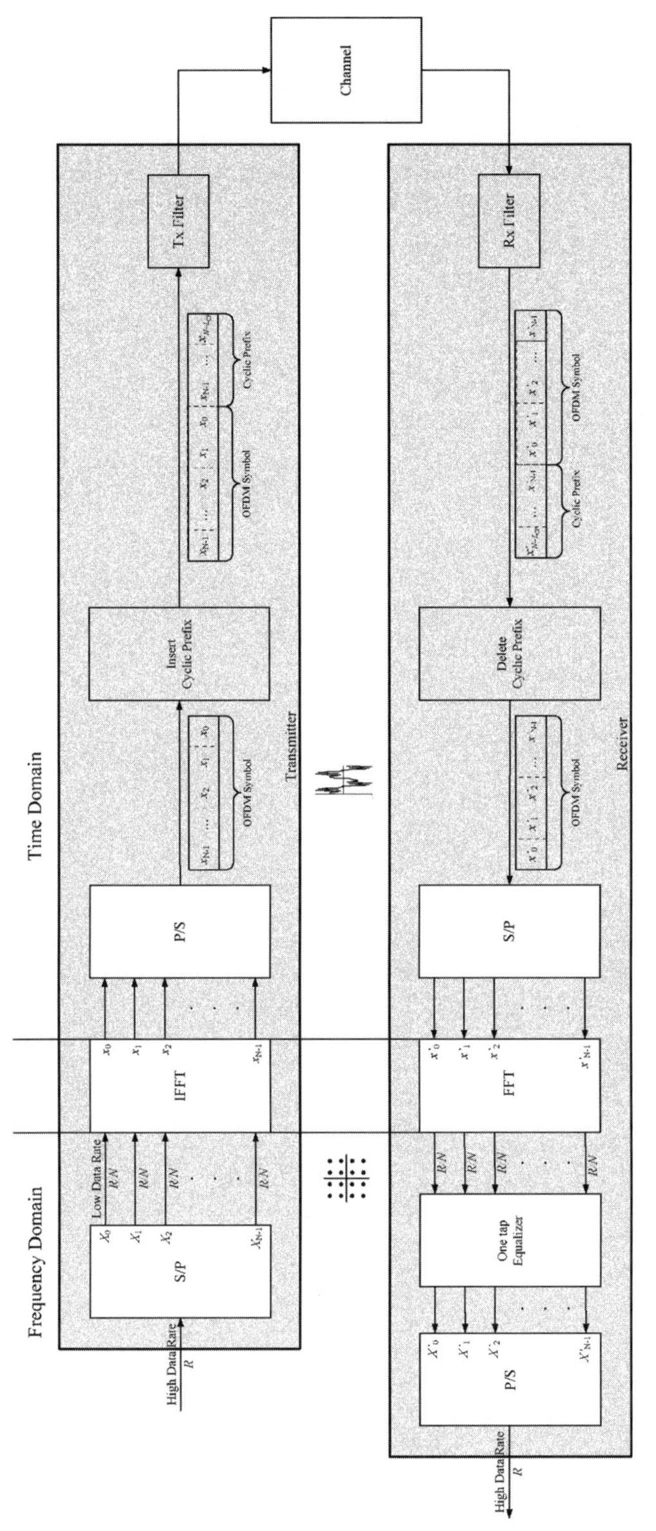

Figure 2.62 OFDM system with cyclic prefix (CP)

Because of this the cyclic prefix length L_{CP} should be chosen as short as possible. The best choice of L_{CP} is to set it exactly equal to the channel length L. However, in wireless communication the channel length L is not constant and the system designer should choose the cyclic prefix length L_{CP} just taking into account some reasonable constraints. Then in some cases it may appear that the channel length during some period of time will be greater than the cyclic prefix length. This of course leads to the appearance both ISI and ICI. If $L_{CP} < L$ the residual ISI for the kth sample of the DFT is given by [15]:

$$ISI_k = \sum_{n=0}^{N-1} \sum_{l=L_{CP}+n+1}^{L-1} h_l \cdot x_{n-l} \cdot \exp\left(-\frac{2\pi j}{N} nk\right) = \sum_{n=0}^{N-1} \sum_{l=L_{CP}+1}^{L-n-1} h_{n+l} \cdot x_{-l} \cdot \exp\left(-\frac{2\pi j}{N} nk\right) \quad (2.121)$$

Notice that here x_{-l} refers to the samples of previously transmitted OFDM symbol. Assuming that $N \geq L - L_{CP} - 1$ and changing the order of summation (2.121) can be rewritten as follows:

$$\begin{aligned} ISI_k &= \sum_{l=L_{CP}+1}^{L-1} x_{-l} \cdot \sum_{m=l}^{L-1} h_m \cdot \exp\left(-\frac{2\pi j}{N}(m-l)k\right) = \\ &= \sum_{l=L_{CP}+1}^{L-1} x_{-l} \cdot \exp\left(\frac{2\pi j}{N} lk\right) \cdot \sum_{m=l}^{L-1} h_m \cdot \exp\left(-\frac{2\pi j}{N} mk\right) = \sum_{l=L_{CP}+1}^{L-1} x_{-l} \cdot \exp\left(\frac{2\pi j}{N} lk\right) \cdot H_{k,l} \end{aligned} \quad (2.122)$$

where $H_{k,l} = \sum_{m=l}^{L-1} h_m \cdot \exp\left(-\frac{2\pi j}{N} mk\right)$ is the DFT of the tail of channel impulse response. Then the ISI power spectral density is given by [15]:

$$\begin{aligned} P_{ISI,k} &= E\left\{ISI_k \cdot ISI_k^*\right\} = \sum_{l=L_{CP}+1}^{L-1} \sum_{m=L_{CP}+1}^{L-1} H_{k,l} \cdot H_{k,m}^* \cdot E\left\{x_{-l} \cdot x_{-m}^*\right\} \cdot \exp\left(\frac{2\pi j}{N} lk\right) \\ &= \sigma_x^2 \cdot \sum_{l=L_{CP}+1}^{L-1} \left|H_{k,l}\right|^2 \end{aligned} \quad (2.123)$$

The ICI appears in case $L_{CP} < L$ because the cyclic prefix is not long enough to mimic the convolution of an extended OFDM symbol with the channel as a cyclic convolution. To do this the length of cyclic prefix should be extended at least up to the length of the channel L. Then the ICI is easy to calculate as the negative of the convolution of hypothetic extension of the cyclic prefix length up to L with the channel [15]:

$$ICI_k = -\sum_{n=0}^{N-1} \sum_{l=L_{CP}+n+1}^{L-1} h_l \cdot x_{(n-l) \bmod N} \cdot \exp\left(-\frac{2\pi j}{N} nk\right) \quad (2.124)$$

The expression (2.124) is very similar to (2.121). The only difference is that in this case x_i refers to the samples of the current OFDM symbol not involving a computation of the samples of the previously transmitted OFDM symbol as in (2.121). Assuming that samples of all OFDM symbols have the same distribution it is possible to say that the ICI power spectral density coincides with $P_{ISI,k}$:

$$P_{ICI,k} = P_{ISI,k} \quad (2.125)$$

Unfortunately the OFDM is quite sensitive to some system impairments, for example, to the frequency offset, which can be caused by non-perfectly synchronized oscillators at the transmitter and the receiver. This leads to a frequency shift of the received signal spectrum. Due to this fact the IDFT and DFT operation in the transmitter and receiver does not correspond exactly to each other anymore. This effect is called the "*DFT leakage*". It is easier to explain it using the OFDM system model without DFT.

Recall the representation of the OFDM as a set of N subcarriers $\psi_n = \exp(j2\pi n f_0 t) = \exp\left(\frac{j2\pi nt}{T}\right), n = 0, \ldots, N-1$ as in (2.104). Consider two subcarriers with indexes k and $k+m$. To restore the data transmitted by these subcarriers the received signal should be multiplied by the complex conjugate of the corresponding subcarriers $\psi_k^* = \exp\left(-\frac{j2\pi kt}{T}\right)$ and $\psi_{k+m}^* = \exp\left(-\frac{j2\pi(k+m)t}{T}\right)$. Now assume that the subcarriers' pulses are generated at the receiver with some frequency offset δ, ($|\delta| < 0.5$), that is instead of pulse ψ_{k+m}^* the offset pulse $\psi_{k+m+\delta}^* = \exp\left(-\frac{j2\pi(k+m+\delta)t}{T}\right)$ is generated. In this case the subcarriers are losing orthogonality and the ICI arises, that is, the energy from one subcarrier leaks into adjacent subcarriers. Then the ICI between subcarriers k and $k+m$ can be written as follows:

$$ICI_{m,\delta} = \int_0^T \psi_k \cdot \psi_{k+m+\delta}^* dt = \int_0^T \exp\left(\frac{j2\pi kt}{T}\right) \cdot \exp\left(-\frac{j2\pi(k+m+\delta)t}{T}\right) dt = \frac{T(1-\exp(-j2\pi\delta))}{j2\pi(m+\delta)} \tag{2.126}$$

The power of interference between subcarriers k and $k+m$ is given by:

$$P_{ICI_{m,\delta}} = ICI_{m,\delta} \cdot ICI_{m,\delta}^* = \frac{T^2 \sin^2(\pi\delta)}{\pi^2(m+\delta)^2} \tag{2.127}$$

In Figure 2.63 the "leakage" of the energy from the kth subcarrier to the $(k+m)$th subcarrier (actually the power of ICI) for the case where there is no frequency shift and for the case $\delta = 0.1$ is plotted. As can be seen from the plot if there were no frequency shift ($\delta = 0$), all the subcarrier energy is concentrated in the frequency corresponding to the kth subcarrier and the energy does not spill into adjacent subcarriers at all. In the case of frequency shift mostly the adjacent subcarriers suffer from the ICI.

The dependence of the ICI power on the frequency offset is depicted in Figure 2.64. As can be seen from the plot the ICI power grows exponentially with the growth of the frequency offset.

The overall ICI energy from $N-1$ subcarriers to the kth subcarrier can be estimated as follows:

$$E_{ICI}(k,\delta) = \sum_{m=-k}^{-1} P_{ICI_{m,\delta}} + \sum_{m=1}^{N-k-1} P_{ICI_{m,\delta}} \tag{2.128}$$

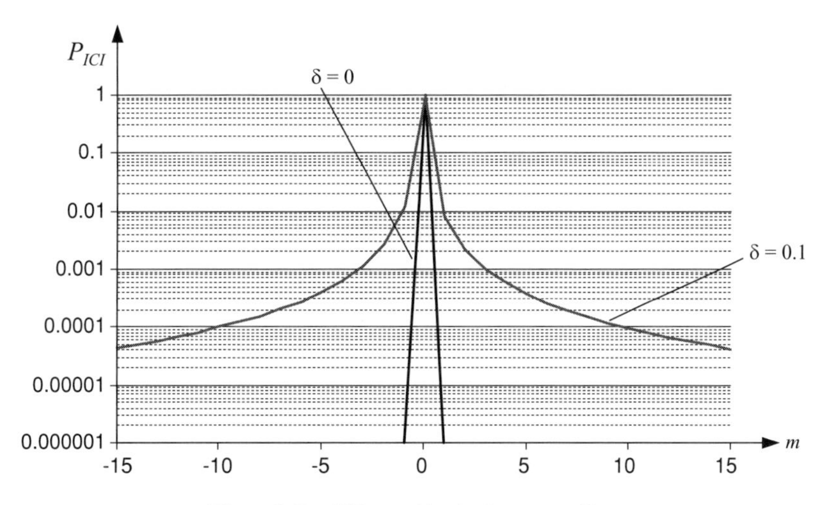

Figure 2.63 ICI caused by the frequency offset

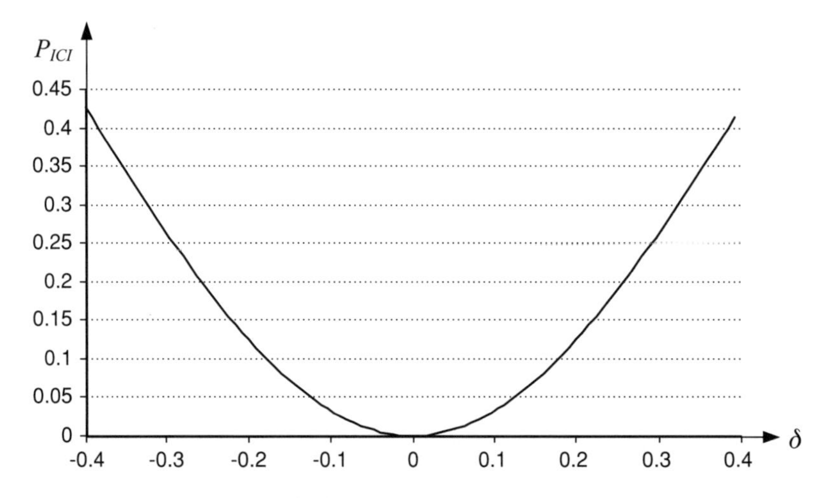

Figure 2.64 ICI power vs. frequency offset

The plot of ICI energy against the position of subcarrier k is depicted in Figure 2.65. Obviously, the ICI energy in this case is less for the subcarriers, which are closer to the border of an OFDM symbol since at this position the subcarrier has less adjacent subcarriers. However, as can be seen from the plot for most subcarriers the overall ICI caused by the frequency offset and originated from over $N - 1$ subcarriers is very similar and in many cases can be represented as an AWGN.

The most popular way to cope with the ICI caused by the frequency offset is to estimate the frequency offset and then to compensate it. Usually it is done with the help of a *phase-locked loop* (*PLL*).

In the mobile radio environment, the relative movement between transmitter and receiver causes Doppler frequency shifts in the received signal. This effect is another source of ICI in an OFDM system. Actually, that means the multipath channel model cannot be considered simply as a filter of length L with the coefficients $h_l, l = 0, \ldots, L - 1$ anymore. Now the channel coefficients are changing with time.

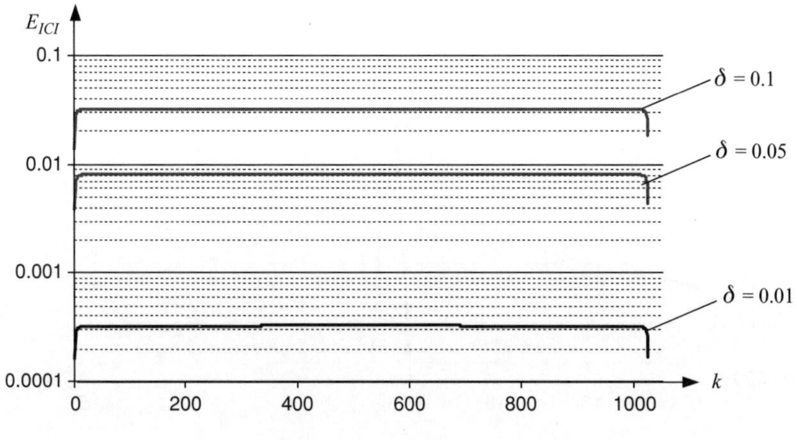

Figure 2.65 Selective mapping

Then (2.113) can be rewritten as:

$$r_n = \sum_{l=0}^{L-1} h_{l,n} \cdot z_{n-l} + w_n, \quad n = 0, \ldots, N + L - 1 \tag{2.129}$$

where $h_{l,n}$ is the value of the channel impulse response at the position l and instant n. Then the demodulated symbols can be written as follows [16]:

$$\hat{X}_k = \sum_{n=0}^{N-1} \sum_{l=0}^{L-1} X_n \cdot H_l(n-k) \cdot \exp\left(-\frac{2\pi j}{N} nl\right) + w_k, \quad k = 0, \ldots, N-1 \tag{2.130}$$

where

$$H_l(n-k) = \frac{1}{N} \sum_{m=0}^{N-1} h_{l,L_{CP}+(m-L_{CP}) \bmod N} \cdot \exp\left(\frac{2\pi j}{N} m(n-k)\right) \tag{2.131}$$

Let us denote:

$$H_k = \sum_{l=0}^{L-1} H_l(0) \cdot \exp\left(-\frac{2\pi j}{N} kl\right) \tag{2.132}$$

and

$$ICI_k = \sum_{n \neq k}^{N-1} \sum_{l=0}^{L-1} X_n \cdot H_l(n-k) \cdot \exp\left(-\frac{2\pi j}{N} nl\right) \tag{2.133}$$

Then (2.130) can be written as follows:

$$\hat{X}_k = X_k \cdot H_k + ICI_k + w_k, \quad k = 0, \ldots, N-1 \tag{2.134}$$

where H_k is the DFT of the tail of channel impulse response and ICI_k is the inter-carrier interference caused by the Doppler shift and w_k is the AWGN. Obviously, if the channel were time-invariant, that is, $h_{l,k} = h_l$, then $ICI_k = 0$ and (2.134) coincides with (2.119). It is shown in [16] that with sufficiently large value of N the inter-carrier interference ICI_k can be represented as a Gaussian random variable with zero mean and variance:

$$C_0^2 = E_s - \frac{E_s}{N^2} \left(N + 2 \sum_{i=1}^{N-1} (N-i) \cdot J_0(2\pi f_D T i) \right) \tag{2.135}$$

where f_D is the maximum Doppler frequency, E_s is the symbol energy and $J_0()$ is the zero-order Bessel function of the first kind.

The most common method to cope with ICI is the frequency domain equalization with the help of training signals. In this case it is not enough to use the simple one-tap equalizer and the receiver implementation complexity significantly increases since it is necessary to use N equalizers. Moreover, the bandwidth efficiency is reduced by training signals.

Another method is the time-domain windowing method comprising the multiplying of the transmitted time-domain signals by a well-designed windowing function. Actually the extension of the OFDM symbol with cyclic prefix is also the time-domain windowing. The drawback of this method is that some additional subcarriers are needed for windowing, which are discarded at the receiver. Because not all of the received signal power is being used in generating data estimates, the method has a reduced overall SNR compared with OFDM without windowing.

The third method is so called *self ICI cancellation* [17]. This method maps the data to be transmitted onto adjacent pairs of subcarriers with a 180-phase difference between them rather than onto single subcarriers. The disadvantage of this method is that it is less bandwidth efficient as two subcarriers are used to transmit one complex value. To some extent the efficiency of this method can be increased if some groups of subcarriers are used for mapping rather than just adjacent pairs [18].

Another serious problem arising for an OFDM system is high *peak to average power ratio* (*PAPR*). If the signal comprises the sum of N sinusoids each of maximum amplitude A, then it is possible that at some point all sinusoids will be added with the maximum amplitudes and the amplitude of resulting signal at this point will be NA. These large peaks disturb out of band energy and therefore increase in-band noise and can cause the saturation in power amplifiers, leading to increasing BER when the signal has to go through the amplifier non-linearity. Of course, it is pretty unlikely that all N subcarriers will be added at their maximum point (especially if N is large enough) but still it is possible. Therefore, it is desirable to reduce the PAPR. The mathematical definition of the PAPR is given by [19]:

$$PAPR = \frac{\max\limits_{t \in [0,T]} |s(t)|^2}{E\left\{|s(t)|^2\right\}} \tag{2.136}$$

where $s(t)$ is the modulated OFDM signal and $E\{\}$ is the expectation. For large values of N assuming that $s(t)$ is the sum of N sinusoids it is possible to apply the central limit theorem to $s(t)$. In accordance with central limit theorem the real and imaginary part of $s(t)$ have the Gaussian distribution. That means the amplitude of the OFDM signal has a Rayleigh distribution with zero mean and a variance of Nv, where v is the variance of one subcarrier. Then the cumulative distribution function for the PAPR per OFDM symbol is given by [19]:

$$\Pr\{PAPR > \gamma\} = 1 - (1 - e^{\gamma})^N \tag{2.137}$$

The expression (2.137) confirms that large peaks occur quite rarely. However, even rare large peaks can distort signal significantly. Assume $N = 1024$ and each subcarrier has the power of 1. Then the maximum PAPR can reach 30 dB and even the average PAPR can be around 15 dB.

One obvious approach to reduce PAPR is clipping. Since the probability of large peaks occurring is quite low it is possible just to clip the signal at some desirable level. However, clipping leads to so called *clipping noise*, which degrades BER, and out of band noise, which reduces the spectral efficiency. Filtering after clipping can reduce the out of band noise but may also cause some peak regrowth [20].

A different approach is to apply some shaped window with good spectral properties to the large signal peak [21]. Since the OFDM signal is multiplied with several of these windows the resulting spectrum is a convolution of the original OFDM spectrum with the spectrum of the applied window. So, ideally, the window should be as narrow band as possible. On the other hand, the window should not be too long in the time domain, because otherwise many signal samples are affected, which in turn increases BER.

Another approach usually called *selective mapping* consists in mapping the data by a set of codes, then taking the IDFT of the obtained signals and picking the one providing least PAPR [22], [23], [24], [25] as is shown in Figure 2.66. Usually the data sequence with evenly distributed ones and zeros provides the signal with lower PAPR. So, the codes used in this method should have this property. However, the drawback of this method is that the use of codes with this property decreases the data rate. Also the side information to restore data from the demodulated sequence should be transmitted, which decreases the spectral efficiency. Moreover, as the number of subcarriers increases this method becomes less attractive since the memory needed to store the codebook and the CPU time needed to find the corresponding sequence grows exponentially with the number of subcarriers.

One more very popular method of decreasing the PAPR is the usage of artificial signals in empty subcarriers together with convex optimization [26], [27]. In literature this method quite often is called *tone reservation*. In many cases when transmitting data with the help of N-point DFT and IFFT, not all N frequencies carry data. Thus, we have a few empty carriers per OFDM symbol. The basic premise of this technique is to add sine waves at these empty carrier frequencies in such a way that the composite OFDM symbol will have a lower PAPR. If a data vector contains zeros on positions $X_j = 0$ for $j \in \{j_1, \ldots j_L\}$, then the transmitter can add any vector $\mathbf{C}$ that satisfies, $C_j = 0$ for $j \notin \{j_1, \ldots j_L\}$ to the data vector

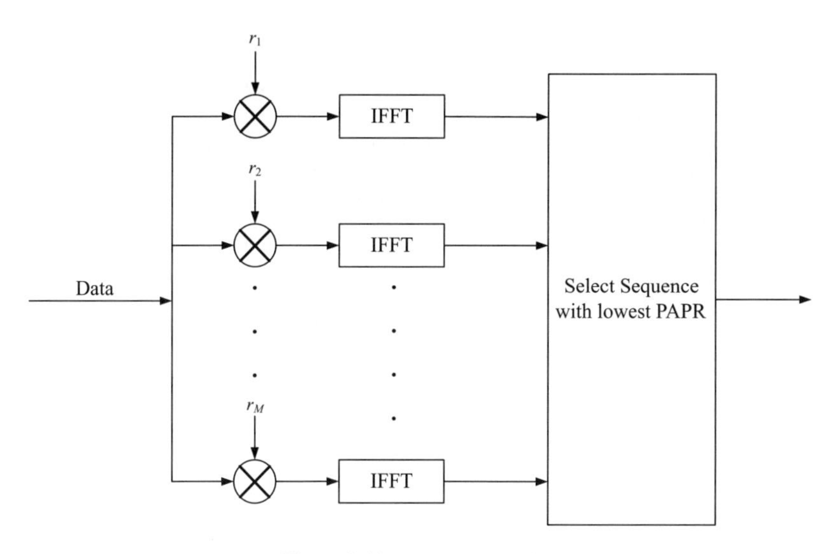

Figure 2.66 Selective mapping

and remove it at the receiver. The IDFT of sum of vectors **X** and **C** results in sum of IDFT of vector **X** and IDFT of vector **C**:

$$IDFT(\mathbf{X} + \mathbf{C}) = \mathbf{x} + \mathbf{c} \tag{2.138}$$

To minimize the PAPR of ($\mathbf{x} + \mathbf{c}$), we must compute the vector $\mathbf{c}_0$ that minimizes the maximum peak value:

$$\mathbf{c}_0 : \max_j \left| x_j + c_{0,j} \right|^2 = \min_{\mathbf{c}} \left\{ \max_j \left| x_j + c_j \right|^2 \right\}, \quad j = 0, \ldots, N-1 \tag{2.139}$$

With this formation, optimizing the time domain signal leads to a convex optimization problem that can be transformed into a linear programming (LP) problem. Solving the LP problem exactly leads to PAPR reduction of 6-10dB, but a simple gradient algorithm could achieve most of this reduction after a few iterations.

References

[1] Trott, M. *The Mathematica Guidebook for Programming.* New York: Springer-Verlag, 2004.
[2] Bracewell, R. *The Fourier Transform and Its Applications*, 3rd edn. New York: McGraw-Hill, 1999.
[3] Proakis, J.G. *Digital Communications*, 3rd edn., McGraw-Hill, 1995.
[4] Nyquist, H. *Certain Topics of Telegraph Transmission Theory.* AIEE Transactions, vol. 47, Apr. 1928, pp. 617–644.
[5] Sklar, B. *Digital Communications. Fundamentals and Applications*, 2nd edn., Prentice Hall, 2001.
[6] Bityutskov, V.I. *Bunyakovskii Inequality*, in Hazewinkel, M., *Encyclopaedia of Mathematics*, Kluwer Academic Publishers, 2001.
[7] Debnath, L. and Mikusinski, P., *Hilbert Spaces with Applications*, 3rd edn., Elsevier, 2005.
[8] Burr. A. *Modulation and Coding for Wireless Communications*, Prentice Hall, 2001.
[9] Doelz, M.L., Heald, E.T., Martin, D.L., *Binary Data Transmission Techniques for Linear Systems.* Proc. IRE, vol. 45, May. 1957, pp. 656–661.
[10] Franco, G.A., Lachs, G., *An Orthogonal Coding Technique for Communications.* IRE Int. Conv. Rec., vol. 9, 1961, pp. 126–133.

[11] Chang, R.W., *Synthesis of Band-Limited Orthogonal Signals for Multichannel Data Transmission*. Bell Syst. Tech. J., vol. 45, Dec. 1966, pp. 1775–1796.
[12] Saltzberg, B.R., *Performance of an Efficient Parallel Data Transmission System*. IEEE Trans. Commun. Tech., vol. 15, Dec. 1967, pp. 805–811.
[13] Weinstein, S.B., Ebert. P.M., *Data Transmission by Frequency-Division Multiplexing Using the Discrete Fourier Transform*. IEEE Trans. Commun., vol. 19, May 1971, pp. 628–634.
[14] Peled, A., Ruiz. A., *Frequency Domain Data Transmission Using Reduced Computational Complexity Algorithms*. IEEE International Conf. on Acoustics, Speech and Signal Proc. (ICASSP'80), 1980, pp. 964–967.
[15] Henkel, W., Tauböck. G., Ödling, P., Börjesson, P.O., Petersen, N., Johansson, A., *The Cyclic Prefix of OFDM/DMT – An Analysis*. IEEE International Zurich Seminar on Broadband Communications, Feb. 2002, pp. 221–223.
[16] Russell, M., Stuber. G.L., *Interchannel Interference Analysis of OFDM in a Mobile Environment*. IEEE Vehicular Technology Conference, Jul. 1995, vol. 2, pp. 820–824.
[17] Zhao, Y. and Häggman, S.G., *Sensitivity to Doppler Shift and Carrier Frequency Errors in OFDM Systems-The Consequences and Solutions*. IEEE Vehicular Technology Conference, Jul. 1996, pp. 2474–2478.
[18] Armstrong, J., *Analysis of new and existing methods of intercarrier interference due to carrier frequency offset in OFDM*. IEEE Trans. Comm., Vol. 47, No 3, Mar. 1999, pp. 365–369.
[19] Muller, S.H., Huber, J.B., *A novel peak power reduction scheme for OFDM*. IEEE Intern. Symposium on Personal, Indoor and Mobile Radio Communications, Sep. 1997, vol. 3, pp. 1090–1094.
[20] Bahai, A.R.S., Singh, M., Goldsmith, A.J., Saltzberg, B.R., *A new approach for evaluating clipping distortion in Multicarrier systems*. IEEE Journal on Selected Areas in Communications, vol. 20, 2002, pp. 3–11.
[21] Van Nee, R., Wild, A., *Reducing the Peak to Average Power Ratio of OFDM*. IEEE Vehicular Technology Conference, Jul. 1998, pp. 2072–2076.
[22] Jones, A.E, Wilkinson, T.A, Barton, S.K, *Block coding scheme for reduction of peak to mean envelope power ratio of multicarrier transmission schemes*. Electronics Letters, Dec. 1994, vol.30, No. 25, pp. 2098–2099.
[23] Eetvelt, P.V., Wade, G. and Tomlinson, M., *Peak to Average Power Reduction for OFDM Schemes by Selective Scrambling*. IEE Electronics Letters, Aug. 1996, vol.32, No. 21, pp. 1963–1964.
[24] Braithwaite, R.N., *Using Walsh code selection to reduce the power variance of band-limited forward link CDMA waveforms*. IEEE Journal on Selected Areas in Communications, vol. 18, 2000, pp. 2260–2269.
[25] Baml, R.W., Fischer, R.F.H. and Huber, J.B., *Reducing the Peak to Average Power Ratio of Multicarrier Modulation by Selected Mapping*. IEE Electronics Letters, vol. 32, No. 22, Sep., 2000, pp. 2056–2057.
[26] Tellado, J., Cioffi, J., *Peak power reduction for multicarrier transmission*. Proceedings Globecom'98, Sydney, Australia, 1998.
[27] Yang Jun, Yang Jiawei, Li Jiandong, *Reduction of the peak-to-average power ratio of the multicarrier signal via artificial signals*. Intern. Conf. on Communication Technology WCC - ICCT 2000., vol. 1, No. 22, 2000, pp. 581–585.

3

Block Codes

Grigorii Kabatiansky[1], Evgenii Krouk[2], Andrei Ovchinnikov[2], and Sergei Semenov[3]
[1] *Institute for Information Transmission Problems, Russian Academy of Sciences, Russia*
[2] *St Petersburg State University of Aerospace Instrumentation, Russia*
[3] *Nokia Corporation, Finland*

This chapter introduces the theory of block codes. Here we will describe mainly the features of block codes which correct the independent errors. The importance of considering this class of codes can be explained by its significance for practice and by the fact that analyses of these codes allows the main methods and results of coding theory to be shown.

3.1 Main Definitions

Let us consider in accordance with Shannon [1] the model of data transmission system represented in Figure 3.1. A data source generates messages $\mathbf{u}_1, \ldots, \mathbf{u}_M$ and a receiver needs to receive them correctly (with high reliability). The data source and the receiver are connected by a channel allowing transmitting symbols from an input alphabet (set) A transmitting in a sequential way. However, due to some noise in the channel the output sequence may differ from the input one. Moreover, in general the input alphabet A and the output alphabet B do not coincide. The probabilistic model of the channel is given by transition probabilities $P(\mathbf{b}|\mathbf{a})$ that an output sequence (of symbols) is $\mathbf{b}$ under condition that an input sequence was $\mathbf{a}$. We restrict our consideration to the case already well explored in coding theory q-ary memoryless channels, for which:

- input and output alphabets coincide;
- cardinal number of the input (and the output) alphabet equals to q;
- statistic characteristics of the output symbol are fully defined by the input symbol (that is, there is no memory in the channel);
- statistic characteristics of the output symbol do not depend on time.

Modulation and Coding Techniques in Wireless Communications Edited by Evgenii Krouk and Sergei Semenov

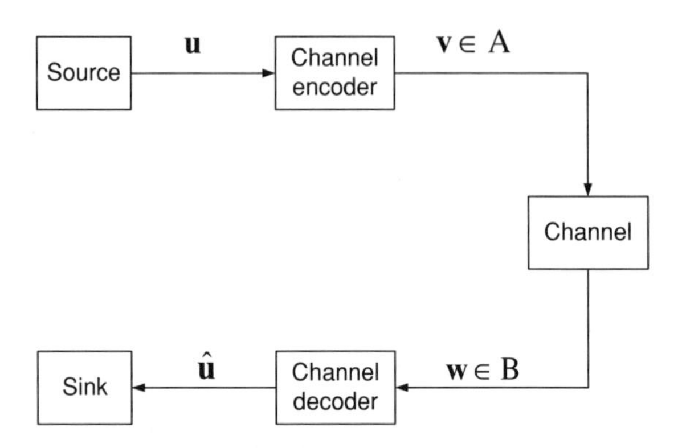

Figure 3.1 Model of data communication system

An important case of this kind of channels is a binary symmetric channel (BSC), where the probability that output (binary) symbol coincides with the input symbol is equal to Q and the probability that output symbols differ from the input symbol is equal to $P = 1 - Q$.

To provide the reliable data transmission messages should be encoded, that is, to each message $\mathbf{u}_i$ corresponds a finite sequence $\mathbf{v}_i = (v_{i1}, \ldots, v_{in_i})$ of symbols of the alphabet A. This sequence is called codeword. The codeword is transmitted over the channel. The set of all codewords is called a code and the mapping $\varphi : \mathbf{u} \to \mathbf{v}$ is called an encoding procedure. If codewords consist of the same number of symbols ($n_i = n$), then a code is called a block code of length n. We assume that an encoding maps different messages to different codewords. Hence a block code is an arbitrary subset V(of cardinality M) of the vector space E_q^n of all q-ary words of length n.

The receiver tries to restore the source message $\mathbf{u}$ relying on the received output sequence $\mathbf{w}$. The corresponding mapping $\Psi : \mathbf{w} \to \hat{\mathbf{u}}$ is called the decoding procedure. This mapping to some extent is the reverse mapping to encoding. Devices, which realize encoding and decoding procedures, are called the encoder and the decoder respectively. Due to the presence of noise in the channel $\mathbf{w}$ may differ from $\mathbf{v}$. In this case message $\Psi(\mathbf{w}) = \hat{\mathbf{u}}$ may differ from source message $\mathbf{u}$ as well. This event is called the decoding error. Since there is reciprocation between the messages and the codewords, then it is possible to consider the mapping $\psi = \varphi \circ \Psi : \mathbf{w} \to \hat{\mathbf{v}} = \varphi(\hat{\mathbf{u}})$ instead of mapping Ψ. Mapping ψ is also called the decoding procedure. It is more convenient to consider the mapping ψ, because the definition of ψ is equivalent to a partition of the set A^n of all words of length n on decision regions Ω_i, so that $\mathrm{A}^n = \bigcup_{i=1}^{M} \Omega_i$; $\Omega_i \cap \Omega_j = \emptyset, i \neq j$, where $\Omega_i = \{\mathbf{w} \in \mathrm{A}^n : \psi(\mathbf{w}) = \mathbf{v}_i\}$. It is intuitively clear that to minimize the probability of decoding error it is necessary to include in Ω_i words of A^n, which are "close enough" to $\mathbf{v}_i$; where the measure of "closeness" should be agreed with the channel in the sense that the closer two words are, the more probable that one word will be received from the output of the channel if another word was fed to the channel input. Such a kind of measure of "closeness" for BSC is the Hamming distance [2].

The *Hamming distance* $d(\mathbf{a}, \mathbf{b})$ between words $\mathbf{a} = (a_1, \ldots, a_n) \in \mathrm{A}^n$ and $\mathbf{b} = (b_1, \ldots, b_n) \in \mathrm{A}^n$ is defined as the number of positions where these words are different.

It is easy to verify that Hamming distance is the metric, that is:

$$\left.\begin{array}{l} d(\mathbf{a}, \mathbf{b}) > 0, \mathbf{a} \neq \mathbf{b}, d(\mathbf{a}, \mathbf{a}) = 0; \\ d(\mathbf{a}, \mathbf{b}) = d(\mathbf{b}, \mathbf{a}); \\ d(\mathbf{a}, \mathbf{c}) \leq d(\mathbf{a}, \mathbf{b}) + d(\mathbf{b}, \mathbf{c}) \end{array}\right\} \tag{3.1}$$

By definition the Hamming distance between transmitted and received words is equal to the number of errors which occurred during transmission over the channel. Therefore we can say that a decoding ψ of a code V corrects t errors, if the decoding result is always correct, on condition that there were no more than t errors during the data transmission over the channel, that is, $\psi(\mathbf{w}) = \mathbf{v}$, if $d(\mathbf{w}, \mathbf{v}) \leq t$. One of the most important decoding procedures is the *minimum distance (MD) decoding*, which for a given received word $\mathbf{w}$ outputs the closest (in Hamming distance) to codeword $\hat{\mathbf{v}}$ (if there are several such codewords, then any of them).

The following notion in many ways characterizes the capability of a code to correct errors. The minimum distance (shortly, distance) $d(V)$ of a code V is the minimum of pairwise Hamming distances between different codewords, that is:

$$d(V) = \min_{\mathbf{v},\mathbf{v}' \in V; \mathbf{v} \neq \mathbf{v}'} d(\mathbf{v}, \mathbf{v}') \tag{3.2}$$

Let us denote a code of length n, cardinal number M and distance d as (n, M, d) code.

It is easy to check that a decoding ψ corrects t errors if and only if for any $i \in 1, \ldots, M$ the decision region Ω_i contains the n-dimensional (hyper)sphere:

$$\mathrm{A}_t^n(\mathbf{v}_i) = \{\mathbf{x} \in \mathrm{A}^n : d(\mathbf{x}, \mathbf{v}_i) \leq t\} \tag{3.3}$$

of radius t and with the centre at the point $\mathbf{v}_i$. Since decision regions Ω_i and Ω_j do not intersect for $i \neq j$, then the corresponding spheres do not intersect either; and this fact, by virtue of properties of metric (3.1), is equivalent to the property: $d(\mathbf{v}_i, \mathbf{v}_j) > 2t$ for all $i \neq j$. So we derive one of the fundamental results of coding theory: code V corrects t errors if and only if $d(V) \geq 2t + 1$.

For many applications it is convenient to extend the definition of encoding procedure by allowing "denial decoding", that is, to define one more decision region Ω_* consisting of received words for which no decision about the transmitted word was done. This kind of decoding is called partial decoding (different from the full decoding described above) and it is the mapping $\psi : \mathrm{A}^n \to \mathrm{A}^n \cup \{*\}$. The most important example of partial decoding is the error detection ψ_0 which refuses to output a codeword if a given received word is not from the code (that is, produce error detection mark *) and in case a received word is a codeword it is assumed that this word was uncorrupted. Thus, $\psi_0(\mathbf{v}_i) = \mathbf{v}_i$ and $\psi_0(\mathbf{w}) = *$ for $\mathbf{w} \notin V$. The generalization of error detection is decoding procedure ψ_t, which corrects no more than t errors. For this method of decoding the decision regions Ω_i coincide with spheres $\mathrm{A}_t^n(\mathbf{v}_i)$ of radius t around $\mathbf{v}_i$ and $\Omega_* = \mathrm{A}^n \backslash \left\{\bigcup_{i=1}^M \mathrm{A}_t^n(\mathbf{v}_i)\right\}$, that is, an error will be detected if the distance between the received word and arbitrary codeword is more than t.

We will say that the decoding procedure ψ of code V is capable of correcting t errors and detecting s errors $s > t$ if the decoding result is the transmitted codeword in cases where no more than t errors occur during transmission, or the decoding result is the error detection in cases where more than t but no more than s errors occur. It is obvious that this is equivalent to the following condition: $\Omega_i = \mathrm{A}_t^n(\mathbf{v}_i)$ and $\mathrm{A}_t^n(\mathbf{v}_i) \cap \mathrm{A}_s^n(\mathbf{v}_j) = 0$ for all $i \neq j$. And the obtained condition, in turn, is equivalent to $d(V) > t + s$. Thus the code V is capable of correcting t errors and detecting s errors $s > t$ if and only if its distance $d(V) \geq t + s + 1$.

Example 3.1 Consider a binary code $V = \{\mathbf{v}_0 = (00000), \mathbf{v}_1 = (11111)\}$ of length 5 and cardinal number 2. The distance of this code is 5. The decision region $\Omega_0 = \mathrm{A}_2^5(\mathbf{v}_0)$ for MD (minimum distance) decoding consists of those binary sequences of length 5 in which the number of zeros is more than the number of ones; and, vice versa, $\Omega_1 = \mathrm{A}_2^5(\mathbf{v}_1)$ consists of sequences in which the number of ones is more than the number of zeros. In other words, the MD decoding in this case is the majority decoding, that is, $\psi_2(\mathbf{w}) = \mathbf{v}_i$, where $i = \mathrm{maj}(w_1, w_2, w_3, w_4, w_5)$. The decoding procedure ψ_1 with the decision regions $\Omega_0 = \mathrm{A}_1^5(\mathbf{v}_0) = \{(00000), (10000), (01000), (00100), (00010), (00001)\}$ and

$\Omega_1 = A_1^5(\mathbf{v}_1) = \{(11111), (01111), (10111), (11011), (11101), (11110)\}$ is capable of correcting one error and detecting two and three errors.

From the above it follows that the greater the code distance, the more errors the code can correct. Thus, one of the main tasks of coding theory is to find optimal codes. An (n, M, d) code is called optimal if for two fixed parameters (of n, M and d) it is impossible to "improve" the third one, that is, to increase the cardinal number, to increase the distance or to decrease the length. Usually this task is formulated as to find the code with maximum cardinal number $M = m_q(n, d)$ for given code length n and distance d. Moreover the solution of this problem helps to find the solution of two other problems ($\min_{V:|V|=M, d(V)=d} n(V)$ and $\max_{V:|V|=M, n(V)=n} d(V)$).

Note that there is little known about optimal codes as well as about the behaviour of the function $m_q(n, d)$ (see Section 3.5).

The capability of a code to correct and/or to detect the errors is connected to redundancy. Since for transmission of M messages over the noiseless channel it is enough to use q-ary k-tuples ($k = \lceil \log_q M \rceil$, where $\lceil x \rceil$ denotes the least integer greater or equal to x) then the value $r = n - k$ symbols is called the redundancy of a code.

The code rate, defined as $R = R(V) = \log_q M \big/ n$, is an important parameter, which characterizes the "slowing down" of the information transmission due to redundancy. The fundamental result of information theory - noisy channel coding theorem by Shannon, states that for any rate R less than the channel capacity C the probability of decoding error for best codes (in fact for almost all codes) tends (exponentially) to zero with increasing of code length; and vice versa, in case $R > C$ the probability of decoding error greater than some constant $a = a(R, C)$ for any code. Thus, for a given acceptable probability of decoding error $P^*_{acc.}$ and code rate R^* ($R^* < C$), it is possible to search for the code with minimum code length of a set of codes $\mathbf{V}$ such that for any code $V \in \mathbf{V}$, $P(V) \leq P^*_{acc.}$ and $R(V) \leq R^*$. Note that this problem is close to the above mentioned extreme problem $\min_{V:|V|=M,\, d(V)=d} n(V)$; and the minimum code length means decreasing of the decoding time, connected to accumulation of all n symbols in the decoder. However, from the practical point of view the complexity of the encoding and decoding procedures is a more important factor.

Until now we considered codes as the arbitrary sets of codewords without any restrictions on a code structure. It is obvious that utilization of codes, defined in such a way, is very restricted. For example, even implementation of an encoding procedure, which usually has much less complexity than decoding, demands the table of mapping messages to codewords to be stored in the memory. In some cases when $k \geq 50$, the size of such a table becomes unacceptably large. Because of this, great attention is given to codes which have some algebraic structure, providing the opportunity to simplify the realization of these codes as well as their construction. The most important class of such kinds of codes is the class of linear codes. To describe these codes we need some information about algebraic structures, stated in the following section.

3.2 Algebraic Structures

In this section we briefly describe such algebraic structures as groups, fields and vector spaces. This knowledge is useful for understanding the following sections. For more detail information see [3] or any comprehensive book on modern algebra. One of the most simple algebraic structures is a *semigroup* defined as a set M with binary operation $\circ$ that assigns to each pair of elements $a, b \in M$ a uniquely defined element, denoted as $a \circ b$. The binary operation $\circ$ should be associative, that is, for any $a, b, c \in M$:

$$(a \circ b) \circ c = a \circ (b \circ c) \tag{3.4}$$

A *group* is defined as a semigroup G which, at first, contains an identity element e such that, for any $a \in G$:

$$a \circ e = e \circ a = a \tag{3.5}$$

and, secondly, for any element $a \in G$, there exists a uniquely *inverse element*, denoted as a^{-1} such that:

$$a \circ a^{-1} = a^{-1} \circ a = e \tag{3.6}$$

If in addition the following condition satisfies:

$$a \circ b = b \circ a \tag{3.7}$$

for all $a, b \in G$, then a group G is said to be *commutative* or *Abelian.*

The binary operation on the group is called (by convention) the multiplication or the addition and is denoted by $\cdot$ or by $+$ respectively. The cardinal number of group G (the number of elements in group) is called the *order of the group* and is denoted by $|G|$. A mapping φ of group G to group G' is called a *homomorphism* if, for all $g_1, g_2 \in G$:

$$\varphi(g_1 \circ g_2) = \varphi(g_1) \circ \varphi(g_2) \tag{3.8}$$

If, moreover, φ is the one-to-one mapping, then it is called the *isomorphism*; and groups **G** and G' are called *isomorphic groups* (that is, algebraic identical).

The important example of a group is the group $S(X)$ of one-to-one mappings of the set X to itself with the superposition of mappings as the binary operation $\circ$, that is, $(f \circ g)(x) = f(g(x)), x \in X$. Let $X = \{1, 2, \ldots, n\}$ be the finite set of n elements, then the group $S(X)$ is called the symmetric group of order n; and its elements, that is, mappings $\sigma : X \to X$ are called *permutations* and denoted as tables:

$$\begin{pmatrix} 1 & 2 & \ldots & n \\ i_1 & i_2 & \ldots & i_n \end{pmatrix}, \quad \text{where} \quad i_k = \sigma(k)$$

For the group G with the "multiplication" it is possible to raise the elements to integer power, that is $g^0 = e, \quad g^i = g \circ g^{i-1}$ for $i > 0$, $g^i = (g^{-1})^{-i}$ for $i < 0$. A group is called *cyclic* if each element is a power of some element a, which is called a generator element, that is, $G = \{a^i : \quad i \in Z\}$. For example, set Z of integers is the cyclic group under the addition with the generator element 1 (or -1). Another example of cyclic group is group Z_q^+ of elements, which are integers (residues) $0, 1, \ldots, q-1$; and the binary operation addition on modulo q is defined as:

$$(i + j) \bmod q = \begin{cases} i + j, & \text{If} \quad i + j < q \\ i + j - q, & \text{otherwise} \end{cases} \tag{3.9}$$

Any cyclic group G is either isomorphic to Z if $|G| = \infty$, or to Z_q^+ where $|G| = q$.

A subset $H \in G$ is called a *subgroup* of G if $h_1 \circ h_2 \in H$ for any $h_1, h_2 \in H$, that is, the set H is a group relative to the binary operation $\circ$.

Example 3.2 The subset $< g >= \{g^i : \quad i \in Z\}$ is always the subgroup, which is called the subgroup generated by the element g; and the order of $< g >$ is said to be the order of element g. If this order is finite, then it is equal to the minimal positive integer n such that $g^n = e$.

A subset $g \circ H = \{g \circ h : h \in H\}$ is called a *left coset* of group G on subgroup H. Any two cosets either coincide or do not intersect each other, that is, they define the partition of G; and all of them have the same order $|H|$. Hence, $|H|$ is the divisor of $|G|$, that is, the order of any subgroup is a divisor of the order of the group. This is the statement of the famous *Lagrange theorem*. Therefore, the order of any element is the divisor of the order of the group; and, for all $g \in G$:

$$g^{|G|} = e \tag{3.10}$$

A subgroup H is said to be *normal* if $g \circ h \circ g^{-1} \in H$, for all $g \in G,\ h \in H$; or that is the equivalent statement, the left coset $g \circ H$ coincides with the *right coset* $H \circ g$ for all g. Obviously any subgroup of a commutative group is normal. For the normal subgroup H the binary operation on the group G induces the binary operation on the set G/H of the cosets, that is, $g_1 H \circ g_2 H = (g_1 \circ g_2)H$; and relevant to this operation the set G/H is the group, called factor-group. For instance, let G be the group Z of all integers with addition as the binary operation; and let H be the subgroup qZ of integers divisible by q. Then the corresponding factor-group is isomorphic to the above-mentioned group Z_q^+.

In some early works on coding theory the so-called *binary group codes* were considered. These codes, defined as arbitrary subgroups of the group Z_2^n of binary sequences of length n with the binary operation of symbol-by-symbol addition by modulo 2. The attempts to transpose these results to the case of an arbitrary finite alphabet of q elements, for example by changing the set E_q^n of q-ary sequences of length n to the group with the binary operation of symbol by symbol addition on modulo q, showed that the group structure is not enough; and the set E_q^n should be regarded (if it is possible) as n-dimension vector space on the field F_q of q elements and the linear subspaces should be chosen as the codes.

Informally saying a field is a set on which it is possible to add, to subtract, to multiply and to divide (by nonzero elements) preserving the "usual" properties of these operations. Well-known examples of fields are the field Q of rational numbers; the field R of real numbers; and the field C of complex numbers.

On the other hand, if we define on the set $\{0, 1\}$ the addition on modulo 2 and conjunction as the binary operations of addition and multiplication respectively, then we obtain the field F_2 consisting of two elements (i.e. $0 + x = x, 1 + 1 = 0, 0 \cdot x = 0, 1 \cdot x = x$). This field plays a very important role in the discrete mathematics, close to the role of the fields R and C in the classic mathematics. To make certain that F_2 is a field despite it being unlike the fields Q, R or C, let us give the formal definition.

A set K of more than one element is called a field if for any two elements $a, b \in K$ there are defined their sum $a + b \in K$ and their product $a \cdot b \in K$ with the following properties:

$$\left.\begin{array}{ll}
1.1 & (a+b)+c = a+(b+c) \\
1.2 & a+b = b+a \\
1.3 & \text{there exists element 0 such that } a+0 = a \\
1.4 & \text{there exists element } (-a) \text{ such that } a+(-a) = 0 \\
2.1 & (a \cdot b) \cdot c = a \cdot (b \cdot c) \\
2.2 & a \cdot b = b \cdot a \\
2.3 & \text{there exists element 1 such that } a \cdot 1 = a, \quad a \neq 0 \\
2.4 & \text{there exists element } a^{-1} \text{ such that } a \cdot a^{-1} = 1, \quad a \neq 0 \\
3 & (a+b) \cdot c = a \cdot c + b \cdot c
\end{array}\right\} \tag{3.11}$$

The axioms of the first section mean that a field is a commutative group relative to addition (see Equations 3.4–3.7); and the axioms of the second section mean that a field without element "0" is a commutative group relative to multiplication.

If we relax the axioms of the second section by excluding conditions 2.2 and 2.4, then we obtain the definition of an associative *ring* with unit. We can define the homomorphism (isomorphism) for the rings, as it was done for groups, with the natural demand that the condition (3.8) should be valid for both operation of addition and multiplication.

A *finite field* of q elements is denoted by F_q or $GF(q)$. In particular, the set of residues by modulo p, where p is the prime number forms the field F_p (or in other notation Z_p). The field F_p consists of integers $0, 1, \ldots, \quad p-1$; and to add or to multiply two elements of F_p means to add (or to multiply) these two elements just as integers, and then to find the remainder after division by p (this remainder is called the residue by modulo p).

If the equation $n \cdot 1_K = 0_K$ $(n \in Z)$ has only zero solution $n = 0$ in a field K, then the field K is said to be a field of zero characteristic. Otherwise, the field K is said to be the field of characteristic p, where p is the minimal positive integer such that $n \cdot 1_K = 0_K$. Thus, p is the order of the element 1_K as an element of the additive group of the field K. It is easy to show that p should be a prime number. A field K of characteristic p contains the subfield $K(1) = \{n \cdot 1_K, n \in Z\}$, which is isomorphic to F_p; and a field K of zero characteristics contains the subfield, which is isomorphic to Q. In particular, a finite field of q elements exist if and only if q is a power of prime number p where p is the characteristic of that field itself, and the field is unique (up to isomorphism).

For any field of characteristic p the following unusual identity ("truncated" Newton identity):

$$(a+b)^p = a^p + b^p \tag{3.12}$$

is true. Moreover, by virtue of Lagrange theorem another useful identity is valid. Namely, for any $a \in F_q, a \neq 0$:

$$a^{q-1} = 1 \tag{3.13}$$

It is known as *Fermat's "small theorem"*. This identity is equivalent to:

$$a^q = a \tag{3.14}$$

for any $a \in F_q$.

Many applications of finite fields in coding theory are based on the fact that it is possible to regard the finite field F_{p^m} as m-dimension vector space on the field F_p; and vice versa.

A set V is called *vector (linear) space* over a field K if:

1. V is an Abelian group under addition.
2. For any $\mathbf{v} \in V$ and $\lambda \in K$, the *multiplication of vector by scalar* (or *scalar multiplication*) $\lambda \cdot \mathbf{v} \in V$ is defined. Moreover,
   2.1 $\lambda \cdot (\mathbf{v}_1 + \mathbf{v}_2) = \lambda \cdot \mathbf{v}_1 + \lambda \cdot \mathbf{v}_2$;
   2.2 $(\lambda_1 + \lambda_2) \cdot \mathbf{v} = \lambda_1 \cdot \mathbf{v} + \lambda_2 \cdot \mathbf{v}$;
   2.3 $(\lambda_1 \cdot \lambda_2) \cdot \mathbf{v} = \lambda_1 \cdot (\lambda_2 \cdot \mathbf{v})$;
   2.4 $1_K \cdot \mathbf{v} = \mathbf{v}$.

As an example of a vector space we can consider so called "n-dimension coordinate (arithmetic) space K^n", the elements of which are the sequences $\mathbf{a} = (a_1, \ldots, a_n), a_i \in K$; and the operations of the addition and the scalar multiplication are defined as follows:

$$\begin{aligned} \mathbf{a} + \mathbf{b} &= (a_1 + b_1, \ldots, a_n + b_n) \\ \lambda \cdot \mathbf{a} &= (\lambda \cdot a_1, \ldots, \lambda \cdot a_n) \end{aligned} \tag{3.15}$$

Vector $\sum_{i=1}^{n} \lambda_i \cdot \mathbf{v}_i$ is called a *linear combination* of vectors $\mathbf{v}_1, \ldots, \mathbf{v}_n$; and λ_i are called coefficients of linear combination. The basis of vector space V over field K is the set of vectors $\mathbf{v}_1, \ldots, \mathbf{v}_m \in V$ such that any vector $\mathbf{x} \in V$ can be represented uniquely as the linear combination:

$$\mathbf{x} = \sum_{i=1}^{m} \lambda_i \cdot \mathbf{v}_i, \quad \lambda_i \in K \tag{3.16}$$

The coefficients λ_i are called *coordinates* of vector $\mathbf{x}$ in the basis $\{\mathbf{v}_i, i = 1, \ldots, m\}$. All bases of given vector space V consist of the same number of vectors, which is referred to as the dimension of vector space V and is denoted by dim V. The vector space V is called m-dimension vector space, where $m = \dim V$.

A set of vectors $\{\mathbf{v}_i\}$ is said to be linear independent if a linear combination of these vectors is equal to 0 only if all coefficients of the linear combination are zeros. Otherwise the set $\{\mathbf{v}_i\}$ is said to be linear dependent. Another definition of the basis is that it is the maximal (in the sense of being the maximal set under inclusion) set of linear independent vectors.

The set of vectors $\{\mathbf{f}_i, i = 1, \ldots, m\}$ with the coefficients f_{ij} in the basis $\mathbf{v}_1, \ldots, \mathbf{v}_n$ (i.e. $\mathbf{f}_i = \sum f_{ij} \cdot \mathbf{v}_j$) is the basis of the space if and only if $\det(f_{ij}) \neq 0$; calculation of the determinant should be done in the field K.

A mapping $A : V \to U$ of vector space V to vector space U is said to be linear if for any $\mathbf{v}, \mathbf{v}', \mathbf{v}'' \in V, \lambda \in K$:

$$\begin{aligned} A(\mathbf{v}' + \mathbf{v}'') &= A(\mathbf{v}') + A(\mathbf{v}'') \\ A(\lambda \cdot \mathbf{v}) &= \lambda \cdot A(\mathbf{v}) \end{aligned} \tag{3.17}$$

A linear mapping is a homomorphism over the field K; and as in the case of groups and rings, the vector spaces V and U are called the isomorphic vector spaces if there exist the one-to-one mapping $A : V \to U$. Since there exists the one-to-one mapping of any vector $\mathbf{v} \in V$ to coordinates of this vector $\mathbf{v}$ in some fixed basis, an n-dimension vector space over the field K is isomorphic to the n-dimension coordinate space. Therefore, all vector spaces of the same dimension over the same field are isomorphic. In particular, any n-dimension vector space over the field F_q consists of $|K^n| = q^n$ vectors.

Let $\mathbf{v}_1, \ldots, \mathbf{v}_n$ be the basis of V and $\mathbf{u}_1, \ldots, \mathbf{u}_m$ be the basis of U, then every linear mapping $A : V \to U$ corresponds to $(m \times n)$ matrix $[a_{ij}]$, $i = 1, \ldots, m; \quad j = 1, \ldots, n$, which coefficients are defined by the following equation:

$$A(\mathbf{v}_j) = \sum_{i=1}^{m} a_{ij} \cdot \mathbf{u}_i, \quad j = 1, \ldots, n \tag{3.18}$$

An important case of linear mapping is a *linear functional* $f : V \to K^1$, which in accordance with (3.18) can be represented as $f(\mathbf{x}) = \sum f_i \cdot x_i$, where $f_i = f(\mathbf{v}_i)$, $\mathbf{x} = \sum x_i \cdot \mathbf{v}_i$. The set of linear functionals forms a vector space V^* under the operations of an addition of functionals and their multiplication by elements of the field K. This vector space is said to be dual to V and it has the same dimension: $\dim V^* = \dim V$.

Another important case is the linear mapping $A : V \to V$, called a *linear operator* over V. In this case $\mathbf{u}_i = \mathbf{v}_i$ (only single basis is used); and the linear operator A corresponds to a square $(n \times n)$ matrix $[a_{ij}]$ such that:

$$A(\mathbf{v}_j) = \sum_{i=1}^{n} a_{ij} \cdot \mathbf{v}_i, \quad j = 1, \ldots, n \tag{3.19}$$

It is possible not only to add linear operators and multiply them by an element of the field, but also to multiply operator by operator (as mappings). In this case the operator $E : E(\mathbf{v}) = \mathbf{v}$, referred to as a unit operator, is the unit (the neutral element under multiplication), since for any operator A, $EA = AE = A$. The matrix of operator E in any basis can be written as:

$$\begin{aligned} \mathbf{E}_n &= [\delta_{ij}] \\ \text{where } \delta_{ij} &= \begin{cases} 1 & \text{if} \quad i = j \\ 0 & \text{if} \quad i \neq j \end{cases} \end{aligned} \tag{3.20}$$

The existence of the inverse operator A^{-1}: $A \cdot A^{-1} = A^{-1} \cdot A = E$ is equivalent to any one of the following conditions:

- $\text{Ker}\, A = \{\mathbf{v} \in V : \quad A(\mathbf{v}) = 0\} = \{0\}$(Nonsingular)
- $\text{Im}\, A = \{A(\mathbf{v}) : \quad \mathbf{v} \in V\} = V$
- $\det(a_{ij}) \neq 0$

The set of nonsingular operators form the group under multiplication, which is called the *general linear group* and denoted by $GL(n, K)$. If a metric is defined on V, then the subset of nonsingular linear operators such that any of them preserves the distance between points of V, forms the subgroup. Such an operator is called *isometry operator*, and the corresponding group is called the isometry group. For Hamming metric this subgroup consists of linear operators $\sigma \cdot \Delta_\Lambda$, where σ is the linear operator of permutation ($\sigma(\mathbf{e}_i) = \mathbf{e}_{\sigma(i)}$); $\Delta_\Lambda(\mathbf{e}_j) = \lambda_j \cdot \mathbf{e}_j$ is the "diagonal" operator; and $\mathbf{e}_i$ is the i th row of matrix E_n, $\mathbf{e}_i$ has only one nonzero component at i th position.

The subset $L \in V$ is said to be a *linear subspace* of the space V if L is the subgroup of V under addition, and for any $\mathbf{l}, \mathbf{l}_1, \mathbf{l}_2 \in L, \quad \lambda \in K$, the following statement is correct: $\lambda \cdot \mathbf{l} \in L$. In other words, L is the linear space under the operations of vector addition and multiplication by scalar, defined on the whole set V. Therefore, the basis $\mathbf{l}_1, \ldots, \mathbf{l}_k$ exists in L, where $k = \dim L$. Thus we obtain the definition of subspace L as follows:

$$L = \left\{ \mathbf{x}; \quad \mathbf{x} = \sum_{i=1}^{k} \lambda_i \cdot \mathbf{l}_i \right\} \tag{3.21}$$

An important fact is that any basis of a subspace could be extended up to the basis of the full space.

A linear subspace can also be described as solutions of some system of linear equations. Define the subspace L^* of the dual space V^*, which consists of all linear functionals such that any of these functionals equal to zero for any vector of L:

$$L^* = \{ f \in V^* : \quad f(\mathbf{l}) = 0, \quad \mathbf{l} \in L \} \tag{3.22}$$

This subspace is said to be dual to L. Then for any basis $f_1, \ldots, f_r$ of space L^*, where $r = \dim L^* = n - \dim L$, L is the set of solutions of system of r linear equations:

$$L = \{ \mathbf{x} \in V : \quad f_i(\mathbf{x}) = 0, \quad i = 1, 2, \ldots, r \} \tag{3.23}$$

Define the scalar product of two vectors $\mathbf{x} = \sum x_i \cdot \mathbf{v}_i$ and $\mathbf{y} = \sum y_i \cdot \mathbf{v}_i$, $x_i, y_i \in K$ as:

$$(\mathbf{x}, \mathbf{y}) = \sum_{i=1}^{n} x_i \cdot y_i \tag{3.24}$$

Then an arbitrary linear functional f can be represented as:

$$f(\mathbf{x}) = (\mathbf{F}, \mathbf{x}) \tag{3.25}$$

where $\mathbf{F} = (f(\mathbf{v}_1), \ldots, f(\mathbf{v}_n))$. The equation (3.25) establishes the isomorphism between V and V^*. In this case dual subspace L^* corresponds to so-called orthogonal complement to L, denoted by $\bar{L}$ and defined as:

$$\bar{L} = \{ \mathbf{x} \in V, \mathbf{l} \in L : \quad (\mathbf{x}, \mathbf{l}) = 0 \} \tag{3.26}$$

Note that for the fields of finite characteristic the "usual" (for fields of zero characteristic, like the field of complex numbers) property $L \cap \bar{L} = 0$ is not correct; and it is possible that L intersect $\bar{L}$. Furthermore, L could belong to $\bar{L}$ and such subspace is called *self-orthogonal*.

Example 3.3 Consider 5-dimensional vector space V of all 5-tuples over the finite field F_2. The operations of vector addition and scalar multiplication are defined according to (3.15) and the scalar product of two vectors is defined by (3.24). The following four vectors form a 2-dimensional subspace L of V:

$$(00000), \quad (00011), \quad (01100), \quad (01111)$$

The orthogonal complement $\bar{L}$ consists of the following eight vectors:

$$\begin{aligned} &(00000), \quad (00011), \quad (01100), \quad (01111), \\ &(10000), \quad (10011), \quad (11100), \quad (11111) \end{aligned}$$

The dimension of $\bar{L}$ is 3. Since $L \cap \bar{L} \neq 0$, $L \subset \bar{L}$, and hence, the subspace L is the self-orthogonal subspace.

For any subspace L of any vector space V, $\bar{\bar{L}} = L$ and $\dim L + \dim \bar{L} = \dim V$. For any linear mapping $A : V \to U$, the dimension of the kernel $\mathrm{Ker}\, A \subset V$ is connected with the dimension of the image $\mathrm{Im}\, A \subset U$ by the following equation:

$$\dim \mathrm{Ker}\, A + \dim \mathrm{Im}\, A = \dim V$$

Now we will pay special attention to the ring of polynomials and its quotient rings. The most important example of associative rings is the ring of integers **Z**. Another very important example, which is very close to the previous one, is the ring $K[x]$ of the polynomials with coefficients from the field K. The elements of the $K[x]$ are the polynomials, that is, the sequences $f = (f_0, f_1, \ldots)$, where $f_i \in K$ and no all coefficients are equal to zero. The maximal m such that $f_m \neq 0$ is called the *degree of the polynomial* and denoted by $\deg f(x)$, where $f(x)$ is more usual representation of the polynomial with one *variable* x:

$$f(x) = f_0 + f_1 x + \ldots + f_m x^m \tag{3.27}$$

If $f_m = 1$ then a polynomial is called *normalized.* Two polynomials can be added and multiplied in accordance with standard formulas:

$$f(x) + g(x) = (f_0 + g_0) + (f_1 + g_1)x + \ldots + (f_i + g_i)x^i + \ldots \tag{3.28}$$

$$f(x) \cdot g(x) = h_0 + h_1 x + \ldots + h_i x^i + \ldots, \text{ where, } h_k = \sum_{i+j=k} f_i \cdot g_j \tag{3.29}$$

and:

$$\deg(f(x) + g(x)) \leq \max(\deg f(x), \deg g(x)), \quad \deg(f(x) \cdot g(x)) = \deg f(x) + \deg g(x) \tag{3.30}$$

Example 3.4 $f(x) = x^4 + 3x^2 + 4, \quad g(x) = 2x^2 + x + 3, \quad K = F_5$

$$f(x) + g(x) = x^4 + (3+2)x^2 + x + (4+3) = x^4 + x + 2$$

$f(x) \cdot g(x) = (1 \cdot 2)x^6 + (1 \cdot 1)x^5 + (1 \cdot 3 + 3 \cdot 2)x^4 + (3 \cdot 1)x^3 + (3 \cdot 3 + 4 \cdot 2)x^2 + (4 \cdot 1)x + 4 \cdot 3 = 2x^6 + x^5 + 4x^4 + 3x^3 + 2x^2 + 4x + 2$

It is easy to verify that zero element and unit element of the ring $K[x]$ are polynomials $\mathbf{0} = (0, \ldots, 0)$ and $\mathbf{1} = (1, 0, \ldots, 0)$ respectively, and $f(x) \cdot g(x) = \mathbf{0}$ if and only if $f(x) = \mathbf{0}$ or $g(x) = \mathbf{0}$. The latter property means that there are no divisors of zero in $K[x]$. Moreover, the ring $K[x]$ as well as the ring Z is an example of the *Euclidean* ring. The commutative ring L is said to be Euclidean if there is a nonnegative integer function $\mu()$ defined on $L \backslash \{0\}$ such that:

1. $\mu(a \cdot b) \geq \mu(a)$ for all $a,\ b \neq 0$ from L;
2. for any $a,\ b \in L, \quad b \neq 0$ there exist $q,\ r \in L$ (the quotient and the reminder) such that

$$a = q \cdot b + r, \text{ where } \mu(r) < \mu(b) \quad \text{or } r = 0.$$

The reminder r is also called the *residue* of a by modulo b and denoted by $a \bmod b$.

For Z this function is $|a|$ and for $K[x]$ $\mu(f(x)) = \deg f(x)$. The property 2) can be realized by the usual algorithm of division of polynomials.

Example 3.5 $a(x) = x^7 - 2x^5 + 4x^3 + 2x^2 - 2x + 2$

$$b(x) = 2x^5 + 3x + 4, \quad K = F_5$$

$$\begin{array}{rllll}
 & & & 3x^2 - 1 & \\
\cline{2-5}
2x^5 + 3x + 4) & x^7 & -2x^5 & +4x^3 + 2x^2 - 2x + 2 & \\
 & x^7 & & +4x^3 + 2x^2 & \\
\cline{2-5}
 & & -2x^5 & -2x \quad +2 & \\
 & & -2x^5 & -3x \; - \; 4 & \\
\cline{3-5}
 & & & \qquad\qquad x + 1 &
\end{array}$$

As the result of the division $a(x)$ by $b(x)$ we obtain the quotient $q(x) = 3x^2 - 1$ and the remainder $r(x) = x + 1$, that is:

$$a(x) = q(x) \cdot b(x) + r(x) = x^7 - 2x^5 + 4x^3 + 2x^2 - 2x + 2 = (3x^2 - 1) \cdot (2x^5 + 3x + 4) + x + 1$$

A subgroup $I \subset L$ of the additive group of the ring L is called an *ideal* of the commutative ring L if for all $a \in L, \quad v \in I$:

$$a \cdot v \in I$$

The set L/I of cosets $\{I + a\}$ is the ring under the operations of addition and multiplication, defined on the ring L: $\{a + I\} + \{b + I\} = \{a + b + I\}$; $\{a + I\} \cdot \{b + I\} = \{a \cdot b + I\}$. The ring L/I is called *quotient ring* or *residue class ring* by modulo of ideal I. The simplest example of an ideal is the *principal ideal* $V = \{a \cdot v : \quad a \in L\}$, generated by an element $v \in L$. Any Euclidian ring is the ring of principal ideals, that is, there are no ideals except the principal ones; and the elements of the quotient ring L/V can be represented as elements $r \in L$ such that $\mu(r) < \mu(v)$ or $r = 0$ if we define operations as follows:

$$\begin{aligned} (r_1 + r_2)_{L/V} &= (r_1 + r_2) \bmod v, \\ (r_1 \cdot r_2)_{L/V} &= (r_1 \cdot r_2) \bmod v \end{aligned} \tag{3.31}$$

For example, for the ring $K[x]$ and the principal ideal generated by $g(x)$, elements of the quotient ring $K[x]/g(x)$ are the polynomials of the degree less than $n = \deg\, g(x)$. These elements can be added as usual polynomials; and the multiplication of polynomials by modulo $g(x)$ is chosen as the operation of multiplication.

It is a well known fact that any positive integer can be uniquely represented as the product of prime numbers. This statement can be generalized to any ring of principal ideals, in particular to the Euclidean rings. Let us restrict our attention to the case of the ring $K[x]$. The polynomial $f(x)$, $\deg\, f(x) \geq 1$ is said to be *irreducible* over the field K if it cannot be represented as a product of two polynomials (with coefficients of K) of nonzero degree.

Example 3.6 $f_1(x) = x^2 - 2$ is the irreducible polynomial over Q, $f_2(x) = x^2 + 1$ is the irreducible polynomial over R, $f_3(x) = x^2 + x + 1$ is the irreducible polynomial over F_2. Notice that the field of the coefficients is significant. For instance, $f_1(x)$ is reducible polynomial over R, $f_2(x)$ can be reduced over C and $f_3(x)$ can be reduced over F_4.

Theorem 3.1 Any polynomial $f(x) \in K[x]$ can be uniquely represented as the product of the element of the field K and the irreducible normalized polynomials.

Notice that for the ring $K[x]$ there are known simple algorithms (with the polynomial complexity) of factorization of polynomials to irreducible polynomials, as distinguished from the case of Z [3].

Consider calculations over finite fields since it is a very important issue for the main part of codes constructed with the help of algebraic coding theory. Let us outline several useful (for calculations) properties of finite fields. First of all, the field F_{p^m} can be represented as the m-dimension vector space over the field F_p, where p is a prime number. That means the addition of the elements in F_{p^m} can be regarded as the addition by modulo p of the m-tuples. Secondly, the multiplicative group of the field F_q consist of $q - 1$ elements and it is the cyclic group, that is, there is at least one *primitive element* $\gamma \in F_q$ such that $a = \gamma^i, \quad 0 \le i < q - 1$, for any $a \in F_q, \quad a \neq 0$; the number i is called the logarithm of a to the base γ and denoted by $\log_\gamma a$. In fact, there are $\varphi(q - 1)$ primitive elements of the field F_q, where $\varphi()$ is the Euler's function. This property allows us to use the "logarithmic" representation of the elements in the process of the multiplication:

$$\log(a \cdot b) = (\log a + \log b) \bmod (q - 1) \tag{3.32}$$

One more useful (but unusual in comparison with fields Q, R and C) property of the finite field F_{p^m} was mentioned above (3.12), (3.14):

$$(a + b)^p = a^p + b^p \tag{3.33}$$

for any $a, b \in F_{p^m}$:

$$(\lambda \cdot a)^p = \lambda \cdot a^p \tag{3.34}$$

for any $a \in F_{p^m}, \ \lambda \in F_p$. Therefore, the mapping $\sigma_p : \ F_{p^m} \to F_{p^m} \ (a \to a^p)$ is a linear operator on F_{p^m} regarded as the m-dimension vector space over the field F_p. Moreover, the mappings $\sigma_{p^i}, \ i = 0, \ 1, \ldots, m - 1$ are the automorphisms of the field F_{p^m} and form the group.

3.3 Linear Block Codes

Let us return to the model of reliable data communication system, described in Section 3.1. A discrete memoryless channel (we restrict consideration to this class of channels) is defined by the crossover probability $p(x/y)$ that is the (conditional) probability of receiving a q-ary symbol y as a channel's output if a q-ary symbol x was transmitted. Let an additive group structure be defined on the q-ary channel alphabet A (for instance, consider A as the group Z_q^+ of residues by modulo q). If the received word $\mathbf{y} = (y_1, \ldots, y_n)$ does not coincide with the transmitted word $\mathbf{x} = (x_1, \ldots, x_n)$ then it is said that the error (or error vector) occurs during the data transmission over the channel. If:

$$P(\mathbf{y}|\mathbf{x}) = P(\mathbf{y} - \mathbf{x}|\mathbf{0}) = \mathbf{P}(\mathbf{y} - \mathbf{x}) \tag{3.35}$$

then such kind of channel is called a *channel with additive noise*.

It is natural to use for a channel with additive noise codes correcting some set of errors E. A code $V \subset \mathrm{A}^n$ can correct a set of errors $E = \{\mathbf{0}, \mathbf{e}_1, \ldots, \mathbf{e}_m\}$ if any equation $\mathbf{v} + \mathbf{e} = \mathbf{v}' + \mathbf{e}'$, where $\mathbf{v}, \mathbf{v}' \in V$, $\mathbf{e}, \mathbf{e}' \in E$ has unique solution $\mathbf{v} = \mathbf{v}'$ and $\mathbf{e} = \mathbf{e}'$. Choice of a set of correctable errors E should depend on the probability distribution of errors $\mathbf{P}$. Since the code V with a set of correctable errors E guarantees the decoding error probability P_e no more than $1 - \sum_{\mathbf{e} \in E} \boldsymbol{P}(\mathbf{e})$ then a suitable set E usually is formed in such a way to include the most probable error patterns. Therefore the problem of construction of corresponding optimal code, that is, the code with maximum cardinal number (or with maximum code rate) should be investigated. Nevertheless, let us note that such a kind of choice of E is not necessarily the best in the sense of maximum code rate for a given decoding error probability P_e.

The *Hamming weight* of vector $\mathbf{x} = (x_1, \ldots, x_n)$, as the number of nonzero components of $\mathbf{x}$. If a group under addition is defined on alphabet A, and a set A^n of all words of length n is regarded as a group denoted by $wt(\mathbf{x})$, is defined under component-wise addition, then the relation between Hamming distance and Hamming weight can be written as follows:

$$d(\mathbf{x}, \mathbf{y}) = wt(\mathbf{x} - \mathbf{y}) \tag{3.36}$$

Consider codes, which are the subgroups of A^n and called *group codes*. To calculate the distance of a group code by virtue of property (3.36) it is enough to find the minimum weight of its nonzero codewords, i.e.:

$$d(V) = \min_{\mathbf{v}\in V, \mathbf{v}\neq 0} wt(\mathbf{v}) \tag{3.37}$$

The group structure on A is not enough to construct good codes in A^n; and the main results of coding theory are obtained in cases where q is a prime power when the alphabet A can be regarded as the finite field F_q, and A^n is regarded as n-dimension vector space F_q^n over F_q. By the definition, a q-ary *linear block* (n, k) *code* is an arbitrary k-dimension subspace of vector space F_q^n.

Since a linear block code is a group code, the equation (3.37) is correct for any linear block code. Notice that in cases where $q = p$, where p is a prime number, the definition of a linear code coincides with the definition of a group code.

Since the number of vectors in arbitrary k-dimension subspace of vector space F_q^n over the field F_q is equal to q^k, the same is the number of messages M, which is possible to transmit by a q-ary (n, k) code. It is convenient to represent these $M = q^k$ messages as k-dimension vectors $\mathbf{u}_i = (u_i^1, \ldots, u_i^k)$, $i = 1, \ldots, M$, from F_q^k, that is, $\{\mathbf{u}_1, \ldots, \mathbf{u}_M\} = F_q^k$.

In the previous section two methods of description of linear subspaces were presented. Let us start with the first of them. Consider an (n, k) code V and $k \times n$ matrix $\mathbf{G}$, where rows are vectors $\mathbf{v}_1, \ldots, \mathbf{v}_k$ forming a basis of the subspace V, that is:

$$\mathbf{G} = [g_{ij}], \text{ where } (g_{i1}, \ldots, g_{in}) = \mathbf{v}_i$$

Matrix $\mathbf{G}$ is called a *generator matrix* of the code. Every (n, k) code has exactly $\prod_{i=0}^{k-1} (q^k - q^i)$ bases and, therefore, the same number of generator matrices. Each generator matrix defines the encoding procedure $\varphi_{\mathbf{G}}: \quad F_q^k \to V$ by the following formula:

$$\varphi_{\mathbf{G}}(\mathbf{u}) = \varphi_{\mathbf{G}}(u_1, \ldots, u_k) = \mathbf{u} \cdot \mathbf{G} = \sum_{i=1}^{k} u_i \cdot \mathbf{v}_i \tag{3.38}$$

which is a linear mapping. Let $\mathbf{G}$ be some generator matrix of an (n, k) code V. Then an arbitrary generator matrix $\mathbf{G}'$ of this code can be represented as $\mathbf{G}' = \mathbf{C} \cdot \mathbf{G}$, where $\mathbf{C}$ is a nonsingular $k \times k$ matrix. Let us split the generator $k \times n$ matrix into matrices $\mathbf{G}_1$ and $\mathbf{G}_2$:

$$\mathbf{G} = [\mathbf{G}_1 | \mathbf{G}_2] \tag{3.39}$$

where $\mathbf{G}_1$ is $k \times k$ matrix, and $\mathbf{G}_2$ is $k \times (n - k)$ matrix. If $\mathbf{G}_1$ is a nonsingular matrix, then matrix $\mathbf{G}'$:

$$\mathbf{G}' = \mathbf{G}_1^{-1} \cdot \mathbf{G} = [\mathbf{I}_k | \mathbf{G}_2'], \quad \mathbf{G}_2' = \mathbf{G}_1^{-1} \cdot \mathbf{G}_2 \tag{3.40}$$

is also a generator matrix of the code V and defines in accordance with (3.38) the encoding procedure:

$$\varphi_{\mathbf{G}}(u_1, \ldots, u_k) = (v_1, \ldots, v_k, v_{k+1}, \ldots, v_{k+r}) = (\mathbf{u}, \ \mathbf{u} \cdot \mathbf{G}_2') \tag{3.41}$$

Such encoding procedure is called a *systematic encoding*, because the first k symbols of any codeword coincide with the corresponding symbols of an uncoded message ($v_i = u_i, \quad i = 1, \ldots, k$). A code with generator matrix $\mathbf{G}'$ is called a *systematic code*. Not every linear code is systematic, because matrix $\mathbf{G}_1$ in (3.39) may appear to be a singular matrix. However, it is always possible to find k linear independent columns of the matrix $\mathbf{G}$ (since rank over the columns coincides with rank over the rows). Therefore, it is possible to transform the code V to a systematic form with by some permutation of coordinates, that is, any (n, k) code is *equivalent* to a systematic one. Hereafter, we often assume that considered (n, k) codes are systematic.

Let code V be a systematic code and matrix $\mathbf{G}'$ has the same form as in (3.40), then V can be defined as aforementioned as the set of solutions of the following system of linear equations:

$$\mathbf{H} \cdot \mathbf{v} = \mathbf{0} \tag{3.42}$$

where $\mathbf{H} = [-\mathbf{G}_2'|\mathbf{I}_r]$. It means that the matrix $\mathbf{H}$ is a generator matrix of the subspace $\bar{V} = \{\mathbf{x} \in L : (\mathbf{x}, \mathbf{v}) = 0, \quad \mathbf{v} \in V\}$, which is called the *dual code*. This statement immediately follows from the substitution of (3.41) in (3.42), which shows that (3.42) is correct for any codeword and from the comparison of the dimensions ($\dim V + \dim \bar{V} = n$). The matrix $\mathbf{H}$ satisfying (3.42) is called a *parity-check matrix* of the code V. The equation (3.42) is the equation of linear dependence between those columns $\mathbf{h}_i$ of matrix $\mathbf{H}$, where $v_i \neq 0$. It leads immediately to the following useful result.

Lemma 3.1 *(Bose criterion). The minimum distance of a code V is no less than d if any $d - 1$ columns of its parity-check matrix* $\mathbf{H}$ *are linear independent.*

It follows from Lemma 3.1 that to construct the code correcting one error, the matrix $\mathbf{H}$ with non-collinear columns should be constructed. For instance, such kind of maximal (in number of columns) matrix $\mathbf{H}$ can be constructed by induction:

$$\mathbf{H}_r = \left[\begin{array}{ccc|cccc} 1 & \ldots & 1 & 0 & 0 & \ldots & 0 \\ & * & & & \mathbf{H}_{r-1} & & \end{array}\right]$$

or, what is the same:

$$\mathbf{H}_r = \left[\begin{array}{ccc|cccc} 1 & \ldots & 1 & 0 & 0 & \ldots & 0 \\ & & & 1 & 1 & & \vdots \\ & & & & & & 0 \\ & F_q^{r-1} & & & F_q^{r-2} & & 1 \end{array}\right] \tag{3.43}$$

The equation (3.43) allows errors to be detected in a very simple manner. Namely, it is enough to calculate vector $\mathbf{S}$ called a *syndrome*:

$$\mathbf{S} = \mathbf{b} \cdot \mathbf{H}^{\mathrm{T}} \tag{3.44}$$

where $\mathbf{b}$ is the received vector, and check if $\mathbf{S}$ is equal to zero or not since $\mathbf{S} = \mathbf{0}$ if and only if $\mathbf{b}$ belongs to the code. Notice that the value of the syndrome depends not only on vector $\mathbf{b}$ but also on the form of the parity-check matrix of the code. This fact we will use later when considering decoding algorithms of linear codes.

Of course, nontrivial code (that is, code which consists of more than one word) cannot correct any errors. In particular, if errors which occurred in the channel, form a codeword, then the received vector $\mathbf{b}$ is a codeword but not the transmitted one. This kind of error cannot even be detected, because the syndrome of the received vector is equal to zero. Let us introduce the concept of a *standard array* to describe errors, which can be corrected and detected by the code.

Let V be an (n, k) linear binary code $(n - k = r)$. Let $\mathbf{v}_0, \mathbf{v}_1, \ldots, \mathbf{v}_{2^k-1}$ be all codewords of the code:

$$V = \{\mathbf{v}_0, \mathbf{v}_1, \ldots, \mathbf{v}_{2^k-1}\}$$

where $\mathbf{v}_0$ is the all-zero word. Let us form the table of 2^k columns and 2^r rows as follows. Any row consists of 2^k vectors. The first row we constrain all codewords with $\mathbf{v}_0$ as the first element of the row. Then we take any n-vector $\mathbf{e}_1$, which does not belong to the code; and the second row consists of elements that are the sum $\mathbf{e}_1 + \mathbf{v}_i$, $i = 0, \ldots 2^k - 1$. Then we choose an element $\mathbf{e}_2$, which does not belong to the first and the second row; and form the third from the sums $\mathbf{e}_2 + \mathbf{v}_i$. We continue this process until all vector space has been exhausted. As the result of this procedure we obtain an array, which is called a standard array:

$$\begin{array}{ccccc} \mathbf{v}_0 & \mathbf{v}_1 & \mathbf{v}_2 & \cdots & \mathbf{v}_{2^k-1} \\ \mathbf{e}_1+\mathbf{v}_0 & \mathbf{e}_1+\mathbf{v}_1 & \mathbf{e}_1+\mathbf{v}_2 & \cdots & \mathbf{e}_1+\mathbf{v}_{2^k-1} \\ \cdots\cdots & \cdots\cdots & \cdots\cdots & \cdots & \cdots\cdots \\ \mathbf{e}_{2^r-1}+\mathbf{v}_0 & \mathbf{e}_{2^r-1}+\mathbf{v}_1 & \mathbf{e}_{2^r-1}+\mathbf{v}_2 & \cdots & \mathbf{e}_{2^r-1}+\mathbf{v}_{2^k-1} \end{array} \tag{3.45}$$

It is obvious that different rows of this array do not contain the same elements. Therefore, the number of rows is equal to 2^r. The syndromes of all vectors in the same row are identical:

$$(\mathbf{e}_i+\mathbf{v}_{j_1})\cdot\mathbf{H}^{\mathrm{T}}=(\mathbf{e}_i+\mathbf{v}_{j_2})\cdot\mathbf{H}^{\mathrm{T}}=\mathbf{e}_i\cdot\mathbf{H}^{\mathrm{T}}$$

and the syndromes of the elements from the different rows are different.

The standard array is the method of writing of the whole n-dimension vector space. Any error vector in the channel can occur, but the code can correct only one received vector from the row of the standard array, because the vectors, placed in the same row has identical syndromes. The rows of the standard array are usually called the *cosets* of the code and the elements in the first column are called *coset leaders*. Any element in the row (in the coset) can be used as the coset leader.

A binary linear code can correct only 2^r vectors, which is significantly less than the overall number of possible error vectors 2^n. However, in most channels the different error vectors have different probability. In any channel it is necessary to choose the most probable error vectors as the coset leaders to realize the decoding on *maximum likelihood*. In particular, in the channel with independent errors the vectors with minimum Hamming weight should be chosen as the coset leaders.

If $\mathbf{e}_0,\ \mathbf{e}_1,\ldots,\mathbf{e}_{2^r-1}$ ($\mathbf{e}_0$ is the all-zero vector) are the coset leaders of code V, then the decoding error probability, provided by this code P_e is:

$$P_e=1-\sum_{i=0}^{2^r-1}P(\mathbf{e}_i) \tag{3.46}$$

where $P(\mathbf{e}_i)$ is the probability of vector $\mathbf{e}_i$ being the error vector in the channel. A code can be used in the channel if:

$$1-P_{e.acc.}<\sum_{i=0}^{2^{r-1}}P(\mathbf{e}_i)$$

where $P_{e.acc.}$ is the acceptable error probability.

To calculate the error probability with the help of formula (3.46) it is necessary to calculate 2^r probabilities, which is, as usual, a highly complex problem. Notice that the coding theorems of information theory shows that there should be a subset of coset leaders among the long enough codes with a code rate less than the channel capacity, which includes the set of most probable channel error vectors. That is, there exist the code, providing an arbitrary small value of error probability P_e. The formula (3.34) defines the exact value of error probability, provided by the code in the channel with independent errors with minimum distance decoding. The estimations of error probability, based on use of minimum distance can be obtained for the case of decoding in hypersphere of radius ν. Let us find out the size of radius ν to provide the decoding in hypersphere to be very close (in the sense of error probability) to the minimum distance decoding.

Let A^n be the set of n-tuples with symbols from the alphabet A; and let E_r be the set of q^r most probable error vectors $\mathbf{e}\in\mathrm{A}^n$. Let V be the $(n,\,k)$-code $(n-k=r)$ over A. Let E_V be the set of leader cosets of code V, and let $P(\mathrm{B})$ be the probability of error vector in the channel and a vector from some set B.

Lemma 3.2 [6]:

$$P(A^n \backslash (E_V \cap E_r)) \le 2P(A^n \backslash E_V)$$

Proof. Since the number of elements in E_r is equal to the number of elements in E_V, then $|E_r \backslash (E_V \cap E_r)| = |E_V \backslash (E_V \cap E_r)|$. Therefore, in accordance with the definition of set E_r:

$$P(E_r \backslash (E_V \cap E_r)) \ge P(E_V \backslash (E_V \cap E_r)) \quad (3.47)$$

Then from the obvious inclusion:

$$A^n \backslash E_V \supseteq E_r \backslash (E_V \cap E_r)$$

and in accordance with (3.47) it follows that:

$$P(E_V \backslash (E_V \cap E_r)) \le P(A^n \backslash E_V) \quad (3.48)$$

From the equation:

$$A^n \backslash (E_V \cap E_r) = (A^n \backslash E_V) \cup (E_V \backslash (E_V \cap E_r))$$

and from the inequality (3.48) obtain:

$$P(A^n \backslash (E_V \cap E_r)) = P(A^n \backslash E_V) + P(E_V \backslash (E_V \cap E_r)) \le 2P(A^n \backslash E_V)$$

Lemma 3.2 shows that decoding only those coset leaders which belong to the set E_r (instead of decoding all error vectors, which can be corrected by the code) leads to the fact that the error probability $P(A^n \backslash (E_V \cap E_r))$ will not exceed the double error probability for decoding on maximum likelihood $2P(A^n \backslash E_V)$.

□

The decoding in hypersphere of radius ν means that the received vector is decoded to the nearest codeword, which is at a distance no more than ν from the received vector. Moreover, the received vector is compared only with coset leaders of weight no more than ν. Therefore, to make it possible that the error probability for decoding in hypersphere does not exceed more than two times the error probability for maximum likelihood decoding, it is necessary to choose the minimum value of ν, satisfying:

$$A_\nu^n(\mathbf{0}) \supseteq E_r \cap E_V$$

where $A_\nu^N(\mathbf{0})$ is the hypersphere of radius ν and with the centre in all-zero vector.

In particular, it is enough if $A_\nu^N(\mathbf{0}) \supseteq E_r$; and for BSC it means that $\left|A_\nu^n(\mathbf{0})\right| = \sum_{i=0}^{\nu} \binom{n}{i} \ge 2^r$.

Notice that given proof does not depend on the errors model, that is, this proof is applicable to any additive channel. Moreover, the proof does not depend on the method of full decoding, that is, the proof is correct for any full decoding algorithm, not only for the maximum likelihood decoding.

3.4 Cyclic Codes

The cyclic codes form the most explored subclass of the linear codes. The greatest number of known good codes are the cyclic codes also. There is the simple encoding procedure for these codes; and simple decoding procedures for many of cyclic codes are also known.

Definition 3.1 A linear code is called a *cyclic code* if every cyclic shift of a codeword is also a codeword.

Thus, if $\mathbf{a}(a_0, a_1, \ldots, a_{n1})$ is the codeword of the cyclic code of length n, then the cyclic shift of this codeword $T(\mathbf{a}) = (a_{n1}, a_0, a_1, \ldots, a_{n2})$ is the codeword of the same code. Let each n-dimension vector $\mathbf{f} = (f_0, f_1, \ldots, f_{n1}), f_i \in K$ correspond to the polynomial $f(x) = f_0 + f_1 x + \ldots + f_{n-1}x^{n-1} \in K[x]$. Then each n-tuple corresponds to the polynomial of degree no more than $n - 1$. Hereafter we will not distinguish vector and the corresponding polynomial.

Let $a(x)$ be the codeword of the cyclic code of length n. Consider the polynomial $xa(x)$ mod$(x^n - 1)$:

$$xa(x) = a_{n-1}x^n + a_{n-2}x^{n-1} + \ldots + a_1x^2 + a_0x$$

and the residue of $xa(x)$ on modulo $(x^n - 1)$ is equal to:

$$xa(x) \bmod (x^n - 1) = a_{n-2}x^{n-1} + \ldots + a_1x^2 + a_0x + a_{n-1} \tag{3.49}$$

The right side of (3.49) is the cyclic shift of codeword $a(x)$. Therefore, $xa(x)$ mod$(x^n - 1)$ is the codeword of the cyclic code. Considering the cyclic shifts of vector $a(x)$: $xa(x)$ mod$(x^n - 1)$, $x^2a(x)$ mod$(x^n - 1)$, and so on, obtain that any polynomial $x^ia(x)$ mod$(x^n - 1)$ is the codeword. Since the cyclic code is the linear code, each linear combination of its codewords also is the codeword, that is, all polynomials:

$$\sum_{i,j} \lambda_i \cdot x^j a(x) \bmod (x^n - 1), \lambda_i \in K \tag{3.50}$$

are the code words. Thus, the set of codewords is an ideal in the ring $K[x]/(x^n - 1)$. As was mentioned above $K[x]/f(x)$ is the ring of principal ideals. Therefore, there exists the element $g(x) \in K[x]/(x^n - 1)$ such that $I =< g(x) >$, that is, this element generates the cyclic code I. It is convenient to choose nonzero normalized polynomial of minimum degree as the element $g(x)$. Then it is easy to verify that any codeword $v(x)$ of the code I can be represented uniquely as:

$$v(x) = m(x) \cdot g(x), \quad \deg m(x) < n - \deg g(x) \tag{3.51}$$

Let us consider the division of $v(x)$ by $g(x)$:

$$v(x) = m(x) \cdot g(x) + r(x)$$

where deg $r(x)<$ deg $g(x)$ or $r(x) = 0$. The first statement can not be correct since in that case $r(x) = v(x) - m(x) \cdot g(x) \in I$, that is, $r(x)$ is the codeword (polynomial) of degree less than degree of $g(x)$; and this fact contradicts the choice of $g(x)$. The fact that the polynomial $m(x) \cdot g(x)$ belongs to the code follows from the properties of an ideal. The uniqueness of the representation (3.51) follows from the fact that there are no divisors of zero in the ring of polynomials. The polynomial $g(x)$ is called the *generator polynomial* of the code. Notice that the generator polynomial $g(x)$ is the divisor of the polynomial $x^n - 1$. Since the degree of the polynomial $x^{n-\deg g(x)}$ $g(x)$ is equal to n, then it can be represented as:

$$x^{n-\deg g(x)}g(x) = x^n - 1 + r(x) \tag{3.52}$$

where $r(x) = (x^{n-\deg g(x)}g(x)) \bmod (x^n - 1)$. In accordance with (3.50) $r(x)$ is the codeword, that is, $g(x)$ is the divisor of $r(x)$. Then from (3.52) it follows that $x^n - 1$ also is divisible by $g(x)$.

We showed that all codewords could be represented as in (3.51); the number of such kind of words is equal to the number of possible choices of the information polynomial $m(x)$, that is, $q^{n-\deg g(x)} = q^k$. The number of information symbols of the code $k = n -$ deg g(x). The generator matrix **G** of the cyclic code can be formed in accordance with (3.51) by the cyclic shifts of $g(x)$:

$$\mathbf{G} = \overbrace{\begin{bmatrix} g_0 & \cdot & \cdot & \cdot & \cdot & g_r & & & \\ & g_0 & & & & & g_r & & \\ & & \cdot & & & & & \cdot & \\ & & & \cdot & & & & & \cdot \\ & & & & g_0 & \cdot & \cdot & \cdot & \cdot & g_r \end{bmatrix}}^{n} \Bigg\} k, \tag{3.53}$$

where $r = n - k$. Any cyclic code is defined by the corresponding generator polynomial $g(x)$, which is the divisor of $x^n - 1$. The opposite is also true, that is, if we choose the polynomial $g(x)$ and form the code from the words of form (3.51) then we obtain the cyclic (n, k) code, where n is such positive integer that $g(x)$ is the factor of $x^n - 1$ and $k = n -$ deg g(x).

The above results can be formulated as the theorem [4]:

Theorem 3.2 Any q-ary cyclic (n, k) code is generated by the normalized polynomial $g(x)$ over $GF(q)$ of degree $n - k$, where $g(x)$ is the factor of $x^n - 1$. And vice versa, any normalized polynomial $g(x)$ over $GF(q)$ of degree $n - k$, where $g(x)$ is the factor of $x^n - 1$, generates the cyclic (n, k) code.

Let polynomial $h(x)$ be:

$$h(x) = \frac{x^n - 1}{g(x)} \tag{3.54}$$

Then the multiplication of any codeword $v(x) = m(x) \cdot g(x)$ by $h(x)$ is equal to:

$$v(x) \cdot h(x) = m(x) \cdot h(x) \cdot g(x) = m(x) \cdot (x^n - 1) = 0 \mod (x^n - 1)$$

This equation defines the parity-check sums for codewords, and the polynomial $h(x)$ is called the *parity polynomial*. The parity-check matrix of the cyclic code can be represented with the help of $h(x)$ as follows:

$$\mathbf{H} = \left.\overbrace{\begin{bmatrix} h_k & \cdot & \cdot & \cdot & \cdot & h_0 & & & & \\ & h_k & & & & & h_0 & & & \\ & & \cdot & & & & & \cdot & & \\ & & & \cdot & & & & & \cdot & \\ & & & & h_k & \cdot & \cdot & \cdot & \cdot & h_0 \end{bmatrix}}^{n}\right\} n - k \tag{3.55}$$

A minimum distance of the cyclic code can be found using the parity-check matrix with the help of the Lemma 3.1.

Example 3.7 Consider the polynomial $g(x) = x^{10} + x^8 + x^5 + x^4 + x^2 + x + 1$. It is easy to verify that the minimal n, for which $x^n - 1$ is divisible by $g(x)$, is equal to 15. Then the polynomial $g(x)$ generates (15, 5) cyclic code over F_2, and $h(x) = \frac{x^{15}-1}{g(x)} = x^5 + x^3 + x + 1$. Therefore:

$$\mathbf{G} = \left.\overbrace{\begin{bmatrix} 1 & 1 & 1 & 0 & 1 & 1 & 0 & 0 & 1 & 0 & 1 & 0 & 0 & 0 & 0 \\ 0 & 1 & 1 & 1 & 0 & 1 & 1 & 0 & 0 & 1 & 0 & 1 & 0 & 0 & 0 \\ 0 & 0 & 1 & 1 & 1 & 0 & 1 & 1 & 0 & 0 & 1 & 0 & 1 & 0 & 0 \\ 0 & 0 & 0 & 1 & 1 & 1 & 0 & 1 & 1 & 0 & 0 & 1 & 0 & 1 & 0 \\ 0 & 0 & 0 & 0 & 1 & 1 & 1 & 0 & 1 & 1 & 0 & 0 & 1 & 0 & 1 \end{bmatrix}}^{15}\right\} 5 \tag{3.56}$$

$$\mathbf{H} = \left.\overbrace{\begin{bmatrix} 1 & 0 & 1 & 0 & 1 & 1 & 0 & 0 & 0 & 0 & 0 & 0 & 0 & 0 & 0 \\ 0 & 1 & 0 & 1 & 0 & 1 & 0 & 0 & 0 & 0 & 0 & 0 & 0 & 0 & 0 \\ 0 & 0 & 1 & 0 & 1 & 0 & 1 & 0 & 0 & 0 & 0 & 0 & 0 & 0 & 0 \\ 0 & 0 & 0 & 1 & 0 & 1 & 1 & 1 & 0 & 0 & 0 & 0 & 0 & 0 & 0 \\ 0 & 0 & 0 & 0 & 1 & 0 & 0 & 1 & 1 & 0 & 0 & 0 & 0 & 0 & 0 \\ 0 & 0 & 0 & 0 & 0 & 1 & 1 & 0 & 1 & 1 & 0 & 0 & 0 & 0 & 0 \\ 0 & 0 & 0 & 0 & 0 & 0 & 0 & 1 & 0 & 1 & 1 & 1 & 0 & 0 & 0 \\ 0 & 0 & 0 & 0 & 0 & 0 & 1 & 0 & 1 & 0 & 1 & 1 & 1 & 0 & 0 \\ 0 & 0 & 0 & 0 & 0 & 0 & 0 & 1 & 0 & 1 & 0 & 0 & 1 & 1 & 0 \\ 0 & 0 & 0 & 0 & 0 & 0 & 0 & 0 & 1 & 0 & 1 & 1 & 0 & 1 & 1 \end{bmatrix}}^{15}\right\} 10 \tag{3.57}$$

The generator and the parity matrices can be reduced to the systematic form:

$$\mathbf{G} = \overbrace{\begin{bmatrix} 1 & 0 & 0 & 0 & 0 & 1 & 1 & 1 & 0 & 1 & 1 & 0 & 0 & 1 & 0 \\ 0 & 1 & 0 & 0 & 0 & 0 & 1 & 1 & 1 & 0 & 1 & 1 & 0 & 0 & 1 \\ 0 & 0 & 1 & 0 & 0 & 1 & 1 & 0 & 1 & 0 & 1 & 1 & 1 & 1 & 0 \\ 0 & 0 & 0 & 1 & 0 & 0 & 1 & 1 & 0 & 1 & 0 & 1 & 1 & 1 & 1 \\ 0 & 0 & 0 & 0 & 1 & 1 & 1 & 0 & 1 & 1 & 0 & 0 & 1 & 0 & 1 \end{bmatrix}}^{15} \Bigg\} 5 \tag{3.58}$$

$$\mathbf{H} = \overbrace{\begin{bmatrix} 1 & 0 & 1 & 0 & 1 & 1 & 0 & 0 & 0 & 0 & 0 & 0 & 0 & 0 & 0 \\ 1 & 1 & 1 & 1 & 1 & 0 & 1 & 0 & 0 & 0 & 0 & 0 & 0 & 0 & 0 \\ 1 & 1 & 0 & 1 & 0 & 0 & 0 & 1 & 0 & 0 & 0 & 0 & 0 & 0 & 0 \\ 0 & 1 & 1 & 0 & 1 & 0 & 0 & 0 & 1 & 0 & 0 & 0 & 0 & 0 & 0 \\ 1 & 0 & 0 & 1 & 1 & 0 & 0 & 0 & 0 & 1 & 0 & 0 & 0 & 0 & 0 \\ 1 & 1 & 1 & 0 & 0 & 0 & 0 & 0 & 0 & 0 & 1 & 0 & 0 & 0 & 0 \\ 0 & 1 & 1 & 1 & 0 & 0 & 0 & 0 & 0 & 0 & 0 & 1 & 0 & 0 & 0 \\ 0 & 0 & 1 & 1 & 1 & 0 & 0 & 0 & 0 & 0 & 0 & 0 & 1 & 0 & 0 \\ 1 & 0 & 1 & 1 & 0 & 0 & 0 & 0 & 0 & 0 & 0 & 0 & 0 & 1 & 0 \\ 0 & 1 & 0 & 1 & 1 & 0 & 0 & 0 & 0 & 0 & 0 & 0 & 0 & 0 & 1 \end{bmatrix}}^{15} \Bigg\} 10 \tag{3.59}$$

To detect the errors in the received word $b(x)$ it is enough to check the condition:

$$b(x) \cdot h(x) = 0 \mod (x^n - 1) \tag{3.60}$$

We show that the generator polynomial of the cyclic code is the factor of $(x^n - 1)$. Therefore, it is necessary to consider all combinations of the factors of the polynomial $(x^n - 1)$ in order to enumerate all cyclic codes of length n. It is well known that if the characteristic p of the field is not the divisor of n, then the polynomial $(x^n - 1)$ can be factored by the irreducible divisors:

$$x^n - 1 = f_1(x) \cdot \ldots f_1|(x)$$

Therefore, it is possible to choose any polynomial of form:

$$g(x) = f_{i_1}(x) \cdot \ldots f_{i_s}(x), \quad i_1 < i_2 < \ldots < i_s, \quad s < l$$

as the generator polynomial. Each of these polynomials corresponds to the code with some values of k and d; and there are $2^l - 2$ nontrivial cyclic codes of length n at all.

Example 3.8 Construct all binary cyclic codes of length 7. The polynomial $x^7 - 1$ can be factored as follows:

$$x^7 - 1 = (x + 1) \cdot (x^3 + x + 1) \cdot (x^3 + x^2 + 1)$$

The corresponding six cyclic codes of length 7 are defined by the following polynomials:

$$g_1(x) = (x + 1), g_2(x) = (x^3 + x + 1), g_3(x) = (x^3 + x^2 + 1)$$
$$g_4(x) = (x + 1) \cdot (x^3 + x + 1), g_5(x) = (x + 1) \cdot (x^3 + x^2 + 1)$$
$$g_6(x) = (x^3 + x + 1) \cdot (x^3 + x^2 + 1)$$

Table 3.1 Parameters of some cyclic codes

n	k	d	Generator polynomial $g(x)$
7	4	3	13 *)
15	11	3	23
	9	3	171
	7	5	721
	5	7	2467
31	26	3	45
	21	5	3551
	16	7	107657
	11	11	5423325
	6	15	313365047
63	57	3	103
	51	5	12471
	45	7	1701317
	39	9	166623567
	30	13	157464165547
	16	23	6331141367235453
	7	31	5231045543503271737
127	120	3	211
	85	13	130704476322273
	71	19	6255010713253127753
	22	47	123376070404722522435445626637647043
255	247	3	435
	187	19	52755313540001322236351
	139	31	461401732060175561570722730247453567445
	47	85	2533542017062646563033041377406233175123334145446045005066024552543173

* All generator polynomials are given in the octal format, for example. $13_8 = 1011 = x + x + 1$.

It is easy to verify (for instance, by enumerating the codewords) that the codes corresponding to these polynomials have the following parameters:

$$\begin{array}{ll} G_1: & k=6, d=2; \\ G_2, G_3: & k=4, d=3; \\ G_4, G_5: & k=3, d=4; \\ G_6: & k=1, d=7 \end{array}$$

The parameters of some binary cyclic codes are listed in Table 3.1.

One of the most important operations for the implementation of the cyclic codes is the calculation of the remainder resulting from dividing one polynomial by another.

This operation can be executed with the help of a tapped filter, that is, a device containing delay elements with taps, adders in field $GF(q)$ and multipliers in field $GF\ (q)$ (Figure 3.2).

The state of the delay element is $s^{(t)} = s$ at the time t if at the moment t we obtain symbol $s \in GF\ (q)$ at the output of this delay element. Let r be the number of delay elements in the filter. Then vector $\mathbf{s}^{(t)} = (s_1^{(t)}, \ldots, s_r^{(t)})$, where $s_i^{(t)}$ is the state of the i th element at time t, is called the state of the filter at time t.

Hereafter we will consider the filters with one input and one output. The input and output signals of the filters will be the sequences of symbols from $GF(q)$.

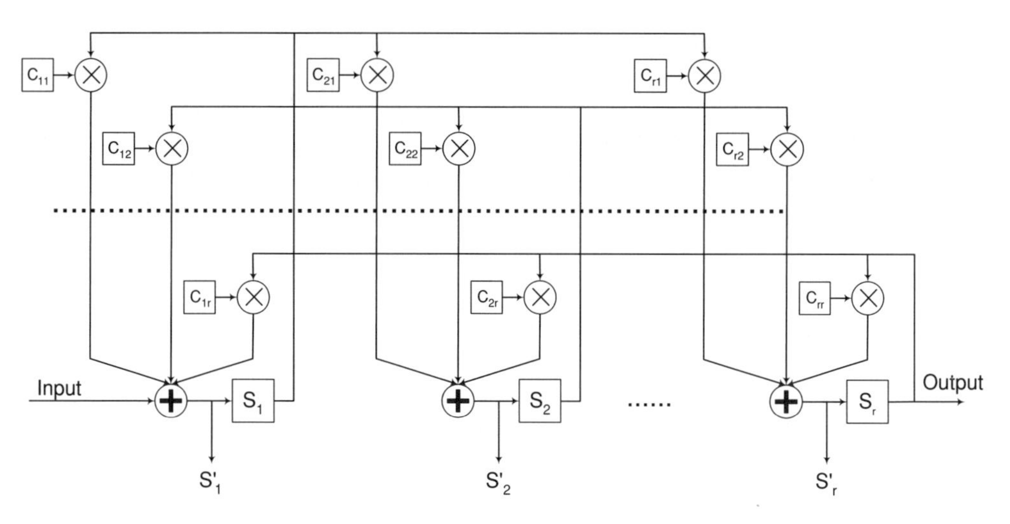

Figure 3.2 General structure of filter

Notice that such a kind of filter is the *linear filter*, that is, the filter response (the output signal) to the sum of input signals is the sum of filter responses to each input signal (sum in $GF(q)$).

Let $\mathbf{s} = (s_1, \ldots, s_r)$ be the preceding and $\mathbf{s}' = (s'_1, \ldots, s'_r)$ be the succeeding states of the filter. Then to define the filter it is necessary to define the following equations:

$$\begin{aligned} s'_1 &= c_{11} \cdot s_1 + \ldots + c_{1r} \cdot s_r; \\ &\ldots\ldots\ldots\ldots\ldots\ldots\ldots\ldots \\ s'_r &= c_{r1} \cdot s_1 + \ldots + c_{rr} \cdot s_r \end{aligned} \tag{3.61}$$

where c_{ij} is the coefficient defined by the structure of the filter; the addition and the multiplication is carried out in $GF\ (q)$. That means the preceding state fully defines the succeeding state in cases where there is no signal at the filter input.

Assume that the input signal appears at the input of the first delay element. Then if the input signal is equal to α, filter from the state $s = (s_1, \ldots, s_r)$ goes to the state $s' = (s'_1, \ldots, s'_r)$ defined by the equations:

$$\begin{aligned} s'_1 &= c_{11} \cdot s_1 + \ldots + c_{1r} \cdot s_r + \alpha; \\ &\ldots\ldots\ldots\ldots\ldots\ldots\ldots\ldots \\ s'_r &= c_{r1} \cdot s_1 + \ldots + c_{rr} \cdot s_r \end{aligned} \tag{3.62}$$

Let us consider the synthesis of the calculator of the remainder. The $(r \times r)$ matrix $\mathbf{C} = [c_{ij}]$ is called the transfer matrix. This matrix defines the filter. The general view of the filter defined by matrix $\mathbf{C}$ is shown in Figure 3.2.

This is put in correspondence to the j th column of matrix $\mathbf{C}$ the polynomial $c_j(x) = \sum_{i=1}^{r} c_{ij} x^{i-1}$.

Theorem 3.3 Let $c_i(x) = x^i \bmod g(x)$ be the i th column of the matrix $\mathbf{C}$, where $g(x)$ is an arbitrary polynomial of degree r. Then if signal α appears at the filter input:

$$s'(x) = \alpha + xs(x) \bmod g(x) \tag{3.63}$$

where $s'(x) = \sum_{i=1}^{r} s'_i x^{i-1}$.

Proof. It follows from (3.62) that:

$$s'(x) = s_1 \cdot c_1(x) + s_2 \cdot c_2(x) + \ldots + s_r \cdot c_r(x) + \alpha$$

Then substituting expressions for $c_i(x)$ obtain:

$$\begin{aligned}
s'(x) &= \alpha + s_1 x \bmod g(x) + s_2 x^2 \bmod g(x) + \ldots + s_r x^r \bmod g(x) = \\
&= (\alpha + s_1 x + s_2 x^2 + \ldots + s_r x^r) \bmod g(x) = (\alpha + x(s_1 + s_2 x + \ldots + s_r x^{r-1})) \bmod g(x) = \\
&= (\alpha + x s(x)) \bmod g(x)
\end{aligned}$$

□

Thus, for the arbitrary polynomial $g(x)$ and the input signal α in Theorem 3.3 the filter goes to the state s' defined by (3.63). Let the initial state of the filter be all-zero and the elements $a_{n-1}, a_{n-2,\ldots,}a_0$, which are the coefficients of the polynomial $a(x) = \sum_{i=0}^{n-1} a_i x^i$, consecutively appear at the input of the filter. Then the filter will consecutively go to the states:

$$\begin{aligned}
&s^{(0)}(x) = 0, \quad s^{(1)}(x) = a_{n-1}, \quad s^{(2)}(x) = (a_{n-2} + a_{n-1}x) \bmod g(x) \\
&s^{(3)}(x) = (a_{n-3} + a_{n-2}x + a_{n-1}x^2) \bmod g(x), \ldots, \\
&s^{(n)}(x) = (a_0 + a_1 x + \ldots + a_{n-1}x^{n-1}) \bmod g(x)
\end{aligned}$$

but $s^{(n)}(x)$ is the remainder resulting from dividing $a(x)$ by $g(x)$.

Example 3.9 $g(x) = x^3 + x + 1, \quad a(x) = x^6 + x^3 + 1, \quad GF(2)$. The transfer matrix is:

$$\mathbf{C} = \begin{bmatrix} 0 & 0 & 1 \\ 1 & 0 & 1 \\ 0 & 1 & 0 \end{bmatrix}$$

The filter defined by the matrix **C** is shown in Figure 3.3. The states of the filter are:

$$\begin{aligned}
s^{(0)}(x) = 0, \quad s^{(1)}(x) = a_{n-1} = a_6 = 1, \quad s^{(2)}(x) = a_5 + a_6 x = x, \quad s^{(3)}(x) = x^2, \\
s^{(4)}(x) = x, \quad s^{(5)}(x) = x^2, \quad s^{(6)}(x) = x + 1, \quad s^{(7)}(x) = x^2 + x + 1
\end{aligned}$$

The remainder resulting from dividing $a(x)$ by $g(x)$ is equal to $x^2 + x + 1$.

All codewords of a cyclic code can be represented in form of (3.51), where $m(x)$ is the information message and $m(x) \cdot g(x)$ is the corresponding codeword. Such encoding procedure corresponds to the generator matrix of form (3.53). There is no necessity to keep the whole matrix (3.53) for the encoding. It is enough to keep only the first row of this matrix, that is, the generator polynomial. Thus, the realization

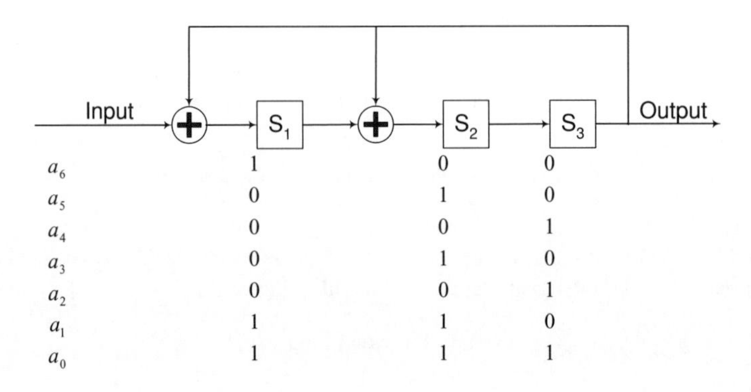

Figure 3.3 Filter for the division by $g(x) = x^3 + x + 1$

of the encoding reduces to the realization of multiplication of two polynomials. However, the encoding procedure:

$$m(x) \to m(x) \cdot g(x) \tag{3.64}$$

defines the nonsystematic code. It is impossible to select the information symbols in the codeword $m(x) \cdot g(x)$; and it is necessary to divide the codeword by $g(x)$ in order to obtain the information message $m(x)$. Here we will show two methods of systematic encoding of the cyclic codes, for which the information symbols are the coefficients of most significant powers of the polynomial corresponding to the codeword.

Let $a(x) = \sum_{i=0}^{n-1} a_i x^i$ be the codeword and $h(x) = \sum_{i0}^{k} h_i x^i$ be the parity polynomial of the (n, k) cyclic code. Consider the product of $a(x)$ by $h(x)$:

$$\begin{aligned} f(x) = a(x) \cdot h(x) = m(x) \cdot g(x) \cdot h(x) = x^n m(x) - m(x) = \\ m_{k-1} x^{n+k-1} + \ldots + m_0 x^n - m_{k-1} - \ldots - m_0 \end{aligned} \tag{3.65}$$

It follows from the definition of the parity polynomial that the coefficients of the polynomial $f(x) = \sum_{i=0}^{n+k-1} f_i x^i$ for x^i, $k \le i \le n-1$ are equal to zero. Now substituting coefficients of $m(x)$ and $h(x)$ to f_i obtain:

$$f_i = \sum_{j=0}^{k} h_j \cdot a_{i-j}, \quad k \le i \le n-1 \tag{3.66}$$

Since $h_k = 1$, then we can derive the following equation:

$$a_{i-k} = -\sum_{j=0}^{k} h_j \cdot a_{i-j}, \quad k \le i \le n-1 \tag{3.67}$$

The equation (3.67) defines the recurrent formula for the sequential calculation of $a_{n-1-k}, a_{n-1-k-1}, \ldots, a_0$ using the information symbols $a_{n-1}, a_{n-2}, \ldots, a_{n-k}$. Thus, (3.67) defines the method of systematic encoding for cyclic code.

The circuit implementing the calculation by the formula (3.67) is shown in Figure 3.4. The operation of the circuit can be described as follows:

Initially gate 1 is switched on and gate 2 is switched off. The information symbols $\alpha_{n-1}, \alpha_{n-2}, \ldots, \alpha_{n-k}$ sequentially are shifted into the register. As soon as k information symbols entered the shift register, gate 1 is switched off and gate 2 is switched on. Symbol α_{n-1} appears at the output of the encoder; and the new symbol is fed to the input. This new symbol is the inverted sum of products of symbols from the register elements by the corresponding coefficients h_i, that is, as follows from Equation 3.67 α_{n-1-k}.

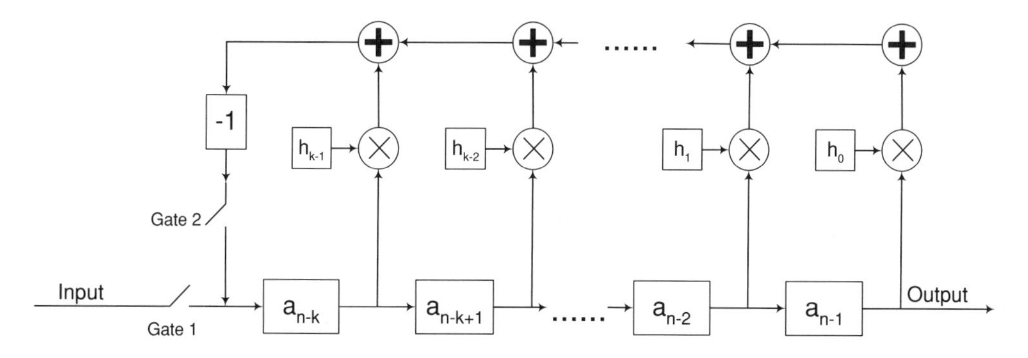

Figure 3.4 The encoder of cyclic code with k delay elements

During the next shifts the symbols $\alpha_{n-2}, \alpha_{n-3}, \ldots$ will appear at the output of encoder; and the symbols $\alpha_{n-2}, \alpha_{n-k-3}, \ldots$ will be fed to the input of the register. After n shifts we obtain the whole codeword at the output of the encoder. Notice that multiplication and addition should be executed in the field $GF(q)$, so we have to have the special devices for these operations.

The encoder considered above uses k delay elements (or k-stage shift register). Let us consider the encoder with $(n-k)$ delay elements. It is obvious that such a kind of encoder will be more economic in case $k > n-k$.

Let us represent the information symbols as the polynomial $\alpha_{n-1}x^{n-1} + \alpha_{n-2}x^{n-2} + \ldots + \alpha_{n-k}x^{n-k}$. In accordance with the algorithm of division:

$$\alpha_{n-1}x^{n-1} + \alpha_{n-2}x^{n-2} + \ldots + \alpha_{n-k}x^{n-k} = m(x) \cdot g(x) + r(x) \tag{3.68}$$

where $\deg r(x) < \deg g(x) = n-k$. It follows from (3.68) that the polynomial:

$$\alpha_{n-1}x^{n-1} + \alpha_{n-2}x^{n-2} + \ldots + \alpha_{n-k}x^{n-k} - r(x)$$

is the codeword. Therefore, it is enough to obtain the remainder resulting from dividing $\alpha_{n-1}x^{n-1} + \alpha_{n-2}x^{n-2} + \ldots + \alpha_{n-k}x^{n-k}$ by $g(x)$ in order to calculate the parity-check symbols. The device implementing this algorithm is shown in Figure 3.5

The device is operating as follows. Initially gate 1 and gate 2 are switched on and gate 3 is switched off. The information symbols $\alpha_{n-1}, \alpha_{n-2}, \ldots, \alpha_{n-k}$ are fed to the input of shift register and appear at the output of the encoder simultaneously. After $n-k$ shifts symbol α_{n-1} appears at the output of the shift register and multiplied by the coefficients $g_{n-k-1}, g_{n-k-2}, \ldots, g_0$ is subtracted from the elements $\alpha_{n-2}, \alpha_{n-3}, \ldots, \alpha_{k-1}$ correspondingly. This step corresponds to the first operation in the algorithm of the division of polynomials. After k shifts gate 1 is switched off. After n shifts from the start of operation the remainder $r(x)$ we obtain the whole codeword at the output of the encoder.

Notice that in the case of binary code the described circuits contain adders on modulo 2 and binary memory elements. The multiplication by "1" is realized by the presence and the multiplication by "0" by the absence of the feedback.

Example 3.10 Consider the (15, 4) binary cyclic code with parity polynomial $h(x) = x^4 + x + 1$. The k-stage encoder is shown in Figure 3.6.

Let $\alpha_{n-1} = \alpha_{n-3} = 1, \quad \alpha_{n-2} = \alpha_{n-4} = 0$. The work of the encoder is explained by Table 3.2.

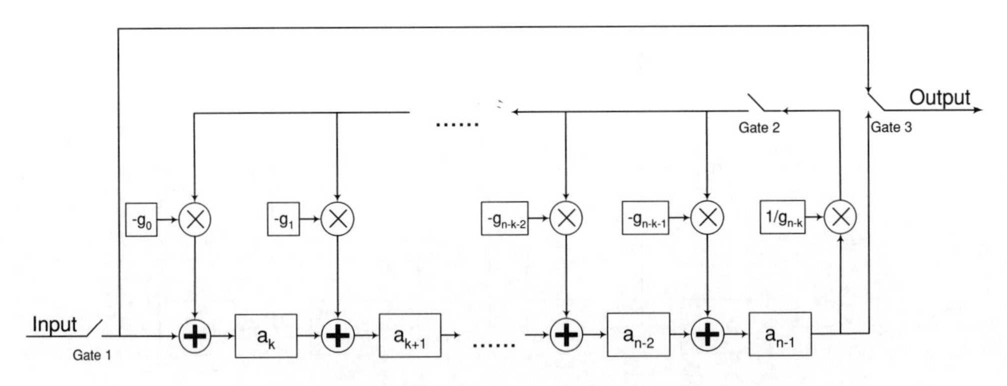

Figure 3.5 The encoder of cyclic code with $(n$-$k)$ delay elements

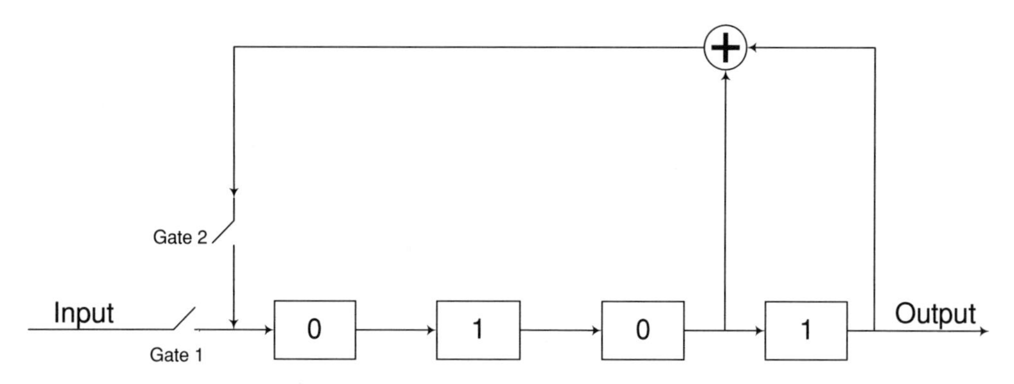

Figure 3.6 The k-stage encoder of the (15, 4) code

Example 3.11 Consider the (15, 11) binary cyclic code with generator polynomial $g(x) = x^4 + x^3 + 1$. The $(n - k)$-stage encoder of this code is shown in Figure 3.7.

Let $\alpha_{14} = \alpha_{13} = \alpha_{10} = \alpha_9 = \alpha_5 = \alpha_4 = 1$, $\alpha_{12} = \alpha_{11} = \alpha_8 = \alpha_7 = \alpha_6 = 0$. The work of the encoder is explained by Table 3.3.

In the example above the division by the polynomial $g(x)$ begins only when the symbol α_{n-1} occupy the last (right) delay element in the shift register. From the other hand it takes an additional four cycles to obtain the calculated parity symbols at the output of the encoder. Hence, the overall encoding time is equal to 19 cycles (the same as for the scheme shown in Figure 3.6). It is possible to decrease the

Table 3.2 The work of the encoder of (15, 4) code

No. of shift	State	Feedback symbol	Output symbol	
0	0	0	0	
1	1	0	0	
2	0	0	0	
3	1	0	0	
4	0	0	0	
5	1	1	1	The result of the encoding
6	1	1	0	
7	1	1	1	
8	1	1	0	
9	0	0	1	
10	0	0	1	
11	0	0	1	
12	1	1	1	
13	0	0	0	
14	0	0	0	
15	1	1	0	
16	1	1	1	
17	0	0	0	
18	1	1	0	
19	0	0	1	

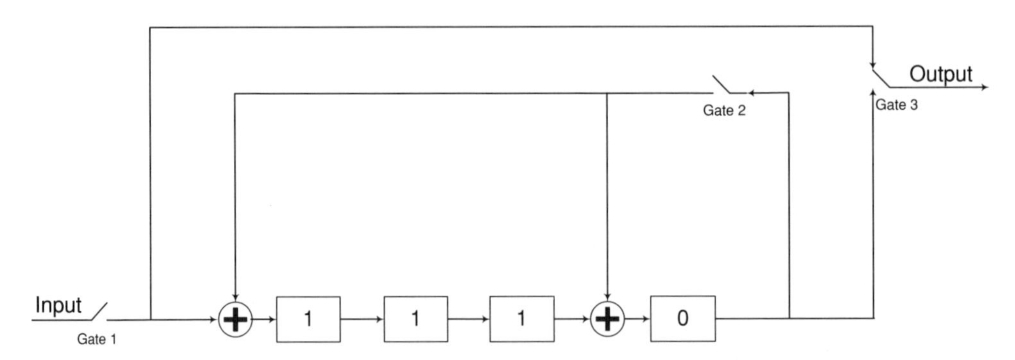

Figure 3.7 The (n-k)-stage encoder of the (15, 11) code

encoding time if we start the division by the generator polynomial simultaneously with feeding the information symbols. Such a kind of encoder is shown in Figure 3.8.

This encoder executes the multiplication of the information polynomial by x^{n-k} (which is equivalent to $(n-k)$ shifts of the information symbols in the shift register) due to feeding the information symbols at point A. The encoding time of the encoder in Figure 3.8 is equal to n cycles.

In order to detect the error with the help of the cyclic code it is enough to calculate the remainder resulting from division of the received word $b(x)$ by the generator polynomial $g(x)$ and to compare it (the remainder) with zero.

The remainder resulting from the division of the received word $b(x)$ by the generator polynomial $g(x)$ is called syndrome. Let us denote the syndrome by $S(x)$. Then:

$$S(x) = b(x) \bmod g(x) \tag{3.69}$$

If the syndrome of the received word is equal to zero, then the received word is regarded as codeword and it is assumed that no error occur during the transmission of this word over the channel (or the undetectable error occur). If the syndrome is not equal to zero the error is detected.

Table 3.3 The work of the encoder of (15, 11) code

No. of shift	State	Feedback symbol
0	0	0
1	1	0
2	1	0
3	0	0
4	0	0
5	0	1
6	1	0
7	0	0
8	0	0
9	0	0
10	0	1
11	0	1
12	1	1
13	1	1
14	1	1
15	1	1

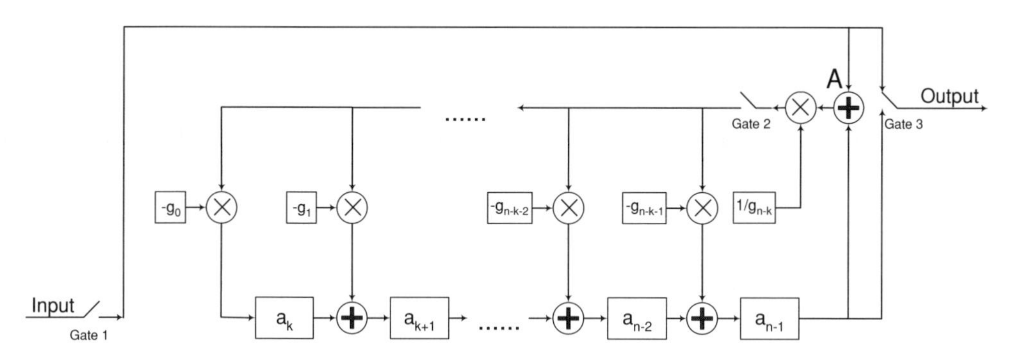

Figure 3.8 The (n-k)-stage encoder of cyclic code with the premultiplication by x^{n-k}

The syndrome can be calculated with the help of $(n-k)$-stage circuit with premultiplication by x^{n-k} used for the calculation of parity-check symbols.

Example 3.12 Consider the circuit for error detection for binary (15, 9) cyclic code with the generator polynomial $g(x) = x^6 + x^5 + x^4 + x^3 + 1$. The circuit is shown in Figure 3.9. In 15 shifts the remainder resulting from division of the received word by the generator polynomial $g(x)$ will be stored in the shift register, that is, the syndrome of the received word. If at least one coefficient of the syndrome is not equal to zero, then the signal of error detection appears at the output of the OR gate.

The correction of the error pattern also can be done with the help of syndrome. In general case to solve this problem it is needed to keep in memory the table, in which every syndrome corresponds to the error pattern.

For cyclic codes the solution is not so complex, because it is possible to keep in memory not all syndromes but only those which correspond to error patterns containing nonzero symbol in the first position. Using the cyclic shifts of the received word $b(x)$ we obtain word $b'(x)$ with the error symbol in the first position and syndrome of this word is stored in memory and can be used to correct the error pattern.

Let us describe the process of the error correction for cyclic code in detail. Let S be the set of syndromes of correctable error patterns with nonzero symbol in the first position. Let us denote by $S^{(\nu)}$ the subset of S, which contain the syndromes of error patterns with symbol ν ($\nu \neq 0, \in GF(q)$) in the first position.

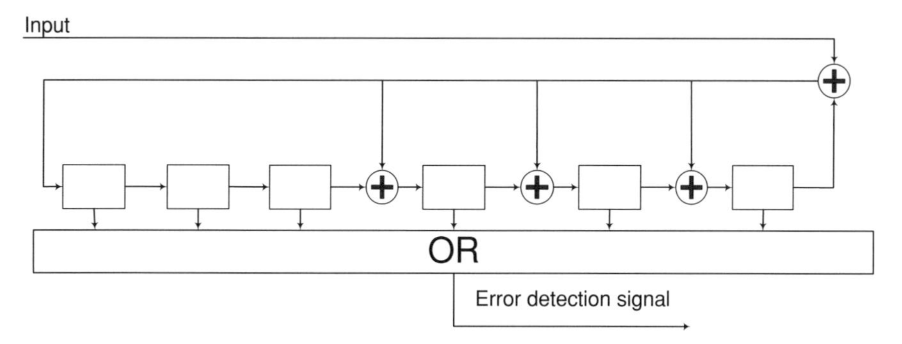

Figure 3.9 The error detector for the (15, 9) code

It is obvious that $S = \bigcup_{v-1}^{q-1} S^{(v)}$. The algorithm of decoding the cyclic code consists of the following steps:

1. For received vector **b** there are computed syndromes S_i $(i = 0, \ldots, n-1)$ of the cyclic shift of vector **b** by i positions.
2. For each S_i the sets $S^{(v)}$ are formed $v = 1, \ldots, q-1$ such that $S_i \in S^{(v)}$. Then the i th element of error pattern e_i is assumed to be equal to v. If for none of v the condition $S_i \in S^{(v)}$ is correct, then e_i is assumed to be zero.
3. When all e_i, $(i = 0, \ldots n = 1)$ are obtained vector **b** is added with computed vector **e**.

Step 2 is usually called the *selection* and the device executing this operation is called *selector*. The most difficult part of realization of the syndrome decoding is the design of the selector. As usual the complexity of this device is very high. Usually the syndrome decoding is used for short codes or in case we tend to correct the small number of errors. For instance, the selector for correction of one error is a very simple device.

The list of cyclic codes is restricted by the fact that for most, but not all, values of n and k there exist the polynomial $g(x)$ of degree $n-k$, which is the factor of $x^n - 1$. Moreover, for some values of n all cyclic codes are inefficient. For example, for all even n the (n, k) cyclic code has the distance $d = 2$ independently of the value of k. The list of codes may be extended if we consider the shortened cyclic codes.

The *shortened cyclic* $(n-i, k-i)$ *code* is the code which is constructed from the cyclic (n, k) code by the rejection of i high-order information symbols of each codeword. The length of the obtained code is $n-i$ and the number of information symbols is $k-i$. The distance of the shortened code is no less than distance of original cyclic code.

The shortened code is not the cyclic code because it contains the words which are not the cyclic shift of some codeword. However, all codewords of the shortened code are divisible by the generator polynomial of the original code, because the codewords of the shortened code are the codewords of the original code, for which the i high-order information symbols are equal to zero. Because of this fact the same circuits as those employed by the original cyclic code can accomplish the encoding of the shortened cyclic code.

It is also possible to use the decoder of the original (n, k) cyclic code to error detection and error correction of the shortened $(n-i, k-i)$ code. It is enough to add i zeros to the word of shortened code to do it. However, the decoder of the original code needs in n shifts to calculate the syndrome; and the decoding of the shortened code can be done with $n-i$ shifts.

Let symbols $\underbrace{0, 0, \ldots, 0}_{i}, b_{n-1-i}, b_{n-2-i}, \ldots, b_o$ consecutively appear at the input of the syndrome calculator. This device is executing the division of polynomial $b(x) = 0 \cdot x^{n-1} + 0 \cdot x^{n-2} + \ldots + 0 \cdot x^{n-i} + b_{n-1-i}x^{n-1-i} + b_{n-2-i}x^{n-2-i} + \ldots + b_o$ by the polynomial $g(x)$; and the operation takes n shifts. However, since the i high-order coefficients of $b(x)$ are equal to zero the feedback signal of the shift register is equal to zero during the first i shifts; and, in fact, there is no operation of the division of polynomials executed during these first i shifts. To avoid this inefficient operation it is possible by preliminary shifting of $b(x)$ i times to correspond to premultiplication of $b(x)$ by x^i on modulo $g(x)$.

Considering the premultiplication by x^{n-k}, which is necessary to compute the syndrome of the original code in n shifts, the calculator of the syndrome of the shortened $(n-i, k-i)$ code needs preliminary multiplication of the received word $b(x)$ by x^{n-k+i}.

Let polynomial $f(x)$ be $f(x) = f_{n-k-1}x^{n-k-1} + f_{n-k-2}x^{n-k-2} + \ldots + f_o x^{n-k+i} \mod g(x)$, then the premultiplication of $b(x)$ by $f(x)$ corresponds to the multiplication by x^{n-k+i}. This operation can be executed with the help of the circuit shown in Figure 3.10.

The syndrome calculator for shortened cyclic code executing the multiplication by $f(x)$ and the division by $g(x)$ is shown in Figure 3.11.

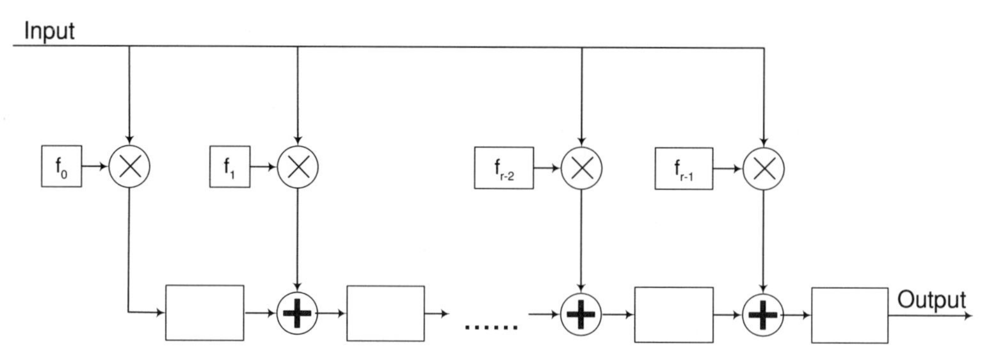

Figure 3.10 The circuit for multiplication by $f(x)$

Example 3.13 Consider the syndrome calculator of (10, 6) code obtained by shortening by five symbols the original cyclic (15, 11) code with the generator polynomial $g(x) = x^4 + x + 1$.

$f(x)x^{15-11+5} \bmod g(x) = x^9 \bmod \left(x^4 + x + 1\right) = x^3 + x$. Figure 3.12 shows the syndrome calculator for this code.

The cyclic codes can be used for burst-error correction. Here we will show the simple burst-error-correcting decoder.

Let $a(x)$ be the transmitted codeword of (n, k) cyclic code, $e(x)$ be the polynomial corresponding to error vector. Then the received word $b(x)$ can be represented as:

$$b(x) = a(x) + e(x) \tag{3.70}$$

Notice that if the burst-error occur on the parity-check symbols of the decoding word, then $\deg e(x) \le n - k - 1$ (the parity-check symbols correspond to less significant digits of the word). Then $S(x) = b(x) \bmod$

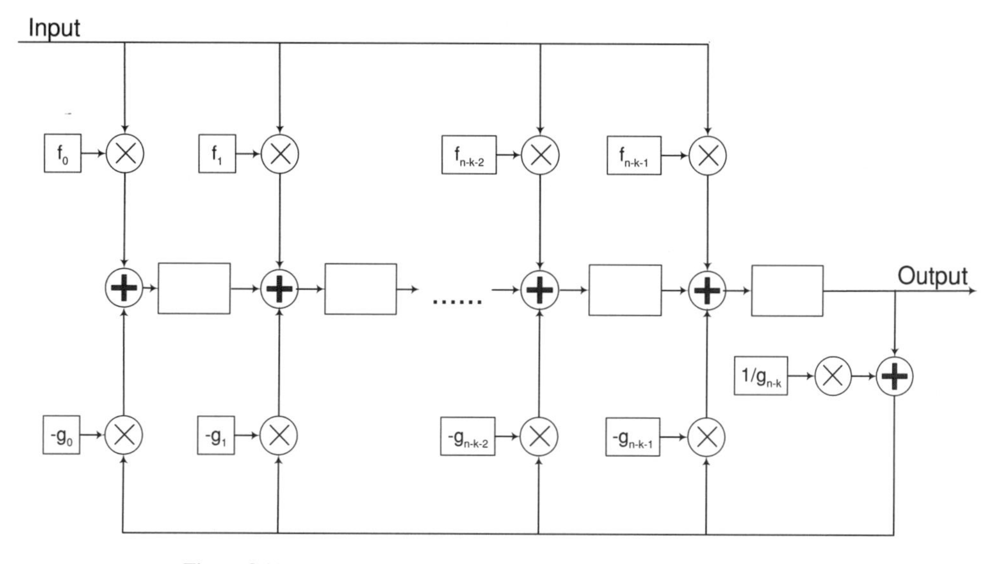

Figure 3.11 The circuit for calculation of shortened cyclic code syndrome

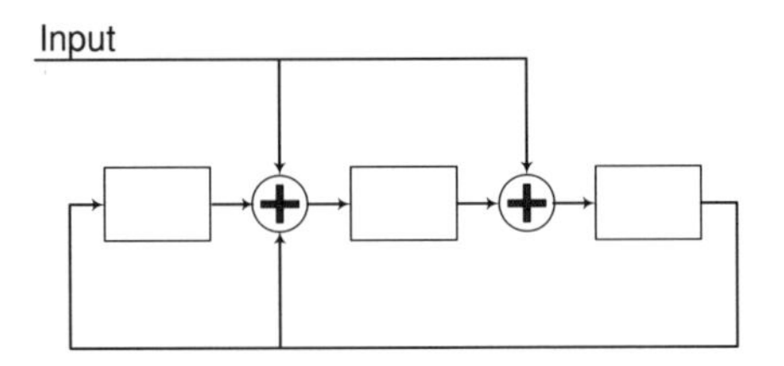

Figure 3.12 The syndrome calculator of shortened (10, 6) code

$g(x) = a(x) \bmod g(x) + e(x) \bmod g(x) = e(x) \bmod g(x) = e(x)$, since $a(x) \bmod g(x) = 0$ and $\deg e(x) < \deg g(x)$. In this case it is enough to calculate the syndrome $S(x)$ and to add it with $b(x)$:

$$a(x) = b(x) + S(x) \tag{3.71}$$

If the burst-error occur on the information symbols it is possible to shift it to the parity-check symbols by shifting the received codeword. Let:

$$b(x) = a(x) + e(x) \cdot x^i$$

where $\deg e(x) < \deg g(x) = n - k$ and i satisfies the relation:

$$\deg\left(e(x) \cdot x^i\right) = i + \deg e(x) \geq n - k$$

Then the polynomial:

$$\left(x^{-i} \cdot b(x)\right) \bmod \left(x^n - 1\right) = \left(x^{-i} \cdot a(x) + e(x)\right) \bmod \left(x^n - 1\right)$$

is the sum of the codeword $a'(x) = \left(x^{-i} \cdot a(x)\right) \bmod \left(x^n - 1\right)$ and the burst-error $e(x)$ on the parity-check symbols positions.

Now by shifting the received word we shift the burst-error on the parity-check symbols positions, and then by adding $\left(x^{-i} \cdot b(x)\right) \bmod \left(x^n - 1\right)$ with the corresponding syndrome we obtain the polynomial $x^{-i} \cdot a(x)$. Then we obtain the codeword $a(x)$ with the help of i cyclic shifts. However, it is not clear on which cyclic shift of the received word $b(x)$ the burst-error appears at the parity-check positions.

Theorem 3.4 Let code corrects the burst-errors of length b or less and let the decoding word $f(x)$ be the sum of the codeword $a(x)$ and burst-error $e(x)$ of length no more than b, $f(x) = a(x) + e(x)$. In order to the degree of $e(x)$ be no more than $b - 1$ it is necessary and sufficient that the degree of the syndrome of word $f(x)$ be no more than $b - 1$.

Proof. The necessity immediately follows from the fact that if $\deg e(x) \leq b - 1 \leq n - k - 1$, then $S(x) = e(x)$.

The sufficiency we will prove by contradiction. Let $\deg e(x) \leq b$ and $\deg S(x) \leq b$. In accordance with definition of $S(x)$:

$$f(x) = m_1(x) \cdot g(x) + S(x)$$

on the other hand:

$$f(x) = m_2(x) \cdot g(x) + e(x)$$

Therefore:

$$e(x) \bmod g(x) = S(x) \bmod g(x) \tag{3.72}$$

The equation (3.72) means that $e(x)$ and $S(x)$ belong to the same coset. But this fact in turn means that the code is capable of correcting only one burst-error from two of length no more than b ($S(x)$ or $e(x)$). And this fact contradicts the condition of the theorem. □

Thus, the algorithm of decoding of burst errors consists in the shift of the decoding word and calculation of the syndrome until the degree of the syndrome become less than b. Then addition of the shifted word and the syndrome results in the shifted codeword. The last step is the reverse cyclic shift of the obtained codeword. The algorithm can be formulated as follows:

1. Calculation of the syndromes $S_i(x) = x^{-i} f(x) \bmod g(x)$, $i = 0, 1 \ldots$ until:

$$\deg S_i(x) \leq b - 1 \tag{3.73}$$

2. If $S_i(x)$ satisfy (3.73) then calculate:

$$a_i(x) = x^{-i} f(x) + S_i(x) \tag{3.74}$$

3. Calculation of the decoding result $\hat{a}(x)$ in form of:

$$\hat{a}(x) x^i a_i(x) \bmod (x^n - 1) \tag{3.75}$$

4. If $S_i(x)$ does not satisfy (3.73) for no one of $i = 0, 1, \ldots, n-1$, then it is assumed that the undecodable error occur.

To calculate the syndromes $S_i(x)$ we need the following result.

Theorem 3.5 The syndrome of the i-fold cyclic shift ($i = \pm 0, \pm 1, \ldots$) of the received word is equal to i-fold cyclic shift of the syndrome of the received word:

$$\left(x^i f(x)\right) \bmod g(x) = x^i (f(x)) \bmod g(x) \tag{3.76}$$

Proof. In accordance with the definition of the syndrome:

$$f(x) = m(x) \cdot g(x) + S(x)$$

Then $\left(x^i f(x)\right) \bmod g(x) = \left(x^i m(x) \cdot g(x) + x^i S(x)\right) \bmod g(x) = x^i S(x) \bmod g(x)$ □

The described algorithm of burst errors correcting can be implemented with the help of the device shown in Figure 3.13. The logic element OR NOT in this decoder has $n - k - b$ inputs that are fed by the output of the syndrome calculator. If all these inputs are equal to zero it means that b less significant digits of the syndrome form the error burst of length b or less, which can be corrected. In this case the feedback of the syndrome calculator is broken with the help of Gate 4 and the shifted word is added with the syndrome. After this the corrected word is shifted $n - i$ times (which is equivalent to i reverse shifts).

Example 3.14 The binary cyclic (15, 9) code with the generator polynomial $g(x) = x^6 + x^5 + x^4 + x^3 + 1$ is capable of correcting burst-errors of length 3. Let $e(x) = x^7 + x^6 + x^5$ be the burst-error of length 3. The syndrome of word $e(x)$ is:

$$S(x) = e(x) \bmod g(x) = x^4 + x$$

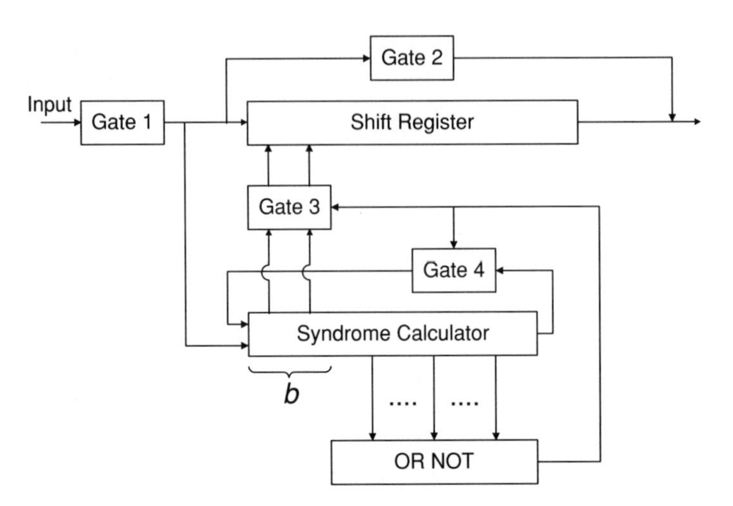

Figure 3.13 The decoder for burst error correction

If we shift this syndrome to the less significant digits, then the contents of the syndrome register will be as follows:

Shift	Content of the syndrome register
0	$x^4 + x$
1	$x^3 + 1$
2	$x^5 + x^4 + x^3$
3	$x^4 + x^3 + x^2$
4	$x^3 + x^2 + x$
5	$x^2 + x + 1$

After five shifts $n - k - b = 15 - 9 - 3 = 3$ most significant digits of the syndrome are equal to zero; and less significant digits form the burst-error. It is possible to correct the error by shifting the decoding word five times to the less significant digits and adding it with the content of the syndrome register on modulo 2. The decoder is shown in Figure 3.14.

3.5 Bounds on Minimum Distance

In this section we recall some well-known bounds on the minimum code distance. Let V be a binary code of length n, consisting of M codewords and capable of correcting t errors. If a code can correct t errors then the minimum distance of this code d satisfies the inequality $d \geq 2t + 1$ or, what the equivalent, the spheres of radius t surrounding all codewords are disjointed. The number of binary vectors of length n at distance exactly i from a given binary vector (of the same length n) equals $\binom{n}{i}$. Then each of these spheres of radius t contains $\sum_{i=0}^{t} \binom{n}{i}$ vectors (assuming that $\binom{n}{0} = 1$). On the other hand, the number of all binary vectors of length n equals to 2^n. This gives the *sphere-packing bound* known also as the *Hamming bound* [5]:

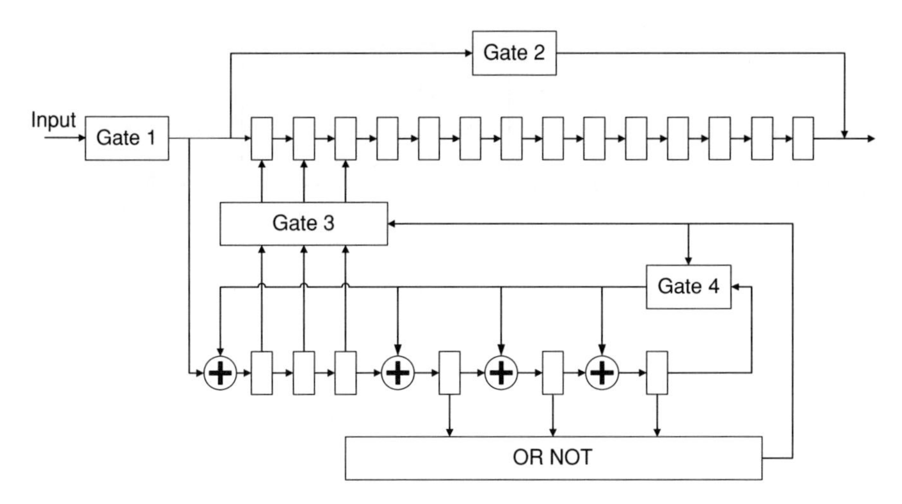

Figure 3.14 The decoder of (15, 9) code

Theorem 3.6 For any binary code of length *n,* consisting of *M* code words and capable of correcting *t*-fold errors:

$$M \cdot \sum_{i=0}^{t} \binom{n}{i} \leq 2^n \tag{3.77}$$

For a q-ary code (3.77) can be written as follows:

$$M \cdot \sum_{i=0}^{t} (q-1)^i \cdot \binom{n}{i} \leq q^n \tag{3.78}$$

For linear q-ary (n, k) code $M = q^k$ and its code rate $R = \frac{k}{n} = \frac{\log_q M}{n}$. Therefore, for an arbitrary code its rate is defined as $R = \frac{\log_q M}{n}$. Taking this into account (3.78) can be represented in the following form:

$$n - k \geq \log \sum_{i=0}^{t} (q-1)^i \cdot \binom{n}{i} \tag{3.79}$$

For $t < \frac{(q-1)n}{q}$ the sum in (3.79) can be upper estimated as $(q-1)^t \cdot \binom{n}{t}$ since:

$$(q-1)^i \cdot \binom{n}{t} > \sum_{i=0}^{t-1} (q-1)^i \cdot \binom{n}{i} \tag{3.80}$$

Applying Stirling's formula to binomial coefficient $\binom{n}{t}$ obtain the following approximation:

$$\binom{n}{t} \approx \binom{n}{n\delta/2} \approx q^{nH_4(\delta/2)} \tag{3.81}$$

where $\delta = \frac{d}{n}$ is the *relative minimum distance*, $0 < \delta < 1$, $H_q(x) = -x \log_q x - (1-x) \log_q (1-x)$ is the q-ary entropy function. Most of the results in this section are given in asymptotic, as $n \to \infty$. In this case it is more convenient to use the *asymptotic code rate* $R(\delta)$ as a function of relative minimum distance,

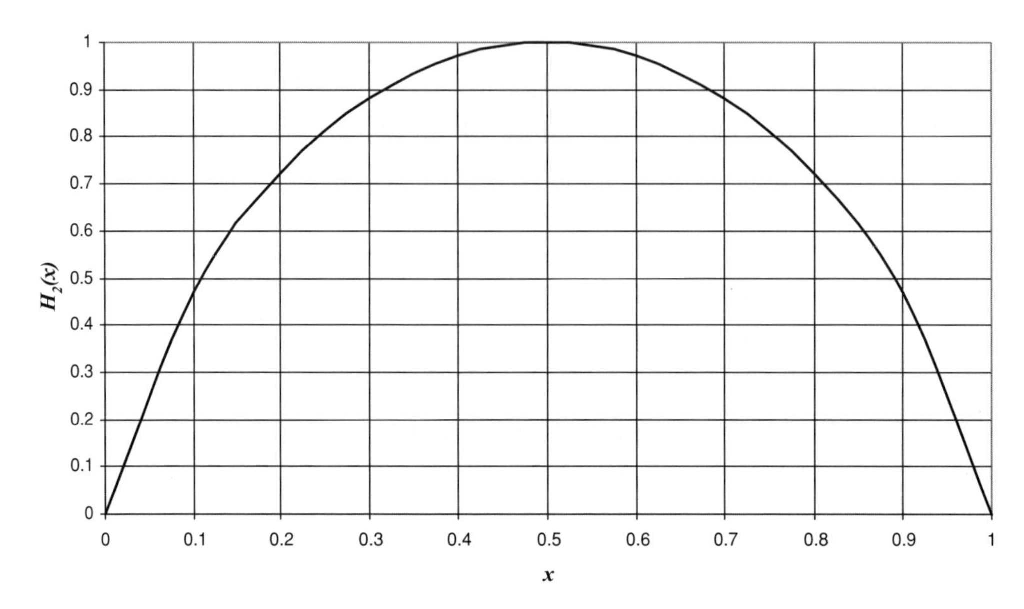

Figure 3.15 Binary entropy function $H_2(x)$

that is, the code rate of code of maximal cardinality as function of $\delta = \frac{d}{n}$ with $n \to \infty$. Altogether it gives that for any q-ary code its asymptotic rate $R(\delta)$:

$$R(\delta) \leq 1 - \frac{\delta}{2} \log_q (q-1) - H_q(\delta/2), \quad 0 < \delta < 1 \tag{3.82}$$

For binary case (3.82) simplifies to the following form known as *asymptotic Hamming bound*:

$$R(\delta) \leq 1 - H_2\left(\frac{\delta}{2}\right), \quad 0 < \delta < 1 \tag{3.83}$$

The binary entropy function is presented in Figure 3.15.

The codes achieving the Hamming bound are called *perfect codes*. Hence spheres of radius t surrounding codewords of a perfect code contain all vectors of the corresponding vector space. It means that a perfect code can correct any error pattern of weight no more than t, and cannot correct any error pattern of weight more than t. For example, binary Hamming codes are perfect codes.

For a code, capable of correcting t errors, the ratio of the overall number of vectors in all spheres of radius t surrounding the code words to the number of vectors of the whole vector space is called the *packing density* of a code. That is, the packing density is the ratio of "volume" of all spheres of radius t surrounding the code words to the "volume" of vector space and the Hamming bound means that the packing density of any code is at most 1. For perfect codes the packing density is equal to 1, that is, a perfect code correcting t errors is capable of packing the whole vector space by the spheres of radius t.

The Hamming bound is an upper bound on the code rate of the code with given (relative) minimum distance. Now we consider the *Gilbert-Varshamov bound*, which is a lower bound that shows an existence of good codes [5], in particular, among linear codes.

Theorem 3.7 If the following inequality holds true:

$$q^{n-k} > \sum_{i=0}^{d-2} (q-1)^i \cdot \binom{n-1}{i} \tag{3.84}$$

then there exists a linear block q-ary code of length n with number of information symbols k that has the minimum distance no less than d.

The proof of the theorem can be found in [5]. The *asymptotic Gilbert-Varshamov bound* has form:

$$R(\delta) \geq 1 - \delta \log_q (q-1) - H_q(\delta), \quad 0 < \delta < \frac{q-1}{q} \tag{3.85}$$

where $\delta = \frac{d}{n} < \frac{q-1}{q}$. It is known that codes satisfying Gilbert-Varshamov bound not only exist, but almost all linear coders satisfying it are asymptotically tight. Therefore it was a very surprising discovery of construction of *algebraic-geometry codes* for which [7]:

$$R(\delta) \geq 1 - \delta - \frac{1}{\sqrt{q} - 1} \tag{3.86}$$

These codes are even asymptotically better than Gilbert-Varshamov bound for $q \geq 49$. Recently very effective algorithms of generation of such codes [8] as well as their decoding (and even beyond of $d/2$ [9]) were discovered and this is what makes this class of codes very attractive for practical applications.

For $q = 2$ (3.85) can be written as:

$$R(\delta) \geq 1 - H_2(\delta), \quad 0 \leq \delta < \frac{1}{2} \tag{3.87}$$

On the other hand, there is a very popular conjecture that for binary codes the Gilbert-Varshamov is asymptotically tight, that is:

$$R(\delta) \stackrel{?}{=} 1 - H_2(\delta), \quad 0 \leq \delta < \frac{1}{2}$$

Now we have the upper Hamming bound (3.82) and the lower Gilbert-Varshamov bound (3.85). The next few theorems address the largest gap between these bounds.

Theorem 3.8 (*Plotkin bound*) The minimum distance d of any q-ary code block code of length n containing M words satisfies the following inequality:

$$d \leq \frac{(q-1)nM}{q(M-1)} \tag{3.88}$$

The proof of Theorem 3.8 can be found in [4]. There is a simple recursion: $A_q(n, d) \leq q^{n-n'} A_q(n', d)$, where $A_q(n, d)$ is the maximal cardinality of a q-ary code of length n and distance d. This recursion together with Theorem 3.8 leads to a more general form of the Plotkin bound: $A_q(n, d) \leq d \cdot q^{n - \frac{q(d-1)}{q-1}}$ or, in asymptotic form:

$$R(\delta) \leq 1 - \frac{q}{q-1}\delta \tag{3.89}$$

as $n \to \infty$. This bound is more tight than Hamming bound for low code rates.

The next step was taken by P. Elias and L. Bassalygo [10], [11]:

Theorem 3.9 (*Elias-Bassalygo bound*) For any q-ary (n, M, d) code the following inequality holds true:

$$R(\delta) \leq 1 - H_q\left(\left(1 - \frac{1}{q}\right) \cdot \left(1 - \sqrt{1 - \frac{q}{q-1}\delta}\right)\right), \quad 0 < \delta < \frac{q-1}{q} \tag{3.90}$$

The proof of Theorem 3.9 can be found in [5].

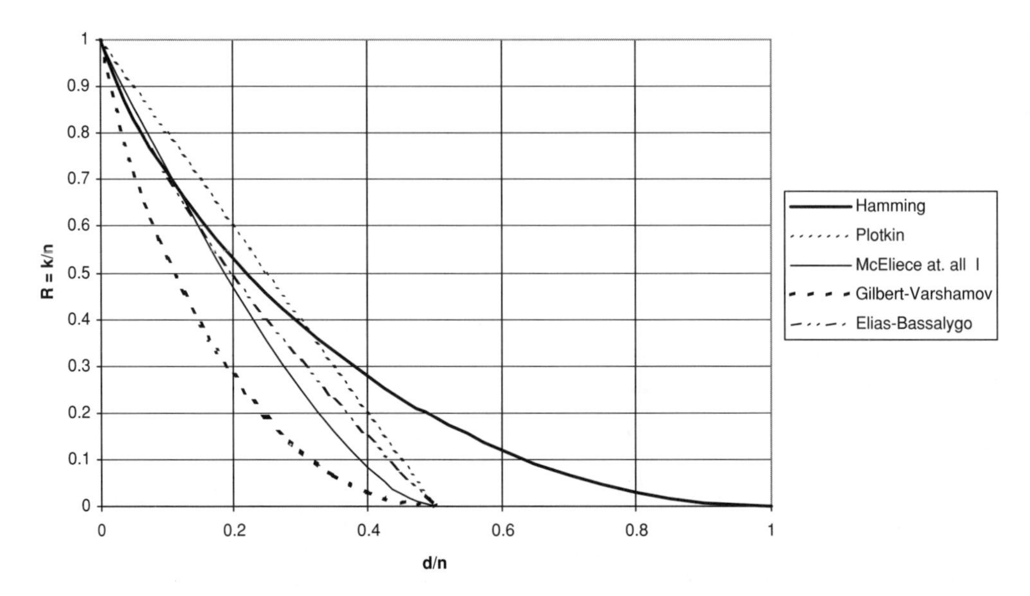

Figure 3.16 Bounds for binary block codes

The Elias-Bassalygo bound is better than Hamming and Plotkin bounds for any $0 < \delta < \frac{q-1}{q}$. At the present time the latest step towards tightening the upper bounds is the *McEliece-Rodemich-Rumsey-Welch bound* derived with the help of Linear Programming method (see [5]). It consists of two parts. Let us consider them as two theorems:

Theorem 3.10 [5, 12] For any binary block code of length n with the minimum distance d the code rate R satisfies the following inequality as $n \to \infty$:

$$R(\delta) \leq H_2\left(\frac{1}{2} - \sqrt{\delta(1-\delta)}\right) \tag{3.91}$$

The discussed bounds for binary block codes are represented in Figure 3.16.

As can be seen from plots in Figure 3.16, for values of $\delta > 0.1$ all codes are between Gilbert-Varshamov and McEliece-Rodemich-Rumsey-Welch I bound. And for low values of δ the Hamming bound is less optimistic than McEliece-Rodemich-Rumsey-Welch I bound. In this region of δ values the second part of McEliece-Rodemich-Rumsey-Welch bound (or McEliece-Rodemich-Rumsey-Welch II bound) can be used:

Theorem 3.11 [5, 12] For any binary block code of length n with the minimum distance d the code rate R satisfies the following inequality as $n \to \infty$:

$$R(\delta) \leq B(\delta) \tag{3.92}$$

where $B(\delta) = \min B(u, \delta), \quad 0 < u \leq 1 - 2\delta,$

$$B(u, \delta) = 1 + h\left(u^2\right) - h\left(u^2 + 2\delta u + 2\delta\right)$$

$$h(x) = H_2\left(\frac{1}{2} - \frac{1}{2}\sqrt{1-x}\right)$$

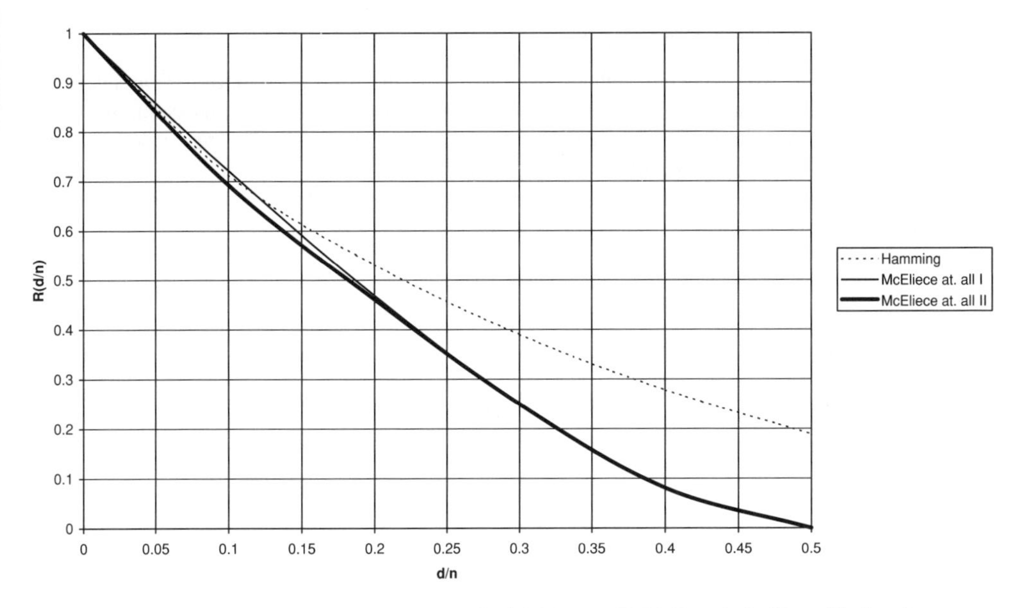

Figure 3.17 First and second McEliece-Rodemich-Rumsey-Welch Bounds for binary block codes

The bound given by (3.92) improves the first McEliece-Rodemich-Rumsey-Welch bound in the region of low values of δ (it coincides with the first bound for $\delta \geq 0.273$). The comparison of first and second McEliece-Rodemich-Rumsey-Welch bound with Hamming bound is depicted in Figure 3.17.

3.6 Minimum Distance Decoding

Usually we take the words minimum distance decoding to mean the procedure of searching the codeword, which is the closest one to the received word. This kind of procedure can be realized by the exhaustive search on the set of codewords or on the set of syndromes of probable error vectors.

The exhaustive search algorithm on the set of codewords consists of a comparison of the received word with all codewords and choice of the closest codeword.

The exhaustive search algorithm on the set of syndromes of error vectors can be realized in the case where we have the table T, in which syndromes correspond to coset leaders. The algorithm is as follows:

1. Calculation of the syndrome **s** of the received word **b**;
2. The search of the syndrome **s** and the corresponding coset leader **e** in the table T;
3. Calculation of decoded vector $\hat{\mathbf{a}}$:

$$\hat{\mathbf{a}} = \mathbf{b} + \mathbf{e}$$

The first algorithm usually needs a greater number of operations, which is related to the need to generate all codewords (to execute q^k encoding procedures) and to compare them with the received word (q^k comparisons). The decoding on the syndromes leads to needing to keep in memory the table with q^r words of length n.

Usually it is hard to implement these algorithms in practice due to the high complexity. Now we will consider those algorithms with less complexity, which, however, can not provide the full decoding, that is, the decoding of all errors, which is possible to correct with the help of given codes (all coset leaders).

3.7 Information Set Decoding

The key concept for many general methods of decoding is the concept of *information set* of the code. It is a known fact that for systematic code first k symbols $a_0, \ldots, a_{k-1}$ fully define the word $a_0, \ldots, a_{k-1}, a_k, \ldots, a_{n-1}$ of the (n, k) code. That means there is only one word in the code where the symbols $a_0, \ldots, a_{k-1}$ occupy the positions $\{0, \ldots k-1\}$.

However, the set of positions $\{0, \ldots, k-1\}$ is not the unique one. For every code there exist many other sets $\{j_1, \ldots, j_k\}$, $0 \le j_1 < \ldots < j_k \le n-1$, such that for any $a_{j_1}, \ldots, a_{j_k}$ there exist only one word in the code where the symbols $a_{j_1}, \ldots, a_{j_k}$ occupy the positions $\{j_1, \ldots, j_k\}$. Hence, the symbols on the positions $\{j_1, \ldots, j_k\}$ also fully define the codeword.

Definition 3.2 The set of positions $\{j_1, \ldots, j_k\}$, $(0 \le j_1 < \ldots < j_k \le n-1)$ is called the information set of the code V if the symbols $a_{j_1}, \ldots, a_{j_k}$ uniquely define the codeword from V.

Let $\mathbf{G}$ be the generator matrix of code V. Let us denote by $\mathbf{G}(\gamma)$ the matrix constructed from the columns of $\mathbf{G}$ enumerated by the elements of set γ. It is obvious that the set γ is the information set if and only if the mapping $f_\gamma : V \to \mathrm{A}^k$, which put in correspondence to the codeword it coordinates with numbers from γ, is one-to-one mapping. This fact is equivalent, as was shown in Section 3.3, to the nonsingularity of the matrix of mapping f_γ, which is the matrix $\mathbf{G}(\gamma)$ in the basis $\{\mathbf{v}_1, \ldots, \mathbf{v}_k\}$. Thus, the following statement is correct:

Lemma 3.3 *The set of positions $\gamma = \{j_1, \ldots j_k\}$ is the information set if and only if the matrix $\mathbf{G}(\gamma)$ is nonsingular.*

If in the received erroneous word there is at least one information set without erroneous symbols (that is, symbols with indexes from this information set do not contain errors), then the transmitted word can be restored on the basis of this information set. In this case the decoding procedure can be regarded as the search of information set, which is free of errors. As this takes place, the issue of the "stop rule" is very important, that is, we should choose the rule in accordance with which it is possible to identify the fact that the information set free of errors is found. Hereafter we will consider the decoding of t-fold errors.

Let us describe now the decoding algorithm based on the information sets. Let $\gamma = \{j_1, \ldots j_k\}$ be the information set of the code V, and let $\mathbf{G}$ and $\mathbf{H}$ be the generator and the parity matrix of this code. Let us denote by $\mathbf{G}_\gamma$ the matrix:

$$\mathbf{G}_\gamma = (\mathbf{G}(\gamma))^{-1} \cdot \mathbf{G} \tag{3.93}$$

It is obvious that the columns of matrix $\mathbf{G}_\gamma$ with numbers $\{j_1, \ldots, j_k\}$ form the identity $(k \times k)$-matrix. Multiplying the vector $(a_{j_1}, \ldots, a_{j_k})$ by matrix $\mathbf{G}_\gamma$ results in the codeword with symbols $a_{j_1}, \ldots, a_{j_k}$ on the positions $\{j_1, \ldots, j_k\}$:

$$\left(a_{j_1}, \ldots, a_{jk}\right) \cdot \mathbf{G}_\gamma = \left(a_{j_1}, \ldots, a_{jk}\right) \cdot \begin{bmatrix} & j_1 & & j_2 & & j_k & \\ \cdots & 1 & \cdots & 0 & \cdots & 0 & \cdots \\ \cdots & 0 & \cdots & 1 & \cdots & 0 & \cdots \\ \cdots & \cdots & \cdots & \cdots & \cdots & \cdots & \cdots \\ \cdots & 0 & \cdots & 0 & \cdots & 1 & \cdots \end{bmatrix} = \begin{pmatrix} & j_1 & & j_2 & & j_k & \\ \cdots & a_{j_1} & \cdots, & a_{j2} & \cdots, & a_{j_k} & , \cdots \end{pmatrix}$$

Let us put matrix $\mathbf{G}_\gamma$ in correspondence to parity matrix:

$$\mathbf{H}_\gamma = (\mathbf{H}(\bar{\gamma}))^{-1} \cdot \mathbf{H} \tag{3.94}$$

where $\bar{\gamma} = \{1, 2, \ldots, n\} \setminus \gamma$, that is, the set of positions, which are not included in γ, and $\mathbf{H}(\bar{\gamma})$ is the matrix formed by the columns of matrix $\mathbf{H}$ with indexes from $\bar{\gamma}$. The columns of matrix $\mathbf{H}_\gamma$ with indexes not included in γ form the identity $(r \times r)$-matrix. Hence, if all nonzero elements of error vector $\mathbf{e}$ are located in the set $\bar{\gamma}$ (that is, γ is free of errors), then the weight of the syndrome:

$$\mathbf{s}_\gamma(\mathbf{e}) = \mathbf{e} \cdot \mathbf{H}_\gamma^{\mathrm{T}} \tag{3.95}$$

is equal to the weight of error vector. Therefore, the algorithm of information set decoding of ν-fold errors, $\nu \leq t = \left\lceil \frac{d-1}{2} \right\rceil$, can be formulated as follows:

Information Set Decoding:

Let $\Gamma = \{\gamma_1, \gamma_2, \ldots, \gamma_l\}$ be the set of information sets of the code. Let us assume that the set Γ contains a reasonable number of information sets to correct the ν-fold errors.

1. Calculation of the syndromes $\mathbf{s}_{\gamma_i}(\mathbf{b})$, where $\mathbf{b} = (b_0, b_1, \ldots, b_{n-1})$ is the received vector, $\mathbf{b} = \mathbf{a} + \mathbf{e}$, $\mathbf{a}$ is the transmitted vector and $\mathbf{e}$ is the error vector, until information set $\gamma = \{j_1, j_2, \ldots, j_k\}$ is found such that the weight of the corresponding syndrome:

$$w\left(\mathbf{s}_\gamma(\mathbf{b})\right) \leq \nu \tag{3.96}$$

2. If the condition (3.96) is satisfied the codeword $\hat{\mathbf{a}}$ is regarded as the decoded word:

$$\hat{\mathbf{a}} = \mathbf{b}(\gamma) \cdot \mathbf{G}_\gamma = (b_{j1}, b_{j2}, \ldots, b_{jk}) \cdot \mathbf{G}_\gamma \tag{3.97}$$

3. If no one information set γ_i satisfies the condition (3.96) the calculation (3.97) is not executed and it is assumed that an uncorrectable error is detected.

Example 3.15 Consider the decoding of the binary (7, 4) code with $d = 3$. Let the generator matrix $\mathbf{G}$ and the parity matrix $\mathbf{H}$ be as follows:

$$\mathbf{G} = \begin{bmatrix} 1 & 0 & 0 & 0 & 1 & 0 & 1 \\ 0 & 1 & 0 & 0 & 1 & 1 & 1 \\ 0 & 0 & 1 & 0 & 1 & 1 & 0 \\ 0 & 0 & 0 & 1 & 0 & 1 & 1 \end{bmatrix}; \quad \mathbf{H} = \begin{bmatrix} 1 & 1 & 1 & 0 & 1 & 0 & 0 \\ 0 & 1 & 1 & 1 & 0 & 1 & 0 \\ 1 & 1 & 0 & 1 & 0 & 0 & 1 \end{bmatrix} \tag{3.98}$$

The set of symbols γ'={0,1,2,6} is the information set of the code and the set $\gamma'' = \{0,1,2,4\}$ is not the information set of the code because the determinant of the matrix $\mathbf{G}(\gamma')$:

$$|\mathbf{G}(\gamma')| = \begin{vmatrix} 1 & 0 & 0 & 1 \\ 0 & 1 & 0 & 1 \\ 0 & 0 & 1 & 0 \\ 0 & 0 & 0 & 1 \end{vmatrix} \neq 0$$

and the determinant of the matrix $\mathbf{G}(\gamma'')$:

$$|\mathbf{G}(\gamma'')| = \begin{vmatrix} 1 & 0 & 0 & 1 \\ 0 & 1 & 0 & 1 \\ 0 & 0 & 1 & 1 \\ 0 & 0 & 0 & 0 \end{vmatrix} = 0$$

In the same way we can verify that $\gamma^{(0)} = \{0, 1, 2, 3\}$, $\gamma^{(1)} = \{3, 4, 5, 6\}$, $\gamma^{(2)} = \{0, 1, 2, 6\}$ are the information sets. For these information sets $\mathbf{H}_{\gamma^{(0)}} = \mathbf{H}$, $\mathbf{G}_{\gamma^{(0)}} = \mathbf{G}$:

$$\mathbf{H}_{\gamma^{(1)}} = (\mathbf{H}(0, 1, 2))^{-1} \cdot \mathbf{H} = \begin{bmatrix} 1 & 1 & 1 \\ 0 & 1 & 1 \\ 1 & 1 & 0 \end{bmatrix}^{-1} \cdot \mathbf{H} = \begin{bmatrix} 1 & 1 & 0 \\ 1 & 1 & 1 \\ 1 & 0 & 1 \end{bmatrix} \cdot \begin{bmatrix} 1 & 1 & 1 & 0 & 1 & 0 & 0 \\ 0 & 1 & 1 & 1 & 0 & 1 & 0 \\ 1 & 1 & 0 & 1 & 0 & 0 & 1 \end{bmatrix}$$

$$= \begin{bmatrix} 1 & 0 & 0 & 1 & 1 & 1 & 0 \\ 0 & 1 & 0 & 0 & 1 & 1 & 1 \\ 0 & 0 & 1 & 1 & 1 & 0 & 1 \end{bmatrix}$$

$$\mathbf{G}_{\gamma^{(1)}} = (\mathbf{G}(3, 4, 5, 6))^{-1} \cdot \mathbf{G} = \begin{bmatrix} 1 & 0 & 1 & 1 & 0 & 0 & 0 \\ 1 & 1 & 1 & 0 & 1 & 0 & 0 \\ 1 & 1 & 0 & 0 & 0 & 1 & 0 \\ 0 & 1 & 1 & 0 & 0 & 0 & 1 \end{bmatrix};$$

$$\mathbf{H}_{\gamma^{(2)}} = (\mathbf{H}(3, 4, 5))^{-1} \cdot \mathbf{H} = \begin{bmatrix} 1 & 1 & 0 & 1 & 0 & 0 & 1 \\ 1 & 1 & 1 & 0 & 1 & 0 & 0 \\ 1 & 0 & 1 & 0 & 0 & 1 & 1 \end{bmatrix};$$

$$\mathbf{G}_{\gamma^{(2)}} = (\mathbf{G}(0, 1, 2, 6))^{-1} \cdot \mathbf{G} = \begin{bmatrix} 1 & 0 & 0 & 0 & 1 & 0 & 1 \\ 0 & 1 & 0 & 0 & 1 & 1 & 1 \\ 0 & 0 & 1 & 0 & 1 & 1 & 0 \\ 0 & 1 & 1 & 0 & 0 & 0 & 1 \end{bmatrix}$$

The set of information sets $\Gamma = \{\gamma^{(0)}, \gamma^{(1)}, \gamma^{(2)}\}$ allows decoding any 1-fold errors in (7, 4) code. Let $\mathbf{a} = (0, 0, 0, 0, 0, 0, 0)$ be the transmitted word and $\mathbf{b} = (0, 0, 0, 1, 0, 0, 0)$ be the received word. The decoding procedure in accordance with the described above algorithm is as follows:

$$\mathbf{s}_{\gamma^{(0)}}(\mathbf{b}) = \mathbf{b} \cdot \mathbf{H}^{\mathrm{T}}_{\gamma^{(0)}} = (0, 1, 1) \Rightarrow w(\mathbf{s}_{\gamma^{(0)}}(\mathbf{b})) = 2 > 1 = \frac{d-1}{2};$$

$$\mathbf{s}_{\gamma^{(1)}}(\mathbf{b}) = \mathbf{b} \cdot \mathbf{H}^{\mathrm{T}}_{\gamma^{(1)}} = (1, 0, 1) \Rightarrow w(\mathbf{s}_{\gamma^{(1)}}(\mathbf{b})) = 2 > 1 = \frac{d-1}{2};$$

$$\mathbf{s}_{\gamma^{(2)}}(\mathbf{b}) = \mathbf{b} \cdot \mathbf{H}^{\mathrm{T}}_{\gamma^{(2)}} = (1, 0, 0) \Rightarrow w(\mathbf{s}_{\gamma^{(2)}}(\mathbf{b})) = 1 = \frac{d-1}{2}$$

With the help of the information set $\gamma^{(2)}$ we can calculate the decoded word:

$$\hat{\mathbf{a}} = (b_0, b_1, b_2, b_6) \cdot \mathbf{G}_{\gamma^{(2)}} = (0000) \cdot \begin{bmatrix} 1 & 0 & 0 & 0 & 1 & 0 & 1 \\ 0 & 1 & 0 & 0 & 1 & 1 & 1 \\ 0 & 0 & 1 & 0 & 1 & 1 & 0 \\ 0 & 0 & 0 & 1 & 0 & 0 & 1 \end{bmatrix} = (0000000) = \mathbf{a}$$

In general the cardinal number of set Γ, that is, the number of information sets, which is required for decoding, increases very quickly with the increasing of the code length and the number of correctable errors. And since each information set γ from Γ needs to keep or calculate matrices $\mathbf{G}_\gamma$ and (or) $\mathbf{H}_\gamma$, then the complexity of the algorithm is growing up. The simplification of the information set decoding is associated with two modifications of this algorithm - *permutation decoding* and decoding with the help of *covering polynomials* (or *covering-set decoding*).

As was defined above the permutation π ($\pi \in S_n$) is the one-to-one self-mapping of set $\{1, 2, \ldots, n\}$. Each permutation $\pi \in S_n$ corresponds to linear operator on space A^n, that is, $\pi(a_1, \ldots, a_n) = (a_{\pi(1)}, \ldots, a_{\pi(n)})$. The arbitrary permutation transfers the word of the code V to some other word (in general this word does not belong to the code V). However, permutations do exist which transfer any codeword to the codeword of the same code. This kind of permutation is said to be preserving code permutation, and the code is said to be invariant relative to this permutation. It is easy to verify that the set of permutations preserving the code V forms the subgroup in the group S_n of all permutations. This subgroup is denoted as Aut$V$.

Example 3.16 Consider the binary linear (3, 2)-code consisting of four words (000), (110), (100), (010). The permutation of the first and the second symbol of any codeword transfers it to the codeword, but the permutation of the second and the third symbol transfer the codeword (110) to the word (101), which do not belong to the code.

Let **G** be the generator matrix and **H** be the parity matrix of the code V:

$$\mathbf{G} = \begin{bmatrix} \mathbf{g}_1 \\ \cdot \\ \cdot \\ \mathbf{g}_k \end{bmatrix}; \quad \mathbf{H} = \begin{bmatrix} \mathbf{h}_1 \\ \cdot \\ \cdot \\ \mathbf{h}_r \end{bmatrix}$$

It is a necessary and sufficient condition for some permutation π to preserve the code V if matrix:

$$\pi(\mathbf{G}) = \begin{bmatrix} \pi(\mathbf{g}_1) \\ \cdot \\ \cdot \\ \pi(\mathbf{g}_k) \end{bmatrix}$$

satisfies the following equation:

$$\pi(\mathbf{G}) \cdot \mathbf{H}^{\mathrm{T}} = \mathbf{0}$$

Let π be the permutation preserving the code V. Then for any vector $\mathbf{b} = \mathbf{a} + \mathbf{e}$, $\mathbf{a} \in \mathrm{V}$:

$$\pi(\mathbf{b}) = \pi(\mathbf{a}) + \pi(\mathbf{e}) = \mathbf{a}' + \mathbf{e}'$$

where $\mathbf{a}'$ is some codeword and $\mathbf{e}'$ is the error vector, which has the same weight as vector $\mathbf{e}$. If the weight of vector $\mathbf{e}$ is no more than t, then the weight of vector $\mathbf{e}'$ also is no more than t.

With the help of permutations preserving the code, the information set decoding can be realized as follows. Let **G** be the generator matrix and **H** be the parity matrix of code V, and let both these matrices be in systematic form, that is, they correspond to the information set $\gamma^{(0)} = \{0,1,\ldots,\mathrm{k}-1\}$. Let Aut$V = \{\pi_1, \ldots, \pi_l\}$ be the set of permutations preserving the code. Let us calculate the syndromes of vectors $\pi_i(\mathbf{b})$ with the help of information set $\gamma^{(0)}$:

$$\mathbf{s}(\pi_i(\mathbf{b})) = \pi_i(\mathbf{b}) \cdot \mathbf{H}^{\mathrm{T}} \tag{3.99}$$

then we calculate the weight of syndromes (3.99). If some permutation π transfer vector $\mathbf{e}$ to vector e$'$, where all nonzero components are located on the positions $\{\mathrm{k}, \ldots, \mathrm{n}-1\}$, then the information set $\gamma^{(0)}$ is free of errors, and the weight of the corresponding syndrome is no more than t. In this case it is enough to use the permutation π^{-1} to vector $\mathbf{a}'$ to restore the transmitted codeword.

Example 3.17 Consider the permutation decoding of (7, 4)-code with $d = 3$, which was defined in Example 3.15. This code is invariant relative to the cyclic permutation T, $T(i) = (i + 1)\mathrm{mod}\ 7$. Really:

$$T(\mathbf{G}) = \begin{bmatrix} 1 & 1 & 0 & 0 & 0 & 1 & 0 \\ 1 & 0 & 1 & 0 & 0 & 1 & 1 \\ 0 & 0 & 0 & 1 & 0 & 1 & 1 \\ 1 & 0 & 0 & 0 & 1 & 0 & 1 \end{bmatrix}$$

$$T(\mathbf{G}) \cdot \mathbf{H}^{\mathrm{T}} = \mathbf{0}$$

Let us decode the word $\mathbf{b} = \mathbf{a} + \mathbf{e} = (1011\ 0\ 0\ 0) + (010\ 0\ 0\ 0\ 0) = (11110\ 0\ 0)$. Calculate $s_{\gamma^{(0)}}(T^i(\mathbf{b}))$ for $i = 0, 1, \ldots, 6$ since $T^7 = \mathrm{I} = T^0$:

$$\mathbf{s}_{\gamma^{(0)}}(\mathbf{b}) = \mathbf{b} \cdot \mathbf{H}^{\mathrm{T}} = (111) \Rightarrow w(\mathbf{s}_{\gamma^{(0)}}(\mathbf{b})) = 3 > 1 = \frac{d-1}{2};$$

$$\mathbf{s}_{\gamma^{(0)}}(T(\mathbf{b})) = T(\mathbf{b}) \cdot \mathbf{H}^{\mathrm{T}} = (110) \Rightarrow w(\mathbf{s}_{\gamma^{(0)}}(T(\mathbf{b}))) = 2 > 1 = \frac{d-1}{2};$$

$$\mathbf{s}_{\gamma^{(0)}}(T^2(\mathbf{b})) = T^2(\mathbf{b}) \cdot \mathbf{H}^{\mathrm{T}} = (011) \Rightarrow w(\mathbf{s}_{\gamma^{(0)}}(T^2(\mathbf{b}))) = 2 > 1 = \frac{d-1}{2}$$

$\mathbf{s}_{\gamma^{(0)}}(T^3(\mathbf{b})) = T^3(\mathbf{b}) \cdot \mathbf{H}^{\mathrm{T}} = (100) \Rightarrow w(\mathbf{s}_{\gamma^{(0)}}(T^3(\mathbf{b}))) = 1 = \frac{d-1}{2}$, and the condition (3.96) is satisfied. With the help of information set $\gamma^{(0)}$ calculate the word $a' = T^3(\hat{\mathbf{a}})$, $\hat{\mathbf{a}}$ is the decoded word:

$$\mathbf{a}' = T^3(\hat{\mathbf{a}}) = (0001) \cdot \mathbf{G} = (0001011)$$

and

$$\hat{\mathbf{a}} = T^{-3}(\mathbf{a}') = (1011000) = \mathbf{a}$$

Another way of “clearing” the information set from errors is the covering of errors in the information set. This method is called the decoding with the help of covering polynomials[1] .

Let θ be the vector (covering polynomial), which coincide with error vector $\mathbf{e}$ on the positions of the information set γ and with zeroes on the other positions. Then for vector $(\mathbf{b}-\theta)$ the information set γ is free of errors, and the weight of corresponding syndrome:

$$\mathbf{s}_{\gamma}(\mathbf{b} - \theta) = (\mathbf{e} - \theta) \cdot \mathbf{H}^{\mathrm{T}}$$

is

$$w(\mathbf{s}_{\gamma}(\mathbf{b} - \theta)) \leq t - w(\theta) \tag{3.100}$$

If we search vectors θ in increasing order of their weights (starting with $\theta_0 = (0\ldots0)$) until some θ^* satisfy (3.100), then it will be possible to restore the transmitted vector with the help of vector $\mathbf{b}^* = (\mathbf{b} - \theta^*)$ and the information set γ (of course, if the weight of error vector does not exceed t).

Example 3.18 Let us use the decoding with the help of covering polynomials for the case considered in Example 3.15. Let the set of covering polynomials be

$\theta_0 = (0000000)$, $\theta_1 = (1000000)$, $\theta_2 = (0100000)$, $\theta_3 = (0010000)$, $\theta_4 = (0001000)$. Consider the information set $\gamma^{(0)}$. Let the received vector be $\mathbf{b} = (0001000) = \mathbf{a} + \mathbf{e} = (0000000) + (0001000)$. Calculate the syndromes:

$$\mathbf{s}_{\gamma^{(0)}}(\mathbf{b} - \theta_0) = (0001000) \cdot \mathbf{H}^{\mathrm{T}} = (011) \Rightarrow w(s_{\gamma^{(0)}}(\mathbf{b} - \theta_0)) = 2 > 1 = \frac{d-1}{2};$$

$$\mathbf{s}_{\gamma^{(0)}}(\mathbf{b} - \theta_1) = (1001000) \cdot \mathbf{H}^{\mathrm{T}} = (110) \Rightarrow w(\mathbf{s}_{\gamma^{(0)}}(\mathbf{b} - \theta_1)) = 2 > 1 = \frac{d-1}{2} - w(\theta_1);$$

$$\mathbf{s}_{\gamma^{(0)}}(\mathbf{b} - \theta_2) = (0101000) \cdot \mathbf{H}^{\mathrm{T}} = (100) \Rightarrow w(\mathbf{s}_{\gamma^{(0)}}(\mathbf{b} - \theta_2)) = 1 > 0 = \frac{d-1}{2} - w(\theta_2);$$

$$\mathbf{s}_{\gamma^{(0)}}(\mathbf{b} - \theta_3) = (0011000) \cdot \mathbf{H}^{\mathrm{T}} = (101) \Rightarrow w(\mathbf{s}_{\gamma^{(0)}}(\mathbf{b} - \theta_3)) = 2 > 0 = \frac{d-1}{2} - w(\theta_3);$$

$$\mathbf{s}_{\gamma^{(0)}}(\mathbf{b} - \theta_4) = (0000000) \cdot \mathbf{H}^{\mathrm{T}} = (000) \Rightarrow w(\mathbf{s}_{\gamma^{(0)}}(\mathbf{b} - \theta_4)) = 0 = \frac{d-1}{2} - w(\theta_4)$$

[1] The term ⟨⟨covering polynomial⟩⟩ well-established in coding theory is not exactly correct. The more proper term is ⟨⟨covering vector⟩⟩ or ⟨⟨covering word⟩⟩.

Then $\theta^* = \theta_4$, $\mathbf{b}^* = (\mathbf{b} - \theta^*) = (0\,0\,0\,0\,0\,0\,0)$, and the transmitted word can be restored with the help of $\gamma^{(0)}$ and vector $\mathbf{b}^*$:

$$\hat{\mathbf{a}} = (b_0^*, b_1^*, \ldots, b_{k-1}^*) \cdot \mathbf{G}_{\gamma^{(0)}} = (0\,0\,0\,0\,0\,0\,0) = \mathbf{a}$$

The best results can be obtained with the joint use of the algorithms considered above. Let $\Gamma = \{\gamma^{(0)}, \ldots, \gamma^{(m)}\}$ be the set of the information sets of code V and $\theta^{(0)}, \ldots, \theta^{(m)}$ be the sets of covering polynomials corresponding to information sets $\gamma^{(0)}, \ldots, \gamma^{(m)}$: $\theta^{(0)} = \{\theta_{00}, \ldots, \theta_{0l_0}\}, \ldots, \theta^{(m)} = \{\theta_{m0}, \ldots, \theta_{ml_m}\}$ with $\theta_{j0} = \mathbf{0}$ and $w(\theta_{j0}) < w(\theta_{j1}) \leq \ldots \leq w(\theta_{jl_j})$, $j = 0, \ldots, m$. The decoding algorithm based on the joint use of information sets and covering polynomials is as follows:

Covering-Set Decoding:

1. Calculate the vector $\tilde{\mathbf{b}}_{ij} = \mathbf{b} - \theta_{ij}$ and the syndrome $\mathbf{s}_{\gamma^{(i)}} = (\tilde{\mathbf{b}}_{ij}) = \tilde{\mathbf{b}}_{ij} \cdot \mathbf{H}^{\mathrm{T}}$ for each pair $\gamma^{(i)}, \theta_{ij}$ ($i = 0, \ldots, m; j = 0, \ldots, l_j$) until the pair i^*, j^* is found, which satisfy the following condition:

$$w(\mathbf{s}_{\gamma^{(i^*)}}(\tilde{\mathbf{b}}_{i^*j^*} - \theta)) \leq t - w(\theta_{i^*j^*}) \tag{3.101}$$

2. If the condition (3.101) is satisfied, calculate:

$$\hat{\mathbf{a}} = \tilde{\mathbf{b}}(\gamma^{(i^*)}) \cdot \mathbf{G}_{\gamma^{(i^*)}}$$

   where $\tilde{\mathbf{b}}(\gamma^{(i^*)})$ is the subvector of vector $\tilde{\mathbf{b}}_{i^*j^*}$ combined from the elements of vector $\tilde{\mathbf{b}}_{i^*j^*}$, which belong to the information set $\gamma^{(i^*)}$
3. If any pair $\gamma^{(i^*)}, \theta_{ij}$ does not satisfy the condition (3.101), then it is assumed that the transmitted word was corrupted by the uncorrectable error.

The set of set Γ and sets of the covering polynomials $\Theta = \{\theta^{(0)}, \ldots, \theta^{(m)}\}$ is called the decoding set and is denoted as $DS = \{\Gamma, \Theta\}$.

Quite often it is convenient to use for decoding not all covering polynomials, but only those with a weight no more than ν, that is, vectors with nonzero elements located on the positions of set γ, and the number of these nonzero elements does not exceed ν. This kind of vectors we denote as $\theta_\gamma(\nu)$.

Example 3.19 Consider (7, 4) code from Example 3.15. Consider $\gamma_0 = \{0, 1, 2, 3\}$. The set of polynomials $\theta_{\gamma^{(0)}}(1)$ consists of five polynomials $\theta_{00} = (0\,0\,0\,0\,0\,0\,0)$, $\theta_{01} = (1\,0\,0\,0\,0\,0\,0)$, $\theta_{02} = (0100000)$, $\theta_{03} = (0010000)$, $\theta_{04} = (0001000)$.

It is easy to verify that if γ is the information set of the code V and π is the permutation preserving the code, then the set $\pi(\gamma)$ is also the information set of the code V.

Example 3.20 Consider the decoding of the (15, 5) Hamming code with distance $d = 7$. It is possible to use the decoding set $DS = \{\Gamma, \Theta(1)\}$ to decode this code, where $\Gamma = \{\gamma^{(0)}, \gamma^{(1)} = T^5(\gamma^{(0)})\}$, $\gamma^{(0)} = \{0, 1, 2, 3, 4\}$, and $T^5(\gamma^{(0)}) = \{5, 6, 7, 8, 9\}$ is the cyclic shift of the information set $\gamma^{(0)}$ by five positions. Let the received word $\mathbf{b}$ be $\mathrm{b} = \mathrm{a} + \mathrm{e} = (001110110010100) + (010100010000000)$, that is, the received word is corrupted in the first, third and seventh position. Calculate $\mathbf{s}_{0j}(\mathbf{b}) = \tilde{\mathbf{b}}_{0j} \cdot \mathbf{H}^{\mathrm{T}}_{\gamma^{(0)}}$ with the help of information set $\gamma^{(0)}$:

$$\tilde{\mathbf{b}}_{00} = \mathbf{b} + \theta_{00} = \mathbf{b} + (0\,0\,0\,0\,0\,0\,0\,0\,0\,0\,0\,0\,0\,0\,0),$$

$$\mathbf{s}_{00}(\mathbf{b}) = \tilde{\mathbf{b}}_{00} \cdot \mathbf{H}^{\mathrm{T}}_{\gamma^{(0)}} = (0011110110) \Rightarrow w(\mathbf{s}_{00}(\mathbf{b})) = 6 > 3 = \frac{d-1}{2} - w(\theta_{00});$$

$$\tilde{\mathbf{b}}_{01} = \mathbf{b} + \theta_{01} = \mathbf{b} + (1\,0\,0\,0\,0\,0\,0\,0\,0\,0\,0\,0\,0\,0\,0\,0),$$

$$\mathbf{s}_{01}(\mathbf{b}) = \tilde{\mathbf{b}}_{01} \cdot \mathbf{H}^{\mathrm{T}}_{\gamma^{(0)}} = (1101000100) \Rightarrow w(\mathbf{s}_{01}(b)) = 4 > 2 = \frac{d-1}{2} - w(\theta_{01});$$

....

$$\tilde{\mathbf{b}}_{05} = \mathbf{b} + \theta_{05} = \mathbf{b} + (000010000000000),$$

$$\mathbf{s}_{05}(\mathbf{b}) = \tilde{\mathbf{b}}_{05} \cdot \mathbf{H}^{\mathrm{T}}_{\gamma^{(0)}} = (1110010011) \Rightarrow w(\mathbf{s}_{05}(\mathbf{b})) = 6 > 2 = \frac{d-1}{2} - w(\theta_{05})$$

Now let us use the information set $\gamma^{(1)} = T^5(\gamma^{(0)})$, and the corresponding matrices $\mathbf{G}_{\gamma^{(1)}} = T^5(\mathbf{G}_{\gamma^{(0)}}$ and $\mathbf{H}_{\gamma^{(1)}} = T^5(\mathbf{H}_{\gamma^{(0)}})$ are:

$$\mathbf{G}_{\gamma^{(1)}} = \begin{bmatrix} 1&0&0&1&0&1&0&0&0&0&1&1&1&0&1 \\ 1&1&0&0&1&0&1&0&0&0&0&1&1&1&0 \\ 1&1&1&1&0&0&0&1&0&0&1&1&0&1&0 \\ 0&1&1&1&1&0&0&0&1&0&0&1&1&0&1 \\ 0&0&1&0&1&0&0&0&0&1&1&1&0&1&1 \end{bmatrix}$$

$$\mathbf{H}_{\gamma^{(1)}} = \begin{bmatrix} 0&0&0&0&0&1&0&1&0&1&1&0&0&0&0 \\ 0&0&0&0&0&1&1&1&1&1&0&1&0&0&0 \\ 0&0&0&0&0&1&1&0&1&0&0&0&1&0&0 \\ 0&0&0&0&0&0&1&1&0&1&0&0&0&1&0 \\ 0&0&0&0&0&1&0&0&1&1&0&0&0&0&1 \\ 1&0&0&0&0&1&1&1&0&0&0&0&0&0&0 \\ 0&1&0&0&0&0&1&1&1&0&0&0&0&0&0 \\ 0&0&1&0&0&0&0&1&1&1&0&0&0&0&0 \\ 0&0&0&1&0&1&0&1&1&0&0&0&0&0&0 \\ 0&0&0&0&1&0&1&0&1&1&0&0&0&0&0 \end{bmatrix}$$

Calculate $\mathbf{s}_{1j}(\mathbf{b}) = \tilde{\mathbf{b}}_{1j} \cdot \mathbf{H}^{\mathrm{T}}_{\gamma^{(1)}}$ with the help of information set $\gamma^{(1)}$:

$$\tilde{\mathbf{b}}_{10} = \mathbf{b} + \theta_{110} = \mathbf{b} + (000000000000000),$$

$$\mathbf{s}_{10}(\mathbf{b}) = \tilde{\mathbf{b}}_{10} \cdot \mathbf{H}^{\mathrm{T}}_{\gamma^{(1)}} = (1101010100) \Rightarrow w(s_{10}(\mathbf{b})) = 5 > 3 = \frac{d-1}{2} - w(\theta_{10});$$

$$\tilde{\mathbf{b}}_{11} = \mathbf{b} + \theta_{11} = \mathbf{b} + (000010000000000),$$

$$\mathbf{s}_{11}(\mathbf{b}) = \tilde{\mathbf{b}}_{11} \cdot \mathbf{H}^{\mathrm{T}}_{\gamma^{(1)}} = (0011100110) \Rightarrow w(\mathbf{s}_{11}(\mathbf{b})) = 5 > 2 = \frac{d-1}{2} - w(\theta_{11});$$

....

$$\tilde{\mathbf{b}}_{13} = \mathbf{b} + \theta_{13} + \mathbf{b} + (000000010000000),$$

$$\mathbf{s}_{13}(\mathbf{b}) = \tilde{\mathbf{b}}_{13} \cdot \mathbf{H}^{\mathrm{T}}_{\gamma^{(1)}} = (0000001010) \Rightarrow w(\mathbf{s}_{13}(\mathbf{b})) = 2 = \frac{d-1}{2} - w(\theta_{13})$$

With the help of the word $\tilde{\mathbf{b}}_{13}$ and the information set $\gamma^{(1)}$ we can restore the word $\hat{\mathbf{a}}$:

$$\hat{\mathbf{a}} = \tilde{\mathbf{b}}_{13}(\gamma^{(1)}) \cdot \mathbf{G}_{\gamma^{(1)}} = (01100) \cdot \mathbf{G}_{\gamma^{(1)}} = (001110110010100) = \mathbf{a}$$

In example the decoding set can be constructed with the help of the permutation T^5 rather than the usage of the information set $\gamma^{(1)}$. In this case the decoding algorithm consists of calculation of $\mathbf{s}_{0\mathrm{j}}(\mathbf{b})$ and then the calculation of $\mathbf{s}_{0\mathrm{j}}(T^5(\mathbf{b}))$.

Actually finding the set Γ is very difficult. A few nontrivial examples are found in [13], [14], and [15]; see also [16]. Therefore, to implement the general information-set decoding algorithm, we have to specify a way of choosing information sets. One obvious suggestion is to take random uniformly distributed k-subsets of set $\{0,1,\ldots, n-1\}$. We call the following algorithm *generalized covering-set decoding* because it, in essence, produces a random covering design and can be regarded as the generalization of the algorithms considered above.

Generalized Covering-Set Decoding:

1. Set $\hat{\mathbf{a}} = \mathbf{0}$.
2. Choose randomly a k-subset γ. Form a list of codewords $M(\gamma) = \{\mathbf{c} \in V | \mathbf{c}(\gamma) = \mathbf{b}(\gamma)\}$.
3. If there is a $\mathbf{c} \in M(\gamma)$ such that:

$$dist(\mathbf{c}, \mathbf{b}) < dist(\hat{\mathbf{a}}, \mathbf{b})$$

4. assign $\hat{\mathbf{a}} \leftarrow \mathbf{c}$.
5. Repeat the last two steps $L_n(k)$ times. Output $\hat{\mathbf{a}}$.

The number of steps $L_n(k)$ needed to execute the algorithm will be discussed later.

An improvement of this algorithm was achieved in two steps in [17], [18]. The idea in [17] is to organize the syndrome table more economically by computing the syndrome separately for the "left" and "right" part of the received vector $\mathbf{b}$.

Suppose for a while that the actual number of errors is t. Let us split the set $\{0,1,\ldots, n-1\}$ into two parts, $l = \{0,1,\ldots, m-1\}$and $r = \{m, m+1,\ldots, n-1\}$, and let $[\mathbf{H}_l|\mathbf{H}_r]$ be the corresponding partition of the parity-check matrix $\mathbf{H}$. Any error vector $\mathbf{e} = (\mathbf{e}_l|\mathbf{e}_r)$ with $\mathbf{e} \cdot \mathbf{H}^{\mathrm{T}} = e_l \cdot H_l^{\mathrm{T}} + e_r \cdot H_r^{\mathrm{T}} = \mathbf{s}$ is a plausible candidate for the decoding output. Assume, in addition, that the number of errors within the subset l equals u, where the numbers u and m are chosen in accordance with the natural restrictions $u \le m$, $t - u \le n - m$. For every possible m-vector $\mathbf{e}_l$, compute the product $\mathbf{s}_l = \mathbf{e}_l \cdot \mathbf{H}_l^{\mathrm{T}}$ and store it, together with the vector $\mathbf{e}_l$, as an entry of the table X_l. Likewise, form the table X_r and look for a pair of entries $(\mathbf{s}_l, \mathbf{s}_r)$ that add up to the received syndrome $\mathbf{s}$. Therefore, for every given $\mathbf{s}_r$ occurring in X_r, we should inspect X_l for the occurrence of $\mathbf{s} - \mathbf{s}_r$. One practical way to do this is to order X_l with respect to the entries $\mathbf{s}_l$.

However, in reality we know neither the number of errors nor their distribution. Therefore, we have to repeat the described procedure for several choices of m and u. In doing so, we may optimize on the choice in order to reduce the total size of memory used for the tables X_l and X_r. For every choice of m there are not more than t different options for the choice of u. Hence, by repeatedly building the tables, though not more than nt times, we shall capture any distribution of t errors. Finally, the entire procedure should be repeated for all $t = 1, 2, \ldots, d$ until we find the error vector that has the "received" syndrome $\mathbf{s}$. Let us give a more formal description of the algorithm.

Split Syndrome Decoding:

Precomputation stage: For every weight t, $1 \le t \le d$ find the point m such that the tables X_l and X_r have an (almost) equal size. Store the pair (m, u) in the set $E(t)$.

1. Compute $\mathbf{s} = \mathbf{b} \cdot \mathbf{H}^{\mathrm{T}}$ and set $t = 1$
2. For every entry of $E(t)$, form the tables X_l and X_r as described.
3. Order X_l with respect to the entries $\mathbf{s}_l$.
4. For every entry of X_r check whether X_l contains the vector $\mathbf{s}_l = \mathbf{s} - \mathbf{s}_r$. If this is found, then output $\hat{\mathbf{a}} = \mathbf{b} - (\mathbf{e}_l|\mathbf{e}_r)$ and STOP.
5. Otherwise, set $t = \mathrm{t} + 1$ and repeat Steps 2–5 while $t < d$.

3.8 Hamming Codes

The Hamming codes are capable of correcting single error. These codes were invented by R.W. Hamming and his paper [19] is considered to be the first work on the coding theory. The code construction can be easily explained with the help of Bose criterion (Lemma 3.1). In accordance with this criterion a linear code is capable of correcting single error if and only if a parity-check matrix **H** of this code does not contain collinear columns. In binary case that means matrix **H** consists of distinct nonzero columns. The total number of different nonzero columns of dimension r is 2^r-1. Hence, any $r \times (2^r - 1)$ matrix H_r combined from these columns defines the linear *Hamming code* with length $n = 2^r-1$, number of information symbols $k = 2^r-r-1$ and minimum distance $d = 3$. Any (not necessarily linear) code of length $n = 2^r-1$ capable of correcting single error has in accordance with Hamming bound (see Section 3.5) the cardinality of no more than $\frac{2^n}{n+1} = 2^{n-r}$ code words. Therefore, the binary Hamming codes are perfect. Recall that codes achieving Hamming bound are called *perfect* and they are optimal. It is known that there are no perfect codes capable of correcting t-fold errors $(t > 1)$ except of binary (23, 12, 7) Golay code, correcting triple errors, and ternary (11, 6, 5) Golay code, correcting double errors [5].

It is convenient to arrange the columns of matrix **H** in such a way that a column $\mathbf{h}_j$ be the binary representation of its own index j. Then to decode the received word **x** (to correct a single error) it is necessary to calculate the syndrome $\mathbf{s} = \mathbf{H} \cdot \mathbf{x} = (s_0, \ldots, s_{r-1})$ and the number $S = s_0 + 2s_1 + \ldots + 2^{r-1}S_{r-1}$ is the index of the corrupted symbol. If $S = 0$, that is, **s** is the all-zero vector then it is assumed that no error occurs. If $n+1$ is not the power of 2 then it is possible to consider the shortened Hamming code, that is, a code defined by the parity-check matrix obtained from matrix $\mathbf{H}_r$ by deletion of some i $(i = 2^r-1-n)$ from 2^r-1 columns, where $r = [\log_2(n + 1)]$. The shortened Hamming codes cannot be improved in the class of linear codes, that is, $k(n, d) = n-[\log_2(n + 1)]$. Also it is known that the shortened Hamming codes of length $n = 2^r-1-i$, $i = 1, 2, 3$ are optimal in the class of all codes [20]. On the other hand for $i = \lambda \cdot 2^r (0\langle\lambda > 0.5)$ and for large n there exist nonlinear codes with $d = 3$, which have about $(1-\lambda)^{-1}$ times more code words than the shortened Hamming codes of the same length. These nonlinear codes asymptotically achieve Hamming bound [21].

Now consider nonbinary codes capable of correcting single errors over the alphabet B the cardinality of which $q = |\mathrm{B}|$ is the power of prime number, that is, $q = p^m$ and $\mathrm{B} = F_q$. Then, as it was pointed above, matrix **H** should not contain the collinear columns. As an example of this kind of matrix with maximal possible number $n_r = \frac{q^r-1}{q-1}$ of r-dimensional columns consider the matrix:

$$\mathbf{H}_{r,q} = \begin{bmatrix} 1 & \ldots & 1 & \vdots & 0 & \ldots & 0 & \vdots & & \vdots & 0 \\ & & & \vdots & 1 & \ldots & 1 & \vdots & & \vdots & \ldots \\ & & & \vdots & & & & \vdots & \ldots & \vdots & 0 \\ & \mathbf{F}_q^{r-1} & & \vdots & & \mathbf{F}_q^{r-1} & & \vdots & & \vdots & 1 \end{bmatrix} \tag{3.102}$$

where $\mathbf{F}_q^l$ is $l \times q^l$ matrix the columns of which are all q^l l-dimensional vectors over the field F_q. Arrange the columns of matrix $\mathbf{F}_q^l$ in such a way that the column with index j be the q-ary representation of number j. A code defined by the matrix $\mathbf{H}_{r,q}$ is called Hamming code. The decoding algorithm of the q-ary Hamming code is as follows:

1. Calculate the syndrome of the received vector **x**: $\mathbf{s} = (s_0, \ldots, s_{r-1}) = \mathbf{H}_{r,q} \cdot \mathbf{x}$.
2. Find nonzero element s_i with minimal index i, then the error value $e = s_i$ and the error position j coincides with the position of column $\mathbf{s}' = (\frac{s_0}{s_i}, \ldots, \frac{s_{r-1}}{s_i})$ in matrix $\mathbf{H}_{r,q}$ (division in the field F_q). If $\mathbf{s} = \mathbf{0}$ it is assumed that no error occurred.

As in the binary case the q-ary Hamming codes are optimal because they reach the Hamming bound $|V| \leq \frac{q^n}{1+(q-1)\cdot n}$, and the shortened Hamming codes are optimal in a class of linear codes. However, if

$q = p^m$ and $m > 1$ then shortened Hamming codes are not optimal already in a class of group codes. In a class of all codes capable of correcting single errors the Hamming codes of length $n = \frac{q^r-1}{q-1} \cdot (1-\lambda)$ have about $(1-\lambda)$ times less code words than optimal codes $(0 < \lambda < 1 - 1/q)$, which as it is proved in [21] reach the Hamming bound asymptotically.

More often the Hamming codes are used in the cyclic representation. Recall that the minimum distance of a binary cyclic code is no less than 3 if and only if the code length n is equal to the smallest positive integer (called *period* of $g(x)$) for which generator polynomial $g(x)$ divides $x^n - 1$. Therefore, if $g(x)$ is the primitive polynomial of degree m we obtain the cyclic code $V_{g(x)}$ of length $n = 2^m - 1$ with the number of parity symbols m and with minimum distance $d \geq 3$. The parity-check matrix of this code contains all possible nonzero m-vectors as columns, that is, it differs from the parity-check matrix $\mathbf{H}_m$ only by a permutation of columns. Thus, a binary Hamming code is equivalent to the cyclic code with the primitive generator polynomial.

For q-ary codes the answer on the corresponding question depends on the fact that numbers $(q-1)$ and m are relatively prime or not. If they are then the q-ary Hamming code of length $n = \frac{q^m-1}{q-1}$ is equivalent to the cyclic code with the generator polynomial $g(x) = f_\beta(x)$, where $\beta \in F_{q^m}$ and the period of β is $\frac{q^m-1}{q-1}$ (i.e. $\beta = \alpha^{q-1}$, where α is the primitive element of F_{q^m}), $f_\beta(x)$ is the minimal polynomial over F_q of β (that is, the monic polynomial of the lowest possible degree from $F_q[x]$ for which β is the root). In other cases the q-ary Hamming code is not equivalent to any cyclic code.

Consider now the construction of near optimal single-error-correcting codes. Very useful for construction of codes with better parameters than shortened Hamming codes is the structure called a *subcode over subset*.

Let alphabet B $(|B| = q)$ be the subset of the alphabet B $(|\mathrm{B}| = Q > q)$ and V be a code of length n with minimum distance d and cardinal number $M = |V|$ over the alphabet B. Then q-ary code $V_B = \{\mathbf{v} = (v_0, \ldots, v_{n-1}) : v_i \in B, \mathbf{v} \in V\}$ consisting of B-ary words of code V is called a subcode of code V over subset B. Obviously, the length of the code V_B is equal to n and the distance is not less than d because V_B is the subcode, that is, $V_B \subset V$. In general it is quite difficult to estimate the cardinal number of the particular code V_B. Instead of this consider the family of q-ary codes $V_{B,\mathbf{b}} = \{V + \mathbf{b}\}_B = \{\mathbf{x} = (x_0, \ldots, x_{n-1})\, x_i \in B, \mathbf{x} - \mathbf{b} \in V\}$. Since code $\{V + \mathbf{b}\}$ has exactly the same parameters n, M, d as code V, the distance of the subcode of this code $d\left(V_{B,\mathbf{b}}\right) \geq d$ and the mean cardinal number of code $\bar{M}$ in this family is equal to:

$$\bar{M} = \frac{1}{Q^n} \cdot \sum_{\mathbf{b} \in E_Q^n} |V_{B,\mathbf{b}}| = \sum_{\mathbf{v} \in V} \sum_{\mathbf{a} \in E_Q^n} 1 = \left(\frac{q}{Q}\right)^n \cdot |V| \tag{3.103}$$

Hence, there exists a code $V_{B,\mathbf{b}}$ in this family, such that $|V_{B,\mathbf{b}}| \geq \bar{M}$.

Example 3.21 [21, 22]. Construct a 4-ary single-error-correcting code of length 6. Let $q = 4$, $Q = 5$ and code V be the 5-ary Hamming code of length 6 with 5^4 code words. It follows from (3.103) that $\bar{M} \geq \frac{4^6}{5^2}$ and hence, there exists a code $\{V + \mathbf{b}\}$ such that its 4-ary subcode has at least $\left[\frac{4^6}{5^2}\right] = 164$ code words. Notice that the shortened 4-ary Hamming code of the same length has only $4^3 = 64$ code words.

With the help of subcode over subset construction it is possible to construct from the Hamming codes over large alphabet the q-ary single-error-correcting codes with packing density asymptotically no less than $\frac{q-1}{\hat{q}-1}$, where $\hat{q}$ is the minimum integer greater than q that can be represented as the power of the prime number.

The other useful structure for the construction of error-correcting codes is the coset. Let $\mathbf{H}$ be the parity-check $r \times n$ matrix of some q-ary $(n, n\text{-}r)$ code C with minimum distance $d = 2t + 1$ (now we again assume that q-ary alphabet is the field F_q). Then all linear combinations of t or less columns of matrix $\mathbf{H}$ are distinct vectors. Denote this set of r-dimensional vectors as $U_{\mathbf{H}}$ ($\mathbf{0} \in U_{\mathbf{H}}$ as a result of linear combination with all-zero coefficients), where $|U_{\mathbf{H}}| = \sum_{i=0}^{t} \binom{n}{i} \cdot (q-1)^i$. Let V be a q-ary code of length r capable of correcting the set of errors $U_{\mathbf{H}}$, that is, it follows from $\mathbf{u} + \mathbf{v} = \mathbf{u}' + \mathbf{v}'$, $\mathbf{u}, \mathbf{u}' \in U_{\mathbf{H}}$, $\mathbf{v}, \mathbf{v}' \in V$ that $\mathbf{u} = \mathbf{u}'$, $\mathbf{v} = \mathbf{v}'$. Then the following statement holds true:

Theorem 3.12 Code $C_V = \{\mathbf{x} = (x_0, \ldots, x_{n-1}) \in E_q^n : \mathbf{H} \cdot \mathbf{x} \in V\}$ is capable of correcting t errors and has $|C_V| = |C| \cdot |V|$ code words.

Proof. Let us prove this statement by reductio ad absurdum. Assume that there exist the error vectors $\mathbf{e}$ and $\mathbf{e}'$: $\|\mathbf{e}\| \le t$, $\|\mathbf{e}'\| \le t$ and vectors $\mathbf{x} \ne \mathbf{x}' \in C_V$ such that $\mathbf{x} + \mathbf{e} = \mathbf{x}' + \mathbf{e}'$. Multiplying both parts of this equation by matrix $\mathbf{H}$ obtain $\mathbf{H} \cdot \mathbf{x} + \mathbf{H} \cdot \mathbf{e} = \mathbf{H} \cdot \mathbf{x}' + \mathbf{H} \cdot \mathbf{e}'$, where $\mathbf{H} \cdot \mathbf{x} + \mathbf{H} \cdot \mathbf{x}' \in V$, $\mathbf{H} \cdot \mathbf{e}, \mathbf{H} \cdot \mathbf{e}' \in U_{\mathbf{H}}$. Then $\mathbf{H} \cdot \mathbf{e} = \mathbf{H} \cdot \mathbf{e}'$, because the code V is capable of correcting errors from the set $U_{\mathbf{H}}$. Recall that all linear combinations of t or less columns of matrix $\mathbf{H}$ are distinct; hence $\mathbf{e} = \mathbf{e}'$, which means that $\mathbf{x} = \mathbf{x}'$ and this statement contradicts the assumption. The code C_V is the union over all $\mathbf{v}$ of solutions of the equation $\mathbf{H} \cdot \mathbf{x} = \mathbf{v}$, $\mathbf{v} \in V$, where r is the rank of matrix $\mathbf{H}$. Since for any $\mathbf{v}$ the number of solutions is the same and equals to $q^{n-r} = |C|$, we have that the overall number of solutions for all $\mathbf{v}$ equals to $|C| \cdot |V|$.

□

For the particular case of single-error-correcting codes it is possible to choose as $U_{\mathbf{H}}$ any homogeneous (that is, $\mathbf{u} \in U_{\mathbf{H}} \Rightarrow \lambda \cdot \mathbf{u} \in U_{\mathbf{H}}$, $\lambda \in F_q$) subset of E_q^n. The corresponding construction is called the method of *homogeneous packing* [21], since it is possible to construct from the code V capable of correcting the homogeneous set of errors $U_{\mathbf{H}}$ the code $V_{\mathbf{H}}$ of length $\frac{|U_{\mathbf{H}}|}{q-1}$ capable of correcting single errors, and it is easy to verify that the packing density in this case does not decrease.

The important example of the homogeneous set is the set $U_{l,N}$ of those vectors from the space $E_q^{N \cdot l}$ for which the nonzero positions are grouped in one of N phased packets of length l, that is, $U_{l,N} = \bigcup_{j=0}^{N-1} \{\mathbf{x} = (x_0, \ldots, x_{N \cdot l-1}) : \ x_i \ne 0, i \in \{j \cdot l, \ldots, j \cdot l + l - 1\}\}$. The code capable of correcting the set of errors $U_{l,N}$ is called the code capable of correcting single phased packets of length l. Since the elements of the field F_{q^l} can be regarded as l-dimensional vectors over field F_q, this correspondence specifies the isomorphism of the spaces $E_{q^l}^N$ and E_q^{Nl} where q-ary code of length $N \cdot l$ capable of correcting single phased packets of length l corresponds the q^l-ary code of length N capable of correcting single errors. The union of this structure with subcodes over subsets and with the method of homogeneous packing gives us the class of asymptotically optimal single-error-correcting codes [21].

Example 3.22 Construct the binary single-error-correcting code of length 18. Consider the 4-ary code of length 6 of Example 3.21 as a binary code of length 12 capable of correcting single phased error packets of length 2 and consisting of $M \ge 164$ code words. From the latter code with the help of cosets construction obtain a binary code of length $n = 6.3$, consisting of $M \cdot 2^6$ code words. This is the single-error-correcting code of length 18 with $2^8.41$ code words which is 1.28 times greater than the number of code words in the shortened Hamming code [21, 22].

As usual for coding theory in cases when q is not a prime power there is a lack of information. During the single-error-correcting codes the best parameters have the codes of length n defined by the following system of equations:

$$\begin{cases} \sum_{i=0}^{n} x_i = \alpha \bmod (2q-1) \\ \sum_{i=0}^{n} i x_i = \beta \bmod p \end{cases} \tag{3.104}$$

where p is a prime number greater than n. These codes are an analog of Varshamov-Tenengoltz codes for correcting single asymmetric errors [23]. The best of the codes defined by the system (3.104) has the number of code words M_{opt} no less than the mean over α and β, that is:

$$M_{opt} \geq \bar{M}_{\alpha,\beta} = \frac{q^n}{p \cdot (2q-1)} \tag{3.105}$$

Since $\lim_{n\to\infty} \frac{p(n)}{n} = 1$, where $p(n)$ is the minimum prime number greater than n, then the packing density of these codes (that is, the ratio of the cardinal number to the value of Hamming bound) with large n is about $\frac{q-1}{1q-1}$, that is, less than $\frac{1}{2}$.

As was mentioned above the structure of subcode over subset allows codes of a large size to be constructed, that is, with packing density asymptotically no less than $\frac{q-1}{\hat{q}-1}$ [21], where $\hat{q}$ is the minimal prime power such that $\hat{q} \geq q$. It is shown in [24] that if there exist perfect (reaching the Hamming bound) q-ary code single-error-correcting code of length $q+1$, then with $n \to \infty$ there exist codes asymptotically reaching the Hamming bound, that is, the packing density of the optimal codes tends to increase the code length in a general q-ary case also.

3.9 Reed-Solomon Codes

The Reed-Solomon (RS) codes invented in 1960 [25] are still one of the most applicable class of codes and despite (or maybe due to) their simplicity are the basis of the new deep generalizations. Lets start the discussion of RS codes from the following simple remark. The rank of a parity-check matrix **H** of the *(n, n-r)* code with minimum distance d is no less than d-1 since any of its d-1 columns are linearly independent. On the other hand, the rank of matrix is no more than the number of rows, and hence for any *(n, n-r)* code the following Singleton bound holds true:

$$d \leq r+1 \tag{3.106}$$

Note that the bound (3.106) holds true for any (not necessary linear) code, namely:

$$M_q(n,d) \leq q^{n-d-1} \tag{3.107}$$

The code reaching the bound (3.107) is an optimal code and it is called a *maximum distance separable (MDS) code*. Thus, in the parity-check $r \times n$ matrix of MDS code any r columns are linearly independent. The example of such kind of matrix is the matrix $\mathbf{H} = \left[h_{ij}\right] = \left[\alpha_j^i\right], i = 0, \ldots, r-1, j = 0, \ldots, n-1$, where $\alpha_0, \ldots, \alpha_{n-1}$ are distinct elements of the field F_q $(n \leq q)$. In other words, the ith row of matrix **H** consists of the values of the term z^i calculated in the points $\alpha_0, \ldots, \alpha_{n-1}$ of the field F_q:

$$\mathbf{H}_r = \begin{array}{c} 1 \\ z \\ \vdots \\ z^{r-1} \end{array} \left\| \begin{bmatrix} 1 & 1 & \ldots & 1 \\ \alpha_0 & \alpha_1 & \ldots & \alpha_{n-1} \\ \ldots & \ldots & \ldots & \ldots \\ \alpha_0^{r-1} & \alpha_1^{r-1} & \ldots & \alpha_{n-1}^{r-1} \end{bmatrix} \right. \tag{3.108}$$

Let us show that any r columns in matrix (3.108) are linearly independent. Lets assume this is not the case, and the columns $\mathbf{h}_{j1}, \ldots, \mathbf{h}_{j_r}$ are linearly dependent. This means $\det\left(\mathbf{h}_{j1}, \ldots, \mathbf{h}_{j_r}\right) = 0$. Hence, the rows of this minor determinant $\mathbf{H}_{j1}, \ldots, {}_{j_r}$ also are linearly dependent.

The linear dependence of rows of this minor determinant with coefficients $\lambda_0, \ldots, \lambda_{r-1}$ means that the polynomial $\lambda_0 + \lambda_1 \cdot z + \ldots + \lambda_{r-1} \cdot z^{r-1}$ of degree less than r has r different roots $\mu_{j1}, \ldots, \mu_{j_r}$, and that is impossible.

It is easy to obtain from the statement about matrix (3.108) that in matrices $\mathbf{H}_{a,r}$ of more general form with the rows defined by the values of term z^j $(j = a, a+1, \ldots, a+r-1)$ on elements $\alpha_j \in F_q \backslash 0$:

$$\mathbf{H}_{a,r} = \begin{bmatrix} \alpha_0^a & \ldots & \ldots & \alpha_{n-1}^a \\ \ldots & \ldots & \ldots & \ldots \\ \ldots & \ldots & \ldots & \ldots \\ \alpha_0^{a+r-1} & \ldots & \ldots & \alpha_{n-1}^{a+r-1} \end{bmatrix} \quad (3.109)$$

any r columns are linearly independent. The codes defined by the matrices (3.109) are called *Reed-Solomon (RS) codes*. More often RS codes are defined as cyclic codes. Let elements $\alpha_0, \ldots \alpha_{n-1}$ form a subgroup. This subgroup should be a cyclic subgroup because the group under multiplication of the whole finite field is the cyclic group. Hence, there exists element $\beta \in F_q$ such that $\alpha_i = \beta^i, i = 0, \ldots, n-1$. Then the matrix (3.109) can be converted to the form:

$$\mathbf{H}_{a,r,\beta} = \begin{bmatrix} 1 & \beta^a & \ldots & \beta_{n-1}^{a(n-1)} \\ 1 & \beta^{a+1} & \ldots & \beta^{(a+1)(n-1)} \\ \ldots & \ldots & \ldots & \ldots \\ 1 & \beta^{a+r-1} & \ldots & \beta_{n-1}^{(a+r-1)(n-1)} \end{bmatrix} \quad (3.110)$$

Since the cyclic shift of any row of matrix $\mathbf{H}_{a,r,\beta}$ (this shift can be obtained by the multiplication of the row by $\beta^{a(n-1)}$) belongs to the linear space of rows of this matrix, the code defined by the matrix $\mathbf{H}_{a,r,\beta}$ is the cyclic code (as well as the corresponding dual code). The condition that vector $\mathbf{v}$ is the codeword of the code defined by the matrix $\mathbf{H}_{a,r,\beta}$:

$$\mathbf{H}_{a,r,\beta} \cdot \mathbf{v} = \mathbf{0}$$

is equivalent to the following condition:

$$v(\beta^j) = 0, \quad j = a, \ldots, a+r-1 \quad (3.111)$$

where $v(x)$ is the polynomial corresponding to vector $\mathbf{v}$. Therefore, the generator polynomial of this code is:

$$g(x) = \text{L.C.M}\left\{f_{\beta^j}(x)\right\}, \quad j = a, \ldots, a+r-1 \quad (3.112)$$

where $f_\gamma(x)$ is the minimal polynomial of the element γ. In the considered case $f_\gamma(x) = x - \gamma$ and therefore:

$$g(x) = \coprod_{j=a}^{a+r-1} \left(x - \beta^j\right) \quad (3.113)$$

The most interesting case is when β is the primitive element of field F_q. Then $n = q-1$ and such kind of codes most often are called RS codes [5]. The code defined by the matrix (3.108) has the length q when $\{\alpha_0, \ldots, \alpha_{n-1}\} = F_q$ and is called *1-extended RS code*. It is possible to extend this kind of code by one symbol more if we add symbol ∞ to the set $\{\alpha_0, \ldots, \alpha_{n-1}\} = F_q$ with formal definition $\infty^j = 0$ for $j = 0, 1, \ldots, r-2$ and $\infty^{r-1} = 1$. The question about possibility of construction of nontrivial MDS codes (that is, $2 < d < n$) of length $n+1$ is the unsolved problem throwing back to the projective geometry. A well-accepted hypothesis maintains that these codes do not exist except in the cases when $q = 2^m, n = q+2$ and either code with distance $d = 4$ or its dual code with distance $d = n$.

Example 3.23 Specify the code of length $q+2$ with the help of systematic encoding that is, match up to the set of information symbols $a_0, \ldots, a_{q-2}$ three parity-check symbols b_0, b_1, b_2 computed in accordance with formula:

$$b_0 = \sum_{i=0}^{q-2} a_i; b_1 = \sum_{i=0}^{q-2} a_i \cdot \alpha^{i+1}; b_2 = \sum_{i=0}^{q-2} a_i \cdot \alpha^{2(i+1)}$$

where α is the primitive element of the field F_q. It is easy to verify that for $q = 2^m$ the distance of this code is 4, that is, this is the nontrivial MDS code of length $q + 2$.

Let V be an (n, n-r) RS code defined by the parity-check matrix (3.109) over the field F_Q, $Q = q^m$. Since the field F_q may be defined as the subfield of the field F_Q consider a subcode V_q of the code V over subfield F_q. This code is no more than q-ary BCH code [5]. It follows from the general relations of subcode over subset that $\left|V_q\right| \geq q^{n-rm}$. In this particular case the given estimation can be improved by considering the field F_Q as an m-dimensional vector space over F_q and the elements of the matrix $\mathbf{H}_{a,r}$ as m-dimensional columns (vectors) over the field F_q. Then the rank (q-ary) of the new matrix $\mathbf{H}_{a,r}^{(q)}$ equals to the number of parity-check symbols of code V_q. It is a known fact [5] that there exists normal basis in the field F_Q, that is, the element γ such that the elements $\gamma, \gamma^q, \ldots, \gamma^{q^{m-1}}$ are the basis of the vector space F_Q over the field F_q. In this basis the "block" of m rows of the matrix $\mathbf{H}_{a,r}^{(q)}$ corresponding to the row $z^{j \cdot q}$ of the matrix $\mathbf{H}_{a,r}$ can be obtained by the cyclic shift (downward) of the rows of the "block" corresponding to the row z^j. From this follows the estimations of a number of parity-check symbols $r^{(q)}$ of the q-ary code V_q of length $n \leq q^m$ and with minimum distance no less than d, which we will write for two particular cases:

$$r^{(q)} \leq m \cdot \left(d - 1 \left\lfloor \frac{d-1}{q} \right\rfloor\right) \quad \text{for } a = 1 \tag{3.114}$$

$$r^{(q)} \leq 1 + m \cdot \left(d - 2 - \left\lfloor \frac{d-2}{q} \right\rfloor\right) \quad \text{for } a = 0 \tag{3.115}$$

Substituting $q = 2$ and $d = 2t + 1$ in (3.114) obtain that the number of parity-check symbols of BCH code of length n capable of correcting t errors is no more than $t \cdot \lfloor \log_2 n \rfloor$. It follows from this statement and from Hamming bound that BCH codes are asymptotically optimal "on redundancy", that is, $\lim\limits_{n\to\infty} \frac{r_{BCH}}{r(n, 2t+1)} = 1$.

Substituting $q = 3$ and $d = 5$ in (3.115) obtain that the number of parity-check symbols of the corresponding 3-ary BCH codes of length n capable of correcting two errors is no more than $1 + 2 \cdot \lfloor \log_3 n \rfloor$. In particular, Hamming bound shows that for $n = 3^n$ these codes are optimal in the class of linear codes. It is known in addition that there is one more class of codes – 4-ary codes capable of correcting two errors [26, 27], which satisfies the Hamming bound for a number of parity-check symbols. There are no known classes of codes with parameters q and t except the mentioned above that satisfies the following equation:

$$\lim_{\substack{n\to\infty \\ q,t=const}} \frac{r}{t \cdot \log_q n} = 1 \tag{3.116}$$

The BCH codes do not satisfy (3.116) with $q < 2$ (except the case $q = 3$, $t = 2$) and for the case $t = 2$ there are exist codes asymptotically better than BCH codes, that is, for these codes the left part of (3.116) has value 7/6 which is better than 1.5 for BCH codes [28]. For an extended discussion of the BCH codes see the next section.

3.10 BCH Codes

For the estimation of the minimum distance the following description of the cyclic codes is very useful. Let $g(x)$ be the generator polynomial of the cyclic (n, k) code V over the field F_q. As before we assume that the code is nontrivial, that is, $2 < d < n$ and, therefore, n is the period of the polynomial $g(x)$, that is, $g(x)|x^n - 1$ and $g(x)$ does not divide $x^{n'} - 1$ for $n' < n$. Let the code length n and the characteristic p of field F_q be relatively prime. Then the polynomial $x^n - 1$ has no repeated roots since $LCD\left(x^n - 1, \frac{d}{dx}(x^n - 1)\right) = 1$. Hence the polynomial $g(x)$ also has no repeated roots because it is the divisor of the polynomial $x^n - 1$. Let m be the minimal positive integer such that n divides $q^m - 1$. Then F_{q^m} is the minimal field containing all roots of $g(x)$. The following theorem called *BCH bound* holds truth.

Theorem 3.13 Let α be an element of the field F_{q^m} and let $\alpha^a, \alpha^{a+1}, \ldots, \alpha^{a+s-1}$ be the roots of the polynomial $g(x)$. Then the minimum distance of the cyclic code V with the generator polynomial $g(x)$ is no less than $s + 1$.

Proof. Any codeword $\mathbf{v} \in V$ can be considered as a polynomial $v(x)$ and the elements $\alpha^a, \alpha^{a+1}, \ldots, \alpha^{a+s-1}$ are the roots of this polynomial because $g(x)$ is the divisor of $v(x)$, that is, $v(\alpha^j) = 0$ for $j = a, \ldots, a\_s - 1$. In accordance with (3.111) that means the codeword $\mathbf{v}$ belongs to the cyclic RS code with the minimum distance $s + 1$.

□

More often this theorem is used for a particular case when α is the primitive element of the field F_{q^m}. The corresponding statement says that the minimum distance of a cyclic code is more than the length of the maximal "set of consecutive roots", that is, the maximum number of consecutive powers $j_0, \ldots, j_{s-1}$, where $\alpha^{j_0}, \ldots, \alpha^{j_{s-1}}$ are the roots of the polynomial $g(x)$. The parameter $s + 1$ usually is called the *designed distance* of a BCH code.

Example 3.24 Consider binary cyclic code V with the generator polynomial $g(x) = x^7 + x^6 + x^5 + x^2 + x + 1$. Find the parameters of this code. To obtain the code length we have to find x^n of minimal power such that the remainder of division x^n by $g(x)$ is equal to 1. The remainder of x^7 divided by $g(x)$ is:

$$x^7 = x^6 + x^5 + x^2 + x + 1 \mod g(x)$$

Now multiply by x on modulo $g(x)$ both left and right part of this equation
$x^8 = x^7 + x^6 + x^3 + x^2 + x = x^6 + x^5 + x^2 + x + 1 + x^6 + x^3 + x^2 + x = x^5 + x^3 + 1 \mod g(x)$
In the same manner we will calculate remainders of division terms x^9, x^{10}, ... by $g(x)$ until the remainder become equal to 1:

$$\left.\begin{aligned}
x^9 &= x^6 + x^4 + x \\
x^{10} &= x^6 + x + 1 \\
x^{11} &= x^6 + x^5 + 1 \\
x^{12} &= x^5 + x^2 + 1 \\
x^{13} &= x^6 + x^3 + x \\
x^{14} &= x^6 + x^5 + x^4 + x + 1 \\
x^{15} &= 1
\end{aligned}\right\} \mod g(x)$$

Thus, the code length is $n = 15$. The number of information symbols k equals to the difference of code length and the degree of the generator polynomial:

$$k = n - \deg g(x) = 15 - 7 = 8$$

Since $g(x)$ is the divisor of $x^n - 1$, all roots of $g(x)$ belong to the field $GF(2^4)$. Let γ be the primitive element of this field such that $\gamma^4 + \gamma + 1 = 0$. It is easy to verify that the roots of $g(x)$ are $\gamma^0 = 1, \gamma^5, \gamma^6, \gamma^9, \gamma^{10}, \gamma^{12}$. The longest set of consequent powers consists of two elements γ^5, γ^6 or γ^9, γ^{10}. It follows from the lower BCH bound that the minimum distance of given code $d \geq 3$. Use the theorem

3.13 with $\alpha = \gamma^5$. Then $\alpha^0, \alpha^1 = \gamma^5, \alpha^2 = \gamma^{10}$ are the roots of the polynomial $g(x)$. Hence, the minimum distance of the code is no less than 4 and the exhaustive search shows that it is the real minimum distance of the code.

Now let us introduce the classical definition of the BCH code [2]. Let n be the divisor of $q^m - 1$ and n is not divisor of $q^i - 1$ for $i < m$. Then the q-ary BCH code with the designed distance d is the cyclic code of length n with the generator polynomial:

$$g(x) = \text{LCM}\left\{\varphi_{\alpha^a}(x), \varphi_{\alpha^{a+1}}(x), \ldots, \varphi_{\alpha^{a+d-2}}(x)\right\} \tag{3.117}$$

where α is the primitive nth root of unit in the field F_{q^m}, $\varphi_\gamma(x)$ is the minimal polynomial of the element γ. If $n = q^m - 1$ and $a = 1$, then the code is called a *primitive* BCH *code*.

Using (3.117), (3.113) and (3.112) it is easy to verify that this definition of BCH code coincides with that given in Section 3.9's definition of BCH code as a q-ary subcode of a q^m-ary RS code. The theorem 3.13 gives only the lower bound of the distance of BCH code and the real minimum distance can be greater than the design distance. The meaning of the design distance is in fact that there exist effective algorithms capable of correcting the corresponding to the design distance number of errors.

3.11 Decoding of BCH Codes

Let V be a cyclic (n, k) BCH code over the field F_q with the design distance $d = 2t + 1$ and $g(x)$ being the generator polynomial of this code. Let $\gamma, \gamma^2, \ldots, \gamma^{2t}$ be the roots of the generator polynomial $g(x)$. Then the parity-check matrix of the code can be written in the form:

$$\mathbf{H} = \begin{bmatrix} 1 & \gamma & \gamma^2 & \ldots & \gamma^{n-1} \\ 1 & \gamma^2 & (\gamma^2)^2 & \ldots & (\gamma^{n-1})^2 \\ \ldots & \ldots & \ldots & \ldots & \ldots \\ 1 & \gamma^{2t} & (\gamma^2)^{2t} & \ldots & (\gamma^{n-1})^{2t} \end{bmatrix} \tag{3.118}$$

Let $b(x)$ be the received word equal to the sum of the codeword $a(x)$ and the error pattern $e(x)$:

$$b(x) = a(x) + e(x) \tag{3.119}$$

and the error polynomial:

$$e(x) = e_0 + e_1 x + \ldots + e_{n-1} x^{n-1} \tag{3.120}$$

contains exactly $\nu \le t$ nonzero coefficients $e_{i_1}, \ldots, e_{i_\nu}$. Multiplying $e(x)$ by $\mathbf{H}$ obtain the syndrome vector with the components:

$$s_j = b(\gamma^j), \quad j = 1, \ldots 2t \tag{3.121}$$

Since $a(\gamma^j) = 0$ for any codeword, (3.121) can be written as follows:

$$s_j = e_{i_1} \cdot (\gamma^j)^{i_1} + e_{i_2} \cdot (\gamma^j)^{i_2} + \ldots + e_{i_\nu} \cdot (\gamma^j)^{i_\nu} \tag{3.122}$$

The error polynomial (3.120) is fully defined by the set of pairs $\left\{e_{i_1}, i_1\right\}, \left\{e_{i_2}, i_2\right\}, \ldots, \left\{e_{i_\nu}, i_\nu\right\}$. Denote e_{i_m} by Y_m and γ^{i_m} by X_m, then:

$$s_j = X_1^j Y_1 + X_2^j Y_2 + \ldots + X_\nu^j Y_\nu, \quad j = 1, \ldots, 2t \tag{3.123}$$

Y_m is called *error-value* and X_m is called *error-location number*. The syndrome components s_j can be calculated directly from the received vector, therefore (3.123) can be regarded as the system of $2t$ nonlinear equations relative to 2ν unknowns $X_1, X_2, \ldots, X_\nu, \quad Y_1, Y_2, \ldots, Y_\nu$. If $t \ge \nu$ system (3.123) has the unique solution that should be found to decode the received word.

However, it is not so easy to solve the system of nonlinear equations directly. To avoid some difficulties with the direct solving of this system consider the polynomial $\sigma(x)$:

$$\sigma(x) = 1 + \sigma_1 x + \ldots + \sigma_\nu x^\nu = (1 - xX_1) \cdot (1 - xX_2) \cdot \ldots \cdot (1 - xX_\nu) \tag{3.124}$$

The polynomial $\sigma(x)$ called *error-location polynomial* is the polynomial of minimal degree such that the roots of this polynomial are values $X_1^{-1}, X_2^{-1}, \ldots, X_\nu^{-1}$, that is, the reciprocals of the error-locations.

If the coefficients $\sigma_1, \ldots, \sigma_\nu$ of the polynomial $\sigma(x)$ are known then we should find the roots of this polynomial to calculate the error-locations. Hence the problem of calculation of the error-locations can be solved in two steps: first, to find the coefficients of the polynomial $\sigma(x)$ and second, to find the roots of $\sigma(x)$.

To find the coefficients $\sigma_1, \ldots, \sigma_\nu$ it is necessary to show the relation of these coefficients (unknown values) with the known syndrome components $s_j, \quad j = 1, \ldots, 2t$. Substituting values $X_1^{-1}, X_2^{-1}, \ldots, X_\nu^{-1}$ in $\sigma(x)$ obtain the system of equations:

$$1 + \sigma_1 \cdot (X_j^{-1}) + \sigma_2 \cdot (X_j^{-1}) + \ldots + \sigma_\nu \cdot (X_j^{-1})^\nu = 0, \quad j = 1, \ldots, \nu \tag{3.125}$$

Multiplying the left and right parts of (3.125) by $X_1^{\nu+1}Y_1, \quad X_2^{\nu+1}Y_2, \ldots, \quad X_\nu^{\nu+1}Y_\nu$ correspondingly obtain:

$$\begin{cases} Y_1 X_1^{\nu+1} + \sigma_1 Y_1 X_1^\nu + \sigma_2 Y_1 X_1^{\nu-1} + \ldots + \sigma_\nu Y_1 X_1 = 0 \\ \ldots\ldots\ldots\ldots\ldots\ldots\ldots\ldots\ldots\ldots \\ Y_\nu X_\nu^{\nu+1} + \sigma_1 Y_\nu X_\nu^\nu + \sigma_2 Y_\nu X_\nu^{\nu-1} + \ldots + \sigma_\nu Y_\nu X_\nu = 0 \end{cases} \tag{3.126}$$

Adding equations (3.126) obtain:

$$(Y_1 X_1^{\nu+1} + \ldots + Y_\nu X_\nu^{\nu+1}) + \sigma_1 (Y_1 X_1^\nu + \ldots + Y_\nu X_\nu^\nu) + \ldots + \sigma_\nu (Y_1 X_1 + \ldots + Y_\nu X_\nu) = 0 \tag{3.127}$$

Taking into account (3.123) we can write (3.127) as follows:

$$s_{\nu+1} + \sigma_1 s_\nu + \ldots + \sigma_\nu s_1 = 0 \tag{3.128}$$

Now multiplying the left and right parts of the equations (3.125) by $X_j^{\nu+2}Y_j, \quad X_j^{\nu+3}Y_j, \ldots$ and with the same manipulations as above obtain the system of equations that defines the relation between the syndrome components and the coefficients of the error-location polynomial:

$$\begin{cases} s_1\sigma_\nu + s_2\sigma_{\nu-1} + \ldots + s_\nu\sigma_1 = -s_{\nu+1} \\ s_2\sigma_\nu + s_3\sigma_{\nu-1} + \ldots + s_{\nu+1}\sigma_1 = -s_{\nu+2} \\ \ldots\ldots\ldots\ldots\ldots\ldots\ldots\ldots \\ s_\nu\sigma_\nu + s_{\nu+1}\sigma_{\nu-1} + \ldots + s_{2\nu-1}\sigma_1 = -s_{2\nu} \end{cases} \tag{3.129}$$

The system (3.129) unlike (3.123) is the system of linear equations and there are well-known methods of solving it.

However, to obtain the system (3.129) it is necessary to know the value of ν. The following theorem allows ν to be obtained.

Theorem 3.14 The determinant of the matrix $\mathbf{M}_\mu$:

$$\mathbf{M}_\mu \begin{bmatrix} s_1 & s_2 & \ldots & s_\mu \\ s_2 & s_3 & \ldots & s_{\mu+1} \\ \ldots & \ldots & \ldots & \ldots \\ s_\mu & s_{\mu+1} & \ldots & s_{2\mu-1} \end{bmatrix}$$

is nonzero $|\mathbf{M}_\mu| \neq 0$ if μ is equal to the number of errors $\mu = \nu$. If $\mu > \nu$ then $|\mathbf{M}_\mu| = 0$.

Proof. Let $X_j = 0$ for $j > \nu$ and let:

$$\mathbf{W}_\mu = \begin{bmatrix} 1 & 1 & \dots & 1 \\ X_1 & X_2 & \dots & X_\mu \\ \dots & \dots & \dots & \dots \\ X_1^{\mu-1} & X_2^{\mu-1} & \dots & X_\mu^{\mu-1} \end{bmatrix}$$

be the Vandermonde matrix. Denote by $\mathbf{D}_\mu$ the diagonal matrix:

$$\mathbf{D}_\mu = \begin{bmatrix} Y_1X_1 & 0 & \dots & 0 \\ 0 & Y_2X_2 & \dots & 0 \\ \dots & \dots & \dots & \dots \\ 0 & 0 & \dots & Y_\mu X_\mu \end{bmatrix}$$

It is easy to verify that in accordance with (3.123):

$$\mathbf{M}_\mu = \mathbf{W}_\mu \cdot \mathbf{D}_\mu \cdot \mathbf{W}_\mu^{\mathrm{T}}$$

where $\mathbf{W}_\mu^{\mathrm{T}}$ is the transposed Vandermonde matrix. Then the determinant of matrix $\mathbf{M}_\mu$ is:

$$\left|\mathbf{M}_\mu\right| = \left|\mathbf{W}_\mu\right| \cdot \left|\mathbf{D}_\mu\right| \cdot \left|\mathbf{W}_\mu^{\mathrm{T}}\right| \tag{3.130}$$

If $\mu = \nu$ then all the elements $X_1, \dots, X_\mu, \quad Y_1, \dots Y_\mu$ differ from zero and $X_1, \dots, X_\mu$ are the distinct elements. Then all determinants in the right part of (3.130) differ from zero and $\left|\mathbf{M}_\mu\right| \neq 0$. If $\mu > \nu$ then $\left|\mathbf{W}_\mu\right| = 0$ and $\left|\mathbf{M}_\mu\right| = 0$. □

With the help of Theorem 3.14 it is possible to find the value of ν in the following way. Consider matrices $\mathbf{M}_\mu$ for $\mu = t, t-1, t-2$ and so on, until μ_0 is found such that $\left|\mathbf{M}_{\mu_0}\right| \neq 0$. Then $\mu_0 = \nu$ is the true number of errors.

For given ν it is possible to solve the system (3.129) (for example, with the help of Gauss method) and obtain the coefficients of the error-location polynomial $\sigma(x)$. Now it is enough to find the roots of $\sigma(x)$ and we can obtain the error-location numbers just by inverting these roots. To find the roots of the polynomial $\sigma(x)$ in the finite field we can just substitute in $\sigma(x)$ all field elements in turn (this procedure for searching the roots of a polynomial is called *Chien's search*).

To calculate error values $Y_1, \dots, Y_\nu$ it is enough to solve the system of first v equations (3.123) after substituting in them the known values of $X_1, \dots, X_\nu$.

Thus, the algorithm of decoding a BCH code, which is usually called *Peterson-Gorenstein-Zierler algorithm* or the direct method of decoding a BCH code, consists of four steps:

- syndrome calculation;
- finding the coefficients of the error-location polynomial;
- finding the roots of the error-location polynomial;
- calculation of error-values.

The formal algorithm is as follows:

1. Calculate the syndrome components $s_j = b(\gamma^j), \quad j = 1, \dots, 2t$, where $b(x)$ is the received polynomial.
2. Consider determinants $|\mathbf{M}_t|, |\mathbf{M}_{t-1}|, \dots$ until find ν such that $|\mathbf{M}_\nu| \neq 0$.
3. Calculate the coefficients of the error-location polynomial $\sigma_1, \dots, \sigma_\nu$:

$$(\sigma_\nu, \dots, \sigma_1) = (-s_{\nu+1}, \dots, -s_{2\nu}) \cdot \left(\mathbf{M}_\nu^{-1}\right)^{\mathrm{T}}$$

4. Find the roots of the error-location polynomial $\sigma(x)$ with the help of Chien's search.

5. Calculate error-values:

$$\begin{bmatrix} Y_1 \\ Y_2 \\ \vdots \\ Y_\nu \end{bmatrix} = \begin{bmatrix} X_1 & X_2 & \dots & X_\nu \\ X_1^2 & X_2^2 & \dots & X_\nu^2 \\ \dots & \dots & \dots & \dots \\ X_1^\nu & X_2^\nu & \dots & X_\nu^\nu \end{bmatrix} \cdot \begin{bmatrix} S_1 \\ S_2 \\ \vdots \\ S_\nu \end{bmatrix}$$

The most computation complexity in this algorithm has step 3 associated with solving the system (3.129). However, (3.129) is not the arbitrary system, the coefficients of the equalities in this system are well structured. All known methods of simplification of the direct decoding are based on this structure. However, these simplified methods do not help to clarify the understanding of the process of decoding a BCH code.

One of the best methods of implementation of step 3 is believed to be the iterative Berlekamp-Massey algorithm. The idea of this algorithm is that the polynomial $\sigma(x)$ is calculated with the help of the sequential approximations $\sigma^{(0)}(x)$,$\sigma^{(1)}(x)$, and so on, until $\sigma^{(\nu)}(x) = \sigma(x)$ and in doing so $\sigma^{(j)}(x)$ is chosen as the refinement of $\sigma^{(j-1)}(x)$.

It follows from (3.129) that the syndrome components can be expressed in the form of the recursive equation:

$$S_j = -\sum_{i=1}^{\nu} \sigma_i \cdot s_{j-i}, \quad j = \nu+1, \dots, 2\nu \tag{3.131}$$

We will say that the polynomial $\sigma^{(j)}(x)$. generates $S_1,\dots S_j$ if the equation:

$$S_j = -\sum_{i=1}^{L} \sigma_i^{(j)} \cdot S_{j-i}$$

holds true, where L is the degree of the polynomial $\sigma^{(j)}(x)$ and $\sigma_i^{(j)}$ are the coefficients of this polynomial.

Then the solution of the system (3.129) is equivalent to the finding of the polynomial $\sigma(x)$of minimal degree that generates the syndrome components $S_1, S_2, \dots, S_{2t}$. The iterative search of $\sigma(x)$ starts with $\sigma^{(0)}(x)$ and is as follows: for given $\sigma^{(j-1)}(x)$ that generates $S_1, S_2, \dots, S_{j-1}$ verify the capability of $\sigma^{(j-1)}(x)$ to generate s_j; if it is so then $\sigma^{(j)}(x)$ is assumed to be equal $\sigma^{(j-1)}(x)$, otherwise $\sigma^{(j)}(x)$ is chosen as:

$$\sigma^{(j)}(x) = \sigma^{(j-1)}(x) + \Delta_j(x)$$

where $\Delta_j(x)$ is the correction polynomial for the jth step. The process is repeated until the polynomial that generates all the syndrome components is found.

Let us present the formal *Berlekamp-Massey algorithm* by searching the polynomial $\sigma(x)$. It contains some auxiliary polynomials $B(x)$ and $T(x)$, which do not have meaningful interpretation. L is the degree of the current polynomial $\sigma(x)$ at the jth step:

1. Set the initial parameters $\sigma^{(0)}(x) = 1, j = 0, L = 0, B(x) = 1$.
2. $j = j + 1$.
3. Calculate the jth discrepancy $\Delta_j = s_j + \sum_{i=1}^{L} \sigma_i^{(j-1)} \cdot S_{j-1}$.
4. Verify if $\sigma^{(j-1)}(x)$ generates s_j or not: compare Δ_j with zero. If $\Delta_j = 0$, then $\sigma^{(j)}(x) = \sigma^{(j-1)}(x)$ and go to step 9.
5. Calculate $T(x) = \sigma^{(j-1)}(x) - \Delta_j \cdot x \cdot B(x)$.
6. Check is it necessary to increase the degree of the current polynomial. Compare $2L$ with $j-1$. If $2Lj-1$ then go to step 8.
7. Calculate new $B(x)$, $\sigma(x)$ and L. $B(x) = \Delta_j^{-1} \cdot \sigma^{(j-1)}(x)$; $\sigma^{(j)}(x) = T(x)$; $L = L-j$. Go to step 10.
8. Calculate $\sigma^{(j)}(x) = T(x)$.

9. Calculate $B(x) = x \cdot B(x)$.
10. Check the condition $j = 2t$. If $j < 2t$ go to step 2.
11. Check the condition $\deg \sigma(x) = L$. If $\deg \sigma(x) = L$ then go to step 13.
12. Stop the Berlekamp-Massey algorithm, $\sigma(x) = \sigma^{(j)}(x)$. Go to the next decoding stage.
13. Stop the Berlekamp-Massey algorithm; the uncorrectable error pattern is detected.

Example 3.25 Consider Berlekamp-Massey algorithm for decoding 16-ary (15, 9) Reed-Solomon code capable of correcting triple errors. Let the transmitted polynomial be $a(x) = 0$, the received polynomial be $b(x) = a(x) + e(x) = e(x) = \alpha x^7 + \alpha^5 x^5 + \alpha^{11} x^2$. $S_1 = \alpha \cdot \alpha^7 + \alpha^5 \cdot \alpha^5 + \alpha^{11} \cdot \alpha^2 = \alpha^{12}$, $s_2 = 1$, $s_3 = \alpha^{14}$, $s_4 = \alpha^{13}$, $s_5 = 1$, $s_6 = \alpha^{11}$. The steps of the Berlekamp-Massey algorithm are listed in Table 3.4.

Table 3.4

j	Δ_j	$T(x)$	$B(x)$	$\sigma^{(i)}(x)$	L
0			1	1	0
1	α^{12}	$1 + \alpha^{12}x$	α^3	$1 + \alpha^{12}x$	1
2	α^7	$1 + \alpha^3 x$	$\alpha^3 x$	$1 + \alpha^3 x$	1
3	1	$1 + \alpha^3 x + \alpha^3 x^2$	$1 + \alpha^{32} x$	$1 + \alpha^3 x + \alpha^3 x^2$	2
4	1	$1 + \alpha^{14} x$	$x + \alpha^3 x^2$	$1 + \alpha^{14} x$	2
5	α^{11}	$1 + \alpha^{14} x + \alpha^{11} x^2 + \alpha^{14} x^3$	$\alpha^4 + \alpha^3 x$	$1 + \alpha^{14} x + \alpha^{11} x^2 + \alpha^{14} x^3$	3
6	0	$1 + \alpha^{14} x + \alpha^{11} x^2 + \alpha^{14} x^3$	$\alpha^4 + \alpha^3 x$	$1 + \alpha^{14} x + \alpha^{11} x^2 + \alpha^{14} x^3$	3

Then $\sigma(x) = 1 + \alpha^{14} x + \alpha^{11} x^2 + \alpha^{14} x^3 = (1 + \alpha^7 x) \cdot (1 + \alpha^5 x) \cdot (1 + \alpha^2 x)$, the roots are α^{-7}, α^{-5}, α^{-2}; and the error-location numbers are α^7, α^5, α^2.

3.12 Sudan Algorithm and Its Extensions

The Sudan algorithm for decoding some low rate Reed-Solomon codes beyond half of their minimum distance d was invented in 1997 [29]. Later Guruswami and Sudan [9] managed to significantly improve this algorithm to make it capable of decoding almost all Reed-Solomon (RS) codes beyond $d/2$ limit. The main idea of these works was based on the fact that if more than $[\frac{d-1}{2}]$ errors have occurred in a received word, then the decoding may not be unique, so the decoder may output a *list* of codewords within a certain distance from the received word. The final decision can be carried out using some additional information, for example "soft" information from the demodulator. Some other authors Roth-Ruckenstein [30], Kötter [31, 32], and Nielsen [33, 34] succeeded in decreasing the complexity of Guruswani-Sudan (GS) algorithm.

Following P. Elias [35] definition a list decoding algorithms of decoding radius T should produce for any received vector $\mathbf{y}$ the list $L_T(\mathbf{y}) = \{\mathbf{c} \in C : \quad d(\mathbf{y}, \mathbf{c}) \leq T\}$ of all vectors $\mathbf{c}$ from a code C, which are at distance at most T apart from vector $\mathbf{y}$. Bounded distance decoding, that is, correcting up to $[\frac{d-1}{2}]$ errors, is a particular case of list decoding when $T = [\frac{d-1}{2}]$ and any list contains no more than one code vector.

Let us consider the problem of list decoding of Reed-Solomon codes, one of the most widely used and well-studied classes of error-correcting codes. There are many ways to define the Reed-Solomon code, one of them was considered in Section 3.9. Here it is convenient to use the "dual" definition:

Consider finite field $GF(q)$ and some set $X = \{x_1, \ldots, x_n\}$ of its distinct elements. The (n, k) RS code consists of all vectors $\mathbf{f} = (f(x_1), \ldots, f(x_n))$, where $f(x) = f_0 + f_1 x + \ldots + f_{k-1} x^{k-1}$ is a polynomial over $GF(q)$ of degree less than k. Since the number of roots of $f(x)$ does not exceed $\deg f(x)$ we have the Hamming weight $wt(\mathbf{f}) \geq n - \deg f(x) \geq n - k + 1$. Hence, the minimum distance of RS code $d \geq n-k+1$, and by the Singleton bound (3.106) we have $d = n-k+1$.

According to the general definition a list decoding algorithm for (n, k) RS code can be reformulated in the following way. For any given received vector $y = (y_1, \ldots, y_n)$ find all polynomials $p(x)$ of degree $\deg p(x) < k$ such that $p(x_s) = y_s$ for at least $n-T$ values x_s. We call such $p(x)$ T-consistent. Denote $\hat{k} = k - 1$, which is more convenient to use in the formulae given below.

The original breakthrough of Sudan's algorithm [29] exploits two very simple mathematical facts:

- a homogenous system of linear equations has nontrivial, that is, nonzero solution if the number of equations is less than the number of variables;
- if a polynomial has more roots than its degree, then this is identically zero polynomial.

Geometrically list decoding of RS code means to find all curves $y - p(x) = 0$, which pass through at least $n - T$ points (x_s, y_s) of the "plane" F_q^2. Consider instead some general algebraic curve:

$$Q(x, y) = \sum q_{ij} x^i y^i$$

which passes through all n points (x_s, y_s), that is:

$$Q(x_s, y_s) = 0, \quad 1 \leq s \leq n \tag{3.132}$$

Definition 3.3 Define weighted degree of $Q(x, y)$ as:

$$\deg_{\{1,\hat{k}\}} Q(x, y) = \max_{\{i,j\}: Q_{ij} \neq 0} i + \hat{k} j$$

This definition becomes clear due to the following

Lemma 3.4 *If $n - T > \deg_{\{1,\hat{k}\}} Q(x, y)$ and $p(x)$ is T-consistent, then $y - p(x)$ divides $Q(x, y)$.*

Proof. Consider a univariate polynomial $g(x) = Q(x, p(x))$. The degree of $g(x)$ does not exceed weighted degree $\deg_{\{1,\hat{k}\}} Q(x, y)$ of $Q(x, y)$. On the other hand, $g(x_s) = 0$ for every s such that $y_s = p(x_s)$. Hence, $g(x)$ has at least $n - T$ roots and under Lemma's condition it leads to $g(x) \equiv 0$. The least means that $(y-p(x)) | Q(x, y)$ or, saying in other words, that $p(x)$ is a root of $Q(x, y)$ considered as a univariate polynomial $\hat{Q}(y)$ on y with coefficients from the ring $F_q[x]$.

□

The next lemma shows that desired $Q(x, y)$ of relatively low weight degree exist.

Lemma 3.5 *For any l such that $l^2 \geq 2\hat{k}n$ there exists $Q(x, y)$ of weighted degree* $\deg_{\{1,\hat{k}\}} Q(x, y) \leq l$ *satisfying* (3.132).

Proof. Consider (3.132) as a system of linear equations for unknown "variables" q_{ij}. Namely:

$$\sum_{i,j} q_{ij} \cdot x_s^i \cdot y_s^j = 0, \quad 1 \leq s \leq n \tag{3.133}$$

Since $i + \hat{k}j \leq l$ there are

$l + 1$ "variables" $q_{i,0}, \quad 0 \leq i \leq l,$

$l + 1 + \hat{k}$ "variables" $q_{i,1}, \quad 0 \leq i \leq l - \hat{k},$

....

$l + 1 - \hat{k}[\frac{1}{\hat{k}}]$ "variables" $q_i, [l/\frac{l}{\hat{k}}], \quad 0 \leq i \leq l - \hat{k}[\frac{1}{\hat{k}}].$

Hence, there are totally M variables, where $M = \sum_{\nu=0}^{[l/\hat{k}} (l + 1 - \nu\hat{k}) = \left([\frac{l}{k} + 1]\right) \cdot (l + 1) - \hat{k} \sum_{\nu=0}^{[l/\hat{k}} \nu = \left([\frac{l}{k} + 1]\right) \cdot \left(l + 1 - \frac{1}{2}\hat{k}.[\frac{l}{k}]\right) > \frac{l^2}{2\hat{k}}$. Therefore, if $\frac{l^2}{2\hat{k}} > n$ then the number of unknown variables is greater than the number of equations and hence, there exist nontrivial, that is, nonzero solution of (3.132).

□

Combining these two lemmas obtain the *original Sudan algorithm*. Namely, for any T and l such that $n{-}T\ l$ and $l^2 \geq 2\hat{k}n$

1) Find a bivariate polynomial $Q(x, y)$ of $\deg_{\{1,\hat{k}\}} Q(x, y) \leq l$ satisfying (3.132). It can be done by solving the linear system (3.133), for instance with the help of Gauss elimination procedure.
2) Find all "roots" $p(x)$ of $\hat{Q}(y)$, that is, all divisors $(y - p(x))|Q(x, y)$.
3) Output only T-consistent roots $p(x)$, that is, those that satisfy $d(y, p(x)) \leq T$.

The best choice $l = [\sqrt{2\hat{k}n}]$ gives the list decoding algorithm with decoding radius $T = n - 1 - [\sqrt{2\hat{k}n}]$.

The main difference between the GS and the original Sudan algorithm is that in the GS algorithm it is not sufficient to say $Q(x_i, y_i) = 0$. It is required that every point (x_i, y_i) is a singularity of Q. Informally, a singularity is a point where the curve given by $Q(x_i, y_i) = 0$ intersects itself. Then in the first phase of GS algorithm the additional constraints will force us to increase the allowed degree of Q. However, we gain in the second phase. In this phase we look for roots of Q and now we know that p passes through many singularities of Q, rather than just points on Q. In such a case we need only half as many singularities as regular points, and this is where the advantage comes from [9].

The singularities of a bivariate polynomial over a finite field can be defined as follows.

Definition 3.4 A polynomial $Q(x, y)$ has a singularity of order r at point (α, β) if the "shifted" polynomial $Q(x + \alpha, y + \beta)$ has no monomials of ordinary total degree ((1; 1) weighted degree) less than r.

To calculate the order of $Q(x, y)$ at point (α, β), we need to be able to express $Q(x + \alpha, y + \beta)$ as a polynomial in x and y. The following propositions, due to H. Hasse [36], tell us one way to do this [37].

Proposition 3.1. If $Q(x) = \sum_i q_i x^i \in F[x]$, then for any $\alpha \in F$, we have:

$$Q(x + \alpha) = \sum_j Q_j(\alpha) x^i \tag{3.134}$$

where

$$Q_j(x) = \sum_i \binom{i}{j} q_i x^{i-j} \tag{3.135}$$

The function $Q_j(x)$ in the left part of equation (3.135) is called the jth *Hasse derivative* of $Q(x)$.

Note that (3.134) is Taylor's series (without reminder) when field F has characteristic 0, since in that case:

$$Q_j(x) = \frac{1}{j!} \cdot \frac{d^j}{dx^j} Q(x)$$

Proposition 3.2. Let $Q(x, y) = \sum_{i,j} q_{i,j} x^i y^i \in F[x, y]$. For any (α, β), we have:

$$Q(x + \alpha, y + \beta) = \sum_{u,v} Q_{u,v}(\alpha, \beta) x^u, y^v \tag{3.136}$$

where

$$Q_{u,v}(x, y) = \sum_{i,j} \binom{i}{u} \binom{j}{v} q_{i,j} x^{i-u} y^{i-v} \tag{3.137}$$

The function $Q_{u,v}(x, y)$ in the left part of (3.137) is called the (u, v)th *Hasse (mixed partial) derivative* of $Q(x, y)$.
Proof. Using the binomial theorem, we express $Q(x+\alpha, y+\beta)$ as a polynomial in x and y:

$$Q(x + \alpha, y + \beta) = \sum_{i,j} q_{i,j} (x + \alpha)^i (y + \beta)^j = \sum_{i,j} q_{i,j} \left(\sum_u \binom{i}{u} x^u \alpha^{i-u} \right) \left(\sum_v \binom{j}{v} y^v \beta^{j-v} \right)$$

$$= \sum_{u,v} x^u y^v \left(\sum_{i,j} \binom{i}{u} \binom{j}{v} q_{i,j} \alpha^{i-u} \beta^{j-v} \right) = \sum_{u,v} Q_{u,v}(\alpha, \beta) x^u y^v$$

□

Corollary. A polynomial $Q(x, y)$ has a root of multiplicity (order) r at a point (α, β) iff all its Hasse derivatives at (α, β) of total order less than r are equal to zero, that is, $Q_{u,v}(\alpha, \beta) = 0$ for all u and v such that $0 \leq u + v \leq r$.

Now we can give a formal description of the *Guruswami-Sudan algorithm.*

The inputs of the algorithm are the code length n, code dimension k, the interpolation points $\{(x_s, y_s)\}$, $s = 1 \ldots n$, and the required root multiplicity r.

1. Compute a non-zero polynomial $Q(x, y)$ of minimal possible $(1, k - 1)$ weighted degree l such that:

$$Q_{j1,j2}(x_i, y_i) = \sum_{j'_1 \geq j_1} \sum_{j'_2 \geq j_2} \binom{j'_1}{j_1} \binom{j'_2}{j_2} q_{j'_1, j'_2}, x^{j'_1 - j_1} y^{j'_2 - j_2} = 0, \quad j_1 + j_2 < r, \quad i = 1 \ldots n \tag{3.138}$$

2. Find all polynomials $p(x) \in GF(q)[x]$ of degree $\deg(p(x)) < k$ such that p is a root of Q, that is, $Q(x, p(x)) = 0$ or $y - p(x)$ is a factor of $Q(x, y)$. For each of these polynomials check if $p(x_i) = y_i$ for at least t values of $i = 1 \ldots n$, and if so, include p in output list.

Obviously, the algorithm can be executed in time polynomial in n since the underlying problems are solving the system of linear equations (3.138) and factorization (or, which is even simpler, finding roots) of a polynomial. The last problem is not a simple one since we need to factorize bivariate polynomials.

The next two lemmas are generalizations of Lemma 3.4.

Lemma 3.6 *If (x_i, y_i) is an input point of the algorithm, and $p(x)$ is a polynomial such that $p(x_i) = y_i$, then $(x - x_i)^r$ divides $g(x) = Q(x, p(x))$.*

Proof. Let $p'(x) = p(x + x_i) - y_i$. Since $p'(0) = 0$, $p'(x) = xp''(x)$ for some polynomial $p''(x)$.

Let $Q^{(i)}(x, y)) = Q(x + x_i; y + y_i)$ and:

$$g'(x) = Q^{(i)}(x, p'(x)) \tag{3.139}$$

Compute $g(x) = Q(x, p(x)) = Q^{(i)}(x - x_i, p(x) - y_i)) = Q^{(i)}(x - x_i, p'(x - x_i)) = g'(x - x_i)$. Since $Q^{(i)}(x, y)$ must not have any coefficients of total degree less than r, substitution of $p'(x) = xp''(x)$ into (3.135) leads to a polynomial $g'(x)$ divisible by x^r. Thus, $(x - x_i)r$ divides $g(x)$.

□

Lemma 3.7 *If $p(x)$ is a polynomial of degree less than k such that $y_i = p(x_i)$ for at least t values of x_i and $rt > l = \deg_{\{1,\hat{k}\}} Q(x, y)$, $(\hat{k} = k - 1)$ then $y - p(x)$ divides $Q(x, y)$.*

Proof. Consider the polynomial $g(x) = Q(x, p(x))$. Obviously, $\deg g(x) \leq l = \deg_{\{1,\hat{k}\}} Q(x, y)$. In accordance with Lemma 3.6, for every i such that $y_i = p(x_i)$, $(x - x_i)^r$ divides $g(x)$. Therefore, $\pi(x) = \prod_{i:p(x_i)=y_i} (x - x_i)^r$ divides $g(x)$. It follows from $\deg \pi(x) \geq rt > l = \deg g(x)$ that $g(x) = 0$, which means that $y = p(x)$ is a root of $Q(x, y)$.

□

The only problem remaining is the selection of such parameters (actually, the only parameter is r) of the algorithm that the polynomial $Q(x, y)$ does exist. In fact, the only requirement is that the number of equations in (3.135), which equals to $n \binom{r+1}{2}$, must be greater than the number of variables (that is, coefficients in the polynomial), which is at least $\frac{l^2}{2\hat{k}}$ (see Lemma 3.5). It leads to a simple quadratic inequality. In [9] authors show that:

$$r = 1 + \left\lfloor \frac{(k-1)n + \sqrt{((k-1)n)^2 + 4(t^2 - (k-1)n}}{2(t^2 - (k-1)n)} \right\rfloor$$

guarantees the existence of polynomial $Q(x, y)$ of small enough degree $(rt - 1)$.

Theorem 3.15 The Guruswami-Sudan algorithm returns all $p(x)$ such that $p(x_i) = y_i$ for at least $n - T > \sqrt{(k-1)n}$ values *of* x_i.

Proof. If $t > \sqrt{(k-1)n}$, one can *always* select r to be large enough to ensure the existence of $Q(x, y)$ of small degree. By Lemma 3.7 all polynomials of a degree less than k such that $y_i = p(x_i)$ for at least t values of x_i divide such a polynomial should thus be discovered at the second step of the algorithm.

□

Since r affects the running time of the algorithm one can set r to be less than is required to handle as many errors as possible. This will significantly reduce the computational complexity at the cost of some performance degradation.

Now let us consider some developments of the GS algorithm.

One of the serious improvements of the GS algorithm deals with invention of the efficient algorithm for finding roots of a bivariate polynomial. It is possible to show [30], that the roots of a bivariate polynomial can be expressed as roots of a number of univariate polynomials. The algorithm exploiting this fact is presented below. This algorithm uses as a subroutine a procedure for finding roots of a univariate polynomial. The algorithm takes as input a bivariate polynomial $Q(x, y)$ and positive integer k, and returns as output ϕ the set of all y-roots of $Q(x, y)$ of degree $\leq k$. The pseudocode of the *Roth-Ruckenstein algorithm* of finding roots of a bivariate polynomial can be written as follows

RECONSTRUCT($Q(x, y), k, i, \phi$)

1. Find the largest r such that x^r divides $Q(x, y)$
2. $M(x, y) := Q(x, y)/x^r$;
3. Find all roots m_j of a univariate polynomial $M(0, y)$;
4. for Each root m_j
5. do $\phi[i] := m_j$;
6. if $i = k - 1$

7. then return ϕ;
8. else $\hat{M}(x, y) := M(x, xy + m_j)$;
9. RECONSTRUCT $(\hat{M}(x, y), k, i + 1, \phi)$.

Another natural extension of the GS algorithm is the case of weighted curve fitting. In the GS algorithm one does not need to make all points to be singularities of the same order. Each point (x_i, y_i) may be assigned with an integer weight w_i and (3.138) should be modified as follows:

$$Q_{j_1,j_2}(x_i, y_i) = \sum_{j_1' \geq j_1} \sum_{j_2' \geq j_2} \binom{j_1'}{j_1} \binom{j_2'}{j_2} q_{j_1' j_2'} x^{j_1'-j_1} y^{j_1'-j_2} = 0, \quad j_1 + j_2 < r \times w_i, \quad i = 1 \ldots n \tag{3.140}$$

for some integer r. Again, one can select such parameters of the algorithm (weights w_i and r) that the polynomial $Q(x, y)$ does exist and the second step of the algorithm returns all polynomials $p(x)$: $\deg p(x) < k$. such that:

$$\sum_{i:p(x_i)=y_i} w_i \geq t \tag{3.141}$$

In [9] it is shown that $t > \sqrt{k \sum_{i=1}^{n} w_i^2}$ is sufficient to solve this problem. Note, that x_i values *need NOT* to be distinct ones and there may be as many interpolation points n as one needs provided the number of constraints is less than the number of coefficients in the polynomial $Q(x, y)$.

This algorithm can be used for the soft decision decoding of Reed-Solomon and some other codes. In [32] authors suggested an algorithm, which can be used for soft-decision decoding of Reed-Solomon codes. As input the algorithm accepts a matrix $\prod = [\pi_{ij}] : \pi_{ij} = \Pr\{y_i = \alpha_j\}$, that is, the cells of the matrix define a-posteriori probability distribution of each symbol in the received word. The algorithm presented below computes integer weights m_{ij}, $i, j = 1 \ldots n$ (multiplicity matrix) for the polynomial reconstruction problem as stated above. The main idea of it is to set greater weights m_{ij} for most probable pairs $(x_i; y_j)$.

The algorithm must be provided with code parameters n, k, $n \times n$ matrix Π and a total number of interpolation points to be generated. The algorithm has not been proved to be optimal (in fact, authors in [32] prove that the problem of construction of an optimal multiplicity matrix M is *NP*-hard), however, with carefully selected parameters it provides significant performance gain [38]. The pseudocode of the algorithm of soft decision decoding of RS code can be presented as follows:

SoftRSDecode(n, k, Π, s)

1. $\Pi^* := \Pi$;
2. $M := 0$;
3. **while** $s < 0$
4. **do**
5. Find a position (i, j) of the greatest entry π_{ij}^* in Π^*;
6. $\pi_{ij}^* := \frac{\pi_{ij}}{m_{ij}+1}$;
7. $m_{ij} := m_{ij} + 1$;
8. $s := s - 1$;
9. Solve the system (3.140) with weights defined by M and find a polynomial $Q(x, y)$;
10. Find its roots $p_i(x) : \deg p_i(x) < k$ and corresponding codewords $Y_i = \{y_1, \ldots, y_n\}$.
11. Select as output the most probable codeword $Y = \arg\max_i \prod_{k=1}^{n} \prod_k y_{ik}$.

One more improvement of the GS algorithm is the Nielsen interpolation algorithm [33, 34]. The algorithm presented in [33] exploits the structure of the system of linear equations (3.138). Here we describe an improved version of the algorithm found in [34].

The main idea of the algorithm is to split the set of all possible solutions of the interpolation problem into a number of disjointed classes and iteratively construct an interpolation polynomial of minimal degree for each class. Finally one can select the smallest one as a solution of the interpolation problem.

Let us introduce *lexicographic monomial ordering* as:

$$x^{\alpha} y^{\beta} \leq lex\, x^{a} y^{b} \Leftrightarrow \alpha < a \vee (\alpha = \alpha \wedge \beta \leq b)$$

Since the GS algorithm requires construction of an interpolation polynomial having minimal possible weighted degree, one has to introduce the *weighted-degree monomial ordering*:

$$f \leq w\text{deg}\, g \Leftrightarrow w\text{deg}_{(1,k)}(f) < w\text{deg}_{(1,k)}(g) \vee (w\text{deg}_{(1,k)}(f) - w\text{deg}_{(1,k)}(g) \wedge f \leq w\text{deg}\, g)$$

Let LT $f(x, y)$ denote the leading term of a polynomial $f(x, y)$ with respect to $\leq w$deg monomial ordering. Then:

$$f(x, y) \leq w\text{deg}\, g(x, y) \Leftrightarrow \text{LT}\, f(x, y) \leq w\text{deg}\, \text{LT}g(x, y)$$

In order to obtain a non-zero solution of the system (3.138) it is sufficient to consider polynomials of form $Q(x, y) = \sum_{i=0}^{N} q_i m_i$, where $N = n\binom{r+1}{2}$ is the total number of equations in the system and m_i are all distinct monomials $x^{j1} y^{j2}$ ordered by their weighted degree:

$$m_0 \leq w\text{deg}\, m_1 \leq w\text{deg} \ldots \leq w\text{deg}\, m_i \leq w\text{deg} \ldots$$

It can be easily shown that:

$$w\text{deg}_{(0,1)}\, m_i \leq \rho_r - 1, \quad i \leq N$$

where

$$\binom{\rho_r}{2} \leq \frac{n\binom{r+1}{2}}{k} < \binom{\rho_r + 1}{2}$$

that is, one can consider only polynomials $Q(x, y)$ with degree in y less than ρ_r. Let us split the set of polynomials $Q(x,y)$: $w\text{deg}_{(0,1)} m_i \leq \rho_r - 1$ into a number of disjointed classes $G_j\{Q(x, y)|w\text{deg}_{(0,1)}(\text{LT}Q) = j\}, \quad k = 0 \ldots \rho_r - 1$. Let us sequentially process constraints (3.138) and at each step for each class G_j construct the minimal with respect to wdeg$_{(1;k)}$ polynomial $Q_j(x, y) \in G_j$ satisfying all processed constraints. Allowed operations are:

1. Add to $Q_j(x, y) \in G_j$ another polynomial $\gamma \cdot Q_{j0}(x, y)$ $Q_{j0}(x, y) \leq w\text{deg}\, Q_j(x, y)$, $Q_{j0}(x, y) \in G_{j0}$, $j \neq j_0, \quad \gamma \in GF(q)$. Clearly, this operation does not increase the order of $Q_j(x, y)$ and keeps it in the same class G_j.
2. Multiply $Q_{j0}(x, y) \in G_{j0}$ by $(x - \delta), \quad \delta \in GF(g)$. This operation introduces the minimal possible increase in degree of the polynomial but keeps it in the same class.

The pseudocode of the Nielsen interpolation algorithm is as follows:

IterativeInterpolation $(n, \{(x_i, y_i), i = 1 \ldots n\}, r, \rho_r)$

1. **for** $i := 0$ **to** $\rho_r - 1$
2. **do** $Q_i(x, y)$: y^t;
3. **for** $i := 1$ **to** n
4. **do for** $\beta := 0$ **to** $r - 1$
5.   **do for** $\alpha := 0$ **to** $r - \beta - 1$
6.   **do** Compute $\Delta_j := \text{coeff}(Q_j(x + x_i, y + y_i), x^{\alpha} y^{\beta}), \quad j = 0 \ldots \rho_r - 1$
7.   Find $j_0 = \arg \min_{j:\Delta j \neq 0} Q_j(x, y)$
8.   **for** $j \neq j_0$
9.   **do** $Q_j(x, y) := Q_j(x, y) - \frac{\Delta_j}{\Delta_{j0}} Q_{j0}(x, y)$;
10. $Q_{j0}(x, y) := Q_{j0}(x, y)(x - x_i)$;
11. **return** $\min_i Q_j(x, y)$

The proof of the algorithm can be found in [33].

The complexity of the algorithm can be estimated as follows. At each step (a) evaluation of Hasse derivatives is performed and (b) polynomials $Q_j(x, y)$ are multiplied by scalar values and summed. From (3.137) one can see that for a bivariate polynomial having s terms $O(s)$ operations are required to compute its Hasse derivative at any point. Number of terms in polynomials $Q_s(x, y)$ grows from 1 at algorithm startup to $O\left(n\binom{r+1}{2}\right)$. Thus, the overall complexity of evaluation of Hasse derivatives is $O\left(n\left(\frac{r(r+1)}{2}\right)^2\rho_r\right) = O\left(n^2\left(\frac{n}{k}\right)^{1/2}r^5\right)$.

Similarly, complexity of manipulations with polynomials at each step is $O(\rho_r, s)$ and the overall complexity is $O\left(n^2\left(\frac{n}{k}\right)^{1/2}r^5\right)$.

3.13 LDPC Codes

Low-density parity-check codes (LDPC-codes) were first suggested by R. Gallager [39, 40], and were investigated further in [41, 42, 43, 44]. Traditionally, LDPC-code is defined by its parity-check matrix **H**, which has the sparse property, that is, its rows and columns have a low number of non-zero elements compared to matrix size. More precisely, we define (n, γ, ρ)-code as a linear code of length n, with parity-check matrix containing the columns of weight γ and the rows of weight ρ.

The parity-check matrix **H** contains:

$$r = n\gamma/\rho \tag{3.142}$$

rows, and therefore, the code rate is lower-limited as [39]:

$$R \geq 1 - \gamma/\rho \tag{3.143}$$

The example of (16, 3, 4)-code is shown in Figure 3.18.

1				1				1				1			
	1				1				1				1		
		1				1				1				1	
			1				1				1				1
1							1			1			1		
	1			1							1			1	
		1			1			1							1
			1			1			1			1			
1						1					1				1
	1						1	1						1	
		1		1					1				1		
			1		1					1		1			

Figure 3.18 Example of LDPC matrix

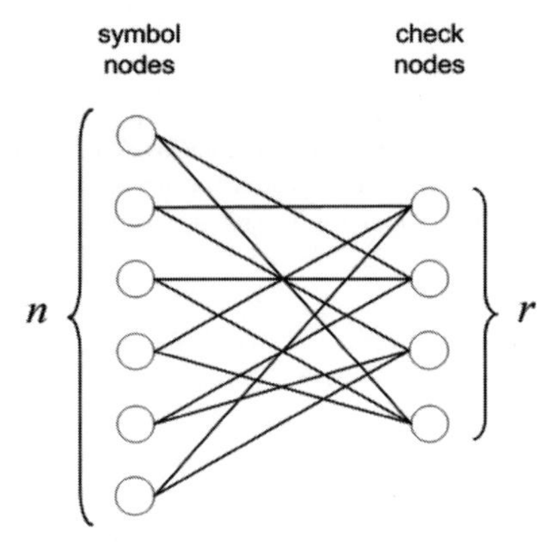

Figure 3.19 LDPC Tanner graph

Besides the traditional definition of a code as a zero-space of its parity-check matrix, LDPC-codes are often defined by means of the incidence graph of **H** matrix (the so-called *Tanner graph* [45]). Such an incidence graph is a bipartite graph with two sets of nodes: n symbol nodes, which correspond to columns, and r check nodes, which corresponds to the rows of the parity-check matrix. The edges of the graph are correspondent to non-zero positions in **H**. An example of such a graph is presented in Figure 3.19.

The quality of LDPC construction is defined by different characteristics: minimum Hamming distance d_0, the minimum length of the cycle in Tanner graph (*girth*) g_0, weights distributions of rows and columns in parity-check matrix.

LDPC-codes with the same number of ones in rows and columns are called regular [39, 41], while the codes with an unequal number of ones are called irregular [46]. Weights distributions can be defined by means of generating functions $\lambda(x)$ and $\rho(x)$ [46]:

$$\begin{aligned} \lambda(x) &= \sum_{i=2}^{d_v} \lambda_i x^{i-1} \\ \rho(x) &= \sum_{i=2}^{d_c} \rho_i x^{i-1} \end{aligned} \tag{3.144}$$

where λ_i is the ratio of parity-check columns with weight i, ρ_i is the ratio of rows in **H** with weight i, and d_v, d_c are the maximum weights of columns and rows, correspondingly. For example in Figure 3.18, $\lambda(x) = x^3$, $\rho(x) = x^4$.

Let us define [46]:

$$\sum_{i \geq 2} \lambda_i / i = \int_0^1 \lambda(x)\, dx$$

Then (3.142) and (3.143) can be written as:

$$r = n \frac{\int_0^1 \rho(x)\, dx}{\int_0^1 \lambda(x)\, dx} \tag{3.145}$$

$$R \geq 1 \frac{\int_0^1 \rho(x)\, dx}{\int_0^1 \lambda(x)\, dx} \tag{3.146}$$

Table 3.5 Properties of regular LDPC codes

	LDPC properties
1.	each row of parity-check matrix **H** contains exactly ρ ones
2.	each column of parity-check matrix **H** contains exactly γ ones
3.	the number of nonzero positions common to any two columns, no more than 1

Usually, constructing the good irregular codes uses probability methods, analysis of such codes is made in asymptotic, while the regular constructions are based on the objects with known properties, and can be analyzed using these properties. Suitably chosen, weights distributions of irregular codes can give gain especially with low SNRs (in AWGN channel), when the quality of the code is defined by its average characteristics. By increasing SNR, the distance properties of the code become determinative in the error probability, and here the gain can be obtained using regular constructions, which allow analyzing their minimum distance, and to construct codes with better spectral properties.

3.13.1 LDPC Constructions

In this section the review and description of some LDPC constructions are presented. These LDPC constructions are those that are well known at the moment and give the lower error probability in the AWGN channel.

The following characteristics can be selected to compare different constructions:

1. minimum distance d_0;
2. girth g_0;
3. flexible selection of parameters: code length n and rate R;
4. error probability in AWGN channel.

The basis of our consideration is the regular constructions. The properties of regular constructions are presented in Table. 3.5. Additional property 3 assures that the girth g_0 is greater than 4.

We shall consider the following LDPC constructions: *Euclidean-geometry codes (EG-LDPC)*, codes based on Reed-Solomon codes (RS-LDPC), generalized Gilbert codes based on Vandermonde matrix (W-LDPC), PEG construction. All these constructions except the latter are regular.

3.13.1.1 LDPC codes based on finite geometries

Error-correcting codes based on finite geometries: projective geometry *PG* or Euclidean geometry *EG*, were described in [5, 47]. However, as low-density codes these codes were also considered in [48, 49], and compare to other known regular LDPC constructions these codes based on finite geometries show good performance in AWGN channel.

The drawback of these constructions is their inflexibility in parameters selection, which is the consequence of finite geometries properties. The decoding of LDPC codes based on finite geometries can be done by the majority-logic decoder [47, 50, 51, 52] with hard or soft decisions. Besides, any common LDPC decoding procedure is suitable for decoding these codes (see also Section 3.13.2).

Here we describe the constructions of Euclidean-geometry low-density codes and give their known basic characteristics and results.

Euclidean-geometry codes are defined as incidence system of geometry $EG(m, q)$, $q = p^s$. Since the number of ones in parity-check matrix of Euclidean-geometry code is small compared to matrix size, this code can be considered as LDPC-code.

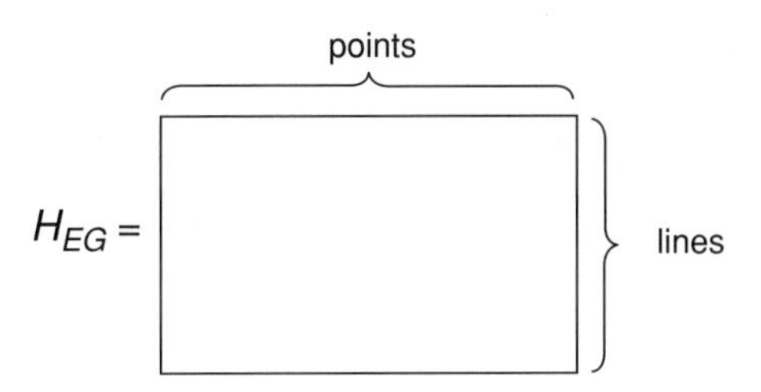

Figure 3.20 Parity-check matrix of *EG*-code

The LDPC-code based on Euclidean geometry with parity-check matrix $\mathbf{H}_{EG}$, is constructed in the following way. The rows of parity-check matrix correspond to lines in Euclidean geometry, while the columns correspond to points in $EG(m, p^s)$. The elements of $\mathbf{H}_{EG}$ are defined from incidence vectors of Euclidean geometry lines (Figure 3.20):

$$\mathbf{H}_{EG}(i, j) = \begin{cases} 1, & \text{if point } j \text{ is on the line } i, \\ 0 & \text{otherwise} \end{cases} \tag{3.147}$$

The parity-check matrix $\mathbf{H}_{EG}$ has:

$$n = q^m \tag{3.148}$$

columns and:

$$r = q^{m-1}(q^m - 1)/(q - 1) \tag{3.149}$$

rows. Each column contains:

$$\gamma = (q^m - 1)/(q - 1) \tag{3.150}$$

ones, and each row contains:

$$\rho = p^s \tag{3.151}$$

ones (all parameters follow from the properties of Euclidean geometry, see (3.A1)-(3.A5)).

Usually, the Euclidean-geometry codes with $p = 2$, not including zero point, are considered [48, 49]. Such codes are sometimes called the EG-codes of type 0, and are cyclic codes [2, 47] with the parameters:

$$n = 2^{ms} - 1$$
$$r = (2^{(ms-1)s} - 1)(2^{(ms)} - 1)/(2^s - 1)$$

The number of information symbols of such codes is estimated in [53].

Besides defining the parity-check matrix as in Figure 3.20, the EG-code with parity-check matrix transposed to (3.147) can be considered. Then the rows are correspondent to geometry points, the columns to geometry lines. In both cases the geometry properties and (3.148)–(3.151) provide the required characteristics of regular LDPC-code, formulated in Table 3.5:

1. each row contains ρ ones;
2. each column contains γ ones;
3. any two columns have no more than one non-zero position in common (since only one line can be passed through two points);
4. any two rows have no more than one non-zero position in common (since two lines intersect at no more than one point).

Additional property 4 means that the code with the parity-check matrix $\mathbf{H}_{EG}^{\mathrm{T}}$ is also a regular LDPC-code satisfying all required parameters.

Since the columns of parity-check matrix (3.147) have no more than one non-zero position in common, any γ columns of parity-check matrix are linearly independent, and hence can not form the zero syndrome. The minimum distance of the code with parity-check matrix (3.147) is estimated as:

$$d_0 \geq \gamma + 1 \tag{3.152}$$

The girth of EG-LDPC codes is:

$$g_0 = 6$$

3.13.1.2 Construction Based on Reed-Solomon Codes

One more scheme based on the incidence system was suggested in [54]. While the combinatorial objects: finite geometries, were used for preceding construction, here the words of Reed-Solomon code are used. The code parameters are selected in such a way to ensure the properties of Table 3.5.

Let us consider the Reed-Solomon codes (RS-codes) over $GF(q)$. RS-codes are the MDS codes, and therefore they have minimum distance $d_{RS} = n - k + 1$, where $n = q - 1$ is code length, k is the number of information symbols. Shortening of the RS-code on $n - \rho$ information symbols gives $(\rho, 2, \rho - 1)$-code C_p of length ρ with two information symbols. All non-zero code words of this code have weight ρ or $\rho - 1$.

Let us choose codeword a of weight ρ and form the sub-space of the code:

$$C_\rho^{(1)} = \{\beta a : \beta \in GF(q)\}$$

The set $C_\rho^{(1)}$ consists of q vectors, and each non-zero vector from $C_\rho^{(1)}$ has weight ρ, and each two vectors differ in all positions.

Let us construct the cosets $C_\rho^{(1)}, \ldots, C_\rho^{(q)}$ of space C_ρ, based on subspace $C_\rho^{(1)}$. Any two vectors in any coset differ in all positions, two vectors from different cosets differ in $\rho - 1$ or ρ positions.

The constructed cosets give the base for the incidence system, from which the parity-check matrix $\mathbf{H}_{RS\text{-}LDPC}$ of LDPC code is constructed. The parity-check matrix consists of γ horizontal stripes-submatrices, $\gamma \in \{1, \ldots, q\}$, and has a form:

$$\mathbf{H}_{RS\text{-}LDPC} = \begin{bmatrix} \mathbf{H}^{(1)} \\ \mathbf{H}^{(2)} \\ \cdots \\ \mathbf{H}^{(r)} \end{bmatrix} \tag{3.153}$$

where $\mathbf{H}^{(t)}$ is defined by coset $C_p^{(t)}$ as follows. Let $a_j^{(t)}(s)$ be the j-th symbol of s-th vector from the set $C_\rho^{(t)}$, $j \in \{0, \ldots, q - 1\}$, $s \in \{1, \ldots, q\}$. Let $c(\alpha)$ be the incidence vector of field element $\alpha \in GF(q)$, that is:

$$c_j(\alpha) = \begin{cases} 1, & \text{if } j = \alpha \\ 0, & \text{otherwise} \end{cases} \tag{3.154}$$

Then $\mathbf{H}^{(t)}$ can be defined as:

$$\mathbf{H}^{(t)} = \begin{bmatrix} c(\alpha_0^{(t)}(1)) & c(\alpha_1^{(t)}(1)) & \cdots & c(\alpha_{q-1}^{(t)}(1)) \\ c(\alpha_0^{(t)}(2)) & c(\alpha_1^{(t)}(2)) & \cdots & c(\alpha_{q-1}^{(t)}(2)) \\ \cdots & \cdots & \cdots & \cdots \\ c(\alpha_0^{(t)}(q)) & c(\alpha_1^{(t)}(q)) & \cdots & c(\alpha_{q-1}^{(t)}(q)) \end{bmatrix} \tag{3.155}$$

or, in other words, it follows from (3.154) and (3.155) that elements of $\mathbf{H}^{(t)}$ are:

$$\mathbf{H}^{(t)}(i,j) = \begin{cases} 1, & \text{if } \alpha^{(t)}_{\lfloor j/q \rfloor}(i) \equiv j \bmod q \\ 0, & \text{otherwise} \end{cases} \quad (3.156)$$

It follows from the properties of shortened RS-code C_p, cosets $C_p^{(t)}$ and construction method (3.153)-(3.156) that if $\alpha^{(t)}_{\lfloor j/q \rfloor} \equiv j \bmod q$ then LDPC-code defined by the parity-check matrix $\mathbf{H}_{RS\text{-}LDPC}$ has the properties formulated in Table 3.5.

In [54] the estimate of minimum distance is given:

$$d_0 \geq \begin{cases} \gamma + 1, & \text{if } \gamma \text{ is odd} \\ \gamma + 2, & \text{if } \gamma \text{ is even} \end{cases} \quad (3.157)$$

which is the same estimate as for finite-geometries constructions relative to the weight γ of columns in parity-check matrix. This estimation is also based on the number of orthogonal parity-checks. In practice the true minimum distance of RS-LDPC code can be much higher than the estimate (3.157).

The parameters of RS-LDPC codes are rather flexible. However, there are no known expressions for the rates of RS-LDPC codes.

3.13.1.3 Gilbert Codes and Their Generalizations

The *Gilbert codes* are low-density parity-check codes (LDPC-codes). They were suggested by Gilbert [55] to correct error bursts. The burst-correcting capability of these codes was considered in [47, 56, 57], where the estimation of maximal correctable burst length was obtained.

A Gilbert code is defined by its parity-check matrix $\mathbf{H}_l$:

$$\mathbf{H}_l = \begin{bmatrix} \mathbf{I}_m & \mathbf{I}_m & \mathbf{I}_m & \dots & \mathbf{I}_m \\ \mathbf{I}_m & \mathbf{C} & \mathbf{C}^2 & \dots & \mathbf{C}^{l-1} \end{bmatrix} \quad (3.158)$$

where $\mathbf{I}_m$ is $(m \times m)$-identity matrix, $\mathbf{C}$ is the $(m \times m)$-matrix of cyclic permutation:

$$\mathbf{C} = \begin{bmatrix} 0 & 0 & 0 & \dots & 0 & 1 \\ 1 & 0 & 0 & \dots & 0 & 0 \\ 0 & 1 & 0 & \dots & 0 & 0 \\ \dots & \dots & \dots & \dots & \dots & \dots \\ 0 & 0 & 0 & \dots & 1 & 0 \end{bmatrix} \quad (3.159)$$

where $l \leq m$ [56]. Clearly, a Gilbert code is $(2, l)$ regular LDPC-code with the $\gamma = 2$ ones in column and $\rho = l$ ones in row. The code length is $n = ml$, the number of redundant symbols is estimated as $r = 2m - 1$ [56].

The minimum distance d_0 of Gilbert code is connected with girth g_0:

$$d_0 = g_0/2 \quad (3.160)$$

The spectral and distance properties of Gilbert codes are estimated using the following statements.

Theorem 3.16 Let $\mathbf{H}_l$ be the matrix of (3.158), $Z_\ell = \{0, 1, \dots \ell - 1\}$ be the set of residues modulo $\ell - 1$. Then, if the sets of integers $\{a_i\}$, $\{b_i\}$ such that equation:

$$\sum_{i=0}^{\omega-1} (-1)^i (a_i - b_i) = 0 \bmod m$$

holds, where:

$$\begin{array}{ll} a_i \in Z_\ell, & b_i \in Z_\ell, \\ a_0 \neq b_0, & a_{\omega-1} \neq b_{\omega-1}, \\ a_i \neq a_{i-1}, & b_i \neq b_{i-1} \end{array}$$

then the code with parity-check matrix $\mathbf{H}_l$ contains the codeword of weight 2ω.

Theorem 3.17 The minimum distance and girth of Gilbert code with $\ell \geq 3$ are:

$$\begin{array}{l} d_0 = 4, \\ g_0 = 8 \end{array}$$

As follows from Theorem 3.17, the Gilbert code has very low minimum distance, and hence, cannot be used to correct the independent errors. However, the generalizations of Gilbert codes can be defined as follows. Consider the parity-check matrix:

$$\mathbf{H}_{s,l} = \begin{bmatrix} \mathbf{I}_m & \mathbf{I}_m & \mathbf{I}_m & \dots & \mathbf{I}_m \\ \mathbf{C}^0 & \mathbf{C}^1 & \mathbf{C}^2 & \dots & \mathbf{C}^{l-1} \\ \mathbf{C}^{i_0^{(3)}} & \mathbf{C}^{i_1^{(3)}} & \mathbf{C}^{i_2^{(3)}} & \dots & \mathbf{C}^{i_{l-1}^{(3)}} \\ \dots & \dots & \dots & \dots & \dots \\ \mathbf{C}^{i_0^{(s)}} & \mathbf{C}^{i_1^{(s)}} & \mathbf{C}^{i_2^{(s)}} & \dots & \mathbf{C}^{i_{l-1}^{(s)}} \end{bmatrix} \tag{3.161}$$

where $\mathbf{H}_{s,l}$ is $s \times l$-matrix, $i_j^{(k)} \in \{0, \dots, m\}$ is the degree of the cyclic permutation matrix $\mathbf{C}$ in j-th block of k-th stripe. Since one of parameters of LDPC-code is the girth, the numbers $i_j^{(k)}$ of any stripe k should not repeat. Then the set $\{i_j^{(k)} : j = 0, \dots, \ell - 1\}$ is defined by the permutation of different residues modulo m.

Construction (3.161) can be the basis for defining regular LDPC codes. Notice that not only the cyclic permutation matrix $\mathbf{C}$ can be used as a block, but any generator of the cyclic group of order not less than l as well.

As an example of such construction consider $\mathbf{H}_{s,l}$, where the degrees of cyclic permutation matrix can be selected corresponding to Vandermonde matrix [2, 47].

In this case we get (γ, ρ) LDPC-code with the parity-check matrix:

$$\mathbf{H}_V = \begin{bmatrix} \mathbf{I}_m & \mathbf{I}_m & \dots & \mathbf{I}_m \\ \mathbf{I}_m & \mathbf{C} & \dots & \mathbf{C}^{\rho-1} \\ \dots & \dots & \dots & \dots \\ \mathbf{I}_m & \mathbf{C}^{\gamma-1} & \dots & \mathbf{C}^{(\gamma-1)(\rho-1)} \end{bmatrix} \tag{3.162}$$

With m prime there is no column having more than one non-zero position in common in this kind of matrix, and the minimum distance of this code can be estimated as:

$$\gamma + 1 \leq d_0 \leq 2m \tag{3.163}$$

The upper bound of minimum distance does not depend on parameters γ and ρ. If we construct the code with such minimum distance, this means that in the given (γ, ρ)-ensemble we provide minimum distance increasing linearly from code length, increasing only m. This task is perspective.

3.13.1.4 PEG Construction

In [58] the empiric procedure is suggested, that constructs the Tanner graph maximizing the girth g_0.

for *j*=0 to $n-1$ **do**
begin
  for *k*=0 to $d_{s_j}-1$ **do**
  begin
    if *k*=0
      $E^0_{s_j} \leftarrow \text{edge}(c_i{}^c s_j)$, where $E^0_{s_j}$ is the first edge incident to S_j and c_i is one check node such that it has the lowest check degree under the current graph setting $E_{S0} \cup E_{s1} \cup \ldots \cup E_{sj-1}$.
    else
      expanding a tree from symbol S_j up to depth ℓ under the current graph setting such that $\overline{N^{\ell}_{S_j}} \neq \emptyset$ but $\overline{N^{\ell \Downarrow 1}_{S_j}} \neq \emptyset$, or the cardinality of $\overline{N^{\ell}_{S_j}}$ stops increasing but is less than m, then $E^k_{s_j} \leftarrow$ edge(c_i, S_j), where $E^k_{S_j}$ is the k-th edge incident to S_j and c_i is one check node picked from the set $\overline{N^{\ell}_{S_j}}$ having the lowest check-node degree.
  end
end

Figure 3.21 PEG construction

PEG construction based on pre-calculated weight distributions of symbol and check nodes in Tanner graph, for example, using "density evolution" procedure. However, the algorithm can use any other distribution as well, including regular distribution.

The algorithm of the constructing graph with given $\lambda(x)$ and $\rho(x)$ bases on iterative edge-by-edge steps, maximizing local girth for given nodes. The result could be in regular or irregular form, depending on weights distribution. In [58] the lower bounds on minimum distance and girth for PEG codes are obtained.

In Figure 3.21 the procedure of constructing the Tanner graph is presented [58]. Let Tanner graph consist of n symbol nodes v_i, $1 \leq i \leq n$, and r check nodes c_j, $1 \leq j \leq r$. Let d_{v_i}, d_{c_j} be the degree of symbol node v_i and check node c_j, correspondingly, where node degree means the number of edges incident to it (this value is defined by weights distributions $\lambda(x)$ and $\rho(x)$), E_{v_i} is the set of edges incident to symbol node v_i, $N^{\ell}_{v_i}$ is the set of check nodes that can be reached from symbol node v_i by ℓ edges or less, $\overline{N^{\ell}_{v_i}}$ is the complementary set of $N^{\ell}_{v_i}$, in other words, $N^{\ell}_{v_i} \cup \overline{N^{\ell}_{v_i}} = V_c$, where V_c is the set of all check nodes in the graph.

In [58] the estimations of girth g_0 and minimum distance d_0 of PEG codes are presented. Let d_v and d_c be the maximal weights of symbol and check nodes in the Tanner graph, respectively. Then the girth is lower bounded as:

$$g_0 \geq 2([t] + 2$$

where

$$t = \frac{\log(rd_c - \frac{rd_c}{d_v} - r + 1)}{\log((d_v - 1)(d_c - 1))} - 1$$

Let us consider Tanner graphs with regular symbol nodes, having the constant degree d_v, and let the graph have the girth g_0. Then the minimum distance of the code defined by such a graph is estimated as:

$$d_0 \geq \begin{cases} 1 + \dfrac{d_v\left((d_v - 1)^{\lfloor (g_0-2)/4 \rfloor} - 1\right)}{d_v - 2} & \text{if } g_0/2 \text{ is odd} \\ 1 + \dfrac{d_v\left((d_v - 1)^{\lfloor (g_0-2)/4 \rfloor} - 1\right)}{d_v - 2} & +(d_v - 1)^{\lfloor (g_0-2)/4 \rfloor} \text{ if } g_0/2 \text{ is even} \end{cases}$$

In PEG construction the main attention is given to the absence of short cycles in the Tanner graph. However, in practice it is not clear how the presence of short cycles can degrade the decoder's performance. In [59] the presence or absence of short cycles is not considered a problem, but how these cycles are connected to each other. The idea of this is that if there are many edges leading from nodes that form a short cycle, then the decoder can work well even in the presence of short cycles. The parameter ACE is suggested, and the procedure of constructing the code, maximizing ACE is considered in [59]. As a result, the irregular codes with improved performance with high SNRs were obtained.

3.13.2 Decoding of LDPC Codes

Decoding algorithms for low-density parity-check (LDPC) codes were first introduced by Gallager in 1963 [39] both for hard and soft decision cases (bit-flip and belief propagation algorithms, respectively). The soft-decision belief propagation iterative algorithm can operate with both probabilities and log-likelihood ratios and it gives good results in AWGN channel as it is shown in [39, 41, 60].

3.13.2.1 Decoding in Discrete Channel (Bit-Flip Decoding)

The idea of decoding in discrete channel (hard-decision decoding) is that for some received symbol c_i any other symbol can be no more than in one parity-check of symbol c_i, because of parity-check matrix sparsity and absence of short cycles in the Tanner graph. In other words, the set of parity-checks is orthogonal on symbol c_i [47, 2, 52]. Then the columns of parity-check matrix have less non-zero positions in common, and hence, unsatisfied parity-check (syndrome position) more probably consists of one erroneous symbol, than of the sum of three or more. This leads to the following decoding procedure.

1. Calculate syndrome from the received word on zero-th iteration, or from the result of the preceding iteration. If the syndrome is zero, or the maximum number of iterations is reached, the procedure is finished.
2. Calculate the number ℓ_i of unsatisfied parity-checks for each symbol c_i.
3. Flip the symbol or symbols with the largest ℓ_i.
4. Go to step 1.

The scheme of decoding algorithm is presented in Figure 3.22. The algorithm is processed iteratively, until the codeword is obtained, or the maximum number of iterations is reached.

The complexity of described decoding procedure is very low, because for every bit the syndrome update is needed, the update complexity consists of few XOR complexities. This decoder also has simple implementation.

3.13.2.2 Decoding in Soft Channel (Belief Propagation Decoding)

For decoding in soft channels (soft-decision decoding) the task is to maximize the conditional probability $P(\mathbf{C}_m/\mathbf{Y})$, where $\mathbf{C}_m$ is the codeword, and $\mathbf{Y}$ is the block of symbols observed on the channel output.

In the case of LDPC codes, decoder gives symbol-by-symbol decisions, and in fact, to make the decision on particular symbol, calculates the likelihood ratio:

$$\mathrm{LR}(c_i) = \frac{P(c_i = 1|\mathbf{Y})}{P(c_i = 0|\mathbf{Y})} \tag{3.164}$$

where c_i is code bit in position i. Instead of the likelihood ratio (3.164) it is more convenient to use the log-likelihood ratio:

$$\mathrm{LLR}(c_i) = \log \frac{P(c_i = 1|\mathbf{Y})}{P(c_i = 0|\mathbf{Y})} \tag{3.165}$$

Values of LR or LLR are often called the symbol reliability.

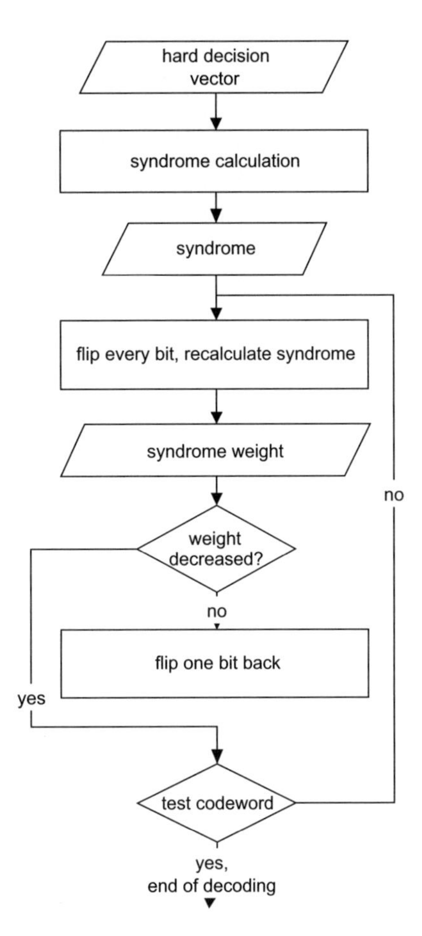

Figure 3.22 Hard decoding of LDPC-code

Algorithm "belief propagation", the standard procedure for decoding LDPC-codes, suggested by Gallager [39, 61, 62], can be described as follows. The LLRs of symbols of received "soft" word is set to correspondent symbol nodes of Tanner graph. Then the decoder processes iterations, each consisting of two stages. During the first, "vertical" stage, each i-th symbol node, $1 \leq i \leq n$, sends to each incident check node j, $1 \leq j \leq \gamma$, some value, called "message", that depends on all values received by i-th symbol node from all incident check nodes besides j-th.

The second, "horizontal" stage, operates similarly, the only difference is that the messages are calculated and sent from check nodes to symbol nodes. The one iteration of decoder is shown in Figure 3.23. Here the message flow between nodes v_1 and c_2 during one iteration is shown. The function f^c denotes message calculation by check node, f^v by symbol node.

After each iteration the algorithm makes a hard decision on every symbol, corresponding to the sign of current message in symbol node. If an obtained hard vector is a code word, or the maximum number of iterations is reached, the algorithm stops. The scheme of algorithm is presented in Figure 3.24.

The maximal number of decoding iterations is selected depending on code length, required error probability and decoding complexity requirements. It is shown in [44] that $\log n$ iterations can be enough for decoding that gives complexity $n \log n$.

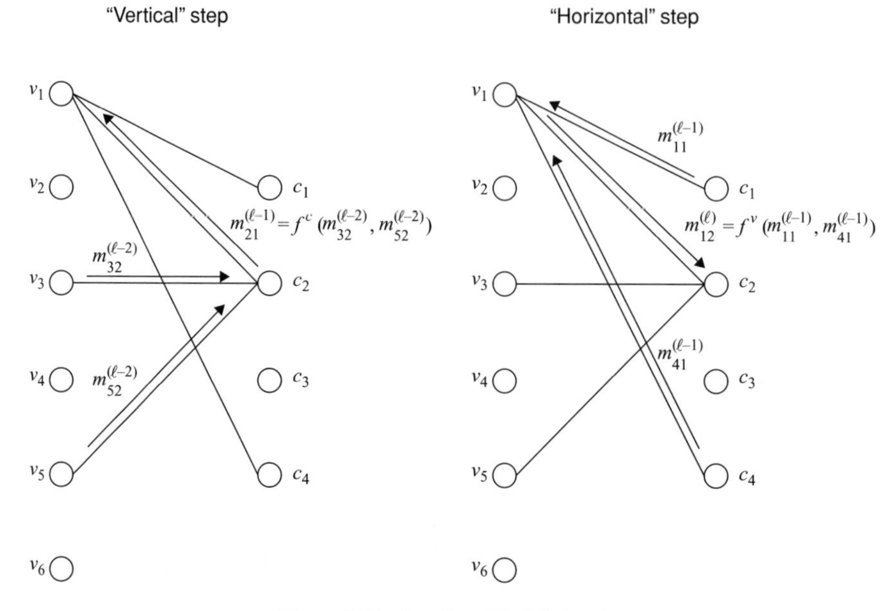

Figure 3.23 Iteration of LDPC-decoder

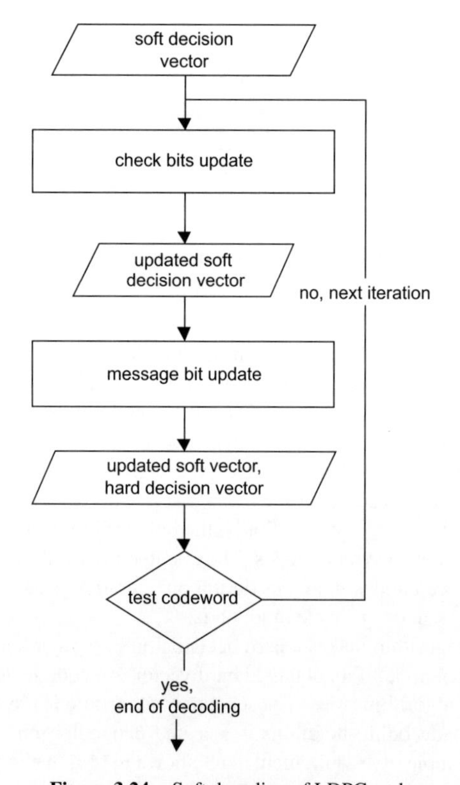

Figure 3.24 Soft decoding of LDPC-code

In practice, the maximal number of iterations performed by decoder is the trade-off between the decoding speed and the error probability requirements. Note that different LDPC constructions may have different convergence speed. The convergence speed of iterative decoder for a given construction is one of the parameters for selecting LDPC-code in a particular communication system.

The complexity of belief propagation decoding is higher than the complexity of hard bit-flip decoding due to sophisticated probabilities update and float number operations. If the belief propagation decoder works in log-likelihood domain it requires LOG() calculation or making a lookup table to avoid exact LOG() calculation.

More detailed information on LDPC codes constructions and decoding algorithms can be found in [63].

References

[1] Shannon, C.E., *A Mathematical Theory of Communication*, Bell System Tech, 1948.

[2] Peterson, W.W., Weldon, E.J., *Error. Correcting Codes*, MIT Press, 1972.

[3] Birkhoff, G., Mac Lane, S., *A Survey of Modern Algebra*, Macmillan, New York, 1941.

[4] Gallager, R.G., *Information Theory and Reliable Communication*. John Wiley, 1968.

[5] MacWilliams, F. J., Sloan, J.J., *The Theory of Error-Correcting Codes*, North-Holland, Amsterdam, 1977.

[6] Evseev, G.S., *Complexity of decoding for linear codes*, Probl. Inform. Transm., vol. 19, no. 1, pp. 3–8 (in Russian) and pp. 1–6 (English translation), 1983.

[7] Tsfasman, M.A., Vladuts, S.G., *Algebraic-Geometry Codes*, Kluwer, Dordrecht, 1991.

[8] Tsfasman, M.A., Vladuts, S.G., Zink, T., *Modular curves, Shimura curves, and Goppa codes better than Varshamov-Gilbert Bound*, Math. nachr., vol 109, pp. 21–28, 1982.

[9] Guruswami, V., Sudan, M., *Improved decoding of Reed–Solomon codes and algebraic geometry codes*, IEEE Trans. Inform. Theory, vol. 45, no. 6, pp. 1757–1767, September 1999.

[10] Bassalygo, L.A, *New upper bounds for error-correcting codes*, Problems of Information Transmission, no. 4, pp. 41–44, 1965.

[11] Shannon, C.E., Gallager, R.G., Berlekamp, E.R., *Lower bounds to error probability for coding on discrete memoryless channels*, Info. and Control, no. 10, pp. 65–103 and 522–552, 1967.

[12] McEliece, R.J., Rodemich, E.R., Rumsey, H.C. Jr., Welch, L.R, *New upper bounds on the rate of a code via Delsarte – MacWilliams inequalities*, IEEE Trans. Inform. Theory, vol. 23, no. 2, pp. 157–166, March 1977.

[13] Gordon, D.M., *Minimal permutation sets for decoding the binary Golay code*, IEEE Trans. Inform. Theory, vol. IT-28, May 1982, pp. 541–543.

[14] Wolfmann, J., *A permutation decoding of the (24, 12, 8) Golay code*, IEEE Trans. Inform. Theory, vol. IT-29, Sept. 1983, pp. 748–751.

[15] Krouk, E.A., Fedorenko, S.V., *Decoding by generalized information sets*, Probl. Inform. Transm., vol. 31, no. 2, pp. 54–61 (in Russian) and pp. 134–139 (English translation), 1995.

[16] Barg, A., *Complexity issues in coding theory*, in Handbook of Coding Theory, vol. 1, V. Pless and W. C. Huffman, Eds. Amsterdam, The Netherlands: Elsevier Science, 1998, pp. 649–754.

[17] Dumer, I., *Two decoding algorithms for linear codes*, Probl. Inform. Transm., vol. 25, no. 1, pp. 24–32 (in Russian) and pp. 17–23 (English translation), 1989.

[18] Dumer, I., *On minimum distance decoding of linear codes*, in Proc. 5th Joint Soviet–Swedish Int. Workshop Information Theory (Moscow, USSR, 1991), pp. 50–52.

[19] Hamming, R.W., *Error Detecting and Error Correcting Codes*, Bell Syst. Tech. J., April 1950, vol. 29, pp. 147–160.

[20] Best, M.R., Brouwer, A.E., *Triply shortened binary Hamming code is optimal*. Discr. Math., vol. 17, pp. 235–245, 1977.

[21] Kabatyanskii, G.A., Panchenko, V.I., *Unit-sphere packings and coverings of the Hamming space*, Problems of Information Transmissions, v. 24, n. 4, pp. 3–16, 1988.

[22] Hamalainen, H., *Two new binary codes with minimum distance three*, IEEE on Information Transmission, vol 34, 1988, p. 885.

[23] Varshamov, R.R., Tenengolts, G.M., *Codes which correct single asymmetric error*, Automation and Remote Control (in Russian), v. 26, n. 2, pp. 286–290, 1965.

[24] Panchenko, V.I., *Packings and coverings over an arbitrary alphabet*, Problems of Information Transmissions, v. 24, n. 4, pp. 93–96, 1988.

[25] Reed, I.S., Solomon, G., *Polynomial codes over certain finite fields*, J. SIAM, 8(1960), pp. 300–304.

[26] Gevorkian, D.N., Avetisian, A.M., Tigranyan, V.A, *On the construction of codes correcting two errors in Hamming metric over Galois fields*, in Vychislitelnia Technika, Kuibishev, no.3, 1975, pp. 19–21 (in Russian).

[27] Dumer, I.I., Zinoviev, V.A., *New maximal code over the Galois field GF(4)*, Problems of Information Transmission, v. 14, n. 3, 1978, pp. 24–34.

[28] Dumer, I.I., *Nonbinary double-error-correcting codes designed by means of algebraic varieties*, IEEE Trans. on of Information Theory, v. 41, n. 6, 1995, pp. 1550–1560.

[29] Sudan, M., *Decoding of Reed–Solomon Codes beyond the Error-Correction Bound*, J. Complexity, v. 13, 1997, pp. 180–193.

[30] Roth, R., Ruckenstein, G., *Efficient decoding of Reed-Solomon codes beyond half the minimum distance*. IEEE Transactions on Information Theory, 46(1), 2000, pp. 246–257.

[31] Kötter, R., *Fast Generalized Minimum-Distance Decoding of Algebraic-Geometry and Reed–Solomon Codes*, IEEE Trans. Inform. Theory, v. 42, n. 3, May 1996, pp. 721–736.

[32] Kötter, R., Vardy, A., *Algebraic soft-decision decoding of Reed-Solomon codes* in Proceedings of 38th Annual Allerton Conference on Communication, Control and Computing, 2000.

[33] Nielsen, R.R., Hoholdt, T., *Decoding Reed-Solomon codes beyond half the minimum distance*. In Proceedings of the International Conference on Coding Theory and Cryptography, Mexico 1998. Springer-Verlag, 1998.

[34] Nielsen, R.R., *List decoding of linear block codes*. PhD thesis, Technical University of Denmark, 2001.

[35] Elias, P., *List decoding for noisy cannels*, Tech. Report 335, Research Lab. of Electronics, MIT, 1957.

[36] Hasse, H., *Theorie der höheren Differentiale in einem algebraishen Funcktionenk örper mit vollkommenem Konstantenkörper nei beliebeger Charakteristic*, J. Reine. Ang. Math., vol. 175, 1936, pp. 50–54.

[37] McEliece, R.J., *The Guruswami-Sudan Decoding Algorithm for Reed-Solomon Codes*, in IPN Progress Report, May 2003.

[38] Gross, W.J., Kschischang, F.R., Koetter, R., Gulak, P., *Simulation results for algebraic soft-decision decoding of Reed-Solomon codes*. In Proceedings of the 21st Biennial Symposium on Communications, June 2002, pp. 356–360.

[39] Gallager, R.G., *Low Density Parity Check Codes*, Cambridge, MA: MIT Press, 1963.

[40] Gallager, R.G., *Low Density Parity Check Codes*. IRE Transactions on Information Theory, Jan. 1962.

[41] MacKay, D., *Good error correcting codes based on very sparse matrices*. IEEE Transactions on Information Theory, 45, Mar. 1999.

[42] MacKay, D., Neal, R., *Near Shannon Limit Performance of Low-Density Parity-Check Codes*. IEEE Transactions on Information Theory, 47(2), Feb. 2001.

[43] Richardson, T.J., Urbanke, R.L., *The capacity of low-density parity-check codes under message-passing decoding*. IEEE Transactions on Information Theory, 47(2), Feb. 2001.

[44] Zyablov, V.V., Pinsker, M.S., *Estimation of the error-correction complexity for Gallager low-density codes*. Problemy Peredachi Informatsii, 11(1):18–28, January-March 1975.

[45] Tanner, M., *A recursive approach to low complexity codes*. IEEE Transactions on Information Theory, IT(27):533–547, Sept. 1981.

[46] Richardson, T.J., Urbanke, R.L., Shokrollahi, M.A., *Design of capacity-approaching irregular low-density parity-check codes*. IEEE Transactions on Information Theory, 47(2), Feb. 2001.

[47] Blahut, R., *Theory and Practice of Error Control Codes*. Addison-Wesley, 1984.

[48] Kou, Y., Lin, S., Fossorier, M.P.C., *Construction of low-density parity-check codes: A geometric approach*. In Proc. 2nd Int. Symp. Turbo Codes and Related Topics, pp. 137–140, Brest, France, Sept. 2000.

[49] Kou, Y., Lin, S., Fossorier, M.P.C., *Low-density parity-check codes based on finite geometries: a rediscovery and new results*, IEEE Trans Inf. Theor., v. 47, n. 7, Nov 2001.

[50] Chen, C.L. *On majority-logic decoding of finite geometry codes*. IEEE Transactions on Information Theory, IT-17(3), May 1971, pp. 332–336.

[51] Kasami, T., Lin, S., *On majority-logic decoding for duals of primitive polynomial codes*. IEEE Transactions on Information Theory, IT-17(3), May 1971, pp. 322–331.

[52] V. Kolesnik and E. Mironchikov. *Decoding of cyclic codes* (In Russian). M.: Svyaz, 1968.

[53] Lin, S., *On the number of information symbols in polynomial codes*. IEEE Transactions on Information Theory, 18(6), Nov. 1972, 785–794.

[54] Djurdjevic, I., Xu, J., Abdel-Ghaffar, K., Lin, S., *A class of low-density parity-check codes constructed based on Reed-Solomon codes with two information symbols*, IEEE Communication Letters, vol. 7, NO. 7, Jul 2003.
[55] Gilbert, E.N., *A problem in binary encoding*. In Proceedings of the Symposium in Applied Mathematics, v. 10, 1960, pp. 291–297.
[56] Krouk, E.A., Semenov, S.V., *Low-density parity-check burst error-correcting codes*. In 2 International Workshop "Algebraic and combinatorial coding theory", Leningrad, 1990, pp. 121–124.
[57] Zhang, W., Wolf, J., *A class of binary burst error-correcting quasi-cyclic codes*. IEEE Transactions on Information Theory, IT-34, May 1988, pp. 463–479.
[58] Hu, X.Y., Eleftheriou, E., Arnold, D-M., *Regular and Irregular Progressive Edge-Growth Tanner Graphs*. IBM Research, Zurich Research Laboratory, 2003.
[59] Tian, T., Jones, C., Villasenor, J., Wesel, R., *Construction of Irregular LDPC Codes with Low Error Floors*. In Proceedings of ICC'2003, Anchorage, Alaska 11–15 May 2003, pp. 3125–3129.
[60] Fossorier, M.P.C., Mihaljevic, M., Imai, H., *Reduced complexity iterative decoding of low-density parity-check codes based on belief propagation*. IEEE Transactions on Communications, 47(5), May 1999.
[61] Lechner, G., *Convergence of Sum-Product Algorithm for Finite Length Low-Density Parity-Check Codes*. Winter School on Coding and Information Theory, Monte Verita, Switzerland, Feb. 24–27, 2003.
[62] Lucas, R., Fossorier, M.P.C., Kou, Y., Lin, S., *Iterative decoding of one-step majority logic decodable codes based on belief propagation*. IEEE Trans. Commun., 48(6), June 2000, pp. 931–937.
[63] Kabatiansky, G., Krouk, E., Semenov, S., *Error Correcting Coding and Security for Data Networks*, J. Wiley, 2005.

4

Convolutional Codes and Turbo-Codes

Sergei Semenov[1], and Andrey Trofimov[2]
[1]*Nokia Corporation, Finland*
[2]*St. Petersburg State University of Aerospace Instrumentation, Russia*

4.1 Convolutional Codes Representation and Encoding

The general structure of the convolutional encoder is represented in Figure 4.1. Every time moment the block of k symbols is fed to the input of the encoder. The encoder has the memory, which keeps the values of not more than $\nu - 1$ previous input blocks ($k \cdot (\nu - 1)$ symbols). The encoder forms n output symbols at a time. The block of n output symbols depends on $k \cdot \nu$ input symbols and each output symbol is the linear combination of information symbols. The number k is called the *number of information symbols*, the number n is called the *number of encoded symbols* and the number ν is known as the *constraint length* of the code. The value k/n is called the *code rate*. The corresponding convolutional code usually is denoted as $(n,\ k,\ \nu)$ code. Note that in some literature the used notation is $(n,\ k,\ \nu - 1)$.

The encoder works as follows: The input data block of k symbols is fed to k shift registers (each symbol to its own register) with the help of demultiplexer. Each shift register consists of no more than $\nu - 1$ delay elements. The outputs of the delay elements are multiplied by some fixed coefficients $g_{ij}^{(l)}$ $(i = 0, \ldots, n - 1;\ j = 0, \ldots, k - 1;\ l = 0, \ldots, \nu - 1)$. These weighted outputs are distributed by n sets and each set generates the output symbol just by adding the members of the set. All multiplications and additions are executed over field K. Thus, the output block of n encoded symbols corresponds to input data block of k symbols. After encoding, the n output symbols are multiplexed into a single sequence. Note that each encoded symbol is obtained with the help of the linear combination of the outputs of the delay elements.

It is obvious that the output symbol depends on the input symbol and on the contents of the shift registers. In this sense a convolutional code is close to a block code. However, in block coding a code block of length n depends on k current data symbols; so code blocks formed independently of each other. In convolutional coding code block of length n depends on current data block of length k and on ν previous data blocks. Thus, any output coded n-tuple depends on previous ones. Unlike a block code

Modulation and Coding Techniques in Wireless Communications Edited by Evgenii Krouk and Sergei Semenov

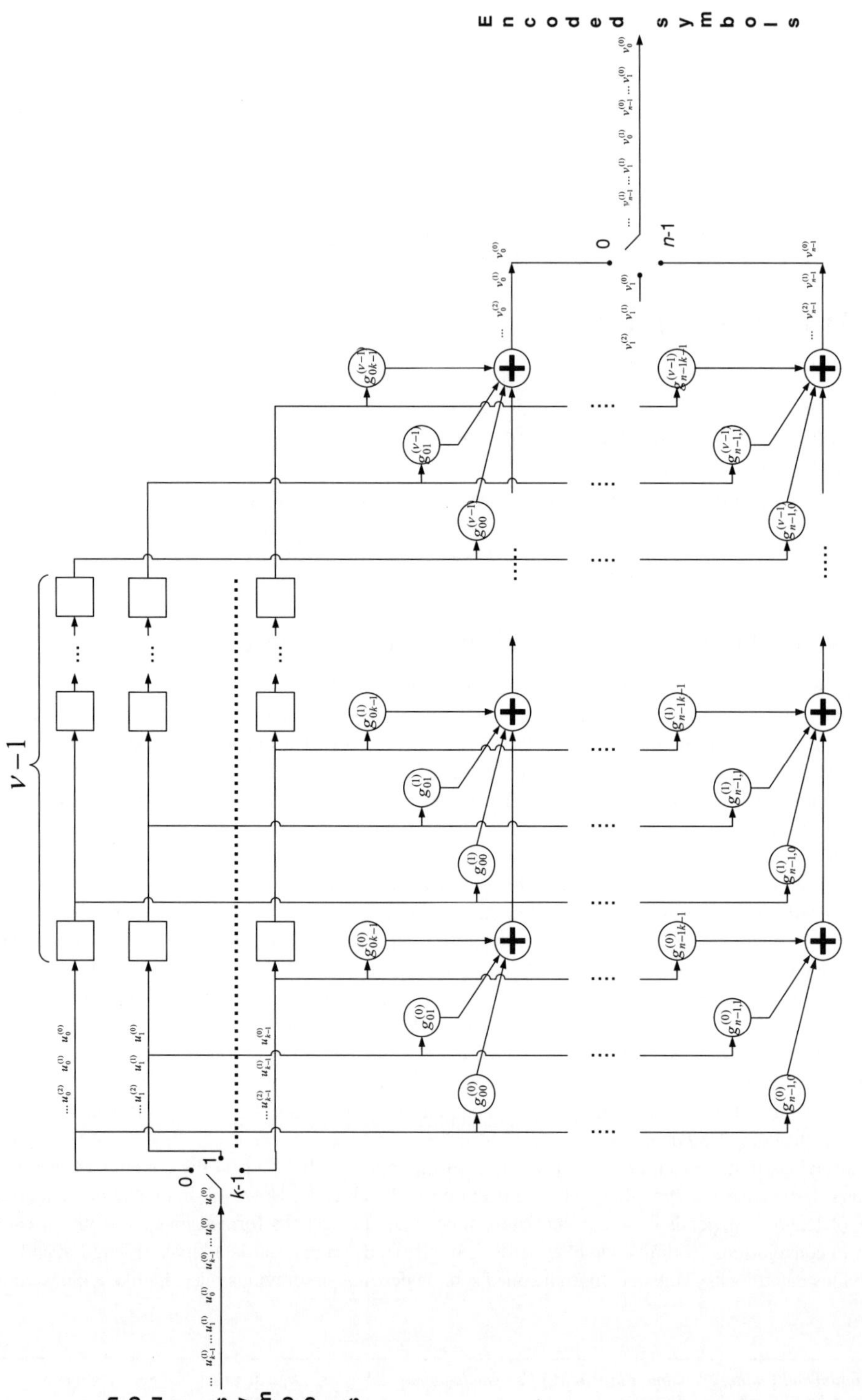

Figure 4.1 A general convolutional encoder

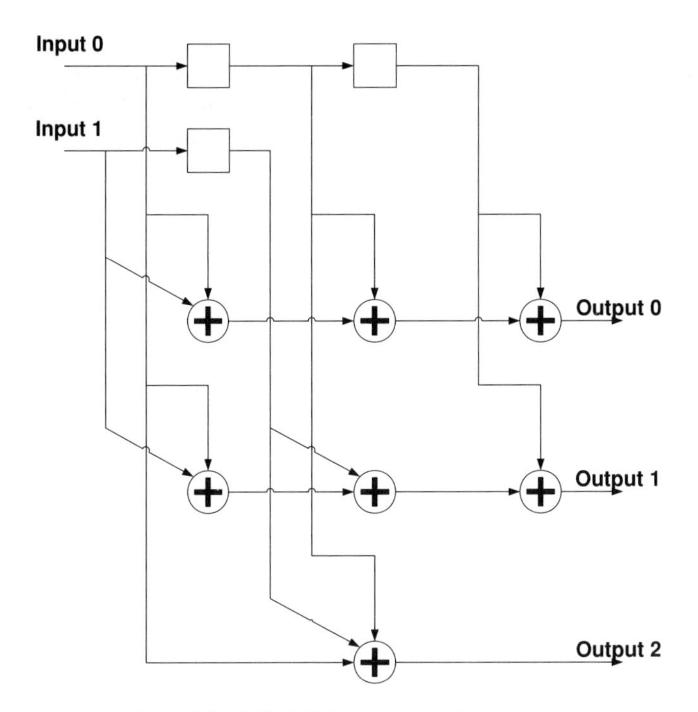

Figure 4.2 A (3, 2, 3) binary convolutional encoder

that has a fixed word length, a convolutional code has no particular word length. We can say that a word of a convolutional code is a semi-infinite sequence.

We can describe the encoder of the convolutional code with the help of its *impulse response*, that is the output of the encoder obtained after feeding the '1' to the encoder input and assuming that the encoder starts from the zero state. It is easy to verify that if the '1' is fed to the jth input then the output sequence (or impulse response) from the ith output will be represented by the sequence of coefficients $\mathbf{g}_{ij} = g_{ij}^{(l)}$, $(l = 0, \ldots, \nu - 1)$.

Example 4.1 Consider a binary (3, 2, 3) convolutional code. The encoder of this code is shown in Figure 4.2. Note that the constraint length of the code is 3. That means the maximum length of the impulse response also is equal to 3. The impulse responses of this encoder can be written as follows:

$\mathbf{g}_{00} = (1, 1, 1)$; Output sequence from the output 0, assuming that input sequence at the input 0 is $(1\,0\,0\ldots)$.

$\mathbf{g}_{01} = (1, 0, 0)$; Output sequence from the output 0, assuming that input sequence at the input 1 is $(1\,0\,0\ldots)$.

$\mathbf{g}_{10} = (1, 0, 1)$; Output sequence from the output 1, assuming that input sequence at the input 0 is $(1\,0\,0\ldots)$.

$\mathbf{g}_{11} = (1, 1, 0)$; Output sequence from the output 1, assuming that input sequence at the input 1 is $(1\,0\,0\ldots)$.

$\mathbf{g}_{20} = (1, 1, 0)$; Output sequence from the output 2, assuming that input sequence at the input 0 is $(1\,0\,0\ldots)$.

$\mathbf{g}_{21} = (0, 1, 0)$; Output sequence from the output 2, assuming that input sequence at the input 1 is $(1\,0\,0\ldots)$.

Let us denote the input symbols, which after the demultiplexer are fed to the jth input of the encoder as u_j, and the encoded symbols at the ith output we denote as v_i. Then the input sequence can be written as $\mathbf{u} = (u_0^{(0)}, u_1^{(0)}, \ldots, u_{k-1}^{(0)}, u_0^{(1)}, u_1^{(1)}, \ldots, u_{k-1}^{(1)}, u_0^{(2)}, u_1^{(2)}, \ldots, u_{k-1}^{(2)}, \ldots)$ and the corresponding output sequence after multiplexing as $\mathbf{v} = (v_0^{(0)}, v_1^{(0)}, \ldots, v_{n-1}^{(0)}, v_0^{(1)}, v_1^{(1)}, \ldots, v_{n-1}^{(1)}, v_0^{(2)}, v_1^{(2)}, \ldots, v_{n-1}^{(2)}, \ldots)$. Now let us consider k input sequences (after the demultiplexer):

$$\mathbf{u}_0 = (u_0^{(0)}, u_0^{(1)}, u_0^{(2)}, \ldots);$$

$$\mathbf{u}_1 = (u_1^{(0)}, u_1^{(1)}, u_1^{(2)}, \ldots);$$

$$\ldots\ldots\ldots\ldots\ldots\ldots$$

$$\mathbf{u}_{k-1} = (u_{k-1}^{(0)}, u_{k-1}^{(1)}, u_{k-1}^{(2)}, \ldots)$$

and n output sequences (before the multiplexer):

$$\mathbf{v}_0 = (v_0^{(0)}, v_0^{(1)}, v_0^{(2)}, \ldots);$$

$$\mathbf{v}_1 = (v_1^{(0)}, v_1^{(1)}, v_1^{(2)}, \ldots)$$

$$\ldots\ldots\ldots\ldots\ldots\ldots$$

$$\mathbf{v}_{n-1} = (v_{n-1}^{(0)}, v_{n-1}^{(1)}, v_{n-1}^{(2)}, \ldots)$$

Then it is easy to verify that the output sequences can be written as the sum of the convolutions of the corresponding input sequence and the impulse response:

$$\begin{aligned} \mathbf{v}_0 &= \sum_{j=0}^{k-1} \mathbf{u}_j * \mathbf{g}_{0j}; \\ \mathbf{v}_1 &= \sum_{j=0}^{k-1} \mathbf{u}_j * \mathbf{g}_{1j}; \\ &\ldots\ldots\ldots\ldots \\ \mathbf{v}_{n-1} &= \sum_{j=0}^{k-1} \mathbf{u}_j * \mathbf{g}_{n-1,j} \end{aligned} \tag{4.1}$$

where $*$ denotes discrete convolution. This feature explains the name convolutional for the codes.

The (4.1) can be rewritten as:

$$v_i^{(f)} = \sum_{j=0}^{k-1}\sum_{l=0}^{\nu-1} u_j^{(f-l)} \cdot g_{ij}^{(l)}, i = 0, \ldots, n-1 \tag{4.2}$$

where $u_j^{(f-l)} = 0$ if $f < l$.

These equations can be written also in the form of a matrix multiplication:

$$\mathbf{v} = \mathbf{u} \cdot \mathbf{G} \tag{4.3}$$

where $\mathbf{v}$ and $\mathbf{u}$ are the semi-infinite sequences and $\mathbf{G}$ is a semi-infinite matrix:

$$\mathbf{G} = \begin{bmatrix} \mathbf{G}^{(0)} & \mathbf{G}^{(1)} & \ldots & \mathbf{G}^{(\nu-1)} & & \ldots \\ & \mathbf{G}^{(0)} & \ldots & \mathbf{G}^{(\nu-2)} & \mathbf{G}^{(\nu-1)} & \ldots \\ & & \ldots & & & \ldots \end{bmatrix} \tag{4.4}$$

and submatrix $\mathbf{G}^{(l)}$ is a $k \times n$ matrix, which can be written as follows:

$$\mathbf{G}^{(l)} = \begin{bmatrix} g_{00}^{(l)} & g_{10}^{(l)} & \cdots & g_{n-1,0}^{(l)} \\ g_{01}^{(l)} & g_{11}^{(l)} & \cdots & g_{n-1,1}^{(l)} \\ \cdots & \cdots & \cdots & \cdots \\ g_{0,k-1}^{(l)} & g_{1,k-1}^{(l)} & \cdots & g_{n-1,k-1}^{(l)} \end{bmatrix}, l = 0, \ldots, \nu-1 \tag{4.5}$$

Matrix $\mathbf{G}$ is called the *generator matrix* of the convolutional code. It is easy to see that each output (encoded) sequence $\mathbf{v}$ can be obtained as the linear combination of rows of the matrix $\mathbf{G}$. Hence the convolutional code defined by the generator matrix $\mathbf{G}$ is the *linear* code.

Example 4.2 Consider the same binary (3, 2, 3) convolutional code as in previous example. The generator matrix $\mathbf{G}$ of this code can be written as:

$$\mathbf{G} = \begin{bmatrix} 1 & 1 & 1 & 1 & 0 & 1 & 1 & 1 & 0 & & & & & & \\ 1 & 1 & 0 & 0 & 1 & 1 & 0 & 0 & 0 & & & & & & \\ & & 1 & 1 & 1 & 1 & 0 & 1 & 1 & 1 & 0 & & & & \\ & & 1 & 1 & 0 & 0 & 1 & 1 & 0 & 0 & 0 & & & & \\ & & & & 1 & 1 & 1 & 1 & 0 & 1 & & & & & \\ & & & & 1 & 1 & 0 & 0 & 1 & 1 & & & & & \end{bmatrix}$$

Let the input sequence $\mathbf{u} = (u_0^{(0)}, u_1^{(0)}, u_0^{(1)}, u_1^{(1)}, u_0^{(2)}, u_1^{(2)}, \ldots)$ be $\mathbf{u} = (1,0, 0,1, 1,1, \ldots)$. Then it can be easily verified that in accordance with (4.1)–(4.5) the output sequence $\mathbf{v} = (v_0^{(0)}, v_1^{(0)}, v_2^{(0)}, v_0^{(1)}, v_1^{(1)}, v_2^{(1)}, v_0^{(2)}, v_1^{(2)}, v_2^{(2)}, \ldots) = (1,1,1, 0,1,1, 1,0,0, \ldots)$.

The convolutional encoder can be described as a device that may take a finite number of states. The state of the encoder is defined by the contents of the shift registers. As was mentioned above the n output symbols are defined by the present k input symbols and by the encoder state. And the input symbols change the encoder state. Such a device is known as a *finite state machine* (FSM). It is convenient to describe the operation of FSM with the help of *state diagram*. The state diagram is a directed graph. The nodes represent the possible states and the branches represent the allowed transitions between states. The branches are labelled with input symbols that cause this transition and with the output symbols, which are emanated during the transition. The total number of the encoder states is equal to $M^{k(\nu-1)}$, where M is the cardinal number of the input alphabet. To be more precise $M^{k(\nu-1)}$ is the upper bound to the total number of the encoder states since not necessarily all k shift registers comprising the encoder have the same length $(\nu - 1)$. Recall that $(\nu - 1)$ is just the maximum number of memory elements in each of the k registers. That means the total number of the encoder states can be represented as $M^{\sum_{i=1}^{k} m_i} \leq M^{k(\nu-1)}$, where m_i is the number of memory elements in ith shift register $(m_i < \nu)$. It is obvious that this method of the code representation is convenient only for the case of the encoder with a small number of states.

Example 4.3 Consider a binary (2, 1, 3) convolutional code. The encoder of this code is shown in Figure 4.3. The state diagram for this code is represented in Figure 4.4. Note that output 0 of this encoder represents just the input symbol. That means the output (encoded) sequence contains the unchanged input

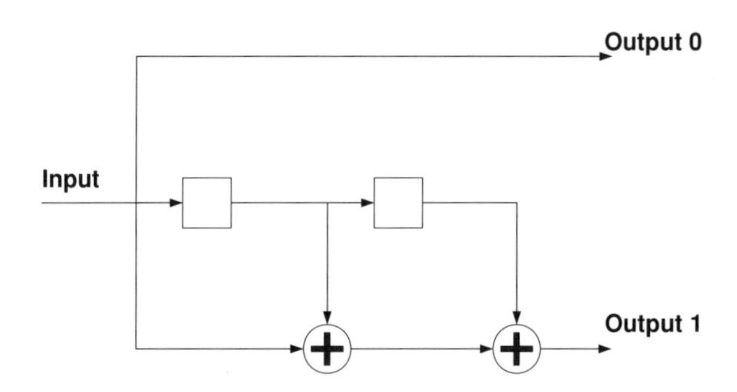

Figure 4.3 A (2, 1, 3) binary systematic convolutional encoder

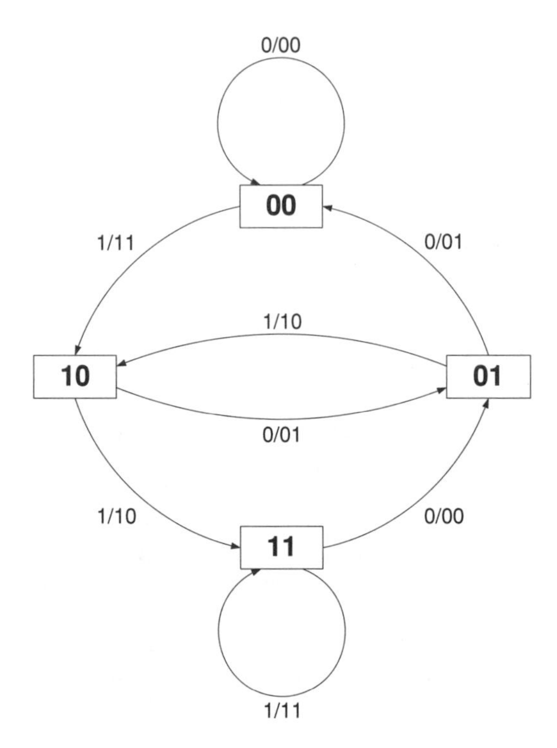

Figure 4.4 A state diagram of the (2, 1, 3) binary systematic convolutional code

(information) sequence. This kind of code is called a *systematic* convolutional code. This code has only four states: 00, 01, 10 and 11. The branches are labelled with the following notation: inS/outS0outS1, where inS is the input symbol that causes this transition and outS0, outS1 are the output symbols at the corresponding outputs.

Another form of the convolutional code representation is the *tree diagram*. The tree diagram represents the encoding process as a tree with the branches corresponding to the transition of the encoder from one possible state to another at a given moment in time. Each node of the tree corresponds to the possible state of the encoder. The number of branches stemming from each node is equal to the number of possible combinations of input symbols, that is, each branch starting from the given node corresponds to one of a few possible combinations of symbols at the input of the encoder. For example, if the cardinal number of the input alphabet is M and the number of the encoder inputs is k, then the number of branches stemming from each node is equal to M^k.

The tree diagram for the code of Example 4.3 for four input bit intervals is shown in Figure 4.5. The upper branch (marked with a solid line) from each node corresponds to input data '0', the lower branch (marked with dashed line) to '1'. The labels on the branches show the corresponding data at the encoder output. The tree diagram helps to visualize the encoding process since it adds the dimension of time to the state diagram (that is, it represents each moment of time with a separate state diagram). Now the encoding procedure can be described by traversing the tree diagram from left to right according to input symbols. An input information sequence defines a specific path through the tree diagram, and an output sequence corresponding to the input information sequence can be easily read from the branch labels of

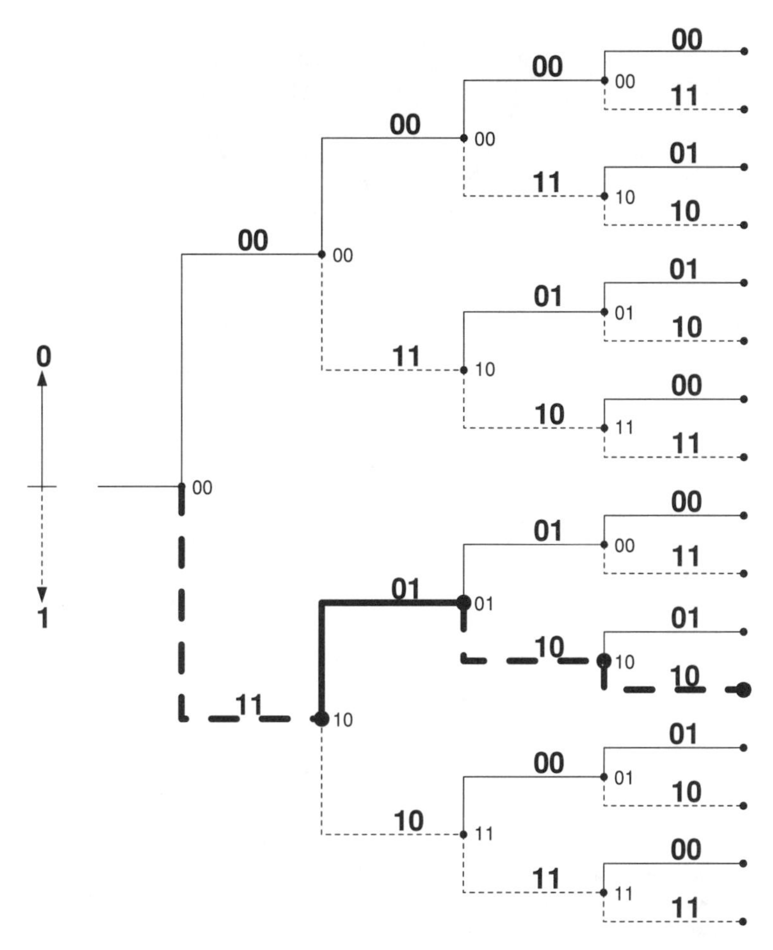

Figure 4.5 A tree diagram of the (2, 1, 3) binary systematic convolutional code

this path. For example, it is easy to see from Figure 4.5 that the input sequence 1011 corresponds to the output sequence 11 01 10 10. Thus any code sequence is represented as a path through the tree.

However, the number of branches in the tree increases exponentially with the length of the input sequence and it is very hard work to draw the tree diagram for the long input sequence. On the other hand it is easy to see that the tree diagram contains a lot of redundant elements. It is enough to compare the parts of the diagram starting at the same state to understand that they are identical. That means the tree diagram can be simplified.

Let us merge all the parts of the tree diagram where the encoder takes the same state at the same time. Now we obtain the diagram in which the number of nodes at any time moment is no more than the number of states. This kind of diagram is called the *trellis diagram*. The trellis diagram provides a more manageable encoder description than the tree diagram does. This compact representation is very helpful for describing the decoding of the convolutional codes, as will be discussed later. The trellis diagram for the code of Example 4.3 is shown in Figure 4.6. Here we use the same convention as for the tree diagram. A solid line denotes the branch generated by an input data '0', and a dashed line denotes the branch generated by an input data '1'. The branches are labelled with the output data. As one can see

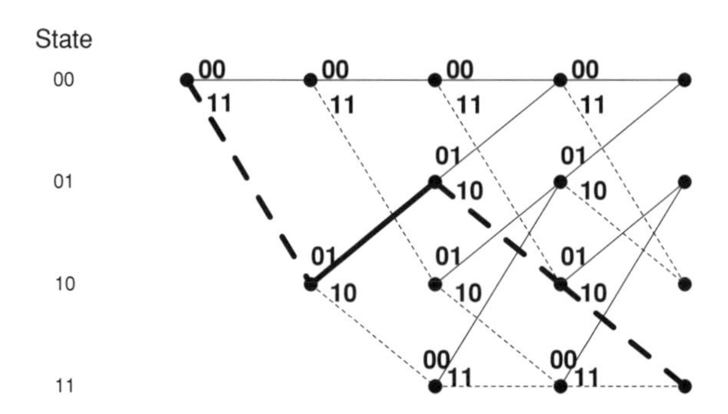

Figure 4.6 A trellis diagram of the (2, 1, 3) binary systematic convolutional code

from Figure 4.6 the structure of the diagram becomes fixed after trellis depth 3 is reached. In general the structure of the trellis becomes fixed after depth ν is reached. With the help of a trellis diagram we can see the correspondence between paths through the diagram and code sequences, while the complexity of the diagram no longer grows exponentially. The bold lines in the trellis in Figure 4.6 show the path corresponding to the input sequence 1011.

With the help of the trellis diagram we can define the most important measure of performance of a convolutional code, which is called *free distance*. It is the analogue of the minimum distance of block code. As we cannot divide the code sequences into code words of the same length, we have to consider the distance between the complete code sequences. The free distance of a convolutional code d_{free} can be defined as:

$$d_{free} = \min\{d(\mathbf{v},\ \mathbf{v}') :\ \mathbf{u} \neq \mathbf{u}'\} \tag{4.6}$$

where $\mathbf{v},\ \mathbf{v}'$ are the code sequences corresponding to the information sequences $\mathbf{u},\ \mathbf{u}'$, respectively. It is assumed that if sequences $\mathbf{u},\ \mathbf{u}'$ are of different length, the necessary number of zeros is appended to the shorter sequence. In other words, free distance is the smallest Hamming distance between any pair of code sequences. Because a convolutional code is a linear code, there is no loss in generality in finding the minimum distance between each of the code sequences and the all-zero sequence. Assuming that the all-zero code sequence can be generated only by the all-zero information sequence (in condition of the all-zero initial state of the encoder), the paths of interest (to compare with all-zero code sequence) are those that diverge from and remerge with the all-zero state. These are the closest sequences that could be confused by the decoder. For example, the free distance of the code of Example 4.3 can be found from trellis in Figure 4.6 by computing the distances between the all-zero path and the paths starting from the left-hand node and returning to the all-zero state later. It is easy to verify that the free distance of this code is 4. For calculating the error-correcting capability of the code we can use the following equation:

$$t = \left\lfloor \frac{d_{free} - 1}{2} \right\rfloor \tag{4.7}$$

where t is the maximum number of errors that the code is capable of correcting, $\lfloor x \rfloor$ means the largest integer not to exceed x. Of course, it does not mean that the convolutional code is capable of correcting no more than t errors on the infinite sequence length. In the same way as block coding we define t as the maximum number of errors that can be corrected by the code on the block length, d_{free} defines the convolutional code error-correcting capability on the length corresponding to the length of loop of minimum weight. In accordance with (4.7) the code of Example 4.3 can correct only one error. It is possible to build a nonsystematic convolutional code with the same parameters (2, 1, 3), which is capable of correcting two errors. In general, making the convolutional code systematic reduces the

Table 4.1 Comparison of free distance for systematic and nonsystematic convolutional codes. Rate $= 1/2$

Constraint length	d_{free} Systematic	d_{free} Nonsystematic
2	3	3
3	4	5
4	4	6
5	5	7
6	6	8
7	6	10
8	7	10

maximum possible free distance for a given constraint length and rate. That means, unlike the block codes, a convolutional nonsystematic code cannot be transformed into a systematic code with the same parameters and error-correcting capability. In Table 4.1 the free distances of systematic and nonsystematic codes of rate $\frac{1}{2}$ are compared [1].

Some convolutional codes cause an infinite number of errors in output sequence after decoding when only a finite number of errors occur during the transmission of a code sequence over the channel. This event is called the *catastrophic error propagation* and this kind of code is called a *catastrophic convolutional code*. This type of code needs to be avoided and can be identified by the state diagram. A state diagram having a loop in which a nonzero information sequence corresponds to all-zero output sequence identifies a catastrophic convolutional code. The examples of such loops are shown in Figure 4.7.

4.2 Viterbi Decoding Algorithm

The best known algorithm of decoding of the convolutional codes was introduced by A. Viterbi in 1967 [2]. Code sequence $\mathbf{v}$ is transmitted over the channel and the received sequence $\mathbf{r}$ can be written as:

$$\mathbf{r} = \mathbf{v} + \mathbf{e} \tag{4.8}$$

where $\mathbf{e}$ is the error sequence. The Viterbi algorithm finds a code sequence $\mathbf{y}$ such that it maximizes the probability $P(\mathbf{r}\,|\mathbf{y})$ that sequence $\mathbf{r}$ is received in condition that sequence $\mathbf{y}$ is transmitted. And in case the weight of error sequence $\mathbf{e}$ does not exceed the code error correcting capability, found sequence $\mathbf{y}$ coincides with the transmitted sequence $\mathbf{v}$. Usually it is more convenient to maximize logarithm of the probability $P(\mathbf{r}\,|\mathbf{y})$ rather than $P(\mathbf{r}\,|\mathbf{y})$ itself. However, since the logarithms are the monotonic

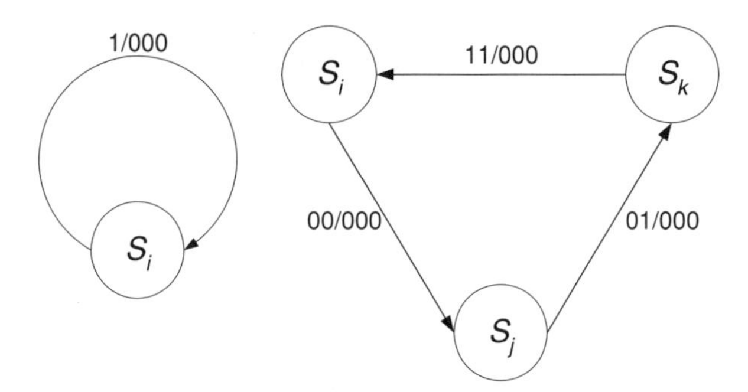

Figure 4.7 Examples of catastrophic convolutional code

functions the code sequence $\mathbf{y}$ that maximizes value $\log P(\mathbf{r}|\mathbf{y})$ maximizes $P(\mathbf{r}|\mathbf{y})$ also. In this case the channel is BSC and the Viterbi algorithm finds the code sequence $\mathbf{y}$ that is closest to the received sequence $\mathbf{r}$ in the sense of minimum Hamming distance. That means the Viterbi algorithm is a *maximum likelihood sequence detection* algorithm. The advantage of the Viterbi algorithm compared with, for example, brute-force maximum likelihood decoding, is that the complexity of the Viterbi algorithm does not depend exponentially on the number of symbols in the code sequence.

As discussed earlier the Viterbi algorithm selects the code sequence $\mathbf{y}$ such that it maximizes the probability $P(\mathbf{r}|\mathbf{y})$ that sequence $\mathbf{r}$ is received in condition that sequence $\mathbf{y}$ is transmitted. This probability is called the *likelihood function*. The channel is assumed to be memoryless, and thus the noise process affects each received symbol independently of all the other received symbols. Since the probability of joint, independent events is equivalent to the product of the probabilities of the individual events, the probability $P(\mathbf{r}|\mathbf{y})$ can be written as:

$$P(\mathbf{r}|\mathbf{y}) = \prod_i P(r_i|y_i) \tag{4.9}$$

where $P(r_i|y_i)$ is a channel transition probability. It is more convenient to use log-likelihood function $\log P(\mathbf{r}|\mathbf{y})$ rather than the likelihood function itself. It follows from (4.9) that:

$$\log P(\mathbf{r}|\mathbf{y}) = \sum_i \log P(r_i|y_i) \tag{4.10}$$

To simplify the manipulation of the summations over the log function, a *symbol metric* $M(r_i|y_i)$ is defined as:

$$M(r_i|y_i) = c_1 \cdot (\log P(r_i|y_i) + c_2) \tag{4.11}$$

where c_1 and c_2 are chosen such that the symbol metric can be well approximated by integers.

From the symbol metric a *path metric* and a *branch metric* are defined as follows:

$$M(\mathbf{r}|\mathbf{y}) = \sum_{j=0}^{L-1} M(\mathbf{r}_j|\mathbf{y}_j) = \sum_{j=0}^{L-1}\left(\sum_{i=0}^{n-1} M(r_i|y_i)\right) \tag{4.12}$$

where $M(\mathbf{r}|\mathbf{y})$ is the path metric, $M(\mathbf{r}_j|\mathbf{y}_j)$ is the branch metric and L is the number of blocks of n symbols in the sequence. In the same way we can define the *partial path metric* $M^j(\mathbf{r}|\mathbf{y})$ as:

$$M^i(\mathbf{r}|\mathbf{y}) = \sum_{j=0}^{i} M(\mathbf{r}_j|\mathbf{y}_j) \tag{4.13}$$

The symbol metric shows the cost of choosing symbol y_i as the estimate of the corresponding symbol r_i. The branch metric indicates the cost of choosing the branch from the trellis, the partial path metric $M^j(\mathbf{r}|\mathbf{y})$ corresponds to the cost of choosing given path $\mathbf{y}$ up to time index j as a part of decoded sequence and finally the path metric shows the total cost of estimating the received sequence $\mathbf{r}$ with the sequence $\mathbf{y}$.

4.2.1 Hard Decision Viterbi Algorithm

Let us consider for simplicity the Viterbi algorithm for BSC (that is, the hard-decision Viterbi algorithm) first. If the symbols are transmitted over the BSC with the crossover probability p the likelihood function for the received sequence $\mathbf{r}$ of length N can be written as:

$$P(\mathbf{r}|\mathbf{y}) = (1-p)^{N-d(\mathbf{r},\mathbf{y})} \cdot p^{d(\mathbf{r},\mathbf{y})} = \prod_{i=0}^{N-1}(1-p)\cdot\left(\frac{1-p}{p}\right)^{-d(r_i,y_i)} \tag{4.14}$$

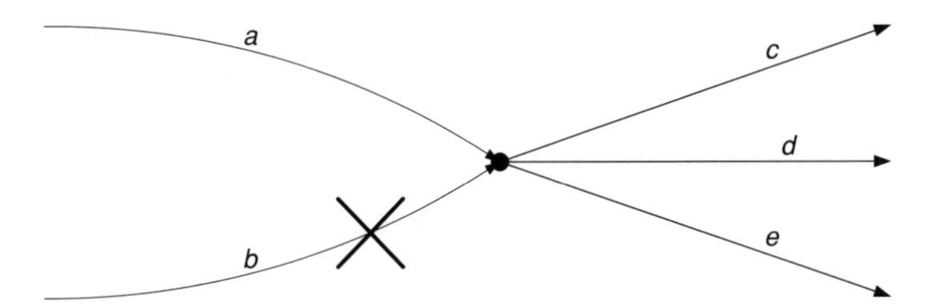

Figure 4.8 Eliminating of one of merged paths in trellis in accordance with the principle of optimality

where $d(\bullet,\ \bullet)$ is the Hamming distance. Then the symbol metric in accordance with (4.11) and (4.14) can be written in the form of:

$$M\left(r_i\,|y_i\right) = c_1 \cdot \left(\log\left((1-p)\cdot\left(\frac{1-p}{p}\right)^{-d(r_i, y_i)}\right) + c_2\right) \tag{4.15}$$

where coefficients c_1 and c_2 can be chosen as follows:

$$c_1 = \left(\log\frac{1-p}{p}\right)^{-1} \tag{4.16}$$

$$c_2 = -\log(1-p) \tag{4.17}$$

Then the symbol metric becomes the Hamming metric and can be written as:

$$M\left(r_i\,|y_i\right) = -d(r_i,\ y_i) \tag{4.18}$$

In this case the problem of finding sequence **y** that maximizes the probability $P\left(\mathbf{r}\,|\mathbf{y}\right)$ can be formulated as the search of optimum path **y** (with the minimum Hamming distance between **y** and **r**) through the trellis. This is equivalent to a dynamic programming problem of finding the path with minimum weight through a weighted graph [3]. Viterbi algorithm is based on the *principle of optimality*. The principle of optimality states that if any two paths in the trellis merge to the same state, one of them can always be discarded in the search for an optimum path, because the path with more weight could not turn out to be the prefix of the optimum path through the trellis. This statement can be illustrated by the picture in Figure 4.8.

The weights of branches $a,\ b,\ c,\ d,\ e$ correspond to Hamming distance between parts of two sequences; that means all the weights are nonzero values. Let $a < b$, then the weights of paths generated by path with weight a, that is, $a+c,\ a+d,\ a+e$ will be less than $b+c,\ b+d,\ b+e$ for all possible values of $c,\ d,\ e$. In this case we can eliminate path with weight b in the search for an optimum path with minimum weight. This principle allows us to consider only the constant number of paths on each stage of the decoding procedure.

Example 4.4 Let us consider the hard-decision Viterbi algorithm on the example of decoding of a nonsystematic (2, 1, 3) binary convolutional code. The encoder of this code is shown in Figure 4.9. The corresponding trellis diagram is represented in Figure 4.10. As one can see from the trellis diagram the free distance of this code is 5. Let the information sequence be $\mathbf{u} = 1100011111$. This information sequence generates the code sequence $\mathbf{v} = 11\ \ 10\ \ 10\ \ 11\ \ 00\ \ 11\ \ 10\ \ 01\ \ 01\ \ 01$.

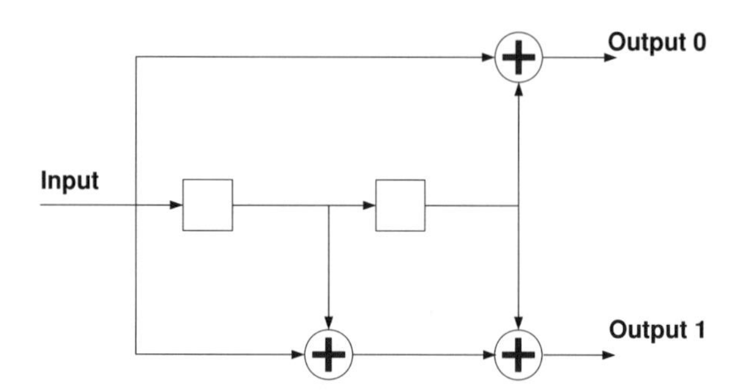

Figure 4.9 A (2, 1, 3) binary nonsystematic convolutional encoder

Assume that the received sequence is $\mathbf{r} = 10\ 00\ 10\ 11\ 00\ 11\ 10\ 01\ 01\ 01$, that is, two errors occur during the transmission of the code sequence over the channel. Let us go through the trellis from the left to the right and search the optimum path that is the closest path to the received sequence $\mathbf{r}$ in the sense of Hamming distance assuming that the initial state of the encoder is all-zero (at the beginning of the encoding of each sequence the encoder should be flushed). The process of decoding is shown in Figure 4.11. At the first stage (Figure 4.11a) we have two partial paths in the trellis: 00 and 11. The partial path is defined as the path from state $S = 0$ at time $j = 0$ to a particular state $S = i$ at time $j \geq 0$. Both of these paths differ from the received symbols 10 in one position. Hence, both of these paths have weight 1. There are no paths merging at the same state at this stage, so we keep all the paths, because we have no choice yet. At the second stage (Figure 4.11b) the number of partial paths is 4. The weight of each partial path (*path metric*) is obtained as the sum of the weight of the previous partial path and the weight of the corresponding branch (*branch metric*). The corresponding symbols of the received sequence $\mathbf{r}$ are 00. Then the branch metric of the branch from state 00 to state 00 is equal to $d(00,\ 00) = 0$, the branch metric of the branch from state 00 to state 10 is $d(00,\ 11) = 2$, the branch metric of the branch from state 10 to state 01 is $d(00,\ 01) = 1$, the branch metric of the branch from state 10 to state 11 is $d(00,\ 10) = 1$. Summing up the branch metric and the path metric of the previous partial path obtain the following path metrics of the corresponding paths: 1, 3, 2, 2. Here we also have no merging paths, so we keep all four paths. At the third stage (Figure 4.11c) we have eight paths. We can calculate the path metrics as in previous stages and then we have two paths merging to each state. Now in accordance with the principle of optimality we can eliminate the path with greater weight. For example, the path merging

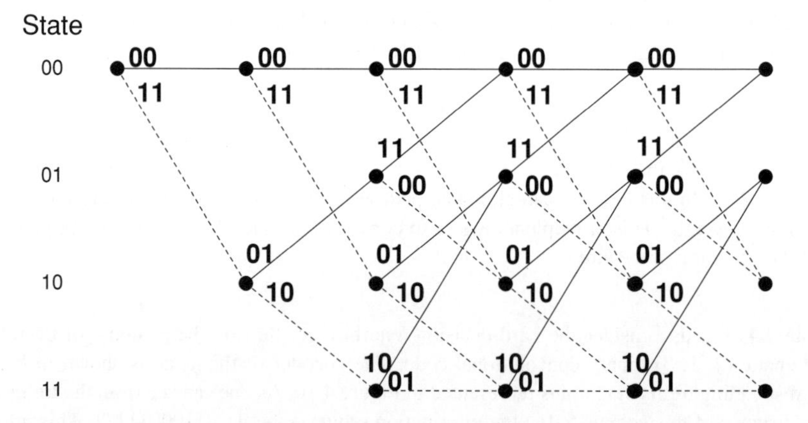

Figure 4.10 A trellis diagram of the (2, 1, 3) binary nonsystematic convolutional code

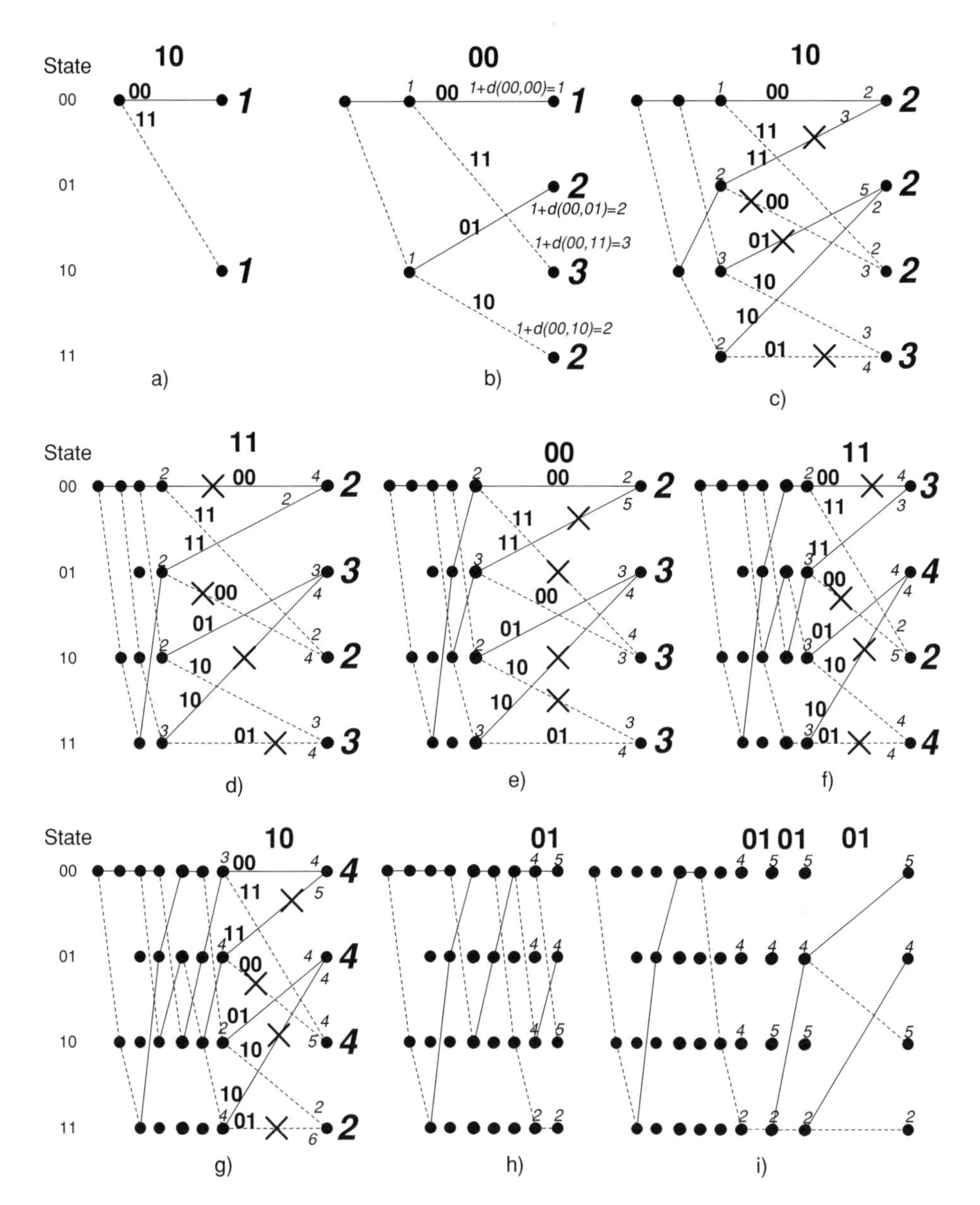

Figure 4.11 The stages of Viterbi decoding of the received sequence $r = 10\ 00\ 10\ 11\ 00\ 11\ 10\ 01\ 01\ 01$ in trellis diagram of the (2, 1, 3) binary nonsystematic convolutional code

to state 00 from state 00 has the path metric 2 and the path merging to the same state 00 from state 01 has the path metric 3 that means the last path should be discarded. The path merging to state 00 from state 00 is called a *survivor path*. After this procedure we have to keep only four paths. Hereafter at each stage we have to keep only four survivor paths.

In cases where there are two paths with the same weight merged to a single node an arbitrary decision about discarding one of these two paths can be made (for example, at the sixth stage in Figure 4.11. If two paths with weight 4 merge to the state 01 and two paths with weight 4 merge to the state 11; in this example we eliminate the lower from the two paths with the same weight). As can be seen from Figure 4.11e already at the fifth stage the path metric of the partial path corresponding to the sequence 11 10 10 11 00 (which is the part of the transmitted sequence **v**) is the best among the four metrics of the survivor paths. Since in our example there are no new errors in the latter symbols it is obvious that the correct path will be found at the end. On the other hand we can see from Figure 4.11 that the survivor paths can differ from each other time. Only at stage 10, as can be seen from Figure 4.11i, do the first eight branches of the survivor paths coincide. At this time we can make a decision about the first eight transmitted symbols since the survivor paths merge. The depth of this merge is an arbitrary value and it depends only on the errors in the channel during the transmission. In any case we can see that the decoding introduces the severe delay, which is much more than a one stage period. In practice it is impossible to wait until the survivor paths merge. Usually the fixed depth of the decoding is defined for the decoder. After reaching this *decoding depth* (the certain number of stages) the decision about the first symbols (in accordance with the path with the best metric) is made. This of course leads to some degradation in performance and the algorithm is no more optimal but suboptimal. On the other hand, if the depth of making the decision is five or seven times the constraint length, the degradation of performance is negligible [4].

Now we can formulate the hard decision Viterbi algorithm as follows [5]:

Hard-Decision *Viterbi Decoding:*

$S_{i,j}$ is the state in the trellis diagram that corresponds to the state S_i at time j. Every state in the trellis is assigned a value denoted $V(S_{i,j})$. L is the decoding depth (or as it is often called the *truncation window length*).

1. (a) Initialize time $j = 0$.
   (b) Initialize $V(S_{0,0}) = 0$ and all other $V(S_{i,j}) = \infty$.
2. (a) Set time $j = j + 1$.
   (b) For all i compute the partial path metrics for all paths going to state S_i at time j . To do this: first, calculate the branch metric, and then add the branch metric to $V(S_{i,j-1})$.
3. (a) For all i set $V(S_{i,j})$ to the best partial path metric going to state S_i at time j.
   (b) If there is a tie for the best partial path metrics, then any one of the tied partial path metric may be chosen.
4. If $j < L$ go to the step 2.
5. (a) Start trace-back through the trellis by following the branches of the best survivor path.
   (b) Store the associated survivor k symbols. These are currently decoded k information symbols.
   (c) Set time $j = 0$; go to step 2. Here is the start of the new truncation window.

Usually it is more convenient to use the code words with fixed length rather than the semi-infinite code sequences. In this case it is possible to add $k \cdot (\nu - 1)$ dummy zeros (so called *tail symbols*) to the end of the information sequence of fixed length before encoding, which forces the encoder to return to the all-zero state and terminates the trellis. This simplifies the work of the decoder, because now it is enough to check only the survivor that ends at the all-zero state. Obviously with using this technique the convolutional code becomes the block code.

4.2.2 *Soft Decision Viterbi Algorithm*

The soft-decision Viterbi algorithm exploits the additional information, which is provided by the soft-decision demodulator and this additional information allows the performance to increase. The algorithm itself is the same as for the hard decisions. The only difference is that Hamming distance is not used as a metric. Generally speaking, the metric used in the algorithm should be defined by the channel. For example, the Euclidean distance is the optimal metric for the Gaussian channel. Let us consider the example of the soft-decision Viterbi decoding for the *discrete memoryless channel* (DMC).

Example 4.5 Let us consider the binary input, 8-ary output DMC represented in Figure 4.12. The transition probabilities $P(r\,|y)$ of this channel are shown in the following table:

$P(r\,\|y)$	$\mathbf{0_4}$	$\mathbf{0_3}$	$\mathbf{0_2}$	$\mathbf{0_1}$	$\mathbf{1_1}$	$\mathbf{1_2}$	$\mathbf{1_3}$	$\mathbf{1_4}$
0	0.439	0.2	0.17	0.1	0.06	0.025	0.005	0.001
1	0.001	0.005	0.025	0.06	0.1	0.17	0.2	0.439

Taking the logarithms obtain the log-likelihood values $\log P(r\,|y)$:

$\log P(r\,\|y)$	$\mathbf{0_4}$	$\mathbf{0_3}$	$\mathbf{0_2}$	$\mathbf{0_1}$	$\mathbf{1_1}$	$\mathbf{1_2}$	$\mathbf{1_3}$	$\mathbf{1_4}$
0	−0.82	−1.61	−1.77	−2.3	−2.81	−3.69	−5.3	−6.91
1	−6.91	−5.3	−3.69	−2.81	−2.3	−1.77	−1.61	−0.82

Let us choose the coefficient $c_2 = -\min_y P(r\,|y)$ and the coefficient $c_1 = 1.35$. Then the symbol metric values $M(r\,|y)$ in accordance with (4.11) can be written as follows:

$M(r\,\|y)$	$\mathbf{0_4}$	$\mathbf{0_3}$	$\mathbf{0_2}$	$\mathbf{0_1}$	$\mathbf{1_1}$	$\mathbf{1_2}$	$\mathbf{1_3}$	$\mathbf{1_4}$
0	8	5	3	1	0	0	0	0
1	0	0	0	0	1	3	5	8

Let the information and encoded sequence be the same as in the previous example $\mathbf{u} = 1100011111$, $\mathbf{v} = 11\ 10\ 10\ 11\ 00\ 11\ 10\ 01\ 01\ 01$. Assume that the received sequence is $\mathbf{r} = 1_4 0_3\ 0_2 0_2\ 0_1 0_4\ 1_1 1_1\ 0_2 0_3\ 1_2 1_3\ 1_2 0_3\ 0_4 1_1\ 0_2 1_1\ 0_1 1_2$. The decoding process for this received sequence is shown in Figure 4.13.

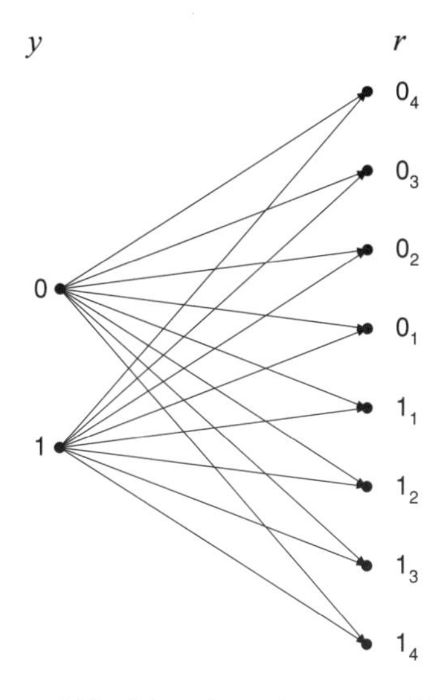

Figure 4.12 Binary input, 8-ary output DMC

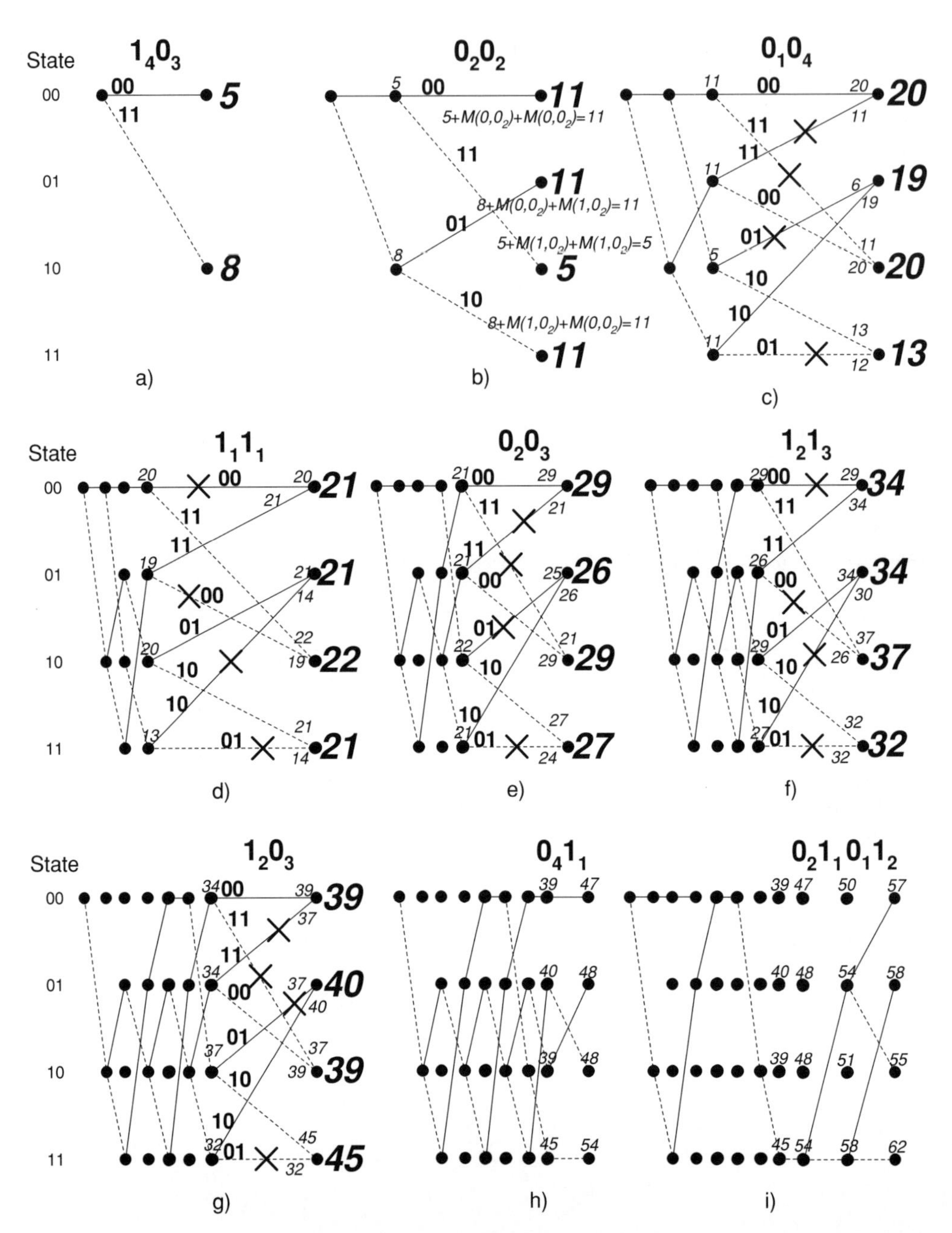

Figure 4.13 The stages of soft-decision Viterbi decoding of the received sequence $r = 1_4 0_3\ 0_2 0_2\ 1_1 0_4\ 1_1 1_1\ 0_2 0_3\ 1_2 1_3\ 1_2 0_3\ 0_4 1_1\ 0_2 1_1\ 0_1 1_2$ in trellis diagram of the (2, 1, 3) binary nonsystematic convolutional code

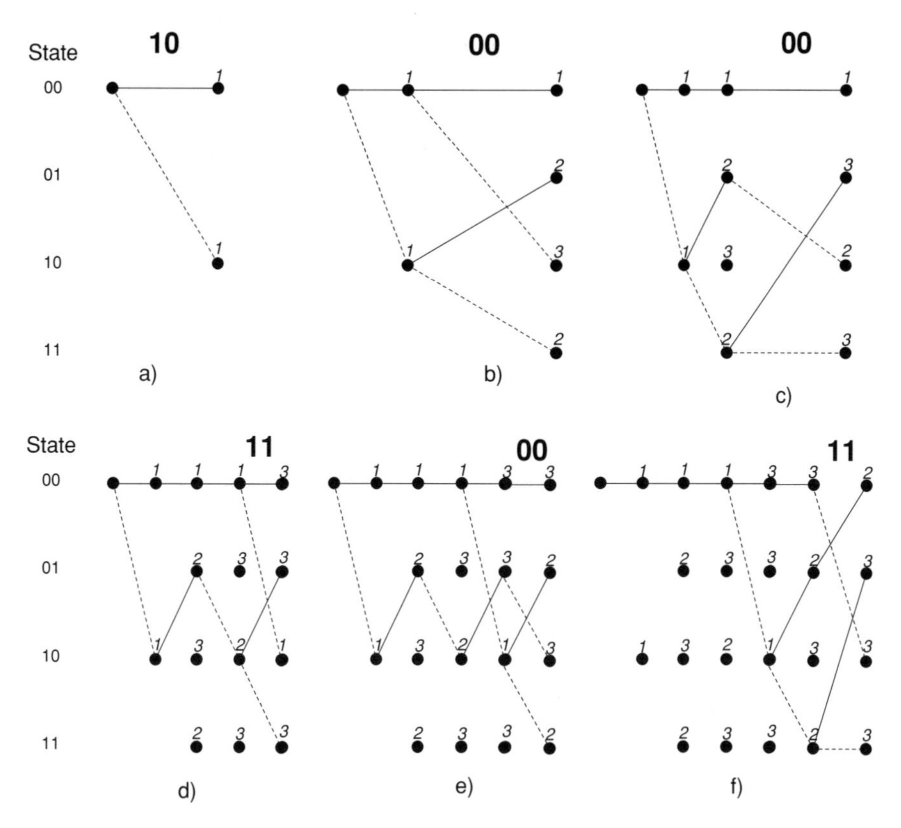

Figure 4.14 The stages of hard-decision Viterbi decoding of the received sequence $r = 10\ 00\ 10\ 11\ 00\ 11\ 10\ 01\ 01\ 01$ in trellis diagram of the (2, 1, 3) binary nonsystematic convolutional code

As one can see from Figure 4.13 the first eight symbols are decoded correctly at stage 10. Notice that if we merge the soft decisions outputs $0_1\ 0_2\ 0_3\ 0_4$ into the hard decision output 0 and the outputs $1_1\ 1_2\ 1_3\ 1_4$ into the hard decision output 1, the hard decision received sequence becomes $\mathbf{r} = 10\ 00\ 00\ 11\ 00\ 11\ 10\ 01\ 01\ 01$. The hard decision decoding of this sequence is represented in Figure 4.14. As expected, in this case the decoder chooses the path that does not coincide with the transmitted sequence (Figure 4.14.f).

Now we can write the soft-decision Viterbi algorithm as follows:

Soft-Decision *Viterbi Decoding:*

1. (a) Initialize time $j = 0$.
   (b) Initialize $\mathrm{V}(S_{0,0}) = 0$ and all other $\mathrm{V}(S_{i,j}) = -\infty$.
2. (a) Set time $j = j + 1$.
   (b) For all i compute the partial path metrics for all paths going to state S_i at time j. To do this: first, calculate the branch metric $M\left(\mathbf{r}_j\left|\mathbf{y}_j\right.\right)$ in accordance with (4.12), and then compute the j-th partial path metric $M^j\left(\mathbf{r}\left|\mathbf{y}\right.\right) = \mathrm{V}(S_{i,j-1}) + M\left(\mathbf{r}_j\left|\mathbf{y}_j\right.\right)$.

3. (a) For all i set $V(S_{i,j})$ to the 'best' partial path metric going to state S_i at time j.
   (b) If there is a tie for the best partial path metrics, then any one of the tied partial path metric may be chosen.
4. If $j < L$ go to step 2.
5. (a) Start trace-back through the trellis by following the branches of the best survivor path.
   (b) Store the associated survivor k symbols. These are currently decoded k information symbols.
   (c) Set time $j = 0$; go to step 2. Here is the start of the new truncation window.

As can be seen the soft-decision algorithm differs from the hard-decision algorithm only by the used metric. Example 4.5 demonstrates the gain, which can be obtained by using the same Viterbi algorithm exploiting the additional information from the soft-decision demodulator. Usually the soft-decision decoding increases the coding gain of a convolutional code by about 2 dB (of course, the actual gain of using soft-decision decoding differs for different channels).

As discussed earlier the number of nodes at each trellis stage is equal to $M^{k(\nu-1)}$, where M is the cardinal number of the input alphabet. At each node of the trellis M^k calculations are needed to perform the Viterbi algorithm. Hence, the complexity of the Viterbi algorithm is the order of $O(M^{k(\nu-1)} \cdot M^k \cdot L)$. This value is significantly less than the complexity of brute-force ML decoding, which can be estimated as $O(M^{kL})$. However, the increase of the number of information symbols k or the constraint length ν leads to the exponential growth of the Viterbi algorithm complexity.

4.3 List Decoding

The list decoding is a suboptimal non-backtracking algorithm, which consists in choosing the several best partial paths at each stage of the decoding process. These partial paths form the list of size L. Unlike the Viterbi algorithm the list decoder considers the extensions of only these best partial paths from the list, not all partial paths. The list decoding algorithm belongs to the class of breadth-first algorithms, as does the Viterbi algorithm. Of course, the list size L should be less than the number of states $M^{\nu-1}$. Obviously the complexity of the list decoding is less than the complexity of the Viterbi algorithm, but because of the fact that some partial paths are not considered, the list decoding is not the optimal algorithm.

Example 4.6 Let us consider the hard-decision list decoding of the received sequence $\mathbf{r} =$ 10 00 10 11 00 11 10 01 01 01 of Example 4.4. The size of the list is three, that is, we will find the extensions of only the three best paths at each decoding stage. The decoding process is shown in Figure 4.15. As can be seen from Figure 4.15 the received sequence is successfully decoded and already at stage 9 (Figure 4.15.h) where the first seven branches of the survivor paths coincide.

So in this particular case we managed to decode correctly the received sequence with less complexity than required by the Viterbi algorithm. Unfortunately the list algorithm as mentioned above is not the optimal one and in some cases it is possible to miss the correct path. The greater the list size the less the probability of missing the correct path.

4.4 Upper Bound on Bit Error Probability for Viterbi Decoding

Let rate of a binary convolutional code be $R = k/n$. Denote a data sequence of length lk, $l = 1, 2, \ldots$, as $\mathbf{u}$ i.e. $\mathbf{u} \in \{0, 1\}^{kl}$, l here is number of k-bit blocks of data sequence. Let $\mathbf{v}(\mathbf{u})$ be code sequence corresponding to the data sequence $\mathbf{u}$, $\mathbf{v}(\mathbf{u}) \in \{0, 1\}^{nl}$. Any path in code trellis consists of intermediate states and transitions connecting the states. Since any data sequence corresponds to a path in code trellis and vice versa we can consider intermediate states of data sequence instead of intermediate states of

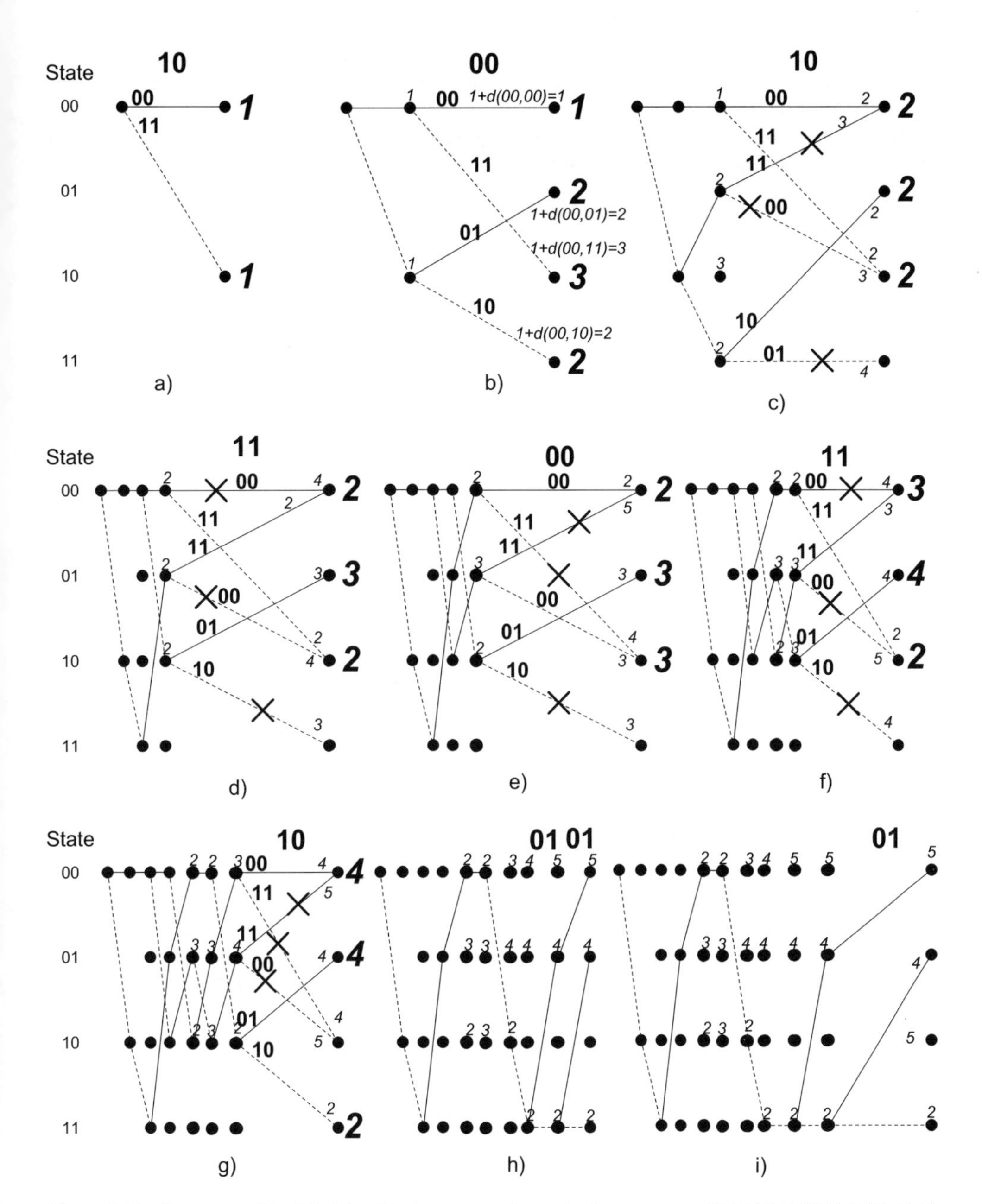

Figure 4.15 The stages of hard-decision List decoding of the received sequence $r = 10\ 00\ 10\ 11\ 00\ 11\ 10\ 01\ 01\ 01$ in trellis diagram of the (2, 1, 3) binary nonsystematic convolutional code

trellis path. Assume that the data sequence $\mathbf{u}$ corresponds to a path in code trellis starting in a state σ_1 and after l transitions coming to trellis state σ_2. Let $\mathbf{u}'$ be another data sequence of the same length kl, such that initial and final state of $\mathbf{u}'$ are σ_1 and σ_2 but all intermediate states of $\mathbf{u}$ and $\mathbf{u}'$ are different. Denote the set of all such data sequences $\mathbf{u}'$ as $E(\mathbf{u})$. Then the classic upper bound on bit error probability based on union bound argumentation [1] can be written as follows:

$$P_b \leq \frac{1}{k} \sum_{l=1}^{\infty} \sum_{\mathbf{u} \in \{0,1\}^{kl}} \sum_{\mathbf{u}' \in E(\mathbf{u})} \Pr\left[\mathbf{v}(\mathbf{u}) \rightarrow \mathbf{v}(\mathbf{u}')\right] P(\mathbf{u}) d_H(\mathbf{u}, \mathbf{u}') \tag{4.19}$$

where $\Pr\left[\mathbf{v}(\mathbf{u}) \rightarrow \mathbf{v}(\mathbf{u}')\right]$ is the error probability for code of two words $\mathbf{v}(\mathbf{u})$ and $\mathbf{v}(\mathbf{u}')$, $P(\mathbf{u})$ is the probability of transmission data sequence $\mathbf{u}$, $d_H(\cdot, \cdot)$ is Hamming distance.

The event $\mathbf{v}(\mathbf{u}) \rightarrow \mathbf{v}(\mathbf{u}')$ occurs when metric value of the correct path $\mathbf{v}(\mathbf{u})$is less than metric value of the incorrect path $\mathbf{v}(\mathbf{u}')$. The probability $\Pr\left[\mathbf{v}(\mathbf{u}) \rightarrow \mathbf{v}(\mathbf{u}')\right]$ can be upper bounded using modified Chernoff bound (see Appendix 4.A) as:

$$\Pr\left[\mathbf{v}(\mathbf{u}) \rightarrow \mathbf{v}(\mathbf{u}')\right] \leq \min_{\beta} K\left(\beta, d_H\left(\mathbf{v}(\mathbf{u}), \mathbf{v}(\mathbf{u}')\right)\right) D_0(\beta)^{d_H\left(\mathbf{v}(\mathbf{u}), \mathbf{v}(\mathbf{u}')\right)} \tag{4.20}$$

where $K(\beta, d)$ and $D_0(\beta)$ are a nonexponential coefficient and exponent of the modified Chernoff bound respectively, and β is an optimization parameter. The minimization here and after is assumed over all *allowed values* of the Chernoff bound parameter β. Note that Chernoff bound coefficient $K(\beta, d)$ monotonically decreases with increase of its second argument.

Since $d_H\left(\mathbf{v}(\mathbf{u}), \mathbf{v}(\mathbf{u}')\right) = w_H\left(\mathbf{v}(\mathbf{u}) + \mathbf{v}(\mathbf{u}')\right)$, where $w_H(\cdot)$ is Hamming weight, and for any linear code $\mathbf{v}(\mathbf{u}) + \mathbf{v}(\mathbf{u}') = \mathbf{v}(\mathbf{u} + \mathbf{u}')$, we have:

$$\Pr\left[\mathbf{v}(\mathbf{u}) \rightarrow \mathbf{v}(\mathbf{u}')\right] < \min_{\beta} K\left(\beta, w_H(\mathbf{v}(\mathbf{u} + \mathbf{u}'))\right) D_0(\beta)^{w_H(\mathbf{v}(\mathbf{u}+\mathbf{u}'))} \tag{4.21}$$

The plus sign means here addition in $GF(2)$. After substitution of right hand part of (4.21) to inequality (4.19) we get:

$$P_b < \min_{\beta} \frac{1}{k} \sum_{l=1}^{\infty} \sum_{\mathbf{u} \in E(\mathbf{0})} K\left(\beta, w_H(\mathbf{v}(\mathbf{u}))\right) D_0(\beta)^{w_H(\mathbf{v}(\mathbf{u}))} w_H(\mathbf{u}) \tag{4.22}$$

where $E(\mathbf{0})$ is a set of data sequences of length kl bits with zero initial and zero final state but nonzero intermediate states. The trellis paths corresponding to data sequences $\mathbf{u} \in E(\mathbf{0})$ are traditionally called *loops* or *detours* with respect to all zero path in trellis. The values $w_H(\mathbf{v}(\mathbf{u}))$ and $w_H(\mathbf{u})$ are called weight of loop and information weight of loop respectively. If we denote $A(l, d, i)$ the number of loops $\mathbf{v}(\mathbf{u})$ of length l edges such that $w_H(\mathbf{v}(\mathbf{u})) = d$ and $w_H(\mathbf{u}) = i$, $l = 1, 2, \ldots$, $d = d_{free}, d_{free} + 1, \ldots$, $i = 1, 2, \ldots$, where d_{free} is the free distance of the convolutional code, then we can transform the inequality (4.22) to:

$$P_b < \min_{\beta} \frac{1}{k} \sum_{l=1}^{\infty} \sum_{d=d_f}^{\infty} \sum_{i=1}^{\infty} i A(l, d, i) K(\beta, d) D_0(\beta)^d \tag{4.23}$$

Recalling that $K(\beta, d)$ monotonically decreases when d grows and making the bound in (4.23) slightly less accurate obtain:

$$P_b < \min_{\beta} \frac{K\left(\beta, d_{free}\right)}{k} \sum_{l=1}^{\infty} \sum_{d=d_f}^{\infty} \sum_{i=1}^{\infty} i A(l, d, i) D_0(\beta)^d \tag{4.24}$$

The well known [1] way to compute the right hand part of (4.24) is based on usage of the *generating function* of weights of loops in code trellis, or *weight enumerator function* of the code.

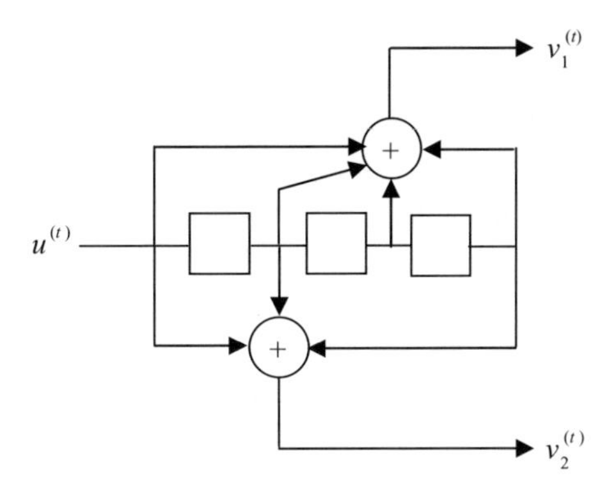

Figure 4.16 Example of rate 1/2 convolutional encoder

It is defined as:

$$T(Z, N) = \sum_{d=d_f}^{\infty} \sum_{i=1}^{\infty} A(d, i) Z^d N^i \quad (4.25)$$

where Z, N are formal variables, and $A(d, i) = \sum_{l=0}^{\infty} A(l, d, i)$ is the number of loops with weight d and information weight i. The generating function can be found from analysis of the convolutional encoder as a finite state machine by the flow graph technique.

Let us consider briefly how to find the expression for generating function $T(Z, N)$. Let rate-1/2 convolutional encoder be as it is depicted in Figure 4.16. The code generators are $\mathbf{g}_1 = [1, 1,1, 1] = 17_8$ and $\mathbf{g}_1 = [1,1,0, 1] = 15_8$. Encoder memory amount is equal to 3, so the number of state is 8. Let us define encoder state transition as $\sigma^{(t-1)} \to \sigma^{(t)}$, where for this example $\sigma^{(t-1)} = (u^{(t-1)}, u^{(t-2)}, u^{(t-3)})$ and $\sigma^{(t)} = (u^{(t)}, u^{(t-1)}, u^{(t-2)})$. We assign the label $N^i Z^d$ to transition $\sigma^{(t-1)} \to \sigma^{(t)}$, where $i = w_H(u^{(t)})$ and $d = w_H([v_1^{(t)}, v_2^{(t)}]$. All states of transition can be arranged as an 8×8 symbolic matrix $\mathbf{F}(Z, N) = [F_{ij}(Z, I)]$, where:

$$F_{ij}(Z, N) = \begin{cases} N^i Z^d, & \text{if transition } i \to j \text{ exists} \\ 0, & \text{otherwise} \end{cases}$$

where indices i, j correspond to the encoder states written as integers in decimal form, $i, j = 0, 1, \ldots 7$. Matrix $\mathbf{F}(D, I)$ for this example is equal to:

$$\mathbf{F}(Z, N) = \begin{bmatrix} 1 & 0 & 0 & 0 & N^2 Z & 0 & 0 & 0 \\ Z^2 & 0 & 0 & 0 & N & 0 & 0 & 0 \\ 0 & Z & 0 & 0 & 0 & NZ & 0 & 0 \\ 0 & Z & 0 & 0 & 0 & NZ & 0 & 0 \\ 0 & 0 & Z^2 & 0 & 0 & 0 & N & 0 \\ 0 & 0 & 1 & 0 & 0 & 0 & NZ^2 & 0 \\ 0 & 0 & 0 & Z & 0 & 0 & 0 & NZ \\ 0 & 0 & 0 & Z & 0 & 0 & 0 & NZ \end{bmatrix}$$

To compute the function $T(Z, N)$ let us present matrix $\mathbf{F}(Z, N)$ as:

$$\mathbf{F}(Z, N) = \begin{bmatrix} 1 & \mathbf{F}_1(Z, N) \\ \mathbf{F}_2(Z, N) & \mathbf{F}_0(Z, N) \end{bmatrix}$$

where $\mathbf{F}_1(Z, N)$ is row vector corresponding to all transitions $0 \to \sigma, \sigma \neq 0$, $\mathbf{F}_2(Z, N)$ is column vector corresponding to all transitions $\sigma \to 0$, $\sigma \neq 0$, and $\mathbf{F}_0(Z, N)$ is matrix vector corresponding to all transitions $\sigma \to \sigma', \sigma, \sigma' \neq 0$. Then the generating function $T(Z, N)$ can be found as [1]:

$$T(Z, N) = \sum_{l=0}^{\infty} \mathbf{F}_1(Z, N)\mathbf{F}_0(Z, N)^l \mathbf{F}_2(Z, N) = \mathbf{F}_1(Z, N)(\mathbf{I} - \mathbf{F}_0(Z, N)^{-1}\mathbf{F}_2(Z, N)$$

It can be shown that for this example:

$$T(Z, N) = \frac{N^2Z^6 + NZ^7 - N^2Z^8}{1 - 2ZN - Z^3N}$$

For evaluation of the bound of bit error probability let us consider the derivative of the function $T(Z, N)$ with respect to formal variable N. The function $F(Z)$ defined as follows:

$$F(Z) = \left.\frac{dT(Z, N)}{dN}\right|_{N=1} = \left.\sum_{d=d_{free}}^{\infty} \sum_{i=1}^{\infty} iA(d, i)Z^d N^{i-1}\right|_{N=1} = \sum_{d=d_{free}}^{\infty} \sum_{i=1}^{\infty} iA(d, i)Z^d = \sum_{d=d_{free}}^{\infty} A(d)Z^d \tag{4.26}$$

where $A(d) = \sum_{i=1}^{\infty} iA(d, i)$, can be used for computation the bound in (4.24). It follows from (4.24), (4.25) and (4.26) that:

$$P_b < \min_{\beta} \frac{K\left(\beta, d_{free}\right)}{k} F\left(D_0(\beta)\right) \tag{4.27}$$

Formulas for $K(\beta, d)$ and $D_0(\beta)$ for some channel models can be found in the Appendix 4A.

In many cases it is not necessary to have the full expression for the function $F(D)$. It is sufficient to use a finite number of terms of the series expansion:

$$F(Z) = A(d_{free})Z^{d_{free}} + Z(d_{free} + 1)Z^{d_{free}+1} + A(d_{free} + 2)Z^{d_{free}+2} + \dots$$

that is, the bound (9) can be computed as:

$$P_b < \min_{\beta} \frac{K\left(\beta, d_{free}\right)}{k} \sum_{d=d_{free}}^{d_{free}+K} A(d)D_0(\beta)^d \tag{4.28}$$

for K large enough; in practically important cases the value $K = 4 \dots 6$ is sufficient for good accuracy of the bound because the terms in the sum in (4.28) are decreasing very fast. For the example given above the series expansion of the generating function is:

$$F(Z) = \frac{2Z^6 - Z^7 - 2Z^8 + Z^9 + Z^{11}}{(1 - 2Z - Z^3)^2} = 2Z^6 + 7Z^7 + 18Z^8 + 49Z^9 + 130Z^{10}$$
$$+333Z^{11} + 836Z^{12} + 2069Z^{13} + 5060Z^{14} + \dots$$

It means that $d_{free} = 6$, and $A(d_{free}) = 2$, $A(d_{free} + 1) = 7$, $A(d_{free} + 2) = 18$, and so on.

Table 4.2 Weight spectrum coefficients for some binary convolutional codes

	$R = 1/3$			$R = 1/2$		
ν	$\mathbf{g}_1, \mathbf{g}_2, \mathbf{g}_3$	d_{free}	$A(d_{free}), \ldots, A(d_{free}+4)$,	$\mathbf{g}_1, \mathbf{g}_2$	d_{free}	$A(d_{free}), \ldots, A(d_{free}+4)$,
2	7,7,5	8	3,0,15,0,58	7,5	5	1,4,12,32,80
3	17,15,13	10	6,0,6,0,58	17,15	6	2,7,18,49,130
4	37,33,25	12	12,0,12,0.56	35,23	7	4,12,20,72,225
5	75,53,47	13	1,8,26,20,19	75,53	8	2,36,32,62,332
6	171,165,133	15	7,8,22,44,22	171,133	10	36,0,211,0,1404
7	367,331,225	16	1,0,24,0,113	371,247	10	2,22,60,148,340
8	557,663,711	18	11,0,32,0,195	753,561	12	12,33,0,281,0,2179

Parameters of good and optimal binary convolutional codes suitable for applications with Viterbi decoding algorithm are known. Some examples are presented in Table 4.2. More examples can be found in [6], [17], [18].

4.5 Sequential Decoding

The sequential decoding algorithms were invented by Wozencraft and then by Fano before the discovery of the Viterbi algorithm. Due to the exponential growth of the Viterbi algorithm complexity with the growth of the constraint length of the code the application of this algorithm is limited to the case of relatively small values of the constraint lengths. Unlike the Viterbi algorithm the complexity of the sequential decoding is essentially independent of constraint length. The sequential decoding algorithms belong to the class of backtracking algorithms known as depth first, because they goes forward in the depth of the code tree tracing a given path as long as the metric indicates that the choice is likely, otherwise they return and start tracing the new path. The Viterbi algorithm on the contrary belongs to the class of algorithms known as breadth first, because it explores all the paths on the given stage of trellis before considering the next stage. It is possible that sequential decoding algorithm misses the best path because it does not explore the whole trellis. Hence, the sequential decoding algorithms are the sub-optimal algorithms.

In sequential decoding we have to compare paths of different lengths. To do this usually the *Fano metric* is used. The hard-decision Fano metric for the path $\mathbf{y}$ can be represented as:

$$M_F(\mathbf{r},\mathbf{y}) = \sum_{j=0}^{L-1} M_F(\mathbf{r}_j,\mathbf{y}_j) \tag{4.29}$$

where L is the path length in n-tuples (branches) and $M_F(\mathbf{r}_j,\mathbf{y}_j)$ is the *Fano branch metric*. The Fano branch metric can be written as:

$$M_F(\mathbf{r}_j,\mathbf{y}_j) = (n - d(\mathbf{r}_j,\mathbf{y}_j)) \cdot a + d(\mathbf{r}_j,\mathbf{y}_j) \cdot b \tag{4.30}$$

where

$$\begin{aligned} a &= \log(1-p) + 1 - R \\ b &= \log p + 1 - R \end{aligned} \tag{4.31}$$

p is crossover probability for BSC and $R = \frac{k}{n}$ is the rate of a convolutional code. The Fano branch metric can be expressed as the sum of *Fano symbol metrics*:

$$M_F(\mathbf{r}_j,\mathbf{y}_j) = \sum_{i=0}^{n-1} M_F(r_i, y_i) \tag{4.32}$$

where the Fano symbol metric $M_F(r_i, y_i)$ is:

$$M_F(r_i, y_i) = \begin{cases} a, & \text{if } r_i = y_i \\ b, & \text{if } r_i \neq y_i \end{cases} \tag{4.33}$$

where a and b are represented in (4.31). The Fano metric increases along the correct path and decreases along the incorrect path.

4.5.1 Stack Algorithm

The idea of the stack algorithm is very simple. The decoder creates the list (stack) of more likely candidates to be the correct path. The list is sorted in such a way that more probable candidate (with greatest Fano metric) is always on the top of the stack. So, when the exploration of the code tree is finished the top path is assumed to be the decoded sequence. The algorithm can be formulated as follows [6]:

Stack Decoding *Algorithm:*

1. Load the stack with the root and the metric zero.
2. Remove the top node and place its successors in the stack according to their metrics.
3. If the top path leads to the end of the tree, then stop and choose the top path to be the decoded sequence; otherwise go to step 2.

Example 4.7 Consider the (2, 1, 3) binary convolutional code of Example 4.4. Let the channel be BSC with the crossover probability $p = 0.05$. Let the information sequence be $\mathbf{u} = 01000$ and the corresponding encoded sequence $\mathbf{v} = 00\ 11\ 01\ 11\ 00$. Assume that two errors occur during the transmission and the received sequence is $\mathbf{r} = 10\ 01\ 01\ 11\ 00$. Let us decode the received sequence $\mathbf{r}$ with the help of stack-algorithm using the Fano metric (for hard-decisions). First of all, let us calculate the symbol Fano metric for given convolutional code and BSC:

$$\begin{aligned} a &= \log(1 - 0.05) + 1 - \frac{1}{2} = 0.449 \\ b &= \log 0.05 + 1 - \frac{1}{2} = -2.496 \end{aligned} \tag{4.34}$$

For convenience we will use more approximate values than in (4.34):

$$\begin{aligned} a &= 0.5 && \text{if } y_i = r_i; \\ b &= -2.5 && \text{if } y_i \neq r_i \end{aligned}$$

Then following the steps of stack-algorithm let us find the decoded sequence by exploring the code tree and putting the obtained values of the Fano path metric in the stack.

1. The initial metric value for the root node is 0.
2. Explore paths 00 and 11. Compare corresponding part of the received sequence 10 with branches (paths) 00 and 11. Both branches differ from the received sequence in 1 bit. That means the Fano metric for each branch is equal to $a + b = 0.5 - 2.5 = -2$. Both paths have metric -2. Path 00 is on the top of the stack.

Path	Path metric
00	−2
11	−2

3. Explore branches 00 and 11, which are the successors of the path 00. The corresponding part of the received sequence is 01. Add corresponding branch metrics (−2 and −2) to the path metric. Put the obtained path metrics (−4 and −4) and the corresponding paths in the stack. Now the path 11 with metric −2 is on the top of the stack.

Path	Path metric
11	−2
00 00	−4
00 11	−4

4. Explore branches 01 and 10, which are the successors of the path 11. The corresponding part of the received sequence is 01. Add corresponding branch metrics (+1 and −5) to the path metric. Put the obtained path metrics (−1 and −7) and the corresponding paths in the stack. Now the path 11 01 with metric −1 is on the top of the stack.

Path	Path metric
11 01	−1
00 00	−4
00 11	−4
11 10	−7

5. Explore branches 11 and 00, which are the successors of the path 11 01. The corresponding part of the received sequence is 01. Add corresponding branch metrics (−2 and −2) to the path metric. Put the obtained path metrics (−3 and −3) and the corresponding paths in the stack. Now the path 11 01 11 with metric −3 is on the top of the stack.

Path	Path metric
11 01 11	−3
11 01 00	−3
00 00	−4
00 11	−4
11 10	−7

6. Explore branches 00 and 11, which are the successors of the path 11 01 11. The corresponding part of the received sequence is 11. Add corresponding branch metrics (−5 and +1) to the path metric. Put the obtained path metrics (−8 and −2) and the corresponding paths in the stack. Now the path 11 01 11 11 with metric −2 is on the top of the stack.

Path	Path metric
11 01 11 11	−2
11 01 00	−3
00 00	−4
00 11	−4
11 10	−7
11 01 11 00	−8

7. Explore the successors of path 11 01 11 11. The corresponding part of the received sequence is 00.

Path	**Path metric**
11 01 00	−3
00 00	−4
00 11	−4
11 01 11 11 01	−4
11 01 11 11 10	−4
11 10	−7
11 01 11 00	−8

8. Explore the successors of path 11 01 00. The corresponding part of the received sequence is 11.

Path	**Path metric**
00 00	−4
00 11	−4
11 01 11 11 01	−4
11 01 11 11 10	−4
11 01 00 01	−5
11 01 00 01	−5
11 10	−7
11 01 11 00	−8

9. Explore the successors of path 00 00. The corresponding part of the received sequence is 01.

Path	**Path metric**
00 11	−4
11 01 11 11 01	−4
11 01 11 11 10	−4
11 01 00 01	−5
11 01 00 01	−5
00 00 00	−6
00 00 11	−6
11 10	−7
11 01 11 00	−8

10. Explore the successors of path 00 11. The corresponding part of the received sequence is 01.

Path	**Path metric**
00 11 01	−3
11 01 11 11 01	−4
11 01 11 11 10	−4
11 01 00 01	−5
11 01 00 01	−5
00 00 00	−6
00 00 11	−6
11 10	−7
11 01 11 00	−8
00 11 10	−9

11. Explore the successors of path 00 11 01. The corresponding part of the received sequence is 11.

Path	Path metric
00 11 01 11	−2
11 01 11 11 01	−4
11 01 11 11 10	−4
11 01 00 01	−5
11 01 00 01	−5
00 00 00	−6
00 00 11	−6
11 10	−7
00 11 01 00	−8
11 01 11 00	−8
00 11 10	−9

12. Explore the successors of path 00 11 01 11. The corresponding part of the received sequence is 00. The path on top of the stack reached the end of the code tree. The decoded path is 00 11 01 11 00, which coincide with the transmitted sequence **v**.

Path	Path metric
00 11 01 11 00	−1
11 01 11 11 01	−4
11 01 11 11 10	−4
11 01 00 01	−5
11 01 00 01	−5
00 00 00	−6
00 00 11	−6
11 10	−7
00 11 01 11 11	−7
00 11 01 00	−8
11 01 11 00	−8
00 11 10	−9

The corresponding partially explored code tree is represented in Figure 4.17.

As can be seen from Figure 4.17 we have to explore about $1/3$ of the code tree in this example to find the decoded sequence. The Fano metric can easily be used for the soft-decisions as well and stack-algorithm does not need any changes except the new metric for soft-decision decoding. The serious drawback of the stack-algorithm is the necessity to keep the long list (stack) of paths-candidates and to sort the stack at each stage.

4.5.2 *Fano Algorithm*

The Fano algorithm differs from the stack algorithm in such a way that it explores only the immediate successors or predecessors of the current path. It never "jumps" as the stack-algorithm. It moves along a certain path while the metric exceeds some threshold. In other cases it returns and explores the next successor of the previous partial path.

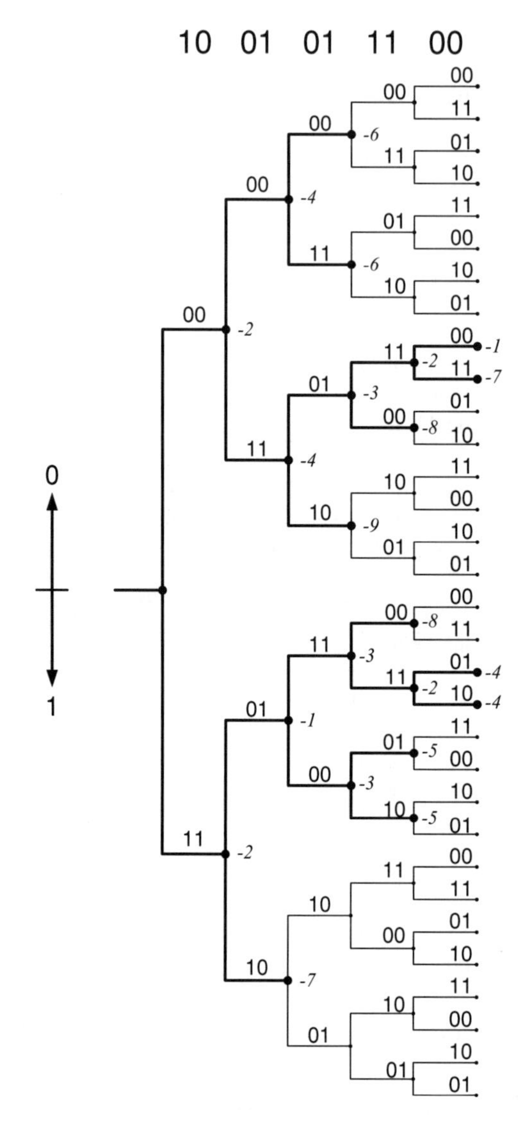

Figure 4.17 The partially explored code tree of Example 4.7. Stack-algorithm

Example 4.8 Consider the decoding of the same received sequence $\mathbf{r} = 10\ 01\ 01\ 11\ 00$ as of Example 4.7 with the help of Fano algorithm. The same Fano symbol metric $a = 0.5$; $b = -2.5$ is used. Let the threshold be -5.

1. Compare branches 00 and 11 with the received symbols 10. Both branches have the same metric -2. The decoder can choose arbitrarily any of them. Let us in this example always choose the moving in the upper part of the code tree if the alternatives are equal. Then the decoder choice is the moving along the path 00. The current path metric is -2.
2. Compare the successors of path 00 branches 00 and 11 with the received symbols 01. Again the branch metrics are equal and the decoder chooses the path 00 00. The current path metric is -4.

3. Compare the successors of path 00 00 branches 00 and 11 with the received symbols 01. The branches again have the equal metrics but now the path metric −6 is less than threshold. That means the decoder returns one step back to the path 00.
4. Explore the next successor of path 00, that is, the branch 11. Now the current path is 00 11 and the current metric is −4.
5. Compare the successors of path 00 11 branches 01 and 10 with the received symbols 01. The branch 01 has the better metric and the new chosen path is 00 11 01 with metric −3.

The decoder continues in this manner and in two steps it finds the decoded sequence 00 11 01 11 00, which is the correct answer. The corresponding code tree is represented in Figure 4.18. The dashed line

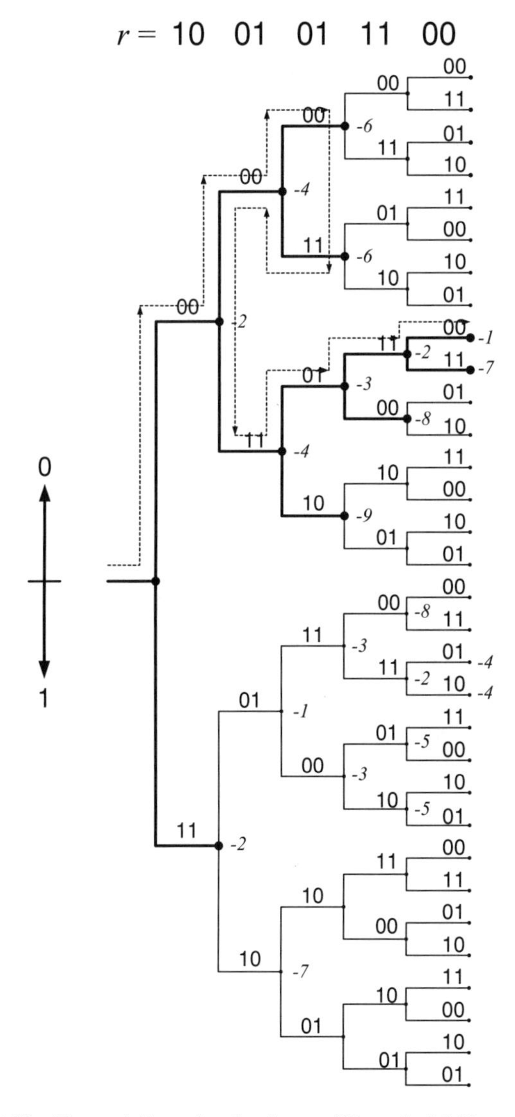

Figure 4.18 The partially explored code tree of Example 4.8. Fano algorithm

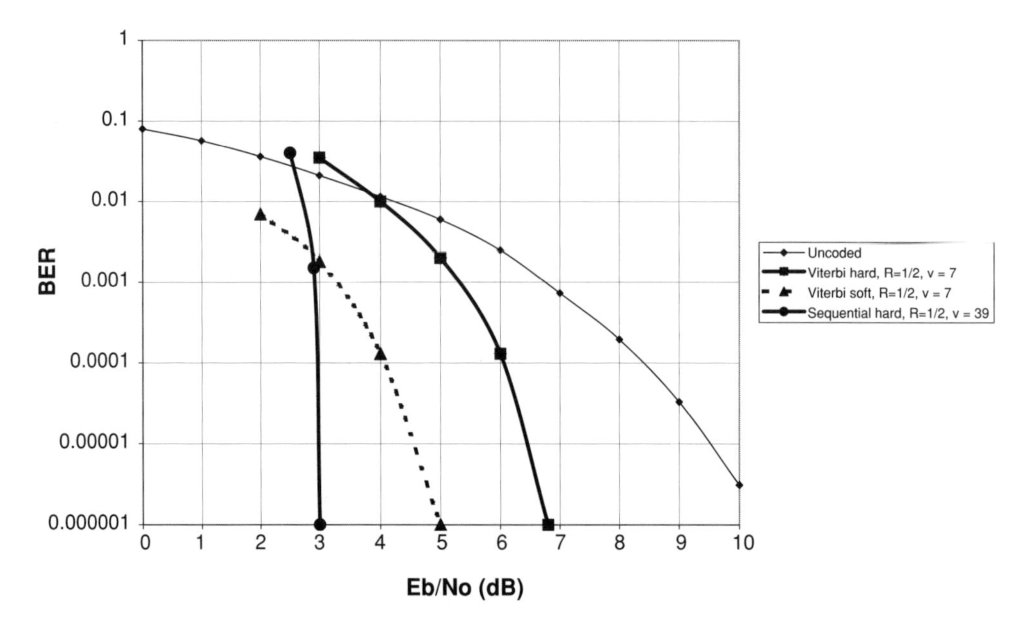

Figure 4.19 Performance comparison of different decoding algorithms for rate 1/2 convolutional codes. BPSK over AWGN channel

shows the moving of the decoder along the tree. In fact, here only the idea of the Fano algorithm is represented. The algorithm itself is more complicated and needs changes in the threshold during the moving along the code tree.

The performance of different decoding algorithms is compared in Figure 4.19 (bit error probability vs. SNR). As one can see from curves in Figure 4.19 the soft decision Viterbi decoding outperforms the hard-decision Viterbi algorithm by 2 dB. The Viterbi decoding of code with the constraint length 7 is compared with the sequential decoding of code with $\nu = 39$ because of significant difference in the decoding complexity of these methods. As can be seen from Figure 4.19 the comparable in complexity Viterbi and sequential decoding algorithms provide significantly different performance. Unfortunately, in real life there is the serious limitation on use of sequential decoding because of the necessity of buffering the input sequence while the algorithm is exploring the code tree. In cases where the input symbol arrival rate exceeds the decoding rate, the buffer will overflow and the data will be lost. And the buffer overflow threshold is a very sensitive function of SNR [5]. This fact restricts the use of sequential decoding algorithms.

4.6 Parallel-Concatenated Convolutional Codes and Soft Input Soft Output Decoding

Parallel-concatenated convolutional codes were introduced by Berrou, Glavieux and Thitimajshima [7]. The encoder of the parallel-concatenated convolutional code (or *turbo-code*) consists of two *recursive systematic convolutional* (RSC) code encoders concatenated in such a way that one of the encoders is fed via an interleaver. The codes are concatenated to employ the principle of iterative decoding. This principle will be discussed later. As mentioned above usually at high values of SNR the systematic

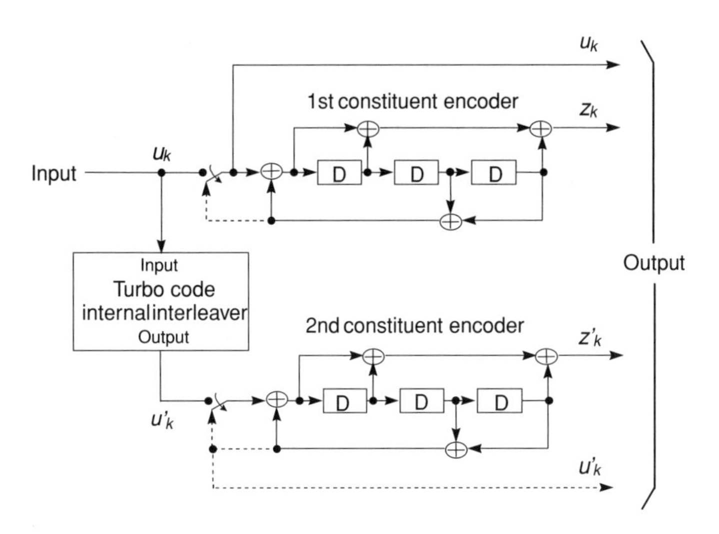

Figure 4.20 Structure of rate 1/3 Turbo encoder (dotted lines apply for trellis termination only)

convolutional codes show worse performance than nonsystematic convolutional codes, but at low SNR values the opposite is true [7].

The structure of the RSC encoder is shown in Figure 4.20. As can be seen from Figure 4.20 the RSC encoder can be represented as infinite impulse response (IIR) filter. In fact, any nonrecursive nonsystematic convolutional code is equivalent to some RSC code in the sense it generates the same set of code sequences. The main difference in nonrecursive nonsystematic and RSC form of the encoder is that for RSC encoder the low weight input sequence could generate an infinite weight output sequence. For example, the unit weight input sequence 00 . . . 00100 . . . will always generate a low weight sequence at the output of the nonrecursive nonsystematic convolutional encoder, which is not the case for the RSC encoder. This is a very important feature for the design of the parallel-concatenated codes. The other important issue in the design of the parallel-concatenated codes is the using of the interleaver. Let us consider the encoder of the turbo code used in 3GPP mobile radio system. The structure of this encoder is shown in Figure 4.20. The output code sequence of the encoder is the result of multiplexing the output sequences of the constituent codes. It is obvious that the weight distribution of the code sequences of the turbo code depends on the way of combining the code sequences of the constituent encoders. For example, the pairing of low weight code sequences of the constituent codes leads to poor performance of the result code. This kind of pairing can be avoided with the help of the interleaver. For the encoder of Figure 4.20 the minimum weight sequences of the constituent encoder is generated by the information sequence . . .0 0 1 0 0 0 1 1 0 0 . . . . The weight of this code sequence is 6. Another information sequence that generates the code sequence of the same weight 6 is . . .0 0 1 1 0 1 0 0 . . . . Due to the interleaver it is not likely that one of these sequences appear at the input of the second constituent encoder if it already appeared at the input of the first encoder. That means the interleaver makes it possible to decrease the probability of combining the two code sequences of minimum weight. As mentioned above, the interleaver cannot help in cases where the low weight code sequence is generated by the information sequence of weight 1, but because of the recursive structure of the constituent encoder of Figure 4.20 the information sequence . . .0 0 1 0 0 . . . generates the code sequence of weight 13. Thus the effect of the recursive structure of the constituent encoder combined with the effect of the interleaver leads to the rise in the distance structure of turbo code. Usually the turbo codes are used for the transmission of the finite length codewords rather than the semi-infinite sequences. In this case some zero tail bits are added at the end of each information sequence, which leads to the flush of the constituent encoders. For the encoder of Figure 4.20 six tail bits are padded after the encoding of information bits. The first three tail bits are used to terminate the first constituent encoder and the second three tail bits are used to terminate the second constituent encoder. For the encoder of Figure 4.20 the flushing is performed by taking the tail bits from the shift register feedback after all information bits are encoded, which corresponds to the lower position of the switches.

An encoder of the type shown in Figure 4.20 is usually employed together with *iterative decoding*. Each iteration of iterative decoding is executed in two phases. Each phase corresponds to the decoding of the codeword of one of two constituent codes. The idea is that additional information about the reliability of the information symbols obtained during decoding of one constituent code should be used in the following phase.

The calculation of likelihood functions used in iterative decoding is based on Bayes' rule:

$$\begin{aligned} P(u = i\,|r) &= \frac{p(r\,|u = i) \cdot P(u = i)}{p(r)}, \quad i = 0, \ldots, M-1 \\ p(r) &= \sum_{i=0}^{M-1} p(r\,|u = i) \cdot P(u = i) \end{aligned} \tag{4.35}$$

where u is the symbol transmitted over the channel, M is the cardinal number of the alphabet, r is the random variable at the channel output, $P(u = i\,|r)$ is the a posteriori probability (APP) of the decision that transmitted symbol $u = i$ conditioned on r, $p(r\,|u = i)$ is the probability density function (pdf) of the random variable r conditioned on the transmitted symbol $u = i$, $P(u = i)$ is the *a priori probability* of occurrence symbol i at the channel input.

We will consider the binary case, that is, $M = 2$. Let the binary symbols 0 and 1 be represented by the voltages -1 and $+1$ respectively. Then the *maximum a posteriori* (MAP) rule states that the decision $(u = +1)$ should be chosen in case $P(u = +1\,|r) > P(u = -1\,|r)$ and otherwise the decision $(u = -1)$ should be chosen in case $P(u = +1\,|r) < P(u = -1\,|r)$. The MAP rule provides the minimum probability of error. The MAP conditions can be written in terms of likelihood ratios:

$$\begin{aligned} &\frac{P(u = +1\,|r)}{P(u = -1\,|r)} > 1, \quad \text{decision } (u = +1) \\ &\frac{P(u = +1\,|r)}{P(u = -1\,|r)} < 1, \quad \text{decision } (u = -1) \end{aligned} \tag{4.36}$$

Using the Bayes' rule the *likelihood ratio* $\frac{P(u=+1|r)}{P(u=-1|r)}$ can be written as follows:

$$\frac{P(u = +1\,|r)}{P(u = -1\,|r)} = \frac{p(r\,|u = +1) \cdot P(u = +1)}{p(r\,|u = -1) \cdot P(u = -1)} \tag{4.37}$$

In practice, of more use is the metric called *log-likelihood ratio* (LLR). It can be obtained by taking the logarithm of likelihood ratio and is denoted by $L(u\,|r)$. Using (4.37) obtain:

$$L(u\,|r) = \ln\left(\frac{p(r\,|u = +1) \cdot P(u = +1)}{p(r\,|u = -1) \cdot P(u = -1)}\right) = \ln\left(\frac{p(r\,|u = +1)}{p(r\,|u = -1)}\right) + \ln\left(\frac{P(u = +1)}{P(u = -1)}\right) \tag{4.38}$$

or

$$L(u\,|r) = L_c(r) + L(u) \tag{4.39}$$

The value $L(u\,|r)$ can be interpreted as the soft decision output of the demodulator; the value $L_c(r)$ as the reliability of the detected symbol, which can be obtained by the measurement of the channel at the receiver input; and $L(u)$ is the a priori LLR of the transmitted symbol (bit).

For the decoder of a systematic code the information is available both from the received information symbols and from the redundant symbols. The information available from the redundant symbols is called the *extrinsic information*. The information from the code stream $L(u)$ is called the *intrinsic information*.

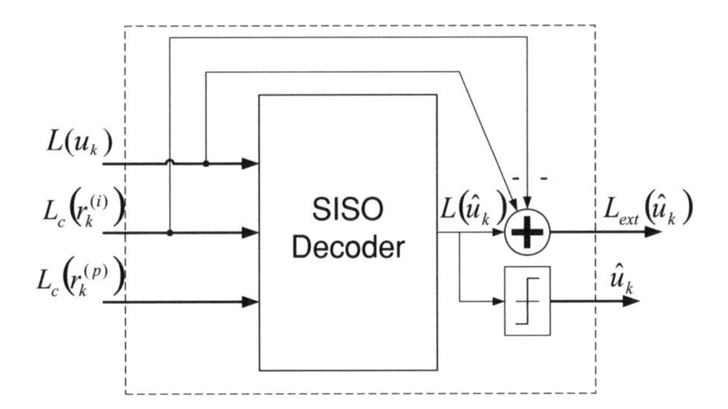

Figure 4.21 A constituent Soft Input/Soft Output decoder

It was shown in [7] that for the systematic codes the extrinsic information does not depend on decoder input. Then it is possible to obtain the soft output of the decoder in the following form:

$$L(\hat{u}) = L(u\,|r) + L_{ext}(\hat{u}) \tag{4.40}$$

where $L_{ext}(\hat{u})$ denotes the extrinsic information. Substituting (4.30) into (4.40) obtain:

$$L(\hat{u}) = L_c(r) + L(u) + L_{ext}(\hat{u}) \tag{4.41}$$

The decoder soft output $L(\hat{u})$ represents both the hard decision itself and the reliability of that hard decision. The hard decision is defined by the sign of $L(\hat{u})$ and the magnitude of $L(\hat{u})$ defines the reliability of the hard decision. We can regard the extrinsic information $L_{ext}(\hat{u})$ as the improvement of the reliability of the received information symbol. The idea of the iterative decoding is to forward to the next phase of decoding only the extrinsic information $L_{ext}(\hat{u})$, since the information from the information symbols is already available to it.

A constituent decoder that accepts a priori information at its input and produces a posteriori information at its output is called a *soft-input/soft-output* (SISO) decoder. The constituent SISO decoder is shown in Figure 4.21. The inputs of the SISO decoder are the information LLRs $L_c(r_k^{(i)})$ that agrees with the transmitted information bits u_k, the parity LLRs $L_c(r_k^{(p)})$ that agrees with the transmitted parity bits z_k or z'_k, and the a priori information $L(u_k)$. The decoder output is $L(\hat{u}_k)$. In accordance with (4.41) we can obtain the extrinsic information by subtracting from the output the inputs $L_c(r_k^{(i)})$ and $L(u_k)$. Also the sign of the output LLR $L(\hat{u}_k)$ gives us the hard decision, that is, the decoded information bits $\hat{u}_k$. Regarding these operations of subtraction and comparison with threshold as the inner operations of the decoder we can say that the outputs of the SISO constituent decoder are the extrinsic information $L_{ext}(\hat{u}_k)$ and the decoded information bits $\hat{u}_k$. This form of the decoder corresponds to the dashed line in Figure 4.21.

The scheme of a standard turbo decoder is shown in Figure 4.22. The first SISO decoder receives the information LLRs $L_c(r_k^{(i)})$ that agrees with the transmitted information bits u_k, the parity LLRs $L_c(r_k^{(p)})$ that agrees with the transmitted parity bits z_k, and the a priori information $L(u_k)$. For the first decoding iteration it is assumed that the a priori information $L(u_k) = 0$, that is, the information bits assumed to be equally likely. The output of the first SISO decoder is the extrinsic information $L_{ext}(\hat{u}_k)$. The second SISO decoder receives the interleaved information LLRs $L'_c(r_k^{(i)})$ that agrees with the interleaved information bits u'_k and the parity LLRs $L'_c(r_k^{(p)})$ that agrees with the transmitted parity bits z'_k. The interleaved extrinsic information obtained from the output of the first SISO decoder is fed to the input of the second SISO decoder and is used as the a priori information. The extrinsic information produced by the second SISO decoder in turn is used after deinterleaving as the input a priori information for

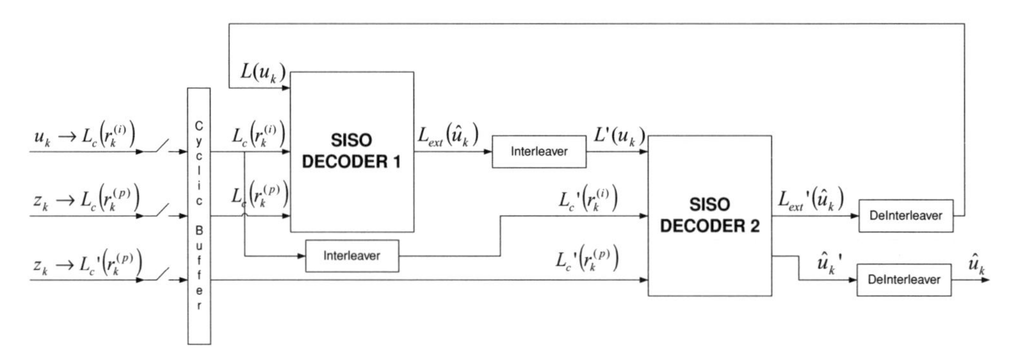

Figure 4.22 A standard turbo decoder

the first SISO decoder in the next decoding iteration. It is extrinsic information that is passed between constituent SISO decoders, rather than the decoded data. As mentioned above usually the constituent encoders of turbo encoder are terminated by the tail bits. Similarly, the constituent decoders of the turbo decoder work on a block-by-block principle. The LLRs corresponding to the block of received symbols (bits) are stored in the cyclic buffers and are fed to the inputs of the constituent SISO decoders at each iteration. Usually several iterations are needed to achieve the required performance. After the last iteration the decoded information bits (hard decisions) $\hat{u}_k$ can be obtained from the output of the second SISO decoder.

In Figure 4.23 the curves of bit error probability vs. SNR for the decoder with a different number of iterations are represented. As one can see, the performance of the turbo decoder increases with increasing the number of iterations.

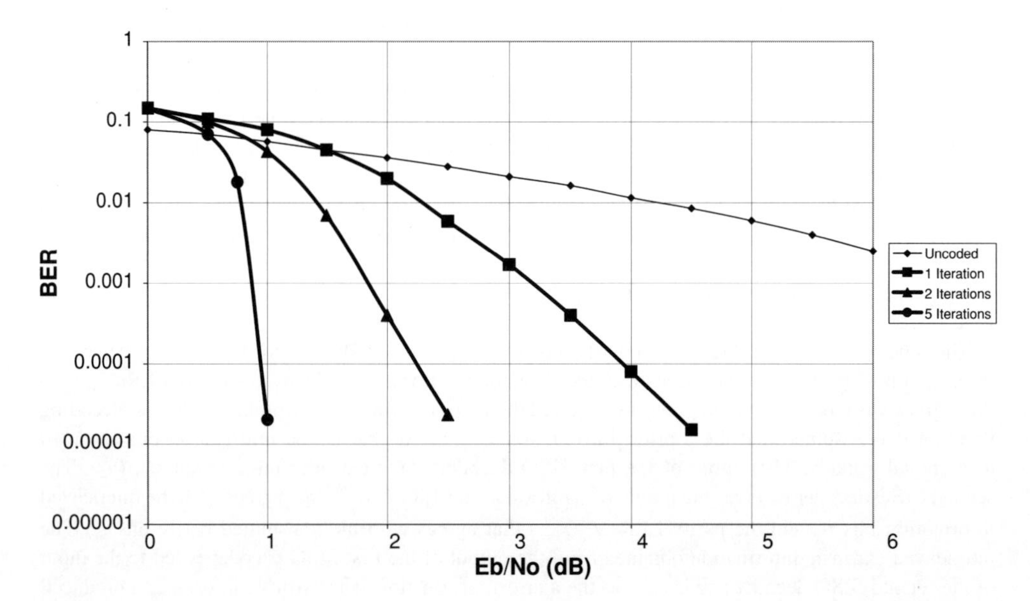

Figure 4.23 Performance of turbo decoder with different number of iterations. BPSK over AWGN, MAP algorithm

4.7 SISO Decoding Algorithms

The main part of the turbo decoder is the SISO constituent decoder. Several different algorithms can be used to implement the SISO decoder. These algorithms can be split into two groups. One group represents the algorithms derived from the Viterbi algorithm and the second group includes algorithms based on the *Maximum A Posteriori* (MAP) *algorithm*. In this section we will follow mostly the [8].

4.7.1 MAP Algorithm and Its Variants

The symbol-by-symbol MAP algorithm was introduced in 1974 by Bahl, Cocke, Jelinek and Raviv [9] for decoding the convolutional codes. Very often the MAP algorithm is called the BCJR algorithm. Unlike the Viterbi algorithm that minimizes the probability of error per sequence, the MAP algorithm minimizes the probability of error per symbol. The implementation of the MAP algorithm is close to the implementation of the Viterbi algorithm performing in forward-backward directions over a block of code symbols. The MAP algorithm finds for each decoded bit u_k the a posteriori LLR $L\left(u_k\left|\mathbf{r}\right.\right)$, where $\mathbf{r}$ is the received sequence. This LLR corresponds to $L(\hat{u})$ in (4.41). We will denote it as $L(\hat{u}_k)$. Let us consider one trellis section of a terminated trellis shown in Figure 4.24. If the previous state $S_{k-1}=i$ and the present state $S_k=j$ are known in a trellis then the input bit u_k, which caused the transition between these two states is also known. Then using the Bayes' rule we can write $L\left(u_k\left|\mathbf{r}\right.\right)$ as follows:

$$L(\hat{u}_k)=L\left(u_k\left|r\right.\right)=\ln\left(\frac{\sum\limits_{(i,j)\Rightarrow u_k=+1} P\left(S_{k-1}=i,\ S_k=j,\ \mathbf{r}\right)}{\sum\limits_{(i,j)\Rightarrow u_k=-1} P\left(S_{k-1}=i,\ S_k=j,\ \mathbf{r}\right)}\right) \tag{4.42}$$

where $(i,j)\Rightarrow u_k=+1$ is the set of transitions from the state $S_{k-1}=i$ to the state $S_k=j$ that can occur if the input bit $u_k=+1$, and similarly for $(i,j)\Rightarrow u_k=-1$ and $P\left(S_{k-1}=i,\ S_k=j,\ \mathbf{r}\right)$ is the joint probability that the previous state $S_{k-1}=i$ and the present state $S_k=j$ conditioned on the received binary sequence $\mathbf{r}$. In the case of continuous output channel $P\left(S_{k-1}=i,\ S_k=j,\ \mathbf{r}\right)$ is the probability density function. The received sequence $\mathbf{r}$ can be split into three sections: the received symbols associated with the current transition $\mathbf{r}_k$, the received sequence prior the current transition $\mathbf{r}_{prev}$ and the received sequence after the current transition $\mathbf{r}_{post}$ as shown in Figure 4.24.

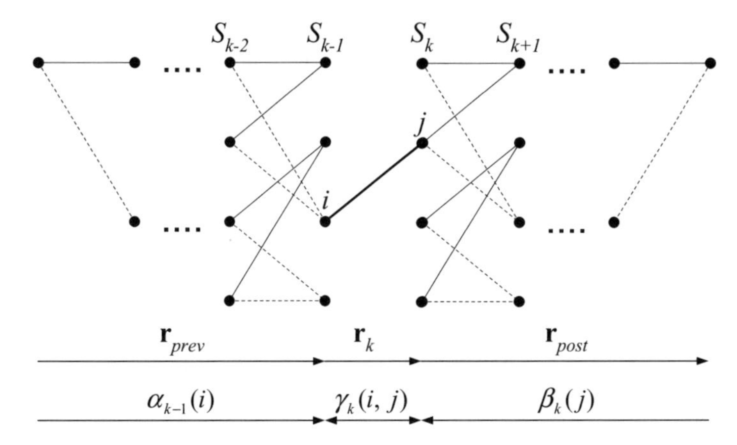

Figure 4.24 A section of MAP decoder trellis

Assuming that the channel is memoryless the properties of the Markov process can be used to write the probability $P\left(S_{k-1}=i,\ S_k=j,\ \mathbf{r}\right)$ as follows:

$$\begin{aligned}&P\left(S_{k-1}=i,\ S_k=j,\ \mathbf{r}\right)=P\left(S_{k-1}=i,\ S_k=j,\ \mathbf{r}_{prev},\ \mathbf{r}_k,\ \mathbf{r}_{post}\right)=\\&P\left(\mathbf{r}_{post}\left|S_k=j\right.\right)\cdot P\left(S_{k-1}=i,\ S_k=j,\ \mathbf{r}_{prev},\ \mathbf{r}_k\right)=\\&P\left(\mathbf{r}_{post}\left|S_k=j\right.\right)\cdot P\left(\left\{S_k=j,\ \mathbf{r}_k\right\}\left|S_{k-1}=i\right.\right)\cdot P\left(S_{k-1}=i,\ \mathbf{r}_{prev}\right)\end{aligned} \tag{4.43}$$

Let us denote the probability that the trellis is in state i at time moment $k-1$ and the received sequence prior to time moment k is $\mathbf{r}_{prev}$ as $\alpha_{k-1}(i)$:

$$\alpha_{k-1}(i)=P\left(S_{k-1}=i,\ \mathbf{r}_{prev}\right) \tag{4.44}$$

the probability that the received sequence after time moment k is $\mathbf{r}_{post}$, conditioned on the trellis is in state j at time moment k as $\beta_k(j)$:

$$\beta_k(j)=P\left(\mathbf{r}_{post}\left|S_k=j\right.\right) \tag{4.45}$$

and the probability of transition from state i at the time moment k-1 to state j at time moment k caused by the received symbols $\mathbf{r}_k$ as $\gamma_k(i,\ j)$:

$$\gamma_k(i,\ j)=P\left(\left\{S_k=j,\ \mathbf{r}_k\right\}\left|S_{k-1}=i\right.\right) \tag{4.46}$$

The value $\gamma_k(i,\ j)$ is called the branch metric associated with the transition $i\rightarrow j$. Substituting (4.43), (4.44), (4.45) and (4.46) into (4.42) obtain:

$$L\left(\hat{u}_k\right)=L\left(u_k\left|\mathbf{r}\right.\right)=\ln\left(\frac{\sum\limits_{(i,j)\Rightarrow u_k=+1}\alpha_{k-1}(i)\cdot\beta_k(j)\cdot\gamma_k(i,\ j)}{\sum\limits_{(i,j)\Rightarrow u_k=-1}\alpha_{k-1}(i)\cdot\beta_k(j)\cdot\gamma_k(i,\ j)}\right) \tag{4.47}$$

The MAP algorithm finds probabilities $\alpha_k(j)$ and $\beta_k(j)$ for all possible states j throughout the trellis, that is, for all values of k; and the branch metric $\gamma_k(i,\ j)$ for all branches in the trellis.

Let us consider the calculation of $\alpha_k(i)$, $\beta_k(i)$ and $\gamma_k(i,\ j)$.

Consider the definition of the branch metric $\gamma_k(i,\ j)$ in (4.46). Using the Bayes' rule we can write (4.46) as follows:

$$\begin{aligned}&\gamma_k(i,\ j)=P\left(\left\{S_k=j,\ \mathbf{r}_k\right\}\left|S_{k-1}=i\right.\right)=P\left(\mathbf{r}_k\left|\left\{S_{k-1}=i,\ S_k=j\right\}\right.\right)\cdot P\left(S_k=j\left|S_{k-1}=i\right.\right)=\\&P\left(\mathbf{r}_k\left|\left\{S_{k-1}=i,\ S_k=j\right\}\right.\right)\cdot P\left(u_k\right)=P\left(\mathbf{r}_k\left|\mathbf{x}_k\right.\right)\cdot P\left(u_k\right)\end{aligned} \tag{4.48}$$

where u_k is the input information bit necessary to cause the transition from state $S_{k-1}=i$ to state $S_k=j$; $P\left(u_k\right)$ is a priori probability of bit u_k.

Obviously the event $\left(S_{k-1}=i,\ S_k=j\right)$ coincides with the event that bit vector $\mathbf{x}_k$ generated by the transition $\left(S_{k-1}=i\right)\rightarrow\left(S_k=j\right)$ was transmitted over the channel. Then (4.48) can be written as:

$$\gamma_k(i,\ j)=P\left(\mathbf{r}_k\left|\left\{S_{k-1}=i,\ S_k=j\right\}\right.\right)\cdot P\left(u_k\right)=P\left(\mathbf{r}_k\left|\mathbf{x}_k\right.\right)\cdot P\left(u_k\right) \tag{4.49}$$

If the encoder forms n output symbols (bits) during each transition the probability $P\left(\mathbf{r}_k\left|\mathbf{x}_k\right.\right)$ can be written in the following way:

$$P\left(\mathbf{r}_k\left|\mathbf{x}_k\right.\right)=\prod_{s=0}^{n-1}P\left(r_{k,s}\left|x_{k,s}\right.\right) \tag{4.50}$$

where $x_{k,s}$ is the transmitted bit and $r_{k,s}$ is the corresponding received symbol. Let us assume the channel model to be flat fading with Gaussian noise. Then the probability that the received symbol is

r conditioned on the transmitted bit x is:

$$P(r|x) = \frac{\sqrt{E_b}}{\sqrt{\pi \cdot N_0}} \cdot e^{-\frac{E_b}{N_0}(r-ax)^2} \tag{4.51}$$

where $\frac{E_b}{N_0}$ is the signal to noise ratio per bit; a is the fading amplitude (for nonfading Gaussian channel $a = 1$). Then the branch metric can be calculated in the following way:

$$\gamma_k(i, j) = P(u_k) \cdot \prod_{s=0}^{n} \frac{\sqrt{E_b}}{\sqrt{\pi \cdot N_0}} \cdot e^{-\frac{E_b}{N_0}\left(r_{k,s} - ax_{k,s}\right)^2} \tag{4.52}$$

The a priori probability $P(u_k)$ can be derived from the input a priori LLR $L(u_k)$:

$$L(u_k) = \ln\left(\frac{P(u_k = +1)}{P(u_k = -1)}\right) \tag{4.53}$$

Solving the (4.53) for $P(u_k = +1)$ or for $P(u_k = -1)$, obtain:

$$\begin{aligned} P(u_k = +1) &= \frac{e^{L(u_k)}}{1 + e^{L(u_k)}}, \\ P(u_k = -1) &= \frac{1}{1 + e^{L(u_k)}} \end{aligned} \tag{4.54}$$

Taking into account that for the calculation of the numerator in (4.47) we need in $P(u_k = +1)$ rather than $P(u_k)$ and for the calculation of the denominator we need in $P(u_k = -1)$ rather than $P(u_k)$, the (4.54) can be used for calculation of the branch metrics in accordance with (4.49) or (4.52).

Consider $\alpha_k(j)$. From the definition of $\alpha_{k-1}(i)$ in (4.44) we can write:

$$\alpha_k(j) = P\left(S_k = j,\ \mathbf{r}_{prev},\ \mathbf{r}_k\right) = \sum_i P\left(S_{k-1} = i,\ S_k = j,\ \mathbf{r}_{prev},\ \mathbf{r}_k\right) \tag{4.55}$$

Using the Bayes' rule and the properties of Markov process we can write (4.55) as follows:

$$\begin{aligned} &\alpha_k(j) = \sum_i P\left(S_{k-1} = i,\ S_k = j,\ \mathbf{r}_{prev},\ \mathbf{r}_k\right) = \\ &\sum_i P\left(\{S_k = j,\ \mathbf{r}_k\} \left| \{S_{k-1} = i,\ \mathbf{r}_{prev}\}\right.\right) \cdot P\left(S_{k-1} = i,\ \mathbf{r}_{prev}\right) = \\ &\sum_i P\left(\{S_k = j,\ \mathbf{r}_k\} | S_{k-1} = i\right) \cdot P\left(S_{k-1} = i,\ \mathbf{r}_{prev}\right) = \sum_i \alpha_{k-1}(i) \cdot \gamma_k(i,\ j) \end{aligned} \tag{4.56}$$

That means the probabilities $\alpha_k(j)$ can be calculated recursively. Assuming that the trellis has the initial state $S_0 = 0$, the initial conditions for this recursion are:

$$\begin{aligned} &\alpha_0(S_0 = 0) = 1 \\ &\alpha_0(S_0 = i) = 0, \quad i \neq 0 \end{aligned} \tag{4.57}$$

This is the forward recursion. Using the same technique it can be shown that the probabilities $\beta_k(i)$ can be calculated with the help of backward recursion:

$$\beta_k(i) = \sum_j \beta_{k+1}(j) \cdot \gamma_{k+1}(i,\ j) \tag{4.58}$$

Assuming the length of information sequence equal to K and that the tail bits put the encoder in the zero state, the conditions for the backward recursion can be written as follows:

$$\begin{aligned} &\beta_K(S_K = 0) = 1 \\ &\beta_K(S_K = i) = 0, \quad i \neq 0 \end{aligned} \tag{4.59}$$

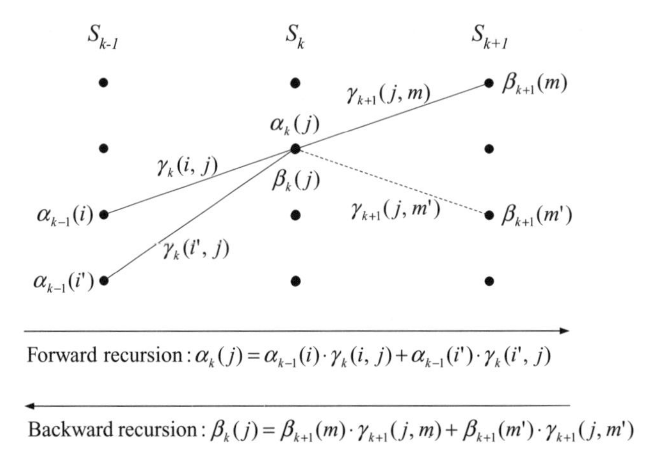

Figure 4.25 Calculation of forward and backward recursion in MAP algorithm

The process of calculating forward and backward recursion in MAP algorithm is illustrated in Figure 4.25.

From the description given above, we see that the MAP decoding of a received sequence **r** to give the a posteriori LLRs $L(u_k|\mathbf{r})$ can be carried out as follows. As the channel values $r_{k,s}$ are received, they and the a priori LLRs $L(u_k)$ are used to calculate $\gamma_k(i, j)$ according to (4.49) or (4.52). Then the forward recursion in accordance with (4.56) and (4.57) can be used to calculate $\alpha_{k-1}(i)$. When all channel values have been received the backward recursion in accordance with (4.58) and (4.59) can be used to calculate $\beta_k(j)$. Finally, all the calculated values $\alpha_{k-1}(i)$, $\beta_k(j)$ and $\gamma_k(i, j)$ are used to calculate a posteriori LLRs $L(u_k|\mathbf{r})$ in accordance with (4.47).

In accordance with (4.41) the a posteriori LLR $L(u_k|\mathbf{r})$ can be represented as follows:

$$L(u_k|\mathbf{r}) = L_c\left(r_k^{(i)}\right) + L(u_k) + L_{ext}(\hat{u}_k) \tag{4.60}$$

where $L_c(r_k^{(i)})$ is the information LLR that agrees with the transmitted information bit u_k and $L_{ext}(\hat{u}_k)$ is the extrinsic information obtained as a result of decoding. As mentioned above we are interested in passing to the next decoder the extrinsic information rather than the a posteriori LLR $L(u_k|\mathbf{r})$. To obtain extrinsic information $L_{ext}(\hat{u}_k)$ we should subtract values of $L_c(r_k^{(i)})$ and $L(u_k)$ from the a posteriori LLR $L(u_k|\mathbf{r})$. In accordance with (4.38) the information LLR $L_c(r_k^{(i)})$ can be represented as:

$$L_c\left(r_k^{(i)}\right) = \ln\left(\frac{p\left(r_k^{(i)}|u_k=+1\right)}{p\left(r_k^{(i)}|u_k=-1\right)}\right) \tag{4.61}$$

Assuming the channel model to be flat fading with Gaussian noise, (4.61) can be written as:

$$L_c\left(r_k^{(i)}\right) = 4\frac{E_b}{N_0}\cdot a\cdot r_k^{(i)} \tag{4.62}$$

where $r_k^{(i)}$ is the received symbol corresponding to the transmitted information bit u_k and a is the fading amplitude. For simplicity (4.62) can be rewritten as:

$$L_c\left(r_k^{(i)}\right) = L_c\cdot r_k^{(i)} \tag{4.63}$$

where

$$L_c = 4\frac{E_b}{N_0} \cdot a \tag{4.64}$$

The MAP algorithm is, in the form described above, extremely complex. However, much work was done to reduce its complexity. Initially the Max-Log-MAP algorithm was proposed by Koch and Bayer [10] and Efranian et al. [11]. This technique simplifies the MAP algorithm by transferring the recursions into the log domain and invoking the approximation to dramatically reduce the complexity. Because of this approximation its performance is suboptimal compared to that of the MAP algorithm. Later Robertson [12] proposed the Log-MAP algorithm, which corrected the approximation used in the Max-Log-MAP algorithm and hence gave a performance almost identical to that of the MAP algorithm, but at a fraction of its complexity.

Max-Log-MAP Algorithm

The Max-Log-MAP algorithm simplifies the calculations of MAP algorithm by transferring the equations into the log arithmetic domain and then using the approximation:

$$\ln\left(\sum_i e^{x_i}\right) \approx \max_i(x_i) \tag{4.65}$$

Then, with $\mathrm{A}_k(i)$, $\mathrm{B}_k(i)$ and $\Gamma_k(i, j)$ defined as follows:

$$\begin{aligned} &\mathrm{A}_k(i) = \ln(\alpha_k(i)),\\ &\mathrm{B}_k(i) = \ln(\beta_k(i)),\\ &\Gamma_k(i, j) = \ln(\gamma_k(i, j)) \end{aligned} \tag{4.66}$$

we can rewrite (4.56) as

$$\begin{aligned} &\mathrm{A}_k(j) = \ln(\alpha_k(j)) = \ln\left(\sum_i \alpha_{k-1}(i)\cdot\gamma_k(i,\ j)\right) = \ln\left(\sum_i e^{\mathrm{A}_{k-1}(i)+\Gamma_k(i,\ j)}\right) \approx\\ &\max_i\,(\mathrm{A}_{k-1}(i) + \Gamma_k(i,\ j)) \end{aligned} \tag{4.67}$$

(4.67) implies that for each path in Figure 4.24 from the previous stage in the trellis to the state $S_k = j$ at the present stage, the algorithm adds a branch metric term $\Gamma_k(i, j)$ to the previous value $\mathrm{A}_{k-1}(i)$ to find a new value $\mathrm{A}_k(j)$ for that path. The new value of $\mathrm{A}_k(j)$ according to (4.67) is then the maximum of the values of the various paths reaching the state $S_k = j$. This can be thought of as selecting one path as the "survivor" and discarding any other paths reaching the state. The value of $\mathrm{A}_k(j)$ should give the natural logarithm of the probability that the trellis is in state $S_k = j$ at stage k, given that the received channel sequence up to this point has been $\mathbf{r}_{prev}$. However, because of the approximation of (4.65) used to derive (4.67), only the Maximum Likelihood (ML) path through the state $S_k = j$ is considered when calculating this probability. Thus, the value of $\mathrm{A}_k(j)$ in the Max-Log-MAP algorithm actually gives the probability of the most likely path through the trellis to the state $S_k = j$, rather than the probability of *any* path through the trellis to state $S_k = j$. This approximation is one of the reasons for the sub-optimal performance of the Max-Log-MAP algorithm compared to the MAP algorithm.

We see from (4.67) that in the Max-Log-MAP algorithm the forward recursion used to calculate $\mathrm{A}_k(j)$ is exactly the same as the forward recursion in the Viterbi algorithm – for each pair of merging paths the survivor is found using two additions and one comparison. Notice that for binary trellises the summation, and maximization, over all previous states $S_{k-1} = i$ in (4.67) will in fact be over only two states, because there will be only two previous states $S_{k-1} = i$ with paths to the present state $S_k = j$. For all other values of S_{k-1} we will have $\gamma_k(i, j) = 0$.

Similarly to (4.67) for the forward recursion used to calculate the $A_k(j)$, we can rewrite (4.58) as:

$$B_k(i) = \ln(\beta_k(i)) \approx \max_j (B_{k+1}(j) + \Gamma_{k+1}(i,\ j)) \tag{4.68}$$

giving the backward recursion used to calculate the $B_k(i)$values. Again, this is equivalent to the recursion used in the Viterbi algorithm except it proceeds backward rather than forward through the trellis.

Using (4.49) and (4.52), we can write the branch metrics in the above recursive (4.67) and (4.68) as:

$$\Gamma_k(i,\ j) = \ln(\gamma_k(i,\ j)) = C + \frac{1}{2}u_k L(u_k) + \frac{L_c}{2}\sum_{s=0}^{n-1} r_{k,s} \cdot x_{k,s} \tag{4.69}$$

where C does not depend on u_k or on the transmitted bits sequence $\mathbf{x}_k$ and so can be considered a constant and omitted. Hence, the branch metric is equivalent to that used in the Viterbi algorithm, with the addition of the a priori LLR term $u_k L(u_k)$. Furthermore, the correlation term $\sum_{s=0}^{n-1} r_{k,s} \cdot x_{k,s}$is weighted by the channel reliability value L_c of (4.64).

Finally, from (4.47), we can write for the a posteriori LLRs $L(\hat{u}_k) = L(u_k\,|\mathbf{r})$ that the Max-Log-MAP algorithm calculates:

$$\begin{aligned} L(\hat{u}_k) = L(u_k\,|\mathbf{r}) = \ln\left(\frac{\sum\limits_{(i,j)\Rightarrow u_k=+1} \alpha_{k-1}(i)\cdot\beta_k(j)\cdot\gamma_k(i,\ j)}{\sum\limits_{(i,j)\Rightarrow u_k=-1} \alpha_{k-1}(i)\cdot\beta_k(j)\cdot\gamma_k(i,\ j)}\right) \approx \\ \max_{(i,j)\Rightarrow u_k=+1} (A_{k-1}(i) + B_k(j) + \Gamma_k(i,\ j)) - \max_{(i,j)\Rightarrow u_k=-1} (A_{k-1}(i) + B_k(j) + \Gamma_k(i,\ j)) \end{aligned} \tag{4.70}$$

This means that in the Max-Log-MAP algorithm for each bit u_k the a posteriori LLR $L(\hat{u}_k) = L(u_k\,|\mathbf{r})$ is calculated by considering every transition from the trellis stage S_{k-1} to the stage S_k. These transitions are grouped into those that might have occurred if $u_k = +1$, and those that might have occurred if $u_k = -1$. For both of these groups the transition giving the maximum value of $A_{k-1}(i) + B_k(j) + \Gamma_k(i,\ j)$ is found, and the a posteriori LLR is calculated based on only these two "best" transitions.

Log-MAP Algorithm

Due to the approximation in (4.65) used in the Max-Log-MAP algorithm it has worse performance in comparison with MAP algorithm. Robertson et al. [12] proposed to use instead of approximation in (4.65) the exact formula:

$$\ln(e^{x_1} + e^{x_2}) = \max(x_1,\ x_2) + \ln(1 + e^{-|x_1 - x_2|}) \tag{4.71}$$

where $\ln(1 + e^{-|x_1 - x_2|})$ can be regarded as a correction factor that will tend to zero as the difference between the arguments increases. Similarly to the Max-Log-MAP algorithm, values for $A_k(j) = \ln(\alpha_k(j))$ and $B_k(j) = \ln(\beta_k(j))$ are calculated using a forward and a backward recursion. However, the maximization in (4.67) and (4.68) is complemented by the correction factor in (4.71). This means that the exact rather than approximate values of $A_{k-1}(i)$ and $B_k(j)$ are calculated. The correction factor can be stored in a look-up table. This means that the Log-MAP algorithm is only slightly more complex than the Max-Log-MAP algorithm, but it gives almost the same performance as the MAP algorithm. Depending on the size of used look-up table the performance of the Log-MAP algorithm can achieve the performance of MAP algorithm.

4.7.2 Soft-In/Soft-Out Viterbi Algorithm (SOVA)

The Soft-In/Soft-Out Viterbi Algorithm (SOVA) was proposed by Hagenauer [13]. The SOVA operates similarly to the Viterbi algorithm except the ML sequence is found with the help of modified metric. The path metrics used in SOVA are modified to take account of a priori information when selecting the ML path through the trellis. As shown above the Max-Log-MAP algorithm also outputs the ML sequence over the whole trellis. Moreover, the recursion defined by (4.67) selects the metric corresponding to the ML path to node j, which in terms of the Viterbi algorithm is the survivor path. Thus the forward recursion performs the same operations as the Viterbi algorithm [14]. Due to this fact the Viterbi algorithm can be modified so that it provides a soft output in the form of the a posteriori LLR $L(\hat{u}_k) = L(u_k|\mathbf{r})$ for each decoded bit.

Consider the state sequence $\mathbf{S}_k^{(m)}$, which gives the states along the surviving path m at stage k in the trellis. The Viterbi algorithm searches for the state sequence $\mathbf{S}^{(m)}$ that maximizes the a posteriori probability $P\left(\mathbf{S}^{(m)}|\mathbf{r}\right)$. By using Bayes' rule the a posteriori probability can be written as:

$$P\left(\mathbf{S}^{(m)}|\mathbf{r}\right) = p\left(\mathbf{r}|\mathbf{S}^{(m)}\right) \cdot \frac{P\left(\mathbf{S}^{(m)}\right)}{p(\mathbf{r})} \tag{4.72}$$

Since the received sequence $\mathbf{r}$ is fixed and does not depend on m it can be discarded. Then we can equivalently maximize:

$$\max_m p\left(\mathbf{r}|\mathbf{S}^{(m)}\right) \cdot P\left(\mathbf{S}^{(m)}\right) \tag{4.73}$$

This maximization is realized in the code trellis, when for each state $\mathbf{S}^{(m)}$ and each stage k, the path with the largest probability $P(\mathbf{S}_k^{(m)}, \mathbf{r}_k)$ is selected. This probability can be calculated by multiplying the branch transition probabilities associated to path m. They are $\gamma_l\left(i^{(m)}, j^{(m)}\right)$ for $1 \le l \le k$ and defined in (4.49). The maximum is not changed if we take the logarithm, and hence the metric computations are the same as described for the forward recursion of the Max-Log-MAP algorithm. The values $\mathrm{A}_k(i)$ and $\mathrm{B}_k(i)$ from (4.67) and (4.68) are additive and the same for all paths m and therefore are irrelevant for the maximization. Let us denote the path m entering state $S_k = j^{(m)}$ at the stage k by $\mathbf{S}_k^{(j^{(m)})}$. Then if the path $\mathbf{S}_k^{(j^{(m)})}$ at the kth stage has the path $\mathbf{S}_{k-1}^{(i^{(m)})}$ as its prefix and assuming the memoryless channel we can choose the following metric $M(\mathbf{S}_k^{(j^{(m)})})$ for the path $\mathbf{S}_k^{(j^{(m)})}$:

$$M\left(\mathbf{S}_k^{(j^{(m)})}\right) = \ln\left(P\left(\mathbf{S}_k^{(j^{(m)})}, \mathbf{r}_k\right)\right) = M\left(\mathbf{S}_{k-1}^{(i^{(m)})}\right) + \ln\left(\gamma_k\left(i^{(m)}, j^{(m)}\right)\right) \tag{4.74}$$

Using (4.69) and omitting the constant C obtain:

$$M\left(\mathbf{S}_k^{(j^{(m)})}\right) = M\left(\mathbf{S}_{k-1}^{(i^{(m)})}\right) + \frac{1}{2}u_k \cdot L(u_k) + \frac{1}{2}\sum_{l=0}^{n-1} L_c \cdot r_{k,l} \cdot x_{k,l} \tag{4.75}$$

where u_k is the information bit and $x_{k,l}$ are the coded bits of path m at the stage k.

This slight modification of the metric of the Viterbi Algorithm in (4.75) with the additional term $u_k \cdot L(u_k)$ included, incorporates the a priori information about the probability of the information bits. The balance between a priori information $L(u)$ and the channel reliability L_c is very important for the SOVA metric. If the channel is very good, $|L_c \cdot r|$ will be larger than $|L(u)|$, and decoding relies on the received channel values. If the channel is bad, as during the deep fade, decoding relies on the a priori information $L(u)$. In iterative decoding this is the extrinsic information from the previous decoding step [15]. If this balance is not achieved, catastrophic effects may result in the degradation of the decoder performance.

Let us now discuss the second modification of the algorithm required, that is, to give soft output.

The modified Viterbi algorithm proceeds in the usual way by calculating the path metrics using (4.75). If the two paths $\mathbf{S}_k^{(j^{(m)})}$ and $\mathbf{S}_k^{(j^{(l)})}$ reaching state $S_k = j$ have metrics $M(\mathbf{S}_k^{(j^{(m)})})$ and $M(\mathbf{S}_k^{(j^{(l)})})$ respectively, and the path $\mathbf{S}_k^{(j^{(m)})}$ is selected as the survivor because of higher metric, then we can define the metric difference Δ_k^j as:

$$\Delta_k^j = M\left(\mathbf{S}_k^{(j^{(m)})}\right) - M\left(\mathbf{S}_k^{(j^{(l)})}\right) \geq 0 \tag{4.76}$$

The probability that the decision is correct $P(\text{correct decision at } S_k = j)$ can be written as follows:

$$P(\text{correct dccision at } S_k = j) = \frac{P\left(S_k^{(j^{(m)})}\right)}{P\left(S_k^{(j^{(m)})}\right) + P\left(S_k^{(j^{(l)})}\right)} \tag{4.77}$$

Using the metric definition in (4.74) we can rewrite (4.77) as:

$$P(\text{correct decision at } S_k = j) = \frac{e^{M\left(S_k^{(j^{(m)})}\right)}}{e^{M\left(S_k^{(j^{(m)})}\right)} + e^{M\left(S_k^{(j^{(l)})}\right)}} = \frac{e^{\Delta_k^j}}{1 + e^{\Delta_k^j}} \tag{4.78}$$

Therefore the LLR that the decision is correct or "soft" value of this binary path decision is Δ_k^j because:

$$\ln \frac{P(\text{correct decision at } S_k = j)}{1 - P(\text{correct decision at } S_k = j)} = \Delta_k^j \tag{4.79}$$

The examples of metric differences in trellis are shown in Figure 4.26. Along the ML path several nonsurviving paths were discarded. Usually all the surviving paths at some stage in the trellis have come from the same path at some point at most δ transitions before the given stage in the trellis. The value δ is set to be five or six times the constraint length of the convolutional code. If the value of the bit u_k associated with the transition from state $S_{k-1} = i$ to state $S_k = j$ on the ML path differs from the value of corresponding bit of the competing path which merged with the ML path at stage $k + \delta$ and if this competing path was chosen by the decoder, then there is a bit error. Thus, when calculating the LLR of the bit u_k, the SOVA must take into account the probability that the paths merging with the ML path from stage k to stage $k + \delta$ in the trellis were incorrectly discarded. This is done by considering the values of the metric difference $\Delta_i^{s_i}$ for all states s_i along the ML path from stage $i = k$ to stage $i = k + \delta$. It is shown in [16] that this LLR can be approximated by:

$$L\left(u_k \,|\mathbf{r}\right) \approx u_k \cdot \min_{\substack{i=k\ldots+\delta, \\ u_k \neq u_k^i}} \Delta_i^{s_i} \tag{4.80}$$

where u_k is the value of the bit given by the ML path and u_k^i is the value of the corresponding bit in the competing path that merged with the ML path and was discarded at the stage i. The minimum is only over

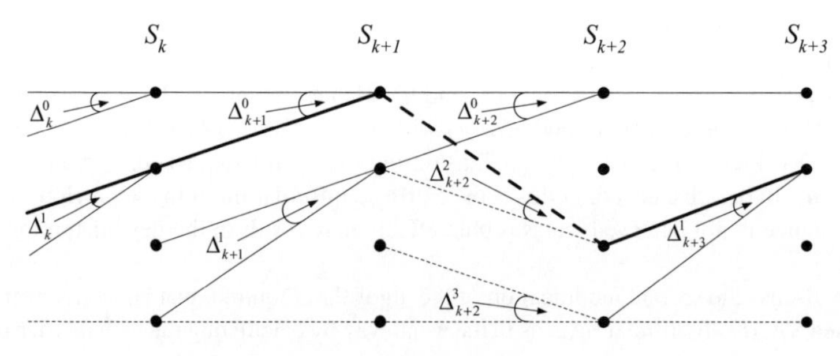

Figure 4.26 A section of SOVA decoder trellis

those nonsurviving paths, which would have led to different values of the bit u_k. Consider the section of trellis shown in Figure 4.26. In this figure solid lines represent transitions taken when the input bit is –1, and dashed lines represent transitions taken when the input bit is $+1$. The bold line marks the ML path. As can be seen from Figure 4.26 the ML path gives a value of –1 for u_k. Assume that $\delta = 3$. Then we can make a decision about u_k at stage S_{k+3}. The other paths merge with ML path at stages S_k, S_{k+1}, S_{k+2}, S_{k+3}, which gives us the metric differences Δ_k^1, Δ_{k+1}^0, Δ_{k+2}^2, Δ_{k+3}^1. However, only the competing paths merging with ML path at stages S_{k+2} and S_{k+3} give the opposite value $+1$ for bit u_k. That means only the minimum of values Δ_{k+2}^2 and Δ_{k+3}^1 should be taken into account during the calculation of soft output for u_k.

The SOVA can be implemented as follows:

Soft Input Soft Output Viterbi Decoding:

1. a) Initialize stage number $k = 0$.
   b) Initialize $M\left(\mathbf{S}_0^{(j)}\right) = 0$ for $j = 0$, $M\left(\mathbf{S}_0^{(j)}\right) = -\infty$ for all $j \neq 0$
2. a) Set $k = k + 1$.
   b) Compute the metric $M\left(\mathbf{S}_k^{(j(m))}\right) = M\left(\mathbf{S}_{k-1}^{(i(m))}\right) + \frac{1}{2}u_k \cdot L(u_k) + \frac{1}{2}\sum_{l=0}^{n-1} L_c \cdot r_{k,l} \cdot x_{k,l}$ for each state in trellis, where
   m denotes the path (branch) number;
   u_k is the information bit of the path m;
   $x_{k,l}$ is the lth bit of n bits for stage k associated with the branch m;
   $r_{k,l}$ is the received value from the channel corresponding to $x_{k,l}$;
   $L_c = 4\frac{E_b}{N_0} \cdot a$ is the channel reliability value;
   $L(u_k)$ is the a priori information. This value is obtained from the previous decoding step. If there was no previous decoding step this value is set to zero.
3. Find $M_k^{(j)} = \max_m M\left(\mathbf{S}_k^{(j(m))}\right)$ for each state j.
4. Store $M_k^{(j)}$ and its associated survivor bit and state paths.
5. Compute $\Delta_k^j = M_k^j - M\left(\mathbf{S}_k^{(j(l))}\right) \geq 0$ for each state j, where $M\left(\mathbf{S}_k^{(j(l))}\right)$ is the metric of the discarded path.
6. Update $\Delta_k^{j_{\min}} = \min_{\substack{i=k-\delta,\ldots,\\ u_k \neq u_k^i}} \Delta_i^{s_i}$ by choosing the minimum metric difference.
7. Go to step 2 until the end of the received sequence.
8. Output the estimated bit sequence $\mathbf{u} = \{u_k\}$ and the corresponding LLRs $\left\{L(\hat{u}_k) = L(u_k\,|\mathbf{r}) = u_k \cdot \Delta_k^{j_{\min}}\right\}$.

As mentioned already earlier the recursion used to find the metric in SOVA is identical to forward recursion in (4.67) in Max-Log-MAP algorithm. Using the notation associated with the Max-Log-MAP algorithm, once a path merges with the ML path, it will have the same value of $B_k(j)$ as the ML path. Hence, as the metric in the SOVA is identical to the $A_k(j)$ values in the Max-Log-MAP, taking the difference between the metrics of the two merging paths in the SOVA algorithm is equivalent to taking the difference between two values of $A_{k-1}(i) + B_k(j) + \Gamma_k(i,\ j)$ in the Max-Log-MAP algorithm, as in (4.70). The only difference is that in the Max-Log-MAP algorithm one path will be the ML path, and the other will be the most likely path that gives a different hard decision for u_k. In the SOVA algorithm again one path will be the ML path, but the other may not be the most likely path that gives a different hard decision for u_k. Instead, it will be the most likely path that gives a different hard decision for u_k and survives to merge with the ML path. Other, more likely paths, which give a different hard decision for the bit u_k to the ML path, may have been discarded before they merge with the ML path, as shown in Figure 4.27. Due to this fact the performance of the SOVA algorithm is slightly worse

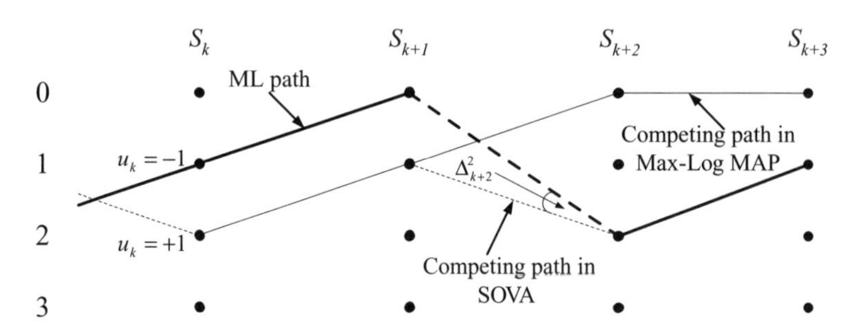

Figure 4.27 Different competing paths in Max-Log-MAP and SOVA algorithm

compared to the performance of the Max-Log-MAP algorithm. However, as pointed out in [12] the SOVA and Max-Log-MAP algorithms will always give the same hard decisions, as in both algorithms these hard decisions are determined by the ML path, which is calculated using the same metric in both algorithms.

In Figure 4.28 the performance of the different SISO decoding algorithms is compared. The simulation results were obtained for QPSK modulation over flat fading channel, the number of iterations in the turbo-decoder was set to 8. As can be seen from Figure 4.28 the Log-MAP algorithm shows almost the same performance as MAP algorithm. The performance of Max-Log-MAP and SOVA appears to be quite close especially in the range of low SNRs, and in the range of high SNRs the Max-Log-MAP outperforms SOVA about 0.5 dB due to the reasons discussed above.

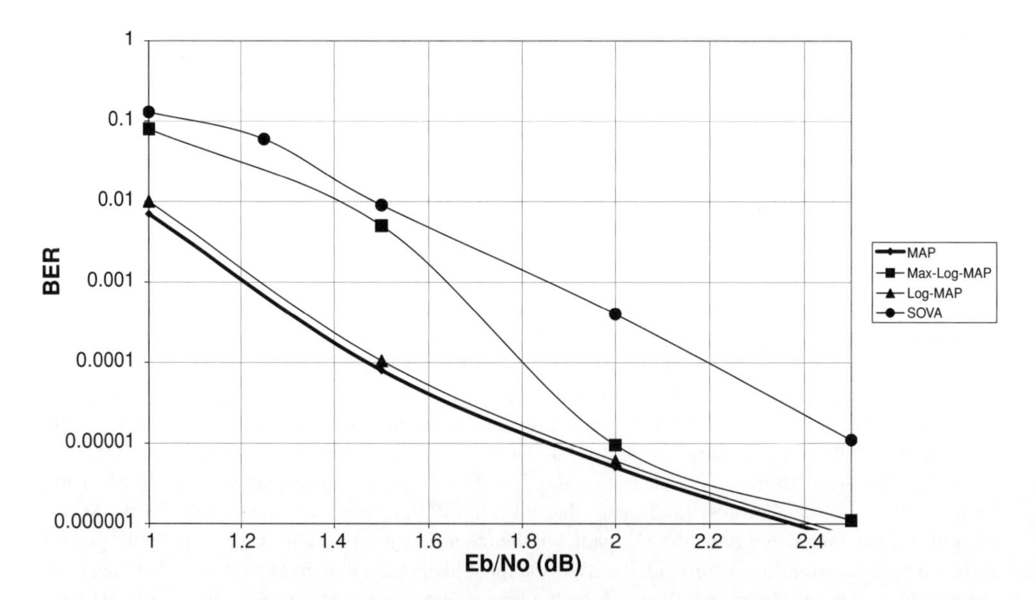

Figure 4.28 Comparison of different SISO decoding algorithms

References

[1] Viterbi, A.J., Omura, J.K. *Principles of Digital Communication and Coding*, McGraw-Hill Book Company, N.Y., 1979.

[2] Viterbi, A.J. "Error Bounds for Convolutional Codes and an Asymptotically Optimum Decoding Algorithm", IEEE Trans. Inf. Theory, IT-13, pp. 260–269, April 1967.

[3] Omura, J.K. "On the Viterbi Decoding Algorithm", IEEE Trans. Inf. Theory, IT-15, pp. 177–179, January 1969.

[4] Proakis, J. G., *Digital Communications*, 3rd ed., New York, McGraw-Hill, 1995.

[5] Rappaport, T. S., *Wireless Communications Principles and Practice*, New Jersey, Prentice-Hall, 1996.

[6] Johannesson, R., Zigangirov, K. Fundamentals of Convolutional Coding. NJ: IEEE, 1999.

[7] . Berrou, C., Glavieux, A. and Thitimajshima, P. Near Shannon limit error-correcting coding: turbo codes. Proceedings IEEE International Conference on Communications, Geneva, Switzerland, 1993, pp. 1064–1070.

[8] Woodard, J.P., Hanzo, L. Comparative Study of Turbo Decoding Techniques: An Overview. IEEE Transactions on Vehicular Technology, vol.49, No.6, November 2000., pp. 2208–2233.

[9] Bahl, L., Cocke, J., Jelinek, F. and Raviv, J. Optimal decoding of linear codes for minimizing symbol error rate. IEEE Transactions on Information Theory, IT-20, March 1974, pp. 284–287.

[10] Koch, W. and Baier, A. "Optimum and sub-optimum detection of coded data disturbed by time-varying inter-symbol interference," IEEE Globecom, pp. 1679–1684, Dec. 1990.

[11] Erfanian, J. A., Pasupathy, S. and Gulak, G. "Reduced complexity symbol detectors with parallel structures for ISI channels," IEEE Trans. Commun., vol. 42, pp. 1661–1671, 1994.

[12] Robertson, P., Villebrun, E. and Hoeher, P. "A comparison of optimal and sub-optimal MAP decoding algorithms operating in the log domain," Proc. Int. Conf. Communications, June 1995, pp. 1009–1013.

[13] Hagenauer, J. and Hoeher, P., "A Viterbi Algorithm with Soft-DecisionOutputs and Its Applications," GLOBECOM 1989, Dallas, Texas, pp.1680–1686, Nov. 1989.

[14] Burr, A. *Modulation and Coding for Wireless Communications*. Prentice Hall. 2001.

[15] Hagenauer, J., Offer, E., and Papke, L., "Iterative Decoding of Binary Block and Convolutional Codes," IEEE Transactions of Information Theory, Vol. 42, No. 2, pp. 429–445, March 1996.

[16] Hagenauer, J., "Source-Controlled Channel Decoding," IEEE Transactions on Communications, Vol. 43, No. 9, pp. 2449–2457, Sept. 1995.

[17] Clark, G.C.C. Jr. and Cain, J.B. *Error-Correcting Coding for Digital Communications*. NY, Plenum Press, 1981.

[18] Lin, S., Costello, D. Error control codes: Fundamentals and Applications, Englewood Cliffs, NJ, Prentice Hall, 1981.

Appendix 4.A

Modified Chernoff Bound and Some Applications

Andrey Trofimov
St. Petersburg State University of Aerospace Instrumentation, Russia

One of the most important problems in analysis of communication systems is error probability evaluation. The exact computation of error probability for coded system is a hard, if not irresolvable, task for many practical situations. That is why the upper bounding of error probability is a common method in system performance estimation. In many cases the error probability can be computed using probability of large deviation of a random variable. One of the most powerful techniques is based on usage of so called Chernoff bound. This bound is very well-known and it is mentioned and referenced practically in any monograph and textbook on coding, information and communication theory.

Let Z be a random value. It is necessary to evaluate the value of probability $P = \Pr[Z \geq 0]$. This probability is equal by definition to:

$$P = \int_0^\infty w_Z(x)dx \tag{4.A1}$$

where $w_Z(\cdot)$ is probability density function of the random value Z. Straightforward computation of P by (4.A1) can be difficult or even not possible. For example, the explicit expression for probability density function $w_Z(\cdot)$ might be unknown. The following inequality gives the expression for common Chernoff bound for probability P:

$$P \leq \overline{\exp(-\beta Z)} \tag{4.A2}$$

where bar denotes expectation, β is a parameter of the bound. It must satisfy the conditions $\beta_0 > \beta \geq 0$, where β_0 is upper bound for parameter following from convergence conditions of the right hand part of (4.A2). The bound (4.A2) turns out to be most useful when the random variable is a sum of n independent random variables. In such a case the bound (4.A2) is *exponentially tight* with respect to n. In this section we consider an advanced approach to estimating the probability in (4.A1) leading to a *non-exponential* refinement of the bound (4.A2).

Modulation and Coding Techniques in Wireless Communications Edited by Evgenii Krouk and Sergei Semenov

Let us introduce step function as:

$$e(x) = \begin{cases} 1, & x \geq 0 \\ 0, & x < 0 \end{cases}$$

then $P = \int_{-\infty}^{\infty} e(x)w_Z(x)dx = \overline{e(Z)}$, where bar denotes expectation over all random variable under the bar. Let $f(x)$ and $\varphi(x)$ be functions such that $f(x)\varphi(x) = e(x)$. It is clear, that $P = \int_{-\infty}^{\infty} f(x)\varphi(x)w_Z(x)dx$.

Let us introduce an invertible integral transform:

$$F(\alpha) = \int_{-\infty}^{\infty} K_1(\alpha, x)f(x)dx$$

and

$$f(x) = \int_{-\infty}^{\infty} K_2(x, \alpha)F(\alpha)d\alpha$$

Here $K_1(\cdot, \cdot)$and $K_2(\cdot, \cdot)$ are kernels of forward and inverse transforms respectively. Then one can write that:

$$P = \int_{-\infty}^{\infty} \varphi(x)\left(\int_{-\infty}^{\infty} K_2(x, \alpha)F(\alpha)d\alpha\right)w_Z(x)dx = \int_{-\infty}^{\infty} F(\alpha)\left(\int_{-\infty}^{\infty} \varphi(x)K_2(x, \alpha)w_Z(x)dx\right)d\alpha$$

or

$$P = \int_{-\infty}^{\infty} F(\alpha)\Phi(\alpha)d\alpha$$

where $\Phi(\alpha) = \overline{\varphi(Z)K_2(Z, \alpha)} = \int_{-\infty}^{\infty} \varphi(x)K_2(x, \alpha)w_Z(x)dx$, $F(\alpha) = \int_{-\infty}^{\infty} K_1(\alpha, x)f(x)dx$, and $f(x)$ and $\varphi(x)$ are some functions, satisfied condition $f(x)\varphi(x) = e(x)$.

Let us choose the Fourier transform as an integral transform, i.e., assume, that $K_1(\alpha, x) = e^{-j\alpha x}$, $K_2(x, a) = (2\pi)^{-1}e^{j\alpha x}$. Let us assign $\varphi(x) = e^{\beta x}$, where $\beta > 0$, and:

$$f(x) = \begin{cases} e^{-\beta x}, & x \geq 0 \\ 0, & x < 0 \end{cases}$$

Evidently, $f(x)\varphi(x) = e(x)$. Then:

$$F(\alpha) = \int_{0}^{\infty} e^{-\beta x}e^{-j\alpha x}dx = \frac{1}{\beta + j\alpha}$$

and

$$\Phi(\alpha) = \frac{1}{2\pi}\int_{-\infty}^{\infty} e^{\beta x}e^{j\alpha x}w_Z(x)dx = \frac{1}{2\pi}\overline{e^{(\beta+j\alpha)Z}} = \frac{1}{2\pi}C_Z(\alpha - j\beta)$$

where

$$C_Z(\omega) = \overline{e^{j\omega Z}} = \int_{-\infty}^{\infty} e^{j\omega x}w_Z(x)dx$$

is a characteristic function of random variable Z. Then the probability P can be found as:

$$P = \frac{1}{2\pi}\int_{-\infty}^{\infty} \frac{C_Z(\alpha - j\beta)}{\beta + j\alpha}d\alpha \tag{4.A3}$$

where $\beta > 0$. In some cases computation or estimation of this integral is more simple than computation of right hand part of the (4.A1); in particular it is possible to apply the following upper bound:

$$P = |P| = \frac{1}{2\pi}\left|\int_{-\infty}^{\infty} \frac{C_Z(\alpha - j\beta)}{\beta + j\alpha} d\alpha\right| \leq \frac{1}{2\pi}\int_{-\infty}^{\infty} \left|\frac{C_Z(\alpha - j\beta)}{\beta + j\alpha}\right| d\alpha = \frac{1}{2\pi}\int_{-\infty}^{\infty} \frac{|C_Z(\alpha - j\beta)|}{\sqrt{\beta^2 + \alpha^2}} d\alpha$$

giving as a result the inequality:

$$P \leq \frac{1}{2\pi}\int_{-\infty}^{\infty} \frac{|C_Z(\alpha - j\beta)|}{\sqrt{\beta^2 + \alpha^2}} d\alpha, \quad \beta > 0 \tag{4.A4}$$

Bound 4.A4 can be optimized by choosing the optimum value of the free parameter β, because its value is arbitrary, $\beta > 0$. In such form the value was given in [1].

The expressions or (4.A3) and (4.A4) are most helpful when the random variable Z is a sum of some other random variables. We consider further examples with application in communication and coding theory. All these examples deal with independent identically distributed (i.i.d.) random variables. But in general the expressions (4.A2) and (4.A3) are applicable to the sum of dependent variables with different distributions.

Example 4.A1 Binary symmetric channel.
Let $Z = \sum_{i=1}^{n} z_i - n/2 = \sum_{i=1}^{n} (z_i - 1/2)$, where z_i, $i = 1, 2, \ldots, n$, are i.i.d. defined as follows:

$$z_i = \begin{cases} 1, & \text{with probability } p, \\ 0, & \text{with probability } 1 - p \end{cases}$$

The value p, $p < 1/2$, is considered as symbol error probability in a binary symmetric channel (BSC), and the event $Z > 0$ corresponds to not more than $n/2$ errors occurring on the length n. In other words P is the probability of error decoding of one of two sequences differing in n positions transmitted over BSC. It is easy to see that:

$$C_Z(\omega) = \overline{e^{j\omega Z}} = \overline{e^{j\omega \sum_{i=1}^{n}(z_i - 1/2)}} = \overline{\prod_{i=1}^{n} e^{j\omega(z_i - /2)}} = \prod_{i=1}^{n} \overline{e^{j\omega(z_i - /2)}} = (pe^{j\omega/2} + (1 - p)e^{-j\omega/2})^n$$

and

$$C_Z(\alpha - j\beta) = (pe^{j\alpha/2}e^{\beta/2} + (1 - p)e^{-j\alpha/2}e^{-\beta/2})^n$$

Since the value of parameter β is arbitrary positive, $\beta > 0$, let us assign it as $\beta_0 = \log((1 - p)/p)$. Then it can be shown that:

$$C_Z(\alpha - j\beta)|_{\beta=\beta_0} = \left(2\sqrt{p(1 - p)}\right)^n \cos^n(\alpha/2)$$

and after substitution into (4.A3) we get:

$$P = \left(2\sqrt{p(1 - p)}\right)^n I_n$$

where

$$I_n = \frac{1}{2\pi}\int_{-\infty}^{\infty} \frac{\cos^n(\alpha/2)}{\beta_0 - j\alpha} d\alpha$$

The remaining is to evaluate the value of this integral. Direct derivation gives:

$$I_n = \frac{1}{2\pi}\int_{-\infty}^{\infty}\frac{\cos^n(\alpha/2)}{\beta_0 - j\alpha}d\alpha = \frac{1}{2\pi}\int_{-\infty}^{\infty}\frac{\cos^n(\alpha/2)(\beta_0 - j\alpha)}{\alpha^2+\beta_0^2}d\alpha =$$

$$= \frac{\beta_0}{2\pi}\int_{-\infty}^{\infty}\frac{\cos^n(\alpha/2)}{\alpha^2+\beta_0^2}d\alpha - \frac{1}{2\pi}j\underbrace{\int_{-\infty}^{\infty}\frac{\alpha\cos^n(\alpha/2)}{\alpha^2+\beta_0^2}d\alpha}_{=0} = \frac{\beta_0}{2\pi}\int_{-\infty}^{\infty}\frac{\cos^n(\alpha/2)}{\alpha^2+\beta_0^2}d\alpha$$

Using the identities (see [2], equations 1.320.5, 1.320.7):

$$\cos^{2n} x = \frac{1}{2^{2n}}\left(\sum_{k=0}^{n-1}2C_{2n}^k\cos 2(n-k)x + C_{2n}^n\right) \tag{4.A5}$$

$$\cos^{2n-1} x = \frac{1}{2^{2n-2}}\left(\sum_{k=0}^{n-1}C_{2n-1}^k\cos(2n-2k-1)x\right) \tag{4.A6}$$

for $n = 2m-1$, $m = 1, 2, \ldots$. we get:

$$I_n = I_{2m-1} = \frac{1}{2\pi}\frac{\beta_0}{2^{2m-2}}\sum_{k=0}^{m-1}C_{2m-1}^k\int_{-\infty}^{\infty}\frac{\cos((\alpha/2)(2m-2k-1))}{\alpha^2+\beta_0^2}d\alpha$$

It is known ([3], equation 859.001), that for $a > 0$, $m \geq 0$:

$$\int_0^{\infty}\frac{\cos mx}{a^2+x^2}dx = \frac{\pi}{2a}e^{-ma} \tag{4.A7}$$

Using the (4.A7) we have:

$$I_{2m-1} = \frac{1}{2^{2m-1}}e^{-\beta_0((2m-1)/2)}\sum_{k=0}^{m-1}C_{2m-1}^k e^{k\beta_0} \leq \frac{C_{2m-1}^{m-1}}{2^{2m-1}}e^{-\beta_0((2m-1)/2)}\sum_{k=0}^{m-1}e^{k\beta_0}$$

$$= \frac{C_{2m-1}^{m-1}}{2^{2m-1}}e^{-\beta_0((2m-1)/2)}\frac{e^{\beta_0 m}-1}{e^{\beta_0}-1} == \frac{C_{2m-1}^{m-1}}{2^{2m-1}}e^{\beta_0/2}\frac{1-e^{-\beta_0 m}}{e^{\beta_0}-1} < \frac{C_{2m-1}^{m-1}}{2^{2m-1}}\frac{e^{\beta_0/2}}{e^{\beta_0}-1}$$

Since $C_{2m-1}^{m-1} = C_{2m}^m/2$, $C_{2m}^m < 4^m/\sqrt{\pi m}$, then:

$$I_{2m-1} < \frac{1}{\sqrt{\pi m}}\frac{e^{\beta_0/2}}{e^{\beta_0}-1} \tag{4.A8}$$

Let us consider the even values of n, $n = 2m$. Using (4.A4), we get:

$$I_{2m} = \frac{1}{2\pi}\frac{\beta_0}{2^{2m}}\left[\sum_{k=0}^{m-1}2C_{2m}^k\int_{-\infty}^{\infty}\frac{\cos(m-k)\alpha}{\alpha^2+\beta_0^2}d\alpha + C_{2m}^m\int_{-\infty}^{\infty}\frac{d\alpha}{\alpha^2+\beta_0^2}\right]$$

and further, using (4.A7),

$$I_{2m} = \frac{e^{-\beta_0 m}}{2^{2m}}\sum_{k=0}^{m-1}C_{2m}^k e^{k\beta_0} + \frac{C_{2m}^m}{2^{2m+1}} \leq \frac{C_{2m}^{m-1}}{2^{2m}}e^{-m\beta_0}\sum_{k=0}^{m-1}e^{k\beta_0} + \frac{C_{2m}^m}{2^{2m+1}} = \frac{C_{2m}^{m-1}}{2^{2m}}e^{-\beta_0 m}\frac{e^{\beta_0 m}-1}{e^{\beta_0}-1} + \frac{C_{2m}^m}{2^{2m+1}} =$$

$$= \frac{C_{2m}^{m-1}}{2^{2m}}\frac{1-e^{-\beta_0 m}}{e^{\beta_0}-1} + \frac{C_{2m}^m}{2^{2m+1}} < \frac{C_{2m}^{m-1}}{2^{2m}}\frac{1}{e^{\beta_0}-1} + \frac{C_{2m}^m}{2^{2m+1}}$$

Since $C_{2m}^{m-1} = (m/(m+1))C_{2m}^m < C_{2m}^m$, and $C_{2m}^m < 4^m/\sqrt{\pi m}$, we have the inequality:

$$I_{2m} < \frac{1}{\sqrt{\pi m}}\left(\frac{1}{e^{\beta_0}-1}+\frac{1}{2}\right) = \frac{1}{2\sqrt{\pi m}}\frac{e^{\beta_0}+1}{e^{\beta_0}-1} \tag{4.A9}$$

Combining (4.A7) and (4.A8) and taking into account that $\beta_0 = \log((1-p)/p)$, we get:

$$I_n < \begin{cases} \dfrac{1}{\sqrt{2\pi(n+1)}}\dfrac{2\sqrt{p(1-p)}}{1-2p} & \text{for odd } n, n = 2m-1, \\ \dfrac{1}{\sqrt{2\pi n}}\dfrac{1}{1-2p} & \text{for even } n, n = 2m \end{cases}$$

Final expression for bound is:

$$P < K(\beta_0, n)D_0(\beta_0)^n \tag{4.A10}$$

where

$$K(\beta_0, n) = \begin{cases} \dfrac{2\sqrt{p(1-p)}}{\sqrt{2\pi(n+1)}(1-2p)} & \text{for odd } n \\ \dfrac{1}{\sqrt{2\pi n}(1-2p)} & \text{for even } n \end{cases} \tag{4.A11a}$$

and

$$D(\beta_0) = 2\sqrt{p(1-p)} \tag{4.A11b}$$

Inequalities (4.A10) and (4.A11) look like the similar bound in [4]. Note, that common Chernoff bound gives inequality $P < (2\sqrt{p(1-p)})^n$ for all n, and the exact formula is:

$$P = \begin{cases} \sum\limits_{i=n/2+1}^{n} C_n^i p^i (1-p)^{n-i} + \dfrac{1}{2}C_n^{n/2} p^{n/2}(1-p)^{n/2}, & n \text{ is even} \\ \sum\limits_{i=(n-1)/2+1}^{n} C_n^i p^i (1-p)^{n-i}, & n \text{ is odd} \end{cases}$$

Figure 4.A1a gives an example of exact value and bounds for $n = 5$ The ratio of bounds and exact value of the value P is shown in Figure 4.A1b.

Example 4.A2 Additive white Gaussian noise channel, binary antipodal signaling, soft decision demodulation.

Let $Z = \sum_{i=1}^n z_i$, where $z_i = n_i - A$, $i = 1, 2, \ldots, n$, n_i are Gaussian i.i.d. with zero mean and variance equal to one. The value of A, $A > 0$, can be expressed using signal to noise ratio E/N_0 as $A = \sqrt{2E/N_0}$. Under these conditions the probability $P = \Pr[Z > 0]$ is the probability of decoding error of two sequences differing in n positions in maximum likelihood decoding in AWGN channel with soft output. It is well-known that exact formula for this probability is $P = Q(\sqrt{2nE/N_0})$. But this expression is difficult to immediately use in, for example, evaluation of union bound for Viterbi decoding based on code weight generating function.

Let us find the characteristic function $C_Z(\omega)$. It is equal to:

$$C_Z(\omega) = \overline{e^{j\omega Z}} = \overline{e^{j\omega \sum_{i=1}^{n} z_i}} = \overline{\prod_{i=1}^{n} e^{j\omega z_i}} = \prod_{i=1}^{n} \overline{e^{j\omega z_i}} = e^{n(-j\omega A - \omega^2/2)}$$

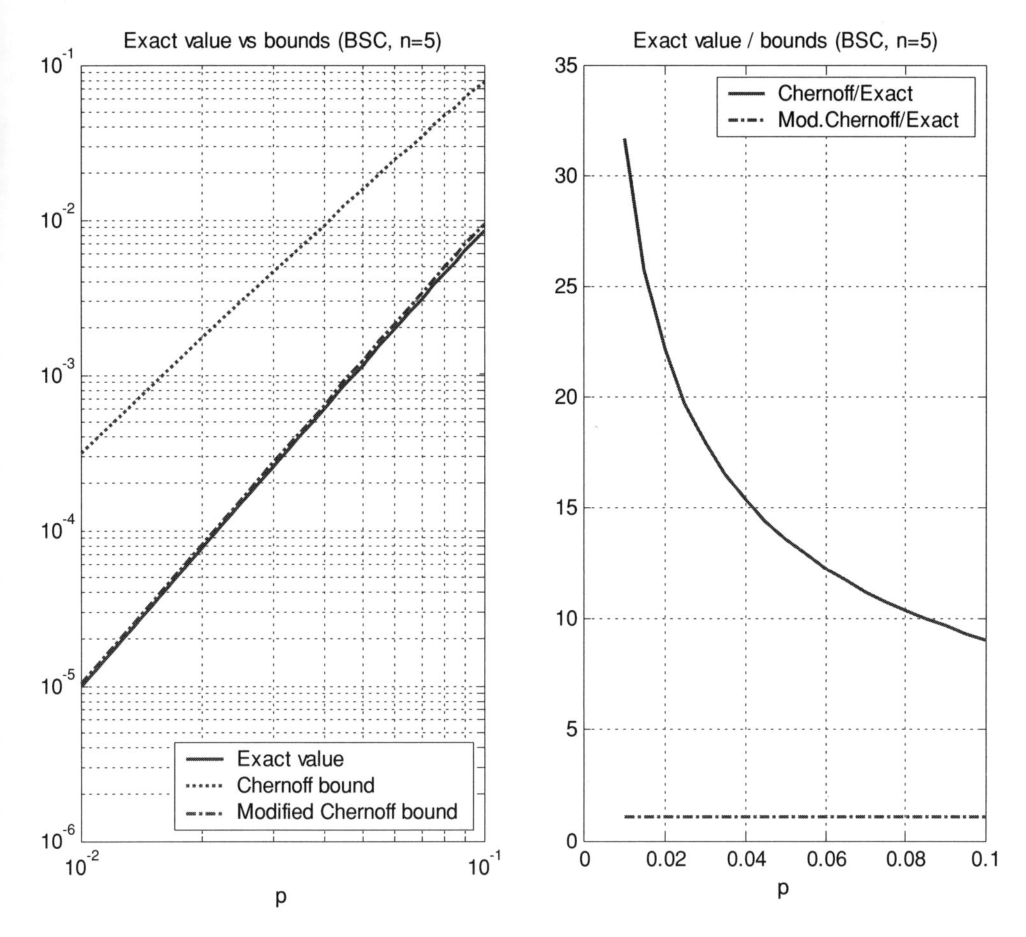

Figure 4.A1 Binary symmetric channel. Chernoff bound vs exact value (a), ratio of exact value and bounds (b)

Therefore $C_Z(\alpha - j\beta) = e^{-n(\beta A+(\alpha^2-\beta^2)/2)} e^{-jn\alpha(A-\beta)}$ и $|C_Z(\alpha - j\beta)| = e^{-n(\beta A+(\alpha^2-\beta^2)/2)}$. After substitution of this expression into (4.A4) we get:

$$P \le \min_{\beta \ge 0} \frac{1}{2\pi} e^{-n(\beta A - \beta^2/2)} \int_{-\infty}^{\infty} \frac{e^{-n\alpha^2/2}}{\sqrt{\beta^2 + \alpha^2}} d\alpha$$

Evidently:

$$\int_{-\infty}^{\infty} \frac{e^{-n\alpha^2/2}}{\sqrt{\beta^2 + \alpha^2}} d\alpha < \frac{1}{\beta} \int_{-\infty}^{\infty} e^{-n\alpha^2/2} d\alpha = \frac{1}{\beta}\sqrt{\frac{2\pi}{n}}$$

and hence:

$$P \le \min_{\beta \ge 0} \frac{1}{\sqrt{2\pi n}\beta} e^{-n(\beta A - \beta^2/2)}$$

Let us set β equal to $\beta_0 = A = \sqrt{2E/N_0}$ (this value gives the minimum of exponent); then:

$$P < K(\beta_0, n)D_0(\beta_0)^n \tag{4.A12}$$

where

$$K(\beta_0, n) = \frac{1}{2\sqrt{\pi n E/N_0}} \tag{4.A13}$$

and

$$D_0(\beta_0) = e^{-E/N_0} \tag{4.A14}$$

Common Chernoff bound in this case is $P < e^{-nE/N_0}$. It should be noted that the bound 4.A12 can be easily obtained using the known bound for function $Q(x)$ for $x > 0$:

$$Q(x) < \frac{1}{\sqrt{2\pi}x}e^{-x^2/2}$$

Therefore the derivation of the bound (4.A12) illustrates the application of the general approach given by formulas 4.A3 and 4.A4.

Example of bounds and exact value of probability P for $n = 5$ are shown in Figure 4.A2a. The ratio of the bounds and exact value of the value P is shown in Figure 4.A2b.

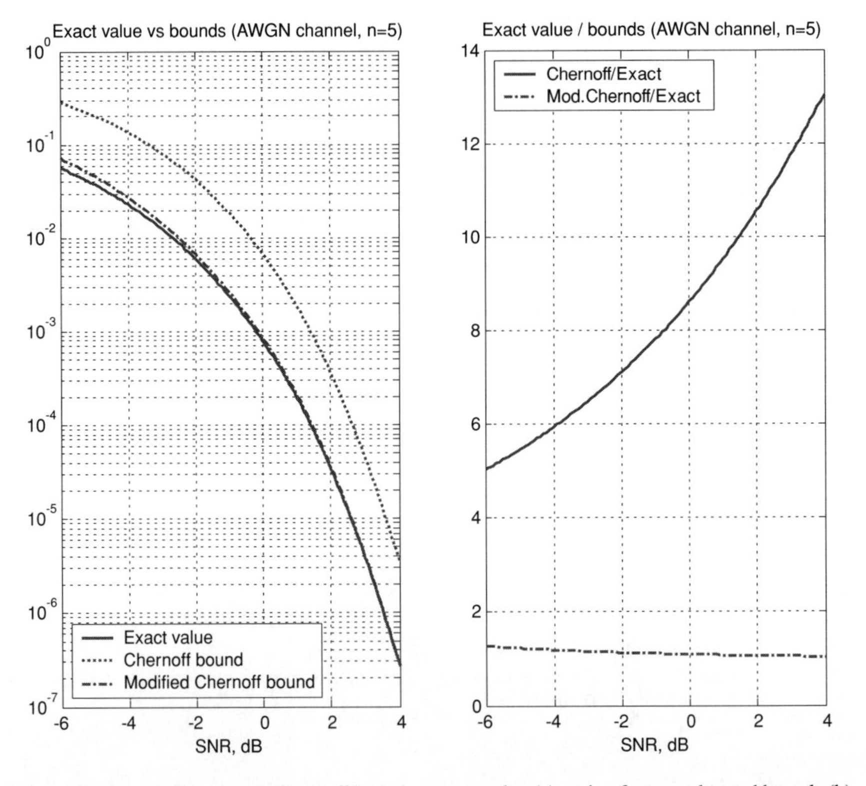

Figure 4.A2 AWGN channel. Chernoff bound vs exact value (a), ratio of exact value and bounds (b)

Example 4.A3 AWGN channel, orthogonal FSK signaling, noncoherent reception, soft decision demodulation.

In this model $Z = \sum_{i=1}^{n} z_i$, where $z_i = n_{3i}^2 + n_{4i}^2 - (n_{2i} + a)^2 - (n_{1i} + b)^2$, $i = 1, 2, \ldots, n$, and $n_{1i}, n_{2i}, n_{3i}, n_{4i}$ are i.i.d. Gaussian random values with zero mean ad variance equal to one. The value of $A = a^2 + b^2$, $A > 0$, and it can be expressed using signal to noise ratio E/N_0 as $A = 2E/N_0$. Under these conditions the probability $P = \Pr[Z > 0]$ is the probability of decoding error of two sequences differing in n positions in maximum likelihood decoding in noncoherent AWGN channel with soft output.

As before, we are starting from derivation of the characteristic function. It is equal to:

$$C_Z(\omega) = \overline{e^{j\omega Z}} = \overline{e^{j\omega \sum_{i=1}^{n} z_i}} = \overline{\prod_{i=1}^{n} e^{j\omega z_i}} = \prod_{i=1}^{n} \overline{e^{j\omega z_i}} = c_z(\omega)^n \tag{4.A15}$$

where

$$c_z(\omega) = \frac{1}{1 - 2j\omega} \frac{1}{1 + 2j\omega} e^{-\frac{j\omega A}{1+2j\omega}}$$

is characteristic function of z_i, and $\mathrm{Re}(j\omega) < 1/2$; this is condition of convergence the integral for calculation of the average $\overline{e^{j\omega z_i}}$.

Let us find the value of $c_z(\alpha - j\beta)$. It can be shown that:

$$c_z(\alpha - j\beta) = \frac{1}{(1 - 2\beta) - j2\alpha} \frac{1}{(1 + 2\beta) + j2\alpha} \exp\left(-\frac{(\beta(1 + 2\beta) + 2\alpha^2)A}{(1 + 2\beta)^2 + 4\alpha^2}\right) \exp(jA\alpha)$$

and $\beta < 1/2$, as it is needed for $\mathrm{Re}(j\omega) < 1/2$. Also there exists the restriction $\beta > 0$. Thus:

$$|c_z(\alpha - j\beta)| = \frac{1}{\sqrt{(1 - 2\beta)^2 + 4\alpha^2}} \frac{1}{\sqrt{(1 + 2\beta)^2 + 4\alpha^2}} \exp\left(-\frac{\beta(1 + 2\beta) + 2\alpha^2}{(1 + 2\beta)^2 + 4\alpha^2} A\right)$$

for $0 < \beta < 1/2$. It is easy to see that the factor in exponent:

$$\frac{\beta(1 + 2\beta) + 2\alpha^2}{(1 + 2\beta)^2 + 4\alpha^2}$$

achieves minimum value of $\beta/(1 + 2\beta)$ at $\alpha = 0$, and it can be shown that:

$$|c_z(\alpha - j\beta)| < \frac{1}{\sqrt{(1 - 2\beta)^2 + 4\alpha^2}} \frac{1}{\sqrt{(1 + 2\beta)^2 + 4\alpha^2}} e^{-\frac{\beta A}{1+2\beta}}$$

Consider the expression $((1 - 2\beta)^2 + 4\alpha^2)((1 + 2\beta)^2 + 4\alpha^2)$ in denominator:

$$\begin{aligned}
&((1 - 2\beta)^2 + 4\alpha^2)((1 + 2\beta)^2 + 4\alpha^2) = (1 - 4\beta^2)^2 + (1 - 2\beta)^2 4\alpha^2 + (1 + 2\beta)^2 4\alpha^2 + (4\alpha^2)^2 = \\
&= (1 - 4\beta^2)^2 + 4\alpha^2(1 - 4\beta + 4\beta^2 + 1 + 4\beta + 4\beta^2) + (4\alpha^2)^2 = \\
&= (1 - 4\beta^2)^2 + (4\alpha^2)2(1 + 4\beta^2) + (4\alpha^2)^2 > \\
&> (1 - 4\beta^2)^2 + 2(1 - 4\beta)^2 4\alpha^2 + (4\alpha^2)^2 = ((1 - 4\beta^2) + 4\alpha^2)^2
\end{aligned}$$

Therefore:

$$|c_z(\alpha - j\beta)| < \frac{1}{(1 - 4\beta^2) + 4\alpha^2} e^{-\frac{\beta A}{1+2\beta}} \tag{4.A16}$$

Substituting right hand part of inequality (4.A16) in (4.A15), and using it in (4.A4) we have:

$$P < \min_{0<\beta<1/2} \frac{e^{-\frac{\beta n A}{1+2\beta}}}{2\pi} \int_{-\infty}^{\infty} \frac{1}{\left((1 - 4\beta^2) + 4\alpha^2\right)^n} \frac{1}{\sqrt{\alpha^2 + \beta^2}} d\alpha < \min_{0<\beta<1/2} \frac{e^{-\frac{\beta n A}{1+2\beta}}}{2\pi\beta} \int_{-\infty}^{\infty} \frac{d\alpha}{\left((1 - 4\beta^2) + 4\alpha^2\right)^n}$$

or

$$P < \min_{0<\beta<1/2} \frac{1}{2\pi\beta} e^{-\frac{\beta n A}{1+2\beta}} I_n$$

where

$$I_n = \int_{\infty}^{\infty} \frac{d\alpha}{\left((1-4\beta^2)+4\alpha^2\right)^n} = \frac{1}{4^n} \int_{-\infty}^{\infty} \frac{d\alpha}{\left((1-4\beta^2)/4+\alpha^2\right)^n}$$

Let us estimate the value of this integral. It is known [3, equation 856.21], that:

$$\int_0^{\infty} \frac{dx}{(a^2+x^2)^n} = \frac{1\cdot 3\cdot 5\cdot \cdots \cdot(2n-3)}{2\cdot 4\cdot 6\cdot \cdots \cdot(2n-2)} \frac{\pi}{2a^{2n-1}} = \frac{C_{2n-2}^{n-1}}{2^{2n-2}} \frac{\pi}{2a^{2n-1}}, \quad a>0 \tag{4.A17}$$

Applying this identity we get:

$$I_n = \frac{1}{4^n} 2 \frac{C_{2n-2}^{n-1}}{2^{2n-2}} \frac{\pi 2^{2n-1}}{2(1-\beta^2)^{(2n-1)/2}} = \frac{2\pi C_{2n-2}^{n-1}}{4^n} \frac{\sqrt{1-\beta^2}}{(1-4\beta^2)^n}$$

then using the equation and bound:

$$C_{2n-2}^{n-1} = \frac{C_{2n}^n n^2}{(2n-1)2n} = \frac{n}{2(2n-1)} C_{2n}^n < \frac{n}{2(2n-1)} \frac{4^n}{\sqrt{\pi n}} \tag{4.A18}$$

we come to:

$$I_n < \frac{2\pi n}{2(2n-1)\sqrt{\pi n}} \frac{\sqrt{1-4\beta^2}}{(1-4\beta^2)^n}$$

and

$$P < \min_{0<\beta<1/2} K(\beta,n) D_0(\beta)^n$$

where

$$K(\beta,n) = \frac{n\sqrt{1-4\beta^2}}{2\beta(2n-1)\sqrt{\pi n}}$$

and

$$D(\beta) = \frac{1}{1-4\beta^2} e^{-\frac{\beta A}{1+2\beta}}$$

Introduce parameter $2\beta = \lambda$. Then the final expression for modified Chernoff bound becomes:

$$P < \min_{0<\lambda<1} \frac{n}{(2n-1)\sqrt{\pi n}} \sqrt{\frac{1-\lambda^2}{\lambda^2}} \left(\frac{1}{1-\lambda^2} e^{-\frac{\lambda}{1+\lambda}\frac{E}{N_0}}\right)^n \tag{4.A19}$$

Common Chernoff bound in this case is:

$$P < \min_{0<\lambda<1} \left(\frac{1}{1-\lambda^2} e^{-\frac{\lambda}{1+\lambda}\frac{E}{N_0}}\right)^n \tag{4.A20}$$

Optimal value of the parameter λ in right hand part of (4.A20) can be found from the equation:

$$\frac{d}{d\lambda}\left(-\frac{\lambda}{1+\lambda}\frac{E}{N_0} - \log(1-\lambda^2)\right) = 0$$

After solving this equation we have:

$$\lambda = \frac{\sqrt{(E/N_0)^2 + 12(E/N_0) + 4} - (E/N_0 + 2)}{4} \tag{4.A21}$$

Value in (4.A21) can be used in bound (4.A19) instead of true optimum value which can be found numerically. The true optimum value depends as it follows from (4.A19) on E/N_0 and on n. The loss of using suboptimal value in (4.A21) instead of optimal is very small.

Exact value of probability P can be found as [5, (1.42)]:

$$P = (e^{-nE/2N_0}/2) \sum_{i=0}^{n-1} \frac{(nE/2N_0)^i}{i!(n+i-1)!} \sum_{k=i}^{j-1} \frac{(k+n-1)!}{(k-i)!2^{k+n-1}}$$

Figure 4.A3a illustrates the example of bounds and exact value of probability P for $n = 5$ for optimal and suboptimal value of the parameter λ. The ratio of bounds and exact value of the value P is shown in Figure 4.A3b.

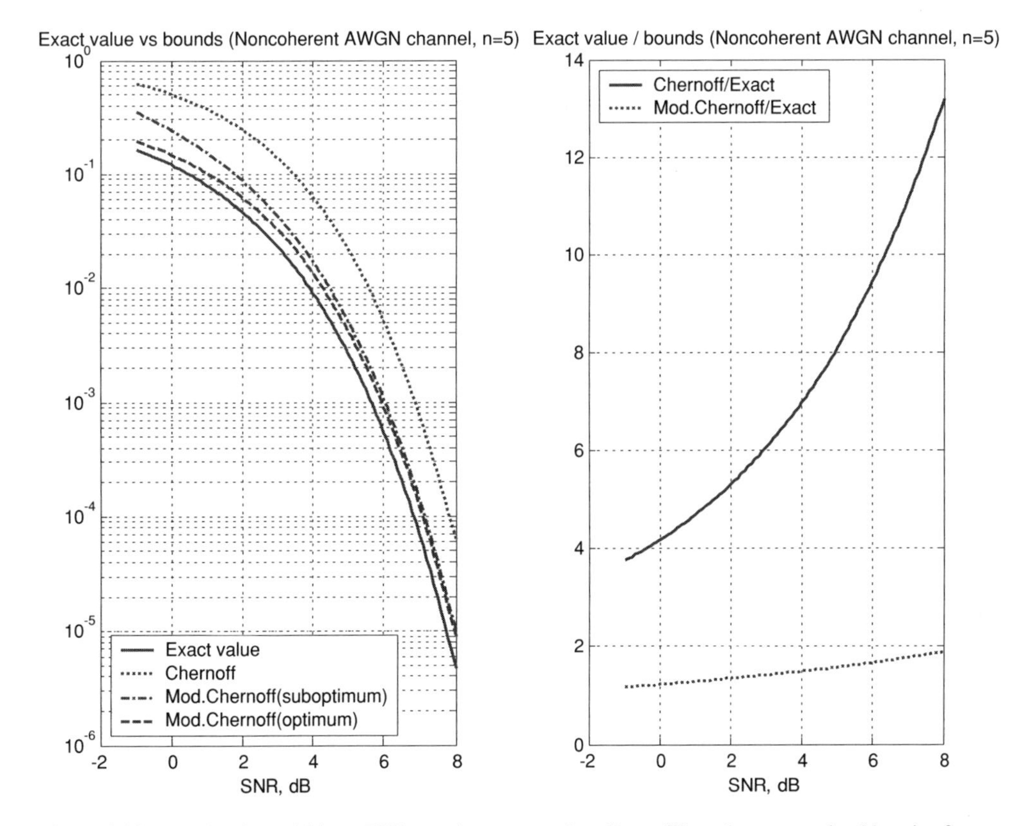

Figure 4.A3 AWGN channel, binary FSK noncoherent reception. Chernoff bound vs exact value (a), ratio of exact value and bounds (b)

Example 4.A4 Rayleigh fading channel, binary FSK orthogonal signaling, noncoherent reception, soft decision demodulation.

In this model $Z = \sum_{i=1}^{n} z_i$, where $z_i = n_{3i}^2 + n_{4i}^2 - (n_{2i} + x)^2 - (n_{1i} + y)^2$, $i = 1, 2, \ldots, n$, $n_{1i}, n_{2i}, n_{3i}, n_{4i}$ are i.i.d. Gaussian random value with zero mean and variance equal to one, x and y are independent Gaussian random variables with zero mean and variance γ. The value of γ is equal to average signal to noise ratio, $\gamma = E/N_0$. The probability $P = \Pr[Z > 0]$ here is the probability of decoding error of two sequences differing in n positions in maximum likelihood decoding in noncoherent Rayleigh fading channel with soft demodulator output.

The characteristic function $C_Z(\omega)$ in this case is:

$$C_Z(\omega) = \overline{e^{j\omega Z}} = \overline{e^{j\omega \sum_{i=1}^{n} z_i}} = \overline{\prod_{i=1}^{n} e^{j\omega z_i}} = \prod_{i=1}^{n} \overline{e^{j\omega z_i}} = c_z(\omega)^n \tag{4.A22}$$

where

$$c_z(\omega) = \frac{1}{1 - 2j\omega} \frac{1}{1 + 2j\omega(\gamma + 1)}$$

is characteristic function of the random variable z_i. Again the condition $\mathrm{Re}(j\omega) < 1/2$ arises here for convergence of the expectation $\overline{e^{j\omega z_i}}$. It follows from (4.A22), that:

$$C_Z(\alpha - j\beta) = \left[\frac{1}{1 - 2j(\alpha - j\beta)} \frac{1}{1 + 2j(\alpha - j\beta)(\gamma + 1)}\right]^n =$$

$$= \left[\frac{1}{(1 - 2\beta) - j2\alpha} \cdot \frac{1}{(1 + 2\beta(\gamma + 1)) + j2\alpha(\gamma + 1)}\right]^n$$

and

$$|C_Z(\alpha - j\beta)| = \frac{1}{(\gamma + 1)^n} \left[\frac{1}{\sqrt{(1 - 2\beta)^2 + 4\alpha^2}} \cdot \frac{1}{\sqrt{\left(\frac{1}{\gamma+1} + 2\beta\right)^2 + 4\alpha^2}}\right]^n$$

Let us assign $\beta = \beta_0 = \gamma/(4(\gamma + 1))$. This value meets the condition $0 < \beta < 1/2$. As a result we have:

$$|C_Z(\alpha - j\beta_0)| = \frac{1}{(\gamma + 1)^n} \frac{1}{\left[\left(\frac{\gamma+2}{2(\gamma+1)}\right)^2 + 4\alpha^2\right]^n} = \frac{1}{4^n(\gamma + 1)^n} \frac{1}{\left[\left(\frac{\gamma+2}{4(\gamma+1)}\right)^2 + \alpha^2\right]^n}$$

From formulas 4.A3 and 4.A4 we get:

$$P \leq \frac{4(\gamma + 1)}{2\pi\gamma} \frac{1}{4^n(\gamma + 1)^n} \int_{-\infty}^{\infty} \frac{d\alpha}{\left[\left(\frac{\gamma+2}{4(\gamma+1)}\right)^2 + \alpha^2\right]^n}$$

Taking into account (4.A16), we have:

$$P < \frac{2C_{2n-2}^{n-1}}{4^n} \frac{(\gamma + 2)}{\gamma} \left[\frac{4(\gamma + 1)}{(\gamma + 2)^2}\right]^n$$

Using (4.A17) and introducing a parameter $p = (\gamma + 2)^{-1}$ we obtain the final expression for modified Chernof bound for this example as:

$$P < K(\beta_0, n) D(\beta_0) \tag{4.A23}$$

where

$$K(\beta_0, n) = \frac{n}{(2n-1)\sqrt{\pi n}} \frac{1}{1-2p} \tag{4.A24a}$$

and

$$D(\beta_0) = 4p(1-p) \tag{4.A24b}$$

Common Chernoff bound for this case:

$$P < [4p(1-p)]^n$$

and the exact value is [6, equation (7.124b)]:

$$P = p^n \sum_{j=0}^{n} C_{n-j+1}^{j} (1-p)^j$$

Figure 4.A4a illustrates the example of bounds and exact value of probability P for $n = 5$. The ratio of bounds and exact value of the value P is shown in Figure 4.A4b.

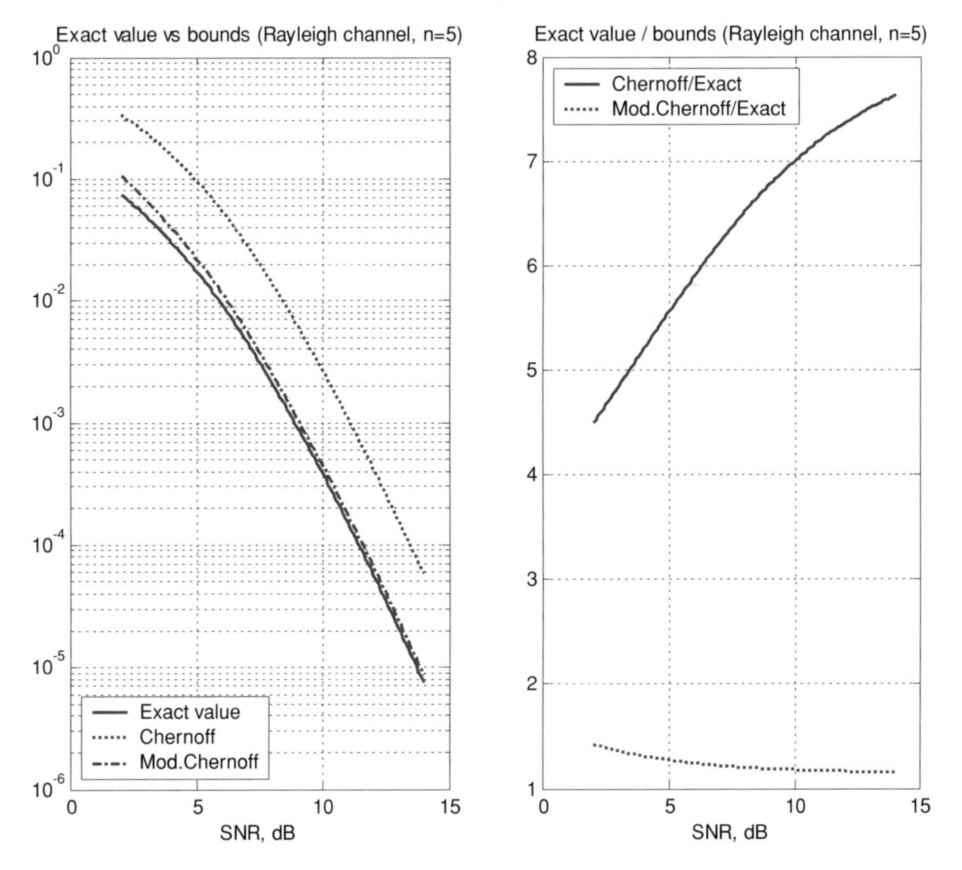

Figure 4.A4 Rayleigh fading channel, binary FSK, noncoherent reception. Chernoff bound vs exact value (a), ratio of exact value and bounds (b)

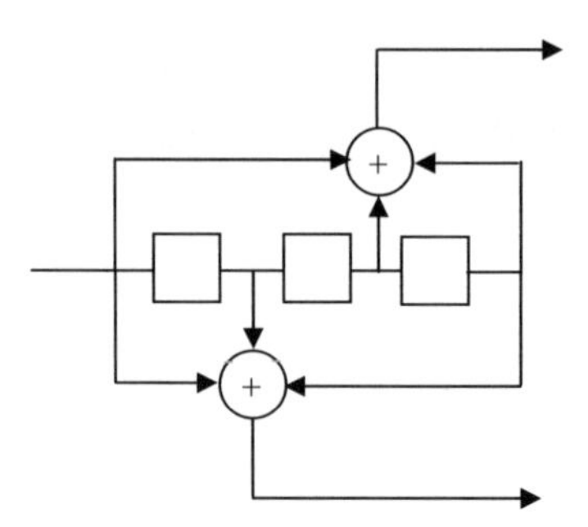

Figure 4.A5 A rate convolutional encoder

To demonstrate usage of the modified Chernof bound let us consider an example of estimation bit error probability for a convolutional code. The Viterbi decoding is assumed for decoding coded data transmitted over AWGN channel. We assume also binary FSK and noncoherent soft demodulation. For definiteness let us assume binary convolutional code with rate $R = 1/2$ and constraint length $\nu = 3$ with encoder shown in Figure 4.A5.

It can be shown [7] that weight generation function for this code is given as:

$$T(Z, N) = \frac{Z^6N^3 - Z^{10}N^3 + Z^6N}{1 - 2Z^2N^3 - 3Z^2N + 2Z^6N^3}$$

and

$$F(Z) = \left.\frac{dT(Z, N)}{dN}\right|_{N=1} = \frac{4N^6 - 20N^8 + 11N^{12} - 3N^{10} + 18N^{15} - 9N^{19}}{(1 - 5N^2 + 2N^6)^2}$$

The bit error probability can be upper bounded using common Chernoff bound as:

$$P_b < F(D_0) \tag{4.A25}$$

where D_0 is the Chernoff bound exponent defined in this case, see (4.A20), as:

$$D_0 = \min_{0<\lambda<1} \frac{1}{1 - \lambda^2} e^{-\frac{\lambda}{1+\lambda}\frac{E}{N_0}} = \frac{1}{1 - \lambda_0^2} e^{-\frac{\lambda_0}{1+\lambda_0}\frac{E}{N_0}}$$

here λ_0 is optimal value of the parameter λ. Modified Chernoff bound leads to the bound:

$$P_b < K(d_f)F(D_0) \tag{4.A26}$$

where, coefficient $K(n)$ is defined as (see (4.A19)),

$$K(n) = \frac{n}{(2n - 1)\sqrt{\pi n}}\sqrt{\frac{1 - \lambda_0^2}{\lambda_0^2}}$$

and d_f is code free distance, $d_f = 6$ for this example. The plots of bounds (4.A25) and (4.A26) and error probability of the uncoded transmission are shown in Figure 4.A6. The difference between common and modified Cheroff bound is about 0.4...0.5 dB in signal to noise ratio, and almost ten times in bit error probability.

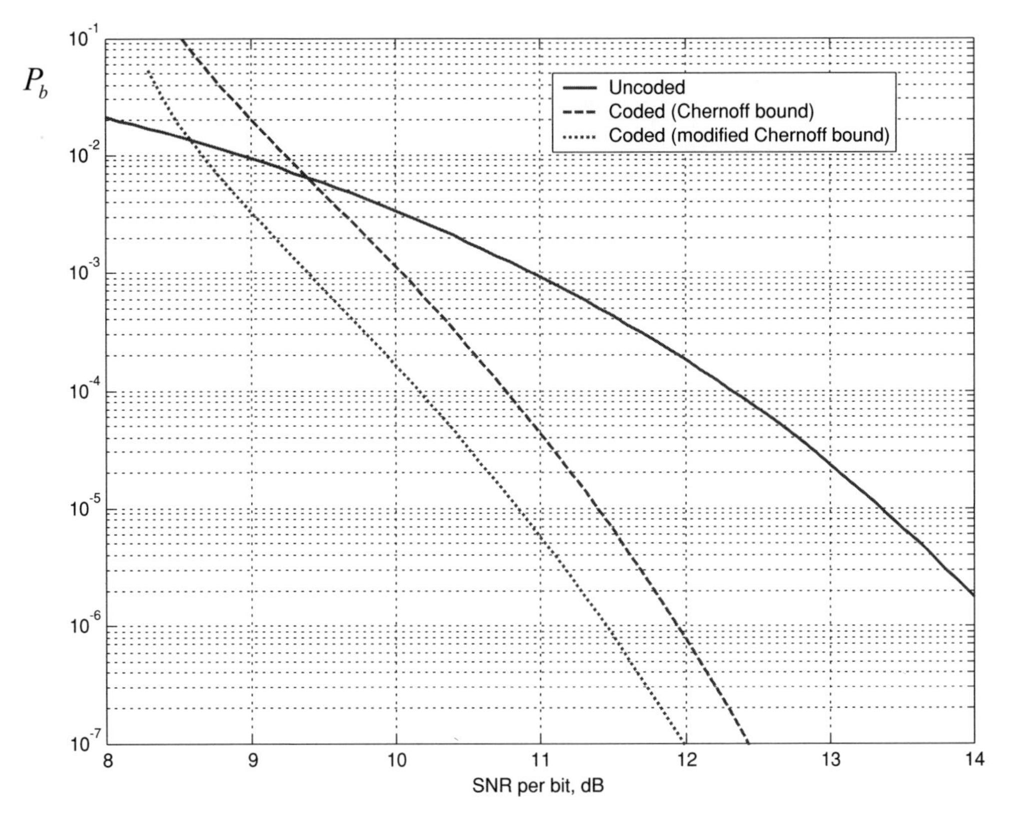

Figure 4.A6 Bounds on bit error probability based on common and modified Chernoff bound for binary convolutional code, $R = 1/2$, $v = 3$, $FSK2$, noncoherent soft output demodulation

References

[1] Prabhu, V.K. "Modified Chernoff for PAM systems for noise and interference". IEEE Trans. Information Theory, vol. IT-28, 1982, N1, pp.95–100.
[2] Gradstein, I.S., Ryzhik, I. *Tables of Integrals, Series and Products*. NY, Academic, 1980.
[3] Dwight, H.B. *Tables of Integrals and Other Mathematical Data*. NY The Macmillan Co, 1961.
[4] Meeberg, L., van de "A Tightened Upper Bound on the Error probability of Binary Convolutional Codes with Viterbi Decoding", IEEE Trans. Information Theory, vol. IT-20, 1974, May, pp.389–390.
[5] Clark, G.C.C. Jr. and Cain, J.B. *Error-Correcting Coding for Digital Communications*. NY, Plenum Press, 1981.
[6] Wozencraft, J. M. and Jacobs, I.M. *Principles of Communication Engineering*, NY, Wiley, 1965.
[7] Wicker, S.B. *Error Control Systems for Digital Communication and Storage*. Upper Saddle River, New Jersey, Prentice Hall, 1995.

5

Equalization

Sergei Semenov
Nokia Corporation, Finland

As discussed in Chapter 2, from the theoretical point of view the transmission of a digital signal requires infinite bandwidth. However, in real life the bandwidth is a shared and very expensive resource. So, in real life we deal only with bandlimited systems. The result of this is that each received pulse is smeared by adjacent pulses. This phenomenon called intersymbol interference (ISI) was described in Chapter 2. The ISI is one of the major obstacles to reliable data transmission over bandlimited channels. One of the ways to eliminate or reduce the ISI is to use for transmission the waveforms obeying Nyquist criterion, which was described in Section 2.1.3. Unfortunately, the use of pulse-shaping allows the ISI to be eliminated only in the case of the ideal channel, that is, $H_C(f) = 1, |f| \leq W$, were $H_C(f)$ is the channel frequency function and W is the channel bandwidth. In this chapter we consider the methods of reducing the ISI in case of nonideal channel. The general name for these methods is *channel equalization* or simply *equalization*. These methods can be roughly divided into two broad categories: *methods based on sequence estimation*, and *methods based on equalization with filters*. For the first group of equalization methods it is assumed that the data stream at the receiver is detected as a sequence rather than each symbol being detected separately. It is quite logical since the ISI, by its nature, introduces dependence of one symbol from the adjacent one. Therefore, each symbol in the received data stream cannot be detected independently from the other symbols. This group of equalization methods includes *maximum-likelihood sequence estimation* (MLSE), which is an optimal method, and some sub-optimal methods, for example, *sphere detection*. The received signal corrupted by the channel in this case is not changed by the equalizer. The equalizer just provides some information to the detection block, which enables the detector to tune the decision thresholds accordingly to the channel conditions. Usually the methods based on sequence estimation provide very good performance (an optimal one for MLSE) but the drawback of these methods is high implementation complexity.

Another approach is based on the attempt to decrease the channel impact to transmitted data by filtering the received signal. The idea is to find the filter that can compensate the distortion inserted to signal by the channel. This approach allows sub-optimal algorithms only to be created but the implementation complexity of these algorithms could be quite low.

One more approach to decreasing the ISI is using diversity. The most known example of this approach is *rake receiver* that can be used for CDMA transmission over a multipath channel. Actually, rake receiver

---

Modulation and Coding Techniques in Wireless Communications Edited by Evgenii Krouk and Sergei Semenov

is quite seldom referred to as an equalizer. However, we consider it in this chapter since in many wireless systems rake receiver is competing with equalization methods based on filtering.

5.1 Equalization with Filtering

The approach based on filtering the received signal is common to a large number of different equalization methods. These methods can be divided into subcategories in accordance with different criteria. The most obvious criterion is the type of filter used for equalization. In line with this criterion it is possible to distinguish *linear equalization* and *nonlinear equalization*. In linear equalization the transversal filters containing only feedforward elements are used and in nonlinear equalization filters containing both feedforward and feedback elements can be used. Another criterion could be eagerness to adapt the equalization filter to changing channel conditions. According to this criterion equalizers can be considered as *preset* or *adaptive*. The preset equalizers are used if it is assumed that the channel is not changing during the transmission time, and correspondingly it is enough to calculate the equalizer filter coefficients once before transmission. The adaptive equalizers are used when the channel is changing during the transmission. In this case the equalizer filter coefficients must be updated with every new sample at the equalizer input or at least once per some small number of samples. Also, there exist so called *automatic equalizers*, which are something intermediate between preset and adaptive equalizers. Automatic equalizers are used in cases when it is assumed that during some short time period the channel coefficients can be considered as constants or when the channel is constant but unknown in advance, then the equalizer coefficients are updated once per this time period. One more criterion is the equalizer filter's resolution. If just one sample is fed to equalizer input per symbol time the equalizer is called *symbol spaced*. If multiple samples are provided for each symbol the equalizer is known as *fractionally spaced* [1]. The latter criterion can be applied to any equalizer independently as it is using filtering or some other equalization method.

It should be taken into account that an equalizer using the filtering approach comprises not only equalizer filter but also the block calculating the filter coefficients. Usually this block is called *tap solver*. If tap solver calculates the filter coefficients based on known *training sequence* (transmitted periodically or in parallel with data as in CDMA) the corresponding equalizer is called *training sequence based equalizer*. If the training sequence is not used for calculation of filter coefficients the corresponding equalizer is called *blind equalizer*.

The scheme of wireless communication systems with filter based equalization is depicted in Figure 5.1. In the description of filter based equalization we will follow mostly [1]. As can be seen this model resembles the system model depicted in Figure 2.14. Assuming that the overall system frequency function is frequency function of raised-cosine filter $H_{RC}(f)$, (2.43) can be written in the following form [1]:

$$H_{RC}(f) = H_{Tx}(f) \cdot H_C(f) \cdot H_{Rx}(f) \cdot H_{Eq}(f) \tag{5.1}$$

where $H_{Tx}(f)$ is the transmitter filter frequency function, $H_C(f)$ is the channel frequency function, $H_{Rx}(f)$ is the receiver filter frequency function and $H_{Eq}(f)$ is the equalizer filter frequency function. However, in real life the channel frequency function $H_C(f)$ usually is not known, and for this reason the transmitter and receiver filters are designed in such a way that their overall frequency function

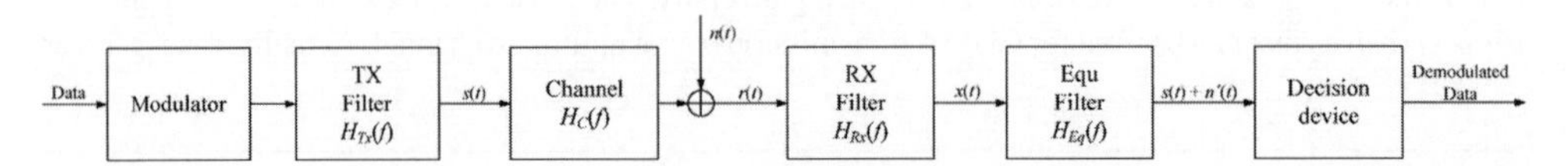

Figure 5.1 Communication system with equalization filter

corresponds to raised-cosine filter:

$$H_{RC}(f) = H_{Tx}(f) \cdot H_{Rx}(f) \tag{5.2}$$

As it was described in Section 2.1.4, in this case transmitter and receiver filters are chosen in a way that both filters have frequency function corresponding to root-raised-cosine filter. Then the equalizing filter frequency function $H_{Eq}(f)$ should be the reversal of the channel frequency function $H_C(f)$ (see (2.47)). In fact, it is correct for zero-forcing equalizers only. For other types of equalizers this dependence is more comprehensive.

Thus, receiver filter takes care of ISI induced by the transmitter filter and equalizer takes care of ISI introduced by the channel. Then, the communication system model from the equalizer task point of view can be simplified to the model depicted in Figure 5.2. Notice that in Figure 5.2 $n'(t)$ comprises coloured Gaussian noise rather than AWGN as in Figure 5.1.

Thus, we can assume that the transmitted data is modulated with raised-cosine pulses and the equalizer task is to reconstruct the original raised-cosine pulse. Quite often the equalizer consists of two filters: a noise-whitening filter, which "whitens" the noise sequence in such a way that the noise samples are uncorrelated; and the equalizing filter itself. Actually, the noise-whitening filter depends only on $H_{Rx}(f)$ and can be determined a priori. Then the noise at the input of the equalizing filter can be considered as an AWGN.

Assume that FIR filter of length $2N + 1$ is used as the equalizing filter. Then the equalization is the linear process and the corresponding equalizer schematic is represented in Figure 5.3.

The delay between filter taps τ corresponds to input samples rate. In this case the equalizer output $y(k)$ is the convolution of input samples $x(k)$ and filter taps $c(i)$ [1]:

$$y(k) = \sum_{i=-N}^{N} x(k-i) \cdot c(i) \tag{5.3}$$

The filter taps must be adjusted by tap solving algorithm in such a way to reduce the effect of ISI caused by the channel. For example, consider the equalizer filter output sequence of length $4N + 1$:

$$y(k), \quad k = -2N, \ldots, 2N \tag{5.4}$$

Then (5.3) can be written in matrix form as follows:

$$\mathbf{y} = \mathbf{X}\mathbf{c} \tag{5.5}$$

where

$$\mathbf{X} = \begin{bmatrix} x(-N) & 0 & 0 & \cdots & 0 & 0 \\ x(-N+1) & x(-N) & 0 & \cdots & 0 & 0 \\ \vdots & \vdots & \vdots & \vdots & \vdots & \vdots \\ x(N) & x(N-1) & x(N-2) & \cdots & x(-N+1) & x(-N) \\ \vdots & \vdots & \vdots & \vdots & \vdots & \vdots \\ 0 & 0 & 0 & \cdots & x(N) & x(N-1) \\ 0 & 0 & 0 & \cdots & 0 & x(N) \end{bmatrix} \tag{5.6}$$

$$\mathbf{y} = \begin{bmatrix} y(-2N) \\ \vdots \\ y(0) \\ \vdots \\ y(2N) \end{bmatrix}, \quad \mathbf{c} = \begin{bmatrix} c(-N) \\ \vdots \\ c(0) \\ \vdots \\ c(N) \end{bmatrix} \tag{5.7}$$

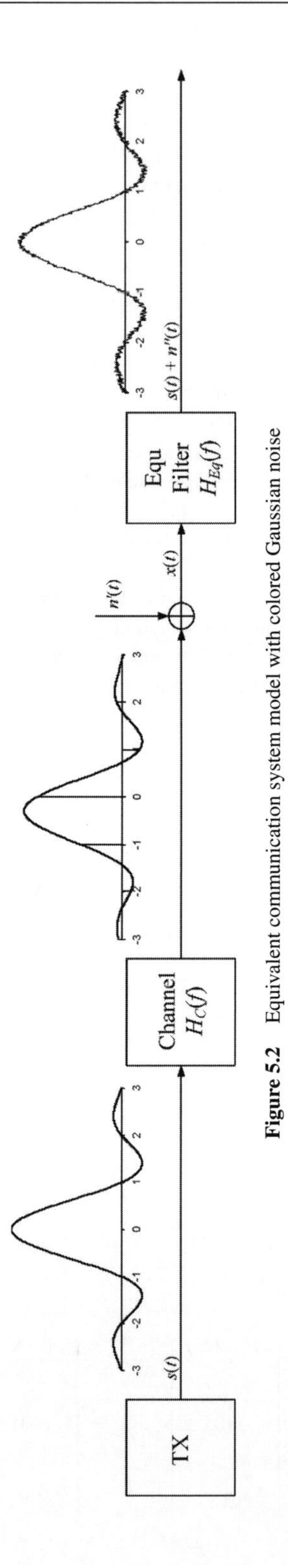

Figure 5.2 Equivalent communication system model with colored Gaussian noise

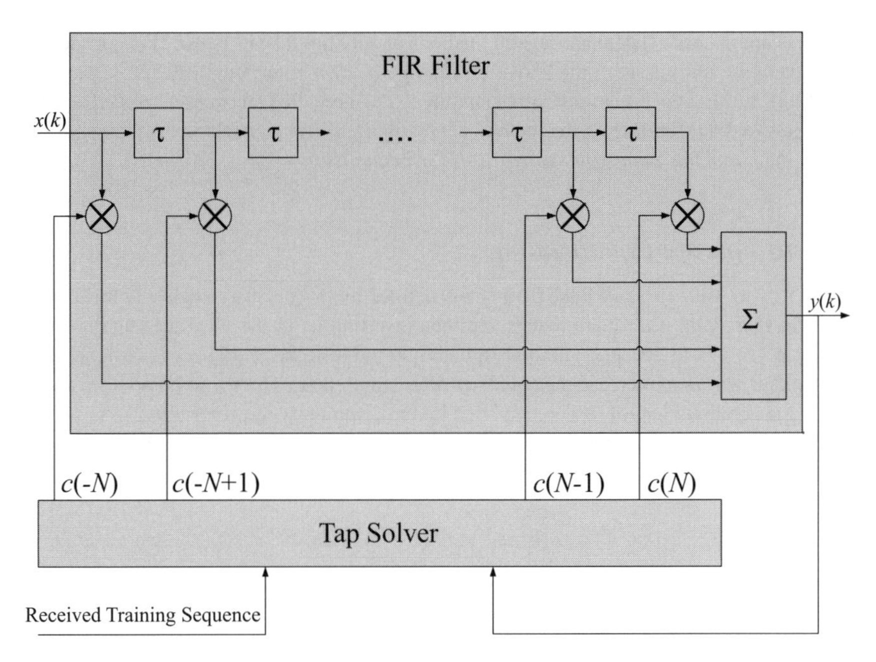

Figure 5.3 Linear equalizer

Actually, the number of samples considered at the output of equalizer (that is, length of vector **y** and number of rows in matrix **X**) does not depend on filter length (that is, length of vector **c** and number of columns in matrix **X**). These values are design parameters of the equalizer. On the one hand, the greater is the filter length, the more accurate the equalizer filter can compensate the ISI introduced by channel. In theory, the equalizer filter must be of infinite length to be able to compensate channel influence completely. Of course, in real life the equalizer filter length is limited by complexity requirements. Moreover, the greater is equalizer filter length, the more noise components are involved in calculation, which in some cases can degrade the equalizer performance. The length of vector **y** depends on channel impulse response length (or simply channel length) but in real life this parameter quite often is chosen to be constant and depends on receiver requirements and system parameters, that is, it is assumed that the equalizer is designed for some channel with a length not exceeding some constant). For example, the greater is the symbol interval the smaller length of vector **y** can be chosen, since in this case only long channel (that is, the channel with length of impulse response comparable with the symbol length) can contribute severe ISI. Very often the special training sequence is used for estimation of channel impulse response. Usually the training sequence comprises sequence of symbols known at the receiver side. Without loss of generality we can assume that delta-function is fed at the channel input instead of training sequence and thus we can estimate the channel impulse response $h(t)$ (for simplicity we will use notation $h(t)$ instead of $h_C(t)$ for channel impulse response and $H(f)$ instead of $H_C(f)$ for channel frequency function). Then the equalizer filter coefficients can be found by substituting in (5.5) the channel matrix **H** instead of matrix **X** (channel matrix **H** is formed in the same way as matrix **X**) and solving the obtained equation:

$$\mathbf{y} = \mathbf{Hc} \tag{5.8}$$

If matrix **H** is square matrix the solution is trivial:

$$\mathbf{c} = \mathbf{H}^{-1}\mathbf{y} \tag{5.9}$$

but usually it is not the case. There are several methods of solving (5.8) if matrix **H** is not square matrix. If the parameters are chosen as in the above example: equalizer filter length is $2N + 1$ and length of vector **y** is $4N + 1$, the set of equations corresponding to (5.8) is called an overdetermined set and matrix **H** cannot be inverted. Depending on the method of solving (5.8) it is possible to distinguish *zero-forcing equalization* and *minimum mean-square error (MMSE) equalization.*

5.1.1 Zero-Forcing Equalization

The simplest way to solve (5.8) with the parameters defined by (5.4) is just to delete N top and N bottom rows of matrix **H** making it a square matrix and then inverting it. That means the length of vector **y** is decreased from $4N + 1$ to $2N + 1$, and correspondingly the resulting equalizer is designed for shorter channels. Another way is to increase the equalizer filter length up to $4N + 1$ to obtain the square matrix **H**. However, the equalizer length can be restricted by the complexity requirements.

Assume that the equalizer filter is of infinite length. Then the cumulative impulse response of channel and equalizer can be written in accordance with (5.3) as follows:

$$q(k) = \sum_{i=-\infty}^{\infty} h(k-i) \cdot c(i) \tag{5.10}$$

And the equalizer output can be written as [2]:

$$y(k) = s(k) \cdot q(0) + \underbrace{\sum_{i \neq k} s(i) \cdot q(k-i) + \sum_{j=-\infty}^{\infty}}_{\text{ISI}} c(j) \cdot n'(k-j) \tag{5.11}$$

where $s(k)$ is the transmitted symbol and $n'(k)$ is the equivalent noise (AWGN filtered by the receiver filter) as depicted in Figure 5.2. The first term in (5.11) is the scaled version of transmitted symbol, the second term comprises ISI and the third term is the noise filtered by the equalizer filter. The peak value of the ISI term is called the *peak distortion*. The peak distortion is given by:

$$D = \sum_{\substack{k=-\infty \\ k \neq 0}}^{\infty} |q(k)| = \sum_{\substack{k=-\infty \\ k \neq 0}}^{\infty} \left| \sum_{j=-\infty}^{\infty} c(j) \cdot h(k-j) \right| \tag{5.12}$$

As can be seen from (5.12) the peak distortion depends on equalizer filter coefficients $c(j)$. The most obvious criterion of calculation of the equalizer filter coefficients is the minimization of peak distortion. In accordance with (5.12) the peak distortion can be reduced to zero if the following condition holds:

$$q(k) = \begin{cases} 1, & k = 0 \\ 0, & k \neq 0 \end{cases} \tag{5.13}$$

If the equalizer filter has the finite length of $2N + 1$ taps that means the tap coefficients in (5.10)–(5.12) $c(j) = 0$ for $|j| > N$. Then the minimization of peak distortion criterion means that tap coefficients must be chosen in such a way that equalizer output is forced to zero at N sample points on either side of the desired pulse [1]:

$$y(k) = \begin{cases} 1, & k = 0 \\ 0, & k = \pm 1, \pm 2, \ldots, \pm N \end{cases} \tag{5.14}$$

Vector **y** corresponding to (5.14) is used for solving (5.9). This kind of solution is called the *zero-forcing solution* and the corresponding equalizer is called *zero-forcing equalizer*. In this case the noise term

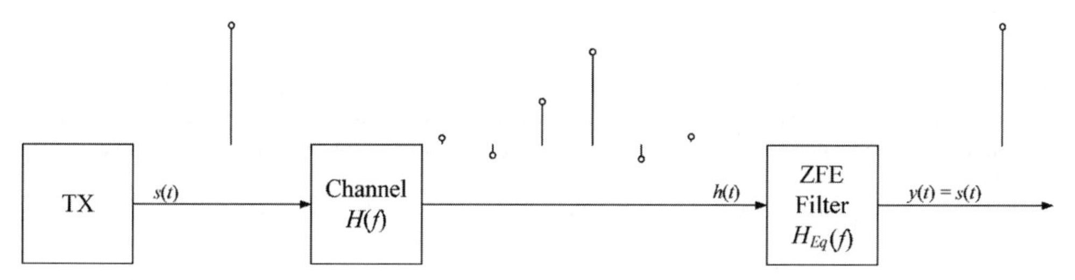

Figure 5.4 System model in accordance with zero-forcing solution (minimizing peak distortion)

in (5.11) is not taken into account, that is, the system model is simplified to one that is depicted in Figure 5.4.

If the equalizer filter length is big enough to be calculated in this way equalizer filter coefficients make the equalizer frequency function $H_{Eq}(f)$ close to reciprocal of channel frequency function $H(f)$:

$$H_{Eq}(f) = \frac{1}{H(f)} \tag{5.15}$$

This gives perfect results with the absence of noise at the equalizer input. If we consider the noise impact the serious problem with the zero-forcing approach arises when spectral nulls in channel frequency function exist, since the gain of equalizer filter in those regions becomes very high. That results in excessive noise enhancement, since the channel noise will also be passing through the equalizer. We can assume that noise $n'(t)$ is the Gaussian noise with zero mean and variance σ_0. Actually the noise at the equalizer filter is coloured noise since it is filtered with the receiver filter, but if the equalizer filter is coupled with the whitening filter we can assume that $n'(t)$ is AWGN. Then the power spectral density of the noise sequence $n''(t)$ at the equalizer output can be written as follows:

$$S_{n''}(\omega) = \frac{N_0}{2} H_{Eq}\left(\exp(j\omega T)\right) |\omega| \leq T \tag{5.16}$$

where N_0 is the noise power spectral density at the equalizer input and the coefficient $\frac{1}{2}$ indicates that power spectral density is double sided. In this case the noise power at the equalizer output can be calculated as:

$$\sigma_n^2 = \frac{N_0}{2} \cdot \frac{T}{2\pi} \int_{-\pi/T}^{\pi/T} H_{Eq}\left(\exp(j\omega T)\right) d\omega = \frac{N_0}{2} \cdot \frac{T}{2\pi} \int_{-\pi/T}^{\pi/T} \frac{1}{H\left(\exp(j\omega T)\right)} d\omega \tag{5.17}$$

Obviously σ_n^2 can take very high values and even can be infinite, for example, in cases when $H\left(\exp(j\omega T)\right)$ is zero over an interval. Actually (5.17) is the IFT of $H_{Eq}(f)$ calculated at $k = 0$. Then (5.17) can be written as:

$$\sigma_n^2 = \frac{N_0}{2} \cdot c(0) \tag{5.18}$$

where $c(0)$ is the corresponding tap of the equalizer filter. Thus, the noise enhancement directly depends on the value of the equalizer filter centre tap (to be more precise the noise enhancement depends on all equalizer filter taps but the central tap contribution dominates), which is not desired since in some cases this may lead to severe performance degradation.

5.1.2 MMSE Equalization

Zero-forcing equalizer removes ISI completely (in the case of infinite equalizer length) but it does not take into account noise, which may lead to serious noise enhancement at the equalizer output. A different solution taking into account noise is the equalization using *minimum mean-square error (MMSE) criterion.*

Without loss of generality we can assume that the transmitted data sequence $\{s(k)\}$ comprises the realization of wide-sense stationary (WSS) random process. Now we will try to choose linear equalizer using the MMSE criterion, that is, the equalizer must minimize the mean-square error (MSE) between the original symbols $\{s(k)\}$ and the equalizer output $\{y(k)\}$:

$$J = E\left\{|\varepsilon(k)|^2\right\} = E\left\{|s(k) - y(k)|^2\right\} \tag{5.19}$$

Assume that the equalizer filter has an infinite length ($N = \infty$). Then the equalizer output can be written as follows:

$$y(k) = \sum_{j=-\infty}^{\infty} c(j) \cdot x(k-j) \tag{5.20}$$

where $x(k)$ is the equalizer input. Recall that the channel can be represented as a FIR filter. Then the equalizer input can be written in the form of:

$$x(k) = \sum_{l=0}^{L-1} h(l) \cdot s(k-l) + n(k) \tag{5.21}$$

where L is the channel length and $n(k)$ is noise. As mentioned above $n(k)$ is the coloured noise, but assuming that the receiver filter is coupled with the whitening filter we can consider $n(k)$ as AWGN with PSD $\frac{N_0}{2}$.

As mentionned above the equalizer filter coefficients should be chosen in such a way to minimize the cost function (5.19). As can be seen from (5.19) cost function is the quadratic function. That means it has only one extreme, and moreover since the function is the square of difference, this extreme is the minimum. Then the minimization can be done in two ways. One way is to take the derivative of J with respect to the equalizer filter coefficients $c(j)$ and equating the result to zero:

$$\nabla_c J = 0 \tag{5.22}$$

Then solving the (5.22) obtain the equalizer filter coefficients that minimize J. Another way is to use the *principle of orthogonality*. Principle of orthogonality claims that the cost function J reaches its minimum value if the corresponding value of the error $\varepsilon(k)$ is orthogonal to each input sample $x(k-i)$ of the equalizer filter for $-\infty < i < \infty$:

$$E\left\{\varepsilon(k) \cdot x^*(k-i)\right\} = E\left\{(s(k) - y(k)) \cdot x^*(k-i)\right\} = 0, \quad -\infty < i < \infty \tag{5.23}$$

In this case error $\varepsilon(k)$ is not correlated with equalizer input. Substituting (5.20) in (5.23) obtain:

$$E\left\{\left(s(k) - \sum_{j=-\infty}^{\infty} c(j) \cdot x(k-j)\right) \cdot x^*(k-i)\right\} = 0, \quad -\infty < i < \infty \tag{5.24}$$

This in turn leads to Wiener-Hopf equation:

$$\sum_{j=-\infty}^{\infty} c(j) \cdot E\left\{x(k-j) \cdot x^*(k-i)\right\} = E\left\{s(k) \cdot x^*(k-i)\right\}, \quad -\infty < i < \infty \tag{5.25}$$

Solution of (5.25) depends only on *second-order statistics* given by autocorrelation:

$$r_{xx}(i-j) = E\{x(k-j)\cdot x^*(k-i)\} \tag{5.26}$$

and *cross-correlation*:

$$r_{sx}(-i) = E\{s(k)\cdot x^*(k-i)\} \tag{5.27}$$

Assuming that data signal can be represented as a random variable with zero mean and substituting (5.21) in (5.26) yields:

$$r_{xx}(i-j) = \sum_{l=0}^{L-1} h^*(l)\cdot h(l+i-j) + \frac{N_0}{2}\delta_{ij} = \begin{cases} r_{hh}(i-j) + \frac{N_0}{2}\delta_{ij}, & |i-j| \le L-1 \\ 0, & \text{otherwise} \end{cases} \tag{5.28}$$

Applying the same assumptions to (5.27) obtain:

$$r_{sx}(-i) = \begin{cases} h^*(-i), & -L+1 \le i \le 0 \\ 0, & \text{otherwise} \end{cases} \tag{5.29}$$

Substituting (5.28) and (5.29) in (5.25) and taking z-transform of resulting equation obtain [2]:

$$C(z)\cdot\left(H(z)H^*(z^{-1}) + \frac{N_0}{2}\right) = H^*(z^{-1}) \tag{5.30}$$

Assuming that noise whitening filter $\frac{1}{H^*(z^{-1})}$ is incorporated in the equalizer filter (5.30) can be written as follows:

$$C'(z)\cdot\left(H(z)H^*(z^{-1}) + \frac{N_0}{2}\right) = 1 \tag{5.31}$$

That means the MMSE equalizer frequency function can be written as:

$$H_{Eq}(f) = \frac{1}{H(f) + \frac{N_0}{2}} \tag{5.32}$$

where $H(f)$ is the channel frequency function.

That means, unlike the zero-forcing equalizer, the MMSE equalizer avoids excessively amplifying the noise, besides equalizing the channel. The main benefit of using the MMSE equalizer is that it does not allow noise enhancement as ZF does, especially in the case when the channel has a spectral null. If noise is negligible, that is, $N_0 \to 0$ the zero-forcing and MMSE equalizer become identical. When $N_0 > 0$ the zero-forcing equalizer eliminates ISI completely but allows significant noise enhancement, MMSE equalizer suppresses noise but some residual ISI and noise will be observed at the output of the MMSE equalizer. However, in general this residual ISI and noise at the output of the MMSE equalizer does not lead to significant performance degradation as noise enhancement does for ZF equalizer.

Now assume that the equalizer filter length is restricted to $2N+1$. Then (5.19) can be written as:

$$J = E\left\{|\varepsilon(k)|^2\right\} = E\left\{\left|s(k) - \sum_{j=-N}^{N} x(k-j)\cdot c(k)\right|^2\right\} = E\left\{\left|s(k) - \mathbf{x}^T\mathbf{c}\right|^2\right\} \tag{5.33}$$

where $\mathbf{x} = [x(k+N), x(k+N-1), \ldots, x(k-N)]^T$, $\mathbf{c} = [c(-N), c(-N+1), \ldots, c(N)]^T$.

To find the optimal values of the equalizer filter $c(j)$ it is possible to apply the principle of orthogonality as it was done above or simply differentiate J with respect to $c(j)$ and set the result to zero. Then obtain the following equation:

$$E\left\{\mathbf{x}\left(s(k) - \mathbf{x}^T\mathbf{c}\right)\right\} = \mathbf{0} \tag{5.34}$$

Rearranging, obtain:

$$\mathbf{r} = \mathbf{Rc} \tag{5.35}$$

where

$$\mathbf{r} = E\left\{s(k)\mathbf{x}\right\} \tag{5.36}$$

is the cross-correlation vector and:

$$\mathbf{R} = E\left\{\mathbf{xx}^H\right\} \tag{5.37}$$

is the auto-correlation matrix. Then the coefficients of the MMSE equalizer filter can be calculated as:

$$\mathbf{c} = \mathbf{R}^{-1}\mathbf{r} \tag{5.38}$$

Another approach to calculation of the MMSE equalizer filter is to consider the equalizer filter as an LMMSE estimator to the Bayesian linear model [3]. Sequence at the equalizer filter input $\mathbf{x}$ can be written in the Bayesian linear model form:

$$\mathbf{x} = \mathbf{Hs} + \mathbf{n} \tag{5.39}$$

where $\mathbf{x} = [x(k+N), x(k+N-1), \ldots, x(k-N)]^T$ is the sequence at the equalizer input, $\mathbf{s} = [s(k+N), s(k+N-1), \ldots, s(k-N), s(k-N-1), \ldots, s(k-N-L+1)]^T$ is the transmitted sequence, $\mathbf{n} = [n(k+N), n(k+N-1), \ldots, n(k-N)]^T$ is the noise vector, and $\mathbf{H}$ is the $((2N+1) \times (2N+L))$ channel matrix:

$$\mathbf{H}^T = \begin{bmatrix} h(0) & 0 & \ddots & 0 \\ h(1) & h(0) & \ddots & 0 \\ \vdots & \vdots & \ddots & \vdots \\ h(L-1) & h(L-2) & \ddots & h(0) \\ 0 & h(L-1) & \ddots & h(1) \\ \vdots & \vdots & \ddots & \vdots \\ 0 & 0 & \ddots & h(L-1) \end{bmatrix}$$

The equalizer output sequence $\mathbf{y}$ is the estimate of the transmitted sequence $\mathbf{s}$. Then in accordance with the Bayesian Gauss-Markov Theorem [3] the LMMSE estimator of $\mathbf{s}$ can be written as:

$$\mathbf{y} = E\left\{\mathbf{s}\right\} + \mathbf{R}_{ss}\mathbf{H}^H\left(\mathbf{HR}_{ss}\mathbf{H}^H + \mathbf{R}_{nn}\right)^{-1}\left(\mathbf{x} - \mathbf{H}E\left\{\mathbf{s}\right\}\right) \tag{5.40}$$

where $\mathbf{R}_{ss} = E\left\{\mathbf{ss}^H\right\}$ is the auto-correlation matrix of the transmitted sequence $\mathbf{s}$ and $\mathbf{R}_{nn} = E\left\{\mathbf{nn}^H\right\}$ is the noise auto-correlation matrix. Assuming the noise being AWGN the noise auto-correlation matrix can be represented as the diagonal matrix:

$$\mathbf{R}_{nn} = E\left\{\mathbf{nn}^H\right\} = \begin{bmatrix} \sigma_0^2 & 0 & 0 & 0 \\ 0 & \sigma_0^2 & \cdots & 0 \\ \cdots & \cdots & \ddots & \vdots \\ 0 & 0 & \cdots & \sigma_0^2 \end{bmatrix} = \sigma_0^2\mathbf{I} \tag{5.41}$$

where $\sigma_0^2 = \frac{N_0}{2}$ is the noise variance.

Quite often the transmitted sequence is masked with the pseudorandom sequence. For example, it is done in wide-band signalling. In this case we can assume **s** to be a random i.i.d. sequence with zero mean $E\{\mathbf{s}\} = 0$ [4], [5]. With this assumption the auto-correlation matrix of the transmitted sequence **s** can be represented also as a diagonal matrix:

$$\mathbf{R}_{ss} = E\left\{\mathbf{s}\mathbf{s}^H\right\} = \begin{bmatrix} \sigma_s^2 & 0 & 0 & 0 \\ 0 & \sigma_s^2 & \cdots & 0 \\ \cdots & \cdots & \ddots & \vdots \\ 0 & 0 & \cdots & \sigma_s^2 \end{bmatrix} = \sigma_s^2\mathbf{I} \tag{5.42}$$

where σ_s^2 is the transmitted signal power. Then (5.40) can be simplified to:

$$\mathbf{y} = \sigma_s^2\mathbf{H}^H\left(\sigma_s^2\mathbf{H}\mathbf{H}^H + \sigma_0^2\mathbf{I}\right)^{-1}\mathbf{x} \tag{5.43}$$

Actually, we are interested in value $y(k)$ only. That means we need only one row of matrix $\sigma_s^2\mathbf{H}^H\left(\sigma_s^2\mathbf{H}\mathbf{H}^H + \sigma_0^2\mathbf{I}\right)^{-1}$ to calculate $y(k)$. Moreover, this row can be regarded as coefficients of the equalizer filter. Then the corresponding values of the equalizer filter coefficients can be calculated as follows [5]:

$$\mathbf{c} = \sigma_s^2\delta_D\mathbf{H}^H\left(\sigma_s^2\mathbf{H}\mathbf{H}^H + \sigma_0^2\mathbf{I}\right)^{-1} \tag{5.44}$$

where δ_D is all zeroes $(1 \times (2N + L))$ vector except for unity in the $(D + 1)$-th position. The delay D is chosen in such a way to improve the equalizer performance [5].

With the assumptions (5.41) and (5.42) and with the best choice of D (5.44) may be derived directly from (5.19). The corresponding system model is depicted in Figure 5.5. Notice that as can be seen from Figure 5.5 the equalizer comprises simply a Wiener filter.

Then the cost function (5.19) may be written as follows:

$$J = E\left\{|\varepsilon(k)|^2\right\} = E\left\{|s(k-D) - y(k)|^2\right\} = E\left\{|\delta_D\mathbf{s} - y(k)|^2\right\} \tag{5.45}$$

Taking into account that $y(k)$ is the result of filtering sequence **x** with filter **c** $y(k) = \mathbf{c}\mathbf{x}$ (5.45) can be written as:

$$J = E\left\{|\delta_D\mathbf{s} - y(k)|^2\right\} = E\left\{|\delta_D\mathbf{s} - \mathbf{c}\mathbf{x}|^2\right\} \tag{5.46}$$

Then substitution of (5.39) into (5.46) gives:

$$J = E\left\{|\delta_D\mathbf{s} - \mathbf{c}\mathbf{x}|^2\right\} = E\left\{|\delta_D\mathbf{s} - \mathbf{c}\mathbf{H}\mathbf{s} - \mathbf{c}\mathbf{n}|^2\right\} \tag{5.47}$$

Expanding (5.47) obtain:

$$J = E\left\{((\delta_D - \mathbf{c}\mathbf{H})\mathbf{s} - \mathbf{c}\mathbf{n})\left(\mathbf{s}^H(\delta_D^T - \mathbf{H}^H\mathbf{c}^H) - \mathbf{n}^H\mathbf{c}^H\right)\right\} \tag{5.48}$$

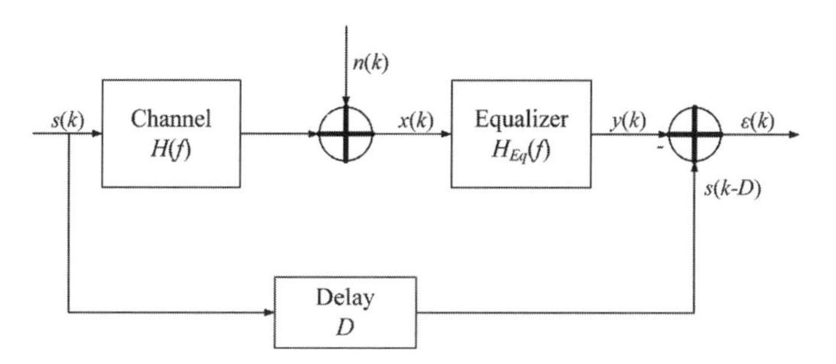

Figure 5.5 System model in accordance with MMSE solution

Then taking in account that data and noise are uncorrelated $E\left\{\mathbf{sn}^H\right\} = 0$ (5.48) is equivalent to:

$$J = (\delta_D - \mathbf{cH})\, E\left\{\mathbf{ss}^H\right\}\left(\delta_D^T - \mathbf{H}^H\mathbf{c}^H\right) + \mathbf{c}E\left\{\mathbf{nn}^H\right\}\mathbf{c}^H \tag{5.49}$$

Substituting (5.41) and (5.42) into (5.49) obtain:

$$J = \sigma_s^2 - \sigma_s^2\delta_D\mathbf{H}^H\mathbf{c}^H - \mathbf{c}\sigma_s^2\mathbf{H}\delta_D^T + \mathbf{c}\left(\sigma_s^2\mathbf{HH}^H + \sigma_0^2\mathbf{I}\right)\mathbf{c}^H \tag{5.50}$$

Denote $\sigma_s^2\mathbf{H}\delta_D^T$ by $\mathbf{r}$ and $\left(\sigma_s^2\mathbf{HH}^H + \sigma_0^2\mathbf{I}\right)$ by $\mathbf{R}$. Then (5.50) is the quadratic function of $\mathbf{c}$:

$$J = \sigma_s^2 - \mathbf{r}^H\mathbf{c}^H - \mathbf{cr} + \mathbf{cRc}^H \tag{5.51}$$

Taking the derivative of J with respect to $\mathbf{c}$ and equating the result to 0 obtain:

$$\nabla_{\mathbf{c}}J = 2\left(\mathbf{c}_{opt}\mathbf{R} - \mathbf{r}^H\right) = 0 \tag{5.52}$$

Then the optimal MMSE equalizer filter coefficients are:

$$\mathbf{c}_{opt} = \mathbf{r}^H\mathbf{R}^{-1} = \sigma_s^2\delta_D\mathbf{H}^H\left(\sigma_s^2\mathbf{HH}^H + \sigma_0^2\mathbf{I}\right)^{-1} \tag{5.53}$$

which coincides with (5.44).

Actually, the same approach can be applied to zero-forcing equalization. It is easy to verify that in this case the zero-forcing equalizer filter coefficients can be obtained as follows:

$$\mathbf{c}_{ZF} = \sigma_s^2\delta_D\mathbf{H}^H\left(\sigma_s^2\mathbf{HH}^H\right)^{-1} \tag{5.54}$$

The (5.44) or (5.53) can be written in another form. It can be done with the help of some matrix identities.

Let us start with the trivial identity:

$$\mathbf{A} + \mathbf{ABA} = \mathbf{A}(\mathbf{I} + \mathbf{BA}) = (\mathbf{I} + \mathbf{AB})\mathbf{A} \tag{5.55}$$

It follows immediately from (5.55) that:

$$\mathbf{A}(\mathbf{I} + \mathbf{BA})^{-1} = (\mathbf{I} + \mathbf{AB})^{-1}\mathbf{A} \tag{5.56}$$

Applying (5.56) to (5.44) obtain:

$$\mathbf{c} = \sigma_s^2\delta_D\left(\sigma_s^2\mathbf{H}^H\mathbf{H} + \sigma_0^2\mathbf{I}\right)^{-1}\mathbf{H}^H \tag{5.57}$$

To find the equalizer filter coefficients with the help of (5.57) it is necessary to invert $(2N + L) \times (2N + L)$ matrix. So, the (5.44) is more popular for calculation of the equalizer filter coefficients since it requires the inversion of $(2N + 1) \times (2N + 1)$ matrix.

The coefficients of the MMSE equalizer filter can also be found iteratively. As mentioned above, the MSE is a quadratic function of $\mathbf{c}$. The gradient of the MSE with respect to $\mathbf{c}$ gives the direction to change equalizer filter coefficients in such a way to increase the MSE. The gradient is described by (5.52). To decrease the MSE, the equalizer filter coefficients $\mathbf{c}$ must be updated in the direction opposite to the gradient. Denote the equalizer filter coefficients calculated at step k by $\mathbf{c}_k$. Then the process of updating the MMSE equalizer filter coefficients can be written as follows:

$$\mathbf{c}_k = \mathbf{c}_{k-1} + \mu\left(\mathbf{r} - \mathbf{c}_{k-1}\mathbf{R}\right) \tag{5.58}$$

where μ is a small positive constant that controls the rate of convergence to the optimal solution. This kind of procedure is called the *steepest descent algorithm*.

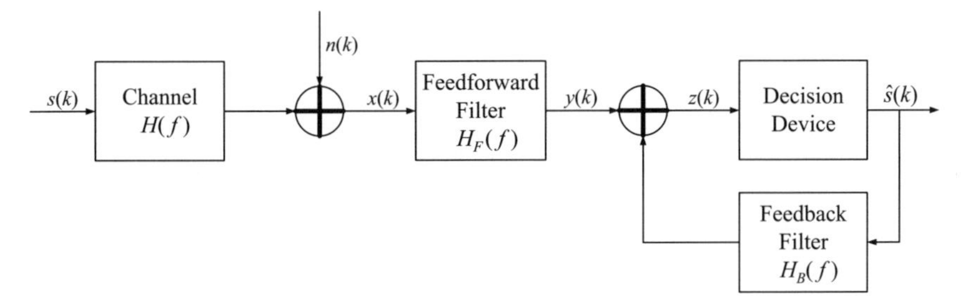

Figure 5.6 Decision feedback equalizer structure

5.1.3 DFE

The linear equalizers usually perform poorly on channels having severe amplitude distortion [1]. Unfortunately quite a few of these channels, especially the radio channels are characterized by this property. In this case a nonlinear equalization may provide better performance. A *decision feedback equalizer* (*DFE*) is a nonlinear equalizer that uses previous decisions to eliminate the ISI on current received signal. Recall the equalizer output written in (5.11). Then the current received symbol can be estimated as follows:

$$\hat{s}(k) = y(k) - \underbrace{\sum_{i \neq k} s(i) \cdot q(k-i)}_{\text{ISI}} \tag{5.59}$$

In real life the previously transmitted symbols $s(i)$ are not known, but it is possible to use previously detected symbols $\hat{s}(i)$ instead of $s(i)$ assuming that they were detected correctly. So, the idea of DFE algorithm is to estimate and subtract the impact of those interfering symbols which have been detected earlier. The DFE consists of two parts: *feedforward filter* and *feedback filter* as it is depicted in Figure 5.6.

Assume that the feedforward equalizer filter has the finite length of $K_1 + 1$ taps and the feedback equalizer filter has the finite length of K_2 taps. Then the equalizer output can be expressed as [2]:

$$z(k) = \sum_{i=-K_1}^{0} c_F(i) \cdot x(k-i) + \sum_{i=1}^{K_2} c_B(i) \cdot \hat{s}(k-i) \tag{5.60}$$

where $c_F(i)$ denotes the coefficients of the feedforward filter and $c_B(i)$ the coefficients of the feedback filter, $\hat{s}(i)$ are previously detected symbols.

The ISI can be split into two parts: *precursor ISI* (involving data ahead of the detection time) and *postcursor ISI* (involving data behind the detection time). The task of the feedforward filter is to decrease the precursor ISI as much as possible, pushing its energy into the postcursor domain. It is also sufficient for the feedforward filter to keep the noise enhancement as small as possible. And the task of the feedback filter is to cancel the postcursor ISI. Notice that the feedforward filtering modifies the postcursor ISI. To be more precise the feedforward filtering adds some more ISI to the postcursor domain since it redistributes the precursor ISI energy into the postcursor region. However, the feedback filter is capable of cancelling even the increased postcursor ISI if the previously detected symbols were detected correctly. The precursor and postcursor ISI before feedforward filtering is depicted in Figure 5.7. The ISI after feedforward filtering is depicted in Figure 5.8. The result of feedforward filtering is the severe decrease of the precursor ISI and the increase of the postcursor ISI. Notice that the detection time after the feedforward filtering is delayed.

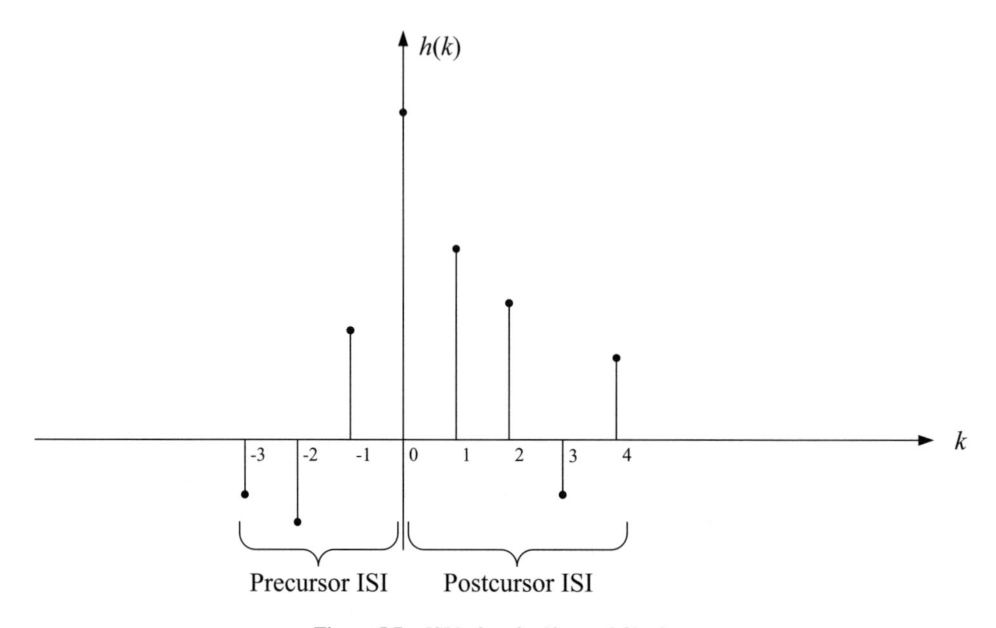

Figure 5.7 ISI before feedforward filtering

Both zero-forcing and MMSE principle can be used for calculation of the feedforward and feedback filter coefficients. Since the MSE criterion provides better results than the peak distortion criterion we will consider here the MMSE solution only.

Assume that all previously detected symbols were detected correctly. That means $\hat{s}(i) = s(i)$. Then (5.60) can be represented in matrix form as follows:

$$z(k) = \mathbf{c}_F\mathbf{x} + \mathbf{c}_B\mathbf{s} \tag{5.61}$$

where $\mathbf{x} = [x(k+K_1), x(k+K_1-1), \ldots, x(k)]^T$, $\mathbf{s} = [s(k-1), s(k-2), \ldots, s(k-K_2)]^T$, $\mathbf{c}_F = [c_F(-K_1), c_F(-K_1+1), \ldots, c_F(0)]$, $\mathbf{c}_B = [c_B(1), c_B(2), \ldots, c_B(K_2)]$. Actually, we can consider feedforward and feedback filters as two parts of one equalizer filter $\mathbf{c} = [\mathbf{c}_F, \mathbf{c}_B]$.

Then the MSE criterion is given by:

$$J = E\left\{|\varepsilon(k)|^2\right\} = E\left\{|s(k) - z(k)|^2\right\} = E\left\{|s(k) - \mathbf{c}_F\mathbf{x} - \mathbf{c}_B\mathbf{s}|^2\right\} \tag{5.62}$$

Differentiating (5.62) with respect to $\mathbf{c}_F$ and $\mathbf{c}_B$ and equating the result to zero obtain:

$$\nabla_{\mathbf{c}_F} J = E\left\{(s(k) - \mathbf{c}_F\mathbf{x} - \mathbf{c}_B\mathbf{s})\left(-\mathbf{x}^H\right)\right\} = -E\left\{s(k)\mathbf{x}^H\right\} + \mathbf{c}_F E\left\{\mathbf{x}\mathbf{x}^H\right\} + \mathbf{c}_B E\left\{\mathbf{s}\mathbf{x}^H\right\}\mathbf{c}_B = \mathbf{0} \tag{5.63}$$

$$\nabla_{\mathbf{c}_B} J = E\left\{(s(k) - \mathbf{c}_F\mathbf{x} - \mathbf{c}_B\mathbf{s})\left(-\mathbf{s}^H\right)\right\} = -E\left\{s(k)\mathbf{s}^H\right\} + \mathbf{c}_F E\left\{\mathbf{x}\mathbf{s}^H\right\} + \mathbf{c}_B E\left\{\mathbf{s}\mathbf{s}^H\right\} = \mathbf{0} \tag{5.64}$$

Then solving the (5.63) and (5.64) obtain the equalizer filter coefficients that minimize J. With the same assumptions that were made in Section 5.1.2 that the data symbols $s(i)$ comprise a random i.i.d. sequence with zero mean, $E\{s(k)\mathbf{s}\} = 0$ and $E\left\{\mathbf{s}\mathbf{s}^H\right\} = \sigma_s^2\mathbf{I}$. Then (5.63) and (5.64) can be simplified to:

$$\begin{cases} E\{s(k)\mathbf{x}\} = \mathbf{c}_F E\left\{\mathbf{x}\mathbf{x}^H\right\} + \mathbf{c}_B E\left\{\mathbf{x}\mathbf{s}^H\right\} \\ \sigma_s^2\mathbf{c}_B = -\mathbf{c}_F E\left\{\mathbf{x}\mathbf{s}^H\right\} \end{cases} \tag{5.65}$$

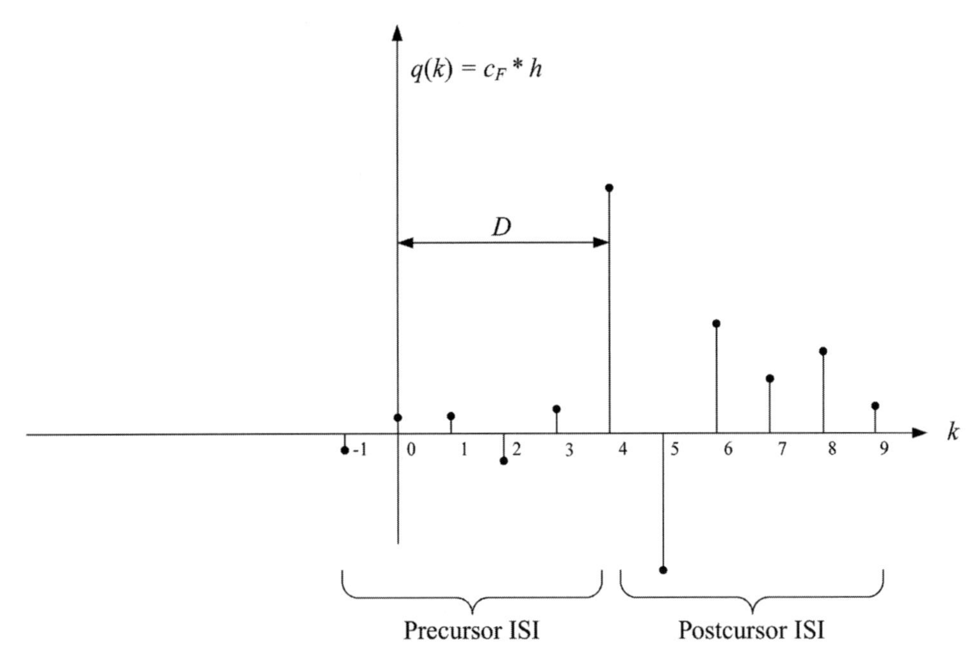

Figure 5.8 ISI after feedforward filtering

Solving system 5.65 obtain:

$$\begin{cases} \mathbf{c}_F = \left(E\left\{\mathbf{x}\mathbf{x}^H\right\} - \sigma_s^{-2} E\left\{\mathbf{x}\mathbf{s}^H\right\} E\left\{\mathbf{s}\mathbf{x}^H\right\}\right)^{-1} E\left\{s(k)\mathbf{x}\right\} \\ \mathbf{c}_B = -\sigma_s^{-2} \mathbf{c}_F E\left\{\mathbf{x}\mathbf{s}^H\right\} \end{cases} \tag{5.66}$$

This solution does not include optimization over delay, which can improve the equalizer performance [4, 5].

System model taking into account the equalizer filter delay is depicted in Figure 5.9.

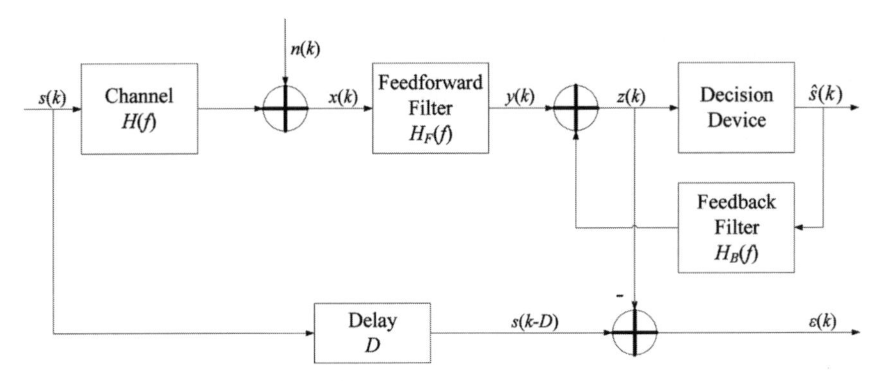

Figure 5.9 System model with DFE

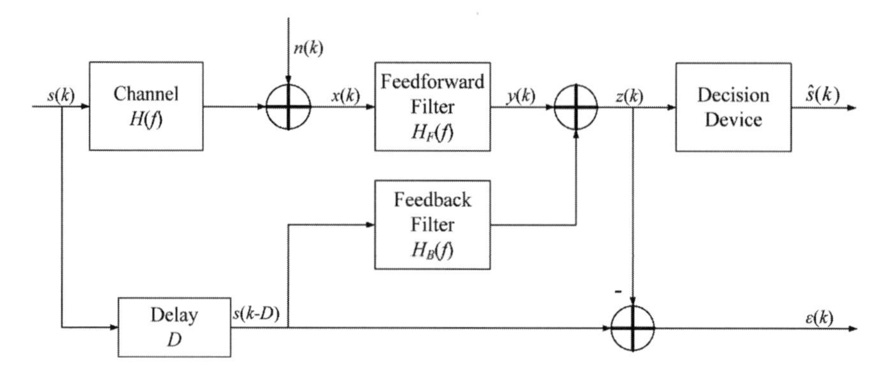

Figure 5.10 Simplified system model with DFE

Now assuming that previously detected symbols were detected correctly the input of the feedback filter $\hat{s}(k)$ can be substituted by transmitted symbols and taking in account the delay D the substitute is:

$$\hat{s}(k) = s(k - D) \tag{5.67}$$

The corresponding simplified system model is depicted in Figure 5.10.

Notice that this simplified model of the DFE became linear. With the assumption (5.67), (5.60) can then be written as:

$$z(k) = \sum_{i=-K_1}^{0} c_F(i) \cdot x(k-i) + \sum_{i=1}^{K_2} c_B(i) \cdot s(k-D-i) \tag{5.68}$$

Expanding (5.68) obtain:

$$\begin{aligned} z(k) &= \sum_{i=-K_1}^{0} c_F(i) \cdot \left(\sum_{l=0}^{L-1} h(l) \cdot s(k-l-i) + n(k-i) \right) + \sum_{i=1}^{K_2} c_B(i) \cdot s(k-D-i) \\ &= \sum_{i=-K_1}^{0} q(i) \cdot s(k-i) + \sum_{i=1}^{K_2} c_B(i) \cdot s(k-D-i) + \sum_{i=-K_1}^{0} c_F(i) \cdot n(k-i) \end{aligned} \tag{5.69}$$

where

$$q(i) = \sum_{l=0}^{L-1} h(l) \cdot c_F(i-l) \tag{5.70}$$

Then the MSE criterion can be written as follows:

$$\begin{aligned} J &= E\left\{|\varepsilon(k)|^2\right\} = E\left\{|s(k-D) - z(k)|^2\right\} = E\left\{|s(k-D) - \mathbf{c}_F\mathbf{x} - \mathbf{c}_B\hat{\mathbf{s}}|^2\right\} \\ &= E\left\{|s(k-D) - \mathbf{c}_F\mathbf{H}\mathbf{s} - \mathbf{c}_B\hat{\mathbf{s}} - \mathbf{c}_F\mathbf{n}|^2\right\} \end{aligned} \tag{5.71}$$

where $\mathbf{s} = [s(k), s(k-1), \ldots, s(k-K_1-L)]^T$, $\hat{\mathbf{s}} = [s(k-D-1), \ s(k-D-2), \ldots, \ s(k-D-K_2)]^T$, $\mathbf{n} = [n(k+K_1), n(k+K_1-1), \ldots, n(k)]^T$.

Differentiating (5.71) with respect to $\mathbf{c}_B$ and equating to zero obtain:

$$\begin{aligned} \nabla_{\mathbf{c}_B} J &= E\left\{(s(k-D) - \mathbf{c}_F\mathbf{H}\mathbf{s} - \mathbf{c}_B\hat{\mathbf{s}} - \mathbf{c}_F\mathbf{n})\left(-\hat{\mathbf{s}}^H\right)\right\} \\ &= -E\left\{s(k-D)\hat{\mathbf{s}}^H\right\} + \mathbf{c}_F\mathbf{H}E\left\{\mathbf{s}\hat{\mathbf{s}}^H\right\} + \mathbf{c}_B E\left\{\hat{\mathbf{s}}\hat{\mathbf{s}}^H\right\} + \mathbf{c}_F E\left\{\mathbf{n}\hat{\mathbf{s}}^H\right\} = \mathbf{0} \end{aligned} \tag{5.72}$$

Taking in account that $\hat{\mathbf{s}} = [s(k-D-1), s(k-D-2), \ldots, s(k-D-K_2)]^T$ does not contain the element $s(k-D)$, $E\left\{s(k-D)\hat{\mathbf{s}}^H\right\} = 0$. Assuming that data and noise are not correlated, $E\left\{\mathbf{n}\hat{\mathbf{s}}^H\right\} = 0$.

In accordance with (5.42) $E\left\{\hat{\mathbf{s}}\hat{\mathbf{s}}^H\right\} = \sigma_s^2\mathbf{I}$. Notice that vectors $\mathbf{s}$ and $\hat{\mathbf{s}}$ may overlap in case $D < K_1 + L + 1$. In fact, delay D cannot be greater than $K_1 + L + 1$ since it is the delay introduced by filtering. Then assuming $K_2 + D < K_1 + L + 1$, $E\left\{\mathbf{s}\hat{\mathbf{s}}^H\right\}$ can be written as follows:

$$E\left\{\mathbf{s}\hat{\mathbf{s}}^H\right\} = \sigma_s^2\mathbf{M} \tag{5.73}$$

where

$$\mathbf{M} = \left[\mathbf{0}_{K_2\times D}\mathbf{I}_{K_2\times K_2}\mathbf{0}_{K_2\times(K_1+L-K_2-D+1)}\right]^T \tag{5.74}$$

Then (5.72) can be reduced to:

$$\nabla_{\mathbf{c}_B} J = \sigma_s^2\left(\mathbf{c}_F\mathbf{H}\mathbf{M} + \mathbf{c}_B\right) = \mathbf{0} \tag{5.75}$$

That means the optimal coefficients of the feedback filter $\mathbf{c}_B$ can be expressed as:

$$\mathbf{c}_B = -\mathbf{c}_F\mathbf{H}\mathbf{M} \tag{5.76}$$

Substitution of (5.76) into (5.71) gives:

$$J = E\left\{\left|s(k-D) - \mathbf{c}_F\mathbf{H}\mathbf{s} + \mathbf{c}_F\mathbf{H}\mathbf{M}\hat{\mathbf{s}} - \mathbf{c}_F\mathbf{n}\right|^2\right\} \tag{5.77}$$

Then differentiating (5.77) with respect to $\mathbf{c}_F$ and equating to zero obtain:

$$\begin{aligned}
\nabla_{\mathbf{c}_F} J &= E\left\{\left(s(k-D) - \mathbf{c}_F\mathbf{H}\mathbf{s} + \mathbf{c}_F\mathbf{H}\mathbf{M}\hat{\mathbf{s}} - \mathbf{c}_F\mathbf{n}\right)\left(-\mathbf{s}^H\mathbf{H}^H + \hat{\mathbf{s}}^H\mathbf{M}^H\mathbf{H}^H - \mathbf{n}^H\right)\right\} \\
&= -E\left\{s(k-D)\left(\mathbf{s}^H\mathbf{H}^H - \hat{\mathbf{s}}^H\mathbf{M}^H\mathbf{H}^H + \mathbf{n}^H\right)\right\} \\
&\quad + \mathbf{c}_F\mathbf{H}E\left\{\mathbf{s}\mathbf{s}^H\right\}\mathbf{H}^H - \mathbf{c}_F\mathbf{H}E\left\{\mathbf{s}\hat{\mathbf{s}}^H\right\}\mathbf{M}^H\mathbf{H}^H + \mathbf{c}_F\mathbf{H}E\left\{\mathbf{s}\mathbf{n}^H\right\} \\
&\quad - \mathbf{c}_F\mathbf{H}\mathbf{M}E\left\{\hat{\mathbf{s}}\mathbf{s}^H\right\}\mathbf{H}^H + \mathbf{c}_F\mathbf{H}\mathbf{M}E\left\{\hat{\mathbf{s}}\hat{\mathbf{s}}^H\right\}\mathbf{M}^H\mathbf{H}^H - \mathbf{c}_F\mathbf{H}\mathbf{M}E\left\{\hat{\mathbf{s}}\mathbf{n}^H\right\} \\
&\quad + \mathbf{c}_F E\left\{\mathbf{n}\mathbf{s}^H\right\}\mathbf{H}^H - \mathbf{c}_F E\left\{\mathbf{n}\hat{\mathbf{s}}^H\right\}\mathbf{M}^H\mathbf{H}^H + \mathbf{c}_F E\left\{\mathbf{n}\mathbf{n}^H\right\} = \mathbf{0}
\end{aligned} \tag{5.78}$$

Taking in account the same assumptions as above (5.78) can be reduced to:

$$\begin{aligned}
\nabla_{\mathbf{c}_F} J &= -E\left\{s(k-D)\mathbf{s}^H\right\}\mathbf{H}^H \\
&\quad + \mathbf{c}_F\sigma_s^2\mathbf{H}\mathbf{H}^H - \mathbf{c}_F\sigma_s^2\mathbf{H}\mathbf{M}\mathbf{M}^H\mathbf{H}^H \\
&\quad - \mathbf{c}_F\sigma_s^2\mathbf{H}\mathbf{M}\mathbf{M}^H\mathbf{H}^H + \mathbf{c}_F\sigma_s^2\mathbf{H}\mathbf{M}\mathbf{M}^H\mathbf{H}^H \\
&\quad + \mathbf{c}_F\sigma_n^2\mathbf{I} \\
&= -E\left\{s(k-D)\mathbf{s}^H\right\}\mathbf{H}^H + \mathbf{c}_F\left(\sigma_s^2\mathbf{H}\mathbf{H}^H - \sigma_s^2\mathbf{H}\mathbf{M}\mathbf{M}^H\mathbf{H}^H + \sigma_0^2\mathbf{I}\right) = \mathbf{0}
\end{aligned} \tag{5.79}$$

Recall that $s(k-D)$ can be written as $\delta_D\mathbf{s}$, where δ_D is all zeroes $(1 \times (K_1 + L + 1))$ vector except for unity in the $(D+1)$-th position. Then the optimal coefficients of the feedforward filter can be calculated as follows:

$$\begin{aligned}
\mathbf{c}_F &= \delta_D\mathbf{H}^H\left(\sigma_s^2\mathbf{H}\mathbf{H}^H - \sigma_s^2\mathbf{H}\mathbf{M}\mathbf{M}^H\mathbf{H}^H + \sigma_0^2\mathbf{I}\right)^{-1} \\
&= \delta_D\mathbf{H}^H\left(\sigma_s^2\mathbf{H}\left(\left(\mathbf{I} - \mathbf{M}\mathbf{M}^H\right)\mathbf{H}^H\right) + \sigma_0^2\mathbf{I}\right)^{-1}
\end{aligned} \tag{5.80}$$

Designating $\mathbf{P} = \mathbf{I} - \mathbf{M}\mathbf{M}^H$ obtain:

$$\begin{aligned}
\mathbf{c}_F &= \delta_D\mathbf{H}^H\left(\sigma_s^2\mathbf{H}\mathbf{P}\mathbf{H}^H + \sigma_0^2\mathbf{I}\right)^{-1} \\
\mathbf{c}_B &= -\mathbf{c}_F\mathbf{H}\mathbf{M}
\end{aligned} \tag{5.81}$$

Comparing (5.81) and (5.44) one can see that the main difference in calculation the filter coefficients of the LMMSE equalizer and the DFE feedforward filter is that for LMMSE $\mathbf{P} = \mathbf{I}$. Actually that means the DFE feedforward filter is designed in such a way that it does not try to suppress ISI on the whole length of data vector $\mathbf{s}$. Matrix $\mathbf{P}$ defines some "window" where only noise rather than ISI is suppressed. This is done to more effectively push the precursor ISI energy to the postcursor region.

The derivation of optimum feedforward and feedback filter coefficients relies on the assumption (5.67). However, in real life there could be errors in data detection. Unfortunately, in this case the postcursor ISI

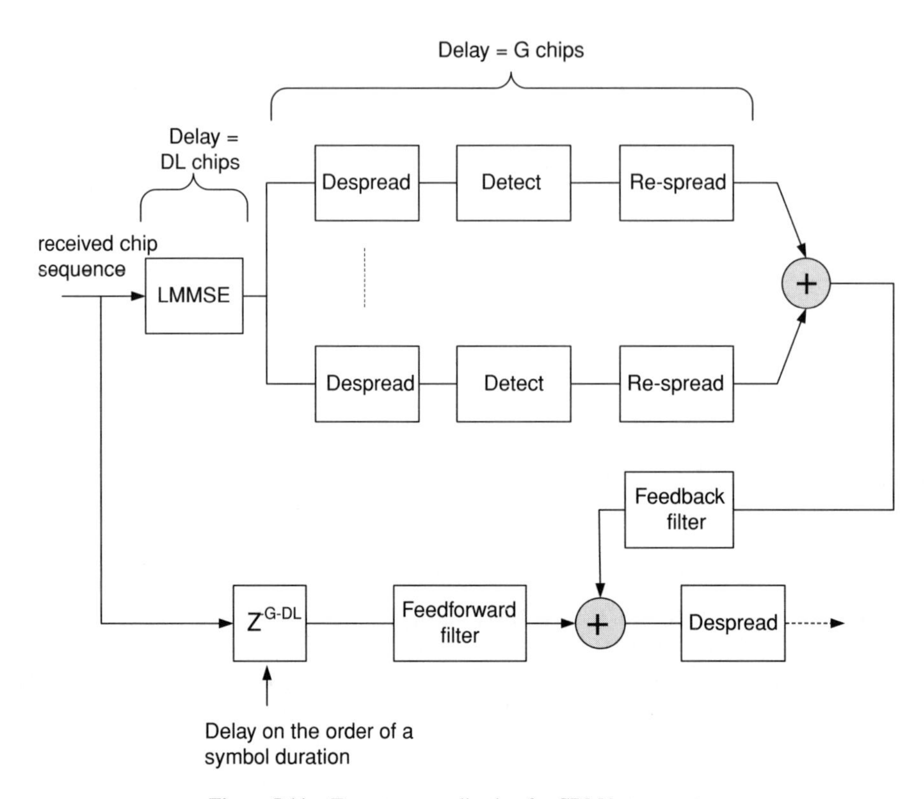

Figure 5.11 Two-stage equalization for CDMA-type system

amplified due to feedforward filtering cannot be cancelled and is propagated to the next data symbols. In this case the DFE faces the phenomena of error propagation, that is, one detection error can lead to a long series of equalization errors.

The usage of DFE in wideband systems where each symbol is spread to the sequence of chips differs from the case of narrowband transmission. Notice that for the linear ZF and LMMSE equalizers there is almost no difference between symbol-level and chip-level transmission due to linearity of these equalizers. DFE is not possible to use without any changes for chip-level transmission since the chip SNR is extremely low and thus there is no sense to use chip decisions for the feedback filter input.

The solution for this problem suggested in [6] can be the use of two-stage equalization when the linear equalizer is used at first stage, and the DFE at the second stage as can be seen in Figure 5.11.

Of course, the feedback delay in this case is increased that makes the implementation requirements tougher. The first stage LMMSE equalizer has delay D_L chips, the detected/recreated chip sequence is delayed by $G + D_L$ chips relative to the input. This delay should be applied at the DFE input to align the two stages. The delay in the DFE feedforward filter D is the separate parameter. That means the DFE output is delayed by $G + D_L + D$ chips relative to the input.

Applying the same technique as above it can be shown that the optimum coefficients of the feedforward and the feedback filters can be calculated as follows [6]:

$$\begin{aligned} \mathbf{c}_F &= \left(\Omega - \mathbf{X}\Phi^{-1}\mathbf{X}^H\right)^{-1}\mathbf{k} \\ \mathbf{c}_B &= -\frac{1}{\sigma_s^2}\mathbf{X}^H\mathbf{c}_F \end{aligned} \tag{5.82}$$

where

$$\mathbf{\Omega} = \sigma_s^2 \mathbf{H}\mathbf{H}^H + \sigma_0^2 \mathbf{I}, \tag{5.83}$$

$$\mathbf{X} = \sigma_s^2 \mathbf{H} \begin{bmatrix} 0 & \cdots & 0 \\ \vdots & \ddots & \vdots \\ 0 & \cdots & 0 \\ 1 & \cdots & 0 \\ \vdots & \ddots & \vdots \\ 0 & \cdots & 1 \\ 0 & \cdots & 0 \\ \vdots & \ddots & \vdots \\ 0 & \cdots & 0 \end{bmatrix} = \sigma_s^2 \mathbf{H} \begin{bmatrix} \mathbf{O}_{D+1 \times K_2} \\ \mathbf{I}_{K_2 \times K_2} \\ \mathbf{O}_{K_1+L-K_2-D-2 \times K_2} \end{bmatrix} = \sigma_s^2 \tilde{\mathbf{H}} \tag{5.84}$$

$$\mathbf{\Phi} = \sigma_s^2 \mathbf{I}_{K_2 \times K_2} \tag{5.85}$$

$$\mathbf{k} = \sigma_s^2 \mathbf{H} \delta_D \tag{5.86}$$

Substituting (5.83)–(5.86) into (5.82) obtain:

$$\begin{aligned} \mathbf{c}_F &= \sigma_s^2 \left(\sigma_s^2 \mathbf{H}\mathbf{H}^H + \sigma_0^2 \mathbf{I} - \sigma_s^2 \tilde{\mathbf{H}} \tilde{\mathbf{H}}^H \right)^{-1} \mathbf{H} \delta_D \\ \mathbf{c}_B &= -\tilde{\mathbf{H}}^H \mathbf{c}_F \end{aligned} \tag{5.87}$$

5.2 Equalization Based on Sequence Estimation

All equalization techniques considered in Section 5.1 employ the observation of a limited number of received symbols (corresponding to the filter length) to make a decision about one symbol. In cases where the channel length is greater than the equalizer filter length these methods cannot provide the optimal solution. In general, the optimum decision criterion requires that the entire sequence of received symbols is observed for making the decision. There are two principles that can provide optimal decision based on observation of the whole received sequence. One is MAP principle and another is *maximum likelihood sequence estimation* (*MLSE*). Both principles were considered in Chapter 4 for decoding the convolutional codes. Here we will consider the application of these principles for the equalization.

5.2.1 MLSE Equalization

Consider the received sequence at the equalizer input as it is described in (5.21) where $k = 0, \ldots, K-1$. Now we try to make the decision on the whole sequence of transmitted symbols $\{s(k)\}_{k=0}^{K-1}$ rather than on one symbol $s(k)$.

Let vector $\mathbf{s} = [s(0), s(1), \ldots, s(K-1)]^T$ represent the transmitted sequence of symbols, vector $\hat{\mathbf{s}} = [\hat{s}(0), \hat{s}(1), \ldots, \hat{s}(K-1)]^T$ designates the sequence of detected symbols, and vector $\mathbf{x} = [x(0), x(1), \ldots, x(K-1)]^T$ be the sequence of received symbols at the equalizer input. Let A be the cardinality of the alphabet from which symbols $s(k)$ are chosen. Then the vector space of the received symbols can be divided in A^K non-overlapping decision regions $R_\mathbf{v}$ in such a way that if the received vector $\mathbf{x}$ belongs to $R_\mathbf{v}$, then vector $\mathbf{v} = [v(0), v(1), \ldots, v(K-1)]^T$ was transmitted. In this case the probability of the correct decision can be calculated as:

$$\Pr\{\hat{\mathbf{s}} = \mathbf{s}\} = \sum_{\mathbf{v}} \Pr\{\mathbf{x} \in R_\mathbf{v} | \mathbf{s} = \mathbf{v}\} \Pr\{\mathbf{s} = \mathbf{v}\} = \sum_{\mathbf{v}} \int_{R_\mathbf{v}} P(\mathbf{u}|\mathbf{v})\, d\mathbf{u} \Pr\{\mathbf{s} = \mathbf{v}\} \tag{5.88}$$

Then the *MLSE criterion* can be written as follows:

$$\hat{\mathbf{s}} = \mathbf{s}_{ML} = \arg\max_{\mathbf{v}} P\left(\mathbf{x}|\,\mathbf{v}\right) \tag{5.89}$$

That means we choose as a detected sequence $\hat{\mathbf{s}}$ the vector $\mathbf{v}$, which maximizes the probability to observe vector $\mathbf{x}$ at the equalizer input.

The probability $P\left(\mathbf{x}|\,\mathbf{v}\right)$ can be represented as:

$$P\left(\mathbf{x}|\,\mathbf{v}\right) = \prod_{k=0}^{K-1} p\left(x(k)|\,\mathbf{v}\right) \tag{5.90}$$

Since we assume that vector $\mathbf{s}$ in (5.21) takes value of $\mathbf{v}$ the PDF $p\left(x(k)|\,\mathbf{v}\right)$ is defined only by the noise components of vector $\mathbf{x}$. Thus (5.90) may be written as follows:

$$P\left(\mathbf{x}|\,\mathbf{v}\right) = \prod_{k=0}^{K-1} \frac{1}{\pi\sigma_0^2} \exp\left(-\frac{1}{\pi\sigma_0^2}\left|x(k) - u(k)\right|^2\right) \tag{5.91}$$

where

$$u(k) = \sum_{l=0}^{L-1} h(l) \cdot v(k-l) \tag{5.92}$$

As explained in Section 4.2 it is more convenient to deal with logarithm of $P\left(\mathbf{x}|\,\mathbf{v}\right)$ rather than with this probability itself. The logarithm of (5.91) may be estimated as:

$$-\ln\left(P\left(\mathbf{x}|\,\mathbf{v}\right)\right) \approx \sum_{k=0}^{K-1} \left|x(k) - u(k)\right|^2 \tag{5.93}$$

Then the MLSE criterion can be expressed as:

$$\mathbf{s}_{ML} = \arg\min_{\mathbf{v}} \sum_{k=0}^{K-1} \left|x(k) - u(k)\right|^2 \tag{5.94}$$

Denote as $a(k)$ the convolution of sequences h and x:

$$a(k) = \sum_{l=0}^{L-1} h(l)x(k+l) \tag{5.95}$$

and as $b(k)$ the cross-convolution of sequence h:

$$b(k) = \begin{cases} \sum_{l=k}^{L-1} h(l)h(l-k), & \text{if} \quad k < L \\ 0, & \text{if } |k| \geq L \end{cases} \tag{5.96}$$

Taking into account that the minimum of function $|x(k) - u(k)|^2$ is defined by the maximum of function $x(k)u(k) - \frac{1}{2}u^2(k)$ it is possible to introduce metric:

$$M_{K-1}(\mathbf{v}) = \sum_{k=0}^{K-1} v(k) \cdot a(k) - \frac{1}{2} \sum_{k=0}^{K-1} \sum_{j=0}^{K-1} v(k) \cdot v(j) \cdot b(k-j) \tag{5.97}$$

which corresponds to the following form of the MLSE criterion:

$$\mathbf{s}_{ML} = \arg\max_{\mathbf{v}} M_{K-1}(\mathbf{v}) \tag{5.98}$$

The metric in (5.97) can be updated in the following way:

$$M_K(\mathbf{v}) = M_{K-1}(\mathbf{v}) + v(K) \cdot \left(a(K) - \frac{1}{2} \left(v(K) \cdot b(0) + \sum_{j=0}^{K-1} v(j) \cdot (b(K-j) + b(j-K)) \right) \right) \tag{5.99}$$

It follows from (5.96) that updated part of (5.99) depends only on L previous symbols of sequence $\mathbf{v}$ $[v(K), v(K-1), \ldots, v(K-L+1)]$, that is, (5.99) can be written as follows:

$$M_K(\mathbf{v}) = M_{K-1}(\mathbf{v}) + m_K(\mathbf{v}) \tag{5.100}$$

where

$$m_K(\mathbf{v}) = v(K) \cdot \left(a(K) - \frac{1}{2} \left(v(K) \cdot b(0) + \sum_{j=K-L+1}^{K-1} v(j) \cdot (b(K-j) + b(j-K)) \right) \right) \tag{5.101}$$

Then at each moment i the equalizer state S_i is defined by the pattern $S_i = [v(i), v(i-1), \ldots, v(i-L+1)]$, that is, the number of equalizer states is equal to A^L. When the new input symbol $x(i+1)$ arrives, the equalizer state changes to $S_{i+1} = [v(i+1), v(i), \ldots, v(i-L+2)]$. Thus, the equalization process can be represented as a solution of the problem of estimating the state of a finite state machine or as an iterative process of search for the path with maximum weight in trellis with A^L states. In the same way as in Chapter 4 designate $m_i(\mathbf{v})$ as branch metric, $M_i(\mathbf{v})$ as partial path metric and $M_{K-1}(\mathbf{v})$ as path metric. Then the Viterbi algorithm can be applied to find the path with maximum path metric $M_{K-1}(\mathbf{v})$ with the complexity $O\left((K-1)A^L\right)$ instead of complexity $O\left(A^{K-1}\right)$ required to find the solution of (5.98) with brute-force method. The example of trellis diagram for $A = \{-1, 1\}$ and $L = 2$ is depicted in Figure 5.12. The solid lines in Figure 5.12 correspond to transitions caused by the new input symbol $x(k) = -1$ and the dashed lines correspond to $x(k) = 1$.

For the detailed description of Viterbi algorithm see Section 4.2. Even though the Viterbi algorithm has a reduced computational complexity in comparison with brute-force algorithm, the complexity of the MLSE equalization is still very high in comparison with the complexity of equalization methods based on filtering. Due to this fact the MLSE equalization methods are rarely used for real-time implementation purposes.

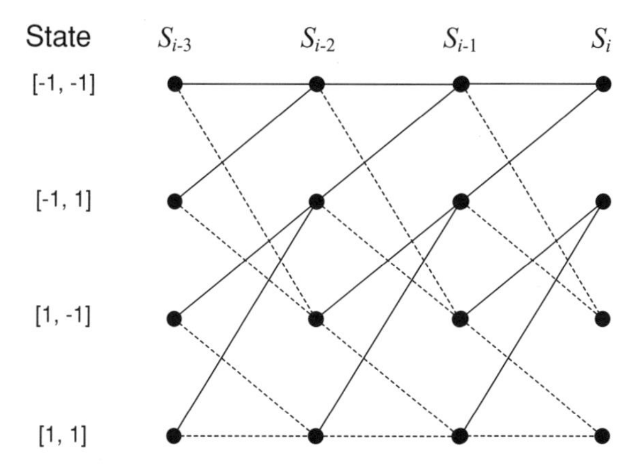

Figure 5.12 Trellis diagram for $A = \{\text{-}1, 1\}$, $L = 2$

5.2.2 Sphere Detection

The *Sphere Detection* (*SD*) algorithm [7]–[8] solves the MLSE criterion by performing a search in a hyper sphere around the received signal, $\mathbf{x}$. To employ SD, first write (5.21) in matrix form as it is done in (5.39):

$$\mathbf{x}' = \mathbf{H}'\mathbf{s}' + \mathbf{n}' \tag{5.102}$$

where $\mathbf{x}' = \left[x'(0), x'(1), \ldots, x'(K+L-1)\right]^T$ is the sequence at the equalizer input, $\mathbf{s}' = \left[s'(0), s'(1), \ldots, s'(K-1)\right]^T$ is the transmitted sequence, $\mathbf{n}' = \left[n'(0), n'(1), \ldots, n'(K+L-1)\right]^T$ is the noise vector, and $\mathbf{H}'$ is the $((K+L) \times K)$ channel matrix:

$$\mathbf{H}' = \begin{bmatrix} h'(0) & 0 & \ddots & 0 \\ h'(1) & h'(0) & \ddots & 0 \\ \vdots & \vdots & \ddots & \vdots \\ h'(L-1) & h'(L-2) & \ddots & h'(0) \\ 0 & h'(L-1) & \ddots & h'(1) \\ \vdots & \vdots & \ddots & \vdots \\ 0 & 0 & \ddots & h'(L-1) \end{bmatrix}$$

Originally the SD algorithm was created for real values, that is, the assumption was that the elements of $\mathbf{s}'$, $\mathbf{x}'$, $\mathbf{H}'$ and $\mathbf{n}'$ are taken from the field of real numbers. However, if we deal with QAM (5.39) can be written in the following way:

$$\mathbf{x} = \mathbf{H}\mathbf{s} + \mathbf{n} \tag{5.103}$$

where

$$\begin{aligned} \mathbf{x} &= \left[\mathrm{Re}(\mathbf{x}')^T \mathrm{Im}(\mathbf{x}')^T\right]^T; \\ \mathbf{H} &= \begin{bmatrix} \mathrm{Re}(\mathbf{H}') & -\mathrm{Im}(\mathbf{H}') \\ \mathrm{Im}(\mathbf{H}') & \mathrm{Re}(\mathbf{H}') \end{bmatrix}; \\ \mathbf{s} &= \left[\mathrm{Re}(\mathbf{s}')^T \mathrm{Im}(\mathbf{s}')^T\right]^T; \\ \mathbf{n} &= \left[\mathrm{Re}(\mathbf{n}')^T \mathrm{Im}(\mathbf{n}')^T\right]^T \end{aligned} \tag{5.104}$$

Then the *MLSE criterion* can be written as follows:

$$\mathbf{s}_{ML} = \arg\min_{\mathbf{v}} \|\mathbf{x} - \mathbf{H}\mathbf{v}\|^2 \tag{5.105}$$

where $\|\cdot\|$ is the Euclidean norm. This problem can be expressed geometrically: given a point $\mathbf{x}$ in $2K$-dimensional space, find the closest point (the point that has the smallest Euclidean distance to $\mathbf{x}$) in a skewed lattice $\Lambda(\mathbf{H})$. The *lattice* generated by $\mathbf{H}$ is:

$$\Lambda(\mathbf{H}) = \left\{\mathbf{H}\mathbf{v} : \mathbf{v} \in R^{2K}\right\} \tag{5.106}$$

The rows of $\mathbf{H}$ are called *basis vectors of lattice* Λ. The solution of (5.105) with the help of brute force algorithm, that is, exhaustive search has the exponential complexity. The SD algorithm solves (5.105) by choosing the candidates only from those points of lattice Λ that belong to a sphere of radius r with the centre in $\mathbf{x}$ as it is shown in Figure 5.13.

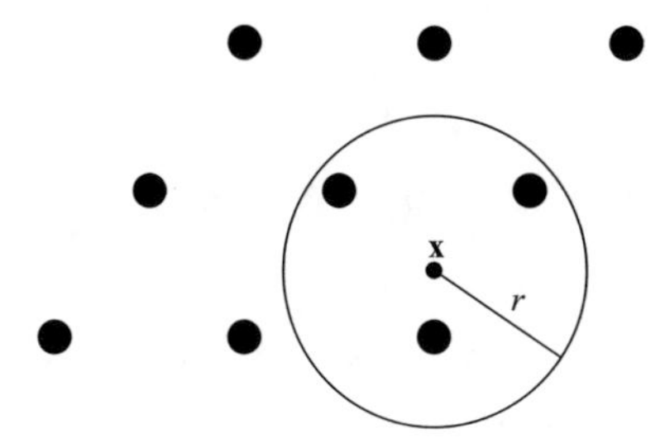

Figure 5.13 Sphere Detection principle

The condition that lattice points belong to a sphere of radius r with the center in $\mathbf{x}$ can be written as follows:

$$\|\mathbf{x} - \mathbf{H}\mathbf{v}\|^2 \leq r^2 \tag{5.107}$$

The channel matrix $\mathbf{H}$ can be decomposed by QR decomposition (QRD) [10] and then (5.104) can be written as:

$$\|\mathbf{x} - \mathbf{Q}\mathbf{R}\mathbf{v}\|^2 \leq r^2 \tag{5.108}$$

where $\mathbf{Q}$ is the $(2(K+L) \times 2(K+L))$ unitary matrix, i.e. $\mathbf{Q}^H\mathbf{Q} = \mathbf{Q}\mathbf{Q}^H = \mathbf{I}$ (in case elements of $\mathbf{Q}$ are real, matrix $\mathbf{Q}$ is called orthogonal matrix $\mathbf{Q}^T\mathbf{Q} = \mathbf{Q}\mathbf{Q}^T = \mathbf{I}$), and $\mathbf{R}$ is the $(2(K+L) \times 2K)$ upper triangular matrix with positive diagonal elements. One of the properties of orthogonal matrix is that it preserves the length, that is, for any orthogonal matrix $\mathbf{U}$ the following equality holds [10]:

$$\|\mathbf{U}\mathbf{a}\| = \|\mathbf{a}\| \tag{5.109}$$

where $\mathbf{a}$ is any vector of corresponding dimension. Since $\mathbf{Q}$ is orthogonal matrix, $\mathbf{Q}^T$ is also orthogonal matrix. Then (5.107) can be written as:

$$\left\|\mathbf{Q}^T\mathbf{x} - \mathbf{R}\mathbf{v}\right\|^2 \leq r^2 \tag{5.110}$$

or as:

$$\|\tilde{\mathbf{x}} - \mathbf{R}\mathbf{v}\|^2 \leq r^2 \tag{5.111}$$

where $\tilde{\mathbf{x}} = \mathbf{Q}^T\mathbf{x}$. Due to the upper triangular form of $\mathbf{R}$, (5.111) defines the set of conditions:

$$\sum_{j=i}^{2K-1} \left| \tilde{x}(j) - \sum_{m=j}^{2K-1} R(j,m) \cdot v(m) \right|^2 \leq r^2, \quad i = 0, \ldots, 2K-1 \tag{5.112}$$

which must satisfy candidate vector components $v(m)$ to belong to considered sphere. Denote the last $2K - m + 1$ components of candidate vector $\mathbf{v}$ by $\mathbf{v}_m^{2K-1} = [v(m), v(m+1), \ldots, v(2K-1)]$. Then the SD algorithm can be represented as a search in depth on the nodes of the tree with the root corresponding to vector $\mathbf{v}_{2K-1}^{2K-1} = [v(2K-1)]$, where layer i of the tree corresponds to the particular value of vector $\mathbf{v}_{2K-i-1}^{2K-1}$ length like it is shown in Figure 5.14.

Each node at level i is connected with $\sqrt{A}$ nodes (A is the transmitted alphabet cardinality) at level $i+1$ since there are $\sqrt{A}$ possible choices of symbol $v(2K-i-1)$ preceding vector $\mathbf{v}_{2K-i}^{2K-1}$. However, appending vector $\mathbf{v}_{2K-i}^{2K-1}$ by new symbol $v(2K-i-1)$ may lead to a situation when the points of lattice $\mathbf{H}\mathbf{v}$ corresponding to the choice of $v(2K-i-1)$ do not belong to the considered sphere anymore. Denote the set of admissible values for each symbol $v(2K-i-1)$ preceding vector $\mathbf{v}_{2K-i}^{2K-1}$ by $T_{2K-i-1}\left(\mathbf{v}_{2K-i}^{2K-1}\right)$. Applying the back-substitution algorithm to (5.112) exploiting the upper triangular form of matrix $\mathbf{R}$,

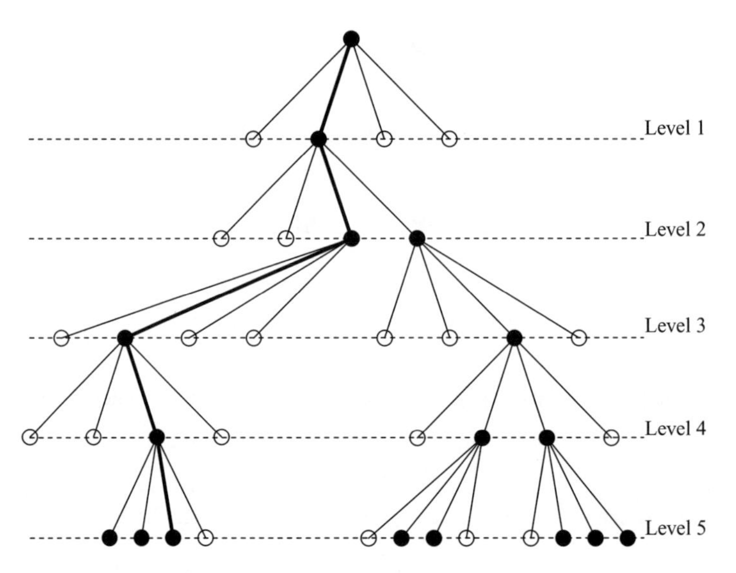

Figure 5.14 Example of Sphere Detection algorithm for $K = 3$, 16-QAM

obtain the set $T_{2K-i-1}\left(\mathbf{v}_{2K-i}^{2K-1}\right)$ as an interval $T_{2K-i-1}\left(\mathbf{v}_{2K-i}^{2K-1}\right) = \left[C_{2K-i-1}\left(\mathbf{v}_{2K-i}^{2K-1}\right), D_{2K-i-1}\left(\mathbf{v}_{2K-i}^{2K-1}\right)\right]$ from which the new symbol $v(2K - i - 1)$ can take the integer values. This interval $T_{2K-i-1}\left(\mathbf{v}_{2K-i}^{2K-1}\right)$ usually is called the *admissible interval*. The values of $C_{2K-i-1}\left(\mathbf{v}_{2K-i}^{2K-1}\right)$ and $D_{2K-i-1}\left(\mathbf{v}_{2K-i}^{2K-1}\right)$ for each level i can be expressed as follows [9], [11]:

$$C_{2K-i-1}\left(\mathbf{v}_{2K-i}^{2K-1}\right) = \left\lceil a_i \left(\tilde{x}(2K-i-1) - \sum_{j=2K-i}^{2K-1} R(2K-i-1, j) \cdot v(j) - \sqrt{r^2 - \sum_{j=2K-i}^{2K-1} \left| \tilde{x}(j) - \sum_{m=j}^{2K-1} R(j,m) \cdot v(m) \right|^2} \right) \right\rceil \tag{5.113}$$

$$D_{2K-i-1}\left(\mathbf{v}_{2K-i}^{2K-1}\right) = \left\lfloor a_i \left(\tilde{x}(2K-i-1) - \sum_{j=2K-i}^{2K-1} R(2K-i-1, j) \cdot v(j) + \sqrt{r^2 - \sum_{j=2K-i}^{2K-1} \left| \tilde{x}(j) - \sum_{m=j}^{2K-1} R(j,m) \cdot v(m) \right|^2} \right) \right\rfloor, \tag{5.114}$$

where $a_i = \frac{1}{R(2K-i-1, 2K-i-1)}$.

If $C_{2K-i-1}\left(\mathbf{v}_{2K-i}^{2K-1}\right) > D_{2K-i-1}\left(\mathbf{v}_{2K-i}^{2K-1}\right)$ or if $\sum_{j=2K-i}^{2K-1} \left| \tilde{x}(j) - \sum_{m=j}^{2K-1} R(j,m) \cdot v(m) \right|^2 > r^2$, then $T_{2K-i-1}\left(\mathbf{v}_{2K-i}^{2K-1}\right) = \{\emptyset\}$, and the node corresponding to this value of symbol $v(2K - i - 1)$ is pruned out of tree since any child of this node corresponds to a vector that does not belong to the considered sphere. In Figure 5.14 the pruned nodes are represented by the unfilled circles. When the algorithm reaches level $2K - 1$ the candidate vector $\mathbf{v}$ belonging to the considered sphere of radius r is found. Then it is possible to calculate the distance between the lattice point $\mathbf{Hv}$ corresponding to the candidate vector and the centre of sphere $\mathbf{x}$. Actually, this distance is equal to the distance between point $\mathbf{Rv}$ and vector $\tilde{\mathbf{x}}$. Then the squared Euclidean distance between the point $\mathbf{Hv}$ and the centre of sphere $\mathbf{x}$ $d^2(\mathbf{x}, \mathbf{Hv})$ can be calculated as follows:

$$d^2(\mathbf{x}, \mathbf{Hv}) = \sum_{j=0}^{2K-1} \left| \tilde{x}(j) - \sum_{m=j}^{2K-1} R(j, m) \cdot v(m) \right|^2 \tag{5.115}$$

Vector $\mathbf{v}$ providing the minimum value of $d^2(\mathbf{x}, \mathbf{Hv})$ is assigned to the result of search vector $\hat{\mathbf{s}}$. If radius r is big enough, vector $\hat{\mathbf{s}}$ is the solution of (5.105). In fact, the SD algorithm searches the points of lattice $\mathbf{Rv}$ belonging to the sphere around centre $\tilde{\mathbf{x}}$ rather than points of lattice $\mathbf{Hv}$ belonging to the sphere around centre $\mathbf{x}$ but since there is one to one correspondence between $\mathbf{Rv}$ and $\mathbf{Hv}$ and between $\tilde{\mathbf{x}}$ and $\mathbf{x}$ it does not impact the search result. If at level $2K - 1$ no point in the sphere is found, the search must be repeated for spheres with greater value of radius r.

In Figure 5.14 the example of SD algorithm for $K = 3$ and 16-QAM is represented. Each I and Q branch of 16-QAM chooses the symbols from the alphabet of four elements, for example $\{-3, -1, 1, 3\}$. Thus each node of the tree has four child nodes. As mentioned above the SD algorithm is depth-first algorithm and after reaching level 5 the search is returning back up to the root node. The distance comparison is done only at bottom level (level 5 in this example). As can be seen from Figure 5.14 eventually only eight lattice points are considered. If there were no restrictions, that is, $r = \infty$ the number of lattice points that should be checked is $4^6 = 4096$. Still the complexity of the SD algorithm is much higher than the complexity of equalization algorithms based on filtering. To be more precise, with the assumption that the radius r is chosen not randomly but taking into account some noise statistics, the complexity of SD algorithm is a random variable with the mean significantly less than the complexity of Viterbi algorithm.

In fact, the complexity of the SD algorithm can be decreased. Notice that the squared Euclidean distance can be calculated recursively with the help of partial squared Euclidean distance:

$$d^2(\mathbf{x}, \mathbf{Hv}_{2K-i-1}^{2K-1}) = d^2(\mathbf{x}, \mathbf{Hv}_{2K-i}^{2K-1}) + \left| \tilde{x}(2K - i - 1) - \sum_{m=2K-i-1}^{2K-1} R(2K - i - 1, m) \cdot v(m) \right|^2 \tag{5.116}$$

Then (5.115) can be written as:

$$d^2(\mathbf{x}, \mathbf{Hv}) = d^2(\mathbf{x}, \mathbf{Hv}_{2K-1}^{2K-1}) \tag{5.117}$$

It means that it is possible to add one more restriction for search process. Unless the bottom level is not reached and at least for one lattice point the squared Euclidean distance to the centre of the sphere is not calculated the algorithm proceeds without any changes, but after that for each partial vector $\mathbf{v}_{2K-i}^{2K-1}$ the partial squared Euclidean distance $d^2(\mathbf{x}, \mathbf{Hv}_{2K-i}^{2K-1})$ is calculated and if it is greater than the stored current squared Euclidean distance the corresponding node is pruned out of the tree even if this node belongs to the set of admissible values.

Another improvement of the SD algorithm is the usage of the *Schnorr-Euchner enumeration* [12]. In the described above SD algorithm the *Pohst enumeration* is used [7], [8]. Pohst enumeration is based on so-called natural spanning of the admissible intervals $T_{2K-i-1}\left(\mathbf{v}_{2K-i}^{2K-1}\right)$ at each level i, that is, the values for symbol $v(2K - i - 1)$ are considered in the following order $C_{2K-i-1}\left(\mathbf{v}_{2K-i}^{2K-1}\right)$, $C_{2K-i-1}\left(\mathbf{v}_{2K-i}^{2K-1}\right) + 1, \ldots, D_{2K-i-1}\left(\mathbf{v}_{2K-i}^{2K-1}\right)$. In Schnorr-Euchner enumeration the symbol $v(2K - i - 1)$ takes values from the admissible interval $T_{2K-i-1}\left(\mathbf{v}_{2K-i}^{2K-1}\right)$ in zigzag order starting from the interval middle point:

$$S_{2K-i-1}\left(\mathbf{v}_{2K-i}^{2K-1}\right) = \operatorname{int}\left(a_i \cdot \left(\tilde{x}(2K - i - 1) - \sum_{m=2K-i}^{2K-1} R(2K - i - 1, m) \cdot v(m)\right)\right) \tag{5.118}$$

where $a_i = \frac{1}{R(2K-i-1,2K-i-1)}$, and function int () denotes the rounding to the nearest integer. To be more precise, the Schnorr-Euchner enumeration implies that symbol $v(2K-i-1)$ takes values from the admissible interval $T_{2K-i-1}\left(\mathbf{v}_{2K-i}^{2K-1}\right)$ as follows:

If

$$\tilde{x}(2K-i-1) - \sum_{m=2K-i}^{2K-1} R(2K-i-1,m)\cdot v(m) - R(2K-i-1,2K-i-1)\cdot S_{2K-i-1}\left(\mathbf{v}_{2K-i}^{2K-1}\right) \geq 0 \tag{5.119}$$

then the order is $S_{2K-i-1}\left(\mathbf{v}_{2K-i}^{2K-1}\right)$, $S_{2K-i-1}\left(\mathbf{v}_{2K-i}^{2K-1}\right)+1$, $S_{2K-i-1}\left(\mathbf{v}_{2K-i}^{2K-1}\right)-1$, $S_{2K-i-1}\left(\mathbf{v}_{2K-i}^{2K-1}\right)+2$, $S_{2K-i-1}\left(\mathbf{v}_{2K-i}^{2K-1}\right)-2, \ldots$.

If

$$\tilde{x}(2K-i-1) - \sum_{m=2K-i}^{2K-1} R(2K-i-1,m)\cdot v(m) - R(2K-i-1,2K-i-1)\cdot S_{2K-i-1}\left(\mathbf{v}_{2K-i}^{2K-1}\right) < 0 \tag{5.120}$$

then the order is $S_{2K-i-1}\left(\mathbf{v}_{2K-i}^{2K-1}\right)$ $S_{2K-i-1}\left(\mathbf{v}_{2K-i}^{2K-1}\right)-1$, $S_{2K-i-1}\left(\mathbf{v}_{2K-i}^{2K-1}\right)+1$, $S_{2K-i-1}\left(\mathbf{v}_{2K-i}^{2K-1}\right)-2$, $S_{2K-i-1}\left(\mathbf{v}_{2K-i}^{2K-1}\right)+2, \ldots$.

One more aspect having great influence on the algorithm complexity is the choice of the sphere radius. The problem of choosing the sphere radius r is two-fold. The infinite radius guarantees that the MLSE solution will be found but in this case the SD algorithm simply becomes the exhaustive search with the exponential complexity. On the other hand, if the sphere radius r is too small it becomes less likely the MLSE solution will be found. Obviously for the sphere of radius r centred at $\mathbf{x}$ the probability that the point $\mathbf{Hs}$ (where $\mathbf{s}$ is the true transmitted data) is given by:

$$\Pr\left(\|\mathbf{x}-\mathbf{Hs}\|^2 \leq r^2\right) = \Pr\left(\|\mathbf{n}\|^2 \leq r^2\right) \tag{5.121}$$

Assuming the AWGN conditions $\|\mathbf{n}\|^2$ is a chi-square random variable with $2K$ degrees of freedom and PDF [13]:

$$f(\lambda) = \frac{\lambda^{K-1}}{(2\sigma_0^2)^K\Gamma(K)}\exp\left(-\frac{\lambda}{2\sigma_0^2}\right) \tag{5.122}$$

where $\Gamma()$ denotes the gamma function. Then the radius r of the sphere centred at $\mathbf{x}$ containing the point $\mathbf{Hs}$ with probability of (5.121) can be calculated by solving the following equation with respect to r:

$$\Pr\left(\|\mathbf{n}\|^2 \leq r^2\right) = \int_0^{r^2} f(\lambda)d\lambda \tag{5.123}$$

A more simple approach is to find some rough estimate of $\mathbf{s}$ with the help of some equalization method with low complexity and then use the distance between the received vector $\mathbf{x}$ and the lattice point $\mathbf{H\hat{s}}$ ($\hat{\mathbf{s}}$ is the estimate of $\mathbf{s}$) as the radius r for the SD algorithm:

$$r = \|\mathbf{x}-\mathbf{H\hat{s}}\| \tag{5.124}$$

Obviously the sphere of this radius must include the point corresponding to the MLSE solution. Usually the zero-forcing solution is used for the calculation of radius r. Then $\hat{\mathbf{s}}$ can be written as:

$$\hat{\mathbf{s}} = \left(\mathbf{H}^T\mathbf{H}\right)^{-1}\mathbf{H}^T\mathbf{x} \tag{5.125}$$

Equation (5.125) can be obtained from (5.54).

Another trick aiming to decrease the SD algorithm complexity is to change the sphere radius r during the search. For example, it is possible to start the search within some sphere of big radius r, but, after finding the first candidate the sphere radius can be decreased up to the distance from the candidate to the sphere centre and the search of other candidates is done inside the sphere of smaller radius and so on.

One more approach quite similar to the one described above was suggested in [14]. The squared distance from the centre of the sphere to the lattice point corresponding to the transmitted sequence **s** can be written as:

$$\|\mathbf{x}-\mathbf{H}\mathbf{s}\|^2 = \left\|\mathbf{H}\mathbf{s} - \left(\mathbf{H}\left(\mathbf{H}^T\mathbf{H}\right)^{-1}\mathbf{H}^T + \mathbf{I} - \mathbf{H}\left(\mathbf{H}^T\mathbf{H}\right)^{-1}\mathbf{H}^T\right)\mathbf{x}\right\|^2 \tag{5.126}$$

where matrix $\mathbf{H}\left(\mathbf{H}^T\mathbf{H}\right)^{-1}\mathbf{H}^T$ defines the projection of **x** to the column space of **H**. Simplifying (5.126) obtain:

$$\|\mathbf{x}-\mathbf{H}\mathbf{s}\|^2 = \left\|\mathbf{H}(\mathbf{s}-\hat{\mathbf{s}}) - \left(\mathbf{I} - \mathbf{H}\left(\mathbf{H}^T\mathbf{H}\right)^{-1}\mathbf{H}^T\right)\mathbf{x}\right\|^2 = \|\mathbf{H}(\mathbf{s}-\hat{\mathbf{s}})\|^2 + \left\|\left(\mathbf{I} - \mathbf{H}\left(\mathbf{H}^T\mathbf{H}\right)^{-1}\mathbf{H}^T\right)\mathbf{x}\right\|^2 \tag{5.127}$$

where $\hat{\mathbf{s}}$ is the zero-forcing solution of (5.125). Under the AWGN conditions the value of $\left\|\left(\mathbf{I} - \mathbf{H}\left(\mathbf{H}^T\mathbf{H}\right)^{-1}\mathbf{H}^T\right)\mathbf{x}\right\|^2$ is defined only by noise. Then one possible choice of radius r is:

$$r^2 = 2\sigma_0^2\alpha K - \left\|\left(\mathbf{I} - \mathbf{H}\left(\mathbf{H}^T\mathbf{H}\right)^{-1}\mathbf{H}^T\right)\mathbf{x}\right\|^2 \tag{5.128}$$

where $2\sigma_0^2 K$ is the expected value of chi-square random variable with $2K$ degrees of freedom and $\alpha \geq 1$ is the scaling factor chosen so that we are reasonably sure that the lattice point corresponding to true s is inside the considered sphere.

Recall that the SD algorithm considered above originally was created for real numbers and can be applied to the complex case only for modulation allowing the decoupling to I and Q branches. In principle, this kind of I and Q decoupling is possible not only for QAM but for other modulations as well (see Chapter 5.2). The problem is that, for example, for M-PSK the distance between projections of signal points to I and Q branches are not equal anymore. That means the calculation of admissible intervals in (5.113)–(5.114) should be adapted to the particular M-PSK. However, it is possible to modify the SD algorithm to handle complex constellations without decoupling the signal to I and Q branches. Here we will follow [14]. For brevity of notation, denote in (5.102) $\mathbf{x} = \mathbf{x}'$, $\mathbf{H} = \mathbf{H}'$, $\mathbf{s} = \mathbf{s}'$, $\mathbf{n} = \mathbf{n}'$. Inequalities (5.105)–(5.111) are valid for both real and complex cases (with changing the transpose by the Hermitian transpose). In a complex case the dimensions of all vectors and matrices are two times less. (5.127) is also valid for the complex case and now can be written as:

$$\|\mathbf{x}-\mathbf{H}\mathbf{s}\|^2 = \left\|\mathbf{H}(\mathbf{s}-\hat{\mathbf{s}}) - \left(\mathbf{I} - \mathbf{H}\left(\mathbf{H}^H\mathbf{H}\right)^{-1}\mathbf{H}^H\right)\mathbf{x}\right\|^2 = \|\mathbf{H}(\mathbf{s}-\hat{\mathbf{s}})\|^2 + \left\|\left(\mathbf{I} - \mathbf{H}\left(\mathbf{H}^H\mathbf{H}\right)^{-1}\mathbf{H}^H\right)\mathbf{x}\right\|^2 \tag{5.129}$$

where

$$\hat{\mathbf{s}} = \left(\mathbf{H}^H\mathbf{H}\right)^{-1}\mathbf{H}^H\mathbf{x} \tag{5.130}$$

As mentioned above under the AWGN conditions $\|(\mathbf{I} - \mathbf{H}(\mathbf{H}^H\mathbf{H})^{-1}\mathbf{H}^H)\mathbf{x}\|^2$ depends only on noise and does not depend on **s**. That means, minimization of $\|\mathbf{H}(\mathbf{s}-\hat{\mathbf{s}})\|^2$ leads to minimization of $\|\mathbf{x}-\mathbf{H}\mathbf{s}\|^2$. Then the problem (5.105) is equivalent to:

$$\mathbf{s}_{ML} = \arg\min_{\mathbf{v}} \|\mathbf{H}(\mathbf{v}-\hat{\mathbf{s}})\|^2 \tag{5.131}$$

If the search on vectors **v** is restricted to sphere of radius r, the condition that the considered lattice points belong to a sphere of radius r with the centre in $\hat{\mathbf{s}}$ can be written in the same way as (5.107):

$$\|\mathbf{H}(\mathbf{v}-\hat{\mathbf{s}})\|^2 = (\mathbf{v}-\hat{\mathbf{s}})^H\mathbf{H}^H\mathbf{H}(\mathbf{v}-\hat{\mathbf{s}}) \leq r^2 \tag{5.132}$$

With the help of Cholesky factorization [2] it is possible to find an upper triangular matrix **U** such that:

$$\mathbf{U}^H\mathbf{U} = \mathbf{H}^H\mathbf{H}, \qquad u_{ii} > 0, u_{ii} \in Z \tag{5.133}$$

Then (5.132) can be written as:

$$\|\mathbf{H}(\mathbf{v}-\hat{\mathbf{s}})\|^2 = (\mathbf{v}-\hat{\mathbf{s}})^H\,\mathbf{U}^H\mathbf{U}(\mathbf{v}-\hat{\mathbf{s}}) = \sum_{i=0}^{K-1} u_{ii}^2\left|v_i - \hat{s}_i + \sum_{j=i+1}^{K-1}\frac{u_{ij}}{u_{ij}}(v_j - \hat{s}_j)\right|^2 \le r^2 \tag{5.134}$$

As can be seen (5.134) is very similar to (5.112) and can be used in the same way. With $i = K-1$ (5.134) is reduced to:

$$u_{K-1,K-1}^2\,|v_{K-1} - \hat{s}_{K-1}|^2 \le r^2 \tag{5.135}$$

which yields:

$$|v_{K-1} - \hat{s}_{K-1}| \le \frac{r}{u_{K-1,K-1}} \tag{5.136}$$

The inequality (5.136) means that the last element of vector v must belong to complex disk of radius $\frac{r}{u_{K-1,K-1}}$ centred in $\hat{s}_{K-1}$. At the same time the element v_{K-1} belongs to the constellation corresponding to modulation used in the transmission. Assume that the PSK was used. Then the search of v_{K-1} value is limited to the intersection of the complex disk with the PSK constellation, that is, with the complex circle. This intersection illustrated in Figure 5.15 is an arc. The angular sweep of this arc can be obtained analytically.

The element v_{K-1} belongs to M-PSK. Then it can be represented in accordance with (2.51) as $v_{K-1} = r_c \cdot \exp(j\phi_{K-1})$, where $\phi_{K-1} \in \left\{0, \frac{2\pi i}{M}, \ldots, \frac{2\pi(M-1)}{M}\right\}$, $r_c > 0$ is the radius of the M-PSK constellation.

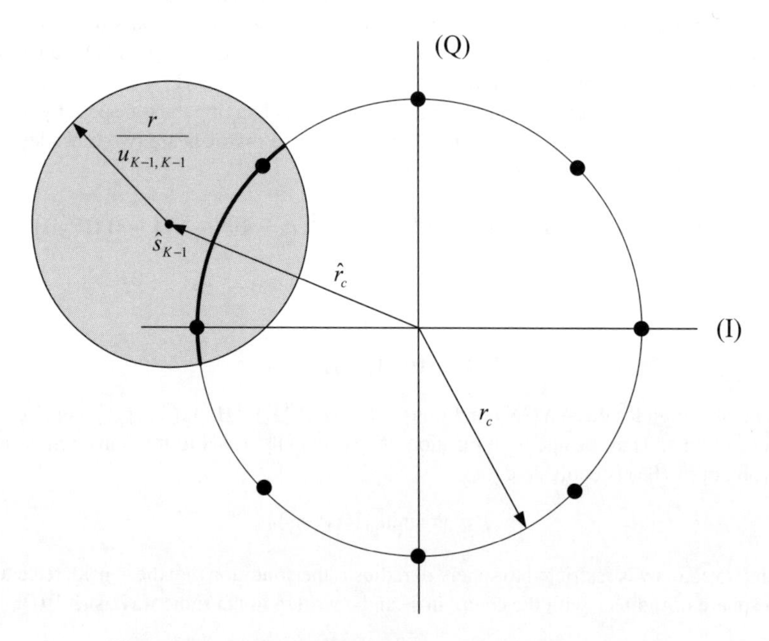

Figure 5.15 Intersection of complex disk and 8-PSK constellation

In the same way it is possible to represent the element $\hat{s}_{K-1} = \hat{r}_c \cdot \exp\left(j\hat{\phi}_{K-1}\right)$, where $\hat{r}_c$ is the radius corresponding to the distorted constellation. Then (5.136) can be written as [14]:

$$|v_{K-1} - \hat{s}_{K-1}|^2 = r_c^2 + \hat{r}_c^2 - 2r_c\hat{r}_c \cos\left(\phi_{K-1} - \hat{\phi}_{K-1}\right) \leq \frac{r^2}{u_{K-1,K-1}^2} \tag{5.137}$$

which yields:

$$\cos\left(\phi_{K-1} - \hat{\phi}_{K-1}\right) \geq \frac{1}{2r_c\hat{r}_c}\left(r_c^2 + \hat{r}_c^2 - \frac{r^2}{u_{K-1,K-1}^2}\right) = \eta \tag{5.138}$$

If $\eta > 1$, then the search disk does not contain any point of the M-PSK constellation. If $\eta < -1$, then all constellation points belong to the search disk. For the case $-1 \leq \eta \leq 1$:

$$\left|\phi_{K-1} - \hat{\phi}_{K-1}\right| \leq \arccos(\eta) \tag{5.139}$$

Assuming $0 \leq \arccos() \leq \pi$, the admissible interval of the constellation points for v_{K-1} is given by [14]:

$$\left\lceil \frac{M}{2\pi}\left(\hat{\phi}_{K-1} - \arccos(\eta)\right)\right\rceil \leq \frac{M}{2\pi}\phi_{K-1} \leq \left\lfloor \frac{M}{2\pi}\left(\hat{\phi}_{K-1} + \arccos(\eta)\right)\right\rfloor \tag{5.140}$$

Now the candidate v_{K-1} is chosen from the range of values for which ϕ_{K-1} satisfy (5.140). The next steps of the algorithm are similar to the ones described above. When v_{K-1} is chosen, the admissible interval for v_{K-2} is chosen by searching the new arc with the help of (5.140) and so on.

Actually, this complex sphere detection algorithm can be applied not only to M-PSK but to QAM also. The difference is that in the QAM case the intersection of complex constellation and complex search disk comprises the bunch of concentric arcs (as shown in Figure 5.16) rather than one arc as in M-PSK case.

The only difference is that it should be used for a few different values of radius r_c corresponding to different concentric circles comprising the QAM constellation. In some cases the usage of this kind of complex SD algorithm is more preferable since it requires the use of vectors and matrices with two times lower dimensions than for the real SD algorithm.

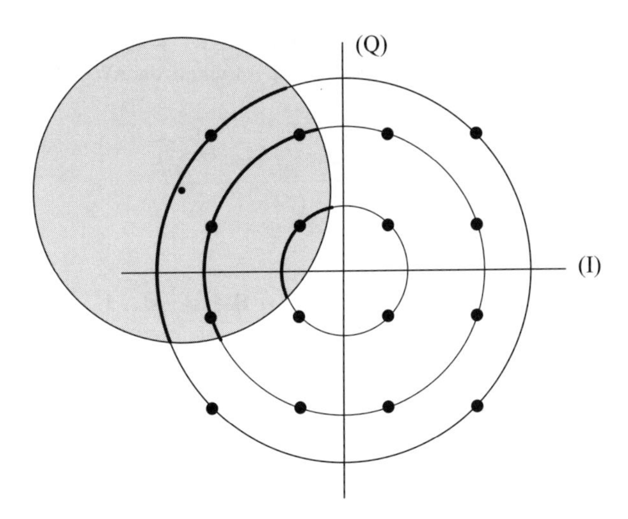

Figure 5.16 Intersection of complex disk and 16-QAM constellation

Many modifications of SD algorithm exist, for example, it is possible to use breadth-first algorithm instead of depth-first algorithm to search the tree of candidates. This kind of modification usually leads to significant decrease of the complexity, but in this case there is no guarantee to find the ML solution and algorithm becomes suboptimal.

5.2.2.1 List Sphere Detection

As mentioned above, except for the MLSE it is possible to use the MAP principle to provide the optimal decision based on observation of the whole received sequence. Without loss of generality assume that the transmitted sequence $\mathbf{s} = \{s_k\}, k = 0, \ldots, K-1$ consists of binary signals, that is, $s_k \in \{-1, +1\}$. If the transmitted signal is not binary, the sequence **s** can be considered as a binary representation of the nonbinary transmitted sequence. Also assume that the Gray coding and the bit-level interleaving are used in the transmitter. Then the binary symbols s_k can be considered as mutually independent. In this case the a posteriori LLR of the bit s_k, conditioned on the received sequence x can be written in accordance with (4.38) as follows:

$$L(s_k|\mathbf{x}) = \ln \frac{P(s_k = +1|\mathbf{x})}{P(s_k = -1|\mathbf{x})} = L_A(s_k) + L_E(s_k) \tag{5.141}$$

where

$$L_A(s_k) = \ln \frac{P(s_k = +1)}{P(s_k = -1)} \tag{5.142}$$

is the a priori LLR of the bit s_k, and $L_E(s_k)$ is the extrinsic LLR. The extrinsic LLR can be represented as [14]:

$$L_E(s_k) = \ln \frac{P(\mathbf{x}|s_k = +1)}{P(\mathbf{x}|s_k = +1)} = \ln \frac{\sum\limits_{\mathbf{s}\in\mathbf{S}_{k,+1}} P(\mathbf{x}|\mathbf{s}) \cdot \exp\left(\frac{1}{2}\mathbf{s}_{[k]}^T \cdot \mathbf{L}_{A,[k]}\right)}{\sum\limits_{\mathbf{s}\in\mathbf{S}_{k,-1}} P(\mathbf{x}|\mathbf{s}) \cdot \exp\left(\frac{1}{2}\mathbf{s}_{[k]}^T \cdot \mathbf{L}_{A,[k]}\right)} \tag{5.143}$$

where $\mathbf{S}_{k,+1}$ is the set of vectors **s** of length K such that the k-th element of any vector from this set is $+1$, i.e., $\mathbf{S}_{k,+1} = \{\mathbf{s} : s_k = +1\}$, correspondingly $\mathbf{S}_{k,-1} = \{\mathbf{s} : s_k = -1\}$, $\mathbf{s}_{[k]}$ is vector of length $K-1$ containing all elements of vector **s** except of the element $s(k)$, $\mathbf{s}_{[k]} = [s(0), s(1), \ldots, s(k-1), s(k+1), \ldots, s(K-1)]$, and $\mathbf{L}_{A,[k]} = [L_A(s_0), L_A(s_1), \ldots, L_A(s_{k-1}), L_A(s_{k+1}), \ldots, L_A(s_{K-1})]$. Assuming the AWGN channel, it follows from (5.39) that:

$$P(\mathbf{x}|\mathbf{s}) = \frac{\exp\left(-\frac{1}{2\sigma_0^2}\|\mathbf{x} - \mathbf{Hs}\|^2\right)}{\left(2\pi\sigma_0^2\right)^K} \tag{5.144}$$

Substituting (5.144) in (5.143) obtain:

$$L_E(s_k) = \ln \frac{\sum\limits_{\mathbf{s}\in\mathbf{S}_{k,+1}} \exp\left(-\frac{1}{2\sigma_0^2}\|\mathbf{x} - \mathbf{Hs}\|^2 + \frac{1}{2}\mathbf{s}_{[k]}^T \cdot \mathbf{L}_{A,[k]}\right)}{\sum\limits_{\mathbf{s}\in\mathbf{S}_{k,-1}} \exp\left(-\frac{1}{2\sigma_0^2}\|\mathbf{x} - \mathbf{Hs}\|^2 + \frac{1}{2}\mathbf{s}_{[k]}^T \cdot \mathbf{L}_{A,[k]}\right)} \tag{5.145}$$

The application of the Max-log approximation (4.65) to (5.145) gives

$$L_E(s_k) \approx \frac{1}{2}\left(\max_{\mathbf{s}\in\mathbf{S}_{k,+1}}\left\{-\frac{1}{\sigma_0^2}\|\mathbf{x} - \mathbf{Hs}\|^2 + \mathbf{s}_{[k]}^T \cdot \mathbf{L}_{A,[k]}\right\} - \max_{\mathbf{s}\in\mathbf{S}_{k,-1}}\left\{-\frac{1}{\sigma_0^2}\|\mathbf{x} - \mathbf{Hs}\|^2 + \mathbf{s}_{[k]}^T \cdot \mathbf{L}_{A,[k]}\right\}\right) \tag{5.146}$$

Thus, the ML estimate must maximize the term $-\frac{1}{\sigma_0^2}\left\|\mathbf{x}-\mathbf{Hs}\right\|^2+\mathbf{s}_{[k]}^T\cdot\mathbf{L}_{A,[k]}$ for both conditions:

$$\mathbf{s}_{ML}=\begin{cases}\arg\max\limits_{\mathbf{v}\in\mathbf{S}_{k,+1}}\left(-\frac{1}{\sigma_0^2}\left\|\mathbf{x}-\mathbf{Hv}\right\|^2+\mathbf{v}_{[k]}^T\cdot\mathbf{L}_{A,[k]}\right)\\ \arg\max\limits_{\mathbf{v}\in\mathbf{S}_{k,-1}}\left(-\frac{1}{\sigma_0^2}\left\|\mathbf{x}-\mathbf{Hv}\right\|^2+\mathbf{v}_{[k]}^T\cdot\mathbf{L}_{A,[k]}\right)\end{cases}$$

The basic idea behind the *List Sphere Detection* (LSD) algorithm is to modify the SD algorithm in such a way to generate the list of candidate vectors $\mathbf{v}$ maximizing both terms in (5.146) and remaining within a sphere of predefined radius r. Actually, the changes needed to transform the SD algorithm to LSD algorithm are not severe: the result of the algorithm is the list of N_{cand} points $\lambda=\{\mathbf{v}_i\}, i=0,\ldots,N_{cand}$ minimizing $\left\|\mathbf{x}-\mathbf{Hv}_i\right\|^2$ rather than just one optimal point $\mathbf{v}_{opt}$ for which $\left\|\mathbf{x}-\mathbf{Hv}_{opt}\right\|^2$ is the minimum one. The point from this list $\lambda=\{\mathbf{v}_i\}, i=0,\ldots,N_{cand}$ maximizing both terms in (5.146) is chosen as the ML estimate $\mathbf{s}_{ML}$. That means (5.146) can be written as follows:

$$L_E(s_k)\approx\frac{1}{2}\left(\max_{\mathbf{s}\in\lambda\cap\mathbf{S}_{k,+1}}\left\{-\frac{1}{\sigma_0^2}\left\|\mathbf{x}-\mathbf{Hs}\right\|^2+\mathbf{s}_{[k]}^T\cdot\mathbf{L}_{A,[k]}\right\}-\max_{\mathbf{s}\in\lambda\cap\mathbf{S}_{k,-1}}\left\{-\frac{1}{\sigma_0^2}\left\|\mathbf{x}-\mathbf{Hs}\right\|^2+\mathbf{s}_{[k]}^T\cdot\mathbf{L}_{A,[k]}\right\}\right) \tag{5.147}$$

The list size N_{cand} must be big enough to make possible the calculation of (5.147). That means the complexity of the LSD algorithm is a little bit greater than the complexity of the SD algorithm, since it requires for each considered point the comparison of the distance between the point and the sphere centre not only with the best one candidate but with the list of N_{cand} candidates to update it. However, this complexity increase is not very severe. Still, the necessity to have a relatively big size of candidates list requires using quite a big sphere radius r that leads to more severe complexity increase of the LSD algorithm in comparison with the SD algorithm.

5.3 RAKE Receiver

Strictly speaking the *RAKE receiver* introduced in [15] cannot be regarded as an equalizer since its ability to cope with ISI is quite limited. Still we consider this technique here since it is very popular in DS-CDMA systems because of low implementation complexity.

In CDMA spread spectrum systems, each symbol is spread with the help of chip sequence into the chips and the chip rate is typically much greater than the flat fading bandwidth of the channel. If the components of the multipath propagation channel are delayed in time by more than one chip duration the multipath channel is seen as a discrete set of reflections. In this case the multipath components can be seen as a source of diversity used by the RAKE receiver. The multipath channel model is represented in Figure 5.17.

The channel impulse response $h(t)$ can be written as follows:

$$h(t)=\sum_{l=1}^{L}h_l\cdot\delta(t-\tau_l) \tag{5.148}$$

where L is the number of multipaths, τ_l is the time delay of the l-th path component, h_l is the gain of the l-th path component, $\delta()$ denotes the delta-function. Then the received signal can be represented as:

$$r(t)=h(t)\cdot s(t)+n(t)=\sum_{l=1}^{L}h_l\cdot s(t-\tau_l)+n(t)=y(t)+n(t) \tag{5.149}$$

where $s(t)$ is the transmitted signal and $n(t)$ is the AWGN. Assume that the channel time delays τ_l and the channel gain coefficients h_l are known. Then in accordance with the matched filter theory (see

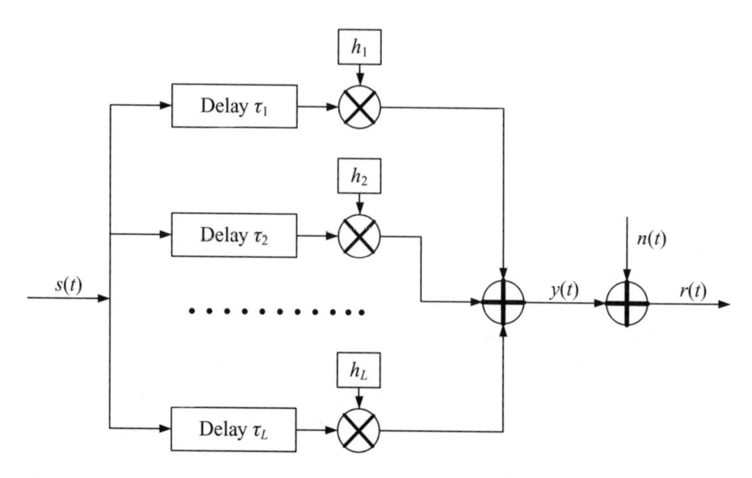

Figure 5.17 Multipath channel model

Section 2.1.4) at the receiver the received signal $r(t)$ must be correlated with the complex conjugate of the expected waveform $y(t)$. Since the number of the transmitted waveforms $s(t)$ is limited and the channel time delays and gain coefficients are known it is possible to generate the limited number of the expected waveforms $y(t)$. Then the decision variable corresponding to the transmitted signal $s(t)$ can be expressed as follows:

$$u = \mathrm{Re}\left\{\int_0^{T_s} r(t)\cdot y^*(t)dt\right\} = \mathrm{Re}\left\{\int_0^{T_s} r(t)\cdot\sum_{l=1}^{L} h_l^*\cdot s^*(t-\tau_l)\,dt\right\}$$

$$= \mathrm{Re}\left\{\sum_{l=1}^{L} h_l^*\cdot\int_0^{T_s} r(t)\cdot s^*(t-\tau_l)\,dt\right\} \tag{5.150}$$

where T_s is the symbol duration. As a result the receiver optimizing the SNR can be implemented with the help of L correlators. Each of these correlators correlates the received signal $r(t)$ with a delayed version of the transmit waveform $s(t)$ and the outputs of the correlators are weighted according to the channel gain coefficients as shown in Figure 5.18.

Each correlator called *finger* extracts from the received signal $r(t)$ only the portion of the signal energy that corresponds to the particular delay. Then the receiver combines the received signals of different paths proportionally to the strength of each path. This kind of combining is called the *Maximal Ratio Combining* (*MRC*) [16]. Since each path undergoes different attenuations, combining them with different weights yields an optimum in the SNR sense solution when noise is white. Thus, the receiver utilizes the multipath channel as the source of diversity. The structure of the receiver is similar to the garden rake and correspondingly the name "RAKE receiver" is used for this kind of receiver. In Figure 5.18 only part of the receiver corresponding just to one particular value of the transmitted signal $s(t)$ is represented. Actually, each finger consists of M correlators, where M is the cardinality of the transmitted alphabet and correspondingly M inputs are fed to the decision device. The finger allocation algorithm estimates the multipath time delays τ_l and provides them to the corresponding fingers. This is done with the help of matched filter. Since the chip sequence used in the DS-CDMA systems is pseudo-noise sequence, the strongest peaks at the output of the matched filter can correspond to the multipath time delays only. Another important block is the channel estimator. It estimates the amplitude and phase of the multipath components h_l. Usually this is done with the help of a special pilot sequence of symbols.

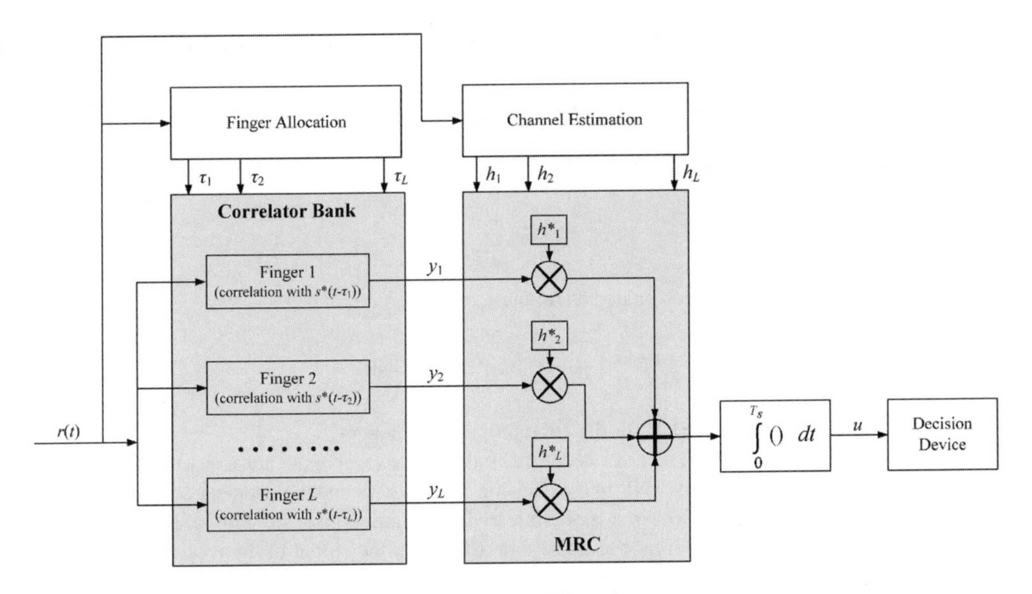

Figure 5.18 RAKE receiver

The transmitted pilot symbols are known at the receiver side beforehand and based on the distortion of these symbols it is possible to estimate the multipath components. Usually in the DS-CDMA systems the pilot symbols are transmitted on the special pilot channel simultaneously with the data. Since in the DS-CDMA systems all channels are split with the help of different spreading codes the multipath components seen on the pilot channel coincide with the multipath components seen on the data channel. Here the pilot channel and the data channel denotes only the type of the transmitted data, do not mix them with the multipath channel.

The performance of the RAKE receiver can be estimated by estimating the SNR at the receiver output. Assume that the transmitted waveforms are ideally orthogonal and that the noise can be represented as an AWGN. Substituting (5.149) into (5.150) obtain:

$$u = \mathrm{Re}\left\{\sum_{l=1}^{L} h_l^* \cdot \sum_{k=1}^{L} h_k \cdot \int_0^{T_s} s\,(t-\tau_k) \cdot s^*\,(t-\tau_l)\,dt\right\} + \mathrm{Re}\left\{\sum_{l=1}^{L} h_l^* \cdot \int_0^{T_s} n(t) \cdot s^*\,(t-\tau_l)\,dt\right\} \tag{5.151}$$

The first term in the right hand side of (5.151) corresponds to the desired signal and the second term corresponds to the noise. Taking into account that $s(t)$ is generated with the help of pseudo-noise sequence with good autocorrelation properties, the autocorrelation function of signal $s(t)$ can be written as follows:

$$R(k-l) = \int_0^{T_s} s\,(t-kT_c) \cdot s^*\,(t-lT_c)\,dt = \begin{cases} 0, & k \neq l \\ T_s, & k = l \end{cases} \tag{5.152}$$

where T_c is the chip duration. Then (5.151) can be written as:

$$u = \mathrm{Re}\left\{\sum_{l=1}^{L} |h_l|^2 \cdot T_s\right\} + \mathrm{Re}\left\{\sum_{l=1}^{L} h_l^* \cdot \int_0^{T_s} n(t) \cdot s^*\,(t-\tau_l)\,dt\right\} \tag{5.153}$$

For the fixed set of channel gain coefficients $\{h_l\}$ the decision variable u is Gaussian with mean value:

$$E\{u\} = T_s \sum_{l=1}^{L} |h_l|^2 \tag{5.154}$$

and variance:

$$\sigma_0^2 = T_s N_0 \sum_{l=1}^{L} |h_l|^2 \tag{5.155}$$

Then the SNR per symbol can be estimated as follows:

$$\mathrm{SNR} = \frac{T_s}{2N_0} \sum_{l=1}^{L} |h_l|^2 = \sum_{l=1}^{L} \mathrm{SNR}_l \tag{5.156}$$

where SNR_l is the instantaneous SNR for the kth multipath component.

As mentioned above the RAKE receiver is optimal only in cases where the noise is AWGN and there is no ISI since it maximizes only the SNR of the decision statistic with finger placement corresponding to the channel impulse response. However, it is possible to use the same simple structure of RAKE receiver to maximize the signal plus interference to noise ratio (SINR) at the output of the receiver. In fact, the interference components of the different RAKE fingers are regarded as coloured Gaussian noise and in this sense it is enough to maximize the SNR almost in the same way as is done in the conventional RAKE. This receiver structure usually is called the *generalized RAKE* or *G-RAKE* [17]. For G-RAKE the finger delays do not match exactly the channel delays anymore. The number of fingers in G-RAKE J is much greater than the number of resolvable channel paths L (the suitable choice is $J \leq 2L$). Choosing the finger delays is a tradeoff between matching to the channel and whitening the coloured noise. The output of J finger correlators $\mathbf{y} = [y_1, y_2, \ldots, y_J]^T$ can be represented as:

$$\mathbf{y} = \mathbf{g} \cdot s(t) + \mathbf{n} \tag{5.157}$$

where $\mathbf{n}$ is the vector of noise samples and $\mathbf{g}$ is the vector corresponding to the channel coefficients filtered with the finger correlators. Now instead of channel gain coefficients $\{h_l\}$ the combining coefficients $\mathbf{w} = [w_1, w_2, \ldots, w_J]^T$ will be used for calculation of a decision statistic u. Then the detector for $s(t)$ can be obtained if the combining coefficients are calculated as follows:

$$\mathbf{w} = \mathbf{R}_n^{-1} \cdot \mathbf{g} \tag{5.158}$$

where $\mathbf{R}_n = E\left\{\mathbf{n}\mathbf{n}^H\right\}$ is the noise covariance matrix. Note that $\mathbf{n}$ comprises the coloured zero-mean Gaussian noise and thus includes not only the white noise components but the interference as well. The performance of the G-RAKE is close to the performance of the LMMSE equalizer. However, if the number of fingers J is close to infinity the optimal performance can be achieved.

5.4 Turbo Equalization

The main idea of the *turbo equalization* is to improve the receiver performance by communicating soft information between the soft-input/soft-output (SISO) equalizer and the SISO decoder. To do this the received signals are jointly equalized and decoded iteratively in a turbo structure. First the principle of the turbo equalization was proposed in [18]. In [18] the SOVA was used for both SISO equalization and SISO decoding. In [19] the MAP algorithm was used for SISO equalizer and decoder. The general principle of turbo equalization is depicted in Figure 5.19.

The turbo equalization structure shown in Figure 5.19 requires multiple iterations between the SISO equalizer and the SISO decoder to obtain the decoded received symbol. The processing is done in accordance with the iterative decoding principle used in decoding of turbo-codes (see Section 4.6). The

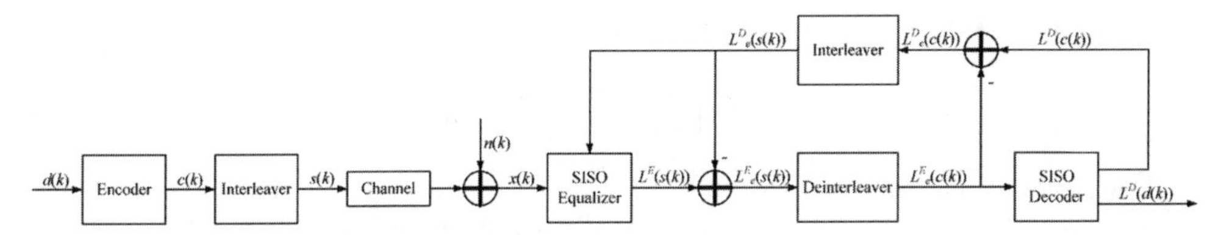

Figure 5.19 Turbo equalization principle

SISO equalization algorithm observes the value $x(k)$ and the a priori LLR $L_e^D\,(s(k))$ obtained from the SISO decoder (at first iteration this a priori LLR is not available) and calculates the symbol's a posteriori LLR $L^E\,(s(k))$. The a priori LLR $L_e^D\,(s(k))$ represents prior information on the occurrence probability of $s(k)$ and is provided by the decoder. Actually, $L_e^D\,(s(k))$ is the extrinsic LLR obtained from the decoder but in an equalizer it is used as the a priori LLR, that is, it can be represented as:

$$L_e^D\,(s(k)) = \ln \frac{P\,(s(k) = +1)}{P\,(s(k) = -1)} \tag{5.159}$$

Treating feedback as a priori information is the essential feature for turbo principle. The equalizer estimates the a posteriori LLR $L^E\,(s(k))$:

$$L^E\,(s(k)) = \ln \frac{P\,(s(k) = +1\,|\mathbf{x})}{P\,(s(k) = -1\,|\mathbf{x})} \tag{5.160}$$

where $\mathbf{x} = [x(0), x(1), \ldots x(K-1)]$ is the received sequence. In accordance with (4.41) the extrinsic LLR:

$$L_e^E\,(s(k)) = \ln \frac{P\,(\mathbf{x}\,|s(k) = +1)}{P\,(\mathbf{x}\,|s(k) = -1)} \tag{5.161}$$

can be obtained by subtracting the a priori LLR $L_e^D\,(s(k))$ from the a posteriori LLR $L^E\,(s(k))$. This extrinsic LLR is deinterleaved using the same deinterleaver pattern as in the encoder, and is passed to SISO decoder, which uses it as the a priori LLR $L_e^E\,(c(k))$. The SISO decoder computes the a posteriori LLR $L^D\,(c(k))$:

$$L^D\,(c(k)) = \ln \frac{P\,(c(k) = +1\,|\mathbf{L})}{P\,(c(k) = -1\,|\mathbf{L})} \tag{5.162}$$

where $\mathbf{L} = [L\,(c(0))\,, L\,(c(1))\,, \ldots L\,(c(K-1))]$ is the sequence of corresponding LLRs.

In the same way as it is described before the extrinsic decoder LLR:

$$L_e^D\,(c(k)) = \ln \frac{P\,(\mathbf{L}\,|c(k) = +1)}{P\,(\mathbf{L}\,|c(k) = -1)} \tag{5.163}$$

can be obtained as follows:

$$L_e^D\,(c(k)) = L^D\,(c(k)) - L_e^E\,(c(k)) \tag{5.164}$$

Recall that the deinterleaved extrinsic equalizer LLR $L_e^E\,(c(k))$ is considered as the a priori LLR for the decoder. Then the extrinsic decoder LLR $L_e^D\,(c(k))$ is interleaved and the obtained LLR $L_e^D\,(s(k))$ is used as the a priori LLR for the equalizer at the next iteration. The iterative equalization and decoding proceed until a stop criterion is met. Then the output of the decoder $L^D\,(d(k))$ is sent to a decision device. The decision device makes the decision based on the simple rule:

$$d(k)=\begin{cases}+1, & L^{D}\left(d(k)\right)\geq 0\\ -1, & L^{D}\left(d(k)\right)<0\end{cases} \tag{5.165}$$

Thus the process of joint iterative equalization and decoding is similar to the process of turbo decoding. Comparing Figures 4.22 and 5.19 one can see that the SISO equalizer simply substitutes one of constituent SISO decoders depicted in Figure 4.22. The SISO decoder and the SISO equalizer exchange extrinsic information between each other using it as the a priori information. Using this a priori information is supposed to improve the quality of the a posteriori information at the output of the equalizer and decoder from iteration to iteration. It is important to feed back only the extrinsic information because the correlation between the a priori information and the previous decisions should be minimized.

As mentioned above it is possible to consider different algorithms for SISO equalization and SISO decoding. The MAP and SOVA algorithms used for SISO equalization are pretty much in line with the description of the corresponding algorithms used for decoding of turbo codes that can be found in Section 4.7. So, here we briefly consider the employing of MAP algorithm for both SISO equalizer and SISO decoder just to show some differences in comparison with the content of Section 4.7.

The LLR $L^{E}\left(s(k)\right)$ can be calculated as follows:

$$L^{E}\left(s(k)\right)=\ln\frac{P\left(s(k)=+1\left|\mathbf{x}\right.\right)}{P\left(s(k)=-1\left|\mathbf{x}\right.\right)}=\ln\frac{\sum\limits_{(S',S),s(k)=+1} p\left(S',S,\mathbf{x}\right)}{\sum\limits_{(S',S),s(k)=-1} p\left(S',S,\mathbf{x}\right)} \tag{5.166}$$

where S' and S denote the state of trellis at level $k-1$ and k respectively as shown in Figure 5.20.

The summation in (5.166) is taken for all transitions in trellis from state S' to state S labeled with $s(k)=+1$ for nominator and with $s(k)=-1$ for denominator. In the same way the LLR for bits $d(k)$ can be computed by the SISO decoder (substituting $\mathbf{x}$ by $\mathbf{L}$ and $s(k)$ by $d(k)$). Also SISO decoder computes LLR for coded bits:

$$L^{D}\left(c(k,i)\right)=\ln\frac{P\left(c(k,i)=+1\left|\mathbf{L}\right.\right)}{P\left(c(k,i)=+1\left|\mathbf{L}\right.\right)}=\ln\frac{\sum\limits_{(S',S),c(k,i)=+1} p\left(S',S,\mathbf{L}\right)}{\sum\limits_{(S',S),c(k,i)=-1} p\left(S',S,\mathbf{L}\right)} \tag{5.167}$$

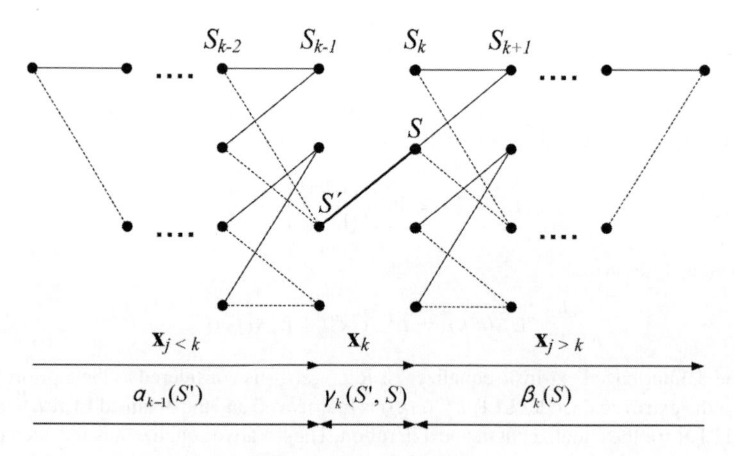

Figure 5.20 MAP trellis

where $c(k, i), i = 0, \ldots, N-1$ denotes the coded bits corresponding to $\mathbf{x}_k$ or what is the same to $d(k)$, N is the code length. The joint probability $p(S', S, \mathbf{x})$ can be represented as follows (see (4.43)–(4.46)):

$$p\left(S', S, \mathbf{x}\right) = \underbrace{p\left(S', \mathbf{x}_{j<k}\right)}_{\alpha_{k-1}(S')} \cdot \underbrace{P\left(S \left| S'\right.\right) \cdot p\left(\mathbf{x}_k \left| S', S\right.\right)}_{\gamma_k(S', S)} \cdot \underbrace{p\left(\mathbf{x}_{j>k} \left| S\right.\right)}_{\beta_k(S)} \tag{5.168}$$

In the same way the probability $p(S', S, \mathbf{L})$ can be calculated. Values $\alpha_{k-1}(S')$ and $\beta_k(S)$ can be calculated recursively:

$$\alpha_k(S) = \sum_{S'} \alpha_{k-1}\left(S'\right) \cdot \gamma_k\left(S', S\right) \tag{5.169}$$

$$\beta_{k-1}\left(S'\right) = \sum_{S} \beta_k(S) \cdot \gamma_k\left(S', S\right) \tag{5.170}$$

For the existing transitions the branch transition probability $\gamma_k(S', S)$ can be represented as follows:

$$\gamma_k\left(S', S\right) = P\left(d(k)\right) \cdot p\left(\mathbf{x}_k \left| S', S\right.\right) \tag{5.171}$$

The only difference in the calculation of metrics for MAP equalizer and MAP decoder is in calculation of (5.171). Taking into account the channel model (5.21) the branch transition probability $\gamma_k(S', S)$ for a transition labelled with $s(i)$ for MAP equalizer can be calculated as [19]:

$$\gamma_i\left(S', S\right) = \exp\left(-\frac{1}{2\sigma_0^2}\left|x(i) - \sum_{l=0}^{L-1} h(l) \cdot s(i-l)\right|^2 + \frac{1}{2} s(i) \cdot L\left(s(i)\right)\right) \tag{5.172}$$

where $L(s(i))$ denotes the a priori LLR used by the MAP equalizer and σ_0 is the variance of AWGN.

For the MAP decoder $\gamma_k(S', S)$ is calculated in the following way [19]:

$$\gamma_k\left(S', S\right) = \exp\left(\sum_{i=0}^{N-1}\left(\frac{1}{2}c(k, i) \cdot L\left(c(k, i)\right)\right) + \frac{1}{2}d(k) \cdot L\left(d(k)\right)\right) \tag{5.173}$$

where $L(d(k))$ is the a priori LLR for bit $d(k)$, $L(c(k, i))$ is the input LLR.

In [20] the turbo equalization with MMSE equalizer or DFE used as SISO equalizer was considered. The MMSE SISO equalizer is depicted in Figure 5.21. It computes the estimates $\hat{s}(k)$ of the transmitted symbols $s(k)$ from the received symbols $x(k)$ by minimizing the cost function $E\left\{|s(k) - \hat{s}(k)|^2\right\}$.

It is assumed that the MMSE SISO equalizer output is the extrinsic LLR $L_e(s(k))$. This extrinsic LLR is calculated using the estimates $\hat{s}(k)$ instead of $x(k)$:

$$L_e\left(s(k)\right) = \ln\frac{P\left(s(k) = +1\left|\hat{\mathbf{s}}\right.\right)}{P\left(s(k) = -1\left|\hat{\mathbf{s}}\right.\right)} - \ln\frac{P\left(s(k) = +1\right)}{P\left(s(k) = -1\right)} = \ln\frac{P\left(\hat{\mathbf{s}}\left|s(k) = +1\right.\right)}{P\left(\hat{\mathbf{s}}\left|s(k) = -1\right.\right)} \tag{5.174}$$

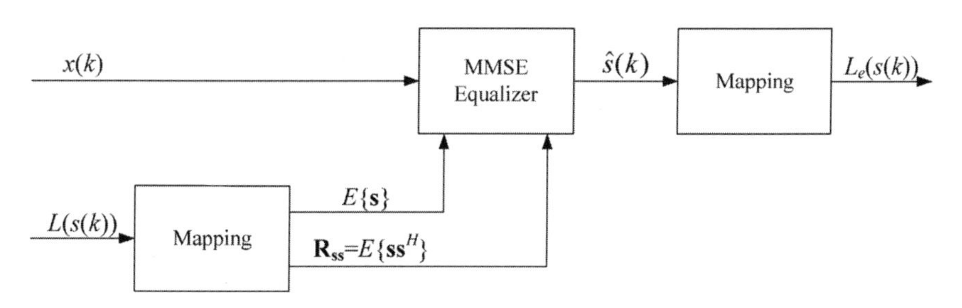

Figure 5.21 SISO MMSE equalizer

where $\hat{\mathbf{s}} = [\hat{s}(0), \hat{s}(1), \ldots \hat{s}(K-1)]$, which requires the knowledge of the distribution $p\,(\hat{s}(k)\,|s(k) = +1\,)$ and $p\,(\hat{s}(k)\,|s(k) = -1\,)$. To perform MMSE equalization, the statistics $E\,\{\mathbf{s}\}$ and $\mathbf{R}_{ss} = E\left\{\mathbf{s}\mathbf{s}^H\right\}$ are required (see Section 5.1.2). After MMSE equalization the PDFs $p\,(\hat{s}(k)\,|s(k) = +1\,)$ and $p\,(\hat{s}(k)\,|s(k) = -1\,)$ can be considered as Gaussian with the parameters $\mu_{+1}(k) = E\,\{\hat{s}(k)\,|s(k) = +1\,\}$, $\sigma_{+1}^2(k) = E\,\{\hat{s}(k)\cdot \hat{s}*(k)\,|s(k) = +1\,\}$ and $\mu_{-1}(k) = E\,\{\hat{s}(k)\,|s(k) = -1\,\}$, $\sigma_{-1}^2(k) = E\,\{\hat{s}(k)\cdot \hat{s}*(k)\,|s(k) = -1\,\}$ [20]:

$$p\,(\hat{s}(k)\,|s(k) = +1\,) \approx \frac{\phi\left(\frac{\hat{s}(k)-\mu_{+1}(k)}{\sigma_{+1}(k)}\right)}{\sigma_{+1}(k)}$$
$$p\,(\hat{s}(k)\,|s(k) = -1\,) \approx \frac{\phi\left(\frac{\hat{s}(k)-\mu_{-1}(k)}{\sigma_{-1}(k)}\right)}{\sigma_{-1}(k)} \tag{5.175}$$

where $\phi\,(x) = \frac{1}{\sqrt{2\pi}}\cdot \exp\left(-\frac{x^2}{2}\right)$. Then the extrinsic LLR $L_e\,(s(k))$ can be calculated as follows:

$$L_e\,(s(k)) = \ln\frac{P\,(\hat{\mathbf{s}}\,|s(k) = +1\,)}{P\,(\hat{\mathbf{s}}\,|s(k) = -1\,)} = \ln\frac{\phi\left(\frac{\hat{s}(k)-\mu_{+1}(k)}{\sigma_{+1}(k)}\right)\cdot \sigma_{-1}(k)}{\phi\left(\frac{\hat{s}(k)-\mu_{-1}(k)}{\sigma_{-1}(k)}\right)\cdot \sigma_{+1}(k)} = \frac{2\hat{s}(k)\cdot \mu_{+1}(k)}{\sigma_{+1}^2(k)} \tag{5.176}$$

The coefficients of the equalizer filter $\mathbf{c}$ can be calculated as it is described in (5.44). Then the parameters of the PDFs $p\,(\hat{s}(k)\,|s(k) = +1\,)$ and $p\,(\hat{s}(k)\,|s(k) = -1\,)$ can be estimated as [20]:

$$\begin{aligned}
&\mu_{+1}(k) = \mathbf{c}\cdot \mathbf{H}\cdot \delta_D^H, \\
&\mu_{-1}(k) = -\mathbf{c}\cdot \mathbf{H}\cdot \delta_D^H, \\
&\sigma_{+1}^2(k) = \sigma_{-1}^2(k) = \mathbf{c}\cdot \mathbf{H}\cdot \delta_D^H - |\mu_{+1}(k)|^2
\end{aligned} \tag{5.177}$$

Substituting (5.177) into (5.176) obtain:

$$L_e\,(s(k)) = \frac{2\hat{s}(k)}{1-\mathbf{c}\cdot \mathbf{H}\cdot \delta_D^H} \tag{5.178}$$

The same approach can be applied if the DFE is used as a SISO equalizer. Taking into account (5.60), the general expression for SISO DFE output can be written as follows:

$$\hat{s}(k) = E\,\{s(k)\} + \sum_{i=-K_1}^{0} c_F(i)\cdot (x(k-i) - E\,\{x(k-i)\}) + \sum_{i=1}^{K_2} c_B(i)\cdot (\hat{s}(k-i) - E\,\{\hat{s}(k-i)\}) \tag{5.179}$$

where $\hat{s}(k-i)$ are previously detected symbols.

Now using the relation between the coefficients of feedforward and feedback filters (5.76) and employing (5.40), (5.179) can be represented as:

$$\hat{s}(k) = E\,\{s(k)\} + \mathbf{c}_F^H\,(\mathbf{x} - \mathbf{H}E\,\{\mathbf{s}\}) \tag{5.180}$$

where $E\,\{\mathbf{s}\} = [\hat{s}(k-K_2), \hat{s}(k-K_2+1), \ldots, \hat{s}(k-1), E\,\{s(k)\}, E\,\{s(k+1)\}, \ldots, E\,\{s(k+K_1)\}]^T$, $\mathbf{x} = [x(k+K_1), x(k+K_1-1), \ldots, x(k)]^T$. The coefficients of the equalizer feedforward filter $\mathbf{c}_F$ are calculated in accordance with (5.81). Then the parameters of the PDFs $p\,(\hat{s}(k)\,|s(k) = +1\,)$ and $p\,(\hat{s}(k)\,|s(k) = -1\,)$ can be estimated as [20]:

$$\begin{aligned}
&\mu_{+1}(k) = \mathbf{c}_F\cdot \mathbf{H}\cdot \delta_D^H, \\
&\mu_{-1}(k) = -\mathbf{c}_F\cdot \mathbf{H}\cdot \delta_D^H, \\
&\sigma_{+1}^2(k) = \sigma_{-1}^2(k) = \mathbf{c}_F\cdot \mathbf{H}\cdot \delta_D^H - |\mu_{+1}(k)|^2
\end{aligned} \tag{5.181}$$

where $\mathbf{H}$ is $(K_1+1)\times(K_1+L+1)$ matrix, δ_D is all zeroes $(1\times(K_1+L+1))$ vector except for unity at the $(D+1)$-th position. The extrinsic LLR is calculated in the same way as for MMSE SISO

Table 5.1 ITU pedestrian A

Tap	Relative delay (ns)	Average Power (dB)
1	0	0
2	110	–9.7
3	190	–19.2
4	410	–22.8

equalizer:

$$L_e(s(k)) = \frac{2\hat{s}(k)}{1 - \mathbf{c}_F \cdot \mathbf{H} \cdot \delta_D^H} \tag{5.182}$$

The same method can be applied also in cases when the turbo coding is used. In this case the turbo decoder is simply used as a SISO decoder.

5.5 Performance Comparison

In this section the performance simulation results for some of the equalization schemes are presented. All simulations were done for WCDMA system under the conditions of fading channel. Three channel profiles were considered: ITU Pedestrian A with the speed 3 km/h, ITU Pedestrian B with the speed 3 km/h and ITU Vehicular A with the speed 30 km/h. The channel profiles are described in Tables 5.1– 5.3.

The following equalizers were used in simulations:

1. Zero-Forcing (ZF) equalizer with 47 taps (taps resolution is 0.5 chip);
2. Linear minimum mean-square error (LMMSE) equalizer with 48 taps (taps resolution is 0.5 chip);
3. Decision feedback equalizer (DFE) represented in Figure 5.11 with 48 taps LMMSE equalizer and feedforward filter with 32 taps and feedback filter with 16 taps;
4. Maximum likelihood sequence estimation (MLSE) equalization with $K = 1024$;
5. List sphere detection (LSD) with $K = 1024$;
6. RAKE receiver with 8 fingers;
7. G-RAKE with 16 fingers;
8. Turbo equalization with MAP SISO equalizer.

The BER results against the *geometry factor G* for different channel profiles are represented in Figures 5.22–5.24. The geometry factor G expresses the distance and quality of the path between the user and Node B. It is defined as:

$$G = \langle \hat{I}_{\text{or}} \rangle / N_0$$

Table 5.2 ITU pedestrian B

Tap	Relative delay (ns)	Average Power (dB)
1	0	0
2	200	–0.9
3	800	–4.9
4	1200	–8.0
5	2300	–7.8
6	3700	–23.9

Table 5.3 ITU vehicular A

Tap	Relative delay (ns)	Average Power (dB)
1	0	0
2	310	–1
3	710	–9
4	1090	–10
5	1730	–15
6	2510	–20

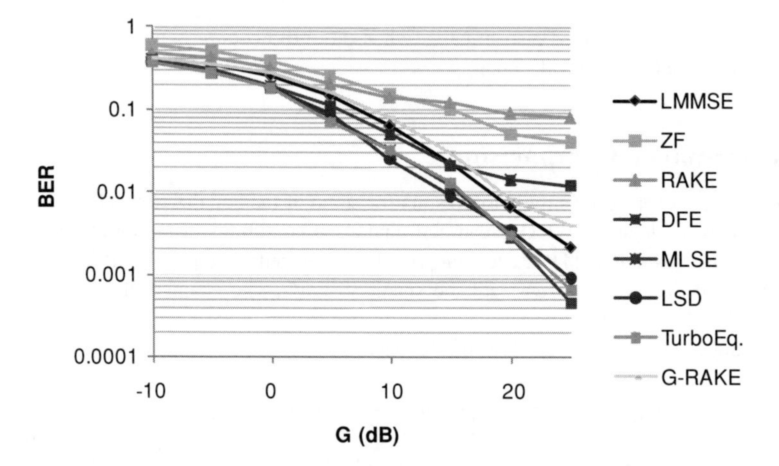

Figure 5.22 Equalizers' performance. Pedestrian A, 3 km/h

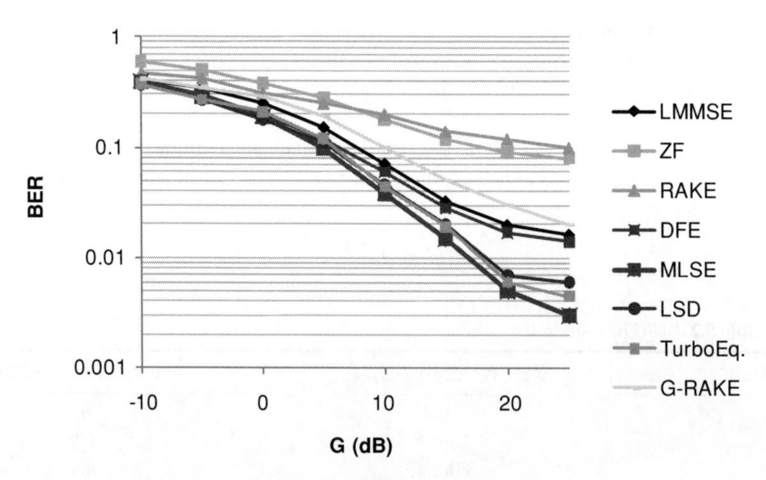

Figure 5.23 Equalizers' performance. Pedestrian B, 3 km/h

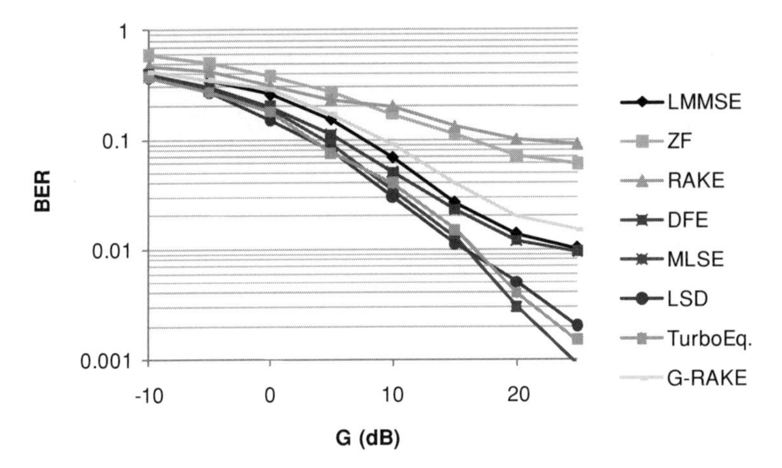

Figure 5.24 Equalizers' performance. Vehicular A, 30 km/h

where $\langle \hat{I}_{or} \rangle$ is the average total power received from the own cell, and N_0 is the noise power which includes both thermal noise and interference from other cells.

As can be seen from Figures 5.22–5.24 the behaviour of different equalizers is quite similar for different channel profiles. The worst performance with low values of G (high noise) shows the ZF equalizer. It is explained by the noise enhancement, which is inherent in ZF equalizer. As expected with high G values ZF equalizer outperforms the RAKE receiver. The LMMSE equalizer significantly outperforms both ZF equalizer and RAKE receiver. This fact is especially important since the complexity of the LMMSE equalization is quite close to the complexity of ZF equalization and not significantly higher than the complexity of RAKE receiver. The performance demonstrated by G-RAKE is quite close to the performance of LMMSE equalizer. It is also an expected result, since the G-RAKE optimization method is quite close to the LMMSE. Slightly better performance than LMMSE is shown by the DFE. This performance gain is especially visible in more demanding channel profiles like Pedestrian B and Vehicular A, while in Pedestrian A with high G values LMMSE equalizer outperforms DFE. This can be explained by the impact of error propagation typical for DFE. As can be seen from plots in Figures 5.22–5.24 the equalization methods based on sequence estimation provide much better performance than methods based on filtering. The best performance is demonstrated by the MLSE equalization. Unfortunately the complexity of this method is extremely high. The similar performance is shown by LSD and turbo equalization with MAP SISO equalizer. However, the complexity of these methods is also quite high.

References

[1] Sklar, B. *Digital Communications. Fundamentals and Applications,* 2nd ed., Prentice Hall, 2001.

[2] Proakis, J.G. *Digital Communications,* 3rd ed., McGraw-Hill, 1995.

[3] Kay, S.M. *Fundamentals of Statistical Signal Processing: Estimation Theory*. Vol. I, Prentice Hall, 2009.

[4] Hooli, K., Latva-Aho, M., and Juntti, M., *Multiple Access Interference Suppression with Linear Chip Equalizers in WCDMA Downlink Receivers,* in Proc. Global Telecommunications Conf., pp. 467–471, Rio de Janero, Brazil, Dec. 5–9 1999.

[5] Krauss, T.P., Zoltowski, M.D. *Oversampling diversity versus dual antenna diversity for chip-levelequalization on CDMA downlink,* in Proceedings of the Sensor Array and Multichannel Signal Processing Workshop, 2000, pp. 47–51.

[6] Mailaender, L. and Proakis, J. G. *Linear-Aided Decision-Feedback Equalization for the CDMA Downlink,* Proceedings Thirty Seventh Asilomar Conference on Signals, Systems and Computers, 2003, pp. 131–135.
[7] Pohst, M. *On the computation of lattice vectors of minimal length, successive minima and reduced basis with applications,*. ACM SIGSAM Bull., vol. 15, pp. 37–44, 1981.
[8] Fincke, U. and Pohst, M. *Improved methods for calculating vectors of short length in a lattice, including a complexity analysis,*. Mathematics of Computation, vol. 44, pp. 463–471, April 1985.
[9] Viterbo, E., Boutros, J., *A Universal Lattice Code Decoder for Fading Channels.* IEEE Trans. Inf. Theory, vol. 45, No 5, July 1999, pp. 1639–1642.
[10] Golub, G.H. Van Loan, C.F., *Matrix Computations* (3rd ed.), Johns Hopkins, 1996.
[11] Damen, M.O., El Gamal, H., Caire, G., *On maximum-likelihood detection and the search for the closest lattice point.* IEEE Trans. Inf. Theory., vol. 49, No 10, Oct. 2003, pp. 2389–2402.
[12] Schnorr, C.P., Euchner, M., *Lattice basis reduction: Improved practical algorithms and solving subset sum problems,* Math. Programming, vol. 66, no. 2, pp. 181–191, Sep. 1994.
[13] Hassibi, B., Vikalo, H., *On the Sphere-Decoding Algorithm I. Expected Complexity.* IEEE Trans. Signal Processing, vol. 53, No. 8, August 2005, pp. 2806–2818.
[14] Hochwald, B., Brink, S.T., *Achieving near-capacity on a multiple antenna channel,* IEEE Trans. Commun., vol. 51, no. 3, Mar. 2003.
[15] Price, R., Green, P.E., *A Communication Technique for Multipath Channels.* Proc. IRE, Mar. 1958, vol. 46, pp. 555–570.
[16] Brennan, D.G., *Linear Diversity Combining Techniques.* Proc. IRE, Jun. 1959, vol. 47, pp. 1075–1102.
[17] Bottomley, T.O.G., Wang, Y. *A generalized rake receiver for interference suppression.* IEEE Journal Selected Areas Communications, Vol. 18, No. 8, Aug. 2000, pp. 1536–1545.
[18] Douilard, C., Jezequel, M., Berrou, C., Picart, A., Didier, P., Glavieux, A., *Iterative correction of intersymbol interference: Turbo-equalization,* European Transactions on Telecommunications, vol. 6, Sep.–Oct. 1995, pp. 507–511.
[19] Bauch, G., Khorram, H., Hagenauer, J., *Iterative equalization and decoding in mobile communications systems,* Proc. European Personal Mobile Commun. Conf., pp. 307–312.
[20] Tuchler, M., Koetter, R., Singer, A.C., *Turbo Equalization: Principles and New Results.* IEEE Trans. On Communications, Vol. 50, No. 5, May 2002, pp. 754–767.

6

ARQ

Evgenii Krouk
St. Petersburg State University of Aerospace Instrumentation, Russia

In closed-loop systems the usage of feedback channel allows the transmission characteristics to be improved. The idea of utilizing the feedback channel is that in case of receiving a very noisy message it is possible not to make a decision, but rather ask the transmitter to repeat the transmission. In the most general cases such organization of communication allows a decision concerning transmitted message to be made only under conditions of noise absence (or very low noise level), repeating the message when the noise level is high. Surely, usage of repeated transmission decreases the overall transmission rate. However, in most cases such rate decreasing is not significant. Communication systems utilizing the feedback channel are called systems with repeat request (ARQ, Automatic Repeat reQuest) or systems with feedback [10, 16, 18].

6.1 Basic ARQ Schemes

6.1.1 Basic Concepts

Among the variety of algorithms for data transmission over the feedback channels the two basic classes of such systems may be distinguished: *systems with decision feedback* and *systems with data feedback*.

In systems with **decision feedback** the decision of repeated message transmission is made by the receiving side of the system as the result of analyzing the received data. In systems with **data feedback** the decision is made by the transmitter based on the information transmitted over the feedback channel. Systems in which the decision of repeated transmission can be made both on receiving and transmitting sides are called combined or *hybrid transmission systems with feedback*.

Consider the simplest model of the system with decision feedback, see Figure 6.1.

Let the transmission be made using the linear (n, k)-code. The code is used in detection mode (this may be, for example, standard CRC code), that is, during the decoding process the belonging of the received word to the code is verified, and the repeat request is made when the received word does not belong to the code. In this case the transmission over the feedback channel consists of the repeat request signal (positive acknowledgment, ACK, or negative acknowledgment, NACK or NAK).

Modulation and Coding Techniques in Wireless Communications Edited by Evgenii Krouk and Sergei Semenov

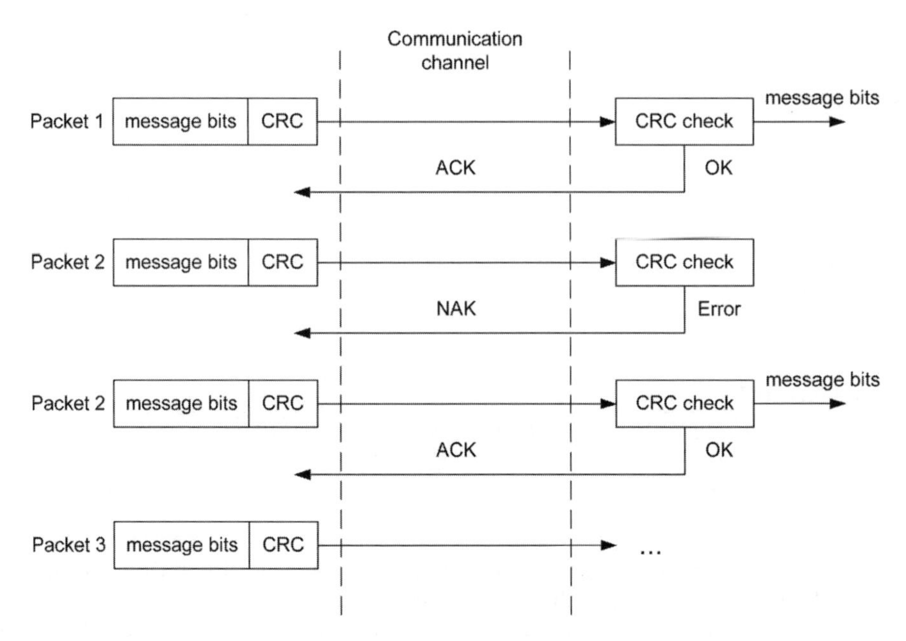

Figure 6.1 Basic ARQ scheme

Denote as P_o the probability that an error detected during the decoding process, let Q be the probability of correct receive, and let P_1 be the probability of erroneous decoding. Then it is clear that:

$$P_o + P_1 + Q = 1 \tag{6.1}$$

Since in the system with feedback the number of symbols obtained by the receiver depends on the channel properties and is a random variable, we define the system transmission rate R as the ratio of mathematical expectation of the number of information symbols obtained by the receiver to the overall number of symbols sent to the forward communication channel. After obtaining every noncorrupted codeword of length n receiver obtained k information symbols, and in the case of receiving the word which does not belong to code, the receiver obtains no information, so the transmission rate is:

$$R = \frac{k}{n}(1 - P_o) \tag{6.2}$$

The probability of error P_e in such a system is the sum of probabilities that after decoding the user will obtain the wrong codeword without repeat request, after one repeat, after second repeat and so on. This probability is:

$$P_e = P_1 + P_1 \cdot P_0 + P_1 \cdot P_0^2 + \ldots = \frac{P_1}{1 - P_0} \tag{6.3}$$

The described simplest model of the system with feedback shows the basic advantages of the feedback channel usage.

Firstly, from (6.3) one can see that with the reasonable repeat request probability $P_o \ll 1$, the error probability is defined by the value of the non-detection probability P_1. Since the detection capability of the code is much higher than the error-correcting capability, much shorter codes may be used in the system with feedback compared to systems without feedback. This in turn allows the complexity of communication system to be significantly decreased.

Secondly, it follows from (6.2) that in the system with feedback the transmission rate is changing together with the change of the detection probability. This is not a special advantage in the channel with constant parameters, when the value of P_o is constant. However, in channels with variable parameters, as is the case of all wireless channels, the mentioned above property of systems with feedback provides adaptation of communication systems to the communication channel.

Considering the simplest model, we did not take into account many effects arising in real systems with feedback. In particular, we did not consider errors occurring in feedback channel. As a rule, there are much less information transmitted over the feedback channel, compared to the forward channel, thus the error probability in the feedback channel is usually much less. Nevertheless, errors in the feedback channel are possible, that is, the situations are possible when the positive acknowledgment (ACK) of successful transmission would be considered as repeat request (NAK) and vice versa. In this case some special errors specific for systems with feedback may occur: some transmissions may be sent twice, and some may be lost.

To combat such deletions and insertions in the systems with feedback the special mechanisms are used.

The problem of forward and feedback channels synchronization also requires investigation. After receiving the erroneous word, the repeat request signal is transmitted over the feedback channel. If the duration of transmission over the feedback channel is comparable to the duration of transmission over the forward channel, then before obtaining the repeated word the receiver may obtain one or more words following the requested. There are several synchronization strategies for forward and feedback channels.

All the strategies of data transmission with repeat requests in the case of detecting the receive errors may be divided into three types:

- Stop and Wait (SAW);
- in case of error repeat last N packets (Go Back N, GBN);
- Selective Repeat (SR).

6.1.2 *Stop-and-Wait ARQ*

As has been noted, the implementation of the simplest system with repeat request involves some difficulties caused by message delay both in forward and feedback channels. During the period of transmission the repeat request signal over the feedback channel there may be some subsequent messages transmitted over the forward channel. The delivery of messages to the user in the initial order requires the buffering. Of course the buffer size must be big enough to allow the whole sequence of messages that came before the successful transmission of repeated message to be stored.

At that, if several retransmissions are needed for successful receive, then the buffer size may become quite large. Equation (6.3) is correct for infinite buffer length.

Stop and Wait (SW) method allows avoiding the buffering (or using a very small buffer size), see Figure 6.2. The idea of the method is that the transmitter is waiting for either positive acknowledgment or repeat request before starting the transmission of the next block. The complexity of Stop and Wait ARQ system implementation is minimal. The receiver should be able to detect errors, but it need not contain the buffer. If the transmitter has sent the packet, it is waiting for the receiver's answer. If the receiver detects no errors, it sends positive acknowledgment (ACK), otherwise the negative acknowledgment (NAK) is sent. In case of ACK reception the transmitter flushes the buffer containing the last sent packet. Otherwise, in case of NAK reception, the transmitter repeats the transmission of the packet containing the buffer.

Clearly, this strategy is not efficient, especially when the delay δ_{RT} of forward and feedback channels is high. Here the delay of forward and feedback channels is the sum of all time period durations between transmission of the last symbol and receiver's positive acknowledgment.

Suppose that the packets of n bits, including k information bits, are transmitted with the physical rate V (bits/s), i.e. $V\delta_{RT}$ bits will be transmitted during time period $n\delta_{RT}$. Then by analogy with (6.2) the transmission rate for Stop and Wait systems will be:

$$R_{SW} = \frac{k}{n}(1 - P_o) \cdot \frac{n}{n + V\delta_{RT}} \tag{6.4}$$

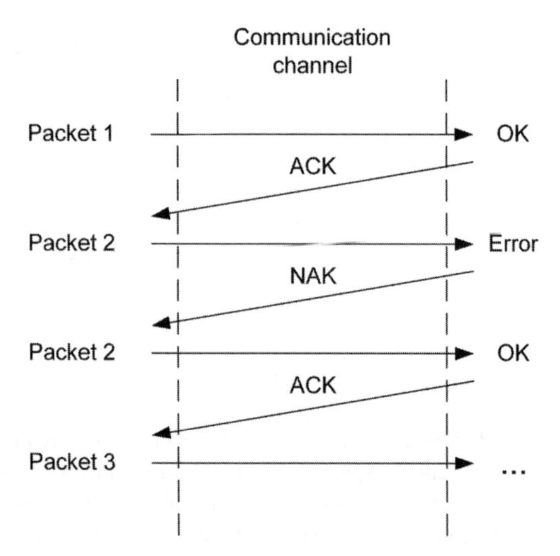

Figure 6.2 Stop-and-Wait ARQ

The value of rate R_{SW} defined by (6.4) becomes lower than that provided by the expression in (6.3) with the increase of the delay in channels and the transmission physical rate. Since the delay of the forward and feedback channels and data transmission rate are usually determined by the channel, the transmission rate in (6.4) may be somewhat improved by means of increasing the packet length. However, this leads to increasing the value P_o, which limits the increasing of n.

Besides the low implementation complexity, the advantage of systems with waiting is low cost of service information which is needed for every transmission. There is no need for packets numbering, since the packets order is inherently preserved. Unfortunately, reliability of transmission of the one bit of acknowledgment is not as simple as it seems. Moreover, due to the protocol's nature it is impossible to collect some determined number of acknowledgments.

The effectiveness of systems with waiting may be significantly improved by means of organization of the system with separated virtual channels.

Multichannel ARQ protocol with waiting. In multichannel ARQ with waiting (MC-SW-ARQ) N Stop and Wait systems are used, see Figure 6.3. Each system is functioning on its own virtual channel, which are independent from each other. For example, the channel separation by time, frequency or code is used. The packets in multichannel ARQ are uniformly distributed over N channels. If the number of channels is selected as:

$$N \geq \left\lceil \frac{V \cdot \delta_{RT}}{k} \right\rceil \tag{6.5}$$

then during the repeat request period in each channel there will be no new messages.

Thus, the rate achieved in multichannel Stop and Wait system is determined by expression (6.3). The main problem in using the multichannel system is the necessity of establishing the correct order of packets incoming. To achieve this, the buffer, intended for packets reordering is still needed at the receive side.

There are different strategies for packets ordering and reordering when distributing packets over the parallel channels. For example, the fixed procedure may be used, when the i-th packet is directed to the channel with number $i \mod N$. Surely, reordering device should always have the information about

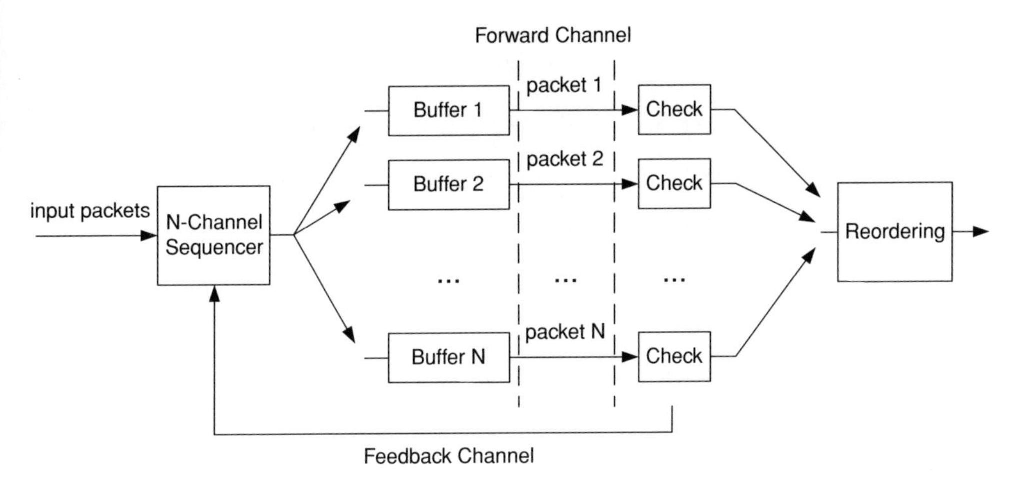

Figure 6.3 Multichannel SW-ARQ

packets distribution at the transmitter side. Differences in ordering techniques affect the amount of memory required by the system and the delay during the messages transmission.

6.1.3 *ARQ with* N *Steps Back (Go Back* N*, GBN)*

The mechanism of Go Back N (GBN) ARQ system does not require confirmation of every data block: the transmission is performed continuously. But if the transmitter receives the message about the lost or erroneous packet, it repeats the transmission starting from the corrupted packet. In this case even successfully transmitted packets are sent repeatedly. Such mechanism is convenient in that it does not require buffering and storing the data at the receiving side, but in case of error the channel load significantly increases. Because of the one erroneous packet correctly received data may be retransmitted many times, see Figure 6.4.

In GBN ARQ system the packets are numbered and sent to channel continuously, that is, the next packet is transmitted before the confirmation of successful receiving of the previous packet is obtained. If the receiver detects an error, the number of erroneous packet is sent back as the signal for repeat transmission. As a result the transmitter goes back to send an erroneous packet and every subsequent packet. If there are buffers for N packets at the transmitter and receiver, then the transmission in forward channel is continuous, if the condition in (6.5) is satisfied.

In the considered scheme the feedback communication system may be in two states:

1. Error is not detected; the probability of this state is $1 - P_o$, transmission rate is $\frac{k}{n}$;
2. Error is detected; the probability of this state is P_o, transmission rate is $\frac{k}{n} \cdot \frac{1}{N}$, since N packets are transmitted for the transmission of one packet. Then, assuming independent packets transmission, the transmission rate in GBN-ARQ scheme is:

$$R_{GBN} = \frac{k}{n} \cdot \frac{1 - P_o}{1 - P_o + P_o \cdot N} = \frac{k}{n} \cdot \frac{1 - P_o}{1 + P_o \cdot (N - 1)} \tag{6.6}$$

It is clear that the GBN-ARQ scheme is not optimal, since some packets will be retransmitted even in case of successful reception. This drawback is addressed in the ARQ scheme with selective repeat.

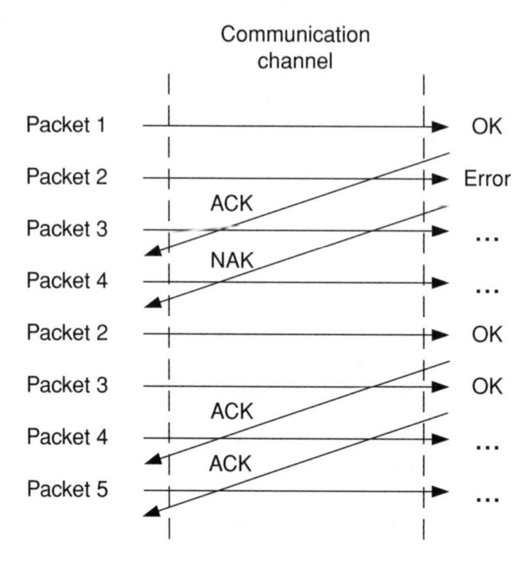

Figure 6.4 Go Back N ARQ

6.1.4 ARQ with Selective Repeat (SR)

When transmitting messages with high rates and on large distances, the value of N in (6.5) may turn out to be very large; this leads to significant decrease of the transmission rate (6.6). The effectiveness of communication may be improved by means of selective repeat request.

In the ARQ scheme with *Selective Repeat (SR)* all the packets have their sequence number, and the receiver is equipped with buffer. Thus only packets which are received with errors should be retransmitted, see Figure 6.5.

The feedback signal leads to retransmission of only one packet, so the transmission rate is determined by (6.3), which makes this scheme preferable compared to SW and GBN systems.

Theoretically, under some conditions the multichannel SAW-ARQ and SR-ARQ may have the same transmission rate. However, in MC-SAW-ARQ the costs of service information (packet numbers,

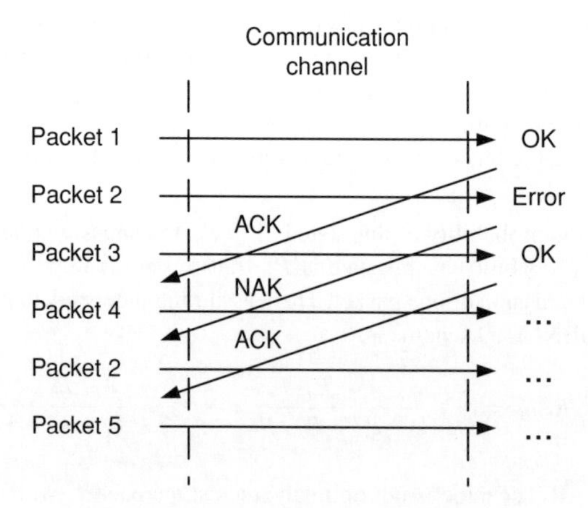

Figure 6.5 Selective Repeat ARQ

acknowledgments, etc.) may not be significant. This is explained by the structure of virtual channels, which guarantees the preserving of the packets order.

In ARQ with selective repeat the persistent failure in transmission of one of the previous packets would prevent the transmission of subsequent packets. In this situation the channel is used ineffectively, since there will be no data sending until the blocking packet has been successfully transmitted.

On the contrary, the persistent denial in transmission of one block in MC-SW-ARQ system does not affect the transmissions being performed in parallel. Surely, the delay in multichannel system will also increase, since the delay in particular channels is possible. However, delayed packets affect the transmission only in their channels, which decreases the rate much less than in SR-ARQ.

The mentioned drawbacks of SR-ARQ comparing to MC-SW-ARQ do not mean the definite comparison result of these ARQ schemes in favour of multichannel systems, since in practice the organization of a multichannel scheme is not always possible.

The SR algorithm allows the transmission of only those packets which are corrupted or lost to be repeated. But in this case the transmitter should store some number of the last sent packets. Nevertheless, since this method is the most efficient regarding the channel resources, it is the basic technique in wireless telecommunications systems.

6.2 Hybrid ARQ

The classical ARQ scheme described in Section 6.1 may not be very efficient in practice, when the noise level is relatively large (when functioning in the area of low- or medium-SNR). In this case the packet transmission would always be erroneous, so the receiver should always request for retransmission, which greatly reduces the transmission throughput or even makes it impractical.

The idea of the Hybrid ARQ (HARQ) scheme is to combine the error-detecting ARQ scheme with the error-correcting FEC scheme. To achieve this, prior to the transmission the message is encoded by some error-correcting code. At the receiving side the decoding attempt is made, after which the ARQ scheme of the receiver should make the decision whether the decoding was successful or not, depending on this decision the ACK or NAK signal is transmitted over the feedback channel.

The detection properties for the ARQ scheme may be provided either by special ARQ detecting procedures like CRC computation or by capabilities of the error-correcting code itself. For example, symbol-by-symbol iterative decoding for some classes of codes (convolutional codes, turbo-codes, LDPC codes) in case of decoding error may lead not to the wrong codeword, but to some word which does not belong to the code (and usually contains less errors than the received word, that is, which is "closer" to the correct codeword than the received word), and this situation may be easily detected.

In the following sections we consider the basic variants of Hybrid ARQ schemes.

6.2.1 Type-I Hybrid ARQ (Chase Combining)

The most straightforward technique of combining the ARQ and FEC is called *Type-I Hybrid ARQ* [3, 4, 14] and is shown in Figure 6.6.

The transmitted packet consists of the message (information) part, and the correspondent redundant parity bits calculated by means of some error-correcting code. After the packet being transmitted and the symbols are demodulated (usually providing soft-decisions of received signals or their log-likelihood ratios, LLRs), they are fed into a FEC decoder. The decoder's output should be checked by some means to determine whether a decoding error occurred. As mentioned at the beginning of Section 6.2, this may be achieved by applying the CRC check, or in some cases this may be directly followed from the syndrome of the decoder's output.

In the case of successful decoding the decoded message bits are sent to the user, the ACK signal is sent over the feedback channel, and the next packet is prepared for transmission. In the case of a decoder's failure, the NAK signal is sent to the transmitter, and the encoded packet is repeated.

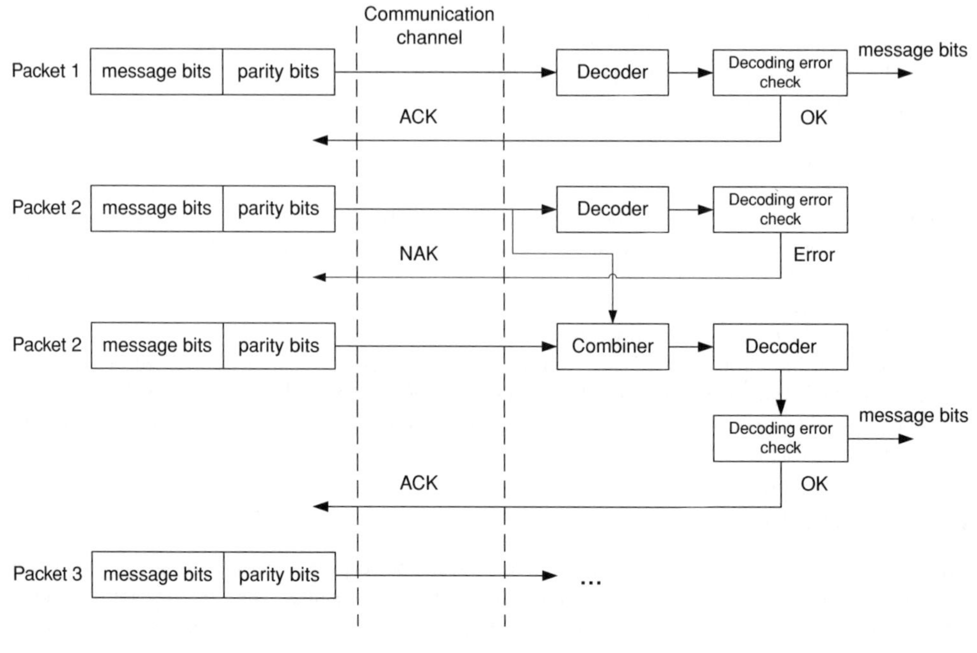

Figure 6.6 Type-I HARQ

When the receiver receives the second response of the same packet (this is the case for Packet 2 in Figure 6.6), it may decode the packet independently on the previously received version (in this case there is no need for a buffer at the receiving side), or the receiver may combine the first received packet with the second one. Usually such combining is made by means of *Chase combining* [2], when the LLRs (soft demodulated decisions) of the correspondent bits from two received copies of transmitted packet are simply added to provide better decoding quality. Due to Chase combining procedure, the Type-I HARQ is sometimes called CC-HARQ. In fact such combining is analogous to diversity combining using MRC scheme, when two received copies of the same transmitted data are combined to provide better signal-to-noise ratio for the received symbols, see Chapter 8.

The main drawback of this HARQ scheme is the low throughput when functioning in the area of relatively large SNRs, when the decoding errors sometimes occurred, but the number of errors at the decoder's output is low and the retransmission of the whole packet is redundant and leads to throughput decreasing.

6.2.2 *Type-II Hybrid ARQ (Full IR)*

The idea of *Type-II HARQ* systems is to use the variable level of redundant bits to ensure the correction of errors at the receiving side using no more parity-check bits than is actually required in current channel conditions [7, 11, 12]. This scheme is depicted in Figure 6.7.

In fact, to ensure the Type-II HARQ scheme, the set of T "nested" error-correcting codes is required. For the fixed number of k information symbols, for the information word m different sequences of parity-check symbols $r_1, r_2, \ldots, r_T$ are calculated, thus providing the set of possible rates $R_1 \geq R_2 \geq \ldots \geq R_T$. The decoding procedure should be possible for any rate from this set.

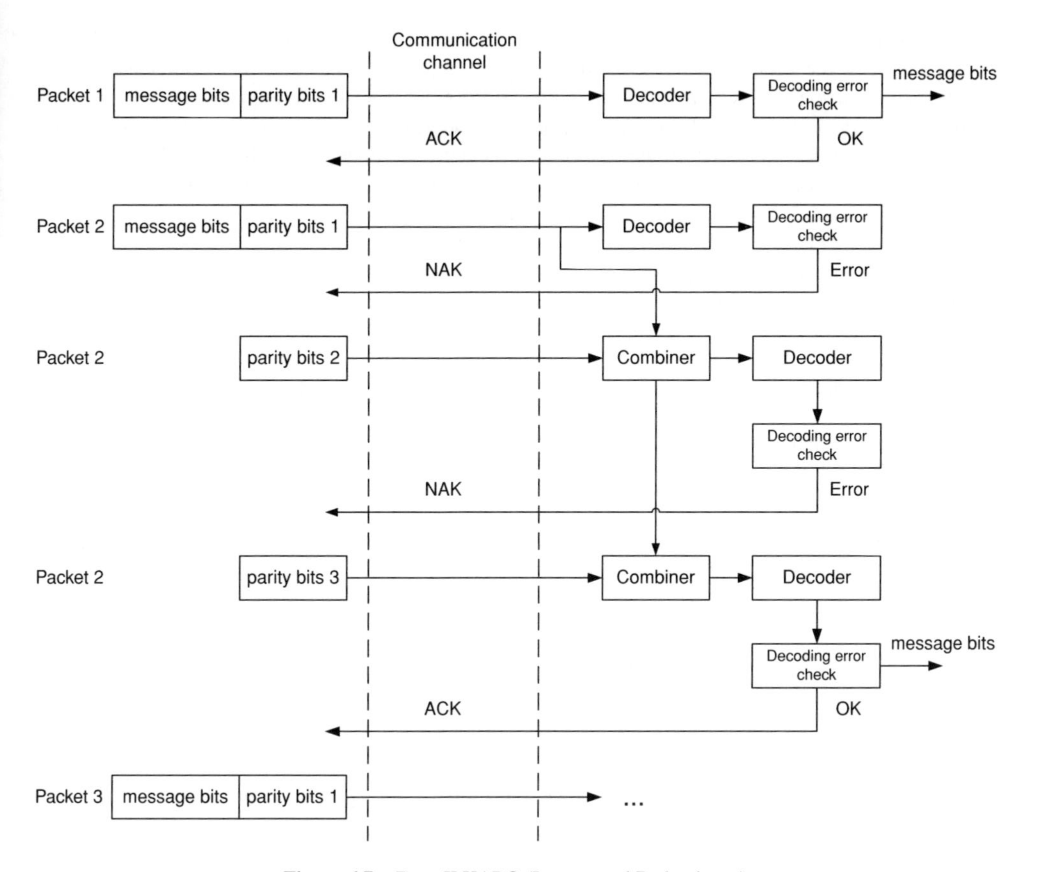

Figure 6.7 Type-II HARQ (Incremental Redundancy)

Then, during the first transmission attempt, only the message bits m and the parity bits r_1 are sent, that is, the code with rate R_1 is used. This code is decoded at the receiver side, and in the case of success the ACK is sent over the feedback channel. Otherwise, the NAK is sent, and during the second attempt only the parity-check bits r_2 are sent (without repeating the message). At the receiver, these parity-check bits are combined with m and r_1, thus obtaining the code with rate $R_2 \leq R_1$ and higher error-correcting capability (the combination and decoding procedures here depend on the type of codes being used). If necessary, during the subsequent retransmissions only the extra (new) redundant bits are sent. So, this scheme is often called the *Incremental Redundancy (IR) HARQ*.

In fact, this scheme provides adaptation of the code rate (and corresponding error-correcting power) to the channel conditions: for better channel state only the small portion of redundant bits are used, and the transmission rate is supported at high level, but if the channel state becomes worse, extra redundant bits are sent and the errors are corrected by cost of rate decreasing.

The most common approaches to construct such rate-compatible families of codes use the codes on trellis (convolutional codes or turbo-codes) [6, 13, 15], or LDPC codes [5, 17, 20]. For these codes it is possible to "separate" the overall redundant bits into portions and to decode the message using only part of the initial code's redundancy.

Let us give an example of throughput estimation for Type-II HARQ scheme for some basic scenario [7, 9]. We assume that the rate-compatible family of convolutional codes for Type-II HARQ transmission is used for BSC or AWGN channel with BPSK or QPSK. The information sequence of k bits is first encoded by the (n, k)-code for error detection, where $n = k + r$. For error correction after the i-th transmission the i-th $(2 + i, 1, m)$-convolutional code C_i from the family is used, where $i = 0$ corresponds to the initial rate-1/2 code with memory n_m and generator polynomials $[G_1(x), G_2(x)]$. The information word m is encoded by the error-detection code into codeword c, let the polynomial $c(x)$ is correspondent to this codeword, then at each subsequent transmission either $c(x)G_1(x)$ or $c(x)G_2(x)$ are sent, each containing $n_0 = n + n_m$ bits. Then, if the decoding attempts for $c(x)G_1(x)$ and $c(x)G_2(x)$ fail, the $c(x)G_1(x)$ is sent again and combined with previously received sequences for decoding with $(3, 1, m)$-code with generator polynomial $[G_1(x), G_2(x), G_1(x)]$. If this decoding is also in error, then the $c(x)G_2(x)$ is sent again and combined with previously received sequences for decoding with $(4, 1, m)$-code with generator polynomial $[G_1(x), G_2(x), G_1(x), G_2(x)]$, and so on. So, the i-th code C_i has the rate $1/(2 + i)$, $i = 1, 2, \dots$ [7].

Suppose that the SR-ARQ scheme is used (Section 6.1.4). Let R_0 be the event {received packet is error-free}, R_u be the event {the received packet contains undetected error} and R_{det} corresponds to {the received packet contains detected error}. Clearly:

$$\Pr\{R_{det}\} = 1 - \Pr\{R_0\} - \Pr\{R_u\} \tag{6.7}$$

If the codeword contains $n_0 = n + n_m = (k + r) + n_m$ bits, where k is the number of information bits, r is the number of parity-check bits and n_m is the number of tail bits for trellis termination, then:

$$\Pr\{R_0\} = (1 - p)^{n_0} \tag{6.8}$$

where p is the bit error probability for BSC channel, or:

$$p = Q\left(\sqrt{\frac{2E_b}{N_0}}\right)$$

for AWGN channel (see (2.73)), where E_b is the transmitted energy per bit, N_0 is the noise power, and Q-function is defined in (2.63):

$$Q(x) = \frac{1}{\sqrt{2\pi}} \int_x^{\infty} e^{\frac{-y^2}{2}} dy$$

If we assume that the probability of non-detected error in channel $\Pr\{P_u\}$ is negligible, then from (6.7) and (6.8) we have:

$$\Pr\{R_{det}\} \approx 1 - \Pr\{R_0\} = 1 - (1 - p)^{n_0}$$

Now consider the sequences obtained after the Viterbi decoding procedure after i-th transmission. Let's define another set of events $D_0^{(i)}$, $D_u^{(i)}$ and $D_{det}^{(i)}$ as {decoded sequence contains no errors}, {decoded sequence contains undetected errors}, and {decoded sequence contains detected errors} correspondingly. Again we assume that $\Pr\{D_u^{(i)}\}$ is negligible. The probability of correct decoding may be limited as:

$$\Pr\{D_0^{(i)}\} \geq (1 - \Pr\{E^{(i)}\})^n$$

where $E^{(i)}$ is decoding error event for Viterbi decoding. Then the probability of detected errors after decoding is:

$$\Pr\{D_{det}^{(i)}\} \approx 1 - \Pr\{D_0^{(i)}\} \geq 1 - (1 - \Pr\{E^{(i)}\})^n \tag{6.9}$$

The value of $\Pr\{E^{(i)}\}$ may be bounded as [19]:

$$\Pr\{E^{(i)}\} \leq \sum_{d=d_{free}^{(i)}}^{\infty} a_d^{(i)} P_d$$

where $d_{free}^{(i)}$ is the free distance of C_i, $a_d^{(i)}$ is distance spectra of C_i, and P_d is the probability of selection the wrong path at distance d. For the BSC channel:

$$P_d = \begin{cases} \sum_{j=(d+1)/2}^{d} \binom{d}{j} p^j (1-p)^{d-j} & d \text{ odd}, \\ \sum_{j=d/2+1}^{d} \binom{d}{j} p^j (1-p)^{d-j} + \frac{1}{2}\binom{d}{d/2} p^{d/2}(1-p)^{d/2} & d \text{ even} \end{cases}$$

and for AWGN channel:

$$P_d = Q\left(\sqrt{\frac{2dE_b}{N_0}}\right)$$

The average number of transmissions $\bar{N}$ is [7]:

$$\begin{aligned} \bar{N} = 1 + \Pr\{R_{det}\} + \Pr\left\{R_{det}, R_{det}, D_{det}^{(1)}\right\} + \Pr\left\{R_{det}, R_{det}, D_{det}^{(1)}, R_{det}, D_{det}^{(2)}\right\} + \\ + \Pr\left\{R_{det}, R_{det}, D_{det}^{(1)}, R_{det}, D_{det}^{(2)}, R_{det}, D_{det}^{(3)}\right\} + \ldots \\ \ldots + \Pr\left\{R_{det}, R_{det}, D_{det}^{(1)}, R_{det}, D_{det}^{(2)}, R_{det}, D_{det}^{(3)}, \ldots, R_{det}, D_{det}^{(i)}\right\} + \ldots \end{aligned}$$

This value may be bounded as:

$$1 + \Pr\{R_{det}\} + \sum_{i=1}^{\infty} \Pr\{R_{det}\}^{i+1} \prod_{j=1}^{i} \Pr\left\{D_{det}^{(j)}\right\} \leq \bar{N} \leq 1 + \Pr\{R_{det}\} + \sum_{i=1}^{\infty} \Pr\left\{D_{det}^{(i)}\right\} \quad (6.10)$$

In fact, the lower and upper bounds on $\bar{N}$ in (6.10) are approximately the same, so we may approximate this value as:

$$\bar{N} \approx 1 + \Pr\{R_{det}\} + \sum_{i=1}^{\infty} \Pr\left\{D_{det}^{(i)}\right\}$$

The throughput η is defined as:

$$\eta = \frac{R}{\bar{N}}$$

where R is code rate, which is in our case $R = k/n_0$, so we have:

$$\eta = \frac{R}{\bar{N}} \approx \left(\frac{k}{k+r+n_m}\right) \cdot \left(\frac{1}{1 + \Pr\{R_{det}\} + \sum_{i=1}^{\infty} \Pr\{D_{det}^{(i)}\}}\right)$$

The drawback of the Incremental Redundancy scheme is the case when the initial transmission containing the message bits fall under very bad channel conditions, and the number of errors occurred is very high. First, in this case extra parity bits may not be helpful, and second, during subsequent transmissions there is no new information about the message part itself, which would allow combining as in Type-I HARQ. This problem is partly overcome in Type-III HARQ.

6.2.3 Type-III Hybrid ARQ (Partial IR)

The main property of *Type-III HARQ* system [1, 8] (which is also sometimes referred to as Partial Incremental Redundancy) is that all transmitted packets are self-decodable. This scheme is shown in Figure 6.8.

With this type of HARQ transmission, all the transmitted packets contain the message part, and in all cases (including retransmission) the decoding attempt of the receiving packet is first made. If the attempt fails, then NAK is sent over the feedback channel (if no retransmissions of this packet are available), or

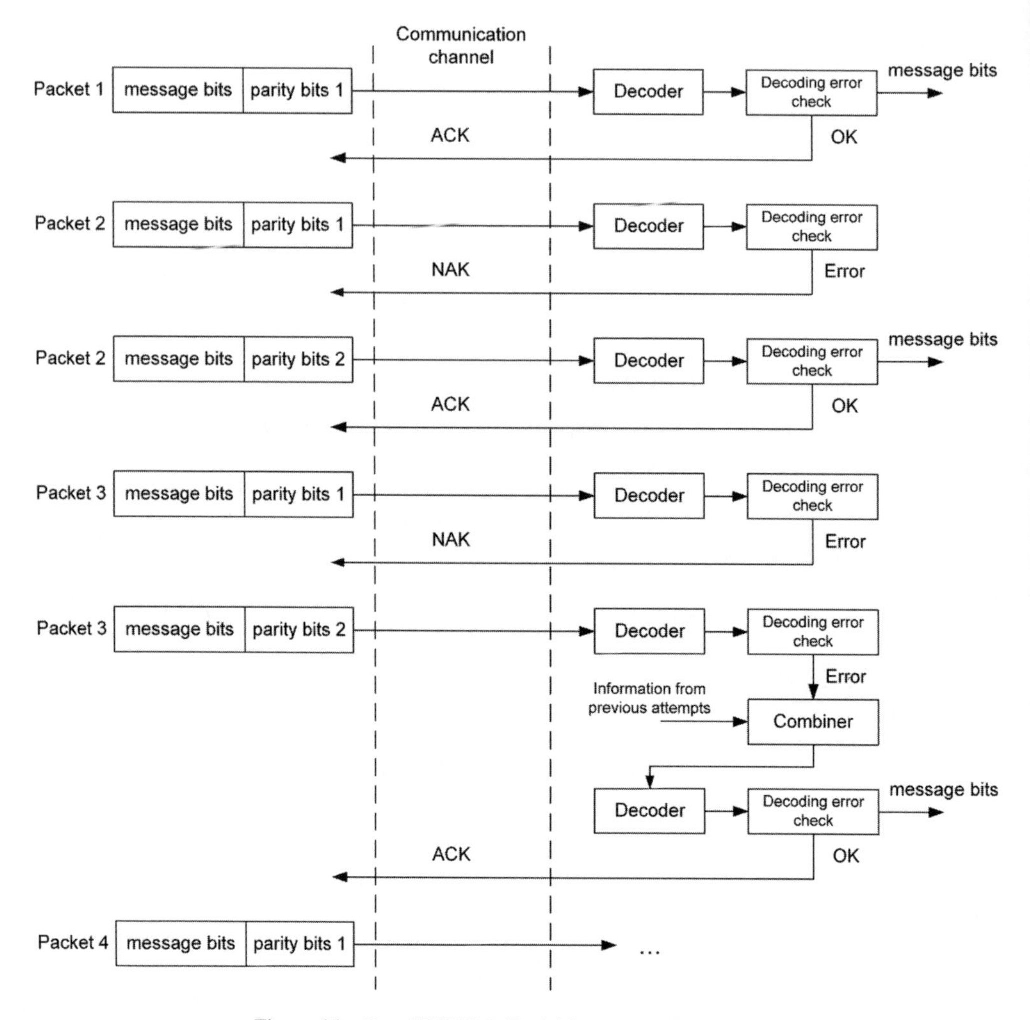

Figure 6.8 Type-III HARQ (Partial Incremental Redundancy)

the packet is consequently combined with the previously received versions and new decoding attempts are made. If all the decoding attempts are failed, the NAK is sent.

In some sense, this type of transmission may be considered as a version of Type-I HARQ, when all the repeated packets contain the same message part, but with different parity-checks, and in some sense, since the parity-checks are changed, this scheme may be considered as Incremental Redundancy scheme.

Since all the packets should transfer information part and parity-check part, there is no rate adaptation to the variable channel conditions in Type-III scheme as compared to Type-II scheme. In the situation when the noise level is relatively high, Type-III HARQ may overcome the Type-II HARQ systems with full Incremental Redundancy, but with the increasing of SNR it become worse in terms of throughput since there is no need for multiple transmission of the same information part.

Again, we will estimate the throughput of the Type-III HARQ scheme for the Gaussian channel using the same family of convolutional codes as in Section 6.2.2 [8]. In this case, q puncturing patterns P_i, $i = 1, 2, \ldots, q$ are used to obtain the codes C_i of rate $R = b/V$ from the original convolutional code of rate $1/V_0$. The combination of j such codes $C^{(j)}$ is obtained using the puncturing matrix $P^{(j)} = P_1 + P_2 + \ldots + P_j$. Each packet of $n_0 = k + r + n_m$ bits, where k is information length, r is the number of redundant bits for error detection and n_m is the memory of the encoder, is encoded with the rate-$1/V_0$ code. Then the coded bits from $C_1, C_2, \ldots$ are sent (until ACK is obtained from the feedback channel) and decoded at the receiver side using the patterns P_1 (after the first transmission), P_2 and, in case of error, $P^{(2)}$ (after the second transmission), and so on. At step $i > 1$, first the decoding attempt for code C_i with pattern P_i is made, and then, in case of error, the combined code $C^{(i)}$ with pattern $P^{(i)}$ is applied. If after q attempts decoding is failed, then the received sequence for C_1 is discarded and the procedure continues with sending the bits from C_1 again, and so on.

Let $F^{(i)}$ be the event {decoding error after i transmissions}, and $D^{(i)}_{det}$ be the event {error detected after decoding in $C^{(i)}$}. Then the average number of transmissions $\bar{N}$ is:

$$\begin{aligned}\bar{N} &= 1 + \Pr\{F^{(1)}\} + \Pr\{F^{(1)}, F^{(2)}\} + \\ &+ \Pr\{F^{(1)}, F^{(2)}, F^{(3)}\} + \ldots \\ &\ldots \Pr\{F^{(1)}, F^{(2)}, F^{(3)}, \ldots, F^{(i)}\} + \ldots\end{aligned} \tag{6.11}$$

Event $F^{(i)}$ is equivalent to joint event $\{D^{(1)}_{det}, D^{(i)}_{det}\}$ for $i < q$ and to joint event $\{D^{(1)}_{det}, D^{(q)}_{det}\}$ for $i \geq q$. Each term in (6.11) may be bounded as:

$$\Pr\{F^{(1)}, \ldots, F^{(i)}\} \leq \begin{cases} \Pr\{D^{(i)}_{det}\}, & i \leq q \\ \Pr\{D^{(q)}_{det}\}^j, & i = jq,\ j = 1, 2, \ldots \\ \Pr\{D^{(q)}_{det}\}^j \Pr\{D^{(i-jq)}_{det}\}, & jq < i < (j+1)q,\ j = 1, 2, \ldots \end{cases} \tag{6.12}$$

From (6.11) and (6.12) obtain:

$$\bar{N} \leq \frac{1 + \sum_{i=1}^{q-1} \Pr\{D^{(i)}_{det}\}}{1 - \Pr\{D^{(q)}_{det}\}}$$

where the value of $\Pr\{D^{(i)}_{det}\}$ may be estimated as in (6.9). Finally, the estimation of throughput for the considered example is:

$$\eta = \frac{R}{\bar{N}} \geq \frac{R(1 - \Pr\{D^{(q)}_{det}\})}{1 + \sum_{i=1}^{q-1} \Pr\{D^{(i)}_{det}\}}$$

References

[1] 3GPP TSG RAN WG1#2 Tdoc R1-99061, *"Hybrid ARQ techniques for efficient support of packet data"*, Panasonic, Japan, Feb. 1999. Available: http://www.3gpp.org/ ftp/ tsg_ran/ WG1_RL1/ TSGR1_02/ Docs/ pdfs/ R1-99061.pdf

[2] Chase, D. *Code combining. A maximum likelihood decoding approach for combining an arbitrary number of noisy packets*, IEEE Transactions on Communications, vol. 33, pp. 385–393, May 1985.

[3] Chockalingam, A., Zorzi, M., Tralli, V. *Wireless TCP performance with link layer FEC/ARQ*, Proceedings, IEEE ICC'1999, vol. 2, pp. 1212–1216, Canada, 1999.

[4] Deng, R. H. *Hybrid ARQ scheme using TCM and code combining*, IEE Electronics Letters, vol. 27, no. 10, pp. 866–868, May 1991.

[5] Ha, J., Kim, J., McLaughlin, S. W. *Rate-compatible puncturing of low-density parity-check codes*, IEEE Trans. on Information Theory, vol. 50, no. 1, pp. 2824–2836, 2004.

[6] Hagenauer, J. *Rate Compatible punctured convolutional codes (RCPC) and their applications*, IEEE Transactions on Communications, vol. 36, no. 4, pp. 389–400, 1988.

[7] Kallel, S. *Analysis of a type-II Hybrid ARQ scheme with code combining*, IEEE Transactions on Communications, vol. 38, no. 8, pp. 1133–1137, Aug. 1990.

[8] Kallel, S. *Complementary punctured convolutional (CPC) codes and their applications*, IEEE Transactions on Communications, vol. 43, no. 6, pp. 2005–2009, June 1995.
[9] Kallel, S, Haccoun, D. *Generalized type II Hybrid ARQ scheme using punctured convolutional coding*, IEEE Transactions on Communications, vol. 38, no. 11, pp. 1938–1946, 1990.
[10] Lin, S., Costello, D. J. *Error control coding*, second ed., Prentice Hall, Englewood Cliffs, NJ, 2004.
[11] Lin, S., Yu, P. S. *A hybrid ARQ scheme with parity retranmission for error control of satellite channels*, IEEE Transactions on Communications, vol. COM-30, no. 7, pp. 1701–1719, 1982.
[12] Mandelbaum, D. M. *An adaptive-feedback coding scheme using incremental redundancy*, IEEE Trans. on Information Theory, vol. 20, pp. 388–389, 1974
[13] Rowitch, D., Milstein, L. *On the performance of hybrid FEC/ARQ systems using rate compatible punctured turbo (RCPT) codes*, IEEE Transactions on Communications, vol. 48, pp. 948–959, 2000.
[14] Schmitt, M. P. *Hybrid ARQ scheme employing TCM and packet combining*, IEE Electronics Letters, vol. 34, no. 18, pp. 1725–1726, September 1998.
[15] Schmitt, M. P. *Improved retransmission strategy for hybrid ARQ schemes employing TCM*, Proceedings, IEEE Wireless Communications and Networking Conference, vol. 3, pp. 1226–1228, September 1999.
[16] Schmitt, M. P. *ARQ systems for wireless communications*, Ph.D. Thesis, Technical University Darmstadt, Darmstadt, 2002.
[17] Sesia, G. C. S., Vivier, G. *Incremental Redundancy Hybrid ARQ Schemes Based on Low-Density Parity-Check Codes*, IEEE Transactions on Communications, vol. 52, no. 8, pp. 1311–1321, Aug. 2004.
[18] Stender, B. *Signaling Aspects for Wireless Communications*, Ph.D. Thesis, University of Ulm, Germany, 2007
[19] Viterbi, A. J. *Convolutional codes and their performance in communication systems*, IEEE Transactions on Communications, vol. COM-19, Oct. 1971.
[20] Yazdani, M., Banihashemi, A. *On construction of rate-compatible low-density parity-check codes*, IEEE Communications Letters, 8 (3), 2004.

7

Coded Modulation

Andrey Trofimov
St. Petersburg State University of Aerospace Instrumentation, Russia

7.1 Principle of Coded Modulation

For uncoded M-ary transmission using signals, belonging to a signal set, or signal alphabet S, $M = |S|$, $M = 2^m$, m is integer, data transmission rate is equal to:

$$R_u = \frac{\log_2 M}{T}, \quad \text{bit/s} \tag{7.1}$$

where T is signalling interval. Hereafter the narrowband signalling is assumed, for example, PSK and QAM as most important cases. For such signal sets the bandwidth W used for transmission does not depend on number of signals M in signal set S, and it can be evaluated as:

$$W = \frac{\alpha}{T} \tag{7.2}$$

where α is a pulse shape dependent constant. In particular, for signals of duration T with a constant envelope $\alpha = 2$, for signals with envelope of sinc($\cdot$), transmitted with period T, $\alpha = 1$. In further consideration we assume that the pulse shape is chosen and remains unchangeable, hence the value of α is constant for all cases of uncoded and coded systems considered in this chapter.

Ratio of transmission rate to bandwidth is called *spectral efficiency*. For uncoded transmission the spectral efficiency is equal to:

$$\eta_u = \frac{R_u}{W}, \quad \text{bit/(s} \cdot \text{Hz)} \tag{7.3}$$

and as it follows from (7.1) and (7.2):

$$\eta_u = \frac{\log_2 M}{\alpha}, \quad \text{bit/(s} \cdot \text{Hz)} \tag{7.4}$$

For coded data transmission the structure of the transmitting part of the system can be presented (with a minor loss of generality negligible for practice) as it is shown in Figure 7.1

Modulation and Coding Techniques in Wireless Communications Edited by Evgenii Krouk and Sergei Semenov

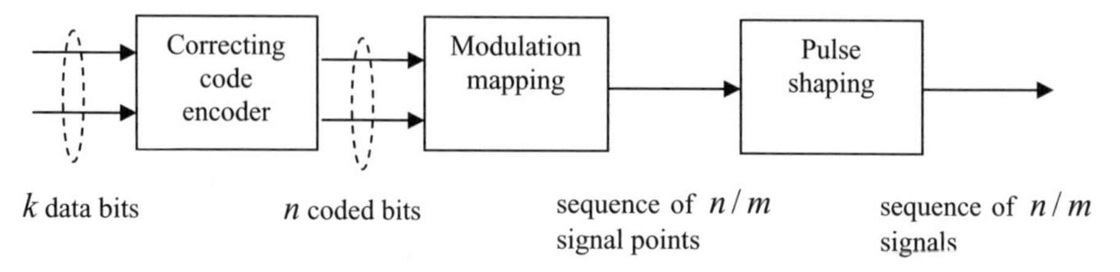

Figure 7.1 General structure of transmitting part of coded system

Let us denote input data block at time t as $\mathbf{u}^{(t)} = (u_0^{(t)}, \ldots, u_{k-1}^{(t)}) \in \{0, 1\}^k$, and coded block at time t as $\mathbf{v}^{(t)} = (v_0^{(t)}, \ldots, v_{n-1}^{(t)}) \in \{0, 1\}^n$. Coded block $\mathbf{v}^{(t)}$ before mapping is divided into (n/m) sub-blocks [1] of length m bits as follows $\mathbf{v}^{(t)} = (\mathbf{v}_0^{(t)}, \ldots, \mathbf{v}_{n/m-1}^{(t)})$, where $\mathbf{v}_l^{(t)} \in \{0, 1\}^m$, $l = 0, 1, \ldots, n/m - 1$. Modulation mapping $\mathrm{M} : \{0, 1\}^m \to S$ can be defined as $\mathbf{s}^{(t)} = s(\mathbf{v}^{(t)})$, where $\mathbf{s}^{(t)} \in S$ [2]. Since the pulse shaping is assumed to be fixed the *set of signal points* (*signal constellation*) and *signal set* are isomorphic. From here on we assume the set S as subset of D-dimensional real Euclidean space, that is, $S \subset \mathrm{R}^D$. For most important in practice examples of PSK and QAM signal space dimension $D = 2$.

Transmission rate R_c for coded transmission can be computed as:

$$R_c = \frac{k}{(n/m)T} = \frac{R \log_2 M}{T}, \quad \text{bit/s} \tag{7.5}$$

where R is the correcting code rate, $R = k/n$. The spectral efficiency η_c for coded transmission can be expressed as:

$$\eta_c = \frac{R_c}{W} = \frac{R \log_2 M}{\alpha}, \quad \text{bit/(s} \cdot \text{Hz)} \tag{7.6}$$

It immediately follows from comparison between (7.6) and (7.4) that $\eta_c = \eta_u$ if $R = 1$, i.e. for nonredundant transmission. It also follows from (7.6) and (7.4) that error correcting coding decreases the spectral efficiency $1/R$ times. In practice that means decreasing data transmission rate and/or increasing the bandwidth. Positive outcome of error correcting coding is an improved reliability of data transmission. For years the trade-off "spectral efficiency – reliability" was considered as an unavoidable feature of error correcting coding.

In [1] G. Ungerboeck has suggested a scheme of *combined design* of error correcting code and modulation mapping of coded subblocks into an *extended signal set*. This combination of coding and modulation does not cause the spectral efficiency to decrease, but it has an improved performance in comparison with uncoded transmission. In other words, the coding in such cases does not cause either bandwidth extension, or slowing down of the transmission rate. This effect is achieved by using an *extended signal set* (extended signal constellation).

Let S_c be the extended signal set with M_c signal points. $M_c = |S_c| > M$. In this case (7.6) can be written down as:

$$\eta_c = \frac{R \log_s M_c}{\alpha}, \quad \text{bit/(s} \cdot \text{Hz)} \tag{7.7}$$

Equating the right hand parts of (7.7) and (7.4) we get the condition for rate R of the error correcting code:

$$R = \frac{\log_2 M}{\log_2 M_c} \tag{7.8}$$

[1] We assume that m is divisor of n.

[2] Distinguish Greek M for mapping and italic M for cardinality of signal set.

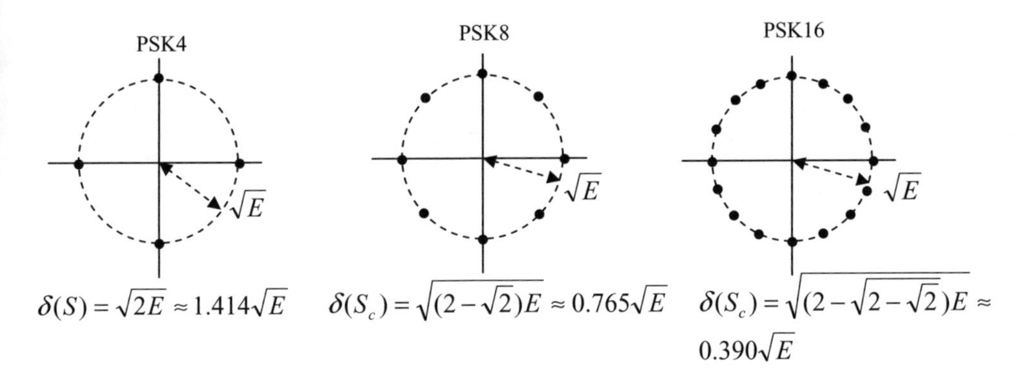

Figure 7.2 Signal constellations for PSK4 (basic signal set) and PSK8 and PSK16 (extended signal sets)

which keeps the spectral efficiency unchangeable in comparison with uncoded transmission. For example, if for uncoded transmission $M = 4$, and for coded transmission $M_c = 8$, then for code with rate $R = \log_2 M / \log_2 M_c = 2/3$ there is no loss of spectral efficiency. If the signal alphabet is extended up to $M_c = 16$ signal points then the code with $R = 1/2$ can be used.

From general principles of coding theory it follows that decreasing of code rate leads to increasing of correcting capability. On the other hand, minimum distance in signal set diminishes with increasing the number of signal points and keeping energy (or average energy) of the signal constant.

Let $d(x, y)$ be the Euclidean distance between any points $x, y \in \mathrm{R}^{\mathrm{D}}$. Let us introduce minimum distance δ in signal set S as:

$$\delta(S) = \min_{x \neq y, x, y, \in S} d(x, y)$$

Examples of signal constellations and their minimum distances for PSK4, PSK8 and PSK16 are shown in Figure 7.2. Signal energy, or Euclidean norm of signal points is denoted as E.

Examples of signal constellations and their minimum distances for QAM4, QAM8 and QAM16 are given in Figure 7.3. Here $\overline{E}$ is average signal energy, or average squared Euclidean norm of signal points.

It follows from Figure 7.2 that the minimum distance decreases 1.848 times and 3.624 times with increasing the signal points in PSK signal constellation from 4 to 8 and to 16 respectively. For QAM constellation, see Figure 7.3, the corresponding factors are 1.158 and 2.236. Nevertheless, it is possible

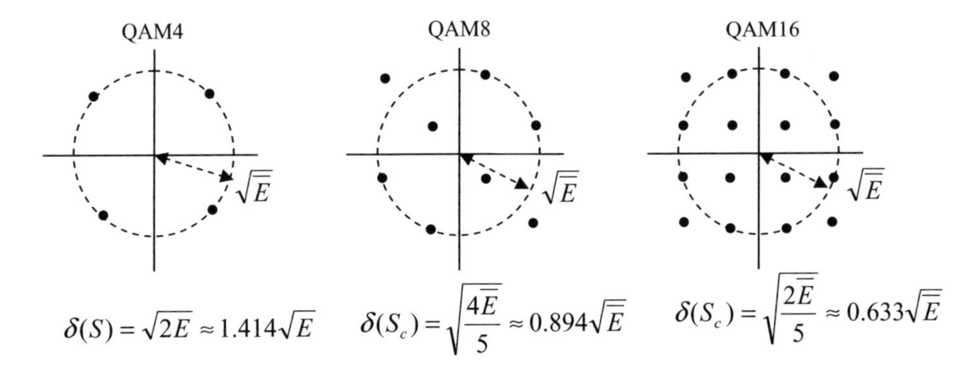

Figure 7.3 Signal constellations for QAM4 (basic signal set) and QAM8 and QAM16 (extended signal sets)

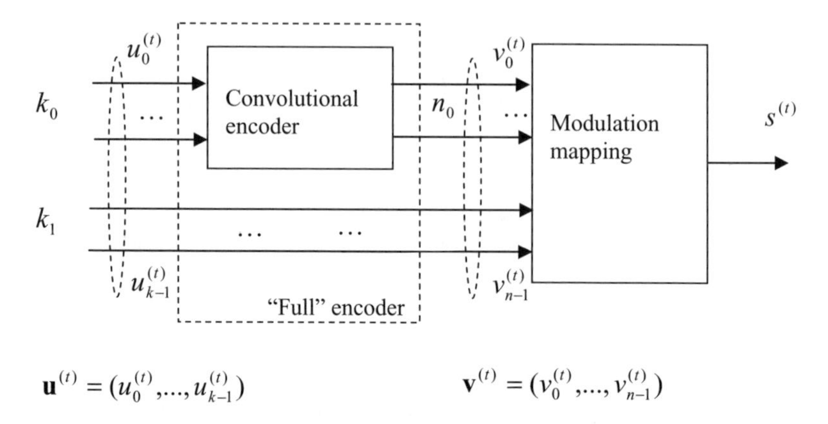

Figure 7.4 Trellis coded modulation scheme

to increase the minimum distance between *sequences of signal points*, using the specially designed correcting code and modulation mapping.

A general scheme of combined *trellis coded modulation (TCM)* using an extended signal set is shown in Figure 7.4.

The scheme consists of encoder of convolutional code of rate k_0/n_0 with memory of ν bits, and modulation memoryless mapper. Encoder produces n_0 coded bits and after combining them with k_1 uncoded bits forms $n = n_0 + k_1$ bits coming into modulation mapping. The overall code rate is k/n, where $k = k_0 + k_1$.

Let $\sigma^{(t)}$ be an encoder state at time t, $\sigma^{(t)} \in \{0, 1\}^\nu$ for all t. State transitions of the encoder can be defined as $\sigma^{(t+1)} = \sigma(\mathbf{u}^{(t)}, \sigma^{(t)})$, where $\sigma(\cdot, \cdot)$ is time-invariant state transition function, and $\mathbf{u}^{(t)}$ is the encoder input at time t as shown in Figure 7.4. Encoded block $\mathbf{v}^{(t)}$ at time t can be defined as $\mathbf{v}^{(t)} = v(\mathbf{u}^{(t)}, \sigma^{(t)})$, where $v(\cdot, \cdot)$ is time-invariant output function. For convolutional encoder these functions are linear over $GF(2)$, that is:

$$\sigma(\mathbf{u}^{(t)} + \mathbf{u}'^{(t)}, \sigma^{(t)} + \sigma'^{(t)}) = \sigma(\mathbf{u}^{(t)}, \sigma^{(t)}) + \sigma(\mathbf{u}'^{(t)}, \sigma'^{(t)}),$$
$$\mathbf{v}^{(t)} = v(\mathbf{u}^{(t)} + \mathbf{u}'^{(t)}, \sigma^{(t)} + \sigma'^{(t)}) = v(\mathbf{u}^{(t)}, \sigma^{(t)}) + v(\mathbf{u}'^{(t)}, \sigma'^{(t)})$$

In this scheme it is assumed that $m = n$ (see Figure 7.1), that is, each coded block is mapped into one of $M_c = 2^n$ signals.

Decoding for TCM is usually implemented as Viterbi algorithm (VA) [2] on the trellis of the encoder with edges labelled according to modulation mapping. The VA accepts the soft decisions from demodulator and searches for the path in trellis minimizing squared Euclidean distance between received sequence and coded signal sequence. The trellis at each level has 2^νnodes, and every node is connected by 2^{k_1} parallel edges with each of 2^{k_0}previous nodes. Any edge corresponds to one signal belonging to extended signal set S_c.

7.1.1 Illustrative Example

In this example we assume transmission with PSK at rate $2/T$ bit/s, where T is signal interval duration. In this example we examine minimum distances between two signal sequences. Let $\mathbf{u}$ and $\mathbf{v}(\mathbf{u})$ be data

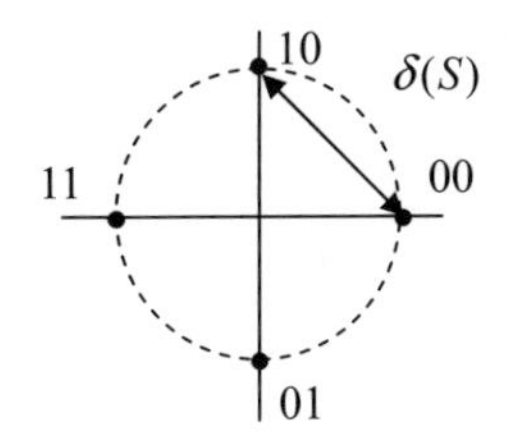

Figure 7.5 Example of modulation mapping for uncoded transmission

sequence and corresponding coded sequence respectively, and $\mathbf{s}(\mathbf{v}(\mathbf{u}))$ be signal sequence at the mapper output.

A. Uncoded transmission. The PSK4 signals conform to transmission rate $2/T$ bit/s. For uncoded transmission $\mathbf{v}(\mathbf{u}) = \mathbf{u}$, and $\min_{\mathbf{u}\neq\mathbf{u}'} d\,(\mathbf{s}(\mathbf{v}(\mathbf{u})), \mathbf{s}(\mathbf{v}(\mathbf{u}'))) = \delta(S)$. Example for modulation mapping for PSK4 (Gray mapping) is shown in Figure 7.5.

It is easy to see that the squared minimum distance between two signal sequences is equal to squared minimum distance between signal points:

$$d^2(\mathbf{s}(\mathbf{v}(\mathbf{u})), \mathbf{s}(\mathbf{v}(\mathbf{u})')) = d^2(\mathbf{s}(\mathbf{u}), \mathbf{s}(\mathbf{u}')) = d^2(s(00), s(10)) = \delta(S) = 2E \tag{7.9}$$

B. Coded transmission. An example of encoder and modulation mapping is shown in Figure 7.6.

The encoder state in time t is $\sigma^{(t)} = (\sigma_0^{(t)}, \sigma_1^{(t)})$, $\sigma_0^{(t)}, \sigma_1^{(t)} \in \{0, 1\}$. It is easy to see from Figure 7.6 that $\sigma_0^{(t)} = u_0^{(t-1)}$, $\sigma_1^{(t)} = \sigma_0^{(t-1)} = u_0^{(t-2)}$. State transition function is defined as $\sigma^{(t+1)} = \sigma(\mathbf{u}^{(t)}, \sigma^{(t)}) = (u_0^{(t)}, \sigma_0^{(t)})$. Output coded block $\mathbf{v}^{(t)} = (v_0^{(t)}, v_1^{(t)}, v_2^{(t)})$ is computed as $\mathbf{v}^{(t)} = v(\mathbf{u}^{(t)}, \sigma^{(t)}) = (\sigma_0^{(t)}, u_0^{(t)} + \sigma_1^{(t)}, u_1^{(t)})$.

Let us consider transmission of the data sequences $\mathbf{u} = (00, 00, 00, \ldots)$ and $\mathbf{u}' = (10, 00, 00, \ldots)$ by this code. Evidently, $\mathbf{v}(\mathbf{u}) = (000, 000, 000, \ldots)$. For data sequence $\mathbf{u}'$ the encoder state transitions and encoder outputs are shown in Table 7.1.

It can be easily computed that squared Euclidean distance between the signal sequences corrresponding to data sequences $\mathbf{u}$ and $\mathbf{u}'$ is:

$$\begin{aligned} d^2(\mathbf{s}(\mathbf{v}(\mathbf{u})), \mathbf{s}(\mathbf{c}(\mathbf{u}'))) &= d^2(s(010), s(000)) + d^2(s(100), s(000)) + d^2(s(010), s(000)) \\ &= 2E + \left(2 + \sqrt{2}\right) E + 2E = \left(6 + \sqrt{2}\right) E \approx 7.414E \end{aligned}$$

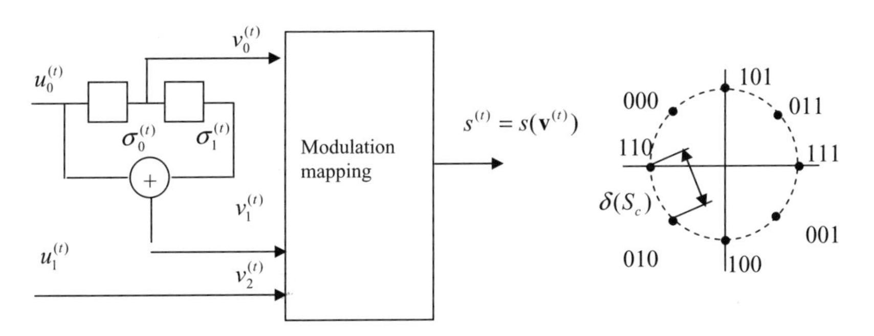

Figure 7.6 Example of trellis coded modulation scheme

Table 7.1 Example of two coded signal sequences

t	$\mathbf{u}^{(t)} = (u_0^{(t)}, u_1^{(t)})$	$\sigma^{(t)} = (\sigma_0^{(t)}, \sigma_1^{(t)})$	$\mathbf{v}^{(t)} = (v_0^{(t)}, v_1^{(t)}, v_2^{(t)})$	$s(\mathbf{v}(\mathbf{u}'))$	$s(\mathbf{0}, \mathbf{0}, \ldots)$
1	1 0	0 0	010		
2	0 0	1 0	100		
3	0 0	0 1	010		
4	0 0	0 0	000		
...	...	...	...		

Comparing this value with (7.9) we see that in this case the first data bit is protected much better.

Trellis for this example is depicted in Figure 7.7. The paths corresponding to $\mathbf{s}(\mathbf{v}(\mathbf{u}))$ and $\mathbf{s}(\mathbf{v}(\mathbf{u}'))$ are shown as bold lines. They differ in the first three transitions and merge together later on.

Free Euclidean distance d_f is a principal parameter for trellis codes and decoding with VA. It can be shown that for the code and modulation mapping shown in Figure 7.6 $d_f^2 = 4E$. That means 3 dB energy gain in comparison with uncoded transmission (see (7.9)).

7.2 Modulation Mapping by Signal Set Partitioning

The Ungerboeck coded modulation scheme is based on a) modulation mapping by set partitioning, and b) especially designed convolutional codes.

Let us consider modulation mapping suggested in [1]. Modulation mapping can be described as a step-by-step procedure. The number of steps is equal to m, $m = \log_2 M_c$. For the Ungerboeck TCM scheme $m = n$, where n is number of output coded symbols. At each step of the procedure the signal subset is divided into two subsets. On the very first stage the signal subset is equal to the whole signal set S_c containing M_c signal points. After last step of partitioning the signal subset consists of a single signal point. Any subset derived on l-th step of partitioning corresponds to a l−bit binary vector. Therefore number of subsets on l-th step is equal to 2^l. Let $\mathbf{v}_l$ be a binary vector of length l, that is, $\mathbf{v}_l \in \{0, 1\}^l$, $l = 1, 2, \ldots, n$. Let $S_c(v_l, l)$ be one of 2^l signal subsets derived on l-th step of partitioning. Then the

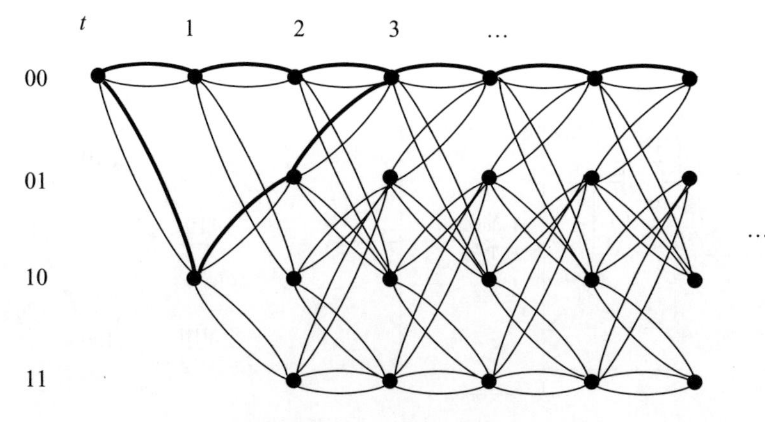

Figure 7.7 Example of code trellis

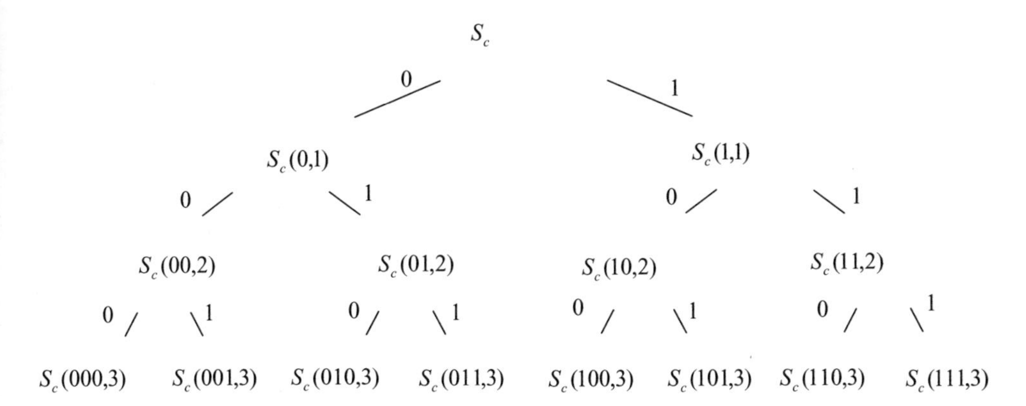

Figure 7.8 Binary partitioning of signal set

partitioning can be defined recurrently as:

$$\begin{cases} S_c = S_c(0,1) \cup S_c(1,1), \\ S_c(\mathbf{v}_l, l) = S([\mathbf{v}_l 0], l+1) \cup S([\mathbf{v}_l 1], l+1), \quad l = 2, 3, \ldots, n-1 \end{cases}$$

where $[\mathbf{v}_l, 0]$, $[\mathbf{v}_l, 1]$ are concatenations of binary vector and binary digit. It is easy to see that the process of partitioning can be presented as a binary tree. An example for $n = 3$ is shown in Figure 7.8

Partitioning must conform to some conditions on the distance between signal points in subsets and distance between signal subsets. Let us define minimum distance between points in subsets of l-th level as:

$$\delta_l = \min_{\mathbf{v} \in \{0,1\}^l} \min_{x,y \in S_c(\mathbf{v},l),\, x \neq y} d(x, y)$$

it is assumed by definition that $\delta_0 = \min_{x,y \in S_c,\, x \neq y} d(x, y) = \delta(S_c)$. Also we define minimum distance between signal subsets on the l-th level as:

$$\Delta_l = \min_{\mathbf{v} \in \{0,1\}^{l-1}} \min_{\substack{x \in S_c([\mathbf{v}\,0],l) \\ y \in S_c([\mathbf{v}\,1],l)}} d(x, y)$$

Then for the Ungerboeck partitioning there exists the chain of inequalities:

$$\delta_0 = \Delta_1 < \delta_1 = \Delta_2 < \ldots < \delta_{n-1} = \Delta_n$$

Figures 7.9 and 7.10 give examples of partitioning for PSK8 and QAM16.

For PSK8 we have the chain of inequalities for distances between signal sets and between signal points $\delta_0 = 2\sqrt{E}\sin(\pi/8) = \Delta_1 < \delta_1 = \sqrt{2E} = \Delta_2 < \delta_2 = 2\sqrt{E} = \Delta_3$.

For QAM16 the minimum distance between signal points $\delta_0 = \sqrt{(2/5)\overline{E}} \approx 0.633\sqrt{\overline{E}}$, where $\overline{E}$ is average energy of signal in QAM16 signal set, and the chain of distances for this case is $\delta_0 = \Delta_1 < \delta_1 = \sqrt{2}\delta_0 = \Delta_2 < \delta_3 = 2\delta_0 = \Delta_3 < \delta_3 = 2\sqrt{2}\delta_0 = \Delta_4$.

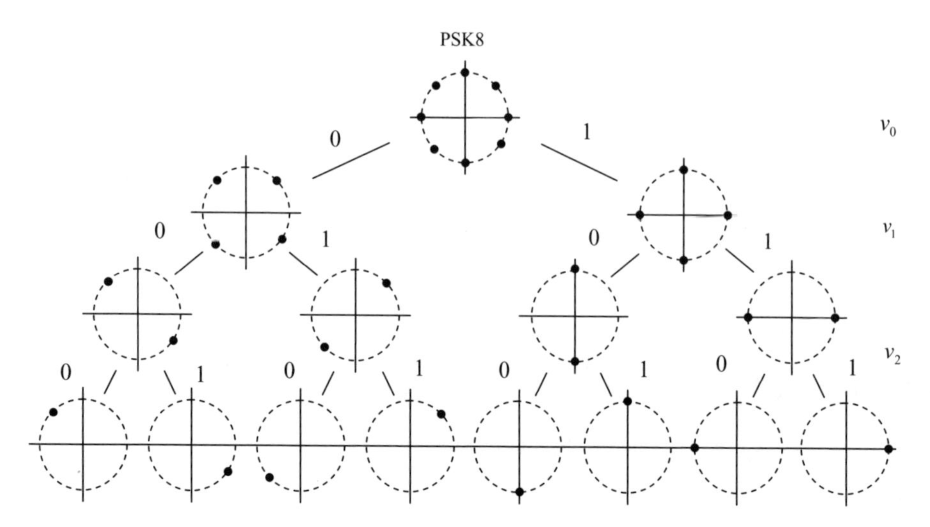

Figure 7.9 Partitioning of signal set PSK8

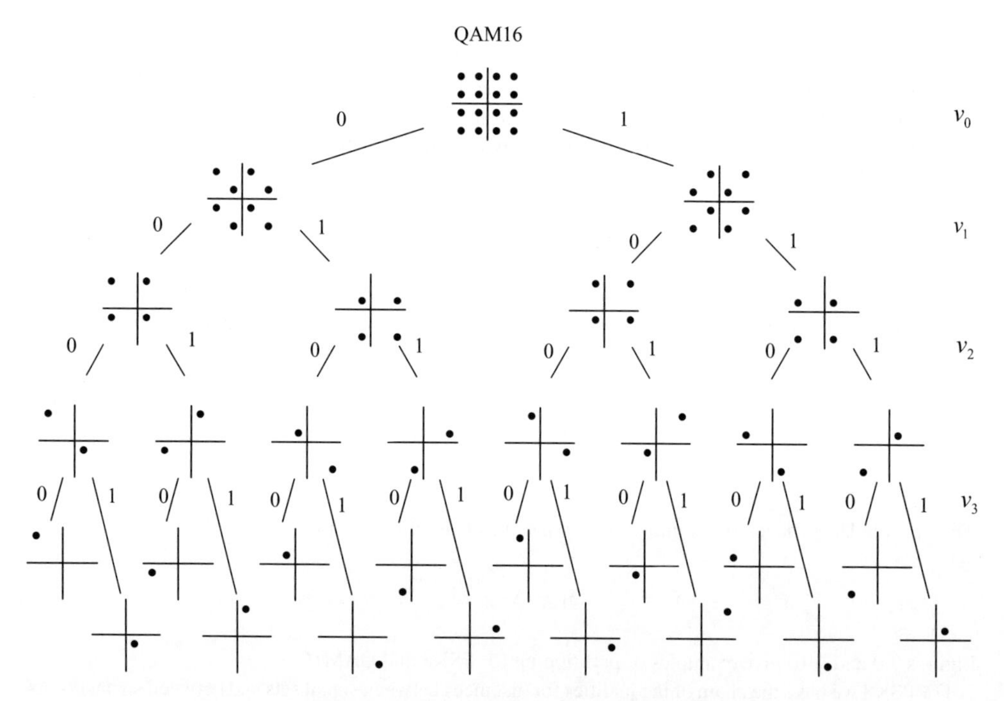

Figure 7.10 Partitioning of signal set QAM16

7.3 Ungerboeck Codes

Trellis coded modulation scheme contains a convolutional encoder of rate k_0/n_0. The encoder of such a code can be defined by polynomial $k_0 \times n_0$ generator matrix $\mathbf{G}_c(z)$:

$$\mathbf{G}_c(z) = \begin{bmatrix} g_{00}(z) & g_{01}(z) & \dots & g_{0n_0-1}(z) \\ g_{10}(z) & g_{11}(z) & \dots & g_{1n_0-1}(z) \\ \dots & \dots & \dots & \dots \\ g_{k_0-1,0}(z) & g_{k_0-1,1}(z) & \dots & g_{k_0-1,n_0-1}(z) \end{bmatrix}$$

where z is a formal variable, and $g_{ij}(z)$ are polynomial of finite field $GF(2)$, $i = 0, 1, \dots, k_0 - 1$, $j = 0, 1, \dots n_0 - 1$, and encoder memory size can be expressed as:

$$\nu = \sum_{i=0}^{k_0-1} \max_j \deg g_{ij}(z)$$

Then the encoding process can be described as the multiplication of vector of formal series corresponding to data sequences by polynomial generator matrix, that is:

$$[\, v_0(z)\ v_1(z) \dots v_{n_0-1}(z) \,] = [\, u_0(z)\ u_1(z) \dots u_{k_0-1}(z) \,]\mathbf{G}_c(z)$$

where $v_j(z) = v_j^{(0)} + v_j^{(1)}z + \dots + v_j^{(t)}z^t + \dots$, $u_i(z) = u_i^{(0)} + u_i^{(1)}z + \dots + u_i^{(t)}z^t + \dots$, are formal series corresponding to data sequence $(u_i^{(0)}, u_i^{(1)}, \dots, u_i^{(t)} \dots)$ and code sequences $(v_j^{(0)}, v_j^{(1)}, \dots, v_j^{(t)}, \dots)$ respectively, $u_i^{(t)}, v_j^{(t)} \in GF(2)$ for all t, $j = 0, 1, \dots, n_0 - 1$, $i = 0, 1, \dots, k_0 - 1$.

For "full" encoder (see Figure 7.4) with k_1 uncoded data sequences the polynomial generator $k \times n$ matrix $\mathbf{G}(z)$ exists in form:

$$\mathbf{G}(z) = \begin{bmatrix} \mathbf{G}_c(z) & \mathbf{0} \\ \mathbf{0} & \mathbf{I} \end{bmatrix} \tag{7.10}$$

and $k = k_0 + k_1$, $n = n_0 + k_1$. Hereafter we use notation $\mathbf{I}$ for identity matrix and $\mathbf{0}$ for zero matrix of a corresponding dimension. It follows from (7.10) that the code sequences $v_0(z), v_1(z), \dots, v_{n_0-1}(z)$ depend on data sequences $u_0(z), u_1(z), \dots, u_{k_0-1}(z)$, and $v_{n_0+j}(z) = u_{k_0+j}(z)$, $j = 1, 2, .., k_1$, and does not depend on $u_0(z), u_1(z), \dots, u_{k_0-1}(z)$.

An equivalent description of the convolutional code can be given by its polynomial parity check $(n - k) \times k$ matrix $\mathbf{H}(z)$, such that $\mathbf{G}(z)\mathbf{H}(z)^T = \mathbf{0}$. For polynomial generator matrix given by (7.10) the parity check matrix can be written as:

$$\mathbf{H}(z) = \begin{bmatrix} \mathbf{H}_v(z) & \mathbf{0} \end{bmatrix}$$

where $\mathbf{H}_c(z)$ is $(n_0 - k_0) \times n_0$ parity check matrix corresponding to $\mathbf{G}_c(z)$, that is, such that $\mathbf{G}_c(z)\mathbf{H}_c(z)^T = \mathbf{0}$. It should be noted that in general elements of matrix $\mathbf{H}_c(z)$ are a rational function of formal variable z over $GF(2)$.

For many examples given in [1] parameters of codes are as follows $n_0 = k_0 + 1$, and the codes are specified by parity check matrices.[3] In this case the dimension of matrix $\mathbf{H}_c(z)$ is $1 \times n_0$, and it can be presented as:

$$\mathbf{H}_c(z) = \begin{bmatrix} \frac{h_1(z)}{h_0(z)} \cdots \frac{h_{k_0}(z)}{h_0(z)} & 1 \end{bmatrix} \tag{7.11}$$

where $h_1(z)$,..,$h_{k_0}(z)$ are feedforward encoder polynomials, and $h_0(z)$ is feedback encoder polynomial. The general scheme of the systematic convolutional code corresponding to the matrix 7.11 is shown in Figure 7.11.

[3] List of some Ungerboeck codes is given below in Tables 7.2–7.5

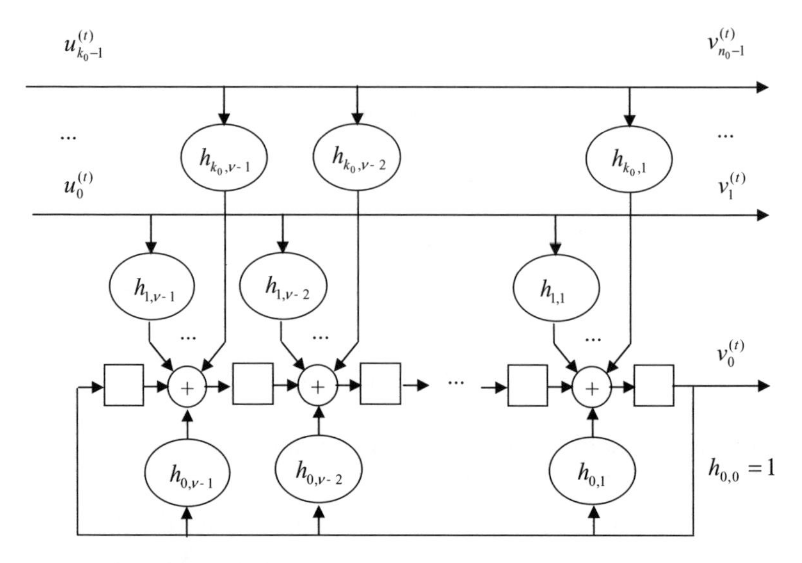

Figure 7.11 Systematic convolutional encoder for Ungerboeck scheme

Example of the encoder for code with the parity check matrix:

$$\mathbf{H}_c(z) = \begin{bmatrix} \frac{z^3}{1+z^5} & \frac{z+z^2}{1+z^5} & 1 \end{bmatrix} \tag{7.12}$$

is shown in Figure 7.12

To make description polynomial generator and parity check matrices more compact we use the common octal notation for polynomial over $GF(2)$. For example, let us consider the parity check matrix in (7.12). The polynomial $z^2 + z$ can be represented by its coefficients as $(1, 1, 0)$ or $110_2 = 6_8$, polynomial z^3 corresponds to $1000_2 = 10_8$, and $1 + z^5$ – $100001_2 = 41_8$. Thus matrix $\mathbf{H}_c(z)$ given in (7.12) can be written in compact octal form as $\begin{bmatrix} 10/41 & 06/41 & 1 \end{bmatrix}$. The generator matrix for this example is:

$$\mathbf{G}_c(z) = \begin{bmatrix} z+z^3 & 1+z+z^2 & z \\ 1+z & z^2 & 0 \end{bmatrix} \leftrightarrow \begin{bmatrix} 12 & 07 & 02 \\ 03 & 04 & 00 \end{bmatrix} \tag{7.13}$$

It can be verified that $\mathbf{G}_c(z)\mathbf{H}_c(z)^T = \mathbf{0}$. The encoder for convolutional code with generator matrix in (7.13) is feedback-free. The code generating by this encoder is equivalent to code with parity check matrix in (7.12).

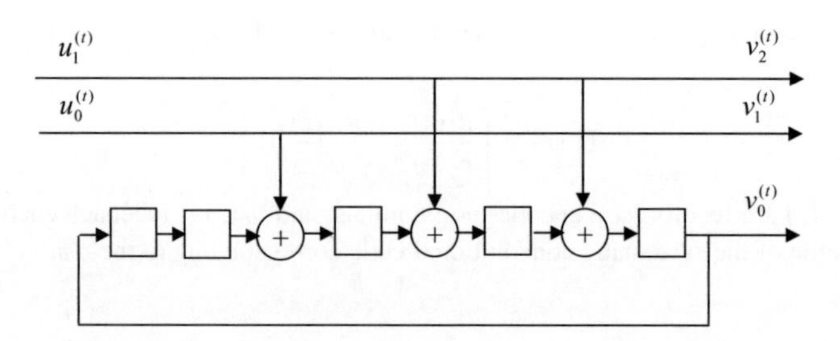

Figure 7.12 Example of systematic convolutional encoder, $\nu = 5$

7.4 Performance Estimation of TCM System

The simple approach to evaluating the error correcting performance for TCM considers minimal (free) Euclidean distance and asymptotic coding gain as a measure of efficiency. Let S be as before a signal set used for uncoded transmission, $|S| = 2^k$, and $\delta(S)$, $\overline{E}(S)$ be the minimum distance between signals in S, and average squared Euclidean norm of signals in S respectively. Let us denote convolutional code of rate k/n used in TCM scheme as V, modulation mapping M, $\mathrm{M} : \{0, 1\}^n \to S_c$, where S_c is extended signal set, $|S_c| = n$, and average squared norm of S_c is $\overline{E}(S_c)$. Let $d_f(C, \mathrm{M}, S_c)$ be free Euclidean distance in trellis for TCM defined by code V, mapping M and signal set S_c. Then the asymptotic coding gain of TCM system in comparison with uncoded system is:

$$G_{dB} = 10\log_{10}\left(\frac{d_f^2(V, \mathrm{M}, S_c)}{\delta^2(S)}\frac{\bar{E}(S)}{\bar{E}(S_c)}\right), \mathrm{dB} \tag{7.14}$$

For code generated by the encoder shown in Figure 7.7, with set partitioning mapping, and $S_c = \mathrm{QAM8}$, $S = \mathrm{QAM4}$ we have $\overline{E}(S_c) = 10$, $\overline{E}(S)= 2$, $d_f^2(V, \mathrm{M}, S_c) = 48$, $\delta^2(S_c) = 4$. Therefore for this code we have $G_{dB} = 10\log_{10}(48 \cdot 2/(4 \cdot 10)) = 3.80\mathrm{dB}$. Some examples of Ungerboeck codes are given in Tables 7.2–7.5. More examples can be found in [1].

A more accurate study includes the (average) number of nearest neighbours of some code sequence (error coefficient, or multiplicity). But for many applications the accuracy of the error probability bounds based only on free distance and number of nearest neighbours may be insufficient.

In this section we consider a technique of estimation of error probability of TCM in the AWGN channel with soft decision. The Viterbi decoding is assumed. The approach is standard for VA performance estimation and it consists of computation of the union bound [2] on error probability using squared Euclidean distance spectra of the trellis code. We follow here the approach published in [3].

7.4.1 Squared Distance Structure of PSK and QAM Constellations

Let s, s' be two signal points, $s, s' \in S_c$. Then the squared Euclidean distance between them may be expressed as follows:

$$d^2(s, s') = n_1 d_1^2 + \ldots + n_p d_p^2 \tag{7.15}$$

where $\mathbf{n} = (n_1, \ldots, n_p)$ is a vector with *integer* coordinates depending on signal pair s, s', and $\mathbf{d}^2 = (d_1^2, \ldots, d_p^2)$ is a constant vector of squared basic distances. The linear combination in (7.15) is called

Table 7.2 Basic parameters of Ungerboeck codes for QAM8

System parameters	Ref. No	Encoder memory, ν	$H(z)$ of Ungerboeck code	$G(z)$ of equivalent code	d_f^2	G_{dB}
$R_c = 2/T$, bit/s $R = 2/3$	1	2	$\begin{bmatrix} 0 & \frac{2}{5} & 1 \end{bmatrix}$	$\begin{bmatrix} 1 & 0 & 0 \\ 0 & 5 & 2 \end{bmatrix}$	32	2.04
S_c =QAM8 $\overline{E}(S_c) = 10$	2	3	$\begin{bmatrix} \frac{04}{11} & \frac{02}{11} & 1 \end{bmatrix}$	$\begin{bmatrix} 4 & 1 & 2 \\ 1 & 2 & 0 \end{bmatrix}$	40	3.01
S =QAM4 $\delta^2(S) = 4$	3	4	$\begin{bmatrix} \frac{16}{23} & \frac{04}{23} & 1 \end{bmatrix}$	$\begin{bmatrix} 7 & 3 & 2 \\ 2 & 7 & 0 \end{bmatrix}$	48	3.80
$\overline{E}(S) = 2$	4	5	$\begin{bmatrix} \frac{10}{41} & \frac{06}{41} & 1 \end{bmatrix}$	$\begin{bmatrix} 12 & 07 & 02 \\ 03 & 04 & 00 \end{bmatrix}$	48	3.80
	5	6	$\begin{bmatrix} \frac{064}{101} & \frac{016}{101} & 1 \end{bmatrix}$	$\begin{bmatrix} 07 & 01 & 02 \\ 07 & 32 & 00 \end{bmatrix}$	56	4.47
	6	7	$\begin{bmatrix} \frac{042}{203} & \frac{014}{203} & 1 \end{bmatrix}$	$\begin{bmatrix} 31 & 02 & 06 \\ 02 & 17 & 00 \end{bmatrix}$	64	5.05

Table 7.3 Basic parameters of Ungerboeck codes for QAM16

System parameters	Ref. No	Encoder memory, ν	$H(z)$ of Ungerboeck code	$G(z)$ of equivalent code	d_f^2	G_{dB}
$R_c = 3/T$, bit/s $R = 3/4$	7	2	$\begin{bmatrix} 0 & \frac{2}{5} & 1 & 0 \end{bmatrix}$	$\begin{bmatrix} 1 & 0 & 0 & 0 \\ 0 & 5 & 2 & 0 \\ 0 & 0 & 0 & 1 \end{bmatrix}$	16	3.01
$S_c = \text{QAM16}$ $\overline{E}(S_c) = 10$ $S = \text{QAM8}$ $\delta^2(S) = 8$	8	3	$\begin{bmatrix} \frac{04}{11} & \frac{02}{11} & 1 & 0 \end{bmatrix}$	$\begin{bmatrix} 4 & 1 & 2 & 0 \\ 1 & 2 & 0 & 0 \\ 0 & 0 & 0 & 1 \end{bmatrix}$	20	3.98
$\overline{E}(S) = 10$	9	4	$\begin{bmatrix} \frac{16}{23} & \frac{04}{23} & 1 & 0 \end{bmatrix}$	$\begin{bmatrix} 7 & 3 & 2 & 0 \\ 2 & 7 & 0 & 0 \\ 0 & 0 & 0 & 1 \end{bmatrix}$	24	4.77
	10	5	$\begin{bmatrix} \frac{10}{41} & \frac{06}{41} & 1 & 0 \end{bmatrix}$	$\begin{bmatrix} 12 & 07 & 02 & 00 \\ 03 & 04 & 00 & 00 \\ 00 & 00 & 00 & 01 \end{bmatrix}$	24	4.77
	11	6	$\begin{bmatrix} \frac{064}{101} & \frac{016}{101} & 1 & 0 \end{bmatrix}$	$\begin{bmatrix} 07 & 01 & 02 & 00 \\ 07 & 32 & 00 & 00 \\ 00 & 00 & 00 & 01 \end{bmatrix}$	28	5.44

integer linear combination. It is easy to see that squared distance representation in (7.15) can be done for any configuration of signal set. In particular, if the number of basic distances $p = M_c(M_c - 1)/2$, that is, equal to number of all possible pairwise distances between signal points, then the representation of (7.15) is valid for any signal set. For QAM and PSK the number p is much less than the number of all possible pairwise distances.

For QAM it is clear that $p = 1$, and:

$$d_1^2 = \delta(S_c)^2 = \begin{cases} \dfrac{6\overline{E}}{M_c - 1}, & M_c = 2^m, \ m \text{ is even} \\ \dfrac{12\overline{E}}{2M_c - 1}, & M_c = 2^m, \ m \text{ is odd} \end{cases} \tag{7.16}$$

where $\overline{E}$ is the average squared Euclidean norm of the QAM signal set.

Table 7.4 Basic parameters of Ungerboeck codes for QAM32

System parameters	Ref. No	Encoder memory, ν	$H(z)$ of Ungerboeck code	$G(z)$ of equivalent code	d_f^2	G_{dB}
$R_c = 4/T$, bit/s $R = 4/5$	12	2	$\begin{bmatrix} 0 & \frac{2}{5} & 1 & 0 & 0 \end{bmatrix}$	$\begin{bmatrix} 1 & 0 & 0 & 0 & 0 \\ 0 & 5 & 2 & 0 & 0 \\ 0 & 0 & 0 & 1 & 0 \\ 0 & 0 & 0 & 0 & 1 \end{bmatrix}$	32	2.80
$S_c = \text{QAM32}$ $\overline{E}(S_c) = 42$ $S = \text{QAM16}$ $\delta^2(S) = 4$ $\overline{E}(S) = 10$	13	3	$\begin{bmatrix} \frac{04}{11} & \frac{02}{11} & 1 & 0 & 0 \end{bmatrix}$	$\begin{bmatrix} 4 & 1 & 2 & 0 & 0 \\ 1 & 2 & 0 & 0 & 0 \\ 0 & 0 & 0 & 1 & 0 \\ 0 & 0 & 0 & 0 & 1 \end{bmatrix}$	40	3.77
	14	4	$\begin{bmatrix} \frac{16}{23} & \frac{04}{23} & 1 & 0 & 0 \end{bmatrix}$	$\begin{bmatrix} 7 & 3 & 2 & 0 & 0 \\ 2 & 7 & 0 & 0 & 0 \\ 0 & 0 & 0 & 1 & 0 \\ 0 & 0 & 0 & 0 & 1 \end{bmatrix}$	48	4.56

Table 7.5 Basic parameters of Ungerboeck codes for PSK8

System parameters	Ref. No	Encoder memory, ν	$H(z)$ of Ungerboeck code	$G(z)$ of equivalent code	d_f^2	G_{dB}
$R_c = 2/T$, bit/s $R = 2/3$ $S_c = \text{PSK8}$ $\overline{E}(S_c) = 1$ $S = \text{PSK4}$ $\delta^2(S) = 2$ $\overline{E}(S) = 1$	1	2	$\begin{bmatrix} 0 & \frac{2}{5} & 1 \end{bmatrix}$	$\begin{bmatrix} 1 & 0 & 0 \\ 0 & 5 & 2 \end{bmatrix}$	4	3.01
	2	3	$\begin{bmatrix} \frac{04}{11} & \frac{02}{11} & 1 \end{bmatrix}$	$\begin{bmatrix} 4 & 1 & 2 \\ 1 & 2 & 0 \end{bmatrix}$	4.59	3.61
	3	4	$\begin{bmatrix} \frac{16}{23} & \frac{04}{23} & 1 \end{bmatrix}$	$\begin{bmatrix} 7 & 3 & 2 \\ 2 & 7 & 0 \end{bmatrix}$	5.17	4.12
	4	5	$\begin{bmatrix} \frac{34}{45} & \frac{16}{45} & 1 \end{bmatrix}$	$\begin{bmatrix} 22 & 01 & 16 \\ 01 & 02 & 00 \end{bmatrix}$	5.76	4.59
	5	6	$\begin{bmatrix} \frac{074}{105} & \frac{036}{105} & 1 \end{bmatrix}$	$\begin{bmatrix} 42 & 01 & 36 \\ 01 & 02 & 00 \end{bmatrix}$	6.00	4.77

For PSK the squared distance between any two signal points s and s' is $d^2(s, s') = 4E \sin^2 \theta(s, s')$, where E is the energy of PSK signals, and $\theta(s, s')$ is the difference of their phases. Evidently, $\theta(s, s') = l(s, s')\pi/M_c$, where $l(s, s')$ is integer, and $l(s, s') \in \{0, 1, \ldots, M_c - 1\}$. Using trigonometric identity $\sin^2 x = (1 - \cos 2x)/2$, we have:

$$d^2(s, s') = 2E\left(1 - \cos\left(2\pi l(s, s')/M_c\right)\right).$$

Using the properties of cosine function we can express the squared distance between two signal points of PSK constellation as given in (7.15) with $p = M_c/4$ squared base distances $d_1^2, d_2^2, \ldots, d_{M_c/4}^2$, where $d_i^2 = 2E\,(1 - \cos(2i\pi/M_c))$, $i = 1, 2, \ldots, M_c/4$, as follows:

$$d^2(s, s') = \begin{cases} 0, & l(s, s') = 0, \\ d_1^2, & 1 \le l(s, s') < M_c/4, \\ 2d_{M_c/4}^2 - d_{M_c/2-l}^2, & M_c/4 \le l(s, s') < M_c/2, \\ 2d_{M_c/4}^2, & l(s, s') = M_c/2, \\ 2d_{M_c/4}^2 - d_{l-M_c/2}^2, & M_c/2 \le l(s, s') < 3M_c/4, \\ d_{M_c-l}^2, & 3M_c/4 \le l(s, s') < M_c \end{cases} \tag{7.17}$$

where $l(s, s') = (M_c/\pi)\,\theta(s, s')$. This rule is valid if the number of signals M_c is a multiple of four. Since in practice $M_c = 2^m$, this fact is not a strong restriction. The equations (7.16) and (7.17) mean that the squared distances between PSK and QAM signals can be computed using a few base distances. For example, for QAM there is only one base distance, and for PSK8 there are two base distances $d_1^2 = 2E(1 - \sqrt{2}/2)$, and $d_2^2 = 2E$.

7.4.2 *Upper Bound on Error Event Probability and Bit Error Probability for TCM*

Evaluation of upper bounds on error probability for trellis coding was considered in [4]–[10]. These bounds can be computed numerically with complexity proportional to complexity of inversion of a $2^{2\nu} \times 2^{2\nu}$ matrix, where ν is the encoder memory length. Some symmetry properties may be used for reducing the complexity of the upper bound computation. It has been shown in [7]–[10] that under some restrictions on the structure of the trellis code the complexity of computation of the upper bound on error probability may be reduced to the complexity of inversion of a $2^\nu \times 2^\nu$ matrix.

In [3] an approach to computation of series expansion of a generating function using computation of the upper bound of error event and bit error probability is presented. The representation of the set of all squared Euclidean distances between code sequences as a countable set is used in this approach. The generating functions of squared Euclidean distances (distance spectra) are obtained by inversion of a square matrix of order 2^ν. The generating functions are defined in terms of one formal variable for QAM, and in terms of $M_c/4$ formal variables for M_c-ary PSK, $M_c = 2^m$, where $m > 2$ is an integer.

Let $Z_1, Z_2, \ldots, Z_p$ and N be formal variables. We also introduce vectors $\mathbf{Z} = (Z_1, Z_2, \ldots, Z_p)$ and $\mathbf{n} = (n_1, \ldots, n_p)$, and notation $\mathbf{Z}^{\mathbf{n}} = Z_1^{n_1} Z_2^{n_2} \ldots Z_p^{n_p}$. Denote by $V(i, j)$ set of code blocks corresponding to encoder transition from state i to state j; we denote such transition as $i \to j$. Let iwt($\mathbf{c}$) be the number of nonzero data bits corresponding to the code block $\mathbf{v}$, or information weight of the vector $\mathbf{v}$. Let $\mathbf{e}$ be a binary vector of length $n = m = \log_2 M_c$, and we define the set of integers $\mathbf{n}(\mathbf{v}, \mathbf{v}+\mathbf{e}) = (n_1, \ldots, n_p)$ such that:

$$d^2(s(\mathbf{v}), s(\mathbf{v}+\mathbf{e})) = n_1 d_1^2 + \ldots + n_p d_p^2$$

Denote a function of the formal variables and vector $\mathbf{e}$:

$$g_{ij}(\mathbf{e}, \mathbf{Z}, N) = 2^{-k} \sum_{\mathbf{v} \in V(i,j)} \mathbf{Z}^{\mathbf{n}(\mathbf{v},\mathbf{v}+\mathbf{e})} N^{\mathrm{iwt}(\mathbf{e})}$$

and $2^\nu \times 2^\nu$ matrix:

$$\mathbf{G}(\mathbf{e}, \mathbf{Z}, N) = [g_{ij}(\mathbf{e}, \mathbf{Z}, N)], \quad i, j = 0, 1, \ldots, 2^\nu - 1$$

Let us introduce the sets $E(l)$ of code sequences corresponding to encoder state transition $0 \to i^{(1)} \to i^{(2)} \to \ldots \to i^{(l-1)} \to 0$, where $i^{(t)} \neq 0$, $1 \leq t \leq l-1$, $l > 1$; by definition we assume that $E(1)$ is a set of nonzero code sequences corresponding to state transition $0 \to 0$. As has been done in [9], we introduce the matrix:

$$\mathbf{G}(\mathbf{Z}, N) = 2^{-\nu} \sum_{l=1}^{\infty} \sum_{\mathbf{e} \in E(l)} \prod_{t=1}^{l} \mathbf{G}(\mathbf{e}^{(t)}, \mathbf{Z}, N)$$

Union upper bounds on the error event probability P_e and bit error probability P_b for AWGN channel may be written as [9]:

$$P_e \leq K \; \mathbf{1}\mathbf{G}(\mathbf{Z}, N)\mathbf{1}^T \big|_{\mathbf{Z}=\mathbf{Z}_0, N=1} \tag{7.18}$$

$$P_b \leq (K/k) \frac{\partial}{\partial N} \mathbf{1}\mathbf{G}(\mathbf{Z}, N)\mathbf{1}^T \Big|_{\mathbf{Z}=\mathbf{Z}_0, N=1} \tag{7.19}$$

where

$$\mathbf{Z}_0 = \left(\exp(-d_1^2/4N_0), \ldots, \exp(-d_p^2/4N_0)\right) \tag{7.20}$$

$$K = Q\left(\frac{d_{\min}}{\sqrt{2N_0}}\right) \exp\left(\frac{d_{\min}^2}{4N_0}\right) \tag{7.21}$$

$$Q(x) = \int_x^{\infty} (2\pi)^{-1/2} \exp(-u^2/2) du$$

and $d_{\min}$ is free Euclidean distance of the trellis code, $\mathbf{1} = (1, \ldots, 1)$ is vector with 2^ν all one's coordinates.

Right hand part of the inequalities in (7.18) and (7.19) can be computed as a value of generating function and its derivative with respect to formal variable N:

$$T(\mathbf{Z}, N) = \mathbf{1}\mathbf{G}(\mathbf{Z}, N)\mathbf{1}^T \tag{7.22}$$

Generally, the function $T(\mathbf{Z}, N)$ may be obtained by inversion of a $2^{2\nu} \times 2^{2\nu}$matrix. Reduction of the dimensionality of the matrix to $2^{\nu} \times 2^{\nu}$may be achieved if some restrictions on the trellis code are imposed. The detailed analysis of conditions providing this complexity reduction is given in [9].

Let the expressions:

$$f(\mathbf{e}, \mathbf{Z}, N) = \sum_{j=0}^{2^{\nu}-1} g_{ij}(\mathbf{e}, \mathbf{Z}, N) \tag{7.23}$$

not depend on i for all $\mathbf{e}$. This means that the matrices $\mathbf{G}(\mathbf{e}, \mathbf{Z}, N)$ are row uniform [9]. Then variation of variable $j$from 0 to $2^{\nu} - 1$ in (7.23) is equivalent to enumeration of signals corresponding to the bundle of branches coming from a node in the trellis, that is, belonging to set $\bigcup_{i'} S(i, i')$, where $S(i, i') = \{s(\mathbf{v}) : \mathbf{v} \in V(i, j)\}$. Let us denote $\mathrm{Is}(\cdot)$ an *isometric* transformation of set of points on the real plane, for instance rotation and (or) translation. If:

$$\bigcup_{i'} S(i, i') = \mathrm{Is}\left(\bigcup_{i'} S(i'', i')\right), \quad i'' \neq i$$

then the sums in (7.23) do not depend on i. It can be shown, as in [3], that for the Ungerboeck TCM these conditions are satisfied. Using notation (7.23) we have:

$$\mathbf{1G}(\mathbf{Z}, N)\mathbf{1}^T = \sum_{l=1}^{\infty} \sum_{\mathbf{e} \in E(l)} \prod_{t=1}^{l} f(\mathbf{e}^{(t)}, \mathbf{Z}, N) \tag{7.24}$$

Denote by $f_{pq}(\mathbf{Z}, N)$ the variable $f(\mathbf{e}, \mathbf{Z}, N)$ corresponding to vector $\mathbf{e}$ generated by encoder transition $p \to q$. Then:

$$\sum_{\mathbf{e} \in E(l)} \prod_{t=1}^{l} f(\mathbf{e}^{(t)}, \mathbf{Z}, N) = \begin{cases} \sum\limits_{i_1} \sum\limits_{i_2} \cdots \sum\limits_{i_{l-1}} \sum\limits_{i_l} f_{0i_1}(\mathbf{Z}, N) f_{i_1 i_2}(\mathbf{Z}, N) \ldots f_{i_{l-1} i_l}(\mathbf{Z}, N) f_{i_l 0}(\mathbf{Z}, N), & l > 1, \\ f_{00}(\mathbf{Z}, N) - 1, \quad l = 1 \end{cases}$$

Substituting right hand part of this equation into (7.24) and the result into (7.22) we get:

$$T(\mathbf{Z}, N) = F_0(\mathbf{Z}, N) + \mathbf{F}_1(\mathbf{Z}, N)\,(\mathbf{I} - \mathbf{F}(\mathbf{Z}, N))^{-1}\,\mathbf{F}_2(\mathbf{Z}, N)^T$$

where

$$F_0(\mathbf{Z}, N) = f_{00}(\mathbf{Z}, N) - 1$$

$$\mathbf{F}_1(\mathbf{Z}, N) = [f_{01}(\mathbf{Z}, N), \ldots, f_{0,\,2^{\nu}-1}(\mathbf{Z}, N)],\ \mathbf{F}_2(\mathbf{Z}, N) = [f_{10}(\mathbf{Z}, N), \ldots, f_{2^{\nu}-1,\,0}(\mathbf{Z}, N)]$$

$$\mathbf{F}(\mathbf{Z}, N) = \begin{bmatrix} f_{11}(\mathbf{Z}, N) & \cdots & f_{1,2^{\nu}-1}(\mathbf{Z}, N) \\ \cdots & \cdots & \cdots \\ f_{2^{\nu}-1,1}(\mathbf{Z}, N) & \cdots & f_{2^{\nu}-1,2^{\nu}-1}(\mathbf{Z}, N) \end{bmatrix}$$

and $\mathbf{I}$ is identity $(2^{\nu} - 1) \times (2^{\nu} - 1)$matrix.

Series expansion for the function $T(\mathbf{Z}, N)$ may be written in the form:

$$T(\mathbf{Z}, N) = \sum_{\mathbf{n}} \sum_{i} A(\mathbf{n}, i)\mathbf{Z}^{\mathbf{n}} N^i$$

where $A(\mathbf{n}, i)$ is the *average* number of the error events having the squared Euclidean norm $(\mathbf{n}, \mathbf{d}^2)$ and information weight i. Denote:

$$A(\mathbf{n}) = \sum_{i=0}^{\infty} A(\mathbf{n}, i)$$

$$B(\mathbf{n}) = \sum_{i=0}^{\infty} i A(\mathbf{n}, i)$$

The sets of numbers $\{A(\mathbf{n})\}$ and $\{B(\mathbf{n})\}$are called the spectra of squared Euclidean distances of the trellis code. It is clear that the numbers $\{A(\mathbf{n})\}$ and $B(\mathbf{n})$ are the coefficients of the series expansion of $T(\mathbf{Z}, 1)$ and $\partial T(\mathbf{Z}, N)/\partial N|_{N=1}$ respectively, that is:

$$T(\mathbf{Z}, 1) = \sum_{\mathbf{n}} A(\mathbf{n})\mathbf{Z}^{\mathbf{n}}$$

$$\left.\frac{\partial T(\mathbf{Z}, N)}{\partial N}\right|_{N=1} = \sum_{\mathbf{n}} B(\mathbf{n})\mathbf{Z}^{\mathbf{n}}$$

It follows from (7.18) and (7.19) that we get:

$$P_e < K \sum_{\mathbf{n}} A(\mathbf{n})\mathbf{Z}_0^{\mathbf{n}} \tag{7.25}$$

$$P_b < \frac{K}{k} \sum_{\mathbf{n}} B(\mathbf{n})\mathbf{Z}_0^{\mathbf{n}} \tag{7.26}$$

where $\mathbf{Z}_0$ and $K$are defined by (7.20) and (7.21).

For practice it is sufficient to compute sums of a truncated series in (7.25) and (7.26), that is, sums of several terms. It means that only a few elements of infinite number of spectral coefficients $\{A(\mathbf{n})\}$ and $\{B(\mathbf{n})\}$ are needed to compute good estimate of probabilities P_e and P_b.

Example 7.1 For encoder shown in Figure 7.6 and QAM8 the values of $\mathbf{F}(\mathbf{Z}, N)$, $\mathbf{F}_1(\mathbf{Z}, N)$, $\mathbf{F}_2(\mathbf{Z}, N)$ and $F(\mathbf{Z}, N)$ are $F_0(\mathbf{Z}, N) = Z^{32}N$, $\mathbf{F}_1(\mathbf{Z}, N) = [\,0\ Z^{16}(N + N^2)\ 0\,]$, $\mathbf{F}_1(\mathbf{Z}, N) = [\,Z^{16}(1 + N)\ 0\ 0\,]$, and:

$$\mathbf{F}(\mathbf{Z}, N) = \begin{bmatrix} 0 & N + Z^{32}N & 0 \\ Z^8 + (Z^8 + 2Z^{40} + Z^{72})N/4 & 0 & (Z^8 + Z^{40})(N + N^2)/2 \\ (Z^8 + Z^{40})(1 + N)/2 & 0 & Z^8N + (Z^8 + 2Z^{40} + Z^{72})N^2/4 \end{bmatrix}$$

Note, that for QAM the formal vector variable $\mathbf{Z}$ becomes a scalar variable Z. The closed expression for $T(Z, N)$ is too complicated to present here (see [3]), but its series expansion can be found, and it is equal to:

$$T(Z, N) = Z^{32}N + Z^{40}N + (9/4)Z^{40}N^2 + (3/2)Z^{40}N^3 + (1/4)Z^{40}N^4 + \ldots$$

The details of computation of the squared Euclidean distance spectra $\{A(\mathbf{n})\}$ and $\{B(\mathbf{n})\}$and results for the computation for some Ungerboeck codes are presented in [3]. Some of the squared Euclidean distance spectra for these codes are listed in Tables 7.6 – 7.7. Tables for other codes see in [3].

Table 7.6 Coefficients of generating function of Ungerboeck codes, QAM-8, $\overline{E}(S_c) = 10$, $d_1^2 = 8$

Ref. No (in Table 7.2)	Encoder memory, ν	$(\mathbf{n}, \mathbf{d}^2)$	$A(\mathbf{n})$	$B(\mathbf{n})$
1	2	32	1.00000	1.00000
		40	5.00000	11.00000
		48	10.25000	37.25000
		56	22.81250	113.93750
		64	52.26563	326.64063
		72	124.83203	915.44922
		80	306.61816	2550.99707
		88	740.32739	6960.78296
		96	1778.09576	18704.04010
		104	4262.43025	49640.95158
2	3	40	2.25000	5.50000
		48	6.25000	23.50000
		56	16.39063	85.40625
		64	43.35156	292.75000
		72	114.04004	944.21680
		80	304.80762	2964.08789
		88	814.87848	9107.28918
		96	2173.81204	27478.49780
		104	5803.64725	81827.93923
		112	15492.44660	241051.55244
3	4	48	4.37500	17.12500
		56	9.98438	45.60937
		64	21.80078	154.10938
		72	77.55469	633.83203
		80	195.86597	1880.13672
		88	556.63696	6282.51172
		96	1602.04042	20208.10577
		104	4415.42456	62449.17698
		112	12517.49614	195725.05378
		120	35194.26880	601372.60482
4	5	48	1.50000	6.25000
		56	5.43750	26.00000
		64	12.53906	84.94531
		72	36.91699	294.37109
		80	109.79541	1037.07568
		88	321.86841	3500.58521
		96	933.30453	11544.05643
		104	2721.82108	37661.62404
		112	7935.07730	121428.97480
		120	23076.82218	387346.71864
5	6	56	2.96875	15.15625
		64	7.71875	41.20313
		72	16.87305	123.41211
		80	49.79883	449.56250
		88	153.05359	1526.53137
		96	456.21996	5199.24472
		104	1326.80325	17065.98297
		112	3916.10218	55776.06466
		120	11599.23146	181249.72367
		128	34147.08199	581842.66684

(Continued)

Table 7.6 (*Continued*)

Ref. No (in Table 7.2)	Encoder memory, ν	$(\mathbf{n}, \mathbf{d}^2)$	$A(\mathbf{n})$	$B(\mathbf{n})$
6	7	64	13.71094	108.30469
		72	0	0
		80	76.07227	832.64795
		88	0	0
		96	779.69955	10961.66870
		104	0	0
		112	6516.27872	112150.75764
		120	0	0
		128	57494.86475	1170203.04167
		136	0	0

Table 7.7 Coefficients of generating function of Ungerboeck codes, PSK8, $E(S_c) = 1, d_1^2 = 0.585787, d_2^2 = 2$

Ref. No (in Table 7.5)	Encoder memory, ν	n_1, n_2	$(\mathbf{n}, \mathbf{d}^2)$	$A(\mathbf{n})$	$B(\mathbf{n})$
1	2	0 2	4.00000	1	1
		1 2	4.58579	4	8
		2 2	5.17157	8	28
		3 2	5.75736	16	76
		4 2	6.34315	32	192
		5 2	6.92893	64	464
		−1 4	7.41421	4	12
		6 2	7.51472	128	1088
		0 4	8.00000	17	66
		7 2	8.10051	256	2496
		1 4	8.58579	54	271
		8 2	8.68629	512	5632
		2 4	9.17157	152	940
		9 2	9.27208	1024	12544
		3 4	9.75736	392	2902
		10 2	9.85786	2048	27648
		4 4	10.34315	960	8288
		11 2	10.44365	4096	60416
		−2 6	10.82843	8	36
		5 4	10.92893	2272	22424
		12 2	11.02944	8192	131072
		−1 6	11.41431	54	301
		6 4	11.51472	5248	58304
		13 2	11.61522	16384	282624
		0 6	12.00000	241	1599
		7 4	12.10051	11904	147040
		14 2	12.20101	32768	606208
		1 6	12.58579	864	6712
		8 4	12.68629	26624	361984
		15 2	12.78680	65536	1294336
		2 6	13.17157	2730	24411
		9 4	13.27208	58880	873856
		16 2	13.37258	131072	2752512
		3 6	13.75736	7948	80560
		10 4	13.85786	129024	2075648

Table 7.7 (*Continued*)

Ref. No (in Table 7.5)	Encoder memory, ν	n_1, n_2	$(\mathbf{n}, \mathbf{d}^2)$	$A(\mathbf{n})$	$B(\mathbf{n})$
		17 2	13.95837	262144	5832704
		−3 8	14.24264	16	100
		4 6	14.34315	21840	247772
		11 4	14.44365	280576	4863488
		−2 8	14.82843	I52	1104
2	3	1 2	4.58579	2.000	5.000
		2 2	5.17157	4.000	17.000
		3 2	5.75736	8.000	50.000
		0 3	6.00000	1.000	1.000
		4 2	6.34315	16.000	132.000
		1 3	6.58579	4.500	12.000
		5 2	6.92893	38.000	328.000
		2 3	7.17157	13.000	60.000
		−1 4	7.41421	2.000	5.000
		6 2	7.51472	64.000	784.000
		3 3	7.75736	34.000	222.500
		0 4	8.00000	9.000	38.000
		7 2	8.10051	128.000	1824.000
		4 3	8.34315	84.000	715.000
		1 4	8.58579	29.000	173.000
		8 2	8.68629	256.000	4160.000
		5 3	8. 92893	200.000	2098.000
		2 4	9.17157	89.500	670.000
		9 2	9.27208	512.000	9344.000
		−1 5	9.41421	3.500	11.000
		6 3	9.51472	464.000	5788.000
		3 4	9.75736	248.750	2283.250
		10 2	9.85786	1024.000	20736.000
		0 5	10.00000	25.000	119.000
		7 3	10.10051	1056.000	15272.000
		4 4	10.34315	653.250	7147.250
		11 2	10.44365	2048.000	45568.000
		1 5	10.58579	105.000	692.500
		8 3	10.68629	2368.000	38960.000
		−2 6	10.82843	4.000	19.000
		5 4	IO.92893	1650.250	21041.000
		12 2	11.02944	4096.000	99328.000
		2 5	11.17157	368.000	3090.500
		9 3	11.27208	5248.000	96800.000
		−1 6	11.41421	28.000	175.000
		6 4	11.51472	4052.250	59128.000
		13 2	11.61522	8192.000	215040.000
		3 5	11.75736	1161.000	11831.500
		10 3	11.85786	11520.000	235456.000
		0 6	12.00000	138.000	1053.000
		7 4	12.10051	9736.250	160221.500
		14 2	12.20101	16384.000	462848.000
		4 5	12.34315	3399.125	40837.125
		11 3	12.44365	25088.000	562816.000
		1 6	12.58579	562.250	5107.250
		8 4	12.68629	22992.250	421656.000

(*Continued*)

Table 7.7 (*Continued*)

Ref. No (in Table 7.5)	Encoder memory, ν	n_1, n_2	$(\mathbf{n}, \mathbf{d}^2)$	$A(\mathbf{n})$	$B(\mathbf{n})$
		−2 7	12.82843	11.000	56.000
		5 5	12.92893	9440.375	130847.000
		2 6	13.17157	2000.000	21347.000
		−1 7	13.41421	100.000	679.500
3	4	2 2	5.17157	2.25000	11.00000
		3 2	5.75736	4.62500	28.25000
		0 3	6.00000	1.00000	1.00000
		4 2	6.34315	6.06250	53.00000
		1 3	6.58579	4.00000	11.00000
		5 2	6.92893	17.65625	183.68750
		2 3	7.17157	6.75000	36.25000
		6 2	7.51472	31.89063	392.06250
		3 3	7.75736	26.37500	197.87500
		0 4	8.00000	8.50000	37.00000
		7 2	8.10051	62.60156	907.64062
		4 3	8.34315	58.37500	542.50000
		1 4	8.58579	16.62500	84.50000
		8 2	8.68629	130.69141	2141.18750
		5 3	8.92893	134.71875	1553.34375
		2 4	9.17157	60.75000	460.00000
		9 2	9.27208	253.69727	4670.18359
		−1 5	9.41421	4.00000	9.00000
		6 3	9.51472	338.68750	4544.59375
		3 4	9.75736	168.21875	1599.50000
		10 2	9.85786	511.91504	10458.68359
		0 5	10.00000	16.50000	90.50000
		7 3	10.10051	749.00781	11549.16406
		4 4	10.34315	461.15625	5193.43750
		11 2	10.44365	1026.59229	22996.80176
		1 5	10.58579	97.37500	683.62500
		8 3	10.68629	1700.80469	29712.09375
		−2 6	10.82843	2.25000	10.00000
		5 4	10.92893	1188.35156	15781.65625
		12 2	11.02944	2044.05493	49908.37695
		2 5	11.17157	306.25000	2679.25000
		9 3	11.27208	3805.22461	73965.12305
		−1 6	11.41421	14.87500	69.25000
		6 4	11.51472	2993.64063	45527.34375
		13 2	11.61522	4098.97913	108271.85034
		3 5	11.75736	1018.65625	11125.59375
		10 3	11.85786	8326.37500	178634.47852
		0 6	12.00000	110.37500	812.50000
		7 4	12.10051	7419.76367	127090.25000
		14 2	12.20101	8192.56793	232756.80591
		4 5	12.34315	3011.50000	38539.18750
		11 3	12.44365	18243.64307	427952.74658
		1 6	12.58579	450.93750	3984.87500
		8 4	12.68629	17837.31836	340673.72266
		15 2	12.78680	16379.63419	498177.88092
		−2 7	12.82843	5.75000	30.25000
		5 5	12.92893	8448.88281	124622.67969

Table 7.7 (*Continued*)

Ref. No (in Table 7.5)	Encoder memory, ν	n_1, n_2	$(\mathbf{n}, \mathbf{d}^2)$	$A(\mathbf{n})$	$B(\mathbf{n})$
		12 3	13.02944	39544.89502	1006541.40?20
		2 6	13.17157	1772.48438	18951.12500
		−1 7	13.41421	93.62500	623.62500
4	5	3 2	5.75736	4.00000	24.00000
		0 3	6.00000	1.00000	1.00000
		4 2	6.34315	4.00000	35.25000
		5 2	6.92893	4.00000	51.12500
		2 3	7.17157	3.00000	14.50000
		6 2	7.51472	20.00000	228.43750
		3 3	7.75736	16.75000	118.50000
		0 4	8.00000	1.00000	2.00000
		7 2	8.10051	36.00000	511.40625
		4 3	8.34315	25.37500	265.00000
		1 4	8.58579	12.00000	68.00000
		8 2	8.68629	52.00000	927.17188
		5 3	8.92893	67.18750	812.31250
		2 4	9.17157	29.25000	232.50000
		9 2	9.27208	132.00000	2367.78906
		6 3	9.51472	198.65625	2670.62500
		3 4	9.75736	66.87500	662.00000
		10 2	9.85786	276.00000	5510.91797
		0 5	10.00000	9.00000	45.00000
		7 3	10.10051	398.29688	6410.01563
		4 4	10.34315	245.50000	2820.62500
		11 2	10.44365	484.00000	11167.10352
		1 5	10.58579	51.25000	364.50000
		8 3	10.68629	888.28906	16160.82813
		5 4	10.92893	662.81250	9093.59375
		12 2	11.02944	1012.00000	24688.94824
		2 5	11.17157	141.00000	1370.75000
		9 3	11.27208	2161.85547	42714.38672
		−1 6	11.41421	12.00000	64.00000
		6 4	11.51472	1595.78125	25448.53125
		13 2	11.61522	2116.00000	55169.68213
		3 5	11.75736	493.43750	5654.81250
		10 3	11.85786	4624.32227	101825.00000
		0 6	12.00000	56.50000	426.50000
		7 4	12.10051	4122.54688	73117.67969
		14 2	12.20101	4052.00000	116067.19019
		4 5	12.34315	1719.62500	22799.06250
		11 3	12.44365	9933.99512	240081.73145
		1 6	12.58579	202.12500	1896.87500
		8 4	12.68629	10327.07031	203600.92188
		15 2	12.78680	8100.00000	247255.74866
		−2 7	12.82843	5.00000	27.50000
		5 5	12.92893	4688.39063	73150.29688
		12 3	13.02944	22255.18994	577434.81543
		2 6	13.17157	906.25000	10135.93750
		9 4	13.27208	24496.43359	535587.00977
		−1 7	13.41421	51.25000	363.50000
		6 5	13.51472	12940.62500	228160.53125
		3 6	13.75736	3360.87500	44505.40625
		0 7	14.00000	248.25000	2321.75000

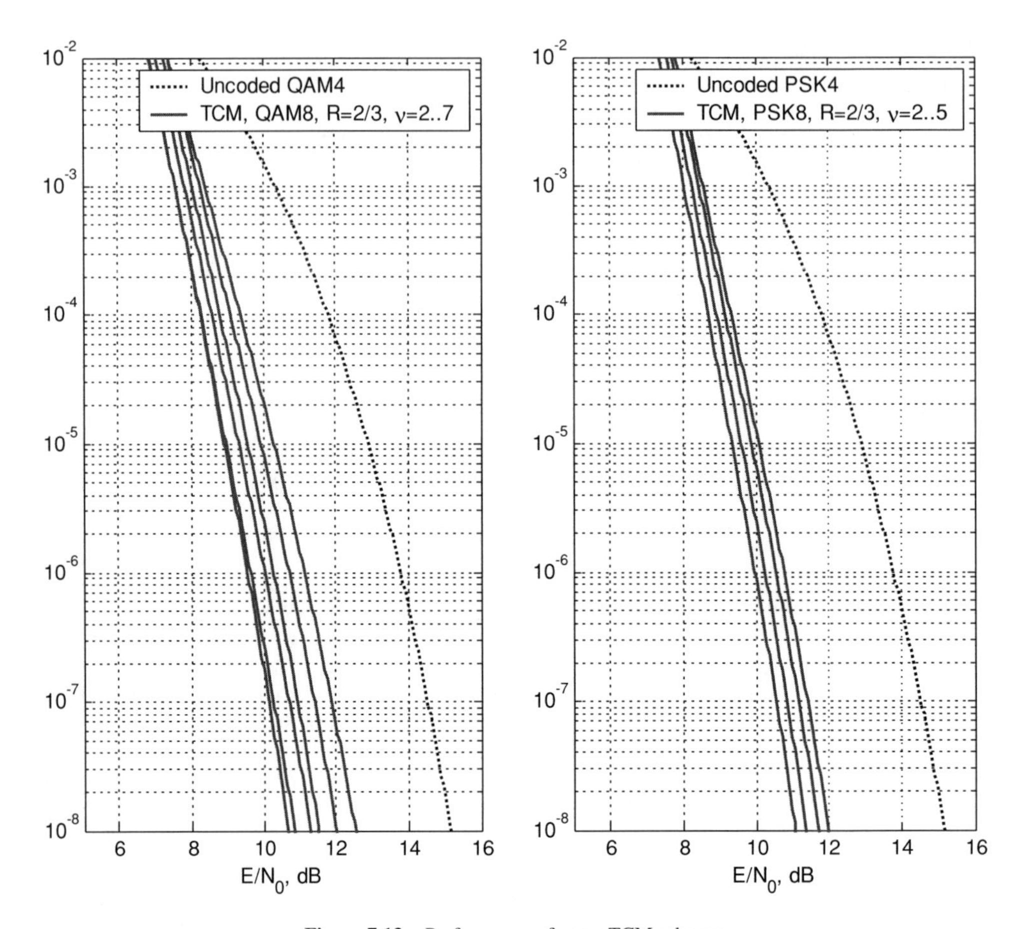

Figure 7.13 Performance of some TCM schemes

It should be noted that results given in the tables are for *equivalent* Ungerboeck codes generated by nonsystematic feedback-free encoders defined by the polynomial generator matrix $G(z)$. The nonsystematic feedback-free representation of the codes is more convenient for computation of the series expansion of the generating function.

Using the distance spectra one can compute the value of upper bound on bit error probability. Figure 7.13 shows plots of bit error probability as a function of signal to noise ratio E/N_0 for PSK and $\overline{E}/N_0$ for QAM in AWGN channel with two-sided noise power spectral density $N_0/2$ for some TCM schemes. This example corresponds to coded and uncoded transmission with the same rate $2/T$ bit/s. The curves are computed according to (7.26), (7.20) and (7.21) using the squared Euclidean distance spectra presented in Tables 7.6 and 7.7. Since the PSK8 and QAM8 are equivalent, the bit error probability for uncoded transmission computed in both cases as:

$$P_b^{(uncoded)} = 2Q(\sqrt{E/N_0}) - Q(\sqrt{E/N_0})^2$$

The plots in Figure 7.13 shows advantages of TCM over uncoded transmission and gives more detailed information about performance, than frequently used estimation based on asymptotic coding gain considerations.

References

[1] Ungerboeck, G. "Channel coding with multilevel/phase signals", IEEE Trans. Inform. Theory, vol. IT-28, pp. 55–66, 1982.

[2] Viterbi, A. J. and Omura, J. K. *Principles of Digital Communication and Coding.* New York: McGraw-Hill, 1979.

[3] Trofimov, A. N. and Kudryashov, B. D. "Distance Spectra and Upper Bounds on Error Probability for Trellis Codes", IEEE Trans. Inform Theory, vol. IT- 41, pp. 561–572, 1995.

[4] Biglieri, E. "High level modulation and coding for nonlinear satellite channels," IEEE Trans. Commun., vol. COM-32, pp. 616–627, 1984.

[5] Divsalar, D., Simon, M., and Yuen, J. H. "Trellis coding with asymmetric modulation," IEEE Trans. Commun., vol. COM-35, pp. 130–141, 1987.

[6] Honig, M. L. "Optimization of trellis codes with multilevel amplitude modulation with respect to an error probability criterion," IEEE Trans. Commun., vol. COM-34, pp. 821–825, 1986.

[7] Zehavi, E. and Wolf, J. K. " On the performance evaluation of trellis codes," IEEE Trans. Inform. Theory, vol. 33, pp. 196–202, 1988.

[8] Biglieri, E. and McLane, P. J. "Uniform distance and error probability properties of TCM schemes," IEEE Trans. Commun., vol. 39, pp. 41–53, 1991.

[9] Liu, Y.-J., Oka, I. and Biglieri, E. "Error probability for digital transmission over nonlinear channels with application to TCM," IEEE Trans. Inform. Theory, vol. 36, pp. 1101–1110, 1990.

[10] Rouanne, M. and Costello, D. J. "An algorithm for computing the distance spectrum of trellis codes," IEEE J. Selected Areas Commun., vol. 7, pp. 929–940, 1989.

8

MIMO

Andrei Ovchinnikov[1] and Sergei Semenov[2]
[1]*St. Petersburg State University of Aerospace Instrumentation, Russia*
[2]*Nokia Corporation, Finland*

8.1 MIMO Channel Model

8.1.1 Fading in Narrowband Channels

In this chapter we consider the transmission via the channels for which the classical model of channel with additive white Gaussian noise (AWGN), with statistically independent errors, mainly caused by the thermal noise at the receiver, is found inadequate and not very useful, since it does not take into account the basic types of errors, arising in such channels, and mainly influencing the receiving quality. Primarily, these are wireless radio channels, in which the effect of interference is more significant than the effect of thermal noise at the receiver. The main reason for communication quality degradation in such channels is the *multipath propagation*, where, because of reflections from obstacles' and objects' motions, the transmitted signal reaches the receiver as a set of interfered responses differing in amplitude, phase, angle of arrival, delay, and so on, that leads to such channel effects as *fading* and *dispersion* [3, 7, 31, 35, 43].

The main reasons for multipath propagation effects, as well as the main parameters characterizing such channels are described in Chapter 1, sections 1.3 and 1.4.

In this chapter, the methods for overcoming the dispersion effects in the channel are not considered in detail, so we will mainly consider the effect of fading in the transmission of narrowband signals. To send a signal over a communication channel the carrier frequency f_c modulated by information symbols is used, and this modulated signal is called a *bandpass signal*. Those symbols which modulate the carrier are called the equivalent baseband (lowpass) signal [7, 31], see Chapter 2 and section 2.1.1.

Let we have the pair of lowpass signals $x^{(p)}(t)$ and $x^{(q)}(t)$ and suppose that the quadrature phase shift keying (QPSK) is used for carrier modulation. In this case, the unmodulated bandpass signal is the sum of *in-phase* carrier and *quadrature* carrier, shifted by $\pi/2$ relative to the in-phase component. It can be easily shown (see (2.3)) that the modulated signal $s(t)$ given complex representation of lowpass signal

Modulation and Coding Techniques in Wireless Communications Edited by Evgenii Krouk and Sergei Semenov

can be written as:

$$s(t) = x^{(p)}(t)\cos(f_c t) - x^{(q)}(t)\sin(f_c t) = \mathrm{Re}\{x(t)e^{2\pi j f_c t}\} \quad (8.1)$$

where f_c is the carrier frequency, and $x(t)$ is the lowpass envelope.

Then, after transmission of signal $s(t)$ via a channel with a fading caused by multipath propagation, the received signal $r(t)$ is equal to:

$$r(t) = \sum_i h_i(t)s[t - \tau_i(t)] \quad (8.2)$$

where $h_i(t)$ is the fading coefficient and $\tau_i(t)$ is the i-th path delay. Here we ignore the additive thermal noise. Applying (8.1) to (8.2), we obtain:

$$r(t) = \mathrm{Re}\left\{\left(\sum_i h_i(t)x\left[t - \tau_i(t)\right]\right)e^{2\pi j f_c[t-\tau_i(t)]}\right\} =$$

$$= \mathrm{Re}\left\{\left(\sum_i h_i(t)e^{-2\pi j f_c \tau_i(t)}x[t - \tau_i(t)]\right)e^{2\pi j f_c t}\right\}$$

Then the correspondent low-frequency signal will have the form:

$$z(t) = \sum_i h_i(t)e^{-2\pi j f_c \tau_i(t)}x[t - \tau_i(t)] \quad (8.3)$$

For simplicity, assume that the unmodulated carrier is transmitted at frequency f_c, that is, $x(t) = 1$ for any t. Then (8.3) simplifies to:

$$z(t) = \sum_i h_i(t)e^{-2\pi j f_c \tau_i(t)} = \sum_i h_i(t)e^{-j\theta_i(t)} \quad (8.4)$$

Thus, the lowpass signal $z(t)$ consists of the sum of time-varying complex vectors with amplitude $h_i(t)$ and phase $\theta_i(t)$. Note that the phase θ_i will change by 2π when the delay τ_i changes by $1/f_c$ only. For example, when the transmitter works on frequency of $f_c = 900$ MHz, the delay $1/f_c$ is 1.1 nanoseconds. This corresponds to a change in the path of spreading signal by 33 cm (in free space). Thus, the phase θ_i in (8.4) may change significantly with relatively small changes in spreading delay.

Vector interpretation of (8.4) helps to graphically describe the effect of fading by means of complex unit circle. In Figure 8.1 the transmitted signal is shown (which can be considered as expected signal), taking into account the presence of the reflected signal $x_i^{(p)}(t) + jx_i^{(q)}(t) = h_i(t)e^{-j\theta_i(t)}$ with in-phase and quadrature components $x^{(p)}$ and $x^{(q)}$, respectively, spread during multipath propagation by the path i. Because of reflections occurring in the path i, the amplitude of the reflected signal decreases, which is taken into account in the $h_i(t)$ coefficient, but because of the propagation distance increasing, the reflected signal is delayed in phase compared to the expected signal. As a result of vector addition of the expected signal and the reflected signal, the resulting signal both attenuated in amplitude and shifted in phase.

In fact, the receiving side is affected not only by the reflected signal, but also the set of signals propagated along different paths. All of them are summed with each other which leads to addition of their in-phase and quadrature components $x_i^{(p)}(t)$ and $x_i^{(q)}(t)$, and after all additions in (8.4) of all the reflected paths the resulting signal $x_r^{(p)}(t)$ and $x_r^{(q)}(t)$ will be obtained. Since the number of such paths is usually large and they are usually assumed independent, then at a fixed moment in time, following the central limit theorem, $x_r^{(p)}(t)$ and $x_r^{(q)}(t)$ will have a Gaussian probability distribution function. When transmitting an unmodulated carrier as in (8.4), the received signal will be:

$$z(t) = r_0(t)e^{-j\theta(t)}$$

where $r_0(t)$ is the resulting envelope amplitude, which equals to the modulus of $z(t)$:

$$r_0(t) = |z(t)| = \sqrt{x_r^{(p)}(t)^2 + x_r^{(q)}(t)^2}$$

and $\theta(t)$ is the resulting phase.

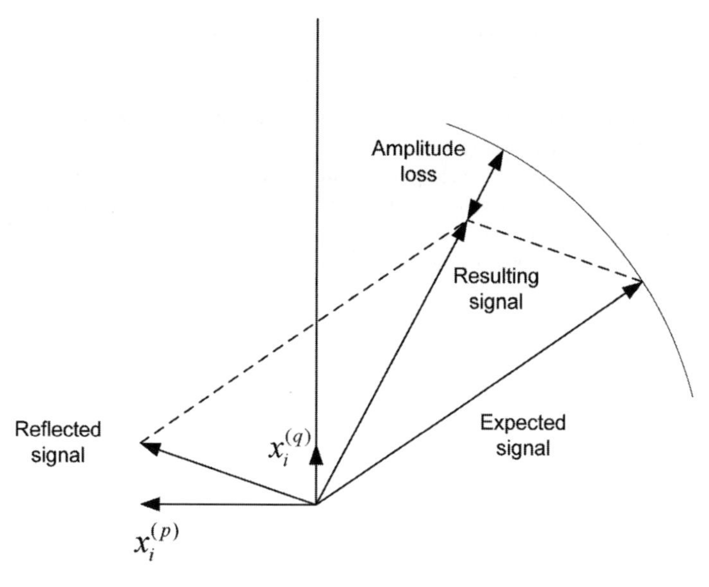

Figure 8.1 The influence of the reflected signal to expected signal

Thus, the amplitude r_0 of the complex envelope will have a Rayleigh distribution (the distribution of the complex Gaussian process amplitude), see section 1.4.1:

$$p(r_0) = \frac{r_0}{\sigma^2} \exp\left(-\frac{r_0^2}{2\sigma^2}\right)$$

where σ^2 is the total power of the multipath signal. Described fading model is called *Rayleigh fading*. If, during the signal propagation, there is the one distinct signal among the set of reflected paths, which is transmitted in direct line of sight and not affected by multipath, then the amplitude of the received signal will have a Ricean distribution, see section 1.4.1:

$$p(r_0) = \frac{r_0}{\sigma^2} \exp\left(-\frac{r_0^2 + s^2}{2\sigma^2}\right) I_0\left(\frac{r_0 s}{\sigma^2}\right)$$

here s^2 is the power of direct sight component, and I_0 designates the modified zero-order Bessel function of the first kind.

If we consider not only the analog signal $r(t)$ at the receiver input, but also the discrete output after the matched filter and sample and hold block, and take into account the effect of thermal noise, which will be considered to have Gaussian distribution with zero mean, we obtain the well-known (discrete) equation for Rayleigh fading channel:

$$y_t = hx_t + \eta_t \tag{8.5}$$

where x_t is the transmitted symbol, y_t is the received symbol, h is the complex Gaussian random variable which describes fading (its real and imaginary parts are Gaussian random variables with zero mean), whose modulus $|\mu|$ is distributed over the Rayleigh distribution law, and η_t is the additive white Gaussian noise.

8.1.2 Fading Countermeasures: Diversity

Here we will consider general approaches usually used to combat the effect of fading described in previous sections. *Diversity* is one of the main means to countermeasure fading in radio channel [24]. The principle of diversity consists of providing two or more channels for the same information signal in such a way that these channels are affected by statistically independent fading. Then, if one of the propagation paths is in deep fade, it is expected that this will not happen with the second path, and therefore, it is possible to maintain a reasonable probability of error. Thus, diversity is used to increase reliability of the received signal. The condition of statistical independence of organized diversed channels is critically important to ensure diversity gain.

In fact, many diversity techniques are simply utilizing the usage of repetition code. Besides diversity methods, it is also necessary to consider methods of *combining* signals, which come from different paths. Diversity methods are classified both by the methods of additional channels organization, and by multiple received signals combining techniques.

The main methods of diversity are diversity in space, frequency and time. These types of diversity are briefly presented in Figure 8.2 [6, 7].

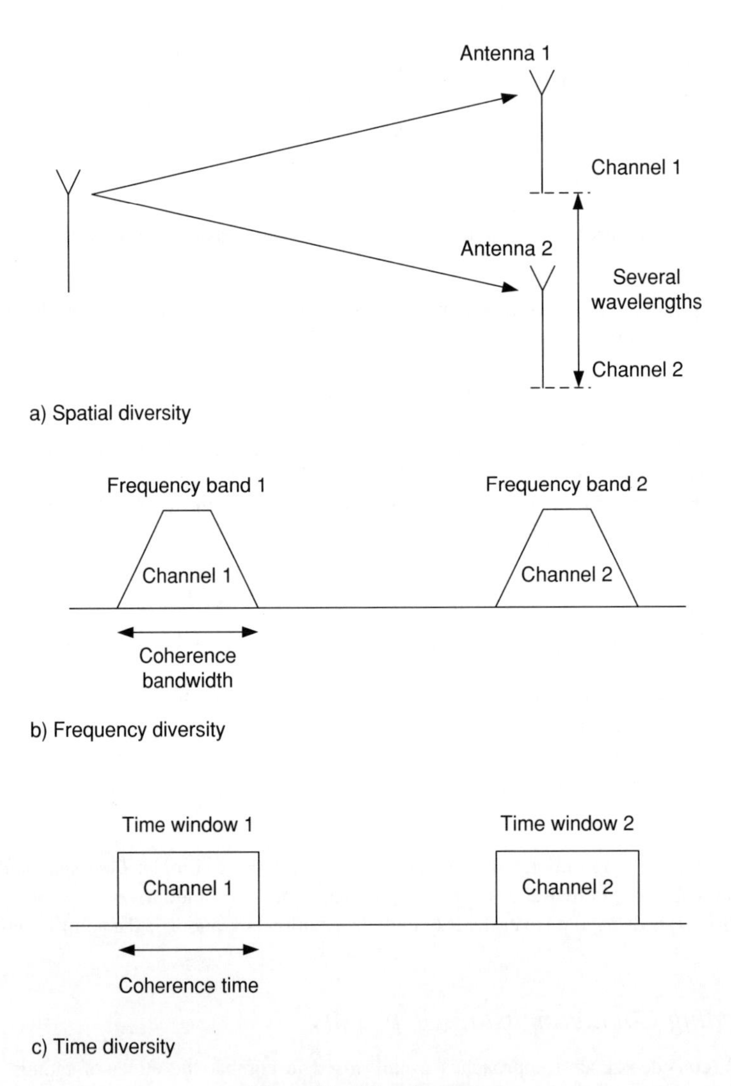

Figure 8.2 Diversity types

1. *Spatial diversity*. In section 8.1.1 it was noted that Rayleigh fading is caused by the summation of transmitted signal responses, differing in phase (and this also causes changes in the amplitude, see Figure 8.1), and that under Rayleigh fading small change in delay (and hence in the distance of signal propagation) may cause a significant change in phase, and hence one can expect that with the spatial diversity of receiving antennas at the distance of several wavelengths (in [31, 36] the distance of about 10 wavelengths is recommended) the fading channels will be completely uncorrelated. In this chapter we will consider MIMO systems based on this diversity type.
2. *Frequency diversity*. If, for some reason, spatial diversity cannot be implemented, then the diversity may be done by frequency. In this case the copies of the signal are transmitted on different frequencies, and again the independence of fading in different frequency channels should be provided, for this purpose the frequencies diversity should be several times larger than the coherence bandwidth $1/\Delta$, where Δ is the delay spread (see section 1.5.2).
3. *Time diversity*. This type of diversity is implemented through the transmission of the signal at different points in time (time slots). To ensure the independence of channels, their diversity in time should be several times larger than the coherence time.

With any type of diversity on the receiving side the various responses of the same signal are obtained, affected by the various (ideally independent) fading. For making decisions concerning the transmitted signal from this redundant set of responses, the received signals must be combined according to some criterion. Below three of the most commonly used methods of combining are listed, each of them may be applied to any of the above diversity methods. Schematically combining techniques are presented in Figure 8.3 [24].

1. *Maximum ratio combining* (MRC). This combining method will be described in detail later; here we only note that this method is optimal in terms of maximum likelihood detection. All received signals are weighted by their personal SNR and then summed. Weighting minimizes the noise occurred as a result of fading (that is, maximizes the SNR). If we suppose that the diversity is performed by n channels and each channel i has the same average SNR $\bar{\gamma}_i = \xi$, then the summation of weighted signals gives the total average SNR $\bar{\gamma}$ [5, 36]:

$$\bar{\gamma} = \sum_{i=1}^{n} \bar{\gamma}_i = \sum_{i=1}^{n} \xi = n\xi$$

   that is, even if in each channel the average SNR ξ is less than some acceptable level (for a given level of error probability), then the maximum ratio combining, nevertheless, may give an acceptable overall average SNR. Summation of signals with this method should be performed in-phase (coherently).
2. *Equal gain combining* (EGC). This combining method is similar to MRC, but the received signals are not weighted (which can be considered as a weighting with equal coefficient 1 for all channels), but immediately summed. This method is used if there is no possibility of tracking the amplitudes of the received signals and their timely change, and of course, it gives a worse result compared to MRC, but nevertheless, it is able to provide diversity gain compared to a system with no diversity. This method also requires coherent summation of signals.
3. *Selection combining* (SC). With selection combining (or with *switched combining*) signals are not actually combined, but diversed channels are switched by the criterion of the strongest signal selection. That is, the signal with maximum SNR is considered as the received signal. It is clear that this combining method does not require coherence of the signals, but it also loses to MRC, because it does not take into account any information except that obtained from one channel which is the best at this moment.

In the future, we will assume that we consider a system with spatial diversity for transmission over frequency-nonselective channel with Rayleigh fading.

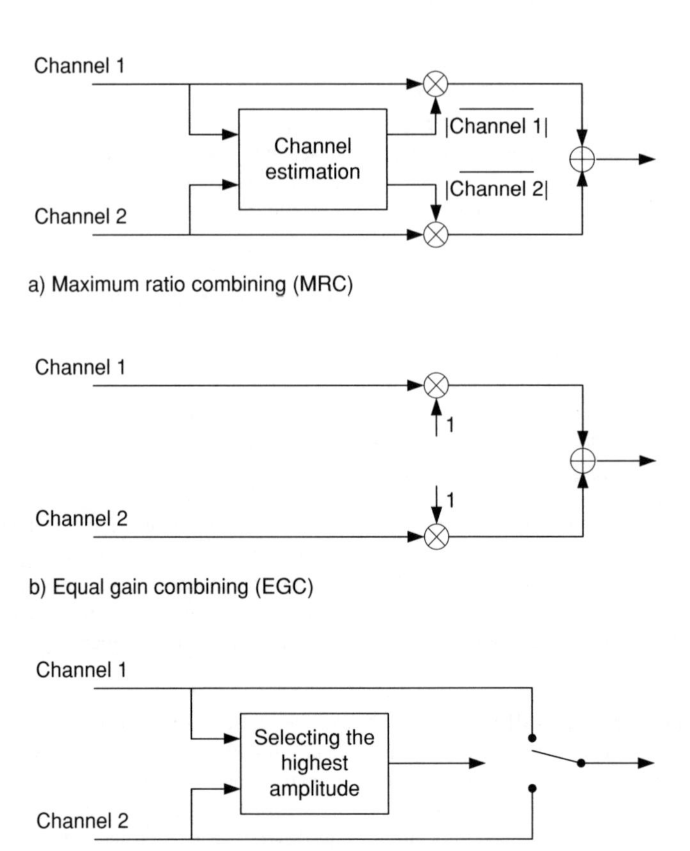

Figure 8.3 Combining methods

8.1.3 MIMO Channel Model

Now, after identifying in section 8.1.1 the discrete-time channel model with Rayleigh fading in (8.5), and describing in section 8.1.2 the methods of diversity and combining as one of the main measures for combat fading, we may proceed to the definition of multiple-input multiple-output (MIMO) system model.

In section 8.1.2 we described spatial diversity, assuming that several *independent* channels are utilized to combat fading, and such channels are obtained by using several receiving antennas and one transmitting antenna. However, this approach may be generalized by considering spatial diversity not only during *receiving* the signal, but during its *transmission*. This requires not only the method for combining different responses of the *same signal* at the receiver, but the organization (possibly nontrivial) of signal transmission by many transmitting antennas. In this section we will consider the channel model for such MIMO systems.

Suppose there are N transmit and M receive antennas. At time t the signals $x_{t,n}$, $n = 1, \ldots, N$ are transmitted simultaneously from N transmit antennas. Each transmitted signal is affected by channel fading, and the response from each signal from each of N transmitting antennas is received at each of M receive antennas. We will consider a frequency-nonselective channel with flat Rayleigh fading (section

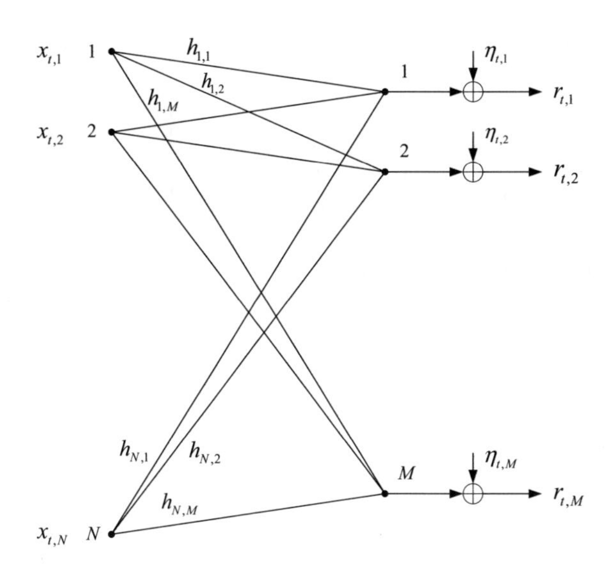

Figure 8.4 Channel with multiple input and multiple output (MIMO)

1.4.4), that is, we will suppose the absence of dispersion in the channel, which leads to intersymbol interference.

For each pair (n, m) of the transmit antenna n and the receive antenna m the transmitted signal is affected by the fading coefficient $h_{n,m}$ (path gain). Then at the antenna m at time t the received signal $r_{t,m}$ is [16]:

$$r_{t,m} = \sum_{i=1}^{N} h_{n,m} x_{t,n} + \eta_{t,m} \tag{8.6}$$

where $\eta_{t,m}$ is the additive Gaussian noise (thermal noise) at the receiving antenna m at time t. Graphically, the model in (8.6) is shown in Figure 8.4.

We will suppose that the channel is quasi-stationary, that is, the fading coefficients $h_{n,m}$ are constant over a certain time period T', and from period to period they are varied independently. Nevertheless, there are models that take into account the correlation between time slots, for example, Jakes model [19].

The value of T' determines the rate of fading. The fading is slow, if the data block is transferred during the time T, less than T'. In this case the coefficients $h_{n,m}$ are constant during the entire transmission period. On the other hand, if $T = T'$, then the coefficients $h_{n,m}$ are changed simultaneously with the transmission of data block.

We may represent the signals transmitted during the time period T from N transmitting antennas using the $(T \times N)$-matrix $\mathbf{X}$:

$$\mathbf{X} = \begin{pmatrix} x_{1,1} & x_{1,2} & \cdots & x_{1,N} \\ x_{2,1} & x_{2,2} & \cdots & x_{2,N} \\ \vdots & \vdots & \ddots & \vdots \\ x_{T,1} & x_{T,2} & \cdots & x_{T,N} \end{pmatrix} \tag{8.7}$$

Similarly, we may represent the received signal in the form of $(T \times M)$-matrix $\mathbf{R}$:

$$\mathbf{R} = \begin{pmatrix} r_{1,1} & r_{1,2} & \cdots & r_{1,M} \\ r_{2,1} & r_{2,2} & \cdots & r_{2,M} \\ \vdots & \vdots & \ddots & \vdots \\ r_{T,1} & r_{T,2} & \cdots & r_{T,M} \end{pmatrix}$$

Assuming $T < T'$, that is, that fading coefficients are constant during the transmission period, these coefficients may also be written in $(N \times M)$ channel matrix $\mathbf{H}$:

$$\mathbf{H} = \begin{pmatrix} h_{1,1} & h_{1,2} & \cdots & h_{1,M} \\ h_{2,1} & h_{2,2} & \cdots & h_{2,M} \\ \vdots & \vdots & \ddots & \vdots \\ h_{N,1} & h_{N,2} & \cdots & h_{N,M} \end{pmatrix} \quad (8.8)$$

Then we can write (8.6) in matrix form:

$$\mathbf{R} = \mathbf{X} \cdot \mathbf{H} + \mathcal{N} \quad (8.9)$$

where $\mathcal{N}$ is the matrix of noise:

$$\mathcal{N} = \begin{pmatrix} \eta_{1,1} & \eta_{1,2} & \cdots & \eta_{1,M} \\ \eta_{2,1} & \eta_{2,2} & \cdots & \eta_{2,M} \\ \vdots & \vdots & \ddots & \vdots \\ \eta_{T,1} & \eta_{T,2} & \cdots & \eta_{T,M} \end{pmatrix}$$

We will assume that the noise $\mathcal{N}$, fading coefficients $\mathbf{H}$ and signals $\mathbf{X}$ are independent of each other. Also assume that the elements of $\mathbf{H}$ (more precisely, their envelopes $|h_{n,m}|$) have the Rayleigh distribution.

Consider the capacity of MIMO channel with N transmitting and M receiving antennas [16, 22, 40]. We will distinguish the case, when the transmitter does not know the state of the channel (the so-called open-loop system), and the case, when the information about the channel (channel matrix $\mathbf{H}$) is available to the transmitter (closed-loop system).

If $\mathbf{K_X}$ is the covariance matrix of input $\mathbf{X}$, and $\mathbf{H}_d$ is the specific (deterministic) realization of the channel matrix $\mathbf{H}$, that the capacity of MIMO channel with N transmitting and M receiving antennas is a random quantity:

$$C(\mathbf{K_X}) = \max_{\text{Tr}(\mathbf{K_X}) \le N} \log_2(\det[\mathbf{I}_M + (\gamma/N)\mathbf{H}_d^H \cdot \mathbf{K_X} \cdot \mathbf{H}_d])\,\text{bit/(s Hz)}$$

here $\mathbf{I}_M$ is $(M \times M)$-identity matrix, γ is SNR at the receiver, $\text{Tr}(\cdot)$ is the trace of matrix, and the operator $(\cdot)^H$ designates Hermitian conjugation, that is, the matrix which is transposed and element-wise complex-conjugate to the original matrix.

In case the channel is not known at the transmitter, one may try uniformly to distribute the input power between the transmit antennas. Then $\mathbf{K_X} = \mathbf{I}_N$, and we obtain the following expression:

$$\hat{C} = \log_2(\det[\mathbf{I}_M + (\gamma/N)\mathbf{H}_d^H \cdot \mathbf{H}_d])\,\text{bit/(s Hz)}$$

which is not exactly the capacity in Shannon's sense, since in some cases the value $\hat{C}$ may be exceeded by means of another power distribution. However, for uncorrelated iid Rayleigh channel the uniform distribution of power is optimal.

If $r = \text{rank}(\mathbf{H}_d)$ and λ_i, $i = 1, 2, \ldots, r$ – nonzero eigenvalues of matrix $\mathbf{H}_d^H \cdot \mathbf{H}_d$, then:

$$\hat{C} = \sum_{i=1}^{r} \log_2[1 + (\gamma/N)\lambda_i]$$

In the case of the closed-loop system, when the matrix $\mathbf{H}_d$ is known to the transmitter, the input power may be redistributed to give more power to channels in a good state and less power to channels in a bad state. To perform this, the procedure of so-called water-filling is known [9]. If the total power γ is distributed on r antennas as powers γ_i, then the capacity of the closed-loop system will be:

$$C = \max_{\sum_{i=1}^{r} \gamma_i = \gamma} \sum_{i=1}^{r} \log_2[1 + \gamma_i \lambda_i]$$

In the case of nondeterministic (random) nature of matrix $\mathbf{H}$ the random capacity (with equal power distribution) may be expressed as:

$$\hat{C} = \log_2(\det[\mathbf{I}_M + (\gamma/N)\mathbf{H}^H \cdot \mathbf{H}])\ \text{bit/(s Hz)} \tag{8.10}$$

It is possible to show that when $M = N$, the capacity increases linearly with the number of antennas.

Note that when considering the SISO (single-input single-output) channel, that is, when $M = N = 1$, the expression in (8.10) becomes:

$$C = \log_2(1 + \gamma |h|^2)$$

which coincides with Shannon's capacity for the Gaussian channel with fading coefficient h.

Now we will consider two particular cases: SIMO (single-input multiple-output) channel with one transmit antenna, $N = 1$, and MISO (multiple-input single-output) channel with one receive antenna, $M = 1$. In both cases, we limit our consideration by the deterministic channels [20].

In the SIMO case, taking into account $M > N$, (8.10) can be written as:

$$C = \log_2(\det[\mathbf{I}_N + (\gamma/N)\mathbf{H}_d^H \cdot \mathbf{H}_d])$$

With $N = 1$, the product $\mathbf{H}_d^H \cdot \mathbf{H}_d = \sum_{i=1}^{M} |h_i|^2$, and then:

$$C = \log_2\left(1 + \sum_{i=1}^{M} |h_i|^2 \gamma\right)$$

In this case knowledge of the channel on the transmitter side does not provide gain.

In MISO case $M = 1$ and $\mathbf{H} \cdot \mathbf{H}_d^H = \sum_{j=1}^{N} |h_j|^2$, hence:

$$C = \log_2\left(1 + \sum_{j=1}^{N} |h_j|^2 (\gamma/N)\right)$$

If the channel is known at the transmitter, and since $\text{rank}(\mathbf{H}_d) = 1$, there is only one nonzero eigenvalue:

$$\lambda = \sum_{j=1}^{N} |h_j|^2$$

so the capacity is:

$$C = \log_2\left(1 + \sum_{j=1}^{N} |h_j|^2 \gamma\right)$$

Thus, with an increasing number of receiving and transmitting antennas in SIMO and MISO channels, correspondingly, the capacity grows logarithmically (in the case of MISO – when the channel matrix is known at the transmitter).

8.2 Space-Time Coding

8.2.1 Maximum Ratio Combining

In this section we will consider in more detail the diversed signal combining with the help of maximum ratio combining technique (MRC) briefly described in section 8.1.2. Strictly speaking, this method is not directly connected to MIMO systems (since it can be applied both in SISO and MIMO systems), consideration of MRC is important for the definition of the space-time coding procedure.

Let us consider the simplest case of spatial diversity (section 8.1.2), when the system comprises one transmit and two receive antennas (Figure 8.2a). In such a way, by means of diversity the receiving antennas in space, the two (independent) channels are organized, on which the same signal x is transmitted, affected by Rayleigh fading according to the model in (8.5) in section 8.1.1.

Assume that the fading coefficient in one channel is described by the value h_1, and in another by the value h_2, which are complex values:

$$\begin{aligned} h_1 &= |h_1| e^{j\theta_1} \\ h_2 &= |h_2| e^{j\theta_2} \end{aligned}$$

In addition to fading in channel, each receive antenna adds to the received signal the thermal noise η_1 and η_2, correspondingly. Then the signals received from two diversed channels are:

$$\begin{aligned} y_1 &= h_1 x + \eta_1 \\ y_2 &= h_2 x + \eta_2 \end{aligned}$$

If we possess the ideal knowledge of the channel (that is, we know the exact values of h_1 and h_2, their estimation is possible, for example, by means of the pilot sequences transmissions [31]), fading can be compensated (optimally from the point of view of signal-to-noise ratio) by multiplying each of the received signals by the value which is a complex conjugate with fading coefficient of the correspondent channel. After this the signals from both channels are combined by additional operation:

$$\begin{aligned} \tilde{x} &= h_1^* y_1 + h_2^* y_2 = \\ &= h_1^* h_1 x + h_1^* \eta_1 + h_2^* h_2 x + h_2^* \eta_2 = \\ &= (|h_1|^2 + |h_2|^2) x + h_1^* \eta_1 + h_2^* \eta_2 \end{aligned} \tag{8.11}$$

Then combined signal $\tilde{x}$ is fed into maximum likelihood detector input. Described combining technique is depicted in Figure 8.5. With such a receiving procedure the maximum likelihood detector coincides with the detector by the minimal Euclidean distance, that is, the decision x' about transmitted symbol may be found by searching the nearest (in Euclidean metrics) symbol to the received combined symbol $\tilde{x}$ from the set of possible transmitted symbols x_i:

$$d_E(\tilde{x}, x') \leq d_E(\tilde{x}, x_i) \quad \text{for all } i$$

where $d_E(a, b)$ is the Euclidean distance between signals a and b.

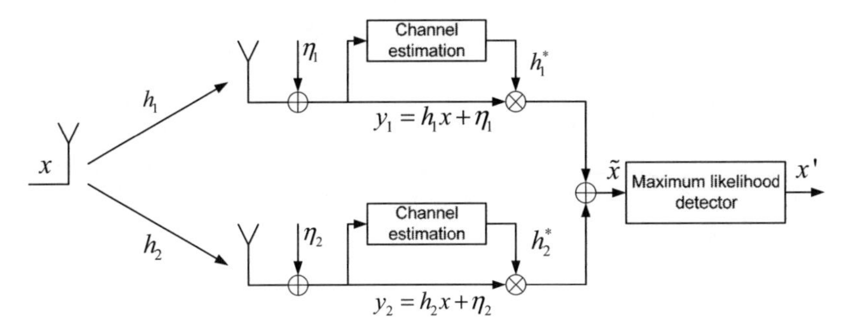

Figure 8.5 MRC technique for two receiving antennas

It may be shown [7, 31], that the symbol error probability for MRC with L receiving antennas is:

$$P_e = \left(\frac{1-\mu}{2}\right)^L \sum_{k=0}^{L-1} \binom{L-1+k}{k} \left(\frac{1+\mu}{2}\right)^k \approx \binom{2L-1}{L} \left(\frac{1}{4E_c/N_0}\right)^L$$

where

$$\mu = \sqrt{(E_c/N_0)/(1 + E_c/N_0)}$$

E_c is the energy of received signal per bit per antenna, and therefore, overall bit energy is $E_b = LE_c$.

8.2.2 *Definition of Space-Time Codes*

Now, when we described in detail the receiving procedure for one transmit and two receive antennas, we may consider a more general case, when diversity is achieved with the help of multiple receive and multiple transmit antennas. The model of such a MIMO channel is described in section 8.1.3. When only one transmit antenna is used, it transmits the dedicated signal x, but in the case of *multiple* transmit antennas we have not only the possibility of combining several responses in one signal (possibly better than each of the received responses), but we may also apply some procedure allowing the system quality to be improved by redistribution of transmitted signals between the transmit antennas at some moment in time. Practically, such a rule describing which signal is to be transmitted on which antenna at the particular time moment (that is, the definition of matrix $\mathbf{X}$ (8.7)), is in fact the definition of the *space-time code*. More precisely, assume that during the time period T we need to transmit k (complex) symbols $x_1, x_2, \ldots, x_k$ using N transmit antennas. If every symbol x_i corresponds to b information bits, then the input to space-time encoder is the block of $k \times b$ bits. These bits form the matrix:

$$\mathcal{G} = \begin{pmatrix} g_{1,1} & g_{1,2} & \cdots & g_{1,N} \\ g_{2,1} & g_{2,2} & \cdots & g_{2,N} \\ \vdots & \vdots & \ddots & \vdots \\ g_{T,1} & g_{T,2} & \cdots & g_{T,N} \end{pmatrix} \tag{8.12}$$

where the elements $g_{i,j}$ of matrix $\mathcal{G}$ are the linear combinations of symbols x_i, $i = 1, \ldots, k$, and their complex conjugates. In fact the code matrix $\mathcal{G}$ is the signal matrix $\mathbf{X}$ in (8.7) for transmission via the MIMO channel (8.9). Since the dimensions of matrix $\mathcal{G}$ are space (antennas) and time, the code obtained by the described procedure is called space-time code. The rate of this code is defined as:

$$R = k/T \tag{8.13}$$

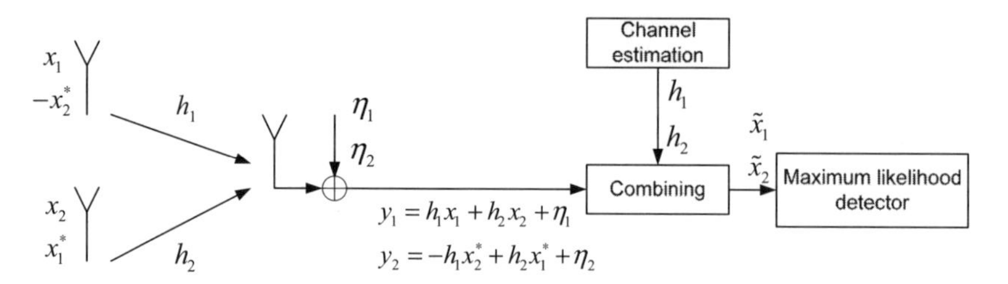

Figure 8.6 Space-time code $\mathbf{G}_2$ with two transmit and one receive antennas

Hence, the rate of the space-time code is limited from above as $R \leq 1$. With $R = 1$ we have the space-time codes with maximal rate. These codes are preferable, since they don't lead to bandwidth expansion.

At the receiving side the responses of transmitted signals are received by M antennas, and after that the combining technique analogous to MRC from section 8.2.1 may be applied. The decoding of space-time codes is described in detail in section 8.3.1.

8.2.3 Space-Time Codes with Two Transmit Antennas

The simplest scheme of space-time coding was proposed by Alamouti [2] for the case of two transmit antennas, $N = 2$. The code matrix was defined as:

$$\mathbf{G}_2 = \begin{pmatrix} x_1 & x_2 \\ -x_2^* & x_1^* \end{pmatrix} \tag{8.14}$$

So, the number of transmitted symbols within one block is also two, $k = 2$. The transmission is made by $T = 2$ time periods, that is, the rate R (see (8.13)) of such code is equal to one. During the first time slot, $T = 1$, the first antenna transmits the symbol x_1, while the second antenna transmits the symbol x_2. During the next time slot, $T = 2$, the first and second antennas transmit $-x_2^*$ and x_1^* correspondingly.

First, we will consider the case when there is only one antenna on the receiving side, $M = 1$, and for this case we will describe the decoding of the space-time code $\mathbf{G}_2$. Then we will generalize the results for the case of the larger number of receive antennas. The receiving scheme for this case is shown in Figure 8.6.

Suppose that the fading coefficients of the channels remain constant within the transmission period, that is, they are equal for both time slots. Taking into account (8.14), the received signals y_1 at time $T = 1$ and y_2 at time $T = 2$ are:

$$\begin{aligned} y_1 &= h_1x_1 + h_2x_2 + \eta_1 \\ y_2 &= -h_1x_2^* + h_2x_1^* + \eta_2 \end{aligned}$$

where η_1 and η_2 is the thermal noise at the receive antenna at time moments $T = 1$ and $T = 2$ correspondingly. To detect the transmitted symbols it is needed to extract the signals x_1 and x_2 from y_1 and y_2. To achieve this, they are fed to a combining block, which also uses the information about the channel. The output of the combining block are the estimates $\tilde{x}_1$ and $\tilde{x}_2$ of the signals x_1 and x_2. For the signal x_1 the estimate is computed as:

$$\begin{aligned} \tilde{x}_1 &= h_1^*y_1 + h_2y_2^* = \\ &= h_1^*h_1x_1 + h_1^*h_2x_2 + h_1^*\eta_1 - h_2h_1^*x_2 + h_2h_2^*x_1^* + h_2\eta_2^* = \\ &= (|h_1|^2 + |h_2|^2)x_1 + h_1^*\eta_1 + h_2\eta_2^* \end{aligned} \tag{8.15}$$

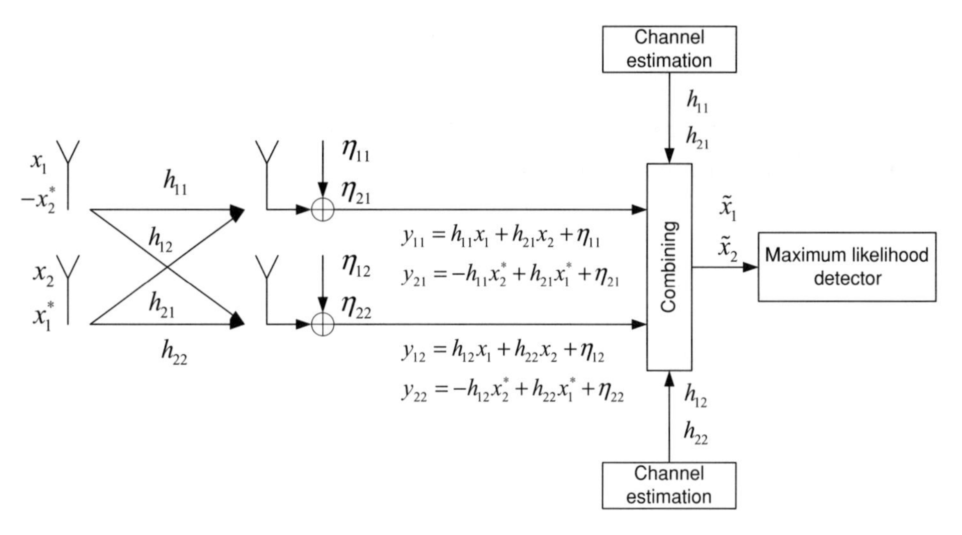

Figure 8.7 Space-time code $\mathbf{G}_2$ with two transmit and two receive antennas

Similarly, for the signal x_2 the estimation is calculated as:

$$
\begin{aligned}
\tilde{x}_2 &= h_2^* y_1 - h_1 y_2^* = \\
&= h_2^* h_1 x_1 + h_2^* h_2 x_2 + h_2^* \eta_1 + h_1 h_1^* x_2 - h_1 h_2^* x_1 - h_1 \eta_2^* = \\
&= (|h_1|^2 + |h_2|^2) x_2 + h_2^* \eta_1 - h_1 \eta_2^*
\end{aligned}
\tag{8.16}
$$

Due to the orthogonality of the code $\mathbf{G}_2$ [38], the second symbol x_2 is cancelled during the symbol x_1 estimation in (8.15), and in (8.16) the x_1 is cancelled in a similar way. Then the estimates $\tilde{x}_1$ and $\tilde{x}_2$ are fed to the input of maximum likelihood detector, which calculates the most probable symbols corresponding to these estimates, as in Figure 8.5.

Now for the same space-time code $\mathbf{G}_2$ we consider the case of two receive antennas, $M = 2$. At the same time we will show how the given receiving methods may be generalized for the case of arbitrary number M of receive antennas [13]. This case is shown in Figure 8.7. Here j in h_{ij}, η_{ij} and y_{ij} denotes the index of receive antenna, i in h_{ij} denotes the index of transmit antenna, and in η_{ij} and y_{ij} i denotes the index of time slot.

Then the output of the first receive antenna is:

$$
\begin{aligned}
y_{11} &= h_{11} x_1 + h_{21} x_2 + \eta_{11} \\
y_{21} &= -h_{11} x_2^* + h_{21} x_1^* + \eta_{21}
\end{aligned}
$$

and the output of the second receive antenna is:

$$
\begin{aligned}
y_{12} &= h_{12} x_1 + h_{22} x_2 + \eta_{12} \\
y_{22} &= -h_{12} x_2^* + h_{22} x_1^* + \eta_{22}
\end{aligned}
$$

These expressions may be generalized as:

$$
\begin{aligned}
y_{1i} &= h_{1i} x_1 + h_{2i} x_2 + \eta_{1i} \\
y_{2i} &= -h_{1i} x_2^* + h_{2i} x_1^* + \eta_{2i}
\end{aligned}
$$

where $i = 1, \ldots, M$, and M is the number of receiving antennas.

Estimates $\tilde{x}_1$ and $\tilde{x}_2$ of the symbols x_1 and x_2 are obtained in combining block as:

$$\tilde{x}_1 = h_{11}^* y_{11} + h_{21} y_{21}^* + h_{12}^* y_{12} + h_{22} y_{22}^* \tag{8.17}$$

$$\tilde{x}_2 = h_{21}^* y_{11} - h_{21} y_{21}^* + h_{22}^* y_{12} - h_{12} y_{22}^* \tag{8.18}$$

Again these expressions may be generalized as:

$$\tilde{x}_1 = \sum_{i=1}^{M} \left(h_{1i}^* y_{1i} + h_{2i} y_{2i}^*\right)$$

$$\tilde{x}_2 = \sum_{i=1}^{M} \left(h_{2i}^* y_{1i} + h_{1i} y_{2i}^*\right)$$

Finally, (8.17) and (8.18) may be simplified as:

$$\tilde{x}_1 = (|h_{11}|^2 + |h_{21}|^2 + |h_{12}|^2 + |h_{22}|^2)x_1 + h_{11}^* \eta_{11} + h_{21} \eta_{21}^* + h_{12}^* \eta_{12} + h_{22} \eta_{22}^*$$

$$\tilde{x}_2 = (|h_{11}|^2 + |h_{21}|^2 + |h_{12}|^2 + |h_{22}|^2)x_2 + h_{21}^* \eta_{11} - h_{11} \eta_{21}^* + h_{22}^* \eta_{12} - h_{12} \eta_{22}^*$$

which gives in generalized form expressions:

$$\tilde{x}_1 = \sum_{i=1}^{M} \left[\left(|h_{1i}|^2 + |h_{2i}|^2\right) x_1 + h_{1i}^* \eta_{1i} + h_{2i} \eta_{2i}^*\right]$$

$$\tilde{x}_2 = \sum_{i=1}^{M} \left[\left(|h_{1i}|^2 + |h_{2i}|^2\right) x_2 + h_{2i}^* \eta_{1i} - h_{1i} \eta_{2i}^*\right]$$

Finally, the estimates $\tilde{x}_1$ and $\tilde{x}_2$ are fed to the input of maximum likelihood detector, as before.

8.2.4 Construction Criteria for Space-Time Codes

In this section we will briefly consider some properties, which may be used for the space-time codes construction, to improve their quality (decrease the error probability) when using them in the MIMO system.

The classical theory of error-correcting codes [4, 29, 30] considers the usage of code's redundancy to achieve the *coding gain*, which in a practical system usually means the decreasing of error probability after the decoding procedure. To achieve this goal, some *metrics* are defined over the set of codewords, allowing computation of the distance $d\left(x_i, x_j\right)$ between any pair of codewords x_i and x_j. The metrics function should satisfy the following properties:

1. $d(x_i, x_j) = d(x_j, x_i)$;
2. $d(x_i, x_j) \geq 0$, and $d(x_i, x_j) = 0$ if and only if $i = j$;
3. $d(x_i, x_j) \leq d(x_i, x_k) + d(x_k, x_j)$ for any codewords x_i, x_j, x_k.

If the metrics is *agreed* with the communication channel, then finding the codeword closest to the received word in terms of distance in given metrics gives the same result as finding the most probable transmitted word for given received word. That is, the minimal distance decoding coincides with the maximum likelihood decoding.

In practice the Hamming metrics or Euclidean metrics are often used. The parameter of the code connected to selected metrics is the *minimal distance* d_0 of the code, that is, the minimal distance from all the distances between all codeword pairs: $d_0 = \min_{i \neq j} d(x_i, x_j)$. Maximization of the parameter d_0 with fixed other code parameters leads to decreasing the decoding error probability and may be used as a criterion for the code construction.

Next we will give some analogous criteria for space-time codes construction. In such codes the redundancy is also used, however, this redundancy allows obtaining not only the coding gain, but also the *diversity gain*. Below we will give estimations on error probability and required parameters of space-time codes connected with this probability. We will assume that the transmission is made over the frequency-nonselective MIMO-channel with Rayleigh fading, fading coefficients of the channel are described by the matrix $\mathbf{H}$ (see (8.8)).

Recall (sections 8.1.3 and 8.2.2) that the codeword $\mathbf{C}^{(i)}$ of the space-time code is defined by the $(T \times N)$-matrix:

$$\mathbf{C}^{(i)} = \begin{pmatrix} c_{1,1}^{(i)} & c_{1,2}^{(i)} & \cdots & c_{1,N}^{(i)} \\ c_{2,1}^{(i)} & c_{2,2}^{(i)} & \cdots & c_{2,N}^{(i)} \\ \vdots & \vdots & \ddots & \vdots \\ c_{T,1}^{(i)} & c_{T,2}^{(i)} & \cdots & c_{T,N}^{(i)} \end{pmatrix}$$

If $P\left(\mathbf{C}^{(i)} \to \mathbf{C}^{(j)}\right)$ denotes the probability that after the decoding procedure the transmitted codeword $\mathbf{C}^{(i)}$ was decoded into codeword $\mathbf{C}^{(j)}$, and $i \neq j$, then the total decoding error probability may be estimated as [16]:

$$P(\text{error}|\mathbf{C}^{(i)} \text{ transmitted}) \leq \sum_{j \neq i} P(\mathbf{C}^{(i)} \to \mathbf{C}^{(j)})$$

Here the summation is made by all codewords besides the transmitted one.

Define the matrix:

$$\mathbf{D}(\mathbf{C}^{(i)}, \mathbf{C}^{(j)}) = \mathbf{C}^{(j)} - \mathbf{C}^{(i)} \tag{8.19}$$

and matrix $\mathbf{A}(\mathbf{C}^{(i)}, \mathbf{C}^{(j)}) = \mathbf{D}(\mathbf{C}^{(i)}, \mathbf{C}^{(j)})^H \cdot \mathbf{D}(\mathbf{C}^{(i)}, \mathbf{C}^{(j)})$, here $(\cdot)^H$ denotes the Hermitian operator. Then, by the theorem of singular decomposition:

$$\mathbf{A}(\mathbf{C}^{(i)}, \mathbf{C}^{(j)}) = \mathbf{V}^H \cdot \mathbf{\Lambda} \cdot \mathbf{V}$$

where $\mathbf{\Lambda} = \text{diag}(\lambda_1, \lambda_2, \ldots, \lambda_N)$, and $\lambda_i, i = 1, \ldots, N$ are the eigenvalues of matrix $\mathbf{A}(\mathbf{C}^{(i)}, \mathbf{C}^{(j)})$. Omitting theoretical background (it may be found in more detail, for example, in [16]), and denoting as $\beta_{i,j}$ the elements of matrix $\mathbf{V} \cdot \mathbf{H}$, the decoding error probability for given pair of codeword $\mathbf{C}^{(i)}$ transmitted over the channel $\mathbf{H}$ and the codeword $\mathbf{C}^{(j)}$ obtained as the result of decoding is [16]:

$$P(\mathbf{C}^{(i)} \to \mathbf{C}^{(j)}|\mathbf{H}) = Q\left(\sqrt{\frac{\gamma}{2}}||(\mathbf{C}^{(j)} - \mathbf{C}^{(i)}) \cdot \mathbf{H}||_F\right) \tag{8.20}$$

where

$$||\mathbf{A}||_F = \sqrt{\text{Tr}(\mathbf{A}^H \cdot \mathbf{A})} = \sqrt{\text{Tr}(\mathbf{A} \cdot \mathbf{A}^H)} \tag{8.21}$$

is the Frobenius norm of matrix $\mathbf{A}$, $\text{Tr}(\mathbf{B})$ is the trace of matrix $\mathbf{B}$, and Q-function is defined in (2.63):

$$Q(x) = \frac{1}{\sqrt{2\pi}} \int_x^{\infty} e^{\frac{-y^2}{2}} dy$$

Calculation of the Frobenius norm in (8.20) gives [16]:

$$||(\mathbf{C}^{(j)} - \mathbf{C}^{(i)}) \cdot \mathbf{H}||_F = \sqrt{\sum_{m=1}^{M} \sum_{n=1}^{N} \lambda_n |\beta_{n,m}|^2}$$

and substituting this into (8.20), we obtain:

$$P(\mathbf{C}^{(i)} \to \mathbf{C}^{(j)}|\mathbf{H}) = Q\left(\sqrt{\frac{\gamma}{2}\sum_{m=1}^{M}\sum_{n=1}^{N}\lambda_n|\beta_{n,m}|^2}\right)$$

Using the expression $Q(x) \le \frac{1}{2}e^{\frac{-x^2}{2}}$ as an upper bound, the following estimation can be made:

$$P(\mathbf{C}^{(i)} \to \mathbf{C}^{(j)}|\mathbf{H}) \le \frac{1}{2}\exp\left(-\frac{\gamma}{4}\sum_{m=1}^{M}\sum_{n=1}^{N}\lambda_n|\beta_{n,m}|^2\right)$$

Since $|\beta_{n,m}|$ have the Rayleigh distribution, the expected value of pairwise error probability can be calculated as:

$$P(\mathbf{C}^{(i)} \to \mathbf{C}^{(j)}) = E[P(\mathbf{C}^{(i)} \to \mathbf{C}^{(j)}|\mathbf{H})] \le \frac{1}{\prod_{n=1}^{N}[1+(\gamma\lambda_n/4)]^M} \tag{8.22}$$

where γ is SNR at the receiver. If $r = \text{rank}\left(\mathbf{A}\left(\mathbf{C}^{(i)}, \mathbf{C}^{(j)}\right)\right) \le N$, we may assume that $\lambda_1 \ge \lambda_2 \ge \cdots \ge \lambda_r > 0$, and $\lambda_{r+1} = \cdots = \lambda_N = 0$. For high values of SNR we may neglect the 1 in the denominator of (8.22) and write down the upper bound for the error probability basing on nonzero eigenvalues:

$$P(\mathbf{C}^{(i)} \to \mathbf{C}^{(j)}) \le \frac{4^{rM}}{\left(\prod_{n=1}^{N}\lambda_n\right)^M \gamma^{rM}} = (G_c\gamma)^{-G_d} \tag{8.23}$$

where G_c is coding gain, connected with the product of eigenvalues $\prod_{n=1}^{r}\lambda_n$, and $G_d = rM$ is diversity gain, or the diversity of space-time code, that is, diversity is defined by the rank of matrix $\mathbf{A}\left(\mathbf{C}^{(i)}, \mathbf{C}^{(j)}\right)$, or equivalently, by the rank of matrix $\mathbf{D}\left(\mathbf{C}^{(i)}, \mathbf{C}^{(j)}\right)$, multiplied by the number M of receive antennas.

It can be seen from (8.23) that the upper bound on error probability may be decreased both by increasing the coding gain and by increasing the diversity gain. Taking into account (8.19), the diversity of the space-time code is maximal and equal to:

$$M \cdot \text{rank}\left[\mathbf{D}\left(\mathbf{C}^{(i)}, \mathbf{C}^{(j)}\right)\right] = NM \tag{8.24}$$

when for all possible codeword pairs $\mathbf{C}^{(i)}$ and $\mathbf{C}^{(j)}$, $i \ne j$, the error matrix $\mathbf{D}\left(\mathbf{C}^{(i)}, \mathbf{C}^{(j)}\right) = \mathbf{C}^{(j)} - \mathbf{C}^{(i)}$ has the full rank N. This requirement is called *rank criterion* [37].

For the codes with full diversity the additional criterion based on coding gain may be used, which is equal in this case to $\prod_{n=1}^{N}\lambda_n$, or equivalently, to the value of determinant of matrix $\mathbf{A}\left(\mathbf{C}^{(i)}, \mathbf{C}^{(j)}\right)$. Define the *coding gain distance* (CGD) between words $\mathbf{C}^{(i)}$ and $\mathbf{C}^{(j)}$ as $\text{CGD}\left(\mathbf{C}^{(i)}, \mathbf{C}^{(j)}\right) = \det\left(\mathbf{A}\left(\mathbf{C}^{(i)}, \mathbf{C}^{(j)}\right)\right)$ (if $r < N$, then the coding gain distance may be determined as the product of nonzero eigenvalues of $\mathbf{A}\left(\mathbf{C}^{(i)}, \mathbf{C}^{(j)}\right)$). This allows the *determinant criterion* as maximization of minimal determinant of matrices $\mathbf{A}\left(\mathbf{C}^{(i)}, \mathbf{C}^{(j)}\right)$ to be determined for all pairs $i \ne j$.

As an example let's consider the Alamouti code in (8.14). Let the signals (s_1, s_2) form the codeword:

$$\mathbf{C} = \begin{pmatrix} s_1 & s_2 \\ -s_2^* & s_1^* \end{pmatrix}$$

and let the signals $\left(s_1', s_2'\right)$ form another codeword:

$$\mathbf{C}' = \begin{pmatrix} s_1' & s_2' \\ -s_2'^* & s_1'^* \end{pmatrix}$$

Then:

$$\mathbf{D}(\mathbf{C}, \mathbf{C}') = \begin{pmatrix} s_1' - s_1 & s_2' - s_2 \\ s_2^* - s_2'^* & s_1'^* - s_1^* \end{pmatrix}$$

The determinant of this matrix is $\det\left[\mathbf{D}(\mathbf{C},\mathbf{C}')\right] = \left|s_1' - s_1\right|^2 + \left|s_2' - s_2\right|^2$. It is equal to zero if and only if $s_1 = s_1'$ and $s_2 = s_2'$. Therefore, with $\mathbf{C} \neq \mathbf{C}'$ the matrix $\mathbf{D}(\mathbf{C},\mathbf{C}')$ is always full-rank, and the Alamouti code satisfies the determinant criterion. Besides, this code provides diversity $2M$ for M receive antennas, and therefore it is the code with full diversity.

In the conclusion of this section we notice that the function $\left\|\mathbf{D}\left(\mathbf{C}^{(i)},\mathbf{C}^{(j)}\right)\right\|_F$ (see (8.21) for the definition of the Frobenius norm) has all the metrics' properties stated above. This metrics may be used for the codewords of space-time code in the same manner as for the classical error-correcting codes the Hamming metrics is used. It may be shown that:

$$P(\mathbf{C}^{(i)} \to \mathbf{C}^{(j)}) \leq \frac{1}{4}\exp\left(-M\left\|\mathbf{D}(\mathbf{C}^{(i)},\mathbf{C}^{(j)})\right\|_F^2 \frac{\gamma}{4}\right)$$

that is, the pairwise probability of wrong decoding is connected with the metrics $\left\|\mathbf{D}\left(\mathbf{C}^{(i)},\mathbf{C}^{(j)}\right)\right\|_F$ – larger value of distance between $\mathbf{C}^{(i)}$ and $\mathbf{C}^{(j)}$ in this metrics correspondent to lower probability of decoding the transmitted codeword $\mathbf{C}^{(i)}$ into $\mathbf{C}^{(j)}$. Thus, another criterion for space-time codes construction may be the maximization of the minimal distance $\left\|\mathbf{D}\left(\mathbf{C}^{(i)},\mathbf{C}^{(j)}\right)\right\|_F$ among all pairs $i \neq j$ [8, 15]. This criterion is called *trace criterion*, since $\left\|\mathbf{D}\left(\mathbf{C}^{(i)},\mathbf{C}^{(j)}\right)\right\|_F^2 = \mathrm{Tr}\left[\mathbf{A}\left(\mathbf{C}^{(i)},\mathbf{C}^{(j)}\right)\right]$.

8.3 Orthogonal Designs

In this section we consider the approaches to construction of the space-time codes, generalizing the scheme described in section 8.2.3. We will distinguish two cases for consideration. In one case the matrix in (8.12) of the space-time code $\mathcal{G}$ is real-valued, and may be used for modulations with real signals. In another case $\mathcal{G}$ consists of complex values (this was the case considered in (8.14)), and consequently it may be used for space-time coding of complex signals. In both cases the construction of the matrix of space-timed code $\mathcal{G}$ is based on the so-called *orthogonal* designs, which we will briefly describe in what follows. The orthogonality property for these designs allows full diversity to be achieved and simple separate decoding of the symbols of space-time code to be performed (see, for example, (8.15) and (8.16)).

8.3.1 Real Orthogonal Designs

First we will consider the case when the number N of transmit antennas and the number M of receive antennas are the same, $N = M$. Define the *orthogonal design* of size N as $N \times N$-matrix $\mathcal{G}$ with real-valued elements $\pm x_1, \pm x_2, \ldots, \pm x_N$, for which the following condition is hold:

$$\mathcal{G}^T\mathcal{G} = \begin{pmatrix} \sum_{i=1}^N x_i^2 & 0 & \cdots & 0 \\ 0 & \sum_{i=1}^N x_i^2 & \cdots & 0 \\ \vdots & \vdots & \ddots & \vdots \\ 0 & 0 & \cdots & \sum_{i=1}^N x_i^2 \end{pmatrix} = \left(\sum_{i=1}^N x_i^2\right)\mathbf{I}_N \tag{8.25}$$

where $\mathbf{I}_N$ is $(N \times N)$-unity matrix.

In fact, satisfying the condition (8.25) means that the scalar product of any two different rows of matrix $\mathcal{G}$ is equal to 0, that is, the rows of this matrix are *orthogonal* to each other. The scalar product of the row by itself evidently should be equal to $\sum_{i=0}^N x_i^2$.

It may be shown [16, 38], that the real orthogonal designs exist only for $N = 2$, 4 and 8. The matrices $\mathcal{G}$ of the correspondent sizes are shown in Figure 8.8. It should be noticed that the matrices in Figure 8.8 are not unique. For any orthogonal design $\mathcal{G}$ and any unitary matrix $\mathbf{U}$ (that is, the matrix for which $\mathbf{U}^T\mathbf{U} = \mathbf{I}$), the matrix $\mathbf{U}\mathcal{G}$ will also be the orthogonal design.

$$\mathcal{G}_2 = \begin{pmatrix} x_1 & x_2 \\ -x_2 & x_1 \end{pmatrix} \qquad \mathcal{G}_4 = \begin{pmatrix} x_1 & x_2 & x_3 & x_4 \\ -x_2 & x_1 & -x_4 & x_3 \\ -x_3 & x_4 & x_1 & -x_2 \\ -x_4 & -x_3 & x_2 & x_1 \end{pmatrix}$$

$$\mathcal{G}_8 = \begin{pmatrix} x_1 & x_2 & x_3 & x_4 & x_5 & x_6 & x_7 & x_8 \\ -x_2 & x_1 & -x_4 & x_3 & -x_6 & x_5 & x_8 & -x_7 \\ -x_3 & x_4 & x_1 & -x_2 & x_7 & x_8 & -x_5 & -x_6 \\ -x_4 & -x_3 & x_2 & x_1 & x_8 & -x_7 & x_6 & -x_5 \\ -x_5 & x_6 & -x_7 & -x_8 & x_1 & -x_2 & x_3 & x_4 \\ -x_6 & -x_5 & -x_8 & x_7 & x_2 & x_1 & -x_4 & x_3 \\ -x_7 & -x_8 & x_5 & -x_6 & -x_3 & x_4 & x_1 & x_2 \\ -x_8 & x_7 & x_6 & x_5 & -x_4 & -x_3 & -x_2 & x_1 \end{pmatrix}$$

Figure 8.8 Orthogonal designs

Now we will show that the real space-time code, determined by matrices from Figure 8.8, will provide the full diversity.

The codeword of the space-time code defined by the matrix $\mathcal{G}$, is the matrix $\mathbf{C} = \mathcal{G}(s_1, \ldots, s_N)$, that is, in fact the matrix $\mathcal{G}$, in which the elements (variables) x_i are replaced by the correspondent signals s_i.

In correspondence with (8.24) the diversity of the space-time code is determined by the rank of matrix $\mathbf{D}\left(\mathbf{C}^{(i)}, \mathbf{C}^{(j)}\right) = \mathbf{C}^{(j)} - \mathbf{C}^{(i)} = \mathcal{G}(s_1^{(j)}, \ldots, s_N^{(j)}) - \mathcal{G}(s_1^{(i)}, \ldots, s_N^{(i)})$ for two different codewords $\mathbf{C}^{(i)}$ and $\mathbf{C}^{(j)}$, $i \neq j$. Evidently:

$$\mathcal{G}(s_1^{(j)}, \ldots, s_N^{(j)}) - \mathcal{G}(s_1^{(i)}, \ldots, s_N^{(i)}) = \mathcal{G}(s_1^{(j)} - s_1^{(i)}, \ldots, s_N^{(j)} - s_N^{(i)})$$

The determinant of orthogonal matrix $\mathcal{G}$ is equal to:

$$\det \mathcal{G} = \det(\mathcal{G}\mathcal{G}^T)^{1/2} = \left[\sum_{i=1}^{N} x_i^2\right]^{N/2}$$

From this:

$$\det(\mathcal{G}(s_1^{(j)} - s_1^{(i)}, \ldots, s_N^{(j)} - s_N^{(i)})) = \left[\sum_{i=1}^{N} |s_i^{(j)} - s_i^{(i)}|^2\right]^{N/2}$$

which is not equal to zero, since the elements $s_k^{(i)}$ and $s_k^{(j)}$ of two different codewords are different at least for one value of $k \in \{1, \ldots, N\}$. Hence, the space-timed codes defined by the orthogonal designs, provide full diversity of NM [38].

Besides the mentioned property of full diversity, the matrices based on orthogonal designs allow the separate decoding of received symbols to be performed, similar to the procedure described for the Alamouti code in section 8.2.3, which simplifies the decoding. The decoding of space-time codes based on orthogonal designs is described in more detail in section 8.3.3.

According to (8.13), the rate of such orthogonal codes is $R = 1$, since the number of transmitted signals $K = N$ and the number of time slots T are the same. Thus, these codes have the maximum achievable rate.

As mentioned above, the real orthogonal designs (for square matrices) exist only for dimensions 2, 4 and 8. Construction of space-time codes for another number of transmit antennas, exploiting all the advantages of orthogonal designs:

1. full diversity;
2. separate decoding of received symbols;
3. high (or maximal) rate;
4. minimal delay (that is, the value of T)

is both of practical and theoretical interest.

The last point (minimal delay) has no sense in consideration of square matrice $\mathcal{G}$, however, in other cases it is desirable to minimize the parameter T, since it affects the decoding delay. Constructions with the least possible T thus are called the "delay optimal".

To obtain the codes with desirable properties, considered orthogonal designs may be generalized [38] by the case when the matrix $\mathcal{G}$ is not square. We will call as *generalized* real orthogonal design the $T \times N$-matrix $\mathcal{G}$ with real elements $\pm x_1, \pm x_2, \ldots, \pm x_K$, for which any two columns are orthogonal, that is, for which the following expression is satisfied:

$$\mathcal{G}^T\mathcal{G} = c\begin{pmatrix} \sum_{i=1}^K x_i^2 & 0 & \cdots & 0 \\ 0 & \sum_{i=1}^K x_i^2 & \cdots & 0 \\ \vdots & \vdots & \ddots & \vdots \\ 0 & 0 & \cdots & \sum_{i=1}^K x_i^2 \end{pmatrix} = c\left(\sum_{i=1}^K x_i^2\right)\mathbf{I}_N \tag{8.26}$$

where $\mathbf{I}_N$ is $(N \times N)$-unity matrix, and c is constant. The rate of the correspondent space-time code, as before, is defined as $R = K/T$. It may be shown that such generalized real orthogonal designs are also providing full diversity and separate decoding.

It was shown [38] that orthogonal real space-time codes with rate $R = 1$ based on generalized real orthogonal designs exist for any N and $T = \min_{\{c,d\geq 0|8c+2^d\geq N\}}\left\{2^{4c+d}\right\}$. Such codes are delay-optimal. Then, for instance, for $N = 3$ transmit antennas obtain $T = K = 4$, for $N = 9$ antennas $T = K = 16$, and so on.

Examples of delay-optimal matrices for real space-time codes with rate $R = 1$ for a different number of transmit antennas are shown in Figure 8.9.

8.3.2 Complex Orthogonal Designs

Now we consider a similar conception of orthogonal design for the case when the transmitted signals are represented by complex modulation.

Let's define the *complex orthogonal design* of size N as $N \times N$-matrix $\mathcal{G}$ with complex elements $\pm x_1, \pm x_2, \ldots, \pm x_N$, their conjugates $\pm x_1^*, \pm x_2^*, \ldots, \pm x_N^*$ and multiplications of these elements by $\pm\sqrt{-1}$, for which the following condition is hold:

$$\mathcal{G}^H\mathcal{G} = \begin{pmatrix} \sum_{i=1}^N |x_i|^2 & 0 & \cdots & 0 \\ 0 & \sum_{i=1}^N |x_i|^2 & \cdots & 0 \\ \vdots & \vdots & \ddots & \vdots \\ 0 & 0 & \cdots & \sum_{i=1}^N |x_i|^2 \end{pmatrix} = \left(\sum_{i=1}^N |x_i|^2\right)\mathbf{I}_N \tag{8.27}$$

where $\mathbf{I}_N$ is $N \times N$-unity matrix.

However, it is proven [38] that such complex orthogonal designs exist only for dimension $N = 2$, and in fact this construction is the basis for the Alamouti code in (8.14). In the same way as the case of real elements, consider the nonsquare matrices with complex elements.

We will nominate as *generalized complex* orthogonal design the $T \times N$ matrix $\mathcal{G}$ with elements obtained from linear combinations of variables $x_1, x_2, \ldots, x_K$ and their conjugates, and in which any

$$\mathcal{G}_{4\times3}=\begin{pmatrix} x_1 & x_2 & x_3 \\ -x_2 & x_1 & -x_4 \\ -x_3 & x_4 & x_1 \\ -x_4 & -x_3 & x_2 \end{pmatrix} \qquad \mathcal{G}_{8\times5}=\begin{pmatrix} x_1 & x_2 & x_3 & x_4 & x_5 \\ -x_2 & x_1 & -x_4 & x_3 & -x_6 \\ -x_3 & x_4 & x_1 & -x_2 & x_7 \\ -x_4 & -x_3 & x_2 & x_1 & x_8 \\ -x_5 & x_6 & -x_7 & -x_8 & x_1 \\ -x_6 & -x_5 & -x_8 & x_7 & x_2 \\ -x_7 & -x_8 & x_5 & -x_6 & -x_3 \\ -x_8 & x_7 & x_6 & x_5 & -x_4 \end{pmatrix}$$

$$\mathcal{G}_{8\times6}=\begin{pmatrix} x_1 & x_2 & x_3 & x_4 & x_5 & x_6 \\ -x_2 & x_1 & -x_4 & x_3 & -x_6 & x_5 \\ -x_3 & x_4 & x_1 & -x_2 & x_7 & x_8 \\ -x_4 & -x_3 & x_2 & x_1 & x_8 & -x_7 \\ -x_5 & x_6 & -x_7 & -x_8 & x_1 & -x_2 \\ -x_6 & -x_5 & -x_8 & x_7 & x_2 & x_1 \\ -x_7 & -x_8 & x_5 & -x_6 & -x_3 & x_4 \\ -x_8 & x_7 & x_6 & x_5 & -x_4 & -x_3 \end{pmatrix}$$

$$\mathcal{G}_{8\times7}=\begin{pmatrix} x_1 & x_2 & x_3 & x_4 & x_5 & x_6 & x_7 \\ -x_2 & x_1 & -x_4 & x_3 & -x_6 & x_5 & x_8 \\ -x_3 & x_4 & x_1 & -x_2 & x_7 & x_8 & -x_5 \\ -x_4 & -x_3 & x_2 & x_1 & x_8 & -x_7 & x_6 \\ -x_5 & x_6 & -x_7 & -x_8 & x_1 & -x_2 & x_3 \\ -x_6 & -x_5 & -x_8 & x_7 & x_2 & x_1 & -x_4 \\ -x_7 & -x_8 & x_5 & -x_6 & -x_3 & x_4 & x_1 \\ -x_8 & x_7 & x_6 & x_5 & -x_4 & -x_3 & -x_2 \end{pmatrix}$$

$$\mathcal{G}_{16\times9}=\begin{pmatrix} x_1 & x_2 & x_3 & x_4 & x_5 & x_6 & x_7 & x_8 & x_9 \\ x_2 & -x_1 & x_4 & -x_3 & x_6 & -x_5 & -x_8 & x_7 & x_{10} \\ x_3 & -x_4 & -x_1 & x_2 & -x_7 & -x_8 & x_5 & x_6 & x_{11} \\ x_4 & x_3 & -x_2 & -x_1 & -x_8 & x_7 & -x_6 & x_5 & x_{12} \\ x_5 & -x_6 & x_7 & x_8 & -x_1 & x_2 & -x_3 & -x_4 & x_{13} \\ x_6 & x_5 & x_8 & -x_7 & -x_2 & -x_1 & x_4 & -x_3 & x_{14} \\ x_7 & x_8 & -x_5 & x_6 & x_3 & -x_4 & -x_1 & -x_2 & x_{15} \\ x_8 & -x_7 & -x_6 & -x_5 & x_4 & x_3 & x_2 & -x_1 & x_{16} \\ x_9 & -x_{10} & -x_{11} & -x_{12} & -x_{13} & -x_{14} & -x_{15} & -x_{16} & -x_1 \\ x_{10} & x_9 & -x_{12} & x_{11} & -x_{14} & x_{13} & x_{16} & -x_{15} & -x_2 \\ x_{11} & x_{12} & x_9 & -x_{10} & x_{15} & x_{16} & -x_{13} & -x_{14} & -x_3 \\ x_{12} & -x_{11} & x_{10} & x_9 & x_{16} & -x_{15} & x_{14} & -x_{13} & -x_4 \\ x_{13} & x_{14} & -x_{15} & -x_{16} & x_9 & -x_{10} & x_{11} & x_{12} & -x_5 \\ x_{14} & -x_{13} & -x_{16} & x_{15} & x_{10} & x_9 & -x_{12} & x_{11} & -x_6 \\ x_{15} & -x_{16} & x_{13} & -x_{14} & -x_{11} & x_{12} & x_9 & x_{10} & -x_7 \\ x_{16} & x_{15} & x_{14} & x_{13} & -x_{12} & -x_{11} & -x_{10} & x_9 & -x_8 \end{pmatrix}$$

Figure 8.9 Generalized orthogonal designs

$$\mathcal{G}_{8\times3} = \begin{pmatrix} x_1 & x_2 & x_3 \\ -x_2 & x_1 & -x_4 \\ -x_3 & x_4 & x_1 \\ -x_4 & -x_3 & x_2 \\ x_1^* & x_2^* & x_3^* \\ -x_2^* & x_1^* & -x_4^* \\ -x_3^* & x_4^* & x_1^* \\ -x_4^* & -x_3^* & x_2^* \end{pmatrix}$$

Figure 8.10 Construction of complex orthogonal design from real design based on (8.29) for $N = 3$

two columns are orthogonal, that is, the following expression is satisfied:

$$\mathcal{G}^H\mathcal{G} = c\begin{pmatrix} \sum_{i=1}^K |x_i|^2 & 0 & \cdots & 0 \\ 0 & \sum_{i=1}^K |x_i|^2 & \cdots & 0 \\ \vdots & \vdots & \ddots & \vdots \\ 0 & 0 & \cdots & \sum_{i=1}^K |x_i|^2 \end{pmatrix} = c\left(\sum_{i=1}^{K} |x_i|^2\right)\mathbf{I}_N \tag{8.28}$$

where $\mathbf{I}_N$ is $(N \times N)$-unity matrix, and c is constant.

As in the case of real signals, the space-time codes constructed on the base of (generalized) complex orthogonal designs, allow full diversity NM and separate decoding of the received signals to be achieved.

Notice that if $\mathcal{G}$ is $(T \times N)$-matrix of real orthogonal space-time code with rate R, and $\mathcal{G}^*$ is the matrix in which the correspondent elements x_i from $\mathcal{G}$ are replaced by x_i^*, then the matrix:

$$\mathcal{G}_{\text{complex}} = \begin{pmatrix} \mathcal{G} \\ \mathcal{G}^* \end{pmatrix} \tag{8.29}$$

with dimension $2T \times N$ defines the complex orthogonal space-time code with rate $R/2$ [16]. Examples of such matrix construction for $N = 3$ and $N = 4$ transmit antennas are given in Figure 8.10 and Figure 8.11.

Since in section 8.3.1 we have noticed that the real orthogonal space-time code with rate $R = 1$ exists for any number N of transmit antennas, hence, for any number N of transmit antennas there exists the complex orthogonal space-time code with rate $R = 1/2$.

$$\mathcal{G}_{8\times4} = \begin{pmatrix} x_1 & x_2 & x_3 & x_4 \\ -x_2 & x_1 & -x_4 & x_3 \\ -x_3 & x_4 & x_1 & -x_2 \\ -x_4 & -x_3 & x_2 & x_1 \\ x_1^* & x_2^* & x_3^* & x_4^* \\ -x_2^* & x_1^* & -x_4^* & x_3^* \\ -x_3^* & x_4^* & x_1^* & -x_2^* \\ -x_4^* & -x_3^* & x_2^* & x_1^* \end{pmatrix}$$

Figure 8.11 Construction of complex orthogonal design from real design based on (8.29) for $N = 4$

$$\mathcal{G}_{16\times 8} = \begin{pmatrix} x_1 & x_2 & x_3 & x_4 & x_5 & x_6 & x_7 & x_8 \\ -x_2 & x_1 & x_4 & -x_3 & x_6 & -x_5 & -x_8 & x_7 \\ -x_3 & -x_4 & x_1 & x_2 & x_7 & x_8 & -x_5 & -x_6 \\ -x_4 & x_3 & -x_2 & x_1 & x_8 & -x_7 & x_6 & -x_5 \\ -x_5 & -x_6 & -x_7 & -x_8 & x_1 & x_2 & x_3 & x_4 \\ -x_6 & x_5 & -x_8 & x_7 & -x_2 & x_1 & -x_4 & x_3 \\ -x_7 & x_8 & x_5 & -x_6 & -x_3 & x_4 & x_1 & -x_2 \\ -x_8 & -x_7 & x_6 & x_5 & -x_4 & -x_3 & x_2 & x_1 \\ x_1^* & x_2^* & x_3^* & x_4^* & x_5^* & x_6^* & x_7^* & x_8^* \\ -x_2^* & x_1^* & x_4^* & -x_3^* & x_6^* & -x_5^* & -x_8^* & x_7^* \\ -x_3^* & -x_4^* & x_1^* & x_2^* & x_7^* & x_8^* & -x_5^* & -x_6^* \\ -x_4^* & x_3^* & -x_2^* & x_1^* & x_8^* & -x_7^* & x_6^* & -x_5^* \\ -x_5^* & -x_6^* & -x_7^* & -x_8^* & x_1^* & x_2^* & x_3^* & x_4^* \\ -x_6^* & x_5^* & -x_8^* & x_7^* & -x_2^* & x_1^* & -x_4^* & x_3^* \\ -x_7^* & x_8^* & x_5^* & -x_6^* & -x_3^* & x_4^* & x_1^* & -x_2^* \\ -x_8^* & -x_7^* & x_6^* & x_5^* & -x_4^* & -x_3^* & x_2^* & x_1^* \end{pmatrix}$$

Figure 8.12 Construction of complex orthogonal design from real design based on (8.29) for $N = 8$

It should be noted that the mentioned above method of construction of the complex code with rate $R/2$ from the real orthogonal code with rate R is not unique [16] and is not the best from the point of view of the delay optimality. For example, consider the matrix shown in Figure 8.12 for $N = 8$, $K = 8$ and obtained from (8.29). Here $T = 16$ and the code rate is $R = 1/2$. At the same time the matrix shown in Figure 8.13 for $N = 8$ also provides the rate of $R = 1/2$, but in this case $K = 4$ and $T = 8$.

In [26] it is shown that for $N > 2$ (that is, for the case of more than two transmit antennas) no complex orthogonal designs exist, providing the space-time code with rate $R = 1$, so the construction used in the Alamouti code, is also unique in this sense.

Besides, in [41] it is shown that for $N > 2$ the generalized complex orthogonal design cannot provide the code with a rate more than 3/4. The matrices for the codes with rate 3/4 for 3 and 4 transmit antennas are shown in Figure 8.14 [38]. One more example of the matrices for codes with rate 3/4 for $N = 3$ and $N = 4$ is given in Figure 8.15 [16].

$$\mathcal{G}_{8\times 8} = \begin{pmatrix} x_1 & x_2 & x_3 & 0 & x_4 & 0 & 0 & 0 \\ -x_2^* & x_1^* & 0 & x_3 & 0 & x_4 & 0 & 0 \\ x_3^* & 0 & -x_1^* & x_2 & 0 & 0 & x_4 & 0 \\ 0 & x_3^* & -x_2^* & -x_1 & 0 & 0 & 0 & x_4 \\ x_4^* & 0 & 0 & 0 & -x_1^* & x_2 & -x_3 & 0 \\ 0 & x_4^* & 0 & 0 & -x_2^* & -x_1 & 0 & -x_3 \\ 0 & 0 & x_4^* & 0 & -x_3^* & 0 & x_1 & x_2 \\ 0 & 0 & 0 & x_4^* & 0 & -x_3^* & -x_2^* & x_1^* \end{pmatrix}$$

Figure 8.13 Another complex orthogonal design for $N = 8$

$$\mathcal{G}_{4\times 3} = \begin{pmatrix} x_1 & x_2 & \frac{x_3}{\sqrt{2}} \\ -x_2^* & x_1^* & \frac{x_2}{\sqrt{2}} \\ \frac{x_3^*}{\sqrt{2}} & \frac{x_3^*}{\sqrt{2}} & \frac{(-x_1 - x_1^* + x_2 - x_2^*)}{2} \\ \frac{x_3^*}{\sqrt{2}} & -\frac{x_3^*}{\sqrt{2}} & \frac{(x_2 + x_2^* + x_1 - x_1^*)}{2} \end{pmatrix}$$

$$\mathcal{G}_{4\times 4} = \begin{pmatrix} x_1 & x_2 & \frac{x_3}{\sqrt{2}} & \frac{x_3}{\sqrt{2}} \\ -x_2^* & x_1^* & \frac{x_2}{\sqrt{2}} & -\frac{x_3}{\sqrt{2}} \\ \frac{x_3^*}{\sqrt{2}} & \frac{x_3^*}{\sqrt{2}} & \frac{(-x_1 - x_1^* + x_2 - x_2^*)}{2} & \frac{(-x_2 - x_2^* + x_1 - x_1^*)}{2} \\ \frac{x_3^*}{\sqrt{2}} & -\frac{x_3^*}{\sqrt{2}} & \frac{(x_2 + x_2^* + x_1 - x_1^*)}{2} & -\frac{(x_1 + x_1^* + x_2 - x_2^*)}{2} \end{pmatrix}$$

Figure 8.14 Design with rate $R = 3/4$ for $N = 3$ and $N = 4$ transmit antennas

For $N > 4$ the problem of calculating the upper bounds on code rates and the problem of construction the (complex) space-time codes with rates in the interval (1/2; 3/4] (or proof of their inexistence) remain the open problems in the general case. However, some examples may be found in [16].

8.3.3 Decoding of Space-Time Codes

Let the space-time code with N transmit, M receive antennas, transmitting in T time slots, is used for transmission over the communication channel.

$$\mathcal{G}_{4\times 3} = \begin{pmatrix} x_1 & x_2 & x_3 \\ -x_2^* & x_1^* & 0 \\ x_3^* & 0 & -x_1^* \\ 0 & x_3^* & -x_2^* \end{pmatrix}$$

$$\mathcal{G}_{4\times 4} = \begin{pmatrix} x_1 & x_2 & x_3 & 0 \\ -x_2^* & x_1^* & 0 & x_3 \\ x_3^* & 0 & -x_1^* & x_2 \\ 0 & x_3^* & -x_2^* & -x_1 \end{pmatrix}$$

Figure 8.15 Other designs with rate $R = 3/4$ for $N = 3$ and $N = 4$ transmit antennas

Denote as $\mathcal{C}$ the space-time code defined by the matrix $\mathcal{G}$. Then the codewords of code $\mathcal{C}$ are the matrices:

$$\mathcal{S} = \begin{pmatrix} s_1^1 & \cdots & s_1^N \\ \vdots & \ddots & \vdots \\ s_T^1 & \cdots & s_T^N \end{pmatrix} \tag{8.30}$$

If the signal s_t^i is transmitted at time moment t from the antenna i, then antenna j receives the signal [38]:

$$r_t^j = \sum_{i=1}^{N} h_{i,j} s_t^i + \eta_t^j$$

where $h_{i,j}$ is the channel transfer coefficient (fading coefficient) from antenna i to antenna j, and η_t^j is the independent complex Gaussian random value.

If the channel transfer coefficients are known, then the maximum likelihood decoding will coincide with the minimum distance decoding in metrics:

$$\mathcal{S} = \arg\min_{\mathcal{S}\in\mathcal{C}} \left\{ \sum_{t=1}^{T} \sum_{j=1}^{M} \left| r_t^j - \sum_{i=1}^{N} h_{i,j} s_t^i \right|^2 \right\} \tag{8.31}$$

over all codewords $\mathcal{S} \in \mathcal{C}$.

Consider as code $\mathcal{C}$ the space-time code with real signals, for which the matrix $\mathcal{G}$ is square, $T = N$, and obtained from orthogonal design (section 8.3.1). Examples of space-time matrices for such codes are shown in Figure 8.8.

As can be seen from Figure 8.8, the first row of matrices $\mathcal{G}$ are formed by symbols $x_1, \ldots, x_N$, while the other rows are the permutations of the first row, with some symbols x_i changing their sign. Denote the permutation correspondent to the row k as π_k, then the element x_i is placed in row k on position $\pi_k(i)$. The sign of this element denote as $\text{sgn}_k(i)$.

Then the received signal may be estimated with maximum likelihood (see maximal ratio combining (8.11) and the Alamouti scheme in (8.15) and (8.16)) as:

$$\tilde{s}_i = \sum_{t=1}^{N} \sum_{j=1}^{M} \text{sgn}_t(i) \cdot r_t^j \cdot h^*_{\pi_t(i),j}$$

Because of the orthogonality of matrix $\mathcal{G}$ the minimization in (8.31) is equivalent to:

$$\mathcal{S} = \arg\min_{\mathcal{S}\in\mathcal{C}} \left\{ \sum_{i=1}^{N} \left(|\tilde{s}_i - s_i|^2 + \left(\sum_{t=1}^{N} \sum_{j=1}^{M} \left|h_{t,j}\right|^2 - 1 \right) |s_i|^2 \right) \right\} \tag{8.32}$$

Again, because of the orthogonality of matrix $\mathcal{G}$ the estimate of signal $\tilde{s}_i$ is independent from estimates of other signals, and hence, minimization in (8.32) may be performed for each term independently:

$$s_i = \arg\min_{s_i} \left\{ |\tilde{s}_i - s_i|^2 + \left(\sum_{t=1}^{N} \sum_{j=1}^{M} \left|h_{t,j}\right|^2 - 1 \right) |s_i|^2 \right\} \tag{8.33}$$

Notice that in the case of using the modulation with equal energy for all symbols, the value:

$$\psi_i = \left(\sum_{t=1}^{N} \sum_{j=1}^{M} \left|h_{t,j}\right|^2 - 1 \right) |s_i|^2$$

is the same for all s_i, and hence it may be removed from (8.33), which gives simply:

$$s_i = \arg\min_{s_i} \left\{ |\tilde{s}_i - s_i|^2 \right\}$$

For the space-time codes with non-square matrix $\mathcal{G}$ (that is, for $T \neq N$) the decoding is practically identical, with evident modification for symbol estimation:

$$\tilde{s}_i = \sum_{t \in \rho(i)}^{N} \sum_{j=1}^{M} \operatorname{sgn}_t(i) \cdot r_t^j \cdot h^*_{\pi_t(i),j} \quad (8.34)$$

where $\rho(i)$ is the set of rows in matrix $\mathcal{G}$, containing s_i.

Decoding of space-time codes with complex signals is similar to the described scheme. We only give examples of decoding schemes for some matrices considered in section 8.3.2 [20, 39].

For the codes obtained from the construction shown by (8.29), the decoded signal is:

$$s_i = \arg\min_{s_i} \left\{ |\tilde{s}_i - s_i|^2 + \left(2 \sum_{t=1}^{N} \sum_{j=1}^{M} \left| h_{t,j} \right|^2 - 1 \right) |s_i|^2 \right\}$$

which is almost the same as (8.33) besides multiplication of $\left| h_{t,j} \right|$ by 2, since for these constructions the constant in (8.28) is evidently equal to $c = 2$.

Estimation of the received signal for these constructions also evidently follows from (8.34) as:

$$\tilde{s}_i = \sum_{t \in \rho(i)}^{N} \sum_{j=1}^{M} \operatorname{sgn}_t(i) \cdot \tilde{r}_t^j(i) \cdot \tilde{h}_{\pi_t(i),j}$$

where

$$\tilde{r}_t^j = \begin{cases} r_t^j, \text{ if } t\text{-th row of } \mathcal{G} \text{ contains } x_i \\ (r_t^j)^*, \text{ if } t\text{-th row of } \mathcal{G} \text{ contains } x_i^* \end{cases}$$

and

$$\tilde{h}_{\pi_t(i),j} = \begin{cases} h^*_{\pi_t(i),j}, \text{ if } t\text{-th row of } \mathcal{G} \text{ contains } x_i \\ h_{\pi_t(i),j}, \text{ if } t\text{-th row of } \mathcal{G} \text{ contains } x_i^* \end{cases}$$

For example, for the codes defined by matrices $\mathcal{G}_{8\times3}$ and $\mathcal{G}_{8\times4}$ in Figure 8.10 and Figure 8.11 correspondently, the estimates of received signals are calculated as shown in Figure 8.16 and 8.17.

$$\tilde{s}_1 = \sum_{j=1}^{M} \left(r_1^j h_{1,j}^* + r_2^j h_{2,j}^* + r_3^j h_{3,j}^* + (r_5^j)^* h_{1,j} + (r_6^j)^* h_{2,j} + (r_7^j)^* h_{3,j} \right)$$

$$\tilde{s}_2 = \sum_{j=1}^{M} \left(r_1^j h_{2,j}^* - r_2^j h_{1,j}^* + r_4^j h_{3,j}^* + (r_5^j)^* h_{2,j} - (r_6^j)^* h_{1,j} + (r_8^j)^* h_{3,j} \right)$$

$$\tilde{s}_3 = \sum_{j=1}^{M} \left(r_1^j h_{3,j}^* - r_3^j h_{1,j}^* - r_4^j h_{2,j}^* + (r_5^j)^* h_{3,j} - (r_7^j)^* h_{1,j} + (r_8^j)^* h_{2,j} \right)$$

$$\tilde{s}_4 = \sum_{j=1}^{M} \left(-r_2^j h_{3,j}^* + r_3^j h_{2,j}^* - r_4^j h_{1,j}^* - (r_6^j)^* h_{3,j} + (r_7^j)^* h_{2,j} - (r_8^j)^* h_{1,j} \right)$$

Figure 8.16 Estimation of received symbols for the code defined by matrix $\mathcal{G}_{8\times3}$ (Figure 8.10)

$$\tilde{s}_1 = \sum_{j=1}^{M} (r_1^j h_{1,j}^* + r_2^j h_{2,j}^* + r_3^j h_{3,j}^* + r_4^j h_{4,j}^* + \\ + (r_5^j)^* h_{1,j} + (r_6^j)^* h_{2,j} + (r_7^j)^* h_{3,j} + (r_8^j)^* h_{4,j})$$

$$\tilde{s}_2 = \sum_{j=1}^{M} (r_1^j h_{2,j}^* - r_2^j h_{1,j}^* - r_3^j h_{4,j}^* + r_4^j h_{3,j}^* + \\ + (r_5^j)^* h_{2,j} - (r_6^j)^* h_{1,j} - (r_7^j)^* h_{4,j} + (r_8^j)^* h_{3,j})$$

$$\tilde{s}_3 = \sum_{j=1}^{M} (r_1^j h_{3,j}^* + r_2^j h_{4,j}^* - r_3^j h_{1,j}^* - r_4^j h_{2,j}^* + \\ + (r_5^j)^* h_{3,j} + (r_6^j)^* h_{4,j} - (r_7^j)^* h_{1,j} - (r_8^j)^* h_{2,j})$$

$$\tilde{s}_4 = \sum_{j=1}^{M} (r_1^j h_{4,j}^* - r_2^j h_{3,j}^* + r_3^j h_{2,j}^* - r_4^j h_{1,j}^* + \\ + (r_5^j)^* h_{4,j} - (r_6^j)^* h_{3,j} + (r_7^j)^* h_{2,j} - (r_8^j)^* h_{4,j})$$

Figure 8.17 Estimation of received symbols for the code defined by matrix $\mathcal{G}_{8\times4}$ (Figure 8.11)

For the codes defined by matrices $\mathcal{G}_{4\times3}$ and $\mathcal{G}_{4\times4}$ in Figure 8.14, the decoding of symbol s_i is performed as:

$$s_i = \arg\min_{s_i} \left\{ |\tilde{s}_i - s_i|^2 + \left(\sum_{t=1}^{N} \sum_{j=1}^{M} |h_{t,j}|^2 - 1 \right) |s_i|^2 \right\}$$

which is similar to (8.33) since for these constructions the constant in (8.28) is equal to $c = 1$. Estimates of the received symbols for these codes are calculated as shown in Figures 8.18 and 8.19 [39].

8.3.4 *Error Probability for Orthogonal Space-Time Codes*

When formulating the criteria of construction of the space-time codes in section 8.2.4 we considered the probability of decoding in wrong space-time codeword $\mathbf{C}^{(j)}$ if the codeword $\mathbf{C}^{(i)}$ had been transmitted over the channel with the transfer coefficients defined by matrix $\mathbf{H}$.

Now consider the analogous probability for the case when the space-time code is represented by the orthogonal designs, more precisely, by generalized complex orthogonal designs defined by (8.28). In

$$\tilde{s}_1 = \sum_{j=1}^{M} \left(r_1^j h_{1,j}^* + (r_2^j)^* h_{2,j} + \frac{(r_4^j - r_3^j) h_{3,j}^*}{2} - \frac{(r_3^j + r_4^j)^* h_{3,j}}{2} \right)$$

$$\tilde{s}_2 = \sum_{j=1}^{M} \left(r_1^j h_{2,j}^* - (r_2^j)^* h_{1,j} + \frac{(r_3^j + r_4^j) h_{3,j}^*}{2} + \frac{(r_4^j - r_3^j)^* h_{3,j}}{2} \right)$$

$$\tilde{s}_3 = \sum_{j=1}^{M} \left(\frac{(r_1^j + r_2^j) h_{3,j}^*}{\sqrt{2}} + \frac{(r_3^j)^* (h_{1,j} + h_{2,j})}{\sqrt{2}} + \frac{(r_4^j)^* (h_{1,j} - h_{2,j})}{\sqrt{2}} \right)$$

Figure 8.18 Estimation of received symbols for the code defined by matrix $\mathcal{G}_{4\times3}$ (Figure 8.14)

$$\tilde{s}_1 = \sum_{j=1}^{M} \left(r_1^j h_{1,j}^* + \left(r_2^j\right)^* h_{2,j} + \frac{(r_4^j - r_3^j)(h_{3,j}^* - h_{4,j}^*)}{2} - \frac{(r_3^j + r_4^j)^*(h_{3,j} + h_{4,j})}{2} \right)$$

$$\tilde{s}_2 = \sum_{j=1}^{M} \left(r_1^j h_{2,j}^* + \left(r_2^j\right)^* h_{1,j} + \frac{(r_3^j + r_4^j)(h_{3,j}^* - h_{4,j}^*)}{2} + \frac{(r_4^j - r_3^j)^*(h_{3,j} + h_{4,j})}{2} \right)$$

$$\tilde{s}_3 = \sum_{j=1}^{M} \left(\frac{(r_1^j + r_2^j)h_{3,j}^*}{\sqrt{2}} + \frac{(r_1^j - r_2^j)h_{4,j}^*}{\sqrt{2}} + \frac{(r_3^j)^*(h_{1,j} + h_{2,j})}{\sqrt{2}} + \frac{(r_4^j)^*(h_{1,j} - h_{2,j})}{\sqrt{2}} \right)$$

Figure 8.19 Estimation of received symbols for the code defined by matrix $\mathcal{G}_{4\times 4}$ (Figure 8.14)

this case the conditional probability in (8.20) may be written as:

$$P(\mathbf{C}^{(i)} \to \mathbf{C}^{(j)}|\mathbf{H}) = Q\left(\sqrt{\frac{\gamma}{2}\mathrm{Tr}\left[\mathbf{H}^H \cdot \left(\mathbf{C}^{(j)} - \mathbf{C}^{(i)}\right)^H \cdot \left(\mathbf{C}^{(j)} - \mathbf{C}^{(i)}\right) \cdot \mathbf{H}\right]}\right)$$

then, taking into account the orthogonality of codewords in (8.28), obtain:

$$\begin{aligned} P(\mathbf{C}^{(i)} \to \mathbf{C}^{(j)}\,|\mathbf{H}) &= Q\left(\sqrt{\frac{\gamma}{2}c\sum_{k=1}^{K}\left|s_k^{(j)} - s_k^{(i)}\right|^2 \mathrm{Tr}\left[\mathbf{H}^H \cdot \mathbf{H}\right]}\right) = \\ &= Q\left(\sqrt{c\frac{\gamma}{2}\sum_{k=1}^{K}\left|s_k^{(j)} - s_k^{(i)}\right|^2 \sum_{n=1}^{N}\sum_{\mathrm{m}=1}^{\mathrm{M}}\left|h_{n,m}\right|^2}\right) \end{aligned} \tag{8.35}$$

Notice that:

$$d_E = \sqrt{\sum_{k=1}^{K}\left|s_k^{(j)} - s_k^{(i)}\right|^2}$$

is the Euclidean distance between transmitted and detected symbols, and thus the expression (8.35) may be rewritten as:

$$P(\mathbf{C}^{(i)} \to \mathbf{C}^{(j)}\,|\mathbf{H}) = Q\left(\sqrt{c\frac{\gamma}{2}d_E^2\sum_{n=1}^{N}\sum_{\mathrm{m}=1}^{\mathrm{M}}\left|h_{n,m}\right|^2}\right) \tag{8.36}$$

Computation of unconditional probability based on conditional from (8.36) requires some additional computational steps, which may be found in [16, 35]. Here we give only the final result of pairwise error probability for orthogonal space-time codes:

$$P(\mathbf{C}^{(i)} \to \mathbf{C}^{(j)}) = \frac{1}{2}\left\{1 - \sqrt{\frac{a}{1+a}}\sum_{i=0}^{MN-1}\binom{2i}{i}\left[\frac{1}{4(1+a)}\right]\right\}$$

where $a = c\frac{\gamma}{4}d_E^2$. The constant c is determined by the structure of the space-time code.

8.4 Space-Time Trellis Codes

8.4.1 Space-Time Trellis Codes

In section 8.3 the space-time codes based on orthogonal designs were considered. These codes exploit the diversity gain, providing at the same time simple decoding methods. In [37] the idea of combining the

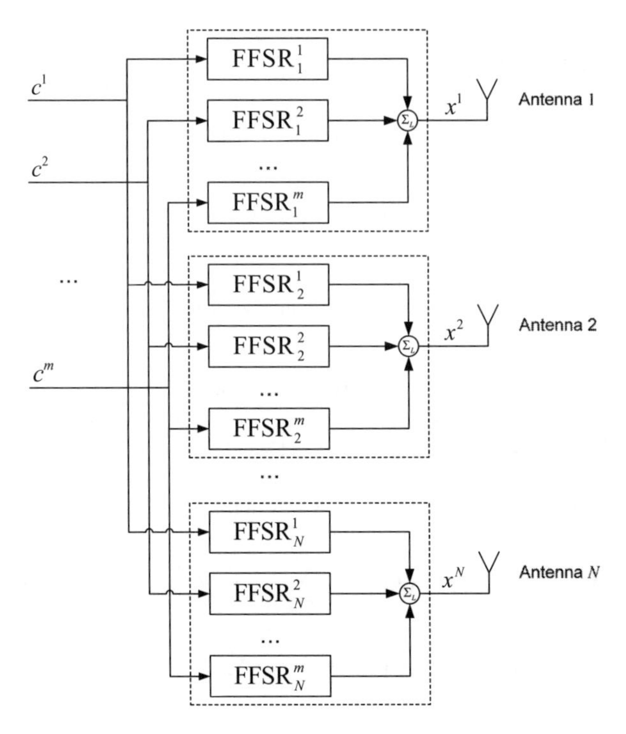

Figure 8.20 Encoder for space-time trellis codes (STTC)

space-time code with the coded modulation scheme is proposed, to obtain also the coding gain, besides diversity gain.

Suppose that in communication system with N transmit antennas the L-PSK modulation is used, and there are $m = \log_2 L$ information bits $c^1, \ldots, c^m$ at the encoder's input. Then at the encoder's output there should be N L-PSK-modulated signals $x^1, \ldots, x^N$ for simultaneous transmission over the N antennas.

Each information bit c^k, $k = 1, \ldots, m$, intended for encoding by space-time trellis code (STTC) for transmission over antenna i, $i = 1, \ldots, N$, is fed to the input of feedforward shift register FFSR_i^k, see Figure 8.20. Here Σ_L denotes addition on modulo L.

Feedforward shift register consists of delay elements, defining its current state, and is functioning by clocks. The register FFSR_i^k with $\nu_k - 1$ delay elements, with symbol c^k at register's input, intended for encoding and transmission over antenna i, is shown in Figure 8.21. When the register is clocked, to calculate the output symbol, the input symbol and the values of delay elements are multiplied by

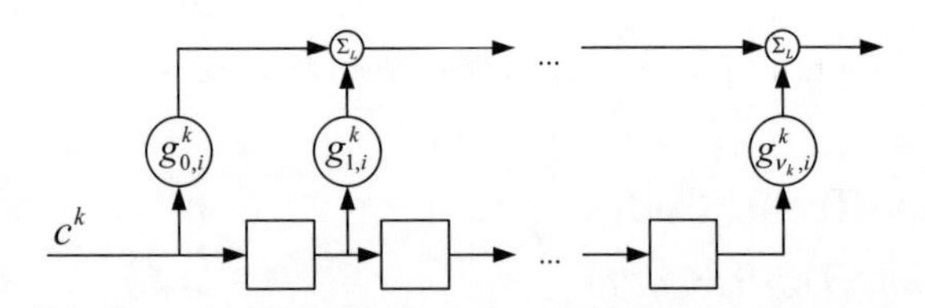

Figure 8.21 Feedforward register FFSR_i^k for encoding symbol c^k and transmission over antenna i

correspondent coefficients (elements of L-PSK signal constellation) $g^k_{j,i}$, $j = 0, \ldots, \nu_k$, and summed on modulo L. The feedforward register may be completely determined by the connection polynomial:

$$G^k_i(D) = g^k_{0,i} + g^k_{1,i}D + \ldots + g^k_{\nu_k,i}D^{\nu_k} \tag{8.37}$$

Since the output symbol is dependent on ν_k preceding input symbols, usage of register adds some "memory" to the system, and in fact the input stream of symbols $c^1, \ldots, c^m$ in Figure 8.20 may be considered as m parallel streams:

$$c^k = (c^k_1, c^k_2, \ldots, c^k_t, \ldots), \qquad k = 1, \ldots, m$$

where t denotes the moment in time. Then the encoder's output x^i_t at time moment t for transmitting antenna i is calculated as:

$$x^i_t = \sum_{k=1}^{m} \sum_{j=0}^{\nu_k} g^k_{j,i} c^k_{t-j} \mod L, \qquad i = 1, 2, \ldots, N$$

Since each register FFSR^k_i may be in 2^{ν_k} states, the total number of states in space-time trellis encoder is 2^ν, where:

$$\nu = \sum_{k=1}^{m} \nu_k$$

The values of ν_k are determined from the value of total encoder's memory as:

$$\nu_k = \left\lfloor \frac{\nu + k - 1}{m} \right\rfloor$$

Transitions between states for different input sequences of length m, and encoder's output for all inputs and states may be traditionally depicted by means of trellis. Expanding the trellis in time, we see that the codeword of space-time codes corresponds to some path through this trellis, and hence, the Viterbi decoding may be used [40].

For example, consider the construction of space-time trellis code for two transmit antennas ($N = 2$) and QPSK modulation ($L = 2$). Then $m = \log_2 L = 2$, and input bits are divided into two parallel streams. Let us represent each of the stream as a polynomial:

$$c^k(D) = c^k_0 + c^k_1 D + c^k_2 D^2 + \ldots, \qquad k = 1, 2$$

where $c^k_j \in \{0, 1\}$, and let feedforward registers for the stream k and antenna i are defined by polynomials $G^k_i(D)$ in (8.37), notice that for QPSK modulation $g^k_{j,i} \in \{0, 1, 2, 3\}$. Then the output of antenna i is:

$$x^i(D) = c^1(D)G^1_i(D) + c^2(D)G^2_i(D) \mod 4$$

or in matrix form:

$$x^i(D) = \begin{bmatrix} c^1(D) & c^2(D) \end{bmatrix} \begin{bmatrix} G^1_i(D) \\ G^2_i(D) \end{bmatrix} \mod 4 \tag{8.38}$$

For the simple example, we take:

$$\begin{array}{ll} G^1_1(D) = 2D, & G^1_2(D) = 2 \\ G^2_1(D) = D, & G^2_2(D) = 1 \end{array}$$

Encoder for such a code is depicted in Figure 8.22. Since $\nu_1 = 1$ and $\nu_2 = 1$, the total memory of the encoder is equal to $\nu = 2$, that is, the encoder may be in $2^2 = 4$ states. At each time moment t two

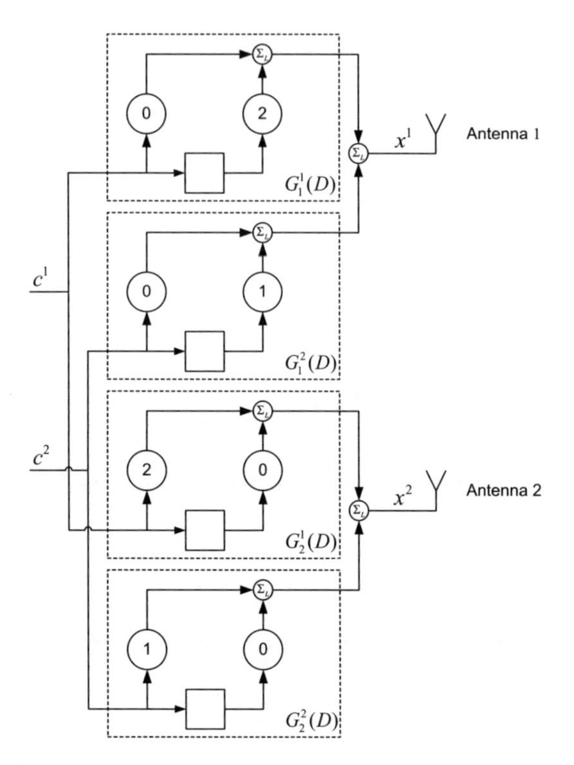

Figure 8.22 Example of STTC for QPSK and 2 transmit antennas

bits c^1 and c^2 are fed to the encoder's input, and the encoder produces two output signals x^1 and x^2 and transitions to the new state. This procedure may be depicted with one trellis section, see Figure 8.23. The label c^1c^2/x^1x^2 corresponds to each edge of the trellis (to each transition from one state to another).

Consider the encoder in state 00 with the input sequence 10 11 01 10 00 .... Then at the decoder's output there will be the sequence 02 23 31 12 20 ..., and this codeword of the space-time code corresponds to the path in the trellis, as shown in Figure 8.24. The decoding task may be formulated as the task of searching in the trellis the path closest (for example, in Euclidean metrics) to the received sequence.

In [40] the tables of STTC code optimal by rank, determinant or trace criteria are given, for different numbers N of transmit antennas and "depth" ν of the trellis. Codes based on rank and determinant criteria, are listed in Table 8.1, and the codes based on trace criterion are listed in Table 8.2. For the stream of information bits c^k the polynomials $G_i^k(D)$ for transmitting antennas 1 to N, defining the correspondent shift registers FFSR_i^k are given as coefficients $[g_{0,i}^k, g_{1,i}^k, \ldots, g_{\nu_k,i}^k]$.

8.4.2 *Space-Time Turbo Trellis Codes*

In section 8.4.1 the space-time codes were considered, using for its construction the shift registers to achieve the coding gain. Based on this construction the *recursive* STTC codes may be defined by adding feedback to the feedforward shift register. This is equivalent to the division of the correspondent connection polynomial $G_i^k(D)$ in (8.37) by the binary irreducible polynomial $q(D)$.

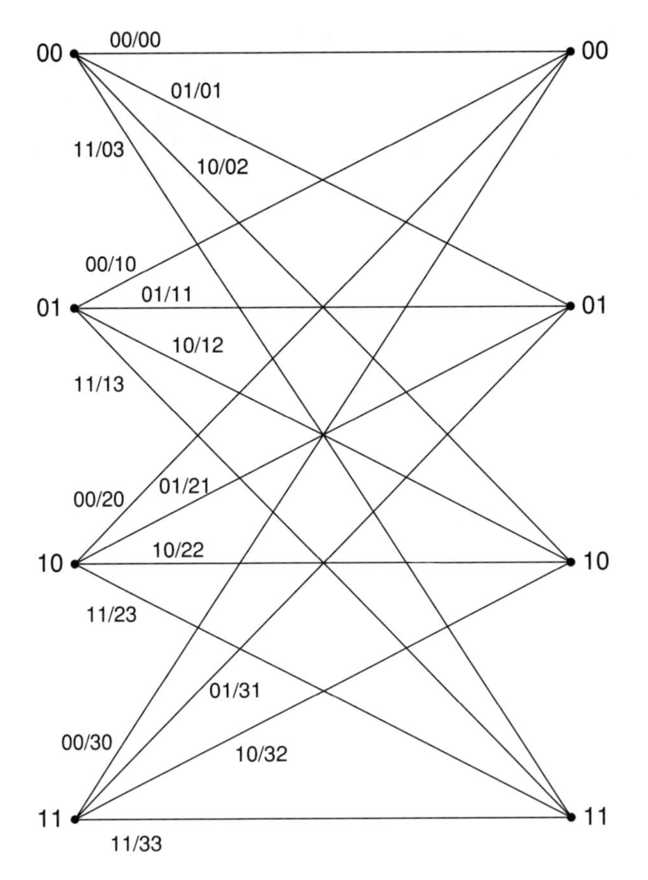

Figure 8.23 STTC trellis for QPSK and 2 transmit antennas

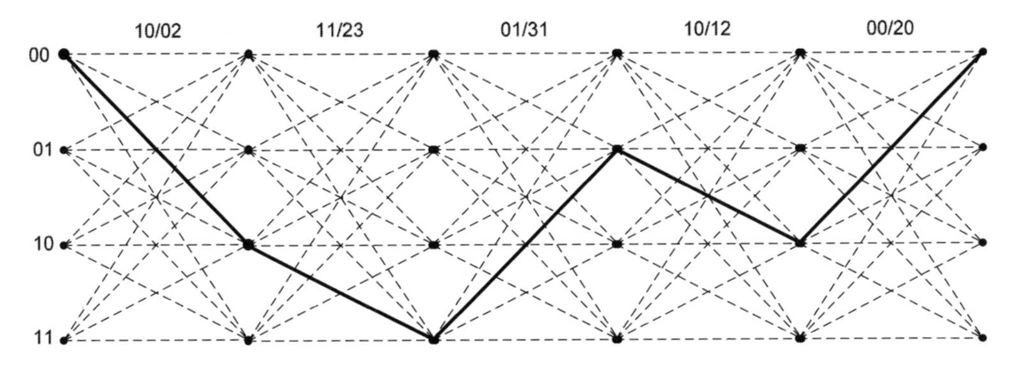

Figure 8.24 Codeword on the trellis

Table 8.1 STTC codes constructed basing on rank and determinant criteria

				FFSR_i^k			
				Antenna i			
Modulation	N	ν	k	1	2	3	4
QPSK	2	2	1	[0,1]	[2,0]		
			2	[2,0]	[2,1]		
		3	1	[0,2]	[2,0]		
			2	[2,1,0]	[1,2,2]		
		4	1	[0,1,2]	[2,2,2]		
			2	[2,1,0]	[0,1,2]		
		5	1	[2,2,0]	[0,3,2]		
			2	[2,1,1,2]	[2,0,2,2]		
		6	1	[1,2,0,2]	[2,2,3,0]		
			2	[2,2,1,0]	[0,0,3,2]		
	3	4	1	[0,0,2]	[0,1,3]	[2,2,1]	
			2	[2,1,2]	[0,2,3]	[0,0,3]	
		5	1	[0,2,0]	[2,0,0]	[1,0,2]	
			2	[3,3,3,2]	[1,2,2,0]	[0,1,2,0]	
		6	1	[1,2,1,2]	[1,1,2,0]	[2,2,0,0]	
			2	[0,0,2,0]	[3,3,2,2]	[0,2,1,2]	
	4	6	1	[0,2,2,2]	[3,3,1,2]	[0,0,1,2]	[2,2,1,0]
			2	[3,2,0,0]	[0,2,0,2]	[2,0,3,2]	[0,2,2,0]
8-PSK	2	3	1	[0,2]	[2,0]		
			2	[0,4]	[4,0]		
			3	[4,5]	[1,4]		
		4	1	[0,4]	[4,0]		
			2	[0,2]	[2,0]		
			3	[2,6,1]	[0,5,4]		
		5	1	[0,4]	[4,4]		
			2	[0,2,2]	[2,2,0]		
			3	[3,0,4]	[5,0,0]		

For the sake of simplicity we consider again QPSK modulation with two transmit antennas and registers with memories ν_1 and ν_2, $\nu_1 \leq \nu_2$. Then the encoding procedure in (8.38) will be as follows:

$$x^i(D) = \begin{bmatrix} c^1(D) & c^2(D) \end{bmatrix} \begin{bmatrix} \dfrac{G_i^1(D)}{q(D)} \\ \dfrac{G_i^2(D)}{q(D)} \end{bmatrix} \mod 4$$

where $q(d) = \sum_{j=0}^{\nu_1} q_j D^j$ is the polynomial with binary coefficients q_j. The encoder structure is similar to Figure 8.20, with the shift register as in Figure 8.25. Here $\oplus$ denotes the summation on modulo 2.

Using the obtained recursive STTC code as the component code, we may construct the scheme of turbo-encoder for space-time coding similar to well-known turbo-encoding schemes which use several component codes and interleavers between them. The scheme of such turbo-STTC code is shown in Figure 8.26 [28, 32].

Table 8.2 STTC codes constructed basing on trace criterion

				$FFSR_i^k$			
				Antenna i			
Modulation	N	v	k	1	2	3	4
QPSK	2	2	1	[0,1]	[2,2]		
			2	[2,2]	[3,0]		
		3	1	[2,2]	[2,1]		
			2	[2,1,0]	[0,2,2]		
		4	1	[1,1,3]	[2,3,2]		
			2	[2,2,2]	[0,2,0]		
		5	1	[0,2,1]	[2,3,2]		
			2	[2,1,2,2]	[2,2,3,0]		
		6	1	[0,3,3,3]	[2,1,3,2]		
			2	[2,2,0,2]	[2,2,0,0]		
	3	2	1	[0,1]	[2,2]	[2,3]	
			2	[2,2]	[3,0]	[3,2]	
		3	1	[2,2]	[2,1]	[2,1]	
			2	[2,1,0]	[0,2,2]	[3,0,2]	
		4	1	[1,1,3]	[2,3,2]	[1,2,1]	
			2	[2,2,2]	[0,2,0]	[2,0,2]	
		5	1	[0,2,1]	[2,3,2]	[2,3,2]	
			2	[2,1,2,2]	[2,2,3,0]	[0,2,1,0]	
		6	1	[0,3,3,3]	[2,1,3,2]	[2,0,2,1]	
			2	[2,2,0,2]	[2,2,0,0]	[0,2,3,1]	
	4	2	1	[0,1]	[2,2]	[2,3]	[0,2]
			2	[2,2]	[3,0]	[3,2]	[2,1]
		3	1	[2,2]	[2,1]	[2,1]	[2,2]
			2	[2,1,0]	[0,2,2]	[3,0,2]	[1,3,1]
		4	1	[1,1,3]	[2,3,2]	[1,2,1]	[1,2,3]
			2	[2,2,2]	[0,2,0]	[2,0,2]	[2,0,2]
		5	1	[0,2,1]	[2,3,2]	[2,3,2]	[2,2,1]
			2	[2,1,2,2]	[2,2,3,0]	[0,2,1,0]	[1,0,0,2]
		6	1	[0,3,3,3]	[2,1,3,2]	[2,0,2,1]	[1,2,2,3]
			2	[2,2,0,2]	[2,2,0,0]	[0,2,3,1]	[2,0,1,2]
8-PSK	2	3	1	[2,3]	[1,4]		
			2	[4,2]	[6,0]		
			3	[0,4]	[4,0]		
		4	1	[2,3]	[4,7]		
			2	[4,6]	[0,6]		
			3	[7,0,4]	[2,7,4]		
		5	1	[0,4]	[4,4]		
			2	[0,2,2]	[2,3,2]		
			3	[4,4,3]	[2,2,7]		
	3	3	1	[2,3]	[1,4]	[3,0]	
			2	[4,2]	[6,0]	[2,4]	
			3	[0,4]	[4,0]	[4,2]	

(*Continued*)

Table 8.2 (*Continued*)

Modulation	N	v	k	FFSR$_i^k$ Antenna i 1	2	3	4
		4	1	[2,3]	[4,7]	[2,2]	
			2	[4,6]	[0,6]	[4,4]	
			3	[7,0,4]	[2,7,4]	[2,6,0]	
		5	1	[0,4]	[4,4]	[0,4]	
			2	[0,2,2]	[2,3,2]	[4,7,7]	
			3	[4,4,3]	[2,2,7]	[6,0,2]	
	4	3	1	[2,3]	[1,4]	[3,0]	[7,5]
			2	[4,2]	[6,0]	[2,4]	[2,4]
			3	[0,4]	[4,0]	[4,2]	[4,0]
		4	1	[2,3]	[4,7]	[2,2]	[2,4]
			2	[4,6]	[0,6]	[4,4]	[4,0]
			3	[7,0,4]	[2,7,4]	[2,6,0]	[0,3,2]
		5	1	[0,4]	[4,4]	[0,4]	[3,3]
			2	[0,2,2]	[2,3,2]	[4,7,7]	[2,1,5]
			3	[4,4,3]	[2,2,7]	[6,0,2]	[5,7,6]

For decoding such turbo-STTC codes the iterative schemes may be applied, analogous to those being used in conventional turbo-codes, for example, using symbol-by-symbol iterative MAP algorithm [40]. The structural scheme of the decoder for M receive antennas is shown in Figure 8.27.

8.5 Differential Space-Time Codes

So far, considering transmission and receiving in multi antenna systems, we assumed that the information about the channel state (CSI, channel side information) is available at the receiver, and this information may be used for decoding. Usually this information is obtained by using channel estimation methods, for example inserting pilot (known a priori) symbols into data traffic. However, in some cases, for example, in channels with fast varying parameters, it is hard to estimate the channel satisfactorily, and the special construction of the receiving scheme is required, which does not use the information about the current channel state.

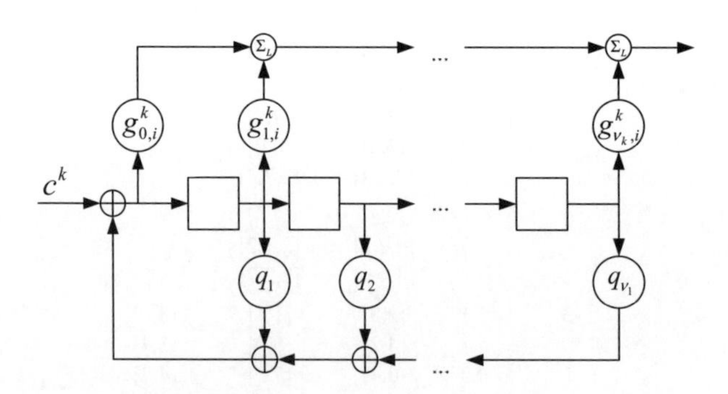

Figure 8.25 Register defined by $G_i^k(D)/q(D)$

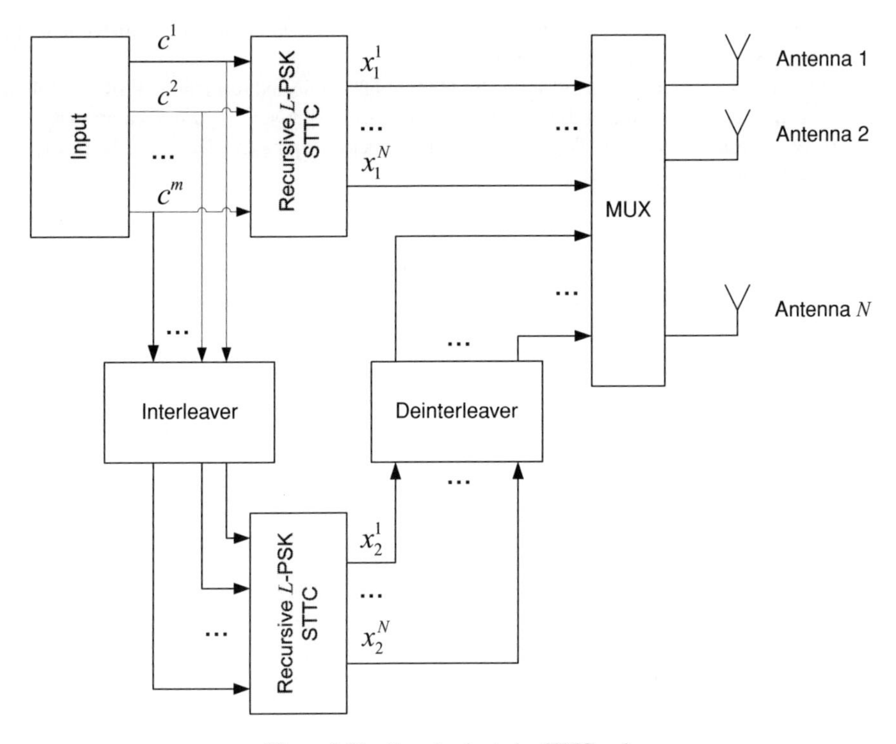

Figure 8.26 Encoder for turbo-STTC code

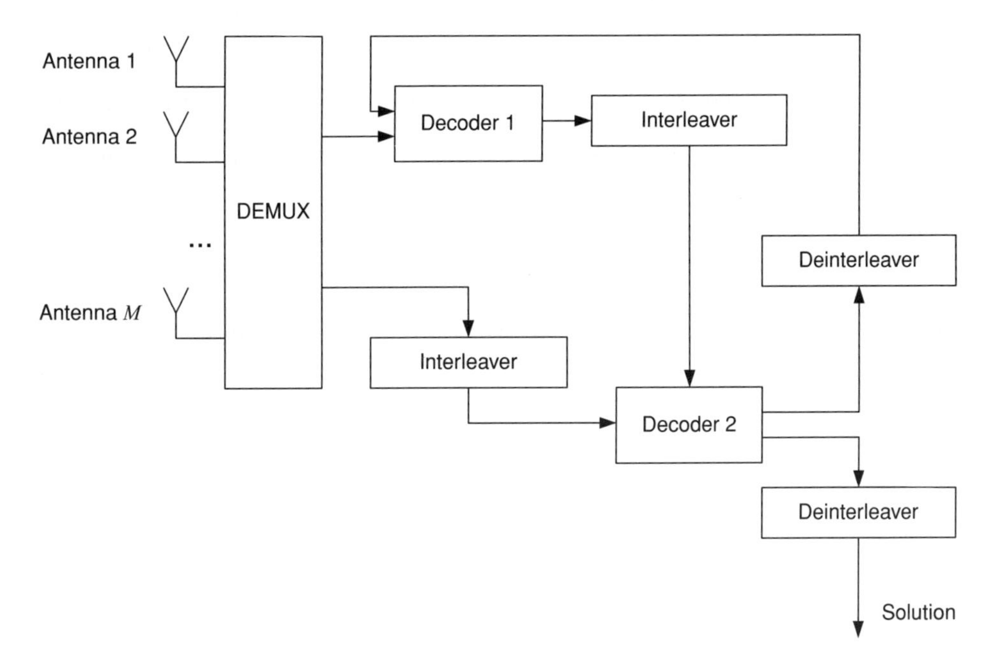

Figure 8.27 Decoder for turbo-STTC code

In the area of space-time codes designing such schemes are based on the so-called differential phase-shift keying (DPSK), which we will consider first.

Suppose that at time moment t we need to transmit the signal s_t, selected from the L-PSK constellation. Instead of transmitting s_t we will send the signal $c_t = c_{t-1}s_t$, setting some a priori known value c_0 as the initial value, for example $c_0 = 1$. The signal c_t will also belong to the L-PSK constellation. Then the received signal would be:

$$r_t = h_t c_t + \eta_t$$

where h_t is transfer coefficient (path gain), and η_t is Gaussian noise. Assuming that the channel state changing is not significant during the transmission time of two adjacent symbols, that is, $h_t \approx h_{t-1} \approx h$, it may be shown [16] that:

$$r_t r_{t-1}^* \approx |h|^2 s_t + \mathcal{N} \tag{8.39}$$

where $\mathcal{N}$ is Gaussian noise. To obtain the estimate $\tilde{s}_t$ of signal s_t we need to find the point in L-PSK constellation, closest to $r_t r_{t-1}^*$, that is:

$$\tilde{s}_t = \arg\min_{s_t} \left| r_t r_{t-1}^* - |h|^2 s_t \right|^2 \tag{8.40}$$

but in L-PSK modulation multiplication of s_t by scaling coefficient $|h|^2$ doesn't change the geometry of the detection region, and hence the problem (8.40) is equivalent to:

$$\tilde{s}_t = \arg\min_{s_t} \left| r_t r_{t-1}^* - s_t \right|^2 \tag{8.41}$$

Obviously, solving (8.41) is independent from the channel state, since it is determined by two adjacent received symbols only.

It may be shown [16] that the usage of such DPSK modulation (that is, ignoring of $|h|^2$ coefficient) leads to reducing the SNR at the receiver (when transmitting over the Rayleigh channel) by approximately 3 dB.

Below we will consider the simple generalization of this approach for the case of space-time coding, when using two transmit antennas.

Suppose that k-th data block encoded by block space-time code (STBC) is represented by the vector of signals:

$$\mathbf{S}^k = \begin{pmatrix} s_1^k \\ s_2^k \end{pmatrix}$$

This vector will be determined by the preceding vector $\mathbf{S}^{k-1}$ and $2m$ input information bits. The signal $\mathbf{S}^0$ will be assumed fixed and known beforehand. Notice that vectors:

$$\mathbf{V}_1(\mathbf{S}^k) = \begin{pmatrix} s_1^k \\ s_2^k \end{pmatrix}, \qquad \mathbf{V}_2(\mathbf{S}^k) = \begin{pmatrix} (s_2^k)^* \\ -(s_1^k)^* \end{pmatrix}$$

form the orthogonal basis. If the vector $\mathbf{S}^k$ is unit-length, then the lengths of $\mathbf{V}_1(\mathbf{S}^k)$ and $\mathbf{V}_2(\mathbf{S}^k)$ will also be equal to one (otherwise these vectors may be normalized).

Next, consider the set $\mathcal{V}$ consisting of 2^{2m} different (2×1)-vectors $\mathbf{P}_1, \mathbf{P}_2, \ldots, \mathbf{P}_{2^{2m}}$ of unit-length. Suppose that the one-to-one mapping $\mathcal{F}$ from $2m$-bit binary sequences to elements of $\mathcal{V}$ is defined. Then the encoding procedure is as follows.

For $2m$ input bits obtain the vector $\mathbf{P}^k = (\mathbf{P}_1^k, \mathbf{P}_2^k)^T \in \mathcal{V}$ by using the $\mathcal{F}$ mapping. Then:

$$\mathbf{S}^k = \mathbf{P}_1^k \mathbf{V}_1(\mathbf{S}^{k-1}) + \mathbf{P}_2^k \mathbf{V}_2(\mathbf{S}^{k-1})$$

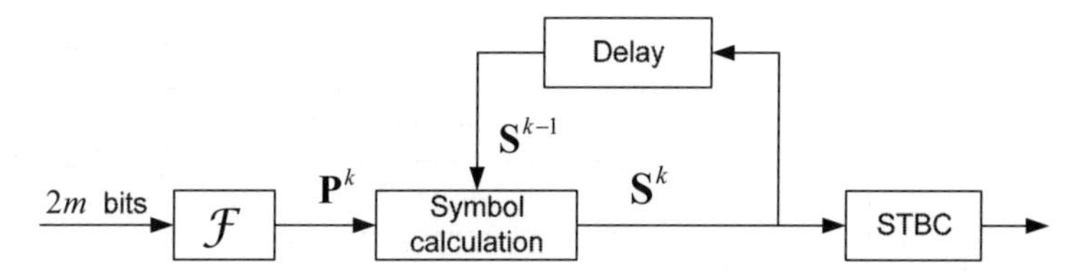

Figure 8.28 Encoder for differential space-time code

Since $\mathbf{V}_1(\mathbf{S}^k)$ and $\mathbf{V}_2(\mathbf{S}^k)$ form the orthogonal basis, then:

$$\mathbf{P}_1^k = [\mathbf{V}_1(\mathbf{S}^{k-1})]^H \cdot \mathbf{S}^k = s_1^k(s_1^{k-1})^* + s_2^k(s_2^{k-1})^*$$
$$\mathbf{P}_2^k = [\mathbf{V}_2(\mathbf{S}^{k-1})]^H \cdot \mathbf{S}^k = s_1^k s_2^{k-1} - s_2^k s_1^{k-1}$$

After that obtained signal $\mathbf{S}^k$ is transmitted by means of STBC based on orthogonal design, for example using the Alamouti code. The encoder scheme is shown in Figure 8.28.

We consider the decoding procedure for the case of one receive antenna. In section 8.2.3 it was shown that the received signals may be written as:

$$r_1^k = h_1 s_1^k + h_2 s_2^k + \eta_1^k$$
$$r_2^k = -h_1(s_2^k)^* + h_2(s_1^k)^* + \eta_2^k$$

where h_1, h_2 are channel transfer coefficients, η_1^k, η_2^k – Gaussian noise. Define the vector $\mathcal{R}$ as:

$$\mathcal{R} = \begin{bmatrix} (r_1^{k-1})^* r_1^k + r_2^{k-1}(r_2^k)^* \\ (r_2^k)^* r_1^{k-1} - r_1^k(r_2^{k-1})^* \end{bmatrix} = (|h_1|^2 + |h_2|^2)P^k + \mathcal{N}$$

In fact the obtained expression reminds (8.39) for DPSK with one transmit antenna. As in the case of DPSK, multiplication of vector $\mathbf{P}^k$ by coefficient $(|h_1|^2 + |h_2|^2)$ does not change the geometry of detection region, and to find the estimate $\tilde{\mathbf{P}}^k$ for the transmitted signal $\mathbf{P}^k$ we need to find in set $\mathcal{V}$ a vector closest to $\mathcal{R}$. Inverse mapping $\mathcal{F}^{-1}$ completes the decoding. The decoder scheme is shown in Figure 8.29.

As for DPSK, the described differential space-time scheme leads to 3 dB loss due to ignoring the scaling coefficient in the detection procedure.

Concrete examples of the set $\mathcal{V}$ and mapping $\mathcal{F}$ selection, allow the decoder to be simplified, and also the generalizations for the cases of more antennas may be found in [12, 14, 16–18].

8.6 Spatial Multiplexing

8.6.1 General Concepts

In sections 8.2 and 8.3 we considered the transmission over MIMO channel by means of the space-time coding. The main task of space-time code is providing the diversity gain in multiantenna transmission and receiving, that is, providing the receiver with different responses of the same signals, which should

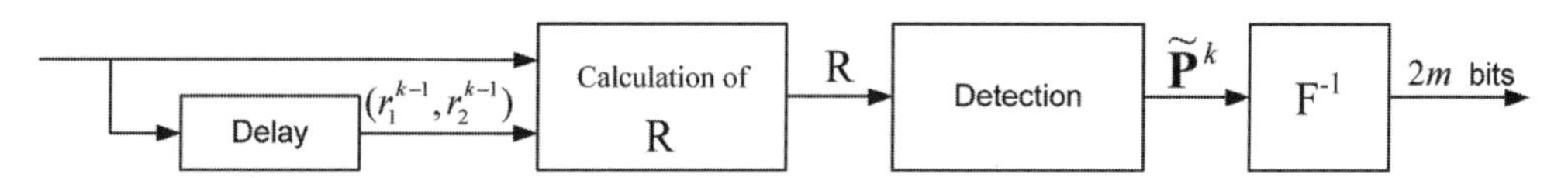

Figure 8.29 Decoder for differential space-time code

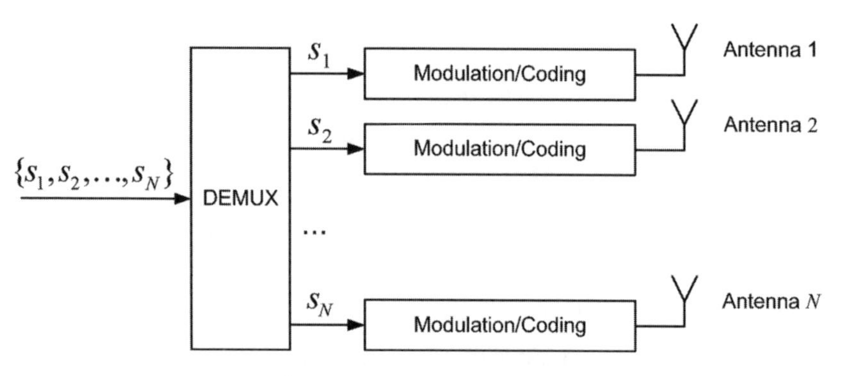

Figure 8.30 Spatial multiplexing

improve the receiving reliability under conditions of transmission over the channels with fading and dispersion. As has been noted in section 8.2, the rate of space-time code in (8.13) is $R = K/T$, that is, it characterizes the number K of symbols which may be transmitted within the period of T time slots. In section 8.3 the orthogonal space-time codes, with rate $R \leq 1$ were considered, that is, the maximal throughput of these codes (with $R = 1$) is one symbol per one time slot (or per one channel use). With this limitation on the coding scheme throughput the orthogonal space-time codes allow the maximal diversity to be achieved. At the same time simple maximum likelihood decoding procedure exists for these codes.

The architecture of the MIMO system may also be used for achieving other goals, for instance, increasing the throughput to more than one symbol per channel use, at the cost of diversity gain decreasing, and hence, increasing of the error probability. For example, this may be acceptable and desirable in applications operating at relatively high SNRs, where error rate is relatively low, but the high transmission rate should be (and may be) achieved.

The possible solution for this problem is the *spatial multiplexing* technique. The idea of this principle is shown in Figure 8.30. Transmitted symbols $s_1, \ldots, s_m$ are split into N parallel streams, which are then modulated (and possibly encoded) and then transmitted over some dedicated antenna or by N antennas simultaneously. From the system architecture point of view the procedures of modulation and coding are not principal, so we will not take them into account in what follows.

If there are M antennas used for receiving, and $\mathbf{s} = (s_1, \ldots, s_N)$ is the vector intended for transmission over the N transmit antennas at some fixed time point then the received vector $\mathbf{r} = (r_1, \ldots, r_M)$ will be:

$$\mathbf{r} = \mathbf{s} \cdot \mathbf{H} + \eta \tag{8.42}$$

where $\mathbf{H}$ is $(N \times M)$-matrix of channel transfer coefficients in (8.8), and η is M-component vector of Gaussian noise.

Then the maximum likelihood decoding (which in Gaussian channel coincides with the minimal Euclid distance decoding) consists in finding the vector $\mathbf{s}$ minimizing the function $d_E\,(\mathbf{r}, \mathbf{s} \cdot \mathbf{H})$, where $d_E\,(\mathbf{a}, \mathbf{b})$ is the Euclidean distance between vectors $\mathbf{a}$ and $\mathbf{b}$. Unlike the case of space-time codes, we do not define the structure of the codewords $\mathbf{s}$ (for example, the structure of orthogonal matrix). Due to this fact the decoding may be performed by the total search through all the modulation signals. If the signal constellation consists of B symbols, then the decoding would require exponential search in B^N possible variants instead of linear search in BN variants required for the codes from section 8.3. In practice such decoding complexity is unacceptable in most cases, and for the decoding the more simple sub-optimal procedures providing acceptable error probability are used.

An example of such simplification is the decoding in the sphere; an approach traditionally exploited in error-correcting coding theory. In this case the search is performed only amongst those codewords,

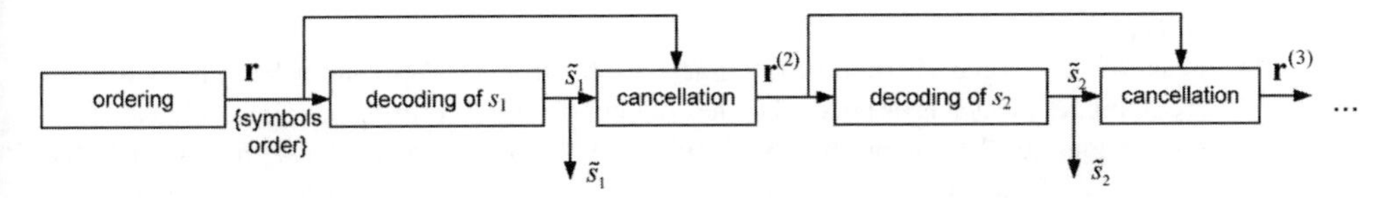

Figure 8.31 V-BLAST decoder (optimal ordering $\{s_1, \ldots, s_N\}$ is assumed)

which are somehow within a defined sphere of some fixed radius and with the centre in the received vector $\mathbf{r}$ rather than by all the codewords. An example of such decoding is sphere detection described in Chapter 5.

Next we will consider some examples of spatial multiplexing based on different versions of space-time architectures *BLAST (Bell Labs Layered Space-Time architectures)* [11].

8.6.2 V-BLAST

First we will describe the architecture of *vertical BLAST (V-BLAST)* [42]. The encoder structure is in fact similar to that shown in Figure 8.30, and received word is described by the expression in (8.42). It may be noted that with the coding scheme as in Figure 8.30, when the transmission is made in fact independently over the N antennas, the decoding task is equivalent to the user detection task when receiving signals from N independent users, so the methods being used in systems with multiple users are applicable here.

Step-by-step decoder described in [42] successively processes the symbols (layers) and is called *SIC* (*successive interference cancellation*). This algorithm works as follows (see Figure 8.31).

1. The codeword symbols $\mathbf{s} = (s_1, \ldots, s_N)$ should be ordered by the decreasing power of the correspondent received signals (decreasing of correspondent SNR). In what follows we will assume that the ordering in consideration coincides with the initial ordering.
2. Suppose that for some $2 \leq i \leq N$ the symbols $s_1, \ldots, s_{i-1}$ are already detected, and their correspondent estimations $\tilde{s}_1, \ldots, \tilde{s}_{i-1}$ are obtained. Then before the detection of symbol s_i the impact of already detected symbols for the received vector $\mathbf{r}$ is taken into account as:

$$\mathbf{r}^{(i)} = \mathbf{r} - \sum_{j=1}^{i-1} \tilde{s}_j \cdot \mathbf{h}^{(j)}$$

   where $\mathbf{h}^{(j)}$ is the j-th column of the channel matrix $\mathbf{H}$. Detection of symbol s_i is done by the vector $\mathbf{r}^{(i)}$. In fact this step resembles the functioning of decision feedback equalizer (DFE, see section 5.1.3). For the detection of symbol s_1 this step is omitted, $\mathbf{r}^{(1)} = \mathbf{r}$ is assumed. It may be shown that in this case the received vector after i-th decoding step is equal to:

$$\mathbf{r}^{(i+1)} = \mathbf{r}^{(i)} - \tilde{s}_i \mathbf{h}^{(i)}$$

   At the next step the detection of correspondent symbol s_i is made by the received vector $\mathbf{r}^{(i)}$.
3. The detection of symbol s_i by vector $\mathbf{r}^{(i)}$ is made with the help of a procedure called interference nulling. In the vector $\mathbf{r}^{(i)}$ the symbols which have been already detected are already taken into account, so the detection is reduced to accounting symbols which are not yet detected, and the impact of which is assumed simple as noise. Such detection may be performed by means of ZF or MMSE equalizers (see section 5.1.2).

(a) *ZF-nulling*
In the case of Zero-Force (ZF) nulling to extract the $s_i \mathbf{h}^{(i)}$ from $\mathbf{r}^{(i)}$ the vector $\mathbf{r}^{(i)}$ may be multiplied by the column $\tilde{\mathbf{h}}^{(i)}$, which is orthogonal to the columns $\mathbf{h}^{(j)}$ for $j = i+1, i+2, \ldots, N$, but which is not orthogonal to the column $\mathbf{h}^{(i)}$. Such column $\tilde{\mathbf{h}}^{(i)}$ may be obtained as i-th column of the nulling matrix $\widetilde{\mathbf{H}}^{+}$, calculated as follows. Replace the rows $1, 2, \ldots, i-1$ of matrix $\mathbf{H}$ by zeros, obtaining the matrix $\widetilde{\mathbf{H}}$ (that is, the matrix not accounting the impact of other antennas), then $\widetilde{\mathbf{H}}^{+}$ will be the pseudo-inverse (Moore-Penrose generalized inverse) to the matrix $\widetilde{\mathbf{H}}$:

$$\widetilde{\mathbf{H}}^{+} = \widetilde{\mathbf{H}}^{H} \cdot (\widetilde{\mathbf{H}} \cdot \widetilde{\mathbf{H}}^{H})^{-1}$$

Then:

$$\mathbf{r}^{(i)} \cdot \tilde{\mathbf{h}}^{(i)} = s_i + \mathcal{N} \cdot \tilde{\mathbf{h}}^{(i)}$$

where $\mathcal{N} \cdot \tilde{\mathbf{h}}^{(i)}$ is Gaussian noise, and therefore the symbol s_i may be detected as symbol $\tilde{s}_i$, closest (in given constellation) to $\mathbf{r}^{(i)} \cdot \tilde{\mathbf{h}}^{(i)}$.

(b) *MMSE-nulling*
In the case of minimum mean-square error (MMSE) nulling the detection is made similarly, but in correspondence with MMSE criterion the nulling matrix $\widetilde{\mathbf{H}}^{+}$ is calculated as:

$$\widetilde{\mathbf{H}}^{+} = \widetilde{\mathbf{H}}^{H} \cdot \left(\frac{\mathbf{I}_N}{\gamma} + \widetilde{\mathbf{H}} \cdot \widetilde{\mathbf{H}}^{H} \right)^{-1}$$

where γ is SNR at the receiver.

To perform the described decoding we should determine the symbols detection order at its first step. This procedure may be different for different equalization methods used for detection. For example, in [42] when using the ZF-equalizer (which is not taking into account the noise during detection, and therefore the noise may be enhanced, but the correspondent noise enhancing factor may be calculated) the ordering is proposed to minimize the maximal noise enhancing factor. In [42] it is proved that the global optimum by this criterion may be achieved, if at each step the locally optimal symbol is selected, that is, the symbol giving the smallest noise enhancing factor.

To overcome the necessity of optimal ordering the algorithm of *parallel interference cancellation* (*PIC*) may be considered [33]. This algorithm may be described as follows.

1. In the absence of SIC ordering all symbols (layers) are detected simultaneously as

$$\tilde{\mathbf{s}} = \mathbf{H}^{+} \cdot \mathbf{r}$$

where $\mathbf{H}^{+} = \mathbf{H}^{H} \cdot (\mathbf{H} \cdot \mathbf{H}^{H})^{-1}$ is pseudo-inverse of the channel matrix $\mathbf{H}$.

2. Interference cancellation in the received vector is performed as

$$\mathbf{r}^{(i)} = \mathbf{r} - \sum_{j \neq i} \tilde{s}_j \cdot \mathbf{h}^{(j)}$$

where $\mathbf{r}^{(i)}$ is the vector of received symbols with interference cancellation in all symbols except i-th, $\mathbf{h}^{(j)}$ is j-th column of the channel matrix $\mathbf{H}$.

3. At the interference nulling step the correspondent nulling matrix $\widetilde{\mathbf{H}}^{+}$ is calculated by ZF or MMSE criterion similar to SIC algorithm, but for its computation during detection the symbol $\tilde{s}_i$ all rows in matrix channel, except i-th, are nulled, and not the rows correspondent to preceding detected symbols, as in SIC algorithm.

PIC algorithm, which is not taking into account ordering, in general loses by the error probability to the SIC algorithm with optimal ordering, but it may be combined with SIC algorithm, which may give advantage compared to basic SIC algorithm.

	Antenna 1	Antenna 2	Antenna 3
$t = 1$	Layer 1		
$t = 2$	Layer 3	Layer 1	
$t = 3$	Layer 2	Layer 3	Layer 1
$t = 4$	Layer 1	Layer 2	Layer 3
$t = 5$	Layer 3	Layer 1	Layer 2
$t = 6$	Layer 2	Layer 3	Layer 1
$t = 7$		Layer 2	Layer 3
$t = 8$			Layer 2

Figure 8.32 Diagonal transmission using D-BLAST layers

8.6.3 D-BLAST

Another variant of BLAST architecture, called *diagonal BLAST (D-BLAST)* [10], differs from the considered V-BLAST by the procedure of mapping the input symbols to the parallel streams being organized. In V-BLAST architecture the input symbols were consecutively distributed over the transmitting antennas from "top" to "down", (see Figure 8.31), which gives the name to the whole system. In the case of D-BLAST the input symbols are distributed over the streams called "layers", and at each consecutive moment of transmission the mapping of these layers to the transmitted antennas is cyclically shifted, forming some diagonal structure.

Let us explain the functioning of the system by example. Let there be $N = 3$ transmit antennas. Then the input symbols stream is split into three layers, and the transmission is made as shown in Figure 8.32.

At first time slot $t = 1$ the symbol of the first layer is transmitted from the first antenna. At the second time slot the symbol of the first layer is transmitted from the second antenna, while the first antenna transmits the symbol of Layer 3 and so on. It may be noticed that not all antennas are taking part in the transmission at $N - 1$ initial and $N - 1$ conclusive time slots. This slightly reduces the throughput of the system.

As can be seen from the Figure 8.32, the symbols belonging to the same layer are transmitted by diagonal in space-time matrix. In fact such a scheme represents the kind of interleaver, and the idea behind it is to avoid grouping of errors into bursts in case one of the antennas is in fading.

Decoding is performed in a very similar way to the decoding in V-BLAST system, but taking into account the diagonality of D-BLAST. First, the decoding of the first (upper) diagonal consisting of N symbols from the Layer 1 is performed. Since at the first time slot there is only one antenna from which transmission is made, the received symbol does not interfere with the other signals from other antennas, and detection may be performed, for example, by means of conventional MRC procedure (see section 8.2.1). At the second time slot the signal from the first antenna (transmitting the symbol from the Layer 3) is considered as noise, and the signal from the second antenna (Layer 1) is decoded as in V-BLAST. At the third time slot the noise from Layers 2 and 3 should be compensated to decode Layer 1. After that N symbols from the diagonal of Layer 3 is decoded, with taking into account that the signals from Layer 2 are not decoded yet and considered as noise, and the signals from Layer 1 are already known and may be compensated as in V-BLAST. Then the decoding proceeds in the same way, "moving" by the diagonals in Figure 8.32 from up to down, compensating received signals by the values of already

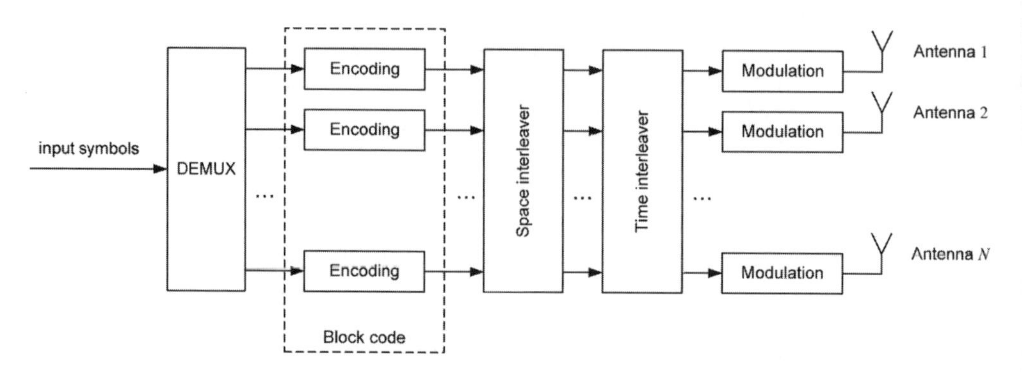

Figure 8.33 Turbo-BLAST encoding scheme

decoded diagonals, which are placed above the current diagonal, and considering the signals from the lower diagonals as noise.

The difference between decoding in V-BLAST and D-BLAST is that in the case of D-BLAST there is no need for the signals sequence ordering to provide detection, since the order here is defined by the interleaving scheme.

8.6.4 Turbo-BLAST

One more variant of BLAST architecture proposed in [34] describes the joint usage of spatial multiplexing with error-correcting coding of obtained streams. Encoding scheme for such *Turbo-BLAST* architecture is shown in Figure 8.33. The data split by N streams are coded with the error-correcting code, and after that are subject to space-time interleaving. In [34] the modification of the scheme shown in Figure 8.32 is considered as space interleaver. This scheme is shown in Figure 8.34, where the distribution of layers over antennas at time slot t is simply the cyclic shift of the distribution at moment $t-1$, as in D-BLAST, but there are no idle antennas at the beginning and ending of transmission. The random interleaving is used as time interleaving.

In [34] it is shown that the proposed scheme allows the artificial channel with time-varying parameters from the quasi-static communication channel with fading to be obtained, in this case space-time

	Antenna 1	Antenna 2	Antenna 3
$t=1$	Layer 1	Layer 2	Layer 3
$t=2$	Layer 3	Layer 1	Layer 2
$t=3$	Layer 2	Layer 3	Layer 1
$t=4$	Layer 1	Layer 2	Layer 3
$t=5$	Layer 3	Layer 1	Layer 2
...	...	...	...

Figure 8.34 Space interleaving in Turbo-BLAST

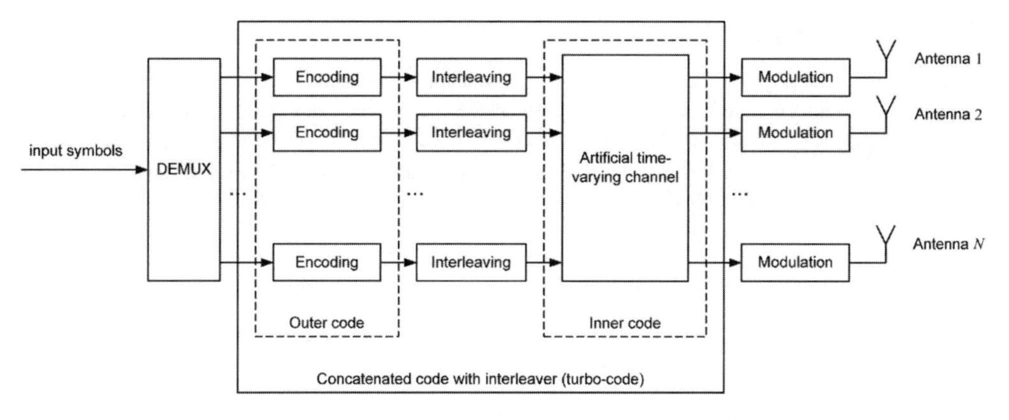

Figure 8.35 Equivalent coding scheme for Turbo-BLAST

interleaving acts as additional channel coding scheme (with rate 1), besides block error-correction coding. Such equivalent coding treatment for Turbo-BLAST is given in Figure 8.35.

In fact we have serially concatenated code, in which outer and inner encoders are separated by interleaver, this is the classical scheme of turbo-code. An iterative decoding scheme may be applied to such code, where the inner decoder is SISO (soft in/soft out) detector, and the outer decoder consists of the decoders of correspondent error-correcting codes. As in usual iterative turbo-decoding schemes, the outer and inner decoders are exchanging information with each other. The scheme of such decoder is given in Figure 8.36.

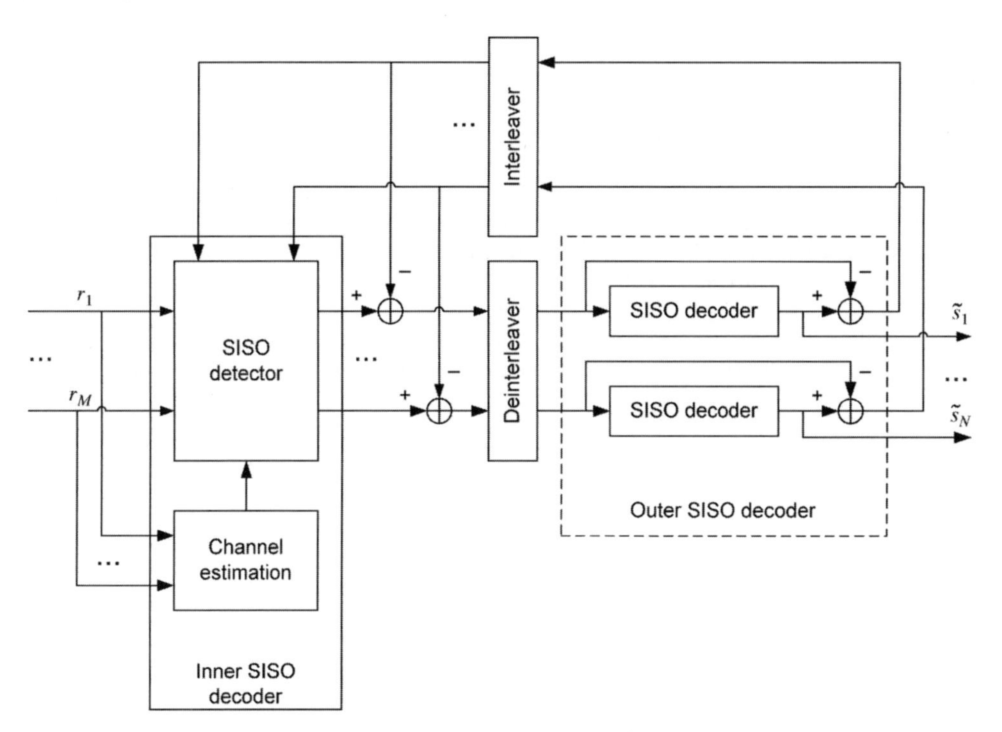

Figure 8.36 Iterative decoder for Turbo-BLAST

8.7 Beamforming

In case of closed-loop system (see section 8.1.3), when the information about the channel is available to the transmitter (this information is sent from the receiver side over the feedback channel), the performance of the system may be improved if the "beams" transmitting the data are "tuned" corresponding to the known (estimated) current channel state. Such procedure is called *beamforming* and may be performed both in analogue domain and with digital data. In this section we will briefly consider "digital beamforming" techniques [16, 27].

Since information about the channel state coming from the feedback channel (*CSI, channel state information*) is usually non-ideal, this may decrease the effect of beamforming, however, in this case combining of space-time coding techniques and digital beamforming methods may be useful [21], see Figure 8.37. Then, instead of sending the codeword $\mathbf{C}$ (represented by matrix) of the space-time code to the transmitting antennas, the following linear transformation will be used:

$$\mathbf{C}' = \mathbf{C} \cdot \mathbf{W} \tag{8.43}$$

where $\mathbf{W}$ is the beamforming matrix. Selection of matrix $\mathbf{W}$ may be done corresponding to different criteria. For example [16, 21], one criterion for the construction may be the minimization of maximal value of pairwise conditional error probability $P(\mathbf{C}_k \to \mathbf{C}_l|\hat{\mathbf{h}})$. This probability is correspondent to the event that after transmission of codeword $\mathbf{C}_k$ of space-time code and given estimation of channel realization $\hat{\mathbf{h}}$ the maximum likelihood decoding makes solution in favour of the codeword $\mathbf{C}_l$. If the upper bound on this probability is denoted as:

$$V(\mathbf{C}_k, \mathbf{C}_l) \geq P(\mathbf{C}_k \to \mathbf{C}_l|\hat{\mathbf{h}})$$

then it may be shown [21] that:

$$\log V(\mathbf{C}_k, \mathbf{C}_l) = f(m_{\mathbf{h}|\hat{\mathbf{h}}}) - \log \det(g(\mathbf{A}(\mathbf{C}_k, \mathbf{C}_l)))$$

where $f(m_{\mathbf{h}|\hat{\mathbf{h}}})$ is some function from conditional mean of the path gains given the partial CSI, that is, determined by the channel and its estimation, and $g(\mathbf{A}(\mathbf{C}_k, \mathbf{C}_l))$ is some function from the matrix $\mathbf{A}(\mathbf{C}_k, \mathbf{C}_l) = (\mathbf{C}_l - \mathbf{C}_k)^H(\mathbf{C}_l - \mathbf{C}_k)$ defined in section 8.2.4, that is, determined by the space-time code and in fact is the determinant criterion.

Then, in the case of absence of CSI, the transmission power is uniformly distributed across all beams. In the case of ideal CSI the first term is dominating in the obtained criterion, and the energy is concentrated in the best beam. In the case of non-ideal CSI the second term is dominating, and matrix $\mathbf{W}$ is optimized to increase the level of spatial diversity (in fact this is achieved by means of determinant criterion, see section 8.2.4).

As an additional example we consider the usage of beamforming in WCDMA/HSDPA standard [1]. In closed-loop mode 1 (CL1) two antennas are used for transmission, each corresponding to some "weight"

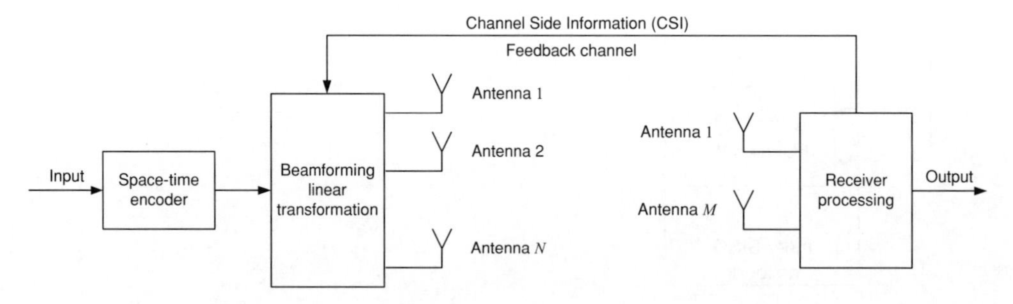

Figure 8.37 Combining of space-time coding and beamforming

w_1 and w_2 [23, 25]. In fact these values are complex values, multiplication by which tunes the phase of the transmitted signals.

Calculation of the tuned phase is made to maximize the received power:

$$P = \mathbf{w}^H \cdot \mathbf{H}^H \cdot \mathbf{H} \cdot \mathbf{w}$$

where $\mathbf{w} = [w_1, w_2]^T$ and $\mathbf{H} = [\mathbf{h}_1, \mathbf{h}_2]$ represent the estimated channel impulse responses for the transmission antennas 1 and 2, of length equal to the length of the channel impulse response. During soft handover, the UE computes the phase adjustment to maximize the total UE received power from the cells in the active set. It can be done by maximizing the following function:

$$P = \mathbf{w}^H \left(\sum_{n=1}^{K} \mathbf{H}_n^H \mathbf{H}_n \right) \mathbf{w}$$

where K is the number of base stations in active set, $\mathbf{H}_i$ is the channel impulse response matrix for cell i. Weight w_1 is constant equal to $1/\sqrt{2}$, and w_2 may take the values of $\frac{1}{2} + j\frac{1}{2}$, $\frac{1}{2} - j\frac{1}{2}$, $-\frac{1}{2} + j\frac{1}{2}$ and $-\frac{1}{2} - j\frac{1}{2}$. Calculated optimal value of w_2 is transmitted over the feedback channel to the base station.

In MIMO 2x2 *Dual-stream Transmit Antenna Adaptive Array* (D-TxAA) mode it is possible to transmit two data streams simultaneously. Each stream is transmitted via both transmit antennas. In the same way as in CL1 the precoding vectors of weighting coefficients are used for each stream. The primary precoding vector (w_1, w_2) is used for transmission of stream 1 and the secondary precoding vector (w_3, w_4) for transmission of stream 2. The precoding weights are defined as follows:

$$\begin{aligned} w_3 &= w_1 = 1/\sqrt{2} \\ w_4 &= -w_2 \\ w_2 &\in \left\{ \frac{1+j}{2}, \frac{1-j}{2}, \frac{-1+j}{2}, \frac{-1-j}{2} \right\} \end{aligned} \tag{8.44}$$

The primary and secondary precoding vectors are orthogonal:

$$w_1 \cdot w_3^* + w_2 \cdot w_4^* = w_3 \cdot w_1^* + w_4 \cdot w_2^* = w_1^2 - w_2^2 = 0 \tag{8.45}$$

Due to this orthogonality it is possible to separate the data streams at the receiver. The precoding weights are calculated in the receiver and reported to base station. The receiver chooses one of four possible combinations of precoding weights (actually value of w_2 defines both primary and secondary precoding vectors) maximizing the throughput for the next transmission. Unlike CL1 the beamforming in MIMO D-TxAA is used in a more comprehensive way since the maximization of the received power of one particular stream does not necessarily lead to maximization of the cumulative throughput of both streams. In the case of MIMO D-TxAA the aim of beamforming is the maximization of SINR in both streams rather than maximization of the received power as in CL1. For a more detailed description of CL1 and MIMO D-TxAA see sections 11.2.8 and 11.3.1.5.

Theoretically D-TxAA scheme can double the throughput provided by one stream transmission by means of spatially multiplexing two data streams. The stream separation at receiver can be done after equalizer based on antenna equalization with the help of postprocessing exploiting the orthogonality of the precoding vectors or immediately at the equalizer output with the help of stream equalization.

Consider the data flow corresponding to one particular channelization code depicted in Figure 8.38. Here symbols s_{11}, s_{12} represent the output of spreader for stream 1 and symbols s_{21}, s_{22} output of spreader for stream 2.

The received signal can be written as follows:

$$\mathbf{r} = \mathbf{H}_1^T \mathbf{x}_1 + \mathbf{H}_2^T \mathbf{x}_2 + \mathbf{n} = \begin{bmatrix} \mathbf{H}_1^T & \mathbf{H}_2^T \end{bmatrix} \begin{bmatrix} \mathbf{x}_1 \\ \mathbf{x}_2 \end{bmatrix} + \mathbf{n} = \begin{bmatrix} \mathbf{H}_1 \\ \mathbf{H}_2 \end{bmatrix}^T \begin{bmatrix} \mathbf{x}_1 \\ \mathbf{x}_2 \end{bmatrix} + \mathbf{n} = \mathbf{H}^T \mathbf{x} + \mathbf{n}$$

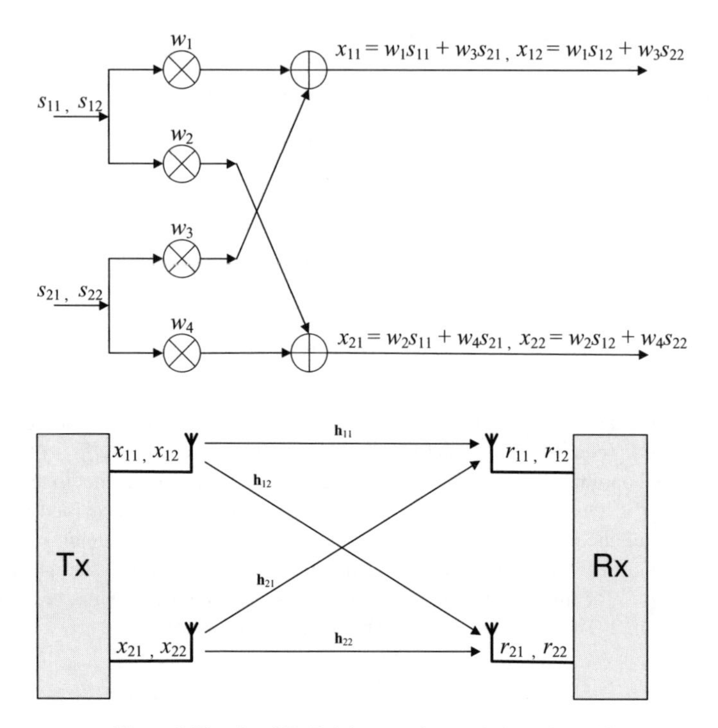

Figure 8.38 2 × 2 D-TxAA transmitter and channel model

where $\mathbf{x}_i$ is the transmitted chip sequence from ith transmit antenna, $\mathbf{n}$ is the noise vector, and $\mathbf{H}_i$ is channel matrix of ith transmit antenna. $\mathbf{H}_i$ is $((F + L) \times 2F)$ matrix, where F is the length of the equalizer filter and L is the channel delay spread in chips.

Taking into account the notation in Figure 8.38 the received signal can be written in the form of:

$$\mathbf{r} = (w_1\mathbf{H}_1 + w_2\mathbf{H}_2)^{\mathrm{T}}\,\mathbf{s}_1 + (w_3\mathbf{H}_1 + w_4\mathbf{H}_2)^{\mathrm{T}}\,\mathbf{s}_2 + \mathbf{n} \tag{8.46}$$

where $\mathbf{s}_i$ is ith stream data. The (8.46) can be written as:

$$\mathbf{r} = \left[\mathbf{H}_1^{\mathrm{T}}\ \mathbf{H}_2^{\mathrm{T}}\right]\mathbf{W}^{\mathrm{T}}\begin{bmatrix}\mathbf{s}_1\\ \mathbf{s}_2\end{bmatrix} + \mathbf{n} = \begin{bmatrix}\mathbf{H}_1\\ \mathbf{H}_2\end{bmatrix}^{\mathrm{T}}\mathbf{W}^{\mathrm{T}}\begin{bmatrix}\mathbf{s}_1\\ \mathbf{s}_2\end{bmatrix} + \mathbf{n}$$

where $\mathbf{W} = \begin{bmatrix}\mathbf{w}_1 & \mathbf{w}_2\\ \mathbf{w}_3 & \mathbf{w}_4\end{bmatrix}$, $\mathbf{w}_i$ is the diagonal matrix given by $\mathbf{w}_i = w_i\mathbf{I}$, I is $(F + L)\times(F + L)$ identity matrix. Now considering matrix:

$$\tilde{\mathbf{H}} = \mathbf{W}\left[\mathbf{H}_1\ \mathbf{H}_2\right] = \left[\tilde{\mathbf{H}}_1\ \tilde{\mathbf{H}}_2\right]$$

as a modified channel matrix it is possible to calculate the equalizer filter coefficients in such a way to make it possible to obtain the separated stream symbols at the equalizer output, that is, provide the stream equalization.

In the case where the equalizer filter is calculated based on matrix $\mathbf{H}$ the estimates of $\begin{bmatrix}\mathbf{x}_1\\ \mathbf{x}_2\end{bmatrix}$ rather than $\begin{bmatrix}\mathbf{s}_1\\ \mathbf{s}_2\end{bmatrix}$ are obtained at the equalizer output as a result of antenna equalization and these values should be

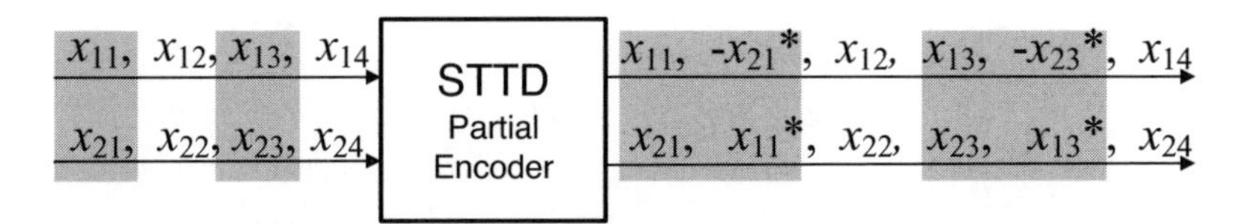

Figure 8.39 Partial STTD Encoder

processed to separate the streams. However, it can be easily done due to the orthogonality of primary and secondary precoding vectors.

Unfortunately, the streams' mutual orthogonality may be easily corrupted under fading channel conditions. In this case the streams would not be separated correctly in both cases using antenna equalization or stream equalization. It is assumed that the usage of adaptive modulation and coding can cope with this problem. However, the use of link adaptation technique in case of loss of streams' orthogonality usually leads to significant decrease of data rate and in some cases the equivalent or even higher throughput can be achieved with one stream transmission using the higher code rate than in MIMO mode which compromises the idea of using the D-TxAA scheme.

One of the possible improvements in this case is to use space-time coding for decreasing the impact of fading. In [44] the modification of D-TxAA scheme consisting in STTD encoding the output of the D-TxAA transmitter was suggested. Actually, it is possible to use the method described by (8.43) in this case but then the matrix **W** used for the transmission would be different from the one described in the standard. However, it is not possible to apply STTD encoding to the whole data since it means two time decrease of the original data rate which immediately leads to performance degradation down to the level provided by one stream transmission. The possible solution is to apply STTD encoding only to the part of the original data flow, that is, only to odd symbols at the output of D-TxAA as shown in Figure 8.39. Of course, the additional STTD encoding in this case limits the theoretically possible throughput of modified scheme to 2/3 of maximum theoretically achieved throughput offered by the original scheme (which is claimed to be 28 Mbps), but this latter value of throughput is not possible to achieve in any case under fading conditions.

Note that taking into account values of precoding beamforming weights in (8.44) the STTD encoding of D-TxAA output keeps the orthogonality of the original streams. The application of STTD encoding on top of D-TxAA scheme makes it impossible to use the stream equalization. Thus, from now on we will consider only antenna equalization.

Now assuming that the data at the output of D-TxAA scheme were encoded with the help of partial STTD encoder represented in Figure 8.39 the output of the despreader after equalization can be written as follows:

$$\begin{array}{ll} d_{11} = h_{11}x_{11} + h_{21}x_{21} + n_{11}, & d_{21} = h_{12}x_{11} + h_{22}x_{21} + n_{21} \\ d_{12} = -h_{11}x_{21}^* + h_{21}x_{11}^* + n_{12}, & d_{22} = -h_{12}x_{21}^* + h_{22}x_{11}^* + n_{22} \\ d_{13} = h_{11}x_{12} + h_{21}x_{22} + n_{13}, & d_{23} = h_{12}x_{12} + h_{22}x_{22} + n_{23} \end{array} \tag{8.47}$$

where d_{ij} is the jth output of despreader corresponding to Rx antenna i, $i \in \{1,2\}$; h_{ij} is the channel coefficient of the path from Tx antenna i to Rx antenna j, i, $j \in \{1,2\}$; n_{ij} is the receiver noise at antenna i, $i \in \{1,2\}$.

In turn the symbols $x_{11}, x_{21}, x_{12}, x_{22}$ shown in Figure 8.38 can be represented as follows:

$$\begin{array}{ll} x_{11} = w_1 s_{11} + w_3 s_{21}, & x_{21} = w_2 s_{11} + w_4 s_{21} \\ x_{12} = w_1 s_{12} + w_3 s_{22}, & x_{22} = w_2 s_{12} + w_4 s_{22} \end{array} \tag{8.48}$$

Assume that equalizer provides perfect channel estimates $\hat{h}_{ij} = h_{ij}$. For the sake of simplicity, we will use only notation h_{ij}. Then the estimates of symbols x_{11} and x_{21} in accordance with [2] can be found from the following equations:

$$\begin{aligned} h_{11}^* d_{11} + h_{12}^* d_{21} + h_{21} d_{12}^* + h_{22} d_{22}^* &= x_{11} \left(|h_{11}|^2 + |h_{12}|^2 + |h_{21}|^2 + |h_{22}|^2 \right) \\ &\quad + h_{11}^* n_{11} + h_{12}^* n_{21} + h_{21} n_{12}^* + h_{22} n_{22}^* \\ h_{21}^* d_{11} + h_{22}^* d_{12} - h_{12} d_{22}^* - h_{11} d_{21}^* &= x_{21} \left(|h_{11}|^2 + |h_{12}|^2 + |h_{21}|^2 + |h_{22}|^2 \right) \\ &\quad + h_{21}^* n_{11} + h_{22}^* n_{12} - h_{12} n_{22}^* - h_{11} n_{11}^* \end{aligned} \tag{8.49}$$

Using the estimates $\hat{x}_{11}$ and $\hat{x}_{21}$ the estimates of symbols s_{11} and s_{21} can be obtained with the help of (8.45) and (8.48) as follows:

$$\begin{aligned} \hat{s}_{11} &= \hat{x}_{11} w_1^* + \hat{x}_{21} w_2^* \\ \hat{s}_{21} &= \hat{x}_{11} w_3^* + \hat{x}_{21} w_4^* \end{aligned} \tag{8.50}$$

The estimates for symbols s_{12} and s_{22} as it follows from (8.47), (8.48) and (8.45) are given by:

$$\begin{aligned} & d_{13} \left(w_1^* h_{22} - w_2^* h_{12} \right) + d_{23} \left(w_2^* h_{11} - w_1^* h_{21} \right) = s_{12} \left(h_{11} h_{22} - h_{12} h_{21} \right) \\ & \quad + n_{13} \left(w_1^* h_{22} - w_2^* h_{12} \right) + n_{23} \left(w_2^* h_{11} - w_1^* h_{21} \right) \\ & d_{13} \left(w_3^* h_{22} - w_4^* h_{12} \right) + d_{23} \left(w_4^* h_{11} - w_3^* h_{21} \right) = s_{22} \left(h_{11} h_{22} - h_{12} h_{21} \right) \\ & \quad + n_{13} \left(w_3^* h_{22} - w_4^* h_{12} \right) + n_{23} \left(w_4^* h_{11} - w_3^* h_{21} \right) \end{aligned} \tag{8.51}$$

Obviously the estimates given by (8.49) and (8.50) are more robust than ones given by (8.51) due to the complex multiplier $(h_{11}h_{22} - h_{12}h_{21})$ in (8.51). Thus the symbols $d_{i(3k+1)}$, $d_{i(3k+2)}$ processed with the help of (8.49) and (8.50) are more reliable than symbols $d_{i(3k+3)}$ obtained with the help of (8.51).

References

[1] 3GPP RAN WG1, *TS25.201: Physical layer – general description*, 3GPP, Tech. Rep., June 2000.
[2] Alamouti, S. *A Simple Transmit Diversity Technique for Wireless Communications*. IEEE Journal on Selected Areas in Communications, 16:1451–1458, Oct. 1998.
[3] Bello, P. *Characterization of randomly time-variant linear channels*. IEEE Transactions on Communication Systems, CS-11: 36–393, 1963.
[4] Blahut, R. *Theory and Practice of Error Control Codes*. Addison-Wesley, 1984.
[5] Brennan, D.G. *Linear diversity combining techniques*. Proc. IRE, vol.47, no.1, pp. 1075–1102, June 1959.
[6] Burr, A. *The multipath problem: an overview*. In IEE Colloquium on Multipath Countermeasures. London, 23 May 1996, Colloquium Digest 1996/120.
[7] Burr, A. *Modulation and Coding for Wireless Communication*. Prentice Hall, 2001.
[8] Chen, Z., Yuan, J., Vucetic, B. *Improved space-time trellis coded modulation scheme on slow fading channels*. Electronic Letters, 37(7):440–441, Mar. 2001.
[9] Cover, T.M., Thomas, J.A. *Elements of Information Theory*. Wiley, 1991.
[10] Foschini, G. *Layered space-time architecture for wireless communication in a fading environment when using multi-element antennas*. Bell Labs Technical Journal, pages 41–59, 1996.
[11] Foschini, G., Chizhik, D., Gans, M., Papadias, C., Valenzuela, R. *Analysis and performance of some basic space-time architectures*. IEEE Journal on Selected Areas in Communications, 21(3):303–320, Apr. 2003.
[12] Ganesan, G., Stoica, P. *Differential modulation using space-time block codes*. IEEE Signal Processing Letters, 9(2):57–60.
[13] Hanzo, L., Liew, T., Yeap, B. *Turbo Coding, Turbo Equalisation and Space-Time Coding*. Wiley-IEEE Press, 2002.
[14] Hochwald, B., Marzetta, T. *Unitary space-time modulation for multiple-antenna communications in Rayleigh flat fading*. IEEE Trans. on Information Theory, 46(2):543–564.
[15] Ionescu, D. *New results on space-time code design criteria*. IEEE Wireless Communications and Networking Conference (WCNC), pages 684–687, 1999.
[16] Jafarkhani, H. *Space-time coding. Theory and Practice*. Cambridge University Press, 2005.

[17] Jafarkhani, H., Tarokh, V. *A differential detection scheme for transmit diversity*. IEEE Journal on Selected Areas in Communications, 18(7):1169–1174.
[18] Jafarkhani, H., Tarokh, V. *Multiple transmit antenna differential detection from generalized orthogonal designs*. IEEE Trans. on Information Theory, 47(6):2626–2631.
[19] Jakes, W. *Microwave Mobile Commumication*. Wiley, 1974.
[20] Jankiraman, M. *Space-Time Codes and MIMO Systems*. Artech House, 2004.
[21] Jongren, G., Skoglund, M., Ottersten, B. *Combining beamforming and orthogonal space-time block coding*. IEEE Trans. on Information Theory, 48(3): Mar. 2002, 611–27.
[22] Kühn, V. *Wireless Communications over MIMO Channels. Applications to CDMA and Multiple Antenna Systems*. Wiley, 2006.
[23] Kurjenniemi, J., Leino, J., Kaipainen, Y., Ristaniemi, T., *Closed Loop Mode 1 Transmit Diversity with High Speed Downlink Packet Access*, ICCT 2003, Beijing, China, Apr, 2003.
[24] Lee, W. *Mobile Communications Engineering*. McGraw Hill, 1982.
[25] Leino, J., Kurjenniemi, J., Rinne, M., *Analysis of Fast Alpha Switching for Closed Loop Mode 1 Transmit Diversity with High Speed Downlink Packet Access*, In Proceedings of IEEE Vehicular Technology Conference (VTC) fall 2004 LA, USA, 26–29 Sept. 2004.
[26] Liang, X, Xia, X. *On the nonexistence of rate-one generalized complex orthogonal designs. IEEE Transactions on Information Theory*, 49(11):2984–2989, Nov. 2003.
[27] Litva, J. and Lo, T. K.-Y. *Digital Beamforming in Wireless Communications*. Artech House Publishers, 1996.
[28] Liu, Y., Fitz, M. *Space-time turbo codes*. 13th Annual Allerton Conf. on Commun. Control and Computing, Sept. 1999.
[29] MacWilliams, F., Sloane, N. *The Theory of Error-Correcting Codes*. North-Holland publishing company, 1977.
[30] Peterson, W., Weldon, E. *Error-Correcting codes*. MIT Press, 1972.
[31] Proakis, J. *Digital communications*. McGraw Hill, 1995.
[32] Robertson, P., Worz, T. *Bandwidth-efficient turbo trellis coded modulation using punctured component codes*. IEEE Journal on Selec. Areas in Communications, 16(2):206–218, Feb. 1998.
[33] Sellathurai, M., Haykin, S. *A simplified diagonal blast architecture with iterative parallel interference cancelation receivers*, IEEE International Conference on Communications, vol. 10, pp. 3067–3071, 2001.
[34] Sellathurai, M., Haykin, S. *TURBO-BLAST for wireless communications: theory and experiments*. IEEE Trans. on Signal Processing, 50(10):2538–2546, Oct. 2002.
[35] Simon, M. *Evaluation of average bit error probability for space-time coding based on a simpler exact evaluation of pairwise error probability*. Int. Jour. Commun. and Networks, 3(3):257–264, Sept. 2001.
[36] Sklar, B. *Digital Communications. Fundamentals and Applications*. Prentice Hall, 2001.
[37] Tarokh, H., Seshadri, V., Calderbank, A. *Space-time codes for high data rate wireless communication: performance analysis and code construction*. IEEE Transactions on Information Theory, 44: 744–765, Mar. 1998.
[38] Tarokh, V., Jafarkhani, H, Calderbank, A. *Space-time block codes from orthogonal designs*. IEEE Transactions on Information Theory, 45: 1456–1467, 1999.
[39] Tarokh, V., Jafarkhani, H, Calderbank, A. *Space-Time Block Coding for Wireless Communications: Performance Results*. IEEE J. Select. Areas Commun., 17(3):451–460, Mar. 1999.
[40] Vucetic, B., Yuan, J. *Space-Time Coding*. Wiley, 2003.
[41] Wang, H., Xia, X. *Upper bounds of rates of space-time block codes from complex orthogonal designs*. IEEE Transactions on Information Theory, 49(10):2788–2796, Oct. 2003.
[42] Wolniansky, P., Foschini, G., Golden, G., Valenzuela, R. *V-BLAST: an architecture for realizing very high data rates over the rich-scattering wireless channel*. International Symposium on Signals, Systems and Electronics (ISSSE), pages 295–300, Sept. 1998.
[43] Zhang, W. *Simulation and modeling of multipath mobile channels*. In Proceedings of the 44th IEEE Vehicular Technology Conference. Stockholm, June, pp. 160–4.
[44] Semenov, S. *Modification of the D-TxAA Scheme for Fading Channel*. In Proceedings of the Fourth Advanced International Conference on Telecommunications 2008. Athens, June 2008, pp. 138–142.

9

Multiple Access Methods

Dmitry Osipov[1], Jarkko Paavola[2], and Jussi Poikonen[2]
[1]Institute for Information Transmission Problems, Russian Academy of Sciences, Russia
[2]Department of Information Technology, University of Turku, Finland

In practice, it is not enough that a single high data rate stream can be transmitted through the air from the transmitter to the receiver. Rather, wireless communication systems must be able to serve tens or even hundreds of users simultaneously with multimedia applications requiring very high data rates. The method allowing several users to communicate simultaneously is called *multiple access* (MA). In this chapter different strategies to implement multiple access are presented.

In a cellular system, the coverage area is divided into parts called cells. Each cell has a base station, which is connected with all user terminals inside the cell. In cellular telecommunication systems, the transmission from the base station to a user terminal (downlink, forward link) and the transmission from a terminal to the base station (uplink, reverse link) must be separated. Traditionally downlink and uplink transmissions use different frequencies or they occur consecutively. These two options are referred to as frequency division duplexing (FDD) and time division duplexing (TDD) [8]. The method allowing several users to communicate simultaneously inside a cell is called a multiple access technique. The capacity of multiple access techniques refers to the maximum number of users K served inside one cell. Here, the term capacity refers specifically to multiple access capacity. From the information theory point of view the maximal rate of data transmitted through the channel is referred to as *channel capacity* to make a distinction from the latter (as established by Claude Shannon [63]). Classical multiple access methods are based on orthogonality, thus they do not inflict *multiple access interference (MAI)* within the communication system. To provide orthogonality signals can be non-overlapping in the time domain or in the frequency domain, but the non-overlapping condition is only sufficient, not necessary. In code-division systems signals overlap in both domains, being orthogonal due to proper code structure. The capacity of orthogonal multiple access methods is $K = WT$, where T is a finite transmission time, and W is the width of the available frequency band. Adding users to systems beyond this limit inflicts MAI. The number of allowed extra users depends on the tolerable amount of interference.

The first generation analog cellular systems used frequency division multiple access (FDMA), where different users communicate on different frequencies. The FDMA is not very convenient from the implementation point-of-view, since K parallel passband filters are required in the receiver and orthogonality is

Modulation and Coding Techniques in Wireless Communications Edited by Evgenii Krouk and Sergei Semenov

difficult to preserve in practice. In the cellular communication environment the cells must be clustered to ensure that the same frequencies are not used in adjacent cells, which would cause inter-cell interference.

Most of the 2G systems used a combination of *time division multiple access (TDMA)* with FDMA. In TDMA, the time axis is divided into slots allocated to different users. Pure TDMA allows all users to have the whole bandwidth at their disposal.

The 3G systems and one 2G system have adopted *code division multiple access (CDMA)* as their multiple access technique. In CDMA users communicate simultaneously within the same frequency band. Different users are separated in the receiver with the help of an individual code signal. At the time of deployment of the 2G systems, CDMA seemed to be too complex from the implementation point-of-view. Thus, only one 2G standard, called IS-95 or cdmaOne, uses CDMA. The rapid development in digital technology has enabled large scale implementations of CDMA in contemporary consumer electronics.

Let us consider the MA problem in more detail. Assume that there are K_0 users that are allowed to transmit in the system under consideration (further on they will be referred to as *subscribers*) and within the scope of consideration K users ($K_0 > K$) are transmitting data via the wireless channel in use (further on they will be referred to as *active users*). Hereinafter two MA system models will be considered. Within the scope of the first model it is assumed that there is a central node called a base station. The base station can communicate with each user and vice versa. It is assumed that users are allowed to communicate directly only with the base station but not with each other. Thus, if a certain user wants to transmit a message to another active user he is to transmit it to the base station. The base station transmits all the messages to the recipients (this model describes many of the contemporary MA systems for example, cellular or satellite systems).

As mentioned, in real life systems the uplink channel and the downlink channel are separated either in the frequency domain or in the time domain. It should be noted that the downlink channel is a one-to-many or broadcasting channel. Thus, all the data streams can be transmitted synchronously. Transmission via the uplink channel is, on the other hand, a many-to-one transmission. So it is very difficult to fully synchronize data streams transmitted by different active users.

On the other hand, it should be noted that it is possible to utilize synchronous CDMA in the uplink. As the cell size is shrinking in the so called hot-spot areas, where extremely high data rates are required, pico cells are utilized. They have coverage areas in the scale of meters or few tens of meters. There, the synchronism can be maintained within a fraction of bit interval also in the uplink due to very small distance between a transmitter and the receiver. Other such quasi-synchronous systems are also proposed for satellite [64] systems, which use a special spreading code to transmit timing information, and microcellular [65] environments, where receiver processing transforms an asynchronous channel into the synchronous equivalent. The synchronous CDMA system developed by Cylink relied on network synchronization for both downlink and uplink [66]. Special orthogonal signals can be constructed to such a quasi-synchronous channel to combat possible small synchronization errors [67, 68].

Within the scope of the second MA model (ad-hoc network) all the users are allowed to communicate with each other. Thus the number of channels that can be potentially requested in an ad hoc network is much greater than in the system with the base station (if the number of active users is the same).

In cellular systems the base station of a certain cell can control activity of all the users in the cell acting as a central node. In ad hoc networks there is no base station and, thus, if control is to be maintained one of the active users is to act as a central node. If for some reason this user exits the system (that is, cannot transmit information to the users in the network under consideration) some other user is to undertake the task. Hereinafter this type of control will be referred to as "decentralized control". Further on we shall assume that any user in the system under consideration is able to transmit to any other.

It should be noted that there are quite a number of MA techniques and the number of techniques resulting from their combination is even greater. Further on only the most fundamental multiple access techniques will be considered. However, in order to give some new insights into the field of multiple access this consideration will be complemented by a short survey of some recently proposed multiple access techniques that can be considered candidates for future implementation in real-life systems. It should be noted that the process of designing real life MA systems and their implementation results in a whole complex of problems (for example, power control, channel estimation and so on). Thus it is worth mentioning that in the course of our consideration of a certain MA method all the theoretical results to be stated are the ones obtained under idealized conditions.

In general, application of any of the abovementioned MA techniques reduces to converting the channel in use into a set of disjoint point-to-point channels (in either frequency domain, time domain or both). By using this approach high spectral efficiency can be obtained but unfortunately the aforesaid approach has certain limitations and drawbacks.

Theoretically speaking it is possible to assign a distinct channel to each subscriber and thus no control is needed in the system under consideration. However, this is possible only if the number of subscribers is few, which certainly is not the case for the majority of modern systems. Thus, in real life systems based on the abovementioned principle control is to be introduced at least in terms of assigning respective channels to the active users, that is, the subscribers who intend to transmit data in the system under consideration will have informed the base station (or the node that acts as one) thereof in advance.

It has been assumed that active users transmit information continuously. However, it turns out that in many real-life systems this is not true. For instance, as has been stated by Viterbi [60], in a two-person telephone conversation each user involved is active from 35% to 50% of time. Note that, since in each channel only one user is allowed to transmit. this channel remains unused throughout all the time period within which the user is not active. Thus, in the previous example the channel allocated to each of the users remains idle for the most of the time. The same holds for various applications where users transmit relatively short messages (for example, commands, requests, authentication information and so on) Therefore classic techniques based on allocating a disjoint portion of a time-frequency resource to each user appear to be very inefficient in terms of channel usage for all applications but for those implying long-term transmission of long messages (for example, multimedia.) To increase the channel usage efficiency of the systems under consideration it is crucial to allocate time and frequency resources that are unused at the moment to the users who are currently active (it is to be noted that switching is possible only if control is introduced into the system under consideration.) This holds even for systems implying long-term transmission since some users may enter the system or new users may exit from the system. Unfortunately, rapid switching between users is practically impossible and, thus, the problem of channel usage efficiency remains one of the major weaknesses of the method under discussion. Moreover, the number of active users' rapid changes (for example, due to active users' high mobility) may result in a very inefficient usage of the control channel.

Another weakness of the method under consideration is its eavesdropping and jamming vulnerability. In the case of a single user transmission it is easy for the intruder to find out which sub-band and/or time slot is assigned to the user. Once this task has been solved the problem of eavesdropping or jamming a signal from the respective user simplifies drastically. As the number of active users grows the task of detecting numbers of time slots and/or sub-bands used by a certain active user becomes more cumbersome. However, in the case of jamming an intruder is to solve a much simpler task – to detect whether this sub-band or time slot is allocated to some active user. In section 9.4 modern MA strategies that are much more jamming and eavesdropping-proof strategies will be considered).

9.1 Frequency Division Multiple Access

The simplest and most intuitive way to solve an MA problem is to allocate a disjoint portion of a time-frequency resource to each user. For instance, in *frequency division multiple access (FDMA)* systems the total bandwidth is subdivided into a set of narrowband subchannels with each user being allowed to transmit or receive only in the subchannels assigned to him (see Figure 9.1).

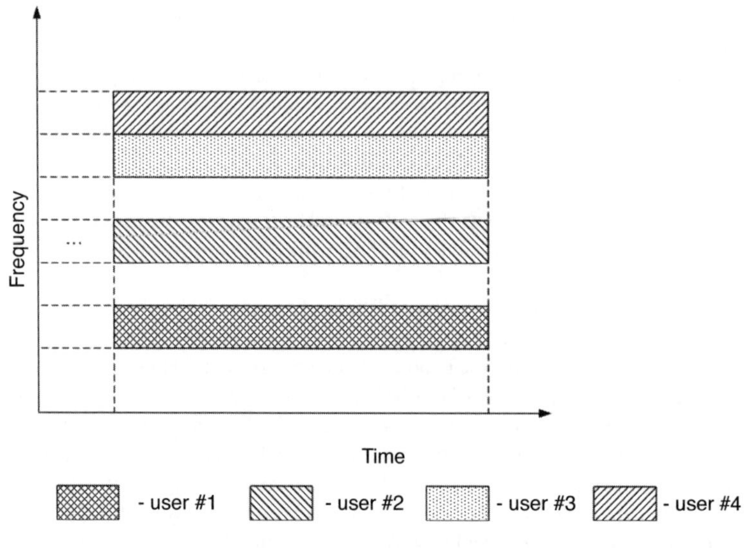

Figure 9.1 FDMA allocation

The basic idea of FDMA is to allocate transmission resources to a number of users in the frequency domain in a non-interfering manner. Thus FDMA is an orthogonal multiple access method (like TDMA), as opposed to non-orthogonal methods described in section 9.3, where controlled interference between users is accepted. Orthogonal MA methods are not capacity-achieving in the information theoretic sense [69], but offer other advantages such as simplicity of realization and flexible system design.

Since the principle of FDMA is to provide a number of users a portion of the total available frequency spectrum, it can easily be applied for continuous-time multiple access. Therefore FDMA was widely used for example in first-generation analog cellular phone systems such as NMT or AMPS [70]. FDMA can be realized simply by communicating with different users by independent band-limited single-carrier transmissions, as described for frequency division multiplexing. Again, this requires the allocation of guard-bands between the users' sub-bands to reduce interference between users due to non-ideal bandpass filtering, possible carrier frequency offsets and Doppler spreading of received signals due to receiver mobility. The use of guard-bands reduces the overall spectral efficiency of a system.

Conventional analog FDMA systems such as the ones mentioned above do not offer a lot of flexibility in the allocation of spectral resources to the users: increasing the number of *frequency slots* available to a user would require the user to have a receiver capable of processing transmissions over disjoint frequency bands, which would require either multiple receivers or complicated wide-band signal processing in the receiver. The flexibility of resource allocation with FDMA can be greatly improved by realizing it using OFDM techniques (see Chapter 2); for such realizations, the term *Orthogonal Frequency Division Multiple Access (OFDMA)* is used. OFDMA will be described in more detail in subsection 9.1.2.

Any practical communication system deals with a certain signal space with finite dimension, which can be quantified for real-valued signal representations as $2WT$, or WT for complex-valued signals where T is a finite transmission time, and W is the width of the available frequency band. In MA systems, this signal space is partitioned and allocated to a number of users. A crucial question in designing large-scale MA systems is how to allow a large number of users to share the limited signal space resource. A solution is to spatially separate the signals of users operating with the same resources. This reuse of the signal space is the basis of cellular networks. Also, multiple-antenna methods such as beamforming can be utilized to mitigate interference between users sharing the same resources in MA systems. In the following, we consider in more detail spectral reuse in FDMA systems.

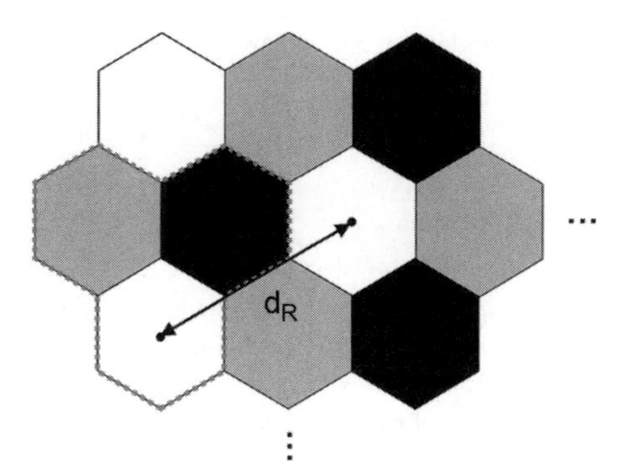

Figure 9.2 Example of spectral reuse assuming a hexagonal tiling of a transmission area. The dashed line indicates a cell cluster with three hexagonal cells (that is, the total spectral resource is distributed into three parts, and thus the reuse factor is 1/3), and d_R indicates the reuse distance

9.1.1 *Spectral Reuse*

Orthogonal MA methods such as TDMA and FDMA are by definition designed so that there is no interference between users operating over different time/frequency intervals. In practice, the delay and Doppler spreads inherent in wireless transmission environments and mobile reception may cause loss of orthogonality and subsequent interference, which is comparable to MAI in non-orthogonal CDMA systems, but not controlled, or directly related to the number of users sharing the resources.

A more relevant interference effect in orthogonal MA systems is that between users sharing the same time/frequency resources. Cellular networks are a practical approach to reducing such interference. The spatial area of a cellular network is partitioned into cell clusters (see Figure 9.2), each containing N cells with one base station each. In an FDMA network, for each cluster of cells, the total available frequency bandwidth is partitioned into N groups of frequency slots, which are allocated to the cells within each cluster. Each cell can accommodate K_U users, where K_U depends on the partitioning of the total spectral resources, that is, N, and on the utilization of frequency bands within the cells (for example, full duplex/half duplex communication). Each cell cluster can thus serve a maximum of $K_U N$ users. In the total FDMA network area spectral reuse is achieved by organizing the cell clusters so that there is maximal attenuation between signals originating from cells using the same frequency slots. Typically this means designing the network for maximal spatial separation between such cells. *Frequency reuse factor* refers to the fraction of the total spectral resources allocated to a single cell (and thus the fraction of the total network area occupied by cells with a given frequency allocation). For simple FDMA networks with omnidirectional BS transmitters the frequency reuse factor is simply *1/N* as shown in Figure 9.2.

For purposes of system analysis, cellular network topologies are often approximated as a two-dimensional tiling of identically shaped cells, for example hexagonal cells to approximate omnidirectional propagation, or diamond-shaped cells to approximate dense urban street grids. A crucial design issue for cellular FDMA networks is then how many cells must be included into each cell cluster. Having more cells in a cluster increases the separation between cells of the same type, but decreases the efficiency of overall spectrum usage within the network. *Reuse distance*, denoted in Figure 9.2 by d_R, can be defined as the distance between the centers (or base-stations) of the two cells utilizing the same frequency resources. Increasing the reuse distance decreases the amount of inter-cell interference experienced by the users, but it also decreases the efficiency of spectrum utilization within the network. The reuse distance should thus be minimized according to the attenuation needed to reach a given average signal-to-interference ratio within the network. This attenuation in turn is also dependent on inter-cell propagation conditions, which are, in modelling cellular networks, typically approximated by an environment-specific path loss exponent.

9.1.2 OFDMA

Orthogonal frequency division multiple access is a MA scheme that utilizes OFDM (see Chapter 2) as the transmission method. OFDMA is a special case of FDMA, which retains the benefits of OFDM versus single-carrier digital transmission (for example, efficient handling of multipath propagation effects), and improves the overall spectral efficiency and flexibility of resource allocation compared to traditional FDMA. This flexibility can be realized not only by assigning unequal numbers of subcarriers to users, but also through user-specific subcarrier distribution, power allocation and bit loading. Furthermore, this flexibility offers a possibility for taking advantage of *multiuser diversity*, where the variation of the frequency-selective fading experienced by the users is taken into account in the subcarrier allocation. Finally, as OFDM signals are by definition arranged into symbols of finite duration, OFDMA allows users to flexibly share both frequency carriers and time slots – in this sense it can be considered a hybrid FDMA/TDMA scheme. In the following we primarily focus on OFDMA resource allocation in the frequency domain.

Assume that in a given time slot, a frequency resource corresponding to K_{SC} OFDM subcarriers is available for downlink data transmission (that is, transmission overhead is excluded), and that this total set of carriers is equally and disjointly allocated to a cell cluster with N cells. If K_U users are then to be provided an equal number of subcarriers within a given cell during the considered time slot, a relevant question is how to allocate the approximately $K_{SC}/(NK_U)$ subcarriers to each of the users. Since the transmission channels between the base station and each of the users differ, generally a good solution in frequency-dependent fading channels would be to allocate the spectral resources so that each user is provided data over the subcarriers corresponding to the best channel conditions, thus taking advantage of the multiuser diversity mentioned above. This requires knowledge of the transmission channels between the base station and the receivers, or *channel state information (CSI)*. Even if CSI is available, optimizing the subcarrier allocation over all possible combinations may not be feasible in real systems. A practical simplification is to optimize the allocation for *chunks* of subcarriers, that is, instead of individual subcarriers, blocks consisting of several consecutive subcarriers are assigned to the users. Scheduling the subcarriers of a single user arbitrarily over the available frequency spectrum is referred to as *distributed* subcarrier allocation, while in *localized* allocation the subcarriers of a single user are adjacent. Figure 9.3 illustrates the two approaches. It should be noted that even if CSI is not available at the transmitter, allocating carriers or chunks of carriers to users according to pseudorandom permutations, that is, using *interleaving*, averages the aggregate channel quality for the users.

Considering downlink transmission for a single user with fixed subcarrier allocation, the total data throughput is generally dependent on the allocation of transmission power to the subcarriers, and the bit loading, or assignment of complex modulation symbols to the subcarriers. The optimal solution for power allocation over varying channel conditions with CSI available is the well-known *water filling* algorithm [69], which has also been generalized as *multi-user water filling* algorithms (see for example [71–72]), which simultaneously address the issue of subcarrier allocation. *Bit loading* algorithms (for example [73]) combine optimization of the power allocation with finding the optimal number of bits (that is, the signal constellation size) to be allocated to each subcarrier when adaptive modulation is used. Efficiency of spectrum usage can be further increased by utilizing *intracell frequency reuse*. This can be implemented for example with *embedded* or *hierarchical* modulation schemes, where high-rate and low-rate signals are superimposed on a single subcarrier, and assigned to users with good and bad channel states, respectively. Figure 9.4 shows an example of embedded modulation using a 64-QAM signal constellation for a given subcarrier. A detailed study of embedded modulation and frequency reuse in OFDMA systems is given in [74].

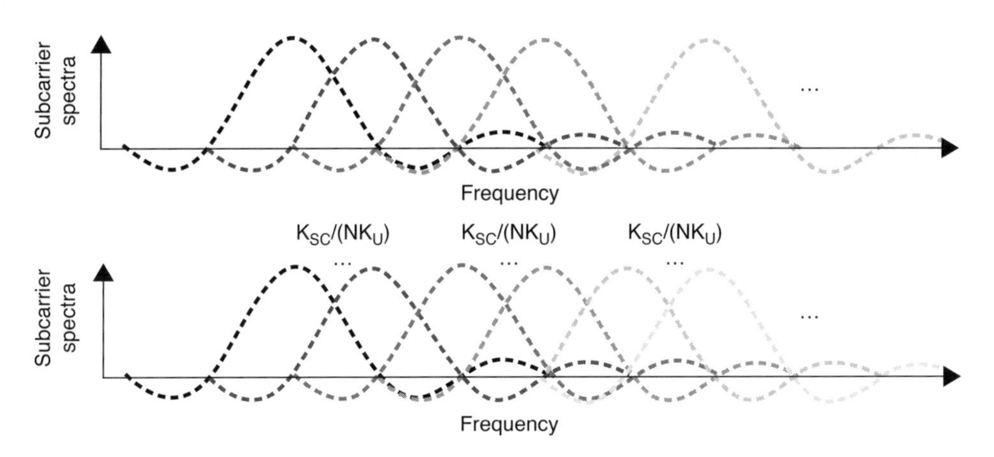

Figure 9.3 Conceptual examples of distributed (upper) and localized (lower) frequency allocation in one OFDMA cell. The black, gray and light gray lines depict theoretical subcarrier spectra for three users. The approximate number of carriers available to each user is indicated in the lower figure

One example of the additional system design flexibility provided by OFDMA compared to earlier FDMA systems is the *fractional frequency reuse* scheme defined in the mobile WiMAX standard [75]. With this reuse scheme users near the base station of each cell are allowed to use the entire available frequency spectrum, as they are less prone to suffer from intercell interference. Users near the cell borders are allocated non-overlapping sets of subcarriers within the cell cluster. This principle is illustrated in Figure 9.5, where frequency reuse for three neighbouring cells is shown.

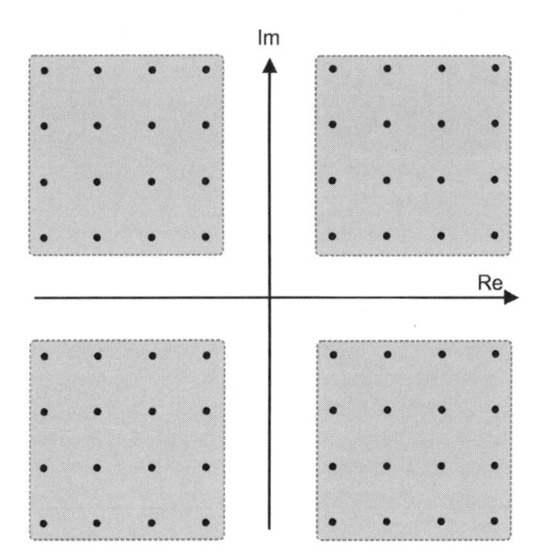

Figure 9.4 Example of embedded modulation. A low resolution signal corresponding to the selection of one of the four groups of modulation symbols (marked with gray squares) is robust against noise and can be transmitted to a user with a poor channel quality. A higher-rate signal is mapped to the modulation symbols within the selected group (marked with dots), which can be transmitted to a user with good channel quality

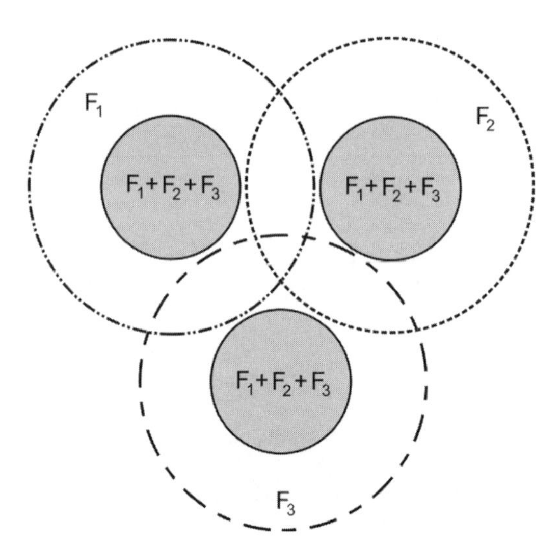

Figure 9.5 Example of fractional frequency reuse. Users on the edges of three cells are assigned non-overlapping portions of the available frequency resource (denoted by F_1, F_2, F_3), while users near the base stations of the cells are assigned the entire available spectrum

9.1.3 SC-FDMA

As described in Chapter 2, one of the disadvantages of OFDM transmission is the large peak-to-average power ratio (PAPR). To reduce signal distortion, the transmitter amplifier should operate in the linear region, which in the presence of a high PAPR results in low transmitter power efficiency. This is especially challenging in the uplink of OFDMA systems with possible mobile users, as power consumption is typically a critical issue for portable terminals. *Single-carrier frequency division multiple access (SC-FDMA)* offers one solution to this problem. SC-FDMA can be considered as a modified OFDMA transmission scheme, where the data (complex modulation symbols) allocated to the subcarriers of a user are precoded using an additional discrete Fourier transform operation as illustrated in Figure 9.6.

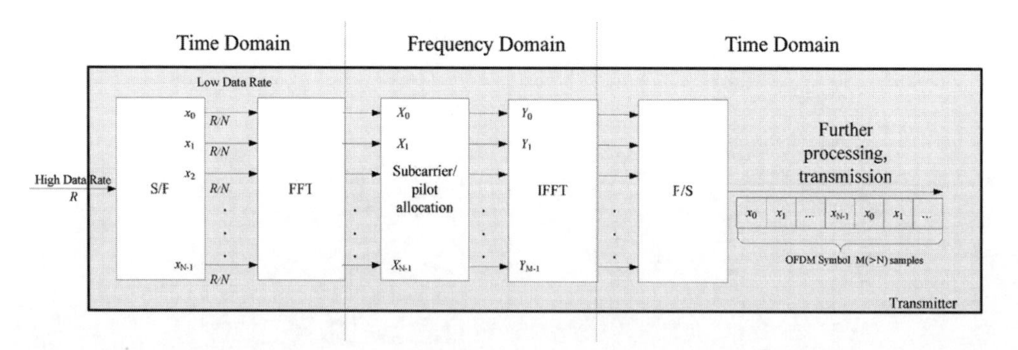

Figure 9.6 SC-FDMA transmitter operations. The input is N modulation symbols. The frequency domain signal with M subcarriers contains the discrete Fourier transform of the input symbols, zero-valued carriers, and pilot symbols. Assuming a distributed subcarrier allocation (referred to as Interleaved SC-FDMA, or IFDMA) results in the shown time-domain signal

As a result, with a distributed subcarrier allocation where the input data is distributed evenly over the available bandwidth, the time domain samples of the transmitted signals are copies of the original complex modulation symbols to be transmitted. This considerably reduces the PAPR. A tutorial presentation on SC-FDMA is given in [76].

9.1.4 WDMA

In optical communications it is practical to describe carriers with their wavelength rather than with their frequency as in radio communications. Wavelength (λ) and frequency (f) have a simple relation as $\lambda = c/f$, where c is the speed of light. As wavelength and frequency are inversely proportional, and since radio and light are both forms of electromagnetic radiation, the two terms can be considered as equivalent. Further details on optical multiplexing can be found, for example in [77].

9.2 Time Division Multiple Access

In *time division multiple access (TDMA)* systems the transmission period is divided into time slots, each user being allowed to transmit or receive only within the slots assigned to him (it is assumed that slots are cyclically repeated in time, that is, if the transmission period is divided into K slots; the slot next to the K-th user slot is again the first user slot). See Figure 9.7.

Data transmission for users of a TDMA system is not continuous but occurs in bursts. Due to burst transmission, synchronization overhead is required in TDMA systems. In addition, guard slots are necessary to separate users. Generally, the complexity of TDMA mobile systems is higher compared with that of the FDMA systems.

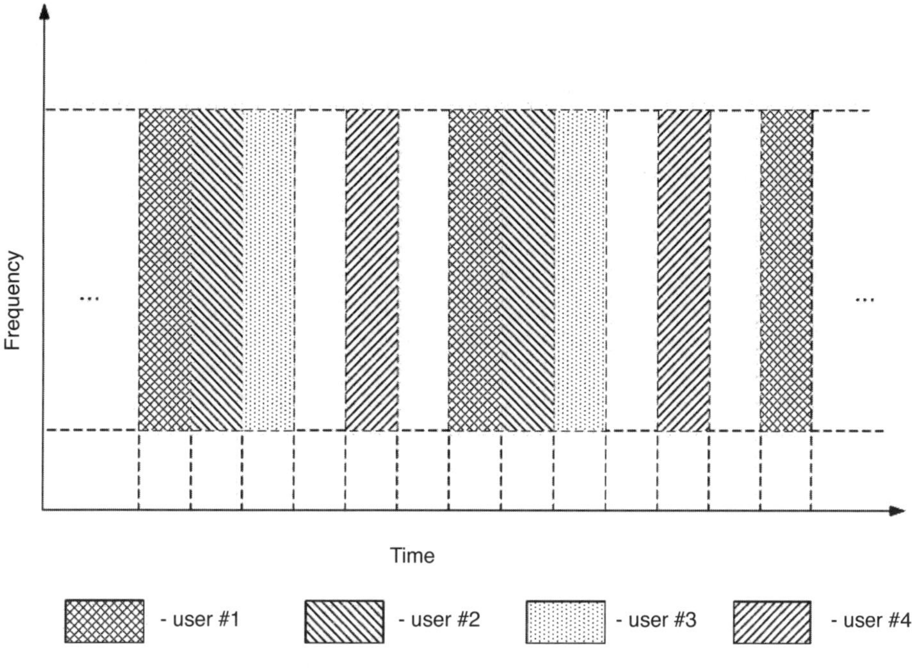

Figure 9.7 TDMA

The practically achievable capacity of pure TDMA is limited by peak power. Peak power, which is the maximal amount of power emitted by a transmitter, should not be dramatically higher than the average transmitted power, since large power variation causes signal distortion due to amplifier nonlinearity in the transmitter. Small duration of time slots entails big power emission, which is impractical in battery powered mobile devices. Also, in cellular environment the frequency re-use restricts the maximum emitted power. For example in GSM the number of time slots is eight. Thus, in order to function in real systems, TDMA must be combined with FDMA.

9.3 Code Division Multiple Access

Spread spectrum system is defined in many sources (for example, [60, 78]) as a system, which employs much wider bandwidth than required by the data rate to be transmitted. This definition, however, is ambiguous. For example, in the GSM cellular standard users reserve the bandwidth of approximately 200 kHz to transmit data at a rate between 9.6-13 kbps, but GSM does not involve any spread spectrum methods in the transmission of user data. Here, we use an older and more universal definition which can be found for example in [16]. There a spread spectrum signal is defined as a signal, whose time-frequency product that is the product of signal duration, T, and its required bandwidth, W, is much greater than one $WT \gg 1$. For plain signals that are not employing spread spectrum methods, the relation in question is $WT \approx 1$.

In principle, spread spectrum systems have numerous merits when compared to systems utilizing plain signals. These merits include good performance in a multipath environment, resistance to narrow-band interference, privacy, possibility for exact time and location measurement, and good electromagnetic compatibility. The most important merit regarding this chapter is the possibility of employing CDMA as a multiple access method. In addition, in a cellular environment CDMA makes universal frequency re-use possible, which increases the capacity and enables the use of soft handover to enhance the performance [60].

Here the two most fundamental code division techniques will be considered: *Direct-Sequence CDMA (DS-CDMA)* and Frequency Hopping CDMA (FH-CDMA). Hereinafter an overview of these two fundamental CDMA techniques will be given. In our treatment of CDMA methods we will for the most part follow monographs by Zigangirov [58] and Ipatov [16].

9.3.1 Direct-Sequence CDMA

Let us assume for simplicity that K active users that are synchronized in transmitter-side with each other are to transmit binary BPSK modulated sequences b_k of length L_m. Before transmitting a signal each user multiplies it by an individual code sequence of length N_c chips. Expression chip is used to distinguish code sequence symbols from information bearing bits. Further we shall assume that the code sequence is also BPSK modulated. Let us further assume that each signature is mapped into a sequence $s_k(t)$ of N_c rectangular pulses of duration T_c and each element of the sequence b_k (or equivalently a rectangular pulse of duration $T = N_c T_c$) is modulated by $s_k(t)$. This operation is shown in Figure 9.8.

Note that the bandwidth of both the signature signal and the resulting signal $W = \frac{1}{T_c}$is N times greater than that of the original signal $W_0 = \frac{1}{T} = \frac{1}{NT_c}$. In other words by multiplying the signal that is to be sent by a spreading sequence we are, so to say, spreading the signal spectrum over a larger bandwidth. Thus, the device implementing the operation is called *a spreader* and that is why DS-CDMA is referred to as spread spectrum communication technique.

The resulting (spread) signal is given by:

$$x_k(t) = \sqrt{P_k} \sum_{i=1}^{L_m} b_k(i) s_k(t - iT) \tag{9.1}$$

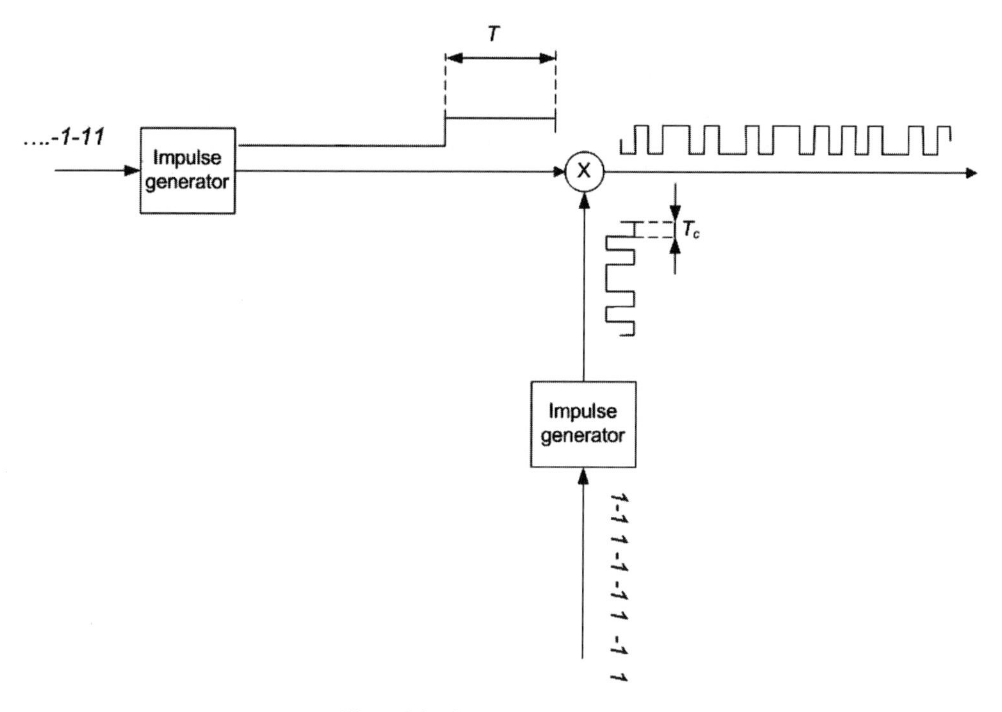

Figure 9.8 Spreading in DS CDMA

where $\sqrt{P_k}$ is energy per bit and the energy of spreading code $s_k(t)$ is normalized to unity. The transmitted signals from all active users are then modulated and sent via the channel in use. General transmitter structure is shown in Figure 9.9.

Let us consider coherent reception of the signal sent by the active user under consideration. Further on we shall assume that both the relative delay of the user under consideration and the phase of the signal from the user under consideration are known to the receiver. At the receiver end the resulting signal is demodulated. The received signal is given by:

$$r(t) = \sum_{k=1}^{K} \sqrt{P_k} \sum_{i=1}^{L_m} b_k(i) s_k(t - iT - \tau_k) + n(t) \tag{9.2}$$

where τ_k denotes random delay caused by impulse response of the channel $h_k(t)$. It is assumed for simplicity that the channel does not cause any other distortions to the signal. CDMA system enables utilization of multipath diversity with a so called RAKE received, which is presented in section 9.3.1.2.

The output of the demodulator of k-th user in i-th bit interval is given by:

$$\begin{aligned} b_k^*(i) &= \int_{iT}^{(i+1)T} r(t) s_k(t) dt \\ &= \sqrt{P_k} b_k(i) + \underbrace{\sum_{l=1, l \neq k}^{K-1} \sqrt{P_l} b_l(i) \int_{iT}^{(i+1)T} s_l(t + iT + \tau_l) s_k(t - iT) dt}_{\text{Multiple access interference}} + \underbrace{\int_{iT}^{(i+1)T} n(t) s_k(t) dt}_{\text{Noise component after receiver filtering}} \end{aligned} \tag{9.3}$$

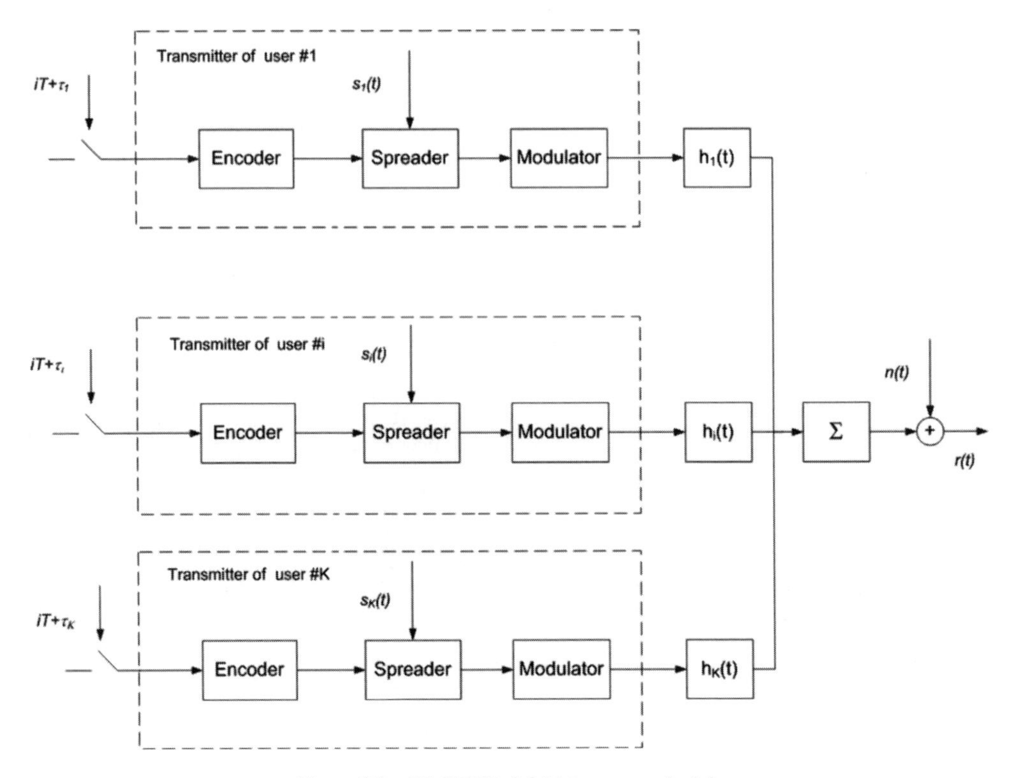

Figure 9.9 DS-CDMA: Multiple access principle

and the estimated bit value for antipodal BPSK is:

$$\hat{b}_k(i) = sign\left(b_k^*(i)\right) \tag{9.4}$$

Note that the value of the component that is introduced by other users' influence depends on the values of cross-correlation function of the spreading sequences in use. Thus spreading sequences are to have small cross-correlation values. Furthermore if multipath propagation is present, the value of the component that is introduced by the inter-symbol interference depends on the values of the spreading sequence's autocorrelation function side lobes. Moreover it should be noted that since one spreading sequence can be used by one user only the number of spreading sequences in the set in use denotes the number of active users that can simultaneously transmit in the system under consideration. Thus it is desirable to obtain sets of spreading sequences having lengths of great powers.

9.3.1.1 Spreading Code Design Aspects

Spreading code design is a problem of minimizing cross-correlation between sequences and minimizing side lobes of code sequence autocorrelation function. In addition, the cross-correlation and autocorrelation values should be known for the cases where perfect code sequences cannot be employed. Comprehensive description on code sequence is found in [85].

If relative delays, τ_l in (9.3), between different transmissions are known in the receiver, orthogonal code sequences can be used, which completely eliminate multiple access interference (MAI). These can be obtained for example with a *Walsh-Hadamard construction* [16]. In UMTS system orthogonal code sequences are called OVSF codes [86] as they have been designed to support various data rates. Unfortunately, total synchronism between active users is difficult to obtain in practice especially in the uplink. Also, the number of orthogonal code sequences is limited by $K = N_c = WT$. Increasing the number of code sequences beyond code sequence length N_c is called overloading or oversaturation. [16, 87, 88].

In asynchronous transmission mode and in the presence of multipath propagation, one of the most popular design criteria is the minmax criterion (see [16]) based on lower bounds for the so-called *total squared correlation (TSC)* of a set of sequences, which is given by:

$$c_{\max} = \max(c_c, c_a) \tag{9.5}$$

where c_c is the maximum value of crosscorrelations and c_a is the maximum value of autocorrelation sidelobe of a certain set of sequences respectively. Let us consider a set of sequences in vector format (that is, each chip is one element of a sequence vector) $\mathbf{S} = \{\mathbf{s}_1, \mathbf{s}_2, \ldots, \mathbf{s}_K\}$ of length N_c satisfying the condition $|\mathbf{s}_k| = 1$. Welch [53] derived the following lower bound for the TSC of such a set:

$$c_{\max}^2 \geq \frac{K-1}{KN_c} \quad \text{for} \quad K > 1 \tag{9.6}$$

(in fact the aforesaid bound is just a corollary of the theorem enunciated by Welch) for $K \gg 1$ this bound simplifies to:

$$c_{\max}^2 \geq \frac{1}{N_c} \tag{9.7}$$

Another bound has been proposed by Sidelnikov [46] for the sequences whose symbols were complex q-th roots of unity for some integer q. For $q = 2$ and $K \geq \frac{N_c}{2}$ this bound simplifies to:

$$c_{\max}^2 \geq \frac{2}{N_c} \tag{9.8}$$

Thus, according to the minmax criteria, TSC of the set of spreading sequence is to approach the Welch bound or at least Sidelnikov bound as N_c grows.

It should be noted that code sequences lying exactly on the Welch bound (Welch Bound Equality (WBE) sequences) have the remarkable property that while they minimize the MAI power, they also maximize the Shannon capacity [89]. However, if the number of active users changes in the system, the capacity property is lost and all code sequences should be re-allocated. Therefore WBE sequences have not been utilized in practical systems [90].

Gold sequences [13] and Kasami sequences [19] have been proposed for spreading in the uplink transmission in modern CDMA standards (for example, Wideband CDMA [54]). The set of Gold sequences of order n (their period is given by $N_c = 2^n - 1$) consists of $K = 2^n + 1$ sequences and the TSC of this set approaches Sidelnikov bound as n (and hence N_c) grows for all $n \neq 0 \bmod 2$. The small set of Kasami sequences order n (their period is given by $N_c = 2^n - 1$ and such sequences exist for all $n = 0 \bmod 2$) consists of $K = 2^{n/2}$ sequences and the TSC of this set approaches Welch bound as N_c grows (note, however, that WCDMA uses the so-called large Kasami set [58] that includes both the Gold sequences and the small set of Kasami sequences as subsets and consists of $K = 2^{(3n)/2}$sequences (if $n = 0 \bmod 4$) and of $K = 2^{(3n)/2} + 2^{n/2}$ sequences (if $n = 2 \bmod 4$).) However, there are other sets of sequences. For instance, Kamaletdinov [17] proposed to complement the small Kasami set with Bent-functions, which results in almost twofold increase of the set cardinality $K = 2^{(n/2)+1}$ and does not affect the correlation properties (note, however, that this set exists for $n = 0 \bmod 4$only). A number of other sets of sequences with good correlation properties has been constructed (see ([18], [25], [26]) including non binary ones ([2], [5], [22], [24], [47]).

9.3.1.2 CDMA Receiver Issues

The simplest form of the CDMA detector is the so-called conventional or single-user detector, which simply calculates the sign of the decision statistic given by (9.4).

This detector works in the same manner as in the case of a single user transmission. Note that the complexity of the single-user receiver is very low. However, it has certain drawbacks. A single-user detector performs well if the power of noise (that is, background noise, other users' interference and inter symbol interference) is much smaller when compared to that of the signal. It has been stated above that other users' interference impact can be reduced by an appropriate spreading sequence choice. However, whatever spreading sequences are used, as the number of active users grows, the impact of the other users' interference increases, which results in substantial degradation of the performance of the single-user detector.

Let us now assume that the power of the signals from some users at the receiver side is sufficiently greater than the power of the signals from other users (for example, because the former are much closer to the receiver than the latter). In that case the interference caused by the signals with greater power can completely swamp the signals with smaller power (the so-called *near-far effect*). The near-far effect can be overcome by the power control introduction. However if the system load is great the single-user detector performance will be poor due to other users' interference impact.

In order to detect signals sent from the active users even in the case of asynchronous transmission (that is, if other users' interference is unavoidable due to spreading sequences' non-orthogonality) a detector is needed that uses all the information obtained from all the matched filters of all the users, that is, a *Multi-user detector (MUD)*. The optimal maximum likelihood multi-user detector (ML MUD) can be derived by rewriting equations corresponding to the outputs of the channel matched filters in matrix notation:

$$\bar{b}^* = R\bar{b} + \bar{n} \tag{9.9}$$

where

$$R = \begin{bmatrix} \rho_{11}(0) & \rho_{12}(0) & \cdots & \rho_{1K}(-2M) \\ \rho_{21}(0) & \rho_{22}(0) & & \rho_{2K}(-2M) \\ \vdots & & & \\ \rho_{11}(1) & \rho_{12}(1) & \cdots & \rho_{1K}(-2M+1) \\ \vdots & & & \vdots \\ \rho_{K1}(2M) & \rho_{K2}(2M) & \cdots & \rho_{KK}(0) \end{bmatrix} \tag{9.10}$$

is the cross-correlation matrix. ML detection is a problem of choosing the transmitted sequence $\bar{y}$ that maximizes the conditional probability to obtain $\bar{b}^*$ at the output of the receiver (the transmitted sequence is chosen from a set of all possible binary sequences of length K, all sequences are considered to be equiprobable):

$$\hat{b}^k_{OMD} = \arg \max_{\bar{y} \in \{-1;1\}} \left\{ p\left(\bar{b}^* \middle| \bar{y}\right) \right\} \tag{9.11}$$

Since all components $\bar{n}$ are normally distributed, (9.11) yields:

$$\hat{b}^k_{OMD} = \arg \min_{\bar{y} \in \{-1;1\}} \left\{ \left(-\bar{y}^T \bar{b}^* + \frac{1}{2} \bar{y}^T R \bar{y} \right) \right\} \tag{9.12}$$

Unfortunately, the complexity of this detector grows exponentially with the number of active users. To reduce the complexity of detecting a number of sub-optimal MUD detectors has been proposed [51].

Suboptimal multiuser receivers aim to cause minimal decrease in the performance of the receiver while providing simple implementation. Suboptimal receivers can be divided into linear [79, 80] and non-linear [81–84] categories, where the latter is also referred to as decision-driven multiuser receiver.

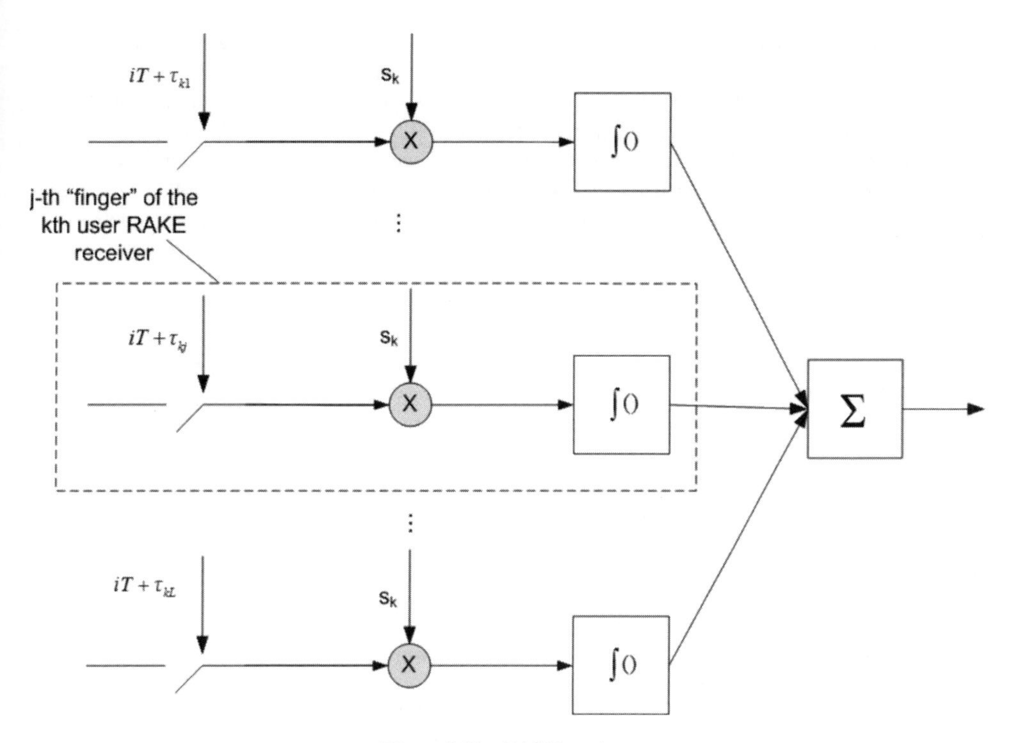

Figure 9.10 RAKE receiver

Developments in the field of multiuser detection have been studied profoundly in [51], which also contains an extensive reference list on the subject.

Let us now make some additional notes on DS-CDMA in multipath channels. The output of the multipath wireless channel is given by the sum of multiple shifted and attenuated replicas of the input signal. Thus, some replicas may add destructively, which results in fading phenomenon. The RAKE receiver decreases the fading impact by receiving replicas of the signal propagating via different paths separately and combining them again in order to maximize the received energy emerging due to multipath propagation. Thus, the RAKE receiver exploits multipath diversity. Let us assume that the transmitted signals' energies and the channel attenuation factors and delays of all paths for all the active users are perfectly known to the receiver, that is, ideal channel and time offset estimation are maintained in the system under consideration (note that although the transmitters of different users can be unsynchronized, the transmitter and the receiver corresponding to each user should be synchronized.) Then the signal can be received by the RAKE receiver (initially proposed in [42].) The RAKE receiver is shown in Figure 9.10.

It is to be noted that in the general model that has been considered above it is assumed that the number of paths via which a signal from a certain user propagates, is different for each user. Moreover it is assumed that the number of paths is known to the receiver and the number of "fingers" of the RAKE receiver is equal to the number of paths. In actual life, however, the number of paths, via which the signal propagates, is unknown to the receiver and furthermore in most real-life applications the actual number of paths can be great, thus making the complexity of the receiver unacceptably high. That is why in real-life systems all the RAKE receivers have the same number of "fingers" (as a rule, RAKE receivers with moderate number of "fingers" are used).

We have considered coherent reception of BPSK modulated signals since it is convenient for explaining DS-CDMA basics. However, in DS-CDMA noncoherent reception (for example, in the downlink if reduced first-order Reed-Muller codes are used) and other modulation techniques (QPSK, DPSK) can be applied (for a comprehensive survey on this matter see [58]).

We have also assumed that the transmitter and the receiver of the user under consideration are synchronized. Moreover, it has been assumed that the receiver has complete information on the received signal (that is, delays, attenuation factors and phase shifts of all the paths; signals energies, initial phases and relative delays of all the signals). In real-life system synchronization and parameters estimation is a very cumbersome task. However, the solution of these problems (primarily initial synchronization) is an indispensable condition for using a DS-CDMA technique in real-life applications.

Potentially, DS-CDMA systems provide a larger *radio channel capacity* (that is, the number of active users that can simultaneously transmit in the system under consideration) than fixed access systems. Moreover, DS-CDMA systems are more adaptable. As has been stated above, if a fixed access system is implemented in order to maximize channel use efficiency even a small change in the system load results in reassignment of sub-bands or slots (or both if the system in question uses mixed strategy). In DS-CDMA, however, a small change in system load results in only a slight change of performance of the active users remaining in the system. However, even if the system load leaps (which can hardly happen in real life applications) the DS-CDMA system is not to be reconfigured. In fixed access methods the base station is to fully control the transmission of all the active users, whereas in many random access systems (for example, slotted ALOHA, see section 9.5) it is to provide feedback broadcasting the results of the reception. In DS-CDMA, however, neither full control nor reception confirmation is needed. Moreover, in fixed access systems the only way to change information transmission rate not requiring resource reallocation is to change the parameters of the outer code in use. In DS- CDMA in addition to different outer codes active users can use spreading sequences with different spreading factors or one user can use several different spreading sequences simultaneously (however, the latter approach is better suited for downlink transmission since in this case all the threads are synchronized).

9.3.2 Frequency-Hopping CDMA

As has been pointed out above, up-to-date real life systems are to be protected against unauthorized users' activity. *Frequency-Hopping code division multiple access (FH-CDMA)* is the fundamental MA concept for solving the task. In conventional frequency hopping the entire available frequency band is divided into Q sub-bands or sub-channels. Following the terminology used in [58] we shall further on refer to the set of all sub-bands available to the user as a *hopset*. Each user's transmitter chooses one of the Q sub-bands (using a specialized sub-band numbers generator) and transmits a signal via the chosen sub-band using a conventional modulation technique (for the most part FSK is used). In the multiple access theory it is common to refer to each change of the sub-band in use as a *hop*. The process of switching between the sub-bands (which can be also interpreted in terms of assigning sub-bands to the users) is called *frequency hopping*. Note that the sequence of sub-bands numbers chosen by the aforesaid generator of the user under consideration can be interpreted as a code sequence.

Let us denote the time duration between transmission of two consecutive symbols as T .The time duration between hops is called *hop duration* or *hopping period* and is denoted by T_c. Frequency hopping can be classified as slow or fast. *Slow frequency hopping* occurs if one or more q-ary symbols are transmitted in the interval between frequency hops. Thus, slow frequency hopping implies that the symbol rate $1/T$ exceeds the hopping rate $1/T_c$. *Fast frequency hopping* occurs if there is more than one frequency hop during one symbol transmission time.

Frequency hopping (assigning sub-bands to the users) can be either coordinated or uncoordinated. Coordinated hopping implies that the code sequence information about other users must be used for the choice of the code. Thus, this approach is suited only for the downlink transmission. The uncoordinated frequency hopping has a number of advantages (it is well protected from unauthorized users' activity, and does not require sophisticated protocols, which are inevitable in a system with coordinated subcarrier assignment) and is well suited for uplink transmission. We shall consider uplink transmission, and, hence, uncoordinated subcarrier assignment strategies. In turn, the uncoordinated subcarrier assignment strategies can be divided into two groups. Methods of the first group exploit code sequences specially designed for multiple access systems and satisfying certain requirements. However, designing a set of sequences well suited for the uplink transmission in real life applications is a cumbersome task. The second group includes methods where subcarrier numbers are assigned to users pseudorandomly. This method is well suited for the uplink transmission.

Let us now consider the reception strategy. It is assumed that the receiver sub-band numbers generator is to be synchronized with that of the transmitter. The latter means that with the reception of a symbol the receiver sub-band numbers generator generates (that is, chooses from the set of numbers of all the sub-bands available to the system subscribers) the number of the very sub-band that has been used by the sender. The reception procedure boils down to two successive procedures: dehopping (that is, extracting the narrowband signal from the subcarrier used by the corresponding user to transmit the symbol of interest) and demodulating.

Since the time of introduction FH CDMA has been considered to be a fundamental technique for designing multiple access systems invulnerable to unauthorized users' activity. As has been mentioned above frequency hopping can be either coordinated or uncoordinated and the uncoordinated frequency hopping can use either special code sequences or pseudorandom code sequences. However, whatever code sequences are used it is assumed the sequences in question are unknown to other users (authorized or unauthorized, that is, nobody but the transmitter and the receiver) and they cannot obtain this information. Thus, frequency hopping is well protected from unauthorized users' activity. However, recently an MA technique that is even better protected in this respect has been proposed. This technique will be considered in section 9.4.3.

9.4 Advanced MA Methods

By combining the techniques considered above a whole range of conjoined MA methods can be obtained. Next some proposed candidates for future MA solutions are presented briefly.

9.4.1 Multicarrier CDMA

DS-CDMA is not just a multiple access technique but also a concept. It is well combined with multicarrier transmission resulting in a whole complex of MA techniques. In MC-CDMA [6], [11], [57] a data stream from each active user is divided into N_p low rate parallel data streams each being spread by its own chip of the N_p-chip spreading sequence and sent via its own subcarrier. Thus, MC-CDMA is a combination of multicarrier transmission (for example, by means of OFDMA) and spreading in the frequency domain. However, this method is better suited for downlink transmission. In DS-MC-CDMA (T) [9] the sequence to be transmitted is first spread in the time domain and then transmitted via multiple subcarriers. Thus, it can be thought of as classical DS-CDMA combined with multicarrier transmission. This technique can be applied for uplink transmission. DS-MC-CDMA (TF) [56] combines spreading in the time domain and spreading in the frequency domain. The approach has certain advantages over both MC-CDMA and DS-CDMA (for example, the complexity of the receiver is lower.) In Multi Tone CDMA (MT-CDMA) [50] the initial data stream is converted into a multiplicity of parallel streams each of the latter being then spread in the time domain so that the spectrum of each subcarrier prior to spreading can satisfy the orthogonality condition with the minimum frequency separation. Therefore the resulting spectrum of each subcarrier no longer satisfies the orthogonality condition. Thus, the MT-CDMA suffers from intersubcarrier overlapping. However, MT-CDMA can use longer codes (the length of the code depends on the number of available subcarriers), which results in radio channel capacity increase (compared to conventional DS-CDMA) and enables better self-interference and MAI combating. For a more detailed overview of multicarrier techniques the reader is encouraged to refer to [14].

9.4.2 Random OFDMA

Random OFDMA has been proposed in [34] and considered in more detail in [37]. Several modifications were considered in [27–36]. In the basic random OFDMA system model, the channel in use is split into N subchannels. Each user is assigned a set of n subchannels randomly chosen from the abovementioned set of Nsubchannels. If several users overlap in certain subchannels these subchannels are considered to be corrupt.

Note that within the scope of the scheme under consideration it is assumed that the set of subchannels that are assigned to a certain user is not to change, that is, this set serves as a certain user's personal identifier. In modern applications, $N = 2^m$ with $m$varying from 12 to 14. Thus though in real-life systems only approximately $\tilde{N} = [0, 8 \cdot N]$ subchannels can be used, the number of available subchannels amounts to several thousands and even for moderate values of n the number of possible sets is much greater than the world population. Therefore it is assumed that the probability of a large number of subchannels in two users' sets overlapping is small. Moreover, it is assumed that the base station knows which users transmit inside its cell (or can obtain this information by analyzing the channel output). Thus, the base station knows all the corrupt subchannels. This information can be used for example, for joint decoding (see [32]). Note, that within the scope of this scheme high order modulation alphabet can be used, thus, enabling high data rates. Besides it is worth mentioning that the maximum number of users in this model is "soft-limited", as in CDMA.

On the basis of simulation results (obtained for the wireless channel model derived in [10] the bandwidth 10 MHz and velocity of mobile users v=50 km/h) it is stated in [37] that the random OFDMA system considered above can perform at least as well as GSM and USDC (IS-45).

A block scheme for the selection of subchannels and also the division into subgroups of subchannels was considered in [34]. Within the scope of the block selection scheme special sequences $\bar{a} = (a_1, a_2, \ldots, a_k)$ have been searched with the property of any value $b_{ij} = \left(a_i - a_j\right) \bmod K$(where $K$is the number of active users) appearing only once. Within the scope of the scheme in question each active user is assigned a set of subchannels $f_i = (s + a_i)$, where s is the number of the respective active user. In this case only three outcomes are possible: two users may either not overlap, overlap in only one subchannel or they may overlap in all the subchannels.

If in the block selection scheme the number n of subchannels per user is greater than k, the total number $N$of subchannels can be divided into $N_{sg} = \frac{n}{k}$ subgroups. Each subgroup must have $N_s > K$subchannels, so that the block selection scheme can still be applied in each subgroup.

In [34] it was shown that the block selection schemes are better for users having less corrupted subchannels (*best users*), but these schemes have a larger number of corrupted subchannels for other users than the completely random scheme for all N subchannels. The results for the random selection of n subchannels from all the Nsubchannels N are very similar to the results with division into n subgroups of $N_s = \frac{n}{k}$ subchannels and random selection of only one subchannel in each subgroup.

A search of better selection schemes based on t-designs [15] and constant weight codes [3] has been performed. However, as is stated in [37], "a t-design with given parameters interesting for the considered system (large N, small n) does not exist." Besides, the construction of very low rate constant weight codes for the given parameters (great codeword length and very small weight) seems impossible.

However, if it is assumed that the central organization has no a priori knowledge about which users will be active in each cell of the system, the best selection strategy is to select randomly n subchannels with equal probability of selecting a given subchannel ($p = \frac{1}{N}$), according to the principle of maximum entropy in statistics, as shown in [37].

9.4.3 DHA-FH-CDMA

The Dynamic Hopset Allocation FH-OFDMA (DHA-FH-OFDMA) has been introduced in [61] and considered in more detail in [41]. Several modifications were considered in [38–40], [62]. Hereinafter we shall consider one of the most promising versions of this concept that has been proposed: DHA-FH-OFDMA with non-coherent threshold detection.

Let us now consider the basics of the scheme under consideration in more detail. Consider a multiple access system where K active users transmit data to the base station through a channel split into Q frequency subchannels; the transmission is asynchronous and uncoordinated (that is, neither of the users has information about the others). It is assumed that all the users transmit binary q-tuples. In the course of the transmission of each consecutive tuple the subchannel number generator assigned to the user under consideration chooses (in a random manner) 2^q subchannels out of Q subchannels. Each tuple (or a part of the tuple) to be transmitted by the aforesaid user within the frame is mapped into the number of the subchannel, via which the signal is to be transmitted.

In what follows we shall assume that in the system under consideration optimal power control is used. The latter means that the powers of all the signals from distinct users are equal at the receiver side. It is assumed that the base station is equipped with the subchannel numbers generator synchronized with that of the active user. The latter means that within the scope of the reception of the respective tuple, the subchannel numbers generator of the base station produces the very same subchannel numbers vector that has been generated by the subchannel numbers generator of the user under consideration . Note, that this assumption is not restrictive since synchronized generators are an essential part of any conventional FH-CDMA system. Thus, we simply replace a generator producing random numbers with a generator producing random vectors.

Within the scope of the reception of a certain tuple sent by the user under consideration the receiver measures the values of power for all the signals received through the subchannels chosen by the subchannel numbers generator of the user under consideration obtaining q statistics in that manner. Thus, the receiver is to decide which subchannel has actually been used by the active user under consideration. To do so the receiver compares each statistic value with a certain threshold (here we assume that the value of the threshold has been chosen in advance and is not to be changed.) If threshold crossing is detected in only one subchannel, the tuple corresponding to the subchannel where the threshold crossing was registered is accepted. Otherwise, an erasure decision is made. If a tuple other than the transmitted one is accepted, we say that an error has occurred. A block diagram of the transmission in a DHA FH OFDMA system is shown in Figure 9.11.

In many respects DHA-FH-OFDMA is similar to FH-CDMA using FSK (let us from now on call it FSK-FH-CDMA). The main difference between them is that no matter how the subchannels are assigned to a certain active user in FSK-FH-CDMA distributed over the entire set of subchannels, the distribution in question is known to the jammer or at least can be detected by him (we assume that all system parameters, for example, total number of subchannels, modulation type, parameters of the outer code and so on, are known to the jammer) and thus the jammer can determine which sub-band has been used for transmission and send a signal (probably having great power) via another subchannel from the same sub-band, causing in that way error or erasure (depending on the receiver strategy in use.) In contrast, in DHA-FH-OFDMA the vector of subchannels that can be used by a certain user is known to the pair "active user – base station" only. Thus, in the system under consideration it is almost impossible for unauthorized users to block the transmission by intentionally sending signals via any of the subchannels chosen by the subchannels number generator of the user under consideration but for the one that has been used by the aforesaid user himself.

Let us now consider the process of the threshold reception procedure that has been analyzed above in terms of two probabilities: probability of a false alarm (that is, probability to detect a signal in the subchannel via which no signal has been transmitted) and probability of a miss (that is, probability not to detect a signal that has been transmitted by the user under consideration via the subchannel in

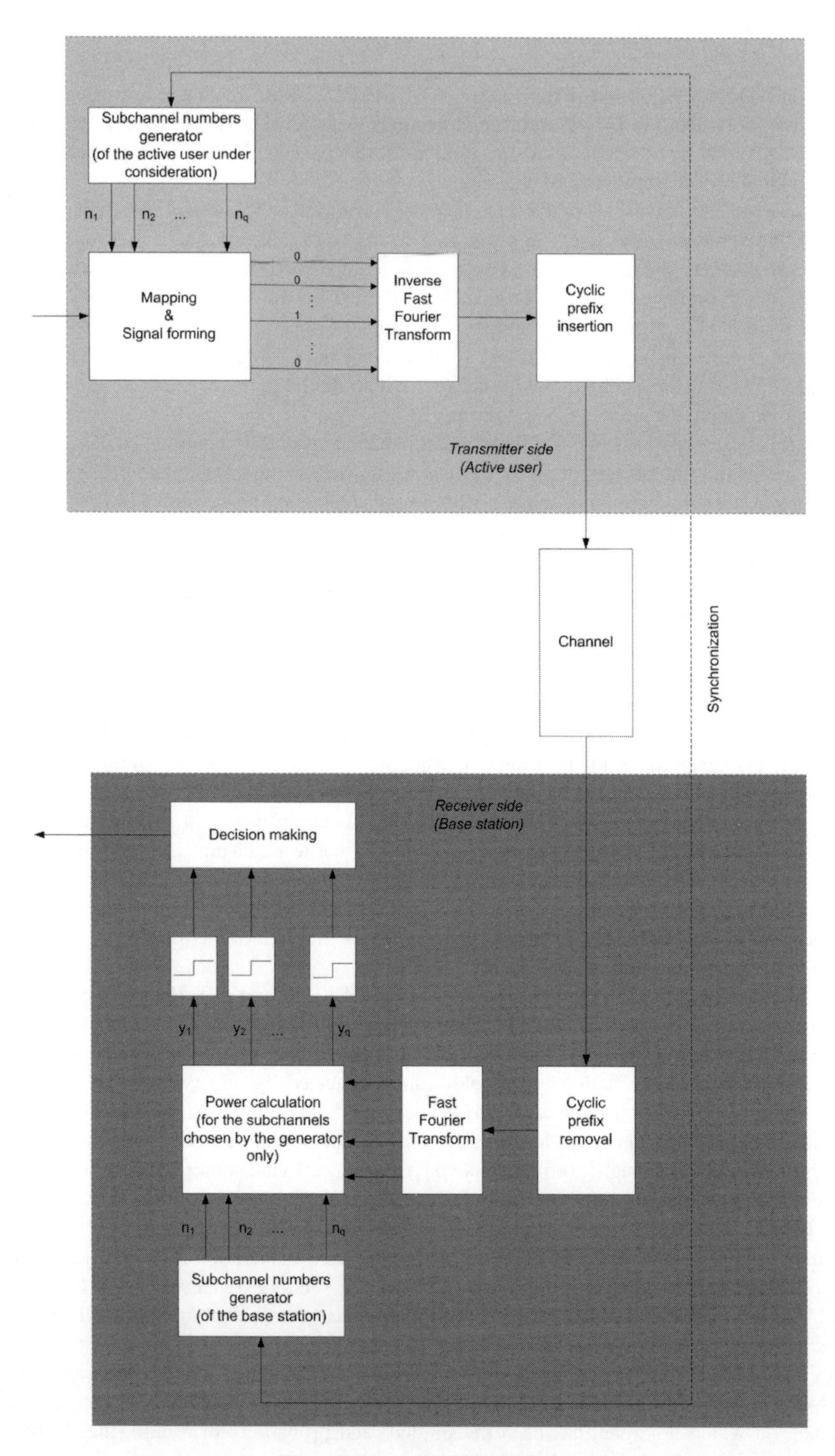

Figure 9.11 DHA-FH-OFDMA

question). Simulations show that for any threshold value that ensures a sufficiently low probability of false alarm (say $p_F < 0.1$) the probability of a miss decreases as the power of the signal transmitted by an interfering user (either authorized (that is, any active user but for the one under consideration) or unauthorized (that is, jammer)) increases (as compared to that of the signal transmitted by the user under consideration). Thus, if the number of active users (either authorized or unauthorized) grows it is the probability of erasure not the probability of error that increases (since error is possible if and only if the signal transmitted via the subchannel used by the user under consideration has not been detected.) Moreover, simulation results show that radio channel capacity of the DHA-FH-OFDMA system amounts to several hundreds of users (if optimal power control is maintained), which is sufficiently greater than that of the conventional FH systems.

9.5 Random Access Multiple Access Methods

As has been mentioned above, in the system where each active user is allocated a disjoint portion of the time-frequency resource each user intending to enter the MA system or to exit it is to state the intention in advance. If the user in question is an active user (that is, has been transmitting data in the system under consideration) the respective uplink and downlink channels can be used (however, in this case a certain part of a frame is to be assigned to control information, which results in rate decrease both in uplink and downlink channels). Assume, however, that the user in question has not yet transmitted data in the system under consideration and wants to enter the system. To declare thereof a special MA method is needed, that is, a method enabling any user to use a certain part of a certain time-frequency resource (or at least try to do so) as soon as he needs to do so, that is, as soon as he has the data to transmit. Such a method can also be made use of in ad hoc networks since in ad hoc networks there is no central node that can assign sub-channels and the number of users (and, respectively, the number of point-to-point sub-channels that can potentially be requested) is unbounded.

This concept is called random access or free-to-all access since all the users try to communicate with the base station (or in the case of an ad hoc network with each other) acting pursuant to a certain statistical strategy. Thus messages from different users may occasionally collide, which results in information losses. In random access systems it is customary to use retransmission to solve this problem. Thus some sort of a feedback is needed (to inform the user, who has transmitted a message, that the transmission has failed) as well as a special strategy enabling the user in question to decide at what instant he is to make another attempt. Hereinafter a brief overview of random access methods will be given. In our treatment of random access methods we will for the most part follow the papers by Gallager [12] and Kleinrock [20].

The simplest random access technique is the so called *slotted ALOHA* ([43]). Let us consider uplink transmission, that is, a set of users each trying to transmit a message (hereinafter following the term used in [12] we shall refer to it as packet). In classic slotted ALOHA the entire time interval, within which the channel is available for uplink transmission (theoretically infinite), is divided into slots, packet transmission being allowed only within a certain slot. Thus, if a packet is generated in the middle of a slot it is to wait until the next slot boundary before it is transmitted. Moreover, it is assumed that:

(a) all the users' transmitters are synchronized;
(b) messages are of fixed length and they fit exactly one time slots, that is, a message is to be neither longer nor shorter (note that this assumption precludes both the possibility of sending short packets to make reservations for long packets and the possibility of carrier sensing, which is to be discussed later);
(c) if more than one transmitter sends a packet in a slot (this situation is referred to as collision) the receiver gets no information about the contents or origins of the transmitted packets;

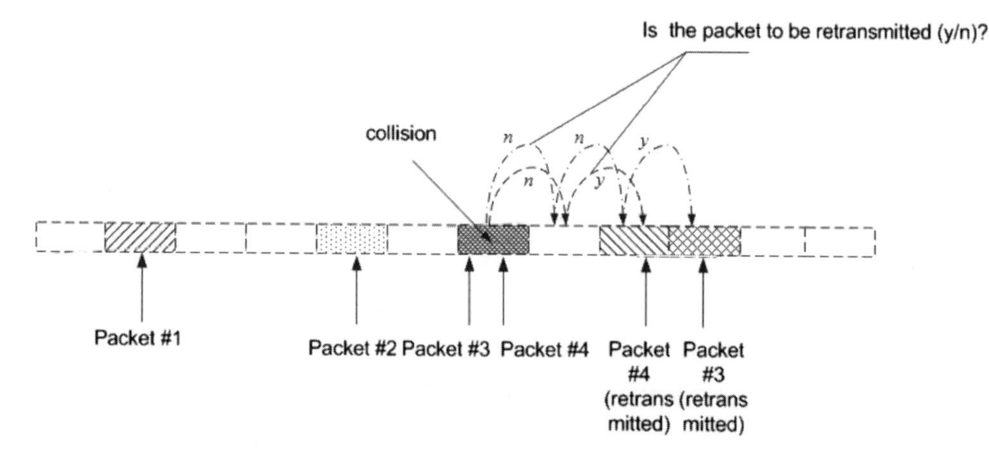

Figure 9.12 Slotted ALOHA

(d) whatever the result of the reception from a certain slot (either no packets have been transmitted, one packet has been transmitted or more than one packet has been transmitted in the slot in question, that is, a collision occurred), each user is informed about it by the beginning of the next slot, that is, it is assumed that there is an immediate error-free feedback from the base station to all the users and the receiver can distinguish between an idle slot and a collision.

If a collision occurs all packets involved are backlogged and are to be retransmitted in the next slot with a certain fixed probability $p(1 > p > 0)$. An example of the transmission in a slotted ALOHA system is shown in Figure 9.12.

Let us assume that the number of users is K, the probability that a user will transmit a packet in some slot is G_i and $S_i(S_i < G_i)$ is the probability of a successful transmission (that is, the probability of the fact that collision hasn't occurred.) Assume further that all K users operate independently and the probability of the users' transmission in the current slot does not depend on the users' transmissions in the previous slot. Then the normalized channel traffic is given by:

$$G = \sum_{i=1}^{K} G_i \tag{9.13}$$

and normalized channel throughput S (expected number of successful packets per packet transmission time) is given by:

$$S = \sum_{i=1}^{K} S_i \tag{9.14}$$

The probability that the packet from the ith user will not collide with the packets from other users is given by:

$$Q_i = \prod_{j=1, j\neq i}^{K} (1 - G_j) \tag{9.15}$$

Note that $S_i = Q_i G_i$

Let us assume that all the users are identical, that is:

$$S_i = \frac{S}{K} \qquad G_i = \frac{G}{K}$$

Then:

$$S = G\left(1 - \frac{G}{K}\right)^{K-1} \tag{9.16}$$

then if $K \to \infty$ we obtain:

$$S(G) = Ge^{-G} \tag{9.17}$$

note that:

$$S_{\max} = \max_{G} S(G) = \frac{1}{e} \tag{9.18}$$

and $G_{\max} = \arg\left(\max_{G} S(G)\right) = 1$

Note that this result has been obtained under the condition $K \to \infty$.

Let us consider the abovementioned assumptions in more detail. Primarily it is to be noted that it has been assumed that a packet can be backlogged only due to collision, that is, a packet being transmitted is not affected by either background noise or channel distortion. However, in real-life applications the transmitted data will also be corrupted by the background noise (in the case of a satellite communication system) or both background noise and fading (for example, in the case of a cellular system). Thus, it is evident that forward error correction coding (FEC) is to be applied. However, distinguishing between an idle slot and a collision (see assumption d.) remains a cumbersome problem (this problem can hardly be solved by applying the existing error correcting codes. Non-coherent threshold decision seems more promising but if the SNR is low the probability of a wrong decision is still substantial).

It is worth mentioning that under some conditions the packet arriving with the largest power at the receiver could be received correctly even in the presence of other packets with lower power, that is, not all packets, which collide on a communications channel, will be lost. Packets attenuated due to channel properties (for example, fading, shadowing) may not be strong enough to corrupt packets with higher power [7, 23]. In communication theory the ability of a base station to receive correctly one packet when two or more packets arrive simultaneously is often referred to as *capture effect*. It is to be noted that we have not considered *capture effect* within the scope of our consideration. Though the problem of capture effect impact on random access techniques performance falls out of the field of the present chapter and will not be considered here (for more information on the subject, please, refer to [43]) we can still conclude that assumption c. is unrealistic.

Within the scope of the foregoing consideration of slotted ALOHA an assumption of immediate feedback has been used. The assumption of immediate feedback is unrealistic, but as is stated in [12], "collision resolution algorithms can usually be modified to deal with delayed feedback" (collision resolution algorithms are to be discussed later.) Finally let us note that slotted ALOHA requires synchronization (see assumption a.)

Hereinabove it has been shown that in slotted ALOHA if all the users operate independently, the probability of the users' transmission in the current slot does not depend on the users' transmissions in the previous slot and all the users are identical, the maximum throughput is $S_{\max} = \max_{G} S(G) = \frac{1}{e}$. In fact this value is the upper bound for classical slotted ALOHA, that is, the system where retransmission of the backlogged packet depends only on the probability of retransmission ($0 < p < 1$). However, the maximum throughput can be further increased if the collision resolution algorithms, for example, the tree algorithms [4] are used. Within the scope of the tree algorithm it is assumed that the system alternates between two modes, the "normal" mode and the "collision resolution" mode. When a collision occurs in some slot all the users go into the "collision resolution" mode. This means that all the users that were not involved in the collision (that is, have not transmitted data in the preceding slot) are not allowed to transmit before the collision resolution. As for the users who were involved in the collision they are all to select (independently) one of the two subsets with equal probability (this procedure is sometimes referred to as "coin flipping"). In the slot following the collision, the users who got into the first subset try to transmit. If another collision occurs, the first subset is further split into two smaller subsets. The first of these subsets is transmitted in the next slot, and if this transmission is successful or idle (that is, the first submit happens to be empty), the second of the subsets is transmitted in the following slot. Thus, whenever the transmission of a subset results in a collision, the subset is split into two. So the split subsets form a tree. When all subsets have been exhausted all the users return to the "normal mode". This approach has been refined by eliminating the "normal mode" (the idea was that if a number of time slots have been spent to resolve the collision it is very likely that within the respective period a number of users ready to transmit packets has arrived; thus it is most likely that the system will return into the "collision resolution" mode after the next slot.) The resulting algorithm has been further improved by Messy who noted that if a certain subset had been split the number of elements in this set was certainly more than one. Thus if the first of the two subsets obtained by splitting the aforesaid subset happens to be empty we can conclude that the second slot contains more than one element. And we need not spend a slot to confirm it. Instead we can proceed directly to resolving collision in the second subset. Further improvements ([49]) resulted in the introduction of the first-come first-served (FCFS) algorithm (Tzybakov and Mikhailov showed in [49] that the optimized version of this algorithm yields $S_{\max} = 0, 4878$).

In **pure ALOHA (unslotted ALOHA)** any user is allowed to start transmission at an arbitrary moment. Historically unslotted ALOHA was the first random access technique that has been developed [1]. In the original pure ALOHA system it has been assumed that if several packets overlap (say packet A overlaps with packet B, which also overlaps with packet C as is shown in Figure 9.13) all packets involved are backlogged and are to be retransmitted at the end of the delay period, its duration being a random value τ, its distribution being given by:

$$p(\tau) = \alpha e^{-\alpha\tau} \tag{9.19}$$

Let T_p be the packet duration. Due to the backoff algorithm that has been considered above, aggregate traffic arrival that results from new arrivals and retransmissions has a Poisson distribution with an average number of arrivals of $\frac{G}{X}$. Collision may occur if at least one packet arrives within the time interval equal

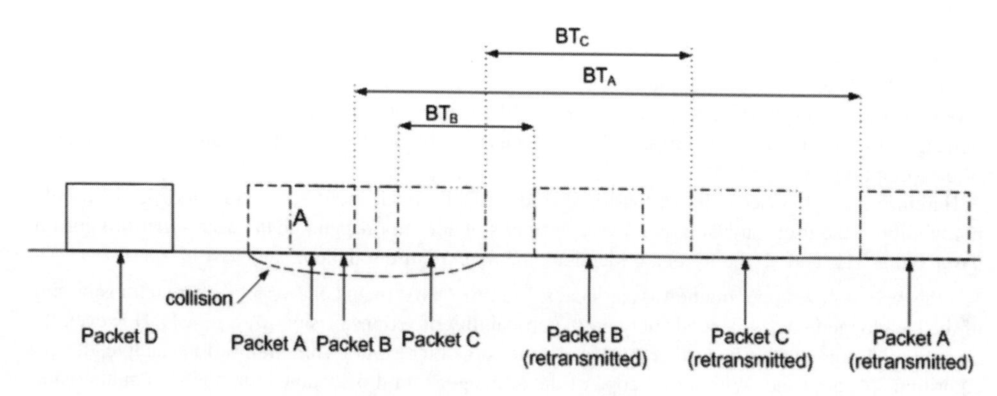

Figure 9.13 Unslotted ALOHA

to $2T_p$. The probability of the arrival of k packets within this interval is given by:

$$p(k) = \frac{(2G)^k}{k!} e^{-2G} \tag{9.20}$$

Thus a normalized channel throughput is given by:

$$S = G \cdot p(0) = G \cdot e^{-2G} \tag{9.21}$$

and

$$S_{\max} = \max_G S(G) = \frac{1}{2e} \tag{9.22}$$

that is, the maximum throughput of unslotted ALOHA is two times smaller than that of the slotted one. On the other hand, unslotted ALOHA does not need synchronization, that is, it is much easier to implement. Moreover, it is to be noted that in unslotted ALOHA packets can be of arbitrary length.

In ALOHA-based systems an active user who has sent a packet via the channel in use can be notified that the aforesaid channel is occupied only if the packet in question has been backlogged. *Carrier Sensing Multiple Access (CSMA)* initially proposed in [21] uses a different approach: it is implied that before sending a packet via the channel in use the user is to sense the medium in order to find out if some user is currently transmitting via the channel in use. If the answer is positive (that is, the channel is occupied), the user in question is not allowed to transmit. CSMA is well suited for ad hoc networks. Note that within the scope of this approach it is implied that any user can receive signals sent by any other active user. However, this is not always the case. Assume that a certain user B cannot receive information from slot A (for example, since signal attenuation is too big and user B simply cannot detect a signal from user A). Thus, he will consider the channel in use to be idle even within the scope of transmission of a signal sent by A and can transmit a packet (transmission protocols are to be considered later) causing, thus, a collision. This problem is known as the *hidden node* problem. However, for the sake of simplicity in what follows we shall not consider the hidden node problem, that is, we shall consider an ad hoc network where each user can receive messages sent by any other user if no collision has occurred; and, thus, carrier sensing always enables the user in question to obtain precise information on the medium state.

Like ALOHA CSMA can be either slotted or unslotted. Hereinafter for the sake of simplicity we shall consider slotted CSMA systems. As has been stated above in CSMA each user senses the channel until it is idle. In 1-persistent CSMA each user that is ready to transmit does so as soon as the channel is found idle. Thus if two or more users were ready to transmit before the channel in use became available a collision would occur. Moreover, since all these users will constantly transmit that results in constant collision. In non-persistent CSMA (np-CSMA) the user senses the medium first. If it is idle the user is to transmit his packet. Otherwise the user in question is to choose a random time period (in the slotted case a random number of slots), wait until this period is over and repeat the preceding steps. A random backoff period results in sufficient collision probability reduction. However, idle periods that are unavoidable in this case lead to efficiency degradation. In p-persistent CSMA a random strategy is applied, that is, if the channel becomes available each user that is ready to transmit is to do so with probability p or defer his transmission by one slot with probability$1 - p$. If the transmission has been deferred by one slot the user under consideration is to repeat the preceding steps at the end of this slot. This technique is considered to be a good tradeoff between 1-persistent CSMA and non-persistent CSMA. However, in wireless ad hoc networks it is common to apply CSMA with collision avoidance (CSMA\CA), which is a slightly modified version of non-persistent CSMA. Note that like other CSMA techniques applied in wireless communications this method only tries to avoid collisions. However, collisions can still occur. Thus, like ALOHA, CSMA needs some sort of feedback. Therefore in CSMA\CA the receiver is to transmit a special acknowledgment message after each transmission (the latter is to be transmitted without sensing the carrier in a special slot, in which no one but the recipient is to transmit.) If the sender fails to receive an acknowledgment message retransmission is applied.

CSMA is known to have a better throughput than ALOHA. However, both concepts are very vulnerable to unauthorized users' activity.

9.6 Conclusions

In this chapter several fundamental multiple access techniques have been considered, namely the fixed access methods, the random access and the code division methods. Also, in order to give some new insights into the field of multiple access this consideration has been complemented by a short survey of several recently proposed multiple access techniques that can be considered candidates for future implementation in real-life systems.

References

[1] Abramson, N., *The ALOHA System - Another Alternative for Computer Communications*, pp. 281–285 in Proceedings of the Fall Joint Computer Conference, 1970.

[2] Boztas, S., Hammons, Jr. A. R., and Kumar, P. V., *4-phase sequences with near-optimum correlation properties*, IEEE Trans. Inform. Theory, vol. 38, pp. 1101–1113, May 1992.

[3] Brouwer, A.E., Shearer, J.B., Sloane, N.J.A., Smith, W.D., *A New Table of Constant WeightCodes*, IEEE Trans. on Information Theory, vol. 36, no. 6, pp. 1334–1380, Nov. 1990.

[4] Capetanakis, J. I., *The multiple access broadcast channel: Protocol and capacity considerations*, Ph.D. dissertation Dept. of Elec. Eng., Mass. Inst. of Tech., Cambridge, MA, Aug. 1977.

[5] Chen, C-C., Yao, K., Umeno, K., and Biglieri, E. *Design of Spread-Spectrum Sequences Using Chaotic Dynamical Systems and Ergodic Theory* IEEE Transactions on circuits and systems—I: Fundamental theory and applications, vol. 48, no. 9, September 2001.

[6] Chouly, A., Brajal, A., Jourdan, S., *Orthogonal Multicarrier Techniques Applied to Direct Sequence Spread Spectrum CDMA Systems*, Proc. of IEEE GLOBECOM '93, Houston, USA, Nov. 1993, pp. 1723–1728.

[7] Cidon, I., Kodesh, H., Sidi, M., *Erasure, Capture, and Random Power Level Selection in Multiple-Access Systems*, IEEE Trans. on Communications, vol. 36, no. 3, pp. 263–271, March 1988.

[8] Cosovic, I., Schnell, M., *Time Division Duplex MC-CDMA for Next Generation Mobile Radio Systems* In proc of the 10th Telecommunications forum TELFOR'2002, Belgrade, Yugoslavia, Nov. 26–28, 2002.

[9] DaSilva, V. M., Sousa, E. S., *Performance of Orthogonal CDMA Codes for Quasi-Synchronous Communication Systems*, Proc. of IEEE ICUPC "93, Ottawa, Canada, Oct. 1993, pp. 995–999.

[10] EC-COST 207, Final report, *Digital land mobile radio communications*, Commission of the European Communities, EUR 12160 EN, Brussels, 1989.

[11] Fazel, K. and Papke, L., *On the Performance of Convolutionally-Coded CDMA/OFDM for Mobile Communications Systems*, in Proc. IEEE PIMRC'93, (Yokohama, Japan), pp. 468–472, September 1993.

[12] Gallager, R., *A Perspective on Multiaccess Channels* IEEE Trans. Information Theory vol. it-31, no. 2, March 1985.

[13] Gold, R., *Optimal binary sequences for spread spectrum multiplexing*, IEEE Trans. Information Theory, vol. IT-13, 1967. pp. 619–621.

[14] Hara, S., Prasad, R., *Overview of Multicarrier CDMA* IEEE Commun. Mag., Dec. 1997, pp. 126–133.

[15] Hall, Jr. M., *Combinatorial Theory* Blaisdell, Waltham, Mass., 1967.

[16] Ipatov, V. P., *Spread spectrum and CDMA: principles and applications*. John Wiley and Sons, 2005.

[17] Kamaletdinov, B. Zh. *An Optimal Ensemble of Binary Sequences Based on the Union of the Ensembles of Kasami and Bent-Function Sequences* Problems of Information Transmission vol. 24, no. 2, pp. 167–169, 1988.

[18] Kamaletdinov, B. Zh. *Optimal Sets of Binary Sequences* Problems of Information Transmission pp. vol. 32, no. 2, 171–175, 1996.

[19] Kasami, T., *Weight distribution formula for some class of cyclic codes*, Coordinated Science Laboratory, University of Illionos, Urbana, Technical Report R-285 (AD632574), 1966.

[20] Kleinrock, L. *On Queueing Problems in Random-Access Communications* IEEE Transactions on Information Theory, vol. it-31, No. 2, March 1985.

[21] Kleinrock, L., Tobagi, F.A., *Packet switching in radio channels: Part I - Carrier Sense Multiple-Access modes and their throughput-delay characteristics*, IEEE Trans. on Communications, vol. 23, no. 12, pp. 1400–1416, Dec. 1975.
[22] Kumar, P. V., Moreno, O., *Prime-phase sequences with periodic correlation properties better than binary sequences*, IEEE Trans. Inform. Theory, vol. 37, pp. 603–616, May 1991.
[23] Linnartz, J.P., Prasad, R., *Near-Far Effect on Slotted Aloha Channels with Shadowing and Capture*, pp. 809–813 in Proceedings IEEE Vehicular Technology Conference, 1989.
[24] Mazzini, G., Setti, G., and Rovatti, R. *Chaotic Complex Spreading Sequences for Asynchronous DS-CDMA* IEEE Transactions on circuits and systems—I: Fundamental theory and applications, vol. 44, no. 10, October 1997.
[25] No, J. S., Kumar, P. V., *A new family of binary pseudorandom sequences having optimal periodic correlation properties and large linear span*, IEEE Trans. Inform. Theory, vol. IT-35, No. 2, pp. 311–319, Mar. 1989.
[26] No, J. S., Yang, K., Chung, H., Song, H.-Y., *New Construction for Families of Binary Sequences with Optimal Correlation Properties*, IEEE Transactions on Information Theory, vol. 43, No. 5, September 1997.
[27] Nogueroles, R., *On multiple access strategies for multicarrier transmission*, Winter School on Coding and Information Theory, Mölle, December 15–18, 1996.
[28] Nogueroles, R., *On Multiple Access Strategies for Multicarrier Transmission in Mobile Radio Systems*, Technical Report ITUU-TR-1997/02, Abt. Informationstechnik, Universität Ulm, February 1997.
[29] Nogueroles, R., Bossert, M. *Comparison of multicarrier FDMA strategies for mobile radio systems*, pp. 143–148 in Proceedings IEEE Communications Society Chapter Germany Workshop Kommunikationstechnik, Schloβ Reisensburg, January 29–31, 1997.
[30] Nogueroles, R., Bossert, M., Donder, A., Zyablov, V. *Improved Performance of a Random OFDMA Mobile Communication System*, pp. 2502–2506 in Proceedings 48th IEEE Vehicular Technology Conference VTC'98, Ottawa, Ontario, Canada, May 18–21, 1998.
[31] Nogueroles, R., Bossert, M., Donder, A., Zyablov, V. *Performance of a Random OFDMA System for Mobile Communications*, pp. 37–43 in Proceedings 1998 International Zurich Seminar on Broadband Communications, Switzerland, February 17–19, 1998.
[32] Nogueroles, R., Bossert, M., Zyablov, V. *Capacity of MC-FDMA in Mobile Communications*, pp. 110–114 in Proceedings 8th IEEE International Symposium on Personal Indoor and Mobile Radio Communications, Helsinki, Finland, September 1–4, 1997.
[33] Nogueroles, R., Bossert, M., Zyablov, V. *Estimation of user capacity in mobile radio multiple access systems based on multicarrier modulation*, pp. 282–283 in Proceedings 4th Int. Symposium on Comm. Theory & Applications, U.K., July 13–18, 1997.
[34] Nogueroles, R., Bossert, M., Zyablov, V. *Multiple access and collision problem in multifrequency transmission systems*, pp. 225–230 in Proceedings Fifth International Workshop on Algebraic and Combinatorial Coding Theory, Bulgaria, June 1–7, 1996.
[35] Nogueroles, R., Bossert, M., Zyablov, V. *Performance of a Novel Mobile Communication System based on Random MC-FDMA*, Proceedings 2. OFDM-Fachgespräch, Braunschweig, September 16–17, 1997.
[36] Nogueroles, R., Bossert, M., Zyablov, V. *Random Access in an OFDMA System*, pp. 20–25 in Proceedings Workshop Advances on Multiuser Communications, DLR Oberpfaffenhofen, February 19–20, 1998.
[37] Nogueroles, R., *Performance Considerations of OFDMA Radio Communication Systems"*, PhD Thesis, Universität Ulm, Ulm, 1998.
[38] Osipov, D. *On the performance of a FH-MC-CDMA system.* In Proc. of the 2nd International Workshop on Multiple Access Communications (MACOM 2009) affiliated with the IEEE International Conference on Communications 2009 (ICC 2009) Dresden, Germany, 14–18 June, 2009.
[39] Osipov, D. *On the probabilistic description of a FH- OFDMA with a MAXP receiver* In Proc. of the XII-th International Symposium on Problems of Redundancy in Information and Control Systems, St.-Petersburg, 26–30 May, 2009.
[40] Osipov, D. S. *Asynchronous OFDMA system with woven turbo coded modulation.* In Proc. of the International Workshop on Multiple Access Communications (MACOM 2008) affiliated with the 15-th International Conference on Telecommunications (ICT 2008), St.-Petersburg, June 16–19, 2008.
[41] Osipov, D. *Multiple access with concatenated coding* (in Russian) Ph. D. Thesis, IITP RAS, Moscow, 2008.
[42] Price, R., Green, P. E., *A communication technique for multipath channels*, Proc. IRE, vol. 46, pp. 555–570.
[43] Roberts, L.G., *ALOHA Packet System with and without Slots and Capture*, Computer Communications Review, vol. 5, no. 2, pp. 28–42, April 1975.

[44] Ruiz, A., Cioffi, J.M., Kasturia, S., *Discrete multiple tone modulation with coset coding for the spectrally shaped channel*, IEEE Trans. on Communications, vol. 40, no. 6, pp. 1012–1029, June 1992.
[45] Sarwate, D. V., *Bounds on crosscorrelation and autocorrelation of sequences*, IEEE Trans. Inform. Theory, vol. IT-25, pp. 720–724, Nov. 1979.
[46] Sidelnikov, V. M., On mutual correlation of sequences," Sov. Math.–Dokl., vol. 12, pp. 197–201, 1971.
[47] Sol'e, P., *A quaternary cyclic code and a family of quadriphase sequences with low correlation properties*, In Coding Theory and Applications, Lecture Notes in Computer Science vol. 388. pp.193–201, 189.
[48] Sun, F.-W., Leib, H., *Optimal Phases for a Family of Quadriphase CDMA Sequences* IEEE Transactions on Information Theory, vol. 43, no. 4, July 1997 pp.1205–1217.
[49] Tsybakov, B. S., Mikhailov, V. A., *Random multiple access of packets: Part and try algorithm*, Problems of Information Transmission, vol. 16, no. 4, 1980., pp. 65–79.
[50] Vandendorpe, L., *Multitone Spread Spectrum Multiple Access Communications System in a Multipath Rician Fading Channel*, IEEE Trans. Vehic. Tech., vol. 44, no. 2, 1995, pp. 327–337.
[51] Verdu, S. *Multiuser detection*, Cambridge University Press, 1998.
[52] Weinstein, S.B., Ebert, P.M., *Data Transmission by Frequency-Division Multiplexing Using the Discrete Fourier Transform*, IEEE Trans. on Communication Technology, vol. 19, no. 5, pp. 628–634, October 1971.
[53] Welch, L. R., *Lower bounds on the maximum cross correlation of signals*, IEEE Trans. Inform. Theory, vol. IT-20, pp. 397–399, May 1974.
[54] *WCDMA for UMTS, Radio Access for Third Generation Mobile Communications*. Second Edition. ed. H. Holma, A. Toskala, New York, Wiley, 2002.
[55] Yang, L. L., Hua, W., Hanzo, L., *Multiuser Detection in Multicarrier CDMA Systems Employing Both Time-Domain and Frequency-Domain Spreading*, In Proceedings of IEEE PIMRC'2003, Beijing China, 7–10 September, 2003.
[56] Yang, L-L., Hanzo, L., *Multicarrier DS-CDMA: A Multiple Access scheme for ubiquitous broadband wireless communications* IEEE Communications Magazine, pp. 116–124, October 2003.
[57] Yee, N., Linnartz, J. P., Fettweis, G., *Multi-carrier CDMA in Indoor Wireless Radio Networks*, in Proc. IEEE PIMRC'93, (Yokohama, Japan), pp. 109–113, September 1993.
[58] Zigangirov, K. Sh. *Theory of Code Division Multiple Access Communication*, IEEE Press, Piscataway, New Jersey, 2004.
[59] Zimmerman, M.S., Kirsch, A.L., *The AN/GSC-10 (KATHRYN) Variable Rate Data Modem for HF Radio*, IEEE Trans. on Communication Technology, vol. 15, no. 2, pp. 197–204, April 1967.
[60] Viterbi, A. J. *CDMA - Principles of Spread Spectrum Communication*. Addison-Wesley, 1995.
[61] Zyablov, V. V., Osipov, D. S., *On the optimum choice of a threshold in a frequency hopping OFDMA system*. Problems of Information Transmission, Vol. 44, No. 2, pp. 91–99, 2008.
[62] Zyablov, V. V., Osipov, D. S., *Equidistant code-based signal-code construction in multiple-access system with concatenated coding* (in Russian) In Proc. of the conference "Information Technologies and Systems" ITAS 09 Bekasovo, 15-18 December, 2009. pp. 145–151.
[63] Shannon, C.E., *A mathematical theory of communication*, Bell Systems Technical Journal, vol. 27, pp. 379–423, 623–656, July–October 1948.
[64] Gaudenzi, R., Elia, C., Viola, R., *Band-limited quasi-synchronous CDMA: A novel satellite access technique for mobile and personal communications systems*, IEEE Journal on Selected Areas on Communication, vol. 10, no. 2, pp. 328–343, February 1992.
[65] Kajiwara, A., Nakagawa, M., *Microcellular CDMA system with a linear multiuser interference canceller*, IEEE Journal on Selected Areas on Communication, vol. 12, no. 4, pp. 605–611, May 1994.
[66] Omura, J., Yang, P., *Spread spectrum S-CDMA for personal communication services*, in Proc. MILCOM 1992, San Diego, United States, October 1992.
[67] DaSilva, V. and Sousa, E. *Multicarrier orthogonal CDMA signals for quasi-synchronous communication systems*, IEEE Journal on Selected Areas on Communication, vol. 12, no. 5, pp. 842–852, June 1994.
[68] Suehiro, N. *A signal design without co-channel interference for approximately synchronized CDMA systems*, IEEE Journal on Selected Areas on Communication, vol. 12, no. 5, pp. 837–841, June 1994.
[69] Cover, T. M., Thomas, J. A. *Elements of Information Theory*, Wiley, New York, 1990.
[70] Goldsmith, A. *Wireless Communications*, Cambridge University Press, New York, 2005.
[71] Münz, G., Pfletschinger, S., Speidel, J. *An Efficient Waterfilling Algorithm for Multiple Access OFDM*, Proc. IEEE Global Telecommunications Conference 2002 (Globecom '02), Taipei, Taiwan, November 2002.

[72] Yu, W. *Multiuser Water-filling in the Presence of Crosstalk*, Information Theory and Applications Workshop, San Diego, CA, U.S.A. Jan-Feb., 2007.

[73] Papandreou, N., Antonakopoulos, T. *A New Computationally Efficient Discrete Bit-Loading Algorithm for DMT Applications*, IEEE Transactions on Communications,vol. 53, no. 5, May 2005.

[74] Pietrzyk, S. *OFDMA for Broadband Wireless Access*, Artech House Publishers, London, 2006.

[75] IEEE, *Air Interface for Fixed and Mobile Broadband Wireless Access Systems*, IEEE Standard, P802.16e/D12, February, 2005.

[76] Myung, H. G., Lim, J., Goodman, D. J. *Single Carrier FDMA for Uplink Wireless Transmission*, IEEE Vehicular Technology Magazine, September 2006.

[77] Cheung, N.K., Nosu, K., Winzer, G. *Guest Editorial / Dense Wavelength Division Multiplexing Techniques for High Capacity and Multiple Access Communication Systems*, IEEE Journal on Selected Areas in Communications, vol. 8 no. 6, August 1990.

[78] Sklar, B. *Digital Communications*, 2nd ed. Prentice Hall, 2001.

[79] Lupas, R. and Verdu, S. *Linear multiuser detectors for synchronous code-division multiple-access channels*, IEEE Transactions on Information Theory, vol. 35, no. 1, pp. 123–136, January 1989.

[80] Madhow, U. and Honig, M. *MMSE interference suppression for direct-sequence spread-spectrum CDMA*, IEEE Transactions on Communications, vol. 42, no. 12, pp. 3178–3188, December 1994.

[81] Varanasi, M. and Aazhang, B. *Near-optimum detection in synchronous code-division multiple access systems*, IEEE Transactions on Communications, vol. 39, no. 5, pp. 725–736, May 1991.

[82] Xie, Z., Short, R. and Rushfort, C. *A family of suboptimum detectors for coherent multiuser communication*, IEEE Journal on Selected Areas on Communication, vol. 8, no. 4, pp. 683–690, May 1990.

[83] Duel-Hallen, A. *A family of multiuser decision-feedback detectors for asynchronous code-division multiple-access channels*, IEEE Transactions on Communications, vol. 43, no. 2, pp. 421–434, February 1995.

[84] Patel, P. and Holtzman, J. *Analysis of a simple successive interference cancellation scheme in a DS/CDMA system*, IEEE Journal on Selected Areas on Communication, vol. 12, no. 5, pp. 796–807, June 1994.

[85] Fan, P. and Darnell, M. *Sequence Design for Communication Applications*. Wiley, 1996.

[86] Walke, B., Seidenberg, P. and Althoff, M. P. *UMTS: The Fundamentals*. Wiley, 2003.

[87] Paavola, J. *Signature Ensembles and Receiver Structures for Oversaturated Synchronous DS-CDMA Systems*, Ph.D. dissertation, University of Turku, 2007.

[88] Vanhaverbeke, F. *Digital communication through overloaded channels*, Ph.D. dissertation, University of Ghent, 2005.

[89] Rupf, M. and Massey, J. L. *Optimum sequence multisets for synchronous code-division multiple-access channels*," IEEE Transactions on Information Theory, vol. 40, no. 4, pp. 1261–1266, July 1994.

[90] Kapur, A., Varanasi, M. and Mullis, C. *On the limitation of generalized Welch-Bound equality signals*, IEEE Transactions on Information Theory, vol. 51, no. 6, pp. 2220–2224, June 2005.

10

Standardization in IEEE 802.11, 802.16

Tuomas Laine[1], Zexian Li[1], Andrei Malkov[1], and Prabodh Varshney[2]
[1]*Nokia Corporation, Finland*
[2]*Nokia, USA*

10.1 IEEE Overview

The IEEE- Standard Association's involvement in electrical standards dates back to 1890, when the *American Institute of Electrical Engineers (AIEE)* proposed a recommendation for the practical unit of self-induction. As a pioneer in voluntary electrical and information technology standards activity, IEEE became a founding member of *American National Standards Institute (ANSI)* in 1918.

In 1963, when the AIEE merged with the *Institute of Radio Engineers (IRE)* to form the IEEE, a formal standards body was established to support standards development. Envisioning the expanded role that standards were to play in the future and their impact on industry, IEEE formed its first Standards Board in 1973. As a standards body, the *IEEE-Standard Association (IEEE-SA)* has responded to changes in the marketplace and as a result, the IEEE-SA of today is quite different and innovative, but still committed to providing the most current, reliable standards knowledge. On average, the IEEE-SA publishes 80 new and revised standards annually and conducts over 245 standards projects ballots, in which a combination of approximately 13,000 individuals participate.

Overview of IEEE Standards Association Organization is shown in Figure 10.1. The IEEE-SA is represented on the IEEE Board of Directors & IEEE Executive Committee. The IEEE-SA is assigned authority for the standardization activities of the IEEE by the IEEE Board of Directors. The IEEE-SA fulfills this assignment by activities such as, but not limited to:

- Encouraging active development of needed standards. This involves, for example, promotion of open and innovative deliberations that result in broad consensus in accordance with due process procedures detailed in the bylaws and operations manuals of the subsidiary boards and committees of the IEEE-SA Board of Governors.
- Building upon the strengths of the standards developing community by involving appropriate interests and outside organizations.

Modulation and Coding Techniques in Wireless Communications Edited by Evgenii Krouk and Sergei Semenov

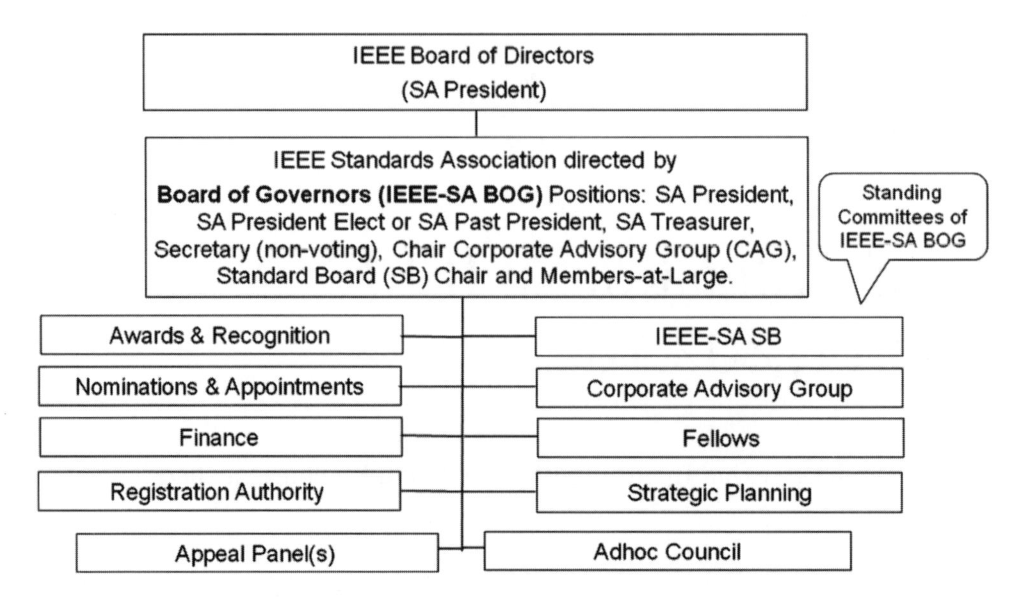

Figure 10.1 IEEE Standards Association (SA) Organization Overview

- Representing IEEE to external bodies on standards matters. This includes providing for cooperation with and IEEE participation in the activities of other organizations consistent with its scope and responsibilities.
- Appointing IEEE members to participate in external bodies on standards matters based on nominations submitted by the IEEE Societies. The IEEE-SA also provides speakers to make presentations at meetings and conferences on subjects related to IEEE's standards interests and to participate in panels on standards-related subjects.

With the approval of the IEEE Board of Directors and as authorized, the IEEE-SA Board of Governors may establish groups:

- To act for the IEEE in product testing or in certification of products or systems to comply with IEEE standards, or
- To offer opinion in the name of the IEEE on the conformance of products or systems to the requirements of IEEE standards for their intended use and safe operation.

Overview of IEEE Standard Association Standard Board Organization is shown in Figure 10.2. The IEEE-SA Standards Board (IEEE-SA SB) is established by the IEEE-SA Board of Governors. The IEEE-SA Standards Board is responsible on an Institute-wide basis for:

- Encouraging and coordinating the development of IEEE standards.
- Reviewing all proposed IEEE standards to determine whether the proposed standards conform to the requirements established by the IEEE-SA Standards Board and whether consensus has been achieved for approval of the proposed standards.
  - Consensus is established when, in the judgment of the IEEE-SA Standards Board, substantial agreement has been reached by directly and materially affected interest categories.
  - Substantial agreement means much more than a simple majority, but not necessarily unanimity.

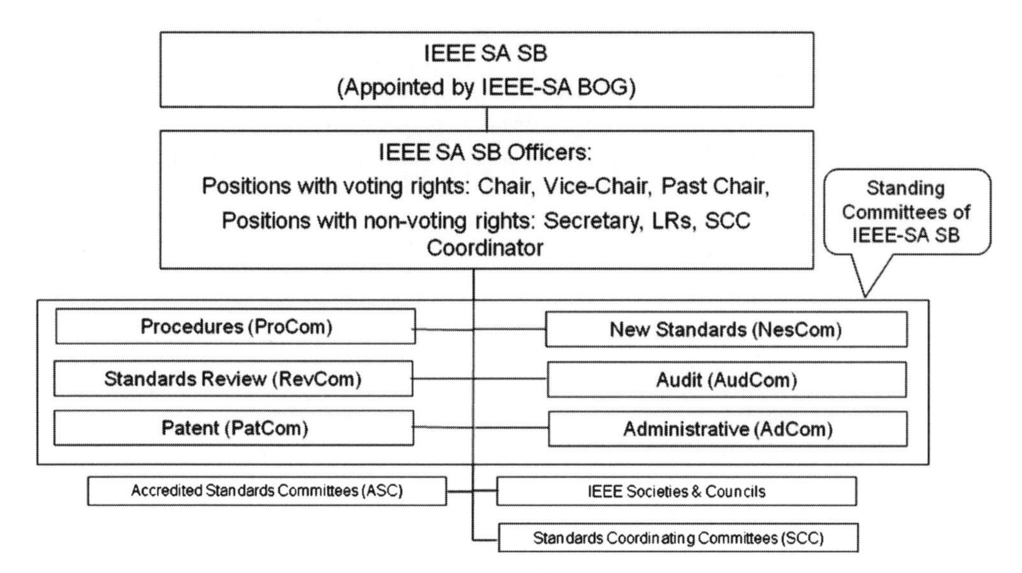

Figure 10.2 IEEE Standard Association Standard Board (SA SB) Organization Overview

- Consensus requires that all views and objections be considered, and that a concerted effort be made toward their resolution.

Matters of standards policy, financial oversight, new directions in standardization, and other standards-related activities in fields of interest to the Institute are the responsibility of the IEEE-SA Board of Governors (BOG).

Procedures Committee (ProCom): This committee is responsible for recommending to the IEEE-SA Standards Board improvements and changes in its bylaws, procedures, and manuals to promote efficient discharge of responsibilities by the IEEE-SA Standards Board and its committees.

New Standards Committee (NesCom): This committee is responsible for ensuring that proposed standards projects are within the scope and purpose of the IEEE, that standards projects are assigned to the proper society or other organizational body, and that interested parties are appropriately represented in the development of IEEE standards. The committee examines Project Authorization Requests (PARs) and makes recommendations to the IEEE-SA Standards Board regarding their approval.

Standards Review Committee (RevCom): This committee is responsible for reviewing proposals for the approval of new and revised standards and for the reaffirmation or withdrawal of existing standards to ensure that the proposals represent a consensus of the members of the official IEEE sponsor balloting group. The committee routinely examines submittals to ensure that all applicable requirements of the *IEEE-SA Standards Board Operations Manual* have been met and make recommendations to the IEEE-SA Standards Board regarding their approval.

Audit Committee (AudCom): This committee provides oversight of the procedures used in the standards-development activities of IEEE Standards Sponsors and review of the procedures used by the Accredited Standards Committees for whom the IEEE serves as (co-)secretariat. This committee also oversees the submission of Sponsor annual reports.

Patent Committee (PatCom): This committee provides oversight for the use of any patents and patent information in IEEE standards. This committee reviews any patent information submitted to the IEEE Standards Department to determine conformity with patent procedures and guidelines.

Administrative Committee (AdCom): The Administrative Committee acts for the IEEE-SA Standards Board between meetings and makes recommendations to the IEEE-SA Standards Board for its disposition at regular meetings.

Standards Coordinating Committees (SCCs): When the scope of a standards activity is too broad to be encompassed in a single IEEE Society, or a society may find itself in a position where it is unable to carry out the work needed to meet an identified need then the IEEE-SA Standards Board shall establish its own committees to perform the required functions.

The IEEE is represented on the Standards Committees (ASCs) in which it has a substantial interest. This provides IEEE with an opportunity to have a direct influence on the development of an American National Standard generated by the ASC.

IEEE 802 is a standard committee sponsored by the Computer Society. The Computer Society is one of the IEEE Societies and councils.

All IEEE standards development is based on projects that have been approved by the IEEE-SA Standards Board, and each project shall be the responsibility of a Sponsor. Projects within IEEE 802, P802.xx, are sponsored by the Computer Society through *LAN/MAN Standard Committee (LMSC)*. The Sponsor accepts responsibility for oversight of any of its assigned standards, including overseeing coordination, balloting, and making annual activity reports to the IEEE-SA Standards Board. Each Sponsor operates in accordance with a written set of policies and procedures (P & P). There are also operating procedures available for Sponsors developing a standard using the entity method of participation, and Sponsors utilize these procedures as the basis for entity standardization.

These projects with P802.xx form a family of IEEE standards dealing with local area networks and metropolitan area networks. The services and protocols specified in IEEE 802 map to the lower two layers (Data Link and Physical) of the seven-layer OSI networking reference model. These standards deal with Ethernet, Token Ring, Bridging & Virtual bridged LANs, WLAN, *Wireless Personal Area Network (WPAN)*: Bluetooth, Zigbee, in or around a body, and so on, Broadband Wireless Access Systems (WiMAX), Media independent handover and *Wireless Regional Area Networks (WRAN)*. Work is organized into a number of Working Groups (WGs) and Technical Advisory Groups (TAGs) operating under the oversight of a sponsor Executive Committee (EC). Current active working groups within IEEE 802 are:

- 802.1 Higher Layer LAN Protocols Working Group
- 802.3 Ethernet Working Group
- 802.11 Wireless LAN Working Group
- 802.15 Wireless Personal Area Network (WPAN) Working Group
- 802.16 Broadband Wireless Access Working Group
- 802.17 Resilient Packet Ring Working Group
- 802.18 Radio Regulatory TAG
- 802.19 Coexistence Working Group
- 802.20 Mobile Broadband Wireless Access (MBWA) Working Group
- 802.21 Media Independent Handover Services Working Group
- 802.22 Wireless Regional Area Networks

10.2 Standard Development Process

Typically a *Study Group (SG)* is formed when a new area is first investigated for standardization. The SG can be within an existing WG or TAG, or it can be independent of the WGs. A new project within an existing group is typically developed by a Sub- Group or Task Group, while a new independent project can lead to the creation of a completely new WG.

Each standard, recommended practice, or guide begins as a group of people with an interest in developing the standard. A *Project Authorization Request (PAR)* is normally submitted for approval within six months of the start of work. In order to avoid duplication, provide for effective management of overall efforts, and expedite approval of final documents, all requests for an initiation of a standards project, in the form of a PAR, shall be approved by the IEEE-SA SB. The IEEE-SA SB has assigned to NesCom the preliminary review of PARs and the responsibility for recommending final approval to the Board. Sponsors are required to submit a PAR at the earliest opportunity when a standards project is contemplated or work is started. PARs that have been submitted by Sponsors to the Secretary of the IEEE-SA Standards Board by the established deadline shall be submitted by the Secretary to the New Standards Committee (NesCom) for review. Unless specifically authorized by the IEEE-SA Standards Board, no proposed standard or revision shall be considered by RevCom without prior approval of the project by the IEEE-SA Standards Board. The lifetime of a PAR shall be four years. In LMSC, new projects require supporting material in the form of 5- Criteria to show that they meet the charter of LMSC. The draft PAR is voted on by the EC, and then it goes to the IEEE Standards Board New Standards Committee (NesCom) which recommends it for approval as an official IEEE Standards Project. Part of the PAR identifies if there will be liaisons with outside standards groups, for instance *International Telecommunication Union (ITU)* for some international standards. The liaisons help avoid conflicts or duplication of effort within an area.

Technical proposals are presented and evaluated either in working group or task group meetings, and a draft standard is written and voted on by the WG. Typically there are six face-to-face meetings every year for working groups. Three of these meetings are called Plenary meetings. In Plenary meetings, all the working groups of IEEE 802 meet and three are called interim meetings. In interim meetings some working groups may meet and some may not. Further, for interim meeting working groups may decide to meet at different locations. When the TG/WG reaches enough consensus on the draft standard, a WG Letter Ballot is done to release it from the WG. It is next approved by the EC and then goes for Sponsor Letter Ballot.

After the Sponsor Letter Ballot has passed and "No" votes are resolved, the draft Standard is sent to the IEEE Standards Board Standards Review Committee (RevCom). All IEEE standards shall be approved by the IEEE-SA Standards Board prior to publication. The IEEE-SA Standards Board has assigned to RevCom the review of standards submittals and the responsibility for recommending final approval to the IEEE-SA Standards Board. Proposed standards, together with the required documentation, that has been submitted by Sponsors to the Secretary of the IEEE-SA Standards Board by the established deadline shall be submitted by the Secretary to the Standards Review Committee (RevCom) for review. Approval by the IEEE-SA Standards Board indicates that the requirements of the *IEEE-SA Standards Board Operations Manual* and these bylaws have been satisfied and, specifically, that the final results of the ballot and statements submitted by other coordinating bodies who participated in the development of the standard indicate that consensus has been achieved and unresolved negative ballots have been considered together with reasons why the comments could not be resolved. Once it is recommended by RevCom and approved by the Standards Board, it can then be published as an IEEE standard.

Some draft standards in LMSC are also sent to ISO at the time they go to Sponsor Letter Ballot. Parallel approval in ISO JTC1/SC6 (Joint Technical Committee 1, Subcommittee 6 - responsible for LANs) may lead to publication as an ISO standard. The process from start to finish can take several years for new standards, and less for revisions or addenda to existing standards.

10.3 IEEE 802.11 Working Group

IEEE 802.11 is considered to be the home of WLAN standardization. Activities in 802 started in 1980. The number "802" was simply the next free number IEEE could assign. In 1987 activities in 802.4L (token bus) were started. 802.11 working group was established in 1990 as outgrowth from 802.4L.

802.11 published its first standard in 1997, standardizing three physical layers, infrared operating at 1 Mbps data rate, Direct sequence spread spectrum and Frequency hopping spread spectrum in 2.4 GHz band using 20 MHz bandwidth with 1 and 2 Mbps data rates.

In 1999, 802.11a was standardized. This standard provided variable data rates from 6 Mbps to 54 Mbps in 5 GHz band using 20 MHz channels. It uses OFDM based air interface. Also in the same year, 802.11b was standardized. This standard uses same air interface (Direct sequence spread spectrum) as base 802.11 standard but increased the data rate to 11 Mbps in 2.4 GHz band. In practice 802.11b has a greater range than 802.11a because of lower frequency in which 802.11b operates but 802.11a offered higher data rates than 802.11b.

802.11c is a network interoperability standard that deals with bridge operation procedures in wireless bridges or access points was published in 2001. It was incorporated in 802.1D in 2004. Also in 2001, 802.11d specified operation in additional regulatory domains. This includes addition of a country information element to beacons, probe requests, and probe responses. The country information elements simplifies the creation of 802.11 wireless access points and client devices that meet the different regulations enforced in various parts of the world.

In 2003, 802.11g introduced OFDM air interface in 2.4 GHz band and increased the data rate to 54 Mbps using 20 MHz channel bandwidth. This standard is backward compatible with 802.11b. Also in the same year 802.11h specified Spectrum and Transmit Power Management Extensions in 5 GHz band to solve problems like interference with satellites and radar. It was originally designed to address European regulations but is now applicable in many other countries. The standard provides *Dynamic Frequency Selection (DFS)* and Transmit Power Control (TPC) to the 802.11 MAC.

In 2004, 802.11i specified security mechanisms for WLAN. It replaced the short authentication and privacy clause of the original standard with a detailed security clause. In the same year 802.11j standard was created for the Japanese market. It allows Wireless LAN operation in the 4.9 to 5 GHz band to conform to the Japanese rules for radio operation for indoor, outdoor and mobile applications.

In 2005, 802.11e defined a set of Quality of Service enhancements for WLAN applications through modifications to the *Medium Access Control (MAC)* layer. This is of importance for delay-sensitive applications, such as Voice over Wireless IP and Streaming Multimedia.

Task group TGma was authorized to incorporate many of the amendments made to base 802.11-1999 standard. REVma or 802.11ma, as it was called, created a single document that merged eight amendments (802.11a, b, d, e, g, h, i, j) with the base standard 802.11-1999. Upon approval on 8 March 2007, 802.11REVma was renamed to the current base standard IEEE 802.11-2007.

In 2008, 802.11k created a standard for a basic set of radio resource measurements. These measurements are required to provide services such as roaming, coexistence, and so on. It is necessary to provide these measurements and other information in order to manage these services. Also in 2008, 802.11r specified fast BSS Roaming/Transition within WLAN networks meeting the real-time handover requirements without compromising the security principles. Further, 802.11y created a standard which enables high powered Wi-Fi equipment to operate on a co-primary basis in the 3650 to 3700 MHz band in the United States, except when near a grandfathered satellite earth station.

Recently in 2009, two new standards were created by 802.11. 802.11n added new features like multiple input multiple output, 40 MHz channel bandwidth and frame aggregation. Using these features 802.11n was able to increase the data rate capability to 600 Mbps (using four spatial stream in 40 MHz channel) from 54 Mbps provided by 802.11a and 802.11g. Also 802.11w created a standard to extended 802.11i to provide protection for selected management frames for example, Disassociation and Deauthentication frames and broadcast/multicast management frames.

10.4 IEEE 802.16 Working Group

IEEE 802.16 is another WG of the IEEE 802 LAN/MAN Standards Committee. Officially the WG starts efforts since 1999 and the first official WG meeting was held in May 1999. The IEEE 802.16 WG on Broadband Wireless Access Standards is responsible for developing standards and recommended practices to support the development and deployment of broadband *Wireless Metropolitan Area Networks (WirelessMAN™)*.

Although the 802.16 family of standards is officially called WirelessMAN in IEEE, it has been commercialized under the name of *"WiMAX" (Worldwide Interoperability for Microwave Access)* by the industry alliance called the WiMAX Forum. The mission of the Forum is to promote and certify compatibility and interoperability of broadband wireless products based on the IEEE 802.16 standards. Here we will focus on IEEE 802.16 WG only.

The first issue of the standard specifics a set of PHY and MAC layer standards was the IEEE Standard 802.16-2001 [1], completed in October 2001 and published on 8 April 2002, defined the WirelessMAN™ air interface specification for *wireless metropolitan area networks (WMANs)*. The intention behind the first release of the standard was to define a technology for *broadband wireless access (BWA)* for fixed users, as an alternative to cabled access networks, such as digital subscriber line (DSL) links. For this reason, the original IEEE 802.16 defines a point-to-multi-point (PMP) network architecture where resources are shared with control from a central node called *base station (BS)* to a set of *subscriber stations (SS)*. From its first release, the *medium access control (MAC)* layer was connection-oriented and supported quality of service (QoS). The standard, as approved in 2001, addresses frequencies from 10 to 66 GHz in *line-of-sight (LOS)* operations using single carrier transmission only. The standard employs QPSK, 16-QAM and 64-QAM as modulation schemes and the modulation scheme can be changed from frame to frame. Moreover, the standard was designed to evolve as a set of air interfaces based on a common MAC protocol, but with physical layer specifications dependent on the spectrum of use and the associated regulations.

In 2003, a new version of the standard, IEEE 802.16a-2003, was published with support for *non-LOS (N-LOS)* operations in frequencies from 2 to 11 GHz, extending the geographical reach of the network.

The subsequent milestone in standard development was IEEE 802.16-2004 [2] which introduced support for two additional physical layers: orthogonal frequency division multiplexing (OFDM) and orthogonal frequency division multiple access (OFDMA). In 2005, a new version of the standard was released to enable combined fixed and mobile operations in licensed bands. The aforementioned standard, IEEE 802.16e-2005 [3], was defined as an amendment to IEEE 802.16-2004 and added several features related to mobile operations and mobile stations (MSs), including power saving, idle mode, handover and an improved OFDMA physical layer. After the 2005 release, the standard development continued to define the *management information base (MIB)* for MAC and PHY (IEEE 802.16f) and the management plane and procedures (IEEE 802.16g), to improve the co-existence for license-exempt operation (IEEE 802.16h), to introduce relay capabilities (IEEE 802.16j), and to refine the MAC and PHY procedures for mobile operations (IEEE 802.16-2009). The latter is also known as the 2009 release, and brings the following major changes: half-duplex mobile terminal operations in OFDMA frequency division duplexing (FDD), *load balancing*, *robust header compression (ROHC)*, enhanced mechanism for resource allocation (for example, persistent allocation), support for location-based services (LBSs) and multicast and broadcast services (MBSs) [10.11]. The clean up for IEEE 802.16-2009 also involved incorporating the IEEE 802.16f and IEEE 802.16g amendments, and removing some stale features, such as the mesh mode.

Targeting to meet the requirement of *IMT-Advanced (IMT-A)* air interface, by the end of 2006, IEEE 802.16 working group (WG) has set up a new task group IEEE 802.16m (TGm). The purpose of IEEE 802.16m TG is to propose an advanced air interface which will include enhancements and extension to IEEE 802.16-2004 and IEEE 802.16e-2005. In addition to this, the new radio interface will be contributed to *ITU Radiocommunication Sector (ITU-R)* as a radio technology proposal for IMT-A. The development of 802.16m is still ongoing, but the RIT proposal submission based on 802.16m was submitted to ITU in 2009.

10.5 IEEE 802.11

10.5.1 Overview and Scope

Wireless Local Area Network (WLAN) technology based on the IEEE 802.11 standard has become very successful during the recent years. There is a huge deployed base of WLAN devices and the shipment rates have been constantly increasing. From a consumer point of view WLAN is becoming a basic connectivity feature even in low price category devices.

The IEEE 802.11 standard specifies the *physical layer (PHY)* and the data link layer (layers 1 and 2, respectively) of the *Open System Interconnection (OSI)* model protocol stack. In this chapter the physical layer (layer 1) features of the most relevant parts of the standard are described. Furthermore, due to the wide scope of the baseline 802.11 specification and the numerous amendments, the focus is mainly on the most recent and commonly available variants of the technology.

The original specification from year 1997 defines three different PHYs: *frequency-hopping spread spectrum (FHSS), direct sequence spread spectrum (DSSS)* and *infrared (IR)*. None of the mentioned PHYs are widely used today. The original specification has since been extended with a large number of amendments over time.

The following PHY related amendments of WLAN are discussed in this chapter:

- 802.11a: Orthogonal frequency division multiplexing (OFDM) PHY specification for the 5 GHz band
- 802.11b: High Rate direct sequence spread spectrum (HR/DSSS) PHY specification
- 802.11g: Extended Rate PHY specification
- 802.11n: High Throughput (HT) PHY specification

Additionally some discussion is included about selected standardization activities that are currently ongoing in the 802.11. Specifically there is discussion about the very high throughput (VHT) task groups TGac and TGad.

10.5.2 Frequency Plan

All the WLAN PHY variants of interest in the context of this discussion are either using the 2.4 GHz *ISM (industrial, scientific and medical)* or the 5 GHz *U-NII (Unlicensed National Information Infrastructure)* radio frequency bands. Both bands are unlicensed in most countries, and they are usually regulated by

Table 10.1 List of available WLAN channels in Europe and North America

Channel	Frequency [GHz]	Europe	North America
1	2.412	X	X
2	2.417	X	X
3	2.422	X	X
4	2.427	X	X
5	2.432	X	X
6	2.437	X	X
7	2.442	X	X
8	2.447	X	X
9	2.452	X	X
10	2.457	X	X
11	2.462	X	X
12	2.467	X	
13	2.472	X	

Table 10.2 Commonly available channels worldwide in 5 GHz band

Band	Channel	Frequency [GHz]
U-NII Lower 5.15–5.25 GHz	36	5.18
	40	5.20
	44	5.22
	48	5.24
U-NII Middle 5.25–5.35 GHz	52	5.26
	56	5.28
	60	5.30
	64	5.32

a national regulatory authority. For the most part regulation of these unlicensed bands is harmonized throughout the world.

In the 2.4 GHz ISM band there are several designated channels that are spaced 5 MHz apart. WLAN signals require wider channel separation, and therefore adjacent channels overlap and cause interference to each other. In the following Table 10.1 there is a list of available 2.4 GHz WLAN channels and indication of commonly used non-overlapping channels in Europe and in North America.

The regulation for the 5 GHz frequency band varies throughout the world. Generally there is a substantially larger amount of unlicensed band available than in the 2.4 GHz band. Typically the channel separation is 20 MHz, which corresponds better to the WLAN signal bandwidth. The list of available channels is country specific and from the user perspective also dependant on the supported channels by the WLAN equipment. In the following Table 10.2 there is a list of commonly available channels worldwide.

10.5.3 Reference Model

The reference model provides an architectural overview of the system, where different parts of the system and the interaction model between the entities can be seen. The two major parts of an 802.11 system are the physical layer and the *medium access control (MAC)* sublayer. They correspond mostly to the OSI reference model layering. Also there is a distinction between the data plane and the management plane. The reference model is illustrated in Figure 10.3.

Each protocol layer provides a *service access point (SAP)* towards upper layers in order to allow the upper layers to use its data plane services. In addition a SAP is defined towards a corresponding management entity for using the control plane services of each protocol layer.

In the data plane the lowest layer specified by 802.11 is the *physical medium dependent (PMD)* sublayer. The PMD sublayer is located directly above the wireless physical medium and it provides services to upper layers for transmitting data frames over the medium using spread spectrum or carrier modulation techniques. Furthermore, the PMD sublayer provides carrier sense indication for detecting activity on the radio channel. This information can be used by the MAC sublayer in order to assess whether the channel is currently idle or busy.

The upper part of the physical layer is the *physical layer convergence procedure (PLCP)* sublayer. It is using the service provided by the PMD sublayer. The main functionality of the PLCP sublayer is frame exchange between the PHY and MAC layers.

Each of the different PHYs defined in the baseline 802.11 and in the amendments is unique in terms of PMD and PLCP details. The overall reference model and SAP definitions still apply for all of them.

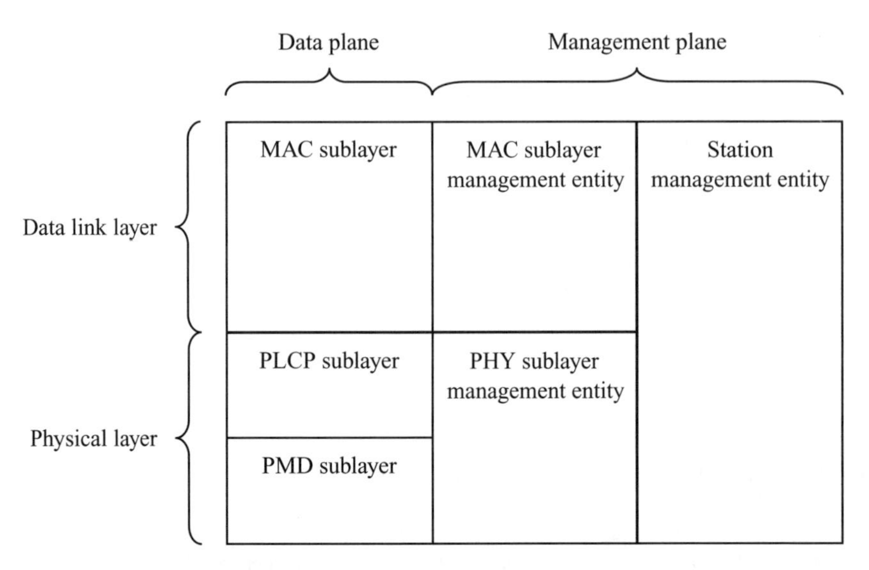

Figure 10.3 802.11 system reference model

10.5.4 Architecture

The most fundamental component of the WLAN architecture is a *station (STA)*, which is connected to the wireless medium. A STA consists of at least a PHY and a MAC. When there are multiple STAs that communicate between each other, they form a *basic service set (BSS)* together. If there is no *access point (AP)* present, the configuration is said to be an *independent BSS (IBSS)*. IBSS is also referred to as WLAN ad-hoc mode. In ad-hoc mode STAs communicate directly towards each other without a centralized forwarder device. When there is an AP present in addition to regular STAs, the configuration is commonly referred to as infrastructure BSS. Currently infrastructure mode is clearly the dominant

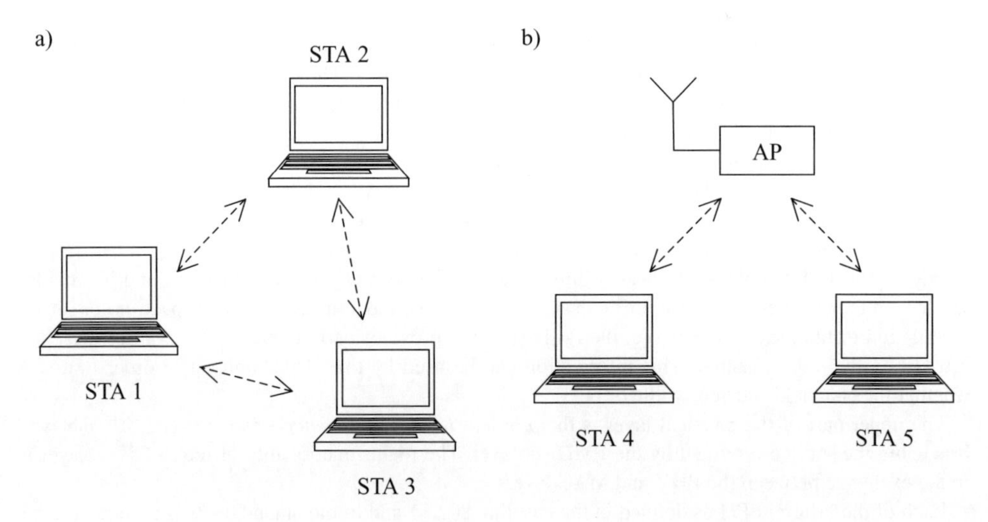

Figure 10.4 Illustration of a) ad-hoc mode and b) infrastructure mode

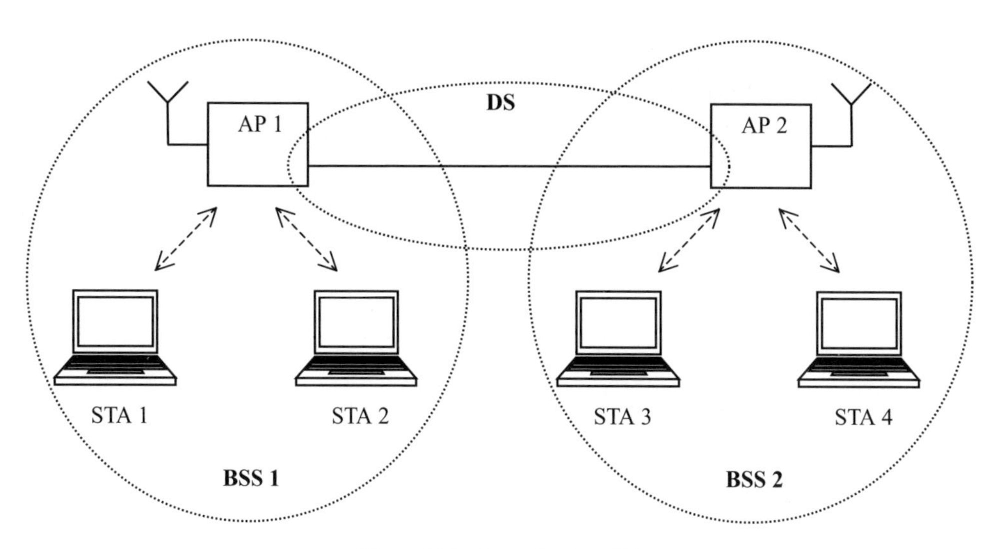

Figure 10.5 Example of an ESS

operational mode in deployed WLAN networks. Ad-hoc mode is rarely used, although many devices have support for it also. Example physical network topologies for both operational modes are shown in Figure 10.4.

Multiple BSSs may form an *extended service set (ESS)* by interconnecting with each other using a *distribution system (DS)*. The DS can for example be based on IEEE 802.3 wired technology. The 802.11 MAC SAP is designed to be compliant with 802.3 MAC SAP. Also the MAC address formats and address spaces between the technologies are compatible. Therefore it is possible to bridge MAC frames directly between these networks. Figure 10.5 depicts an example ESS configuration.

10.5.5 802.11a

The 802.11a amendment defines an OFDM PHY for the 5 GHz band. Originally the specification was released in 1999 and it was incorporated to the main 802.11 standard in the 2007 revision [8]. It allows data rates up to 54 Mbit/s. Today the 802.11a WLAN variant is used mainly in corporate environments, because consumer class devices with 5 GHz capabilities have become available much later than 2.4 GHz devices.

The 802.11a signal bandwidth is specified as 20 MHz. The OFDM uses 52 subcarriers with 0.3125 MHz spacing. At the target maximum bitrate the chosen number of subcarriers provides a practical compromise that allows a sufficient symbol guard interval while keeping the complexity and cost at a reasonable level. A guard interval is required between symbols to remove inter-symbol interference from the received signal. The required guard interval length depends on the delay spread of a typical usage environment for the technology.

Forty-eight of the total 52 subcarriers are used for data and four are used as pilot subcarriers. Every data subcarrier is modulated using binary or quadrature phase shift keying (BPSK or QPSK) or using 16- or 64-quadrature amplitude modulation (16-QAM or 64-QAM). The OFDM symbol time is 4 μs, which corresponds to the symbol rate of 250 000 symbol/s.

As forward error correction (FEC) mechanism 802.11a uses convolutional coding. The allowed coding rates are 1/2, 2/3, or 3/4. Coding is implemented across all subcarriers, which helps to even out the effects of narrowband interference or narrowband fading to the overall OFDM bit error ratio (BER). Decoding

Table 10.3 802.11a modulation and coding schemes

Modulation	Coding rate	Data rate [Mbit/s]
BPSK	1/2	6
BPSK	3/4	9
QPSK	1/2	12
QPSK	3/4	18
16-QAM	1/2	24
16-QAM	3/4	36
64-QAM	2/3	48
64-QAM	3/4	54

by the commonly used Viterbi algorithm is recommended. Table 10.3 lists all possible modulation and coding schemes (MCS) for 802.11a and their corresponding data rates.

The 802.11a *PLCP protocol data unit (PPDU)* consists of the PLCP preamble, the SIGNAL field and the data. The frame format is illustrated in Figure 10.6.

The OFDM physical protocol unit starts with a preamble, which consists of 12 OFDM symbols. The main purpose of the preamble is to synchronize the transmitter and receiver to have common timing. The first part of the preamble is 10 short symbols each having 0.8 μs duration. They are transmitted without a guard interval. This is referred to as the short training sequence. It is used for signal detection, gain control, possible antenna selection and coarse synchronization. Following the short training sequence there is a long training sequence consisting of two 3.2 μs symbols preceded by a guard interval of 1.6 μs. It is used for channel estimation and fine frequency acquisition.

After the preamble there is a SIGNAL field that is encoded within a single OFDM symbol using BPSK modulation and convolutional coding at 1/2 coding rate. SIGNAL includes the RATE and LENGTH fields. RATE defines the MCS used for the rest of the data unit. LENGTH is an integer value representing the number of octets in *PLCP service data unit (PSDU)*. Additionally SIGNAL includes a parity bit for error detection and tail bits that are set to zero. Like all other 802.11a OFDM symbols the duration of the SIGNAL is 4 μs including a 0.8 μs guard interval.

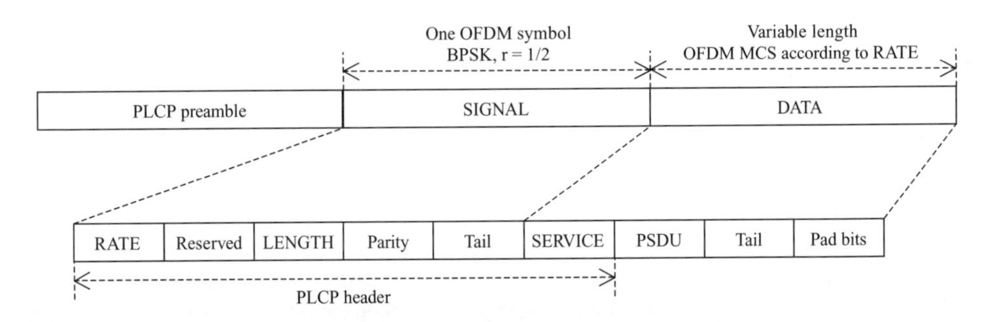

Figure 10.6 802.11a PPDU frame format

10.5.6 802.11b

The 802.11b amendment extends the DSSS PHY of the original standard with increased throughput up to 11 Mbit/s. Like the 802.11a amendment, it was originally released in 1999 and it was incorporated

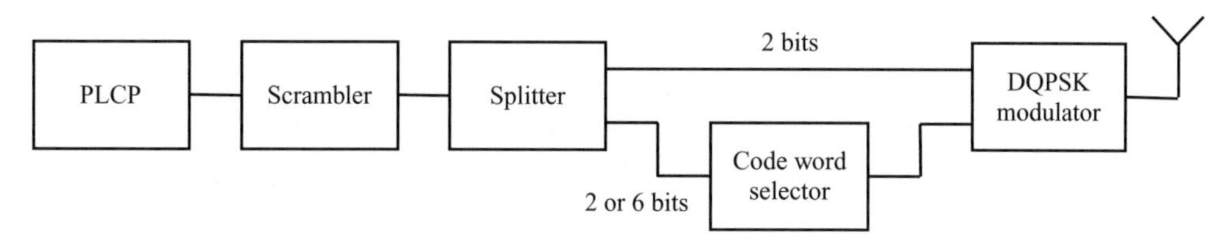

Figure 10.7 802.11b transmitter

to the main 802.11 standard in the 2007 revision [8]. It uses the unlicensed 2.4 GHz frequency band for communication. Since its release it has become one of the most common WLAN variants worldwide, and it is still supported by most WLAN capable devices today.

The DSSS PHY in the original 802.11 specification defines data rates of 1 and 2 Mbit/s. The information signal is spread to wider bandwidth using static repeating chip sequences at a much higher chip rate than the information bit rate. 11-bit code words are transmitted 1 million times per second. Each code word carries one or two information bits depending on the data rate.

Encoding more bits into a single 11-bit code word is no longer practical using regular phase shift keying techniques. Detecting finer phase shifts at the receiver especially in the presence of inter-symbol interference requires significantly more advanced and expensive technology. Therefore the higher data rates of 5.5 and 11 Mbit/s use an alternative encoding method. *Complementary code keying (CCK)* transmits chip stream in symbols of eight chips at the rate of 11 Mchip/s. Depending on the data rate either 4 or 8 bits are encoded to a single code word. Sophisticated transforms are used to derive the code words from the data.

With both 5.5 and 11 Mbit/s data rates the two first bits of a transmitted data block are used for determining the phase shift of the modulated code word. The rest of the bits in the data block select the codeword to be modulated. The total size of a data block is 4 or 8 bits corresponding to 5.5 and 11 Mbit/s data rates respectively. A simplified transmitter structure is shown in Figure 10.7.

All PPDU bits are scrambled before transmission by the DSSS PMD. The purpose of scrambling is to randomize long sequences of 0s or 1s within the data. The receiver can descramble the data after demodulation and de-spreading.

The High Rate DSSS PHY defines also a PPDU format specific to the PHY. The frame consists of a PCLP preamble, PLCP header and PSDU. Two different preambles and headers are defined. The mandatory long preamble and header interoperate with the original 802.11 DSSS PHY. The short preamble and header are added in the 802.11b amendment as optional and are not interoperable with the legacy DSSS PHY. The frame format is illustrated in Figure 10.8.

The first part of the preamble is the SYNC field, which is a string of scrambled 1s. The receiver uses it to detect the incoming signal and synchronize timing. Following SYNC comes the 16-bit *start of frame delimiter (SFD)*. SFD marks the start of the PPDU frame.

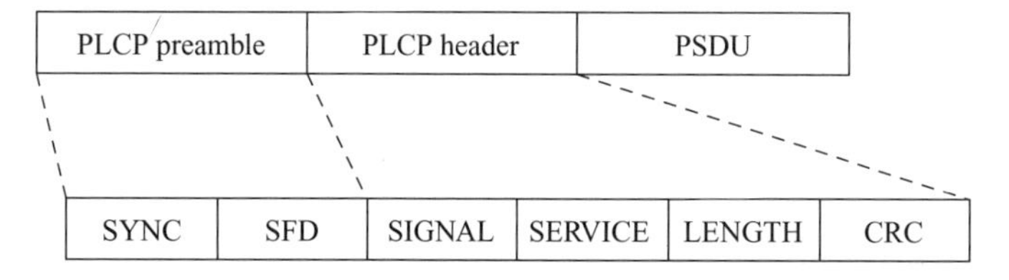

Figure 10.8 802.11b PPDU format

Table 10.4 Long and short PPDU comparison

	Long PPDU	Short PPDU
PLCP preamble modulation	1 Mbit/s DBPSK	1 Mbit/s DBPSK
PLCP preamble length	144 bits, 144 μs	72 bits, 72 μs
PLCP header modulation	1 Mbit/s DBPSK	2 Mbit/s DQPSK
PLCP header length	48 bits, 48 μs	48 bits, 24 μs
PSDU bitrates	1, 2, 5.5 and 11 Mbit/s	2, 5.5 and 11 Mbit/s

The PLCP header consists of the SIGNAL, SERVICE, LENGTH and CRC fields. The 8-bit SIGNAL field indicates the PHY data rate used for the PSDU. The integer value multiplied by 100 kbit/s equals the PSDU rate. The SERVICE field includes indication of the used modulation method and indication of the LENGTH field extension. The LENGTH field is a 16-bit integer value representing the duration of the PSDU in microseconds. CRC is used for error detection. If an error is detected, the receiving MAC can choose to reject the frame.

The short PPDU format provides an enhanced throughput compared to the long PPDU format, but since the two formats are not compatible, the short PPDU format required all devices in the network to support 802.11b. The main differences are listed in Table 10.4.

10.5.7 802.11g

The 802.11g amendment defines an *extended rate PHY (ERP)* for the DSSS PHY in the original specification and the high rate DSSS PHY in the 802.11b amendment. It supports data rates up to 54 Mbit/s in the 2.4 GHz unlicensed frequency range. The specification was originally released in 2003 and it was incorporated to the main 802.11 standard in the 2007 revision [8]. The standard has been very widely adopted all over the world.

An ERP system supports data rates specified in the DSSS and high rate DSSS PHY specifications, which makes it backwards compatible with the mentioned systems. Furthermore, the radio part implements all mandatory modes of 802.11a using the 2.4 GHz frequency band and channelization plan. Table 10.5 includes the data rates and modulation schemes that are used in different ERP modes.

An ERP system must support three different PPDU frame formats: the long and short DSSS frame formats specified in 802.11b and the OFDM frame format specified in 802.11a. Additionally there are

Table 10.5 ERP modes

Data rate	Modulation scheme
1 Mbit/s	DBPSK + DSSS
2 Mbit/s	DQPSK + DSSS
5.5 Mbit/s	DQPSK + DSSS/CCK
11 Mbit/s	DQPSK + DSSS/CCK
6 Mbit/s	OFDM
9 Mbit/s	OFDM
12 Mbit/s	OFDM
18 Mbit/s	OFDM
24 Mbit/s	OFDM
36 Mbit/s	OFDM
48 Mbit/s	OFDM
54 Mbit/s	OFDM

optional frame formats specified in 802.11g. The receiver distinguishes the PHY mode using the PLCP preamble and demodulates the rest of the PPDU accordingly.

The maximum throughput of 802.11g is comparable to 802.11a. However, if there are legacy DSSS mode devices present in the network, the aggregate throughput is decreased significantly. The reason is lower air time utilization efficiency DSSS mode due to the relatively long PLCP preamble and header duration.

10.5.8 802.11n

802.11n [10.9] is an amendment to the year 2007 revision of the original 802.11 specification. The target has been to significantly increase the network throughput of the preceding WLAN standards. The amendment specifies a new *high throughput (HT)* OFDM PHY in order to achieve the goal. The introduced enhancements in the PHY layer include multi-antenna MIMO techniques and 40 MHz channel bandwidth. In the MAC layer the new mechanisms include for example frame aggregation. The features of 802.11n are capable of supporting data rates up to 600 Mbit/s.

The standard was ratified by the IEEE in October 2009. However, there have already been products available for several years before ratification that have been implemented using a draft version of the standard. The Wi-Fi Alliance has been granting certifications for interoperability based on the draft standards since 2007. The certified devices will remain compatible with the final version of the standard.

An HT system may operate either in the 2.4 GHz or the 5 GHz frequency band. When operating in the 5 GHz range the HT system is backwards compatible with the 802.11a specification and similarly in the 2.4 range it is compliant with 802.11g. Support of both bands within a single device is optional.

Similarly to the previous OFDM PHYs, 802.11n OFDM subcarriers are modulated using BPSK, QPSK, 16-QAM or 64-QAM. The subcarrier spacing is identical to the previous OFDM PHYs, but the maximum allowed number of data subcarriers is increased from 48 to 52 in a 20 MHz channel.

The signal bandwidth may be either 20 MHz or 40 MHz. In practice the 40 MHz channel provides approximately double throughput compared to the 20 MHz channel. However, the availability of contiguous 40 MHz channels may be limited due to interference. The problem is prevalent especially in the 2.4 GHz range, which has a low number of non-overlapping channels and typically high amount of devices present. The support for 40 MHz channel bandwidth is optional for all devices.

802.11n supports different MIMO techniques such as *spatial division multiplexing (SDM)*, receive diversity and beamforming. Support for two spatial streams in 20 MHz channel is mandatory for AP devices. All other MIMO features are optional.

MIMO SDM in 802.11n is capable of supporting up to four spatial streams. Current implementations support typically at most two spatial streams. In practice the number is limited by implementation complexity and cost. SDM efficiency is also dependant on the spatial separation of the antennas. Achieving sufficient separation for significant performance increase may not be possible especially on small form factor devices. In the mandatory MCSs all streams use *equal modulation (EQM)*. Optionally *unequal modulation (UEQM)* for individual streams is also supported.

Other optional features that directly affect maximum achievable throughput by means of decreasing the physical protocol overhead are short *guard interval (GI)* and the greenfield frame format. The GI may be shortened from 800 ns to 400 ns with certain preconditions in environments that have sufficiently low delay spread. Typically the short GI is applied only after the highest MCS has been reached by rate adaptation. The greenfield frame format omits some unnecessary parts in the physical protocol headers in case there are no non-HT devices present in the network.

There are a large number of combinations of the previously described parameters, which are supported by the standard. Table 10.6 shows the relationships of MCS parameters that are related to data rate. The table includes only a subset of all MCSs including only EQM modes for simplicity. MCS indexes 0-7 are mandatory for non-AP devices and MCS indexes 0-15 for AP devices.

Table 10.6 Selected 802.11n MCSs

				Data rate [Mbit/s]			
				20 MHz channel		40 MHz channel	
MCS Index	Modulation	Coding rate	Spatial streams	800 ns GI	400 ns GI	800 ns GI	400 ns GI
0	BPSK	1/2	1	6.5	7.2	13.5	15
1	QPSK	1/2	1	13.0	14.4	27	30
2	QPSK	3/4	1	19.5	21.7	40.5	45
3	16-QAM	1/2	1	26.0	28.9	54	60
4	16-QAM	3/4	1	39.0	43.3	81	90
5	64-QAM	2/3	1	52.0	57.8	108	120
6	64-QAM	3/4	1	58.5	65.0	121.5	135
7	64-QAM	5/6	1	65.0	72.2	135	150
8	BPSK	1/2	2	13.0	14.4	27	30
9	QPSK	1/2	2	26.0	28.9	54	60
10	QPSK	3/4	2	39.0	43.3	81	90
11	16-QAM	1/2	2	52.0	57.8	108	120
12	16-QAM	3/4	2	78.0	86.7	162	180
13	64-QAM	2/3	2	104.0	115.6	216	240
14	64-QAM	3/4	2	117.0	130.0	243	270
15	64-QAM	5/6	2	130.0	144.4	270	300
...	...	...	...	...	...	...	...
31	64-QAM	5/6	4	260.0	288.9	540.0	600.0
...	...	...	...	...	...	...	...

In order to transmit multiple spatial streams the original data stream is first divided into multiple independent streams. Each spatial stream is then transmitted using a discrete transmit chain including analog and RF parts for all antennas. A corresponding number of antennas are required at the receiver in order to recover all the spatial streams. MIMO SDM can potentially increase data rate as a function of the number of spatial streams. However, the actual gain from spatial multiplexing is inversely proportional to the amount of spatial correlation between the streams. In addition to spatial separation, multipath signal propagation can reduce spatial correlation.

In a system that has more receive antennas than the number of spatial streams, a technique known as receive diversity may be utilized to improve robustness. The signal from different receive antennas may be combined in a way that substantially increases SNR. The method is referred as maximal ratio combining (MRC).

One optional feature of 802.11n is the usage of space-time block code (STBC). It is a coding technique that is used in order to improve reliability of data transfer by exploiting antenna diversity of a MIMO system. In STBC multiple encoded copies of the same data are transmitted from different antennas. The copies are combined at the receiver with an increased probability to correctly decode the original data. The STBC encoder divides the spatial streams further into space-time streams. If STBC is used in 802.11n, the number of space-time streams is always one or two larger than the number of spatial streams. The maximum number of space-time streams is four.

Low density parity check (LDPC) code is an advanced coding method that has been introduced into 802.11n as an optional feature. Convolutional coding is still the only mandatory coding method. LDPC is a very efficient code and it can provide coding gain of a few decibels over convolutional code. Therefore it improves robustness of an 802.11n system, while maintaining the decoding complexity at a relatively low level.

Transmit beamforming (TxBF) is an optional capability that is provided in the 802.11n standard. In a MIMO system TxBF can be achieved by means of applying weights to the different transmit signals in such a way that reception is improved. It is possible to utilize TxBF for sending data and MRC for receiving data in a single multi-antenna device. An AP is a typical example of a device that benefits from such configuration. TxBF can provide relatively good performance improvements, but implies also increased hardware complexity.

The TxBF method defined in 802.11n is based on knowledge of the propagation environment. To obtain the knowledge the channel needs to be sounded by transmitting a sounding packet between the two devices performing the TxBF. The transmit weighting matrix is calculated using the channel information. Commonly the *singular value decomposition (SVD)* method is used to solve the matrix. For simplicity, the details of SVD are not included in this discussion. If the communication channel is assumed to be reciprocal, implicit channel information feedback can be used. Implicit feedback means that channel sounding is done in the opposite direction as the TxBF. Explicit feedback on the other hand means that channel sounding and TxBF are done in the same direction. The explicit feedback method does not require reciprocal communication conditions, but there is overhead of sending the channel sounding results back to the transmitter. The TxBF SVD needs to be calculated and applied in digital baseband to each individual OFDM subcarrier before the RF and analog parts of the transmit chain.

During the 802.11n development one requirement has been to provide some level of backwards compatibility with legacy 802.11a and 802.11g equipment. For this purpose the following three different PPDU frame formats have been defined.

- Non-HT format
- HT-mixed format
- HT-greenfield format

Support for the non-HT format and the HT-mixed format is mandatory, but the HT-greenfield format is optional. Packets of the non-HT format are legacy packets structured according to either 802.11a or 802.11g specification, depending on the used frequency band. In the HT-mixed format the PPDU contains both a legacy preamble and an HT preamble. Although 802.11a/g devices are only able to decode the legacy part of the PPDU, they will obtain information about the PPDU duration, which is utilized for the MAC protocol. The HT-mixed format therefore enables coexistence of 802.11n devices and 802.11a/g devices. The HT-greenfield format is intended for use in environments that contain exclusively 802.11n devices. Because it contains only the HT preamble, it provides the highest throughput of the different PPDU formats, but does not support optimal coexistence with 802.11a/g devices.

10.5.9 Future Developments

Currently there are ongoing standardization activities in the 802.11 to further increase the throughput from 802.11n data rates. Originally the work has started in the *very high throughput (VHT)* study group even before the final version of 802.11n standard was ratified. Initially many different ideas were proposed to the group as the basis for the VHT work. Eventually two separate task groups emerged from the VHT study group. Each one of them is concentrating on different solutions. The 802.11ac task group (TGac) is developing a VHT WLAN standard for operation below 6 GHz frequency bands. The 802.11ad task group (TGad) is working on a 60 GHz VHT WLAN standard.

The target aggregate throughput for multiple STAs in TGac is at least 1 Gbit/s measured at the MAC SAP interface. For single STA the target is 500 Mbit/s. There have been various PHY proposals for achieving the target. Some potential PHY techniques include using 80 MHz contiguous or even 160 MHz non-contiguous RF carrier, using higher order OFDM MCSs with 256-QAM and increasing the maximum number of spatial streams to eight in multi-user MIMO transmission.

TGad is targeting for even higher data rates. For a single user the minimum target is 1 Gbit/s measured at the MAC SAP. There is a large amount of unlicensed spectrum available worldwide in the 60 GHz frequency band (typically between 57-66 GHz), which usually includes multiple 2 GHz RF carriers. The wide spectrum allows very high throughput, but the millimeter wave propagation has also some strict limitations. Path loss in 60 GHz is of a higher magnitude than in 2.4 or 5 GHz. To overcome the high path loss, directional antennas can be used both in transmitter and receiver side. Because the wavelength is relatively short, the antennas are also substantially smaller than with traditional WLAN devices. Consequently multi-element antenna arrays with a large number of antennas become a feasible solution. With directional antennas the multipath propagation is reduced, which makes MIMO SDM techniques less effective. Both single carrier and multi-carrier modulation methods are considered for 60 GHz. Physical layer coding is likely to build on 802.11n mechanisms. The options have different benefits and the selection will be based on requirements derived from usage models.

10.6 IEEE 802.16x

In this section, we will focus on the standards with mobility support, in particular, IEEE 802.16e and IEEE 802.16m were selected.

10.6.1 Key PHY Features of the IEEE 802.16e

In the following we give the key PHY features of mobile WiMAX technology (PHY and MAC are taken from IEEE 802.16e) and provide short descriptions.

Scalable OFDMA

OFDMA is the multiple access technique for mobile WiMAX. OFDMA is the Orthogonal Frequency Division Multiplexing (OFDM) based multiple access scheme and has become the de-facto single choice for modern broadband wireless technologies adopted in other competing technologies such as 3GPP's Long Term Evolution (LTE) DL. OFDMA demonstrates superior performance in nonline-of-sight (N-LOS) multi-path channels with its relatively simple transceiver structures and allows efficient use of the available spectrum resources by time and frequency subchannelization.The simple transceiver structure of OFDMA also enables feasible implementation of advanced antenna techniques such as MIMO with reasonable complexity. Lastly, OFDMA employed in mobile WiMAX is scalable in the sense that by flexibly adjusting FFT sizes and channel bandwidths with fixed symbol duration and subcarrier spacing, it can address various spectrum needs in different regional regulations in a cost-competitive manner.

TDD

The mobile WiMAX Release 1 Profile has only TDD as the duplex mode even though the baseline IEEE Standards contains both TDD and Frequency Division Duplex (FDD). Even though future WiMAX Releases will have FDD mode as well, TDD is in many ways better positioned for mobile Internet services than FDD.

First of all, Internet traffic is asymmetric typically with the amount of downlink traffic exceeding the amount of uplink traffic; thus, conventional FDD with the same downlink and uplink channel bandwidth does not provide the optimum use of resources. With TDD products, operators are capable of adjusting downlink and uplink ratios based on their service needs in the networks.

Advanced Antenna Techniques (MIMO and BF)

Various advanced antenna techniques have been implemented in the mobile WiMAX Release 1 profile to enable higher cell and user throughputs and improve coverage. As a matter of fact, mobile WiMAX was the first commercially available cellular technology that actually realized the benefits of MIMO techniques promised by academia for years. With its downlink and uplink MIMO features, both operators and end-users enjoy up to twice the data rates of Single-Input Single-Output (SISO) rate, resulting in up to 37 Mbps for downlink and 10 Mbps for uplink sector throughput using just 10 MHz TDD channel bandwidth.

Mobile WiMAX also enhances the cell coverage with its inherent *beamforming (BF)* techniques. Coupled with TDD operation, its powerful BF mechanism allows base stations to accurately form a channel matching beam to a terminal station so that uplink and downlink signals can reach reliably from and to terminals at the cell edge, thus effectively extending the cell range.

Full Mobility Support

Full mobility support is one of the main strength of the mobile WiMAX products. The baseline standard of mobile WiMAX was designed to support vehicles at highway speed with appropriate pilot design and Hybrid Automatic Repeat Request (HARQ), which helps to mitigate the effect of fast channel and interference fluctuation. The systems can detect the mobile speed and automatically switch between different types of resource blocks, called subchannels, to optimally support the mobile user. Furthermore, HARQ helps to overcome the error of link adaptation in fast fading channels and to improve overall performance with its combined gain and time diversity.

Frequency Reuse One and Flexible Frequency Reuse

From the operators' perspective, securing greater frequency spectrum for their services is always costly. Naturally it is in their best interest if a technology allows decent performance in the highly interference-limited conditions with frequency reuse one. Mobile WiMAX technology was designed to meet this goal in a respectable way with its cell-specific subchannelization, low rate coding and power boosting and de-boosting features. It also enables real-time application of flexible frequency reuse where frequency reuse one is applied to terminals close to the cell centre whereas a fraction of frequency is used for terminals at the cell edge, thereby reducing heavy co-channel interference.

The PHY transmission chains of OFDMA are illustrated in Figure 10.9. The blocks are the same with the small difference that OFDMA PHY includes a repetition block. The modulation is one of the four digital modulations described in the previous chapter: BPSK, QPSK, 16-QAM or 64-QAM. The modulated symbols are then transmitted on the OFDM orthogonal subcarriers.

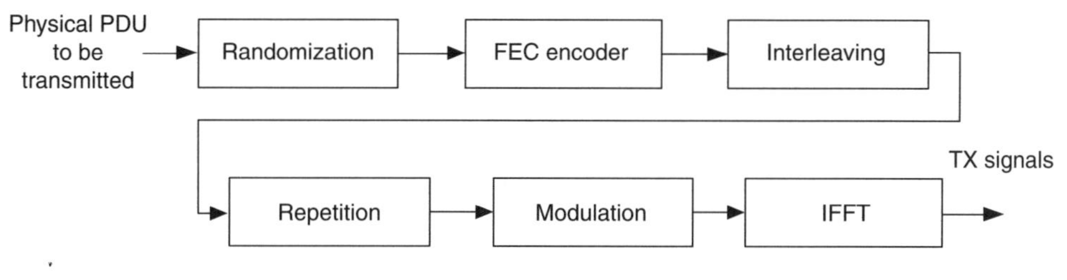

Figure 10.9 OFDMA PHY Chain

10.6.2 IEEE 802.16m

Targeted at IMT-Advanced (IMT-A), IEEE 802.16 working group (WG) had set up a new task group IEEE 802.16m by the end of 2006. The purpose of IEEE 802.16m TG is to propose an advanced air interface which will include enhancements and extension to IEEE 802.16-2004 [1] and IEEE 802.16e-2005 [2]. In addition to this, the new radio interface has been contributed to ITU-R as a radio technology proposal for IMT-A. The IEEE announced on 6th October 2009 that they had submitted a candidate radio interface technology for IMT-Advanced standardization in the Radio Communication Sector of the International Telecommunication Union (ITU-R). The proposal [4] documented that IEEE 802.16m met ITU-R's challenging and stringent requirements in all four IMT-Advanced "environments": Indoor, Microcellular, Urban, and High Speed.

In this section we consider the layer 1 (L1) processing and procedures defined in 802.16m. In both the downlink and uplink direction, OFDMA is employed as a multiple access scheme. 802.16m supports both time division duplex (TDD) and frequency division duplex (FDD) modes, including half-FDD (H-FDD) MS operation. Considering WirelessmMAN-OFDM system,[1] the legacy WiMAX system which is built on top of 802.16e, it is very possible that TDD based system will be deployed first. Therefore, in this chapter we will focus on TDD features only.

10.6.2.1 Frame Structure

10.6.2.1.1 Basic Frame Structure

The OFDM parameters defined in 802.16m are specified in Table 10.7.

The 802.16m basic frame structure is illustrated in Figure 10.10. Each 20 ms superframe is divided into four equally-sized 5 ms radio frames. The first subframe of each superframe contains the *super frame header (SFH)*. The SFH is divided into *primary SFH (P-SFH)* and *secondary SFH (S-SFH)*. The P-SFH has a fixed size and it is transmitted in every superframe, whereas the S-SFH has a variable size. *Advanced Preamble (A-Preamble)* is transmitted at the beginning of every radio frame.

When using the same OFDMA parameters as in Table 10.7 with the channel bandwidth of 5 MHz, 10 MHz, or 20 MHz, each 5 ms radio frame further consists of eight subframes for CP ratio G = 1/8 and 1/16. With the channel bandwidth of 8.75 and 7 MHz, each 5 ms radio frame further consists of seven and six subframes, respectively, for G = 1/8 and 1/16. In the case of G = 1/4, the number of subframes per frame is one less than that of other CP lengths for each bandwidth case. A subframe shall be assigned for either DL or UL transmission. There are four types of subframes:

1. Type-1 subframe consists of six OFDMA symbols,
2. Type-2 subframe consists of seven OFDMA symbols,
3. Type-3 subframe which consists of five OFDMA symbols, and
4. Type-4 subframe which consists of nine OFDMA symbols. This type shall be applied only to UL subframe for the 8.75MHz channel bandwidth when supporting the legacy 802.16e frames.

The basic frame structure is applied to both FDD and TDD duplexing schemes, including H-FDD MS operation. The number of switching points in each radio frame in TDD systems is two, where a switching point is defined as a change of directionality, that is, from DL to UL or from UL to DL.

In a TDD frame with DL to UL ratio of D:U, the 1st contiguous D subframes and the remaining U subframes are assigned for DL and UL, respectively, where $D + U = 8$ for 5, 10 and 20 MHz channel bandwidths, $D + U = 7$ for 8.75 MHz channel bandwidth, and $D + U = 5$ for 7 MHz channel bandwidth. The ratio of $D{:}U$ shall be selected from one of the following values: 8:0, 6:2, 5:3, 4:4, or 3:5 for 5, 10

[1] As in IEEE 802.16m Specification, the legacy WiMAX system is called WirelessMAN-OFDMA system.

Table 10.7 OFDMA parameters

		Nominal Channel Bandwidth (MHz)	5	7	8.75	10	20
		Over-sampling Factor	28/25	8/7	8/7	28/25	28/25
		Sampling Frequency (MHz)	5.6	8	10	11.2	22.4
		FFT Size	512	1024	1024	1024	2048
		Sub-Carrier Spacing (kHz)	10.94	7.81	9.77	10.94	10.94
		Useful Symbol Time T_u (μs)	91.4	128	102.4	91.4	91.4
Cyclic Prefix (CP) $T_g=1/8\ T_u$		Symbol Time T_s (μs)	102.857	144	115.2	102.857	102.857
	FDD	Number of OFDM symbols per Frame	48	34	43	48	48
		Idle time (μs)	62.857	104	46.40	62.857	62.857
	TDD	Number of OFDM symbols per Frame	47	33	42	47	47
		TTG + RTG (μs)	165.714	248	161.6	165.714	165.714
Cyclic Prefix (CP) $T_g=1/16\ T_u$		Symbol Time T_s (μs)	97.143	136	108.8	97.143	97.143
	FDD	Number of OFDM symbols per Frame	51	36	45	51	51
		Idle time (μs)	45.71	104	104	45.71	45.71
	TDD	Number of OFDM symbols per Frame	50	35	44	50	50
		TTG + RTG (μs)	142.853	240	212.8	142.853	142.853
Cyclic Prefix (CP) $T_g=1/4\ T_u$		Symbol Time T_s (μs)	114.286	160	128	114.286	114.286
	FDD	Number of OFDM symbols per Frame	43	31	39	43	43
		Idle time (μs)	85.694	40	8	85.694	85.694
	TDD	Number of OFDM symbols per Frame	42	30	38	42	42
		TTG + RTG (μs)	199.98	200	136	199.98	199.98

and 20 MHz channel bandwidths, and 3:2 or 2:3 for 7 MHz channel bandwidth and 5:2, 4:3, or 3:4 for 8.75 MHz channel bandwidth. In each frame, the *Transmission Transaction Guard (TTG)* and *Receiving Transmission Guard (RTG)* shall be inserted between the DL and UL switching points.

Figure 10.11 illustrates one TDD frame structure example ($D{:}U = 5{:}3$) with the channel bandwidth 5, 10 and 20 MHz where each radio frame is composed of eight subframes for G = 1/8. Each superframe starts with a preamble used for physical synchronization. There are two types of preambles: *Primary*

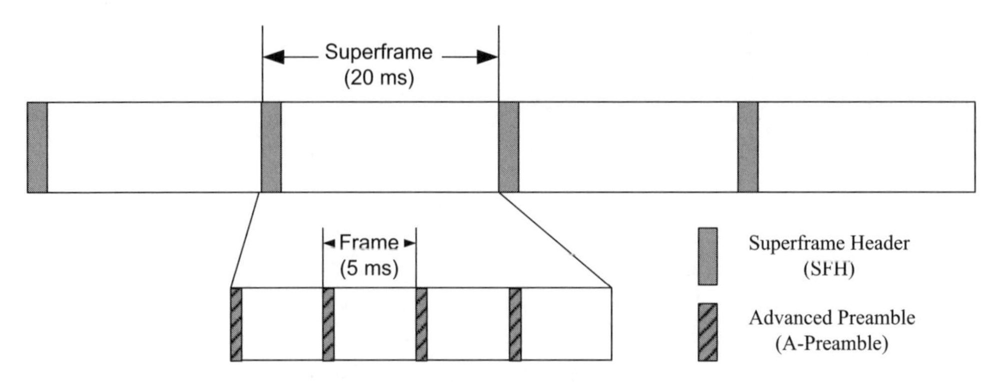

Figure 10.10 Basic Frame Structure

A-Preamble (PA-Preamble) and *Secondary A-Preamble (SA-Preamble)*. A PA-Preamble is transmitted on the first subframe of the second radio frame, while a SA-Preamble is transmitted at the beginning of all the other radio frames.

10.6.2.1.2 TDD Frame Structure for Supporting Legacy 802.16e MSs

According to [3], IEEE 802.16m shall provide continuing support and interoperability for legacy WirelessMAN-OFDMA equipments which are based on 802.16e, including MSs and BSs. Specifically, the features, functions and protocols enabled in IEEE 802.16m shall support the features, functions and protocols employed by WirelessMAN-OFDMA legacy equipment. The following are backward compatibility requirements:

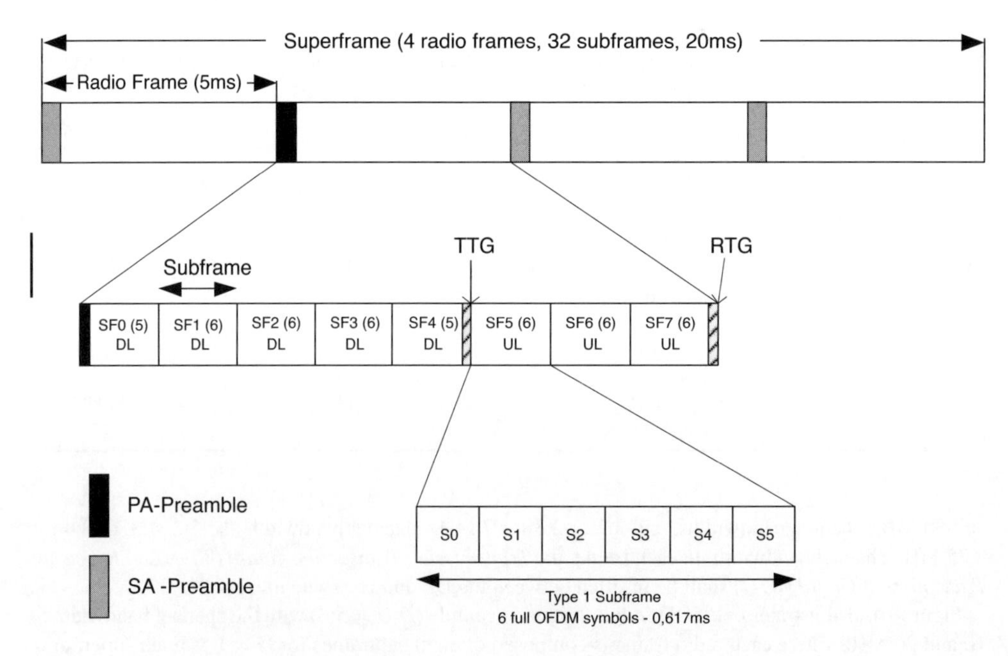

Figure 10.11 Frame structure example for 5/10/20 MHz Mode (G=1/8)

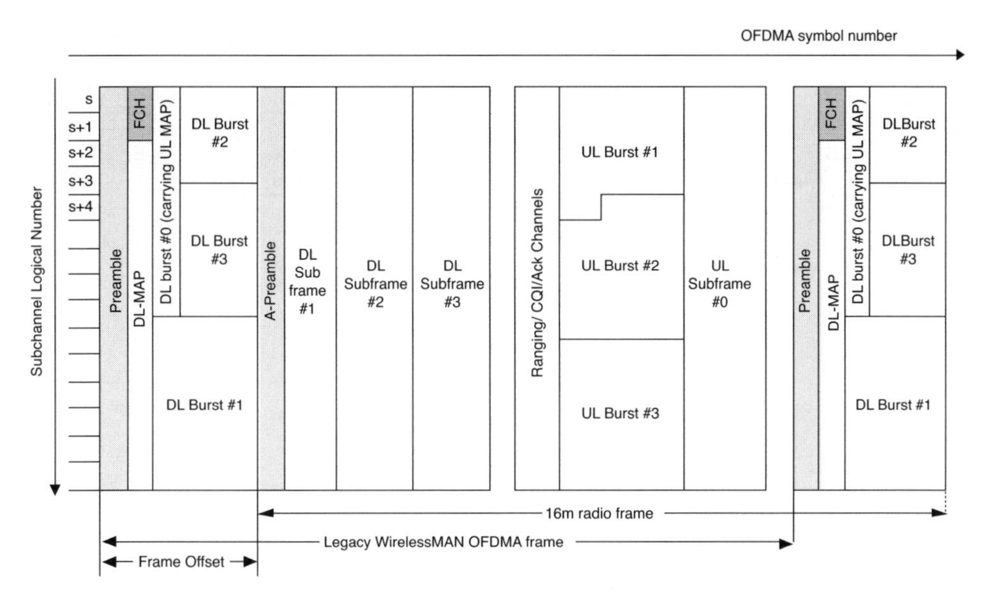

Figure 10.12 TDD frame configuration for supporting the WirelessMAN-OFDMA operation with UL TDM

- An IEEE 802.16m MS shall be able to operate with WirelessMAN-OFDMA BS, at a level of performance equivalent to that of a WirelessMAN-OFDMA MS.
- Systems based on IEEE 802.16m and the WirelessMAN-OFDMA System shall be able to operate on the same RF carrier, with the same channel bandwidth; and should be able to operate on the same RF carrier with different channel bandwidths.
- An IEEE 802.16m BS shall support a mix of IEEE 802.16m and WirelessMAN-OFDMA MSs when both are operating on the same RF carrier. The system performance with such a mix should improve with the fraction of IEEE 802.16m MSs attached to the BS.
- An IEEE 802.16m BS shall support handover of a WirelessMAN-OFDMA MS to and from a WirelessMAN-OFDMA BS and to and from IEEE 802.16m BS, at a level of performance equivalent to handover between two WirelessMAN-OFDMA BSs.
- An IEEE 802.16m BS shall be able to support a WirelessMAN-OFDMA MS while also supporting IEEE 802.16m MSs on the same RF carrier, at a level of performance equivalent to that a WirelessMAN-OFDMA BS provides to a WirelessMAN-OFDMA MS.

In order to support WirelessMAN-OFDMA MSs, 802.16m introduces a special frame structure which is reported in Figures 10.12 and 10.13 for bandwidths of 5, 10, and 20 MHz. In DL direction, legacy WirelessMAN-OFDMA and a new system are multiplexed in a time division fashion and the beginning of legacy and new frames are offset by a fixed number of subframes. This means that the legacy DL subframe is located at the end of the 802.16m radio frame and the corresponding legacy UL subframe is located in the subsequent 16m radio frame. In the UL direction, legacy and new system can multiplexed either in time division (Figure 10.12) or frequency division mode (Figure 10.13).

In UL TDM mode, a subset of UL subframes is dedicated to the WirelessMAN-OFDMA operation to enable one or more WirelessMAN-OFDMA UL time zones. The subset includes the 1st WirelessMAN-OFDMA UL time zone to support the transmission of the ranging channel, feedback channel and ACK channel. Data bursts from the WirelessMAN-OFDMA MSs shall not be transmitted in the UL subframes

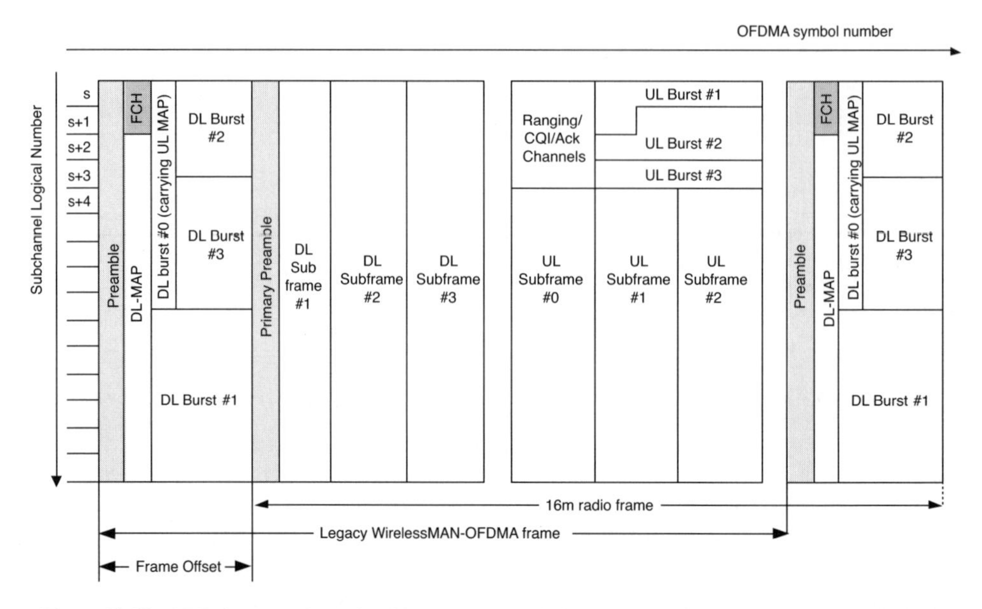

Figure 10.13 TDD frame configuration for supporting the WirelessMAN-OFDMA operation with UL FDM

for operation of 802.16m Air Interface. Those UL subframes shall be indicated as a UL time zone in the UL-MAP message.

While in FDM mode, a group of subcarriers (subchannels), spanning the entire UL transmission, is dedicated to the WirelessMAN-OFDMA operation. The remaining subcarriers, denoted the UL subframes, are dedicated to 802.16m operation. Figure 10.13 illustrates an example frame configuration for supporting the WirelessMAN-OFDMA operation when FDM mode is used in UL. Data bursts from the WirelessMAN-OFDMA MSs shall not be transmitted in the UL subchannels group for operation of 16m. The UL subchannels group for operation of the WirelessMAN-OFDMA shall be indicated by the UL allocated subchannels bitmap TLV or the UL AMC Allocated physical bands bitmap TLV defined in the UCD message.

From the legacy WirelessMAN-OFDMA MSs point of view, the 802.16m DL (UL) subframes are just as a separate DL(UL) zone. This enables legacy MS to correctly detect the beginning and the end of each legacy DL and UL subframe.

10.6.2.1.3 Frame Structure Supporting Wider Bandwidth

One of the main new features introduced in 16m is multiple RF carriers support at the same time. When the multi-carrier feature is supported, the system may define and utilize additional RF carriers to improve the user experience and QoS or provide services through additional RF carriers configured or optimized for specific services. In the case of multi-carrier, a common MAC entity controls a PHY spanning over multiple frequency channels as shown in Figure 10.14. The RF channels can be of the same or different bandwidths (for example, 5, 10 and 20 MHz), be on contiguous or non-contiguous frequency bands. The RF channels may be of different duplexing modes, for example, FDD, TDD, or a mix of bidirectional and broadcast only carriers.

In general, the same frame structure is used for each carrier in multi-carrier mode operation. Figure 10.15 illustrates the example of the frame structure to support multi-carrier operation.

These additional RF carriers are the secondary carriers. Each MS is controlled through an RF carrier which is the primary carrier. Hence, from the MS point of view, a carrier can be either primary or

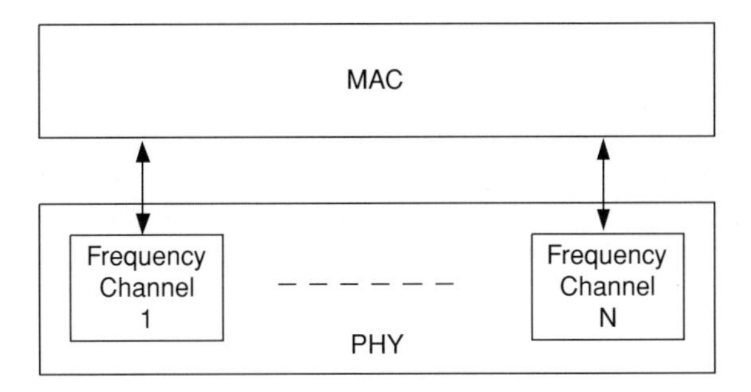

Figure 10.14 Multicarrier support protocol structure

secondary carrier. A MS has one and only one primary carrier which is used to exchange data and control information with the BS. A MS may use another carrier(s) to exchange data with the BS, but the control information is normally transmitted through the primary carrier. Figure 10.15 shows the frame structure for multi-carrier operations with legacy support. In Figure 10.15, RF carrier 2 has to support legacy operations, whereas RF carrier 0 and 1 work in 16m mode only.

IEEE 802.16m supports relay operation as well. More specifically, the two hop data transmission between BS and MS using an intermediate *relay station (RS)*.

When a RS is deployed it uses the same OFDMA signal parameters as its serving BS. The BS frame is divided into 16m Access zone and 16m Relay zone. The 16m Access zone position precedes the 16m Relay zone position inside the frame. The duration of the 16m Access zone and 16m Relay zone may be different in DL and UL directions. The zone configuration of 16m Access zone and 16m Relay zone is informed to the ARS by the 802.16m base station.

The BS frame 16m Access zone consists of 16m DL Access zone and 16m UL Access zone, and 16m Relay zone consists of 16m DL Relay zone and 16m UL Relay zone. The BS frame 16m Access zone is

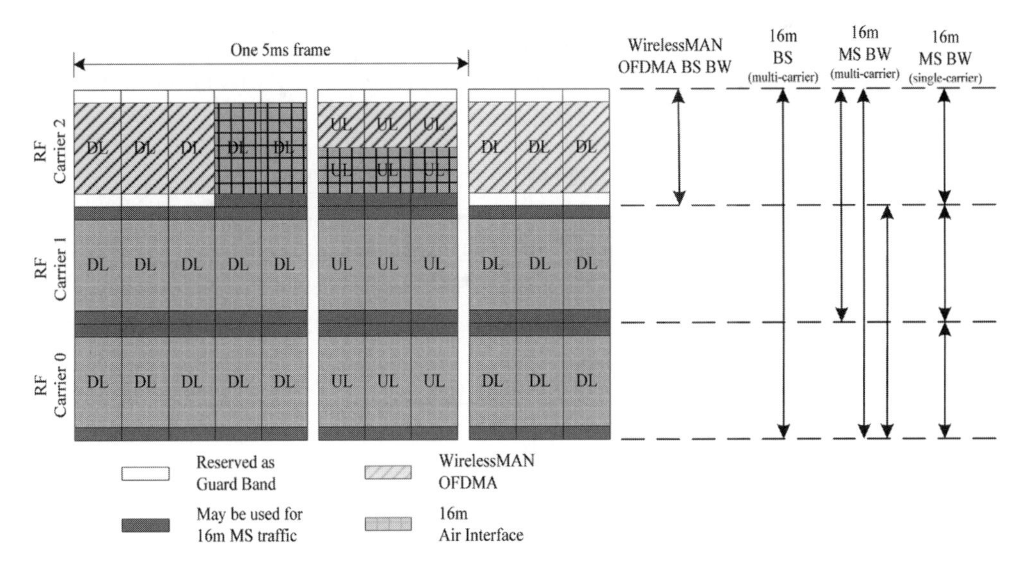

Figure 10.15 Example of frame structure to support multi-carrier operation in WirelessMAN-OFDMA support mode

used for communication with MSs only. The BS frame 16m Relay zone is used for communication with RSs and may be used for communication with MSs. In the 16m DL Relay zone the BS transmits to its subordinate RS and in the 16m UL Relay zone the BS receives transmissions from its subordinate RS.

The RS frame 16m Access zone consists of 16m DL Access zone and 16m UL Access zone and 16m Relay zone consists of 16m DL Relay zone and 16m UL Relay zone. The RS frame 16m Access Zone is used for communication with MSs only. In the 16m DL Relay zone the RS receives transmissions from its superordinate BS and in the 16m UL Relay zone the RS transmits to its superordinate BS.

10.6.2.2 Control Structure (PA-Preamble, Control Channels)

Control channels are used to convey information essential for system operation. Control information is transmitted hierarchically over different time scales from the superframe level to the subframe level. In this section, we will address Advanced Preamble (A-Preamble, that is, synchronization channel), DL control channel and UL control channel.

It should be pointed out that there is no need for 16m MS to decode WirelessMAN-OFDM signaling for network access.

10.6.2.2.1 DL Control Structure

10.6.2.2.1.1 Advanced Preamble There are two types of Advanced Preamble (A-Preamble): primary advanced preamble (PA-Preamble) and secondary advanced preamble (SA-Preamble) defined in 802.16m. One PA-Preamble symbol and three SA-Preamble symbols are transmitted within one superframe. The location of the A-Preamble symbol is specified as the first symbol of each frame. PA-Preamble is located at the first symbol of second frame in a superframe while SA-Preamble is located at the first symbol of the remaining three frames.

Figure 10.16 depicts the location of A-Preamble symbols.

10.6.2.2.1.2 Primary Advanced Preamble (PA-Preamble) PA-Preamble is transmitted within fixed bandwidth and the length of PA-Preamble sequence is 216. PA-Preamble carries the information of

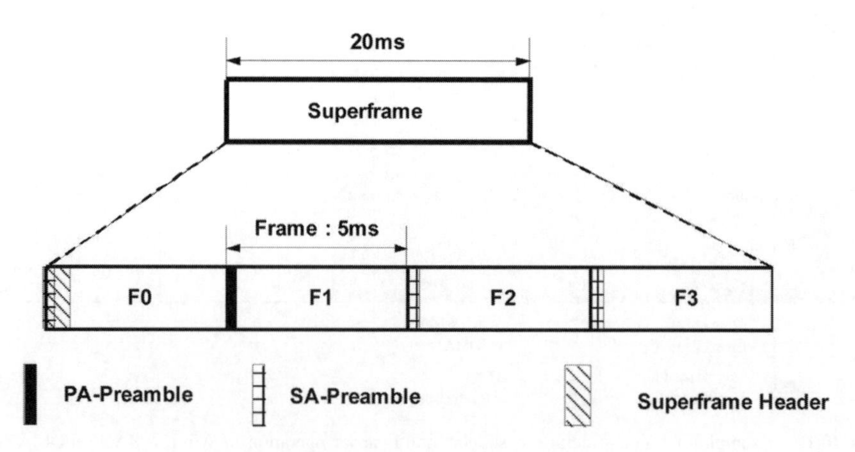

Figure 10.16 The location of A-Preamble symbol

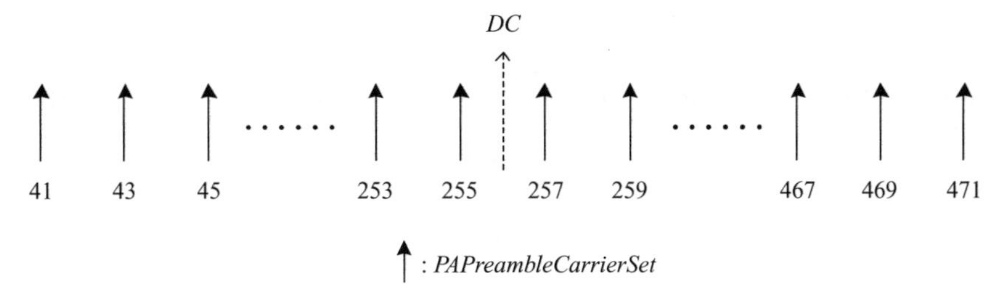

Figure 10.17 Subcarriers occupied by PA-Preamble

system bandwidth and system multi-carrier configuration information. When the subcarrier index 256 is reserved for DC, the allocation of subcarriers is accomplished by the following equation:

$$PAPreambleCarrierSet = 2 \cdot k + 41$$

where
PAPreambleCarrierSet specifies all subcarriers allocated to the PA-Preamble, and
k is a running index from 0 to 215.

Figure 10.17 depicts the symbol structure of PA-Preamble in the frequency domain.

Among the PA-Preamble sequences, 10 sequences are used for fully-configured RF carriers and one PA-Preamble sequence is reserved for partially-configured carriers. Also PA-Preamble carriers system bandwidth information which can be used by MS to learn about the BW of SA-Preamble which is the same as system BW.

10.6.2.2.1.2.1 Secondary Advanced Preamble (SA-Preamble) The lengths of SA-Preamble sequences, N_{SAP}, are 144, 288, and 576 for 512-FFT, 1024-FFT, and 2048-FFT, respectively. The allocation of subcarriers is accomplished by the following equation, when the subcarrier indexes 256, 512, and 1024 are reserved for DC for 512-FFT, 1024-FFT, and 2024-FFT, respectively:

$$SAPreambleCarrierSet_n = n + 3 \cdot k + 40 \cdot \frac{N_{SAP}}{144} + \left\lfloor \frac{2 \cdot k}{N_{SAP}} \right\rfloor \tag{10.1}$$

where
SAPreambleCarrierSet$_n$ specifies all subcarriers allocated to the specific SA-Preamble,
n is the index of the SA-Preamble carrier-set 0, 1 and 2 representing segment ID,
k is a running index 0 to N_{SAP} -1 for each FFT sizes

Each segment uses an SA-Preamble composed of a carrier-set out of the three available carrier-sets in the following manner:

- Segment 0 uses SA-Preamble carrier-set 0.
- Segment 1 uses SA-Preamble carrier-set 1.
- Segment 2 uses SA-Preamble carrier-set 2.

Each cell ID has an integer value *IDcell* from 0 to 767. The IDcell is defined by segment index and an index per segment as follows:

$$IDcell = 256 \cdot n + Idx$$

where
n is the index of the SA-Preamble carrier-set 0, 1 and 2 representing segment ID,
Idx is a running index 0 to 255.

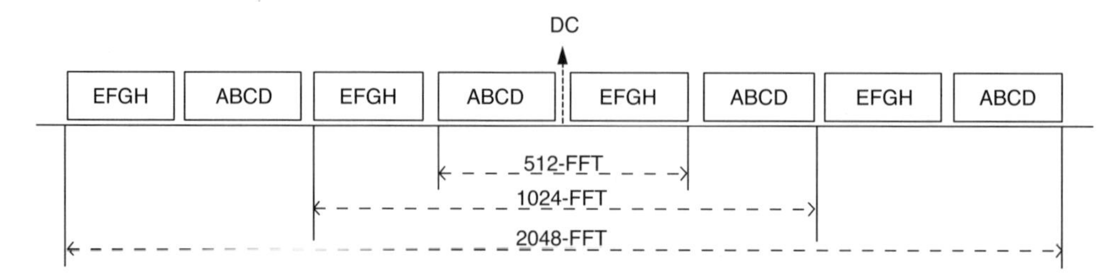

Figure 10.18 Allocation of sequence sub-blocks for each FFT

SA-Preamble sequences are partitioned and each partition is dedicated to a specific base station type like macrocell BS, Macro Hotzone BS, Femto BS. The base station types are categorized into macro BS and non-macro BS cells by hard partition with 255 sequences (85 sequences per segment * 3 segments) dedicated for macro BS. The non-macro BS information is broadcast in a hierarchical structure, which is composed of S-SFH Sub-Packet 3 and AAI_SCD message.

For the 512-FFT size, the 288-bit SA-Preamble sequence is divided into eight main blocks, namely, A, B, C, D, E, F, G, and H. The length of each block is 36 bits. Each segment ID has different sequence blocks. For the 512-FFT size, A, B, C, D, E, F, G, and H are modulated and mapped sequentially in ascending order onto the SA-Preamble subcarrier-set corresponding to segment ID as shown in Figure 10.5. For higher FFT sizes, the basic blocks (A, B, C, D, E, F, G, H) are repeated in the same order. For instance in the 1024-FFT size, E, F, G, H, A, B, C, D, E, F, G, H, A, B, C, D are modulated and mapped sequentially in ascending order onto the SA-Preamble subcarrier-set corresponding to segment ID.

A circular shift is applied over three consecutive sub-carriers after applying subcarrier mapping. Each subblock has common offset. The circular shift pattern for each subblock is:

[2,1,0. . . . . ., 2,1,0, ..., 2,1,0, 2,1,0, DC, 1,0,2, 1,0,2, ..., 1,0,2, ...1,0,2] where the shift is right circular.

10.6.2.2.1.3 Superframe Header (SFH) Superframe Header (SFH) is transmitted in the first subframe in every superframe. It carries essential system parameters and system configuration information. The SFH is divided into two parts: Primary Superframe Header (P-SFH) and Secondary Superframe Header (S-SFH). P-SFH is transmitted in every superframe while it is not necessary to transmit S-SFH in every superframe.

The physical processing block diagram for P-SFH and S-SFH is shown in Figure 10.19. It should be pointed out that the diagram is applicable to both P-SFH and S-SFH.

TBCC channel coding is used for both P-SFH and S-SFH. The encoded bit sequence is modulated using QPSK. Two-stream SFBC with two Tx antennas is used for P-SFH and S-SFH transmission. For more than 2-Tx antenna configurations, the P-SFH and S-SFH are transmitted using 2-stream SFBC with precoding, which is decoded by the *Mobile Station (MS)* without any information on the precoding and antenna configuration.

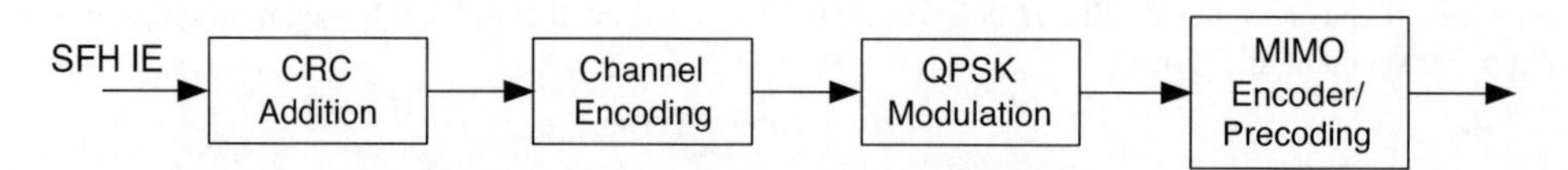

Figure 10.19 Physical processing block diagram for SFH

The P-SFH information element (IEs) contains essential system information and it is mapped to the P-SFH. The format of the P-SFH *information element (IE)* is: LSB of super frame number, information related to S-SFH including: S-SFH change count and change bitmap, S-SFH size extension, MCS indication of S-SFH, S-SFH scheduling information.

The S-SFH IE is mapped to the S-SFH. Essential system parameters and system configuration information belonging to the S-SFH are categorized into three S-SFH subpacket IEs such as SP1, SP2 and SP3. These SPs are transmitted in different timing and periodicity. The periodicity of SP (T_{SP}) is determined with $T_{SP1} < T_{SP2} < T_{SP3}$. Each S-SFH subpacket IE is of a fixed size. S-SFH SP1 IE contains information for network re-entry. S-SFH SP2 IE contains information for initial network entry and network discovery. S-SFH SP3 IE contains the rest of system parameters.

10.6.2.2.1.4 Initial Network Entry In this part, we can have a quick look at the initial network entry procedure since the procedure is mainly related to DL control information. The procedure for initialization of an AMS shall be divided into the following steps:

(a) Scan for DL channel (mainly Primary and Secondary Advanced Preamble) and establish DL PHY synchronization with the BS.

On initialization or after signal loss, the MS shall acquire the DL PHY synchronization by Primary and Secondary A-Preamble.

The MS shall have nonvolatile storage in which the last operational parameters are stored and, when the MS needs to acquire DL PHY synchronization, it may at first try synchronization using the stored DL channel information. But if the trial fails, the MS shall begin to scan the possible channels of the DL frequency band of operation until it finds a valid DL signal.

(b) Obtain DL/UL parameters (from P-SFH/S-SFH IEs and so on) and establish DL MAC synchronization.

For initial network entry, once the MS has achieved DL PHY synchronization with a BS, the MS shall decode the P-SFH and S-SFH in order to obtain the necessary system information containing the DL and UL parameters for the initial network entry. Based on the network information such as the NSP list, the MS shall decide whether to continue the network entry process with this BS or to scan for other BSs. If _bar= this cell does not allow access of new MS and the MS shall select a different cell to restart network entry procedure.

If the MS succeeds to decode the essential system information, the MS is DL MAC synchronized with the BS.

(c) Perform ranging and automatic adjustment.

Ranging is the process of acquiring the correct timing offset, frequency offset and power adjustments so that the MS's transmissions are aligned with the BS, and they are received within the appropriate reception thresholds.

After DL synchronization, the MS shall attempt to perform initial ranging with the BS. If the ranging procedure is successfully completed, the MS is UL synchronized with the BS and obtains Temporary Station ID (TSTID) from the BS.

(d) Negotiate basic capability.

Right after completion of ranging, the MS informs the BS of its basic capabilities by transmitting station basic capability request message.

Upon receipt of the station basic capability request message from the MS, the BS determines whether it could allow or could support the requested feature set or MAC and/or PHY protocol revisions. The BS will send back the basic capability response message to indicate the common supported features.

(e) Perform MS authorization and key exchange.
If authorization support is enabled in basic capability negotiation, the BS and the MS shall perform authorization and key exchange and key agreement procedure. If authorization support is disabled, the step of MS authorization and key exchange shall be skipped.
(f) Perform registration, and setup default service flows.
Registration is the process by which the MS is allowed to enter into the network. After authorization and key exchange are finished, the MS informs the BS of its capabilities and requests the registration for entry into the network.

The example procedure for initialization of an MS is shown in Figure 10.20. This figure shows the overall flow between the stages of initialization in an MS. This figure does not include error paths and is shown simply to provide an overview of the process.

10.6.2.2.1.5 Advanced MAP (A-MAP) The *Advanced MAP (A-MAP)* carries unicast service control information. Unicast service control information consists of user-specific control information and

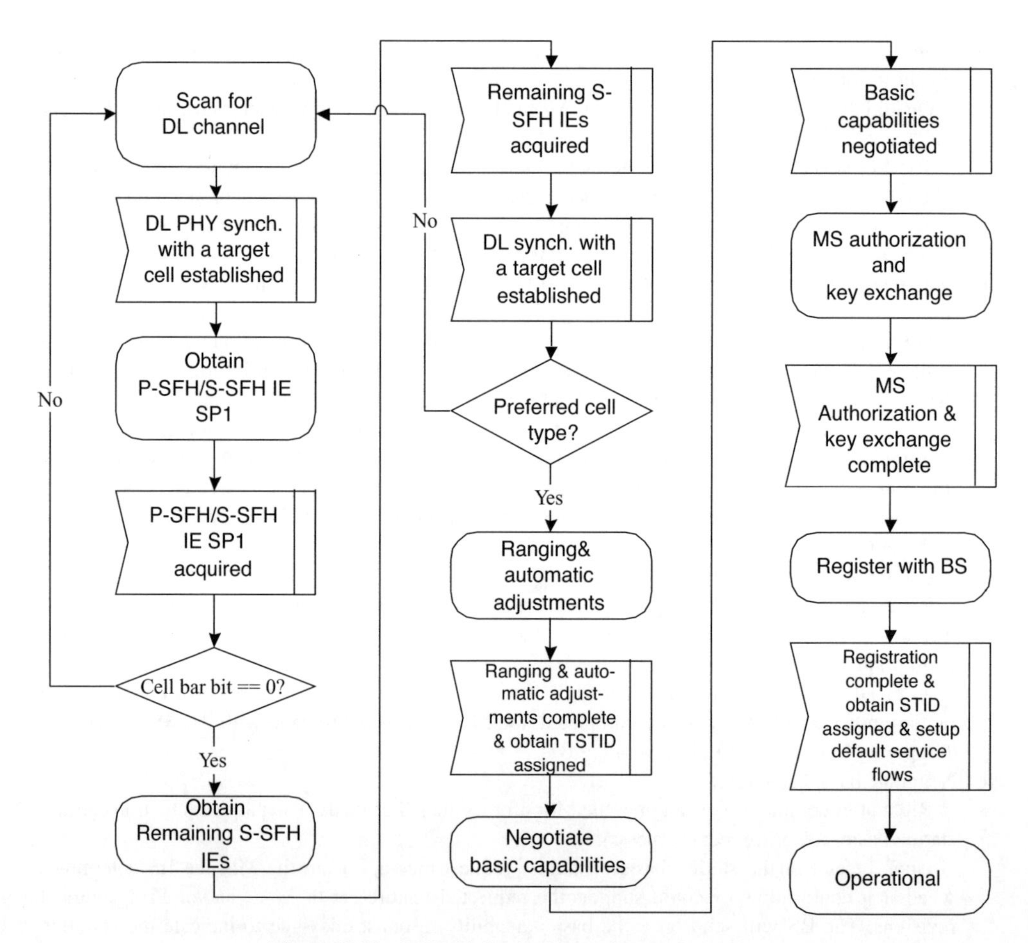

Figure 10.20 MS Initialization Network Entry Procedure

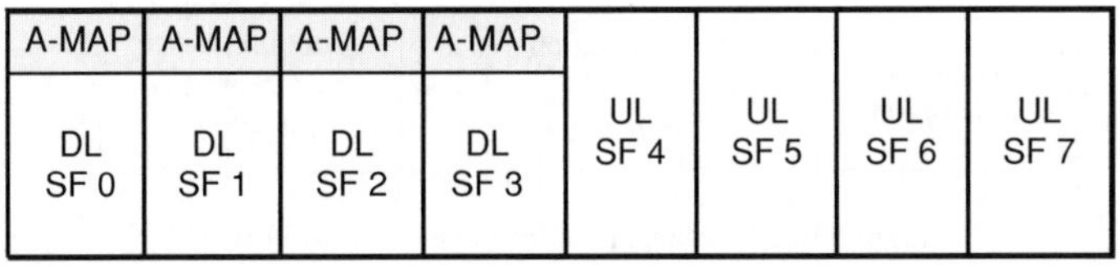

A-MAP	A-MAP	A-MAP	A-MAP	UL SF 4	UL SF 5	UL SF 6	UL SF 7
DL SF 0	DL SF 1	DL SF 2	DL SF 3				

Figure 10.21 Example of locations of A-MAP region in a TDD system (DL:UL ratio=4:4)

non-user-specific control information. User-specific control information is further divided into assignment information, HARQ feedback information, and power control information, and they are transmitted in the assignment A-MAP, HARQ feedback A-MAP, and power control A-MAP respectively.

All the A-MAPs share a region of physical resources called A-MAP region. A-MAP regions shall be present in all DL unicast subframes. When default TTI is used, DL data allocations corresponding to an A-MAP region occupy resources in the *Advanced Air Interface (AAI)* subframe where the A-MAP region is located.

A-MAP consists of both user-specific control information and non-user-specific control information. User-specific control information is further divided into assignment information, HARQ feedback information, and power control information, and they are transmitted in the assignment A-MAP, HARQ feedback A-MAP, and power control A-MAP, respectively. All the A-MAPs share a region of physical resources called A-MAP region. An example of A-MAP region location in TDD with 4:4 subframe DL:UL split is illustrated in Figure 10.21.

In the DL subframes where the A-MAP regions can be allocated, each frequency partition may contain an A-MAP region. The A-MAP region occupies the first few distributed LRUs in a frequency partition. The structure of an A-MAP region is illustrated as an example in Figure 10.22. The resources occupied by

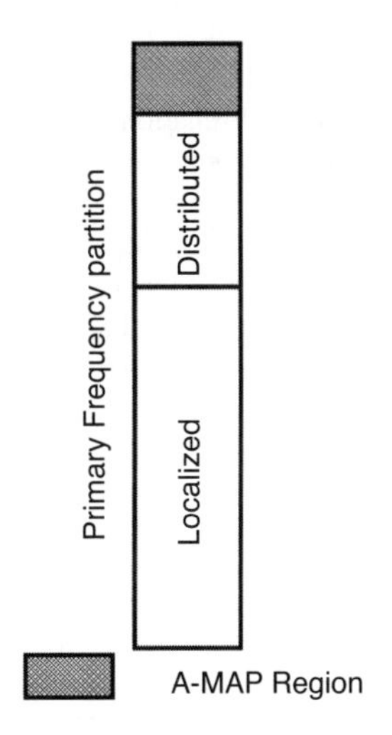

Figure 10.22 Example location of an A-MAP region

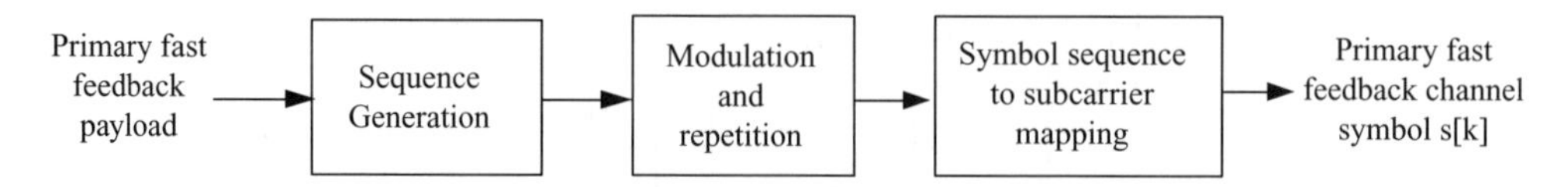

Figure 10.23 Mapping of information in the PFBCH

each A-MAP physical channel may vary depending on the system configuration and scheduler operation. The A-MAP region consists of a number of distributed LRUs.

10.6.2.3 UL Control Channels

The UL control channels carry multiple types of control information in order to support different air interface procedures. Information contained in the control channels is classified into:

1. Channel quality feedback
2. MIMO feedback
3. HARQ feedback (ACK/NACK)
4. Uplink synchronization signals
5. Bandwidth request
6. E-MBS feedback
7. Frequency partition information

10.6.2.3.1 Fast Feedback Control Channel

The UL fast feedback channel carries channel quality feedback and MIMO feedback and BW RE-Qindicators. There are two types of UL fast feedback control channels: *primary fast feedback channel (PFBCH)* and *secondary fast feedback channels (SFBCH)*. The UL fast feedback channel starts at a pre-determined location, with the size defined in a DL broadcast control message.

The UL PFBCH carries 4 to 6 bits of information, providing wideband and narrowband channel quality feedback and MIMO feedback. It is used to support robust feedback reports. The UL SFBCH carries narrowband CQI and MIMO feedback information. The number of information bits carried in the SFBCH ranges from 7 to 24. The process of composing the PFBCH and SFBCH are illustrated in Figure 10.23 and Figure 10.24.

The DRUs occupied by fast feedback channel are permuted by UL tile permutation to form distributed LRUs for both data and control resource/channel. A UL *feedback mini-tile (FMT)* is defined as two contiguous subcarriers by six OFDM symbols. The UL feedback control channels are formed by applying the UL mini-tile permutation to the LRUs allocated to the control resource. The fast feedback channels are comprised of three RFMTs.

The primary fast feedback channel is comprised of three RFMTs.

The secondary fast feedback channel has the same physical control channel structure as the primary fast feedback channel. The secondary fast feedback channels are comprised of three RFMTs.

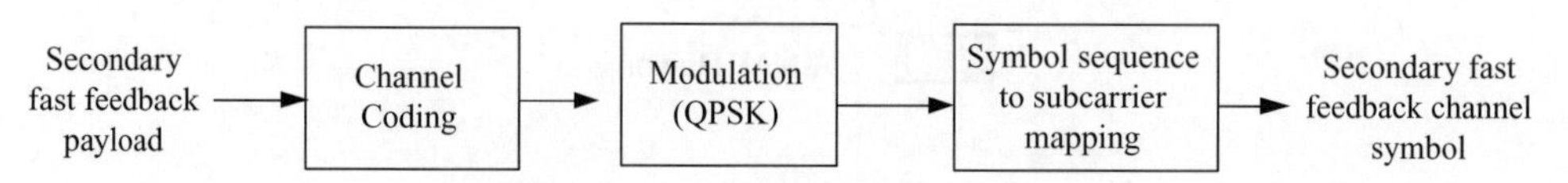

Figure 10.24 Mapping of information in the SFBCH

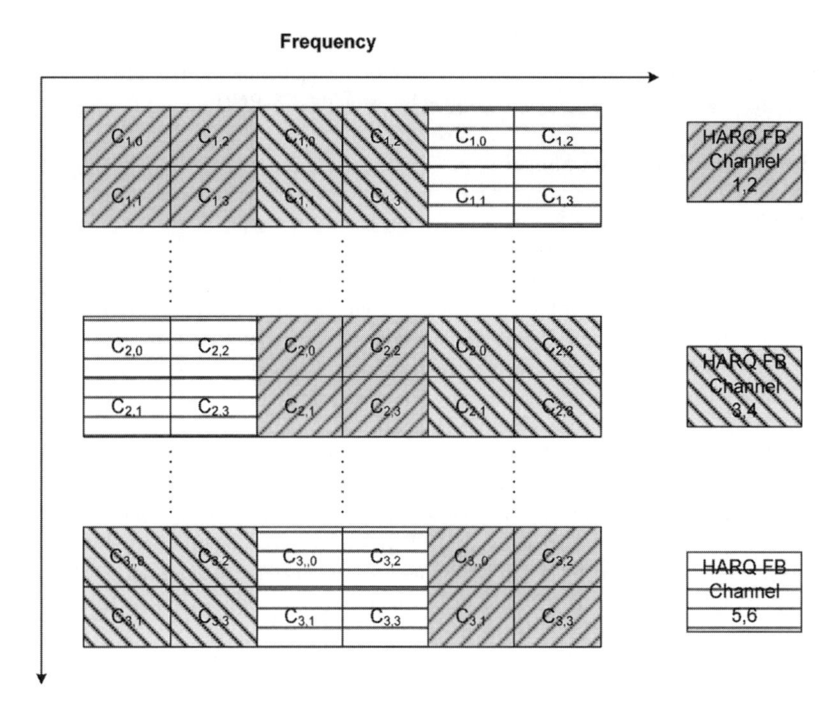

Figure 10.25 2 × 2 HMT Structure

10.6.2.3.2 HARQ Feedback Control Channel

HARQ feedback (ACK/NACK) is used to acknowledge DL transmissions. Multiple codewords in MIMO transmission can be acknowledged in a single ACK/NACK transmission.

Each UL HARQ feedback resource consists of three RFMTs. A total resource of three distributed 2×6 RFMTs supports six UL HARQ feedback channels. The 2×6 RFMTs are further divided into UL HARQ mini-tiles (HMT). A UL HARQ mini-tile has a structure of two subcarriers by two OFDM symbols as illustrated in Figure 10.25.

10.6.2.3.3 Sounding Channel

Uplink channel sounding provides the means for the BS to determine UL channel response for the purpose of UL closed-loop MIMO transmission and UL scheduling. In TDD systems, the BS can also use the estimated UL channel response to perform DL closed-loop transmission to improve system throughput, coverage and link reliability. In this case BS can translate the measured UL channel response to an estimated DL channel response when the transmitter and receiver hardware of BS and AMS are appropriately calibrated.

The sounding channel occupies a single OFDMA symbol in the UL sub-frame, which is located in the beginning of the UL sub-frame. Each UL sub-frame can only contain one sounding channel. Multiple sub-frames in a 5-ms radio frame can be used for sounding. The number of subcarriers for the sounding in a *Physical Resource Units (PRU)* is 18 adjacent subcarriers.

10.6.2.3.4 UL Ranging Channel

The UL *ranging channel (RCH)* is used for UL synchronization. The UL ranging channel can be further classified into *non-synchronized ranging channel (NS-RCH)* and *synchronized ranging channel (S-RCH)* for non-synchronized and synchronized MS. The ranging channel for synchronized MSs is used for periodic ranging and the ranging channel for non-synchronized MSs is used for initial ranging. An MS cannot transmit any other uplink burst or uplink control channel signal in the advanced air-interface (AAI) subframe where it transmits a ranging signal by using the NS-RCH.

10.6.2.3.5 Bandwidth Request Channel

Bandwidth request information is transmitted using contention based random access on this control channel. The bandwidth request (BR) channel contains resources for the AMS to send a BR preamble and an optional quick access message.

In the LZone with PUSC, a BW REQ tile is defined as four contiguous subcarriers by six OFDM symbols. The number of BW REQ tiles per BW REQ channel is three. Each BW REQ tile carries a BW REQ preamble only. In the Mzone, a BW REQ tile is defined as six contiguous subcarriers by six OFDM symbols. Each BW REQ channel consists of three distributed BW-REQ tiles. Each BW REQ tile carries a BW REQ preamble and a quick access message. The AMS may transmit the access sequence only and leave the resources for the quick access message unused.

Each BR channel shall comprise of three distributed BR tiles for frequency diversity. A BR tile in the M-Zone is defined as six contiguous subcarriers by six OFDM symbols. The BWREQ tile is made up of two parts – a preamble portion and a data portion. The preamble portion transmits the BR preamble on a resource that spans four subcarriers by six OFDM symbols. The data portion of the BWREQ tile spans two contiguous subcarriers by six OFDM symbols and transmits the quick access message for the three-step BR. For the three-step BR, 16 bits of BW request information is constructed from 12 bits of STID and 4 bits of pre-defined BR information.

10.6.2.4 DL PHY Structure

Each downlink subframe consists of Physical Resource Units (PRUs). The size of one PRU is equal to P_{sc} subcarriers by N_{sym} OFDMA symbols. P_{sc} is equal to 18 and N_{sym} is equal to the size of the subframe, that is, to five, six or seven OFDMA symbols.

PRUs are mapped to Logical Resource Units (LRU_{FP}s). This mapping is done in several steps (see Figure 10.26):

1. Mapping of the PRUs to the subband and miniband PRUs (PRU_{sb} and PRU_{mb});
2. Permutation of the PRU_{mb}s. Permutated PRU_{mb}s are denoted by $PPRU_{mb}$s;
3. Mapping of the PRU_{sb}s and $PPRU_{mb}$s to the Frequency Partitions (PRU_{FP}s);
4. Mapping of the PRU_{FP}s to the Contiguous and Distributed Resource Units (CRU_{FP}s and DRU_{FP}s);
5. Mapping of the CRU_{FP}s to the contiguous LRU_{FP}s ($CLRU_{FP}$s);

Mapping of the DRU_{FP}s to the distributed LRU_{FP}s ($DLRU_{FP}$s) by subcarrier permutation.

10.6.2.4.1 Mapping of the PRUs to the Subband and Miniband PRUs

One subband comprises $N1=4$ PRUs; one miniband comprises $N2=1$ PRU. A number of the allocated subbands, denoted by K_{sb}, are defined by *Downlink Subband Allocation Count (DSAC)* transmitted in the SFH. The maximum number of subbands depends on the FFT size (see Table 10.8). The minimum number of subbands is equal to 0.

PRUs, which have not been allocated to subbands, are allocated to minibands. PRU_{SB} denotes PRU allocated to subband, PRU_{MB} denotes PRU allocated to miniband. There are $L_{SB}=N1 \cdot K_{sb}$ PRU_{SB}s and $L_{MB}=N_{pru}- L_{SB}$ PRU_{MB}s. PRUs are permuted by the *outer permutation*. The choice of the outer permutation scheme depends on the FFT size and the actual number of the allocated subbands. Figure 10.27 shows outer permutation for the case when the FFT size is equal to 512 and the number of subbands is equal to 4.

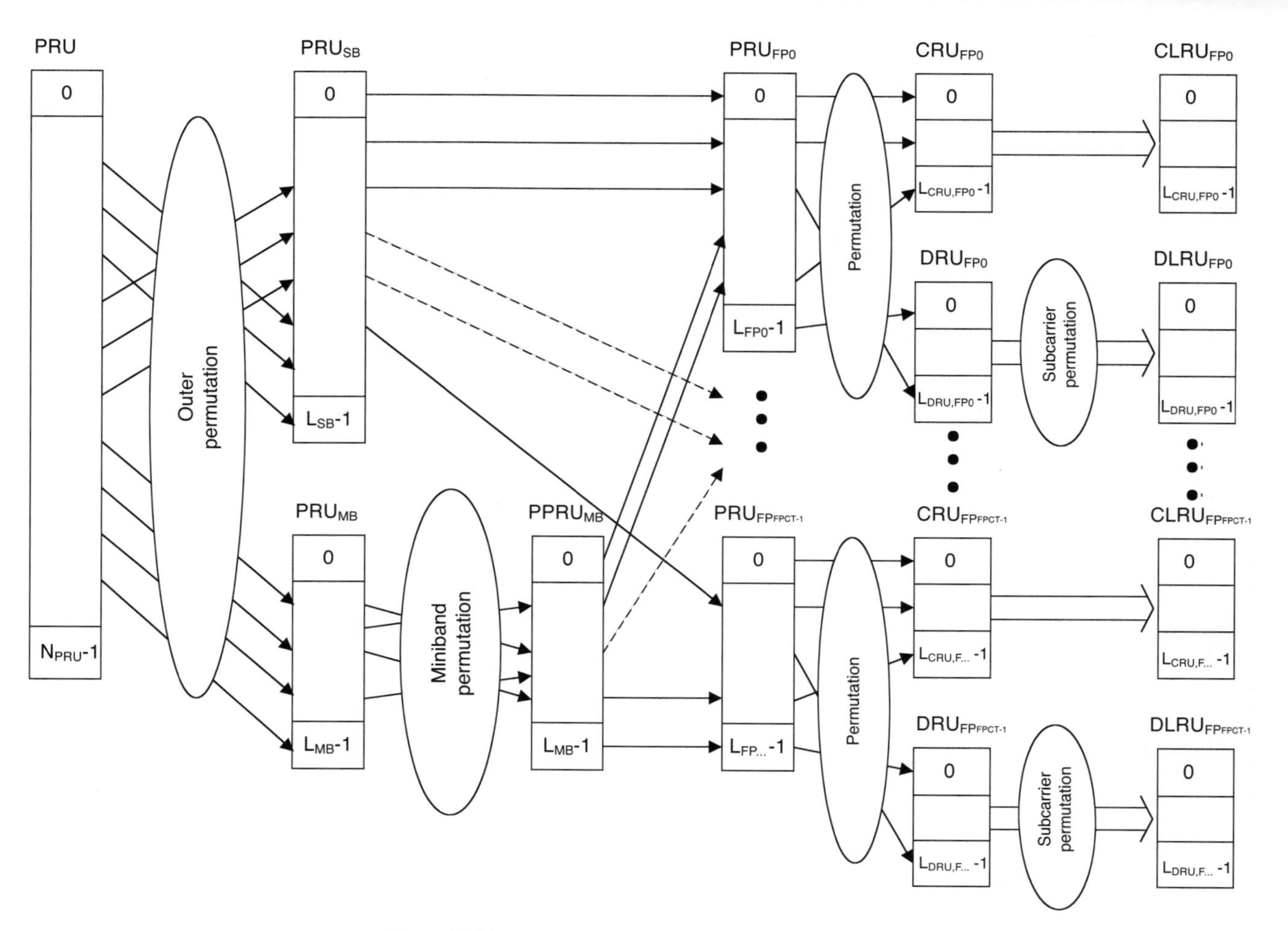

Figure 10.26 General structure of the PRU to LRU mapping

Table 10.8 The maximum number of subbands and the number of PRUs

FFT size	512	1024	2048
The maximum number of subbands	4	10	21
The number of PRUs (N_{pru})	24	48	96

10.6.2.4.2 Permutation of the PRU_{mb}s

PRU_{MB}s undergo miniband permutation. Figure 10.27 shows miniband permutation for eight minibands.

10.6.2.4.3 Mapping of the PRU_{sb}s and PRU_{mb}s to the Frequency Partitions

PRU_{SB}s and PRU_{MB}s are mapped to the frequency partitions. The *Downlink Frequency Partition Configuration (DFPC)* is transmitted in SFH. It defines the number of frequency partitions (*FPCT*) and the Frequency Partition Size (FPS_i). The frequency partitions are enumerated from *0* to *FPCT-1*. The maximum number of the frequency partitions is equal to four. FPS_i defines the size of the frequency partition *i* in PRUs. FPS_0 is equal to *0* for some DFPCs.

The Downlink Frequency Partition Subband Count (*DFPSC*), transmitted in SFH, defines the number of subbands allocated to FP_i for $i > 0$. The number of subbands allocated to FP_0 ($K_{SB,FP0}$) is calculated from K_{SB}, *FPCT* and *DFPSC*. The number of minibands for each frequency partition is calculated from FPS_i and $K_{SB,FPi}$. Figure 10.27 shows frequency partitioning for $FPCT = 4$, $FPS_i = 6$, $0 \le i \le 3$, $DFPSC = 1$.

10.6.2.4.4 Mapping of the PRU_{FP}s to the Contiguous and Distributed Resource Units

PRU_{FPi}s are mapped to the *contiguous and distributed recourse units (CRUs and DRUs)*. There are subband and miniband-based CRUs. The number of the subband-based CRUs in FP_0 ($DCAS_{SB,0}$), measured in subbands, is signal in SFH. The number of the miniband-based CRUs in FP_0 is defined by $DCAS_{MB,0}$, signaled in SFH, and a look-up table. The number of the subband and miniband-based CRUs in FP_i, $i > 0$, is calculated from $DCAS_i$, signaled in SFH, and $K_{SB,FPi}$. Those PRU_{FP}s, which are not allocated to CRUs, are allocated to DRUs.

When PRU_{FP}s are mapped to CRUs and DRUs the mapping to DRUs and to miniband-based CRUs is done with permutation. This permutation is done by means of the permutation sequence, which is generated by random sequence generator. The initial state of the random sequence generator is determined by *IDcell*.

Figure 10.27 shows mapping of the PRU_{FP}s to CRUs and DRUs for $DCAS_{SB,0} = 1$ and $DCAS_{MB,0} = 0$.

10.6.2.4.5 Mapping of the CRUs and DRUs to the Contiguous and Distributed Logical Resource Units

The subband-based CRUs are directly mapped to the subband contiguous LRUs (SLRUs) and the miniband-based CRUs are directly mapped to the miniband contiguous LRUs (NLRUs).

The *distributed LRUs (DLRUs)* are obtained by subcarrier permutation of the DRUs. The permutation is performed on the pairs of data subcarriers. It spreads the subcarriers of one DRU across the whole distributed resource allocation.

10.6.2.5 UL PHY Structure

The UL physical structure is generally similar to the DL physical structure. The major differences are:

- There is one additional size of the PRU corresponding to type-4 AAI subframe. N_{sym} is equal to 9 in that case.
- The minimum unit for forming the distributed logical resource unit (DLRU) is a tile.

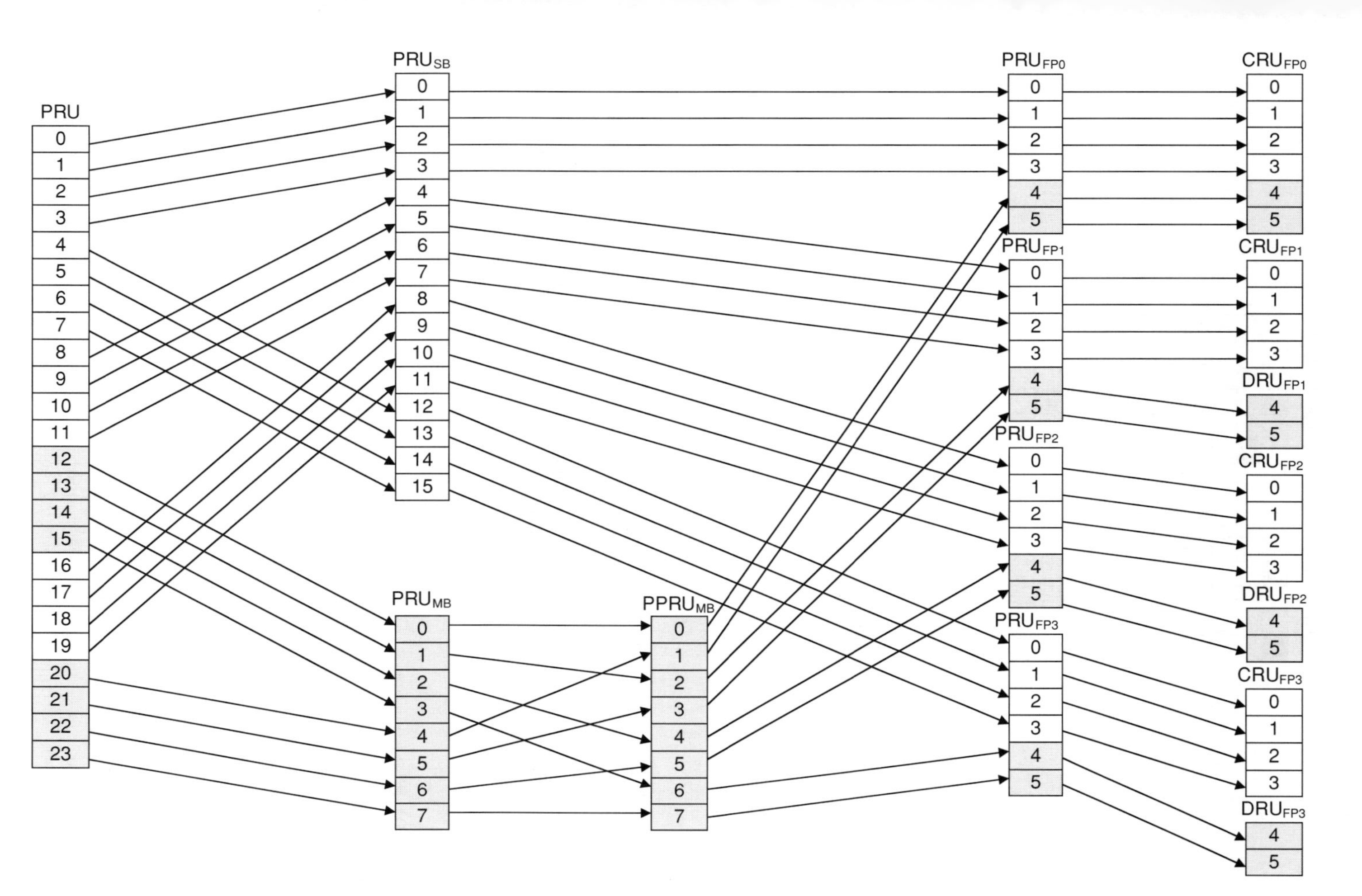

Figure 10.27 Frequency partition for $BW = 5\ MHz$, $K_{sb} = 4$, $FPCT = 4$, $FPS_0 = FPS_i = 6$, $DFPSC = 1$

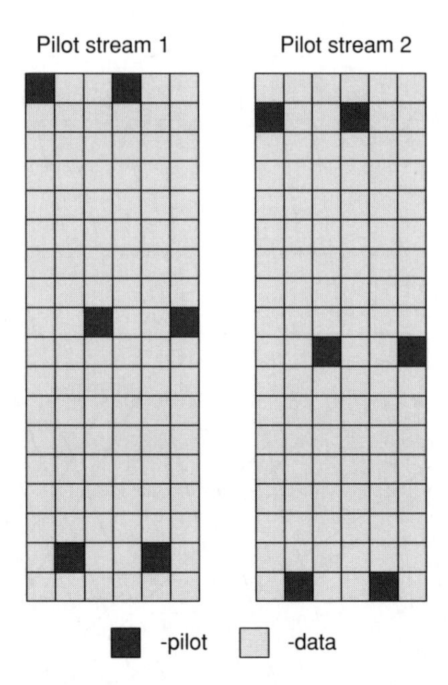

Figure 10.28 Pilot pattern for 1 DL data stream

UL DLRUs may be further divided into data, bandwidth request, and feedback regions.

The PRU to *Logical Resource Unit (LRU)* mapping for UL differs from the PRU to LRU mapping for DL in how Distributed Resource Units (DRUs) are used to form Distributed Logical Resource Units (DLRUs). In the case of DL the minimum unit for permutation algorithm is a pair of subcarriers, in the case of UL the minimum unit is a tile. The tile size is $6\ x\ N_{sym}$. N_{sym} depends on the AAI subframe type.

10.6.2.6 DL Pilot Structure

16m supports common and dedicated pilot. The common pilot can be used by all MSs, the dedicated pilot is intended to be used only by MS allocated to the specific resource allocation. The common and dedicated pilots use unified pilot patterns. Pilot patterns are designed for up to eight data streams. Figure 10.28 – 10.30 show pilot structure for 1, 2 and 4 DL data streams. Subcarrier index is increasing from top to bottom and symbol index is increasing from left to right.

10.6.2.7 UL Pilot Structure

UL pilot is dedicated pilot and it is intended to be used only by MS allocated to the specific resource allocation. UL pilot can be precoded or beamformed in the same way as date subcarriers. They are defined for up to four streams. The pilot patterns for UL CLRUs are the same as for DL. The pilot patterns for UL DLRU are shown in Figure 10.31 and Figure 10.32.

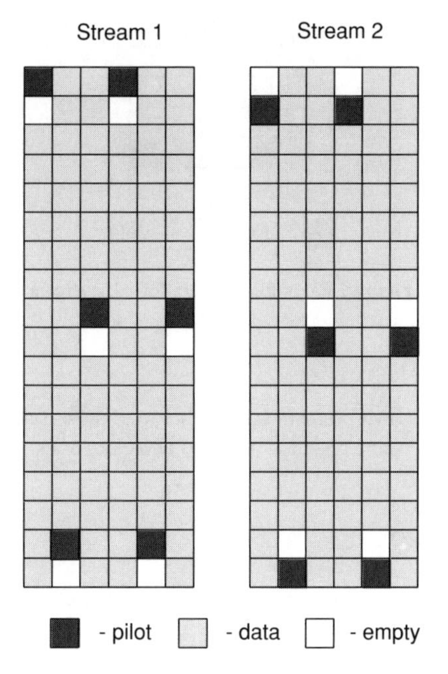

Figure 10.29 Pilot pattern for 2 DL data streams

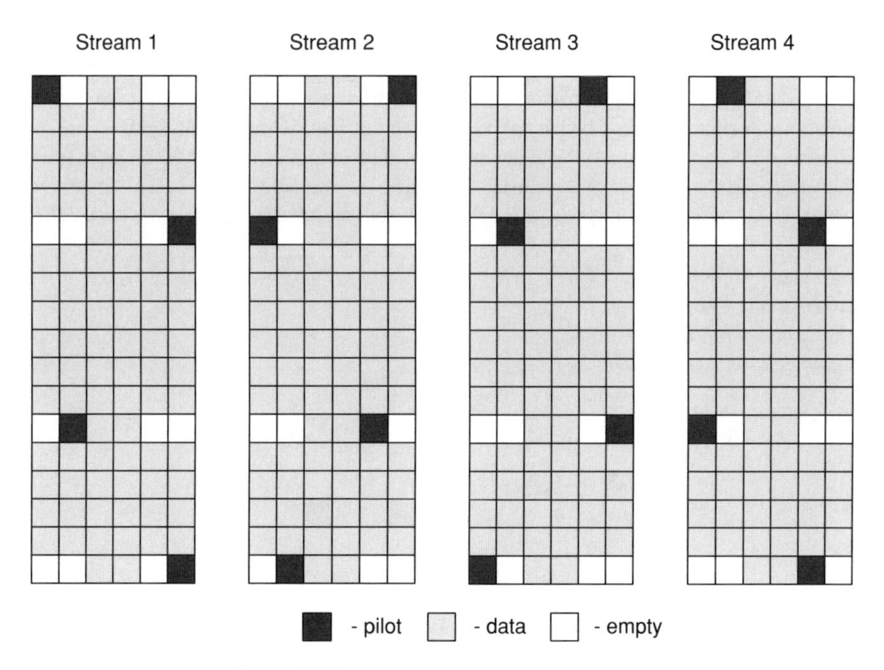

Figure 10.30 Pilot pattern for 4 DL data streams

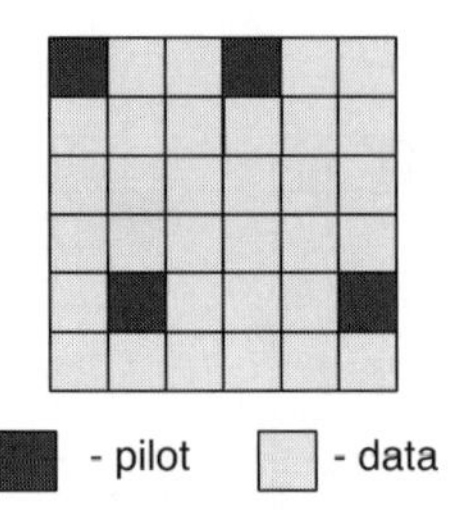

Figure 10.31 Pilot pattern for 1 UL stream

10.6.2.8 Downlink MIMO

802.16m supports several DL MIMO transmission schemes. Those schemes can be classified by the number of users allocated to the same resource unit into two categories:

- Single-user MIMO (SU-MIMO)
- Multi-user MIMO (MU-MIMO).

Depending on the precoding type DL MIMO schemes can be classified into another two categories:

- DL MIMO with adaptive precoding or Closed-loop MIMO (CL MIMO)
- DL MIMO with non-adaptive precoding or Open-loop MIMO (OL MIMO).

Figure 10.33 shows the architecture of the DL MIMO at a transmission side. The MIMO encoder maps MIMO layers into MIMO streams. The MIMO precoder maps MIMO streams into Tx antennas.

The input of the MIMO encoder is a vector $\mathbf{s}$ of the length M_t, where M_t is a number of MIMO streams. For SU-MIMO all M_t symbols are scheduled to the same AMS and belong to the same coding and modulation information path, which is called MIMO layer. For MU-MIMO M_t symbols are scheduled to M_t different AMSs and belong to the different layers. The output of the MIMO encoder is an $M_t x N_F$ matrix, where N_F is the number of subcarriers occupied by one MIMO block. This matrix is called MIMO *space-time coding* (STC) matrix. 802.16m supports four *MIMO encoding formats* (*MEF*):

- *Space-frequency block coding (SFBC)*
- *Vertical encoding (VE)*
- *Horizontal encoding (HE)*
- *Conjugate data repetition (CDR).*

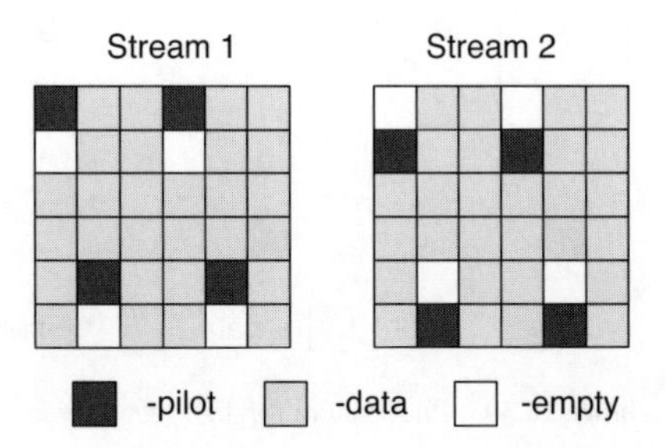

Figure 10.32 Pilot pattern for 2 UL streams

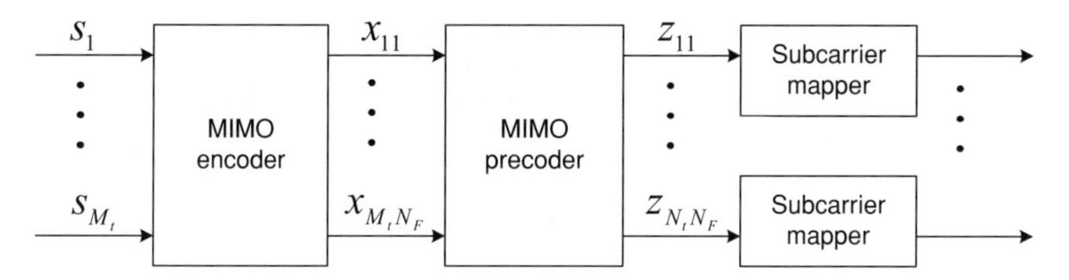

Figure 10.33 DL/UL MIMO architecture at a transmission side

Table 10.9 shows the input of the MIMO encoder ***s*** and the output of the MIMO encoder **x** for different MEFs.

802.16m supports six DL MIMO modes given in Table 10.10. Mode 0 and 5 are Tx diversity modes, while Mode 1 – 4 are *spatial multiplexing* (SM) modes.

One MIMO block occupies two subcarriers in spatial multiplexing modes and one in Tx diversity modes.

The number of MIMO layers at the input of the MIMO encoder is equal to one for SU-MIMO and it is equal to the number of streams for MU-MIMO.

The number of streams should not exceed the number of receive antennas at AMS for MIMO Mode 1 and 2.

The number of supported transmit antennas at BS (N_t) can be equal to 2, 4 or 8.The number of streams (M_t) varies from 1 to 8.

The last column of the Table 10.10 shows which combinations of MIMO modes and DL permutations are allowed. Note, that MIMO Mode 5 is supported only inside so-called Open-Loop (OL) Region. (Table 10.10 shows MIMO modes outside OL region.)

MIMO precoding is defined by $N_t x M_t$ precoding matrix $\mathbf{W}(k)$, where k is the index of the physical subcarrier where $\mathbf{W}(k)$ is applied.

With *non-adaptive precoding* the precoding matrix is selected from the base codebook or a subset of the base codebook. The base codebooks are defined for 2, 4 and 8 transmit antennas.

Table 10.11 and Table 10.12 illustrate correspondingly the rank-1 and rank-2 base code books for two transmit antennas. The precoding matrix is changing every $N_1 P_{SC}$ contiguous physical subcarriers for SFBC with VE and every N_1 PRUs with HE.

Table 10.9 MIMO encoding formats

MIMO encoding format	Input	Output	Note
SFBC	$s = \begin{bmatrix} s_1 \\ s_2 \end{bmatrix}$	$x = \begin{bmatrix} s_1 & -s_2^* \\ s_2 & s_1^* \end{bmatrix}$	
VE	$s = \begin{bmatrix} s_1 \\ \cdots \\ s_M \end{bmatrix}$	$x = \begin{bmatrix} s_1 \\ \cdots \\ s_M \end{bmatrix}$	$s_1 \ldots s_M$ belong to the same layer
HE	$s = \begin{bmatrix} s_1 \\ \cdots \\ s_M \end{bmatrix}$	$x = \begin{bmatrix} s_1 \\ \cdots \\ s_M \end{bmatrix}$	$s_1 \ldots s_M$ belong to the different layers
CDR	$s = s_1$	$x = \begin{bmatrix} s_1 & s_1^* \end{bmatrix}$	

Table 10.10 DL MIMO modes

Mode index	Description	MEF	Number of transmit antennas N_t	Number of streams M_t	Supported permutation
0	OL SU-MIMO	SFBC	2 4 8	2 2 2	DLRU; NLRU
1	OL SU-MIMO	VE	2 4 8	1-2 1-4 1-8	DLRU, M_t=2; NLRU, M_t<=4; SLRU
2	CL SU-MIMO	VE	2 4 8	1-2 1-4 1-8	NLRU, M_t=1; SLRU
3	OL MU-MIMO	HE	2 4 8	2 2-4 2-4	SLRU
4	CL MU-MIMO	HE	2 4 8	2 2-4 2-4	NLRU; SLRU
5	OL SU-MIMO	CDR	2 4 8	1 1 1	Not supported outside OL region

With *adaptive precoding* the precoding matrix is derived from the feedback of the AMS. The feedback can be:

- codebook – based
- sounding – based.

If the codebook based feedback is used then AMS selects preferred matrix from the base codebook and sends *Preferred Matrix Index (PMI)*, which represents an entry of the base codebook to the BS. The BS

Table 10.11 MIMO base codebook of rank-1 for two transmit antennas

	$\begin{bmatrix} c_1 & c_2 \end{bmatrix}^T$
0	$\begin{bmatrix} 0.7071 & -0.7071 \end{bmatrix}^T$
1	$\begin{bmatrix} 0.7071 & -0.5-0.5i \end{bmatrix}^T$
2	$\begin{bmatrix} 0.7071 & -0.7071i \end{bmatrix}^T$
3	$\begin{bmatrix} 0.7071 & 0.5-0.5i \end{bmatrix}^T$
4	$\begin{bmatrix} 0.7071 & 0.7071 \end{bmatrix}^T$
5	$\begin{bmatrix} 0.7071 & 0.5+0.5i \end{bmatrix}^T$
6	$\begin{bmatrix} 0.7071 & 0.7071i \end{bmatrix}^T$
7	$\begin{bmatrix} 0.7071 & -0.5+0.5i \end{bmatrix}^T$

Table 10.12 MIMO base codebook of rank-2 for two transmit antennas

	$\begin{bmatrix} c_{11} & c_{12} \\ c_{21} & c_{22} \end{bmatrix}^T$
0	$\begin{bmatrix} 0.7071 & -0.7071 \\ 0.7071 & 0.7071 \end{bmatrix}^T$
1	$\begin{bmatrix} 0.7071 & -0.5-0.5i \\ 0.7071 & 0.5+0.5i \end{bmatrix}^T$
2	$\begin{bmatrix} 0.7071 & -0.7071i \\ 0.7071 & 0.7071i \end{bmatrix}^T$
3	$\begin{bmatrix} 0.7071 & 0.5-0.5i \\ 0.7071 & -0.5+0.5i \end{bmatrix}^T$

calculates precoder using PMI. There are three modes of the codebook based feedback and depending on the mode the different type of codebooks is used:

- The base mode: the base codebook.
- The transformation mode: the transformation codebook.
- The differential mode: the differential and the base codebook.

The base codebook is the same codebook as for non-adaptive precoding.

The transformation codebook is obtained from the base codebook according to (10.2):

$$\widetilde{v}_i = \frac{Rv_i}{\| Rv_i \|} \tag{10.2}$$

where v_i is the i-th codeword of the base codebook, $\widetilde{v}_i$ is the i-th codeword of the transformed codebook and **R** is $\mathbf{N_t}$ by $\mathbf{N_t}$ correlation matrix. The channel correlation matrix is computed periodically at AMS according to (10.3) and is fed back to the BS:

$$R = E(H_{ij}^H H_{ij}) \tag{10.3}$$

where H_{ij} is the correlation matrix for the i-th OFDM symbol and j-th subcarrier. The averaging is done over a number of OFDM symbols and all subcarriers. The correlation matrix is a conjugate-symmetrical. Because of that only the upper triangular elements are fed back to the BS. They are normalized and quantized before transmission. The transformation codebook is used only for rank-1 MIMO.

In the differential mode the precoder V(t) is computed from the base or from the differential codebook. The feedback matrix D(t) is selected from the base codebook at the start/reset of the feedback, where t is the feedback index. At the start of the feedback this index is set to 0 and it is reset to 0 at $t = T_{max}+1$. At start/reset of the feedback the precoder can be calculated from the feedback matrix alone: V(0) = D(0). At $t = 1,\ldots,T_{max}$ the precoder is computed from the rotation matrix $Q_{V(t)}$ and the feedback matrix D(t): V(t)= $Q_{V(t-1)}$D(t). The dimension of D(t) is $N_t x M_t$ .The rotation matrix $Q_{V(t)}$ is computed from the precoder V(t). It is a unitary matrix. Its dimension is $N_t x N_t$ .The feedback matrix D(t) is selected from the differential codebook. Table 10.13 illustrates the differential codebook for $N_t = 2, M_t = 1$ and Table 10.14 for $N_t = 2, M_t = 2$.

Table 10.13 Differential codebook for $N_t = 2, M_t = 1$

1	$\begin{bmatrix} 1 & 0 \end{bmatrix}^T$
2	$\begin{bmatrix} cos(15^\circ) & sin(15^\circ) \end{bmatrix}^T$
3	$\begin{bmatrix} cos(15^\circ) & sin(15^\circ)\, exp(j120^\circ) \end{bmatrix}^T$
4	$\begin{bmatrix} cos(15^\circ) & sin(15^\circ)\, exp(-j120^\circ) \end{bmatrix}^T$

10.6.2.9 MIMO Feedback

16m UL fast feedback channel carries channel quality feedback and MIMO feedback. MIMO feedback information includes: STC rate, subband selection, MIMO stream index, Quantized correlation matrix, PMI report, CQI and preferred MIMO feedback mode. In order to support both OL and CL MIMO operations, eight MIMO feedback modes have been defined in 16m specification. When allocating a feedback channel, the MIMO feedback mode shall be indicated to the MS, and the MS will feedback information accordingly. Depending on the MIMO feedback format, the MIMO feedback information is mapped to PFBCH or to SFBCH.

10.6.2.10 Uplink MIMO

The architecture of the UL MIMO is illustrated by Figure 10.33. It is similar to the DL MIMO. Single-user MIMO and multi-user MIMO, which is called collaborative spatial multiplexing, is supported. Unlike DL MIMO in the case of the UL MU-MIMO all symbols of the vector **s** at the input of the MIMO encoder belong to the same layer and horizontal encoding is not supported. 802.16m supports two MIMO encoding formats (MEF) for UL:

- Space-frequency block coding (SFBC)
- Vertical encoding (VE).

802.16m supports five UL MIMO modes given in Table 10.15.

One MIMO block occupies two subcarriers in mode 0 and one subcarrier in all other modes.

The number of supported transmit antennas N_t is equal to 1, 2, or 4.The minimum number of receive antennas N_r is equal to 2. The number of streams M_t varies from 1 to 4.

The last column of the Table 10.15 shows what combinations of MIMO modes and UL permutations are allowed.

Table 10.14 Differential codebook for $N_t = 2, M_t = 2$

1	$\begin{bmatrix} 1 & 0 \\ 0 & 1 \end{bmatrix}^T$
2	$\begin{bmatrix} cos(15^\circ) & sin(15^\circ) \\ sin(15^\circ) & -cos(15^\circ) \end{bmatrix}^T$
3	$\begin{bmatrix} cos(15^\circ) & sin(15^\circ)\, exp(j120^\circ) \\ sin(15^\circ)\, exp(j120^\circ) & -cos(15^\circ) \end{bmatrix}^T$
4	$\begin{bmatrix} cos(15^\circ) & sin(15^\circ)\, exp(-j120^\circ) \\ sin(15^\circ)\, exp(-j120^\circ) & -cos(15^\circ) \end{bmatrix}^T$

Table 10.15 UL MIMO modes

Mode index	Description	MEF	Number of transmit antennas N_t	Number of streams M_t	Supported permutation
0	OL SU-MIMO	SFBC	2 4	2 2	DLRU; NLRU
1	OL SU-MIMO	VE	1 2 4	1 1-2 1-4	DLRU, $M_t <= 2$; NLRU, SLRU
2	CL SU-MIMO	VE	2 4	1-2 1-4	DLRU, $M_t <= 2$; NLRU, SLRU
3	OL MU-MIMO	VE	1 2 4	1 1-2 1-3	DLRU, $M_t = 1$; NLRU, SLRU
4	CL MU-MIMO	VE	1 2 4	1 1-2 1-3	DLRU, $M_t = 1$; NLRU, SLRU

10.6.2.11 Channel Coding and Modulation

Channel coding procedure (Figure 10.34) starts with CRC (Cyclic Redundancy Code/Check) addition and partitioning of the data burst into FEC blocks. CRC is appended to the data burst if it is partitioned into more than one FEC block. Before partitioning data burst is padded if necessary to the nearest specified burst size large than the data burst in question. CRC is defined by the polynomial in (10.4):

$$D^{16} + D^{12} + D^{5} + 1 \tag{10.4}$$

Randomization is done by *Pseudorandom Binary Sequence (PRBS)* generator illustrated in Figure 10.35. It is defined by the polynomial in (10.5):

$$D^{15} + D^{14} + 1 \tag{10.5}$$

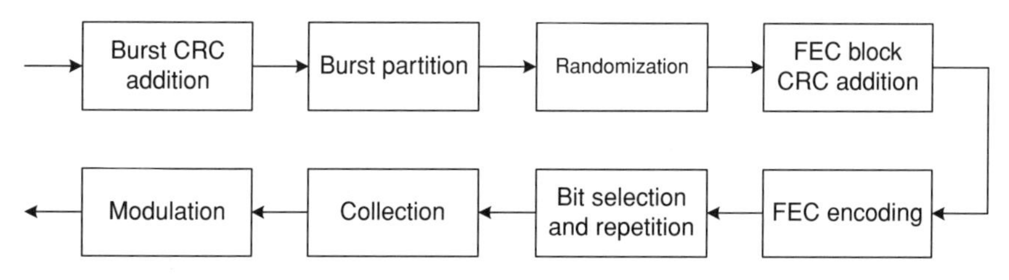

Figure 10.34 Channel coding

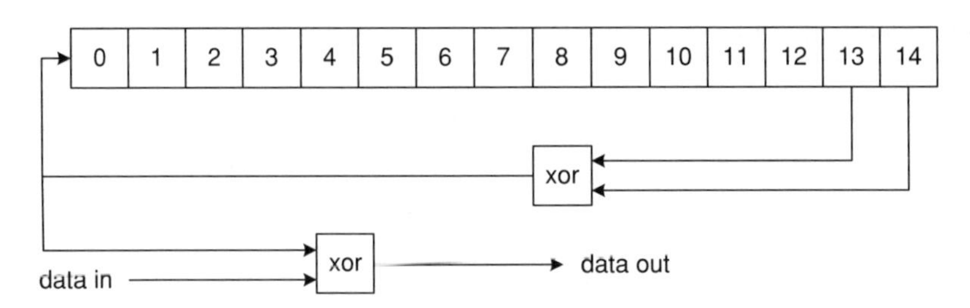

Figure 10.35 Data randomizer

PRBS is initialized by initial vector 011011100010101, written from LSB to MSB, for each FEC block. Randomization is followed by addition of the CRC to every FEC block. CRC is defined by the polynomial in (10.4).

Double binary *CRSC (Circular Recursive Systematic Convolutional)* code is used. *Convolutional turbo code (CTC)* encoder, illustrated on **Figure 10.36**, is defined by the following polynomials:

- Feedback branch:

$$1 + D + D^3 \tag{10.6}$$

- Y parity bits:

$$1 + D^2 + D^3 \tag{10.7}$$

- W parity bits:

$$1 + D^3 \tag{10.8}$$

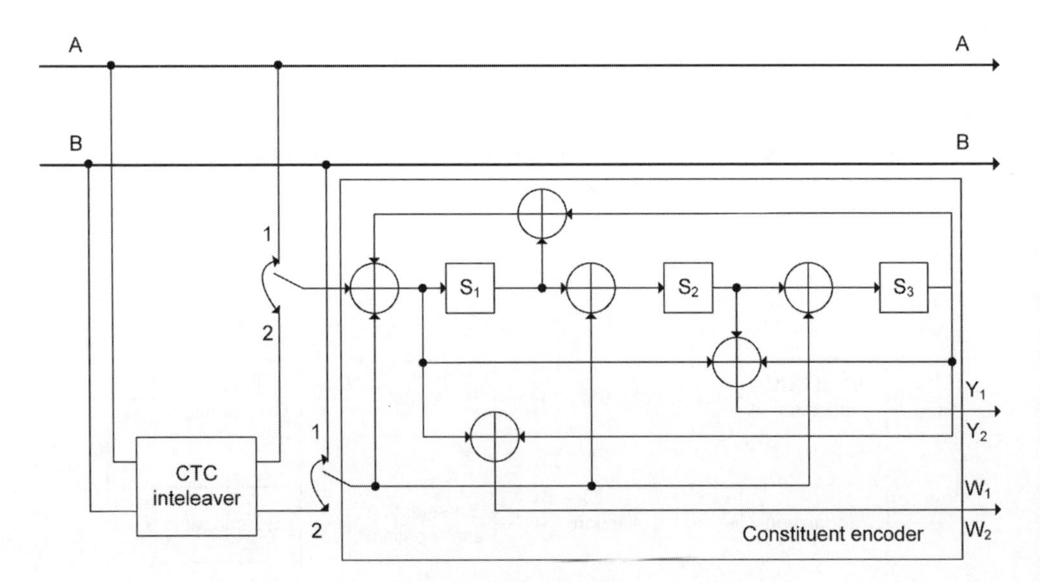

Figure 10.36 CTC encoder

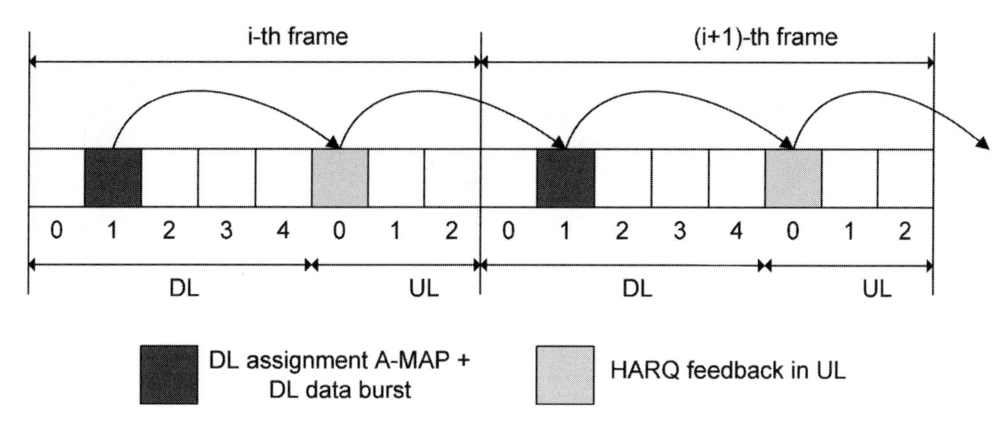

Figure 10.37 Timing and signaling for DL HARQ

The structure of the CTC encoder includes two switches, which work in a synchronized way. First, both switches are in position 1. The constituent encoder is fed by $2N$ bits in a natural order. Then, both switches change position to 2, and the constituent encoder is fed by $2N$ interleaved bits. The output of the CTC encoder is fed into the bit separation block in the following order:

$$A_0 \ldots A_{N-1} B_0 \ldots B_{N-1} Y_{1,0} \ldots Y_{1,N-1} Y_{2,0} \ldots Y_{2,N-1} W_{1,0} \ldots W_{1,N-1} W_{2,0} \ldots W_{2,N-1} \tag{10.9}$$

The bit separation block demultiplexes those bits into subblocks A, B, Y_1, Y_2, W_1, W_2. $A_0 \ldots A_{N-1}$ go to subblock A, $B_0 \ldots B_{N-1}$ go to subblock B, etc. The six subblocks are separately interleaved according to the procedure described in [2]. Next, interleaved subblocks are multiplexed into four blocks.

Collection block collects selected bits from each FEC block for H-ARQ transmission.

10.6.2.12 H-ARQ

The hybrid ARQ (H-ARQ) is based on the N-channel stop-and-wait (SAW) protocol (see Chapter 6). For single carrier the maximum number of channels per MS is 16. For multi-carrier this number can be larger. The maximum number of retransmissions is 4 or 8; 4 is a default value.

Figure 10.37 illustrates timing and signaling for DL HARQ. MS receives DL assignment A-MAP and data burst in 1st sub-frame of i-th DL frame and tries to decode that data burst. After decoding failure MS sends negative acknowledgment to BS in UL sub-frame 0 of the same frame. BS retransmits data in the same DL subframe of the next frame. Transmission format and resource allocation may be different. This process is repeated until BS receives positive acknowledgment or the maximum number of retransmissions is reached.

Figure 10.38 illustrates timing and signaling for UL HARQ. MS receives UL assignment A-MAP in 1st sub-frame of i-th DL frame and transmits UL data burs in subframe 0 of the same frame. BS tries to decode that burst and after decoding failure it sends a new UL assignment A-MAP and negative acknowledgment in the same DL subframe of the next frame. This process is repeated until BS successfully decodes the data and sends positive acknowledgment to MS or the maximum number of retransmissions is reached.

H-ARQ uses Incremental Redundancy (IR) for soft bits combining. Chase combining is supported as a special case of IR.

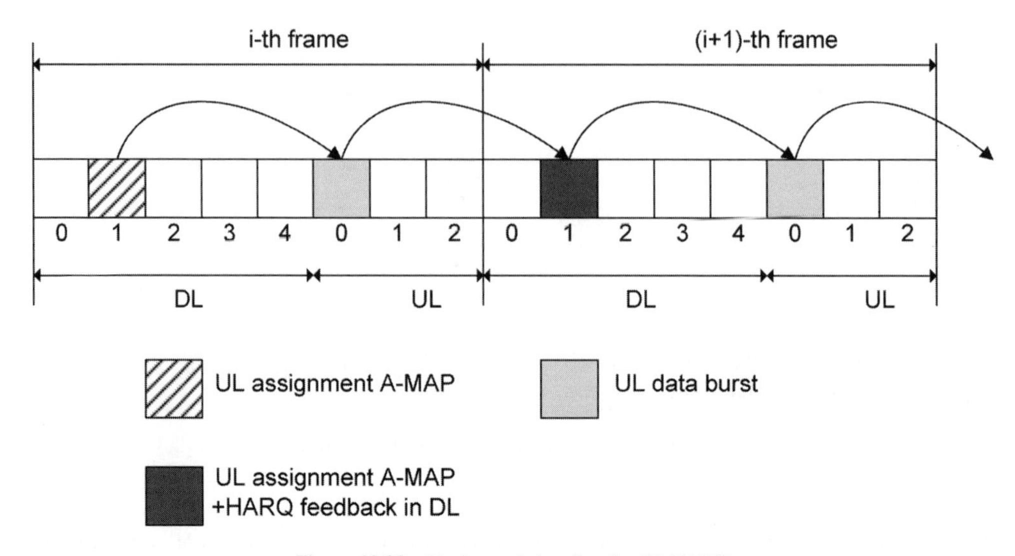

Figure 10.38 Timing and signaling for UL HARQ

References

[1] IEEE 802.16-2001, IEEE Standard for Local and Metropolitan Area Networks — Part 16: Air Interface for Fixed Broadband Wireless Access Systems, Apr. 8, 2002.
[2] IEEE P802.16-2004. Air Interface for Fixed Broadband Wireless Access Systems, October 2004.
[3] IEEE P802.16e-2005, Air Interface for Fixed and Mobile Broadband Wireless Access Systems – Amendment 2: Physical and Medium Access Control Layers for Combined Fixed and Mobile Operation in Licensed Bands and Corrigendum1," February 2006.
[4] IEEE 802.16m System Requirements Document (SRD), Draft, 2010-01-04, http://wirelessman.org/tgm/.
[5] IEEE 802.16m Evaluation Methodology Document (EMD), Draft, 2009-01-15, http://wirelessman.org/tgm/.
[6] IEEE 802.16m System Description Document (SDD), Draft, 2009-09-24, http://wirelessman.org/tgm/.
[7] IEEE P802.16m: Air Interface for Fixed and Mobile Broadband Wireless Access Systems. D3, Dec. 2009, http://wirelessman.org/tgm/.
[8] IEEE 802.11™-2007, IEEE Standard for Information Technology—Telecommunications and information exchange between systems—Local and metropolitan area networks—Specific requirements, Part 11: Wireless LAN Medium Access Control (MAC) and Physical Layer (PHY) Specifications.
[9] IEEE 802.11n™-2009, IEEE Standard for Information Technology—Telecommunications and information exchange between systems—Local and metropolitan area networks—Specific requirements, Part 11: Wireless LAN Medium Access Control (MAC) and Physical Layer (PHY) specifications, Amendment 5: Enhancements for Higher Throughput.
[10] IEEE 802.16-2009, IEEE Standard for Local and metropolitan area networks Part 16: Air Interface for Broadband Wireless Access Systems, May 2009.

11

Standardization in 3GPP

Asbjørn Grøvlen[1], Kari Hooli[2], Matti Jokimies[3], Kari Pajukoski[2], Sergei Semenov[3], and Esa Tiirola[2]
[1]*Nokia, Denmark*
[2]*Nokia Siemens Network, Finland*
[3]*Nokia Corporation, Finland*

In this chapter we consider the layer 1 (L1) procedures in more recent standards of 3GPP (the 3rd Generation Partnership Project). Unlike the first and second generation cellular systems oriented mostly to the voice communications, third generation systems are designed first of all for data communications. This fact reflects the common trend in communications: the shift from voice to data. This trend clearly already showed itself in wired communications. Nowadays only a small fraction of landlines traffic is used for voice transmission, and this fraction is constantly decreasing due to development of VoIP technique. We can now see the beginning of the same process in cellular wireless systems. This process needs a significant increase in data rates, which poses some challenges, taking into account the hard conditions of an air interface channel. However, the proper use of known modulation and coding schemes allows this problem to be solved.

11.1 Standardization Process and Organization

11.1.1 General

The standardization organization specifying the standards of the *Universal Mobile Telecommunications System (UMTS)* and the *Long Term Evolution (LTE)* is *Third Generation Partnership Project (3GPP)*. As a major difference to most telecommunications standardization bodies, 3GPP is not a legal entity, but a partnership project of regional *standards developing organizations (SDOs)*. Each company or other entity participating in the standardization activities of 3GPP has to be a member of one of the regional SDOs, or *Organizational Partners (OPs)*, as they are called in 3GPP [2].

The coordination of 3GPP activities is carried out by the *Program Coordination Group (PCG)*. The PCG is responsible for overall time-frame and management of overall work progress. The specifications

Modulation and Coding Techniques in Wireless Communications Edited by Evgenii Krouk and Sergei Semenov

of 3GPP are created in the *Technical Specification Groups (TSGs)*. The actual technical details and the text for standards are developed in the *working groups (WGs)* of the TSGs. TSG *RAN (Radio Access Network)* is responsible for defining the protocols and procedures of the air interface between the base station and the *user equipment (UE)* for UMTS and LTE. The organization of 3GPP and specifically the organization of TSG RAN are described in detail in the next section [3].

In 3GPP terminology, the radio access for UMTS is called *Universal Telecommunications Radio Access (UTRA)*, and for LTE, it is called *Enhanced UTRA (E-UTRA)* [4]. Typically, the terms *UTRAN (UTRA network)* and *E-UTRAN (E-UTRA network)* are used in the RAN specifications to denote the networks, as a contrast to the UE. However, this distinction is not always clear in 3GPP documentation, and sometimes the terms UTRAN and E-UTRAN refer to the whole radio access level system, including the UE.

11.1.2 Organization of 3GPP

As mentioned above, 3GPP consists of several TSGs. The current TSGs and their main responsibilities are as follows:

- TSG CT; *Core Network and Terminals*
  - protocols of core networks and their counterparts in terminals
  - terminal aspects mainly related to smart cards (for example, to *subscriber identity module (SIM)*)
- TSG GERAN; *GSM EDGE Radio Access*
  - radio access level aspects of GSM/GPRS/EGPRS networks and terminals
- TSG RAN; Radio Access Network
  - radio access level aspects of UTRA and E-UTRA networks and terminals
- TSG SA; *Service and System Aspects*
  - services, architecture, security, codecs and telecom management

Even though the air interface between the base station and the UE is under the responsibility of TSG RAN and its WGs, the other TSGs and their WGs also create standards impacting the radio access level functionalities in UTRA and E-UTRA. For example: Handovers and cell reselections from GERAN to UTRAN and to E-UTRAN are specified by WGs of TSG GERAN. Radio access level encryption standards are the responsibility of WG3 of TSG SA [3].

11.1.3 Organization of TSG RAN

Because TSG RAN is the 3GPP TSG having the main responsibility for the technologies covered by this book, the structure of this TSG is addressed in more detail. TSG RAN consists of five WGs; their primary responsibility areas are listed below:

- RAN WG 1 (RAN1); Radio layer 1
- RAN WG 2 (RAN2); Radio layer 2 and radio layer 3 RR (Radio Resource)
- RAN WG3 (RAN3); Iu, Iub, Iur, S1, and X2 interfaces, and UTRAN/E-UTRAN architecture
- RAN WG4 (RAN4); Radio performance, radio frequency parameters, and base station conformance
- RAN WG5 (RAN5); Mobile terminal conformance testing

In addition to the five WGs, TSG RAN has also an ad hoc group (AHG), RAN AHG1, for International Telecommunications Union (ITU) work. This group coordinates TSG RAN work related to ITU-R.

The detailed specification responsibilities of the WGs are as follows:

- RAN1
  - physical channel structures
  - mapping of the transport channels onto physical channels
  - physical layer multiplexing
  - channel coding
  - error detection
  - spreading and modulation
  - physical layer procedures
  - definition of measurements and their provision by physical layer to upper layers
- RAN2
  - radio interface layer 2 protocols:
    - *medium access control (MAC)*
    - *radio link control (RLC)*
    - *packet data convergence protocol (PDCP)*
  - *radio resource control (RRC)* on layer 3
- RAN3
  - interfaces between radio access level network elements: Iub, Iur, X2
  - interfaces between core network and radio access network: Iu, S1
  - overall UTRAN and E-UTRAN architecture
- RAN4
  - minimum requirements for transmission and reception parameters, and for channel demodulation
  - base station test procedures
  - requirements for other RAN level radio devices than UEs and base stations (for example, repeaters)
- RAN5
  - UE conformance test specifications, based on requirements specified by other groups (also by other WGs than RAN WGs)

As can be seen above, the most relevant 3GPP WGs for the air interface layer 1 between the UE and the base station are RAN4 and RAN1. Also, several of the items specified by RAN2 are relevant, especially the MAC and RRC functionalities controlling L1. The conformance test specifications deal with layer 1 items, but they do not introduce any new requirements in addition to those in the specifications created by other WGs [4]. More details on the RAN specifications are provided in a later sub-section, after the standardization process and the 3GPP releases are presented.

11.1.4 Standardization Process

The outcomes of the standardization process are the *technical specifications (TSs)* and *technical reports (TRs)*. In general, the functionality is specified in normative parts of TSs, while TRs and informative sections of TSs provide further information for background, or they serve as guidelines for implementations, or they are used as temporary documents for 3GPP internal purposes [5].

Typically, new features are handled first in a study item, where various alternatives are evaluated. Also the feasibility of the feature is assessed, and the results are documented in a TR. If the feature has been evaluated on the TSG level to be feasible, the actual work item for the feature is started. Finally, the work results in new TSs, or *change requests (CR)* to existing TSs.

The above-mentioned workflow in 3GPP may consist of three stages, at least in the case of major new items. In stage 1, the service aspects are addressed. In this stage, the RAN WGs dealing with air interface layer 1 aspects are usually not involved. In stage 2, the architectural aspects are standardized, and the existing stage 2 TSs are updated, or a new stage 2 TS is created. In this stage, air interface protocol aspects are also included. Most stage 2 documents serve also as overall descriptions of the features. In the final stage, that is in the stage 3, the TSs specifying the functionalities in the detailed level are created. The stages were specified originally by CCITT (Comité Consultatif International Téléphonique et Télégraphique), which was later transformed to ITU Telecommunications standardization sector (ITU-T) [5], [6]

It should be noted that the cover pages of the 3GPP TSs carry the following disclaimer: "The present document has not been subject to any approval process by the 3GPP Organizational Partners and shall not be implemented". Thus, in principle the 3GPP TSs are not the formal standards for implementation, but only the corresponding standards, published by the OPs, should be implemented. For example, the 3GPP specification TS 36.101 is transposed by European Telecommunications Standards Institute (ETSI) to the ETSI specification RTS/TSGR-0436101. This is done by changing the cover page, headers, and similar information, but without altering the technical content. In practice, however, only the 3GPP standards are referenced instead of the OP standards. However, in formal regulatory requirements, the reference is primarily made to the regional documents published by the OPs.

The TRs intended for 3GPP internal use are not transposed by SDOs to their publications. These TRs, which are often an outcome of a study item, are numbered as .8xx (for example, TR 25.820), in contrast to the published TRs, which have numbering in form of .9xx (for example, TR 36.913) [5].

11.1.5 3GPP Releases

New features are introduced in new releases of 3GPP specifications, allowing the older releases to be maintained and upgraded without impact on the new features.

The first release of 3GPP was Release '99, which specified the baseline of UMTS. The next major layer 1 air interface features in RAN specifications after Release '99 were HSDPA in Release 5, and HSUPA in Release 6. These two were further enhanced to HSPA + in Release 7. LTE was introduced in Release 8, and Release 10 is expected to introduce LTE-Advanced (LTE-A) [6].

The numbering of the first release is according to the yearly release numbering of GSM in ETSI. ETSI specified the GSM/GPRS/EDGE radio access systems (now under 3GPP TSG GERAN), and the GSM/GPRS core network. The latter also became the basis of the 3GPP UMTS core network. Hence, it was natural to adopt initially the release numbering of ETSI in 3GPP, too.

After Release '99, the pace of yearly releases was not applied, and the intended Release '00 was split into two: Release 4 containing minor updates, and Release 5 introducing better support for packet-based traffic. The numbering of the releases was taken from the most significant digit of the TS and TR version number. The versions of TSs and TRs for approved Release '99 specifications had version number in the form of 3.x.x, while the intended Release '00 had version numbers in the form of 4.x.x (the lower version numbers were reserved for drafts). Therefore, the new numbering of the releases was started from number four.

Even though new items are introduced in new releases, the association between features and releases is somewhat ambiguous: The first release, where a feature is included, depends on whether a specific functionality or the whole service is addressed. For example, the LTE MBMS functionalities on layer 1 are specified to a great extent already in Rel-8, even though the whole MBMS feature is introduced in Rel-9.

New versions of 3GPP specifications can be approved in TSG meetings, which are held four times a year in the case of TSG RAN (in March, June, September, and December). The version levels are designated by the month of the meeting; for example, 2009-03, 2009-06 and so on. (The actual month of the TSG meeting is sometimes the preceding month, but the numbering is not altered.) Several releases are maintained in parallel. For example, in TSG RAN meeting of September 2009, that is, on version level 2009-09, updates were approved on one or more RAN specifications of Releases 4, 5, 6, 7, 8 and 9 [7].

11.1.6 Frequency Bands and 3GPP Releases

The requirements on frequency bands can be release independent. That is, the frequency bands may not be restricted to the 3GPP release where they are introduced, and the successors of the release. This means that a UE, implemented according to a specific release, may support frequency bands that are introduced in later releases. In such cases, extra requirements, not included in the base release of the UE, are imposed on the UE. However, these requirements are typically limited to the radio frequency level specifications of the new band, and to the signaling of the new band numbers [8].

11.1.7 RAN Specifications

The UTRA TSs and TRs created in TSG RAN are prefixed with "25.", that is, they belong to the "series 25" specifications. The RAN5 UTRA test specifications have a different prefix "34.", because the predecessor of RAN5 was under a different TSG (under TSG T for terminals, now closed). The E-UTRA TSs and TRs are prefixed with "36.". The first number after the dot (or the prefix in case of UTRA test specifications) denotes the WG responsible for the TS as follows:

- RAN1: 25.2xx, 36.2xx
- RAN2: 25.3xx, 36.3xx
- RAN3: 25.4xx, 36.4xx
- RAN4: 25.1xx, 36.1xx
- RAN5: 34.xxx, 36.5xx

There are, however, several exceptions. For example, the specification TS 25.317 is under the responsibility of RAN4, even though the numbering refers to RAN2.

The TRs are numbered differently. As mentioned earlier, in the case of TRs the first number after the dot is 8 or 9, depending on the type of the TR. Hence, the RAN TRs on UTRA have numbers 25.8xx, 25.9xx, 34.8xxx, and 34.9xx, and those on E-UTRA 36.8xx and 36.9xx.

To some extent, the numbering of E-UTRA TSs is on a par with the numbering of UTRA TSs. For example, UTRA multiplexing and channel coding for FDD is specified in TS 25.212, and the corresponding specification for E-UTRA is TS 36.212. However, this numbering is not applied consistently, for example, the specification for UTRA L1 procedures is TS 25.214, while the E-UTRA L1 procedures are defined in TS 36.213. Additionally, in UTRA, RAN1 and RAN4 TSs are often different for TDD and FDD, while in E-UTRA both modes are included in the same specification, due to small differences between the modes in E-UTRA. For example, TS 25.101 specifies UTRA UE radio transmission and reception for FDD, and TS 25.102 for TDD; in E-UTRA TS 36.101 specifies this functionality for both modes.

The specifications of the groups most relevant to layer 1, that is, those of RAN1, RAN2, and RAN4, can be found in [9], [10], and [11] respectively. The complete list of 3GPP specification series can be found in [12].

11.2 3G WCDMA

The *Wideband Code Division Multiple Access (WCDMA)* is the cellular system transmission protocol that utilizes direct-sequence code division multiple access (DS-CDMA) on a common wideband (5MHz)

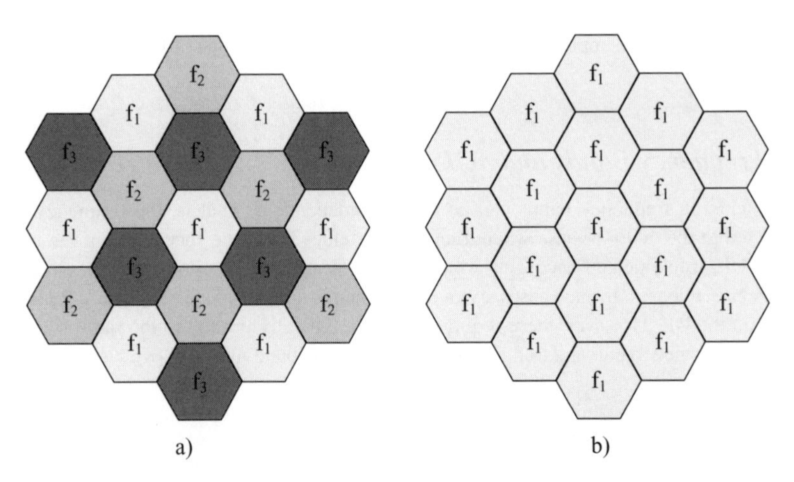

Figure 11.1 *a*) Frequency reuse 3; *b*) Frequency reuse 1

carrier. This system is already widely spread in many countries. Nowadays it is the main representative of the third generation (3G) cellular systems. One of main drivers of the WCDMA design was the addition of flexible multimedia capabilities to the voice transmission. WCDMA has two modes characterized by the duplex method: FDD (frequency division duplex) and TDD (time division duplex), for operating with paired and unpaired bands, respectively [1]. At least for now the deployment of the TDD mode is geographically more limited. So, in this chapter we will concentrate on FDD mode features mostly.

11.2.1 WCDMA Concept. Logical, Transport and Physical Channels

WCDMA is a DS-CDMA system, that is, the data information bits are spread over a bandwidth, which is much greater than the bandwidth corresponding to the original data rate. It is done by multiplying the original data with quasi-random chip sequence called spreading code (see Chapter 9). On one hand, this leads to a need to have a greater amount of such a valuable resource as frequency spectrum; but on the other hand, the DS-CDMA technique has advantages such as increasing the resistance to interference, jamming and interception. Also, usage of DS-CDMA technique simplifies the network planning, since DS-CDMA system is frequency reuse 1 system. The *frequency reuse* parameter shows how many different frequencies should use the neighbouring cells. If the frequency reuse is greater than 1, the adjacent cells must use the different frequency carriers that should be taken into account in the network planning. The example of network with the frequency reuse 3 is depicted in Figure 11.1a. The example of network planning with frequency reuse 1 is depicted in Figure 11.1b.

In WCDMA systems the adjacent base stations use the same frequency band since the different base stations and users are separated by codes rather than frequencies. The different base stations use different scrambling codes (also known as long codes). The transmitted data is divided into *uplink* and *downlink* data. The uplink comprises the data transmitted from mobile station (MS) or *user equipment* (*UE*) to the base station (BS) or *Node B*. And the downlink comprises the data transmitted in the reverse direction from Node B to UE. In the FDD mode the uplink and downlink are carried by the frequencies separated by 5 MHz. In the TDD mode the same 5 MHz bandwidth is time shared between uplink and downlink. The chip rate of the system is 3.84 Mcps. The chip sequence is transmitted by frames of length 10 ms (38400 chips/frame) and each frame is divided into 15 slots (2560 chips/slot). The spreading factors used in uplink vary from 4 to 256 and in the downlink from 4 to 512. Thus, the respective modulation symbol rates vary from 960 k symbols/s to 15 k symbols/s in the downlink and from 960 k symbols/s to

7.5 k symbols/s in the uplink. This is done to provide the support to different quality of service (QoS) requirements. The channels from the same source are separated with the help of orthogonal variable spreading factor (OVSF) channelization codes. The scrambling codes used in the downlink to separate different cells are Gold codes with a 10-ms period (38400 chips). The actual length of used Gold codes is 2^{18}-1 chips. In the uplink the scrambling codes are used to separate the different users. The same Gold codes can be used in the uplink with a 10-ms period, or short codes with a 256-chip period belonging to the family of extended S(2) codes. The scrambling codes are not spreading codes, that is, they are not spreading the data over the greater bandwidth. It is done only with the help of channelization codes. The scrambling code is used on top of OVSF channelization codes.

The data within uplink and downlink is split into different *logical channels* depending on the aim of the data. Logical channels define the type of data that is transferred. A general classification of logical channels is into two groups:

- **Control Channels** (for the transfer of control information);
- **Traffic Channels** (for the transfer of user information).

Thus, the logical channels define the content of the transmitted information, that is, the logical channel describes *what* is transferred over the radio interface.

Another concept is used to describe *how* and with what characteristics data are transferred over the radio interface. This logical structure is *transport channel*. The physical layer (L1) is required to support variable bit rate transport channels to offer bandwidth-on-demand services, and to be able to multiplex several services to one connection [1]. Any control or traffic channel is transmitted via some transport channel. In other words, transport channels specify the corresponding required transmission characteristics by the physical layer (L1).There exists two types of transport channels:

- **Common transport channels.** These channels are used in case there is a need for inband identification of the UEs when particular UEs are addressed.
- **Dedicated transport channels.** These channels are used when the UEs are identified by the physical channel, that is, code and frequency for FDD and code, time slot and frequency for TDD.

The logical channel can be mapped to one transport channel or to several different transport channels, that is, the information corresponding to a particular logical channel can be transferred via different transport channels. In other words the transport channels are used as a resource for logical channels.

Physical channels define the exact physical characteristics of the radio channel. The logical channels are mapped onto transport channels, and the transport channels in turn are mapped onto physical channels. Physical channels are defined by a specific carrier frequency, scrambling code, channelization code (optional), time start & stop (giving a duration) and, on the uplink, relative phase (0 or $\pi/2$).

11.2.2 Logical and Transport Channels

The configuration of logical channel types is depicted in Figure 11.2 [13]

Broadcast Control Channel (BCCH) is a downlink channel for broadcasting system control information. It bears the common for all UEs in the cell information from Node B to UEs.

Paging Control Channel (PCCH) is a downlink channel that transfers paging information. This channel is used when the network does not know the location cell of the UE, or, the UE is in the cell connected state (utilizing UE sleep mode procedures).

Common Control Channel (CCCH) is a bi-directional channel for transmitting control information between network and UEs. This channel is commonly used by the UEs having no RRC connection with

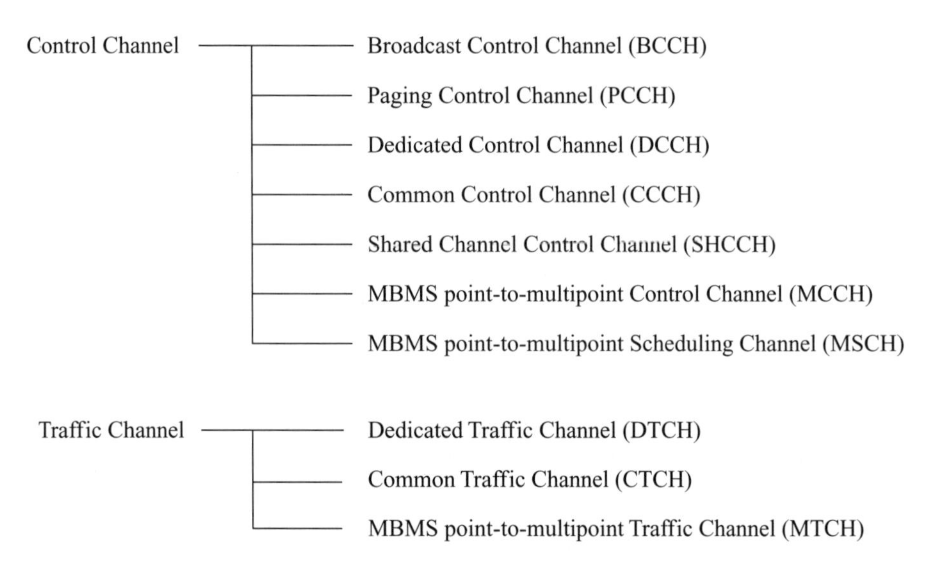

Figure 11.2 Logical channels structure. Reproduced by permission of © 2008 3GPP™

the network and by the UEs using common transport channels when accessing a new cell after cell reselection.

Dedicated Control Channel (DCCH) is a point-to-point bi-directional channel that transmits dedicated control information between a UE and the network.

Shared Channel Control Channel (SHCCH) is a bi-directional channel that transmits control information for uplink and downlink shared channels between network and UEs. This channel is for TDD only.

MBMS point-to-multipoint Control Channel (MCCH) is a point-to-multipoint downlink channel used for transmitting control information from the network to the UE. This channel is only used by UEs that receive Multimedia Broadcast Multicast Service (MBMS). MBMS is a broadcast service that uses multicast distribution in the core network instead of point-to-point links for each UE.

MBMS point-to-multipoint Scheduling Channel (MSCH) is a point-to-multipoint downlink channel used for transmitting scheduling control information, from the network to the UE. This channel is only used by UEs that receive MBMS.

Dedicated Traffic Channel (DTCH) is a point-to-point channel, dedicated to one UE, for the transfer of user information. A DTCH can exist in both uplink and downlink.

Common Traffic Channel (CTCH) is a point-to-multipoint unidirectional channel for transfer of dedicated user information for all or for a group of specified UEs.

MBMS point-to-multipoint Traffic Channel (MTCH) is a point-to-multipoint downlink channel used for transmitting traffic data from the network to the UE. This channel is only used for MBMS.

The configuration of transport channel types is depicted in Figure 11.3. The channels not belonging to Release'99 channels are highlighted.

Common transport channel types are:

- **Random Access Channel(s) (RACH).** It is a contention based uplink channel used for transmission of relatively small amounts of data, for example, for initial access or nonreal-time dedicated control or traffic data. The amount of data borne by this channel is rather small, so it has limited data field, and data rate offered by this type of channel is low.

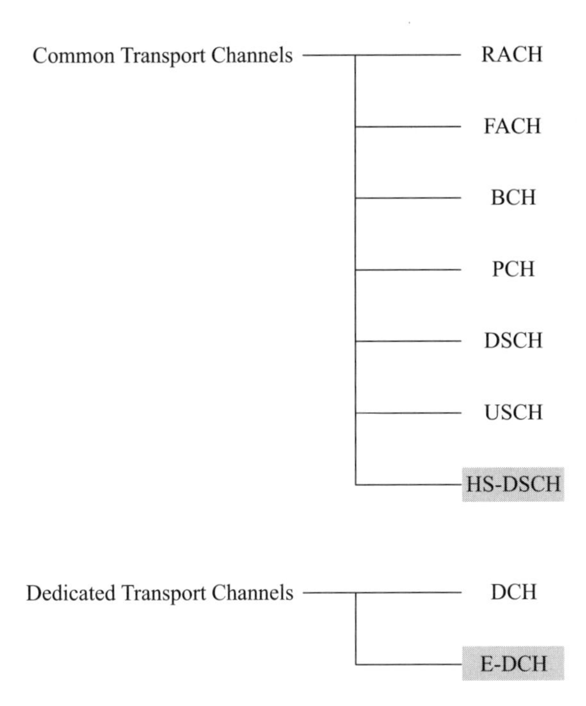

Figure 11.3 Transport channels structure

- **Forward Access Channel(s) (FACH).** It is a common downlink channel without closed-loop power control used for transmission of relatively small amount of data. In addition FACH is used to carry broadcast and multicast data. There can be few FACHs in the cell. At least one of the FACHs must have the data rate low enough to be received by all UEs in the cell, in this case other FACHs can have higher data rates. Due to this the data rate in this type of channel can change fast (each 10ms).
- **Broadcast Channel (BCH).** It is a downlink channel used for broadcast of system- and cell-specific information into an entire cell. The typical information transferred via this channel is the available random access codes and access slots in the cell. The strict requirement of this kind of channel is that it should be received by any UE in the entire coverage area of the cell. The UE cannot be registered in the cell if it is not capable of receiving BCH. The data rate provided by this channel is low and fixed to make it possible for low-end terminals to receive easily the information borne by BCH.
- **Paging Channel (PCH).** It is a downlink channel used for broadcast of control information into an entire cell allowing efficient UE sleep mode procedures. Currently identified information types are paging and notification. It should be broadcast in the entire coverage area of the cell.
- **Downlink Shared Channel(s) (DSCH).** This type of channel is used in TDD only. It is a downlink channel shared by several UEs carrying dedicated control or traffic data. It should be possible to use beamforming and slow power control for this channel.
- **Uplink Shared channel (USCH)**. This type of channel is used in TDD only. It is an uplink channel shared by several UEs carrying dedicated control or traffic data. It should be possible to use beamforming and power control for this channel.
- **High Speed Downlink Shared Channel (HS-DSCH)**. This channel exists in downlink only and is used to carry the packet data. This channel is part of HSDPA concept and is not used in Release'99. The channel should provide high data rates. To achieve this goal, the link adaptation technique is applied to the data transferred with the channel. Also it should be possible to use beamforming and HARQ for this channel.

Dedicated transport channel types are:

- **Dedicated Channel (DCH).** It is a channel dedicated to one UE used in uplink or downlink. The DCH is transmitted over the entire cell or over only a part of the cell using for example, beamforming antennas.
- **Enhanced Dedicated Channel (E-DCH).** It is a channel dedicated to one UE used in uplink only. The E-DCH is subject to Node-B controlled scheduling and HARQ. E-DCH is a part of HSUPA concept and is not used in Release'99.

The transport channels are multiplexed at the physical layer (L1). Due to this operation there exists one more transport channel. It is defined as a single output data stream from the transport channel multiplexing and is denoted as **Coded Composite Transport Channel (CCTrCH)**.

The transport channels form the interface between physical layer (L1) and data link layer (L2). A UE can have simultaneously one or several transport channels in the downlink, and one or more transport channel in the uplink. The basic unit of exchange between layers L1 and L2 is a *transport block*. The number of bits in a transport block (TB) is called the *transport block size (TBS)*. The CRC is attached to each transport block. Actually, the set of transport blocks can be received at the same time instance using the same transport channel. This set is called the *transport block set*. The transport block size is always fixed within a given transport block set, that is, all transport blocks within a transport block set are equally sized. The *transport block set size (TBSS)* is defined as a number of bits in a transport block set. The periodicity at which a transport block set is transferred by the physical layer on the radio interface is called the *transmission time interval (TTI)*, that is, it is the time between consecutive deliveries of data between L2 and L1 [14]. For the Release'99 channels the TTI is always a multiple of 10ms, the length of one radio frame. For the HS-DSCH the TTI is much shorter and comprises 2ms. The example of delivering transport block sets via different transport channels is represented in Figure 11.4.

Each transport channel is associated with one or several *transport formats*. A transport format is defined as a combination of TBS, TBSS, and type of error protection scheme applied to the data. The list of transport formats associated with a particular transport channel is called a *transport format set (TFS)*. The physical layer multiplexes one or several transport channels, and for each transport channel, there exists a TFS. Nevertheless, at a given point in time, not all combinations of transport formats may be used at L1 but only a subset, which is called the *transport format combination (TFC)*. TFC is defined as an authorized combination of the combination of currently valid transport formats that can be submitted simultaneously to the L1 for transmission on a Coded Composite Transport Channel of a UE, that is, containing one transport format from each transport channel. The TFC is represented by a *transport format combination indicator (TFCI)*. There is a one-to-one correspondence between a certain value of the TFCI and a certain Transport Format Combination. The TFCI is used in order to inform the receiving side of the currently valid TFC, and hence how to decode, de-multiplex and deliver the received data

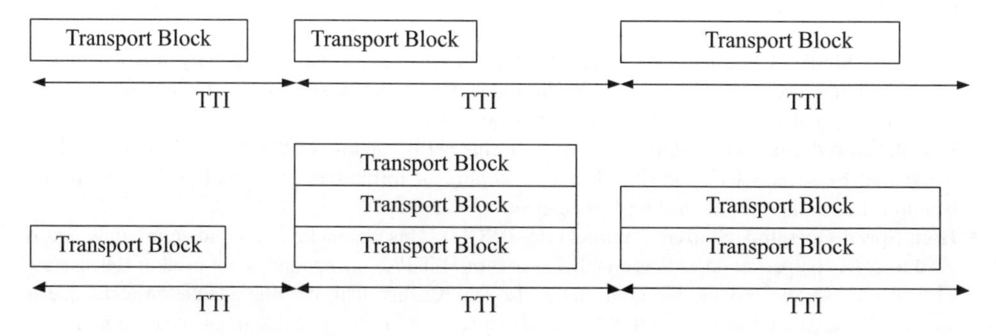

Figure 11.4 Transport block sets and transmission time interval (TTI)

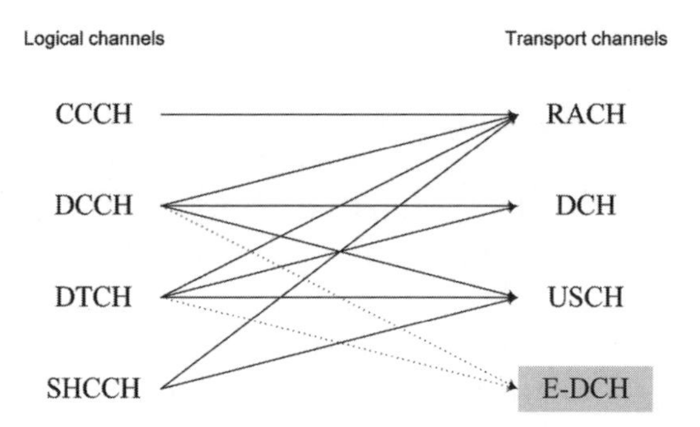

Figure 11.5 Mapping of logical channels onto transport channels. Uplink

on the appropriate transport channels. The TFCI is not used for the HS-DSCH and E-DCH. At each delivery of transport block sets the L1 builds the TFCI from the transport formats of all parallel transport channels of the UE. Through the detection of the TFCI the receiving side is able to identify the TFC. And from the assigned transport format combinations, the receiving side has all the information it needs in order to decode the information.

The mapping of logical channels onto transport channels in uplink is depicted in Figure 11.5, and the mapping of logical channels onto transport channels in downlink is depicted in Figure 11.6. In these pictures channels not belonging to the Release'99 are highlighted.

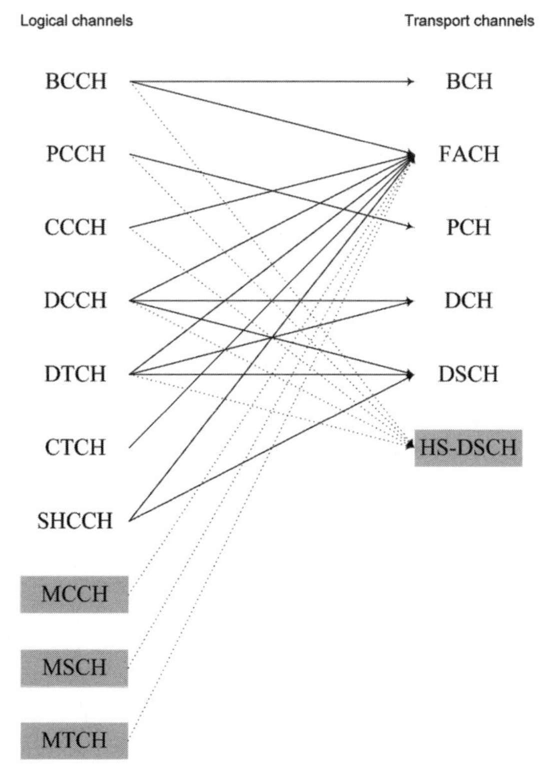

Figure 11.6 Mapping of logical channels onto transport channels. Downlink

Transport Channels	Physical Channels
DCH	Dedicated Physical Data Channel (DPDCH)
	Dedicated Physical Control Channel (DPCCH)
	Fractional Dedicated Physical Channel (F-DPCH)
E-DCH	E-DCH Dedicated Physical Data Channel (E-DPDCH)
	E-DCH Dedicated Physical Control Channel (E-DPCCH)
	E-DCH Absolute Grant Channel (E-AGCH)
	E-DCH Relative Grant Channel (E-RGCH)
	E-DCH Hybrid ARQ Indicator Channel (E-HICH)
RACH	Physical Random Access Channel (PRACH)
	Common Pilot Channel (CPICH)
BCH	Primary Common Control Physical Channel (P-CCPCH)
FACH	Secondary Common Control Physical Channel (S-CCPCH)
PCH	
	Synchronisation Channel (SCH)
	Acquisition Indicator Channel (AICH)
	Paging Indicator Channel (PICH)
	MBMS Notification Indicator Channel (MICH)
HS-DSCH	High Speed Physical Downlink Shared Channel (HS-PDSCH)
	HS-DSCH-related Shared Control Channel (HS-SCCH)
	Dedicated Physical Control Channel (uplink) for HS-DSCH (HS-DPCCH)

Figure 11.7 Mapping of transport channels onto physical channels

11.2.3 Physical Channels

The mapping of transport channels onto physical channels is depicted in Figure 11.7. The channels that do not belong to Release'99 channels are highlighted.

Before mapping onto physical channels the data from Transport block or Transport block set is encoded to offer transport services over the radio transmission link. The data from different transport channels are multiplexed. It is done to provide a continuous data stream, since the continuous data stream allows the channel capacity to be used in a more efficient way. And only after multiplexing the transport channels data are mapped onto physical channels. The process of encoding and mapping is described by the *channel coding scheme*. Channel coding scheme is a combination of error detection, error correcting, rate matching, interleaving and transport channels mapping onto physical channels [16]. The coding/multiplexing steps for uplink and downlink differ only in some details but in any case include *Cyclic Redundancy Check (CRC)* attachment, channel coding, interleaving, rate matching, channel multiplexing, and mapping to physical channels. As was already mentioned in section 11.1.2 the transport channels data after multiplexing forms Coded Composite Transport Channel (CCTrCH). Thus, in fact in most cases only the CCTrCH is mapped to physical channels directly. All other transport channels are mapped onto physical channels via CCTrCH.

11.2.3.1 Uplink Multiplexing

The uplink multiplexing is depicted in Figure 11.8 [16]. This kind of multiplexing is applied to DCH and RACH.

The first operation is CRC attachment to the transport block data. CRC provides the possibility of error detection for each transport block. The size of the CRC is 24, 16, 12, 8 or 0 bits. The actual CRC size that should be used for each TrCH is signalled in TFCI. The process of CRC attachment in more detail is described in section 11.2.4.1.

After the CRC attachment, the next step is the transport blocks concatenation/segmentation. All transport blocks in a TTI are serially concatenated. It is done to fit the transport blocks data to the size of the code block. If the number of bits in a TTI is larger than the maximum size of a code block in question, then code block segmentation is performed after the concatenation of the transport blocks. In other words, if the transport block size is less than the code block size, then several transport blocks are concatenated, and if the size of transport block is greater than the size of code block, the transport block is split into several code blocks. The maximum size of the code blocks depends on whether convolutional coding or turbo coding is used for the TrCH. The maximum code block size for convolutional coding is 504 bits, for turbo coding 5114 bits. The outputs of this procedure are the code blocks of the same length.

After the transport blocks concatenation/segmentation the channel encoding is applied to each code block. The following channel coding schemes can be applied:

- convolutional coding with coding rate 1/3 or 1/2;
- turbo coding with coding rate 1/3.

The coding schemes will be considered in detail in section 11.2.4.

The function of radio frame equalization is to ensure that data can be divided into equal sized blocks when transmitting over more than a single 10 ms radio frame [1] This is done by padding the necessary number of bits until the data can be divided into data segments of the same size.

The first interleaving is called inter-frame interleaving. It is used if the TTI size is greater than the size of radio frame (10 ms). The TTI size can be 10, 20, 40 or 80 ms. If TTI consists of more than one radio frame, then the first interleaving is applied to the data stream. The input bit sequence is written into the R1 $\times$ C1 matrix row by row, where C1 is the number of radio frames in TTI (1, 2, 4 or 8) and R1 is the number of bits in radio frame for transport channel i:

$$\begin{bmatrix} x_{i,1} & x_{i,2} & x_{i,3} & \cdots & x_{i,C1} \\ x_{i,(C1+1)} & x_{i,(C1+2)} & x_{i,(C1+3)} & \cdots & x_{i,(2\times C1)} \\ \vdots & \vdots & \vdots & \cdots & \vdots \\ x_{i,((R1-1)\times C1+1)} & x_{i,((R1-1)\times C1+2)} & x_{i,((R1-1)\times C1+3)} & \cdots & x_{i,(R1\times C1)} \end{bmatrix} \quad (11.1)$$

Then the columns of this matrix are permutated. After permutation of the columns, the bits are denoted by y_{ik}:

$$\begin{bmatrix} y_{i,1} & y_{i,(R1+1)} & y_{i,(2\times R1+1)} & \cdots & y_{i,((C1-1)\times R1+1)} \\ y_{i,2} & y_{i,(R1+2)} & y_{i,(2\times R1+2)} & \cdots & y_{i,((C1-1)\times R1+2)} \\ \vdots & \vdots & \vdots & \cdots & \vdots \\ y_{i,R1} & y_{i,(2\times R1)} & y_{i,(3\times R1)} & \cdots & y_{i,(C1\times R1)} \end{bmatrix} \quad (11.2)$$

To obtain the interleaved output bit sequence of the ith transport channel the data is read column by column from the inter-column permuted R1 $\times$ C1 matrix.

The radio frame segmentation is applied when the transmission time interval is longer than 10 ms. In this case the input bit sequence is segmented and mapped onto several consecutive radio frames. Due to

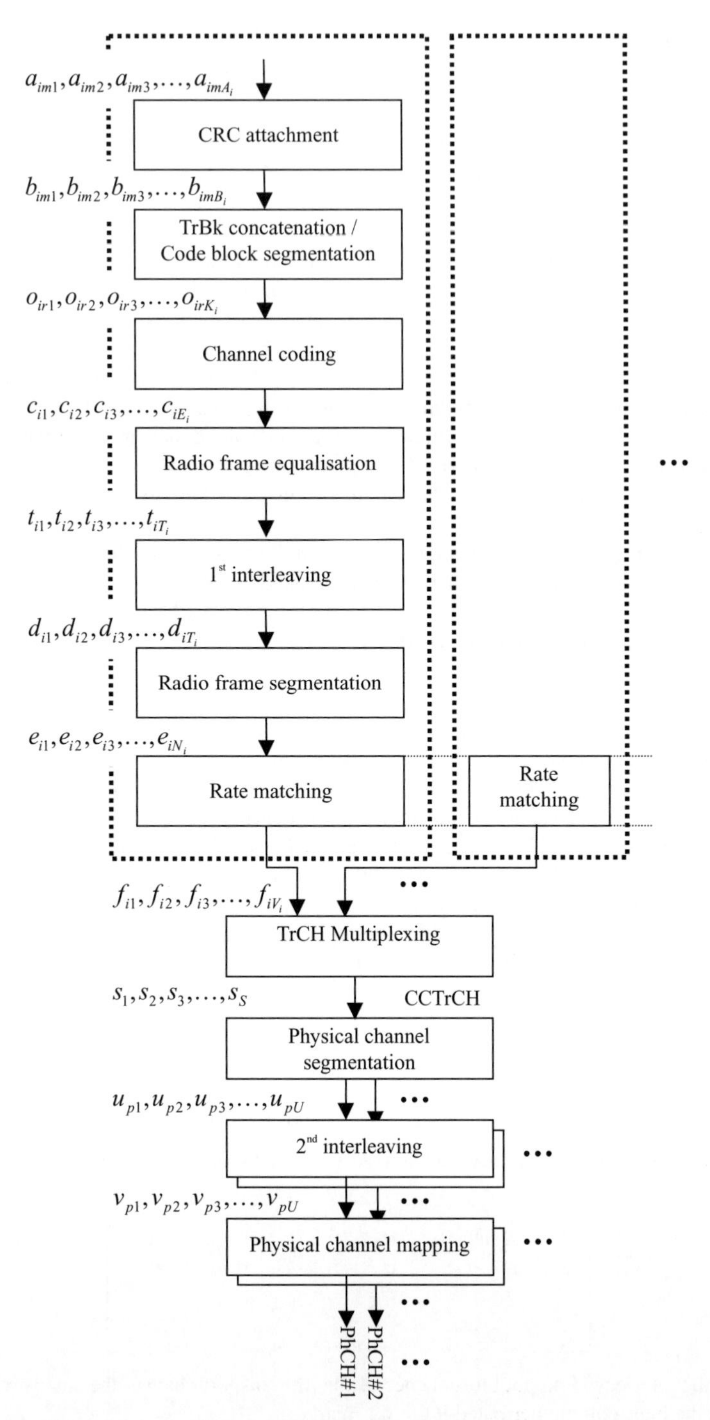

Figure 11.8 Transport channel multiplexing structure for uplink. Reproduced by permission of © 2008 3GPP™

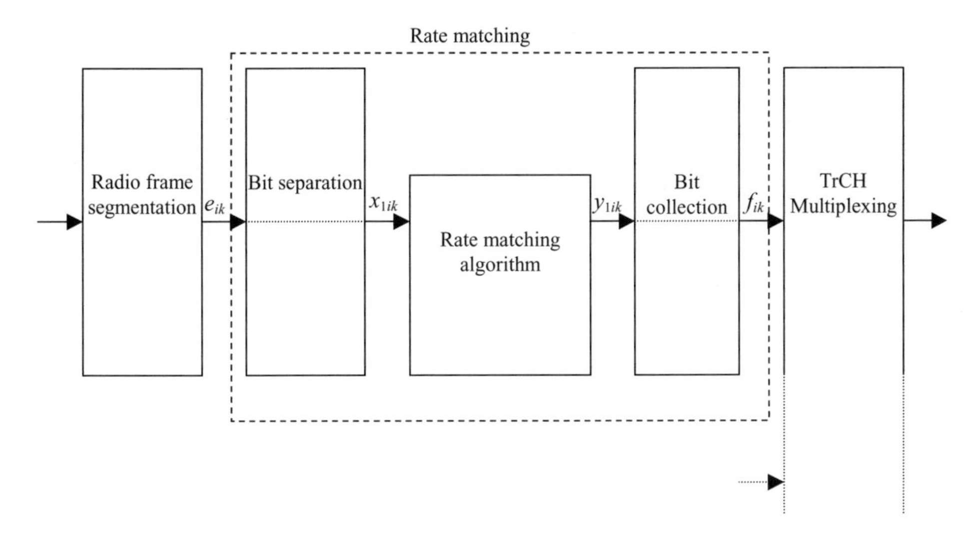

Figure 11.9 Rate matching for convolutionally encoded TrCHs and for turbo encoded TrCHs with repetition in uplink. Reproduced by permission of © 2008 3GPP™

the radio frame size equalization in the UL the input bit sequence length is guaranteed to be an integer multiple of radio frame size.

Rate matching is used to match the number of bits to be transmitted to the number of bits available on a single frame. Rate matching means that bits on a transport channel are repeated or punctured. The number of bits on a transport channel can vary between different TTIs. And in accordance with this change the bits are repeated or punctured to ensure that the total bit rate after transport channels multiplexing is identical to the total channel bit rate of the allocated dedicated physical channels. The rate matching procedure is very complicated and contains a lot of options. Here we consider only the core algorithm. The rate matching for uplink is illustrated in Figure 11.9 [16] and Figure 11.10 [16]

Before starting the rate matching some rate matching parameters should be calculated. First of all to match the number of bits in the transport channel to number of bits in allocated physical channels, the number of bits to be punctured or repeated should be calculated. Also the initial parameters for the rate matching pattern determination algorithm should be calculated (for details see section 4.2.7.1 in [16]). As can be seen from Figures 11.9 and 11.10, the first block in rate matching is *bit separation*. Actually this procedure is active only for puncturing of turbo encoded transport channels. For convolutionally encoded transport channels and for turbo encoded transport channels with repetition the bit separation function and bit collection functions are transparent, that is:

$$x_{1,i,k} = e_{i,k}, \quad k = 1, 2, \ldots, X_i; \quad X_i = N_i \tag{11.3}$$

where N_i is the number of bits in a radio frame before rate matching on TrCH i.

The goal of bit separation is to separate the bit sequence input to the rate matching block into three sequences: systematic bits, first parity bits, and second parity bits. The reason for this is that, as can be seen from Figure 11.10, the systematic bits of turbo encoded transport channel shall not be punctured. Only parity bits may be punctured. To be more precise, the three output sequences at the output of bit separation function contain the following bits.

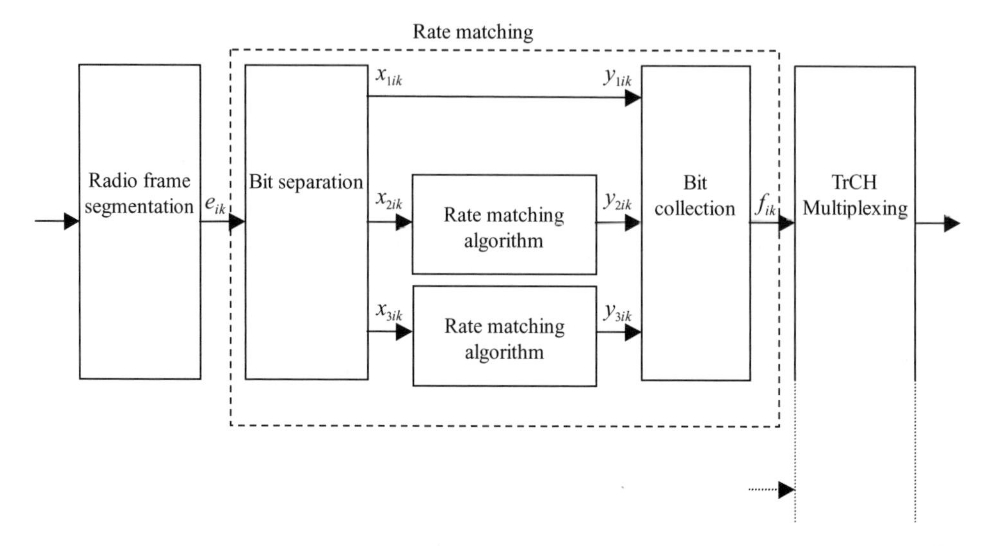

Figure 11.10 Puncturing of turbo encoded TrCHs in uplink. Reproduced by permission of © 2008 3GPP™

The first sequence contains:

- All of the systematic bits that are from turbo encoded TrCHs.
- From 0 to 2 first and/or second parity bits that are from turbo encoded TrCHs. These bits come into the first sequence when the total number of bits in a block after radio frame segmentation is not a multiple of three.
- Some of the systematic, first parity and second parity bits that are for trellis termination.

$$x_{1,i,k} = e_{i,3(k-1)+1+(\alpha_1+\beta_{n_i}) \bmod 3}, \quad k = 1, 2, \ldots, X_i; \quad X_i = \lfloor N_i/3 \rfloor \tag{11.4}$$

where α_1 is the offset due to 1st interleaving listed in Table 11.1 [16], n_i is the radio frame number for TrCH i (the bit separation is different for different radio frames in the TTI), β_{n_i} is the offset for the radio frame number n_i listed in Table 11.2 [16].

The second sequence contains:

- All of the first parity bits that are from turbo encoded TrCHs, except those that go into the first sequence when the total number of bits is not a multiple of three.
- Some of the systematic, first parity and second parity bits that are for trellis termination.

$$x_{2,i,k} = e_{i,3(k-1)+1+(\alpha_3+\beta_{n_i}) \bmod 3}, \quad k = 1, 2, \ldots, X_i; \quad X_i = \lfloor N_i/3 \rfloor \tag{11.5}$$

Table 11.1 TTI dependent offset needed for bit separation

TTI (ms)	α_1	α_2	α_3
10, 40	0	1	2
20, 80	0	2	1

Reproduced by permission of © 2008 3GPP™

Table 11.2 Radio frame dependent offset needed for bit separation

TTI (ms)	β_0	β_1	β_2	β_3	β_4	β_5	β_6	β_7
10	0	NA	NA	NA	NA	NA	NA	NA
20	0	1	NA	NA	NA	NA	NA	NA
40	0	1	2	0	NA	NA	NA	NA
80	0	1	2	0	1	2	0	1

The third sequence contains:

- All of the second parity bits that are from turbo encoded TrCHs, except those that go into the first sequence when the total number of bits is not a multiple of three.
- Some of the systematic, first parity and second parity bits that are for trellis termination.

$$x_{3,i,k} = e_{i,3(k-1)+1+(\alpha_3+\beta_{n_i}) \bmod 3}, \quad k = 1, 2, \ldots, X_i; \quad X_i = \lfloor N_i/3 \rfloor \tag{11.6}$$

The second and third sequences shall be of equal length, whereas the first sequence can contain from 0 to 2 more bits. Puncturing is applied only to the second and third sequences.

The core of rate matching procedure is the *rate matching pattern determination algorithm.* This algorithm determines the positions of bits that should be punctured or repeated.

The inputs of the algorithm are bit sequence $x_{i1}, x_{i2}, \ldots, x_{iX_i}$, where i is the TrCH number and the parameters X_i, e_{ini}, e_{plus}, and e_{minus} that was calculated before starting the rate matching procedure.

The rate matching rule is as follows:

```
if puncturing is to be performed
        e = e_ini          – initial error between current and desired puncturing ratio
        m = 1              – index of current bit
        do while m <= X_i
                e = e – e_minus              – update error
                if e <= 0 then               – check if bit number m should be punctured
                        set bit x_i,m to δ where δ∉{0, 1}
                        e = e + e_plus – update error
                end if
                m = m + 1                    – next bit
        end do
else
        e = e_ini          – initial error between current and desired puncturing ratio
        m = 1              – index of current bit
        do while m <= X_i
                e = e – e_minus              – update error
                do while e <= 0              – check if bit number m should be repeated
                        repeat bit x_i,m
                        e = e + e_plus – update error
                end do
                m = m + 1                    – next bit
        end do
end if
```

A repeated bit is placed directly after the original one.

This algorithm provides fast rate matching pattern calculation for any TFCI and for all possible values of number of bits in transport channel and the number of bits in allocated physical channels. The algorithm is suboptimal in the sense that in some cases better puncturing schemes than provided by the algorithm exist. On the other hand, storing all possible puncturing or repetition schemes requires too much memory.

Bit collection is the inverse function of the bit separation. Also, the process of puncturing is performed in bit collection. As can be seen from the rate matching pattern determination algorithm the bits to be punctured are substituted by non-binary symbols δ. During bit collection procedure these bits are removed. The bits after collection are denoted by $z_{i,1}, z_{i,2}, \ldots, z_{i,Y_i}$. After bit collection, the bits indicated as punctured are removed and the bits are then denoted by $f_{i,1}, f_{i,2}, \ldots, f_{i,V_i}$, where i is the TrCH number and $V_i = N_i + \Delta N_i$, where N_i is the number of bits in a radio frame before rate matching on TrCH i. and ΔN_i, if it is positive is the number of bits that should be repeated, if negative it is the number of bits that should be punctured in each radio frame on TrCH i. The relations between $y_{b,i,k}$, $z_{i,k}$, and f_{ik} are given below.

For turbo encoded TrCHs with puncturing ($Y_i = X_i$):

$$z_{i,3(k-1)+1+(\alpha_1+\beta_{n_i}) \bmod 3} = y_{1,i,k}, \quad k = 1, 2, \ldots, Y_i \tag{11.7}$$

$$z_{i,3\lfloor N_i/3\rfloor+k} = y_{1,i,\lfloor N_i/3\rfloor+k}, \quad k = 1, 2, \ldots, N_i \bmod 3 \tag{11.8}$$

When (N_i mod 3) = 0 (11.8) is not needed:

$$z_{i,3(k-1)+1+(\alpha_2+\beta_{n_i}) \bmod 3} = y_{2,i,k}, \quad k = 1, 2, \ldots, Y_i \tag{11.9}$$

$$z_{i,3(k-1)+1+(\alpha_3+\beta_{n_i}) \bmod 3} = y_{3,i,k}, \quad k = 1, 2, \ldots, Y_i \tag{11.10}$$

After the bit collection, bits $z_{i,k}$ with value δ, where $\delta \notin \{0, 1\}$, are removed from the bit sequence. Bit $f_{i,1}$ corresponds to the bit $z_{i,k}$ with smallest index k after puncturing, bit $f_{i,2}$ corresponds to the bit $z_{i,k}$ with second smallest index k after puncturing, and so on.

For convolutionally encoded TrCHs and turbo encoded TrCHs with repetition:

$$z_{i,k} = y_{1,i,k}, \quad k = 1, 2, \ldots, Y_i \tag{11.11}$$

When repetition is used, $f_{i,k} = z_{i,k}$ and $Y_i = V_i$. When puncturing is used, $Y_i = X_i$.

Every 10 ms, one radio frame from each transport channel is delivered to the transport channel multiplexing. These radio frames are serially multiplexed into a coded composite transport channel (CCTrCH).

When more than one physical channel is used, physical channel segmentation divides the bits among the different physical channels. Each physical channel corresponds to some spreading code. The segmentation is done serially.

The second interleaving is an intra-frame interleaving. It performs the bits permutation inside the 10 ms radio frame. It comprises the bits input to a matrix with padding, the inter-column permutation for the matrix and bits output from the matrix with pruning. The bits input to the interleaver are denoted by $u_{p,1}, u_{p,2}, u_{p,3}, \ldots, u_{p,U}$, where p is the physical channel number and U is the number of bits in one radio frame for one physical channel. The input bit sequence $u_{p,1}, u_{p,2}, u_{p,3}, \ldots, u_{p,U}$ is written into the R2 $\times$ C2 matrix row by row. The number of columns of the matrix C2 $= 30$, R2 is the minimum integer R2 such that $U \leq$ R2 $\times$ C2. The interleaver matrix can be written as follows:

$$\begin{bmatrix} y_{p,1} & y_{p,2} & y_{p,3} & \cdots & y_{p,\mathrm{C2}} \\ y_{p,(\mathrm{C2}+1)} & y_{p,(\mathrm{C2}+2)} & y_{p,(\mathrm{C2}+3)} & \cdots & y_{p,(2\times\mathrm{C2})} \\ \vdots & \vdots & \vdots & \vdots & \vdots \\ y_{p,((\mathrm{R2}-1)\times\mathrm{C2}+1)} & y_{p,((\mathrm{R2}-1)\times\mathrm{C2}+2)} & y_{p,((\mathrm{R2}-1)\times\mathrm{C2}+3)} & \cdots & y_{p,((\mathrm{R2}\times\mathrm{C2})} \end{bmatrix} \tag{11.12}$$

Table 11.3 Inter-column permutation pattern for 2nd interleaving

Number of columns C2	Inter-column permutation pattern < P2(0), P2(1), ..., P2(C2-1) >
30	<0, 20, 10, 5, 15, 25, 3, 13, 23, 8, 18, 28, 1, 11, 21, 6, 16, 26, 4, 14, 24, 19, 9, 29, 12, 2, 7, 22, 27, 17>

where $y_{p,k} = u_{p,k}$ for $k = 1, 2, \ldots, U$ and if R2 × C2 > U, the dummy bits are padded for $k = U + 1, U + 2, \ldots,$ R2 × C2. These dummy bits are pruned away from the output of the matrix after the inter-column permutation. Then the matrix columns are permutated and the output of the interleaver is the bit sequence read out column by column from the inter-column permuted R2 × C2 matrix. The permutation of matrix columns is performed with the help of the pattern $\langle P2(j)\rangle_{j\in\{0,1,\ldots,C2-1\}}$ that is shown in Table 11.3 [16], where P2(j) is the original column position of the j-th permuted column.

The output is pruned by deleting dummy bits that were padded to the input of the matrix before the inter-column permutation.

The bits from the output of the second interleaving are mapped to the physical channels. The physical channels used during a radio frame are either completely filled with bits that are transmitted over the air or not used at all. The number of bits mapped to each physical channel is exactly the number of chips bearing by this channel divided by the corresponding *spreading factor (SF)*, for example, one radio frame (10 ms) comprises 38400 chips, if the SF = 256, exactly 150 bits should be mapped to one radio frame of this particular physical channel.

11.2.3.2 Uplink DPCCH and DPDCH

The *uplink Dedicated Physical Data Channel (uplink DPDCH)* is used to carry the DCH transport channel, that is, to transmit the user data. There may be zero, one, or several uplink DPDCHs on each radio link. The *uplink Dedicated Physical Control Channel (uplink DPCCH)* is used to carry control information generated at L1. The L1 control information consists of known pilot bits to support channel estimation for coherent detection, *transmit power-control (TPC)* commands, *feedback information (FBI)*, and an optional transport format combination indicator (TFCI). The transport format combination indicator informs the receiver about the instantaneous transport format combination of the transport channels mapped to the simultaneously transmitted uplink DPDCH radio frame. There is one and only one uplink DPCCH on each radio link. The frame structure of the uplink DPDCH and the uplink DPCCH is depicted in Figure 11.11 [15]. Each radio frame of length 10 ms is split into five subframes, each of three slots, each of length $T_{slot} = 2560$ chips, corresponding to one power-control period. The DPDCH and DPCCH are always frame aligned with each other.

The parameter k in Figure 11.11 determines the number of bits per uplink DPDCH slot. It is related to the spreading factor (SF) of the DPDCH as $SF = \frac{256}{2^k}$. The DPDCH spreading factor may range from 4 to 256.

The spreading factor of the uplink DPCCH is always equal to 256, that is, there are 10 bits per uplink DPCCH slot. The FBI bits are used to support techniques requiring feedback from the UE for operation of closed loop mode transmit diversity. The use of the FBI bits is described in detail in section 11.2.6. Multi-code operation is possible for the uplink dedicated physical channels. When multi-code transmission is used, several parallel DPDCH are transmitted using different channelization codes. However, there is only one DPCCH per radio link. The maximum user data rate on a single DPDCH is derived from the maximum channel bit rate, which is 960 Kbits/s without channel coding with spreading factor 4. If the higher data rates are needed, up to six parallel DPDCH can be transmitted [1]. Due to this the maximum channel bit rate can be raised up to 5740 Kbits/s (see Table 11.4).

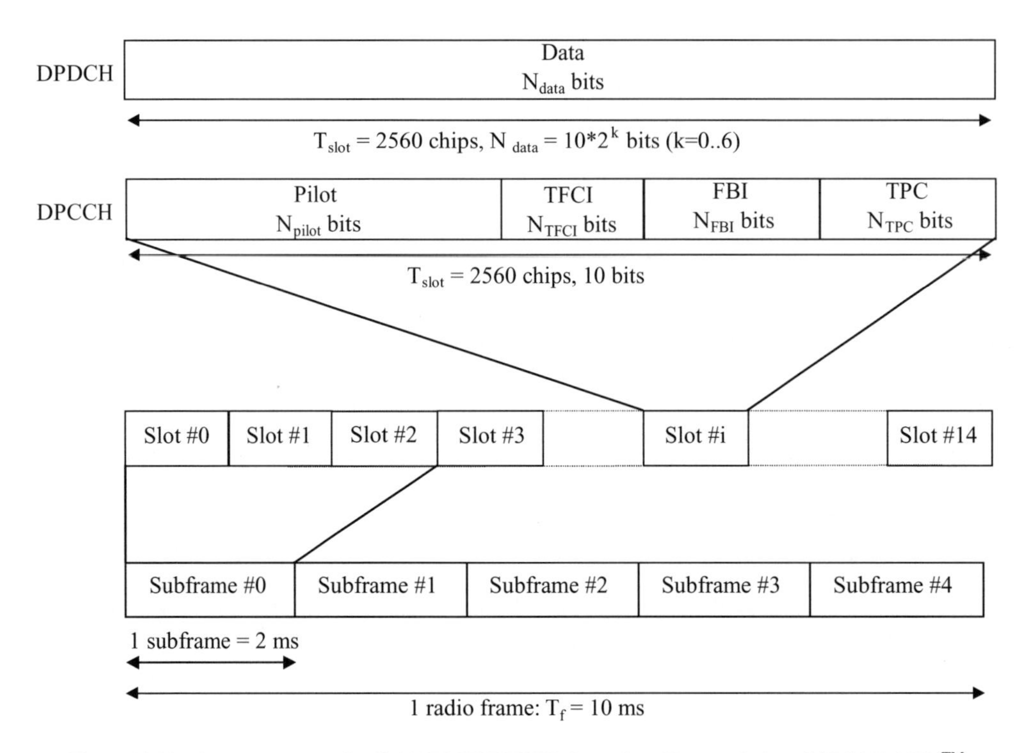

Figure 11.11 Frame structure of uplink DPDCH/DPCCH. Reproduced by permission of © 2008 3GPP™

11.2.3.3 Physical Random Access Channel (PRACH)

The *Physical Random Access Channel (PRACH)* is used to carry the RACH. The random-access transmission is based on a Slotted ALOHA approach [17] with fast acquisition indication. The UE can start the random-access transmission only at the beginning of a set of pre-defined time intervals, denoted *access slots*. There are 15 access slots per two frames and they are spaced 5120 chips apart, see Figure 11.12 [15]. Information on what access slots are available for random-access transmission is given by higher layers.

Table 11.4 Uplink DPDCH data rates

DPDCH SF	Number of parallel channels (spreading codes)	DPDCH channel bit rate (Kbits/s)	Maximum user data rate (with coding rate $^1/_2$) (Kbits/s)
256	1	15	7.5
128	1	30	15
64	1	60	30
32	1	120	60
16	1	240	120
8	1	480	240
4	1	960	480
4	4	5740	2.3 Mbits/s

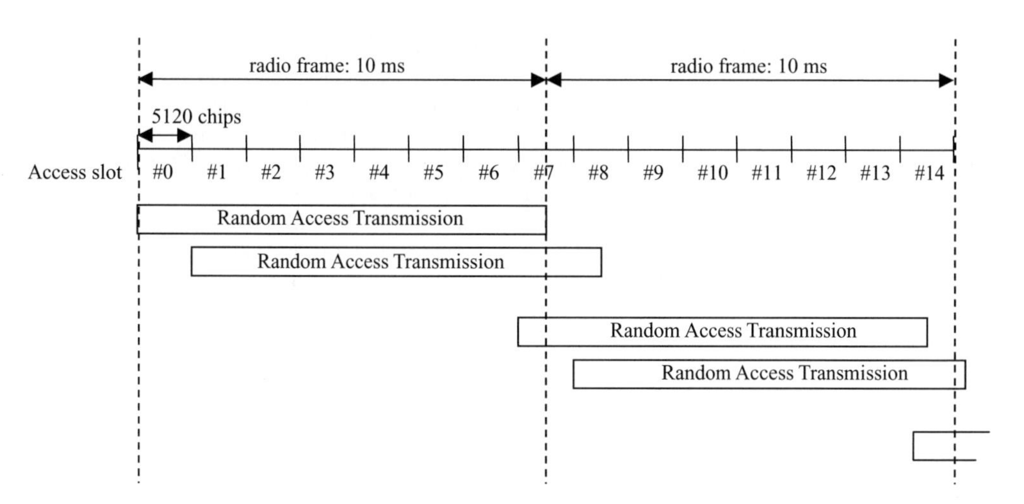

Figure 11.12 RACH access slot numbers and their spacing. Reproduced by permission of © 2008 3GPP™

The random-access transmission consists of one or several *preambles* of length 4096 chips and a *message* of length 10 ms or 20 ms. Random access is the mechanism through which the UE performs the initial access to the system. As the UE is not yet connected to the system, no uplink codes have been allocated, nor is there any control of the uplink power. To choose the correct power UE sends preambles with randomized signatures with increasing power in a "ramping" procedure. When the power is sufficient, the BS transmits an acquisition indicator to request transmission of the actual message from the terminal. This ramping procedure is very fast and together with the signature randomization leads to an efficient, high-capacity RACH. Each preamble is of length 4096 chips and consists of 256 repetitions of a signature of length 16 chips. Totally there are 16 available signatures. When the preamble is detected by BS it sends the acknowledgment with the Acquisition Indicator Channel (AICH). Then UE transmits the message part of the random-access transmission. Figure 11.13 shows the structure of the random-access message part radio frame [15].

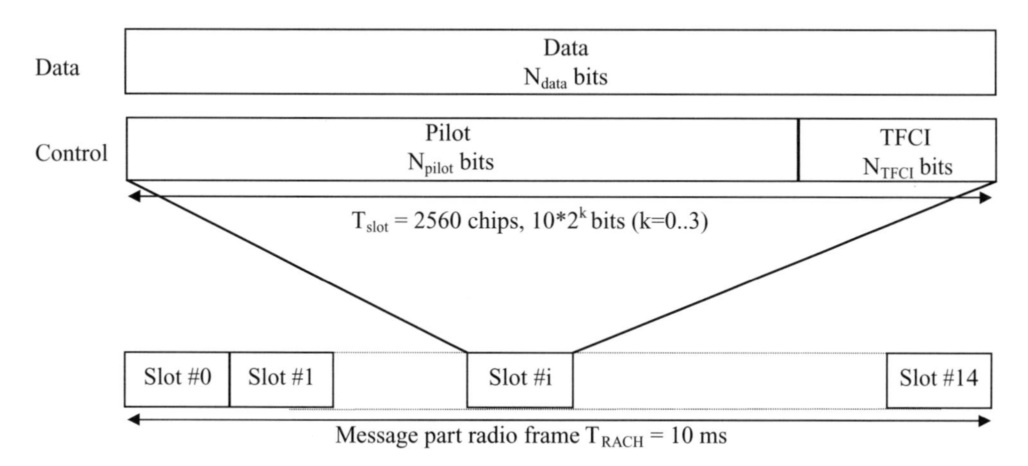

Figure 11.13 Structure of the random-access message part radio frame. Reproduced by permission of © 2008 3GPP™

The 10 ms message part radio frame is split into 15 slots, each of length $T_{slot} = 2560$ chips. Each slot consists of two parts, a data part to which the RACH transport channel is mapped and a control part that carries L1 control information. The data and control parts are transmitted in parallel. A 10 ms message part consists of one message part radio frame, while a 20 ms message part consists of two consecutive 10 ms message part radio frames. The data part consists of $10 \cdot 2^k$ bits, where $k = 0,1,2,3$. This corresponds to a spreading factor of 256, 128, 64, and 32 respectively for the message data part. The control part consists of eight known pilot bits to support channel estimation for coherent detection and two TFCI bits. This corresponds to a spreading factor of 256 for the message control part. The total number of TFCI bits in the random-access message is 15*2 = 30. The TFCI of a radio frame indicates the transport format of the RACH transport channel mapped to the simultaneously transmitted message part radio frame. In case of a 20 ms PRACH message part, the TFCI is repeated in the second radio frame.

11.2.3.4 Downlink Multiplexing

The downlink multiplexing is depicted in Figure 11.14 [16]. This kind of multiplexing is applied to BCH, FACH and PCH.

The greatest part of downlink multiplexing procedures coincides with the uplink ones. The main difference is that the rate matching in downlink is done instead of radio frame size equalization in the uplink. The rate matching procedures in uplink and downlink also differs, but the aim of rate matching in downlink is the same as in uplink. Rate matching procedure for downlink is illustrated in Figures 11.15 [16] and 11.16 [16].

The rate matching pattern determination algorithm is the same for both uplink and downlink. The difference is in calculation of the initial parameters for rate matching pattern determination algorithm (for details see section 4.2.7.2 in [16]) and in bit separation and bit collection procedures. The bits input to the rate matching are denoted by $c_{i1}, c_{i2}, \ldots, c_{iE_i}$, where i is the TrCH number and E_i is the number of bits input to the rate matching block. In the same way as in uplink in downlink the bit separation function is transparent for convolutionally encoded TrCHs and for turbo encoded TrCHs with repetition as it is shown in Figure 11.15, that is:

$$x_{1,i,k} = c_{i,k}, \quad k = 1, 2, \ldots, X_i; \quad X_i = E_i \tag{11.13}$$

For turbo encoded TrCHs with puncturing:

$$x_{1,i,k} = c_{i,3(k-1)+1}, \quad k = 1, 2, \ldots, X_i; \quad X_i = E_i/3 \tag{11.14}$$

$$x_{2,i,k} = c_{i,3(k-1)+2}, \quad k = 1, 2, \ldots, X_i; \quad X_i = E_i/3 \tag{11.15}$$

$$x_{3,i,k} = c_{i,3(k-1)+3}, \quad k = 1, 2, \ldots, X_i; \quad X_i = E_i/3 \tag{11.16}$$

Like in uplink the internal variables in bit collection procedure corresponding to bits after collection are denoted by $z_{i,1}, z_{i,2}, \ldots, z_{i,Y_i}$. After bit collection, the bits indicated as punctured are removed and the bits are then denoted by $g_{i,1}, g_{i,2}, \ldots, g_{i,G_i}$, where i is the TrCH number and $G_i = N_i^{TTI} + \Delta N_i^{TTI}$, where N_i^{TTI} is the number of bits in a transmission time interval before rate matching on TrCH i, and ΔN_i^{TTI} if positive is the number of bits to be repeated in each TTI on TrCH i, if negative is the number of bits to be punctured in each TTI on TrCH i. For convolutionally encoded TrCHs and turbo encoded TrCHs with repetition the bit collection procedure is transparent:

$$z_{i,k} = y_{1,i,k}, \quad k = 1, 2, \ldots, Y_i \tag{11.17}$$

For turbo encoded TrCHs with puncturing ($Y_i = X_i$):

$$z_{i,3(k-1)+1} = y_{1,i,k}, \quad k = 1, 2, \ldots, Y_i \tag{11.18}$$

$$z_{i,3(k-1)+2} = y_{2,i,k}, \quad k = 1, 2, \ldots, Y_i \tag{11.19}$$

$$z_{i,3(k-1)+3} = y_{3,i,k}, \quad k = 1, 2, \ldots, Y_i \tag{11.20}$$

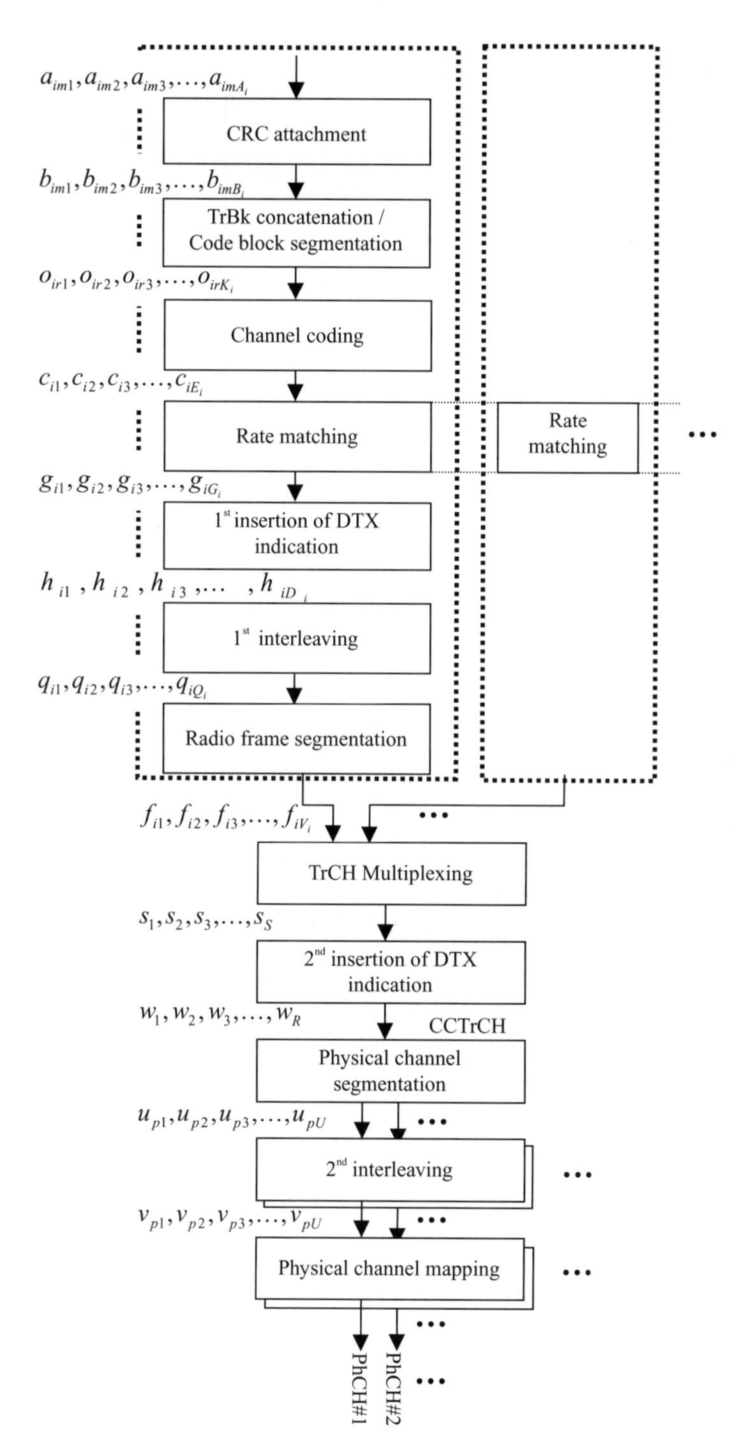

Figure 11.14 Transport channel multiplexing structure for downlink. Reproduced by permission of © 2008 3GPP™

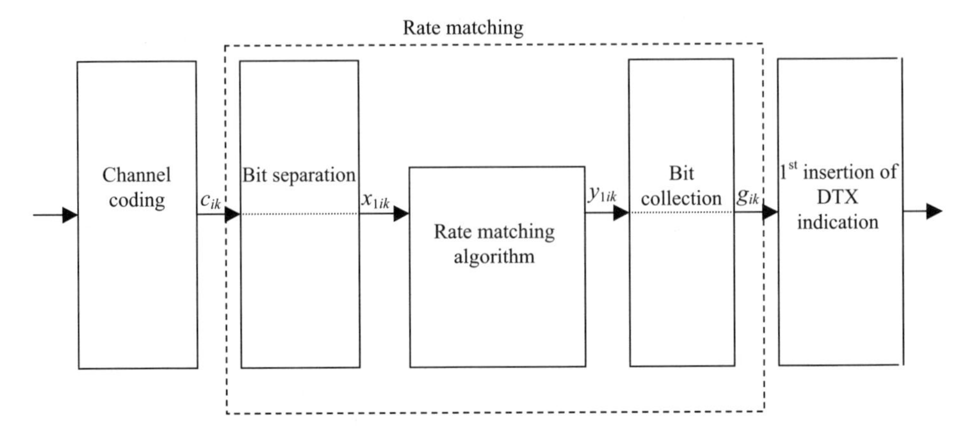

Figure 11.15 Rate matching for convolutionally encoded TrCHs and for turbo encoded TrCHs with repetition in downlink

After the bit collection, bits $z_{i,k}$ with value δ, where $\delta \notin \{0, 1\}$, are removed from the bit sequence. Bit $g_{i,1}$ corresponds to the bit $z_{i,k}$ with smallest index k after puncturing, bit $g_{i,2}$ corresponds to the bit $z_{i,k}$ with second smallest index k after puncturing, and so on. As can be seen from comparison of (11.3)–(11.6) with (11.13)–(11.16) and (11.7)–(11.11) with (11.17)–(11.20), the principle of bit separation and bit collection procedures is quite similar for both uplink and downlink.

Another step that differs between the downlink multiplexing chain and that of the uplink one is the insertion of *discontinuous transmission (DTX)* indication bits. In the downlink the transmission is interrupted if the number of bits is lower than maximum (that is, DTX is used to fill up the radio frame with bits). The insertion point of DTX indication bits depends on whether fixed or flexible positions of the transport channels in the radio frame are used. It is up to the network to decide for each CCTrCH whether fixed or flexible positions are used during the connection. DTX indication bits only indicate when the transmission should be turned off, they are not transmitted.

In downlink as in uplink it is possible to use more than one physical channel to transmit the data. In this case the physical channel segmentation is applied. The interleaving procedures in downlink are done in the same way as in uplink.

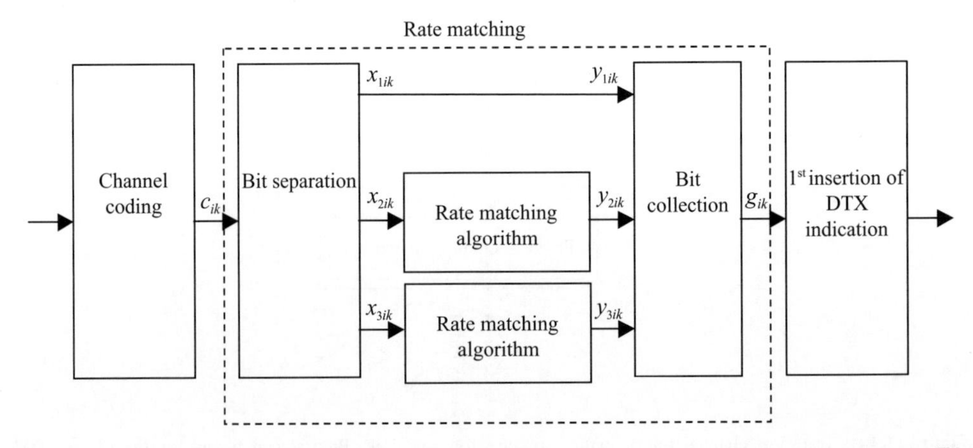

Figure 11.16 Puncturing of turbo encoded TrCHs in downlink. Reproduced by permission of © 2008 3GPP™

The difference of the downlink physical channel mapping from the uplink one is that in downlink, the physical channels do not need to be completely filled with bits that are transmitted over the air. For example, the bits corresponding to DTX indicators are mapped to the DPCCH/DPDCH fields but are not transmitted over the air.

11.2.3.5 Downlink DPCH

Within one *Downlink Dedicated Physical Channel (downlink DPCH)*, dedicated data generated at Layer 2 and above, that is, the dedicated transport channel (DCH), is transmitted in time-multiplex with control information generated at Layer 1 (known pilot bits, TPC commands, and an optional TFCI). The downlink DPCH can thus be seen as a time multiplex of a downlink DPDCH and a downlink DPCCH, compared to uplink DPDCH/DPCCH. The frame structure of the downlink DPCH is represented in Figure 11.17 [15]. Each frame of length 10 ms is split into 15 slots, each of length $T_{slot} = 2560$ chips, corresponding to one power-control period.

The parameter k in Figure 11.17 determines the total number of bits per downlink DPCH slot. It is related to the spreading factor (SF) of the physical channel as $SF = 512/2^k$. The spreading factor may thus range from 512 down to 4. The possible data rates are listed in Table 11.5.

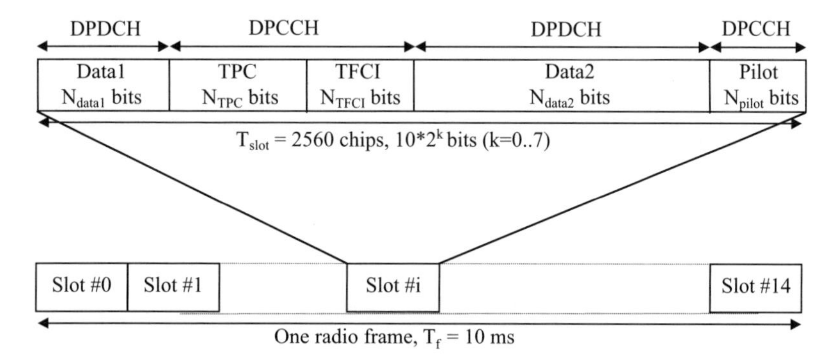

Figure 11.17 Frame structure of downlink DPCH. Reproduced by permission of © 2008 3GPP™

Table 11.5 Downlink DPCH data rates

DPCH SF	Number of parallel channels (spreading codes)	DPCH channel symbol rate (Ksymbols/s)	DPDCH channel bit rate (Kbits/s)	Maximum user data rate (with coding rate ½) (Kbits/s)
512	1	7.5	3–6	1–3
256	1	15	12–246	6–12
128	1	30	42–51	20–24
64	1	60	90	45
32	1	120	210	105
16	1	240	432	215
8	1	480	912	456
4	1	960	1872	936
4	3	2880	5616	2.3 Mbits/s

There are basically two types of downlink Dedicated Physical Channels; those that include TFCI (for example, for several simultaneous services) and those that do not include TFCI (for example, for fixed-rate services). It is the network that determines if a TFCI should be transmitted.

11.2.3.6 Common Pilot Channel (CPICH)

The CPICH is a fixed rate (30 kbps, SF = 256) downlink physical channel that carries a pre-defined bit sequence. This bit sequence is used in the UE receiver for channel estimation. Also CPICH can be used for the measurements during the handover and cell selection. Figure 11.18 [15] shows the frame structure of the CPICH.

In cases where transmit diversity is used the CPICH shall be transmitted from both antennas using the same channelization and scrambling code. In this case, the pre-defined bit sequence of the CPICH is different for Antenna 1 and Antenna 2. These sequences are orthogonal and of even length of no less than 4 bits. There is no transport channel mapped to CPICH.

There are two types of Common pilot channels, the Primary and Secondary CPICH. They differ in their use and the limitations placed on their physical features.

The *Primary Common Pilot Channel (P-CPICH)* has the following characteristics:

- The same channelization code is always used for the P-CPICH;
- The P-CPICH is scrambled by the primary scrambling code;
- There is one and only one P-CPICH per cell;
- The P-CPICH is broadcast over the entire cell.

The *Secondary Common Pilot Channel (S-CPICH)* has the following characteristics:

- An arbitrary channelization code of SF = 256 is used for the S-CPICH;
- The S-CPICH is scrambled by either the primary or a secondary scrambling code;
- There may be zero, one, or several S-CPICH per cell;
- The S-CPICH may be transmitted over the entire cell or only over a part of the cell;

Usually the S-CPICH is used for antenna beamforming in CL1 or MIMO mode.

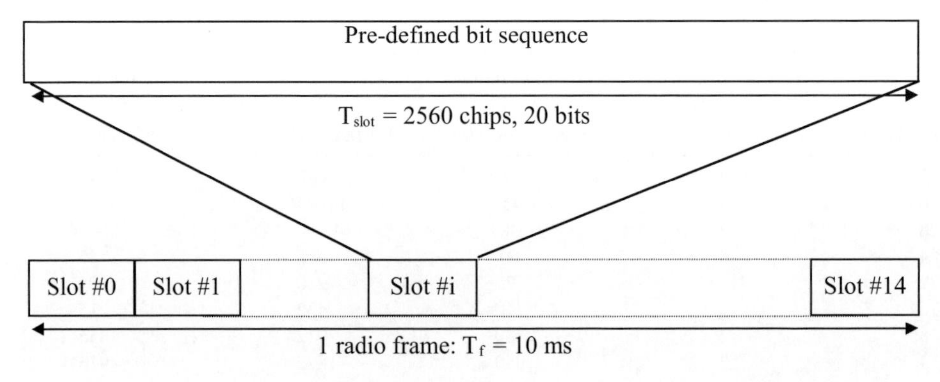

Figure 11.18 Frame structure of CPICH. Reproduced by permission of © 2008 3GPP™

11.2.3.7 Synchronisation Channel (SCH)

The *Synchronisation Channel (SCH)* is a downlink signal used for cell search. The SCH consists of two sub channels, the Primary and Secondary SCH. The UE uses these channels to find the cell. The 10 ms radio frames of the Primary and Secondary SCH are divided into 15 slots, each of length 2560 chips. Figure 11.19 [15] illustrates the structure of the SCH radio frame.

The Primary SCH consists of a modulated code of length 256 chips called the Primary Synchronization Code (PSC). PSC is transmitted once every slot, in Figure 11.19 this code is denoted as c_p. The PSC is the same for every cell in the system. The Secondary SCH consists of repeatedly transmitting sequence of 15 modulated codes of length 256 chips. These codes are called the Secondary Synchronization Codes (SSC) and are transmitted in parallel with the Primary SCH. The SSC is denoted $c_s^{i,k}$ in Figure 11.19, where $i = 0, 1, \ldots, 63$ is the number of the scrambling code group, and $k = 0, 1, \ldots, 14$ is the slot number. Each SSC is chosen from a set of 16 different codes of length 256. This sequence on the Secondary SCH indicates which of the code groups the cell's downlink scrambling code belongs to. The primary and secondary synchronization codes are modulated by the symbol a shown in Figure 11.19, which indicates the presence/ absence of STTD encoding on the P-CCPCH. If P-CCPCH is STTD encoded $a = 1$, otherwise $a = -1$. Transmit diversity, in the form of Time Switched Transmit Diversity (TSTD), can be applied to the SCH. Figure 11.20 [15] illustrates the structure of the SCH transmitted by the TSTD scheme. In even numbered slots both PSC and SSC are transmitted on antenna 1, and in odd numbered slots both PSC and SSC are transmitted on antenna 2.

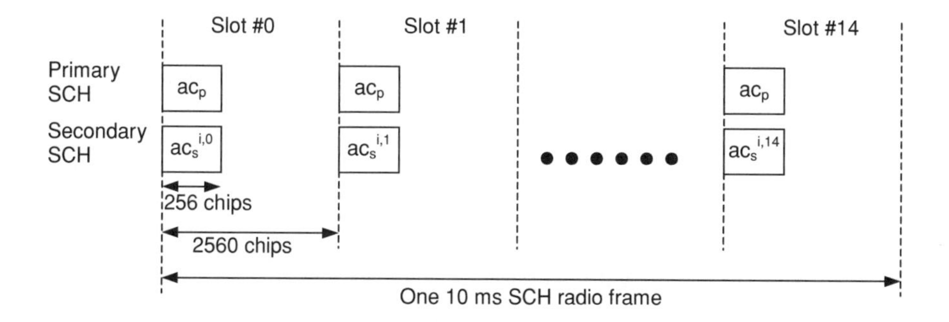

Figure 11.19 Structure of Synchronization Channel (SCH). Reproduced by permission of © 2008 3GPP™

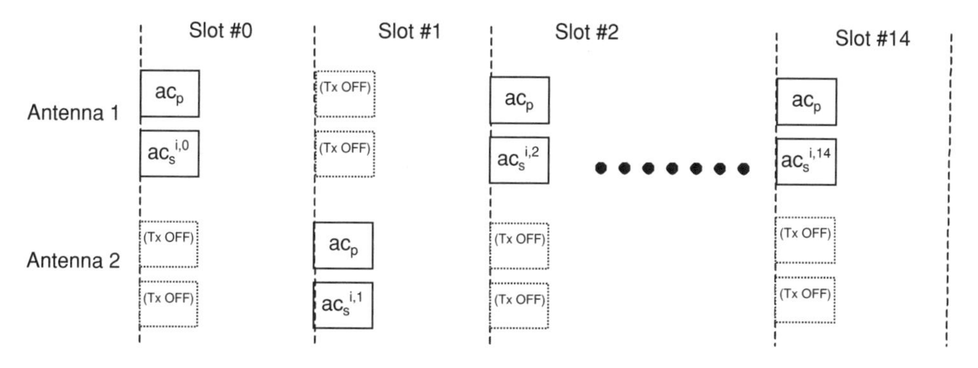

Figure 11.20 Structure of SCH transmitted by TSTD scheme. Reproduced by permission of © 2008 3GPP™

11.2.3.8 Primary Common Control Physical Channel (P-CCPCH)

The Primary CCPCH is a fixed rate (30 kbps, SF = 256) downlink physical channels used to carry the BCH transport channel. Figure 11.21 [15] shows the frame structure of the Primary CCPCH. The frame structure differs from the downlink DPCH in that no TPC commands, no TFCI and no pilot bits are transmitted. The Primary CCPCH is not transmitted during the first 256 chips of each slot. Instead, Primary SCH and Secondary SCH are transmitted during this period.

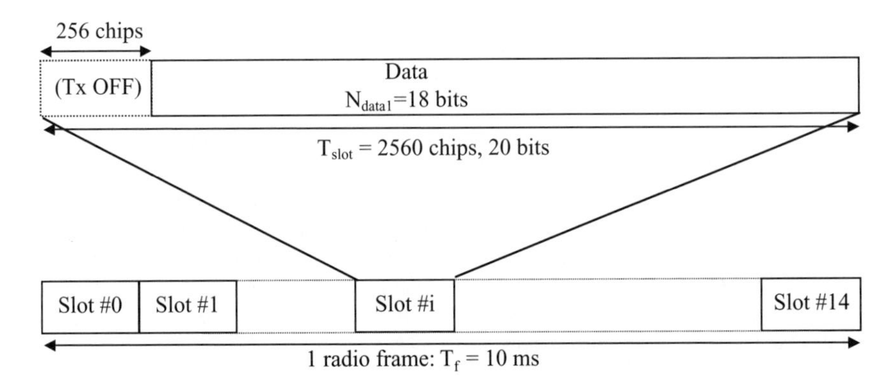

Figure 11.21 Frame structure for P-CCPCH. Reproduced by permission of © 2008 3GPP™

The data rate of P-CCPCH is kept low because any UE in the cell should be able to receive and decode it. If P-CCPCH decoding fails, the UE cannot access the system, since in this case UE is unable to obtain the critical system parameters such as random access codes or channelization codes used for other common channels [1]. It is possible to use the transmit diversity for P-CCPCH transmission. In this case the P-CCPCH is to be transmitted using open loop transmit diversity and the data bits of the P-CCPCH are STTD encoded.

11.2.3.9 Secondary Common Control Physical Channel (S-CCPCH)

The Secondary CCPCH is used to carry the FACH and PCH. There are two types of Secondary CCPCH: those that include TFCI and those that do not include TFCI. It is the network that determines if a TFCI should be transmitted, hence making it mandatory for all UEs to support the use of TFCI. The set of

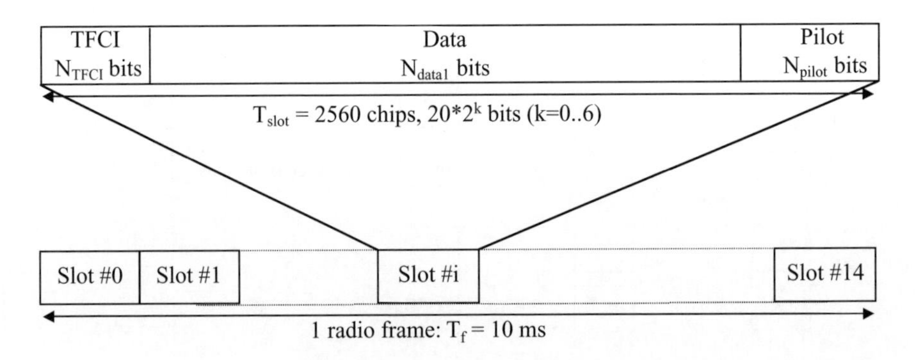

Figure 11.22 Frame structure of S-CCPCH

possible rates for the Secondary CCPCH is the same as for the downlink DPCH. The frame structure of the Secondary CCPCH is shown in Figure 11.22 [15].

The parameter k in Figure 11.22 determines the total number of bits per downlink Secondary CCPCH slot. It is related to the spreading factor (SF) of the physical channel as $SF = 256/2^k$. The spreading factor range is from 256 down to 4. The FACH and PCH can be mapped to the same or to separate Secondary CCPCHs. If FACH and PCH are mapped to the same Secondary CCPCH, they can be mapped to the same frame. The main difference between a CCPCH and a downlink dedicated physical channel is that a CCPCH is not inner-loop power controlled. The main difference between the Primary and Secondary CCPCH is that the transport channel mapped to the Primary CCPCH (BCH) can only have a fixed predefined transport format combination, while the Secondary CCPCH supports multiple transport format combinations using TFCI. In cases of using the transmit diversity the S-CCPCH is to be transmitted using open loop transmit diversity, and the data and TFCI bits of the S-CCPCH are STTD encoded.

11.2.3.10 Acquisition Indicator Channel (AICH)

The *Acquisition Indicator Channel (AICH)* is a fixed rate (SF = 256) physical channel used to carry Acquisition Indicators (AI). Acquisition Indicator AI_s corresponds to signature s sent on the PRACH. With Acquisition Indicator AI_s the BS confirms that the PRACH preamble corresponding to signature s was successfully decoded. Figure 11.23 [15] illustrates the structure of the AICH. The AICH consists of a repeated sequence of 15 consecutive *access slots* (AS), each of length 5120 chips. Each access slot consists of two parts, an *Acquisition-Indicator* (AI) part consisting of 32 real-valued signals $a_0, \ldots, a_{31}$ and a part of duration 1024 chips with no transmission that is not formally part of the AICH. The part of the slot with no transmission is reserved for possible future use by other physical channels.

The real-valued signals $a_0, a_1, \ldots, a_{31}$ in Figure 11.23 are given by:

$$a_j = \sum_{s=0}^{15} AI_s b_{s,j} \tag{11.21}$$

where AI_s, taking the values + 1, −1, and 0, is the acquisition indicator corresponding to signature s and the sequence $b_{s,0}, \ldots, b_{s,31}$ is given by Table 11.6. If the signature s is not a member of the set of available signatures for all the Access Service Class (ASC) for the corresponding PRACH, then AI_s shall be set to 0. If an Acquisition Indicator is set to + 1, it represents a positive acknowledgment. If an Acquisition Indicator is set to -1, it represents a negative acknowledgment. The real-valued signals, a_j, are spread and modulated in the same fashion as bits when represented in $\{+1, -1\}$ form.

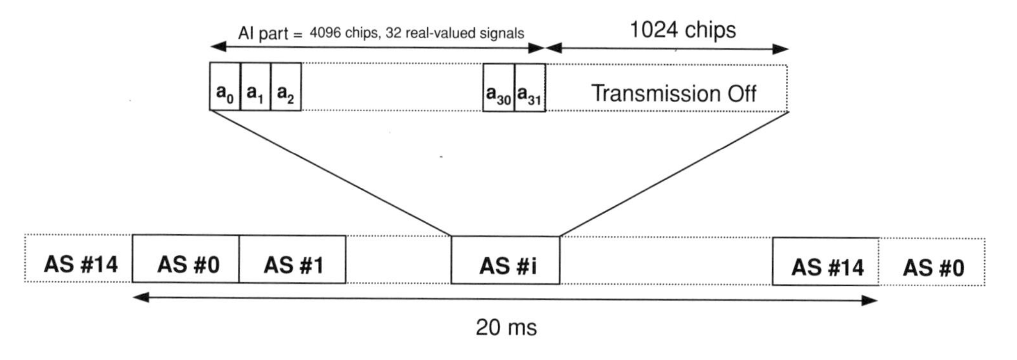

Figure 11.23 Frame structure of AICH. Reproduced by permission of © 2008 3GPP™

Table 11.6 AICH signature patterns

s	$b_{s,0}, b_{s,1} \ldots, b_{s,31}$																														
0	1	1	1	1	1	1	1	1	1	1	1	1	1	1	1	1	1	1	1	1	1	1	1	1	1	1	1	1	1	1	1	1
1	1	1	−1	−1	1	1	−1	−1	1	1	−1	−1	1	1	−1	−1	1	1	−1	−1	1	1	−1	−1	1	1	−1	−1	1	1	−1	−1
2	1	1	1	1	−1	−1	−1	−1	1	1	1	1	−1	−1	−1	−1	1	1	1	1	−1	−1	−1	−1	1	1	1	1	−1	−1	−1	−1
3	1	1	−1	−1	−1	−1	1	1	1	1	−1	−1	−1	−1	1	1	1	1	−1	−1	−1	−1	1	1	1	1	−1	−1	−1	−1	1	1
4	1	1	1	1	1	1	1	1	−1	−1	−1	−1	−1	−1	−1	−1	1	1	1	1	1	1	1	1	−1	−1	−1	−1	−1	−1	−1	−1
5	1	1	−1	−1	1	1	−1	−1	−1	−1	1	1	−1	−1	1	1	1	1	−1	−1	1	1	−1	−1	−1	−1	1	1	−1	−1	1	1
6	1	1	1	1	−1	−1	−1	−1	−1	−1	−1	−1	1	1	1	1	1	1	1	1	−1	−1	−1	−1	−1	−1	−1	−1	1	1	1	1
7	1	1	−1	−1	−1	−1	1	1	−1	−1	1	1	1	1	−1	−1	1	1	−1	−1	−1	−1	1	1	−1	−1	1	1	1	1	−1	−1
8	1	1	1	1	1	1	1	1	1	1	1	1	1	1	1	1	−1	−1	−1	−1	−1	−1	−1	−1	−1	−1	−1	−1	−1	−1	−1	−1
9	1	1	−1	−1	1	1	−1	−1	1	1	−1	−1	1	1	−1	−1	−1	−1	1	1	−1	−1	1	1	−1	−1	1	1	−1	−1	1	1
10	1	1	1	1	−1	−1	−1	−1	1	1	1	1	−1	−1	−1	−1	−1	−1	−1	−1	1	1	1	1	−1	−1	−1	−1	1	1	1	1
11	1	1	−1	−1	−1	−1	1	1	1	1	−1	−1	−1	−1	1	1	−1	−1	1	1	1	1	−1	−1	−1	−1	1	1	1	1	−1	−1
12	1	1	1	1	1	1	1	1	−1	−1	−1	−1	−1	−1	−1	−1	−1	−1	−1	−1	−1	−1	−1	−1	1	1	1	1	1	1	1	1
13	1	1	−1	−1	1	1	−1	−1	−1	−1	1	1	−1	−1	1	1	−1	−1	1	1	−1	−1	1	1	1	1	−1	−1	1	1	−1	−1
14	1	1	1	1	−1	−1	−1	−1	−1	−1	−1	−1	1	1	1	1	−1	−1	−1	−1	1	1	1	1	1	1	1	1	−1	−1	−1	−1
15	1	1	−1	−1	−1	−1	1	1	−1	−1	1	1	1	1	−1	−1	−1	−1	1	1	1	1	−1	−1	1	1	−1	−1	−1	−1	1	1

11.2.3.11 Paging Indicator Channel (PICH)

The *Paging Indicator Channel (PICH)* is a fixed rate (SF = 256) physical channel used to carry the paging indicators (PI). The PICH is associated either with an S-CCPCH to which a PCH transport channel is mapped, or with a HS-SCCH associated with the HS-PDSCH(s) to which a HS-DSCH transport channel carrying paging messages is mapped. A UE, once registered to a network, has been allocated a paging group. Paging Indicator (PI) shows that there are paging messages for this paging group. Once a PI has been detected, the UE decodes the next PCH frame transmitted on the S-CCPCH to see whether there was a paging message intended for it. Figure 11.24 [15] illustrates the frame structure of the PICH. One PICH radio frame of length 10 ms consists of 300 bits ($b_0, b_1, \ldots, b_{299}$). Of these, 288 bits ($b_0, b_1, \ldots, b_{287}$) are used to carry paging indicators. The remaining 12 bits are not formally part of the PICH and shall not be transmitted (DTX). The part of the frame with no transmission is reserved for possible future use.

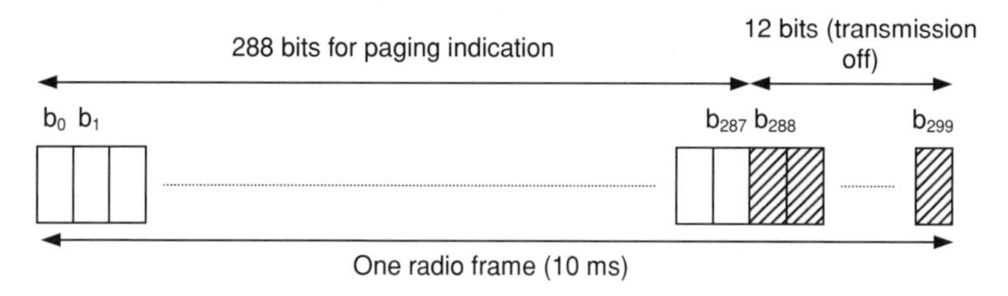

Figure 11.24 Frame structure of PICH. Reproduced by permission of © 2008 3GPP™

In each PICH frame, Np paging indicators $\{P_0, \ldots, P_{Np-1}\}$ are transmitted, where Np = 18, 36, 72, or 144. If a paging indicator in a certain frame is set to "1" it is an indication that UEs associated with this paging indicator and PI should read either the corresponding frame of the associated S-CCPCH, or the corresponding subframes of the associated HS-SCCH. When transmit diversity is employed for the PICH, STTD encoding is used on the PICH data.

Figure 11.25 [15] illustrates the timing between a PICH frame and its associated single S-CCPCH frame, that is, the S-CCPCH frame that carries the paging information related to the paging indicators in the PICH frame. A paging indicator set in a PICH frame means that the paging message is transmitted on the PCH in the S-CCPCH frame starting τ_{PICH} chips after the transmitted PICH frame.

The physical channels not belonging to Release'99 will be described in Section 11.3.

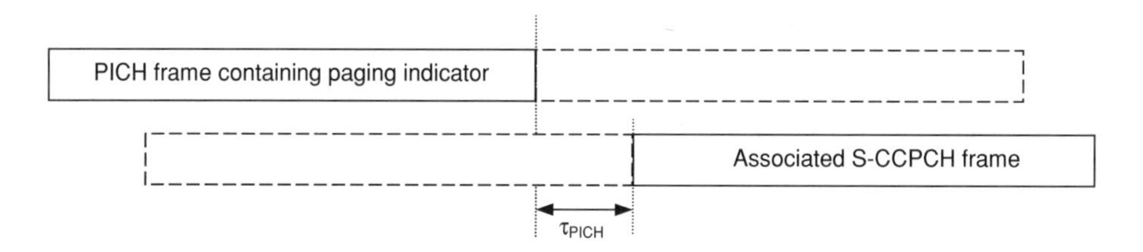

Figure 11.25 Timing relation between PICH frame and associated S-CCPCH frame

11.2.4 Coding, Spreading and Modulation

11.2.4.1 CRC Attachment

Error detection is provided on transport blocks through a Cyclic Redundancy Check (CRC). The size of the CRC is 24, 16, 12, 8 or 0 bits and it is signalled from higher layers what CRC size that should be used for each transport channel. The entire transport block is used to calculate the CRC parity bits for each transport block. The parity bits are generated by one of the following cyclic generator polynomials:

$$g_{CRC24}(D) = D^{24} + D^{23} + D^6 + D^5 + D + 1 \tag{11.22}$$

$$g_{CRC16}(D) = D^{16} + D^{12} + D^5 + 1 \tag{11.23}$$

$$g_{CRC12}(D) = D^{12} + D^{11} + D^3 + D^2 + D + 1 \tag{11.24}$$

$$g_{CRC8}(D) = D^8 + D^7 + D^4 + D^3 + D + 1 \tag{11.25}$$

Denote the bits in a transport block delivered to layer 1 by $a_{im1}, a_{im2}, \ldots, a_{imA_i}$, and the parity bits by $p_{im1}, p_{im2}, \ldots, p_{imL_i}$. A_i is the size of a transport block of TrCH i, m is the transport block number, and L_i is the number of parity bits. L_i can take the values 24, 16, 12, 8, or 0 depending on what is signalled from higher layers.

The encoding is performed in a systematic form, which means that in $GF(2)$, the polynomial:

$$a_{im1}D^{A_i+23} + a_{im2}D^{A_i+22} + \ldots + a_{imA_i}D^{24} + p_{im1}D^{23} + p_{im2}D^{22} + \ldots + p_{im23}D^1 + p_{im24}$$

yields a remainder equal to 0 when divided by $g_{CRC24}(D)$, polynomial:

$$a_{im1}D^{A_i+15} + a_{im2}D^{A_i+14} + \ldots + a_{imA_i}D^{16} + p_{im1}D^{15} + p_{im2}D^{14} + \ldots + p_{im15}D^1 + p_{im16}$$

yields a remainder equal to 0 when divided by $g_{CRC16}(D)$, polynomial:

$$a_{im1}D^{A_i+11} + a_{im2}D^{A_i+10} + \ldots + a_{imA_i}D^{12} + p_{im1}D^{11} + p_{im2}D^{10} + \ldots + p_{im11}D^1 + p_{im12}$$

yields a remainder equal to 0 when divided by $g_{CRC12}(D)$ and polynomial:

$$a_{im1}D^{A_i+7} + a_{im2}D^{A_i+6} + \ldots + a_{imA_i}D^{8} + p_{im1}D^{7} + p_{im2}D^{6} + \ldots + p_{im7}D^1 + p_{im8}$$

yields a remainder equal to 0 when divided by $g_{CRC8}(D)$.

If no transport blocks are input to the CRC calculation ($M_i = 0$), no CRC attachment shall be done. If transport blocks are input to the CRC calculation ($M_i \neq 0$) and the size of a transport block is zero ($A_i = 0$), CRC shall be attached, that is, all parity bits equal to zero.

11.2.4.2 Channel Coding

There are two types of coding scheme, which are used in WCDMA: convolutional coding and turbo coding. It is possible to use convolutional codes with rate 1/2 and 1/3 and the turbo code with rate 1/3. The constrained length of the used convolutional codes is 9. The coders of convolutional codes with rate 1/2 and 1/3 are depicted in Figure 11.26 [16].

Output from the rate 1/3 convolutional coder shall be done in the order output 0, output 1, output 2, output 0, output 1, output 2, output 0, ..., output 2. Output from the rate 1/2 convolutional coder shall be done in the order output 0, output 1, output 0, output 1, output 0, ..., output 1. 8 tail bits with binary value 0 are added to the end of the code block before encoding. The initial value of the shift register of the coder shall be "all 0" when starting to encode the input bits. The number of encoded bits Y at the output of convolutional coder for code rate 1/2 is:

$$Y = 2 \cdot K + 16 \tag{11.26}$$

and for code rate 1/3 is:

$$Y = 3 \cdot K + 24 \tag{11.27}$$

where K is the number of bits in the input code block.

The scheme of Turbo coder is a Parallel Concatenated Convolutional Code (PCCC) with two 8-state constituent encoders and one Turbo code internal interleaver. The coding rate of Turbo coder is 1/3. The structure of Turbo coder is illustrated in Figure 11.27 [16].

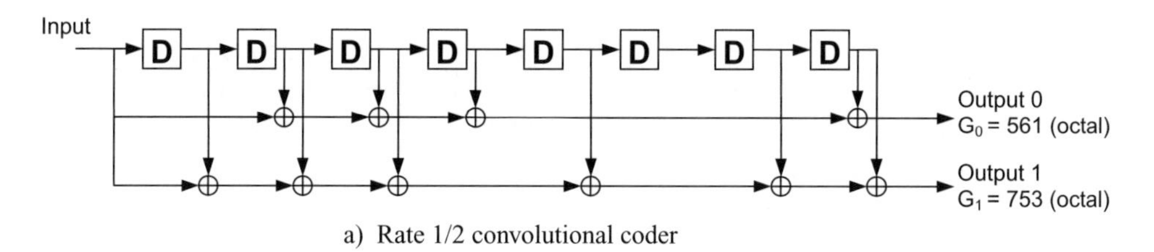

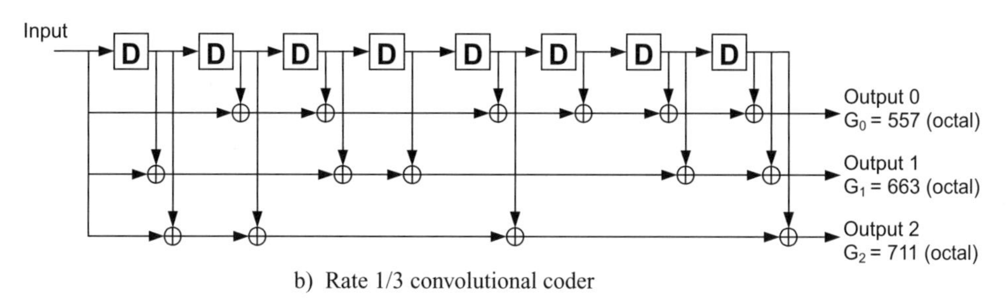

Figure 11.26 Rate 1/2 and rate 1/3 convolutional coders. Reproduced by permission of © 2008 3GPP™

The transfer function of the 8-state constituent code for PCCC is:

$$G(D) = \left[1, \frac{g_1(D)}{g_0(D)}\right] \tag{11.28}$$

where

$$\begin{aligned} g_0(D) &= 1 + D^2 + D^3 \\ g_1(D) &= 1 + D + D^3 \end{aligned} \tag{11.29}$$

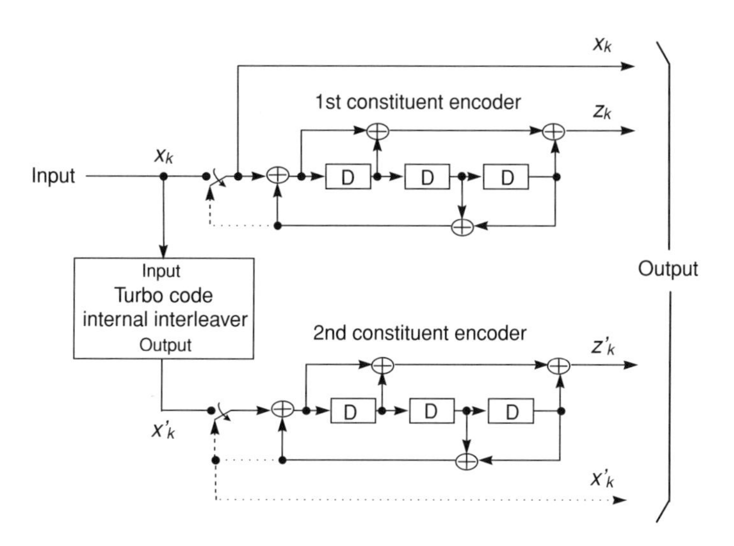

Figure 11.27 Structure of rate 1/3 Turbo coder (dotted lines apply for trellis termination only). Reproduced by permission of © 2008 3GPP™

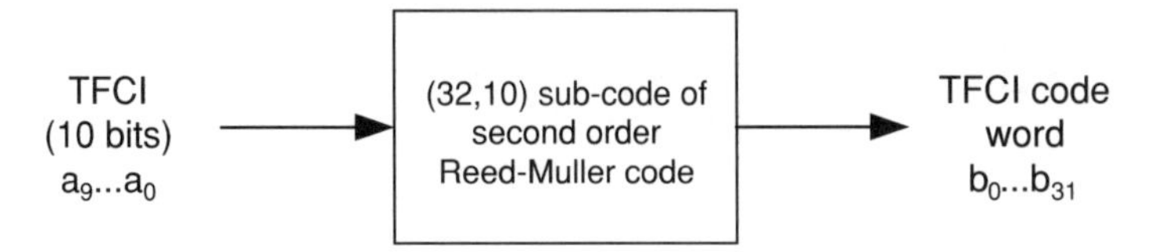

Figure 11.28 TFCI information coding. Reproduced by permission of © 2008 3GPP™

The initial value of the shift registers of the 8-state constituent encoders shall be all zeros when starting to encode the input bits. Output from the Turbo coder is $x_1, z_1, z'_1, x_2, z_2, z'_2, \ldots, x_K, z_K, z'_K$, where $x_1, x_2, \ldots, x_K$ are the bits input to the Turbo coder, that is, both first 8-state constituent encoder and Turbo code internal interleaver, and K is the number of bits, and $z_1, z_2, \ldots, z_K$ and $z'_1, z'_2, \ldots, z'_K$ are the bits output from first and second 8-state constituent encoders, respectively. The information bits x_k are called *systematic bits*, the bits at the output of the first constituent encoder z_k are called *parity 1 bits*, and the bits at the output of the second constituent encoder z'_k are called *parity 2 bits*.

Table 11.7 Basis code words for (32,10) TFCI code

i	$M_{i,0}$	$M_{i,1}$	$M_{i,2}$	$M_{i,3}$	$M_{i,4}$	$M_{i,5}$	$M_{i,6}$	$M_{i,7}$	$M_{i,8}$	$M_{i,9}$
0	1	0	0	0	0	1	0	0	0	0
1	0	1	0	0	0	1	1	0	0	0
2	1	1	0	0	0	1	0	0	0	1
3	0	0	1	0	0	1	1	0	1	1
4	1	0	1	0	0	1	0	0	0	1
5	0	1	1	0	0	1	0	0	1	0
6	1	1	1	0	0	1	0	1	0	0
7	0	0	0	1	0	1	0	1	1	0
8	1	0	0	1	0	1	1	1	1	0
9	0	1	0	1	0	1	1	0	1	1
10	1	1	0	1	0	1	0	0	1	1
11	0	0	1	1	0	1	0	1	1	0
12	1	0	1	1	0	1	0	1	0	1
13	0	1	1	1	0	1	1	0	0	1
14	1	1	1	1	0	1	1	1	1	1
15	1	0	0	0	1	1	1	1	0	0
16	0	1	0	0	1	1	1	1	0	1
17	1	1	0	0	1	1	1	0	1	0
18	0	0	1	0	1	1	0	1	1	1
19	1	0	1	0	1	1	0	1	0	1
20	0	1	1	0	1	1	0	0	1	1
21	1	1	1	0	1	1	0	1	1	1
22	0	0	0	1	1	1	0	1	0	0
23	1	0	0	1	1	1	1	1	0	1
24	0	1	0	1	1	1	1	0	1	0
25	1	1	0	1	1	1	1	0	0	1
26	0	0	1	1	1	1	0	0	1	0
27	1	0	1	1	1	1	1	1	0	0
28	0	1	1	1	1	1	1	1	1	0
29	1	1	1	1	1	1	1	1	1	1
30	0	0	0	0	0	1	0	0	0	0
31	0	0	0	0	1	1	1	0	0	0

Reproduced by permission of © 2008 3GPP™

After all information bits are encoded the trellis termination is performed by taking the tail bits from the shift register feedback. Tail bits are padded after the encoding of information bits. The first three tail bits shall be used to terminate the first constituent encoder (upper switch of Figure 11.27 in lower position) while the second constituent encoder is disabled. The last three tail bits shall be used to terminate the second constituent encoder (lower switch of Figure 11.27 in lower position) while the first constituent encoder is disabled. The transmitted bits for trellis termination shall then be: $x_{K+1}, z_{K+1}, x_{K+2}, z_{K+2}, x_{K+3}, z_{K+3}, x'_{K+1}, z'_{K+1}, x'_{K+2}, z'_{K+2}, x'_{K+3}, z'_{K+3}$.

The special type of coding is used for the TFCI information. The TFCI is encoded using a (32, 10) sub-code of the second order Reed-Muller code. The coding procedure is as shown in Figure 11.28 [16].

The code words of the (32, 10) sub-code of second order Reed-Muller code are linear combination of 10 basis code words. The basis code words are represented in Table 11.7 [16].

If the TFCI consist of less than 10 bits, it is padded with zeros to 10 bits, by setting the most significant bits to zero. The length of the TFCI code word is 32 bits. The output code word bits b_i are given by:

$$b_i = \sum_{n=0}^{9} \left(a_n \times M_{i,n}\right) \bmod 2, \quad i = 0, 1, \ldots, 31 \tag{11.30}$$

11.2.4.3 Spreading and Modulation

Spreading is applied to the physical channels. It consists of two operations. The first is the channelization operation, which transforms every data symbol into a number of chips, thus increasing the bandwidth of the signal. The number of chips per data symbol is called the *Spreading Factor (SF)*. Orthogonality between the different spreading factors is achieved by the tree-structured orthogonal codes. The second operation is the scrambling operation, where a scrambling code is applied to the spread signal. Scrambling is used for cell separation in the downlink and user separation in the uplink. For the Release'99 channels the QPSK modulation is used. The QPSK symbols are represented in I and Q format. The data bits are mapped onto so-called I- and Q-branches. In uplink the I-Q/code multiplexing is used and in the downlink the conventional QPSK modulation is in use.

11.2.4.3.1 Uplink Spreading and Modulation

In the uplink the *Orthogonal Variable Spreading Factor (OVSF) codes* are used as a channelization codes. These codes provide the spreading, that is, increase the signal bandwidth and preserve the orthogonality between a user's different physical channels. The OVSF codes allow preserving the orthogonality across the different symbol rates. The OVSF codes tree is represented in Figure 11.29 [17].

In Figure 11.29 the channelization codes are uniquely described as $C_{ch,SF,k}$, where SF is the spreading factor of the code and k is the code number, $0 \le k \le$ SF-1. The generation method for the channelization code is defined as follows:

$$\begin{aligned}
&C_{ch,1,0} = 1,\\
&\begin{bmatrix} C_{ch,2,0} \\ C_{ch,2,1} \end{bmatrix} = \begin{bmatrix} C_{ch,1,0} & C_{ch,1,0} \\ C_{ch,1,0} & -C_{ch,1,0} \end{bmatrix} = \begin{bmatrix} 1 & 1 \\ 1 & -1 \end{bmatrix},\\
&\begin{bmatrix} C_{ch,2^{(n+1)},0} \\ C_{ch,2^{(n+1)},1} \\ C_{ch,2^{(n+1)},2} \\ C_{ch,2^{(n+1)},3} \\ \vdots \\ C_{ch,2^{(n+1)},2^{(n+1)}-2} \\ C_{ch,2^{(n+1)},2^{(n+1)}-1} \end{bmatrix} = \begin{bmatrix} C_{ch,2^n,0} & C_{ch,2^n,0} \\ C_{ch,2^n,0} & -C_{ch,2^n,0} \\ C_{ch,2^n,1} & C_{ch,2^n,1} \\ C_{ch,2^n,1} & -C_{ch,2^n,1} \\ \vdots & \vdots \\ C_{ch,2^n,2^n-1} & C_{ch,2^n,2^n-1} \\ C_{ch,2^n,2^n-1} & -C_{ch,2^n,2^n-1} \end{bmatrix}
\end{aligned} \tag{11.31}$$

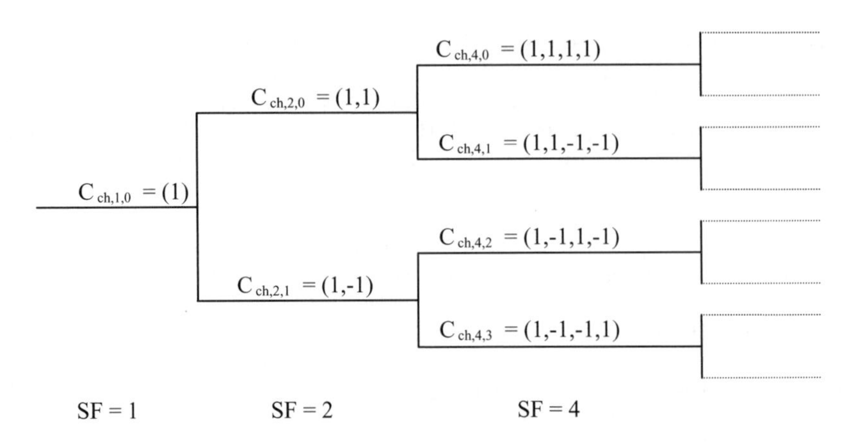

Figure 11.29 Code-tree for generation of Orthogonal Variable Spreading Factor (OVSF) codes. Reproduced by permission of © 2008 3GPP™

The scrambling is used on top of spreading. The scrambling operation does not increase the bandwidth. It is used to separate the data of different UEs. The relation between the spreading and scrambling codes is depicted in Figure 11.30. In the uplink, either short or long scrambling codes are used. The short codes are used to ease the implementation of advanced multiuser receiver techniques; otherwise, long spreading codes can be used. Short codes are S(2) codes of length 256 and long codes are Gold sequences of length 2^{41}, but the latter are truncated to form a cycle of a 10-ms frame. There are 2^{24} long and 2^{24} short uplink scrambling codes.

The long scrambling code is built from two constituent long sequences. The long scrambling sequences $c_{long,1,n}$ and $c_{long,2,n}$ are constructed from position wise modulo 2 sum of 38400 chip segments of two binary m-sequences generated by means of two generator polynomials of degree 25. Let x, and y be the two m-sequences respectively. The x sequence is constructed using the primitive (over $GF(2)$) polynomial $X^{25} + X^3 + 1$. The y sequence is constructed using the polynomial $X^{25} + X^3 + X^2 + X + 1$. The resulting sequences thus constitute segments of a set of Gold sequences.

The sequence $c_{long,2,n}$ is a 16777232 chip shifted version of the sequence $c_{long,1,n}$. Let $n_{23} \ldots n_0$ be the 24 bit binary representation of the scrambling sequence number n with n_0 being the least significant bit.

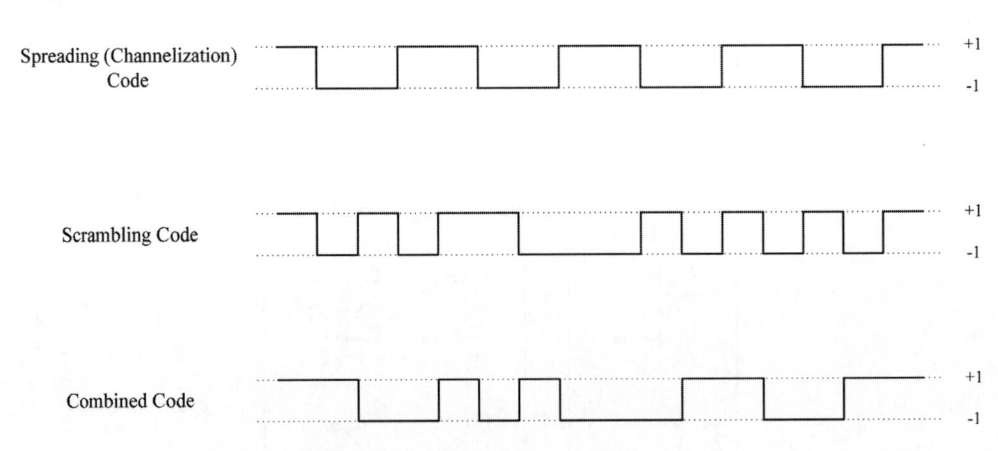

Figure 11.30 Spreading and Scrambling code

The x sequence depends on the chosen scrambling sequence number n and is denoted x_n. Furthermore, let $x_n(i)$ and $y(i)$ denote the ith symbol of the sequence x_n and y, respectively.

The m-sequences x_n and y are constructed as follows.

Initial conditions are:

$$\begin{aligned} &x_n(0) = n_0, \quad x_n(1) = n_1, \ldots, x_n(22) = n_{22}, \quad x_n(23) = n_{23}, \quad x_n(24) = 1 \\ &y(0) = y(1) = \ldots = y(23) = y(24) = 1 \end{aligned} \tag{11.32}$$

Recursive definition of subsequent symbols is:

$$\begin{aligned} &x_n(i+25) = x_n(i+3) + x_n(i) \bmod 2, \quad i = 0, \ldots, 2^{25} - 27 \\ &y(i+25) = y(i+3) + y(i+2) + y(i+1) + y(i) \bmod 2, \quad i = 0, \ldots, 2^{25} - 27 \end{aligned} \tag{11.33}$$

Then the binary Gold sequence z_n is defined as:

$$z_n(i) = x_n(i) + y(i) \bmod 2, \quad i = 0, \ldots, 2^{25} - 2 \tag{11.34}$$

The real valued Gold sequence Z_n is defined by:

$$Z_n(i) = \begin{cases} +1 & if\, z_n(i) = 0 \\ -1 & if\, z_n(i) = 1 \end{cases} \quad \text{for}\, i = 0, 1, \ldots, 2^{25} - 2 \tag{11.35}$$

Now, the real-valued long scrambling sequences $c_{long,1,n}$ and $c_{long,2,n}$ are defined as follows:

$$\begin{aligned} &c_{long,1,n}(i) = Z_n(i), \quad i = 0, \ldots, 2^{25} - 2 \\ &c_{long,2,n}(i) = Z_n((i + 16777232) \bmod (2^{25} - 1)), \quad i = 0, \ldots, 2^{25} - 2 \end{aligned} \tag{11.36}$$

The generator of the long scrambling sequences $c_{long,1,n}$ and $c_{long,2,n}$ is depicted in Figure 11.31 [17].

Finally, the complex-valued long scrambling sequence $C_{long,\,n}$, is defined as:

$$C_{long,n}(i) = c_{long,1,n}(i)\left(1 + j(-1)^i c_{long,2,n}(2\lfloor i/2 \rfloor)\right) \tag{11.37}$$

where $i = 0, 1, \ldots, 2^{25} - 2$ and $\lfloor\rfloor$ denotes rounding to nearest lower integer.

The short scrambling sequences $c_{short,1,n}(i)$ and $c_{short,2,n}(i)$ are defined from a sequence from the family of periodically extended S(2) codes.

Let $n_{23}n_{22}\ldots n_0$ be the 24 bit binary representation of the code number n.

The nth quaternary S(2) sequence $z_n(i)$, $0 \leq n \leq 16777215$, is obtained by modulo 4 addition of three sequences, a quaternary sequence $a(i)$ and two binary sequences $b(i)$ and $d(i)$, where the initial loading of the three sequences is determined from the code number n. The sequence $z_n(i)$ of length 255 is generated

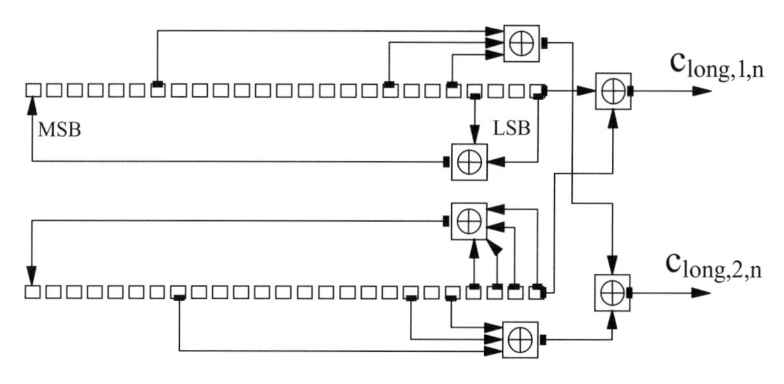

Figure 11.31 Uplink long scrambling sequence generator. Reproduced by permission of © 2008 3GPP™

Table 11.8 Mapping from $z_n(i)$ to $c_{\text{short},1,n}(i)$ and $c_{\text{short},2,n}(i)$

$z_n(i)$	$c_{\text{short},1,n}(i)$	$c_{\text{short},2,n}(i)$
0	+1	+1
1	−1	+1
2	−1	−1
3	+ 1	−1

according to the following relation:

$$z_n(i) = a(i) + 2b(i) + 2d(i) \bmod 4, \quad i = 0, 1, \ldots, 254 \tag{11.38}$$

where the quaternary sequence $a(i)$ is generated recursively by the polynomial $g_0(x) = x^8 + 3x^5 + x^3 + 3x^2 + 2x + 3$ as:

$$\begin{aligned} &a(0) = 2n_0 + 1 \bmod 4 \\ &a(i) = 2n_i \bmod 4, \quad i = 1, 2, \ldots, 7 \\ &a(i) = 3a(i-3) + a(i-5) + 3a(i-6) + 2a(i-7) + 3a(i-8) \bmod 4, \quad i = 8, 9, \ldots, 254 \end{aligned} \tag{11.39}$$

and the binary sequence $b(i)$ is generated recursively by the polynomial $g_1(x) = x^8 + x^7 + x^5 + x + 1$ as:

$$\begin{aligned} &b(i) = n_{8+i} \bmod 2, \quad i = 0, 1, \ldots, 7 \\ &b(i) = b(i-1) + b(i-3) + b(i-7) + b(i-8) \bmod 2, \quad i = 8, 9, \ldots, 254 \end{aligned} \tag{11.40}$$

and the binary sequence $d(i)$ is generated recursively by the polynomial $g_2(x) = x^8 + x^7 + x^5 + x^4 + 1$ as:

$$\begin{aligned} &d(i) = n_{16+i} \bmod 2, \quad i = 0, 1, \ldots, 7 \\ &d(i) = d(i-1) + d(i-3) + d(i-4) + d(i-8) \bmod 2, \quad i = 8, 9, \ldots, 254 \end{aligned} \tag{11.41}$$

The sequence $z_n(i)$ is extended to length 256 chips by setting $z_n(255) = z_n(0)$.

The mapping from $z_n(i)$ to the real-valued binary sequences $c_{\text{short},1,n}(i)$ and $c_{\text{short},2,n}(i)$, $i = 0, 1, \ldots, 255$ is defined in Table 11.8.

Finally, the complex-valued short scrambling sequence $C_{short,\,n}$, is defined as:

$$C_{short,n}(i) = c_{short,1,n}(i \bmod 256)\left(1 + j(-1)^i c_{short,2,n}(2\lfloor i \bmod 256/2 \rfloor)\right) \tag{11.42}$$

where $i = 0, 1, 2, \ldots$ and $\lfloor\rfloor$ denotes rounding to nearest lower integer. The short scrambling sequence generator is depicted in Figure 11.32 [17].

In the uplink the QPSK modulation is used, but it is used in the form of the I-Q/code multiplexing. That means the physical channels are multiplexed in such a way that data bits of one physical channel after spreading are treated as I branch of the QPSK modulated symbol and the bits of other physical channel after spreading are treated as Q branch. It is done to avoid the discontinuous transmission in uplink. If the time multiplexing of physical channels (for example, DPDCH and DPCCH) would be used it will lead to the shutting down of the transmitter during the discontinuous transmission (DTX) period when no information bits are transmitted as shown in Figure 11.33a, which in turn causes the power pulsation in audio frequency. This power pulsation would cause the audible interference in the middle of the telephony voice frequency band [1]. To avoid this effect the I-Q/code multiplexing is used. In this case two physical channels are transmitted in parallel, one on I branch and another on Q branch and due to this fact the transmission became continuous as can be seen in Figure 11.33b.

With the I-Q/code multiplexing the power levels of the DPDCH and DPCCH are typically different, especially as data rates increase [1]. Due to this fact the constellation of the I-Q/code multiplexing is pumped in one of the dimensions as shown in Figure 11.34a.

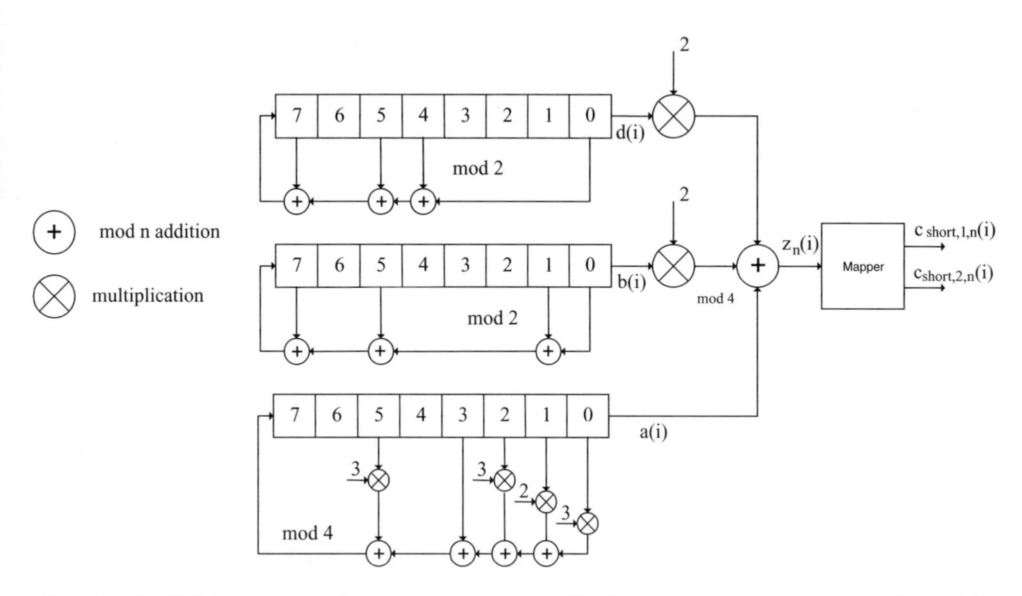

Figure 11.32 Uplink short scrambling sequence generator for 255 chip sequence. Reproduced by permission of © 2008 3GPP™

This constellation distortion can be overcome with the help of complex scrambling. As can be seen from (11.37) and (11.42) $C_{long,n}(i)$ or $C_{short,n}(i)$ can take one of four possible values: $1 + j$, $1 - j$, $-1 + j$, $-1 - j$. That means after multiplication the I-Q/code modulated signal by the scrambling code the resulting constellation is the constellation of the I-Q/code modulated signal rotated by $\pm 45^\circ$ as it is depicted in Figure 11.34b. Moreover, the complex scrambling codes are formed in such a way that the rotations between consecutive chips within one symbol period are limited to $\pm 90^\circ$. The full $\pm 180^\circ$ rotation can happen only between consecutive symbols. Thus, the possible constellation points during

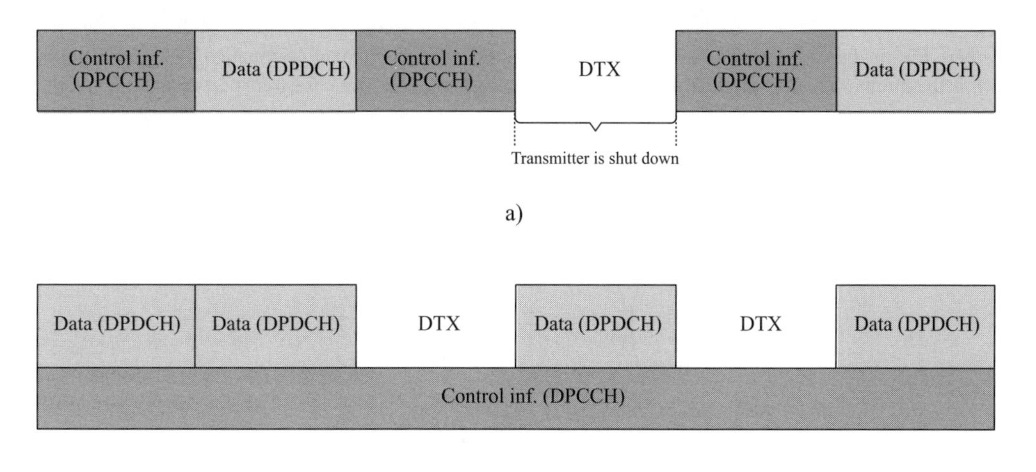

Figure 11.33 a) Time channel multiplexing; b) I-Q/code channel multiplexing

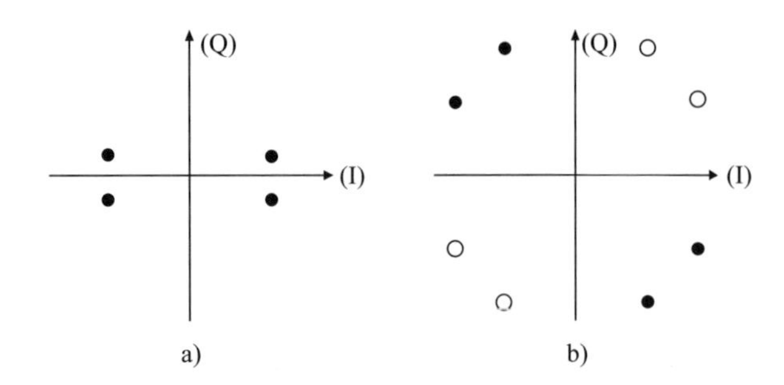

Figure 11.34 Constellation of the I-Q/code multiplexing a) before complex scrambling; b) after complex scrambling

one symbol period are only circles connected with solid lines or circles connected with dashed lines depicted in Figure 11.35.

Then the resulting constellation shown in Figure 11.35 represents the rotated QPSK modulation. Moreover, it is quite close to $\pi/4$-QPSK modulation. This complex scrambling helps to decrease the peak-to-average power ratio (PAPR) in comparison with the original I-Q/code multiplexing, which makes the I-Q/code multiplexing with complex scrambling much more efficient. And the efficiency remains constant irrespective of the power difference between DPDCH and DPCCH [1]. Figure 11.36 [17] shows the uplink spreading of DPCCH and DPDCHs, which are spread by different channelization codes. One DPCCH and up to six parallel DPDCHs can be transmitted simultaneously. After channelization, the real-valued spread signals are weighted by gain factors, which are different for DPCCH and DPDCHs but are the same for all DPDCHs. The gain factors β_c and β_d may be signalled by higher layers or may be computed for certain TFCs, based on the signalled settings for a reference TFC. The calculation is performed as follows.

Let $\beta_{c,ref}$ and $\beta_{d,ref}$ denote the signalled gain factors for the reference TFC. Further, let $\beta_{c,j}$ and $\beta_{d,j}$ denote the gain factors used for the j:th TFC. Also let L_{ref} denote the number of DPDCHs used for the reference TFC and L_j denote the number of DPDCHs used for the j:th TFC. Then define the variable:

$$K_{ref} = \sum_i RM_i \cdot N_i \tag{11.43}$$

where RM_i is the semi-static rate matching attribute for transport channel i (signalled by higher layers), N_i is the number of bits output from the radio frame segmentation block for transport channel i, and the

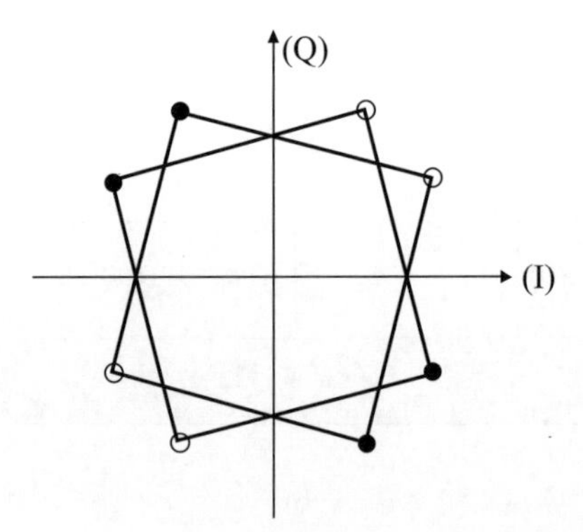

Figure 11.35 Constellation of the I-Q/code multiplexing after complex scrambling

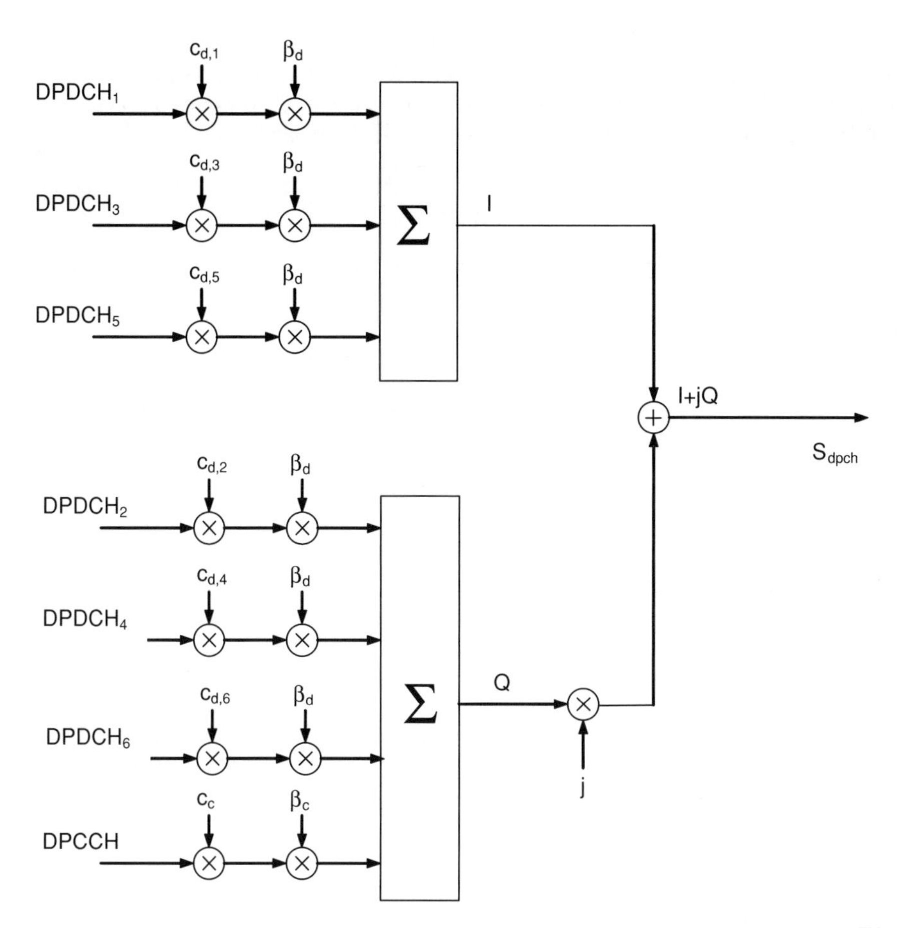

Figure 11.36 Spreading for uplink DPCCH/DPDCHs. Reproduced by permission of © 2008 3GPP™

sum is taken over all the transport channels i in the reference TFC. Similarly, define the variable:

$$K_j = \sum_i RM_i \cdot N_i \tag{11.44}$$

where the sum is taken over all the transport channels i in the j:th TFC. The variable A_j, called the nominal power relation is then computed as:

$$A_j = \frac{\beta_{d,ref}}{\beta_{c,ref}} \cdot \sqrt{\frac{L_{ref}}{L_j}} \sqrt{\frac{K_j}{K_{ref}}} \tag{11.45}$$

The gain factors for the j:th TFC are then computed as follows:

- If $A_j > 1$, then $\beta_{d,j} = 1.0$ and $\beta_{c,j}$ is the largest quantized β-value, for which the condition $\beta_{c,j} \leq 1 / A_j$ holds. Since $\beta_{c,j}$ may not be set to zero, if the above rounding results in a zero value, $\beta_{c,j}$ shall be set to the lowest quantized amplitude ratio of 1/15.
- If $A_j \leq 1$, then $\beta_{d,j}$ is the smallest quantized β-value, for which the condition $\beta_{d,j} \geq A_j$ holds and $\beta_{c,j} = 1.0$.

The gain factors used during a compressed frame for a certain TFC are calculated from the nominal power relation used in normal (non-compressed) frames for that TFC. Let A_j denote the nominal power relation for the j:th TFC in a normal frame. Further, let $\beta_{c,C,j}$ and $\beta_{d,C,j}$ denote the gain factors used for the j:th TFC when the frame is compressed. The variable $A_{C,j}$ is computed as:

$$A_{C,j} = A_j \cdot \sqrt{\frac{15 \cdot N_{pilot,C}}{N_{slots,C} \cdot N_{pilot,N}}} \tag{11.46}$$

where $N_{pilot,C}$ is the number of pilot bits per slot when in compressed mode, and $N_{pilot,N}$ is the number of pilot bits per slot in normal mode. $N_{slots,C}$ is the number of slots in the compressed frame used for transmitting the data.

The gain factors for the j:th TFC in a compressed frame are computed as follows:

- If $A_{C,j} > 1$, then $\beta_{d,C,j} = 1.0$ and $\beta_{c,C,j}$ is the largest quantizedβ-value, for which the condition $\beta_{c,C,j} \leq 1 / A_{C,j}$ holds. Since $\beta_{c,C,j}$ may not be set to zero, if the above rounding results in a zero value, $\beta_{c,C,j}$ shall be set to the lowest quantized amplitude ratio of 1/15.
- If $A_{C,j} \leq 1$, then $\beta_{d,C,j}$ is the smallest quantizedβ-value, for which the condition $\beta_{d,C,j} \geq A_{C,j}$ holds and $\beta_{c,C,j} = 1.0$.

The quantized β-values are defined in Table 11.9 [17].

In the case that no DPDCH is configured, the gain factor β_c is equal to 1. During a compressed frame, the gain factor $\beta_{c,C,j}$ is also equal to 1.

After transforming the signal from real to complex, it is then scrambled by the complex-valued scrambling code, which can be long or short.

As mentioned in section 11.2.3.3 the PRACH comprises preamble part and message part. The PRACH preamble part consists of a complex-valued code $C_{pre,n,}$. It is built from a preamble scrambling code $S_{r\text{-}pre,n}$ and a preamble signature $C_{sig,s}$ as follows:

$$C_{pre,n,s}(k) = S_{r-pre,n}(k) \cdot C_{sig,s}(k) \cdot \exp\left(j\left(\frac{\pi}{4} + \frac{\pi}{2}k\right)\right), \quad k = 0, 1, \ldots, 4095 \tag{11.47}$$

Table 11.9 The quantization of the gain parameters

Signalled values for β_c and β_d	Quantized amplitude ratios β_c and β_d
15	1.0
14	14/15
13	13/15
12	12/15
11	11/15
10	10/15
9	9/15
8	8/15
7	7/15
6	6/15
5	5/15
4	4/15
3	3/15
2	2/15
1	1/15
0	Switch off

where $k = 0$ corresponds to the chip transmitted first in time and $S_{r\text{-}pre,n}$ and $C_{sig,s}$ are defined below. The scrambling code for the PRACH preamble part $S_{r\text{-}pre,n}$ is constructed from the long scrambling sequences. There are 8192 PRACH preamble scrambling codes in total.

The n:th preamble scrambling code, $n = 0, 1, \ldots, 8191$, is defined as:

$$S_{r-pre,n}(i) = \cdot c_{long,1,n}(i), \quad i = 0, 1, \ldots, 4095 \tag{11.48}$$

where the sequence $c_{long,1,n}$ is defined in (11.36). The 8192 PRACH preamble scrambling codes are divided into 512 groups with 16 codes in each group. There is a one-to-one correspondence between the group of PRACH preamble scrambling codes in a cell and the primary scrambling code used in the downlink of the cell. The kth PRACH preamble scrambling code within the cell with downlink primary scrambling code m, $k = 0, 1, 2, \ldots, 15$ and $m = 0, 1, 2, \ldots, 511$, is $S_{r\text{-}pre,n}(i)$ as defined in (11.48) with $n = 16 \times m + k$.

The preamble signature corresponding to a signature s consists of 256 repetitions of a length 16 signature $P_s(n)$, $n = 0, \ldots, 15$. This is defined as follows:

$$C_{sig,s}(i) = \cdot P_s(i \bmod 16), \quad i = 0, 1, \ldots, 4095 \tag{11.49}$$

The signature $P_s(n)$ is from the set of 16 Hadamard codes of length 16.

Figure 11.37 [17] illustrates the principle of the spreading and scrambling of the PRACH message part, consisting of data and control parts. The binary control and data parts to be spread are represented by real-valued sequences, that is, the binary value "0" is mapped to the real value + 1, while the binary value "1" is mapped to the real value –1. The control part is spread to the chip rate by the channelization code c_c, while the data part is spread to the chip rate by the channelization code c_d.

After channelization, the real-valued spread signals are weighted by gain factors β_c for the control part and β_d for the data part. At every instant in time, at least one of the values β_c and β_d has the amplitude 1.0. After the weighting, the stream of real-valued chips on the I- and Q-branches are treated as a complex-valued stream of chips. This complex-valued signal is then scrambled by the complex-valued scrambling code $S_{r\text{-}msg,n}$. The 10 ms scrambling code is aligned with the 10 ms message part radio frames, that is, the first scrambling chip corresponds to the beginning of a message part radio frame.

As can be seen from Figures 11.36 and 11.37 the output of the spreading and scrambling processes is the chip sequence S. The modulation of the complex-valued chip sequence generated by the spreading process is shown in Figure 11.38 [17].

The output of the modulation process is the sequence of QPSK symbols. The modulating chip rate is 3.84 Mcps. The transmit pulse shaping filter is a root-raised cosine (RRC) with roll-off $\alpha = 0.22$ in the

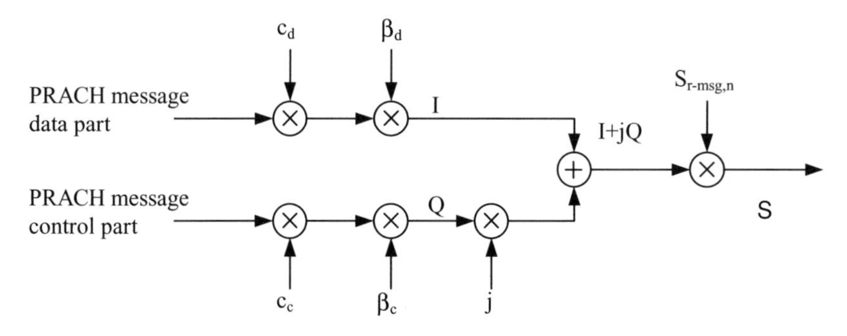

Figure 11.37 Spreading of PRACH message part. Reproduced by permission of © 2008 3GPP™

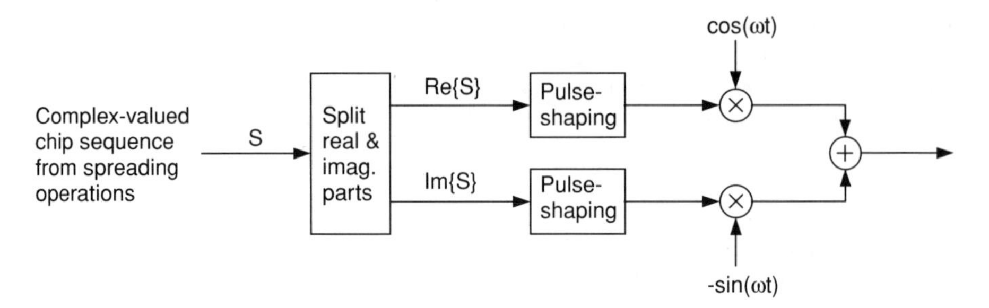

Figure 11.38 Uplink modulation. Reproduced by permission of © 2008 3GPP™

frequency domain. The impulse response of the chip impulse filter $RC_0(t)$ is:

$$RC_0(t) = \frac{\sin\left(\pi \frac{t}{Tc}(1-\alpha)\right) + 4\alpha \frac{t}{Tc}\cos\left(\pi \frac{t}{Tc}(1+\alpha)\right)}{\pi \frac{t}{Tc}\left(1-\left(4\alpha \frac{t}{Tc}\right)^2\right)} \tag{11.50}$$

where the roll-off factor the chip duration T_C is:

$$T_C = \frac{1}{3.84 \cdot 10^6} \approx 0.26042\mu s \tag{11.51}$$

11.2.4.3.2 Downlink Spreading and Modulation

Figure 11.39 illustrates the spreading operation for all downlink physical channels except SCH. The spreading operation includes a modulation mapper stage successively followed by a channelization stage, an IQ combining stage and a scrambling stage. All the downlink physical channels are then combined as shown in Figure 11.39 [17].

In the downlink for all Release'99 channels except SCH the normal QPSK modulation with time multiplexing is used. The high order modulation (16QAM and 64QAM) is used only for HSDPA. The audible interference generated with DTX is not a relevant issue in the downlink since the common channels have continuous transmission in any case [1]. The I and Q branches have the equal power and thus the scrambling operation does not provide the same effect as in uplink. For all downlink channels except AICH, E-HICH and E_RGCH the input digits are mapped to real-valued symbols as follows: the

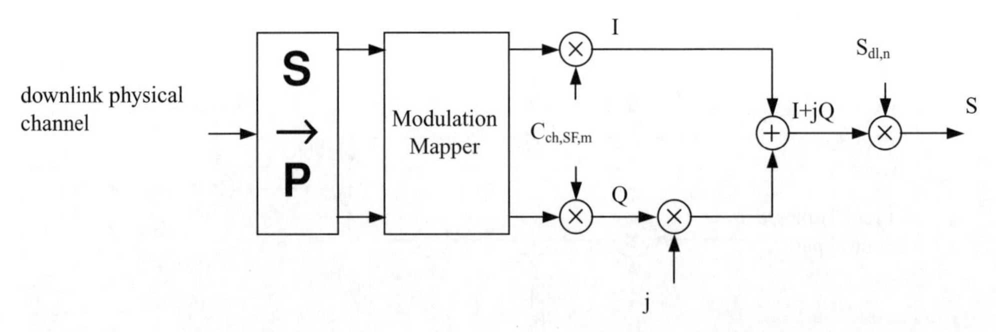

Figure 11.39 Spreading for all downlink physical channels except SCH. Reproduced by permission of © 2008 3GPP™

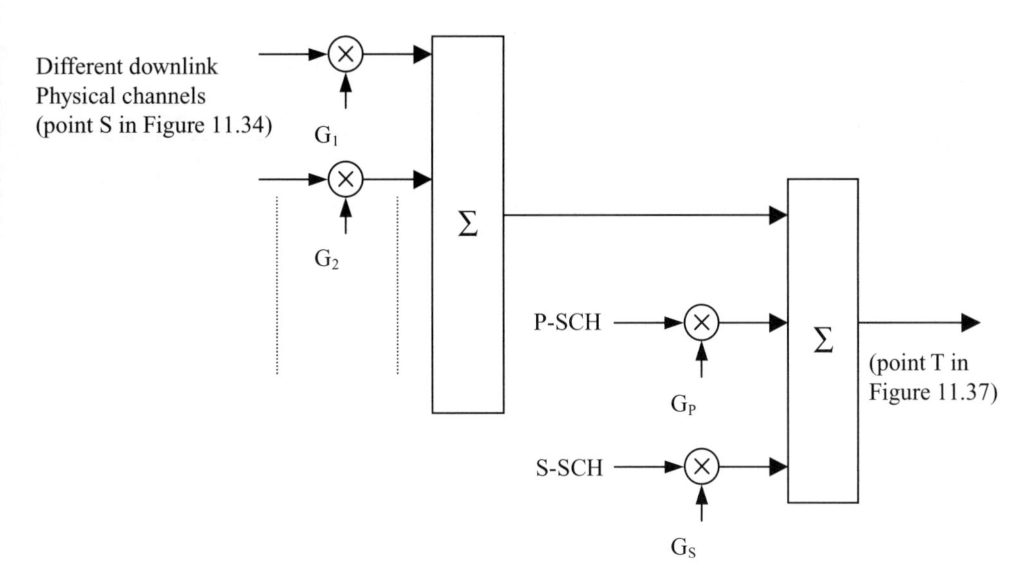

Figure 11.40 Combining downlink physical channels. Reproduced by permission of © 2008 3GPP™

binary value "0" is mapped to the real value + 1, the binary value "1" is mapped to the real value –1 and "DTX" is mapped to the real value 0. Each pair of two consecutive real-valued symbols is first converted from serial to parallel and mapped to an I and Q branch. The even and odd symbols are mapped in the modulation mapper to the I and Q branch respectively.

For all physical channels (except SCH) the I and Q branches are spread to the chip rate by the same real-valued channelization code $C_{ch,SF,m}$, that is, the output for each input symbol on the I and the Q branches is a sequence of SF chips corresponding to the channelization code chip sequence multiplied by the real-valued symbol. The channelization code sequence shall be aligned in time with the symbol boundary.

The channelization codes are the same codes as used in the uplink, namely Orthogonal Variable Spreading Factor (OVSF) codes that preserve the orthogonality between downlink channels of different rates and spreading factors.

Figure 11.40 [17] illustrates how different downlink channels are combined. Each complex-valued spread channel, corresponding to point S in Figure 11.39, may be separately weighted by a weight factor G_i. The complex-valued P-SCH and S-SCH may be separately weighted by weight factors G_p and G_s. All downlink physical channels shall then be combined using complex additions.

A total of $2^{18} - 1 = 262{,}143$ scrambling codes, numbered 0...262,142 can be generated. However, not all the scrambling codes are used. The scrambling codes are divided into 512 sets each of a primary scrambling code and 15 secondary scrambling codes.

The primary scrambling codes consist of scrambling codes $n = 16 \cdot i$, where $i = 0, \ldots, 511$. The ith set of secondary scrambling codes consists of scrambling codes $16 \cdot i + k$, where $k = 1, \ldots, 15$. There is a one-to-one mapping between each primary scrambling code and 15 secondary scrambling codes in a set such that ith primary scrambling code corresponds to ith set of secondary scrambling codes. The set of primary scrambling codes is further divided into 64 scrambling code groups, each consisting of 8 primary scrambling codes. The jth scrambling code group consists of primary scrambling codes $16 \cdot 8 \cdot j + 16 \cdot k$, where $j = 0, \ldots, 63$ and $k = 0, \ldots, 7$.

Each cell is allocated one and only one primary scrambling code. The primary CCPCH, primary CPICH, PICH, MICH, AICH and S-CCPCH carrying PCH shall always be transmitted using the primary

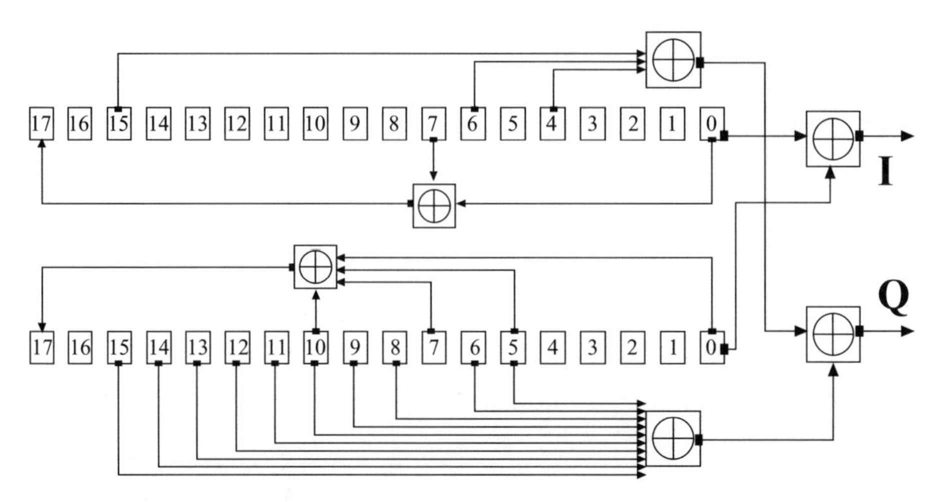

Figure 11.41 Downlink scrambling code generator. Reproduced by permission of © 2008 3GPP™

scrambling code. The other downlink physical channels may be transmitted with either the primary scrambling code or a secondary scrambling code from the set associated with the primary scrambling code of the cell.

The scrambling code sequences are constructed by combining two real sequences into a complex sequence. Each of the two real sequences are constructed as the position wise modulo 2 sum of 38400 chip segments of two binary *m*-sequences generated by means of two generator polynomials of degree 18. The resulting sequences thus constitute segments of a set of Gold sequences. The scrambling codes are repeated for every 10 ms radio frame. Let x and y be the two sequences respectively. The x sequence is constructed using the primitive (over $GF(2)$) polynomial $1 + X^7 + X^{18}$. The y sequence is constructed using the polynomial $1 + X^5 + X^7 + X^{10} + X^{18}$. The sequence depending on the chosen scrambling code number n is denoted z_n, in the sequel. Furthermore, let $x(i)$, $y(i)$ and $z_n(i)$ denote the ith symbol of the sequence x, y, and z_n, respectively.

The *m*-sequences x and y are constructed as follows:

The x sequence is initialized with $x(0) = 1$, $x(1) = x(2) = \ldots = x(16) = x(17) = 0$, and y sequence is initialized with $y(0) = y(1) = \ldots = y(16) = y(17) = 1$. The recursive definition of subsequent symbols is:

$$\begin{aligned} &x(i+18) = x(i+7) + x(i) \bmod 2, \quad i = 0, \ldots, 2^{18} - 20 \\ &y(i+18) = y(i+10) + y(i+7) + y(i+5) + y(i) \bmod 2, \quad i = 0, \ldots, 2^{18} - 20 \end{aligned} \tag{11.52}$$

The nth *Gold code sequence* z_n, $n = 0,1,2,\ldots,2^{18} - 2$, is then defined as:

$$z_n(i) = x\left((i+n) \bmod (2^{18} - 1)\right) + y(i) \bmod 2, \quad i = 0, \ldots, 2^{18} - 2 \tag{11.53}$$

These binary sequences are converted to real valued sequences Z_n by the following transformation:

$$Z_n(i) = \begin{cases} +1 & if z_n(i) = 0 \\ -1 & if z_n(i) = 1 \end{cases} \; for i = 0, 1, \ldots, 2^{18} - 2 \tag{11.54}$$

Finally, the nth complex scrambling code sequence $S_{dl,n}$ is defined as follows:

$$S_{dl,n}(i) = Z_n(i) + j \cdot Z_n\left((i + 131072) \bmod (2^{18} - 1)\right), \quad i = 0, \ldots, 38399 \tag{11.55}$$

The downlink scrambling code generator is depicted in Figure 11.41 [17].

The synchronization channel (SCH) is not under the cell scrambling code. The UE should be able to synchronize to the cell with the help of SCH before knowing the downlink scrambling code. This is done with the help of special synchronization codes as described in section 11.2.3.7. Since the scrambling codes are not used for SCH, the synchronization codes must have the same properties as scrambling codes or at least close to them, taking into account the small length of synchronization codes. The *primary synchronization code (PSC)*, C_{psc} is constructed as a so-called generalized hierarchical *Golay sequence*. The PSC is furthermore chosen to have good aperiodic auto correlation properties. The PSC is generated by repeating the sequence a:

$$a =< x_1, x_2, x_3, \ldots, x_{16} >=< 1, 1, 1, 1, 1, 1, -1, -1, 1, -1, 1, -1, 1, -1, -1, 1 > \quad (11.56)$$

modulated by a Golay complementary sequence, and creating a complex-valued sequence with identical real and imaginary components. The PSC C_{psc} is defined as:

$$C_{psc} = (1 + j) \cdot < a, a, a, -a, -a, a, -a, -a, a, a, a, -a, a, -a, a, a > \quad (11.57)$$

The 16 *secondary synchronization codes (SSCs)*, $\{C_{ssc,1}, \ldots, C_{ssc,16}\}$, are complex-valued with identical real and imaginary components, and are constructed from position wise multiplication of a Hadamard sequence and a sequence z, defined as:

$$z =< b, b, b, -b, b, b, -b, -b, b, -b, b, -b, -b, -b, -b, -b > \quad (11.58)$$

where

$$b =< x_1, x_2, x_3, x_4, x_5, x_6, x_7, x_8, -x_9, -x_{10}, -x_{11}, -x_{12}, -x_{13}, -x_{14}, -x_{15}, -x_{16} > \quad (11.59)$$

and $x_i, i = 1, \ldots, 16$, are same as in (11.56). The *Hadamard sequences* are obtained as the rows of a matrix H_8 constructed recursively by:

$$H_0 = (1)$$
$$H_k = \begin{pmatrix} H_{k-1} & H_{k-1} \\ H_{k-1} & -H_{k-1} \end{pmatrix}, \quad k \geq 1 \quad (11.60)$$

The rows are numbered from the top starting with row 0 (the all ones sequence).

Denote the nth Hadamard sequence h_n as a row of H_8 numbered from the top, $n = 0, 1, 2, \ldots, 255$, in the sequel. Furthermore, let $h_n(i)$ and $z(i)$ denote the ith symbol of the sequence h_n and z, respectively where $i = 0, 1, 2, \ldots, 255$ and $i = 0$ corresponds to the leftmost symbol.

The kth SSC, $C_{ssc,k}$, $k = 1, 2, 3, \ldots, 16$ is then defined as:

$$C_{SSC,k} = (1 + j) \cdot < h_m(0) \cdot z(0), \quad h_m(1) \cdot z(1), \quad h_m(2) \cdot z(2), \ldots, h_m(255) \cdot z(255) > \quad (11.61)$$

where $m = 16 \times (k - 1)$ and the leftmost chip in the sequence corresponds to the chip transmitted first in time.

The 64 secondary SCH sequences are constructed such that their cyclic shifts are unique, that is, a non-zero cyclic shift less than 15 of any of the 64 sequences is not equivalent to some cyclic shift of any other of the 64 sequences. Also, a nonzero cyclic shift less than 15 of any of the sequences is not equivalent to itself with any other cyclic shift less than 15. The SCH code words are modulated with symbol $a = \pm 1$ showing presence or absence of STTD encoding on the BCH as described in section 11.2.3.7.

The modulating chip rate for downlink is the same as in uplink: 3.84 Mcps. And the modulation of the complex-valued chip sequence generated by the spreading process is exactly the same as shown for uplink in Figure 11.38. The pulse shaping filter used in downlink is the same as in uplink.

11.2.5 Cell Search

Before starting any kind of communication with the network, a UE must first synchronize to the signals sent from BS. The initial search procedure in WCDMA is used to identify the scrambling code used by the BS that has the lowest path loss coefficient of the received signal, among all the other base stations. During the cell search, the UE searches for a cell and determines the downlink scrambling code and common channel frame synchronization of that cell. If the UE will search for all possible base station codes at all possible times it can take a lot of time and thus can seriously increase the time from power on the phone till making the call. The WCDMA is the asynchronous system and thus cannot use the GPS support in the search procedure as in synchronous systems like the IS-95/Cdma2000. The grouping of the codes helps to minimize the search time and keep the complexity of the search at low level. The cell search is typically carried out in three steps [18]:

Step 1: Slot synchronization
During the first step of the cell search procedure the UE uses the SCH's primary synchronization code to acquire slot synchronization to a cell. This is typically done with a single matched filter (or any similar device) matched to the primary synchronization code which is common to all cells. The slot timing of the cell can be obtained by detecting peaks in the matched filter output.

Step 2: Frame synchronization and code-group identification
During the second step of the cell search procedure, the UE uses the SCH's secondary synchronization code to find frame synchronization and identify the code group of the cell found in the first step. This is done by correlating the received signal with all possible secondary synchronization code sequences, and identifying the maximum correlation value. Since the cyclic shifts of the sequences are unique the code group as well as the frame synchronization is determined.

Step 3: Scrambling-code identification
During the third step of the cell search procedure, the UE determines the exact primary scrambling code used by the found cell. The primary scrambling code is typically identified through symbol-by-symbol correlation over the CPICH with all codes within the code group identified in the second step. After the primary scrambling code has been identified, the Primary CCPCH can be detected. And the system- and cell specific BCH information can be read.

In case the UE has received information about which scrambling codes to search for, steps 2 and 3 above can be simplified.

11.2.6 Power Control Procedures

In WCDMA, power control is employed in both the uplink and the downlink. In uplink the power control is used for each UE and in downlink for each physical channel.

The main target of the uplink power control is to cope with the *near-far problem*. The near-far problem arises in cases where no power control is used and all UEs in the cell are transmitting with equal power levels as shown in Figure 11.42a.

If the UEs transmit using equal fixed power levels (Tx1 Power = Tx2 Power = Tx3 Power), the cell would be dominated by users closest to BS and the faraway users signals would not be heard by the BS due to the path-loss variation of users with different distances from the BS (Rx1 Power < Rx2 Power < Rx3 Power). This phenomenon is called near-far problem. Actually the fading variation can lead to the fact that the path-loss of the UEs with the same distance from the BS varies differently. From the BS point of view the best situation is that the received power levels of all UEs in the cell are equal regardless

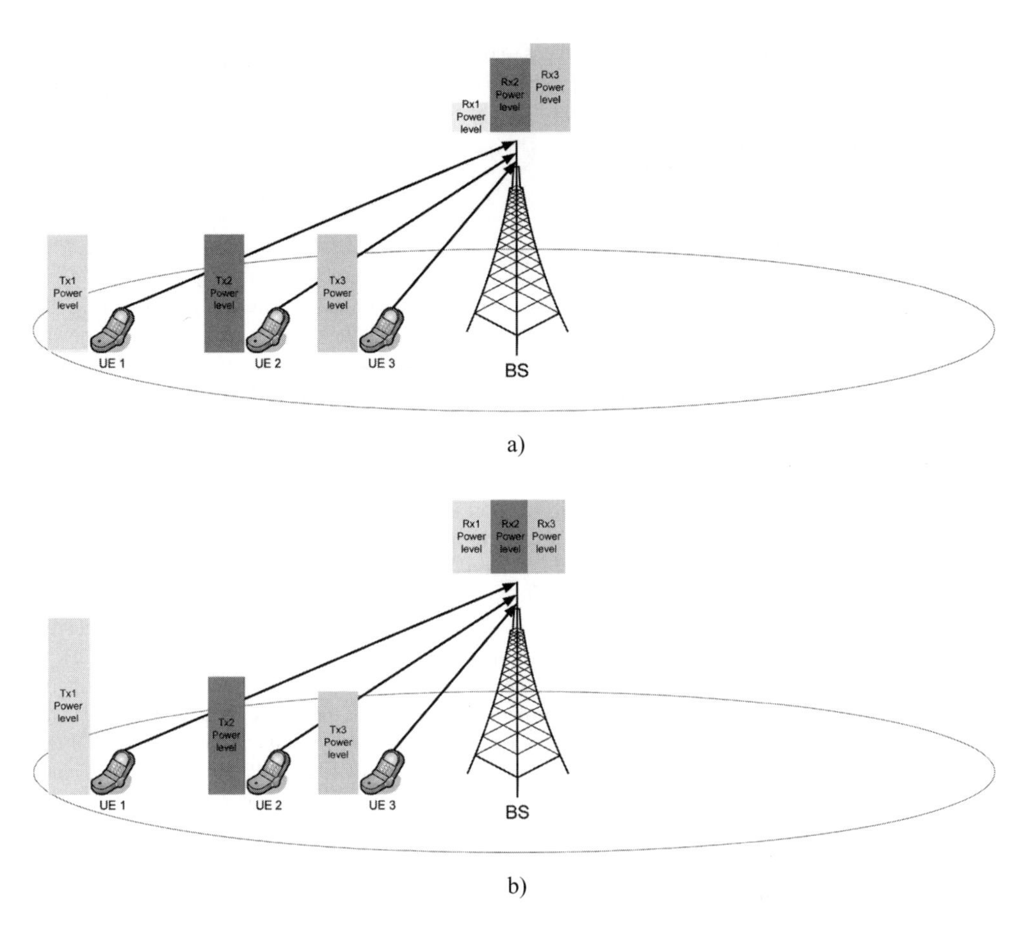

Figure 11.42 a) No power control, near-far problem; b) Uplink power control

of their distance from the BS (Rx1 Power $\approx$ Rx2 Power $\approx$ Rx3 Power) as it is shown in Figure 11.42b. The UEs, which are far away from the BS should transmit with considerably higher power than the UEs close to the BS. In this case the BS receiver is able to decode the UE's transmissions in an optimal way that leads to optimization of cell capacity. It can be reached with the help of uplink power control. In real life not only the distance from the BS but also fading conditions of the each UE should be taken into account to provide the efficient power control. Thus, the main measure of the signal quality is the signal to interference ratio (SIR) rather than the signal power. Another reason for using uplink power control is the mitigation of the intra-cell interference, that is, the interference from other UEs in the cell. Also the uplink power control helps to save the UE's transmit energy in case the UE is close to BS and thus optimize the battery energy consumption.

The near-far problem is also visible in downlink. Due to the signal attenuation the UEs at the cell border experience higher interference than that near to the BS. They have a high level of interfering signals from other BS. In downlink the main goal of power control is to keep the signal at a minimal required level in order to decrease the interference to users in other cells (*intra-cell interference*). One of the main drivers of the development of the power control algorithms is the difficult propagation conditions of the multipath fading channel causing the interference. In uplink the different users are separated with the different orthogonal scrambling codes and in downlink the different base stations also use the different orthogonal scrambling codes. Thus, in an ideal case no intra-cell or inter-cell interference should occur. However, in real life the multipath channel propagation conditions destroy the orthogonality of the codes both in uplink and downlink, and the problem of inter-cell and intra-cell arise. In this case the efficient power control is vital for system functionality.

There are two basic types of power control:

- Open Loop Power Control
- Closed Loop Power Control

In the *Open Loop power control* the UE estimates the transmission signal strength by measuring the received power level of the pilot signal from the BS in the downlink, and adjusts its transmission power level in a way that is inversely proportional to the pilot signal power level. The stronger the received pilot signal, the lower the UE transmitted power. This is a simple and efficient method, but the problem is that the interference estimation is done on the received signal, and the transmitted signal probably uses the different frequency (in the case of FDD), which differs from the received frequency by the system's duplex offset. As uplink and downlink fast fading (on different frequency carriers) do not correlate, this method gives the power values only as an average. This method is more applicable to TDD mode, where both the uplink and downlink use the same frequency and thus their fading processes are strongly correlated. However, Open Loop power control is used in FDD mode also, but only to provide initial power setting of the UE at the beginning of the connection. In this case the UE estimates the received power level of the pilot signal in downlink, and adjusts the transmission power as inversely proportional to the received pilot power level. The information about the allowed power parameters UE receives from the BCCH.

In the *Closed Loop power control* technique the measurements are done on the other end of the connection (in the BS for the uplink power control and in the UE for the downlink power control) and the results are then sent back to UE or BS transmitter correspondingly so that it can adjust its transmitted power. This method gives much better results that the Open Loop power control, but it can hardly react to quick changes in the channel conditions. The Closed Loop power control is used when the radio connection has already been established.

In WCDMA the Closed Loop power control is used in the form of two techniques:

- *Inner Closed Loop power control;*
- *Outer Closed Loop power control.*

The inner (also called fast) closed loop power control adjusts the transmitted power in order to keep the received SIR equal to a given target. The inner closed loop power control algorithm measures the received SIR and sends commands to the transmitter (that is, the UE in case of uplink) for the transmitted power update. The obtained SIR estimate, noted SIR_{est}, is then used by the receiver to generate power control (PC) commands [18]. The transmitted power is updated at each time slot resulting in 1500 Hz command rate. It is increased or decreased by a fixed value:

- if $SIR_{est} > SIR_{target}$ then the transmit power control (TPC) command to transmit is "0", requesting a transmit power decrease;
- if $SIR_{est} < SIR_{target}$ then the TPC command to transmit is "1", requesting a transmit power increase.

The basic step size in the case of uplink is equal to 1 or 2 dB. In downlink the basic step size is 1 dB (mandatory) and 0.5 dB (optional). There exists another inner closed loop algorithm in accordance to which the transmitted power updates each five time slots which simulates smaller power update steps. The power control range in uplink is 30 dB and in downlink is 80 dB.

The outer closed loop PC is used for long term quality control. The setting of the SIR_{target} for the inner closed loop PC is done by the outer loop PC in order to match the required BLER as shown in Figure 11.43. Outer loop PC update frequency is 10-100 Hz [1]. The BLER target is a function of the service

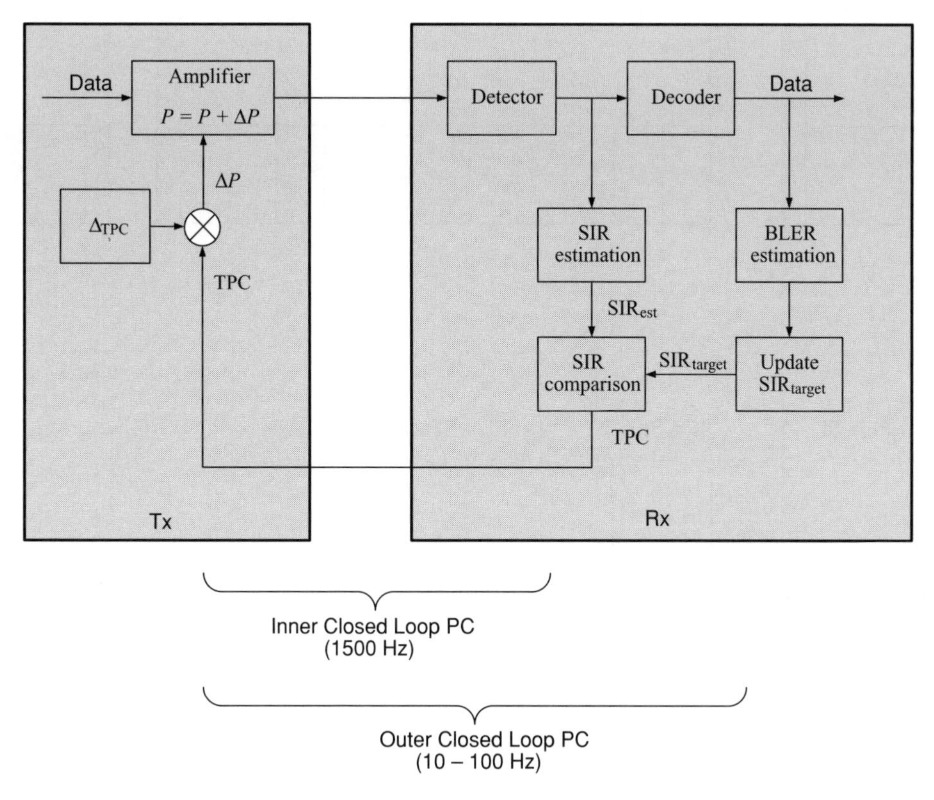

Figure 11.43 Inner and Outer Closed Loop Power Control

that is carried. Ensuring that the lowest possible SIR target is used results in greater network capacity. For the uplink the SIR_{target} is set by the Radio Network Controller (RNC) and the control is located in the BS. For the downlink the outer loop PC is located in UE, and the initial control parameters are set by network.

By adjusting the SIR_{target} for the inner closed loop PC the receiver is able to achieve the maximum target BLER providing the necessary quality of service.

11.2.7 Handover Procedures

Handover is the process in which a UE closes the existing connection in the current cell and establishes a new connection in the new cell or new cell sector. With the help of handover it is possible to provide a continuity of mobile services to a user travelling over cell boundaries. For a user having an ongoing communication and crossing the cell edge, it is preferable to use the radio resources in the new cell because the signal strength perceived in the "old" cell weakens as the user crosses the boundary between the old and the new cells. There are different types of handover:

- Hard handover;
- Soft handover;
- Softer handover;
- Inter- Radio Access Technology (RAT) handover.

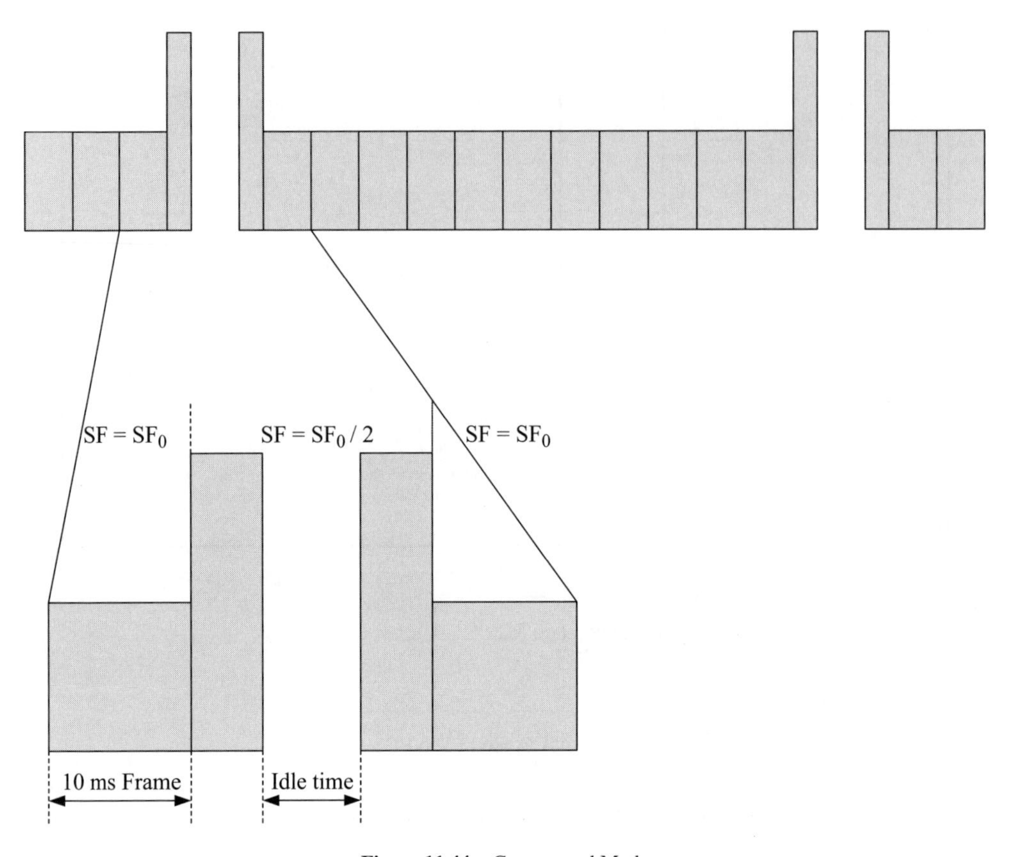

Figure 11.44 Compressed Mode

Hard handover means that all the old radio links in the UE are removed before the new radio links are established. Hard handover is used to change the radio frequency band of the connection between the UE and the network. In practice, each UMTS operator can use several bands of approximately 5MHz. Moreover, in principle Node B is capable of supporting multiple carrier frequencies. Due to this the situation occurs when the UE must switch to another carrier frequency. Usually this happens when the current cell is congested or when the UE is penetrating the cell with different carrier frequency. Hard handover procedure can be initiated by the network or by the UE. Generally it would be initiated by the network using one of the Radio Bearer Control messages. In the case of UE being initiated, it would happen if the UE performs a Cell Update procedure and that Cell Update reaches the RNC on a different frequency [19]. In order to complete inter-frequency handovers, an efficient method is needed for making measurements on other frequencies while still having the connection running on the current frequency. One method is providing the measurements with the help of dual receiver. That means the UE should contain two receiver branches, each of which is capable of receiving the signal independently of another branch. Then, one receiver branch is receiving the current frequency and simultaneously another receiver branch is switched to another frequency for measurements. The advantage of the dual receiver approach is that there is no break in the current frequency connection. Of course, not all UEs are dual mode terminals. So, another option is to use the compressed mode for measurements. The compressed mode is depicted in Figure 11.44.

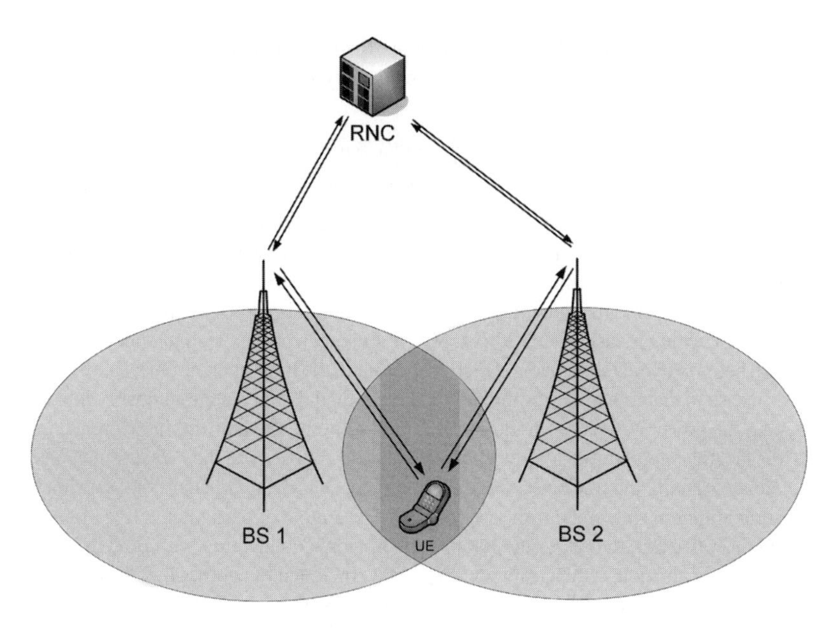

Figure 11.45 Soft handover

The information normally transmitted during a 10-ms frame is compressed either by code puncturing or by decreasing the spreading factor, leaving idle time for measurements on other frequencies.

A *soft handover* occurs when the mobile station is in the overlapping coverage area of two adjacent cells. Then the UE can add and remove radio links in such a manner that the UE always keeps at least one radio link to network. This can be performed on the same carrier frequency only. In this case the UE may simultaneously communicate via a number of radio links towards different cells. With reference to Soft handover, the *Active Set* is defined as the set of radio links simultaneously involved in the communication between the UE and UTRAN (that is, the UTRA cells currently assigning a downlink DPCH to the UE constitute the active set) [20]. In soft handover the UE combines more than one radio link to improve the reception quality. On the other hand, the uplink signals from different base stations are combined to obtain good quality data from the UE. The soft handover is represented in Figure 11.45. In the UE the downlink signals received from the two different base stations are combined using maximum ratio combining (MRC) processing. In the uplink direction the received signals can no longer be combined in the base station but are routed to the RNC. In the RNC the uplink signals from different base stations are combined by means of selective combining, that is, the two signals are compared on a frame-by-frame basis and the best candidate is selected after each interleaving period; that is, every 10, 20, 40 or 80ms. The outer loop power control algorithm measures the SNR of received uplink signals, and based on this information the frame with the best quality is selected. The transmitting power of the UE is controlled by that Node B, to which the lowest propagation loss exists.

The soft handover procedure is composed of a number of single functions:

- Measurements;
- Filtering of Measurements;
- Reporting of Measurement results;
- The Soft Handover Algorithm;
- Execution of Handover.

The main measurement for making handover decisions is the pilot SNR (CPICH Ec/No). It is defined as the received energy per chip divided by the power density in the band [21]. It can be calculated as follows:

$$E_c/N_0 = \frac{\mathit{CPICH\ RSCP}}{\mathit{UTRACarrier\ RSSI}} \tag{11.62}$$

where CPICH RSCP (Received Signal Code Power) is the received power on one code measured on the Primary CPICH, and UTRA Carrier RSSI is the received wide band power, including thermal noise and noise generated in the receiver, within the bandwidth defined by the receiver pulse shaping filter [21]. Before sending the measurements to the network they are filtered in the UE. It is done to average out the effect of fast fading causing measurement errors. Measurement errors can lead to unnecessary handovers. And the filtering helps to avoid these errors. On the other hand, the long filtering period can cause delays in the handover. In case the handover is significantly delayed, the UE can penetrate deeply into the adjacent cell still obeying the power control from the old BS, thus bringing the harmful interference into the new cell. Due to this fact the length of the filter is chosen as a trade-off between the measurement accuracy and the handover delay. Usually the filtering time of 200ms is chosen. UE is constantly measuring the CPICH Ec/No for several neighbouring cells. If the value of CPICH Ec/No of particular cell is high enough, this cell is included in *Active Set*, that is, in the list of cells having a connection with the UE. If the value of CPICH Ec/No is not high enough the corresponding cell is included in another list: *Monitoring Set*. Both sets are constantly updating. Based on the cell measurements, the Soft Handover function evaluates if any cell should be added to (Radio Link Addition), removed from (Radio Link Removal), or replaced in (Combined Radio Link Addition and Removal) the Active Set; performing what is known as "Active Set Update" procedure [20]. Only the cells from the Active Set can be involved in soft handover. Other important information needed for handover is timing information. Since the UTRAN is the asynchronous network, the timing of different BS is not synchronized. To allow UE the combining of data obtained from different base stations the transmissions have to be adjusted in time. To do this UE measures the relative timing difference between cell j and cell i, defined as $T_{CPICHRxj}$ - $T_{CPICHRxi}$, where $T_{CPICHRxj}$ is the time when the UE receives one Primary CPICH slot from cell j, $T_{CPICHRxi}$ is the time when the UE receives the Primary CPICH slot from cell i that is closest in time to the Primary CPICH slot received from cell j [21]. The measurements are reported to network. This report constitutes the basic input to the *Soft Handover Algorithm*. The example of Soft Handover Algorithm is represented in Figure 11.46. At the beginning the UE is connected to the BS1. The UE is measuring the pilot SNR of BS1, BS2 and BS3 and all three corresponding cells are included in Monitoring Set. As the CPICH Ec/No of BS2 and BS3 is not high enough they are not included in ActiveSet constituting at first only cell 1. As the UE penetrates the cell 2 the CPICH Ec/No of BS2 is increasing and when this measurement becomes greater than (*Best_Ss* - *As_Th* + *As_Th_Hyst*) during the *time-to-trigger* ΔT, the cell 2 is added to Active Set and the corresponding report is sent to network. Here *As_Th* is the threshold for macro diversity, *Best_Ss* is the best measured cell present in the Active Set, and *As_Th_Hyst* is the *hysteresis*. The hysteresis parameter is used to limit the amount of event-triggered reports. The hysteresis ensures that the event is not reported until the difference is equal to the hysteresis value. The result of usage of the time-to-trigger is that the report is triggered only after the conditions for the event have existed for this specified time-to-trigger. Thus, the time-to-trigger ΔT allows avoiding the "ping-pong" effect. As UE leaves cell 1, the CPICH Ec/No of BS1 decreases correspondingly, and cell 1 becomes the worst measured cell present in the Active Set (with the CPICH Ec/No of BS1 denoted as *Worst_Old_Ss*). When the measured CPICH Ec/No of BS3 is greater than (*Worst_Old_Ss* + *As_Rep_Hyst*) for a period of ΔT and if Active Set is full, cell 3 replaces cell 1 in Active Set. Here, *As_Rep_Hyst* is the replacement hysteresis. Then, as UE is moving towards BS2, the measured CPICH Ec/No of BS3 decreases, and when it is below (*Best_Ss* - *As_Th* - *As_Th_Hyst*) for a period of ΔT, the worst cell (in this case cell 3) is removed from the Active Set.

To execute the soft handover the UE must obtain from the new cell involved in soft handover what channelization code(s) are used for the new radio link. The channelization codes from different cells are

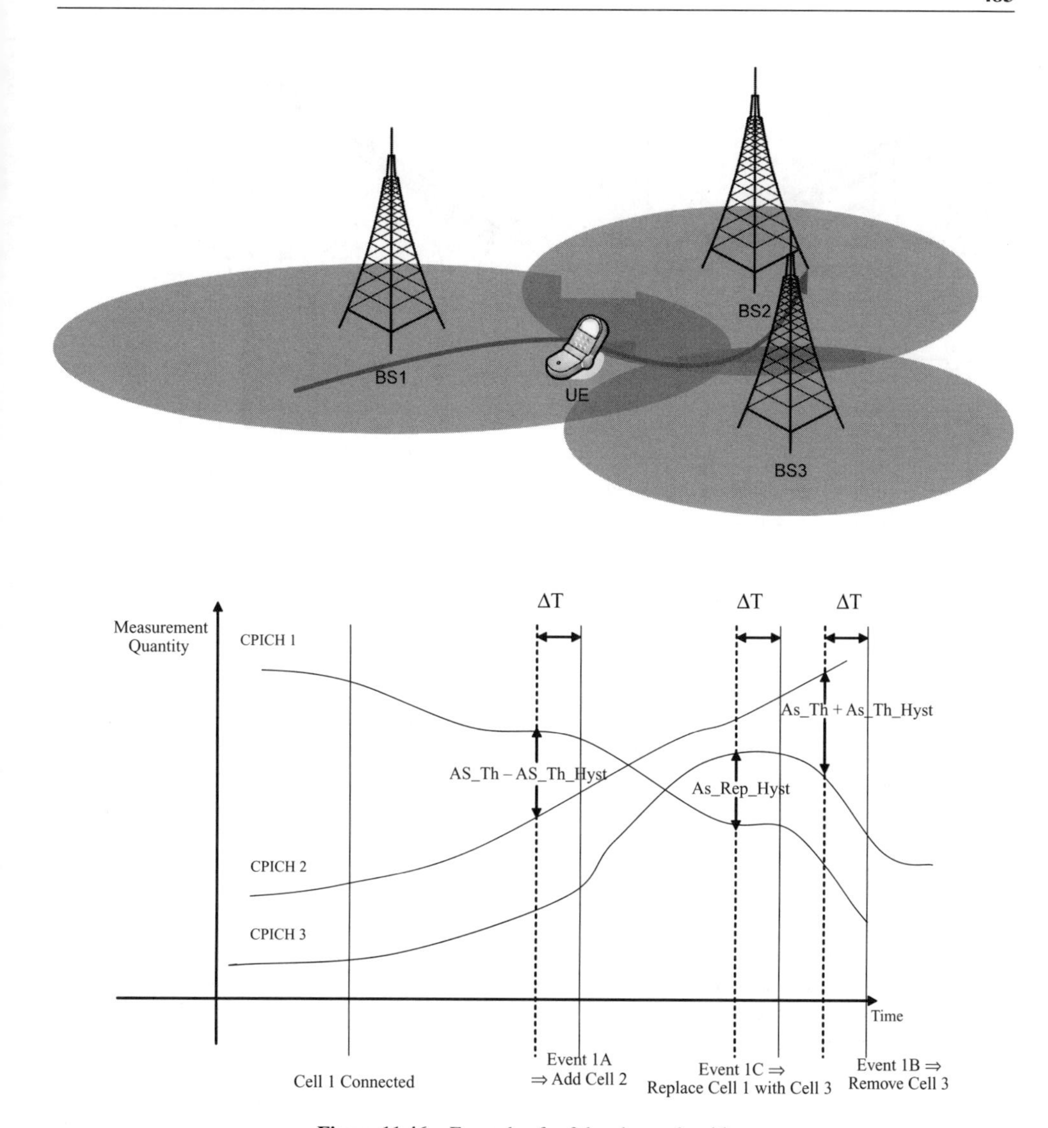

Figure 11.46 Example of soft handover algorithm

not required to be the same as they are under different scrambling codes. Also the UE should calculate the difference between timing of old and new cells involved in soft handover. Before connecting to a new base station, the UE must measure a time difference related to the new base station and the already connected base stations. This measurement is reported to the network, which adjusts the transmission timing of the new base station accordingly. The relative timing information of the new cell, in respect to the timing UE is experiencing from the existing connections (as measured by the UE at its location). The timing difference is reported by UE to network via old cell and the network makes this information available to the new cell as shown in Figure 11.47 [20]. Based on this, the new Node-B can determine what should be the timing of the transmission initiated in respect to the timing of the common channels (CPICH) of the new cell [20].

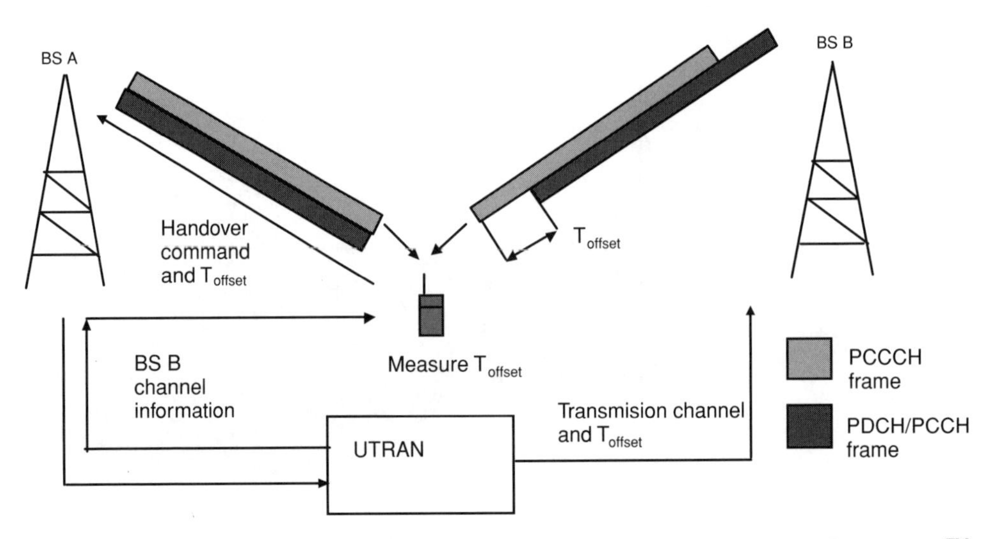

Figure 11.47 Information needed for Soft Handover Execution. Reproduced by permission of © 2007 3GPP™

At the start of soft handover, the uplink dedicated physical channel transmitted by the UE, and the downlink dedicated physical channel transmitted by the Node-B will have their radio frame number and scrambling code phase counted up continuously, and they will not be affected by the soft handover. User data carried on both uplink and downlink will continue without any interruption [20].

A *softer handover* occurs when the UE is in the overlapping coverage area of two adjacent sectors of the same cell as shown in Figure 11.48. Actually, it is not necessary for UE to be in the overlapping coverage area of two sectors. Quite often the additional radio link with the adjacent sector of the same Node B can be established due to the fact that the signal sent from the UE reaches the Node B from the adjacent sector because of reflections on buildings or natural barriers as shown in Figure 11.48. In this case the UE is capable of supporting two simultaneous connections to the network using different air interface channels in the same manner as in soft handover. The difference is that in the case of softer handover these connections are provided via the same Node B. Due to this fact it is possible to use the MRC not only in downlink like in soft handover but in uplink also.

In the downlink combining it should be taken into account that the Node B uses different scrambling codes to separate the different sectors it serves. That means the signals received from different cell sectors must be despread differently before combining them together. Thus the downlink combining is exactly the same as in the case of soft handover.

During the soft or softer handover, transmission is performed through few Node Bs or few sectors of the same Node B. Due to this fact the probability of data correct reception increases, and thus the transmitting power can be decreased keeping the same BER as with one link transmission. Assume a desired BER level is 10^{-6}. This level can be achieved with the help of just one radio link that fulfils the given BER. On the other hand, the same BER can be achieved with the help of two radio links, each of them providing BER of only 10^{-3}. It follows from the fact that the receiver has two copies of data received from two different radio links and sources of noise in these two links are independent of each other. Then the overall error event is a result of two independent error events in each of links. Thus, the overall BER is obtained as a product of BER in link 1 and BER in link 2 ($10^{-3} \times 10^{-3} = 10^{-6}$). Increasing the acceptable link BER allows decreasing the transmission power, which in turn leads to decreasing the interference in the system. This phenomenon is called *soft handover gain*. On the other hand, there is an increase in interference in the system when UEs and base stations are transmitting

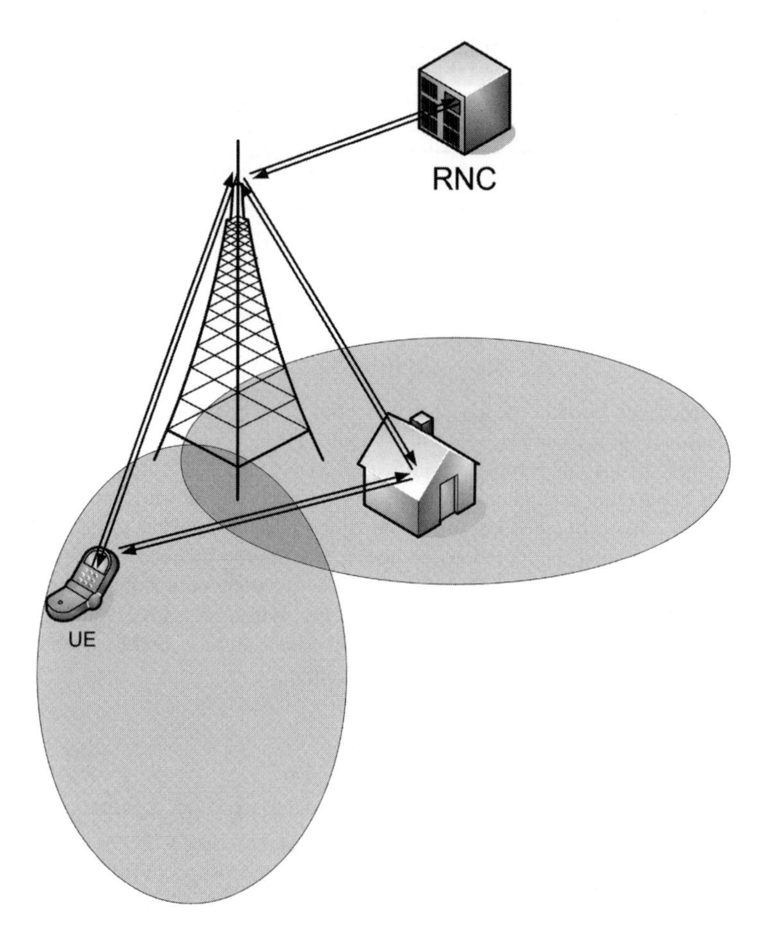

Figure 11.48 Softer handover

additional signals over the air interface compared to a situation when every radio network connection consists of one link. Thus, it is necessary to limit the amount of soft handovers in the system to the level at which the soft handover gain prevails over the additional interference due to multiple link connections.

Inter-RAT handover usually occurs when UE reaches the end of coverage area for UMTS services. In this case UE can handover to some other radio access technology, for example, GSM (in case UE is capable of receiving GSM or other RAT signal). To perform the inter-RAT handover UE must provide measurements on downlink physical channels belonging to another radio access technology than WCDMA, for example, GSM. In order for the UE to perform handover from UTRA FDD mode to GSM without simultaneous use of two receiver chains, a UE can perform measurements by using idle periods in the downlink transmission, where such idle periods are created by using the downlink compressed mode as defined in [16]. The compressed mode is under the control of the UTRAN and the UTRAN signals appropriate configurations of compressed mode pattern to the UE. For some measurements uplink compressed mode is also needed, depending on UE capabilities and measurement objects. Alternatively, independent measurements not relying on the compressed mode, but using a dual receiver approach can be performed, where the GSM receiver branch can operate independently of the UTRA FDD receiver branch [20]. The UE is able to obtain the information from the synchronization bursts in the synchronization

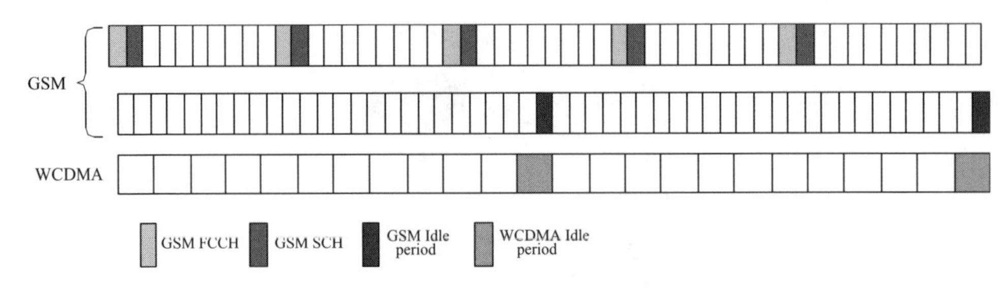

Figure 11.49 WCDMA and GSM timing

frame on GSM carrier with the aid of a frequency correction (FCCH) burst. The timing relation between WCDMA channels and GSM channels is shown in Figure 11.49.

In order to reduce the size of certain size critical messages in UMTS, a network may download/pre-define one or more radio configurations in a mobile via system information. A predefined radio configuration mainly consists of radio bearer- and transport channel parameters. A network knowing that the UE has suitable predefined configurations stored can then refer to the stored configuration requiring only additional parameters to be transferred [20]. A default configuration is a set of radio bearer and transport channel parameters. Inter-RAT handover is always performed as hard handover. Node B must inform dual mode UE of existing GSM frequencies in the area. In turn, GSM network is required to indicate WCDMA spreading codes for easy cell identification.

11.2.8 Transmit Diversity

To increase the reliability of the downlink received data it is possible to use the transmit diversity. There are two types of transmit diversity, which can be used in WCDMA: *open loop* and *closed loop transmit diversity*. Open loop transmit diversity includes *space-time block coding based transmit diversity (STTD)* and *time switched transmit diversity (TSTD)*. STTD can be applied to all downlink physical channels except SCH. TSTD can be applied to SCH only. Closed loop transmit diversity in principle constitutes two modes: *closed loop mode 1 (CL1)* and *closed loop mode 2* (*CL2*). However, it was decided to use only CL1. CL1 can be applied to DPCH or HS-PDSCH (if HS-PDSCH is not used in MIMO mode). The STTD encoder is depicted in Figure 11.50.

Actually, STTD is slightly modified Alamouti scheme having the same properties as the original one. The STTD is optional for network, that is, for Node B but the support of STTD is mandatory for each UE. The usage of TSTD for SCH is depicted in Figure 11.20. In case of using TSTD the signal is never transmitted via both antennas simultaneously. The main difference between open loop and closed loop techniques is that the open loop transmit diversity methods do not take in account the current channel

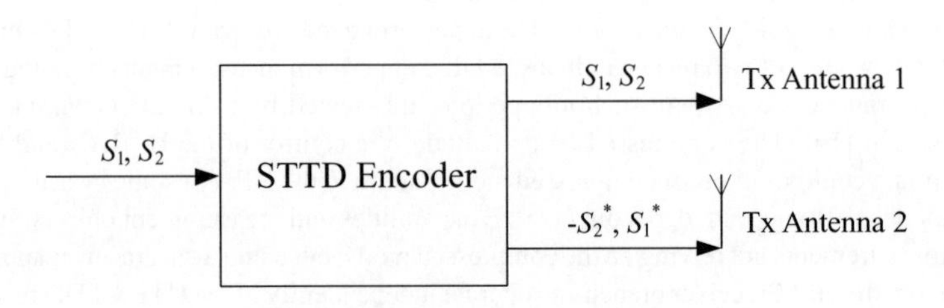

Figure 11.50 STTD Encoder

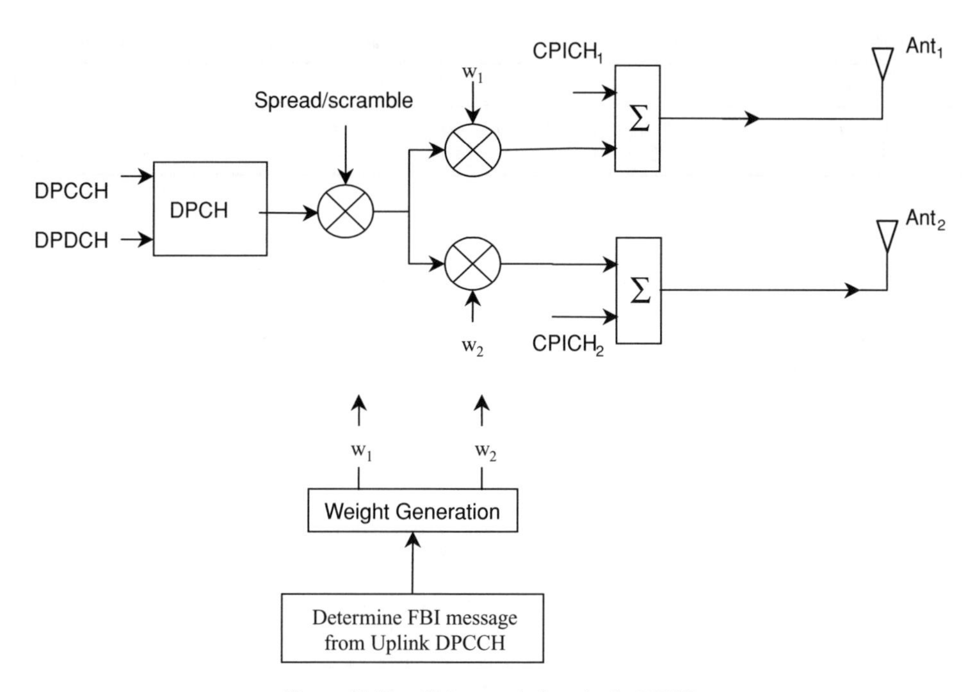

Figure 11.51 CL1 transmit diversity for DPCH

state. Due to this fact the open loop methods do not use the feedback from the receiver to transmitter employed by closed loop techniques.

The general transmitter structure to support closed loop mode transmit diversity for DPCH transmission is shown in Figure 11.51 [15]. Channel coding, interleaving and spreading are done as in non-diversity mode. The spread complex valued signal is fed to both TX antenna branches, and weighted with antenna specific weight factors w_1 and w_2. The weight factor w_1 is a constant scalar and the weight factor w_2 is complex valued signal. The weight factor w_2 (actually the corresponding phase adjustment) is determined by the UE, and signalled to the UTRAN using the FBI field of uplink DPCCH. The weight factors w_1 and w_2 control the Node B beam-forming [15].

The computation of feedback information w_1, w_2 can be accomplished by for example, solving for weight vector **w** that maximizes:

$$P = \mathbf{w}^H \cdot \mathbf{H}^H \cdot \mathbf{H} \cdot \mathbf{w} \tag{11.63}$$

where $\mathbf{w} = [w_1, w_2]^T$ and $\mathbf{H} = [\mathbf{h}_1, \mathbf{h}_2]$ represent the estimated channel impulse responses for the transmit antennas 1 and 2, of length equal to the length of the channel impulse response. Without loss of generality it is possible to assume w_1 is a constant. Then the problem simplifies to search value of w_2 maximizing (11.63). To simplify this task even more, it is possible to limit the choice of w_2 by some set of precalculated values.

During soft handover, the UE computes the phase adjustment to maximize the total UE received power from the cells in the active set. It can be done by maximizing the following function:

$$P = \mathbf{w}^H \cdot (\mathbf{H}_1^H \cdot \mathbf{H}_1 + \mathbf{H}_2^H \cdot \mathbf{H}_2 + \ldots) \cdot \mathbf{w} \tag{11.64}$$

where $\mathbf{H}_i$ is the channel impulse response matrix for cell i.

Table 11.10 Phase adjustments, ϕ_i, corresponding to feedback commands for the slots i of the UL radio frame

Slot #		0	1	2	3	4	5	6	7	8	9	10	11	12	13	14
FSM	0	0	$\pi/2$	0	$\pi/2$	0	$\pi/2$	0	$\pi/2$	0	$\pi/2$	0	$\pi/2$	0	$\pi/2$	0
	1	π	$-\pi/2$	π	$-\pi/2$	π	$-\pi/2$	π	$-\pi/2$	π	$-\pi/2$	π	$-\pi/2$	π	$-\pi/2$	π

The UE uses the CPICH transmitted both from transmit antenna 1 and antenna 2 to calculate the phase adjustment to be applied at UTRAN access point to maximize the UE received power. In each slot, UE calculates the optimum phase adjustment, ϕ, for antenna 2, which is then quantized into ϕ_Q having two possible values as follows:

$$\phi_Q = \begin{cases} \pi, & \text{if } \pi/2 < \phi - \phi_r(i) \le 3\pi/2 \\ 0, & \text{otherwise} \end{cases} \tag{11.65}$$

where

$$\phi_r(i) = \begin{cases} 0, & i = 0, 2, 4, 6, 8, 10, 12, 14 \\ \pi/2, & i = 1, 3, 5, 7, 9, 11, 13 \end{cases} \tag{11.66}$$

If $\phi_Q = 0$, a command '0' is send to Node B using the *Feedback Signalling Message (FSM)* bits transmitted in the FBI field of uplink DPCCH slot(s) as it is shown in Figure 11.11. Correspondingly, if $\phi_Q = \pi$, command '1' is sent to Node B. Thus, the feedback bit corresponds to either the real or imaginary part of the current weight factor w_2 in consecutive slots. As can be seen, the CL1 feedback information load is quite small: 1 bit per slot or 15 bits per 10 ms frame (this does not take into account the encoding of FSM bits). Due to rotation of the constellation at UE the Node B interprets the received commands according to Table 11.10 which shows the mapping between phase adjustment, ϕ_i, and received feedback command for each uplink slot [18].

The weight w_1 for Tx antenna 1 is constant $w_1 = 1/\sqrt{2}$. The weight w_2 is then calculated by averaging the received phases with the help of sliding window over two consecutive slots as follows:

$$w_2 = \frac{\sum_{i=n-1}^{n} \cos(\phi_i)}{2} + j\frac{\sum_{i=n-1}^{n} \sin(\phi_i)}{2} \tag{11.67}$$

Due to the filtering it is possible to obtain four different values of possible phase rotation using only one feedback bit. Thus, there are four possible values of the weight w_2 and at each slot the weight w_2 remains the same as in previous slot or jumps to the neighbouring value, as shown in Figure 11.52.

At frame borders the averaging operation is slightly modified. Upon reception of the FB command for slot 0 of a frame, the average is calculated based on the command for slot 13 of the previous frame and the command for slot 0 of the current frame, that is, ϕ_i from slot 14 is not used. It is done to keep the same sequence of phase adjustments as inside the frame. Otherwise, at frame border there will be possible jumps of phase adjustment by π.

Due to uplink FBI bit errors, the situation could arise when the Node B uses the beamforming weights different from the requested by UE. This causes degradation to the performance firstly because the antenna weights used at Node B are not optimal and secondly because wrong antenna weights may be used in combining process at UE. The latter problem may lead to error floor in performance. To mitigate this problem, *antenna verification* can be applied, which can make use of antenna specific pilot patterns of the dedicated physical channel. For the CL1 different orthogonal dedicated pilot symbols in the DPCCH are sent on the two different antennas. Using the DPCH the UE can estimate the channel coefficients for the Tx antenna 2. Then, the weight w_2 used by Node B can be recovered in UE by comparing the common and dedicated channel estimates. The phase adjustment ϕ_0 is assumed to be

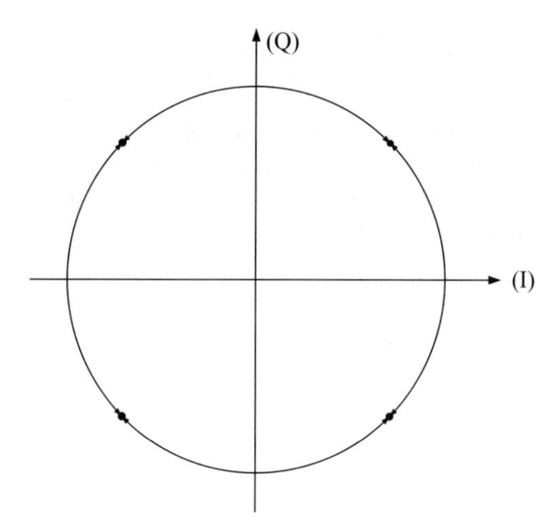

Figure 11.52 Possible transitions of CL1 weight w_2 for Tx antenna 2

$\phi_0 = 0$ if the following inequality holds on:

$$2\sum_{i=1}^{Npath} \frac{1}{\sigma_i^2}\left\{\sqrt{2}\,\mathrm{Re}(\gamma h_{2,i}^{(d)} h_{2,i}^{(p)*})\right\} > \ln\left(\frac{\Pr(\phi_0 = \pi)}{\Pr(\phi_0 = 0)}\right) \tag{11.68}$$

otherwise $\phi_0 = \pi$. The phase adjustment ϕ_1 is assumed to be $\phi_1 = -\pi/2$ if:

$$-2\sum_{i=1}^{Npath} \frac{1}{\sigma_i^2}\left\{\sqrt{2}\,\mathrm{Im}(\gamma h_{2,i}^{(d)} h_{2,i}^{(p)*})\right\} > \ln\left(\frac{\Pr(\phi_1 = \pi/2)}{\Pr(\phi_1 = -\pi/2)}\right) \tag{11.69}$$

and $\phi_1 = \pi/2$ otherwise. For even slots the value of ϕ_0 is calculated and for odd slots value of ϕ_1. Here, $h_{2,i}^{(p)}$is the ith estimated channel tap of antenna 2 using the CPICH, $h_{2,i}^{(d)}$is the ith estimated channel tap of

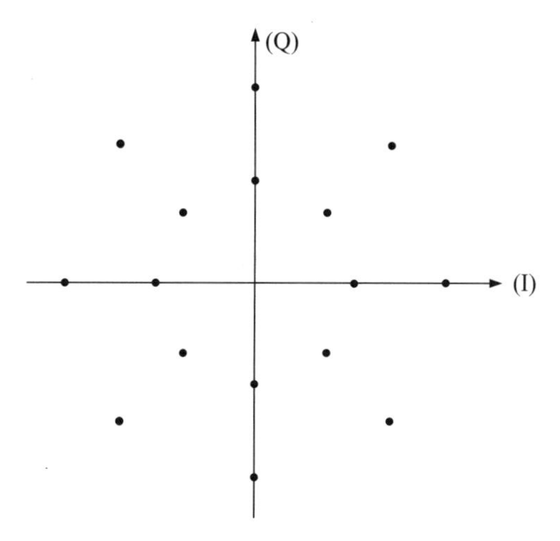

Figure 11.53 Possible states of CL2 weight w_2 for Tx antenna 2

antenna 2 using the DPCCH, γ^2 is the ratio of DPCH Pilot SNIR to CPICH SNIR, $\sigma_i^2$is the noise plus interference power on the ith path. Then, the weight w_2 used by Node B can be calculated in accordance with (11.67).

In closed loop mode 2 (CL2) transmit diversity not only the phase rotation but also the amplitude is adjusted. The possible states of the weight w_2 are represented in Figure 11.53.

With accurate channel estimates CL2 outperforms CL1 due to better resolution of weight coefficient. However, CL2 was removed from 3GPP specifications.

11.3 3.5G HSDPA/HSUPA

The voice service and SMS were the driving force of cellular networks development until recently. However, nowadays this trend is changing. There is a huge demand among subscribers for high-speed data services. Operators are also trying to offer data services to improve ARPU (Average Revenue Per User) and hence profitability. With the help of the high-speed data services operators can offer consumers video on demand, television, streaming music, as well as the ability to record and send short movies. For business users, the most interesting feature is the high speed Internet access from a laptop or mobile phone anywhere, which very often is not possible to do via WiFi connection, since a WiFi coverage is very fragmentary. As can be seen from Tables 11.4 and 11.5, Release'99 channels are able in principle to provide the data rates both in uplink and downlink up to 2 Mbits/s. However, in real life the data rate for Release'99 WCDMA usually is about 384 Kbits/s for downlink and 128 Kbits/s for uplink and the typical round trip delay for packet data is about 120 – 200 ms. The high-speed data service is characterized by the fact that typically data arrives in bursts, which poses the problem of fast changing need in radio resources. Moreover, the low delays are required for many services (for example, for video). The Release'99 WCDMA is not capable of meeting all these requirements. The real answer for the high-speed data services demand was the Release 5 of 3GPP specifications (finalized in early 2002) introducing *high-speed downlink packet access* (*HSDPA*) technology. Almost at the same time, similar works on a method to increase the uplink transmission speed have been started, and Release 6 introducing the enhanced uplink technology: *high-speed uplink packet access* (*HSUPA*) was finalized in early 2005. Both of these technologies, similar to GPRS and EDGE in GSM, allow the packet data transmission speed to be increased. This increase in transmission speed in comparison with Release'99 WCDMA can be up to several times. Quite often both technologies are referred to as *high-speed packet access* (*HSPA*).

11.3.1 HSDPA

The HSDPA technology used in WCDMA significantly improves the quality of data transmission on downlink. The improvement in the transmission quality is possible due to the increase in the speed of the transmission and the decrease in delays. Theoretically, the transmission speed can achieve up to 21 Mbps in non-MIMO mode and up to 28 Mbps in MIMO mode. HSDPA operates on the basis of the concept of shared transmission channel. HSDPA introduces one new transport channel:

- **HS-DSCH** – High Speed Downlink Shared Channel – a transport channel shared by all of the users of a cell;

and three new physical channels to the WCDMA system:

- **HS-PDSCH** – High Speed Physical Downlink Shared Channel – a physical channel used to carry the HS-DSCH,

- **HS-SCCH** – High Speed Shared Control Channel – a downlink physical channel used to carry downlink signalling related to HS-DSCH transmission,
- **HS-DPCCH** – High Speed Dedicated Physical Control Channel – an uplink physical channel that carries uplink feedback signalling related to downlink HS-DSCH transmission.

The HS-DSCH employs 2 ms TTI, which helps to decrease the delays and to share the radio resources much faster than in Release'99. Another difference from Release'99 transport channels is using the high order modulations: 16-QAM and 64-QAM except QPSK. The high order modulations definitely allow the transmission data rate to be increased, but more important for this goal is the usage of *adaptive modulation and coding* (*AMC*) or *fast link adaptation*. AMC allows choosing the modulation and coding scheme based on information about channel conditions provided by the UE to the Node B. Due to short 2 ms TTI the modulation and coding can be changed quite quickly that enables the maximum possible effect to be achieved from using the high order modulation. One more very important technology provided by HSDPA is *Hybrid Automatic Repeat-reQuest* (*HARQ*). The HARQ procedure in HSDPA is done at L1 not involving higher layers, which also helps to decrease the delays. And, of course, one of the main HSDPA features is *fast scheduling*. Typically the high-speed data has a bursty nature, that is, the amount of required radio resources varies rapidly. In Release'99 WCDMA the radio resources (transmission power and channelization codes) are allocated in static manner and the reallocation process is quite slow and involves higher layers. This contradicts the bursty nature of high-speed data. Unlike Release'99, in HSDPA the shared HS-DSCH resources (time slots and channelization codes) can be allocated to the different users on a rapid basis due to the short TTI (2 ms) and the fact that a fast scheduling algorithm is implemented as part of Node B functionality and there is no need to involve higher layers (for example, RNC) in the process of scheduling. Of course, the usage of AMC and fast scheduling leads to some fluctuation in data rate for the user, but packet-switched data services can tolerate some jitter in data rate.

11.3.1.1 Link Adaptation

Usually the fading causes significant variation of the signal to interference and noise ratio (SINR) of the signal received by a UE. In one particular cell the SINR of the received signal may vary over time by as much as 30-40dB. In order to improve system characteristics (such as capacity and coverage reliability) on one hand and optimize the user performance on the other hand, the downlink signal transmitted from Node B to a particular UE should be modified in such a way to take into account the signal quality variation. This process of adaptation of the transmitted signal to the channel conditions is called *link adaptation*. In Release'99 WCDMA channels the fast power control is used for link adaptation as shown in Figure 11.54a. In HSDPA the transmit power is kept constant. Instead the AMC is used for link adaptation. That means, in HSDPA the data rate is adjusted to channel conditions instead of changing the signal power as depicted in Figure 11.54b. This solution is more beneficial for the reduction of inter-cell interference. For example, if the UE is in deep fading, even significant increase of signal power will not help, but rather introduce the significant amount of interference into the system.

The spreading factor for HS-DSCH is fixed (SF = 16). Thus, the data rate can be adjusted to the channel conditions by changing modulation, coding rate or number of channelization codes allocated to particular UE. It is possible to use three modulation schemes in HSDPA: QPSK, 16-QAM and 64-QAM. In good channel conditions high order modulation (16-QAM or 64-QAM) can be used that immediately increases the data rate in comparison with QPSK two or three times correspondingly. For channel coding the turbo-encoder with constant rate 1/3 represented in Figure 11.27 is used. However, with the help of rate matching (that is, with the help of repetition or puncturing) the coding rate can be changed in the range from 0.14 to 1. From 1 to 15 channelization codes with SF = 16 can be used for HSDPA. These channelization codes are considered as a common resource, which can be shared between UEs. The allocation of channelization codes is done every TTI (2 ms) as shown in Figure 11.55.

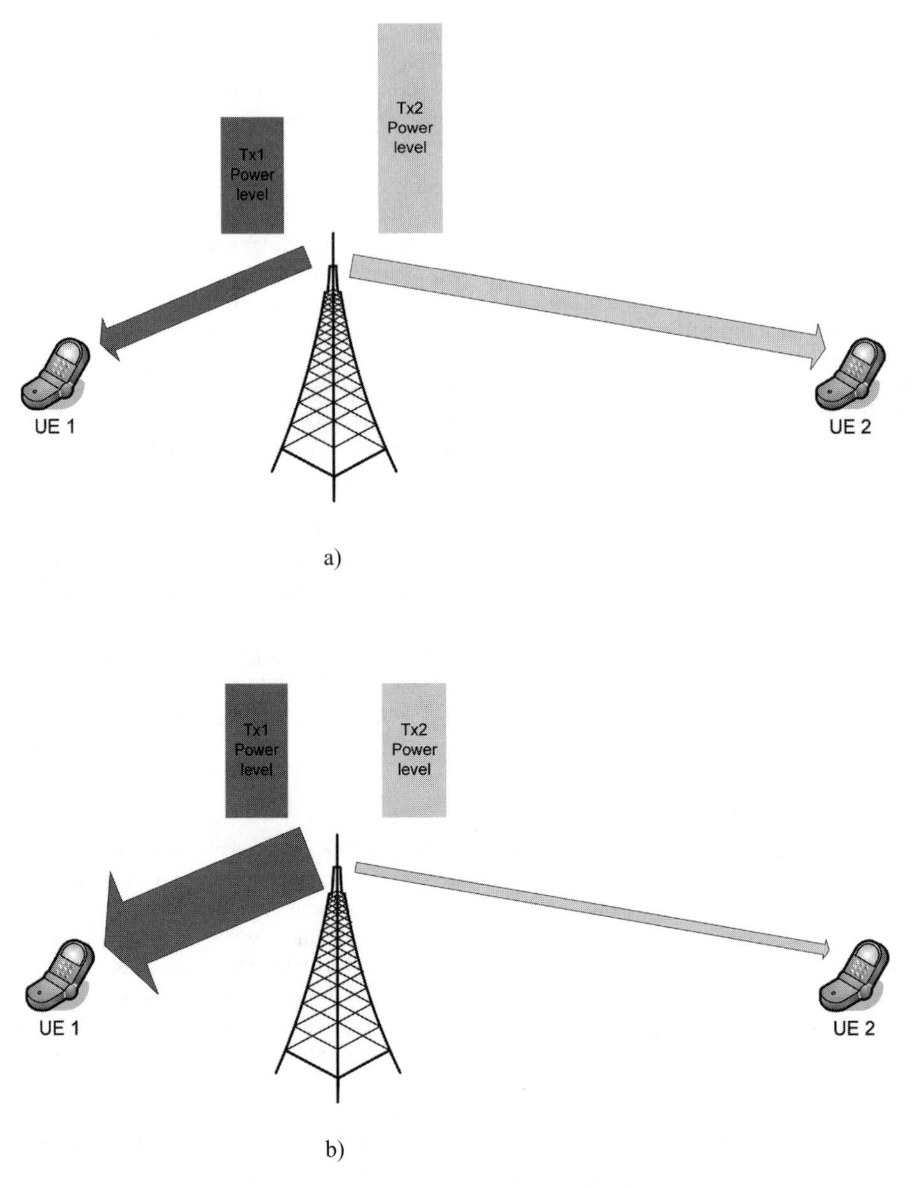

Figure 11.54 Link Adaptation: a) Power Control; b) Adaptive Modulation and Coding (AMC)

The choice of modulation and coding scheme is also done every 2 ms. Thus, the data rate can be adjusted to channel conditions quite rapidly. The basement for data rate selection is the *channel quality indicator (CQI)* report received from UE every TTI. The CQI report calculated in UE in fact is the quantized estimate of the received signal SINR. Taking into account the CQI report from the UE, Node B can adapt the data rate for this particular UE. Usually, the link adaptation is a two-step algorithm containing inner and outer loops, for example, as a power control algorithm does. The link adaptation algorithm based on AMC also consists of inner and outer loop algorithms. The *inner loop link adaptation*

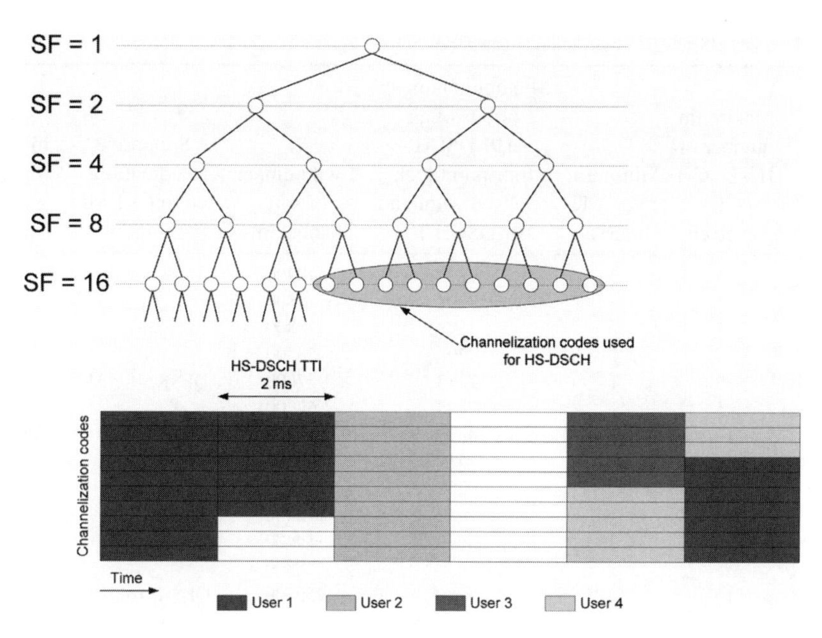

Figure 11.55 Channelization codes allocation for HSDPA users

is based on CQI reports. The main part of this algorithm (calculation of CQI report) as mentioned above is operating in UE. The inner loop algorithm allows Node B to react very rapidly to the change of radio channel quality. However, the UEs in the cell can have totally different receiver architecture, and due to this fact there are some offsets in relative UE performance. For example, some quantization errors may lead to reporting the biased CQI value. Thus, the inner loop algorithm may introduce some bias in CQI reports. And it is the task of *outer loop link adaptation algorithm* to compensate for any bias introduced by the inner loop algorithm. The outer loop algorithm is based on acknowledgment/not-acknowledgment feedback signals (Ack/Nack) sent by UE. The algorithm is operating at Node B. The outer loop algorithm controls the residual BLER after the first retransmission. It is done to take into account the gain from HARQ. The idea is to keep the BLER at the level of 10% (this level of BLER, intended for system operations, usually is called the *operating point*). If during some time the BLER is lower than 10%, Node B can provide higher data rate than that of UE requested by CQI report. And if BLER is higher than operating point, Node B can provide lower data rate than requested by UE. Also the link adaptation algorithm should take into account the UE capability to process the data, which depends on UE receiver implementation (amount of available memory and efficiency of detection and decoding algorithms). This capability is reflected in UE category, which is reported by UE to Node B. The UE categories are listed in Table 11.11 [22].

Inter-TTI interval > 1 means that terminal cannot receive data in consecutive TTIs. It must wait for 2 ms (TTI interval $= 2$) or 4 ms (TTI interval $= 3$) after each received TTI as shown in Figure 11.56.

Thus, the UE category defines the maximum possible data rate that the UE is capable of processing. For example, the UE of category 11 can provide a maximum data rate about 0.9 Mbps: the maximum transport block size (TBS) is 3630 bits and the TTI interval is 2, that is, it can receive 3630 bits per 4 ms $\left(\frac{3630}{0.004} = 0.9075\text{Mbps}\right)$. Actually, that means the UE of category 11 can use quite simple detection algorithm, since the used modulation is QPSK and the maximum coding rate is 3/4. The maximum coding rate can be calculated as follows. The HSDPA spreading factor is 16, and the number of chips per 2 ms TTI is $3.84 \cdot 10^6 \times 0.002 = 7680$, that is, the amount of transmitted symbols per channelization

Table 11.11 FDD HS-DSCH physical layer categories

HS-DSCH category	Maximum number of HS-DSCH codes received	Minimum inter-TTI interval	Maximum number of bits of an HS-DSCH transport block received within an HS-DSCH TTI	Total number of soft channel bits	Supported modulations without MIMO operation	Supported modulations simultaneous with MIMO operation
Category 1	5	3	7298	19200		
Category 2	5	3	7298	28800		
Category 3	5	2	7298	28800		
Category 4	5	2	7298	38400		
Category 5	5	1	7298	57600	QPSK, 16QAM	
Category 6	5	1	7298	67200		Not applicable (MIMO not supported)
Category 7	10	1	14411	115200		
Category 8	10	1	14411	134400		
Category 9	15	1	20251	172800		
Category 10	15	1	27952	172800		
Category 11	5	2	3630	14400	QPSK	
Category 12	5	1	3630	28800		
Category 13	15	1	35280	259200	QPSK, 16QAM, 64QAM	
Category 14	15	1	42192	259200		
Category 15	15	1	23370	345600	QPSK, 16QAM	
Category 16	15	1	27952	345600		
Category 17	15	1	35280	259200	QPSK, 16QAM, 64QAM	–
			23370	345600	–	QPSK, 16QAM
Category 18	15	1	42192	259200	QPSK, 16QAM, 64QAM	–
			27952	345600	–	QPSK, 16QAM
Category 19	For future use; supports the capabilities of category 17 in this version of the protocol					
Category 20	For future use; supports the capabilities of category 18 in this version of the protocol					

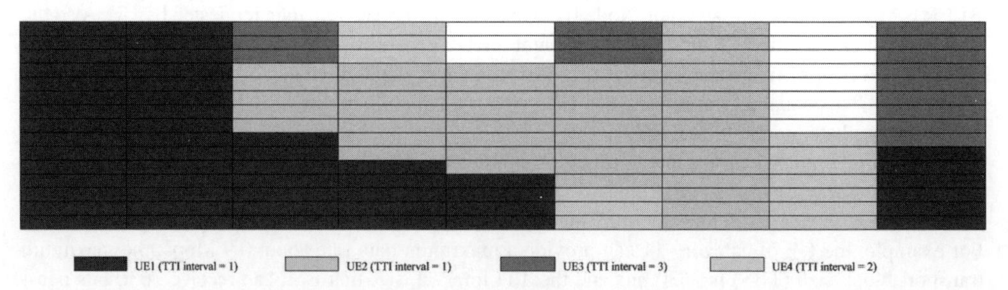

Figure 11.56 Channelization codes allocation for HSDPA users with different TTI interval values

code per TTI is $\frac{7680}{16} = 480$. Taking into account that QPSK bears 2 bits per symbol, obtain 960 bits per channelization code per TTI. Since the maximum number of channelization codes for category 11 is 5, obtain $960 \times 5 = 4800$ transmitted bits per TTI. That gives us the coding rate $\frac{3630}{4800} \approx 0.756$. For UE of category 1 the TTI interval = 3 and the maximum TBS = 7298, that is, the UE of this category can provide maximum data rate 1.2 Mbps using 16-QAM and coding rate 3/4. The UE of category 16 using 16-QAM modulation and coding rate 0.97 with allocation of 15 channelization codes and with TTI = 1 can provide maximum data rate 14 Mbps. And the maximum theoretical data rate in non-MIMO mode can be provided by the UE of category 18: 21 Mbps with 64-QAM, coding rate 0.98 and 15 channelization codes. Of course, in the latter case the UE receiver must have more memory and exploit more comprehensive detection algorithms, and therefore this type of UE should be more expensive. On the other hand, the higher data rates provided by these UEs should stimulate end users to buy more expensive phones with advanced receivers. And this is a clear benefit for the network operators.

11.3.1.2 Fast Scheduling

HS-DSCH is the shared channel, that is, its resources are shared between all UEs in the cell, and the scheduling of these resources to users plays a critical role in HSDPA performance. In each TTI, the scheduler must decide what share of HSDPA resources (channelization codes) should be allocated for each UE. It is possible to use different strategies of scheduling. However, all of them have the inherited contradiction between fairness and total cell throughput. The *Round Robin scheduling* allocates an even number of channelization codes and time slots for each UE independently of UE channel conditions. This leads to a lower user throughput at the cell border compared to that close to the Node B. Another option is the *Fair Throughput scheduling*. This type of scheduling means that the resources are allocated for UEs in such a way to provide the same throughput for all UEs, regardless of their radio channel quality. This implies that UEs with poorer radio channel quality must be assigned a larger amount of resources to balance for their throughput. This scheduling strategy gives a fair user throughput distribution at the cost of a lower cell throughput. On the other hand, the UE channel quality measurements, which enable the fast fading process of UEs to be tracked, gives the possibility of serving UEs on the top of their most favourable conditions, thus introducing the gain commonly known as *multi-user diversity*. This type of scheduling called *Maximum C/I* (carrier to interference ratio), obviously maximizes the cell throughput, but it is done at the cost of fairness because UEs under worse radio conditions are allocated lower amount of resources. The Maximum C/I scheduling is depicted in Figure 11.57. The multi-user diversity increases with increase of channel variations and with the increase of number of UEs in the cell. Thus, in the sense of cell capacity or cell throughput, fading is desirable and can be exploited. The benefit of the strategy exploiting the multi-user diversity depends on the ability of the Node B to obtain recent and accurate channel quality measurements of the fast fading process. So, the low latency of CQI reports is a key factor for this strategy. Due to the short TTI and the fact that scheduling is performed directly in Node B not involving higher layers in the decision process, this requirement is met in HSDPA.

Actually, as can be seen from Figure 11.56, it is not necessary to allocate all resources in current TTI to one UE under the best channel conditions. Some share of channelization codes may be allocated to other UEs. This may prevent the starvation of UEs under poor radio propagation conditions (usually the users at the cell edge). This type of scheduling called *Proportional Fair scheduling* is the compromise between the cell capacity provided by the Maximum C/I scheduling and the fairness [23], [24]. It takes into account the channel quality as well as previously achieved UE throughput. Since the Proportional Fair scheduling provides competitive cell capacity as well as fairness it is also used in HSDPA together with Maximum C/I scheduling. The exact scheduling strategy is not specified in 3GPP and depends on base station manufacturer.

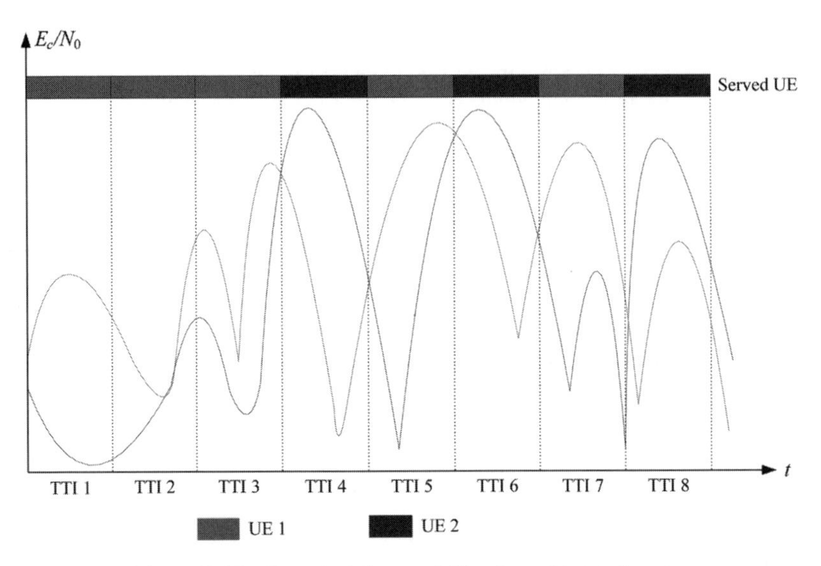

Figure 11.57 Fast scheduling, exploiting the multi-user diversity

11.3.1.3 H-ARQ

In fast fading conditions the AMC alone cannot guarantee the reliability of data transmission. The reasons for this are the data rate selection granularity and the fact that CQI values were chosen and corresponding thresholds were tabulated under some particular channel conditions and for the current channel conditions these CQI tables may show some bias as mentioned in section 11.3.1.1. The hybrid ARQ allows the UE not only to rapidly request the retransmission of the erroneous packet but also to add the redundancy in retransmission only when needed. Thus, the H-ARQ mechanism helps the outer loop link adaptation algorithm to compensate for the errors of inner loop link adaptation. H-ARQ autonomously adapts to the instantaneous channel conditions and is insensitive to the measurement error and delay. Combining AMC with H-ARQ leads to the best of both worlds: AMC provides the coarse data rate selection, while H-ARQ provides for fine data rate adjustment based on channel conditions. The conventional ARQ mechanism implies that the erroneously received packet is just discarded and substituted by its retransmitted copy. Hybrid ARQ uses soft combining mechanism. That means the retransmitted packets are combined with the original packet to improve the decoding probability. Diversity time gain is thus obtained. It is possible to use different types of soft combining in HSDPA. The simplest one is *Chase combining* [15]. In this case the retransmission is identical, that is, exactly the same bits are retransmitted in each retransmission attempt for the packet. Then the soft bits obtained from the two or more identical transmissions are summed up, and the result is fed into the decoder. The idea is that each copy of the packet is weighted by the SINR and therefore the simple summation of the soft bits gives the optimal combination [25]. In more complicated schemes, so-called *incremental redundancy* (IR) is used: the retransmissions are not necessarily identical with the original one but can contain a different number of redundancy bits for error correction. In H-ARQ of type II, the retransmissions need not be decodable alone, that is, they can, for example, contain only parity bits of the channel turbo-encoder. If all transmissions can in principle be decoded separately (self-decodable retransmission), that is, if each retransmission contains all the systematic bits of the data, the scheme is called H-ARQ of type III. So, in principle the original transmission can consist of the systematic bits only, and if the original transmission fails, then some of parity 1 or parity 2 bits can be added to the systematic bits in the retransmissions (type III H-ARQ) or

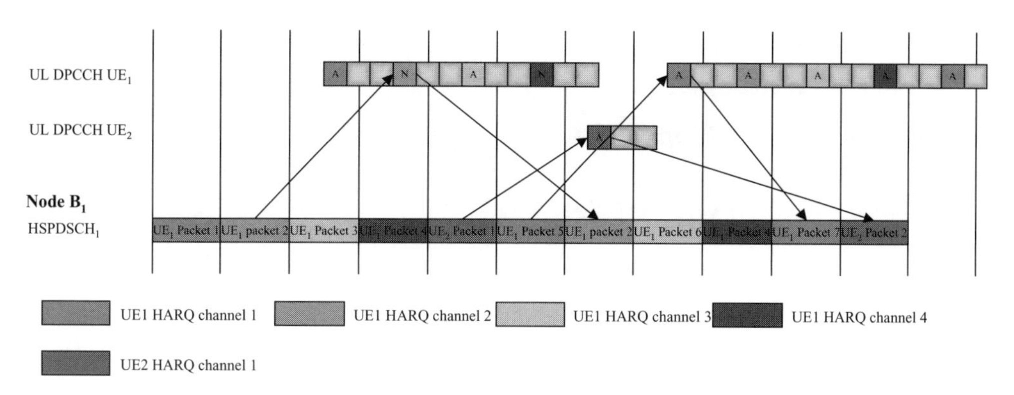

Figure 11.58 Principle of *N*-channel stop-and-wait HARQ ($N = 4$). Reproduced by permission of © 2001 3GPP™

retransmission can consist of some part of systematic bits and parity bits or even only parity bits can constitute the retransmission (type II H-ARQ), that is, the redundancy can be added incrementally.

The ARQ protocol used for HSDPA H-ARQ is *N-channel stop-and-wait* (*SAW*) protocol. As mentioned in Chapter 6, stop-and-wait (SAW) is one of the simplest forms of ARQ requiring very little overhead. However, it has one severe drawback: acknowledgments are not instantaneous and therefore after every transmission, the transmitter must wait to receive the acknowledgment prior to transmitting the next block. This is a well-known problem with SAW ARQ. In the interim, the channel remains idle and system capacity goes wasted. In a slotted system, the feedback delay will waste at least half the system capacity while the transmitter is waiting for acknowledgments. As a result, at least every other timeslot must go idle even on an error free channel. The *N*-channel SAW ARQ differs from the simple SAW in such a way that the transmitted data is split in *N* independent time-multiplexed channels and the simple SAW ARQ mechanism is applied to each of these channels. Thus, while the transmitter is waiting an acknowledgment for one channel the data is transmitted over other channels as shown in Figure 11.58 [26].

The example depicted in Figure 11.58 shows a sequence of events when packets 1-7 are being transmitted to UE1 and packet 1 to UE2 using *N*-channel SAW HARQ with $N = 4$. Packets are transmitted using four parallel ARQ processes for UE1 and one ARQ process for UE2, each using stop-and-wait principle. Each packet is acknowledged during the transmission of other packets so that the downlink channel can be kept occupied all the time if there are packets to transmit. *N*-channel SAW HARQ supports asynchronous transmission: different users can be scheduled freely without waiting completion of a given transmission. Receiver needs to know which HARQ process the packet belongs to, and this information is explicitly signalled on HS-SCCH. Also, the information about the content of the current packet (is it initial transmission or retransmission) and the *redundancy version (RV)* is signalled on HS-SCCH. This helps the receiver to perform the combining and decoding of the transport block(s) properly in case of signalling errors in uplink. Transmission for a given user is assumed to continue when the channel is again allocated. The *N*-channel SAW HARQ used in HSDPA is asynchronous. It can be seen from Figure 11.58: after four packets to UE1, a packet is transmitted to UE2 and the transmission to UE1 is delayed by one TTI. Also, there are five packets to UE1 between packets to UE2. The processing times should be defined such that continuous transmission to a UE is possible. The maximum number of HARQ processes (channels) is $N = 8$. Due to UE and Node B processing delay requirements minimum value of *N* is set to 6.

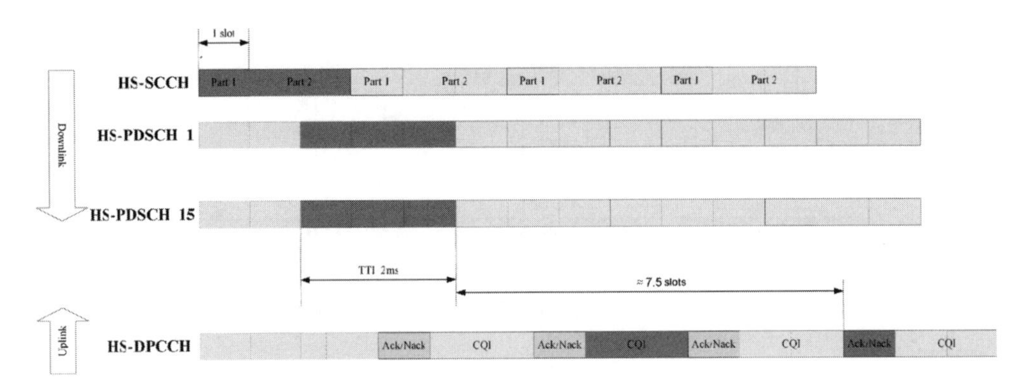

Figure 11.59 HSDPA physical channels timing

11.3.1.4 HSDPA Channels

The timing of HSDPA channels is depicted in Figure 11.59.

First the data carried by HS-SCCH is received. The HS-SCCH TTI (2 ms) consists of two parts. Part 1 occupying the first slot of TTI carries the most critical information needed for HS-DCSH decoding: the set of channelization codes and modulation scheme used for the current HS-DSCH TTI. This information must be decoded before receiving the HS-DSCH TTI. Part 2 of HS-SCCH TTI contains the information that is not so critical and can be decoded later. HS-DSCH, consisting of up to 15 physical channels, (HS-PDSCHs) is delayed relative to HS-SCCH by two slots. That means the UE receiver must decode the HS-SCCH Part 1 information during 1 slot (666 μs). After HS-DSCH decoding the feedback information is sent by UE over HS-DPCCH. The first slot of HS-DPCCH subframe (2 ms) is occupied by acknowledgment information (Ack/Nack) and two other slots comprise CQI and *Precoding Control Indication (PCI)*. PCI is used only in MIMO mode. However, Ack/Nack and CQI/PCI sent in one subframe does not necessarily refer to the same HS-DSCH TTI. The transmit timing of the start of the HS-DPCCH subframe is delayed by about 7.5 slots relative to the end of HS-DSCH TTI. Due to this fact Ack/Nack cannot be sent earlier than 7.5 slots after the end of HS-DSCH TTI. On the other hand, this fact does not prevent UE to send the CQI/PCI report earlier as shown in Figure 11.60.

Node B should have the most recent information about channel state, and thus the CQI/PCI report corresponding to the old channel state can be substituted by the most recent one. In this case the delay between HS-DSCH TTI and CQI report is only about 2.5 slots. Of course, it depends on the ability of the UE receiver to generate the CQI/PCI report during 2.5 slots.

Now consider the HSDPA channels in more detail.

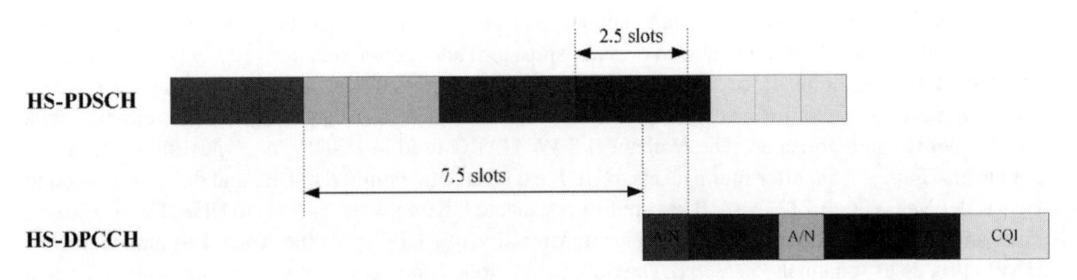

Figure 11.60 CQI/PCI report delay

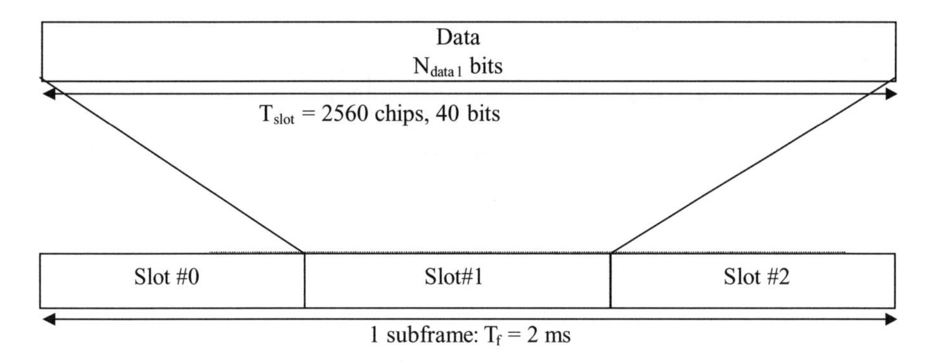

Figure 11.61 Subframe structure for the HS-SCCH. Reproduced by permission of © 2008 3GPP™

The HS-SCCH is a fixed rate (60 kbps, SF = 128) downlink physical channel used to carry downlink signalling related to HS-DSCH transmission. Figure 11.61 [15] illustrates the sub-frame structure of the HS-SCCH.

QPSK is always used for HS-SCCH transmission.

The High Speed Physical Downlink Shared Channel (HS- PDSCH) is used to carry the High Speed Downlink Shared Channel (HS-DSCH).

An HS-PDSCH corresponds to one channelization code of fixed spreading factor SF = 16 from the set of channelization codes reserved for HS-DSCH transmission. Multi-code transmission is allowed, which translates to UE being assigned multiple channelization codes in the same HS-PDSCH subframe, depending on its UE capability.

The subframe and slot structure of HS-PDSCH are shown in Figure 11.62 [15].

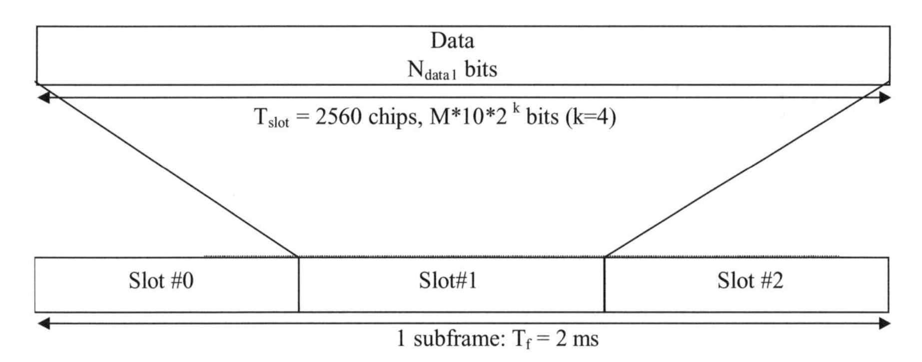

Figure 11.62 Subframe structure for the HS-PDSCH. Reproduced by permission of © 2008 3GPP™

Table 11.12 HS-DSCH fields

Slot format #i	Channel Bit Rate (kbps)	Channel Symbol Rate (ksps)	SF	Bits/ HS-DSCH subframe	Bits/ Slot	Ndata
0(QPSK)	480	240	16	960	320	320
1(16QAM)	960	240	16	1920	640	640
2(64QAM)	1440	240	16	2880	960	960

Reproduced by permission of © 2008 3GPP™

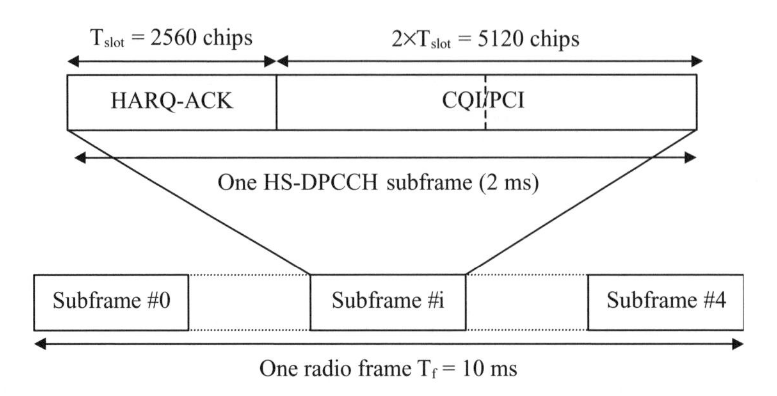

Figure 11.63 Sub-frame structure for the HS-DPCCH

An HS-PDSCH may use QPSK, 16QAM or 64QAM modulation symbols. In Figure 11.62, M is the number of bits per modulation symbols, that is, $M = 2$ for QPSK, $M = 4$ for 16QAM and $M = 6$ for 64QAM. All relevant Layer 1 information is transmitted in the associated HS-SCCH, that is, the HS-PDSCH does not carry any Layer 1 information. The HS-PDSCH slot formats are shown in Table 11.12 [15].

Figure 11.63 [15] illustrates the frame structure of the HS-DPCCH. The HS-DPCCH carries uplink feedback signalling related to downlink HS-DSCH transmission. The HS-DSCH-related feedback signalling consists of Hybrid-ARQ Acknowledgment (HARQ-ACK) and Channel-Quality Indication (CQI) and in this-case the UE is configured in MIMO mode of Precoding Control Indication (PCI) as well [16]. Each sub frame of length 2 ms (3·2560 chips) consists of three slots, each of length 2560 chips. The HARQ-ACK is carried in the first slot of the HS-DPCCH sub-frame. The CQI, and in case the UE is configured in MIMO mode the PCI as well, are carried in the second and third slot of a HS-DPCCH sub-frame. There is at most one HS-DPCCH on each radio link. The HS-DPCCH can only exist together with an uplink DPCCH.

The spreading factor of the HS-DPCCH is 256, that is, there are 10 bits per uplink HS-DPCCH slot. QPSK is always used for HS-DPCCH transmission.

One more channel related to HSDPA is *fractional dedicated physical channel* (F-DPCH). This channel partially substitutes DPCH in cases where only packet services are active in the downlink. In such a case,

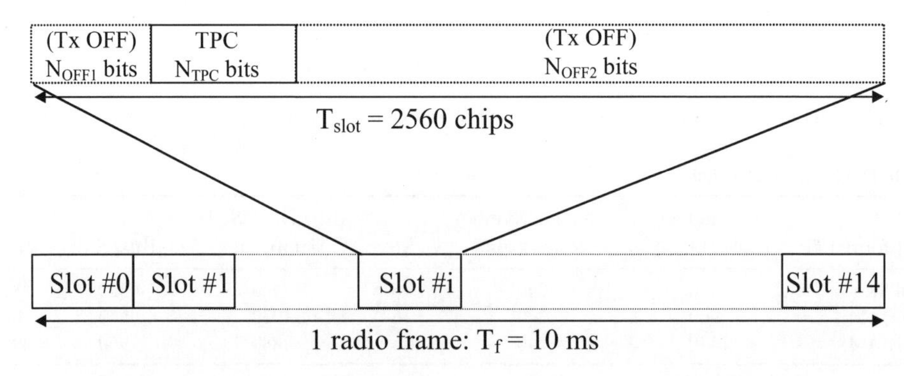

Figure 11.64 Frame structure of F-DPCH. Reproduced by permission of © 2008 3GPP™

Table 11.13 F-DPCH fields

Slot Format #i	Channel Bit Rate (kbps)	Channel Symbol Rate (ksps)	SF	Bits/ Slot	N_{OFF1} Bits/Slot	N_{TPC} Bits/Slot	N_{OFF2} Bits/Slot
0	3	1.5	256	20	2	2	16
1	3	1.5	256	20	4	2	14
2	3	1.5	256	20	6	2	12
3	3	1.5	256	20	8	2	10
4	3	1.5	256	20	10	2	8
5	3	1.5	256	20	12	2	6
6	3	1.5	256	20	14	2	4
7	3	1.5	256	20	16	2	2
8	3	1.5	256	20	18	2	0
9	3	1.5	256	20	0	2	18

especially with lower data rates, the downlink DCH introduces too much overhead and can also consume too much code space if looking for a large number of users using a low data rate service – like VoIP [27]. F-DPCH carries transmission power control information (TPC commands) only. It is a special case of downlink DPCCH. Figure 11.64 [15] shows the frame structure of the F-DPCH. Each frame of length 10 ms is split into 15 slots, each of length $T_{slot} = 2560$ chips, corresponding to one power-control period.

The exact number of bits of the OFF periods and of the F-DPCH fields (N_{TPC}) is described in Table 11.13 [15]. Each slot format corresponds to a different set of OFF periods within the F-DPCH slot.

In compressed frames, F-DPCH is not transmitted in downlink transmission gaps given by transmission gap pattern sequences signalled by higher layers. In the case of STTD for F-DPCH, the TPC bits are not STTD encoded and the same bits are transmitted with equal power from the two antennas.

11.3.1.4.1 HS-SCCH Multiplexing and Coding

There are three types of HS-SCCH: type 1 HS-SCCH is used when UE is not configured in MIMO mode and not in HS-SCCH-less mode; type 2 HS-SCCH is used for so called HS-SCCH-less mode, which will be discussed later; and type 3 HS-SCCH is used only when UE is configured in MIMO mode.

The following information is transmitted by means of the HS-SCCH type 1 physical channel.

- Channelization-code-set information (7 bits): $x_{ccs,1}, x_{ccs,2}, \ldots, x_{ccs,7}$
- Modulation scheme information (1 bit): $x_{ms,1}$
- Transport-block size information (6 bits): $x_{tbs,1}, x_{tbs,2}, \ldots, x_{tbs,6}$
- Hybrid-ARQ process information (3 bits): $x_{hap,1}, x_{hap,2}, x_{hap,3}$
- Redundancy and constellation version (3 bits): $x_{rv,1}, x_{rv,2}, x_{rv,3}$
- New data indicator (1 bit): $x_{nd,1}$
- UE identity (16 bits): $x_{ue,1}, x_{ue,2}, \ldots, x_{ue,16}$

Channelization-code-set information over HS-SCCH is mapped in the following manner: the OVSF codes shall be allocated in such a way that they are positioned in sequence in the code tree. That is, for *P* channelization codes at offset *O* the following codes are allocated:

$$C_{ch,16,O}, \ldots, C_{ch,16,O+P-1} \quad (11.70)$$

Transmitting just two numbers *P* and *O* requires 8 bits, since *P* can take values from 1 to 15 and *O* can take values from 0 to 15. However, taking into account that values of *P* and *O* are not independent, it is possible to transmit values of P and O with the help of only 7 bits (actually 6.5 bits). Bits $x_{ccs,1}$, $x_{ccs,2}$, $x_{ccs,3}$ indicate the code group number:

$$x_{ccs,1}, x_{ccs,2}, x_{ccs,3} = \min(P - 1, 15 - P) \tag{11.71}$$

and bits $x_{ccs,4}$, $x_{ccs,5}$, $x_{ccs,6}$, $x_{ccs,7}$, indicate code offset. If 64QAM is not configured for the UE, or if 64QAM is configured but QPSK is used for the HS-DSCH transmission, then:

$$x_{ccs,4}, x_{ccs,5}, x_{ccs,6}, x_{ccs,7} = |O - 1 - [P/8] \cdot 15| \tag{11.72}$$

Otherwise (that is, if 64QAM is configured for the UE and 16QAM or 64QAM is used for HS-DSCH transmission):

$$x_{ccs,4}, x_{ccs,5}, x_{ccs,6}, x_{ccs,dummy} = |O - 1 - [P/8] \cdot 15| \tag{11.73}$$

where $x_{ccs,dummy}$ is a dummy bit that is not transmitted on HS-SCCH. Furthermore:

$$x_{ccs,7} = \begin{cases} 0 \ \ if \ \ 16QAM \\ 1 \ \ if \ \ 64QAM \end{cases} \tag{11.74}$$

The value of $x_{ms,1}$ is derived from the modulation and given by the following:

$$x_{ms,1} = \begin{cases} 0 \ \ if \ \ QPSK \\ 1 \ \ otherwise \end{cases} \tag{11.75}$$

Thus, bits $x_{ccs,7}$ and $x_{ms,1}$ define the modulation scheme.

Transport-block size information (6 bits) $x_{tbs,1}, x_{tbs,2}, \ldots, x_{tbs,6}$ is the unsigned binary representation of the transport block size.

Hybrid-ARQ process information (3 bits) $x_{hap,1}, x_{hap,2}, x_{hap,3}$ is the unsigned binary representation of the HARQ process identifier.

The redundancy version (RV) parameters r, s and constellation version parameter b are coded jointly to produce the value X_{rv}. X_{rv} is alternatively represented as the sequence $x_{rv,1}, x_{rv,2}, x_{rv,3}$ where $x_{rv,1}$ is the MSB. This is done according to Tables 11.14 [16] and 11.15 [16] according to the modulation mode used:

New data indicator $x_{nd,1}$ is set to "1" for the first transmission and "0" for the retransmission.

The UE identity is the HS-DSCH Radio Network Identifier (H-RNTI) comprising 16 bits. The H-RNTI is allocated by controlling RNC to each UE establishing a HS-DSCH channel. H-RNTI shall be unique within the cell carrying the HS-DSCH. The H-RNTI shows to which UE the current HS-DSCH TTI is intended.

Figure 11.65 [16] illustrates the overall coding chain for HS-SCCH type 1.

Table 11.14 RV coding for 16QAM and 64QAM

X_{rv} (value)	s	r	b
0	1	0	0
1	0	0	0
2	1	1	1
3	0	1	1
4	1	0	1
5	1	0	2
6	1	0	3
7	1	1	0

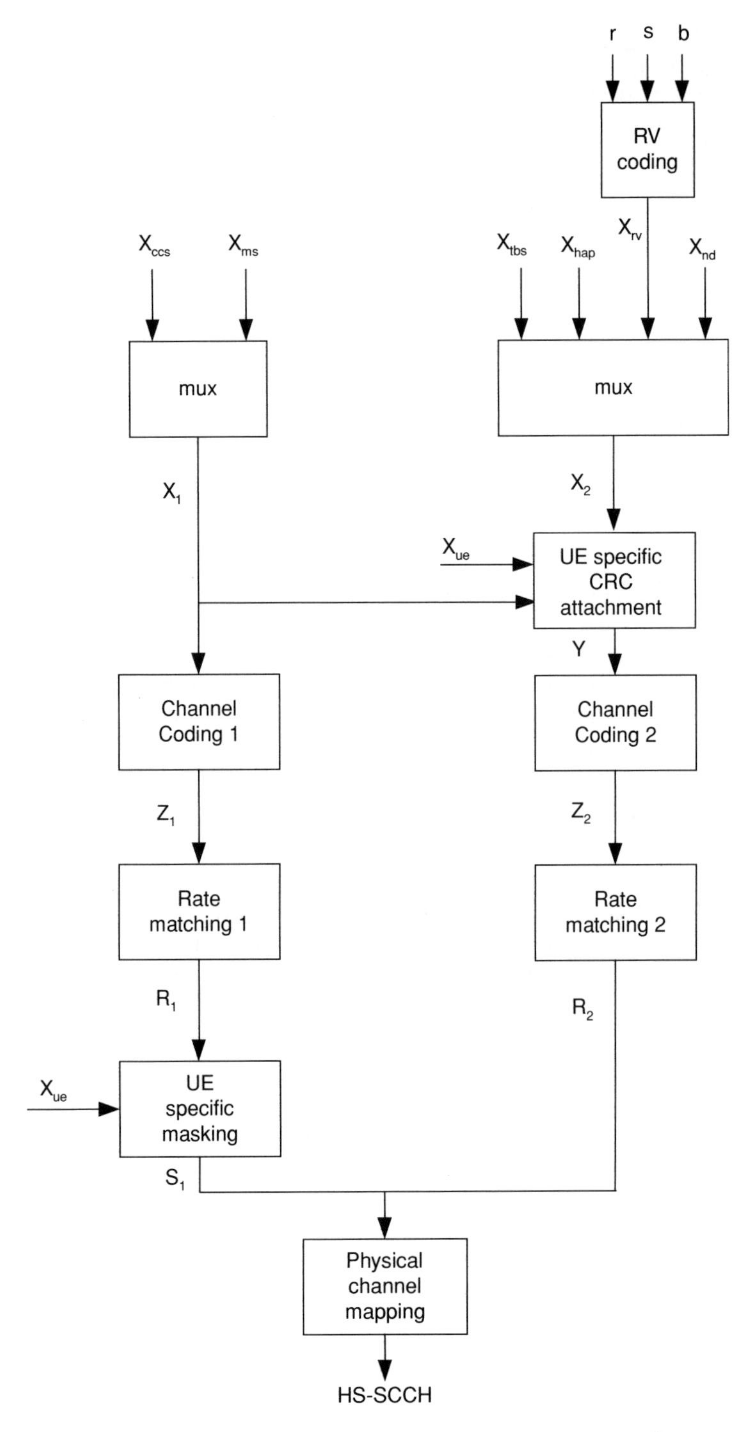

Figure 11.65 Coding chain for HS-SCCH type 1. Reproduced by permission of © 2008 3GPP™

Table 11.15 RV coding for QPSK

X_{rv} (value)	s	r
0	1	0
1	0	0
2	1	1
3	0	1
4	1	2
5	0	2
6	1	3
7	0	3

The channelization code-set information $x_{ccs,1}, x_{ccs,2}, \ldots, x_{ccs,7}$ and modulation-scheme information $x_{ms,1}$ are multiplexed together. This gives a sequence of bits $x_{1,1}, x_{1,2}, \ldots, x_{1,8}$ where:

$x_{1,i} = x_{ccs,i}$ $\quad i = 1,2,\ldots,7$
$x_{1,i} = x_{ms,i\text{-}7}$ $\quad i = 8$

This bit sequence comprises the data transmitted in Part 1 of the HS-SCCH sub-frame. This data is most important for HS-DSCH reception and should be decoded first.

The transport-block-size information $x_{tbs,1}, x_{tbs,2}, \ldots, x_{tbs,6}$, Hybrid-ARQ-process information $x_{hap,1}$, $x_{hap,2}, x_{hap,3}$, redundancy-version information $x_{rv,1}, x_{rv,2}, x_{rv,3}$ and new-data indicator $x_{nd,1}$ are multiplexed together. This gives a sequence of bits $x_{2,1}, x_{2,2}, \ldots, x_{2,13}$ where:

$x_{2,i} = x_{tbs,i}$ $\quad i = 1,2,\ldots,6$
$x_{2,i} = x_{hap,i\text{-}6}$ $\quad i = 7,8,9$
$x_{2,i} = x_{rv,i\text{-}9}$ $\quad i = 10,11,12$
$x_{2,i} = x_{nd,i\text{-}12}$ $\quad i = 13$

This data comprises Part 2 of the HS-SCCH sub-frame.

From the sequence of bits $x_{1,1}, x_{1,2}, \ldots, x_{1,8}, x_{2,1}, x_{2,2}, \ldots, x_{2,13}$ a 16 bits CRC is calculated. This gives a sequence of bits $c_1, c_2, \ldots, c_{16}$. This sequence of bits is then masked with the UE Identity $x_{ue,1}, x_{ue,2}, \ldots,$ $x_{ue,16}$ and then appended to the sequence of bits $x_{2,1}, x_{2,2}, \ldots, x_{2,13}$ to form the sequence of bits $y_1, y_2, \ldots,$ y_{29}, where:

$$\begin{aligned} y_i &= x_{2,i} & i &= 1, 2, \ldots, 13 \\ y_i &= (c_{i-13} + x_{ue,i-13}) \bmod 2 & i &= 14, 15, \ldots, 29 \end{aligned} \tag{11.76}$$

After multiplexing Part 1 and Part 2 bits are encoded separately. Rate 1/3 convolutional coding, as described in section 11.2.4.2, is applied to the sequence of bits $x_{1,1}, x_{1,2}, \ldots, x_{1,8}$. This gives a sequence of bits $z_{1,1}, z_{1,2}, \ldots, z_{1,48}$. The same encoding is applied to the sequence of bits $y_1, y_2, \ldots, y_{29}$. This gives a sequence of bits $z_{2,1}, z_{2,2}, \ldots, z_{2,111}$.

From the input sequence $z_{1,1}, z_{1,2}, \ldots, z_{1,48}$ the bits $z_{1,1}, z_{1,2}, z_{1,4}, z_{1,8}, z_{1,42}, z_{1,45}, z_{1,47}, z_{1,48}$ are punctured to obtain the output sequence $r_{1,1}, r_{1,2} \ldots r_{1,40}$.

From the input sequence $z_{2,1}, z_{2,2}, \ldots, z_{2,111}$ the bits $z_{2,1}, z_{2,2}, z_{2,3}, z_{2,4}, z_{2,5}, z_{2,6}, z_{2,7}, z_{2,8}, z_{2,12}, z_{2,14}, z_{2,15},$ $z_{2,24}, z_{2,42}, z_{2,48}, z_{2,54}, z_{2,57}, z_{2,60}, z_{2,66}, z_{2,69}, z_{2,96}, z_{2,99}, z_{2,101}, z_{2,102}, z_{2,104}, z_{2,105}, z_{2,106}, z_{2,107}, z_{2,108}, z_{2,109},$ $z_{2,110}, z_{2,111}$ are punctured to obtain the output sequence $r_{2,1}, r_{2,2} \ldots r_{2,80}$.

The rate matched bits $r_{1,1}, r_{1,2} \ldots r_{1,40}$ shall be masked in an UE specific way using the UE identity $x_{ue,1}$, $x_{ue,2}, \ldots, x_{ue,16}$, to produce the bits $s_{1,1}, s_{1,2} \ldots s_{1,40}$.

Intermediate codeword bits b_i, $i = 1,2\ldots,48$, are defined by encoding the UE identity bits using the rate $\frac{1}{2}$ convolutional coding described in section 11.1.4.2. Eight bits $b_1, b_2, b_4, b_8, b_{42}, b_{45}, b_{47}, b_{48}$ out of the resulting 48 convolutionally encoded bits are punctured to obtain the 40 bit UE specific scrambling sequence $c_1, c_2, \ldots, c_{40}$. The mask output bits $s_{1,1}, s_{1,2}\ldots s_{1,40}$ are calculated as follows:

$$s_{1,k} = (r_{1,k} + c_k) \bmod 2 \quad \text{for}\, k = 1, 2, \ldots, 40 \tag{11.77}$$

The sequence of bits $s_{1,1}, s_{1,2}, \ldots, s_{1,40}$ comprising Part 1 is mapped to the first slot of the HS-SCCH sub frame. The sequence of bits $r_{2,1}, r_{2,2}, \ldots, r_{2,80}$ comprising Part 2 is mapped to the second and third slot of the HS-SCCH sub-frame.

There are some specific cases when so-called HS-SCCH-less operation is used. For example, such services as VoIP and gaming require relatively little bandwidth per user and thus the number of simultaneous users can be high. That means the amount of signalling is increasing, since there must exist the HS-SCCH for each user, but the amount of data transmitting over HS-DSCH to each user is rather small. Thus, more users mean more signalling overhead which decreases overall available bandwidth for user data. HS-SCCH-less operation aims at reducing this overhead. In HS-SCCH-less mode operation, HS-SCCH is not transmitted for the initial transmission of data on HS-PDSCH. And UE is trying to blindly decode the data on HS-PDSCH with predefined control information (in this case the HS-DSCH data can only have one of four transport block sizes and is always modulated using QPSK and the redundancy and constellation version is always pre-defined to 0). If the UE is unable to blindly decode the initial transmission successfully then the data shall be retransmitted for a maximum of two times. HS-SCCH is transmitted only during the retransmissions. The HS-SCCH that is used for 1st and 2nd retransmission is called HS-SCCH type 2.

When HS-SCCH-less operation is used, the first transmission of an HS-DSCH transport block with 24 bit CRC shall be sent without an associated HS-SCCH. In this case, the UE shall use the following signalling values in order to attempt to decode the transport block:

- Channelization-code-set information: *Configured by higher layers*
- Modulation scheme information: *QPSK*
- Transport-block size information: *Each of four possible sizes configured by higher layers*
- Redundancy and constellation version: $X_{rv} = 0$
- UE identity: *Configured by higher layers*

In case of HS-SCCH-less operation for the initial HS-DSCH transmission UE will not send NACK on HS-DPCCH if it is unable to decode the data. If the Node B does not detect an ACK for a HS-DSCH transmission in HS-SCCH-less mode then the Node B decides that it was a NACK and it will retransmit the data using HS-SCCH Type 2. During second and third transmission, the following information is transmitted by means of the HS-SCCH type 2 physical channel.

- Channelization-code-set information (7 bits): $x_{ccs,1}, x_{ccs,2}, \ldots, x_{ccs,7}$
- Modulation scheme information (1 bit): $x_{ms,1}$
- Special Information type (6 bits): $x_{type,1}, x_{type,2}, \ldots, x_{type,6}$
- Special Information (7 bits): $x_{info,1}, x_{info,2}, x_{info,3}, x_{info,4}, x_{info,5}, x_{info,6}, x_{info,7}$
- UE identity (16 bits): $x_{ue,1}, x_{ue,2}, \ldots, x_{ue,16}$

The channelization code-set bits $x_{ccs,1}, x_{ccs,2}, \ldots, x_{ccs,7}$ are coded in accordance with (11.71), where the value of P shall be set to either 1 or 2.

The value of $x_{ms,1}$ shall be set to '0' (QPSK).

The Special Information type $x_{type,1}, x_{type,2}, \ldots, x_{type,6}$ shall be set to "111110" to indicate HS-SCCH less operation.

The Special Information bits $x_{info,1}, x_{info,2}, \ldots, x_{info,7}$ comprise:

- Transport-block size information (2 bits): $x_{info,1}, x_{info,2} = x_{tbs,1}, x_{tbs,2}$
- Pointer to the previous transmission (3 bits): $x_{info,3}, x_{info,4}, x_{info,5} = x_{ptr,1}, x_{ptr,2}, x_{ptr,3}$
- Second or third transmission (1 bit): $x_{info,6} = x_{sec,3}$
- Reserved (1 bit): $x_{info,7} = x_{res,1}$

The Transport-block size information (2 bits) $x_{tbs,1}, x_{tbs,2}$ is the unsigned binary representation of a reference to one of the four Transport-block sizes and the associated number of HS-PDSCH codes for the first transmission configured by higher layers.

Pointer to the previous transmission (3 bits) $x_{ptr,1}, x_{ptr,2}, x_{ptr,3}$ is the unsigned binary representation of s, such that the previous transmission of the same transport block started (6 + s) sub-frames before the start of this transmission.

Second or Third transmission (1 bit) indicates whether this is the second or third transmission. If $x_{sec,1}$ = '0', this is a second transmission. If $x_{sec,1}$ = '1', this is a third transmission.

The redundancy version X_{rv} for the second and third transmissions shall be equal to 3 and 4 respectively.

The UE identity is the same as for HS-SCCH type 1.

Figure 11.66 [16] illustrates the overall coding chain for HS-SCCH type 2.

The channelization code-set information $x_{ccs,1}, x_{ccs,2}, \ldots, x_{ccs,7}$ and modulation-scheme information $x_{ms,1}$ are multiplexed together. This gives a sequence of bits $x_{1,1}, x_{1,2}, \ldots, x_{1,8}$ where:

$$x_{1,i} = x_{ccs,i} \qquad i = 1,2,\ldots,7$$
$$x_{1,i} = x_{ms,i\text{-}7} \qquad i = 8$$

The Special Information type $x_{type,1}, x_{type,2}, \ldots, x_{type,6}$, and Special Information $x_{info,1}, x_{info,2}, x_{info,3}, x_{info,4}, x_{info,5}, x_{info,6}, x_{info,7}$ are multiplexed together. This gives a sequence of bits $x_{2,1}, x_{2,2}, \ldots, x_{2,13}$ where:

$$x_{2,i} = x_{type,i} \qquad i = 1,2,\ldots,6$$
$$x_{2,i} = x_{info,i\text{-}6} \qquad i = 7,8,\ldots,13$$

The sequence of bits $y_1, y_2, \ldots, y_{29}$, is calculated according to (11.76).

Channel coding, rate matching, UE specific masking and physical channel mapping are performed in the same way as for HS-SCCH type 1.

HS-SCCH type 3 is used when the UE is configured in MIMO mode. The coding chain of HS-SCCH type 3 will be considered later together with MIMO mode.

In some cases HS-SCCH is used for transmitting orders from Node B to UE. The typical orders are to activate or deactivate the *discontinuous downlink reception* operation (DRX) and *discontinuous uplink DPCCH transmission* operation (DTX).

No HS-PDSCH is associated with HS-SCCH orders.

The following information is transmitted by means of the HS-SCCH order physical channel.

- Order type (3 bits): $x_{odt,1}, x_{odt,2}, x_{odt,3}$
- Order (3 bits): $x_{ord,1}, x_{ord,2}, x_{ord,3}$
- UE identity (16 bits): $x_{ue,1}, x_{ue,2}, \ldots, x_{ue,16}$

For an HS-SCCH order, in case of HS-SCCH type 1

- $x_{ccs,1}, x_{ccs,2}, \ldots, x_{ccs,7}, x_{ms,1}$ shall be set to '11100000'
- $x_{tbs,1}, x_{tbs,2}, \ldots, x_{tbs,6}$ shall be set to '111101'
- $x_{hap,1}, x_{hap,2}, x_{hap,3}, x_{rv,1}, x_{rv,2}, x_{rv,3}$ shall be set to $x_{odt,1}, x_{odt,2}, x_{odt,3}, x_{ord,1}, x_{ord,2}, x_{ord,3}$
- $x_{nd,1}$ is reserved

where $x_{odt,1}, x_{odt,2}, x_{odt,3}, x_{ord,1}, x_{ord,2}, x_{ord,3}$.

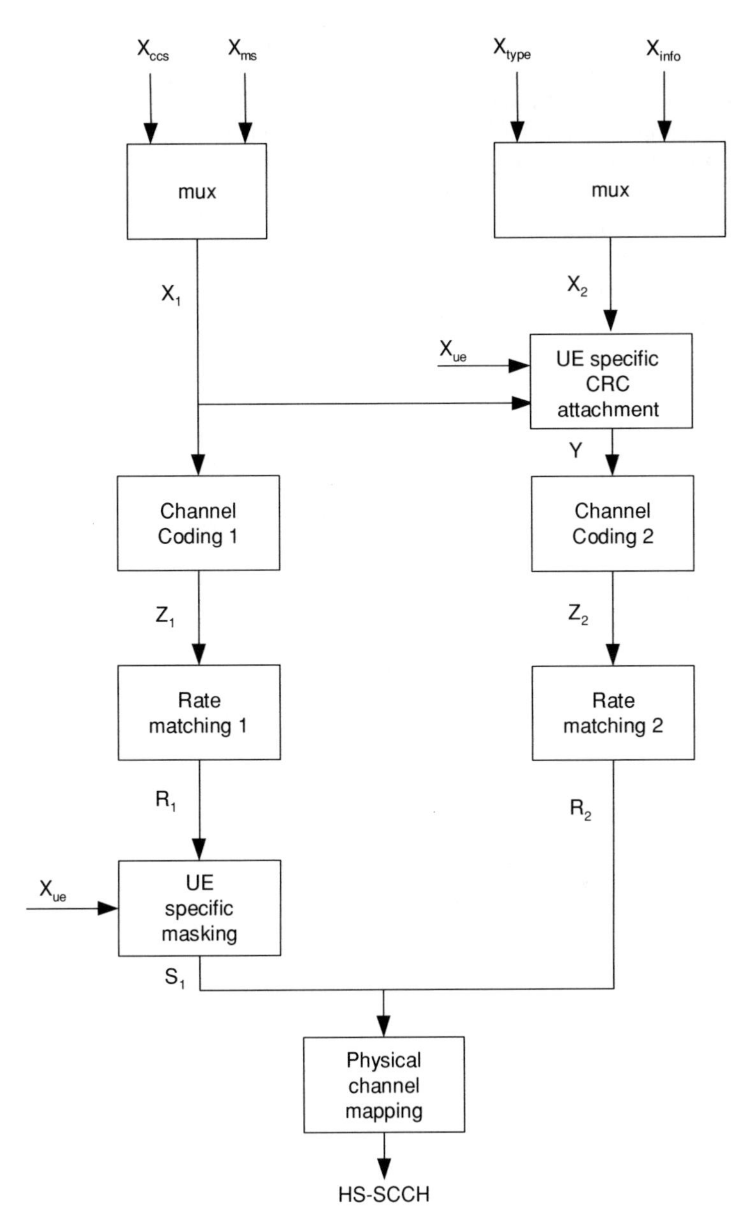

Figure 11.66 Coding chain for HS-SCCH type 2. Reproduced by permission of © 2008 3GPP™

If Order type $x_{odt,1}$, $x_{odt,2}$, $x_{odt,3}$ = '000', then the orders for activation and deactivation of DTX and DRX are transmitted. For this Order type, $x_{ord,1}$, $x_{ord,2}$, $x_{ord,3}$ comprise:

- DRX order activation (1 bit): $x_{ord,1} = x_{drx,1}$
- DTX order activation (1 bit): $x_{ord,2} = x_{dtx,1}$
- Reserved (1 bit): $x_{ord,3} = x_{res,1}$

If $x_{drx,1}$ = '0', then the HS-SCCH order is a DRX De-activation order.
If $x_{drx,1}$ = '1', then the HS-SCCH order is a DRX Activation order.
If $x_{dtx,1}$ = '0', then the HS-SCCH order is a DTX De-activation order.
If $x_{dtx,1}$ = '1', then the HS-SCCH order is a DTX Activation order.
The multiplexing and coding of HS-SCCH orders is done in the same way as for HS-SCCH type 1.

11.3.1.4.2 HS-DSCH Multiplexing and Coding

One transport block of HS-DSCH data is transmitting every transmission time interval (TTI). The TTI is 2 ms which is mapped to a radio sub-frame of three slots. Each sub-frame has 16 channel codes accessible, 15 out of which can be used for the users' data. Thus, each sub-frame includes up to 15 accessible parallel physical channels (HS-PDSCHs) for sending data. The physical channels may be either allocated to just one UE or divided between several UEs.The coding steps for HS-DSCH are shown in Figure 11.67 [16].

First of all the CRC is attached to the transport block. The CRC length is 24 bits. There are two methods of CRC attachment for HS-DSCH. According to the first method simply 24 bit CRC is calculated and attached to transport block. In accordance with method 2 from the sequence of bits $a_1, a_2, a_3, \ldots, a_A$, where A is the size of the HS-DSCH transport block, a CRC of length the 24 bit $c_k, k = 1, \ldots, 24$ is calculated first. This sequence of bits is then masked with the UE Identity $x_{ue,1}, x_{ue,2}, \ldots, x_{ue,16}$ and then appended to the sequence of bits $a_1, a_2, a_3, \ldots, a_A$ to form the sequence of bits $b_1, b_2, b_3, \ldots, b_B$, where $B = A + 24$, and:

$$\begin{aligned} b_k &= a_k && k = 1, 2, \ldots, A \\ b_k &= c_{k-A} && k = A+1, \ldots, A+8 \\ b_k &= (c_{k-A} + x_{ue,k-A-8}) \bmod 2 && k = A+9, \ldots, A+24 \end{aligned} \tag{11.78}$$

The bits output from the HS-DSCH CRC attachment are scrambled in the bit scrambler. The bits input to the bit scrambler are denoted by $b_{im,1}, b_{im,2}, b_{im,3}, \ldots, b_{im,B}$, where B is the number of bits input to the HS-DSCH bit scrambler The bits after bit scrambling are denoted $d_{im,1}, d_{im,2}, d_{im,3}, \ldots, d_{im,B}$.

Bit scrambling is defined by the following relation:

$$d_{im,k} = \left(b_{im,k} + y_k\right) \bmod 2, \quad k = 1, 2, \ldots, B \tag{11.79}$$

and y_k results from the following operation:

$$\begin{aligned} y'_\gamma &= 0, \quad -15 < \gamma < 1 \\ y'_\gamma &= 1, \quad \gamma = 1 \\ y'_\gamma &= \left(\sum_{x=1}^{16} g_x \cdot y'_{\gamma-x}\right) \bmod 2 \quad 1 < \gamma \leq B \end{aligned} \tag{11.80}$$

where $g = \{g_1, g_2, \ldots, g_{16}\} = \{0, 0, 0, 0, 0, 0, 0, 0, 0, 0, 1, 0, 1, 1, 0, 1\}$, $y_k = y'_k$, $k = 1,2,\ldots,B$.

The code block segmentation for HS-DSCH is trivial since there is a maximum of one transport block. The output bits from the code block segmentation function are $o_{ir1}, o_{ir2}, o_{ir3}, \ldots o_{irK}$. Channel coding for the HS-DSCH transport channel shall be done with the general method described in section 11.2.4.2 with the following specific parameters. There will be a maximum of one transport block, i = 1. The rate 1/3 turbo coding shall be used. The value of Z = 5114 for turbo coding shall be used.

The hybrid ARQ functionality matches the number of bits at the output of the channel coder to the total number of bits of the HS-PDSCH set to which the HS-DSCH is mapped. The hybrid ARQ functionality is controlled by the redundancy version (RV) parameters. The exact set of bits at the output of the hybrid ARQ functionality depends on the number of input bits, the number of output bits, and the RV parameters.

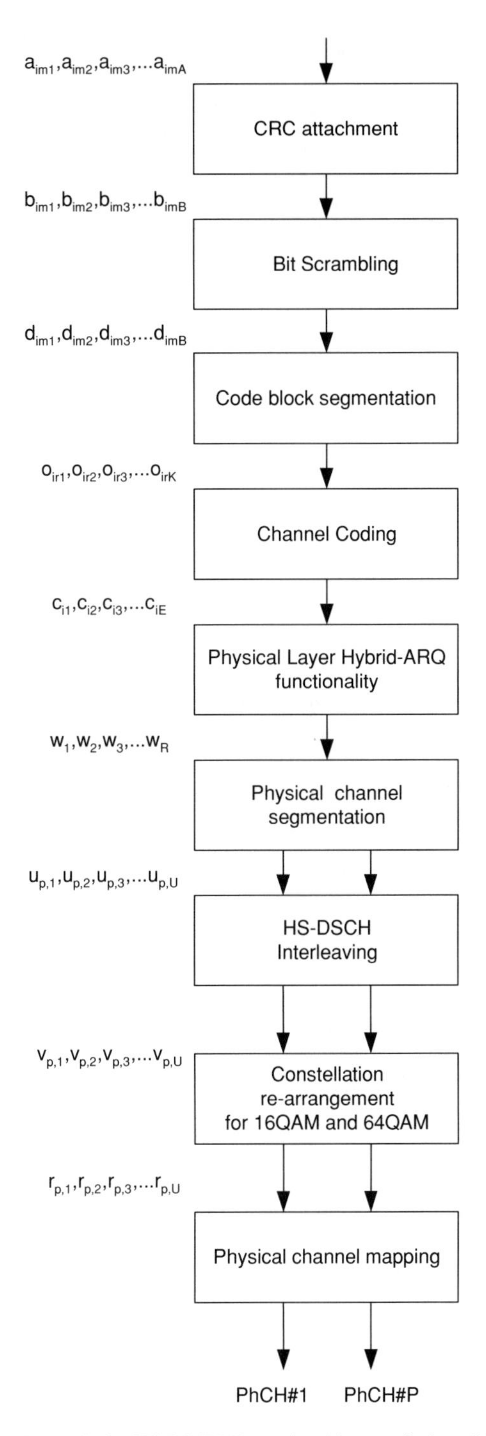

Figure 11.67 Coding chain for HS-DSCH. Reproduced by permission of © 2008 3GPP™

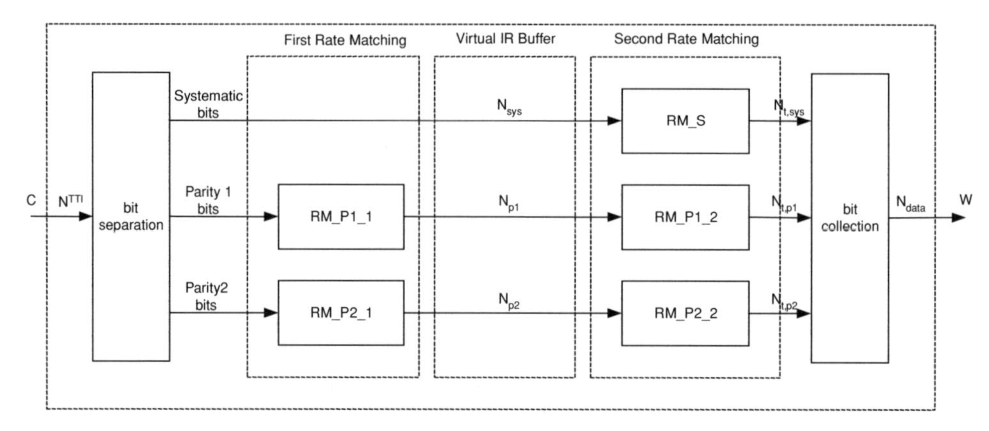

Figure 11.68 HS-DSCH H-ARQ functionality. Reproduced by permission of © 2008 3GPP™

The hybrid ARQ functionality consists of two rate-matching stages and a virtual buffer as shown in Figure 11.68 [16].

The first rate matching stage matches the number of input bits to the virtual IR buffer, information about which is provided by higher layers. Note that, if the number of input bits does not exceed the virtual IR buffering capability, the first rate-matching stage is transparent.

The second rate matching stage matches the number of bits after first rate matching stage to the number of physical channel bits available in the HS-PDSCH set in the TTI.

The parameters of the second rate matching stage depend on the value of the RV parameters s and r. The parameter s can take the value 0 or 1 to distinguish between transmissions that prioritize systematic bits ($s = 1$) and non systematic bits ($s = 0$). The parameter r (range 0 to $r_{max} - 1$) changes the initial error variable e_{ini} which defines the work of rate matching algorithm in the case of puncturing. In case of repetition both parameters r and s change the initial error variable e_{ini}. For details of calculation parameters for rate matching pattern determination algorithm see sections 4.5.4.2 and 4.5.4.3 in [16].

The rate matching pattern determination algorithm used in both first rate matching and second rate matching is the same as that used for Rel'99 channels described in section 11.2.3.1.

The HARQ bit collection is achieved using a rectangular interleaver of size $N_{row} \times N_{col}$.

The number of rows and columns are determined from:

$$N_{row} = 6 \quad \text{for 64 QAM}, \quad N_{row} = 4 \quad \text{for 16QAM} \quad \text{and} \quad N_{row} = 2 \quad \text{for QPSK}$$
$$N_{col} = N_{data}/N_{row}$$

where N_{data} is used as defined in Table 11.12.

Data is written into the interleaver column by column, and read out of the interleaver row by row starting from the first column.

$N_{t,sys}$ is the number of transmitted systematic bits. Intermediate values N_r and N_c are calculated using:

$$N_r = \left\lfloor \frac{N_{t,sys}}{N_{col}} \right\rfloor \quad \text{and} \quad N_c = N_{t,sys} - N_r \cdot N_{col}$$

If $N_c = 0$ and $N_r > 0$, the systematic bits are written into rows $1 \ldots N_r$.

Otherwise systematic bits are written into rows $1 \ldots N_r + 1$ in the first N_c columns and, if $N_r > 0$, also into rows $1 \ldots N_r$ in the remaining N_{col}-N_c columns.

The remaining space is filled with parity bits. The parity bits are written column wise into the remaining rows of the respective columns. Parity 1 and 2 bits are written in alternating order, starting with a parity 2 bit in the first available column with the lowest index number.

In the case of 64QAM for each column the bits are read out of the interleaver in the order row 1, row 2, row 3, row 4, row 5, row 6. In the case of 16QAM for each column the bits are read out of the interleaver in the order row 1, row 2, row 3, row 4. In the case of QPSK for each column the bits are read out of the interleaver in the order row 1, row 2.

When more than one HS-PDSCH is used, physical channel segmentation divides the bits among the different physical channels. The bits input to the physical channel segmentation are denoted by w_1, w_2, w_3,. . .w_R, where R is the number of bits input to the physical channel segmentation block. The number of PhCHs is denoted by P.

The bits after physical channel segmentation are denoted $u_{p1}, u_{p2}, u_{p3}, \ldots, u_{pU}$, where p is PhCH number and U is the number of bits in one radio sub-frame for each HS-PDSCH, that is, $U = \frac{R}{P}$. The relation between w_k and $u_{p,k}$ is given below.

For all modes, some bits of the input flow are mapped to each code until the number of bits on the code is U.

Bits on first PhCH after physical channel segmentation:

$$u_{1,k} = w_k \quad k = 1, 2, \ldots, U$$

Bits on second PhCH after physical channel segmentation:

$$u_{2,k} = w_{k+U} \quad k = 1, 2, \ldots, U$$

. . .

Bits on the Pth PhCH after physical channel segmentation:

$$u_{P,k} = w_{k+(P-1)\times U} \quad k = 1, 2, \ldots, U$$

The interleaving for FDD is done as shown in Figure 11.69, separately for each physical channel. The bits input to the block interleaver are denoted by $u_{p,1}, u_{p,2}, u_{p,3}, \ldots, u_{p,U}$, where p is PhCH number and U is the number of bits in one TTI for one PhCH. For QPSK $U = 960$, for 16QAM $U = 1920$ and for

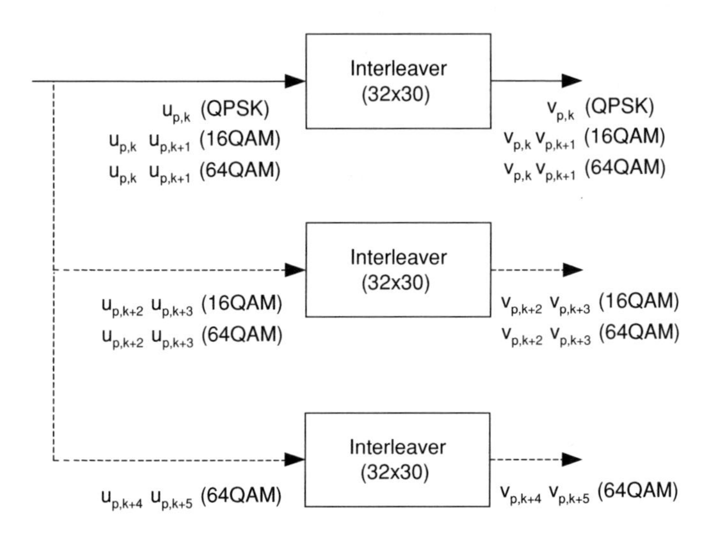

Figure 11.69 Interleaver structure for HS-DSCH

Table 11.16 Constellation re-arrangement for 16QAM

constellation version parameter *b*	Output bit sequence	Operation
0	$v_{p,k}v_{p,k+1}v_{p,k+2}v_{p,k+3}$	None
1	$v_{p,k+2}v_{p,k+3}v_{p,k}v_{p,k+1}$	Swapping MSBs with LSBs
2	$v_{p,k}v_{p,k+1}\overline{v_{p,k+2}v_{p,k+3}}$	Inversion of the logical values of LSBs
3	$v_{p,k+2}v_{p,k+3}\overline{v_{p,k}v_{p,k+1}}$	Swapping MSBs with LSBs and inversion of logical values of LSBs

64QAM $U = 2880$. The basic interleaver is as the 2nd interleaver described by (11.12). The interleaver is of fixed size: R2 = 32 rows and C2 = 30 columns.

For 16QAM, there are two identical interleavers of the same fixed size R2 × C2 = 32 × 30. The output bits from the physical channel segmentation are divided two by two between the interleavers: bits $u_{p,k}$ and $u_{p,k+1}$ go to the first interleaver and bits $u_{p,k+2}$ and $u_{p,k+3}$ go to the second interleaver. Bits are collected two by two from the interleavers: bits $v_{p,k}$ and $v_{p,k+1}$ are obtained from the first interleaver and bits $v_{p,k+2}$ and $v_{p,k+3}$ are obtained from the second interleaver, where k mod 4 = 1.

For 64QAM, there are three identical interleavers of the same fixed size R2 × C2 = 32 × 30. The output bits from the physical channel segmentation are divided two by two between the interleavers: bits $u_{p,k}$ and $u_{p,k+1}$ go to the first interleaver, bits $u_{p,k+2}$ and $u_{p,k+3}$ go to the second interleaver and bits $u_{p,k+4}$ and $u_{p,k+5}$ go to the third interleaver. Bits are collected two by two from the interleavers: bits $v_{p,k}$ and $v_{p,k+1}$ are obtained from the first interleaver, bits $v_{p,k+2}$ and $v_{p,k+3}$ are obtained from the second interleaver and bits $v_{p,k+4}$ and $v_{p,k+5}$ are obtained from the third interleaver, where k mod 6 = 1.

Quite often the bit-mapping onto the signal constellation causes the variations in bit reliabilities, that is, some bits suffer more than others from the symbol error in cases where high order modulation is used. This in turn causes the packetizing of bit errors. To avoid this problem the *constellation rearrangement* is used. That means the different mapping rules are used for different (re)transmissions of the same packet. This significantly improves the receiver error rate performance. The mapping rule used for the current transmission is defined by *constellation version parameter b*. The constellation rearrangement function only applies to 16QAM and 64QAM modulated bits. In the case of QPSK it is transparent. Table 11.16 [16] describes the operations that produce the different rearrangements for 16QAM. The bits of the input sequence are mapped in groups of four so that $v_{p,k}$, $v_{p,k+1}$, $v_{p,k+2}$, $v_{p,k+3}$ are used, where k mod 4 = 1. The output bit sequences map to the output bits in groups of four, that is, $r_{p,k}$, $r_{p,k+1}$, $r_{p,k+2}$, $r_{p,k+3}$, where k mod 4 = 1.

Table 11.17 [16] describes the operations that produce the different rearrangements for 64QAM. The bits of the input sequence are mapped in groups of six so that $v_{p,k}$, $v_{p,k+1}$, $v_{p,k+2}$, $v_{p,k+3}$, $v_{p,k+4}$, $v_{p,k+5}$

Table 11.17 Constellation re-arrangement for 64QAM

constellation version parameter *b*	Output bit sequence	Operation
0	$v_{p,k}v_{p,k+1}v_{p,k+2}v_{p,k+3}v_{p,k+4}v_{p,k+5}$	None
1	$v_{p,k+4}v_{p,k+5}\overline{v_{p,k+2}v_{p,k+3}}v_{p,k}v_{p,k+1}$	Swapping MSBs and LSBs. Inversion of Middle SBs
2	$v_{p,k+2}v_{p,k+3}\overline{v_{p,k+4}v_{p,k+5}}v_{p,k}v_{p,k+1}$	Left circular shift of pair of SBs. Inversion of Middle SBs
3	$v_{p,k}v_{p,k+1}\overline{v_{p,k+2}v_{p,k+3}}v_{p,k+4}v_{p,k+5}$	Inversion of Middle SBs

are used, where k mod 6 = 1. The output bit sequences map to the output bits in groups of six, that is, $r_{p,k}$, $r_{p,k+1}$, $r_{p,k+2}$, $r_{p,k+3}$, $r_{p,k+4}$, $r_{p,k+5}$, where k mod 6 = 1.

After the constellation rearrangement the obtained bit sequence is mapped to physical channels HS-PDSCHs. The bits input to the physical channel mapping are denoted by $r_{p,1}, r_{p,2}, \ldots, r_{p,U}$, where p is the physical channel number and U is the number of bits in one radio sub-frame for one HS-PDSCH. The bits $r_{p,k}$ are mapped to the PhCHs so that the bits for each PhCH are transmitted over the air in ascending order with respect to k.

11.3.1.4.3 HS-DPCCH Multiplexing and Coding

The data transmitted over HS-DPCCH is HARQ acknowledgment and Channel Quality Indicator (CQI). In cases where the UE is configured in MIMO mode the Precoding Control Indicator (PCI) is also transmitted over HS-DPCCH. The coding/multiplexing for HS-DPCCH is defined separately for cases when the UE is or is not configured in MIMO mode.

The general coding flow when the UE is not configured in MIMO mode is shown in Figure 11.70 [16]. This is done in parallel for the HARQ-ACK and CQI as the flows are not directly multiplexed but are transmitted at different times.

The case of HS-DPCCH coding/multiplexing when UE is configured in MIMO mode will be considered later together with MIMO scheme.

Two forms of channel coding are used, one for the channel quality indication (CQI) and another for HARQ acknowledgment (HARQ-ACK).

The Ack/Nack information is not necessarily sent in each subframe, for example, in cases of discontinuous transmission. In this case Node B should detect not only ACK or NACK but also "no transmission" (DTX), which adds complexity to the detection algorithm. Especially undesirable is the case when DTX is detected by Node B as ACK, since it will be followed by sending the new user data to UE, while UE does not expect this. Of course it is possible to cope with this problem by increasing the transmit power on the HS-DPCCH for ACK/NACK fields but obviously it would cause the increase of UE energy consumption. To avoid requiring that the UE transmit at a power level that for most time intervals would

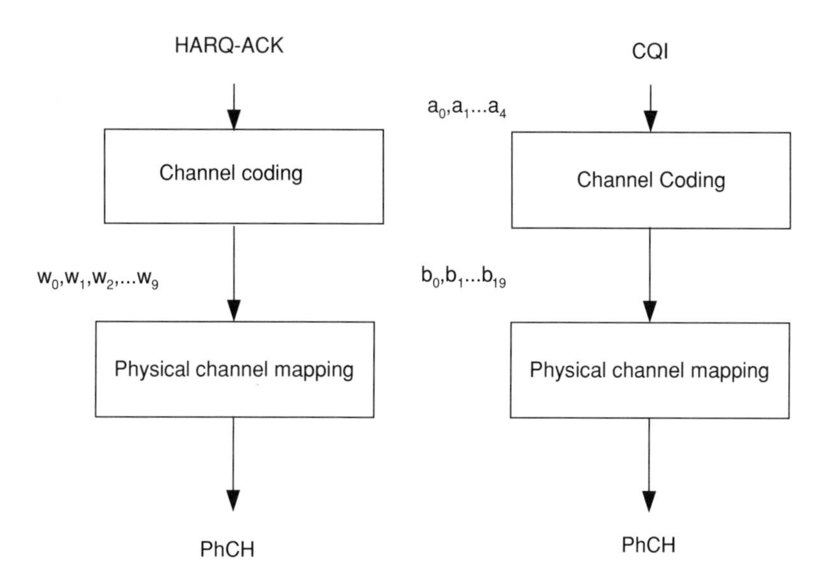

Figure 11.70 Coding for HS-DPCCH when the UE is not configured in MIMO mode. Reproduced by permission of © 2008 3GPP™

Table 11.18 Channel coding of HARQ-ACK when the UE is not configured in MIMO mode

HARQ-ACK message to be transmitted	w_0	w_1	w_2	w_3	w_4	w_5	w_6	w_7	w_8	w_9
ACK	1	1	1	1	1	1	1	1	1	1
NACK	0	0	0	0	0	0	0	0	0	0
PRE	0	0	1	0	0	1	0	0	1	0
POST	0	1	0	0	1	0	0	1	0	0

be unnecessarily high, it is possible to send the special messages HARQ Preamble and HARQ Postamble instead of Ack/Nack information in place of HARQ acknowledgment messages. These messages mark the start and the end of discontinuous transmission (DTX). When a UE receives signalling information intended for this UE on HS-SCCH, the UE would transmit a HARQ Preamble in the subframe preceding the subframe allocated to the HARQ ACK/NACK, unless an ACK or NACK corresponding to the previous HS-PDSCH packet is to be transmitted in this particular subframe. In addition, the UE would transmit a HARQ Postamble in the subframe following the HARQ ACK/NACK, unless another HS-DSCH packet follows immediately and is successfully decoded. Thus, after transition of UE from DTX mode to active mode the first message received by Node B from the UE would be the HARQ Preamble, and then Node B can expect only ACK or NACK in the consequent subframes. Correspondingly when UE is entering the DTX mode the HARQ Postamble would be the last HARQ message sent to Node B. After this the Node B is expecting only HARQ Preamble message. Such a procedure would keep the Node B from detecting DTX as ACK in the HARQ ACK/NACK subframe, allowing a substantial reduction in the

Table 11.19 Basis code words of (20,5) code

i	$M_{i,0}$	$M_{i,1}$	$M_{i,2}$	$M_{i,3}$	$M_{i,4}$
0	1	0	0	0	1
1	0	1	0	0	1
2	1	1	0	0	1
3	0	0	1	0	1
4	1	0	1	0	1
5	0	1	1	0	1
6	1	1	1	0	1
7	0	0	0	1	1
8	1	0	0	1	1
9	0	1	0	1	1
10	1	1	0	1	1
11	0	0	1	1	1
12	1	0	1	1	1
13	0	1	1	1	1
14	1	1	1	1	1
15	0	0	0	0	1
16	0	0	0	0	1
17	0	0	0	0	1
18	0	0	0	0	1
19	0	0	0	0	1

required power. The HARQ acknowledgment message to be transmitted shall be coded to 10 bits as shown in Table 11.18 [16]. The output is denoted $w_0, w_1, \ldots w_9$.

The channel quality indication is coded using a (20,5) code. The code words of the (20,5) code are a linear combination of the 5 basis code words (generation matrix) denoted $M_{i,n}$ defined in the Table 11.19 [16].

The CQI values 1 ... 30 (CQI value of 0 is forbidden) are converted from decimal to binary to map them to the channel quality indication bits (1 0 0 0 0) to (1 1 1 1 1) respectively. The channel quality indication bits are a_0, a_1, a_2, a_3, a_4 (where a_0 is LSB and a_4 is MSB). The output codeword bits b_i are given by:

$$b_i = \sum_{n=0}^{4} (a_n \times M_{i,n}) \bmod 2 \tag{11.81}$$

where $i = 0, \ldots, 19$.

The HS-DPCCH physical channel mapping function shall map the input bits b_k directly to physical channel so that bits are transmitted over the air in ascending order with respect to k.

11.3.1.4.4 HSDPA Spreading and Modulation

The HS-SCCH and HS-PDSCHs are spread in the same way as all downlink channels as described in section 11.2.4.2.2. The HS-SCCH is modulated with QPSK and the HS-PDSCH can be modulated with QPSK, 16-QAM or 64-QAM.

In the case of 16-QAM, a set of four consecutive binary symbols $n_k, n_{k+1}, n_{k+2}, n_{k+3}$ (with k mod $4 = 0$) is serial-to-parallel converted to two consecutive binary symbols ($i_1 = n_k, i_2 = n_{k+2}$) on the I branch and two consecutive binary symbols ($q_1 = n_{k+1}, q_2 = n_{k+3}$) on the Q branch and then mapped to 16-QAM by the modulation mapper as defined in Table 11.20.

In the case of 64-QAM, a set of six consecutive binary symbols $n_k, n_{k+1}, n_{k+2}, n_{k+3}, n_{k+4}, n_{k+5}$ (with k mod $6 = 0$) is serial-to-parallel converted to three consecutive binary symbols ($i_1 = n_k, i_2 = n_{k+2}, i_3 = n_{k+4}$) on the I branch and three consecutive binary symbols ($q_1 = n_{k+1}, q_2 = n_{k+3}, q_3 = n_{k+5}$) on the Q branch and then mapped to 64-QAM by the modulation mapper as defined in Table 11.21.

Table 11.20 16-QAM modulation mapping

$i_1q_1i_2q_2$	I branch	Q branch
0000	0.4472	0.4472
0001	0.4472	1.3416
0010	1.3416	0.4472
0011	1.3416	1.3416
0100	0.4472	−0.4472
0101	0.4472	−1.3416
0110	1.3416	−0.4472
0111	1.3416	−1.3416
1000	−0.4472	0.4472
1001	−0.4472	1.3416
1010	−1.3416	0.4472
1011	−1.3416	1.3416
1100	−0.4472	−0.4472
1101	−0.4472	−1.3416
1110	−1.3416	−0.4472
1111	−1.3416	−1.3416

Table 11.21 64-QAM modulation mapping

$i_1q_1i_2q_2\ i_3q_3$	I branch	Q branch	$i_1q_1i_2q_2\ i_3q_3$	I branch	Q branch
000000	0.6547	0.6547	100000	−0.6547	0.6547
000001	0.6547	0.2182	100001	−0.6547	0.2182
000010	0.2182	0.6547	100010	−0.2182	0.6547
000011	0.2182	0.2182	100011	−0.2182	0.2182
000100	0.6547	1.0911	100100	−0.6547	1.0911
000101	0.6547	1.5275	100101	−0.6547	1.5275
000110	0.2182	1.0911	100110	−0.2182	1.0911
000111	0.2182	1.5275	100111	−0.2182	1.5275
001000	1.0911	0.6547	101000	−1.0911	0.6547
001001	1.0911	0.2182	101001	−1.0911	0.2182
001010	1.5275	0.6547	101010	−1.5275	0.6547
001011	1.5275	0.2182	101011	−1.5275	0.2182
001100	1.0911	1.0911	101100	−1.0911	1.0911
001101	1.0911	1.5275	101101	−1.0911	1.5275
001110	1.5275	1.0911	101110	−1.5275	1.0911
001111	1.5275	1.5275	101111	−1.5275	1.5275
010000	0.6547	−0.6547	110000	−0.6547	−0.6547
010001	0.6547	−0.2182	110001	−0.6547	−0.2182
010010	0.2182	−0.6547	110010	−0.2182	−0.6547
010011	0.2182	−0.2182	110011	−0.2182	−0.2182
010100	0.6547	−1.0911	110100	−0.6547	−1.0911
010101	0.6547	−1.5275	110101	−0.6547	−1.5275
010110	0.2182	−1.0911	110110	−0.2182	−1.0911
010111	0.2182	−1.5275	110111	−0.2182	−1.5275
011000	1.0911	−0.6547	111000	−1.0911	−0.6547
011001	1.0911	−0.2182	111001	−1.0911	−0.2182
011010	1.5275	−0.6547	111010	−1.5275	−0.6547
011011	1.5275	−0.2182	111011	−1.5275	−0.2182
011100	1.0911	−1.0911	111100	−1.0911	−1.0911
011101	1.0911	−1.5275	111101	−1.0911	−1.5275
011110	1.5275	−1.0911	111110	−1.5275	−1.0911
011111	1.5275	−1.5275	111111	−1.5275	−1.5275

The I and Q branches are then both spread to the chip rate by the same real-valued channelization code $C_{ch,16,m}$. The channelization code sequence shall be aligned in time with the symbol boundary. The sequences of real-valued chips on the I and Q branch are then treated as a single complex-valued sequence of chips. This sequence of chips from all multi-codes is summed and then scrambled (complex chip-wise multiplication) by a complex-valued scrambling code $S_{dl,n}$. The scrambling code is applied aligned with the scrambling code applied to the P-CCPCH.

The scrambling and the channel combining for HS-SCCH and HS-PDSCH is the same as for all downlink channels described in section 11.2.4.2.2.

Figure 11.71 [17] illustrates the spreading operation for the HS-DPCCH.

The HS-DPCCH shall be spread to the chip rate by the channelization code c_{hs}. After channelization, the real-valued spread signals are weighted by gain factor β_{hs}

The β_{hs} values are derived from the quantized amplitude ratios A_{hs} which are translated from Δ_{ACK}, Δ_{NACK} and Δ_{CQI} signalled by higher layers.

The translation of Δ_{ACK}, Δ_{NACK} and Δ_{CQI} into quantized amplitude ratios $A_{hs} = \beta_{hs}/\beta_c$ is shown in Table 11.22 [17].

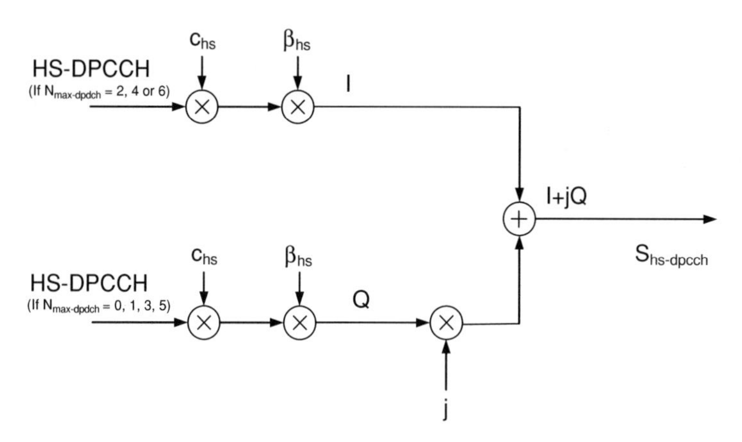

Figure 11.71 Spreading for uplink HS-DPCCH. Reproduced by permission of © 2008 3GPP™

HS-DPCCH shall be mapped to the I branch in case $N_{max\text{-}dpdch}$ is 2, 4 or 6, and to the Q branch otherwise ($N_{max\text{-}dpdch} = 0$, 1, 3 or 5).

The HS-DPCCH shall be spread with code c_{hs} as specified in Table 11.23 [17].

HS-DPCCH is modulated with QPSK. The modulation of uplink channels is described in section 11.2.4.2.1.

Table 11.22 The quantization of the power offset

Signalled values for Δ_{ACK}, Δ_{NACK} and Δ_{CQI}	Quantized amplitude ratios $A_{hs} = \beta_{hs}/\beta_c$
9	38/15
8	30/15
7	24/15
6	19/15
5	15/15
4	12/15
3	9/15
2	8/15
1	6/15
0	5/15

Reproduced by permission of © 2008 3GPP™

Table 11.23 Channelization code of HS-DPCCH

$N_{max\text{-}dpdch}$ (as defined in subclause 4.2.1)	Channelization code c_{hs}
0	$C_{ch,256,33}$
1	$C_{ch,256,64}$
2,4,6	$C_{ch,256,1}$
3,5	$C_{ch,256,32}$

Reproduced by permission of © 2008 3GPP™

11.3.1.4.5 HSDPA Channels Interaction

Scheduler in Node B evaluates every 2 ms for each UE the following factors:

- Channel conditions (with the help of UE CQI reports);
- HARQ buffer status;
- Time from last transmission;
- Retransmissions pending (with the help of ACK/NACK reports);

Based on this information scheduler decides which UE (or UEs) should be served in the current TTI. For each serving UE Node B identifies channelization codes, modulation scheme, transport block size, UE capability limitations, and so on. Then Node B starts to transmit the HS-SCCH information (see section 11.3.1.4.1) 2 slots before corresponding HS-DSCH TTI. There is a set of active control channels (HS-SCCHs). The maximum size of this set is 4. If there were data for the particular UE in the previous TTI the same HS-SCCH as in previous TTI should be used, otherwise any of the four HS-SCCH could be chosen for the transmission. In cases where the UEs are only time multiplexed (only one UE can be served in each TTI) only one HS-SCCH is needed. UE monitors the active set of HS-SCCHs given by network (maximum 4 HS-SCCHs). The word "monitor" here means that UE is trying to decode the Part 1 of HS-SCCH information. Recall that Part 1 information is masked with UE identity. Due to this fact only the UE to which the HS-SCCH information is intended is capable of decoding Part 1 information. If the decoded "channelization-code-set information" and decoded "modulation scheme information" correspond to an indication of an HS-SCCH order, the UE decodes the HS-SCCH order and processes it discarding the HS-DSCH data sent in the current TTI.. Otherwise the UE starts receiving the HS-PDSCHs indicated by this consistent Part 1 control information and decoding Part 2 HS-SCCH information. After decoding H-ARQ parameters from Part 2 UE can determine to which H-ARQ process data belong and whether it should be combined with the data in H-ARQ buffer. If the H-ARQ information is not included in the set configured by upper layers, or if the HS-SCCH CRC is not OK, the UE shall discard the information received on this HS-SCCH and on the HS-PDSCHs. If the information received on HS-SCCH is not discarded, the UE shall transmit a HARQ Preamble in the slot allocated to HARQ-ACK in HS-DPCCH subframe preceding the HS-DPCCH subframe, in which the ACK/NACK corresponding to the current HS-DSCH TTI should be transmitted, unless an ACK or NACK is to be transmitted in this preceding sub-frame as a result of previous HS-DSCH transmission. And accordingly if there will not be ACK/NACK information in the next HS-DPCCH subframe after the subframe corresponding to the current TTI, a HARQ Postamble should be transmitted in this subframe. After decoding the HS-PDSCHs information the HS-DSCH CRC is checked and depending on the CRC result ACK or NACK is sent in the corresponding HS-DPCCH subframe. As mentioned earlier, there is a fixed 7.5 slot delay from the end of HS-DSCH TTI to the start of ACK/NACK transmission in HS-DPCCH. Due to this fact the Node B knows for which packet the ACK/NACK information is related to. If Node B continues to transmit data for the same UE in next TTI the same HS-SCCH that was used in previous TTI should be used for signalling.

There is no soft handover for HSDPA. When a change of serving HS-DSCH cell happens, the UE will flush all the buffers at the handover time and move to listen to the new base station as instructed in the downlink RRC signalling. Respectively, at the moment the handover takes place the Node B also flushes packets still in its buffers, including possibly unfinished HARQ processes [27].

11.3.1.5 HSDPA MIMO

HSDPA MIMO was approved in standardization as a work item in RAN#12 (June 2001). For several years no HSDPA MIMO schemes were standardized. The reason for this was that system performance benefits of HSDPA MIMO appeared to be moderate in a typical WCDMA deployment scenario (macrocell, small penetration of MIMO UE's) while WCDMA MIMO implementation added considerable complexity to the receiver.

However some system performance benefit of HSDPA MIMO was observed in scenarios where high signal-to-noise ratios are achievable. Besides, from operator's point of view, HSDPA MIMO provided a solution to increase HSDPA peak bit rate beyond 14 Mbps in 5MHz bandwidth. Due to this fact in RAN#30 (Dec 2005), it was decided to continue HSDPA MIMO work. Several proposals were presented from different companies. In RAN1#45 (May 2006), an agreement was reached and the MIMO technique for FDD was selected.

The method selected in standardization for HSDPA MIMO (FDD mode) is the so-called 2 × 2 *Dual-stream Transmit Antenna Adaptive Array* (*D-TxAA*) technique. 2 × 2 D-TXAA uses two antennas for transmission and two antennas for reception. The system is *dual stream*, meaning that two data streams are transmitted in parallel. Actually, D-TxAA scheme is an extension of closed loop TX diversity to dual stream transmission. In dual stream transmission, two data streams (or payload packets) are encoded and modulated separately, but transmitted on the same orthogonal spreading code(s). Each of the data streams is transmitted over both antennas applying different antenna weights in an attempt to orthogonalize the streams. The idea is that by applying the correct weight vectors at the transmitter (and receiver) both streams could be separated. This is done using for the primary stream a precoding weight vector orthogonal to the one applied to the secondary stream. Besides, it is rather obvious that the ability to separate the two streams of data is also related to the spatial diversity provided by the propagation channel. The more uncorrelated the antennas are, the better the two streams can be recovered.

D-TxAA is a closed loop system: the phase adjustments applied to the TX antennas are based on the feedback commands sent by the terminal. The antenna weights are chosen by the UE and transmitted to the Node B in the HS-DPCCH. Since the antenna weights for each stream are orthogonal the knowledge of one pair is enough for deriving the other. The antenna weights are signalled on the HS-SCCH in the downlink in order to avoid antenna verification in the terminal. In MIMO mode dynamic scheduling of single and dual stream transmission is possible. Single operation mode is turned on when it is more advantageous (for example, from cell or link throughput maximization point of view) to transmit with full power on one beam. In single stream mode one data stream is encoded and transmitted similarly as with Release 99 CL1 transmit diversity but by using the D-TxAA specific feedback mechanism, being the main difference the rate of feedback updates (1 slot in CL1 vs. 1 TTI in MIMO).

Four HSDPA MIMO categories (15, 16, 17 and 18) have been defined and added to 3GPP specifications (see Table 11.11). Actually, the only difference between category 15 and 17 and between category 16 and 18 is that categories 17 and 18 support 64-QAM modulation scheme but only in cases where the UE is not configured in MIMO mode. Thus, in MIMO mode category 17 coincides with category 15 and category 18 with category 16. All MIMO categories support 15 codes channelization codes. It should be noted that in MIMO mode always exactly 15 channelization codes are used. A MIMO UE shall support up to 15 codes also in non-MIMO HSDPA operation. Peak data rate for categories 15 and 17 is 23.3 Mbps and the maximum code rate is limited to 0.82. Peak data rate for categories 16 and 17 is 27.95 Mbps with code rate of 0.97. A MIMO capable UE can be signalled to operate in "MIMO mode" by RRC. When not in MIMO mode it would operate as a regular non-MIMO UE.

Figure 11.72 [18] depicts the 2 × 2 D-TxAA scheme

Channel coding, interleaving and spreading are done as in non-MIMO mode. The Node B scheduler decides whether to transmit one or two transport blocks to a UE in one TTI. The same number of codes (15) is used for both streams. The spread complex valued signals are fed to both TX antenna branches, and weighted with precoding weights w_1, w_2, w_3 and w_4. The precoding weights w_1 and w_3 are constant real valued scalars and the precoding weights w_2 and w_4 are variable complex valued scalars. The precoding

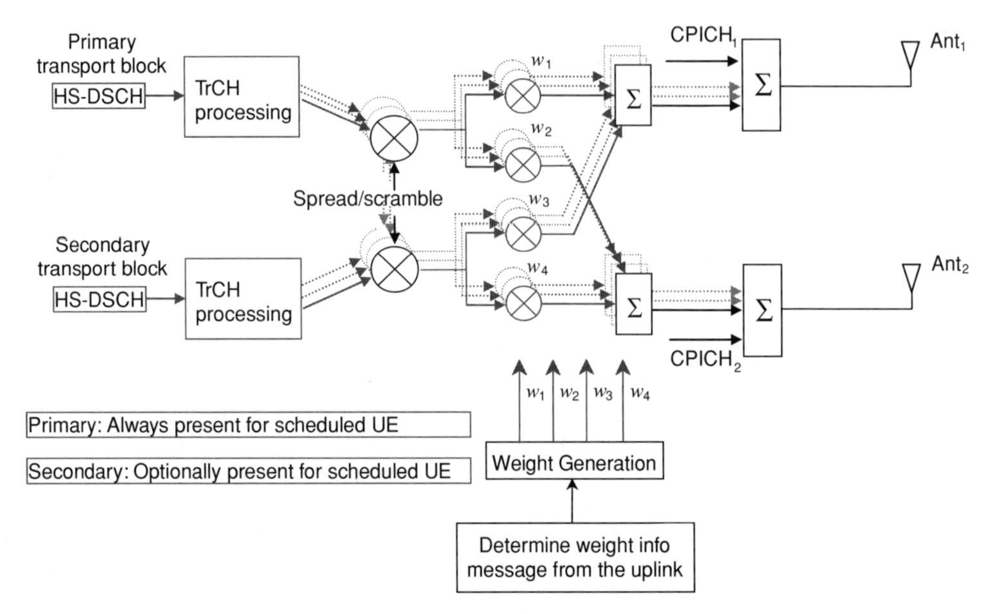

Figure 11.72 D-TxAA HSDPA MIMO scheme. Reproduced by permission of © 2008 3GPP™

weights are defined as follows:

$$\begin{aligned} w_3 &= w_1 = 1/\sqrt{2} \\ w_4 &= -w_2 \\ w_2 &\in \left\{ \frac{1+j}{2}, \frac{1-j}{2}, \frac{-1+j}{2}, \frac{-1-j}{2} \right\} \end{aligned} \tag{11.82}$$

The system can be used in two modes: *single stream mode* and *dual stream mode*. In MIMO mode dynamic scheduling of single and dual stream transmission is possible. The precoding vector (w_1, w_2) is called the *primary precoding vector* and is used for transmitting the primary transport block. The precoding vector (w_3, w_4) is called secondary precoding vector and is used for transmitting the secondary transport block. In cases where the system is in single stream mode only the primary transport block is transmitted and correspondingly only the primary precoding vector (w_1, w_2) is used. Taking into account that the precoding weight w_1 is a constant the single stream mode just repeats the concept of CL1 mode. The difference is that in the MIMO single stream mode the feedback comprises 2 bits, thus there is no need to calculate the feedback information from two consecutive slots like in CL1 (see section 11.2.8). If the UTRAN schedules two transport blocks to a UE in one TTI (dual stream mode), it uses two orthogonal precoding vectors to transmit the two transport blocks. It is easy to verify that the precoding vectors (w_1, w_2) and (w_3, w_4) are orthogonal:

$$w_1 \cdot w_3^* + w_2 \cdot w_4^* = w_1^2 - w_2^2 = 0 \tag{11.83}$$

The UE uses the CPICH to separately estimate the channels seen from each antenna. One of the antennas will transmit the Antenna 1 modulation pattern of the P-CPICH as defined in section 5.3.3.1 of [15]. The other antenna will transmit either the Antenna 2 modulation pattern of the P-CPICH or the Antenna 1 modulation pattern of a S-CPICH. The Pilot configuration in support of MIMO operation of HS-DSCH in the cell is signalled by higher layers. Unfortunately, both methods which use P-CPICH or S-CPICH as a phase reference for Antenna 2 have some drawbacks. In cases using P-CPICH with the Antenna

2 modulation pattern the legacy (non-MIMO) UEs must switch to STTD mode, which leads to some loss in performance. In cases using S-CPICH as Antenna 2 phase reference, S-CPICH becomes one additional source of interference for other channels. To decrease the impact of S-CPICH interference to other channels it is possible to decrease the power of S-CPICH in comparison with the P-CPICH power. This power offset can be up to -6 dB. The power offset is signalled to UE. The usage of S-CPICH as a phase reference for Antenna 2 is regarded as a main case.

The UE determines a preferred primary precoding vector $\left(w_1^{\text{pref}}, w_2^{\text{pref}}\right)$ and signals it to the Node B. The signalled information about the preferred primary precoding vector is termed *precoding control indication* (*PCI*). The precoding weights are calculated in the receiver and reported to NodeB once per TTI. The aim of the PCI calculation algorithm is to choose one of four possible combinations of precoding weights (actually value of w_2 defines both primary and secondary precoding vectors) maximizing the throughput for the next TTI. The PCI is signalled to the Node B together with channel quality indication (CQI) as a composite PCI/CQI report. The UE transmits the composite PCI/CQI report to the Node B using the CQI field on the HS-DPCCH. Based on the composite PCI/CQI reports, the Node B scheduler decides whether to schedule one or two transport blocks to a UE in one TTI and what tranport block size(s) and modulation scheme(s) to use for each of them.

The Node B signals to the UE the precoding weight w_2 applied on the HS-PDSCH sub-frame using the precoding weight indication bits of part 1 of the corresponding HS-SCCH sub-frame. The precoding weight adjustment of each HS-PDSCH is done at the HS-PDSCH sub-frame border [18].

There are two types of PCI/CQI reports: type A and type B. These types are defined as follows [18]:

> *Type A:* CQI reports that indicate the supported transport format(s) for the number of simultaneously transmitted transport blocks that the UE prefers according to the current channel conditions assuming that the preferred primary precoding vector as indicated by the PCI value signalled in the same HS-DPCCH sub-frame would be applied at the Node-B for the primary transport block and in cases where two transport blocks are preferred the precoding vector orthogonal to the preferred primary precoding vector would be applied for the secondary transport block. This type of CQI report contains information on either one transport format or a combination of two transport formats depending on what is currently the preferred number of transport blocks (either 1 or 2).
>
> *Type B:* CQI reports that indicate the supported transport format for a single transmitted transport block according to the current channel conditions assuming that the preferred primary precoding vector as indicated by the PCI value signalled in the same HS-DPCCH sub-frame would be applied at the Node-B for the primary transport block and that no secondary transport block is transmitted.

Usually UE sends a few Type A PCI/CQI reports and then a few Type B reports. The number of Type A and Type B reports to be sent is defined by three constants obtained from the higher layers [18]:

CQI feedback cycle k

The number N of dynamic single/dual CQI reports out of a sequence of M CQI reports in cases where the UE is configured in MIMO mode: N_cqi_typeA, M_cqi, respectively.

The CQI feedback cycle k can take values 16, 32 or 64 ms [19].

The ratio of the params N_cqi_typeA/M_cqi can take values 1/2, 2/3, 3/4, 4/5, 5/6, 6/7, 7/8, 8/9, 9/10, 1/1 [19], that is, no less than 50% of all reports should be Type A reports.

Each TTI the UE decides which type of PCI/CQI report should be sent in accordance with the following rule [18].

For $k = 0$, the UE shall not transmit a composite PCI/CQI value.

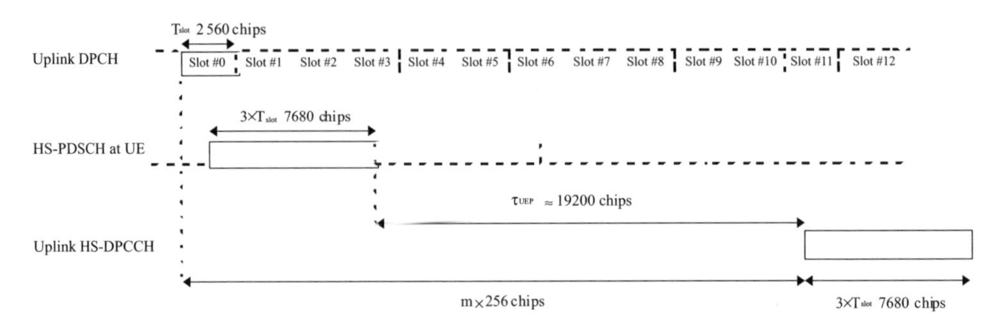

Figure 11.73 Timing structure at the UE for HS-DPCCH control signaling

For $k > 0$ when the UE is not in DTX or DRX mode, the UE shall transmit a composite PCI/CQI value in each subframe that starts $m \times 256$ chips after the start of the associated uplink DPCCH frame with m fulfilling:

$$(5 \times CFN + \lceil m \times 256chip/7680chip \rceil) \bmod k' = 0 \tag{11.84}$$

where $k' = k/(2ms)$, k is the multiple of 2ms time, CFN denotes the connection frame number for the associated DPCH and the set of five possible values of m is calculated as described in section 7.7 in [15], $m \in \{101, \ldots, 250\}$. The corresponding timing diagram is represented in Figure 11.73 [15].

Actually it means that the PCI/CQI composite report should be sent in every HS-DPCCH TTI but since the observation time for PCI and CQI calculation is not restricted it does not prevent us from calculation of PCI on the basis of CPICH signals immediately preceding the mentioned above HS-DPCCH TTI, as depicted in Figure 11.74.

When the relation:

$$\left\lfloor \frac{5 \times CFN + \lceil m \times 256chip/7680chip \rceil}{k'} \right\rfloor \bmod M\_cqi < N\_cqi\_typeA \tag{11.85}$$

holds, the UE shall report a type A CQI value. Otherwise the UE shall report a type B CQI value.

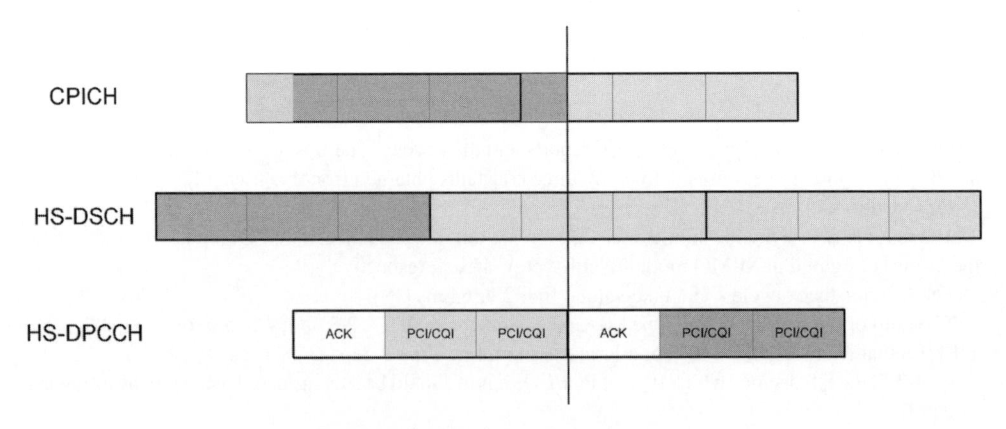

Figure 11.74 PCI reporting timing diagram

The UE shall repeat the transmission of the composite PCI/CQI value derived above over the next *(N_cqi_transmit – 1)* consecutive HS-DPCCH sub frames in the slots respectively allocated to CQI as defined in [15]. The UE does not support the case of $k' < N\_cqi\_transmit$, where *N_cqi_transmit* is the repetition factor of CQI, signalled to the UE and the Node B from higher layers.

That means even being in dual mode and sending Type A report UE should take into account that the reported PCI value can be used for single stream transmission.

The CQI encoding for Type B reports follows the existing Rel-5 mapping. CQI can take values in the range 1-30 (5 bits). The CQI encoding for Type A reports follows a new mapping. In this case CQI can take values in the range 0-255 (8 bits) and are constructed as follows:

$$CQI = \begin{cases} 15 \times CQI_1 + CQI_2 + 31 & \text{when 2 transport blocks are preferred by the UE} \\ CQI_S & \text{when 1 transport block is preferred by the UE} \end{cases} \quad (11.86)$$

where CQI_1 corresponds to the transport block that would be transmitted with the preferred primary precoding vector and CQI_2 corresponds to the transport block that would be transmitted with the precoding vector orthogonal to the preferred primary precoding vector. CQI_S corresponds to the transport block that would be transmitted with the preferred primary precoding vector in cases where a single stream transmission is preferred by the UE. CQI_S takes values in the range 1 – 30, and CQI_1, CQI_2 take values in the range 0 – 14.

For the purpose of CQI reporting, the UE shall assume a total received HS-PDSCH power of:

$$P_{HSPDSCH} = P_{CPICH} + \Gamma(\text{dB}) \quad (11.87)$$

where the total received power is assumed to be evenly distributed among 15 HS-PDSCH codes, and the measurement power offset Γ is signalled by higher layers, P_{CPICH} denotes the sum of the powers received from the set of CPICH(s). The measurement power offset Γ that is signalled by higher layers should be adjusted by the Node B such that it reflects the power that would be available for HS-PDSCH transmission relative to the combined transmit power of the set of CPICH(s) that shall be used for HS-PDSCH demodulation in case the Node B would have 15 OVSF codes available for HS-PDSCH transmission such that it results in the same power per OVSF code as the current power and code resources actually available for HS-PDSCH transmission at the Node B would allow.

When deriving the CQI value, the UE assumes that the Node B would be using a uniform power allocation across 15 channelization codes. In cases where the UE reports a CQI for two transport blocks, it is assumed by the UE that the Node B uses an equal power per channelization code for both of the two transport blocks. If the Node B does not transmit equal power per channelization code, it should not assume that the reported transport block sizes can be received with the specified block error probabilities.

The CQI tables can be found in [18] subclause 6A.2.3.

The UE calculates the preferred precoding vectors to be applied at Node B to maximize the aggregate transport block size that could be supported under current channel conditions. The information on whether one or two transport blocks are preferred is part of the CQI reporting. The precoding weight w_2 is mapped to PCI values as shown in Table 11.24 [18] (recall that the precoding weight w_2 defines both primary and secondary precoding vectors).

11.3.1.5.1 HS-SCCH type 3 Multiplexing and Coding

HS-SCCH type 3 is used when the UE is configured in MIMO mode. If one transport block is transmitted on the associated HS-PDSCH(s) or an HS-SCCH order is transmitted, the following information is transmitted by means of the HS-SCCH type 3 physical channel:

- Channelization-code-set information (7 bits): $x_{ccs,1}, x_{ccs,2}, \ldots, x_{ccs,7}$
- Modulation scheme and number of transport blocks information (3 bits): $x_{ms,1}, x_{ms,2}, x_{ms,3}$
- Precoding weight information (2 bits): $x_{pwipb,1}, x_{pwipb,2}$
- Transport-block size information (6 bits): $x_{tbspb,1}, x_{tbspb,2}, \ldots, x_{tbspb,6}$

Table 11.24 Mapping of preferred precoding weight w_2^{pref} to PCI values

w_2^{pref}	PCI value
$\frac{1+j}{2}$	0
$\frac{1-j}{2}$	1
$\frac{-1+j}{2}$	2
$\frac{-1-j}{2}$	3

- Hybrid-ARQ process information (4 bits): $x_{hap,1}, x_{hap,2}, \ldots, x_{hap,4}$
- Redundancy and constellation version (2 bits): $x_{rvpb,1}, x_{rvpb,2}$
- UE identity (16 bits): $x_{ue,1}, x_{ue,2}, \ldots, x_{ue,16}$

For an HS-SCCH order,

- $x_{ccs,1}, x_{ccs,2}, \ldots, x_{ccs,7}, x_{ms,1}, x_{ms,2}, x_{ms,3}, x_{pwipb,1}, x_{pwipb,2}$ shall be set to '111000000000'
- $x_{tbspb,1}, x_{tbspb,2}, \ldots, x_{tbspb,6}$ shall be set to '111101'
- $x_{hap,1}, x_{hap,2}, x_{hap,3}, x_{hap,4}, x_{rvpb,1}, x_{rvpb,2}$ shall be set to $x_{odt,1}, x_{odt,2}, x_{odt,3}, x_{ord,1}, x_{ord,2}, x_{ord,3}$

where $x_{odt,1}, x_{odt,2}, x_{odt,3}, x_{ord,1}, x_{ord,2}, x_{ord,3}$ are defined in Section 11.3.1.4.1.

If two transport blocks are transmitted on the associated HS-PDSCHs, the following information is transmitted by means of the HS-SCCH type 3 physical channel:

- Channelization-code-set information (7 bits): $x_{ccs,1}, x_{ccs,2}, \ldots, x_{ccs,7}$
- Modulation scheme and number of transport blocks information (3 bits): $x_{ms,1}, x_{ms,2}, x_{ms,3}$
- Precoding weight information for the primary transport block (2 bits): $x_{pwipb,1}, x_{pwipb,2}$
- Transport-block size information for the primary transport block (6 bits): $x_{tbspb,1}, x_{tbspb,2}, \ldots, x_{tbspb,6}$
- Transport-block size information for the secondary transport block (6 bits): $x_{tbssb,1}, x_{tbssb,2}, \ldots, x_{tbssb,6}$
- Hybrid-ARQ process information (4 bits): $x_{hap,1}, x_{hap,2}, \ldots, x_{hap,4}$
- Redundancy and constellation version for the primary transport block (2 bits): $x_{rvpb,1}, x_{rvpb,2}$
- Redundancy and constellation version for the secondary transport block (2 bits): $x_{rvsb,1}, x_{rvsb,2}$
- UE identity (16 bits): $x_{ue,1}, x_{ue,2}, \ldots, x_{ue,16}$

Figure 11.75 [16] illustrates the overall coding chain for HS-SCCH type 3. Note that some information shown is not present if only one transport block is transmitted on the associated HS-PDSCH(s).

The channelization code-set bits $x_{ccs,1}, x_{ccs,2}, \ldots, x_{ccs,7}$ are coded in the same way as described in section 11.2.1.4.1. If two transport blocks are transmitted on the associated HS-PDSCH(s), the same set of channelization codes shall be used for both transport blocks.

In accordance with Release-7 3GPP specifications only QPSK and 16-QAM modulation schemes can be used for HS-DSCH transmitting in MIMO mode. The usage of 64-QAM in MIMO mode is expected in Release-8. The number of transport blocks transmitted on the associated HS-PDSCH(s) and the modulation scheme information are jointly coded as shown in Table 11.25 [16]:

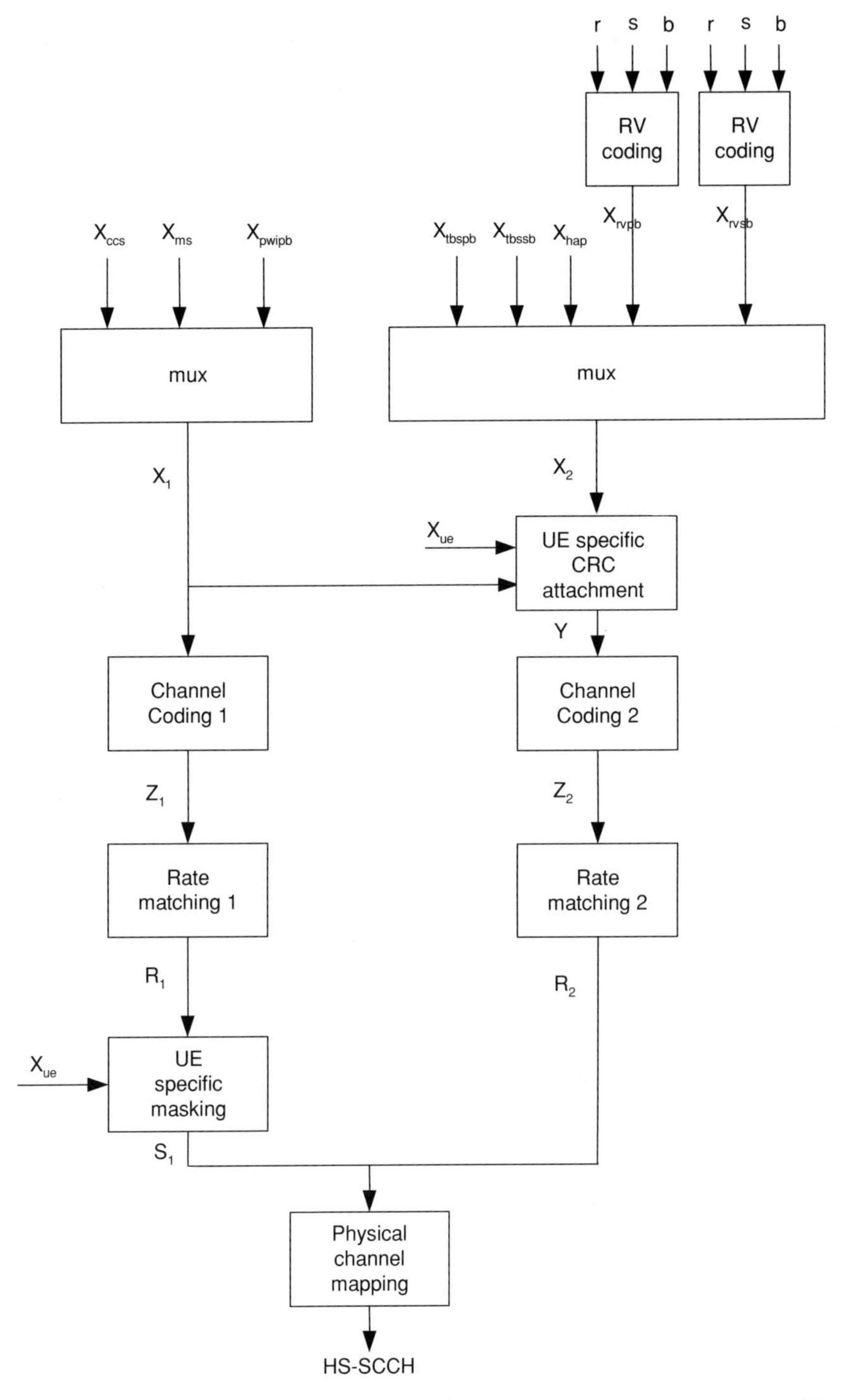

Figure 11.75 Coding chain for HS-SCCH type 3. Reproduced by permission of © 2008 3GPP™

Table 11.25 Mapping of x_{ms}

$x_{ms\text{-}stb,1}$, $x_{ms\text{-}stb,2}$, $x_{ms\text{-}stb,3}$	Modulation for primary transport block	Modulation for secondary transport block	Number of transport blocks
111	16QAM	16QAM	2
110	16QAM	QPSK	2
100	16QAM	n/a	1
011	QPSK	QPSK	2
000	QPSK	n/a	1

The precoding weight information for the primary transport block $x_{pwipb,1}$, $x_{pwipb,2}$ is derived from the precoding weight factor w_2 as defined in Table 11.24, according to Table 11.26 [16].

The transport-block size information $x_{tbspb,1}, x_{tbspb,2}, \ldots, x_{tbspb,6}$ is the unsigned binary representation of the transport block size index for the primary transport block. If two transport blocks are transmitted on the associated HS-PDSCH(s), the transport-block size information $x_{tbssb,1}, x_{tbssb,2}, \ldots, x_{tbssb,6}$ is the unsigned binary representation of the transport block size index for the secondary transport block.

If two transport blocks are transmitted on the associated HS-PDSCH(s), the mapping relationship between the H-ARQ processes and the transport blocks is such that when the H-ARQ-process with identifier HAP_{pb} is mapped to the primary transport block, the HARQ-process with the identifier given by $(HAP_{pb} + N_{proc}/2) \bmod (N_{proc})$ shall be mapped to the secondary transport block, where N_{proc} is the number of HARQ processes configured by higher layers. The combination of H-ARQ-processes is indicated by the H-ARQ process information (4 bits) $x_{hap,1}, x_{hap,2}, x_{hap,3}, x_{hap,4}$ which are the unsigned binary representation of HAP_{pb}. If only one transport block is transmitted on the associated HS-PDSCH(s), the above mapping is ignored and the H-ARQ process information $x_{hap,1}, x_{hap,2}, x_{hap,3}, x_{hap,4}$ is the unsigned binary representation of the H-ARQ process identifier.

For each of the primary transport block and a secondary transport block if two transport blocks are transmitted on the associated HS-PDSCH(s), the redundancy version (RV) parameters r, s and constellation version parameter b are coded jointly to produce the values X_{rvpb} and X_{rvsb} respectively. The transmitted sequences $x_{rvpb,1}, x_{rvpb,2}$ and $x_{rvsb,1}, x_{rvsb,2}$ are the binary representations of X_{rvpb} and X_{rvsb}, respectively. For the primary transport block if only one transport block is transmitted on the associated HS-PDSCH(s), the redundancy version (RV) parameters r, s and constellation version parameter b are coded jointly to produce the value X_{rvpb}. The transmitted sequence $x_{rvpb,1}$, $x_{rvpb,2}$ is the binary

Table 11.26 Mapping of precoding weight information for primary transport block

w_2	$x_{pwipb,1}$, $x_{pwipb,2}$
$\frac{1+j}{2}$	00
$\frac{1-j}{2}$	01
$\frac{-1+j}{2}$	10
$\frac{-1-j}{2}$	11

Table 11.27 RV coding for 16QAM for HS-SCCH type 3

	$N_{sys}/N_{data} < 1/2$			$N_{sys}/N_{data} \geq 1/2$		
X_{rvpb} or X_{rvsb} (value)	s	r	b	s	r	b
0	1	0	0	1	0	0
1	1	1	1	0	1	1
2	1	0	2	0	0	0
3	1	0	3	1	0	2

representation of X_{rvpb}. Joint coding of parameters r, s and constellation version parameter b is done according to Tables 11.27 [16] and 11.28 [16] according to the modulation mode used. If $X_{rvpb} = 0$ or $X_{rvsb} = 0$, the UE shall treat the corresponding transport block as an initial transmission.

The channelization-code-set information $x_{ccs,1}, x_{ccs,2}, \ldots, x_{ccs,7}$, modulation-scheme and number of transport blocks information $x_{ms,1}, x_{ms,2}, x_{ms,3}$ and precoding weight information $x_{pwipb,1}, x_{pwipb,2}$ are multiplexed together. This gives a sequence of bits $x_{1,1}, x_{1,2}, \ldots, x_{1,12}$, where:

$$x_{1,i} = x_{ccs,i} \qquad i = 1, 2, \ldots, 7$$
$$x_{1,i} = x_{ms,i-7} \qquad i = 8, 9, 10$$
$$x_{1,i} = x_{pwipb,i-10} \qquad i = 11, 12$$

If one transport block is transmitted on the associated HS-PDSCH(s), the transport-block-size information $x_{tbspb,1}, x_{tbspb,2}, \ldots, x_{tbspb,6}$, H-ARQ-process information $x_{hap,1}, x_{hap,2}, \ldots, x_{hap,4}$ and redundancy-version information $x_{rvpb,1}, x_{rvpb,2}$ are multiplexed together. This gives a sequence of bits $x_{2,1}, x_{2,2}, \ldots, x_{2,12}$ where:

$$x_{2,i} = x_{tbs,i} \qquad i = 1, 2, \ldots, 6$$
$$x_{2,i} = x_{hap,i-6} \qquad i = 7, 8, \ldots, 10$$
$$x_{2,i} = x_{rv,i-10} \qquad i = 11, 12$$

If two transport blocks are transmitted on the associated HS-PDSCHs, the transport-block-size information for the primary transport block $x_{tbspb,1}, x_{tbspb,2}, \ldots, x_{tbspb,6}$, transport-block-size information for the secondary transport block $x_{tbssb,1}, x_{tbssb,2}, \ldots, x_{tbssb,6}$, Hybrid-ARQ-process information $x_{hap,1}, x_{hap,2}, \ldots, x_{hap,4}$, redundancy-version information for the primary transport block $x_{rvpb,1}, x_{rvpb,2}$, and redundancy-version information for the secondary transport block $x_{rvsb,1}, x_{rvsb,2}$ are multiplexed together. This gives

Table 11.28 RV coding for QPSK for HS-SCCH type 3

	$N_{sys}/N_{data} < 1/2$		$N_{sys}/N_{data} \geq 1/2$	
X_{rvpb} or X_{rvsb} (value)	s	r	s	r
0	1	0	1	0
1	1	1	0	1
2	1	2	0	3
3	1	3	1	2

a sequence of bits $x_{2,1}, x_{2,2}, \ldots, x_{2,20}$ where:

$$\begin{aligned} x_{2,i} &= x_{tbspb,i} && i = 1, 2, \ldots, 6 \\ x_{2,i} &= x_{tbssb,i-6} && i = 7, 8, \ldots, 12 \\ x_{2,i} &= x_{hap,i-12} && i = 13, 14, \ldots, 16 \\ x_{2,i} &= x_{rvpb,i-16} && i = 17, 18 \\ x_{2,i} &= x_{rvsb,i-18} && i = 19, 20 \end{aligned}$$

If one transport block is transmitted on the associated HS-PDSCH(s), from the sequence of bits $x_{1,1}, x_{1,2}, \ldots, x_{1,12}, x_{2,1}, x_{2,2}, \ldots, x_{2,12}$ a 16-bit CRC is calculated. This gives a sequence of bits $c_1, c_2, \ldots, c_{16}$.

This sequence of bits is then masked with the UE Identity $x_{ue,1}, x_{ue,2}, \ldots, x_{ue,16}$ and then appended to the sequence of bits $x_{2,1}, x_{2,2}, \ldots, x_{2,12}$ to form the sequence of bits $y_1, y_2, \ldots, y_{28}$, where:

$$\begin{aligned} y_i &= x_{2,i} && i = 1, 2, \ldots, 12 \\ y_i &= (c_{i-12} + x_{ue,i-12}) \bmod 2 && i = 13, 14, \ldots, 28 \end{aligned}$$

If two transport blocks are transmitted on the associated HS-PDSCHs, from the sequence of bits $x_{1,1}, x_{1,2}, \ldots, x_{1,12}, x_{2,1}, x_{2,2}, \ldots, x_{2,20}$ a 16-bit CRC is calculated. This gives a sequence of bits $c_1, c_2, \ldots, c_{16}$. here

This sequence of bits is then masked with the UE Identity $x_{ue,1}, x_{ue,2}, \ldots, x_{ue,16}$ and then appended to the sequence of bits $x_{2,1}, x_{2,2}, \ldots, x_{2,20}$ to form the sequence of bits $y_1, y_2, \ldots, y_{36}$, where:

$$\begin{aligned} y_i &= x_{2,i} && i = 1, 2, \ldots, 20 \\ y_i &= (c_{i-20} + x_{ue,i-20}) \bmod 2 && i = 21, 22, \ldots, 36 \end{aligned}$$

Rate 1/3 convolutional coding, as described in section 11.2.4.2, is applied to the sequence of bits $x_{1,1}, x_{1,2}, \ldots, x_{1,12}$. This gives a sequence of bits $z_{1,1}, z_{1,2}, \ldots, z_{1,60}$ for HS-SCCH Part 1.

The same rate 1/3 convolutional coding is applied to the sequence of bits $y_1, y_2, \ldots, y_{28}$ if one transport block is transmitted on the associated HS-PDSCH(s), or to the sequence of bits $y_1, y_2, \ldots, y_{36}$ if two transport blocks are transmitted on the associated HS-PDSCHs. This gives a sequence of bits for HS-SCCH Part2 $z_{2,1}, z_{2,2}, \ldots, z_{2,108}$ if one transport block is transmitted or $z_{2,1}, z_{2,2}, \ldots, z_{2,132}$ if two transport blocks are transmitted.

From the input sequence $z_{1,1}, z_{1,2}, \ldots, z_{1,60}$ the bits $z_{1,1}, z_{1,2}, z_{1,4}, z_{1,6}, z_{1,8}, z_{1,12}, z_{1,15}, z_{1,18}, z_{1,21}, z_{1,24}, z_{1,37}, z_{1,40}, z_{1,43}, z_{1,46}, z_{1,49}, z_{1,53}, z_{1,55}, z_{1,57}, z_{1,59}, z_{1,60}$ are punctured to obtain the output sequence $r_{1,1}, r_{1,2} \ldots r_{1,40}$.

If one transport block is transmitted on the associated HS-PDSCH(s), from the input sequence $z_{2,1}, z_{2,2}, \ldots, z_{2,108}$ the bits $z_{2,1}, z_{2,2}, z_{2,3}, z_{2,4}, z_{2,5}, z_{2,6}, z_{2,7}, z_{2,8}, z_{2,12}, z_{2,14}, z_{2,15}, z_{2,24}, z_{2,42}, z_{2,48}, z_{2,63}, z_{2,66}, z_{2,93}, z_{2,96}, z_{2,98}, z_{2,99}, z_{2,101}, z_{2,102}, z_{2,103}, z_{2,104}, z_{2,105}, z_{2,106}, z_{2,107}, z_{2,108}$ are punctured to obtain the output sequence $r_{2,1}, r_{2,2} \ldots r_{2,80}$.

If two transport blocks are transmitted on the associated HS-PDSCHs, from the input sequence $z_{2,1}, z_{2,2}, \ldots, z_{2,132}$ the bits $z_{2,1}, z_{2,2}, z_{2,3}, z_{2,4}, z_{2,5}, z_{2,6}, z_{2,7}, z_{2,8}, z_{2,10}, z_{2,11}, z_{2,13}, z_{2,14}, z_{2,16}, z_{2,19}, z_{2,22}, z_{2,25}, z_{2,28}, z_{2,31}, z_{2,34}, z_{2,37}, z_{2,40}, z_{2,43}, z_{2,46}, z_{2,49}, z_{2,55}, z_{2,61}, z_{2,72}, z_{2,78}, z_{2,84}, z_{2,87}, z_{2,90}, z_{2,93}, z_{2,96}, z_{2,99}, z_{2,102}, z_{2,105}, z_{2,108}, z_{2,111}, z_{2,114}, z_{2,117}, z_{2,119}, z_{2,120}, z_{2,122}, z_{2,123}, z_{2,125}, z_{2,126}, z_{2,127}, z_{2,128}, z_{2,129}, z_{2,130}, z_{2,131}, z_{2,132}$ are punctured to obtain the output sequence $r_{2,1}, r_{2,2} \ldots r_{2,80}$.

The UE specific masking is done in the same way as described in section 11.3.1.4.1, (11.77). The UE specific masking gives the output bits $s_{1,1}, s_{1,2} \ldots s_{1,40}$ comprising HS-SCCH Part 1. The bit sequence $r_{2,1}, r_{2,2} \ldots r_{2,80}$ comprises HS-SCCH Part 2.

The HS-SCCH orders in MIMO mode are coded in the same way as in non-MIMO mode.

11.3.1.5.2 HS-DSCH in MIMO Mode

As can be seen from Figure 11.72 the multiplexing, coding, spreading and modulation of data streams in MIMO mode is done independently as described in section 11.3.1.4.2. Also, the streams can be processed independently in the receiver. In case the receiver uses *successive interference cancellation*

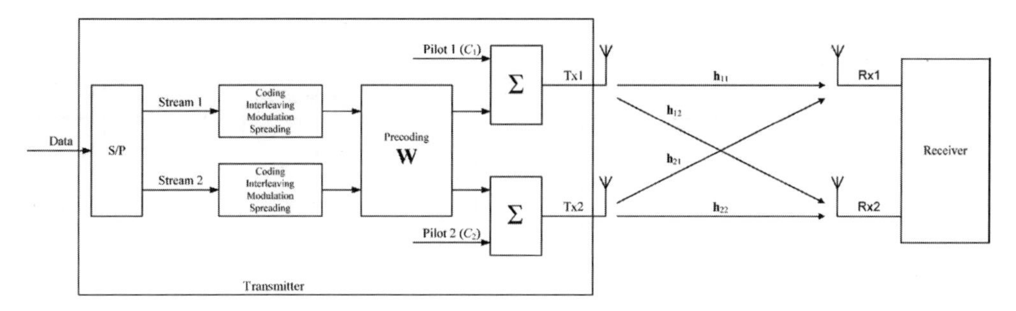

Figure 11.76 HSDPA MIMO System model

(*SIC*) or turbo-equalization the result of detection or decoding of one stream can be used for the detection of another stream but the decoding of each stream is done independently. Thus, the decoding chain of the UE receiver in MIMO mode simply can be doubled. However, the main difference of the UE receiver in MIMO mode from the receiver in non-MIMO mode is that the receiver should be able to separate the received streams. It could be done in different ways. In Figure 11.76 the MIMO system model is depicted.

Here precoding consists in multiplication of data streams by matrix $\mathbf{W} = \begin{bmatrix} w_1 \; w_3 \\ w_2 \; w_4 \end{bmatrix}$, and vector $\mathbf{h}_{ij} = \left[h_{ij}(0), h_{ij}(1), \ldots, h_{ij}(L-1)\right]$ comprises channel coefficients of channel from the transmitter antenna Tx_i to the receiver antenna Rx_j. The most straightforward method of stream separation is first to equalize the channel and despread the data, then the data from the despreader output should be multiplied by the inverse matrix $\mathbf{W}^{-1}$. Taking into account (11.83) it is easy to verify that $\mathbf{W}^{-1} = \begin{bmatrix} w_1^* \; w_2^* \\ w_3^* \; w_4^* \end{bmatrix}$. This method usually is called *antenna equalization*, since the equalizer in this case does not use any information about streams' precoding. The schematic of the UE receiver using antenna equalization is depicted in Figure 11.77.

Another method consists in separating the streams with the help of equalizer. Actually it is possible to consider the HS-DSCH precoding as part of channel. That means, the channel matrix:

$$\mathbf{H} = \begin{bmatrix} \mathbf{h}_{11} & \mathbf{h}_{21} \\ \mathbf{h}_{12} & \mathbf{h}_{22} \end{bmatrix} \tag{11.88}$$

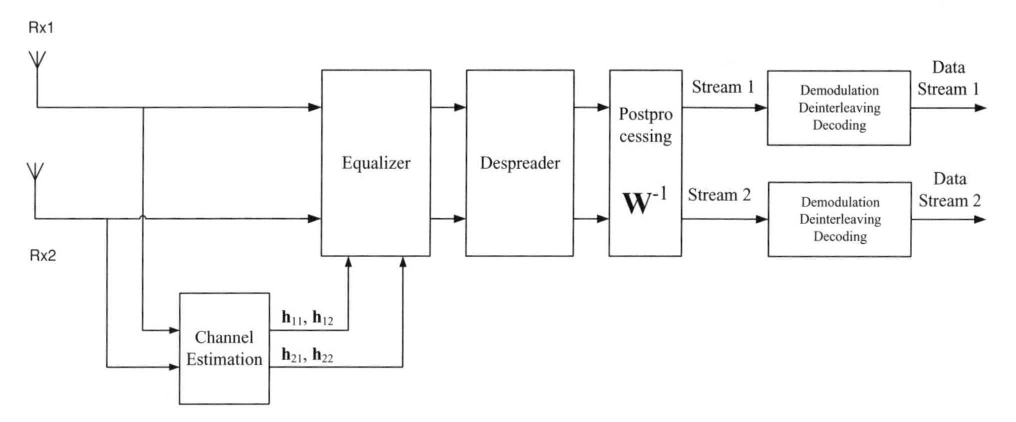

Figure 11.77 Antenna equalization

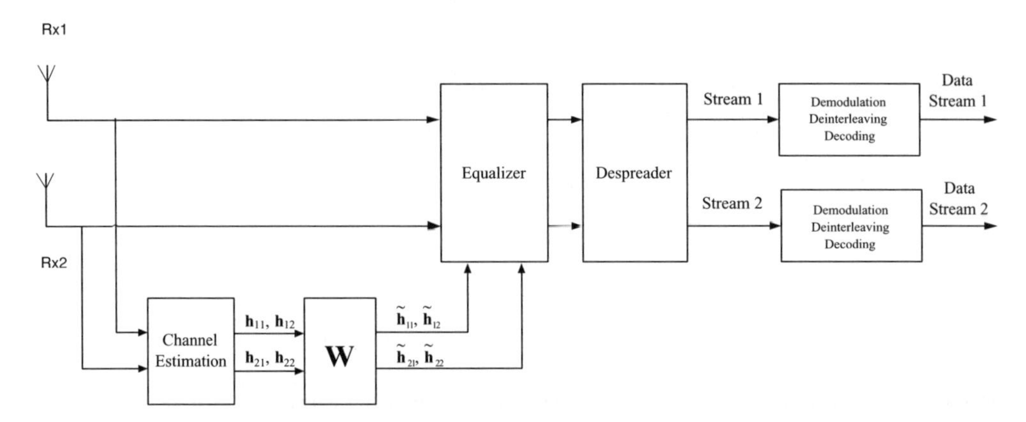

Figure 11.78 Stream equalization

can be substituted by the matrix of "superchannel" including both precoding and channel itself:

$$\tilde{\mathbf{H}} = \begin{bmatrix} \tilde{\mathbf{h}}_{11} & \tilde{\mathbf{h}}_{21} \\ \tilde{\mathbf{h}}_{12} & \tilde{\mathbf{h}}_{22} \end{bmatrix} = \begin{bmatrix} \mathbf{h}_{11}w_1 + \mathbf{h}_{21}w_2 & \mathbf{h}_{11}w_3 + \mathbf{h}_{21}w_4 \\ \mathbf{h}_{12}w_1 + \mathbf{h}_{22}w_2 & \mathbf{h}_{12}w_3 + \mathbf{h}_{22}w_4 \end{bmatrix} \tag{11.89}$$

To make it possible for equalizer to equalize the "superchannel" the channel estimates should be modified in accordance with (11.89). Then the streams are already separated at the equalizer output (to be more precise, at the output of the despreader, since the equalizer output comprises the mix of all downlink channels). This kind of HS-DSCH processing is denoted as *stream equalization*. The schematic of the receiver implementing stream equalization is depicted in Figure 11.78.

In the ideal case when the channel estimates exactly coincide with the channel coefficients both methods should provide the same performance. However, the complexity of the receiver implementing stream equalization is higher due to the fact that other downlink channels except the HS-DSCH in this case should be equalized with different equalizer, since they are not precoded.

11.3.1.5.3 HS-DPCCH Multiplexing and Coding in MIMO Mode

Multiplexing and coding for HS-DPCCH when UE is configured in MIMO mode differs from the case when UE is not in MIMO mode. The precoding control indication (PCI) and channel quality indication (CQI) are coded jointly comprising the composite PCI/CQI report. The H-ARQ acknowledgment is coded separately from PCI/CQI report.

The H-ARQ message in MIMO mode occupies the same 10 bits as in non-MIMO mode but in MIMO mode the H-ARQ message should represent the increased number of events in comparison with non-MIMO mode. In non-MIMO mode there are only four possible events that should be mapped to H-ARQ message: acknowledgment (ACK), non-acknowledgment (NACK), preamble (PRE) and postamble (POST). In MIMO dual stream mode instead of ACK and NACK there are four possible combinations of ACKs and NACKs for two streams. The HARQ acknowledgment message to be transmitted is encoded to 10 bits in accordance with Table 11.29 [16]. The encoded bits are denoted $w_0, w_1, \ldots, w_9$.

When the UE is configured in MIMO mode, two types of CQI reports shall be supported by the UE. Type A CQI reports use values $0, \ldots, 255$ and type B CQI reports use values $1, \ldots, 30$, respectively. In cases where a type A CQI shall be reported, the CQI values 0 .. 255 are converted from decimal to binary to map them to the channel quality indication bits (0 0 0 0 0 0 0 0) to (1 1 1 1 1 1 1 1), respectively. The channel quality indication bits are cqi_0, cqi_1, cqi_2, cqi_3, cqi_4, cqi_5, cqi_6, cqi_7 (where cqi_0 is LSB and cqi_7 is MSB). In cases where a type B CQI shall be reported, the CQI values $1, \ldots, 30$ are converted from decimal to binary to map them to the channel quality indication bits (1 0 0 0 0) to (1 1 1 1 1), respectively. The channel quality indication bits are cqi_0, cqi_1, cqi_2, cqi_3, cqi_4 (where cqi_0 is LSB and cqi_4 is MSB).

Table 11.29 Channel coding of HARQ-ACK when the UE is configured in MIMO mode

HARQ-ACKmessage to be transmitted		w_0	w_1	w_2	w_3	w_4	w_5	w_6	w_7	w_8	w_9
HARQ-ACK in response to a single scheduled transport block											
ACK		1	1	1	1	1	1	1	1	1	1
NACK		0	0	0	0	0	0	0	0	0	0
HARQ-ACK in response to two scheduled transport blocks											
Response to primary transport block	Response to secondary transport block										
ACK	ACK	1	0	1	0	1	1	1	1	0	1
ACK	NACK	1	1	0	1	0	1	0	1	1	1
NACK	ACK	0	1	1	1	1	0	1	0	1	1
NACK	NACK	1	0	0	1	0	0	1	0	0	0
PRE/POST indication											
PRE		0	0	1	0	0	1	0	0	1	0
POST		0	1	0	0	1	0	0	1	0	0

According to the PCI definition, the range of possible PCI values is $0, \ldots, 3$. The PCI values $0, \ldots, 3$ are converted from decimal to binary to map them to the precoding control indication bits (0 0) to (1 1) respectively. The precoding control indication bits are pci_0, pci_1 (where pci_0 is LSB and pci_1 is MSB).

Two formats for composite PCI/CQI information words are possible depending on the type of the reported CQI value. The two formats shall be constructed according to the scheme depicted in Figure 11.79 [16].

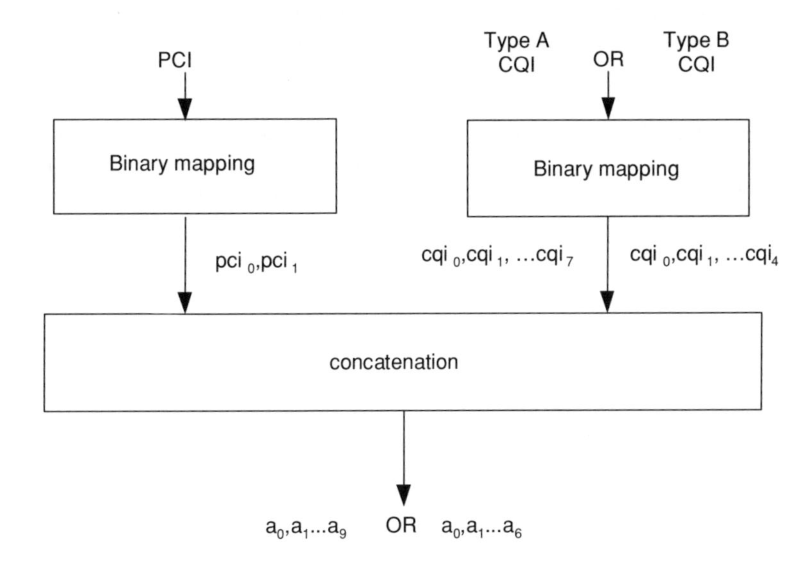

Figure 11.79 Composite (PCI/CQI) report multiplexing

Table 11.30 Basis sequences for channel encoding of composite PCI/CQI reports

i	$M_{i,0}$	$M_{i,1}$	$M_{i,2}$	$M_{i,3}$	$M_{i,4}$	$M_{i,5}$	$M_{i,6}$	$M_{i,7}$	$M_{i,8}$	$M_{i,9}$	$M_{i,10}$
0	1	0	0	0	0	0	0	0	0	0	0
1	0	1	0	0	0	0	0	0	0	0	0
2	0	0	0	1	0	0	0	0	0	0	0
3	0	0	0	0	1	0	0	0	0	0	0
4	0	0	0	0	0	1	0	0	0	0	0
5	0	0	0	0	0	0	0	1	0	0	0
6	0	0	0	0	0	0	0	0	1	0	1
7	0	0	0	0	0	0	0	0	0	1	1
8	1	0	1	0	0	0	1	1	1	0	1
9	1	1	0	1	0	0	0	1	1	1	1
10	0	1	1	0	1	0	0	0	1	1	1
11	1	0	1	1	0	1	0	0	0	1	0
12	1	1	0	1	1	0	1	0	0	0	0
13	1	1	1	0	1	1	0	1	0	0	0
14	0	1	1	1	0	1	1	0	1	0	1
15	0	0	1	1	1	0	1	1	0	1	0
16	0	0	0	1	1	1	0	1	1	0	1
17	1	0	0	0	1	1	1	0	1	1	1
18	0	1	0	0	0	1	1	1	0	1	0
19	1	1	1	1	1	1	1	1	1	1	1

In cases where a type A CQI shall be reported, the precoding control indication bits pci_0, pci_1, and the channel quality indication bits cqi_0, cqi_1, cqi_2, cqi_3, cqi_4, cqi_5, cqi_6, cqi_7 are concatenated to the composite precoding control indication and channel quality indication bits according to the relation:

$$(a_0\, a_1\, a_2\, a_3\, a_4\, a_5\, a_6\, a_7\, a_8\, a_9) = (pci_0\, pci_1\, cqi_0\, cqi_1\, cqi_2\, cqi_3\, cqi_4\, cqi_5\, cqi_6\, cqi_7) \tag{11.90}$$

In case a type B CQI shall be reported, the precoding control indication bits pci_0, pci_1, and the channel quality indication bits cqi_0, cqi_1, cqi_2, cqi_3, cqi_4 are concatenated to the composite precoding control indication and channel quality indication bits according to the relation:

$$(a_0\, a_1\, a_2\, a_3\, a_4\, a_5\, a_6) = \left(pci_0\, pci_1 cqi_0\, cqi_1\, cqi_2\, cqi_3\, cqi_4\right) \tag{11.91}$$

In cases where a type A CQI needs to be reported, the composite precoding control indication and channel quality indication is coded using a (20,10) code. The code words of the (20,10) code are a linear combination of the 10 basis code words denoted $M_{i,n}$ defined in Table 11.30 [16].

The output codeword bits b_i are given by:

$$b_i = \sum_{n=0}^{9} (a_n \times M_{i,n}) \bmod 2 \tag{11.92}$$

where $i = 0, \ldots, 19$.

In case a type B CQI needs to be reported, the composite precoding control indication and channel quality indication is coded using a (20,7) code defined in Table 11.30. The code words of the (20,7) code are a linear combination of the basis sequences denoted $M_{i,n}$ defined in Table 11.30 for $n \in \{0, 1, 3, 4, 5, 7, 10\}$. The output codeword bits b_i in this case are given by:

$$b_i = \left(\sum_{n=0}^{1} (a_n \times M_{i,n}) + \sum_{n=2}^{4} (a_n \times M_{i,n+1}) + a_5 \times M_{i,7} + a_6 \times M_{i,10} \right) \bmod 2 \tag{11.93}$$

where $i = 0, \ldots, 19$.

The bits w_k $(k = 0, \ldots, 9)$ are mapped directly to the slot 1 of HS-DPCCH physical channel and bits b_k $(k = 0, \ldots, 19)$ are mapped directly to the slots 2 and 3 of HS-DPCCH physical channel.

In general, the main drawback of MIMO is the inter-stream interference. In the ideal case the streams are perfectly orthogonal and there is no inter-stream interference, but the impact of the multi-path fading channel destroys the orthogonality of two streams, which cannot be totally compensated by the equalization. Due to this fact for MIMO the inter-stream interference dominates all other interference sources in real life. That is why the dual stream mode does not provide the doubling of performance in comparison with the single stream mode in most cases.

11.3.1.6 Dual Cell HSDPA

Dual Cell HSDPA (*DC-HSDPA*) is an optional feature of the Release 8 in 3GPP. DC-HSDPA has been introduced to 3GPP Release 8 specifications in order to increase the peak bit rate. DC-HSDPA makes the increase at the cost of doubling the bandwidth as shown in Figure 11.80. That means, at Node B similarly with MIMO the data intended for a particular UE is multiplexed into two streams, one of which is transmitted on carrier frequency 1 and the second stream is transmitted on carrier frequency 2. This solution is supported by the fact that UMTS licenses are often issued as 10 or 15 MHz paired spectrum allocations.

A UE in DC-HSDPA operation is able to simultaneously receive HSDPA traffic over two downlink carrier frequencies transmitted in the same frequency band from a single serving sector and to transmit on one uplink carrier frequency. The Release-8 considers only one uplink carrier for DC-HSDPA. The uplink carrier for a DC-HSDPA UE is not strictly tied to one of the two downlink carriers. The DC-HSDPA can be especially efficient at the cell edges where the channel conditions are not favourable and existing techniques such as MIMO cannot be used. As it was mentioned in section 11.3.1.5 the main problem of MIMO transmission scheme in most cases is the inter-stream interference. Unlike MIMO the interference between the carriers in DC-HSDPA is not significant. Due to this, in recent times the operators are much more interested in the implementation of DC-HSDPA rather than MIMO. The additional benefit of DC-HSDPA in comparison with MIMO is that the implementation of DC-HSDPA requires less additional hardware. At least there is no need in additional antennas.

One of the carriers called *anchor carrier* has all the downlink physical channels. There are no restrictions of channel operation on the anchor carrier. The other *supplementary carrier* is reserved only for HSDPA channels (HS-SCCH and HS-PDSCH). The only additional channel on the secondary carrier which the UE needs to demodulate is the common pilot channel (CPICH). On the supplementary carrier, the UE can only monitor DL HSDPA related channels [28]. In Release-8 the carriers are assumed to be adjacent to each other. The two data streams do not have any common L1 signalled parameters.

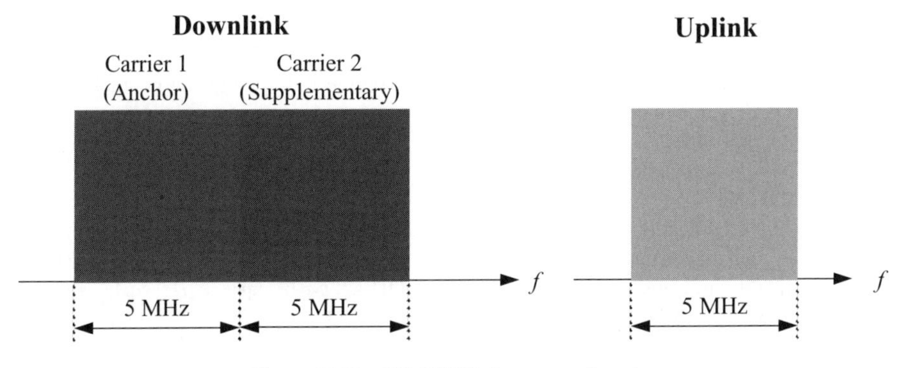

Figure 11.80 DC-HSDPA frequency allocation

Table 11.31 DC-HSDPA FDD HS-DSCH physical layer categories

HS-DSCH category	Maximum number of HS-DSCH codes received	Minimum inter-TTI interval	Maximum number of bits of an HS-DSCH transport block received within an HS-DSCH TTI NOTE 1	Total number of soft channel bits	Supported modulations without MIMO operation or dual cell operation	Supported modulations simultaneous with MIMO operation and without dual cell operation	Supported modulations with dual cell operation
Category 21	15	1	23370	345600	–	–	QPSK,
Category 22	15	1	27952	345600			16QAM
Category 23	15	1	35280	518400			QPSK,
Category 24	15	1	42192	518400			16QAM, 64QAM

The number of channelization codes and modulation are defined individually for both carriers without extra limitations. All the Release-8 modulations (QPSK, 16QAM and 64QAM) are supported also in DC-HSDPA mode of operation. Parallel transmission with 64-QAM modulation each carrier can theoretically provide a downlink peak data rate of 43.2 Mbps in 10MHz without the support of MIMO. The simultaneous use of MIMO and DC-HSDPA operation is not considered in Release-8. The two streams have separate HARQ processes so that the retransmissions need to use the same carrier as the initial transmission.

There are four new UE capability classes for DC-HSDPA added to 3GPP specifications (21, 22, 23 and 24) shown in Table 11.31.

Both the anchor and supplementary carriers can have disjointed HS-SCCH channels. In this case the coding of HS-SCCH does not need to be changed. In order to not restrict the scheduler, the UE should preferably monitor up to four HS-SCCH codes on each carrier, as in the single carrier case, assuming HS-SCCH is transmitted in both carriers [28]. The UE shall be able to receive 1 HS-SCCH on anchor carrier and 1 HS-SCCH on supplementary carrier simultaneously. This approach provides flexibility to scheduling. A new HS-SCCH-order is introduced to switch on and off the DC-HSDPA mode [29]. All the orders can be signalled on both carriers.

In the uplink DC-HSDPA uses only a single carrier. The CQI and ACK/NACK information from both carriers are multiplexed into one common HS-PDCCH. Due to independent CQIs measurements on both carriers an efficient joint scheduling in the Node B based on the link quality can be applied. Channel coding of HARQ-ACK info is defined in Table 11.32 [29]

The composite CQI report is constructed from two individual CQI reports that are represented by CQI1 and CQI2. CQI1 corresponds to the serving HS-DSCH cell and CQI2 corresponds to the secondary serving HS-DSCH cell.

Each constituent CQI report uses values 1, . . . , 30. The individual CQI values are converted from decimal to binary to map them to the channel quality indication bits (1 0 0 0 0) to (1 1 1 1 1) respectively.

The channel quality indication bits corresponding to CQI1 are $cqi1_0$, $cqi1_1$, $cqi1_2$, $cqi1_3$, $cqi1_4$ (where $cqi1_0$ is LSB and $cqi1_4$ is MSB) and those corresponding to CQI2 are $cqi2_0$, $cqi2_1$, $cqi2_2$, $cqi2_3$, $cqi2_4$ (where $cqi2_0$ is LSB and $cqi2_4$ is MSB).

Table 11.32 Channel coding of HARQ-ACK in case Secondary Cell Active as no 0

HARQ-ACK message to be transmitted		w_0	w_1	w_2	w_3	w_4	w_5	w_6	w_7	w_8	w_9
HARQ-ACK when UE detects a single scheduled transport block on the secondary HS-DSCh cell											
ACK		1	1	1	1	1	1	1	1	1	1
NACK		0	0	0	0	0	0	0	0	0	0
HARQ-ACK when UE detects a single scheduled transport blocks on the secondary serving HS-DSCh cell											
ACK		1	1	1	1	1	0	0	0	0	0
NACK		0	0	0	0	0	1	1	1	1	1
HARQ-ACK when UE detects a single scheduled transport blocks on the secondary serving HS-DSCh cell											
Response to transport block from serving HS-DSCH cell	Response to transport block from secondary serving HS-DSCH cell										
ACK	ACK	1	0	1	0	1	0	1	0	1	0
ACK	NACK	1	1	0	0	1	1	0	0	1	1
NACK	ACK	0	0	1	1	0	0	1	1	0	0
NACK	NACK	0	1	0	1	0	1	0	1	0	1
PRE/POST indication											
PRE		0	0	1	0	0	1	0	0	1	0
POST		0	1	0	0	1	0	0	1	0	0

The two individual CQI reports are concatenated to form the composite channel quality indication according to the relation:

$$(a_0\, a_1\, a_2\, a_3\, a_4\, a_5\, a_6\, a_7\, a_8\, a_9) = (cqi1_0\, cqi1_2\, cqi1_2\, cqi1_3\, cqi1_4\, cqi2_0\, cqi2_1\, cqi2_2\, cqi2_3\, cqi2_4) \tag{11.94}$$

The coding for CQI uses the same (20, 10) code which was defined for MIMO type A CQI encoding in Table 11.30.

DC-HSDPA provides the performance gain not only in comparison with a single carrier case, but also in comparison with independent use of two single carriers. The reasons for cell capacity gain are the dynamic multiplexing of users rather than static and the possibility of using joint scheduling. And of course DC-HSDPA allows double the instantaneous data rates in comparison with single carrier case by assigning all the code and power resources to a single user in a TTI, which is not possible for two independent single carriers. Taking into account that the interference between two carriers is not significant it is much easier to provide doubling of data rates with DC-HSDPA than with MIMO HSDPA. Figure 11.81 [28] shows the capacity gain as a function of the number of users per sector. As can be seen from Figure 11.81, DC-HSDPA gain is more pronounced at low loads. At two users per sector, the gain in sector throughput is 25%. At 32 users per sector, it is 7%.

The obvious way of future development of DC-HSDPA is using more than two carriers. This kind of technology is usually referred to as *multi-cell HSDPA* (*MC-HSDPA*). Since the data rate increases with the bandwidth, this could significantly improve peak data rates as well as the system capacity. Of course,

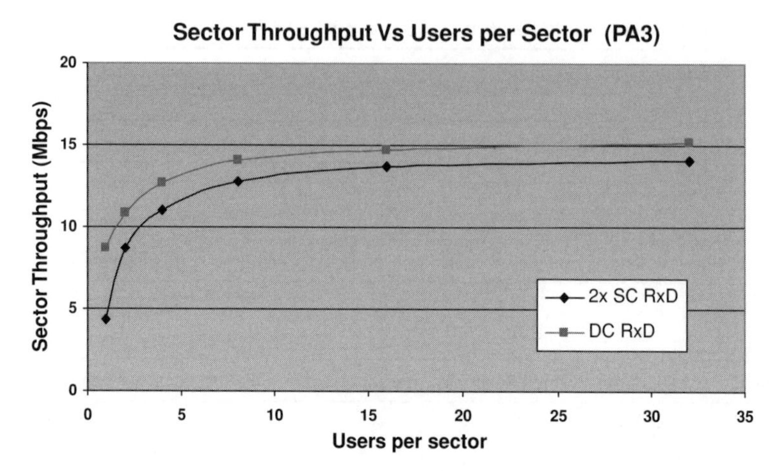

Figure 11.81 Capacity gain from DC-HSDPA over 2xSC-HSDPA. Reproduced by permission of © 2008 3GPP™

this gain is not for free, the used bandwidth should be increased correspondingly. Another possibility to improve the HSDPA performance is to allow the simultaneous use of DC-HSDPA (or even MC-HSDPA) and MIMO. Of course, the latter solution requires significant increase of implementation complexity but the implementation of both DC-HSDPA and MIMO is quite common and consists mostly in scaling the number of parallel decoding chains in UE.

11.3.2 HSUPA

The HSUPA technology, like HSDPA in downlink, allows significantly improve the quality of data transmission on uplink. When the work on the Rlease-5 3GPP specifications introducing the HSDPA was started it became clear that the data rates provided by uplink DCH are not sufficient. The new technology aiming to improve the uplink data rates HSUPA was introduced in the Release-6. The basic principles of HSUPA, allowing the data rates to be increased and the delays decreased, are very similar to those of HSDPA. HSUPA like HSDPA uses fast scheduling, which is implemented in Node B and fast H-ARQ. Also it is possible to use 2 ms TTI in HSUPA. However, AMC and high order modulation are not used in HSUPA in Release-6 and Release-7. The reason for this is that high order modulations require more energy per bit to be transmitted and in UE such resource as energy is quite limited because of the battery life period. Nevertheless, most probably the use of 16-QAM in HSUPA will be included in Release-8 specifications. Theoretically, the data rate provided by HSUPA can achieve up to 5.76 Mbps in non-MIMO mode. The use of MIMO mode for HSUPA is not considered in Release-7 but it may be included in specifications later (most probably in Release-9). HSUPA introduces one new uplink transport channel:

- **E-DCH** – Enhanced Dedicated Channel – a transport channel bearing the uplink packet data;

and five new physical channels:

- **E-DPDCH** – E-DCH Dedicated Physical Data Channel – a physical channel carrying the E-DCH data,
- **E-DPCCH** – E-DCH Dedicated Physical Control Channel – a physical channel carrying the control data for E-DCH,
- **E-AGCH** – E–DCH Absolute Grant Channel – a downlink physical channel carrying the uplink E-DCH absolute grant used for scheduling control,

- **E-RGCH** – E-DCH Relative Grant Channel – a dedicated downlink physical channel carrying the uplink E-DCH relative grants used for scheduling control,
- **E-HICH** – E-DCH Hybrid ARQ Indicator Channel – a dedicated downlink physical channel carrying the uplink E-DCH hybrid ARQ acknowledgment indicator.

The main difference between HSUPA and HSDPA is that HSDPA is based on shared transport channel (HS-DSCH) and HSUPA of course as an uplink technology is based on dedicated transport channel (E-DCH). In spite of many common features with HS-DSCH, E-DCH is still the development of uplink DCH. Each UE has its own E-DCH and naturally the data transmitted over this channel cannot be shared as in HS-DSCH. Like all uplink channels E-DCH is under the power control. In the same way as uplink DCH the E-DPDCH and E-DPCCH are code-multiplexed and transmitted simultaneously in time. Due to this, the greater part of the UE's power is assigned to E-DPDCH, the higher the pay-load bit rate achievable on that channel but the less power is left for E-DPCCH and the less reliable is the signalling in the link. In uplink DCH the ratio between the power of DPDCH and DPCCH is set to a constant. In HSUPA the ratio between the power of E-DPDCH and E-DPCCH is controlled by the Node B. Thus, by adjusting this ratio the scheduler can control the E-DCH data rate. If for HSDPA the shared resource comprises the transmitter power (in one particular TTI the whole transmitted power of Node B can be allocated just to one UE) and the number of channel codes (all codes can be allocated to one UE or shared between several UEs in one TTI), for HSUPA the shared resource is the interference from UEs or that is the same, the total received power at Node B receiver.

Like HS-DSCH, the E-DCH also can employ 2 ms TTI, which helps to decrease the delays. Actually, E-DCH can use both 10 ms and 2 ms TTIs. The reason for keeping 10 ms TTIs is that the transmission with 2 ms TTIs when UE is in poor channel conditions (that is, close to the cell edge) requires much more energy than the transmission with 10 ms TTIs. Unlike HSDPA, HSUPA supports soft handover. The support of soft handover is quite important for E-DCH. First of all, when the UE is in soft handover mode the E-DCH is under the simultaneous power control from neighbouring cells that helps to decrease the interference. Moreover, the soft handover introduces a macro-diversity in transmitted uplink information.

In the same way as for HSDPA the different UE categories are defined for HSUPA. The HSUPA UE categories are listed in Table 11.33 [22].

Table 11.33 FDD E-DCH physical layer categories

E-DCH category	Maximum number of E-DCH codes transmitted	Minimum spreading factor	Support for 10 and 2 ms TTI EDCH	Maximum number of bits of an E-DCH transport block transmitted within a 10 ms E-DCH TTI	Maximum number of bits of an E-DCH transport block transmitted within a 2 ms E-DCH TTI
Category 1	1	SF4	10 ms TTI only	7110	–
Category 2	2	SF4	10 ms and 2 ms TTI	14484	2798
Category 3	2	SF4	10 ms TTI only	14484	–
Category 4	2	SF2	10 ms and 2 ms TTI	20000	5772
Category 5	2	SF2	10 ms TTI only	20000	–
Category 6	4	SF2	10 ms and 2 ms TTI	20000	11484
Category 7	4	SF2	10ms and 2 ms TTI	20000	22996

NOTE: When four codes are transmitted in parallel, two codes shall be transmitted with SF2 and two with SF4

UEs of categories 1 to 6 support QPSK only. Category 7 is not used in Release-7 and is reserved for later releases. UEs of category 7 must support both QPSK and 16-QAM. As can be seen from Table 11.33 the maximum data rate supported by UE of category 6 is 5.7 Mbps. In future UE of category 7 should double this data rate.

11.3.2.1 H-ARQ

The hybrid ARQ scheme used in HSUPA is similar to one used in HSDPA. In the same way as in HSDPA the soft combining of retransmissions with the original transmission is used also in HSUPA. Of course, in the case of HSUPA the (re)transmitted data is buffered and combined in Node B. Like in HSDPA both methods of combining Chase combining and incremental redundancy can be used. And like in HSDPA the N-channel SAW ARQ is implemented in HSUPA. However, there are some differences in H-ARQ implementation for HS-DSCH and E-DCH. Mainly these differences centre around timing. First of all, as mentioned above, two different TTI types (10 ms and 2 ms) can be used in HSUPA. In case of 10 ms TTI, four HARQ processes are configured, and in case of 2 ms TTI, eight HARQ processes are configured. Unlike HSDPA, the retransmissions and ACKs/NACKs in HSUPA are synchronous. That means in HSUPA there is no need to signal the number of (re)transmitted H-ARQ process implicitly like in HSDPA. And the round trip time is 40 ms for 10 ms TTI and 16 ms for 2 ms TTI. The ACK/NACK is sent in downlink over E-HICH. Depending on the length of E-DCH TTI the length of the ACK/NACK varies from 2 to 8 ms. The timing diagram for the case when E-DCH TTI is 10 ms is depicted in Figure 11.82a and for the case E-DCH TTI = 2 ms is depicted in Figure 11.82b.

For each data packet, the UE sends a *retransmission sequence number (RSN)* on E-DPCCH. For the original transmission the RSN value is 0, for the first retransmission it is 1 and is incremented with each consecutive retransmission. RSN together with frame number (or with sub-frame number in case of 2 ms TTI) defines the redundancy version (RV). The RV is used in a similar way as in HSDPA for data combining. In order to simplify the H-ARQ procedure and to limit the size of the RSN field in E-DPCCH

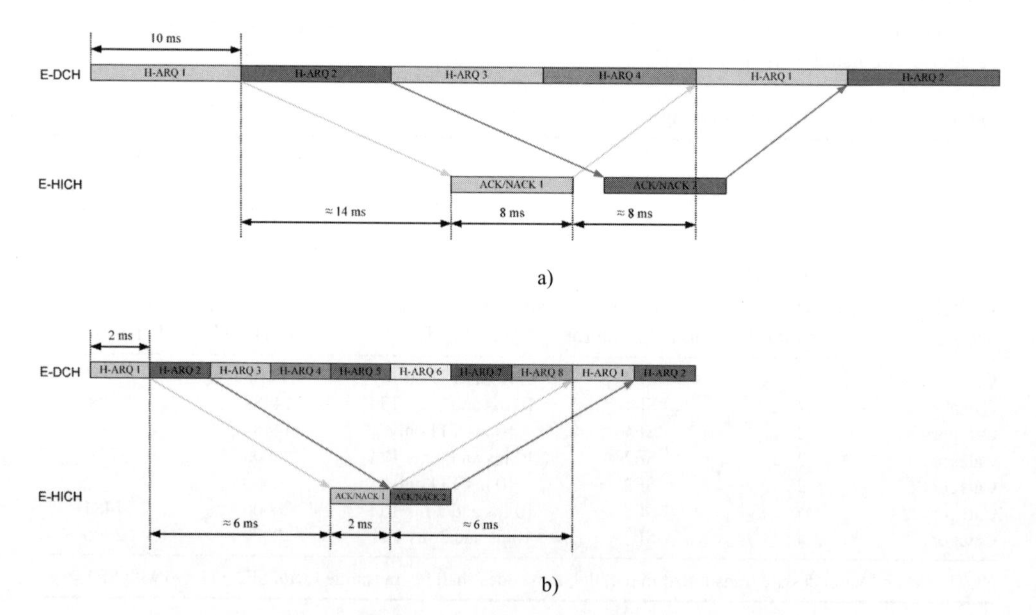

Figure 11.82 HSUPA H-ARQ: a) E-DCH TTI 10 ms, b) E-DCH TTI 2 ms

Table 11.34 Relation between RSN value and E-DCH RV Index

RSN Value	*Code rate* <1/2 E-DCH RV Index	1/2 ≤ *Code rate* E-DCH RV Index
0	0	0
1	2	3
2	0	2
3	$[\lfloor TTIN/N_{ARQ} \rfloor \bmod 2] \times 2$	$\lfloor TTIN/N_{ARQ} \rfloor \bmod 4$

the RSN value is restricted to the maximum of 3. That means, after the 3rd retransmission, the RSN is not incremented anymore and for all retransmissions after 3rd one the RSN = 3. The applied E-DCH RV index specifying the used RV depends on the RSN, on code rate, and if RSN = 3 also on the TTIN (TTI number). For 10 ms TTI the TTI number is equal to the connection frame number (CFN):

$$TTIN = CFN \tag{11.95}$$

for 2 ms TTI:

$$TTIN = 5 \cdot CFN + subframe\ number \tag{11.96}$$

where the subframe number counts the five TTIs which are within a given CFN, starting from 0 for the first TTI to 4 for the last TTI. The relation between RSN and RV index is described in Table 11.34 [16]

N_{ARQ} is the number of Hybrid ARQ processes.

However, if it is signalled by higher layers the UE can use only E-DCH RV index 0 independently of the RSN, that is, use Chase combining only.

11.3.2.2 Fast Scheduling

In general the fast scheduling is the feature that was brought to HSUPA from HSDPA. The task of both downlink and uplink scheduler is to control the shared resources in the served cell. The difference is that in HSDPA the shared resources controlled by the downlink scheduler are time slots and the number of channelization codes, while in HSUPA the shared resource controlled by the uplink scheduler is the uplink interference. If some UE needs a higher data rate, it should be allowed to transmit the E-DCH with higher power. On the other hand, that means this UE will contribute more interference to the cell. Thus, the uplink scheduler grants maximum allowed transmit power ratios to each UE. This effectively limits the uplink interference.

Like in HSDPA the scheduler is placed directly in Node B rather than in RNC, which decreases the delays. Moreover, moving the scheduling algorithm from the RNC to the Node B provides tighter control of the uplink interference which in turn may result in increased capacity and improved coverage. The main difference between the scheduling mechanisms in HSDPA and HSUPA is that the scheduling in HSUPA operates on a request-grant principle where the UE requests permission to send packets and the scheduler decides when and how many UEs will be allowed to do so. A request for transmission contains data about the state of the transmission buffer, the queue at the UE and its available power margin. The available uplink power determines the possible data rate. A *serving grant* represents the maximum E-DPDCH to DPCCH power ratio the UE may use in the next transmission. The serving grant is updated every TTI. The selection of TFC for E-DCH is done based on the serving grant. The scheduling mechanism is based on absolute and relative grants. The *absolute grants* are used to initialize the scheduling process and provide absolute transmit power ratios to the UE. The *relative grants* are used to increase or decrease the resource limitation compared to the previously used value.

Each absolute grant and relative grant is associated with a specific uplink E-DCH TTI, that is, HARQ process. This association is implicitly based on the timing of the E-AGCH and E-RGCH. The timing is tight enough that this relationship is un-ambiguous. When UE receives an absolute grant on the E-AGCH of the serving E-DCH cell it sets the serving grant value to the received value of the absolute grant. The absolute grant can be applied to all H-ARQ processes or per H-ARQ process. This is defined by 1 bit indicator included in absolute grant information sent on E-AGCH. However, if the E-DCH is configured with 10ms TTI, the absolute grant value can be applied only to all H-ARQ processes [30] If the UE has received an absolutc grant, it will ignore any serving cell E-RGCH commands for one H-ARQ cycle (40ms in the case of 10 ms TTI, 16ms in the case of 2 ms TTI). Relative grants can be sent from the serving cell and from non-serving neighbouring cells. A *serving relative grant* is transmitted on the E-RGCH from the serving cell. The serving relative grant allows the Node B scheduler to incrementally adjust the serving grant of UEs under its control. By definition, there can only be one serving relative grant command received at any one time. This indication can take three different values: "UP", "DOWN" or "HOLD". Non-serving relative grant transmits the E-RGCH from a non-serving E-DCH radio link. The *non-serving relative grant* allows neighbouring Node Bs to adjust the transmitted rate of UEs that are not under their control in order to avoid overload situations. By definition, there could be multiple nonserving relative grant commands received by MAC at any time. This indication can take two different values, "DOWN" or "HOLD" [30]. A relative grant is interpreted relative to the UE power ratio in the previous TTI for the same hybrid ARQ process as the transmission which the relative grant will affect as shown in Figure 11.83 [31].

If no data was transmitted at the same hybrid ARQ process in the previous TTI, the UE shall ignore the relative grant.

The handling of the relative grant signalling is based on the *scheduling grant (SG)* table configured by higher layers and shown in Tables 11.35 [20] and 11.36 [30].

The serving grant is updated based on the scheduling grant Table 11.35 or 11.36 as should be configured by higher layers. When UE receive the relative grant it first must determine the index in the configured SG-table close to the EDPDCH/DPCCH power ratio used for the previous TTI on this H-ARQ process. This index is called SG_{LUPR} (LUPR = Last Used Power Ratio). For interpretation of the "UP" serving relative grant, the network can configure the terminal with so-called "*3-index-step threshold*" and "*2-index-step threshold*" values. Then the serving grant is calculated as follows:

If SG_{LUPR} < "3-index-step threshold": Serving Grant = $SG[MIN(SG_{LUPR} + 3, 37)]$.

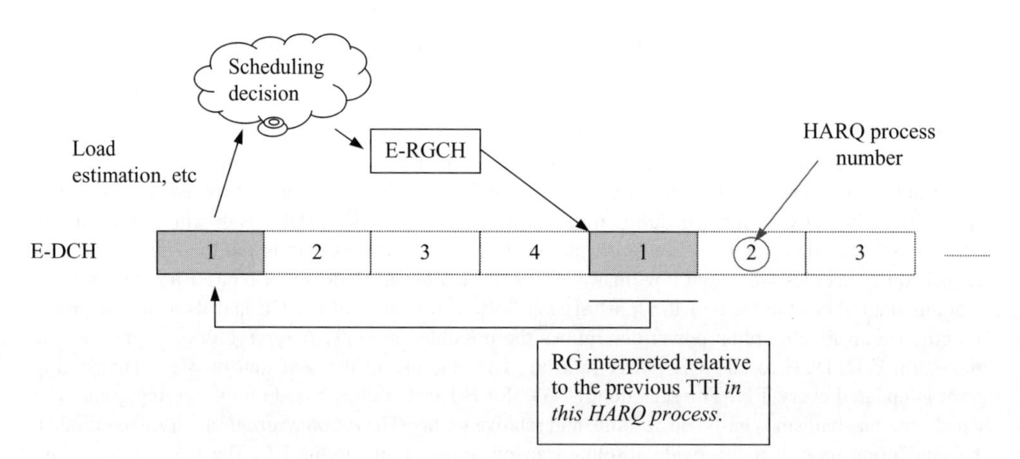

Figure 11.83 Timing relation for Relative Grant. Reproduced by permission of © 2006 3GPP™

Table 11.35 Scheduling Grant Table 1 (SG-table)

Index	Scheduled Grant
37	$(168/15)^2*6$
36	$(150/15)^2*6$
35	$(168/15)^2*4$
34	$(150/15)^2*4$
33	$(134/15)^2*4$
32	$(119/15)^2*4$
31	$(150/15)^2*2$
30	$(95/15)^2*4$
29	$(168/15)^2$
28	$(150/15)^2$
27	$(134/15)^2$
26	$(119/15)^2$
25	$(106/15)^2$
24	$(95/15)^2$
23	$(84/15)^2$
22	$(75/15)^2$
21	$(67/15)^2$
20	$(60/15)^2$
19	$(53/15)^2$
18	$(47/15)^2$
17	$(42/15)^2$
16	$(38/15)^2$
15	$(34/15)^2$
14	$(30/15)^2$
13	$(27/15)^2$
12	$(24/15)^2$
11	$(21/15)^2$
10	$(19/15)^2$
9	$(17/15)^2$
8	$(15/15)^2$
7	$(13/15)^2$
6	$(12/15)^2$
5	$(11/15)^2$
4	$(9/15)^2$
3	$(8/15)^2$
2	$(7/15)^2$
1	$(6/15)^2$
0	$(5/15)^2$

If "3-index-step threshold" $< =$ SG_{LUPR} $<$ "2-index-step threshold": Serving Grant $=$ $SG[MIN(SG_{LUPR} + 2, 37)]$.

If "2-index-step threshold" $< =$ SG_{LUPR}: Serving_Grant $= SG[MIN(SG_{LUPR} + 1, 37)]$.

If the UE received a serving relative grant or non-serving relative grant "DOWN", in both cases the serving grant is calculated as follows:

Serving Grant $= SG[MAX(SG_{LUPR} - 1, 0)]$.

This algorithm allows, with the help of E-RGCH commands, the UE to be directed to lower power ratio levels than can be obtained by using E-AGCH commands only. Actually, the absolute grants that are signalled on E-AGCH are a subset of the SG table.

Table 11.36 Scheduling Grant Table 2 (SG-table)

Index	Scheduled Grant
37	$(377/15)^2 \times 4$
36	$(336/15)^2 \times 4$
35	$(237/15)^2 \times 6$
34	$(212/15)^2 \times 6$
33	$(237/15)^2 \times 4$
32	$(168/15)^2*6$
31	$(150/15)^2*6$
30	$(168/15)^2*4$
29	$(150/15)^2 \times 4$
28	$(134/15)^2 \times 4$
27	$(119/15)^2 \times 4$
26	$(150/15)^2 \times 2$
25	$(95/15)^2 \times 4$
24	$(168/15)^2$
23	$(150/15)^2$
22	$(134/15)^2$
21	$(119/15)^2$
20	$(106/15)^2$
19	$(95/15)^2$
18	$(84/15)^2$
17	$(75/15)^2$
16	$(67/15)^2$
15	$(60/15)^2$
14	$(53/15)^2$
13	$(47/15)^2$
12	$(42/15)^2$
11	$(38/15)^2$
10	$(34/15)^2$
9	$(30/15)^2$
8	$(27/15)^2$
7	$(24/15)^2$
6	$(21/15)^2$
5	$(19/15)^2$
4	$(17/15)^2$
3	$(15/15)^2$
2	$(13/15)^2$
1	$(12/15)^2$
0	$(11/15)^2$

Network can configure each UE with a primary E-RNTI and a secondary E-RNTI. This can be used in cases when there are a group of UEs which only occasionally send large amounts of uplink data. Then the network can assign to this group of UEs a secondary E-RNTI in order to control this group together. Only if a specific UE needs to send more uplink data, the network would assign an individual primary grant to this UE with the primary E-RNTI. The CRC for primary and secondary E-RNTI differs and due to this fact the absolute grant sent on E-AGCH on serving cell to the group of UEs or to some specific UE can be unambiguously identified by the UE.

In order to schedule the resources correctly the scheduling algorithm needs the information about the amount of system resources needed by the UE and the amount of resources it can actually make use of. The UE sends the scheduling information as part of the MAC-e Protocol Data Unit (PDU) over the E-DPDCH. The scheduling information is transmitted periodically when there is a free space in E-DPDCH due to the quantization of the transport block sizes that can be supported or the transmission of this information can be triggered by some events (for example, when SG becomes too small to allow the reliable data transmission but the E-DCH buffers are not empty). The scheduling information includes the following fields:

- Highest priority Logical channel ID (HLID). The HLID field identifies unambiguously the highest priority logical channel with available data. If multiple logical channels exist with the highest priority, the one corresponding to the highest buffer occupancy will be reported. The length of the HLID is 4 bits.
- Total E-DCH Buffer Status (TEBS). The TEBS field identifies the total amount of data available across all logical channels for which reporting has been requested by the RRC and indicates the amount of data in number of bytes that is available for transmission and retransmission in RLC layer. The length of this field is 5 bits. The values taken by TEBS are shown in Table 11.37 [31].

Table 11.37 TEBS Values

Index	TEBS Value (bytes)
0	$TEBS = 0$
1	$0 < TEBS \leq 10$
2	$10 < TEBS \leq 14$
3	$14 < TEBS \leq 18$
4	$18 < TEBS \leq 24$
5	$24 < TEBS \leq 32$
6	$32 < TEBS \leq 42$
7	$42 < TEBS \leq 55$
8	$55 < TEBS \leq 73$
9	$73 < TEBS \leq 97$
10	$97 < TEBS \leq 129$
11	$129 < TEBS \leq 171$
12	$171 < TEBS \leq 228$
13	$228 < TEBS \leq 302$
14	$302 < TEBS \leq 401$
15	$401 < TEBS \leq 533$
16	$533 < TEBS \leq 708$
17	$708 < TEBS \leq 940$
18	$940 < TEBS \leq 1248$
19	$1248 < TEBS \leq 1658$
20	$1658 < TEBS \leq 2202$
21	$2202 < TEBS \leq 2925$
22	$2925 < TEBS \leq 3884$
23	$3884 < TEBS \leq 5160$
24	$5160 < TEBS \leq 6853$
25	$6853 < TEBS \leq 9103$
26	$9103 < TEBS \leq 12092$
27	$12092 < TEBS \leq 16062$
28	$16062 < TEBS \leq 21335$
29	$21335 < TEBS \leq 28339$
30	$28339 < TEBS \leq 37642$
31	$37642 < TEBS$

Table 11.38 HLBS Values

Index	HLBS values (%)
0	$0 < \text{HLBS} \leq 4$
1	$4 < \text{HLBS} \leq 6$
2	$6 < \text{HLBS} \leq 8$
3	$8 < \text{HLBS} \leq 10$
4	$10 < \text{HLBS} \leq 12$
5	$12 < \text{HLBS} \leq 14$
6	$14 < \text{HLBS} \leq 17$
7	$17 < \text{HLBS} \leq 21$
8	$21 < \text{HLBS} \leq 25$
9	$25 < \text{HLBS} \leq 31$
10	$31 < \text{HLBS} \leq 37$
11	$37 < \text{HLBS} \leq 45$
12	$45 < \text{HLBS} \leq 55$
13	$55 < \text{HLBS} \leq 68$
14	$68 < \text{HLBS} \leq 82$
15	$82 < \text{HLBS}$

- Highest priority Logical channel Buffer Status (HLBS). The HLBS field indicates the amount of data available from the logical channel identified by HLID, relative to the highest value of the buffer size range reported by TEBS when the reported TEBS index is not 31, and relative to 50000 bytes when the reported TEBS index is 31. The length of HLBS is 4 bits. The values taken by HLBS are shown in Table 11.38 [31]. In cases where the TEBS field is indicating index 0 (0 byte), the HLBS field shall indicate index 0.
- UE Power Headroom (UPH). The UPH field indicates the ratio of the maximum UE transmission power and the corresponding DPCCH code power. The length of UPH is 5 bits.

One more parameter reported by UE to scheduler is the so called "*happy bit*". Happy bit indicates that UE requests additional uplink resources. The happy bit is a single bit field taking two values: "Not Happy" and "Happy" indicating respectively whether the UE could use more resources or not. Unlike scheduling information happy bit is included in the control data transmitted on E-DPCCH.

11.3.2.3 HSUPA Channels

The example of timing of HSUPA channels for E-DCH TTI = 10 ms is depicted in Figure 11.84, and for E-DCH TTI = 2 ms in Figure 11.85.

E-DPDCH carrying E-DCH data and E-DPCCH carrying E-DCH control information are always transmitted simultaneously, except the case of DTX when E-DPCCH is transmitted without E-DPDCH. After E-DPDCH decoding the feedback information (ACK/NACK) is sent by Node B to UE over E-HICH. The transmit timing of the start of the E-HICH frame (in case of 10 ms TTI) or sub-frame (in case of 2 ms TTI) depends on E-DCH TTI length and downlink DPCH delay.

When the E-DCH TTI is 10 ms the E-HICH frame offset relative to P-CCPCH shall be $\tau_{\text{E-HICH,n}}$ chips with:

$$\tau_{E\text{-}HICH,n} = 5120 + 7680 \times \left\lfloor \frac{\left(\tau_{DPCH,n}/256\right) - 70}{30} \right\rfloor \tag{11.97}$$

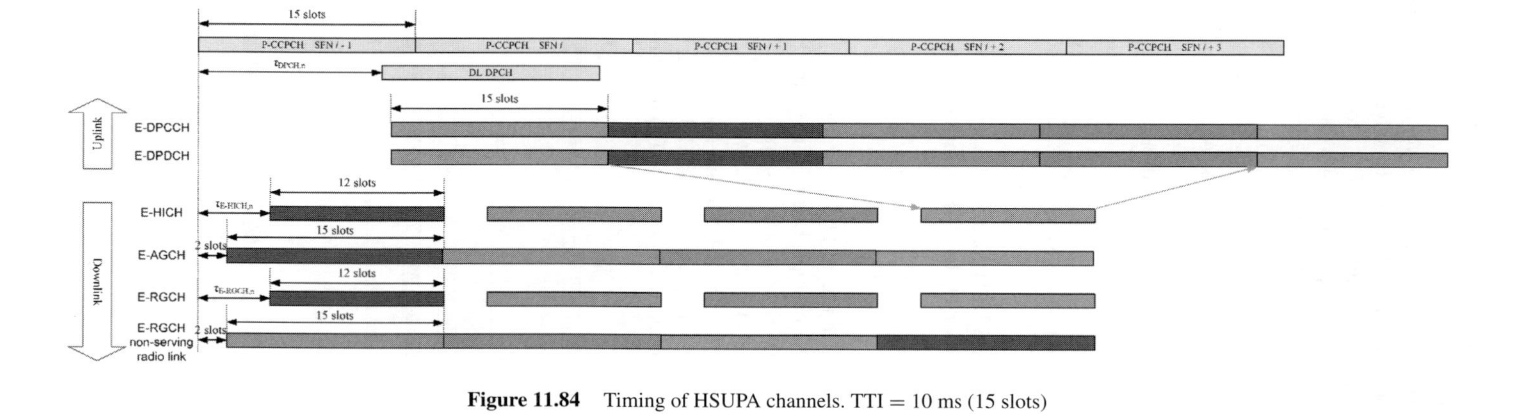

Figure 11.84 Timing of HSUPA channels. TTI = 10 ms (15 slots)

where $\tau_{DPCH,n}$ is the downlink DPCH offset relative to P-CCPCH, $\tau_{DPCH,n}$ is a multiple of 256 chips, that is, $\tau_{DPCH,n} = \mathrm{T_n} \times 256$ chips, $\mathrm{T_n} \in \{0, 1, \ldots, 149\}$.

The P-CCPCH, on which the cell *system frame number (SFN)* is transmitted, is used as timing reference for all the physical channels. For each cell in the E-DCH active set, the UE shall associate the control data received in the E-HICH frame associated with SFN *i* to the data transmitted in the E-DPDCH frame associated with SFN *i*-3. This can be seen in the example depicted in Figure 11.84: the E-HICH frame associated with SFN *i* +3 carries the ACK/NACK information about the E-DCH frame associated with SFN *i*. In the example depicted in Figure 11.84 $\tau_{DPCH,n} = 129 \times 256$ chips and correspondingly $\tau_{E\text{-}HICH,n} = 5$slots. Then the delay between the end of the E-DCH frame (TTI) and the beginning of the corresponding E-HICH frame comprises about 14 ms. Actually, this value can vary between 14 and 16 ms.

When the E-DCH TTI is 2 ms the E-HICH frame offset relative to P-CCPCH can be calculated in accordance with the following formula:

$$\tau_{E\text{-}HICH,n} = 5120 + 7680 \times \left\lfloor \frac{(\tau_{DPCH,n}/256) + 50}{30} \right\rfloor \tag{11.98}$$

In this case, the UE shall associate the E-DCH control data received in sub-frame *j* of the E-HICH frame associated with SFN *i* to sub-frame *t* of the E-DPDCH frame associated with SFN *i*-*s* where:

$$\begin{aligned} s &= 1 - \lfloor j/3 \rfloor \\ t &= (j+2) \bmod 5 \end{aligned} \tag{11.99}$$

The example for this case is depicted in Figure 11.85. For this example $\tau_{DPCH,n} = 9 \times 256$ chips and $\tau_{E\text{-}HICH,n} = 5$ slots. That means, for this example E-HICH sub-frame 3 associated with SFN *i* carries the feedback information referring to E-DCH sub-frame 0 of frame associated with SFN *i* ($s = 0, t = 0$). It is easy to verify that in case E-DCH TTI = 2 ms the delay between the end of the E-DCH sub-frame (TTI) and the beginning of the corresponding E-HICH sub-frame can vary from 6 to 8 ms. That means in case of 2 ms E-DCH TTI the processing time of E-DCH data is much more limited than in case of 10 ms E-DCH TTI. This is the price of decreasing the H-ARQ round trip time with 2 ms E-DCH TTI.

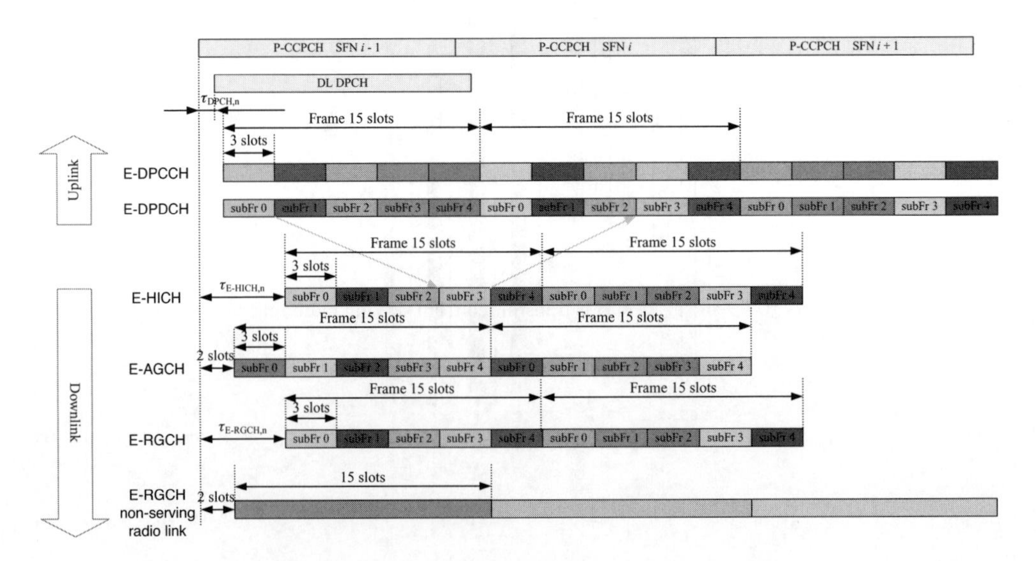

Figure 11.85 Timing of HSUPA channels. TTI = 2 ms (3 slots)

In order to initialize the serving grant the absolute grant is sent over E-AGCH. The E-AGCH frame offset relative to P-CCPCH is always equal to two slots. The timing of the E-AGCH depends on whether 2 ms TTI or 10 ms TTI is used.

If E-DCH TTI = 10 ms, the absolute grant received in the E-AGCH frame associated with SFN i must be used for the E-DCH transmission in the E-DPDCH frame associated with SFN $i+1+s$ where:

$$s = \left\lceil \frac{100 - \left(\tau_{DPCH,n}/256\right)}{150} \right\rceil \tag{11.100}$$

For the example depicted in Figure 11.84 $s = 0$ and as can be seen from this timing diagram the absolute grant received in SFN i is used for the transmission of the E-DCH frame associated with SFN $i+1$.

If E-DCH TTI = 2 ms, the absolute grant received in sub-frame j of the E-AGCH frame associated with SFN i corresponds to E-DCH transmission in sub-frame t of the E-DPDCH frame associated with SFN $i+s$ where:

$$s = \left\lfloor \frac{\left\lceil \frac{30j+100-\left(\tau_{DPCH,n}/256\right)}{30} \right\rceil}{5} \right\rfloor \tag{11.101}$$

and

$$t = \left\lceil \frac{30j + 100 - \left(\tau_{DPCH,n}/256\right) - 150s}{30} \right\rceil \tag{11.102}$$

In the example depicted in Figure 11.85, sub-frame 0 of the E-AGCH frame associated with SFN i is used for the E-DCH transmission in sub-frame 4 of the E-DPDCH frame associated with SFN i ($s = 0$, $t = 4$) as it is shown in timing diagram. The delay between reception of the absolute grant and start of the corresponding E-DCH (sub)frame varies from 5.6 ms to 15.5 ms in case of 10 ms E-DCH TTI and from 3.6 ms to 5.5 ms in case of 2 ms E-DCH TTI.

Every TTI the serving grant is updated with the help of relative grant sent over E-RGCH. The relative grant usually is sent from serving radio link, but in cases when it is necessary to decrease the interference in the neighbouring cell it is possible the relative grant can be sent from non-serving radio link either. The E-RGCH frame offset relative to P-CCPCH coincides with the frame offset of E-HICH if E-RGCH is transmitted from serving radio link set or with E-AGCH frame offset if E-RGCH is transmitted from non-serving radio link set. The timing of the E-AGCH depends on whether 2 ms TTI or 10 ms TTI is used, on downlink DPCH frame offset and on whether E-RGCH is transmitted from serving radio link set or from nonserving radio link.

If the E-DCH TTI is 10 ms and E-RGCH is transmitted from serving radio link set the E-RGCH frame offset relative to P-CCPCH is:

$$\tau_{E-RGCH,n} = 5120 + 7680 \times \left\lfloor \frac{\left(\tau_{DPCH,n}/256\right) - 70}{30} \right\rfloor \tag{11.103}$$

For each cell which belongs to the serving E-DCH radio link set, the relative grant received in the E-RGCH frame associated with SFN i is used in the E-DCH transmission in the E-DPDCH frame associated with SFN $i+1$. The delay between reception of an absolute grant and start of the corresponding E-DCH frame varies in this case from 5.6 ms to 7.5 ms. The example for this case is presented in Figure 11.84.

If 2 ms E-DCH TTI is used and E-RGCH is transmitted from serving radio link set the E-RGCH frame offset relative to P-CCPCH is:

$$\tau_{E-RGCH,n} = 5120 + 7680 \times \left\lfloor \frac{(\tau_{DPCH,n}/256) + 50}{30} \right\rfloor \quad (11.104)$$

In this case the relative grant received in sub-frame j of the E-RGCH frame associated with SFN i is used for E-DCH transmission in sub-frame j of the E-DPDCH frame associated with SFN i +1. The delay between reception of a relative grant and start of the corresponding E-DCH sub-frame in this case varies from 3.6 ms to 15.5 ms. The example for this case is presented in Figure 11.85.

If E-RGCH is transmitted from nonserving radio link set the E-RGCH frame offset relative to P-CCPCH is always 2 slots and 15 slots transmission is used for E-RGCH.

If 10 ms E-DCH TTI is used and E-RGCH is transmitted from non-serving radio link the relative grant received in the E-RGCH frame associated with SFN i shall be used for E-DCH transmission in the E-DPDCH frame associated with SFN i +1 + s where:

$$s = \left\lceil \frac{160 - (\tau_{DPCH,n}/256)}{150} \right\rceil \quad (11.105)$$

The delay between reception of a relative grant and start of the corresponding E-DCH frame in this case varies from 8.7 ms to 19.5 ms. The example for this case is presented in Figure 11.84. In this example the relative grant received from nonserving radio link in SFN i is used for E-DCH transmission in SFN i + 2.

If 2 ms E-DCH TTI is used and E-RGCH is transmitted from nonserving radio link the relative grant received in the E-RGCH frame associated with SFN i shall be used for E-DCH transmission in sub-frame t of the E-DPDCH frame associated with SFN i +1 + s where:

$$s = \left\lfloor \frac{\left\lceil \frac{160-(\tau_{DPCH,n}/256)}{30} \right\rceil}{5} \right\rfloor \quad (11.106)$$

and

$$t = \left\lceil \frac{160 - (\tau_{DPCH,n}/256) - 150s}{30} \right\rceil \quad (11.107)$$

The delay between reception of a relative grant and start of the corresponding E-DCH sub-frame in this case varies from 3.8 ms to 10.8 ms. The example for this case is presented in Figure 11.85. In this example the relative grant received from nonserving radio link in SFN i is used for E-DCH transmission in sub-frame 1 of frame corresponding to SFN i + 2.

When a downlink F-DPCH is configured, in all mentioned above cases $\tau_{DPCH,n} = \tau_{F-DPCH,n}$.

Now consider the HSUPA channels in more detail.

The E-DPDCH is used to carry the E-DCH transport channel. There may be zero, one, or several E-DPDCH on each radio link.

The E-DPCCH is a physical channel used to transmit control information associated with the E-DCH. There is at most one E-DPCCH on each radio link.

E-DPDCH and E-DPCCH are always transmitted simultaneously, except for the following cases when E-DPCCH is transmitted without E-DPDCH:

- when E-DPDCH but not E-DPCCH is DTXed due to power scaling, or
- during the n_{dtx} E-DPDCH idle slots if $n_{max} > n_{tx1}$.

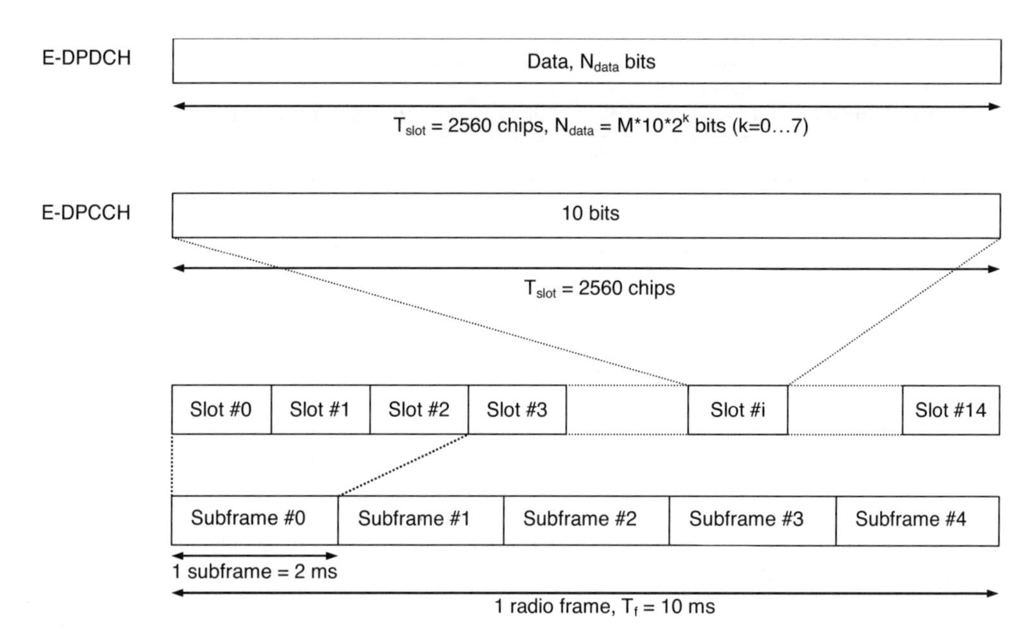

Figure 11.86 E-DPDCH frame structure

E-DPCCH shall not be transmitted in a slot unless DPCCH is also transmitted in the same slot.

Figure 11.86 [15] shows the E-DPDCH and E-DPCCH (sub)frame structure. Each radio frame is divided in five sub-frames, each of length 2 ms; the first sub-frame starts at the start of each radio frame and the fifth subframe ends at the end of each radio frame.

An E-DPDCH may use BPSK or 4PAM modulation symbols. In Figure 11.86, M is the number of bits per modulation symbol, that is, $M = 1$ for BPSK and $M = 2$ for 4PAM.

The E-DPDCH slot formats, corresponding rates and number of bits are specified in Table 11.39 [15]. The E-DPCCH slot format is listed in Table 11.40 [15].

The E-DCH Absolute Grant Channel (E-AGCH) is a fixed rate (30 kbps, SF = 256) downlink physical channel carrying the uplink E-DCH absolute grant. Figure 11.87 [15] illustrates the frame and sub-frame structure of the E-AGCH.

Table 11.39 E-DPDCH slot formats

Slot Format #i	Channel Bit Rate (kbps)	Bits/Symbol M	SF	Bits/ Frame	Bits/ Subframe	Bits/Slot N_{data}
0	15	1	256	150	30	10
1	30	1	128	300	60	20
2	60	1	64	600	120	40
3	120	1	32	1200	240	80
4	240	1	16	2400	480	160
5	480	1	8	4800	960	320
6	960	1	4	9600	1920	640
7	1920	1	2	19200	3840	1280
8	1920	2	4	19200	3840	1280
9	3840	2	2	38400	7680	2560

Table 11.40 E-DPCCH slot formats

Slot Format #i	Channel Bit Rate (kbps)	SF	Bits/ Frame	Bits/ Subframe	Bits/Slot N_{data}
0	15	256	150	30	10

An E-DCH absolute grant shall be transmitted over one E-AGCH sub-frame or one E-AGCH frame. The transmission over one E-AGCH sub-frame and over one E-AGCH frame shall be used for UEs for which E-DCH TTI is set to 2 ms and 10 ms respectively. The scope of the absolute grant tells the UE whether the absolute grant is valid for a specific H-ARQ process or for all H-ARQ processes.

The E-DCH Relative Grant Channel (E-RGCH) is a fixed rate (SF = 128) dedicated downlink physical channel carrying the uplink E-DCH relative grants. Figure 11.88 illustrates the structure of the E-RGCH. A relative grant is transmitted using 3, 12 or 15 consecutive slots and in each slot a sequence of 40 ternary values is transmitted. The 3 and 12 slot duration shall be used on an E-RGCH transmitted to UEs for which the cell transmitting the E-RGCH is in the E-DCH serving radio link set and for which the E-DCH TTI is respectively 2 and 10 ms. The 15 slot duration shall be used on an E-RGCH transmitted to UEs for which the cell transmitting the E-RGCH is not in the E-DCH serving radio link set.

The sequence $b_{i,0}, b_{i,1}, \ldots, b_{i,39}$ transmitted in slot i in Figure 11.88 is given by:

$$b_{i,j} = a \cdot C_{ss,40,m(i),j} \tag{11.108}$$

where a is the relative grant and $C_{ss,40,m(i),j}$ is the orthogonal signature sequence. In a serving E-DCH radio link set, the relative grant a is set to $+1$ (UP), 0 (HOLD), or -1 (DOWN) and in a radio link not belonging to the serving E-DCH radio link set, the relative grant a is set to 0 (HOLD) or -1 (DOWN). This differentiation has been made to allow only the serving radio link set to increase the uplink power. Non-serving radio links can only decrease the uplink power in case too high interference. The orthogonal signature sequences $C_{ss,40,m(i)}$ are given by Table 11.41 [15] and the index $m(i)$ in slot i is given by Table 11.42 [15]. The E-RGCH signature sequence index l in Table 11.42 is given by higher layers.

In cases where STTD-based open loop transmit diversity is applied for E-RGCH, STTD encoding is applied to the sequence $b_{i,j}$. as described in section 11.1.8.

The E-DCH Hybrid ARQ Indicator Channel (E-HICH) is a fixed rate (SF = 128) dedicated downlink physical channel carrying the uplink E-DCH hybrid ARQ acknowledgment indicator. The structure of the E-HICH coincides with structure of the E-RGCH and is represented in Figure 11.88. A hybrid ARQ acknowledgment indicator is transmitted using 3 or 12 consecutive slots and in each slot a sequence of

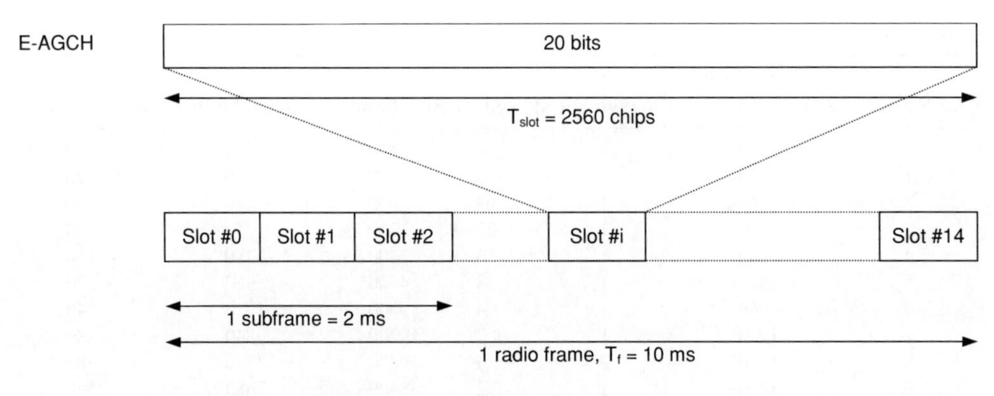

Figure 11.87 E-AGCH (sub)frame structure

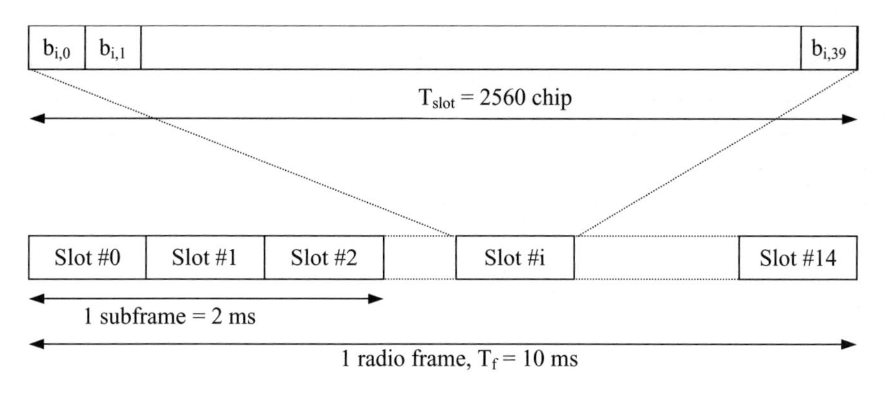

Figure 11.88 E-RGCH and E-HICH structure

40 binary values is transmitted. The 3 and 12 slot duration shall be used for UEs which E-DCH TTI is set to 2 ms and 10 ms respectively.

The sequence $b_{i,0}, b_{i,1}, \ldots, b_{i,39}$ transmitted in slot i in Figure 11.88 is given by (11.108), where a is the hybrid ARQ acknowledgment indicator. In a radio link set containing the serving E-DCH radio link set, the hybrid ARQ acknowledgment indicator a is set to + 1 or –1, and in a radio link set not containing the serving E-DCH radio link set the hybrid ARQ indicator a is set to + 1 or 0. The orthogonal signature sequences $C_{ss,40,m(i)}$ are given by Table 11.41 [15] and the index $m(i)$ in slot i is given by Table 11.42 [15]. The E-HICH signature sequence index l is given by higher layers.

In cases where STTD-based open loop transmit diversity is applied for E-HICH, STTD encoding according to section 11.1.8 is applied to the sequence $b_{i,j}$.

11.3.2.3.1 E-DPCCH Multiplexing and Coding

The following information is transmitted by means of the E-DPCCH:

- Retransmission sequence number (RSN): $x_{rsn,1}, x_{rsn,2}$
- E-TFCI: $x_{tfci,1}, x_{tfci,2}, \ldots, x_{tfci,7}$
- "Happy" bit: $x_{h,1}$

Figure 11.89 [16] illustrates the overall coding chain for E-DPCCH.

To indicate the redundancy version (RV) of each HARQ transmission and to assist the Node B soft buffer management a two bit retransmission sequence number (RSN) is signalled from the UE to the Node B. The Node B can avoid soft buffer corruption by flushing the soft buffer associated to one HARQ process in case more than three consecutive E-DPCCH transmissions on that HARQ process can not be decoded or the last received RSN is incompatible with the current one.

The relation between RSN and applied E-DCH RV index are specified in Table 11.34.

The UE shall set $x_{h,1}$ to '1' if value of happy bit is "Happy" and to '0' if value of happy bit is "Not Happy".

The E-TFCI $x_{tfci,1}, x_{tfci,2}, \ldots, x_{tfci,7}$, the retransmission sequence number $x_{rsn,1}, x_{rsn,2}$ and the "happy" bit $x_{h,1}$ are multiplexed together. This gives a sequence of bits $x_1, x_2, \ldots, x_{10}$ where:

$$x_k = x_{h,1} \qquad k = 1$$
$$x_k = x_{rsn,4-k} \qquad k = 2, 3$$
$$x_k = x_{tfci,11-k} \qquad k = 4, 5, \ldots, 10$$

Table 11.41 E-RGCH and E-HICH signature sequences

$C_{ss,40,0}$	−1	−1	−1	1	−1	1	−1	−1	1	1	−1	−1	1	−1	1	1	−1	1	1	−1	−1	−1	−1	−1	−1	−1	−1	1	−1	1	−1	−1	1	1	1	1	1	−1	−1	−1
$C_{ss,40,1}$	−1	1	1	−1	−1	1	1	1	−1	−1	1	−1	1	1	−1	−1	−1	−1	1	1	1	−1	−1	−1	−1	1	−1	−1	−1	−1	−1	−1	−1	1	1	−1	1	1	−1	−1
$C_{ss,40,2}$	−1	−1	−1	1	−1	1	1	1	−1	−1	−1	−1	1	−1	−1	1	1	−1	−1	1	1	−1	1	1	1	−1	−1	1	1	1	−1	1	−1	−1	−1	−1	−1	−1	−1	−1
$C_{ss,40,3}$	1	−1	−1	−1	−1	−1	−1	1	1	1	−1	1	−1	1	−1	1	−1	−1	1	1	−1	1	−1	−1	1	1	−1	1	−1	−1	1	1	−1	−1	1	−1	−1	−1	−1	−1
$C_{ss,40,4}$	1	1	1	−1	−1	1	−1	1	−1	−1	1	1	1	−1	1	1	1	1	1	1	−1	1	1	1	−1	−1	−1	1	1	−1	1	−1	1	−1	1	1	−1	1	−1	−1
$C_{ss,40,5}$	−1	1	−1	−1	1	1	1	−1	1	1	−1	1	1	1	−1	1	1	1	−1	−1	1	−1	−1	1	−1	1	−1	1	−1	−1	1	−1	1	−1	−1	−1	−1	1	1	−1
$C_{ss,40,6}$	1	1	−1	−1	−1	1	1	−1	1	1	−1	−1	1	−1	−1	−1	−1	1	1	−1	1	1	1	−1	1	−1	1	−1	1	−1	−1	1	1	−1	1	−1	−1	1	−1	1
$C_{ss,40,7}$	−1	1	−1	1	1	1	−1	−1	−1	−1	−1	1	1	1	1	−1	−1	−1	1	−1	−1	−1	1	−1	1	1	−1	−1	1	1	1	1	−1	−1	1	1	−1	1	1	−1
$C_{ss,40,8}$	1	1	−1	1	1	−1	1	1	1	1	−1	−1	−1	−1	1	−1	1	−1	1	1	1	1	−1	1	−1	−1	−1	−1	−1	1	−1	−1	−1	−1	1	1	−1	1	1	−1
$C_{ss,40,9}$	−1	1	−1	−1	−1	−1	1	−1	−1	−1	−1	1	−1	−1	1	1	1	−1	1	−1	1	−1	−1	1	1	−1	1	1	−1	−1	1	1	−1	1	1	1	1	1	−1	1
$C_{ss,40,10}$	−1	1	1	−1	1	1	−1	1	1	1	1	−1	1	−1	1	1	−1	−1	−1	1	−1	−1	−1	−1	1	−1	1	1	−1	−1	−1	1	−1	−1	−1	1	−1	1	1	1
$C_{ss,40,11}$	−1	1	−1	−1	−1	−1	−1	1	1	1	−1	−1	−1	1	1	−1	1	1	−1	1	−1	−1	1	1	1	1	−1	−1	1	−1	−1	1	1	1	−1	1	1	1	−1	−1
$C_{ss,40,12}$	−1	−1	−1	−1	1	−1	1	1	−1	−1	−1	−1	−1	1	1	1	−1	1	1	1	1	−1	1	−1	−1	1	1	1	1	−1	−1	−1	1	−1	1	1	−1	−1	1	1
$C_{ss,40,13}$	1	1	1	1	−1	−1	1	−1	−1	−1	1	−1	−1	1	1	1	−1	1	−1	−1	1	1	−1	−1	1	1	−1	1	−1	1	−1	1	1	−1	−1	1	−1	1	−1	−1
$C_{ss,40,14}$	−1	1	1	1	−1	−1	−1	−1	1	1	1	−1	−1	1	−1	1	1	−1	1	−1	−1	−1	1	1	−1	1	1	1	1	1	−1	−1	−1	−1	1	−1	−1	1	−1	1
$C_{ss,40,15}$	−1	−1	1	1	−1	1	1	1	1	1	1	1	1	1	−1	1	1	1	1	1	1	−1	−1	1	1	1	1	−1	−1	1	1	1	1	−1	1	1	−1	−1	−1	1
$C_{ss,40,16}$	1	−1	−1	−1	−1	1	−1	−1	−1	−1	−1	−1	1	1	1	−1	1	−1	−1	−1	−1	1	−1	1	−1	1	1	−1	−1	−1	−1	−1	−1	−1	−1	1	−1	−1	−1	1
$C_{ss,40,17}$	1	−1	1	−1	1	1	1	−1	1	1	1	−1	1	1	1	1	1	−1	1	−1	1	1	1	1	1	1	−1	1	1	−1	−1	1	−1	1	1	1	1	−1	1	−1
$C_{ss,40,18}$	1	1	−1	1	−1	1	1	1	1	1	−1	1	1	1	1	1	−1	−1	−1	1	1	1	1	−1	−1	1	1	1	1	1	1	−1	−1	1	−1	1	1	1	−1	1
$C_{ss,40,19}$	1	1	−1	1	1	1	−1	1	−1	−1	−1	−1	1	1	−1	1	1	1	1	1	−1	1	−1	1	1	1	1	1	−1	1	−1	1	1	1	1	−1	1	1	1	1
$C_{ss,40,20}$	1	1	1	−1	1	1	−1	1	−1	1	−1	1	−1	−1	−1	1	−1	−1	1	−1	1	−1	−1	1	1	1	1	−1	1	1	−1	−1	1	1	−1	1	−1	−1	−1	−1
$C_{ss,40,21}$	−1	1	1	−1	−1	−1	−1	1	−1	1	−1	−1	1	−1	−1	1	1	−1	−1	−1	1	1	1	−1	−1	1	−1	−1	−1	1	1	1	1	−1	1	1	1	−1	1	1
$C_{ss,40,22}$	−1	−1	−1	1	−1	−1	−1	1	−1	1	1	−1	1	1	−1	−1	−1	−1	1	−1	1	1	−1	1	1	−1	−1	1	1	−1	1	−1	1	1	−1	1	−1	1	1	1
$C_{ss,40,23}$	1	−1	−1	−1	−1	1	1	1	1	−1	1	1	−1	−1	−1	−1	1	−1	−1	−1	−1	−1	1	−1	1	1	−1	1	−1	1	−1	−1	1	1	1	1	−1	1	1	1
$C_{ss,40,24}$	−1	−1	−1	1	1	1	−1	−1	1	−1	1	−1	−1	−1	−1	1	1	−1	1	1	1	1	1	−1	1	1	1	−1	−1	−1	1	−1	1	−1	−1	1	1	1	−1	−1
$C_{ss,40,25}$	−1	1	−1	−1	1	−1	−1	−1	1	−1	1	1	1	−1	−1	−1	−1	1	1	1	1	1	1	1	−1	1	−1	1	−1	1	−1	1	−1	1	−1	1	−1	−1	−1	1
$C_{ss,40,26}$	−1	−1	1	1	1	1	1	1	−1	1	−1	1	−1	−1	1	−1	−1	−1	1	−1	−1	1	1	1	−1	1	−1	1	−1	−1	−1	1	1	−1	−1	−1	1	1	−1	1
$C_{ss,40,27}$	1	−1	1	−1	−1	1	−1	1	1	−1	−1	−1	−1	1	−1	−1	−1	1	1	−1	1	−1	1	1	−1	−1	1	1	−1	1	1	1	−1	−1	−1	1	1	1	1	−1
$C_{ss,40,28}$	1	1	−1	1	1	1	−1	1	1	−1	1	−1	−1	1	1	1	−1	−1	−1	−1	1	−1	1	1	−1	−1	−1	−1	−1	−1	1	1	1	1	1	−1	−1	−1	−1	1
$C_{ss,40,29}$	−1	1	−1	−1	−1	1	−1	−1	−1	1	1	1	−1	1	1	−1	−1	−1	−1	1	1	1	1	1	1	−1	1	1	−1	1	−1	−1	1	−1	1	−1	1	−1	1	−1
$C_{ss,40,30}$	−1	1	1	−1	1	−1	1	1	1	−1	−1	−1	1	1	1	−1	1	−1	1	−1	−1	1	1	−1	1	−1	1	1	−1	1	1	−1	1	1	−1	−1	−1	−1	−1	−1
$C_{ss,40,31}$	−1	1	−1	−1	−1	1	1	1	1	−1	1	−1	−1	−1	1	1	−1	1	1	−1	−1	1	−1	1	1	1	−1	−1	1	1	1	−1	−1	−1	−1	−1	1	−1	1	1
$C_{ss,40,32}$	1	1	1	1	−1	−1	1	−1	1	−1	−1	1	1	1	−1	1	−1	−1	1	1	−1	−1	1	1	1	−1	−1	−1	−1	−1	−1	−1	1	−1	−1	1	1	−1	1	1
$C_{ss,40,33}$	−1	−1	−1	−1	1	−1	1	1	1	−1	1	1	1	1	−1	1	−1	−1	−1	−1	−1	1	−1	1	−1	−1	1	−1	1	1	−1	1	1	−1	1	1	1	1	−1	−1
$C_{ss,40,34}$	1	−1	−1	−1	1	−1	−1	1	−1	1	1	1	1	1	1	1	1	1	1	−1	1	−1	1	−1	1	−1	−1	−1	−1	1	−1	−1	−1	−1	−1	−1	1	1	−1	1
$C_{ss,40,35}$	−1	−1	1	1	−1	−1	−1	1	1	−1	−1	1	1	−1	1	1	−1	1	−1	−1	1	1	1	1	1	1	1	−1	−1	−1	−1	−1	−1	1	1	−1	−1	1	1	−1
$C_{ss,40,36}$	−1	1	1	1	1	1	−1	1	1	−1	−1	1	−1	1	−1	−1	1	1	−1	−1	1	1	−1	−1	1	−1	−1	1	1	−1	−1	−1	−1	−1	1	1	1	−1	−1	1
$C_{ss,40,37}$	1	−1	1	−1	1	−1	−1	−1	1	−1	−1	−1	1	−1	1	−1	−1	−1	−1	1	1	−1	−1	1	1	1	−1	1	1	1	1	−1	1	−1	1	−1	1	1	−1	1
$C_{ss,40,38}$	−1	−1	1	−1	1	1	1	−1	−1	1	−1	−1	−1	1	−1	1	−1	1	−1	1	−1	1	1	1	1	−1	−1	−1	−1	1	1	−1	−1	1	1	1	−1	1	−1	1
$C_{ss,40,39}$	−1	−1	1	−1	−1	1	−1	−1	1	−1	−1	1	−1	1	1	1	1	−1	1	1	1	1	−1	−1	−1	−1	−1	−1	1	1	−1	1	1	1	−1	−1	−1	1	1	1

Table 11.42 E-HICH and E-RGCH signature hopping pattern

	Row index $m(i)$ for slot i		
Sequence index l	i mod 3 = 0	i mod 3 = 1	i mod 3 = 2
0	0	2	13
1	1	18	18
2	2	8	33
3	3	16	32
4	4	13	10
5	5	3	25
6	6	12	16
7	7	6	1
8	8	19	39
9	9	34	14
10	10	4	5
11	11	17	34
12	12	29	30
13	13	11	23
14	14	24	22
15	15	28	21
16	16	35	19
17	17	21	36
18	18	37	2
19	19	23	11
20	20	39	9
21	21	22	3
22	22	9	15
23	23	36	20
24	24	0	26
25	25	5	24
26	26	7	8
27	27	27	17
28	28	32	29
29	29	15	38
30	30	30	12
31	31	26	7
32	32	20	37
33	33	1	35
34	34	14	0
35	35	33	31
36	36	25	28
37	37	10	27
38	38	31	4
39	39	38	6

Channel coding of the E-DPCCH is done using a sub-code of the second order Reed-Muller code. Coding is applied to the output $x_1, x_2, \ldots, x_{10}$ from the E-DPCCH multiplexing, resulting in:

$$z_i = \sum_{n=0}^{9} (x_{n+1} \times M_{i,n}) \bmod 2 \quad i = 0, 1, \ldots, 29 \tag{11.109}$$

where $M_{i,n}$ is the ith bit of nth codeword from the generator matrix of Reed-Muller code. The generator matrix of (32, 10) Reed-Muller code is listed in Table 11.7.

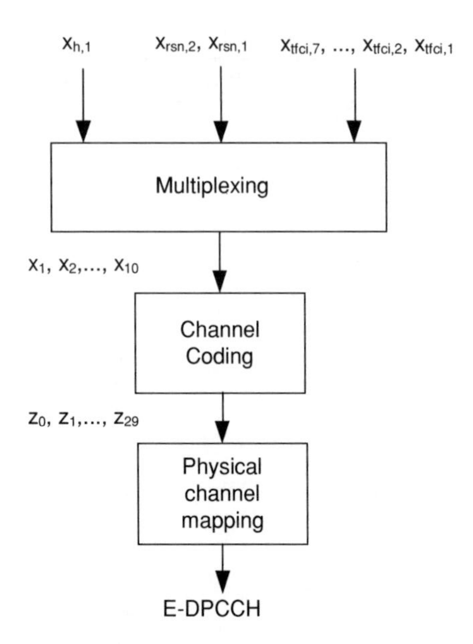

Figure 11.89 Coding chain for E-DPCCH. Reproduced by permission of © 2008 3GPP™

The sequence of bits $z_0, z_1, \ldots, z_{29}$ output from the E-DPCCH channel coding is mapped to the corresponding E-DPCCH sub frame. The bits are mapped so that they are transmitted over the air in ascending order with respect to i. If the E-DCH TTI is equal to 10 ms the sequence of bits is transmitted in all the E-DPCCH sub frames of the E-DPCCH radio frame.

For compressed frames in the uplink and the case when E-DCH TTI length is 10 ms, the bits mapped to the E-DPCCH idle slots shall not be transmitted.

11.3.2.3.2 E-DCH Multiplexing and Coding

Figure 11.90 [16] shows the processing structure for the E-DCH transport channel mapped onto a separate CCTrCH. Data arrives to the coding unit in form of a maximum of one transport block once every transmission time interval (TTI), that is, the number of transport blocks per TTI and the number of transport channels is always one.

CRC attached to the E-DCH transport channel always has the length of 24 bits and is calculated as described in section 11.1.4.1.

Code block segmentation for the E-DCH transport channel is done in the same way as described in section 11.1.3.1 with some specific restrictions: there is a maximum of one transport block, and only turbo coding with block size 5114 shall be used.

The channel coding for E-DCH is performed with the help of turbo coder of rate 1/3 depicted in Figure 11.27.

The hybrid ARQ functionality matches the number of bits at the output of the channel coder to the total number of bits of the E-DPDCH set to which the E-DCH transport channel is mapped. The hybrid ARQ functionality is controlled by the redundancy version (RV) parameters. The H-ARQ functionality is depicted in Figure 11.91 [16].

First of all the parameters for the rate matching pattern determination algorithm are calculated (for details see section 4.8.4.1 and section 4.5.4.3 in [16]).

The parameters of the rate matching stage depend on the value of the RV parameters s and r. The s and r combinations corresponding to each RV allowed for the E-DCH are listed in the Table 11.43.

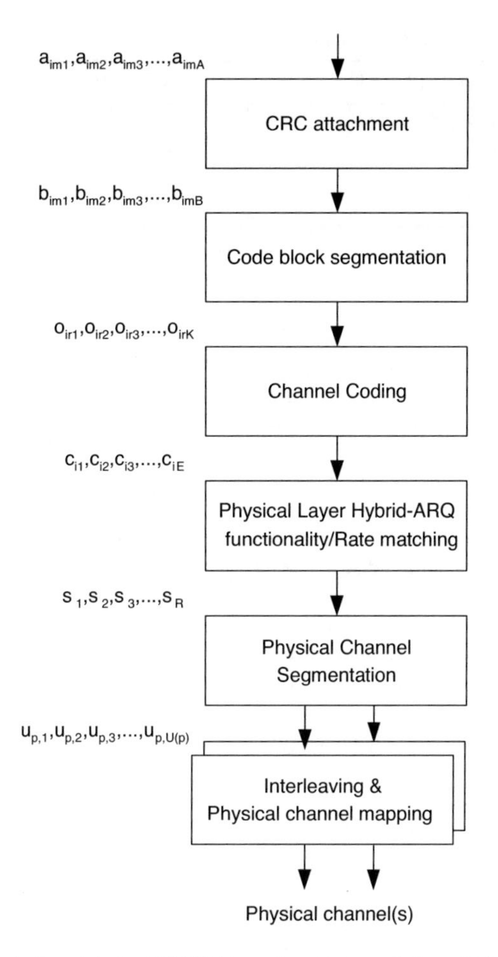

Figure 11.90 Coding chain for E-DCH. Reproduced by permission of © 2008 3GPP™

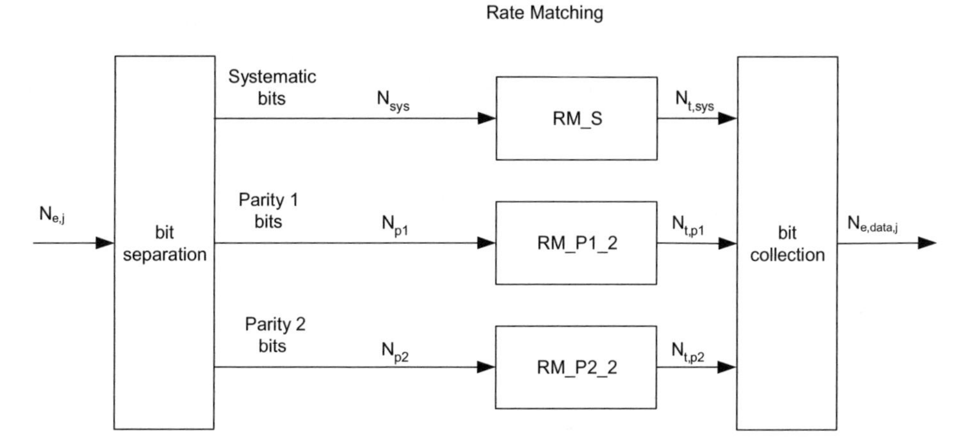

Figure 11.91 E-DCH hybrid ARQ functionality. Reproduced by permission of © 2008 3GPP™

Table 11.43 RV for E-DCH

E-DCH RV Index	s	r
0	1	0
1	0	0
2	1	1
3	0	1

Parameter s defines if systematic bits are prioritized in rate matching algorithm. If $s = 1$ then the redundancy version is "self-decodable", that is, it contains all systematic bits, if $s = 0$ the corresponding redundancy version may contain only part of systematic bits or parity bits only. Parameter r defines the rate matching pattern. The HARQ bit separation function is performed in the same way as bit separation for turbo encoded TrCHs with puncturing described in section 11.2.3.1. The rate matching pattern determination algorithm described in section 11.2.3.1 is used for rate matching. The HARQ bit collection function is performed in the same way as bit collection for turbo encoded TrCHs with puncturing described in section 11.2.3.1.

After rate matching the physical channel segmentation is done. When more than one E-DPDCH is used, physical channel segmentation distributes the bits among the different physical channels. The bits input to the physical channel segmentation are denoted by $s_1, s_2, s_3, \ldots, s_R$, where R is the number of bits input to the physical channel segmentation block. The number of PhCHs is denoted by P.

The bits after physical channel segmentation are denoted $u_{p,k}$ where p is the PhCH number. $U(p)$ is the number of physical channel bits in one E-DCH TTI for the pth E-DPDCH. The relation between s_k and $u_{p,k}$ is given below.

Bits on first PhCH after physical channel segmentation:

$$u_{1,k} = s_k \qquad k = 1, 2, \ldots, U(1) \tag{11.110}$$

Bits on p^{th} PhCH after physical channel segmentation are given by:

$$u_{p,k} = s_{k+\sum_{q=1}^{p-1} U(q)}, \qquad k = 1, 2, \ldots, U(p) \tag{11.111}$$

The interleaving for E-DCH is done as shown in Figure 11.92 [16], separately for each physical channel. The bits input to the block interleaver are denoted by $u_{p,1}, u_{p,2}, u_{p,3}, \ldots, u_{p,U}$, where p is PhCH number and $U = U(p)$ is the number of bits in one TTI for one PhCH.

The basic interleaver is as the second interleaver described in section 11.2.3.1. However, for 4PAM, there are two identical interleavers of the same size R2 × 30, where R2 is the minimum integer fulfilling $\lceil U/2 \rceil \leq R2 \times 30$. The output bits from the physical channel segmentation are divided one by one between the interleavers: bit $u_{p,k}$ goes to the first interleaver and bit $u_{p,k+1}$ goes to the second interleaver.

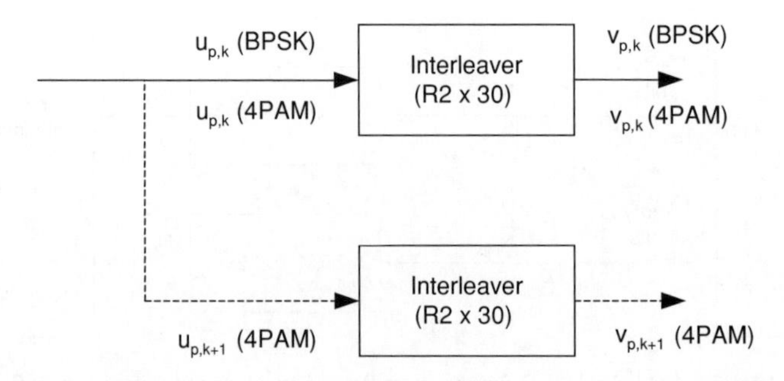

Figure 11.92 Interleaver structure for E-DCH. Reproduced by permission of © 2008 3GPP™

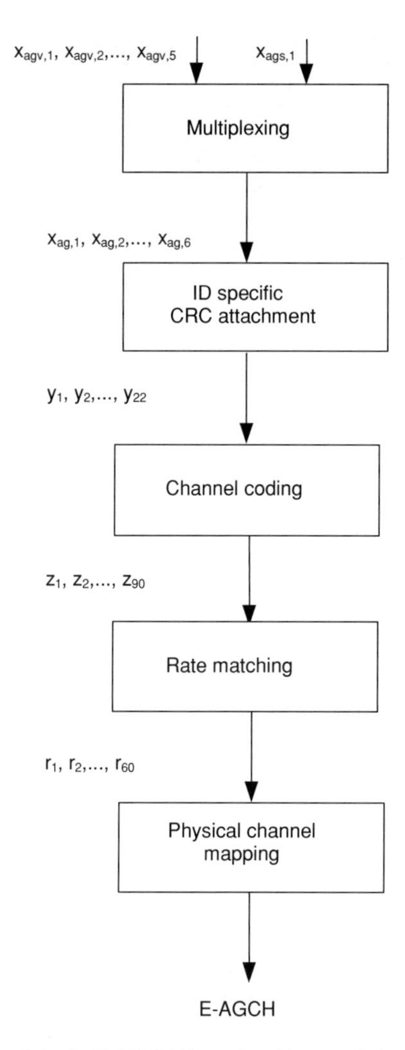

Figure 11.93 Coding chain for E-AGCH. Reproduced by permission of © 2008 3GPP™

Bits are collected one by one from the interleavers: bit $v_{p,k}$ is obtained from the first interleaver and bit $v_{p,k+1}$ is obtained from the second interleaver, where k mod 2 = 1.

The bits input to the physical channel mapping are denoted $v_{p,1}, v_{p,2}, \ldots, v_{p,U(p)}$. The bits $v_{p,k}$ are mapped to the physical channels such that the bits for each physical channel are transmitted over the air in ascending order with respect to k.

11.3.2.3.3 E-AGCH Multiplexing and Coding

The absolute grant channel (E-AGCH) is bearing the following information:

- Absolute Grant Value: $x_{agv,1}, x_{agv,2}, \ldots, x_{agv,5}$
- Absolute Grant Scope: $x_{ags,1}$

Figure 11.93 [16] illustrates the overall coding chain for the E-AGCH.

Table 11.44 Mapping of Absolute Grant Value

Absolute Grant Value	Index
$(168/15)^2 \times 6$	31
$(150/15)^2 \times 6$	30
$(168/15)^2 \times 4$	29
$(150/15)^2 \times 4$	28
$(134/15)^2 \times 4$	27
$(119/15)^2 \times 4$	26
$(150/15)^2 \times 2$	25
$(95/15)^2 \times 4$	24
$(168/15)^2$	23
$(150/15)^2$	22
$(134/15)^2$	21
$(119/15)^2$	20
$(106/15)^2$	19
$(95/15)^2$	18
$(84/15)^2$	17
$(75/15)^2$	16
$(67/15)^2$	15
$(60/15)^2$	14
$(53/15)^2$	13
$(47/15)^2$	12
$(42/15)^2$	11
$(38/15)^2$	10
$(34/15)^2$	9
$(30/15)^2$	8
$(27/15)^2$	7
$(24/15)^2$	6
$(19/15)^2$	5
$(15/15)^2$	4
$(11/15)^2$	3
$(7/15)^2$	2
ZERO_GRANT*	1
INACTIVE*	0

The Absolute Grant Value information is specified in Table 11.44 [16] and Table 11.45 [16]. Based on higher layer signalling, either Table 11.44 or Table 11.45 is selected. Obviously the Tables 11.44 and 11.45 are the subsets of Tables 11.35 and 11.36.

The value of $x_{ags,1}$ is set to '1' if the scope of the absolute grant is "Per HARQ process" and to '0' if the scope is "All HARQ processes".

The Absolute Grant Value information $x_{agv,1}, x_{agv,2}, \ldots, x_{agv,5}$ and the Absolute Grant Scope information $x_{ags,1}$ are multiplexed together. This gives a sequence of bits $x_{ag,1}, x_{ag,2}, \ldots, x_{ag,6}$ where:

$$x_{ag,k} = x_{agv,k} \qquad k = 1, 2, \ldots, 5$$
$$x_{ag,k} = x_{ags,7-k} \qquad k = 6$$

From the sequence of bits $x_{ag,1}, x_{ag,2}, \ldots, x_{ag,6}$ a 16 bit CRC is calculated. That gives the sequence of CRC bits $c_1, c_2, \ldots, c_{16}$. This sequence of bits is then masked with *E-DCH Radio Network Identifier (E-RNTI)* $x_{id,1}, x_{id,2}, \ldots, x_{id,16}$. The E-RNTI plays the same role in HSUPA as H-RNTI in HSDPA: it is the unique identifier of the UE. Then the masked CRC is appended to the sequence of bits $x_{ag,1}, x_{ag,2}, \ldots,$

Table 11.45 Alternative Mapping of Absolute Grant Value

Absolute Grant Value	Index
$(377/15)^2 \times 4$	31
$(237/15)^2 \times 6$	30
$(168/15)^2*6$	29
$(150/15)^2*6$	28
$(168/15)^2*4$	27
$(150/15)^2 \times 4$	26
$(134/15)^2 \times 4$	25
$(119/15)^2 \times 4$	24
$(150/15)^2 \times 2$	23
$(95/15)^2 \times 4$	22
$(168/15)^2$	21
$(150/15)^2$	20
$(134/15)^2$	19
$(119/15)^2$	18
$(106/15)^2$	17
$(95/15)^2$	16
$(84/15)^2$	15
$(75/15)^2$	14
$(67/15)^2$	13
$(60/15)^2$	12
$(53/15)^2$	11
$(47/15)^2$	10
$(42/15)^2$	9
$(38/15)^2$	8
$(34/15)^2$	7
$(30/15)^2$	6
$(27/15)^2$	5
$(24/15)^2$	4
$(19/15)^2$	3
$(15/15)^2$	2
ZERO_GRANT*	1
INACTIVE*	0

$x_{ag,6}$ to form the sequence of bits $y_1, y_2, \ldots, y_{22}$ where:

$$\begin{aligned} y_i &= x_{ag,i} & i &= 1, 2, \ldots, 6 \\ y_i &= (c_{i-6} + x_{ag,i-6}) \bmod 2 & i &= 7, 8, \ldots, 22 \end{aligned} \tag{11.112}$$

In the next step rate 1/3 convolutional coding, as described in section 11.2.4.2 is applied to the sequence of bits $y_1, y_2, \ldots, y_{22}$, resulting in the sequence of bits $z_1, z_2, \ldots, z_{90}$.

Then from the input sequence $z_1, z_2, \ldots, z_{90}$ the bits z_1, z_2, z_5, z_6, z_7, z_{11}, z_{12}, z_{14}, z_{15}, z_{17}, z_{23}, z_{24}, z_{31}, z_{37}, z_{44}, z_{47}, z_{61}, z_{63}, z_{64}, z_{71}, z_{72}, z_{75}, z_{77}, z_{80}, z_{83}, z_{84}, z_{85}, z_{87}, z_{88}, z_{90} are punctured to obtain the output sequence $r_1, r_2, \ldots, r_{60}$.

After that the sequence of bits $r_1, r_2, \ldots, r_{60}$ is mapped to the corresponding E-AGCH sub frame. The bits r_k are mapped so that they are transmitted over the air in ascending order with respect to k. If the E-DCH TTI is equal to 10 ms the same sequence of bits is transmitted in all the E-AGCH sub frames of the E-AGCH radio frame.

Table 11.46 Mapping of RG value

Command	RG Value (serving E-DCH RLS)	RG Value (other radio links)
UP	+1	not allowed
HOLD	0	0
DOWN	−1	−1

11.3.2.3.4 E-RGCH and E-HICH Multiplexing

The relative grant channel (E-RGCH) is bearing the relative grant commands. The relative grant command is mapped to the relative grant value as described in Table 11.46 [16].

The coding of the relative grant value is performed in accordance with (11.108).

The E-DCH H-ARQ acknowledgment indicator channel (E-HICH) is bearing the ACK/NACK command. The ACK/NACK command is mapped to the H-ARQ acknowledgment indicator as described in Table 11.47 [16].

The coding of the H-ARQ acknowledgment indicator is performed in accordance with (11.108).

11.3.2.3.5 HSUPA Spreading and Modulation

Figure 11.94 illustrates the spreading operation for the E-DPDCHs and the E-DPCCH.

The E-DPCCH shall be spread to the chip rate by the channelization code c_{ec}. The kth E-DPDCH, denominated E-DPDCH$_k$, shall be spread to the chip rate using channelization code $c_{ed,k}$.

After channelization, the real-valued spread E-DPCCH and E-DPDCH$_k$ signals shall respectively be weighted by gain factor β_{ec} and $\beta_{ed,k}$. The power of the E-DPCCH and the E-DPDCH(s) is set in relation to the DPCCH. For this purpose, gain factors are used for scaling the uplink channels relative to each other. The E-DPCCH gain factor computation depends on the transmitted E-TFC at a given TTI.

In non compressed frames, if E-TFCI$_i$ is smaller than or equal to E-TFCI$_{ec,boost}$, where E-TFCI$_i$ denotes the E-TFCI of the i:th E-TFC, the E-DPCCH gain factor, β_{ec} is calculated according to:

$$\beta_{ec} = \beta_c \cdot A_{ec} \tag{11.113}$$

where β_c value is signalled by higher-layers or calculated as described in section 11.1.4.3.1, A_{ec} is the quantized amplitude ratio which is translated from $\Delta_{\text{E-DPCCH}}$ signalled by higher layers. The translation of $\Delta_{\text{E-DPCCH}}$ into quantized amplitude ratios $A_{ec} = \beta_{ec}/\beta_c$ is specified in Table 11.48 [17]. The E-TFCI$_{ec,boost}$ value is signalled by higher layers.

When E-TFCI $>$ E-TFCI$_{ec,boost}$, in order to provide an enhanced phase reference, the value of β_{ec} shall be derived as follows. Firstly, the unquantized E-DPCCH gain factor for the i:th E-TFC, $\beta_{ec,i,uq}$, is

Table 11.47 Mapping of HARQ Acknowledgement

Command	HARQ acknowledgement indicator
ACK	+1
NACK (RLSs not containing the serving E-DCH cell)	0
NACK (RLS containing the serving E-DCH cell)	−1

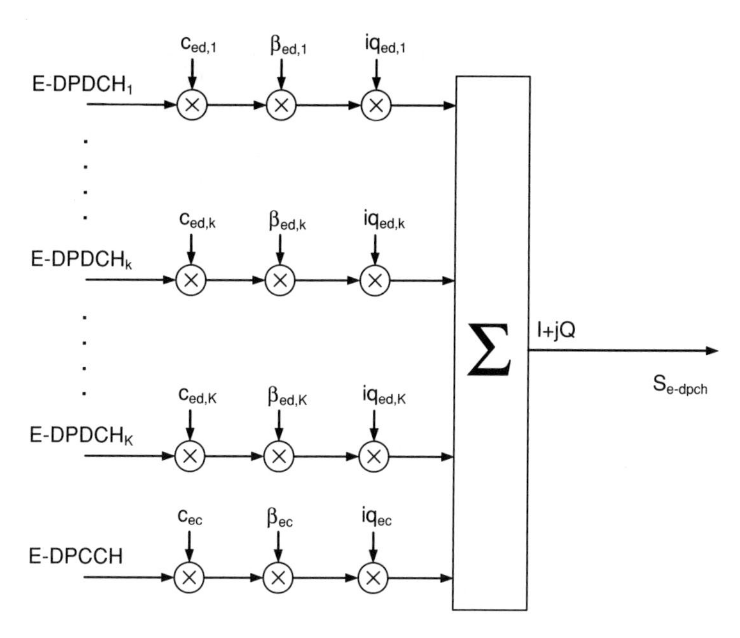

Figure 11.94 Spreading for E-DPDCH/E-DPCCH

calculated according to:

$$\beta_{ec,i,uq} = \beta_c \cdot \sqrt{\max\left(A_{ec}^2, \frac{\sum_{k=1}^{k_{\max,i}} \left(\frac{\beta_{ed,i,k}}{\beta_c}\right)^2}{10^{\frac{\Delta_{T2TP}}{10}}} - 1\right)} \tag{11.114}$$

where Δ_{T2TP} is the traffic to total pilot power offset, configured by higher layers as specified in Table 11.49 [17], $\beta_{ed,i,k}$ is the E-DPDCH beta gain factor for the i:th E-TFC on the kth physical channel and $k_{max,i}$ is the number of physical channels used for the i:th E-TFC.

Table 11.48 Quantization for $\Delta_{\text{E-DPCCH}}$ for E-TFCI $\leq$ $E\text{-}TFCI_{ec,boost}$

Signalled values for $\Delta_{\text{E-DPCCH}}$	Quantized amplitude ratios $A_{ec} = \beta_{ec}/\beta_c$
8	30/15
7	24/15
6	19/15
5	15/15
4	12/15
3	9/15
2	8/15
1	6/15
0	5/15

Table 11.49 Δ_{T2TP}

Signalled values for Δ_{T2TP}	Power offset values Δ_{T2TP} [dB]
6	16
5	15
4	14
3	13
2	12
1	11
0	10

If $\beta_{ec,i,uq}$ is less than the smallest quantized value of Table 11.50 [17], then the E-DPCCH gain factor β_{ec} is set such that β_{ec}/β_c is the smallest quantized value of Table 11.50. Otherwise, β_{ec} is selected from Table 11.50, such that 20*log10(β_{ec}/β_c) is the nearest quantized value to 20*log10($\beta_{ec,i,uq}/\beta_c$).

During compressed frames, the E-DPCCH gain factor β_{ec} should be scaled. This is done in order to avoid the E-DPCCH power being increased by the offset that is applied to the DPCCH during compressed frames. The reason for applying this offset is that the uplink DPCCH slot formats that have TFCI bits contain fewer pilot bits than the formats for normal (non-compressed) mode. This is due to the fact that the number of TFCI bits shall always be the same during a frame to ensure robust transport format detection. Therefore, in order to keep the same channel quality the energy of the pilot must be kept equal, and the power of the DPCCH must therefore be increased by the factor $N_{pilot,C}/N_{pilot,N}$, where $N_{pilot,C}$ is the number of pilot bits per slot on the DPCCH in compressed frames, and $N_{pilot,N}$ is the number of pilot bits per slot in non-compressed frames.

Table 11.50 Quantization for β_{ec}/β_c for E-TFCI > $E\text{-}TFCI_{ec,boost}$

Quantized amplitude ratios β_{ec}/β_c	E-DPDCH modulation schemes which may be used in the same subframe
239/15	4PAM
190/15	4PAM
151/15	4PAM
120/15	BPSK, 4PAM
95/15	BPSK, 4PAM
76/15	BPSK, 4PAM
60/15	BPSK, 4PAM
48/15	BPSK, 4PAM
38/15	BPSK, 4PAM
30/15	BPSK, 4PAM
24/15	BPSK, 4PAM
19/15	BPSK, 4PAM
15/15	BPSK, 4PAM
12/15	BPSK, 4PAM
9/15	BPSK
8/15	BPSK, 4PAM
6/15	BPSK, 4PAM
5/15	BPSK

During compressed frames where the E-DCH TTI is 2msec and if E-TFCI$_i$ is smaller than or equal to E-TFCI$_{ec,boost}$, the E-DPCCH gain factor, β_{ec} is calculated according to:

$$\beta_{ec} = \beta_{c,C,j} \cdot A_{ec} \cdot \sqrt{\frac{N_{pilot,C}}{N_{pilot,N}}} \tag{11.115}$$

and if E-TFCI$_i$ is greater than E-TFCI$_{ec,boost}$ according to:

$$\beta_{ec,i,uq} = \beta_{c,C,j} \cdot \sqrt{\max\left(A_{ec}^2, \frac{\sum_{k=1}^{k_{\max,i}} \left(\frac{\beta_{ed,i,k}}{\beta_c}\right)^2}{10^{\frac{\Delta_{T2TP}}{10}}} - 1\right)} \cdot \sqrt{\frac{N_{pilot,C}}{N_{pilot,N}}} \tag{11.116}$$

where $\beta_{c,C,j}$ is calculated as described in section 11.2.4.3.1.

If a 10 ms TTI is used in a compressed mode, the E-DPCCH gain factor β_{ec} must be additionally scaled by $\sqrt{\frac{15}{N_{slots,C}}}$, where $N_{slots,C}$ is the number of non DTX slots in the compressed frame. This is done to take into account that there are fewer slots available for transmission during the compressed mode.

Then during compressed frames where the E-DCH TTI is 10msec and if E-TFCI$_i$ is smaller than or equal to E-TFCI$_{ec,boost}$, the E-DPCCH gain factor, β_{ec} is calculated according to:

$$\beta_{ec} = \beta_{c,C,j} \cdot A_{ec} \cdot \sqrt{\frac{15 \cdot N_{pilot,C}}{N_{slots,C} \cdot N_{pilot,N}}} \tag{11.117}$$

and if E-TFCI$_i$ is greater than E-TFCI$_{ec,boost}$ according to:

$$\beta_{ec,i,uq} = \beta_{c,C,j} \cdot \sqrt{\max\left(A_{ec}^2, \frac{\sum_{k=1}^{k_{\max,i}} \left(\frac{\beta_{ed,i,k}}{\beta_c}\right)^2}{10^{\frac{\Delta_{T2TP}}{10}}} - 1\right)} \cdot \sqrt{\frac{15 \cdot N_{pilot,C}}{N_{slots,C} \cdot N_{pilot,N}}} \tag{11.118}$$

The E-DPDCH gain factor β_{ed} may take a different value for each E-TFC and H-ARQ offset. There can be one or several E-DPDCH(s) configured. The gain factor for each E-DPDCH is defined by the E-TFC and H-ARQ power offset. The H-ARQ offsets are used for support of different H-ARQ profiles created by higher layers for different data flows. The E-DPDCH gain factor β_{ed} for each E-DPDCH is computed based on reference gain factor(s) $\beta_{ed,ref}$ of E-TFC(s) signalled as reference E-TFC(s). At least one E-TFC of the set of E-TFCs configured by the network shall be signalled as a reference E-TFC. The gain factors may vary on radio frame basis or sub-frame basis depending on the E-DCH TTI used. Furthermore, the setting of gain factors is independent of the inner loop power control.

For each reference E-TFC, a reference gain factor $\beta_{ed,ref}$ is calculated according to:

$$\beta_{ed,ref} = \beta_c \cdot A_{ed} \tag{11.119}$$

The quantized amplitude ratio $A_{ed} = \beta_{ed,ref}/\beta_c$ is translated from $\Delta_{\text{E-DPDCH}}$ signalled by higher layers. The translation of $\Delta_{\text{E-DPDCH}}$ into quantized amplitude ratios A_{ed} is specified in Table 11.51 [17] for the case when E-TFCI $\leq$ E-TFCI$_{ec,boost}$ and Table 11.52 [17] for the case when E-TFCI $>$ E-TFCI$_{ec,boost}$.

The gain factor β_{ed} of an E-TFC is compute based on the signalled settings for its corresponding reference E-TFC. It is possible to use two methods of calculation the gain factor β_{ed}: *E-DPDCH power extrapolation formula* or *E-DPDCH power interpolation formula*. In cases where E-DPDCH power extrapolation formula is used only one reference E-TFC is chosen from the list of signalled as reference E-TFC(s). And corresponding to the chosen reference E-TFC value of reference gain factor $\beta_{ed,ref}$ is used for calculation of β_{ed}. In cases where E-DPDCH power interpolation formula is used two reference

Table 11.51 Quantization for $\Delta_{\text{E-DPDCH}}$ for E-TFCI $\leq$ $E\text{-}TFCI_{ec,boost}$

Signalled values for Δ E-DPDCH	Quantized amplitude ratios $A_{ed} = \beta_{ed}/\beta_c$	E-DPDCH modulation schemes which may be used in the same subframe
29	168/15	BPSK
28	150/15	BPSK
27	134/15	BPSK
26	119/15	BPSK
25	106/15	BPSK
24	95/15	BPSK
23	84/15	BPSK
22	75/15	BPSK
21	67/15	BPSK
20	60/15	BPSK
19	53/15	BPSK, 4PAM
18	47/15	BPSK, 4PAM
17	42/15	BPSK, 4PAM
16	38/15	BPSK, 4PAM
15	34/15	BPSK, 4PAM
14	30/15	BPSK, 4PAM
13	27/15	BPSK, 4PAM
12	24/15	BPSK, 4PAM
11	21/15	BPSK, 4PAM
10	19/15	BPSK, 4PAM
9	17/15	BPSK
8	15/15	BPSK
7	13/15	BPSK
6	12/15	BPSK
5	11/15	BPSK
4	9/15	BPSK
3	8/15	BPSK
2	7/15	BPSK
1	6/15	BPSK
0	5/15	BPSK

E-TFCs is chosen from the list and two corresponding values of reference gain factors $\beta_{ed,ref}$: primary and secondary is needed for calculation of β_{ed}. Whether E-DPDCH power extrapolation formula or E-DPDCH power interpolation formula is used to compute the gain factor β_{ed} is signalled by higher layers.

Let $E\text{-}TFCI_{ref,m}$ denote the E-TFCI of the m:th reference E-TFC, where $m = 1,2,\ldots,M$ and M is the number of signalled reference E-TFCs and $E\text{-}TFCI_{ref,1} < E\text{-}TFCI_{ref,2} < \ldots < E\text{-}TFCI_{ref,M}$. Let $E\text{-}TFCI_i$ denote the E-TFCI of the i:th E-TFC.

For the i:th E-TFC:

If E-DPDCH power extrapolation formula is configured
if $E\text{-}TFCI_i \geq E\text{-}TFCI_{ref,M}$, the reference E-TFC is the M:th reference E-TFC.
if $E\text{-}TFCI_i < E\text{-}TFCI_{ref,1}$, the reference E-TFC is the 1st reference E-TFC.
if $E\text{-}TFCI_{ref,1} \leq E\text{-}TFCI_i < E\text{-}TFCI_{ref,M}$,
the reference E-TFC is the m:th reference E-TFC such that
$E\text{-}TFCI_{ref,m} \leq E\text{-}TFCI_i < E\text{-}TFCI_{ref,m+1}$.

Table 11.52 Quantization for $\Delta_{\text{E-DPDCH}}$ for E-TFCI > *E-TFCI_{ec,boost}*

Signalled values for $\Delta_{\text{E-DPDCH}}$	Quantized amplitude ratios $A_{ed} = \beta_{ed}/\beta_c$	E-DPDCH modulation schemes which may be used in the same subframe
31	377/15	4PAM (applicable only for SF2 code in a 2xSF2 + 2xSF4 configuration)
30	336/15	4PAM (applicable only for SF2 code in a 2xSF2 + 2xSF4 configuration)
29	299/15	4PAM
28	267/15	BPSK (applicable only for SF2 code in a 2xSF2 + 2xSF4 configuration), 4PAM
27	237/15	BPSK (applicable only for SF2 code in a 2xSF2 + 2xSF4 configuration), 4PAM
26	212/15	BPSK, 4PAM
25	189/15	BPSK, 4PAM
24	168/15	BPSK, 4PAM
23	150/15	BPSK, 4PAM
22	134/15	BPSK, 4PAM
21	119/15	BPSK, 4PAM
20	106/15	BPSK, 4PAM
19	95/15	BPSK, 4PAM
18	84/15	BPSK, 4PAM
17	75/15	BPSK, 4PAM
16	67/15	BPSK, 4PAM
15	60/15	BPSK, 4PAM
14	53/15	BPSK, 4PAM
13	47/15	BPSK, 4PAM
12	42/15	BPSK, 4PAM
11	38/15	BPSK
10	34/15	BPSK
9	30/15	BPSK
8	27/15	BPSK
7	24/15	BPSK
6	21/15	BPSK
5	19/15	BPSK
4	17/15	BPSK
3	15/15	BPSK
2	13/15	BPSK
1	11/15	BPSK
0	8/15	BPSK

Else If E-DPDCH power interpolation formula is configured

if $E\text{-}TFCI_i \geq E\text{-}TFCI_{ref,M}$,

the primary and secondary reference E-TFCs are the $(M$-1):th and M:th reference E-TFCs respectively.

if $E\text{-}TFCI_i < E\text{-}TFCI_{ref,1}$,

the primary and secondary reference E-TFCs are the 1st and 2nd reference E-TFCs respectively.

if $E\text{-}TFCI_{ref,1} \leq E\text{-}TFCI_i < E\text{-}TFCI_{ref,M}$,

the primary and secondary reference E-TFCs are the m:th and $(m + 1)$:th reference E-TFCs respectively, such that

$E\text{-}TFCI_{ref,m} \leq E\text{-}TFCI_i < E\text{-}TFCI_{ref,m+1}$.

End If

Table 11.53 HARQ offset Δ_{harq}

Signalled values for Δ_{harq}	Power offset values Δ_{harq} [dB]
6	6
5	5
4	4
3	3
2	2
1	1
0	0

The reference gain factor cannot be directly used for scaling the E-DPDCHs, since except for the number of data bits contained in reference E-TFC, the number of E-DPDCHs used for transmission and the H-ARQ profile should be also taken in account. Due to this, for the *i*th E-TFC, the temporary variable $\beta_{ed,i,harq}$ is computed.

When E-DPDCH power interpolation formula is configured, the temporary variable $\beta_{ed,i,harq}$ is calculated in accordance with:

$$\beta_{ed,i,harq} = \beta_{ed,ref}\sqrt{\frac{L_{e,ref}}{L_{e,i}}}\sqrt{\frac{K_{e,i}}{K_{e,ref}}} \cdot 10^{\left(\frac{\Delta harq}{20}\right)} \tag{11.120}$$

where $L_{e,ref}$ is the number of E-DPDCHs used for the reference E-TFC and $L_{e,i}$ is the number of E-DPDCHs used for the *i*th E-TFC, $K_{e,ref}$ is the transport block size of the reference E-TFC and $K_{e,i}$ is the transport block size of the *i*th E-TFC and the H-ARQ offset Δ_{harq} used for support of different H-ARQ profiles are configured by higher layers as specified in Table 11.53 [17].

If SF2 is used, $L_{e,ref}$ and $L_{e,i}$ are the equivalent number of physical channels assuming SF4. That means, in 2 x SF2 case $L_{e,ref}$ and $L_{e,i}$, should be 4 instead of 2, and in 2 x SF2 + 2 x SF4 case $L_{e,ref}$ and $L_{e,i}$ should be 6 instead of 4. Due to this, the calculated $\beta_{ed,i,harq}$ must be scaled by $\sqrt{2}$ in case SF = 2.

When E-DPDCH power interpolation formula is configured, the temporary variable $\beta_{ed,i,harq}$ is computed as:

$$\beta_{ed,i,harq} = \sqrt{\frac{L_{e,ref,1}}{L_{e,i}}} \cdot \sqrt{\left(\left(\frac{\frac{L_{e,ref,2}}{L_{e,ref,1}}\beta^2_{ed,ref,2} - \beta^2_{ed,ref,1}}{K_{e,ref,2} - K_{e,ref,1}}\right)\left(K_{e,i} - K_{e,ref,1}\right) + \beta^2_{ed,ref,1}\right)} \cdot 10^{\left(\frac{\Delta harq}{20}\right)} \tag{11.121}$$

where $\beta_{ed,ref,1}$ and $\beta_{ed,ref,2}$ denote the reference gain factors of the primary and secondary reference E-TFCs respectively, $L_{e,ref,1}$ and $L_{e,ref,2}$ are the number of E-DPDCHs used for the primary and secondary reference E-TFCs respectively, $L_{e,i}$ is the number of E-DPDCHs used for the *i*th E-TFC, $K_{e,ref,1}$ and $K_{e,ref,2}$ are the transport block sizes of the primary and secondary reference E-TFCs respectively, $K_{e,i}$ is the transport block size of the *i*th E-TFC. If SF2 is used, $L_{e,ref,1}$, $L_{e,ref,2}$ and $L_{e,i}$ are the equivalent number of physical channels assuming SF4. There is one exception in calculation of (11.121): if the following inequality holds:

$$\left(\frac{\frac{L_{e,ref,2}}{L_{e,ref,1}}\beta^2_{ed,ref,2} - \beta^2_{ed,ref,1}}{K_{e,ref,2} - K_{e,ref,1}}\right)\left(K_{e,i} - K_{e,ref,1}\right) + \beta^2_{ed,ref,1} \leq 0 \tag{11.122}$$

the temporary variable $\beta_{ed,i,harq}$ should be set to 0.

Table 11.54 Quantization for $\beta_{ed,k}/\beta_c$ for E-TFCI $\leq$ *E-TFCI$_{ec,boost}$*

Quantized amplitude ratios $\beta_{ed,k}/\beta_c$	E-DPDCH modulation schemes which may be used in the same subframe
168/15	BPSK
150/15	BPSK
134/15	BPSK
119/15	BPSK
106/15	BPSK
95/15	BPSK
84/15	BPSK
75/15	BPSK
67/15	BPSK
60/15	BPSK
53/15	BPSK, 4PAM
47/15	BPSK, 4PAM
42/15	BPSK, 4PAM
38/15	BPSK, 4PAM
34/15	BPSK, 4PAM
30/15	BPSK, 4PAM
27/15	BPSK, 4PAM
24/15	BPSK, 4PAM
21/15	BPSK, 4PAM
19/15	BPSK, 4PAM
17/15	BPSK
15/15	BPSK
13/15	BPSK
12/15	BPSK
11/15	BPSK
9/15	BPSK
8/15	BPSK
7/15	BPSK
6/15	BPSK
5/15	BPSK

For the *i*th E-TFC, the unquantized gain factor $\beta_{ed,k,i,uq}$ for the *k*th E-DPDCH shall be set to $\sqrt{2} \times \beta_{ed,i,harq}$ if the spreading factor for E-DPDCH$_k$ is 2 and to $\beta_{ed,i,harq}$ otherwise. Then, value of $\beta_{ed,k}$ shall be computed by quantizing the unquantized gain factor $\beta_{ed,k,i,uq}$ in accordance with Table 11.54 [17] for the case when E-TFCI $\leq$ *E-TFCI$_{ec,boost}$* and Table 11.55 [17], for the case when E-TFCI $>$ *E-TFCI$_{ec,boost}$*.

The gain factor applied to E-DPDCH is adjusted as a result of compressed mode operation in the following cases:

- E-DCH transmissions that overlap a compressed frame
- For 10 ms E-DCH TTI case, retransmissions that do not themselves overlap a compressed frame, but for which the corresponding initial transmission overlapped a compressed frame.

The gain factors used during a compressed frame for a certain E-TFC are calculated from the nominal power relation used in normal (non-compressed) frames for that E-TFC. When the frame is compressed, the gain factor used for the *i*:th E-TFC is derived from $\beta_{ed,C,i}$ as described below.

Table 11.55 Quantization for $\beta_{ed,k}/\beta_c$ for E-TFCI > $E\text{-}TFCI_{ec,boost}$

Quantized amplitude ratios $\beta_{ed,k}/\beta_c$	E-DPDCH modulation schemes which may be used in the same subframe
377/15	4PAM (applicable only for SF2 code in a 2xSF2 + 2xSF4 configuration)
336/15	4PAM (applicable only for SF2 code in a 2xSF2 + 2xSF4 configuration)
299/15	4PAM
267/15	BPSK (applicable only for SF2 code in a 2xSF2 + 2xSF4 configuration), 4PAM
237/15	BPSK (applicable only for SF2 code in a 2xSF2 + 2xSF4 configuration), 4PAM
212/15	BPSK, 4PAM
189/15	BPSK, 4PAM
168/15	BPSK, 4PAM
150/15	BPSK, 4PAM
134/15	BPSK, 4PAM
119/15	BPSK, 4PAM
106/15	BPSK, 4PAM
95/15	BPSK, 4PAM
84/15	BPSK, 4PAM
75/15	BPSK, 4PAM
67/15	BPSK, 4PAM
60/15	BPSK, 4PAM
53/15	BPSK, 4PAM
47/15	BPSK, 4PAM
42/15	BPSK, 4PAM
38/15	BPSK
34/15	BPSK
30/15	BPSK
27/15	BPSK
24/15	BPSK
21/15	BPSK
19/15	BPSK
17/15	BPSK
15/15	BPSK
13/15	BPSK
11/15	BPSK
8/15	BPSK

When the E-DCH TTI is 2 ms, $\beta_{ed,C,i}$ shall be calculated as follows:
If E-DPDCH power extrapolation formula is configured:

$$\beta_{ed,C,i} = \beta_{c,C,j} \cdot A_{ed} \cdot \sqrt{\frac{L_{e,ref}}{L_{e,i}}} \cdot \sqrt{\frac{K_{e,i}}{K_{e,ref}}} \cdot 10^{\left(\frac{\Delta harq}{20}\right)} \cdot \sqrt{\frac{N_{pilot,C}}{N_{pilot,N}}} \tag{11.123}$$

where $\beta_{c,C,j}$ is calculated for the *j*:th TFC as described in section 11.2.4.3.1, where $N_{pilot,C}$ is the number of pilot bits per slot on the DPCCH in compressed frames, and, $N_{pilot,C}$ is the number of pilot bits per slot on the DPCCH in compressed frames, and $N_{pilot,N}$ is the number of pilot bits per slot in non-compressed frames.

If E-DPDCH power interpolation formula is configured:

$$\beta_{ed,C,i} = \beta_{c,C,j} \cdot \sqrt{\frac{L_{e,ref,1}}{L_{e,i}}} \cdot \sqrt{\left(\left(\frac{\frac{L_{e,ref,2}}{L_{e,ref,1}} A^2_{ed,ref,2} - A^2_{ed,ref,1}}{K_{e,ref,2} - K_{e,ref,1}}\right)(K_{e,i} - K_{e,ref,1}) + \beta^2_{ed,ref,1}\right)} \cdot 10^{\left(\frac{\Delta harq}{20}\right)} \cdot \sqrt{\frac{N_{pilot,C}}{N_{pilot,N}}} \tag{11.124}$$

with the exception that $\beta_{ed,C,i}$ is set to 0 if inequality:

$$\left(\frac{\frac{L_{e,ref,2}}{L_{e,ref,1}} A^2_{ed,ref,2} - A^2_{ed,ref,1}}{K_{e,ref,2} - K_{e,ref,1}}\right)(K_{e,i} - K_{e,ref,1}) + \beta^2_{ed,ref,1} \leq 0 \tag{11.125}$$

holds.

When the E-DCH TTI is 10 ms and the current frame is compressed, $\beta_{ed,C,i}$ shall be calculated as follows:

If E-DPDCH power extrapolation formula is configured:

$$\beta_{ed,C,i} = \beta_{c,C,j} \cdot A_{ed} \cdot \sqrt{\frac{L_{e,ref}}{L_{e,I,i}}} \cdot \sqrt{\frac{K_{e,i}}{K_{e,ref}}} \cdot 10^{\left(\frac{\Delta harq}{20}\right)} \cdot \sqrt{\frac{15 \cdot N_{pilot,C}}{N_{slots,I} \cdot N_{pilot,N}}} \tag{11.126}$$

If E-DPDCH power interpolation formula is configured:

$$\beta_{ed,C,i} = \beta_{c,C,j} \cdot \sqrt{\frac{L_{e,ref,1}}{L_{e,i}}} \cdot \sqrt{\left(\left(\frac{\frac{L_{e,ref,2}}{L_{e,ref,1}} A^2_{ed,ref,2} - A^2_{ed,ref,1}}{K_{e,ref,2} - K_{e,ref,1}}\right)(K_{e,i} - K_{e,ref,1}) + \beta^2_{ed,ref,1}\right)} \cdot 10^{\left(\frac{\Delta harq}{20}\right)} \cdot \sqrt{\frac{15 \cdot N_{pilot,C}}{N_{slots,I} \cdot N_{pilot,N}}} \tag{11.127}$$

with the exception that $\beta_{ed,C,i}$ is set to 0 if inequality:

$$\left(\frac{\frac{L_{e,ref,2}}{L_{e,ref,1}} A^2_{ed,ref,2} - A^2_{ed,ref,1}}{K_{e,ref,2} - K_{e,ref,1}}\right)(K_{e,i} - K_{e,ref,1}) + \beta^2_{ed,ref,1} \leq 0 \tag{11.128}$$

holds.

When the E-DCH TTI is 10 ms and the current frame is not compressed, but is a retransmission for which the corresponding first transmission was compressed, the gain factor used for the kth E-DPDCH for the ith E-TFC is derived from $\beta_{ed,R,i}$ as follows:

If E-DPDCH power extrapolation formula is configured:

$$\beta_{ed,R,i} = \beta_{ed,ref} \sqrt{\frac{L_{e,ref}}{L_{e,I,i}}} \sqrt{\frac{K_{e,i}}{K_{e,ref}}} \sqrt{\frac{15}{N_{slots,I}}} \cdot 10^{\left(\frac{\Delta harq}{20}\right)} \tag{11.129}$$

Table 11.56 IQ branch mapping for E-DPDCH

$N_{max\text{-}dpdch}$	HS-DSCH configured	$E\text{-}DPDCH_k$	$iq_{ed,k}$
0	No/Yes	$E\text{-}DPDCH_1$	1
		$E\text{-}DPDCH_2$	j
		$E\text{-}DPDCH_3$	1
		$E\text{-}DPDCH_4$	j
1	No	$E\text{-}DPDCH_1$	j
		$E\text{-}DPDCH_2$	1
1	Yes	$E\text{-}DPDCH_1$	1
		$E\text{-}DPDCH_2$	j

If E-DPDCH power interpolation formula is configured:

$$\beta_{ed,R,i} = \cdot\sqrt{\frac{L_{e,ref,1}}{L_{e,i}}} \cdot \sqrt{\left(\left(\frac{\frac{L_{e,ref,2}}{L_{e,ref,1}}\beta^2_{ed,ref,2} - \beta^2_{ed,ref,1}}{K_{e,ref,2} - K_{e,ref,1}}\right)\left(K_{e,i} - K_{e,ref,1}\right) + \beta^2_{ed,ref,1}\right)}\sqrt{\frac{15}{N_{slots,I}}} \cdot 10^{\left(\frac{\Delta harq}{20}\right)} \tag{11.130}$$

with the exception that $\beta_{ed,R,i}$ is set to 0 if:

$$\left(\frac{\frac{L_{e,ref,2}}{L_{e,ref,1}}\beta^2_{ed,ref,2} - \beta^2_{ed,ref,1}}{K_{e,ref,2} - K_{e,ref,1}}\right)\left(K_{e,i} - K_{e,ref,1}\right) + \beta^2_{ed,ref,1} \le 0 \tag{11.131}$$

holds.

The quantization is applied in the same way as described above.

After weighting, the real-valued spread signals shall be mapped to the I branch or the Q branch according to the iq_{ec} value for the E-DPCCH and to $iq_{ed,k}$ for $E\text{-}DPDCH_k$ and summed together.

The E-DPCCH shall always be mapped to the I branch, that is, $iq_{ec} = 1$.

The IQ branch mapping for the E-DPDCHs depends on the actual number of configured DPDCHs, denoted $N_{max\text{-}dpdch}$ (it is equal to the largest number of DPDCHs from all the TFCs in the TFC set) and on whether an HS-DSCH is configured for the UE; the IQ branch mapping shall be as specified in Table 11.56 [17].

The modulation of the obtained complex-valued chip sequence is done as shown in Figure 11.38.

The downlink HSUPA channels are spread and modulated as it is described in section 11.1.4.3.2.

11.3.2.4 Operation During Compressed Mode

When E-DCH TTI length is 2 ms, the UE shall not transmit E-DCH data in a TTI which fully or partly overlaps with an uplink transmission gap. For sub-frames within compressed frames that are not overlapping with uplink transmission gaps, an adaptation of the gain factors for E-DPCCH and E-DPDCH is done as described in section 11.3.2.3.5.

However, it is not possible to apply the same solution for the case of 10 ms E-DCH TTI, since it can ruin the H-ARQ processing in the case of frequent compressed frames. When E-DCH TTI length is 10 ms, the UE is allowed to transmit E-DCH data in frames overlapping with uplink transmission gaps, but not in all slots of the frame. Some slots in the frame in this case are idle. The gain factor of E-DPCCH and E-DPDCH is increased according to the amount of non-idle slots as described in section 11.3.2.3.5.

If an initial transmission overlaps with a compressed frame the starting slot of the consecutive E-DPDCH idle slots within the E-DCH TTI is n_{first}, and n_{last} is the final idle slot within the 10 ms E-DCH TTI. The number of transmitted slots n_{txI} is given by $n_{txI} = 14 + n_{first}\text{-}n_{last}$.

If the initial transmission occurs in a non-compressed uplink frame, $n_{txI} = 15$.

In case the retransmission occurs in a compressed frame the maximum number of slots available for the retransmission is given by $n_{max} = 14 + n_{first}\text{-}n_{last}$. Otherwise the maximum number of slots available for the retransmission n_{max} is 15.

If the initial transmission was compressed and in the retransmission more than n_{txI} slots are available for transmission ($n_{max}>n_{txI}$), the last $n_{dtx} = n_{max}\text{-}n_{txI}$ available slots of the E-DPDCH retransmission frame should be E-DPDCH idle slots.

This is illustrated in examples in Figure 11.95. In the example depicted in Figure 11.95a the compressed frame overlaps with the initial transmission, and slots from 2 to 6 are idle in the initial transmission.

There is no compressed mode during retransmission but since there were five idle slots in the initial transmission, the same amount of slots should be idle at the end of retransmission. In the example depicted in Figure 11.95b the compressed mode occurs both during the initial transmission and the retransmission. However, the number of slots transmitted in the initial transmission $n_{txI} = 8$ is less than the maximum number of slots available for the retransmission $n_{max} = 11$. Because of this $n_{dtx} = n_{max}\text{-}n_{txI} = 3$ are not transmitted at the end of the retransmission frame.

The E-DPCCH slots are not DTXed. The idle slots in the E-DPCCH frames correspond only compressed mode idle time as shown in Figure 11.95.

The behaviour of downlink HSUPA channels: E-HICH, E-RGCH and E-AGCH during compressed mode on the DPCH or F-DPCH is similar to the behaviour of E-DPCCH:

- A UE shall decode E-HICH, E-RGCH or E-AGCH transmissions to the UE using all the slots which do not overlap a downlink transmission gap.
- The UE may discard E-HICH, E-RGCH or E-AGCH slots which overlap a downlink transmission gap.

11.3.2.5 Operation During Soft Handover

Unlike HSDPA, HSUPA supports soft and softer handovers. The behaviour of E-DCH during the soft handover is very similar to the behaviour of DCH described in section 11.2.7. However, during soft handover HSUPA uses its own active set, which differs from the active set used by Release'99 channels. The maximum number of Node Bs that can be included in HSUPA active set is four rather than six for Release'99 channels. Actually, soft handover is very important for HSUPA since it helps to decrease the interference originating from the other cell UEs.

One feature helping to decrease the interference and to increase the reliability of the transmitted uplink data involves all Node Bs from the active set in the H-ARQ process. That means all Node Bs included in active set rather than only serving Node B can send ACK/NACK signal to the UE. However, as can be seen from Table 11.47 NACK signal from the non-serving Node B is represented by '0'. Therefore, in reality non-serving Node B can feedback only ACK signal. It can help in situations when the UE is near the cell edge, then it is quite probable that the transmission of a packet to serving cell fails while the non-serving Node B can receive this packet correctly. Then this packet will be received by RNC, which considers the transmission successful and sends the packet to network without waiting the delivery of this packet from the serving cell, that is, the packets at RNC are reordered, as shown in Figure 11.96. After receiving ACK from any Node B belonging to the active set UE will stop the retransmissions. This helps somehow to decrease the interference from this particular UE in the neighbouring cell, since this procedure decreases the packet transmission time. And taking into account that being at the serving cell

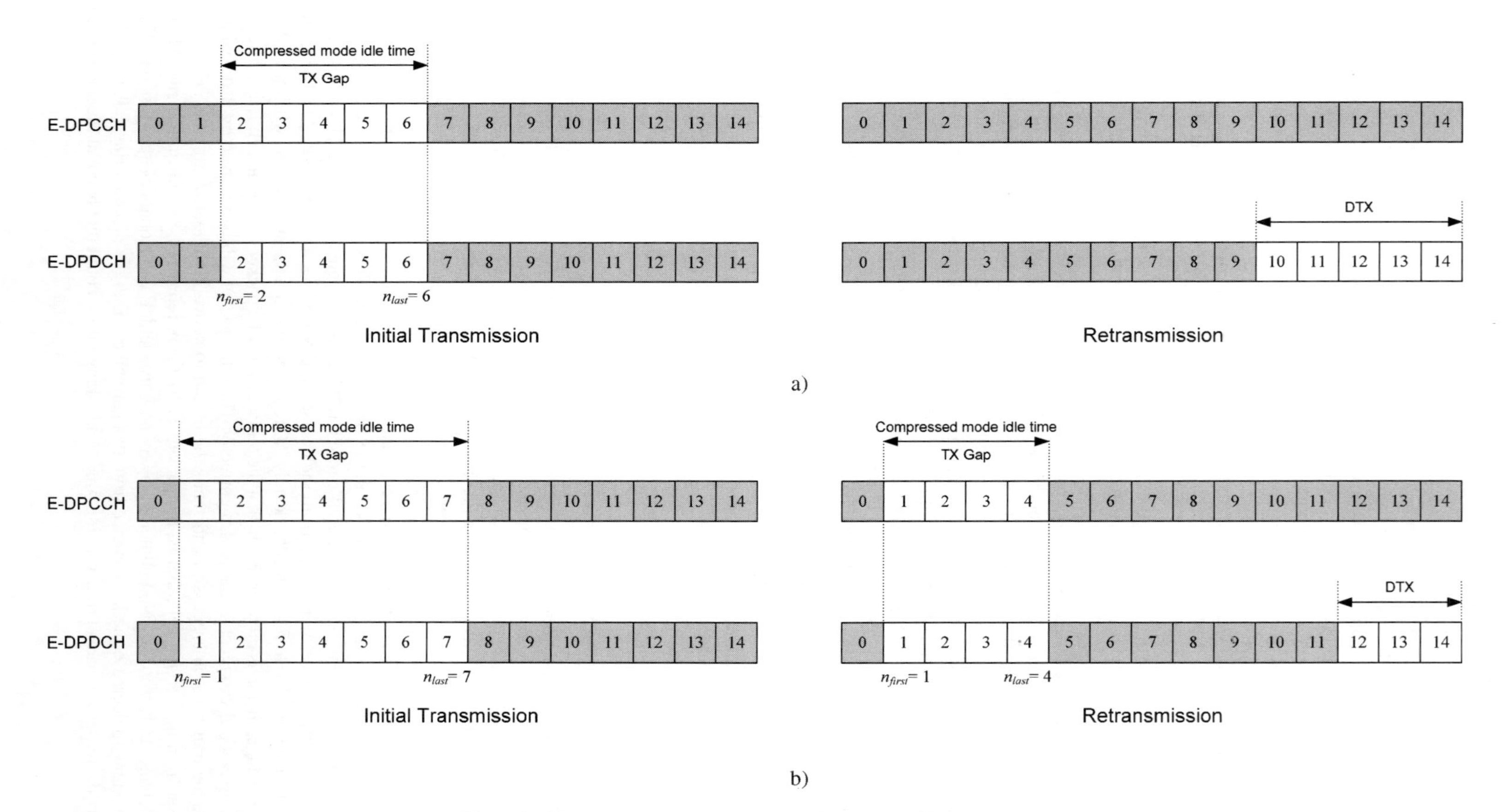

Figure 11.95 E-DPDCH during compressed mode. E-DCH TTI = 10 ms

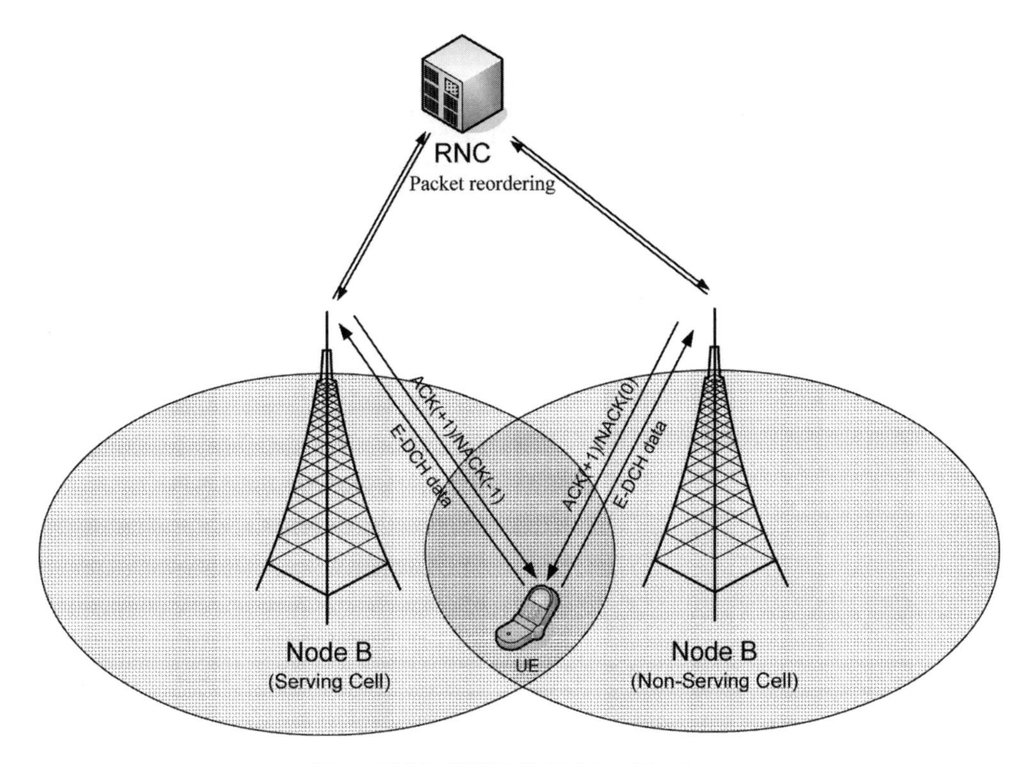

Figure 11.96 HSUPA H-ARQ in soft handover

edge the UE have to transmit at a higher power to compensate for being farther away from the serving Node B, the interference causing by this UE is greater than average. So, it is enough if just any Node B from the active set received the correct version of packet from the UE to deliver this packet to network. On the other hand, the retransmission of the packet can be triggered only by the NACK signal from the serving cell. That allows traffic to be kept as low as possible. In case the UE is in softer handover, the ACK/NACK signals from different sectors are soft combined, and NACK from non-serving sector is mapped to '-1' rather than to '0' as belonging to the same radio link.

The second field, where handover mechanism is very important for HSUPA is the resource allocation. The uplink resource is power (or what is the same interference). As it was mentioned in section 11.3.2.2 the HSUPA scheduling is provided with the help of absolute and relative grants. The very important fact for soft handover is that it is possible to send the relative grants to UE not only from serving cell but from non-serving cells either. For example, if a UE served in low loaded cell asks for more resources by sending the value of Happy Bit = 'Not Happy', it may be given a grant that allows it to transmit at high data rates, maximizing the benefit to that user as shown in Figure 11.97. However, if the UE then moves toward a high loaded cell, its high power transmission can bring additional severe interference to that cell. The serving Node B knows nothing about the interference in the nonserving cell and cannot decrease the UE transmitting power. The solution of this problem is that nonserving Node B can send to UE the relative grant 'DOWN' requesting the UE's transmit power to be reduced as shown in Figure 11.97. The UE can increase data only in cases where the 'UP' relative grant was received from serving Node B and no 'DOWN' relative grant was detected from any Node B belonging to the active set.

The nonserving Node B can send only two types of relative grants: 'HOLD' or 'DOWN'. It is only the serving cell that can allow UE to increase the transmit power by sending relative grant 'UP' or absolute

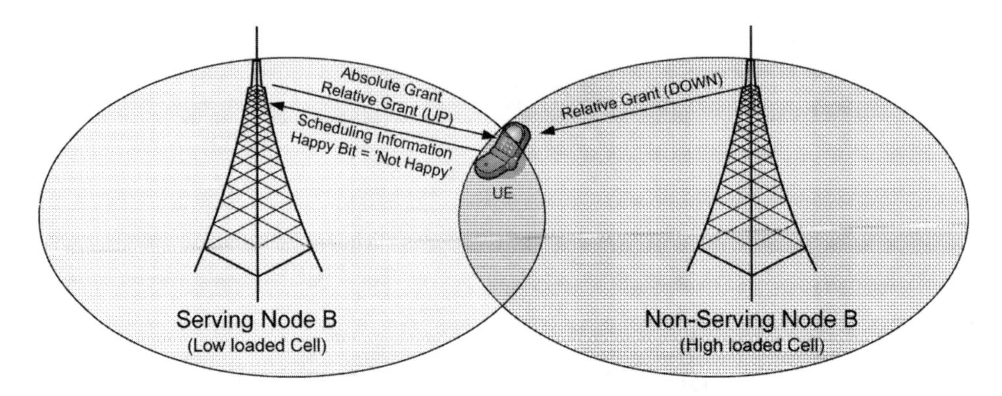

Figure 11.97 HSUPA scheduling in soft handover

grant with high power value. This mechanism allows different Node Bs to manage the uplink air interface without involving the RNC, which makes the scheduling more effective and much faster. When the UE is in softer handover the relative grants from the E-DCH serving radio link set are the same (all of them can send all three possible types of relative grants: 'UP', 'HOLD' and 'DOWN') and shall be soft combined into one single relative grant information.

11.3.3 CPC

Continuous Packet Connectivity (*CPC*) was designed to improve the performance of delay-critical small bit rate services like VoIP. Actually, it is not some new WCDMA technology like HSDPA or HSUPA. CPC is the collection of features applied to HSDPA or HSUPA aiming at optimizing WCDMA for packet switched connections. CPC eliminates the need for continuous transmission and reception when data is not exchanged. The main components of CPC are

- Discontinuous uplink transmission (*uplink gating*);
- Discontinuous downlink reception;
- HS-SCCHless HSDPA.

The benefits of implementing CPC are obvious. First of all, it eliminates the situation when a lot of connected but inactive UEs in the cell consumes significant part of resources and create severe interference without transmission any data. The elimination of this scenario allows increasing the cell capacity, that is, more users can be connected. Another benefit is the decreasing of energy consumption. For UE it is especially important since it allows battery life to be extended and therefore, to increase the talk time. Also, implementing of CPC improves the connection set up time.

The uplink gating or DTX allows a lot of resources to be saved. Without uplink gating the DPCCH control channel is always transmitted to support the power control, even if there is no data for transmission. The comparison of transmitted power levels of uplink channels without and with uplink gating is represented in Figure 11.98. The example of uplink DCH with TTI = 20 ms transmitting the circuit-switched voice with data rate 12.2 Kbps is depicted in Figure 11.98a. In this case both DPCCH and DPDCH are transmitted in every TTI to provide the required CS voice service quality. In Figure 11.98b the example of E-DCH with 10 ms TTI bearing data for VoIP is depicted. In this case only E-DPCCH is transmitted constantly. Since VoIP does not require high data rate there is no need to transmit the E-DPDCH in every TTI. The same example for 2 ms TTI is depicted in Figure 11.98 c). In case of 2 ms E-DCH TTI it is possible to transmit the E-DPDCH even less frequently as in case of 10 ms E-DCH

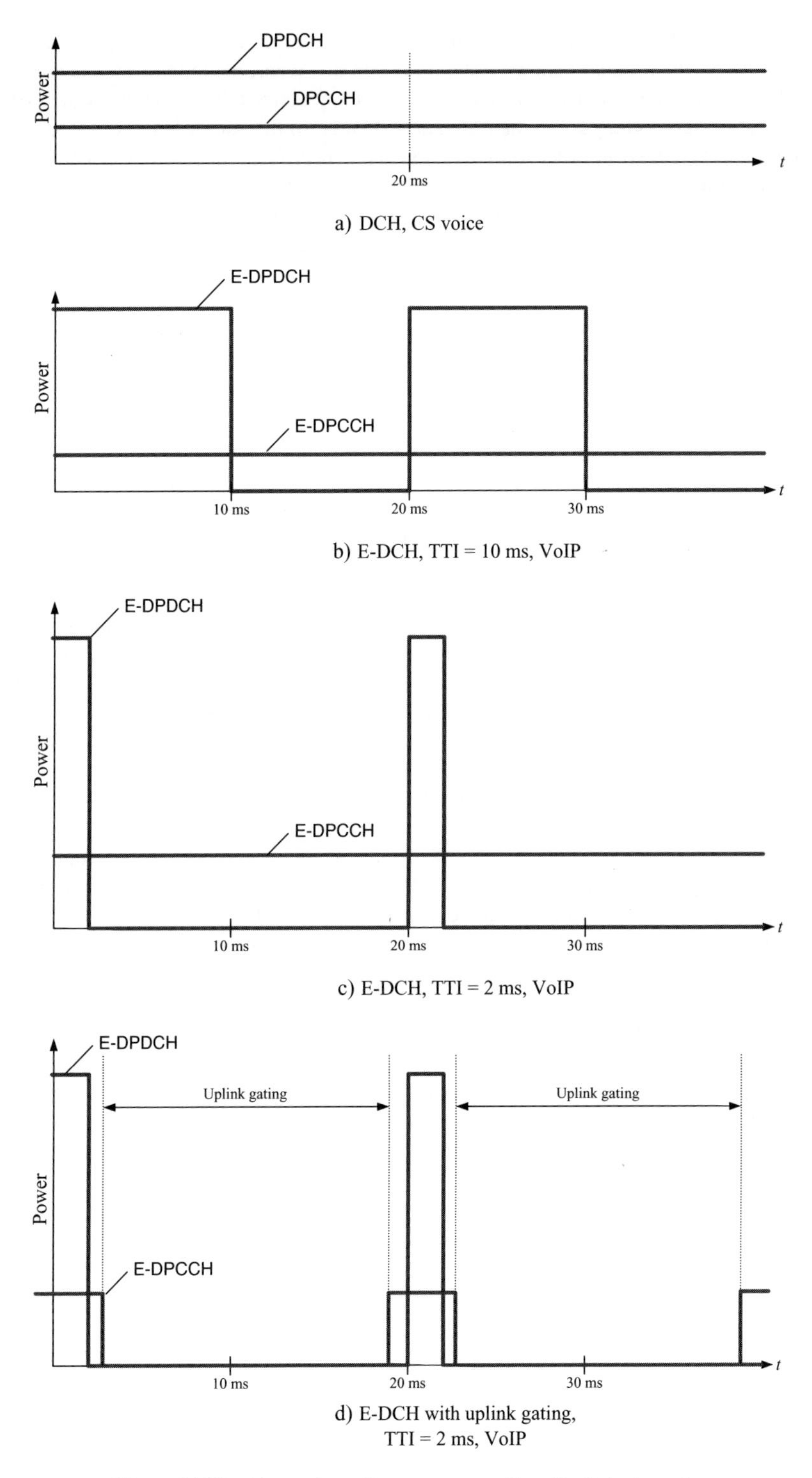

Figure 11.98 Comparison of power consumption of uplink channels

TTI. However, E-DPCCH is still transmitted in every TTI. Uplink gating allows to get rid of necessity for E-DPCCH to be always on, as shown in Figure 11.98 d). According to the CPC the control signals are transmitted on the uplink control channels according to predefined discontinuous patterns during the inactive state of corresponding uplink data channels (that is, in the state when there is no packets for transmission on E-DPDCH) in order to maintain signal synchronization and power control loop with less control signalling. The DTX pattern in example in Figure 11.98 d) consists mostly of idle slots and includes only few active slots before and after data transmission on E-DPDCH. Uplink gating is applied not only to HSUPA channels, but also to normal DPCCH in addition to the uplink control channels of HSDPA. However, the uplink gating for DPCCH has more restrictions than for E-DPCCH. If a UE will start a transmission of E-DPCCH and E-DPDCH on an E-DCH TTI, the UE shall start the DPCCH transmission 2, 4 or 15 slots prior to the E-DCH TTI and continue the DPCCH transmission during the E-DCH TTI and 1 slot after the E-DCH TTI. The uplink gating can be activated/deactivated by higher layers or by serving Node B with fast L1/L2 signalling.

Discontinuous downlink reception (downlink DRX) of CPC is configured by the RNC and allows the UE to restrict the downlink reception times in order to reduce power consumption. When the downlink DRX is enabled, the UE is not required to receive physical downlink channels except for several specific situations. Actually, that means the UE is allowed to power off the radio receiver for a predefined period called DRX period. The DRX cycle comprises active period during which the receiver is on and DRX period during which the receiver is off. During DRX, the UE must just check once per DRX cycle if new data arrives. However, due to the power control, slots corresponding to uplink transmissions must always be received. Downlink DRX is useless without the uplink gating because due to uplink power control feedback continuous downlink transmission and reception is required if uplink transmission is continuous.

In addition, CPC includes an HS-SCCH less operation, which is a special mode of HSDPA operation for reducing HS-SCCH overhead for an H-ARQ process. This feature also allows the UE power consumption to be reduced. Under this mode, the initial HS-DSCH transmission of the HARQ process corresponding to the small size of transport blocks on pre-defined HS-DSCH is performed without the accompaniment of HS-SCCH signaling, and the H-ARQ retransmissions corresponding to the original HS-DSCH transmission are accompanied with the HS-SCCH Type 2 signalling if the retransmissions are needed. This kind of signalling allows the control channel overhead to be eliminated from small packets sent over HSDPA. In this case the UE must use blind transport format detection for the initial transmission based on predefined transport block size and channel coding set. If blind transport format detection is successful, the UE reports ACK; otherwise, the UE reports nothing and waits for retransmission initiated by the Node-B. In order to combine the original transmission with the subsequent retransmissions, the HS-SCCH transmits required control signals of physical channel coding set, transport block size, UE identity, and the pointer notifying the UE of the TTI where the previous transmission has been performed as it is described for HS-SCCH Type 2 in section 11.3.1.4.1. In addition, the UE can report ACK or NACK for the retransmission, and the retransmission is restricted to two times. The first and second retransmissions can be asynchronous with respect to the first transmission, and with respect to each other. The accompanying HS-SCCH follows the same timing relationship with the HS-PDSCH transmission as legacy transmissions do.

All these CPC features allow the UE power consumption to be decreased significantly. Some examples of power saving gain provided by CPC are represented in Table 11.57.

Table 11.57 Examples of power saving gain with CPC

Service	Power saving gain with CPC
Voice call	36 %
Browsing	32 %
Online Data Application	50 %

Table 11.57A Special subframe configuration

UE Category	DL peak data rate [Mbit/s]	Number of RX antennas	UL peak data rate [Mbit/s]
1	10	2	5
2	51	2	25
3	102	2	51
4	151	2	51
5	300	4	75

As can be seen from Table 11.57 the UE energy consumption can be decreased up to 50% due to CPC implementation. Also, the cell capacity of the cell using CPC can be almost doubled in comparison to the cell supporting only Release'99 channels.

These facts provide wide support to faster implementation of CPC and it is included in 3GPP Release 7 specifications.

11.4 4G LTE

Long Term Evolution (LTE) is the next step in the development of cellular services. LTE is positioned as 4G standard. In future the radio access for LTE E-UTRA should replace the UTRA. The orginal requirements for LTE were peak data rate of 100 Mbps in the downlink and 50 Mbps in the uplink and with scalable bandwidth from 1.25 MHz to 20 MHz in order to suit the needs of different network operators with different bandwidth allocations, and also allow operators to provide different services based on spectrum. LTE supports both frequency division multiplexing (FDD) and frequency division multiplexing (TDD) to allow operation in both paired and unpaired spectrum allocation. LTE is also expected to improve spectral efficiency and decrease the RAN latency to 10 ms [32]. The key technologies allowing LTE to provide high performance are OFDMA and MIMO.

The actual supported peak data rates are higher than the requirements and are specified for the given UE categories defined in [42]. Table 11.57A shows the number maximum UL and DL data rates that can be achieved for the different categories.

11.4.1 LTE Downlink

11.4.1.1 Frame Structure

To ensure good radio access technology interoperability(handovers to other radio systems),the LTE frame structure is based on the UMTS radio frame of 10 ms. Each radio frame consists of 10 subframes of length 1 ms and in normal subframes there are two 0.5 ms slots in each subframe. The smallest time unit that can be scheduled to one user is one subframe. The frame structure is shown in Figure 11.99.

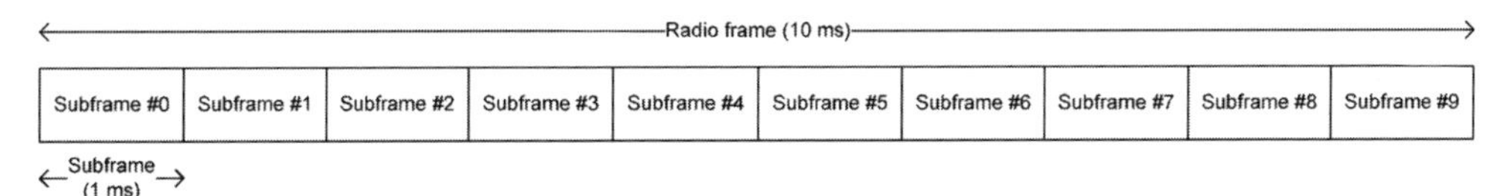

Figure 11.99 LTE frame structure[1]

[1]The 3GPP specifications separate the FDD and TDD frame structures as frame structure Type 1 and Type 2, respectively

Table 11.58 Uplink-downlink configurations

Uplink-downlink configuration	Downlink-to-Uplink Switch-point periodicity	Subframe number									
		0	1	2	3	4	5	6	7	8	9
0	5 ms	D	S	U	U	U	D	S	U	U	U
1	5 ms	D	S	U	U	D	D	S	U	U	D
2	5 ms	D	S	U	D	D	D	S	U	D	D
3	10 ms	D	S	U	U	U	D	D	D	D	D
4	10 ms	D	S	U	U	D	D	D	D	D	D
5	10 ms	D	S	U	D	D	D	D	D	D	D
6	5 ms	D	S	U	U	U	D	S	U	U	D

D: Downlink subframe S: Special subframe U: Uplink subframe

11.4.1.1.1 TDD Frame Structure and the Special Subframe

Since downlink and uplink transmissions in FDD are separated in frequency, all subframes are either uplink or downlink DL, but in TDD the radio frame needs to be split into downlink and uplink frames and this is done on a subframe level. Since the data load in the uplink and downlink may be different, the number of subframes used for uplink and downlink transmissions can be varied, shifting resources between downlink and uplink. For the same reason, the switching point periodicity can be configured to be either 5 ms or 10 ms. Table 11.58 shows the allowed configurations the TDD frame structure with the associated downlink and uplink percentages.

In a TDD system, problems with interference will arise at a base station if the transmitted signals from the neighbouring base stations interfere with the received signals from the UEs. Strict synchronization of the DL and UL transmissions across cells is therefore needed, but even with tight synchronization interference can still be a problem when a delayed signal from a distant base station arrives after the base station has switched from downlink transmission to uplink reception. In addition, a small switching time is needed in both UE and base station to do the actual switch from uplink to downlink and vice versa. Therefore, a guard period with no transmission is needed and this is the purpose of the special subframe which always occupies subframe #1 and additionally subframe #6 in the case of 5 ms downlink to uplink switch point periodicity. Figure 11.100 shows the special subframe.

The special subframe consists of three parts, *DwPTS*, guard period and *UpPTS*. The DwPTS is the downlink part of the subframe and can contain downlink transmissions of control and data. The UpPTS can be used for transmission of uplink sounding and short RACH. In order to allow for deployments with varying site distance, the length of the guard period can be configured by varying the length of DwPTS and UpPTS. The possible lengths are shown in Table 11.59.

A switching time is also needed when switching from uplink to downlink, but this is provided by advancing the UL transmission timing to the DL receiving time as shown in Figure 11.100A.

11.4.1.1.2 Slot Structure

LTE multiplexing is based on OFDM with a sub-carrier spacing is 15 kHz in LTE. In an OFDM system, the OFDM symbol consists of two parts, the data part and the cyclic prefix. The cyclic prefix part is a

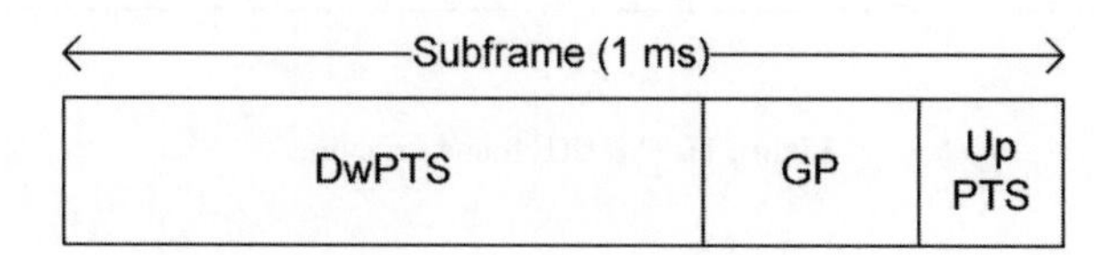

Figure 11.100 LTE special subframe structure

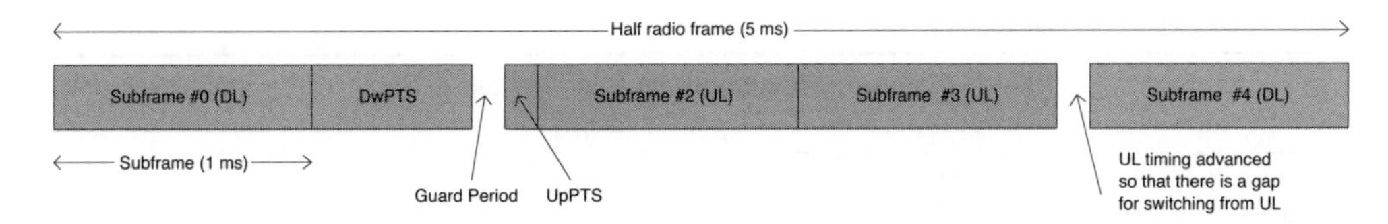

Figure 11.100A Example of TDD switching with uplink downlink configuration 1 and special subframe configuration 7

copy of the last part of the data symbol and is put in front of the data symbol to create a periodic signal over the full OFDM symbol. The purpose of the cyclic prefix is to avoid intersymbol interference caused by multipath propagation. Therefore, the cyclic prefix length should be larger than the delay spread of the channel, but the cyclic prefix length is also overhead and reduces the spectral efficiency of the system. In order to support a large delay spread range while maintaining a reasonable overhead for the more typical deployments, LTE supports two cyclic prefix lengths, normal cyclic prefix and extended cyclic prefix. Since the slot length is fixed at 0.5 ms, the number of OFDM symbols varies based on the cyclic prefix length. The numerology is shown in Table 11.60.

One OFDM sub-carrier in one OFDM symbol is called a *Physical Resource Element*. In order to facilitate simple resource allocation, a *resource block* is defined as 12 resource elements in the frequency domain across the slot as in Figure 11.101.

11.4.1.1.3 Subframe Structure

One subframe consists of two slots and is the smallest time unit that can be scheduled to one UE. The subframe contains two parts, the control region and the data (PDSCH) region. The subframe starts with the control region which can be dynamically varied from one to four OFDM symbols (four OFDM symbols only for bandwidths with less than 10 RBs). The data region occupies the remaining OFDM symbols of the subframe. The subframe structure is shown in Figure 11.102.

11.4.1.1.4 Transmission Bandwidths

LTE is designed to support varying bandwidths able to support different operator's spectrum allocation. For LTE Release 8 the transmission bandwidths listed in Table 11.61 are supported.

The physical layer specifications actually support any bandwidth from 6 to 110 resource blocks, but only the listed transmission bandwidths are supported by LTE Release 8 UEs.

In order for the UE to find a cell without knowing the transmission bandwidth, all procedures used in the initial access are based on the narrowest bandwidth of six resource blocks. To facilitate this,

Table 11.59 Special subframe configuration

Special subframe configuration	DwPTS [symbols]	UpPTS [symbols]	GP [symbols]
0	3		10
1	9		4
2	10	1	3
3	11		2
4	12		1
5	3	2	9
6	9		3
7	10		2
8	11		1

Table 11.60 Downlink numerology

Configuration	Cyclic prefix length [μs]	Number of OFDM symbols per slot N_{symb}^{DL}	Number of resource elements in a resource block
Normal cyclic prefix	4.8	7	84
Extended cyclic prefix	16.7	6	72

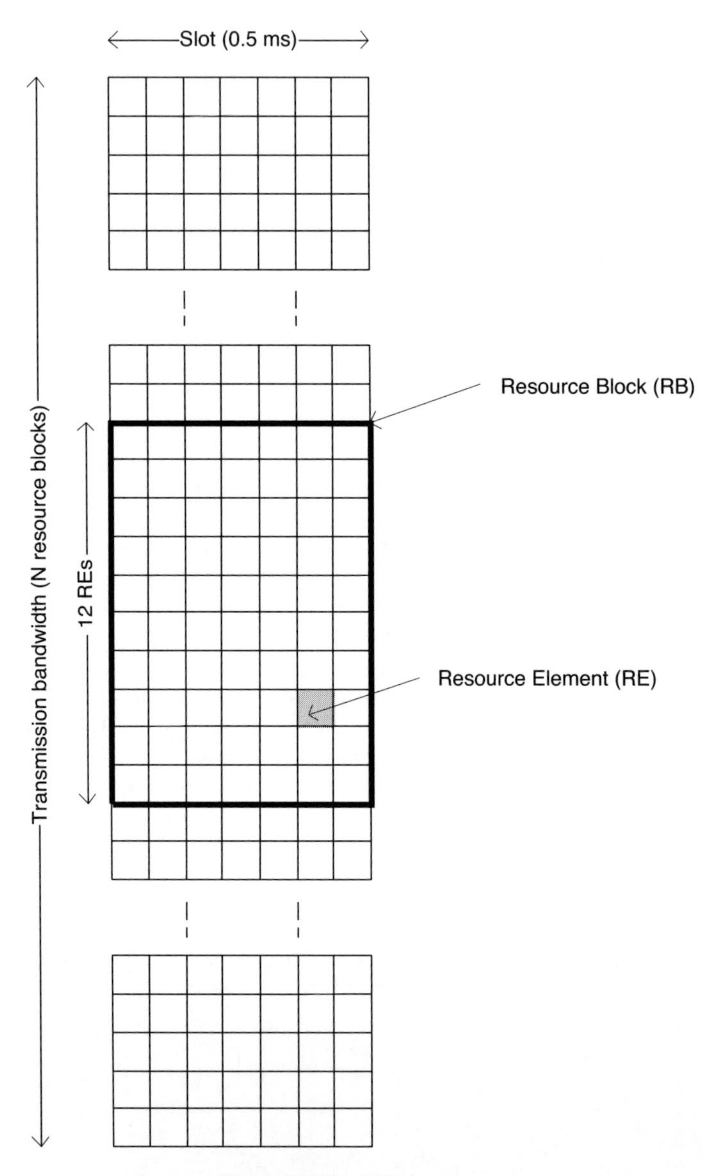

Figure 11.101 LTE slot structure

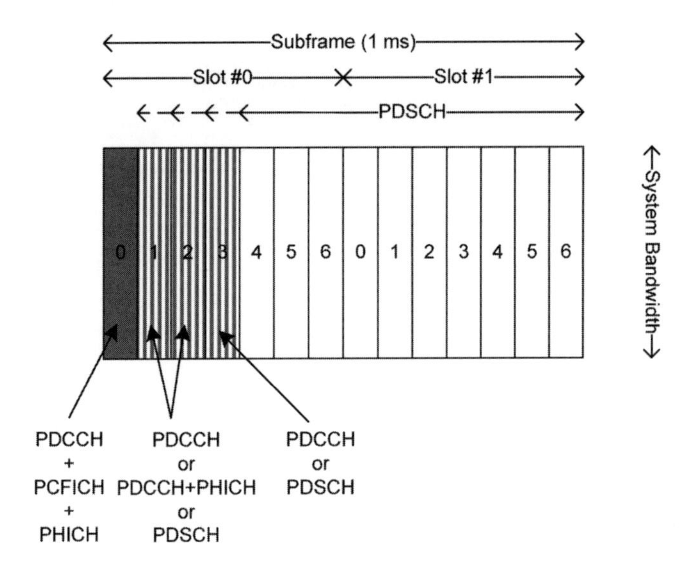

Figure 11.102 LTE subframe structure

the synchronization signals and the primary broadcast channel, which contains information about the downlink transmission bandwidth, only span the six middle resource blocks regardless of the number of resource blocks in the transmission bandwidth.

11.4.1.1.5 LTE Timing

The timing for the downlink transmissions and the associated ACK/NACK transmission is shown in Figure 11.102A.

When the UE receives a transmission on the PDSCH, an ACK or NACK shall be transmitted for this HARQ process (see on the PUCCH three subframes after the end of the PDSCH subframe and this is the time available for the UE for processing the PDSCH and preparing to transmit the PUCCH. The transmissions from different UE are time aligned at the base station and this timing is controlled by the base station by sending timing advance commands to the UE that is just by the UE to adjust its timing. As can be seen from the figure, the timing advance will reduce the available processing time for the UE. The HARQ timing in DL LTE is adaptive, meaning that the BS may transmit a retransmission or a new transmission for a HARQ process at any time after having received the NACK or ACK and the base station processing time will be whatever this delay is.

11.4.1.2 Transport Channels

LTE has the following transport channels:

- *Downlink Shared Channel (DL-SCH)*. Carries user data and system information
- *Broadcast Channel (BCH)*. Delivers the Master Information Block (MIB) of the system information

Table 11.61 LTE transmission bandwidths

Channel bandwidth [MHz]	1.4	3	5	10	15	20
Transmission bandwidth in number of resource blocks	**6**	**15**	**25**	**50**	**75**	**100**

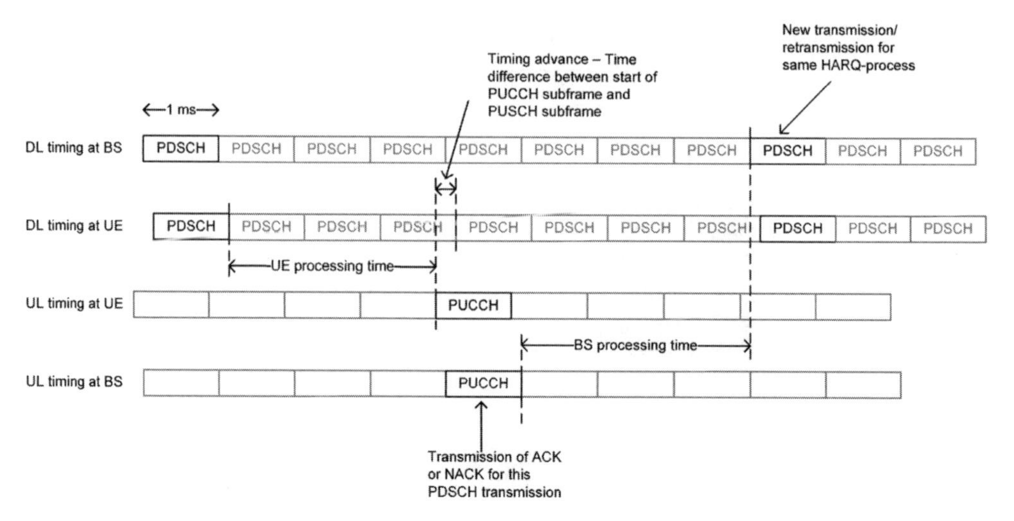

Figure 11.102A The timing for the downlink transmissions and the associated ACK/NACK transmission

- *Paging Channel* (*PCH*). Used for paging UEs
- *Multicast Channel* (*MCH*). Used for Multimedia Broadcast and Multicast Services (MBMS) transmission.

In LTE, the channel encoding and rate matching are done on the transport channel and these functions are common for DSCH, PCH and MCH. BCH has its own processing.

11.4.1.2.1 Transport Block

A transport block is delivered to the physical layer from higher layers. Transport block size can take a given number of sizes and higher layers make sure that the data is padded to a supported transport block size. A 24-bit CRC is added to the transport block to allow for error detection. At the receiving end, the UE finds the transport block size from the signalled number of resource blocks and the modulation and coding scheme using a specified look-up table.

11.4.1.2.2 Channel Coding

The transport channels are encoded using turbo code shown in Figure 11.27. The constituent codes are the same as used in UTRA, but the interleaver has been changed. From an implementation perspective there are clear benefits of being able to generate the interleaver on the fly and thus a polynomial interleaver was chosen. The polynomial used in the interleaver is:

$$\Pi(i) = \left(f_1 \cdot i + f_2 \cdot i^2\right) \bmod K \tag{11.132}$$

where K is the code block size and 188 possible code block sizes between 40 and 6144 bits have been defined with associated values for f_1 and f_2.

In case the transport block size is larger than the maximum code block size, code block segmentation is performed where the transport block is divided into equal sized code blocks that are individually encoded. For each code block a 24-bit CRC is added.

11.4.1.2.2.1 HARQ and Rate Matching Selective repeat Hybrid ARQ (HARQ) is applied to the DL-SCH transport channel. This means that retransmissions are performed in the physical layer, significantly reducing the latency incurred by incorrectly received transmissions. HARQ allows for combining the

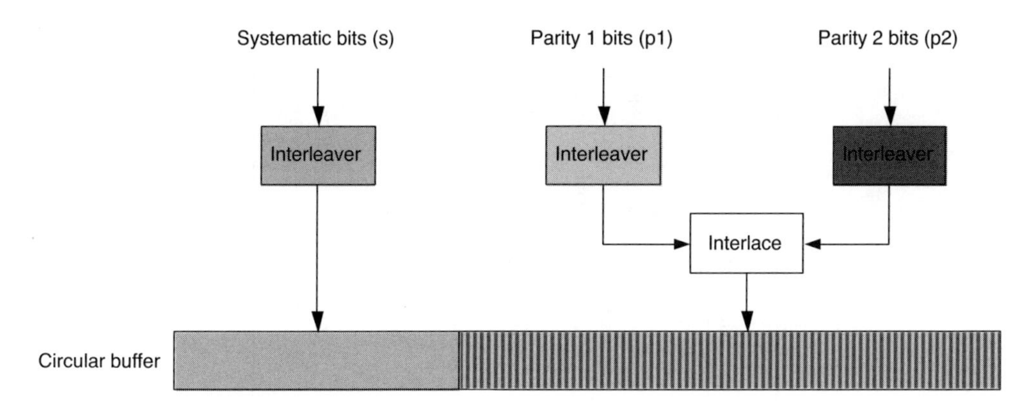

Figure 11.103 LTE circular buffer rate matching

received samples from different transmissions before decoding, improving the link performance over pure ARQ. In the downlink, the HARQ is adaptive both when it comes to timing of the retransmissions and resources and coding rates used, with synchronized transmission time for the ACK or NACK in the UL as shown in section 11.4.1.1.5. The maximum number of HARQ processes is eight.

Adaptive link adaptation in LTE means that the coding rate can be adjusted to the UE's channel conditions. For a given data packet size, the eNB scheduler finds the appropriate coding rate and assigns the necessary number of resource block needed. The purpose of the rate matching is to take the output of the fixed rate channel encoder and output the number of bits that is available for transmission on the PDSCH.

In LTE a *circular buffer rate matching* scheme is deployed. The systematic bits are interleaved and written to the rate matching buffer first. The parity bits are first interleaved and then interlaced and written to the rate matching buffer. The rate matching algorithm then reads out as many bits from the buffer as needed for the transmission. The buffer is circular in the sense that after reaching the end of the buffer, the reading continues from the start of the buffer again.

The circular buffer allows for simple implementation of Incremental Redundancy (IR) HARQ. In IR, the transmitted bits can be different for the retransmissions than for the initial transmission ensuring that after a certain number of transmissions all the channel bits will have been transmitted, effectively reducing the code rate for each transmission. Different redundancy versions are generated by varying the starting address in the circular buffer. The circular buffer rate matching scheme is depicted in Figure 11.103.

11.4.1.3 Physical Channels

A summary of the physical channels is shown in Table 11.62.

11.4.1.3.1 Physical Broadcast Channel (PBCH)

The PBCH carries the broadcast transport channel (BCH) and is used by the UE in the initial access to get the master information block (MIB). The MIB contains essential information for initial access to the system: the DL transmission bandwidth, the used configuration for the PHICH, and the eight most significant bits of the system frame number (SFN). In addition, the PBCH implicitly gives the full SFN and the number the antenna ports used in the transmission of common control information. The two least significant bits of the SFN can be detected from the PBCH timing since the PBCH TTI is 40 ms. The scrambling of the BCH CRC depends on the number of antenna ports configured in the base station

Table 11.62 LTE Physical channels

Physical channel	Abbreviation	Content	Transport channel	Coding	Modulation	Mapping to resource elements
Physical Downlink Shared Channel	PDSCH	User data and system information	DSCH, PCH	Turbo coding	QPSK, 16QAM, 64QAM	Frequency first mapping starting from the second, third or fourth OFDM symbol of the subframe
Physical Broadcast Channel	PBCH	Broadcast information	BCH	Convolutional coding	QPSK	6 center resource blocks in the first four OFDM symbols of the second slot in sub-frame #0.
Physical Control Format Indicator Channel	PCFICH	Gives the length of PDCCH in OFDMS symbols	NA	Block code (1/16)	QPSK	Spread in frequency in the first OFDM symbol of the subframe
Physical Downlink Control Channel	PDCCH	Allocates DL and UL	NA	Convolutional coding	QPSK	Interleaved across the 1-3 first OFDM symbols of the subframe
Physical Hybrid ARQ Indicator Channel	PHICH	ACK/NACK from UL transmissions	NA	Repetition code (1/3)	BPSK	First or three first OFDM symbols of the subframe
Physical Multicast Channel	PMCH	For MBMS transmission	MCH	Turbo coding	QPSK, 16QAM, 64QAM	Frequency first mapping starting from the third OFDM symbol of the subframe

and by doing the CRC check for each of the three possible configurations, the number of ports can be detected. The PBCH is mapped to the centre 72 resource elements of the first four OFDMS symbols of the second slot in subframe #0. Once the PBCH has been decoded, the UE receives the system information from the broadcast channel (BCCH) giving the UE all the needed information to attach to the cell.

11.4.1.3.2 Physical Control Format Indicator Channel (PCFICH)

The number of OFDM symbols used for the control channels (PDCCH) can be varied dynamically to allow for shifting resources between control and data based on the needed control channel capacity. The PCFICH gives the information about the number of OFDM symbols used.

11.4.1.3.3 Physical Downlink Control Channel (PDCCH)

The PDCCH is used for allocating resources in the downlink and the uplink and for providing power control commands to a group of UEs.

11.4.1.3.3.1 PDCCH Content The content of the PDCCH varies depending on whether it is a downlink assignment, uplink grant or power control commands. In addition, different PDSCH transmission modes require different information to be conveyed, resulting in different content. These different formats are called downlink control information (DCI) formats. The following information is transmitted in most DCI formats:

- Resource block assignment – indicates the resource allocation for PDSCH or PUSCH
- HARQ process number (DL only)
- Modulation and coding scheme – joint indication of modulation and transport block size
- Redundancy version (implicit in UL) – changes the sent channel bits for incremental redundancy
- New data indicator – indicates if this is the initial transmission or are retransmission
- Transmit power control command for the scheduled PUSCH(UL) or PUCCH (DL)
- TPC command for PUCCH
- Downlink Assignment Index (TDD only)

Table 11.63 shows the more specific information provided by the different formats.

11.4.1.3.3.2 PDCCH Encoding The PDCCH coding rate variesd based on the number of control channel elements (CCE) it is mapped to. A control channel element is 72 bits and one PDCCH can be rate matched to 1, 2, 4 or 8 control channel elements resulting in a wide coding rate range.

The coding used for the PDCCH is a convolutional code with constraint length 7 and coding rate 1/3. The encoding is as shown in Figure 11.104.

Tail-biting is used to avoid the additional overhead for conventional convolution codes caused by the additional terminating bits needed to terminate the code to a known final state. The initial state is set to be the same as the final state and this is done by setting the shift register initial value to the last 6 input bits. Since this state is not known by the decoder, additional processing is needed to find the initial state. One example of finding the initial state is to run the decoding process twice, where the second run is started from the final state of the first run. This increases the decoding complexity, but since tail biting also gives the same performance for a code with shorter constraint length, the overall complexity is not increased compared to conventional convolutional codes.

The rate matching of PDCCH uses also a circular buffer in a similar manner as for DL-SCH. Multiple PDCCH are interleaved and mapped to one, two or three OFDM symbols depending on the value indicated by the PCFICH.

Table 11.63 DCI formats

DCI format		Number of transport blocks	Resource allocation type	Information provided
0	UL	1	2	Frequency hopping indication Indication of the demodulation reference signal sequence to be used CQI request – used by the network to request an aperiodic CQI report
1	DL	1	0 or 1	
1A	DL	1	2	Can allocate distributed PDSCH To reduce the UE blind decoding, format 1A is always the same size format 0 and a bit indicates which of the two that is being sent. Format 1A is always decoded by the UE
1B	DL	1	2	Same as 1A with additional precoding information for closed loop TxD
1C	DL	1	2[2]	Compact assignment with reduced MCS set
1D	DL	1	2	Same as 1B with an additional power offset information for use in MU-MIMO
2	DL	2	0 or 1	Additional information for second codeword Precoding information for closed loop spatial multiplexing
2A	DL	2	0 or 1	Same as format 2, but with reduced precoding information for large delay CDD
2B	DL	2	0 or 1	Same as format 2A, but with additional information about the antenna port used for UE-specific reference signals.
3	–	–	–	Transmission of TPC commands for PUCCH and PUSCH with 2-bit power adjustments Same size as format 0/1A.
3A	–	–	–	Transmission of TPC commands for PUCCH and PUSCH with single bit power adjustments Same size as format 0/1A.

11.4.1.3.3.3 PDCCH Decoding Since multiple PDCCHs are interleaved together and the UE has limited a priori knowledge of which PDDCH contains information, the UE needs in every subframe to blindly decode multiple PDCCHs. Based on the transmission mode (Table 11.65) the UE is configured to receive certain DCI formats for the given PDCCH code rates. In addition, the search space has been limited by defining two search spaces, the common search space, searched by all UEs, and a UE specific search space that UEs finds from its Cell Radio Network Temporary Identity(C-RNTI). The C-RNTI is the unique identifier of a UE in a cell. Based on the limitations given by these two search spaces, a maximum 44 decoding attempts must be made before covering the full search space. The search spaces of different UEs will overlap, so basing the detection of a transmission for a certain UEs purely on a successful decoding is not sufficient. To prevent any ambiguity the PDCCH CRC has been scrambled with the C-RNTI and if the incorrect C-RNTI is used in the descrambling, the CRC will fail. If the CRC checks correct, the UE knows that the transmission is for it and can proceed with the appropriate actions.

[2]Only distributed transmissions

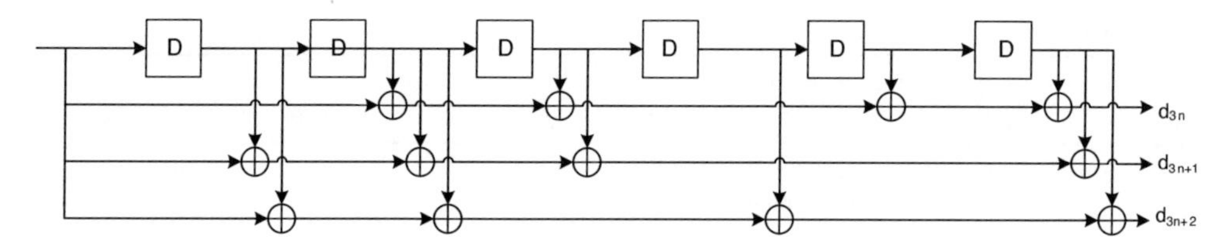

Figure 11.104 PDCCH convolutional encoder

In addition to the C-RNTI, additional RNTIs have been defined to allow identification of special cases. The possible RNTIs are given in the Table 11.64.

11.4.1.3.3.4 PDSCH Resource Allocation In order to provide flexible resource allocation for PDSCH while keeping the associated signalling overhead on PDCCH at a reasonable level, three types of allocation schemes are supported:

0. The resources assignment is for groups of consecutive resource block where the assignment information contains a bitmap where each bit indicates that a group has been assigned to the UE. The number of RBs in a group is from 1-4 and depends on the transmission bandwidth
1. The resource assignment is for a subset of the available resource blocks where the assignment information contains the allocated subset, a possible frequency domain shift of the subset and a bitmap allocating the resource blocks within the subset. The subsets are defined by the specification. Note that this allocation scheme cannot allocate the full bandwidth to one UE
2. The resource assignment is for consecutive resource blocks and the assignment information contains the first resource block assigned and the number of resource blocks assigned

The difference between the resource allocation schemes is shown in Figure 11.105.

In order to increase the frequency diversity for transmissions requiring a small number of resource blocks while not requiring the high signalling overhead incurred by resource allocations types 0 and 1, a special resource allocation mode for type 2 is provided where the allocated resource blocks are distributed in the frequency domain using a specified function and the resource allocation is different in the first and second slot of the subframe creating slot-based frequency hopping within a subframe.

Table 11.64 RNTI types

RNTI	Use	Search space	DCI Formats
C-RNTI	UE specific assignments	Common and UE specific	0, 1, 1A, 1B, 1D, 2, 2A
SPS C-RNTI	Initiation and stopping of SPS	Common and UE specific	0, 1, 1A, 2, 2A
P-RNTI	Indicates PDSCH allocation for paging (PCH)	Common	1A, 1C
RA-RNTI	Random access response	Common	1A, 1C
SI-RNTI	Indicates PDSCH allocation for BCCH	Common	1A, 1C
TPC-PUCCH-RNTI	PUCCH power control commands	Common	3, 3A
TPC-PUSCH-RNTI	PUSCH power control commands	Common	3, 3A

Table 11.65 Transmission modes

Transmission mode	Transmission scheme for PDSCH	Fallback mode	DCI formats	UE feedback
1	Single-antenna port transmission with common reference signals	None	1, 1A	CQI
2	Transmit diversity	None	1, 1A	CQI
3	Large delay CDD	Transmit diversity	1A, 2A	CQI + RI
4	Closed-loop spatial multiplexing	Transmit diversity	1A, 2	CQI + PMI + RI
5	Multi-user MIMO	Transmit diversity	1A	CQI + PMI
6	Closed-loop spatial multiplexing using a single transmission layer	Transmit diversity	1A	CQI + PMI
7	Single-antenna port transmission with UE-specific reference signals	Single-antenna port transmission with common reference signals if the number of antenna ports is one and transmit diversity otherwise	1, 1A	CQI
8	Dual layer transmission	Single-antenna port transmission with common reference signals if the number of antenna ports is one and transmit diversity otherwise	1, 1A	CQI + PMI + RI or CQI + RI

11.4.1.3.4 MIMO Operation

Introduction of high capacity spatial multiplexing is one of the main features of LTE. LTE supports operation with one, two or four *antenna ports*. These ports are virtual, meaning that they do not necessarily correspond to the physical number of antennas. An example of this is beamforming, where the number of physical antennas would be eight, but the UE would only see one, two or four antenna ports and any beamforming is then a matter of base station implementation. A cell is configured with a given number of antenna ports and the UEs processing differs. All UEs support reception of LTE transmissions from 1, 2, or 4 antenna ports at the base, but the maximum number of spatial layers a UE is a function of the UE category and depends on the number of antennas the UE has.

LTE MIMO is based on linear precoding, that is, the mapping from spatial transmission layer to the antenna port, where precoding matrixes have been specified to support the multiple MIMO transmission modes included and these precoding matrixes are assumed in the feedback design for the closed loop mode and in the transmission. In addition, from Release-9 onwards it is also possible to do spatial multiplexing using UE-specific reference signals and then any precoding is possible, but the UE feedback is still based on the standardized precoding matrixes.

The UE is configured to a certain transmission mode as shown in Table 11.65. For each transmission mode there is a fallback mode that provides robustness to the communication link and associated feedback provided to the network by the UE.

Single-antenna port transmission As the name indicates, the PDSCH is in this case transmitted from a single-antenna port. This can either be using cell-specific reference signals or UE-specific reference signals. In single-antenna port transmission, any precoding is the same for the PDSCH as for the reference signals and hence transparent to the UE.

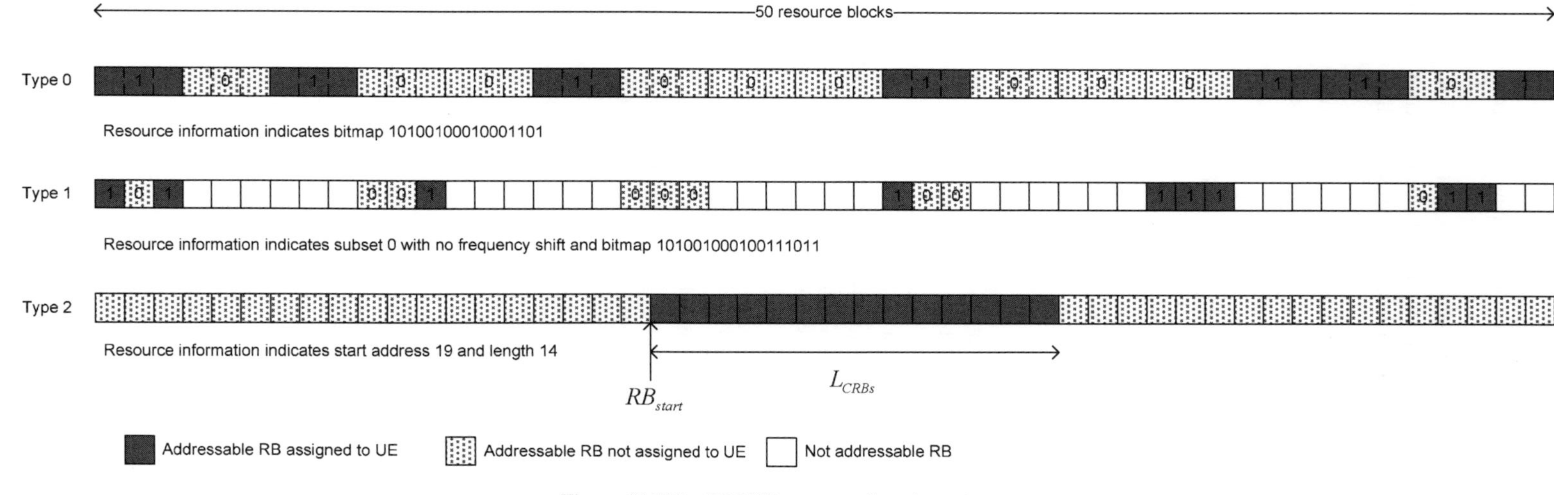

Figure 11.105 PDSCH resource allocation schemes

Transmit diversity For transmit diversity the precoding is based on the *frequency shift transmit diversity(FSTD)*

Large delay CDD In large delay cyclic delay diversity (CDD), a time difference is applied to the transmissions from the different antenna ports. The precoding used is the same as for closed loop spatial multiplexing, but the mode is open loop and the base station cycles through the codebook entries.

Closed-loop spatial multiplexing In closed-loop spatial multiplexing, the UE feeds back the recommended precoding vector from the specified codebook to the network and normal network operation is then to use this recommended vector in the transmission. The network can limit the operation in this transmission scheme to a single layer (Transmission mode 6).

MU-MIMO This mode is used to spatially multiplex two UEs to the same resource (multi-user MIMO) where each user gets a one layer transmission. The precoding is as for closed-loop spatial multiplexing.

Dual layer transmission Dual layer transmission was introduced in LTE release 9 and provides spatial multiplexing using UE specific reference signals. As for single layer transmission, any precoding is transparent since the reference signals and the transmitted data layers have the same precoding.

As shown in Table 11.65, the different modes have different levels of feedback provided by the UE to the network. The feedback provided is:

- Channel Quality Indication (CQI) reporting is similar in LTE as in HSDPA and the UE provides an indication of the channel quality it sees by signalling the combination of transport block size and modulation it estimates it can receive with a block error rate not exceeding 10%. The CQI can be reported for the whole received bandwidth or individual CQIs per subband.
- *Rank Indication (RI)* indicates the number of transmission layers the UE estimates it can receive.
- *Precoding Matrix Indicator (PMI)* indicates to the network what the UE estimates as the best precoding vector to be used for the transmission. This can be for the whole received bandwidth or individual for subbands.

11.4.1.3.5 Physical Channel Multiplexing

Figure 11.106 shows the multiplexing and mapping of the physical channels to the frame structure for FDD. For TDD, the physical channel mapping is the same, but the synchronization signal mapping is different as the primary synchronization signal is mapped to the third OFDM symbol of DwPTS and the secondary synchronization signal is mapped to the last OFDM of the first slot of subframe #0.

11.4.1.4 Physical Signals

11.4.1.4.1 Synchronization Signals

The *primary synchronization signal (PSS)* is used by the UE for initial time and frequency synchronization. There are three possible sequences used for PSS. Once the UE has identified a possible candidate frequency and timing from the PSS, it correlates the incoming signal with the 170 possible secondary synchronization signals sequences. There are 510 combinations of primary and secondary synchronization signals and this gives the physical layer ID for the cell. The cell ID gives the reference signals used in the cell and the UE can then proceed with decoding the PBCH followed by the reception of the system information from the broadcast channel giving the UE all the needed information to attach to the cell.

11.4.1.4.2 Reference Signals

In LTE, different reference signal types are provided to aim channel estimation, timing estimation and measurements:

Cell-specific reference signals: Cell-specific reference signals are used for demodulation of common channels and for PDSCH if UE-specific reference signals are not configured for the UE. It is possible to

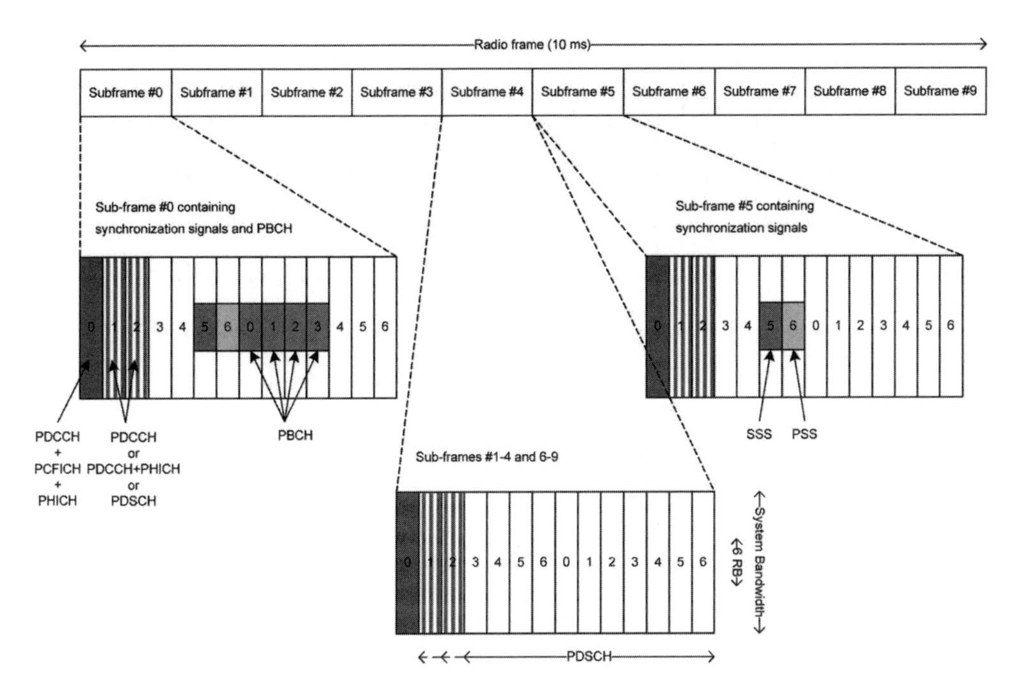

Figure 11.106 Multiplexing and mapping of the physical channels for FDD

configure a cell with one, two or four antenna ports. The cell-specific reference signals are also used for timing estimation and neighbour cell measurements. In order to allow high quality channel estimates, a resource element used for transmission of cell-specific reference signal for a given antenna port is not used for transmission ("nulled") in the other configured antenna ports.

UE-specific reference signals: When cell-specific reference signals are used, the codebook used in the precoding must be known for the UE do the demodulation. In order to allow non-codebook based precoding, for example, higher order beamforming, UE specific reference signals are provided and the UE assumes in the demodulation that the data has been subject to the same precoding as these reference signals. Since common channels still have to be transmitted to the whole cell, UE specific reference signals cannot replace cell specific reference signals, but must be provided in addition.

MBSFN reference signals: In MBMS transmissions, multiple cells broadcast the same synchronized signal and the received signal at the UE antenna is then the sum of the transmitted signals from all the base stations participating in the MBSFN transmission. To the UE, such a transmission appears as if if it is coming from a single base station in a similar way as any multipath transmission. In order to simplify the demodulation, the base stations must also transmit the same reference signals. Hence, cell specific reference signals cannot be used and special reference signal port is therefore introduced, where the reference signal density has been optimized for the long delay spread typically seen in MBSFN transmissions.

Positioning reference signals: LTE Release 9 supports a UE-based positioning method based on Observed Time Difference of Arrival(OTDOA) Special reference signals can be provided by the network to enhance the time difference measurement.

An example of LTE reference signals is shown in Figure 11.107. Here, the base station is configured with two cell-specific antenna ports and in addition, the UE is using Release 9 UE specific reference signals for demodulation of PDSCH.

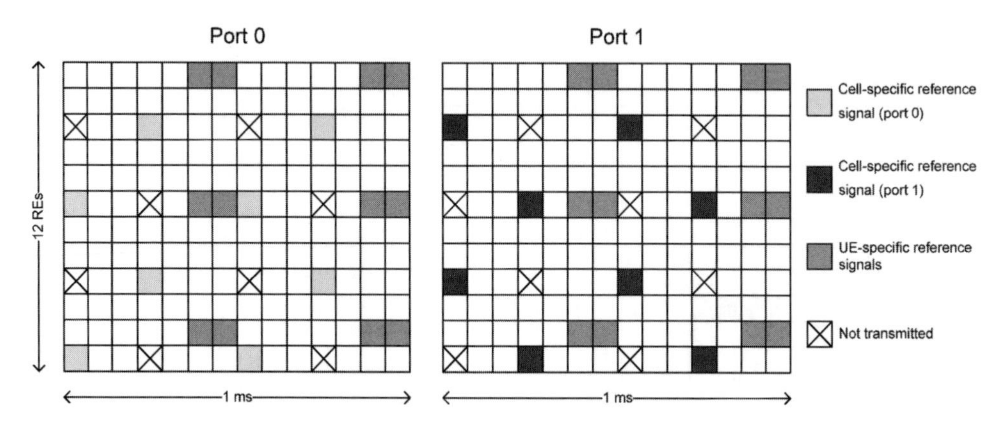

Figure 11.107 LTE reference signals

11.4.2 LTE Uplink

This section deals with *single-carrier frequency division multiple access* (*SC-FDMA*) which is the radio technology selected for LTE Uplink. Special emphasis is put on the reasons and consequences of the selection of SC-FDMA for LTE UL.

It is clear that wide area (WA) coverage was an optimization point for LTE radio. It is noted that in WA environment, Tx power of user equipment is much smaller than that of *evolved Node B* (*eNB*). This indicates that WA coverage is limited by UL direction. The importance of power consumption is another factor which differentiates the UL and DL. Thus, selection of the transmit waveform is such that power efficiency is being maximized as the key performance issue in UL. OFDMA is known to be spectrum efficient but it suffers from high peak to average power ratio (PAPR) which decreases power efficiency of the transmitter.

As an outcome of the standardization different radio technologies have been selected for LTE: OFDMA for DL and SC-FDMA for UL. Selection of the multiple access scheme, including selection of the waveform of transmitted signal, impacts in other parts of the physical layer, such as reference signal and control/data channel multiplexing design. Regardless of the different multiple access scheme, most of the building blocks used in LTE system are common for UL and DL. These methods include the frame structure, channel coding, inter-block-interference free frequency domain equalization and so on.

11.4.2.1 SC-FDMA Basics

The basic SC-FDMA transmission scheme is based on low peak to average power ratio single-carrier transmission with cyclic prefix to achieve uplink inter-user orthogonality and to enable efficient frequency-domain equalization at the receiver side [33]. The block diagram of UL SC-FDMA transceiver chain is shown in Figure 11.108.

The UE can exploit frequency domain generation of SC-FDMA signal, as presented in Figure 11.109. Frequency domain realization of SC-FDMA is called *DFT-S-OFDMA*. The SC-FDMA with frequency domain generation has high degree of commonality with DL OFDMA. The frequency domain generation enabling zero roll-off results in higher spectrum efficiency compared to earlier SC- based radio standards like WCDMA. The modulation methods specified for *Physical Uplink Data Channel* (*PUSCH*) are QPSK, 16-QAM and 64-QAM.

Different UEs are separated by means of Frequency Division Multiplexing in SC-FDMA. Each scheduled UE occupies a variable bandwidth resource, which is contiguous in the frequency domain.

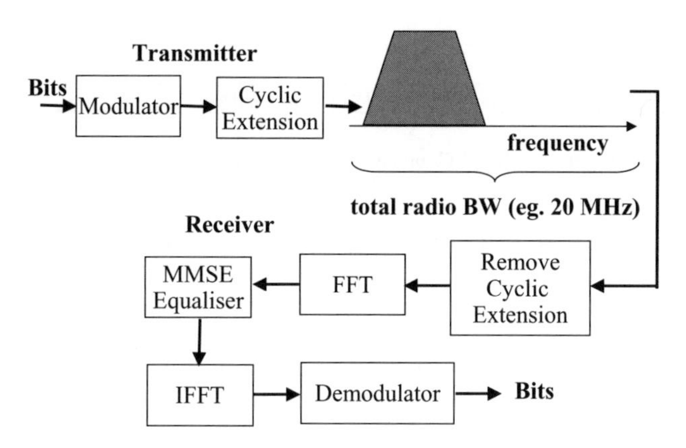

Figure 11.108 Block diagram of LTE UL transceiver chain

The frequency and time domain resources are organized into physical resource blocks and subframes. A variable number of adjacent resource blocks can be allocated to an UE. One physical resource block consists of 12 subcarries corresponding to 180 kHz frequency band and 1 ms subframe, which is further divided into two slots each 0.5 ms. The slot structure is shown in Figure 11.110. Further, the slot is divided into either six or seven SC-FDMA blocks, depending on the length of cyclic prefix (normal/extended). In data channel (physical uplink shared channel, PUSCH), the middlemost block is reserved to *demodulation reference signals* (*DMRS*).

The frequency selectivity of the radio channel can be exploited by channel dependent frequency domain scheduling or frequency hopping between slots. The frequency hopping between slots and/or subframes is typically used for control channels, whereas frequency dependent scheduling is applied for data channel. The channel dependent scheduling in supported by sounding reference signal transmitted in the last SC-FDMA block of the subframe.

11.4.2.2 Comparison Between OFDMA and SC-FDMA

The multiple access selection was based on the careful benchmarking between OFDMA and SC-FDMA. OFDMA represents a spectrum efficient system and allows usage of simple 1-tap Equalizer. The biggest drawback of OFDMA is that it suffers from high peak-to-average power ratio (PAPR). The higher the PAPR, the higher *output power back-off* (*OBO*) is needed to avoid clipping of transmitted signal which in turn result in higher *Adjacent Carrier Leakage Ratio* (*ACLR*) and increased Error Vector Magnitude (EVM) [34]. Thus, high OBO decreases power efficiency of the transmitter, which is very critical for mobile terminals with low form factor from both a battery drain and heat dissipation point of view.

The *Cubic Metric* (*CM*) for both OFDMA and SC-FDMA are shown in Figure 11.111. The CM represents the required OBO [40]. The output power difference between OFDMA and SC-FDMA is then

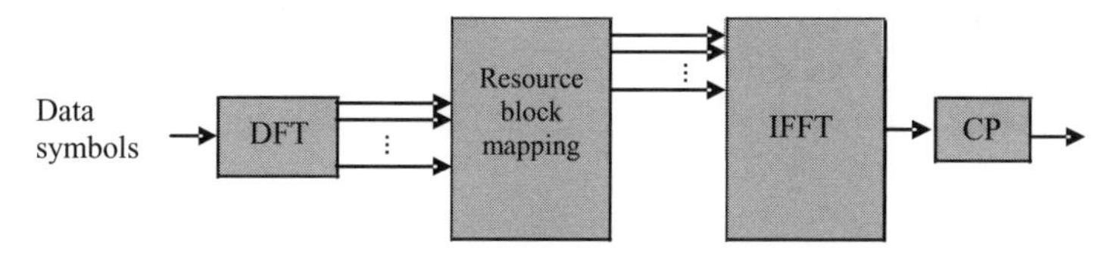

Figure 11.109 Frequency domain generation of SC-FDMA signal

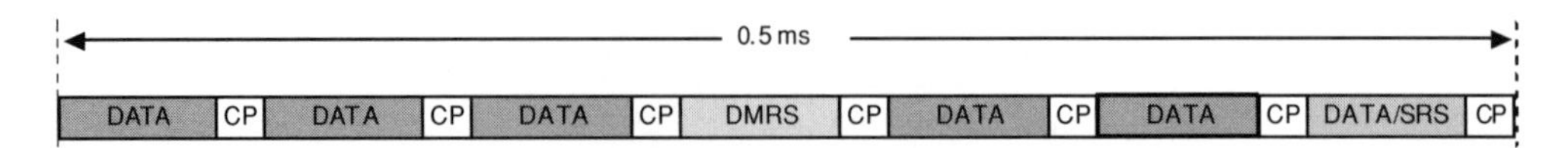

Figure 11.110 LTE uplink slot structure for short cyclic prefix

2.4 dB with QPSK modulation and 1.6 dB with 16-QAM, respectively. The OBO difference translates into corresponding difference in UL link budget.

The link performance for SC-FDMA and OFDMA is shown in Figure 11.112. The simulations are done in Typical Urban (TU) radio channel in 5 MHz bandwidth. The perfect channel state information has been assumed. The results show that SC-FDMA performs equally or slightly worse than OFDMA with QPSK using baseline *Frequency Domain Equalizer* (*FDE*). The difference in link performance varies between 0 dB and 0.4 dB and is dependent on the coding rate. In the case of 16-QAM OFDMA performs about 0.7 dB better than SC-FDMA using simple FDE. As depicted in Figure 11.112 turbo equalizer improves the performance of SC-FDMA over FDE by about 0.2 to 1 dB, resulting that the performance of SC-FDMA with turbo equalizer becomes better than that of OFDMA.

Based on given aspects, it can be said that SC-FDMA is the most preferable technology for wide area uplink, as it has lower OBO than OFDMA. Furthermore, SC-FDMA allows usage of iterative reception technique to achieve better radio link performance, even better than that of OFDMA. The price to pay is higher receiver complexity, but this is considered as a feasible and economical choice in the uplink direction.

11.4.2.3 Uplink Reference Signals

Two types of reference signals are defined in LTE uplink: *demodulation reference signals* (*DM RS*) and *sounding reference signals* (*SRS*). Demodulation reference signal are used for channel estimation and, thus, they enable coherent detection and demodulation of data. Periodic sounding reference signal transmissions provide base station channel state information (CSI) on uplink channel, also when terminal does not have data to transmit. With SRS period longer than channel coherence time, the obtained CSI can

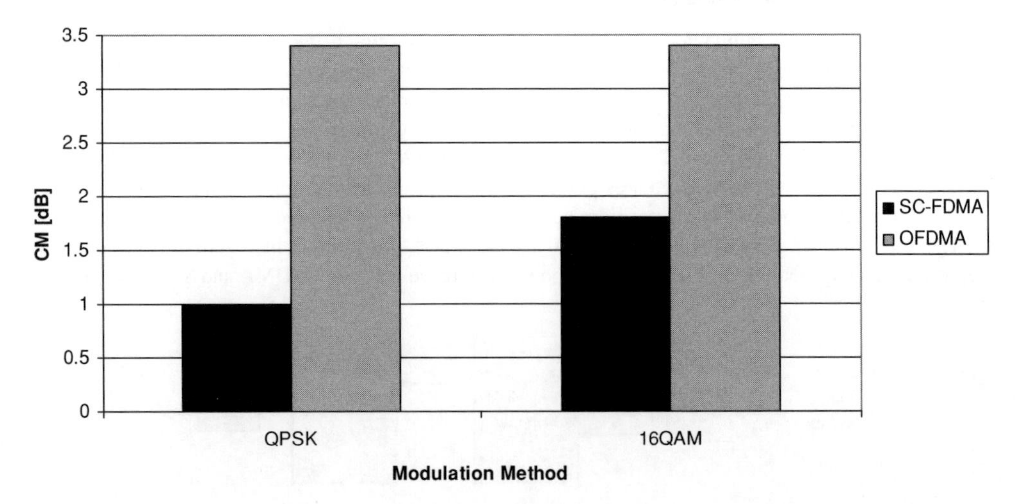

Figure 11.111 CM comparison between SC-FDMA and OFDMA

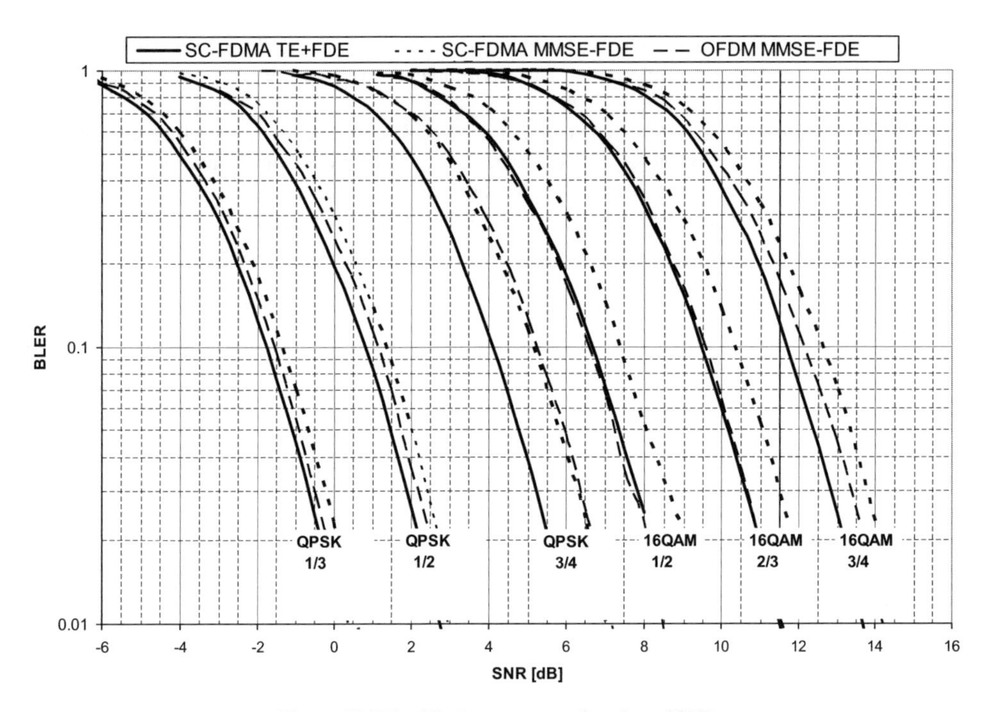

Figure 11.112 Block error rate as function of SNR

be used for adjusting uplink timing or transmission power control in absence of other uplink transmission. With shorter SRS period, the CSI can be used for packet scheduling in frequency domain.

To maintain uplink single carrier property, time division multiplexing is used between data and reference signals. On PUSCH, each slot contains one DM RS SC-FDMA symbol at the centre of the slot, as shown in Figure 11.113. Thus, acceptable channel estimation performance is achieved even at high terminal velocities while keeping DM RS overhead reasonable. DM RS overhead is 14.3% and 16.6% for normal and extended cyclic prefix lengths, respectively. On PUCCH, correct detection of carried control data is equally important for all terminals, also for those ones on cell edge. Control data transmitted on PUCCH is also time-critical so there are no possibilities for retransmissions. On the other hand, carried data amount is considerably smaller than on PUSCH. Consequently, the DM RS overhead is larger on PUCCH than on PUSCH to achieve reasonable balance between the signal energies used for channel estimation and data transmission. On PUCCH, DM RS overhead ranges from 14.3% up to 42.9%, depending on the PUCCH format and cyclic prefix length.

When configured, sounding reference signal is transmitted on the last SC-FDMA symbol of subframe. When the last SC-FDMA symbol is reserved for SRS transmission, no PUSCH data is transmitted on that symbol in the cell. Base station can reserve the last subframe symbol for SRS transmission in every

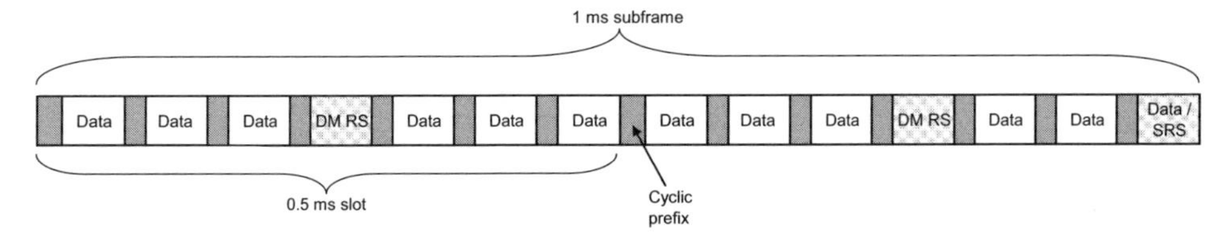

Figure 11.113 Uplink reference signals in normal cyclic prefix subframe

subframe or in some of the subframes, for example, once per 10 ms frame. Thus SRS overhead can be adjusted according to the cell configuration and environment. Overhead ranges from less than 1% up to 7%.

Reference signal sequences are designed so that they have constant amplitude in frequency domain. This considerably simplifies channel estimation especially in frequency domain. Consequently, sequences have ideal periodic autocorrelation. This means that autocorrelation is zero for all other lags except zero lag. As a result, a set of orthogonal reference signals is obtained with cyclic shifts, that is, by circularly shifting the sequence. The set of orthogonal reference signals, or cyclic shifts, is used to code division multiplex several terminals on the same physical resource block, that is, on PUCCH as discussed in section 11.4.2.4. The same sequences are used for DM RS, SRS, as well as on PUCCH.

11.4.2.4 Uplink Control Signaling

In order to benefit from single carrier property low Cubic Metric property needs to be maintained not only when transmitting data on PUSCH, but also when transmitting uplink controls signals with or without simultaneous UL data allocation. Hence, separate control signalling schemes have been specified for these two cases.

PUSCH carries the UL L1/L2 control signals in cases when the UE has been scheduled for data transmission on shared data channel. PUSCH is capable to transmit control signals with a large range of supported signalling sizes. Data and different control fields such as ACK/NACK and CQI are separated by means of time division multiplexing by mapping them into separate modulation symbols prior to the DFT.

Physical Uplink Control Channel (*PUCCH*) is a shared frequency/time resource reserved exclusively for UEs transmitting only L1/L2 control signals. PUCCH has been optimized for a large number of simultaneous UEs with relatively small number of control signaling bits per UE. Due to the single carrier limitations, simultaneous transmission of PUCCH and PUSCH is not allowed in LTE Rel-8.

11.4.2.4.1 PUCCH

PUCCH has special requirements in wide-area coverage optimized system. First of all, quality of service requirement for delay critical control signalling is high. Hence, utilization of frequency diversity is of high importance since PUCCH benefits neither from the fast link adaptation nor the HARQ. On the other hand, channel estimation limits the performance of UL control signalling in the cell-edge conditions. The best trade-off between the channel estimation performance and frequency diversity is obtained using two narrowband frequency domain clusters in a frequency hopping manner. At the same time, concurrent transmission of multiple clusters increasing the cubic metric is avoided.

PUCCH is transmitted in a frequency region located on the edges of the system bandwidth. In order to exploit frequency diversity while maintaining the single-carrier property slot-based frequency hopping symmetrically over the centre frequency is always used on PUCCH. Different UEs are separated on PUCCH by means of Frequency Division Multiplexing (FDM) and Code Division Multiplexing (CDM). FDM is used only between the narrowband resource blocks (180 kHz) whereas code division multiplexing is used inside the PUCCH resource block. CDM is a feasible choice for UE multiplexing in LTE UL since it allows longer time duration for the control signalling message and maximizes the wide area coverage. PUCCH utilizes two separate and LTE-specific ways to realize CDM [36]:

- CDM by means of cyclic shifts of sequences with suitable zero-autocorrelation properties (that is, CDM inside the SC-FDMA block)
- CDM by means of block-wise spreading with the orthogonal cover sequences (that is, CDM between multiple SC-FDMA blocks)

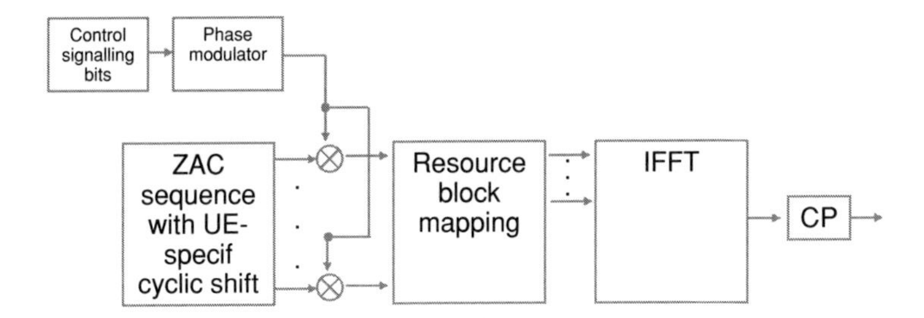

Figure 11.114 Block diagram of sequence modulator

Figure 11.114 shows the block diagram of sequence modulation scheme used in LTE PUCCH. Cyclically shifted *zero-autocorrelation* (*ZAC*) sequences take care of both user multiplexing (CDM) and conveying the control information on PUCCH. The same sequences are used also as reference signal sequences allowing coherent detection. BPSK and QPSK modulation schemes are supported on PUCCH which provides one or two control signalling bits per modulated sequence.

Block-wise spreading increases the multiplexing capacity of PUCCH by a factor of spreading factor (SF) used. The principle of block-wise spreading is shown in Figure 11.115, which illustrates the block spreading operation made for ACK/NACK reference signal (RS) sequence transmitted on PUCCH. Separate block spreading operation is made for the reference signal and ACK/NACK data parts but for simplicity, block processing related to reference symbol (RS) part is neglected in Figure 11.115. PUCCH utilizes Walsh-Hadamard codes as block spreading codes with SF = 4 and SF = 2, whereas DFT codes are used when SF equals to three.

11.4.2.4.2 PUSCH

Figure 11.116 shows the principle of control and data multiplexing within the SC-FDMA symbol (block) on PUSCH. In order to maintain the single carrier properties of transmitted signal data and different control symbols are multiplexed prior to the DFT by means of TDM multiplexing. The data part of PUSCH is punctured by the number of control symbols allocated in the given subframe. Data and different control fields (ACK/NAK, CQI/PMI, Rank Indicator) are coded and modulated separately

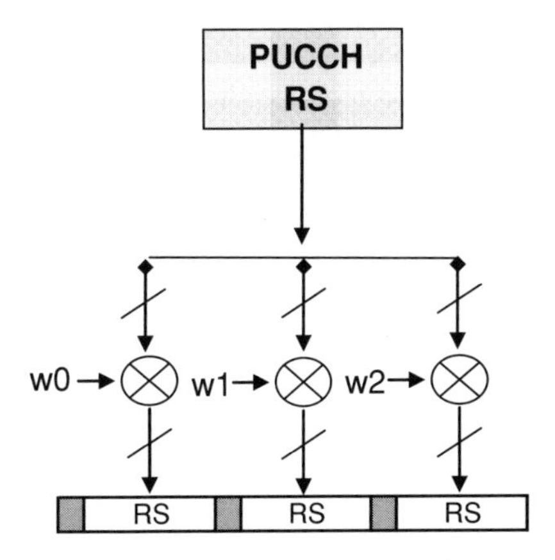

Figure 11.115 Principle of block spreading applied for ACK/NACK reference signal, spreading factor SF = 3

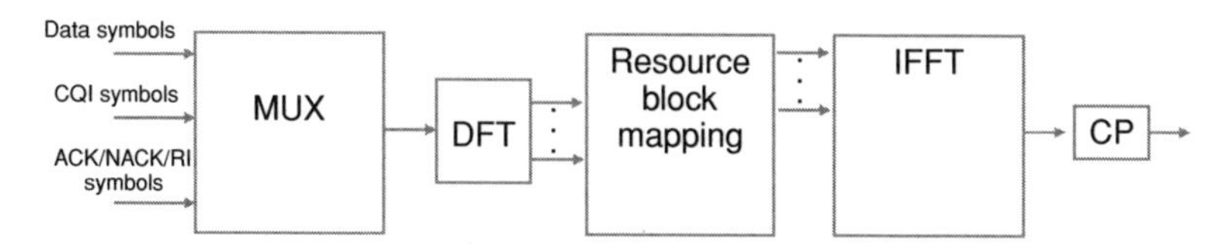

Figure 11.116 Principle of data and control modulation on PUSCH

before multiplexing them into the same SC-FDMA symbol block. Different coding rates for control are achieved by occupying different number of symbols for each control field [37].

CQI/PMI transmitted on PUSCH utilizes the same modulation scheme as the data part. ACK/NACK and *Rank Indicator* (*RI*) are transmitted in such that the coding, scrambling and modulation maximize the Euclidean distance at the symbol level. This means a modulation symbol used for ACK/NACK carries at most two bits of coded control information regardless of the PUSCH modulation scheme. The outermost constellation points having the highest transmission power are used to signal the ACK/NACK and RI in the case of 16-QAM and 64-QAM. This selection provides a small power gain for ACK/NACK and RI symbols, compared to PUSCH data using higher order modulation.

11.4.2.5 RACH

In several situations, UE needs to transmit data to the network but the base station is not aware of the need and has not allocated any uplink resources to the terminal. Such uplink random access is needed when the terminal has been idle and wants to establish radio resource control connection with the network, for example,to access the Internet or to make or receive a phone call. Random access is also needed when the terminal has been idle for some time and all PUCCH resources allocated for the terminal have been released, or when uplink timing synchronization needs to be re-established, for example,during handover between cells.

Random access preamble is transmitted on *Random Access Channel* (*RACH*). There are several requirements that are characteristic to the random access channel:

- RACH is a contention based channel and there is a possibility of collisions, that is, multiple terminals transmit simultaneously the same preamble. It is desirable to minimize collision rate, since collisions will increase random access delay and consume radio resources.
- Random access preamble needs to be detected reliably, that is, missed detection rate needs to be minimized. If preamble is missed, the terminal will transmit another preamble at higher transmission power, if possible. However, repeated transmissions will increase random access delay as well as RACH load and collision probability. RACH detection sensitivity will also set, together with performance of other channels, limits to the maximum cell range.
- False alarm rate needs to be minimized. In the false alarm, base station detects, due to noise and interference, a preamble that was never sent. Due to false detection, base station will transmit random access response at the downlink and reserve uplink resources for the consecutive response from the terminal. Thus, every false alarm will consume unnecessarily radio resources.
- Timing of received preamble needs to be estimated accurately. Terminal is synchronized to the received downlink signal when it transmits preamble. However, terminal's distance from the base station is unknown and the propagation delay both in downlink and uplink cannot be estimated nor compensated for. Thus, preamble reception is unsynchronized and the base station needs to estimate the corresponding timing from the received preamble. Based on the timing estimate, the base station can set the timing advance for the terminal, thus, compensating for propagation delays and synchronizing the consecutive uplink transmissions with the timing of transmissions from other terminals.

- Radio resources need to be used efficiently.
- RACH receiver needs to have acceptable computational complexity.

The requirements are clearly contradictory; for example, collision and missed detection rates can be easily reduced by increasing the radio resources used, and the compromise between missed detection and false alarm rates is a classical statistical decision process problem.

On each RACH subframe, 64 different preamble sequences can be transmitted. Thus, there are up to 64 random access opportunities in each RACH subframe. In this way, the number of random access opportunities can be increased, thus, decreasing preamble collision probability without increasing the amount of radio resources reserved by RACH. Of course, such code division multiple access sets further requirements for the preamble sequences. *Zadoff-Chu sequences* [33] are used as preamble sequences since they have several properties favourable for random access preamble:

- There is a large number of Zadoff-Chu sequences available, which allows for reasonable reuse factors between cells together with different RACH subframe configuration options. In LTE, 838 Zadoff-Chu sequences are used for preamble sequences.
- Zadoff-Chu sequences have ideal periodic cross-correlation and auto-correlation properties, thus, minimizing interference between different preambles as well as simplifying preamble detection and timing estimation.
- They have also low cubic metric, which translates to reasonable back-off at the power amplifier and, thus, more efficient transmission.

Since the Zadoff-Chu sequences have ideal auto-correlation properties, it is possible to derive multiple preamble sequences from a single Zadoff-Chu sequence by using cyclic shifts, as illustrated in Figure 11.117. Interference between cyclic shifts is close to zero, so the use of cyclic shifts as preambles reduces intra-cell interference. Of course, one needs to reserve a sufficient window for the two-way propagation delay for each cyclic shift used as a preamble sequence, so the number of preamble sequences obtained from a single Zadoff-Chu sequence is inversely proportional to the configured cell range.

In addition to preamble sequence, random access preamble contains also cyclic prefix (CP). With the CP, the received signal can be correlated with the preamble sequence for all potential propagation delays in frequency domain as shown in Figure 11.117. Duration of a preamble sequence is 800 μs,

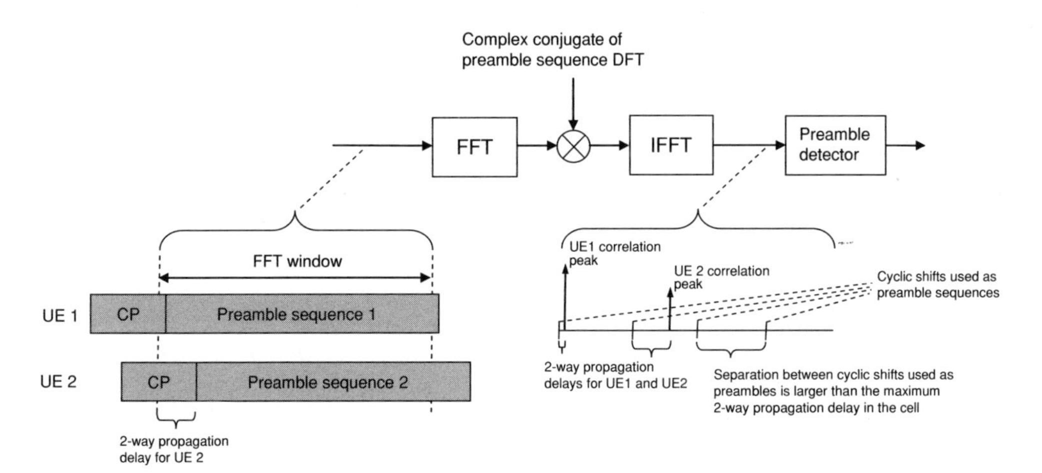

Figure 11.117 Detection of random access preambles based on cyclic shifts

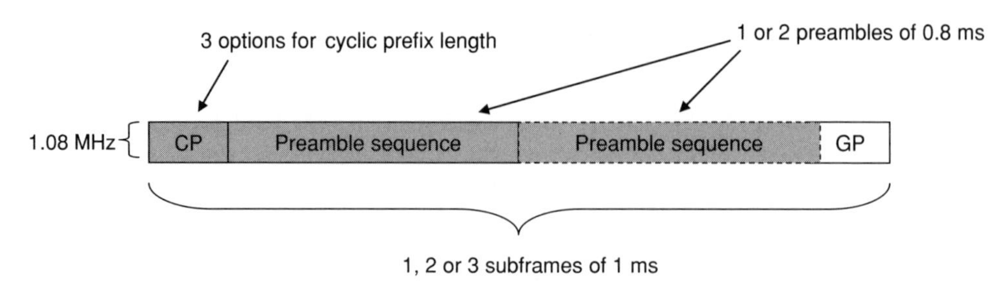

Figure 11.118 Random access preamble formats

and 1.08 MHz are reserved for RACH. The preamble sequence duration is increased to maximize the signal energy available for preamble detection while limiting the whole random access preamble within single 1 ms subframe. Preamble bandwidth is a reasonable compromise between increased frequency diversity and timing estimation accuracy, driving for larger bandwidths, and radio resource consumption and noise bandwidth, both favouring narrower bandwidths. Random access preamble is followed by a silent guard period (GP). The purpose of the GP is simply to avoid a delayed preamble from a cell edge terminal to overlap and, thus, interfere with the PUSCH transmission from other terminals in the following subframe. The length of GP corresponds roughly to the length of cyclic prefix.

LTE supports four preamble formats for both FDD and TDD and one format specific for TDD in order to provide reasonable RACH configurations to various environments. The preamble structure for the four formats suitable for both FDD and TDD is presented in Figure 11.118. Preamble formats contain three different CP lengths, accommodating preamble to different round-trip propagation delays and, thus, cell ranges. With short CP of roughly 0.1 ms, two-way propagation delays up to roughly 15 km cell ranges are supported and the RACH preamble still fits within single subframe. Longer cyclic prefixes are needed in larger cells; CP of roughly 0.2 ms can support 30 km cells ranges, while the longest CP of 684 μs can support two-way propagation delays up to 100 km, that is set as LTE maximum cell range. In two preamble formats with longer CP, RACH detection is also improved by repeating the preamble sequence twice; longer transmission time simply increases signal energy available for RACH detection. However, the prize to be paid is the increased consumption of subframes; one preamble for larger cells reserves two or three subframes. Clearly the benefit of multiple preamble formats is that RACH can be configured to support large cells while radio resource usage can be optimized in typical cells with a preamble of single subframe.

11.4.2.6 LTE Uplink Evolution (LTE-Advanced)

LTE Rel-8 is approaching the commercial deployment phase. As a part of continuous system evolution, the standardization work is now focusing on new LTE-Advanced features towards LTE Rel-10. The main new features in UL side are *single user MIMO (SU-MIMO)* and component carrier aggregation.

11.4.2.6.1 SU-MIMO

As uplink MIMO techniques, LTE is already in support for *multi-user MIMO* (*MU-MIMO*) and antenna selection transmit diversity. In multi-user MIMO, uplink transmissions from multiple terminals are spatially multiplexed by scheduling and received at the base station with suitable MIMO receiver. In LTE-Advanced, *SU-MIMO* transmission with up to four transmit antennas per terminal is supported. Three transmission modes will be specified for SU-MIMO terminals; single antenna port, transmit diversity, and spatial multiplexing modes.

In single antenna port mode, transmitted signal will appear as single antenna transmission to the base station. Single antenna port mode can be used as a fall-back mode for SU-MIMO terminals. As such, it has clear benefits simplifying the overall system design. For example, the base station does not need to have knowledge on the number of transmit antennas at the terminal immediately after random access thanks to single antenna port mode. Additionally, single antenna port mode needs anyway to be implemented at the terminal so that it can access LTE Rel-8 cells that do not support SU-MIMO in the UL. Transmit diversity mode will be used for reliable transmission of critical control signals on PUCCH.

Spatial multiplexing will be controlled by the base station. It will dynamically configure the number of spatial layers to be multiplexed, a.k.a. transmission rank. Transmission rank can be configured from one layer up to four layers. Precoding will also be configured by the base station. The terminal will periodically send channel sounding signal in the uplink. Based on the channel estimates, the base station will select the most suitable precoding matrix from a fixed precoding codebook and signal the corresponding index to the terminal. Also in the precoding codebook design, the cubic metric is taken into account and CM preserving precoding will likely be used.

11.4.2.6.2 CC-Aggregation

Carrier aggregation, where two or more component carriers are aggregated, is considered for LTE-Advanced, in order to support wider transmission bandwidths of, for example, up to 100 MHz and also for spectrum aggregation. The principle of carrier aggregation is shown in Figure 11.119, which shows aggregation of four component carriers. The physical layer specification shall support carrier aggregation for both contiguous and non-contiguous component carriers. Both cases are covered in the example depicted in Figure 11.119.

UL multiple access scheme (waveform) supporting wider bandwidth is N x DFT-S-OFDMA [38], [39]. This means that LTE-Advanced UL continues on the SC-FDMA track. The reason for selecting N x DFT-S-OFDMA is that it will provide full support for component carrier specific MCS, HARQ and power control. The selected approach provides also the best backwards compatibility between LTE Rel-8 and enables that majority of existing control signaling can be directly borrowed from LTE Rel-8. The block diagram of considered transmitter is shown in Figure 11.120 [41]. There is one transport block per component carrier (in the case without spatial multiplexing). The transport block is turbo-encoded and modulated and after the modulator, there is a separate DFT for each component carrier. The multiple IFFTs shown in the figure may be replaced by one large IFFT on condition that the component carriers are located on the common 15 kHz raster. It is also noted that support for multiple UL component carriers is an UE category feature.

Besides the multi-carrier extension it has also been decided to support clustered DFT-S-OFDMA (non-contiguous PRB allocation) within each component carrier. Furthermore, it has been decided that

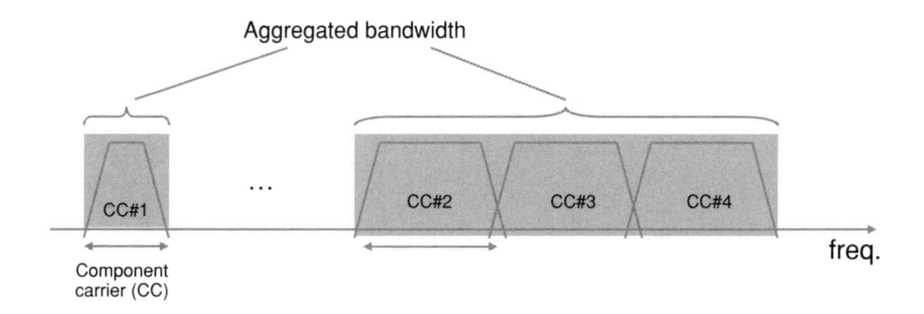

Figure 11.119 Carrier aggregation principle

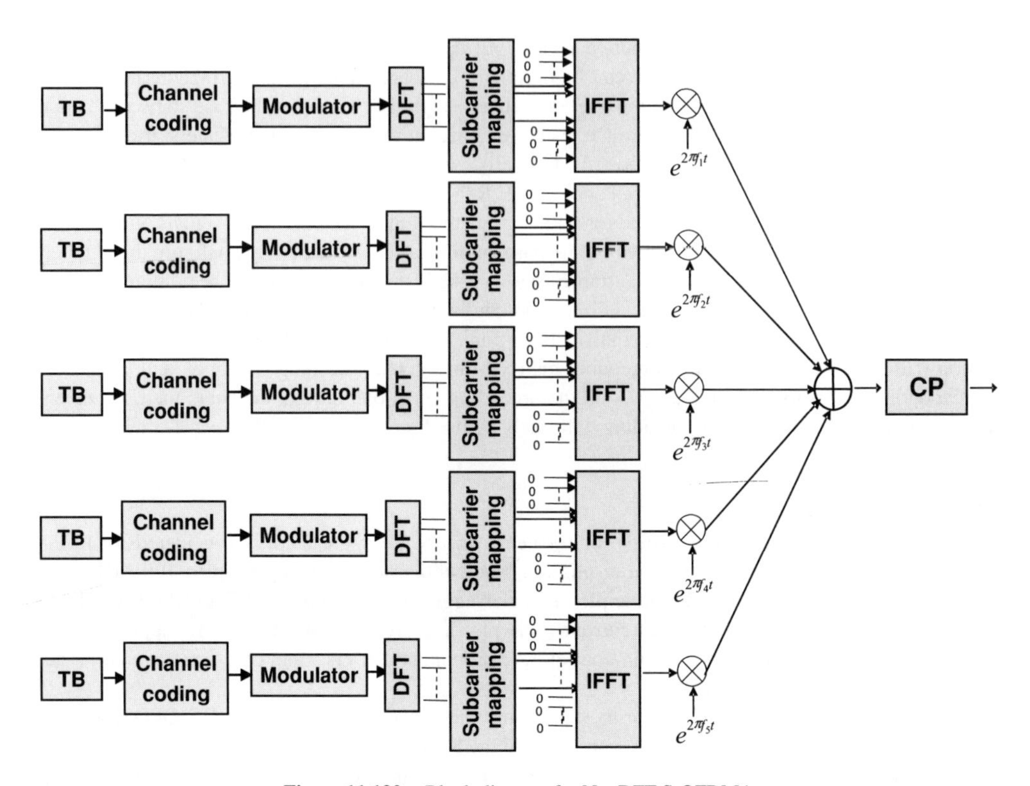

Figure 11.120 Block diagram for N x DFT-S-OFDMA

on top of Rel-8 operation, control-data decoupling (simultaneous PUCCH and PUSCH transmission) is supported.

To summarize, we note that possibility for low-CM transmission is fully maintained in LTE-Advanced including also SU-MIMO. The additional options increasing the CM compared to Rel-8 are designed to enable higher bandwidth (CC aggregation), increased scheduling flexibility (clustered DFT-S-OFDMA) and data-control separation. CM of these options is still reasonable and well below that of OFDMA. Furthermore, the new features are fully backwards compatible and can be made with rather minor standardization efforts.

References

[1] Holma, H., and A. Toskala (eds.), *WCDMA for UMTS*, Chichester, England: John Wiley & Sons, Ltd., 2000.

[2] http://www.3gpp.org/ftp/Inbox/2008_web_files/3GPP_Scopeando310807.pdf, retrieved 18. Dec. 2009

[3] http://www.3gpp.org/Specification-Groups, retrieved 18. Dec. 2009

[4] http://www.3gpp.org/RAN, retrieved 18. Dec. 2009

[5] 3GPP, *TR 21.900 Technical Specification Group working methods, v 8.3.0, 2008.* © 2008. 3GPP™ TSs and TRs are the property of ARIB, ATIS, CCSA, ETSI, TTA and TTC who jointly own the copyright in them. They are subject to further modifications and are therefore provided to you "as is" for information purposes only. Further use is strictly prohibited.

[6] http://www.3gpp.org/releases, retrieved 18 Dec. 2009

[7] 3GPP, *RP-091353, Minutes of the 45th 3GPP TSG RAN meeting, 2009*

[8] 3GPP, *TS 25.307 Requirements on User Equipments (UEs) supporting a release-independent frequency band, v. 8.4.0, 2009.* © 2009. 3GPP™ TSs and TRs are the property of ARIB, ATIS, CCSA, ETSI, TTA and TTC

who jointly own the copyright in them. They are subject to further modifications and are therefore provided to you "as is" for information purposes only. Further use is strictly prohibited.
[9] http://www.3gpp.org/ftp/Specs/html-info/TSG-WG–r1.htm, retrieved 18. Dec. 2009
[10] http://www.3gpp.org/ftp/Specs/html-info/TSG-WG–r2.htm, retrieved 18. Dec. 2009
[11] http://www.3gpp.org/ftp/Specs/html-info/TSG-WG–r4.htm, retrieved 18. Dec. 2009
[12] http://www.3gpp.org/specification-numbering, retrieved 18. Dec. 2009
[13] 3GPP, *TS 25.301 Radio Interface Protocol Architecture*, v. 7.4.0, 2008. © 2008. 3GPP™ TSs and TRs are the property of ARIB, ATIS, CCSA, ETSI, TTA and TTC who jointly own the copyright in them. They are subject to further modifications and are therefore provided to you "as is" for information purposes only. Further use is strictly prohibited.
[14] 3GPP, *TS 25.302 Services provided by the physical layer*, v. 7.6.0, 2008. © 2008. 3GPP™ TSs and TRs are the property of ARIB, ATIS, CCSA, ETSI, TTA and TTC who jointly own the copyright in them. They are subject to further modifications and are therefore provided to you "as is" for information purposes only. Further use is strictly prohibited.
[15] 3GPP, *TS 25.211 Physical channels and mapping of transport channels onto physical channels (FDD)*, v. 7.6.0, 2008. © 2008. 3GPP™ TSs and TRs are the property of ARIB, ATIS, CCSA, ETSI, TTA and TTC who jointly own the copyright in them. They are subject to further modifications and are therefore provided to you "as is" for information purposes only. Further use is strictly prohibited.
[16] 3GPP, *TS 25.212 Multiplexing and channel coding (FDD)*, v. 7.9.0, 2008. © 2008. 3GPP™ TSs and TRs are the property of ARIB, ATIS, CCSA, ETSI, TTA and TTC who jointly own the copyright in them. They are subject to further modifications and are therefore provided to you "as is" for information purposes only. Further use is strictly prohibited.
[17] 3GPP, *TS 25.213 Spreading and modulation (FDD)*, v. 7.6.0, 2008. © 2008. 3GPP™ TSs and TRs are the property of ARIB, ATIS, CCSA, ETSI, TTA and TTC who jointly own the copyright in them. They are subject to further modifications and are therefore provided to you "as is" for information purposes only. Further use is strictly prohibited.
[18] 3GPP, *TS 25.214 Physical layer procedures (FDD)*, v. 7.9.0, 2008. © 2008. 3GPP™ TSs and TRs are the property of ARIB, ATIS, CCSA, ETSI, TTA and TTC who jointly own the copyright in them. They are subject to further modifications and are therefore provided to you "as is" for information purposes only. Further use is strictly prohibited.
[19] 3GPP, *TS 25.331 Radio Resource Control (RRC); Protocol Specification*, v. 7.10.0, 2008. © 2008. 3GPP™ TSs and TRs are the property of ARIB, ATIS, CCSA, ETSI, TTA and TTC who jointly own the copyright in them. They are subject to further modifications and are therefore provided to you "as is" for information purposes only. Further use is strictly prohibited.
[20] 3GPP, *TS 25.922 Radio resource management strategies*, v. 7.1.0, 2007. © 2007. 3GPP™ TSs and TRs are the property of ARIB, ATIS, CCSA, ETSI, TTA and TTC who jointly own the copyright in them. They are subject to further modifications and are therefore provided to you "as is" for information purposes only. Further use is strictly prohibited.
[21] 3GPP, *TS 25.215 Physical layer - Measurements (FDD)*, v. 7.4.0, 2007. © 2007. 3GPP™ TSs and TRs are the property of ARIB, ATIS, CCSA, ETSI, TTA and TTC who jointly own the copyright in them. They are subject to further modifications and are therefore provided to you "as is" for information purposes only. Further use is strictly prohibited.
[22] 3GPP, *TS 25.306 UE Radio Access capabilities*, v. 7.9.0, 2009. © 2009. 3GPP™ TSs and TRs are the property of ARIB, ATIS, CCSA, ETSI, TTA and TTC who jointly own the copyright in them. They are subject to further modifications and are therefore provided to you "as is" for information purposes only. Further use is strictly prohibited.
[23] Holtzman J.M. CDMA Forward Link Waterfilling Power Control. Vehicular Technology Conference, 2000. VTC 2000 Spring. Volume 3. pp. 1663–1667.
[24] Jalali A. *et al.* Data Throughput of CDMA-HDR a High Efficiency-High Data Rate Personal Communication Wireless System. Vehicular Technology Conference, 2000. VTC 2000 Spring. Volume 3. pp. 1854–1858.
[25] Chase, D., Code combining: A maximum-likelihood decoding approach for combining an arbitrary number of noisy packets, IEEE Trans. Commun., Vol. COM -33, pp. 385–393, May 1985.
[26] 3GPP, *TR 25.848 Physical layer aspects of UTRA High Speed Downlink Packet Access*, v. 4.0.0, 2001. © 2001. 3GPP™ TSs and TRs are the property of ARIB, ATIS, CCSA, ETSI, TTA and TTC who jointly own

the copyright in them. They are subject to further modifications and are therefore provided to you "as is" for information purposes only. Further use is strictly prohibited.

[27] Holma, H., Toskala, A (eds.), *HSDPA/HSUPA for UMTS*, Chichester, England: John Wiley & Sons, Ltd., 2006.

[28] 3GPP, *TR 25.825 Dual-Cell HSDPA operation*, v. 1.0.0, 2008. © 2008. 3GPP™ TSs and TRs are the property of ARIB, ATIS, CCSA, ETSI, TTA and TTC who jointly own the copyright in them. They are subject to further modifications and are therefore provided to you "as is" for information purposes only. Further use is strictly prohibited.

[29] 3GPP, *TS 25.212 Multiplexing and channel coding (FDD)*, v. 8 DRAFT.

[30] 3GPP, *TS 25.321 Medium Access Control (MAC) protocol specification*, v. 7.12.0, 2009. © 2009. 3GPP™ TSs and TRs are the property of ARIB, ATIS, CCSA, ETSI, TTA and TTC who jointly own the copyright in them. They are subject to further modifications and are therefore provided to you "as is" for information purposes only. Further use is strictly prohibited.

[31] 3GPP, *TS 25.309 FDD Enhanced Uplink; Overall Description*, v. 6.6.0, 2006. © 2006. 3GPP™ TSs and TRs are the property of ARIB, ATIS, CCSA, ETSI, TTA and TTC who jointly own the copyright in them. They are subject to further modifications and are therefore provided to you "as is" for information purposes only. Further use is strictly prohibited.

[32] 3GPP, *TR 25.913 Requirements for Evolved UTRA (E-UTRA) and Evolved UTRAN (E-UTRAN) (Release 9)*, V9.0.0, 2009 © 2009. 3GPP™ TSs and TRs are the property of ARIB, ATIS, CCSA, ETSI, TTA and TTC who jointly own the copyright in them. They are subject to further modifications and are therefore provided to you "as is" for information purposes only. Further use is strictly prohibited.

[33] *R1-050251, Uplink Considerations for UTRAN LTE*, Nokia

[34] *R1-050639, Impact of the transmitter back-off to the uplink range*, Nokia

[35] Chu, D.C., Polyphase codes with good periodic correlation properties, IEEE Transactions on Information Theory, vol. 18, pp. 531–532, July 1972

[36] *R1-062841, Multiplexing of L1/L2 Control Signalling when UE has no data to transmit*, Nokia

[37] *R1-062840, TDM based Multiplexing Schemes between L1/L2 Control and UL Data*, Nokia

[38] *R1-081842, LTE-A Proposals for evolution*, Nokia Siemens Networks, Nokia

[39] *R1-083732, Comparison between SC-FDMA and OFDMA for LTE-Advanced Uplink*, Nokia Siemens Networks, Nokia

[40] *R1-060023, Cubic Metric in 3GPP-LTE*, Motorola Jan 2006

[41] *R1-082609, Uplink Multiple access for LTE-Advanced Nokia Siemens Networks*, Nokia Siemens Networks, Nokia

[42] 3GPP, *TS 36.306 User Equipment (UE) radio access capabilities (Release 9)*, V9.0.0 2009-09 © 2009. 3GPP™ TSs and TRs are the property of ARIB, ATIS, CCSA, ETSI, TTA and TTC who jointly own the copyright in them. They are subject to further modifications and are therefore provided to you "as is" for information purposes only. Further use is strictly prohibited.

12

CDMA2000 and Its Evolution

Andrei Ovchinnikov
St. Petersburg State University of Aerospace Instrumentation, Russia

12.1 Development of 3G CDMA2000 Standard

With code division access each user is assigned some code, modulation by which disperses the user's data over the bandwidth. The data from all users are overlaid with each other, but because of the correlation properties of modulation sequences the receiver may extract the signals of particular users from the overall received signal [5, 6]. The methods of multiple access with code division were considered in Chapter 9.

The CDMA standards exist both for second generation networks (2G) and for third generation networks (3G). CDMA standards of the second generation are called cdmaOne and include IS-95A and IS-95B standards.

CDMA is also the basis for 3G standards: two main standards of third generation networks based on CDMA: these are CDMA2000 and WCDMA standards. In this chapter we consider the standards of the CDMA2000 family. Below is a brief description of IS-95 and IS-2000 standards.

12.1.1 IS-95 Family of Standards (cdmaOne)

This telecommunication standard, which uses CDMA, was developed by Qualcomm and titled IS-95 (Interim Standard 95, also the title cdmaOne is used). This is the second generation standard (2G) for working in frequency bandwidth used in the USA by analog communication systems (AMPS). The CdmaOne standard contains a full description of the wireless system, and includes IS-95A and IS-95B revisions of CDMA TIA/EIA IS-95 standard. CdmaOne provides the set of services, including cellular communication, wireless local loop and PCS (personal communication service) [2].

The first cellular standard TIA/EIA IS-95 [4] (Telecommunications Industry Association/Electronic Industries Association Interim Standard — 95) was published in July 1993. IS-95A revision was published in May 1995 and formed the basis for many commercial second-generation CDMA systems all over the world. IS-95A describes the structure of the channels with a bandwidth of 1.25 MHz, power control, call processing, hand-offs and registration techniques.

Modulation and Coding Techniques in Wireless Communications Edited by Evgenii Krouk and Sergei Semenov

Basic characteristics of this standard are listed below [6].

- Each channel is spread to bandwidth of 1.25 MГц and filtered to limit the spectrum.
- The transmission rate of elementary signals (chips) is 1.2288 millions signals per second. Nominal transmission rate in RS1 mode (Rate Set 1) is 9.6 Kbit/s, it is also possible to use improved rate mode RS2 to achieve a rate of 14.4 Kbit /s.
- To modulate the signal the binary phase-shift keying (BPSK) is used, with signal spreading by means of QPSK method.
- The convolutional error-correcting code with rate $1/2$ and constraint length $K = 9$ is used for forward error correction with the Viterbi decoding.
- In downlink (forward) channel the base station performs time-division of users with a time interval of 20 ms.
- For signal receiving the RAKE-receiver is used, demodulating three of the strongest components of multipath signal in mobile station and four components in base station. To provide spatial diversity the two antennas are used.
- Base station uses 64 channels for transmission. For channel separation the orthogonal code multiplexing is used.
- Power control allows the base station to equalize the levels of signals from users which are at a different distance from the base station. This allows the power of transmitted signal to be minimized and the interference to be decreased.

IS-95B revision provides a transmission rate of 64 Kbit/s in line-switching mode and 115.2 Kbit/s in batch mode of the CDMA network. Because of the achievable rates the IS-95B standard is referred to the set of 2.5G standards.

Advantages of cdmaOne standard when implemented in the cellular network are

- Capacity increase 8 to 10 times compared to the AMPS analog system and 4 to 5 times compared to the GSM system.
- Improved call quality, with better and more consistent sound as compared to AMPS systems.
- Simplified system planning through the use of the same frequency in every sector of every cell.
- Enhanced privacy.
- Improved coverage characteristics, allowing for the possibility of fewer cell sites.
- Increased talk time for portables.
- Bandwidth on demand.

12.1.2 IS-2000 Family of Standards

The networks of third generation (3G) are the systems, providing high quality of voice communications, and also high speed of data transmission, including access to Internet, working of mobile applications and transferring mobile content. The program IMT-2000 (International Mobile Telecommunications-2000), developed by International Telecommunication Union (ITU), defines technical requirements and standards, as well as frequency bandwidth usage for third generation systems, involved in the program.

According to ITU requirements, third generation networks of IMT-2000 should provide improved capacity and spectrum efficiency compared to 2G systems. Besides, they should support data transmission on rates of 144 kbps in outdoor and 2 Mbps in indoor conditions.

Based on these requirements, in 1999 ITU approved five standards for IMT-2000. Three of them (CDMA2000, TD-SCDMA, WCDMA) are based on CDMA.

CDMA2000 is the family of standards and includes:

- CDMA2000 1x
- CDMA2000 1xEV-DO
  - CDMA2000 1xEV-DO Rel 0
  - CDMA2000 1xEV-DO Rev A
  - CDMA2000 1xEV-DO Rev B

Standards of 1x*EV-DO (evolution data-only, or data-optimized)* family became the development of CDMA2000 1x standard, oriented on improving the data transmission. Also another direction of development is CDMA2000 *EV-DV (evolution data/voice)* standard, in which both data and voice transmission are improved. However, in practice the direction EV-DO turned to became more attractive to telecoms operators, since these standards do not require backward compatibility, network equipment cost is lower, and once completed there was not enough EV-DV equipment to match the market requirements, while at the same time EV-DO equipment was accessible. As the result, in March 2005 Qualcomm froze the development of EV-DV chipsets and focused on development of the EV-DO products.

CDMA2000 uses the advantages of CDMA technology, and also contributes new improvements such as Orthogonal Frequency Division Multiplexing (OFDM and OFDMA), improved methods for signal control and transmission, improved countermeasures against interference, quality of service (QoS) management, and also the novel "antenna" techniques like Multiple-Input Multiple-Output (MIMO) and *Space Division Multiple Access (SDMA)*. All of these significantly improve data transmission rate and services quality, at the same time significantly increasing the capacity of the network and decreasing costs.

The main properties of CDMA2000 are:

- Voice quality.
- High-speed broadband data connectivity.
- Low end-to-end latency.
- Efficient use of spectrum.
- Support for advanced mobile services: CDMA2000 1xEV-DO enables the delivery of a broad range of advanced services, such as high-performance VoIP, push-to-talk, video telephony, multimedia messaging, multicasting and online gaming.
- All-IP: CDMA2000 technologies are compatible with IP and ready to support network convergence.
- Flexibility: CDMA2000 systems have been designed for urban as well as remote rural areas for fixed wireless, wireless local loop (WLL), limited mobility and full mobility applications in multiple spectrum bands.
- Application, user and flow-based QoS.
- Improved security and privacy.

In the following sections we briefly consider the main properties of CDMA2000 and its evolution CDMA2000 1xEV-DO.

12.1.2.1 CDMA2000 1x

CDMA2000 1x (IS-2000) was accepted by International Telecommunications Union (ITU) as IMT-2000 standard in November 1999. This was the first practically deployed IMT-2000 system in the world (October 2000).

CDMA2000 1x supports data transmission at rates up to 307 kbps in one channel with 1.25 MHz bandwidth.

Main properties of CDMA2000 1x are:

- Voice capacity.
- High-speed data transmission: Release 0 supports bi-directional peak data rates of up to 153 kbps and an average of 60-100 kbps in commercial networks in a 1.25 MHz channel. Release 1 can deliver peak data rates of up to 307 kbps.
- Applications: Supports circuit-switched voice, short messaging service (SMS), multimedia messaging service (MMS), games, GPS-based location services, music and video downloads.

Mobile phones of CDMA2000 1x are backwards compatible with cdmaOne standard.

The following basic methods of CDMA are used in CDMA2000 1x and backwards compatible with cdmaOne:

- Direct Sequence Spread Spectrum Multiple Access — to improve spectral efficiency (system capacity).
- Orthogonal Code Channelization — for user separation on the downlink (mitigates interference).
- Random Access — to efficiently share radio access resources among all users.
- Fast Uplink Power Control — to resolve the near-far field effect (reduce interference).
- Rake Receivers — to resolve and benefit from multipath interference and support soft handoffs.
- Soft Handoff — to handoff users between base stations.
- Softer Handoff — to handoff users between base station sectors.
- Soft Handoff (SHO) Active Set — to provide seamless service with increased spectral efficiency.
- Single Frequency Re-use — to increase overall network capacity.
- Downlink Slotted Paging — to extend the battery life of mobile devices.
- Blind Rate Detection — to enable variable rate decoding without additional overhead.
- Downlink Reference Channel — to share a common pilot to increase capacity.
- Downlink Channel Structure — to simplify system implementation and efficiency by separating channels with Walsh codes.
- Scrambling — to provide communications privacy.
- Speech Regulated Vocoders — to reduce interference and increase system capacity.

Besides the already mentioned technologies from cdmaOne in CDMA2000 new technologies are used:

- Variable Length Orthogonal Codes — to support variable data rates.
- Uplink Complex Spreading — to increase data rates and network capacity.
- Fast Downlink Power Control — to reduce transmit power usage and increase capacity.
- Data Rate Configurable Channels — to support applications that use a variety of data rates.
- Dual-Event Downlink Paging — to further extend the battery life of mobile devices.
- Uplink Channel Structure — to multiplex control and data channels.
- Reserve Mode Random Access — to access the network more efficiently.
- Parallel Turbo Codes — to improve capacity through more efficient forward error correction.
- Coherent Uplink Detection — to improve data rates and coverage.
- Continuous Uplink Operation — to increase transmission range and capacity, while reducing interference to hearing aids and other devices.

Further development of CDMA2000 standard is CDMA2000 1xEV-DO presented in the next subsection.

12.1.2.2 CDMA2000 1xEV-DO

CDMA2000 1xEV-DO (Evolution-Data Optimized) standard was approved as IMT-2000 standard in 2001 and provides peak transmission rates above 2 Mbps. Variations of this standard include CDMA2000 1xEV-DO Release 0, CDMA2000 1xEV-DO Revision A and CDMA2000 1xEV-DO Revision B.

CDMA2000 1xEV-DO Release 0 (Rel. 0) standard proposes data transmission on rates up to 2.4 Mbps. Its key features are:

- Broadband data: Provides a peak data rate of 2.4 Mbps in the forward link and 153 kbps in the reverse link in a single 1.25 MHz FDD carrier. In commercial networks, Rel. 0 delivers an average throughput of 300-700 kbps in the forward link and 70-90 kbps in the reverse link.
- Supports IP-based network connectivity and software applications.
- Applications: Supports broadband data applications, such as broadband Internet or VPN access, MP3 music downloads, 3D gaming, TV broadcasts, video and audio downloads.

Besides technologies used in CDMA2000 1x, in CDMA2000 1xEV-DO Rel. 0 the following improvements are used:

- High-speed packet-switched downlink channelization structure — bundling downlink resources into a packet data channel to enable high-speed data rate transmissions by combining all of the available Walsh codes and power.
- Adaptive modulation and coding schemes — to optimize the delivery of packets based on changes in the radio environment.
- Adaptive packet data scheduling — to rapidly adapt to changes in the radio link.
- Hybrid ARQ — to acknowledge correct receipt of data and retransmit erroneous data.
- Incremental redundancy feedback in the Downlink — to increase the effective data rate in the uplink by terminating the transmission of a packet early if it is decoded earlier than expected.
- Rate control — to rapidly adjust to changes in the radio environment in downlink and to efficiently control the transmission of mobile devices in uplink.
- Downlink multiple user separation — to efficiently assign the downlink channel to users.
- Downlink transmission signaling — to indicate the downlink modulation and coding.
- Uplink rate detection — to enable correct decoding of uplink data traffic.
- Short transmission time intervals (TTI) — to accelerate the transmission of packets.

Rev. A is an evolution of CDMA2000 1xEV-DO Rel. 0 that increases peak rates on reverse and forward links to support a wide variety of symmetric, delay-sensitive, real-time, and concurrent voice and broadband data applications. It also incorporates OFDM technology to enable multicasting (one-to-many) multimedia content delivery.

Rev. A allows users to send large files, emails with attachments, high resolution photographs and personal videos from their mobile devices. With its low network latency, service tiering with Quality of Service (QoS) and IP-based broadband architecture, Rev. A is able to support time-sensitive applications, such as Voice over IP (VoIP), Push-to-Talk (PTT) and video telephony. Rev. A was launched in October 2006.

Key features of Rev. A include:

- Improved broadband speeds — Provides a peak data rate of 3.1 Mbps in the forward link and 1.8 Mbps in the reverse link in a 1.25 MHz FDD carrier. In commercial networks, Rev. A achieves average throughput of 450-800 kbps in the forward link and 300-400 kbps in the reverse link.

- Higher spectral efficiency — Supports 1.2 times Rel. 0 forward link sector capacity and 3.4 times reverse link sector capacity. Increased rate quantization on both forward and reverse link enables more efficient use air link resources, better network utilization and lower cost of delivery.
- Increased capacity — On both the forward and reverse link, Rev. A allows operators to support more users and it improves the cost of delivering voice, data and multimedia services.
- Symmetry — Symmetry is important for applications where users send packets of data as often as they receive them, such as receiving and sending email with attachments.
- Low latency — The average latency of Rev. A is below 50 milliseconds, making it ideal for delay-sensitive applications.
- Advanced quality of service (QoS) mechanisms that support the prioritization and delivery of individual packets based on the type of application or user profile. These mechanisms ensure a consistent, high-quality user experience.
- All-IP — Like 1xEV-DO Rel. 0, All-IP Rev. A network provides operators with service flexibility and higher bandwidth efficiencies.
- Advanced services — Enables the enhanced performance of real-time broadband, symmetric data link, and delay sensitive services such as VoIP, push-to-talk (PTT), push-to-media (PTM), video conferencing, multicasting, and 3D gaming with multiple players.
- Backward compatibility — Rev. A networks support existing Rel. 0 applications and devices.

In addition to the air interface techniques used in CDMA2000 1x and 1xEV-DO Rel. 0, the following new high-speed packet-switched uplink techniques are incorporated into CDMA2000 1xEV-DO Rev. A:

- Fast uplink rate control — to efficiently control the transmission of mobile devices.
- Hybrid ARQ in uplink — to acknowledge correct receipt of data and retransmit erroneous data.
- Incremental redundancy feedback in uplink — to increase the effective data rate in the downlink by terminating the transmission of a packet early if it is decoded earlier than expected.
- Uplink channelization — to enable better control of the uplink data flows.
- Short Transmission Time Interval (TTI) — to accelerate the transmission of packets

The Revision B (Rev. B) is an evolutionary step of Rev. A that consists of a software upgrade that aggregates multiple EV-DO Rev. A channels to provide higher performance for multimedia delivery, bi-directional data transmissions and VoIP-based concurrent services. The Rev. B standard was published by the 3GPP2 under document number 3GPP2 C.S0024-B [1] and by the TIA and Electronics Industry Association as TIA/EIA/IS-856-B.

Rev. B builds on the efficiencies of Rev. A by introducing the concept of dynamically scalable bandwidth. Through aggregation of multiple 1.25 MHz Rev. A channels, Rev. B enables data traffic to flow over more than one carrier and hence improve user data rates, latencies on both the forward and reverse link. Peak data rates are proportional to the number of carriers aggregated. When 15 channels are combined within a 20 MHz bandwidth, Rev. B delivers peak rates of 46.5 Mbps in forward link and 27 Mbps in the reverse link. With the 64-QAM scheme, the peak data rate in the forward link increase in a single 1.25 MHz carrier to 4.9 Mbps, an aggregated 5 MHz will deliver up to 14.7 Mbps and within 20 MHz of bandwidth up to 73.5 Mbps. By increasing the bandwidth, an operator can support more users per sector or lower their cost per megabyte to encourage longer usage. To achieve this performance, the 1.25 MHz carriers do not have to be adjacent to one another, thus giving operators the flexibility to combine blocks of spectrum from different bands. This is a unique benefit of Rev. B that is not available to WCDMA/HSDPA [2].

In addition to supporting mobile broadband data and OFDM-based multicasting, the lower latency characteristics of Rev. B improve the performance of delay-sensitive applications such as VoIP, push-to-talk, video telephony, concurrent voice and multimedia and multiplayer online gaming.

Table 12.1 Radio-configuration parameters of CDMA Reverse Channel

Radio Config.	Associated Spreading Rate	Data rates, Forward Error Correction, and general characteristics
1	1	1200, 2400, 4800, and 9600 bps data rates with $R = 1/3$, 64-ary orthogonal modulation
2	1	1800, 3600, 7200, and 14400 bps data rates with $R = 1/2$, 64-ary orthogonal modulation
3	1	1200, 1350, 1500, 2400, 2700, 4800, 9600, 19200, 38400, 76800, and 15300 bps data rates with $R = 1/4$, 307200 bps data rate with $R = 1/2$, BPSK modulation with a pilot
4	1	1800, 3600, 7200, 14400, 28800, 57600, 115200, and 230400 bps data rates with $R = 1/4$, BPSK modulation with a pilot
5	3	1200, 1350, 1500, 2400, 2700, 4800, 9600, 19200, 38400, 76800, and 15300 bps data rates with $R = 1/4$, 307200 and 614400 bps data rate with $R = 1/3$, BPSK modulation with a pilot
6	3	1800, 3600, 7200, 14400, 28800, 57600, 115200, 230400, and 460800 bps data rates with $R = 1/4$, 1036800 bps data rates with $R = 1/2$, BPSK modulation with a pilot

12.2 Reverse Channel of Physical Layer in CDMA2000 Standard

12.2.1 Reverse Channel Structure

CDMA2000 standard [3] described interoperability of the base station and user mobile stations. The transmission channel from base station to mobile station is referred to in the standard as reverse channel, while the transmission channel from mobile station to base station is referred to as forward channel. In this section we will describe the reverse channel.

Transmission over the reverse channel may be performed in different modes. First, the transmission mode may be defined by the so-called spreading rate, which may be denoted by the number 1 or 3. With spreading rate 1 (SR1) the transmission is performed on the rate 1.2288 Mcps, using direct spread of the one carrier. With spreading rate 3 (SR3) and transmission over the CDMA forward channel either three carriers with rate 1.2288 Mcps (Multi-Carrier Forward Channel) or one carrier with direct spread on rate 3.6864 Mcps (Direct-Spread Forward Channel) may be used. During the transmission over the CDMA reverse channel the one carrier with direct spreading of spectrum is used on rate 3.6864 Mcps.

Then, for SR1 mode the four radio-configurations (RC) are defined. Two more radio-configurations are defined for SR3 mode. Parameters of radio-configurations are listed in Table 12.1.

The Reverse Channel structure of the CDMA2000 standard is shown in Figure 12.1, taking into account different SR and RC modes. The channel consists of the following subchannels.

- **Access Channel**. The Access Channel is used for short signaling message exchanges such as call originations, responses to pages, and registrations.
- **Reverse Traffic Channel**. A traffic channel on which data and signaling are transmitted from a mobile station to a base station.
  - **Reverse Fundamental Channel**. A channel which carries higher level data and control information.
  - **Reverse Supplemental Code Channel**. A channel which operates in conjunction with some other channels to provide higher data rate services.
  - **Reverse Pilot Channel**. A reverse pilot channel provides a phase reference for coherent demodulation and may provide a means for signal strength measurement.
  - **Reverse Dedicated Control Channel**. A channel used for the transmission of higher-level data and control information from a mobile station to a base station.

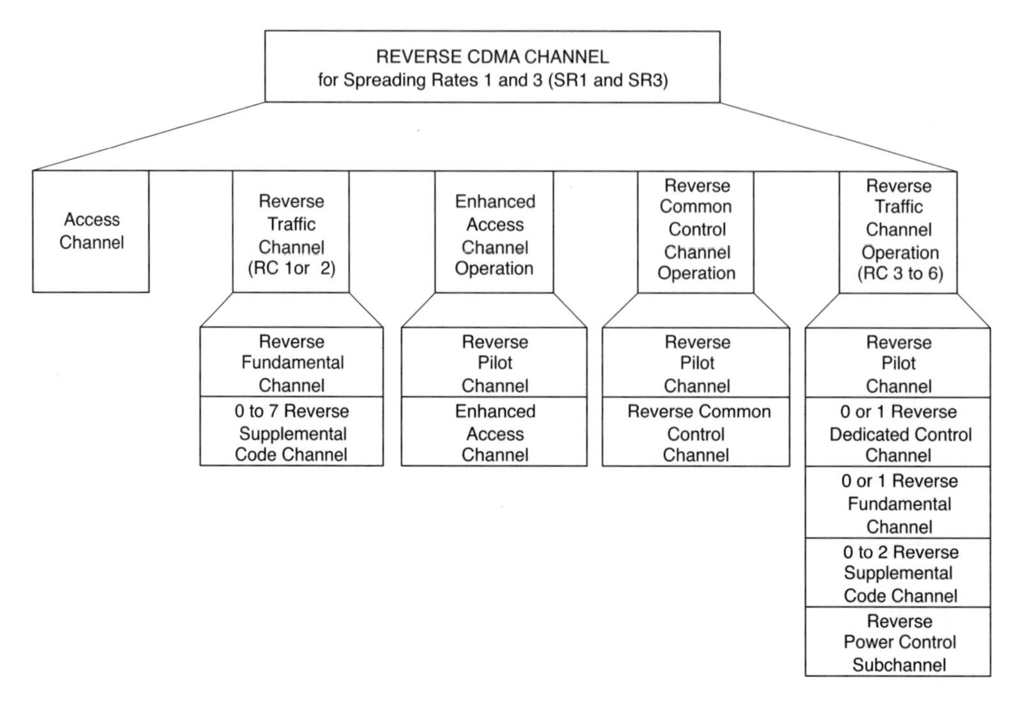

Figure 12.1 Reverse Channel structure

  - **Reverse Supplemental Channel**. A channel which operates in conjunction with some other channels to provide higher data rate services.
  - **Reverse Power Control Subchannel**. A subchannel used by the mobile station to control the power of a base station.
- **Enhanced Access Channel** — used for transmission of short messages, such as signaling, MAC messages, response to pages, and call originations. It can also be used to transmit moderate-sized data packets.
- **Reverse Common Control Channel** — used for the transmission of digital control information from one or more mobile stations to a base station.

We will describe only SR1 mode in radio-configurations RC1 and RC3. In fact, from the functional point of view the other radio-configurations do not differ from these. The Access Channel structure is depicted in Figure 12.2.

The structure of Reverse Traffic Channel (which consists of Reverse Fundamental Channel and Reverse Supplemental Channel) for radio-configuration 1 is depicted in Figure 12.3.

The structure of Enhanced Access Channel and Reverse Common Control Channel are shown in Figures 12.4 and 12.7. The structure of Reverse Traffic Channel for radio-configuration 3 is shown in Figures 12.5–12.7. The structure of Reverse Pilot Channel is also presented in Figure 12.7.

12.2.2 Forward Error Correction (FEC)

Cdma2000 standard uses the convolutional codes as error-correcting codes. To provide FEC coding in Reverse Supplemental Channel the turbo-codes may also be used.

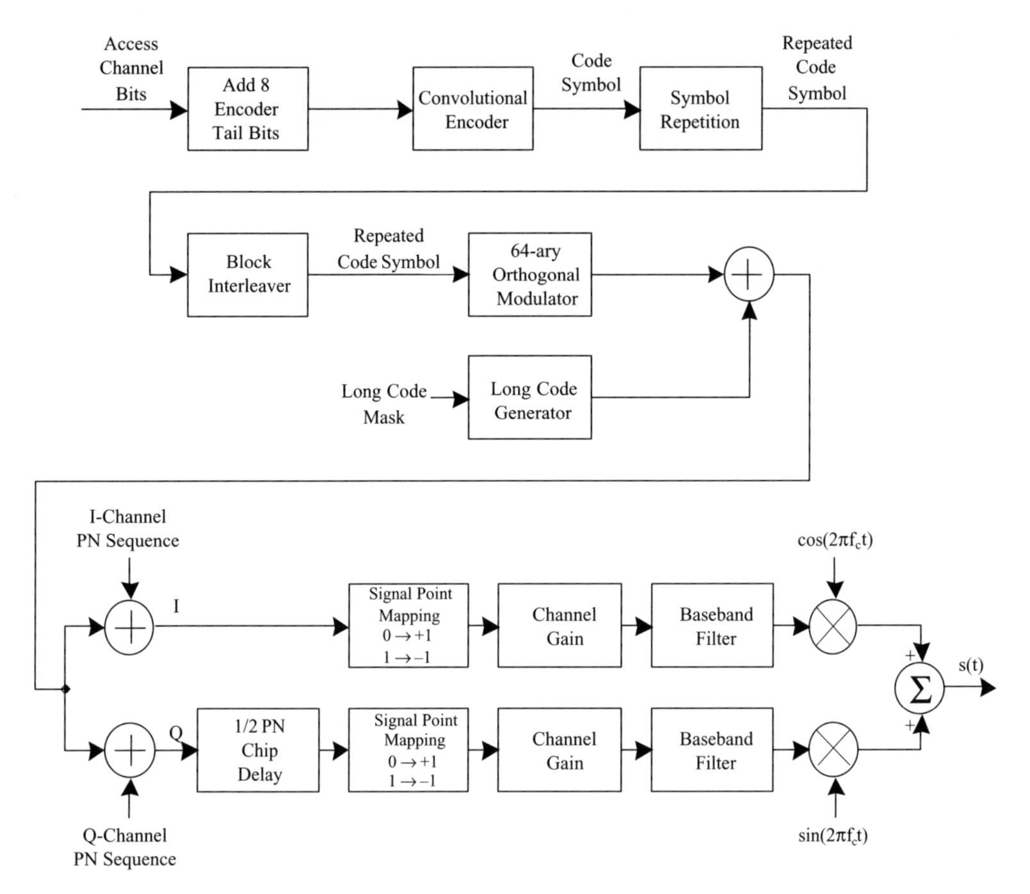

Figure 12.2 Access Channel structure (SR1 mode)

12.2.2.1 The Convolutional Codes in CDMA2000 Standard

The convolutional codes used in CDMA2000 standard have code rates of 1/2, 1/3 and 1/4 and the constraint length of 9. At the initial moment of time all the registers are set to zero, the structure of the code (encoder scheme) with rate 1/4 is shown in Figure 12.8, while the same structures for the rate 1/3 and rate 1/2 codes are described in Figure 12.9 and Figure 12.10 correspondingly.

12.2.2.2 The Turbo-Codes in CDMA2000 Standard

Let K denotes the length of encoded physical layer packet, and turbo-codes is (N_{out}, N_{turbo}, R)-code with codelength N_{out}, information message length N_{turbo} and code rate R.

The information symbols of encoded message are the symbols of physical layer packet. The turbo-code calculates $N_{out} = N_{turbo}/R$ codeword symbols, and also $6/R$ tail symbols.

The encoder of the turbo-code consists of two constituent systematic recursive convolutional encoders, connected in parallel with an interleaver between. To achieve the target rates the outputs of constituent encoders may be punctured or repeated.

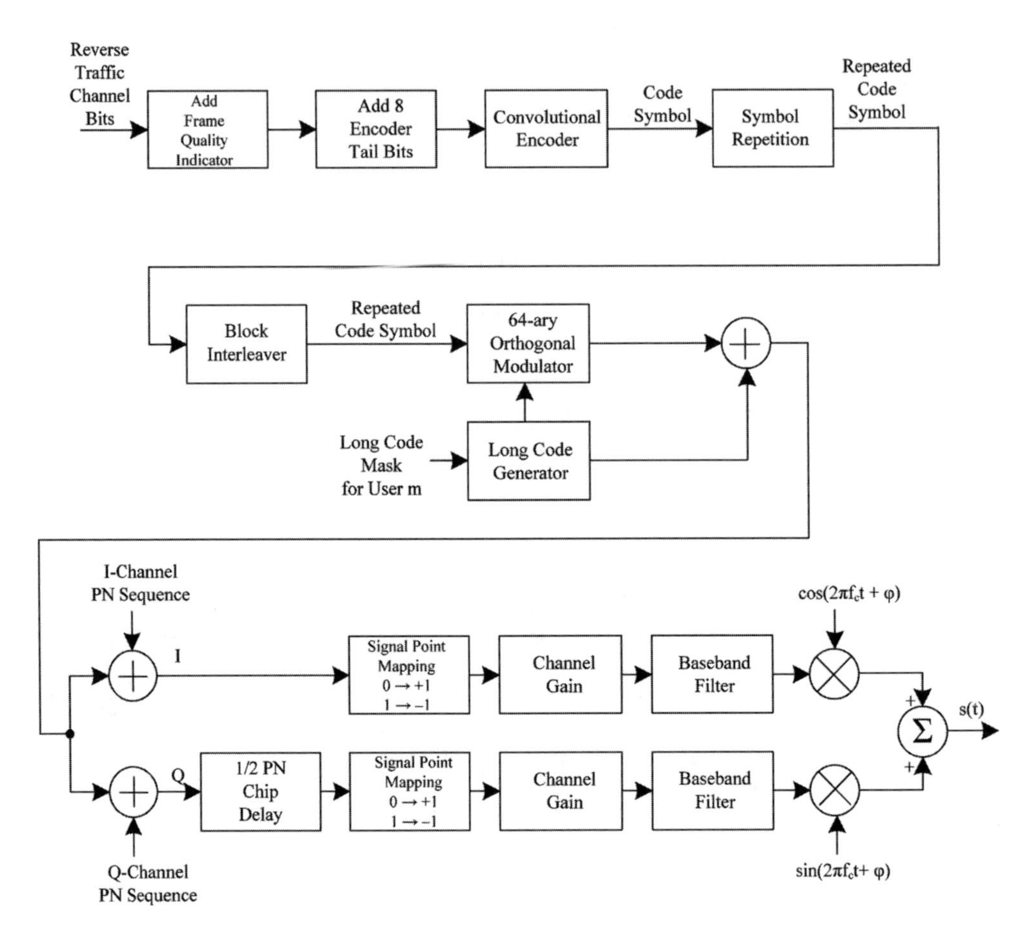

Figure 12.3 Reverse Traffic Channel structure, RC1 mode

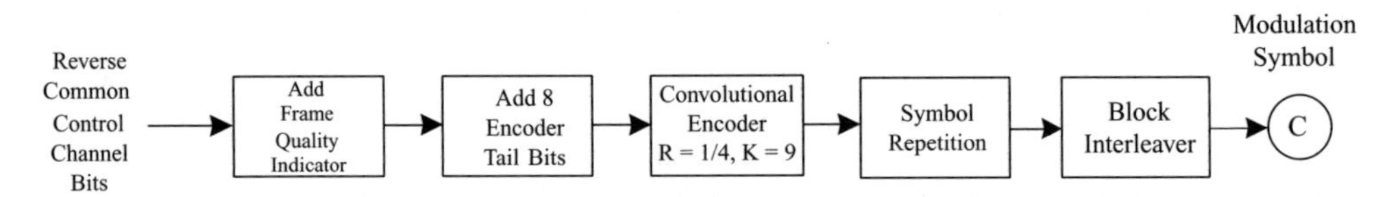

Figure 12.4 Enhanced Access Channel and Reverse Common Control Channel structure

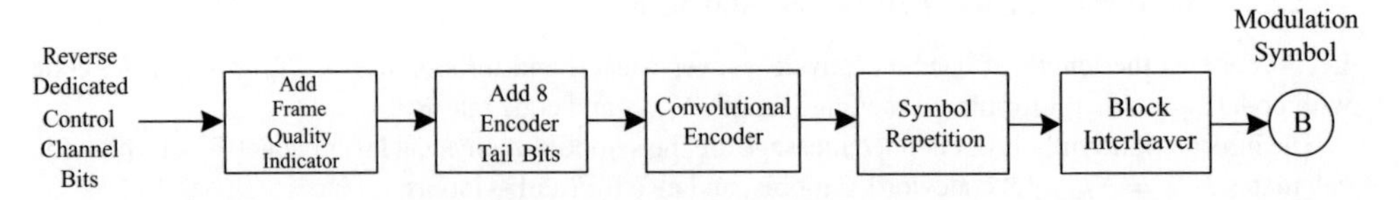

Figure 12.5 Reverse Traffic Channel structure (Reverse Dedicated Control Channel), RC3 mode

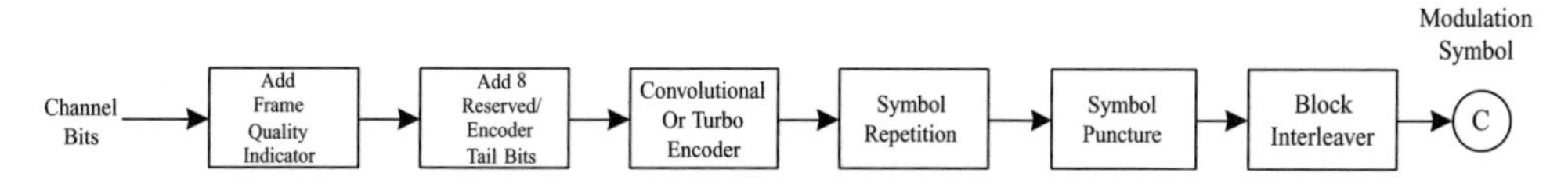

Figure 12.6 Reverse Traffic Channel structure (Reverse Fundamental Channel and Reverse Supplemental Channel), RC3 mode

The transfer function of each constituent encoder is defined by:

$$G(D) = \left[1 \quad \frac{1 + D + D^3}{1 + D^2 + D^3} \quad \frac{1 + D + D^2 + D^3}{1 + D^2 + D^3} \right]$$

The encoder scheme is shown in Figure 12.11. In the initial moment of time the elements of registers are filled with zeros. Then the registers are clocked N_{turbo} times with switches in "up" position to obtain the symbols of the codeword. The outputs of constituent encoders are punctured in correspondence with Table 12.2, where "0" corresponds to symbol puncturing, and "1" corresponds to symbol output.

Then the switches are set into "down" position and $6/R$ tail symbols are calculated: first constituent encoder are clocked three times (the second constituent encoder does not work), then the second constituent encoder are clocked three times (first constituent encoder does not work). The outputs of encoders are punctured as shown in Table 12.3 and in the case of using codes with rate $R = 1/4$ and $R = 1/3$ the X outputs are repeated.

That is, with $R = 1/2$ the output of first three clocks is XY_0, the output of second three clocks is $X'Y_0'$. With $R = 1/3$ the output of first three clocks is XXY_0, the output of second three clocks is $X'X'Y_0'$. With $R = 1/4$ the output of first three clocks is XXY_0Y_1, the output of second three clocks is $X'X'Y_0'Y_1'$. The turbo-interleaver scheme is shown in Figure 12.12. The interleaved symbols are numbered from 0 to $N_{turbo} - 1$, the output of the turbo-interleaver are the position numbers (addresses) from this set. The procedure of addresses calculation is performed as follows.

1. Calculate the parameter $n = \lceil \log_2 N_{turbo} \rceil - 5$, i.e. the address of N_{turbo} may be presented as $n + 5$ bits.
2. Initialize the $(n + 5)$-bit counter to 0.
3. Add 1 to n most significant bits (MSB) of counter $(i_{n+4}, \ldots, i_5)$, keep only n least significant bits (LSB) from the resulted value.
4. Extract the n-bit value from the row of Table 12.4, defined by five LSB bits $(i_4, \ldots, i_0)$ of the counter.
5. Multiply the values obtained on steps 3 and 4 and keep only n LSB of the result.
6. Bit-reverse the five LSB of the counter as $(i_0, \ldots, i_4)$.
7. From a tentative address, in which MSB are defined by the value obtained on step 6, and LSB are defined by the value obtained on step 5.
8. Consider the tentative address as output address, if its value less than N_{turbo}, otherwise ignore the value.
9. Increment the counter by 1 and repeat the steps 3-8, until all N_{turbo} output addresses will be found.

12.2.3 Codeword Symbols Repetition

The output of error-correction encoders are repeated by the factor dependent on channel being used and radio-configuration, as shown in Figures 12.2–12.6.

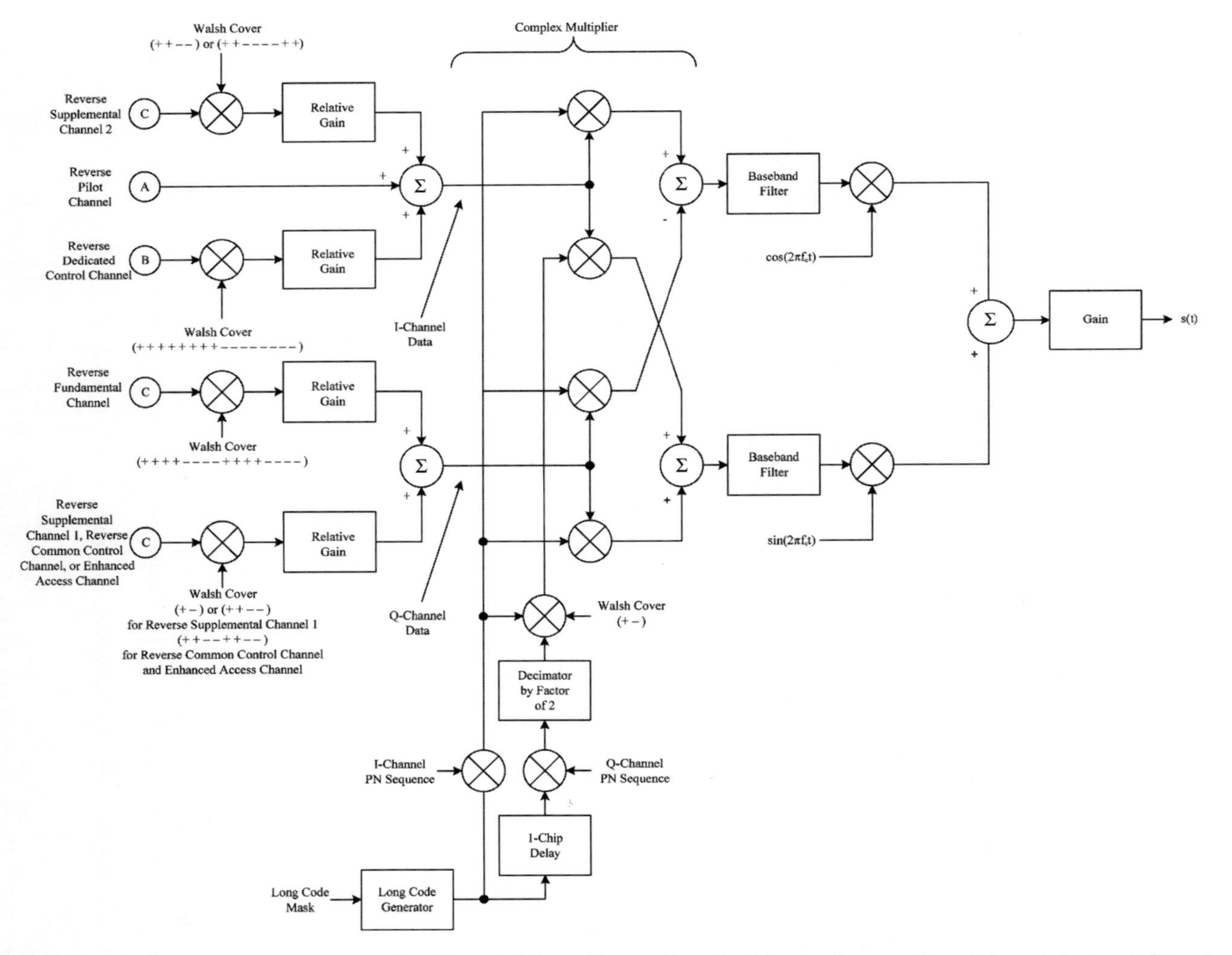

Figure 12.7 Reverse Traffic Channel, Reverse Pilot Channel, Enhanced Access Channel and Reverse Common Control Channel structure, RC3 mode

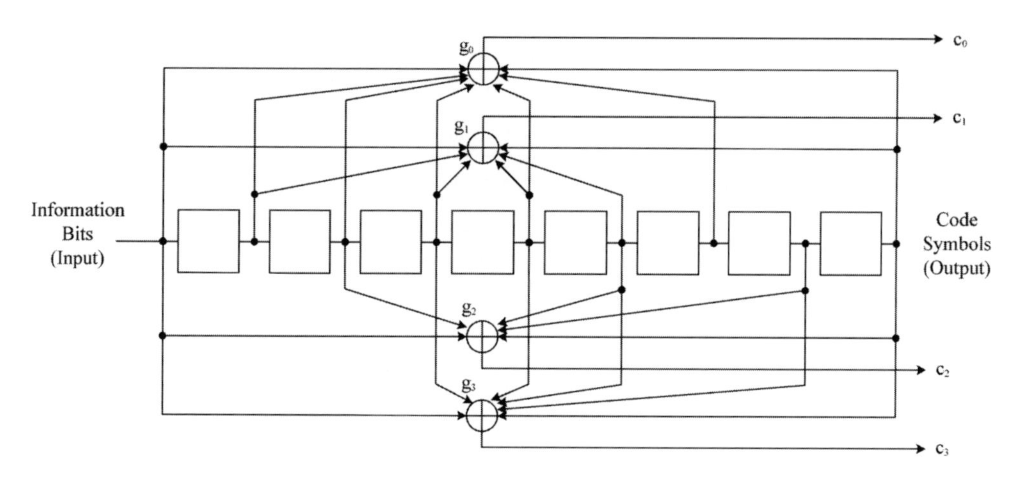

Figure 12.8 Rate 1/4 convolutional encoder structure

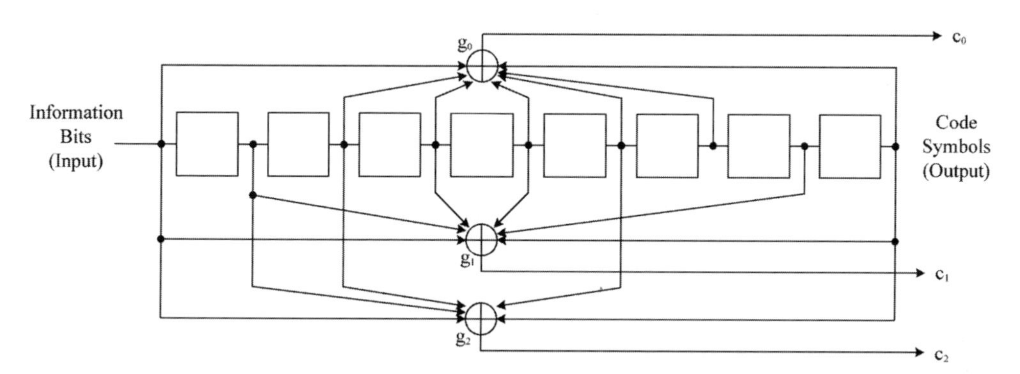

Figure 12.9 Rate 1/3 convolutional encoder structure

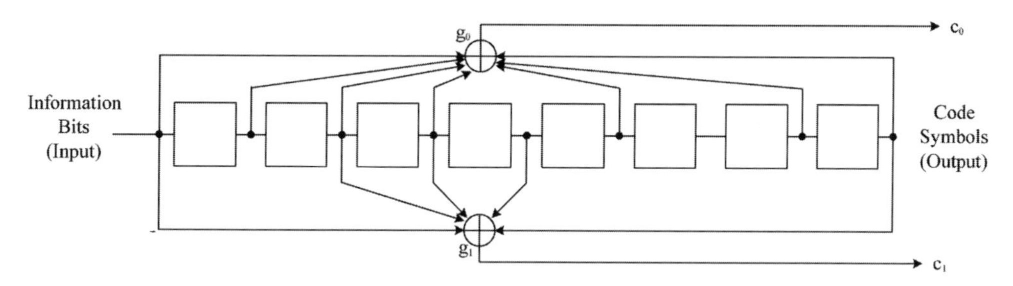

Figure 12.10 Rate 1/2 convolutional encoder structure

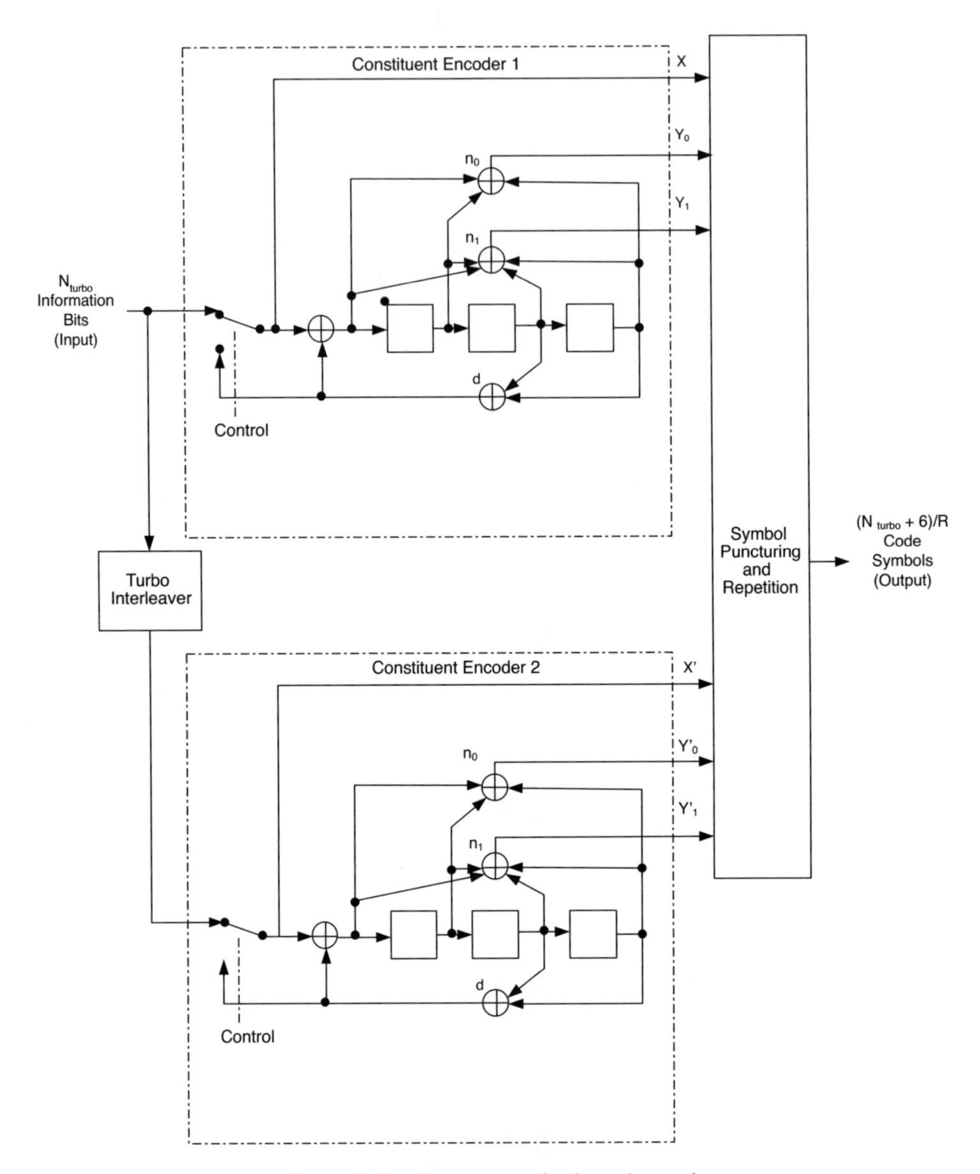

Figure 12.11 The structure of turbo-code encoder

12.2.4 Puncturing

In some radio-configurations after the symbols repetition one more puncturing procedure is performed, see Figure 12.6.

12.2.5 Block Interleaving

After procedures of codeword symbols repetition and puncturing the data is block interleaved, as shown in Figures 12.2–12.6.

Table 12.2 Puncturing the codeword symbols of turbo-code

	Code Rate		
Output	1/2	1/3	1/4
X	11	11	11
Y_0	10	11	11
Y_1	00	00	10
X'	00	00	00
Y'_0	01	11	01
Y'_1	00	00	11

12.2.6 Orthogonal Modulation and Orthogonal Spreading

When using the radio-configuration RC1 the output of block interleaving is processed with 64-ary orthogonal modulation. When using the radio-configuration RC3 the output of block interleaving is processed with orthogonal spreading by using Walsh coverings, see Figures 12.2, 12.3 and 12.7.

12.2.7 Direct Sequence Spreading and Quadrature Spreading

For spreading the sequences obtained after 64-ary orthogonal modulation in radio-configuration RC1, or after Walsh covering in radio-configuration RC3, the so-called long codes are used, as well as short pseudorandom sequences for in-phase and quadrature channels.

The long codes have period $2^{42}-1$. To obtain the long code, the feedback shift register is used, defined by the polynomial:

$$p(x) = x^{42} + x^{35} + x^{33} + x^{31} + x^{27} + x^{26} + x^{25} + x^{22} + x^{21} + x^{19} \\ + x^{18} + x^{17} + x^{16} + x^{10} + x^{7} + x^{6} + x^{5} + x^{3} + x^{2} + x + 1$$

The output of this shift register is masked by means of component-wise multiplication by symbols of masking sequences, as shown in Figure 12.13. The long code mask depends on the channel type, public *Electronic Serial Number (ESN)*, unique for each mobile station, and also private *Voice Privacy Mask (VPM)*, providing the secrecy of transmission.

For spreading rate SR3 the generation of long code uses the procedure of generation of the long code for spreading rate SR1 as its basis, as shown in Figure 12.14. As a result the long code with chip rate 3.6864 Mcps is obtained instead of 1.2288 Mcps for SR1 mode.

Table 12.3 Puncturing the tail symbols of turbo-code

	Code Rate		
Output	1/2	1/3	1/4
X	111 000	111 000	111 000
Y_0	111 000	111 000	111 000
Y_1	000 000	000 000	111 000
X'	000 111	000 111	000 111
Y'_0	000 111	000 111	000 111
Y'_1	000 000	000 000	000 111

Table 12.4 Table for turbo-interleaving calculations

Table Index	n = 4	n = 5	n = 6	n = 7	n = 8	n = 9	n = 10
0	5	27	3	15	3	13	1
1	15	3	27	127	1	335	349
2	5	1	15	89	5	87	303
3	15	15	13	1	83	15	721
4	1	13	29	31	19	15	973
5	9	17	5	15	179	1	703
6	9	23	1	61	19	333	761
7	15	13	31	47	99	11	327
8	13	9	3	127	23	13	453
9	15	3	9	17	1	1	95
10	7	15	15	119	3	121	241
11	11	3	31	15	13	155	187
12	15	13	17	57	13	1	497
13	3	1	5	123	3	175	909
14	15	13	39	95	17	421	769
15	5	29	1	5	1	5	349
16	13	21	19	85	63	509	71
17	15	19	27	17	131	215	557
18	9	1	15	55	17	47	197
19	3	3	13	57	131	425	499
20	1	29	45	15	211	295	409
21	3	17	5	41	173	229	259
22	15	25	33	93	231	427	335
23	1	29	15	87	171	83	253
24	13	9	13	63	23	409	677
25	1	13	9	15	147	387	717
26	9	23	15	13	243	193	313
27	15	13	31	15	213	57	757
28	11	13	17	81	189	501	189
29	3	1	5	57	51	313	15
30	15	13	15	31	15	489	75
31	5	13	33	69	67	391	163

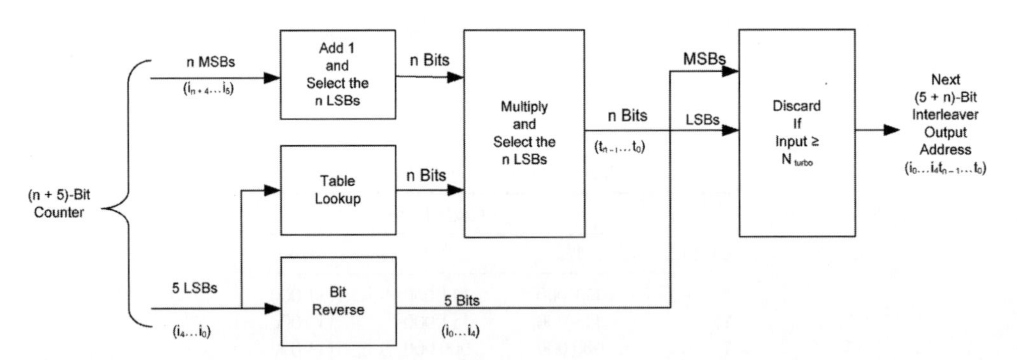

Figure 12.12 The scheme of turbo-interleaver functioning

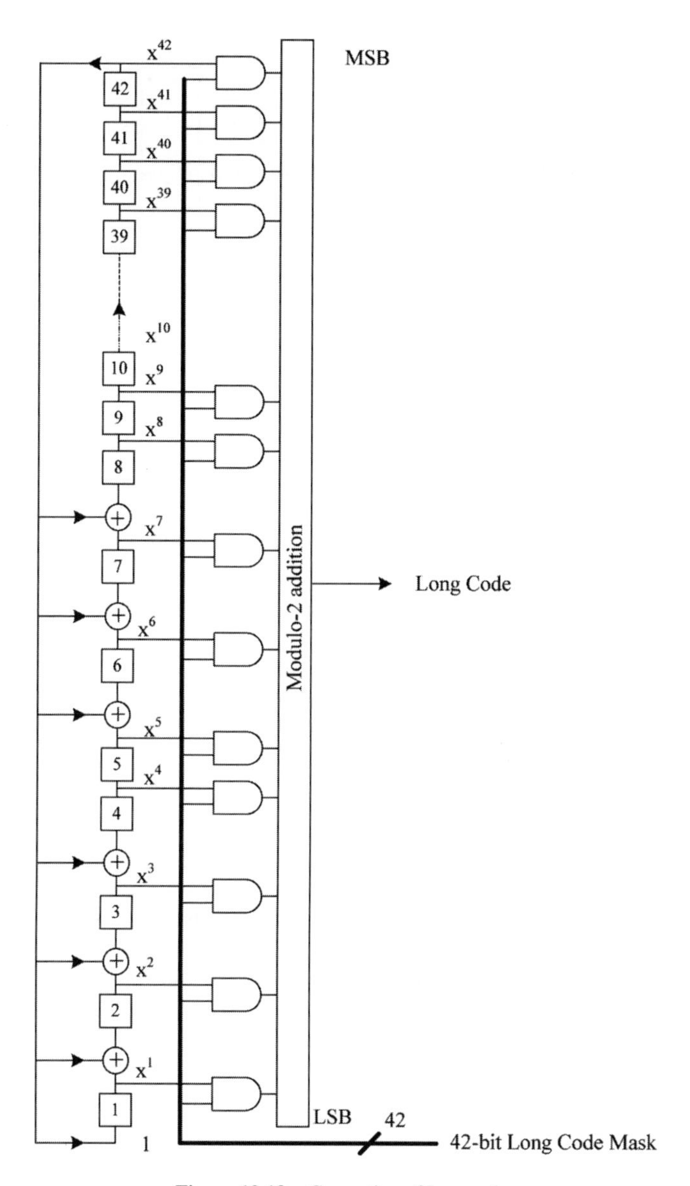

Figure 12.13 Generation of long codes

The short pseudonoise (PN) sequences P_I and P_Q have the period 215 for SR1 mode and 3×2^{15} for SR3 mode. Usage of such short sequences allows the correlation between in-phase and quadrature channels to be avoided.

The sequences for SR1 mode is obtained based on linear feedback shift registers (LFSR), defined by the polynomials:

$$P_I(x) = x^{15} + x^{13} + x^9 + x^8 + x^7 + x^5 + 1 \tag{12.1}$$

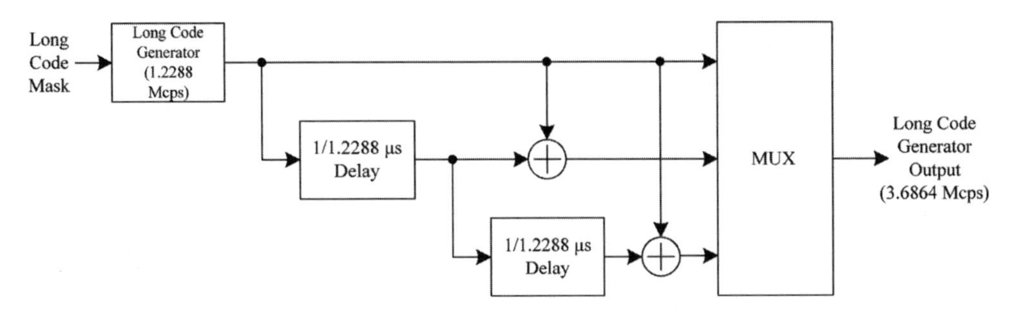

Figure 12.14 Generation of long codes, SR3 mode

and

$$P_Q(x) = x^{15} + x^{12} + x^{11} + x^{10} + x^6 + x^5 + x^4 + x^3 + 1 \quad (12.2)$$

Such LFSRs have the maximal period of $2^{15} - 1$, and to obtain the sequences with the period of 2^{15}, in the only case when in these sequences there is the series of 14 zeros, the one additional (15th) zero is added to this series.

The sequences for SR3 mode are obtained basing on the LFSR defined by the polynomial:

$$P(x) = x^{20} + x^9 + x^5 + x^3 + 1 \quad (12.3)$$

but with different initial conditions for in-phase and quadrature channels. The output of such LFSR has the maximal period of $2^{20} - 1$, and to obtain the period of 3×2^{15} the output sequence is truncated.

12.2.8 *Frame Quality Indicator*

The frames of all channels in Figure 12.1, excluding Reverse Power Control, Reverse Pilot and Access Channels (but including Enhanced Access Channel), apart from the information bits and bits dedicated to encoder tail bits, contain the cyclic redundancy check (CRC) which is called *Frame Quality Indicator*. Depending on the channel type and transmission rate the CRC of lengths 16, 12, 10, 8 and 6 bits are used. For this purpose the following generator polynomials $g(x)$ are used:

$$\begin{aligned}
g(x) &= x^{16} + x^{15} + x^{14} + x^{11} + x^6 + x^5 + x^2 + x + 1\\
g(x) &= x^{12} + x^{11} + x^{10} + x^9 + x^8 + x^4 + x + 1\\
g(x) &= x^{10} + x^9 + x^8 + x^7 + x^6 + x^4 + x^3 + 1\\
g(x) &= x^8 + x^7 + x^4 + x^3 + x + 1\\
g(x) &= x^6 + x^5 + x^2 + x + 1
\end{aligned} \quad (12.4)$$

For example, the shift register for 16-bit Frame Quality Indicator is depicted in Figure 12.15. The other registers may be constructed by the correspondent generator polynomials in a similar way.

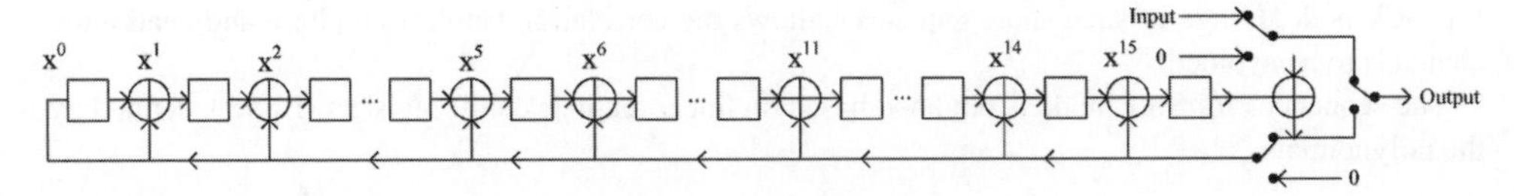

Figure 12.15 16-bit Frame Quality Indicator

Table 12.5 Radio-configuration parameters of CDMA Forward Channel

Radio Configuration	Associated Spreading Rate	Data rates, Forward Error Correction, and general characteristics
1	1	1200, 2400, 4800, and 9600 bps data rates with R = 1/2, BPSK pre-spreading symbols
2	1	1800, 3600, 7200, and 14400 bps data rates with R = 1/2, BPSK pre-spreading symbols
3	1	1500, 2700, 4800, 9600, 19200, 38400, 76800, and 153600 bps data rates with R = 1/4, QPSK pre-spreading symbols, OTD allowed
4	1	1500, 2700, 4800, 9600, 19200, 38400, 76800, 153600, and 230400 bps data rates with R = 1/2, QPSK pre-spreading symbols, OTD allowed
5	1	1800, 3600, 7200, 14400, 28800, 57600, 115200, and 230400 bps data rates with R = 1/4, QPSK pre-spreading symbols, OTD allowed
6	3	1500, 2700, 4800, 9600, 19200, 38400, 76800, 153600, and 307200 bps data rates with R = 1/6, QPSK pre-spreading symbols, DS non-OTD, DS OTD, or MC modes.
7	3	1500, 2700, 4800, 9600, 19200, 38400, 76800, 153600, 307200, and 614400 bps data rates with R = 1/3, QPSK pre-spreading symbols, DS non-OTD, DS OTD, or MC modes.
8	3	1800, 3600, 7200, 14400, 28800, 57600, 115200, 230400, and 460800 bps data rates with R = 1/4 (20 ms) or 1/3 (5 ms), QPSK pre-spreading symbols, DS non-OTD, DS OTD, or MC modes.
9	3	1800, 3600, 7200, 14400, 28800, 57600, 115200, 230400, 460800, and 1036800 bps data rates with R = 1/2 (20 ms) or 1/3 (5 ms), QPSK pre-spreading symbols, DS non-OTD, DS OTD, or MC modes.

12.3 Forward Channel of Physical Layer in CDMA2000 Standard

12.3.1 Forward Channel Structure

CDMA Forward Channel connects the base station to the mobile station. Similar to Reverse Channel, in the Forward Channel of CDMA2000 standard there are two spreading rates SR1 and SR3, for which are defined the radio-configurations 1 to 5 and 6 to 9 correspondingly. The parameters of these radio-configurations are presented in Table 12.5.

As can be seen from Table 12.5, Forward Channel may function in three modes (see also section 12.2.1): direct-spreading (DS) with *orthogonal transmit diversity (OTD)*, DS in non-OTD mode, and multi-carrier (MC) mode.

OTD mode is the technique of transmission in Forward Channel, when the symbols are arranged among multiple antennas and spread with a unique Walsh or quasi-orthogonal function associated with each antenna.

In multi-carrier (MC) transmission the $N > 1$ adjacent 1.2288 Mcps direct-spread RF carriers are used. Interleaved data is demultiplexed onto each of the N adjacent carriers.

In the following, when describing the structure of different Forward Channel subchannels, we will basically limit the consideration by non-OTD mode. Generalization by the case of transmission over multiple antennas (carriers) requires correspondent (usually trivial) demultiplexing and does not affect the basic functional blocks described below.

The general structure of the Forward Channel is shown in Figure 12.16. Forward Channel consists of the following subchannels.

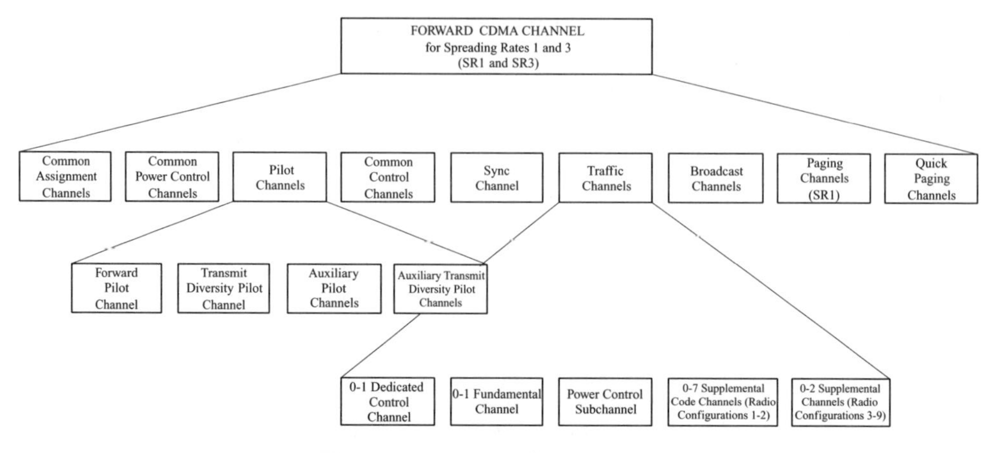

Figure 12.16 Forward Channel Structure

- **Common Assignment Channel**. A forward common channel used by the base station to acknowledge a mobile station accessing the Enhanced Access Channel.
- **Common Power Control Channel**. A forward common channel which transmits power control bits (that is, common power control subchannels) to multiple mobile stations.
- **Pilot channels**.
  - **Forward Pilot Channel**. An unmodulated, direct-sequence spread spectrum signal transmitted continuously by each CDMA base station. The Pilot Channel allows a mobile station to acquire the timing of the Forward CDMA Channel, provides a phase reference for coherent demodulation, and provides means for signal strength comparisons between base stations for determining when to handoff.
  - **Transmit Diversity Pilot Channel**. Used to support forward link transmit diversity. The pilot channel and the transmit diversity pilot channel provide phase references for coherent demodulation of forward link CDMA channels which employ transmit diversity.
  - **Auxiliary Pilot Channel**. An auxiliary pilot channel is required for forward link spot beam and antenna beam forming applications, and provides a phase reference for coherent demodulation of those forward link CDMA channels associated with the auxiliary pilot.
  - **Auxiliary Transmit Diversity Pilot Channel**. A transmit diversity pilot channel associated with an auxiliary pilot channel.
- **Common Control Channel**. A control channel used for the transmission of digital control information from a base station to one or more mobile stations.
- **Sync Channel**. A channel which transports the synchronization message to the mobile station.
- **Traffic Channel**. One or more channels used to transport user and signaling traffic from the base station to the mobile station.
  - **Dedicated Control Channel**. A channel used for the transmission of higher-level data, control information, and power control information from a base station to a mobile station.
  - **Fundamental Channel**. A channel which carries a combination of higher-level data and power control information.
  - **Power Control Subchannel**. A subchannel used by the base station to control the power of a mobile station when operating on the Reverse Traffic Channel.
  - **Supplemental Code Channel**. A channel which operates in conjunction with other channels to provide higher data rate services, and on which higher-level data is transmitted.

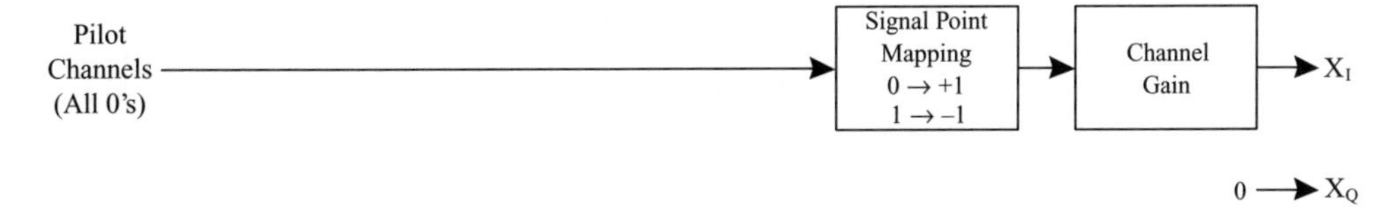

Figure 12.17 Pilot Channel structure

- **Supplemental Channel**. A channel which operates in conjunction with other channels to provide higher data rate services, and on which higher-level data is transmitted.

- **Broadcast Channel**. A channel used for transmission control information and pages from a base station to a mobile station.
- **Paging Channel**. A channel used for transmission control information and pages from a base station to a mobile station.
- **Quick Paging Channel**. An uncoded, spread, and On-Off-Keying (OOK) modulated spread spectrum signal sent by a base station to inform mobile stations operating in the slotted mode during the idle state whether to receive the Forward Common Control Channel or the Paging Channel starting in the next Forward Common Control Channel or Paging Channel frame.

The structures of the channels are shown in Figures 12.17–12.28.

The structure of the Forward Traffic Channel in Figure 12.26 is shown only for radio-configuration RC3, from the functional point of view it is similar to other configurations.

Connection between outputs in Figures 12.17–12.26 and inputs of I and Q mapping in Figure 12.28 is defined by demultiplexing scheme as shown in Figure 12.27.

The description of basic functional blocks of Forward Channel is given below.

12.3.2 Forward Error Correction

As in the case of transmission over the Reverse Channel, the convolutional and turbo-codes may be used to provide an error-correction coding in the Forward Channel.

12.3.2.1 Convolutional Coding

In the Forward Channel the convolution codes with constraint length $K = 9$ and rate 1/2, 1/3, 1/4 and 1/6 are used, depending on radio-configuration, spreading rate and subchannel being used. The codes on rates 1/2, 1/3 and 1/4 are completely similar to those being used in Reverse Channel and described in section 12.2.2.1.

The structure of the convolutional code with rate 1/6 is presented in Figure 12.29.

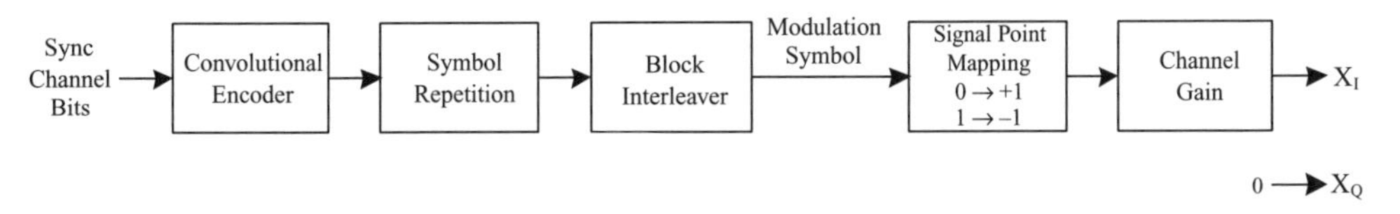

Figure 12.18 Sync Channel structure

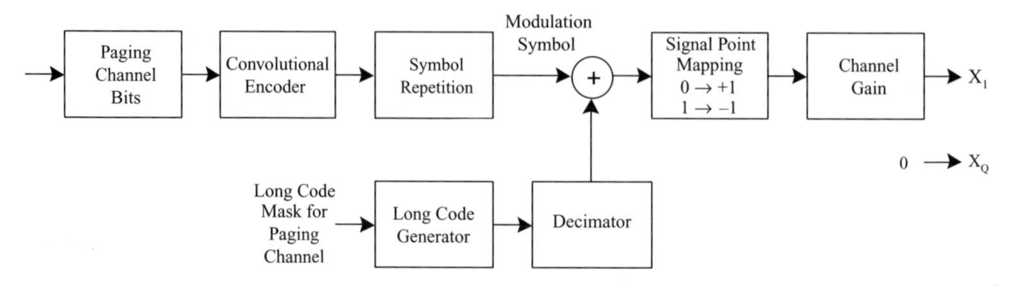

Figure 12.19 Paging Channel structure

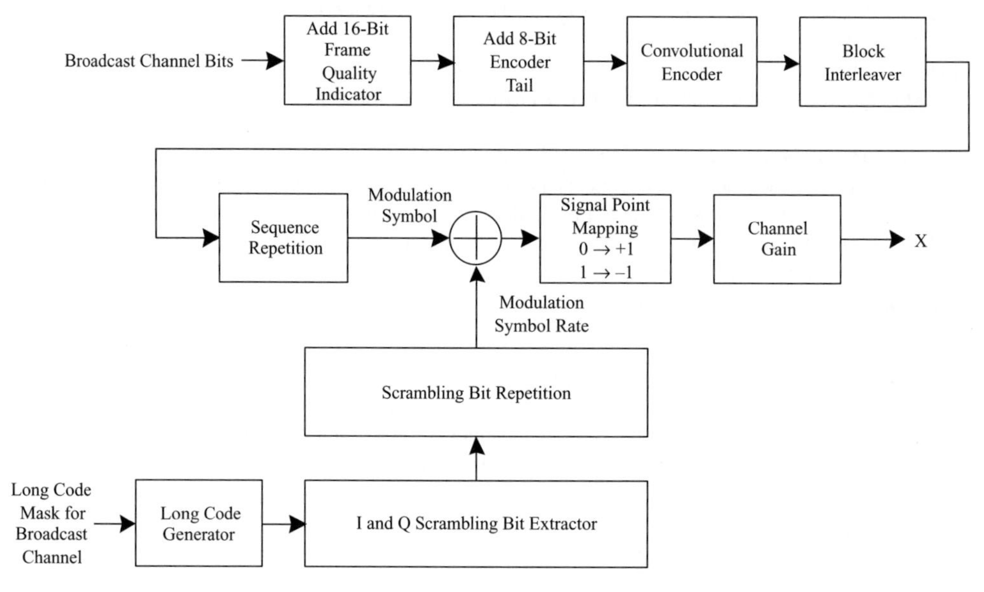

Figure 12.20 Broadcast Channel structure

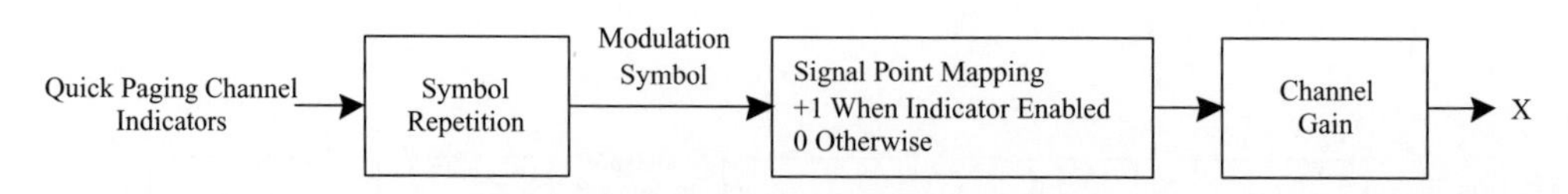

Figure 12.21 Quick Paging Channel structure

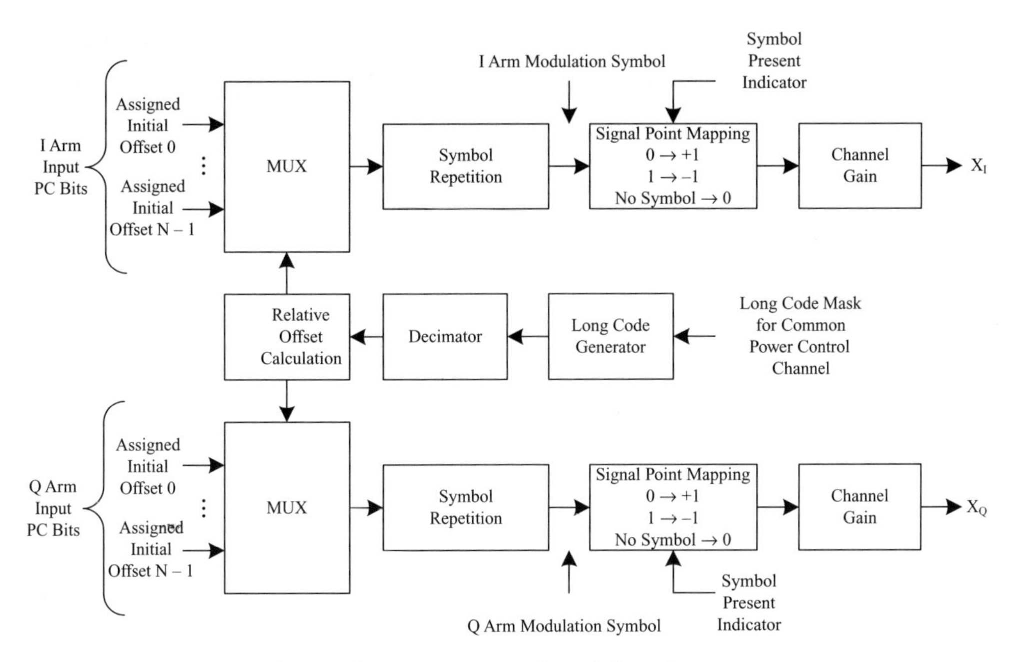

Figure 12.22 Common Power Control Channel structure

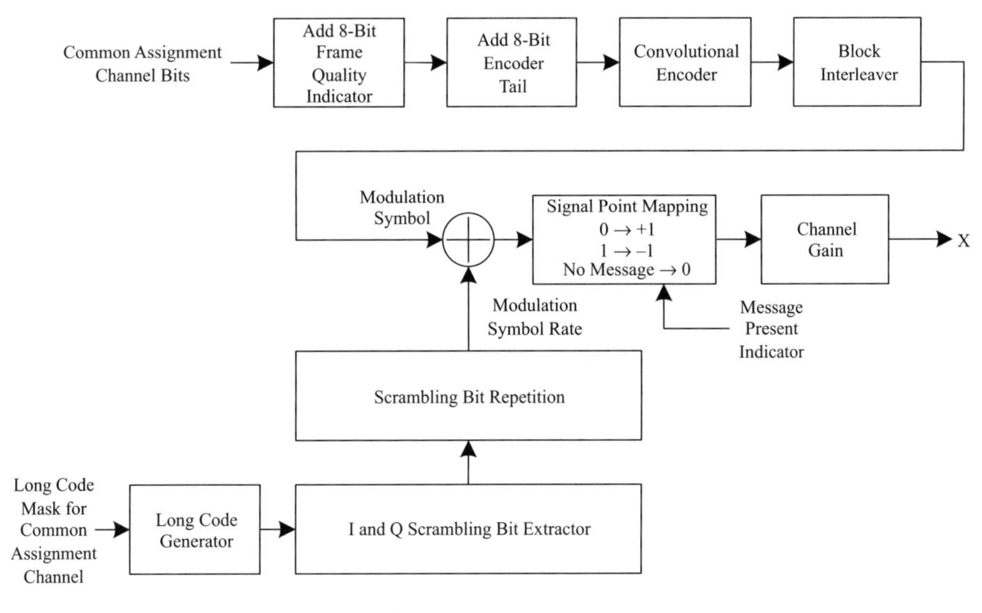

Figure 12.23 Common Assignment Channel structure

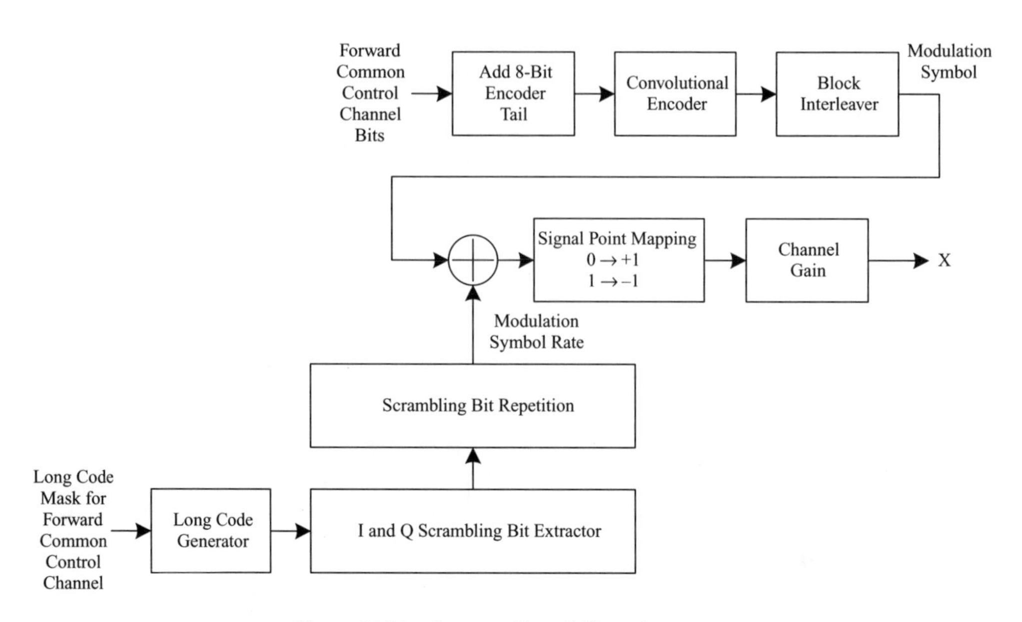

Figure 12.24 Common Control Channel structure

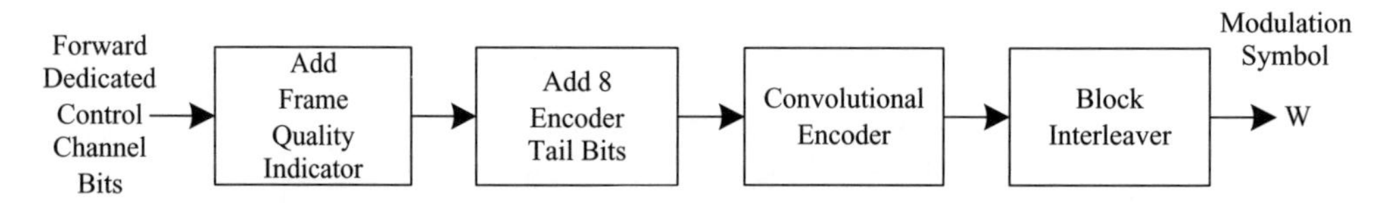

Figure 12.25 Dedicated Control Channel structure

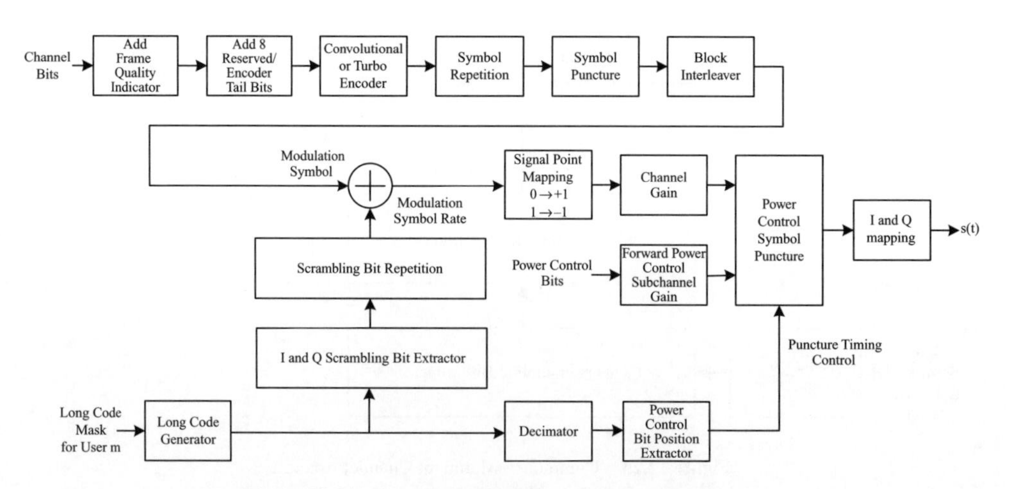

Figure 12.26 Forward Traffic Channel structure, RC3 mode

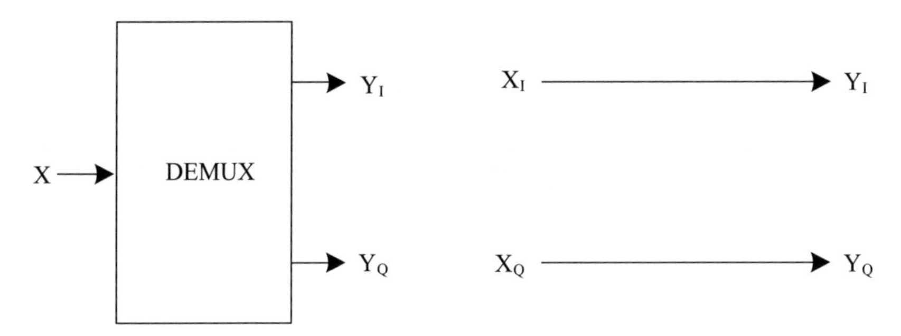

Figure 12.27 Demultiplexer structure

12.3.2.2 Turbo-Coding

In the Forward Channel the turbo-codes with rates 1/2, 1/3 and 1/4 are used. The structure of the encoders, as well as interleaving and puncturing schemes are completely the same as used in Reverse Channel and described in section 12.2.2.2.

12.3.3 Codeword Symbols Repetition

The output of error-correction encoders are repeated by the factors dependent on channel being used and radio-configuration.

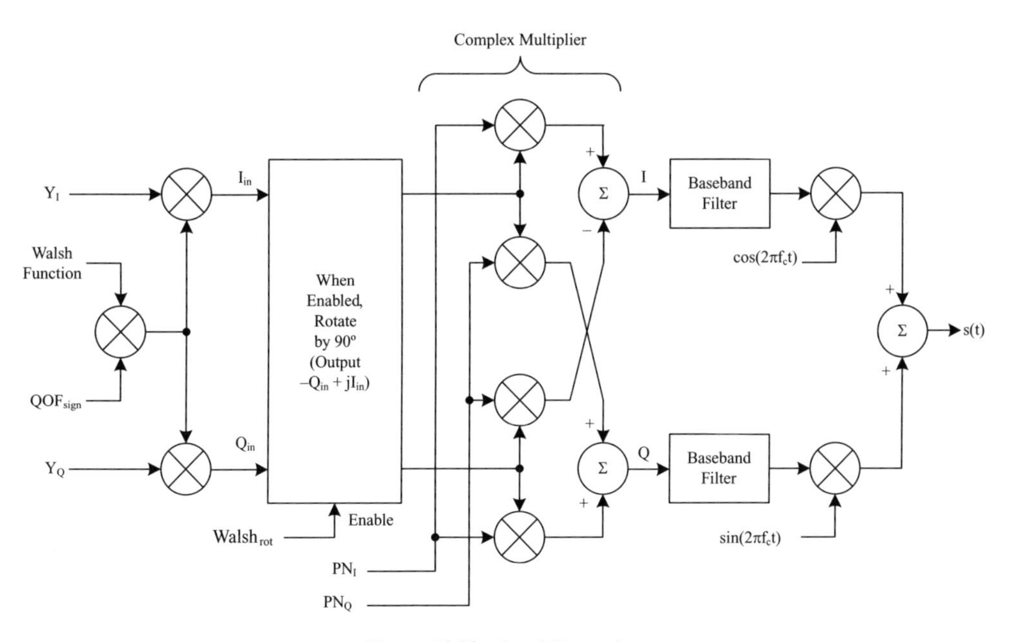

Figure 12.28 I and Q mapping

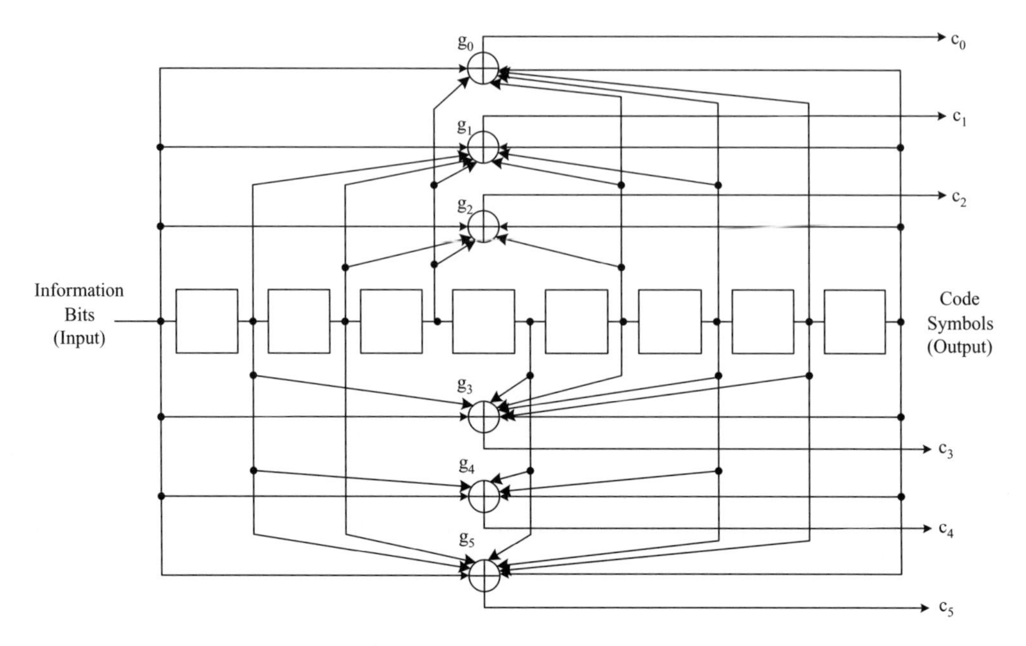

Figure 12.29 Rate 1/6 convolutional encoder structure

12.3.4 Puncturing

In some radio-configurations after the symbols repetition one more puncturing procedure is performed, see Figure 12.26.

12.3.5 Block Interleaving

After procedures of codeword symbols repetition and puncturing the data is block interleaved, as shown in Figures 12.17–12.26. The parameters of interleaving depend on the radio-configuration and the mode being used (DS or MC).

12.3.6 Sequence Repetition

When transmitting over the Broadcast Channel (see Figure 12.20) the sequence repetition is applied. Depending on the transmission rate the sequences are repeated twice, four times or not repeated at all.

12.3.7 Data Scrambling

When transmitting over the Paging, Broadcast, Common Assignment, Forward Common Control and Forward Traffic Channels the data are scrambled, see Figures 12.19, 12.20, 12.23, 12.24, and 12.26. For scrambling the PN-sequence is used, obtained from the long code.

12.3.8 Orthogonal and Quasi-Orthogonal Spreading

When using the radio-configurations RC1 and RC2 the Walsh functions are used for orthogonal spreading. When using the radio-configurations RC3 to RC9 the Walsh functions or quasi-orthogonal functions are used. The maximal length of Walsh sequence is 128 bits for SR1 and 256 bits for SR3.

Quasi-orthogonal functions (QOF) are obtained from Walsh functions masked by multiplication on QOF_{sign} sequences and "rotation", defined by $Walsh_{rot}$ sequence, see Figure 12.28.

12.3.9 Quadrature Spreading

The procedure of quadrature spreading is shown in Figure 12.28.

For the quadrature spreading the PN-sequences are used, analogous to short codes described in Section 12.2.7. For SR1 or MC SR3 modes of the Forward Channel the registers similar to shift registers (12.1) and (12.2) of Reverse Channel's SR1 mode are used.

For DS SR3 mode of the Forward Channel the shift register, similar to (12.3) of Reverse Channel's SR3 mode is used.

12.3.10 Frame Quality Indicator

The frames of Broadcast, Common Assignment, Forward Dedicated Control, Forward Fundamental, Forward Supplemental and Forward Supplemental Code Channels, apart from the information bits and bits dedicated to encoder tail bits, contain the cyclic redundancy check (CRC) which is called Frame Quality Indicator. Depending on the channel type and transmission rate the CRC of lengths 16, 12, 10, 8 and 6 bits are used. For this purpose the same generator polynomials as for Reverse Channel are used, see (12.4) in section 12.2.8.

The shift register for 16-bit Frame Quality Indicator is depicted in Figure 12.15. The other registers may be constructed by the correspondent generator polynomials similarly.

12.4 Architecture Model of CDMA2000 1xEV-DO Standard

Next we will consider the development of CDMA2000 standard – the CDMA2000 1xEV-DO standard. In this development, besides the transmission of the voice information, one of the main tasks is the improvement of basic CDMA2000 standard for digital data transmission. This is given the widely used title of this standard as EV-DO — "evolution-data optimized".

Consider the model shown in Figure 12.30. It includes the access terminal, access network and air interface. The access terminal is the user's device allowing user to access the network. Access network is the analogue of base station, that is, equipment allowing organizing access points for interoperation with user (in fact organizing the "last mile" between network and user's end point). The subparts of access network are the sectors, which are distinguished by their identifiers. The CDMA2000 standard describes an air interface between access terminal and access network.

The standard describes seven levels of the standard OSI model: Application Layer, Stream Layer, Session Layer, Connection Layer, Security Layer, MAC Layer, and Physical Layer. For each of these layers the default protocols are defined, as well as non-default protocols.

In the following we will briefly describe the physical layer of CDMA2000 1xEV-DO Revision B standard [1]. For the physical layer the standard defines four subtypes: the one default subtype (subtype 0) and three non-default subtypes (subtype 1, subtype 2 and subtype 3).

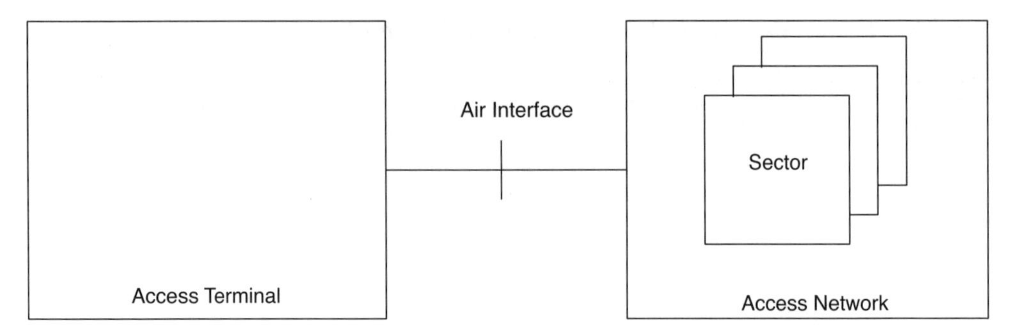

Figure 12.30 Architecture reference model

12.4.1 Structure of Physical Layer Packet

The length of the packet of physical layer may be equal to 256, 512, 1024, 2048, 3072 or 4096 bit, depending on the transmission channel and physical layer subtype being used. Each physical layer packet contains one or more packets from the MAC layer. The lengths of physical layer packets and the number of contained MAC-packets are listed in the Table 12.6.

The structure of the physical layer packet is shown in Figure 12.31. Here the MAC layer packet is the packet obtained from the higher (MAC) layer of the network, FCS (frame check sequence) is the CRC code for integrity check, TAIL is the field for encoder's tail bits, which should be always equal to 0.

12.4.2 FCS Computation

FCS computation is performed by the following scheme (Figure 12.32), using the linear feedback shift register, defined by the polynomial:

$$g(x) = x^{16} + x^{12} + x^5 + 1$$

In the beginning, all register cells are initialized with zeros. Switches in Figure 12.32 are set in "up" position. The input is the symbols of the first field of physical layer packet (in Figure 12.31 this is the MAC layer packet), which are summed with the register's output and defined the feedback values (during this time the symbols themselves are output without changing). Then the switches are set down, the register is clocked 16 more times and the FCS field of the physical layer packet is filled by the register's output. The TAIL field remains unchanged.

Table 12.6 The lengths of physical layer packets and the number of transmitted MAC-packets

Channel	PHY subtype	Packet length	Number of MAC-packets
Control	0, 1	1024	1
Access	0	256	1
	1	256, 512, 1024	1
Forward traffic	0, 1	1024, 2048, 3072, 4096	1,2,3,4
Reverse traffic	0, 1	256, 512, 1024, 2048, 4096	1

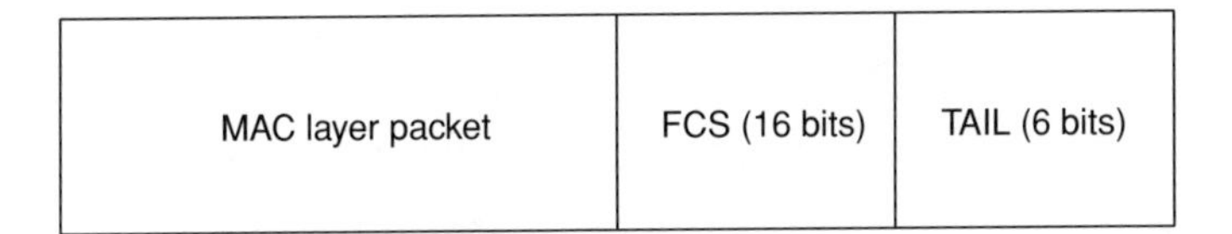

Figure 12.31 The structure of the physical layer packet

12.5 Access Terminal of the CDMA2000 1xEV-DO Standard

12.5.1 Power Control

The management of the access terminal's output power is performed by combining the open-loop power control and closed-loop power control.

Open-loop power control is performed by the access terminal itself, taking into account the power of the signal received from the forward communication channel. Access terminal defines the power for signal transmission over the pilot channel, and from this value the output powers for the transmission over the other channels are defined.

As the result of the open-loop power control, the access terminal obtains the estimate for the output power, which is then corrected by means of the closed-loop power control. With closed-loop power control the access terminal takes into account the commands transmitted from the access network over the special channel named Reverse Power Control Channel (RPC Channel), see section 12.6.4. This channel is the part of the forward channel from the access network to the access terminal, see section 12.6.1.

The access network transmits the signals '0' and '1' over the RPC channel, which give the command to the access terminal to increase or decrease the transmission power, correspondingly. Power increasing and decreasing is performed by the fixed value RPCStep, which is the parameter of Reverse Traffic Channel in MAC Protocol. So, the access network may control the power of signals from the access terminals based on the quality of signals received from them. This may be required, for instance, to equalize the power of the received signals from the access terminals which are on different distances from the base station, or transmitting under different channel conditions.

12.5.2 Reverse Channel Structure

Reverse Channel is the channel connecting the access terminal to the access network. The structure of the Reverse Channel for physical layer subtype 0 and 1 is shown in Figure 12.33.

The main subchannels of the Reverse Channel are listed and described below.

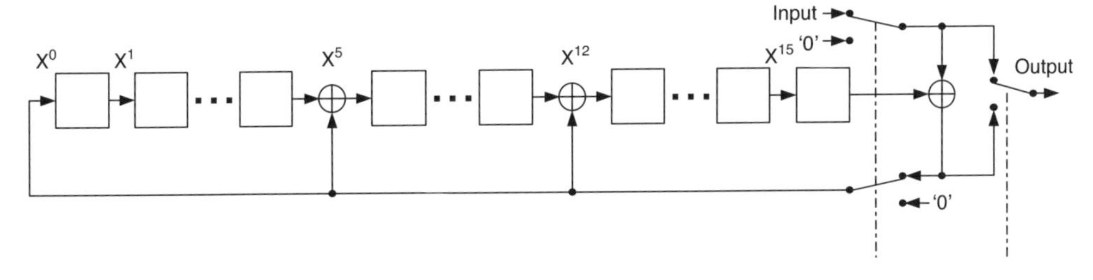

Figure 12.32 Calculation of the FCS check field for the physical layer packet

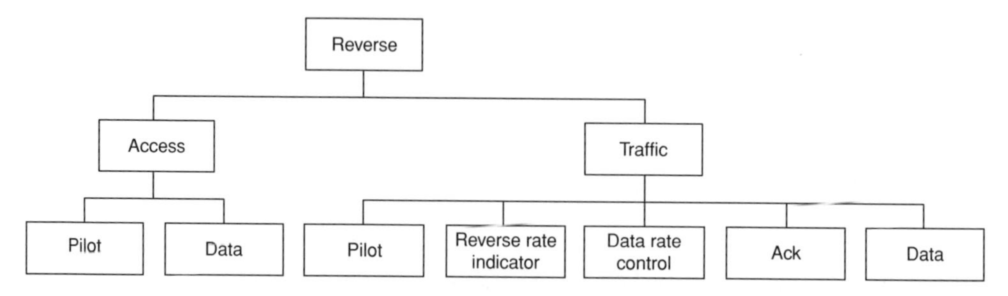

Figure 12.33 Reverse Channel structure

- **Access channel** — intended for initializing the connection between the access terminal and access network or to respond to an access terminal directed message. For each sector of the access network there is the correspondent access channel, identified by the long code.
  - **Pilot channel** — intended for pilot signals transmission.
  - **Data channel** — intended for data transmission.
- **Traffic channel** — used for connecting the access terminal to the access network. Prior to the access terminal authentication the traffic channel is used as non-dedicated resource. After successful authentication the channel may be used as dedicated resource for given access terminal.
  - **Pilot channel** — intended for pilot signals transmission.
  - **Data channel** — intended for data transmission.
  - **Reverse rate indicator (RRI) channel** — indicates the transmission rate in the Reverse Traffic Channel.
  - **Data rate control (DRC) channel** — indicates the rate on which the access terminal may receive data from the Forward Traffic Channel, and also the sector from which the access terminal would like to receive the Forward Traffic Channel.
  - **Acknowledgment (ACK) channel** — indicates to the access network on success or failure in receiving from the Forward Traffic Channel.

The general structures of the Access Channel and Reverse Traffic Channel are shown in Figures 12.34 and 12.35. The features of these channels are considered in Sections 12.5.4 and 12.5.5, and then the common blocks are described.

12.5.3 Modulation Parameters and Transmission Rates

Modulation parameters for the Access Channel and Reverse Traffics Channel are listed in Table 12.7.

When using the subtype 0 of the physical layer (default subtype) the transmission from the access terminal is performed on the rate of 9.6 kbps. In case of using the subtype 1 the transmission rate may achieve 9.6, 19.2 or 38.4 kbps, correspondent to the parameters of MAC protocol.

When transmitting over the Traffic Channel the transmission rate may be equal to 9.6, 19.2, 38.4, 76.8 or 153.6 kbps, correspondent to the parameters of MAC protocol.

12.5.4 Access Channel

The structure of the Access Channel is shown in Figure 12.34. The rates and packet sizes indicated in the first row for the Data Channel packets, are correspondent to the physical layer subtype 0 (default

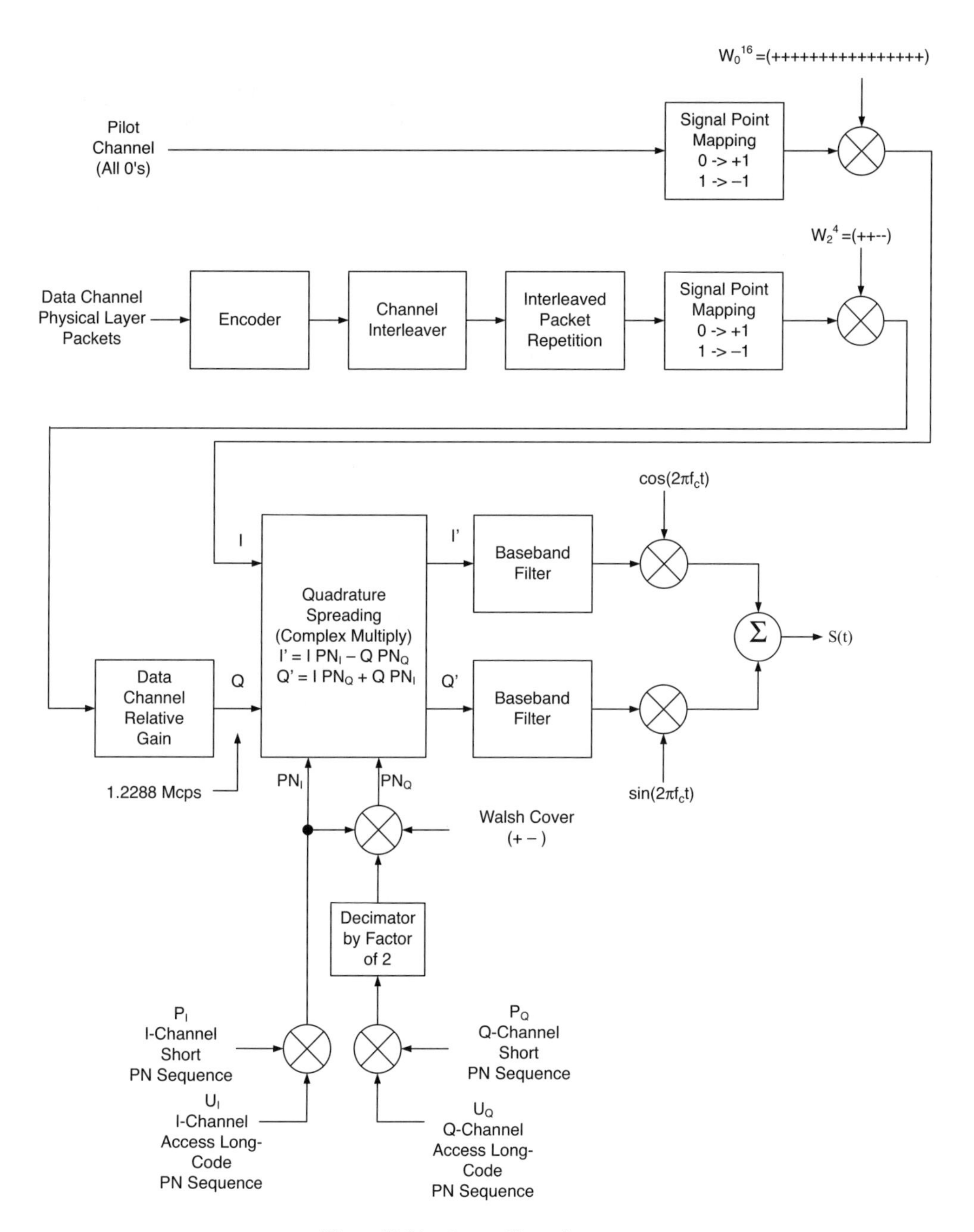

Figure 12.34 Access Channel structure

Table 12.7 Modulation parameters

Parameter	Data Rate (kbps)				
	9.6	19.2	38.4	76.8	153.6
Reverse Rate Index	1	2	3	4	5
Bits per Physical Layer Packet	256	512	1,024	2,048	4,096
Physical Layer Packet Duration (ms)	26.66...	26.66...	26.66...	26.66...	26.66...
Code Rate	1/4	1/4	1/4	1/4	1/4
Code Symbols per Physical Layer Packet	1,024	2,048	4,096	8,192	8,192
Code Symbol Rate (ksps)	38.4	76.8	153.6	307.2	307.2
Interleaved Packet Repeats	8	4	2	1	1
Modulation Symbol Rate (ksps)	307.2	307.2	307.2	307.2	307.2
Modulation Type	BPSK	BPSK	BPSK	BPSK	BPSK
PN Chips per Physical Layer Packet Bit	128	64	32	16	8

subtype), while the subtype 1 also supports parameters indicated in the second and third rows (see section 12.5.3).

To establish the connection over the Access Channel the access probe is done, which means the transmission of the preamble and then — one or more packets of the physical layer Access Channel. The preamble consists of transmission only the symbols of the pilot channel (over in-phase component channel), and during the transmission of the Access Channel packet both the symbols of Pilot Channel (over in-phase component channel) and Data Channel (over quadrature component channel) are sent.

Over the Pilot Channel unmodulated signals ‘0’ are transmitted. Transmission over the Pilot Chanel is performed continuously, the signals are sent over the in-phase channel with covering by Walsh sequence $W_0^{16} = (++++++++++++++++)$.

During the access probe, after preamble transmission one or more Access Channel physical layer packets are sent over the quadrature channel with Walsh covering $W_2^4 = (++--)$.

12.5.5 Reverse Traffic Channel

The structure of the Reverse Traffic Channel is shown in Figure 12.35.

The Pilot Channel, DRC Channel, ACK Channel and Data Channel are orthogonally spread with Walsh functions of length 4, 8 or 16.

The frame of the Reverse Channel Traffic has the duration of $26.66\ldots\mu s$. The frame consists of 16 slots of duration $1.66\ldots\mu s$, each slot contains 2048 chips.

Next we will describe the functioning of different channels shown in Figure 12.35.

12.5.5.1 Pilot Channel

Access terminal transmits unmodulated signals ‘0’ over the Pilot Channel. Transmission over the Pilot Channel is made continuously, the pilot signals are time-division multiplexed with RRI Channel. The transmission is made over the in-phase channel with Walsh covering by sequence $W_0^{16} = (++++++++++++++++)$. The transmission over the Pilot Channel and RRI channel is performed with equal power.

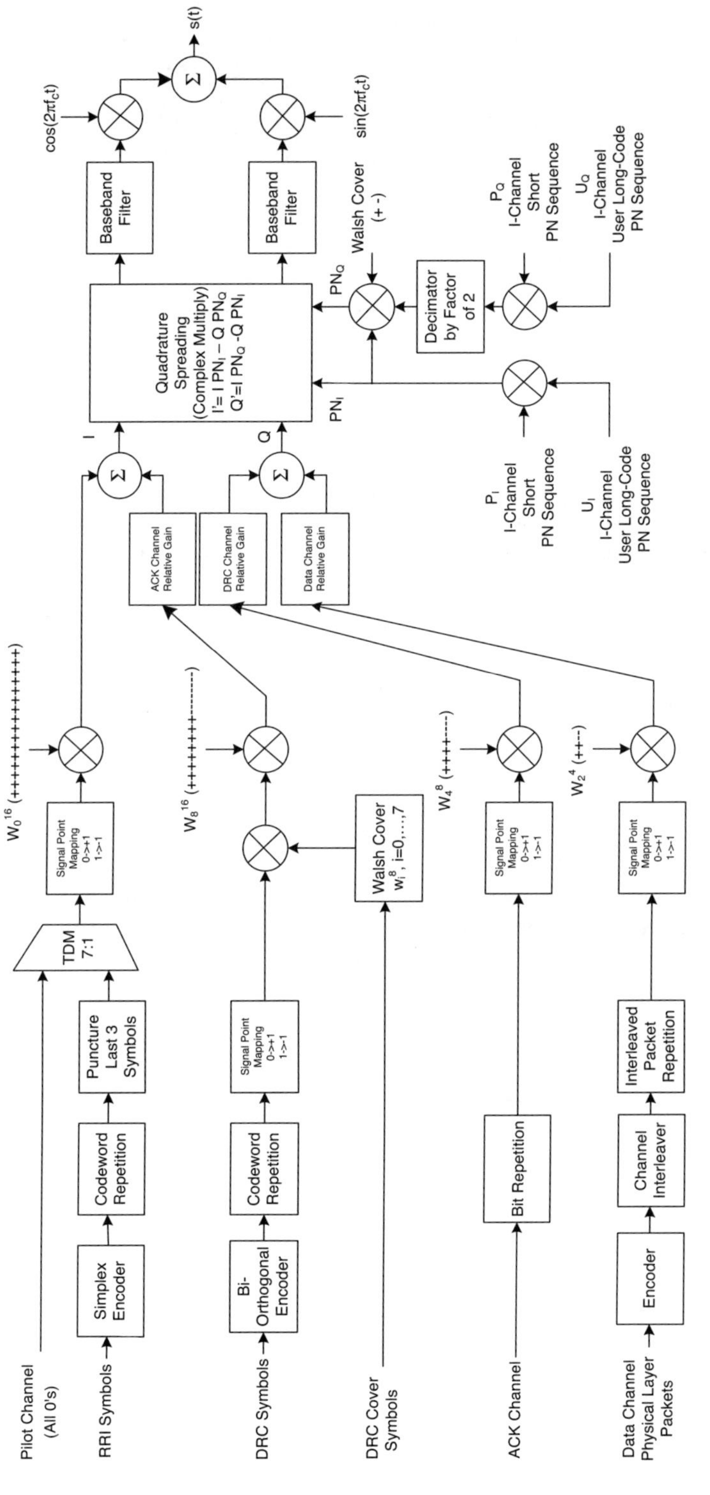

Figure 12.35 Reverse Traffic Channel structure

Table 12.8 RRI symbols encoding

Data Rate (kbps)	RRI Symbol	RRI Codeword
0	000	0000000
9.6	001	1010101
19.2	010	0110011
38.4	011	1100110
76.8	100	0001111
153.6	101	1011010
Reserved	110	0111100
Reserved	111	1101001

12.5.5.2 RRI Channel

With the RRI Channel the access terminal indicates the transmission rate over the Data Channel. The transmission rate of each 16-slot physical layer packet is encoded by 3 bits, then 3-bit symbol is encoded into 7-bit, as shown in Table 12.8. Obtained codeword is repeated 37 times, which gives 259-bit sequences, from which the three last bits are punctured (see Figure 12.35).

Obtained 256 RRI channel bits are time-division multiplexed (TDM, time-division multiplexing) with the symbols of Pilot Channel as shown in Figure 12.36. Then the multiplexed Pilot and RRI Channels are spread by sequence W_0^{16} and transmitted over the I-channel, see section 12.5.5.1.

12.5.5.3 DRC Channel

The Data Rate Control (DRC) Channel is used by the access terminal to indicate the selected servicing sector to the access network, and also to indicate the requested data transmission rate for the Forward Traffic Channel. The DRC Channel is transmitted over the Q-channel, as shown in Figure 12.35.

The requested rate is mapped into 4-bit DRC-value, which is then encoded into 8-bit biorthogonal codeword in correspondence with Table 12.9.

Obtained 8-bit codeword is repeated twice for each slot. Each bit of the repeated codeword is spread by 8-bit Walsh sequence W_i^8 as shown in Table 12.10, where i is defined by the DRCCover parameter of MAC protocol for the Forward Traffic Channel. Next, each bit of 8-bit Walsh sequence is spread by the sequence W_8^{16} (see Figure 12.35).

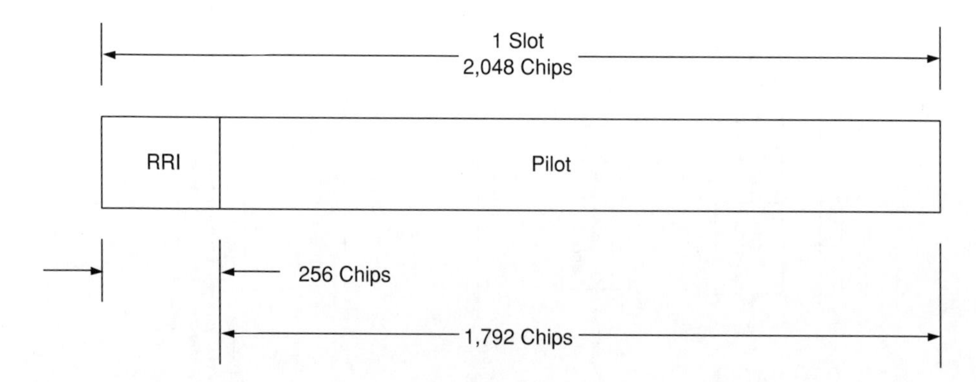

Figure 12.36 TDM for Pilot and RRI Channels of the Reverse Traffic Channel

Table 12.9 DRC symbols encoding

Rate (kbps)	DRC value	Codeword
Null rate	0×0	00000000
38.4	0×1	11111111
76.8	0×2	01010101
153.6	0×3	10101010
307.2	0×4	00110011
307.2	0×5	11001100
614.4	0×6	01100110
614.4	0×7	10011001
921.6	0×8	00001111
1228.8	0×9	11110000
1228.8	$0 \times a$	01011010
1843.2	$0 \times b$	10100101
2457.6	$0 \times c$	00111100
Invalid	$0 \times d$	11000011
Invalid	$0 \times e$	01101001
Invalid	$0 \times f$	10010110

The DRC Channel may function either constantly or periodically (gated functioning), under control of the MAC protocol Forward Traffic Channel. In this case the slots for which the DRC Channel is on, are called active.

12.5.5.4 ACK Channel

The ACK Channel is used for informing the access network about success or failure in receiving the packet from the Forward Traffic Channel. The symbols of ACK Channel are modulated by BPSK, where '0' corresponds to successful receive acknowledgement (ACK), and '1' corresponds to error (negative acknowledgement, NAK). If the physical layer packet of the Forward Traffic Channel is transmitted in slot n, then the correspondent bit of ACK Channel is transmitted in slot $n + 3$ of the Reverse Traffic Channel. The transmission over the ACK Channel takes the half of the slot, is spread by Walsh sequence W_4^8 (see Figure 12.35) and is performed by the I-channel.

12.5.5.5 Data Channel

The transmission over the Data Channel is performed on rates 9.6, 19.2, 38.4, 76.8 or 153.6 kbps. The data are encoded, block interleaved, repeated and orthogonally spread by Walsh sequence W_2^4.

Table 12.10 8-bit Walsh sequences

$W_0{}^8$	0000 0000
$W_1{}^8$	0101 0101
$W_2{}^8$	0011 0011
$W_3{}^8$	0110 0110
$W_4{}^8$	0000 1111
$W_5{}^8$	0101 1010
$W_6{}^8$	0011 1100
$W_7{}^8$	0110 1001

Table 12.11 Turbo-code parameters

Data Rate (kbps)	9.6	19.2	38.4	76.8	153.6
Reverse Rate Index	1	2	3	4	5
Bits per Physical Layer Packet	256	512	1,024	2,048	4,096
Number of Turbo Encoder Input Symbols	250	506	1,018	2,042	4,090
Turbo Encoder Code Rate	1/4	1/4	1/4	1/4	1/2
Encoder Output Block Length (Code Symbols)	1,024	2,048	4,096	8,192	8,192

12.5.6 Encoding

Revers Traffic Channel and Access Channel are encoded with turbo-code. Depending on the size of physical layer packet the codes with rates 1/2 or 1/4 are used, the parameters of the codes are listed in Table 12.11. The turbo-code encoder is similar to the encoder from CDMA2000 standard, described in section 12.2.2.2 for the rates 1/2 and 1/4.

The turbo-interleaving scheme is also similar to the procedure described in section 12.2.2.2, but uses Table 12.12 instead of Table 12.4.

Table 12.12 Table for turbo-interleaving calculation

Table Index	n = 3 Entries	n = 4 Entries	n = 5 Entries	n = 6 Entries	n = 7 Entries
0	1	5	27	3	15
1	1	15	3	27	127
2	3	5	1	15	89
3	5	15	15	13	1
4	1	1	13	29	31
5	5	9	17	5	15
6	1	9	23	1	61
7	5	15	13	31	47
8	3	13	9	3	127
9	5	15	3	9	17
10	3	7	15	15	119
11	5	11	3	31	15
12	3	15	13	17	57
13	5	3	1	5	123
14	5	15	13	39	95
15	1	5	29	1	5
16	3	13	21	19	85
17	5	15	19	27	17
18	3	9	1	15	55
19	5	3	3	13	57
20	3	1	29	45	15
21	5	3	17	5	41
22	5	15	25	33	93
23	5	1	29	15	87
24	1	13	9	13	63
25	5	1	13	9	15
26	1	9	23	15	13
27	5	15	13	31	15
28	3	11	13	17	81
29	5	3	1	5	57
30	5	15	13	15	31
31	3	5	13	33	69

12.5.7 Channel Interleaving and Repetition

The output of the error-correcting encoder is subject to channel interleaving as follows. If the symbols at the output of interleaver are numbered from 0 to $2^L - 1$, and if the i-th symbol at the interleaver's output has bit representation $i = (b_{L-1}, \ldots, b_0)$, then this symbol is the symbol from interleaver's input with number $j = (b_0, \ldots, b_{L-1})$.

When transmitting on rates below 76.8 kbps the sequences of codeword symbols are repeated the number of times listed in Table 12.7, to achieve the modulation symbol rate of 307.2 ksps.

12.5.8 Quadrature Spreading

The powers of ACK Channel, DRC Channel and Data Channel should be scaled relative to the Pilot Channel, this procedure is performed in "ACK Channel Relative Gain", "DRC Channel Relative Gain" and "Data Channel Relative Gain" blocks depicted in Figure 12.35, and also in "Data Channel Relative Gain" block in Figure 12.34. After scaling is finished, the stage of quadrature spreading should be performed.

Pilot Channel and ACK Channel are combined and form the in-phase channel I-Channel, Data Channel and DRC Channel are forming the quadrature channel Q-Channel similarly. Quadrature spreading is equivalent to complex multiplication by the pseudo-noise (PN) sequences PN_I and PN_Q. These sequences are obtained from the correspondent so-called long codes and short codes.

The procedure of obtaining the sequences PN_I and PN_Q is shown in Figures 12.34 and 12.35, where for performing the operation of sequences multiplication the binary sequences are mapped into bipolar, so that '0' is mapped to '1', and '1' is mapped to '−1'. During the decimation procedure every second symbol is destroyed and replaced by the preceding symbol, so there are the pairs of similar symbols at the output of decimator. After combining with the Walsh sequence (+−) the first symbol remains unchanged, while the second changes its sign.

The procedure similar to that being described in section 12.2.7 is used to obtain the long PN-sequences U_I and U_Q. To achieve this, the same PN-sequence is used, which is then masked by two different masks M_I and M_Q defined by the MAC layer protocol. For the Access Channel the long code identifies the sector, and masks M_I and M_Q are dependent on SectorID parameter. In case of Reverse Traffic Channel the long codes identify different users, and masks M_I and M_Q are dependent on ATI (Access Terminal Identifier) parameter.

To obtain the long code, the same common register is used as described in section 12.2.7 for the CDMA2000 standard. This shift register is defined by the polynomial:

$$p(x) = x^{42} + x^{35} + x^{33} + x^{31} + x^{27} + x^{26} + x^{25} + x^{22} + x^{21} + x^{19} \\ + x^{18} + x^{17} + x^{16} + x^{10} + x^7 + x^6 + x^5 + x^3 + x^2 + x + 1$$

The output of this register is masked with the symbols of masking sequences by means of component-wise multiplication, as shown in Figure 12.37.

The short codes are obtained from the registers defined by polynomials:

$$P_I(x) = x^{15} + x^{10} + x^8 + x^7 + x^6 + x^2 + 1$$

for in-phase component and:

$$P_Q(x) = x^{15} + x^{12} + x^{11} + x^{10} + x^9 + x^5 + x^4 + x^3 + 1$$

for quadrature component, which are similar to correspondent polynomials for spreading rate 1 in CDMA2000 standard, see section 12.2.7.

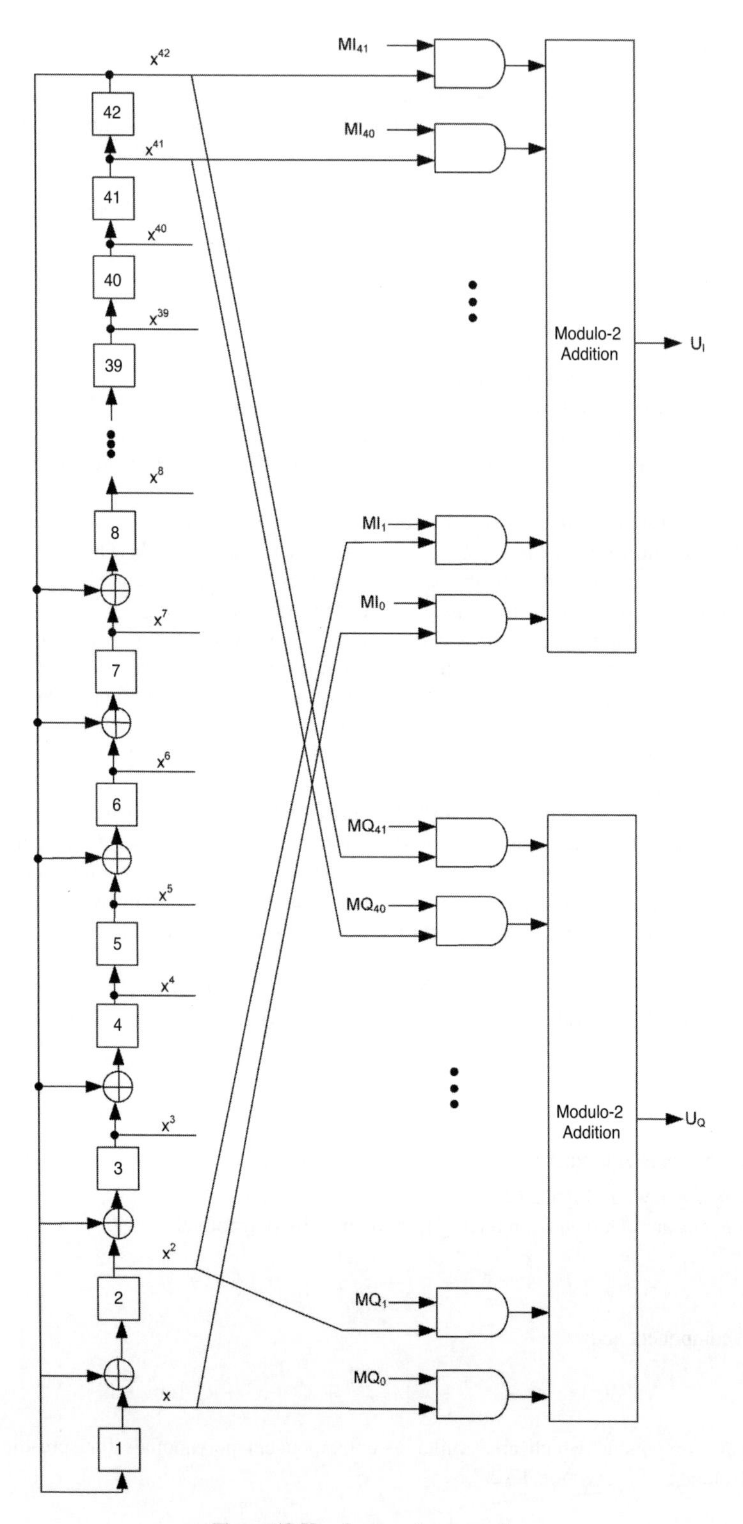

Figure 12.37 Long codes generation

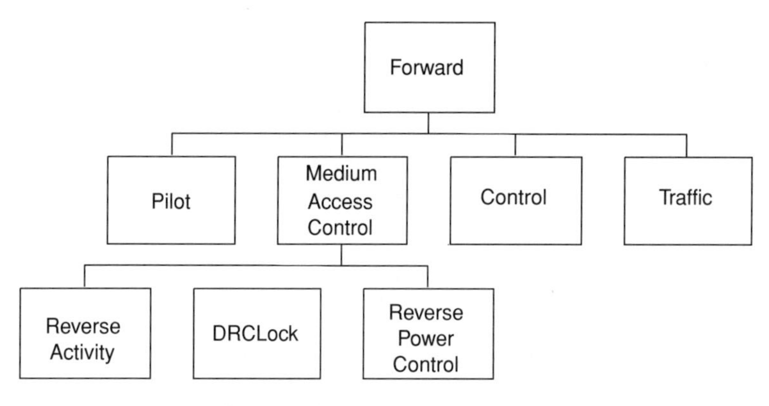

Figure 12.38 Forward Channel structure

At the beginning of each long codes period the long codes generator should be initialized by the value `0x24B91BFD3A8`.

12.6 Access Network of the CDMA2000 1xEV-DO Standard

There is the transmitter in each sector of the access network. The channel from the transmitter to the user station (access terminal) is called the forward channel. The structure of the CDMA2000 1xEV-DO Forward Channel and its basic functional blocks are described in this section.

12.6.1 Forward Channel Structure

Forward Channel is the channel connecting the access network to the access terminal. The structure of the Forward Channel is shown in Figure 12.38.

The main subchannels of the Forward Channel are listed and described below.

- **Pilot channel** — the portion of the Forward Channel that carries the pilot.
- **Control channel** — the channel that carries data to be received by all access terminals monitoring the Forward Channel.
- **Traffic channel** — used for data transmission to particular access terminal. Prior to the access terminal authentication the traffic channel is used as non-dedicated resource. After successful authentication the channel may be used as dedicated resource for given access terminal.
- **MAC Channel** — the portion of the Forward Channel dedicated to Medium Access Control activities. The Forward MAC Channel consists of the Reverse Power Control (RPC), DRCLock, and Reverse Activity (RA) Channels.
  - **Reverse Activity channel** — indicates activity level on the Reverse Channel.
  - **DRCLock channel** — indicates to the access terminal whether or not the access network can receive the DRC Channel and Reverse Link Channel sent by the access terminal.
  - **Reverse Power Control (RPC) channel** — controls the power of the Reverse Channel for one particular access terminal.

The structure of the Forward Channel is shown in Figure 12.39. As can be seen from Figure 12.39, the channels included in Forward Channel (see Figure 12.38) are time-multiplexed with each other. The

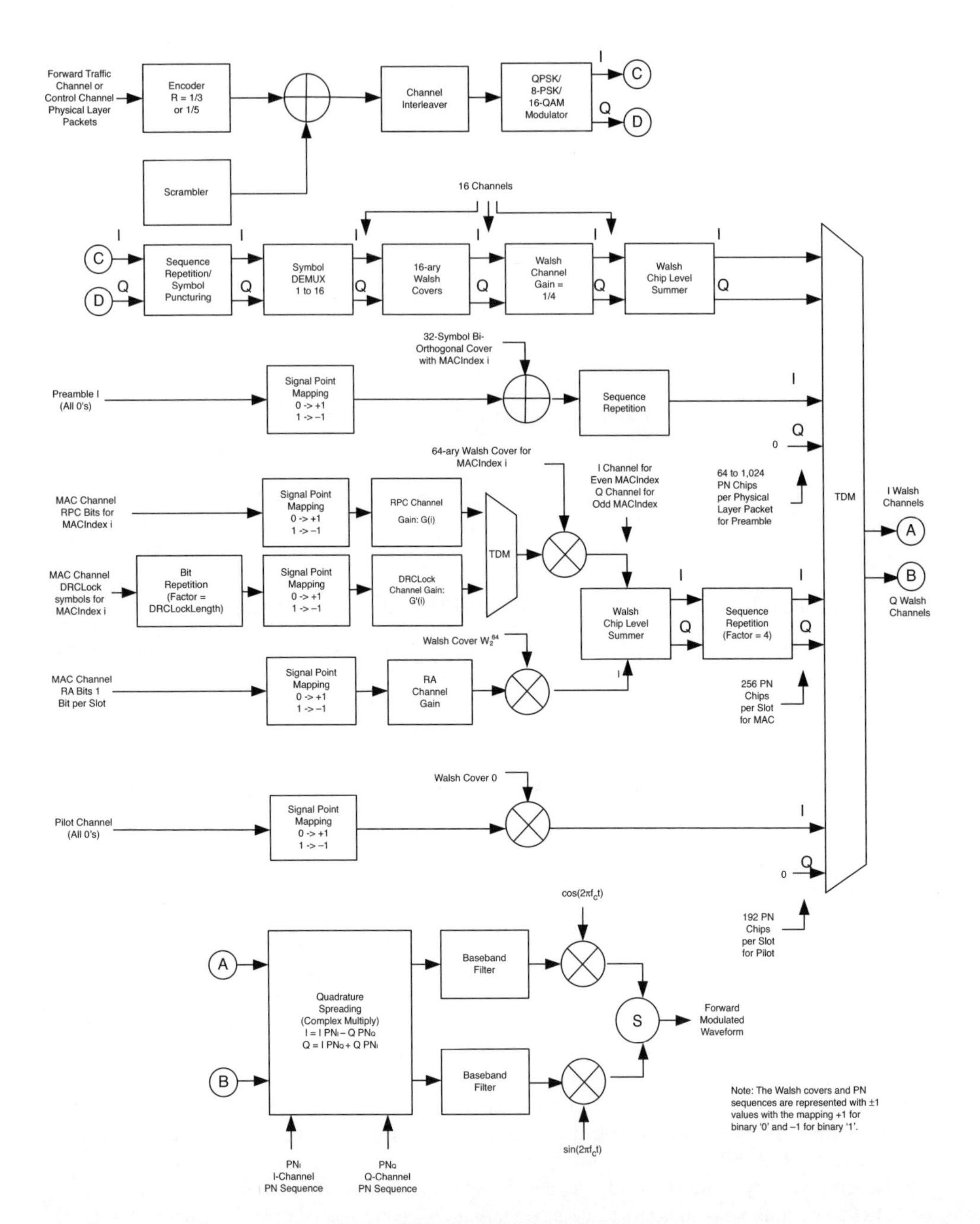

Figure 12.39 Forward Channel structure

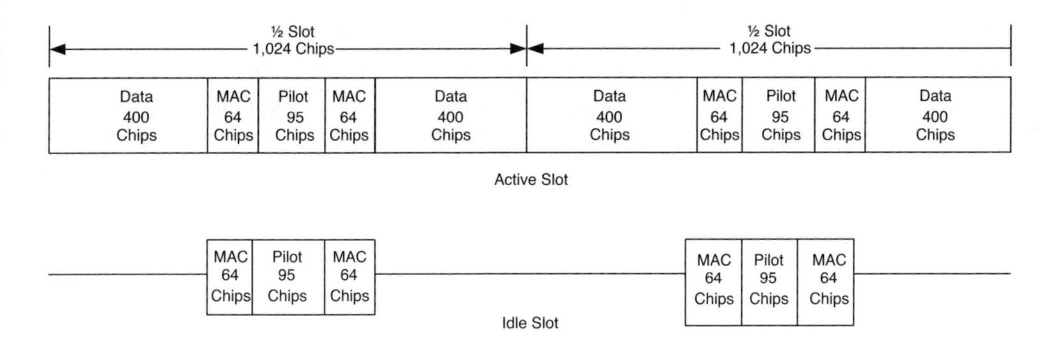

Figure 12.40 Forward Channel slot structure

transmission over the Forward Channel is performed in slots of 2048 chips. Pilot Channel, MAC Channel and Traffic or Control Channel are time-multiplexed within each slot, as shown in Figure 12.40.

The functional blocks of the channels within the Forward Channel structure, shown in Figure 12.39, are described below.

12.6.2 Modulation Parameters and Transmission Rates

Modulation parameters and possible transmission rates for the Forward Traffic Channel and Control Channel are listed in Table 12.13.

The symbols of MAC Channel are modulated with BPSK, as shown in Figure 12.39 and described further in section 12.6.4.

12.6.3 Pilot Channel

The Pilot Channel transmits unmodulated signal, containing '0', over in-phase component. This channel is used for synchronization and other functions which may be needed by the access terminal in sector's coverage area.

12.6.4 Forward MAC Channel

The Forward MAC Channel consists of three subchannels — Reverse Power Control (RPC) Channel, DRCLock Channel and Reverse Activity (RA) Channel. The RPC and DRCLock Channels are time-multiplexed and transmitted via the same MAC-channel.

Symbols, which are transmitted over the channel are modulated with BPSK and covered by Walsh sequences W_i^{64}, where i may take values from 0 to 63. The selection of particular sequence is defined by the MACIndex parameter. For even values of MACIndex (0, 2, . . ., 62) the Walsh sequences $W_0^{64}, \ldots, W_{31}^{64}$ are used, and correspondent subchannels are transmitted over in-phase component, while for odd values of MACIndex (1, 3, . . ., 63) the Walsh sequences $W_{32}^{64}, \ldots, W_{63}^{64}$ are used, and correspondent subchannels are transmitted over quadrature component.

The MAC-channel symbols after Walsh covering are repeated four times within a slot in blocks of 64 chips, just before the pilot symbols and after them, see Figure 12.40.

Table 12.13 Modulation parameters of the Forward Traffic Channel and Control Channel

Data Rate (kbps)	Number of Value per Physical Layer Packet				
	Slots	Bits	Code Rate	Modulation Type	TDM Chips (Preamble, Pilot, MAC, Data)
38.4	16	1,024	1/5	QPSK	1,024 3,072 4,096 24,576
76.8	8	1,024	1/5	QPSK	512 1,536 2,048 12,288
153.6	4	1,024	1/5	QPSK	256 768 1,024 6,144
307.2	2	1,024	1/5	QPSK	128 384 512 3,072
614,4	1	1,024	1/3	QPSK	64 192 256 1,536
307.2	4	2,048	1/3	QPSK	128 768 1,024 6,272
614.4	2	2,048	1/3	QPSK	64 384 512 3,136
1,228.8	1	2,048	1/3	QPSK	64 192 256 1,536
921.6	2	3,072	1/3	8-PSK	64 384 512 3,136
1,843.2	1	3,072	1/3	8-PSK	64 192 256 1,536
1.228,8	2	4,096	1/3	16-QAM	64 384 512 3,136
2,457.6	1	4,096	1/3	16-QAM	64 192 256 1,536

12.6.5 Control Channel

Control Channel transmits the broadband messages and access-terminal-directed messages. The messages are transmitted over the Control Channel on rates 76.8 kbps or 38.4 kbps. Modulation parameters for the Control Channel are the same as for the Forward Traffic Channel. Determining the transmission mode (Control Channel or Forward Traffic Channel) is made by the value of MACIndex (the value of MACIndex, corresponding to Control Channel, equals 2 for the rate of 76.8 kbps and 3 for the rate of 38.4 kbps).

12.6.6 Forward Traffic Channel

12.6.6.1 Forward Traffic Channel Preamble

Preamble, being transmitted with Forward Traffic Channel or Control Channel, serves for establishing the synchronization on the access terminal side. Preamble consists of all-'0' and is transmitted over the in-phase component only. Preamble sequence is covered by Walsh sequences $W^{32}_{i/2}$ and their bitwise complements $\overline{W^{32}_{(i-1)/2}}$, where $i \in [0, 63]$ is MACIndex parameter (see section 12.6.4).

Depending on the transmission rates the preamble sequence may be repeated several times (from 2 to 32).

12.6.6.2 Encoding

For encoding the Forward Traffic Channel the turbo-codes with rates 1/3 and 1/5 are used.

The physical layer packets of the Forward Traffic Channel are encoded similar to encoding in Reverse Channel (see section 12.5.6), that is, the TAIL field symbols are omitted, other symbols are encoded, and $6/R$ tail symbols are added.

Encoding parameters are listed in Table 12.14.

Encoder structure is similar to that described in section 12.5.6 and shown in Figure 12.11. To achieve the required rates the encoder output is subject to repetition and puncturing. The puncturing scheme for the codeword symbols is given in Table 12.15, and the puncturing scheme for the tail symbols is similar

Table 12.14 Encoding parameters for Forward Traffic Channel

Data Rate (kbps)	Values per Physical Layer Packet				
	Slots	Bits	Turbo Encoder Input Bits	Code Rate	Turbo Encoder Output Symbols
38.4	16	1,024	1,018	1/5	5,120
76.8	8	1,024	1,018	1/5	5,120
153.6	4	1,024	1,018	1/5	5,120
307.2	2	1,024	1,018	1/5	5,120
614.4	1	1,024	1,018	1/3	3,072
307.2	4	2,048	2,042	1/3	6,144
614.4	2	2,048	2,042	1/3	6,144
1,228.8	1	2,048	2,042	1/3	6,144
921.6	2	3,072	3,066	1/3	9,216
1,843.2	1	3,072	3,066	1/3	9,216
1,228.8	2	4,096	4,090	1/3	12,288
2,457.6	1	4,096	4,090	1/3	12,288

Table 12.15 Puncturing the codeword symbols of turbo-code

Output	Code Rate	
	1/3	1/5
X	1	1
Y_0	1	1
Y_1	0	1
X'	0	0
Y'_0	1	1
Y'_1	0	1

to those given in Table 12.3, where the first column (for rate 1/2) corresponds to the rate-1/3 case in Forward Channel, and the second column (for rate 1/4) corresponds to the rate of 1/5.

Besides the puncturing procedure, the repetition is also used during the last six clocking for calculating the tail bits. For the rate 1/5 the output of the first three clocks is $XXY_0Y_1Y_1$, the output of the next three clocks is $X'X'Y'_0Y'_1Y'_1$. For the rate 1/3 the output of the first three clocks is XXY_0, the output of the next three clocks is $X'X'Y'_0$.

Turbo-interleaver is similar to that being described in section 12.5.6 and shown in Figure 12.12. For the physical layer packets lengths of 1024, 2048 and 4096 the value of parameter n is equal to 5, 6 and 7 correspondingly. For the packet length 3072 the value of n is set to 7. For turbo-interleaving calculation Table 12.12 is used (columns, correspondent to the values of $n = 5$, 6 and 7).

12.6.6.3 Scrambling

The decoder's output is subject to scrambling for randomizing the resulting value. The scrambling procedure concludes in a summation of encoded symbols with the pseudorandom sequence modulo 2. PN sequence is obtained from the feedback shift register defined by the polynomial:

$$h(D) = D^{17} + D^{14} + 1$$

and shown in Figure 12.41.

The initial state of the shift register is defined by the binary values $[1111111r_5r_4r_3r_2r_1r_0d_3d_2d_1d_0]$. The values of $r_5r_4r_3r_2r_1r_0$ are defined by the 6-bit value of MACValue parameter, and the values of $d_3d_2d_1d_0$ are defined by the transmission rate, as shown in Table 12.16.

12.6.6.4 Channel Interleaving

Interleaving of scrambled symbols at the decoder's output is performed in two steps. First step is symbol reordering; the second step is symbol permuting.

During the reordering stage the scrambled data stream is split into several subsequences, which are then reordered in some way, as described below.

When using the code with rate 1/5 the scrambled symbols are demultiplexed in five parallel streams, we denote it as U, V_0, V_1, V'_0 and V'_1, so that the first symbol at scrambler's output goes to the stream U, the second goes to V_0 and so on (the sixth symbol will be written in U again). Then the streams are ordered as $UV_0V'_0V_1V'_1$.

When using the code with rate 1/3 the scrambled symbols are demultiplexed in three parallel streams U, V_0 and V'_0. Then the streams are ordered as $UV_0V'_0V_1V'_1$.

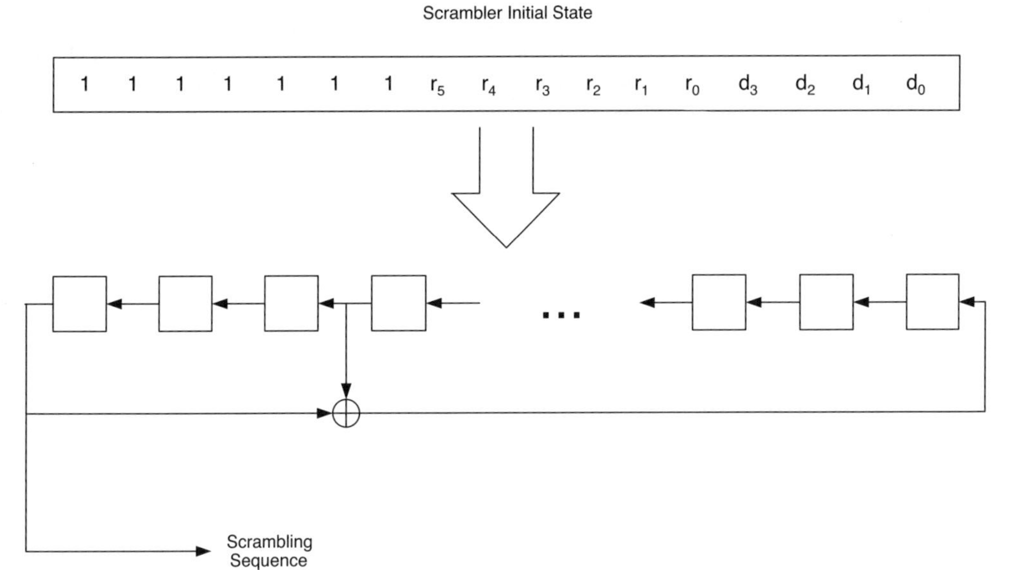

Figure 12.41 Scrambling

Next, the symbol permuting step is performed. After symbol reordering we obtain the sequence U, followed by sequences V_0 and V_0' (denote it as sequence V_0/V_0'), and then, in the case of rate-1/5 code, there are the sequences V_1 and V_1' (denote it as sequence V_1/V_1'). Then for each of the sequences U, V_0/V_0' and V_1/V_1' the following procedure is performed.

1. Write the whole reordered sequence in two-dimensional matrix with K rows and M columns. Parameters K and M are defined by the physical layer packet size and given in Table 12.17.
2. Number the columns of obtained matrix by index j, taking the values from 0 to $M - 1$. Cyclically shift each column of the matrix down by $j \bmod K$ positions for block U and by $\lfloor j/4 \rfloor \bmod K$ positions for blocks V_0/V_0' and V_1/V_1'.

Table 12.16 Parameters $d_3d_2d_1d_0$ for scrambler initializing

Data Rate (kbps)	Slots per Physical Layer Packet	d_3	d_2	d_1	d_0
38.4	16	0	0	0	1
76.8	8	0	0	1	0
153.6	4	0	0	1	1
307.2	2	0	1	0	0
307.2	4	0	1	0	1
614.4	1	0	1	1	0
614.4	2	0	1	1	1
921.6	2	1	0	0	0
1,228.8	1	1	0	0	1
1,228.8	2	1	0	1	0
1,843.2	1	1	0	1	1
2,457.6	1	1	1	0	0

Table 12.17 Parameters for symbol permuting

Physical layer packet size	U block interleaver parameters		V_0/V'_0 and V_1/V'_1 block interleaver parameters	
	K	M	K	M
1,024	2	512	2	1,024
2,048	2	1,024	2	2,048
3,072	3	1,024	3	2,048
4,096	4	1,024	4	2,048

3. Reorder the columns of the matrix, that is, swap the columns with numbers in bit-reversed order to each other.
4. Read the matrix into serial sequence by columns from top to bottom and from the leftmost column.

12.6.6.5 Modulation

At the output of channel interleaver the modulation QPSK, 8-PSK or 16-QAM is used, depending on transmission rate, in correspondence with Table 12.13.

12.6.6.6 Repetition and Puncturing

Since the number of symbols at modulator's output for each physical layer packet and the number of symbols required for data TDM chips may differ in both sides, the sequences of modulated symbols, if necessary, are repeated or punctured, as shown in Table 12.18.

Table 12.18 Repetition and puncturing parameters

Data Rate (kbps)	Values per Physical Layer Packet					
	Number of Slots	Number of Bits	Number of Modulation Symbols Provided	Number of Modulation Symbols Needed	Number of Full Sequence Transmissions	Number of Modulation Symbols in Last Partial Transmission
38.4	16	1,024	2,560	24,576	9	1,536
76.8	8	1,024	2,560	12,288	4	2,048
153.6	4	1,024	2,560	6,144	2	1,024
307.2	2	1,024	2,560	3,072	1	512
614.4	1	1,024	1,536	1,536	1	0
307.2	4	2,048	3,072	6,272	2	128
614.4	2	2,048	3,072	3,136	1	64
1,228.8	1	2,048	3,072	1,536	0	1,536
921.6	2	3,072	3,072	3,136	1	64
1,843.2	1	3,072	3,072	1,536	0	1,536
1,228.8	2	4,096	3,072	3,136	1	64
2,457.6	1	4,096	3,072	1,536	0	1,536

12.6.6.7 Demultiplexing

Both in-phase and quadrature channels are demultiplexed at the modulator's output into 16 parallel streams, so that first in-phase (quadrature) modulated symbol goes to first stream, second goes to second, 17th goes again to first stream and so on. At the output of demultiplexor each parallel stream contains modulated symbols at rate 76.6 ksps.

12.6.6.8 Walsh Channels Forming

For each of the 16 demultiplexed stream (more precisely, for each pair in-phase/quadrature stream) is correspondent one of 16 different Walsh channels W_k^{16}, $k = 0, \ldots, 15$. To maintain the constant transmission power, the symbols of Walsh channels are scaled by the coefficient $1/\sqrt{16} = 1/4$. Then the symbols of 16 Walsh channels are summed component-wise.

12.6.7 Time-Division Multiplexing

The chips of the Forward Traffic Channel (or Control Channel) are time-division multiplexed with the preamble, Pilot Channel and MAC-channel chips, as shown in Figure 12.39. The Walsh chip rate is fixed on 1.2288 Mcps. Diagrams of time-division multiplexing for different cases are shown in Figures 12.42–12.45. Multiplexing parameters are given in Table 12.19.

12.6.8 Quadrature Spreading

Spectrum spreading in the Forward Channel is performed by means of maximal-period linear feedback shift registers (LFSR) producing PN-sequences P_I and P_Q for in-phase and quadrature components,

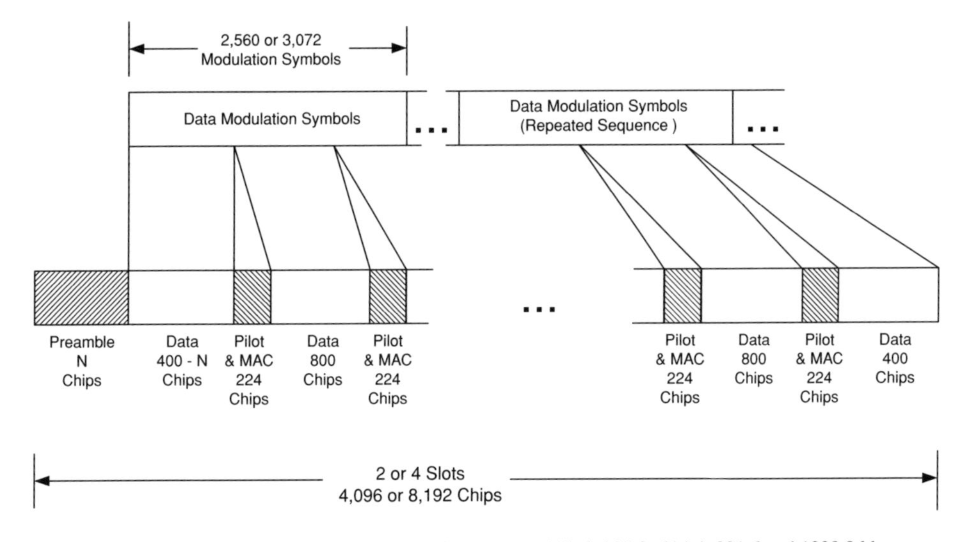

Figure 12.42 TDM for multislot transmission on rates 153.6, 307.2, 614.4, 921.6 and 1228.8 kbps

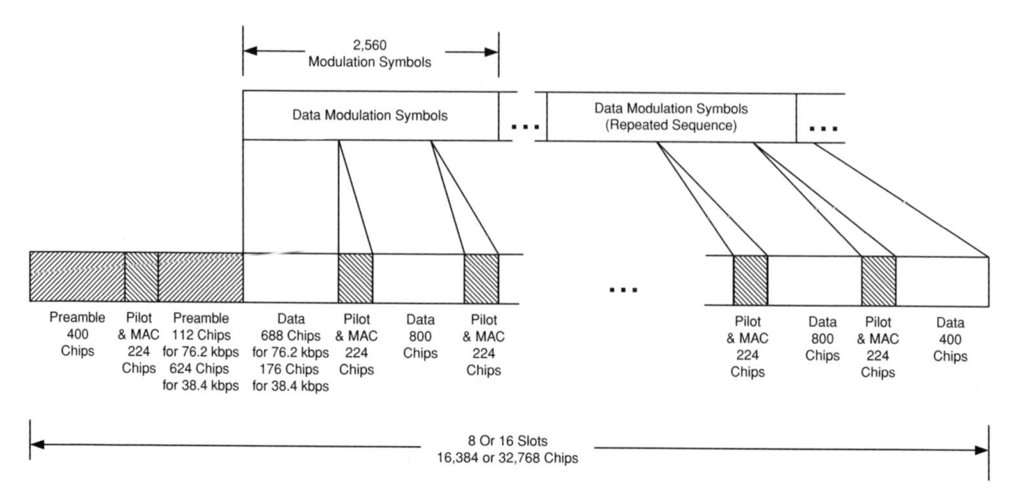

Figure 12.43 TDM for transmission on rates 38.4 and 76.8 kbps

correspondingly, with the period of 2^{15}. These registers are defined by polynomials:

$$P_I(x) = x^{15} + x^{10} + x^8 + x^7 + x^6 + x^2 + 1$$

for in-phase component and:

$$P_Q(x) = x^{15} + x^{12} + x^{11} + x^{10} + x^9 + x^5 + x^4 + x^3 + 1$$

for quadrature component.

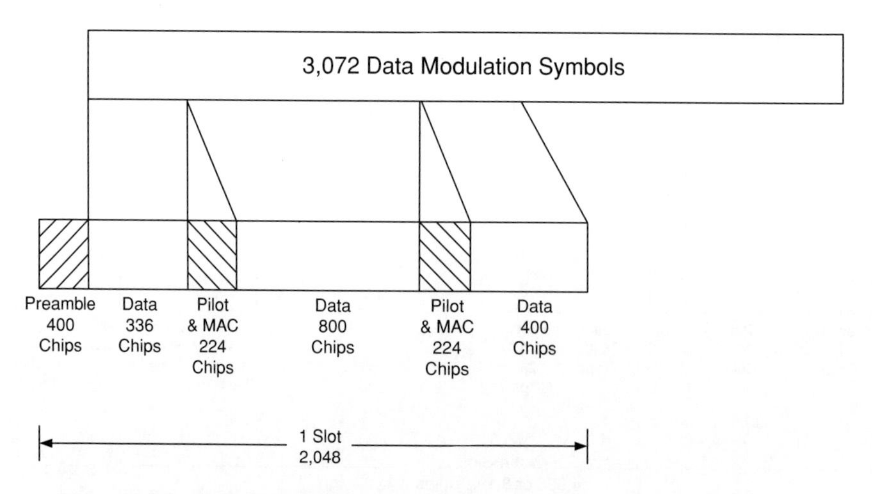

Figure 12.44 TDM for single slot transmission on rates 1.2288, 1.8432 and 2.4576 Mbps

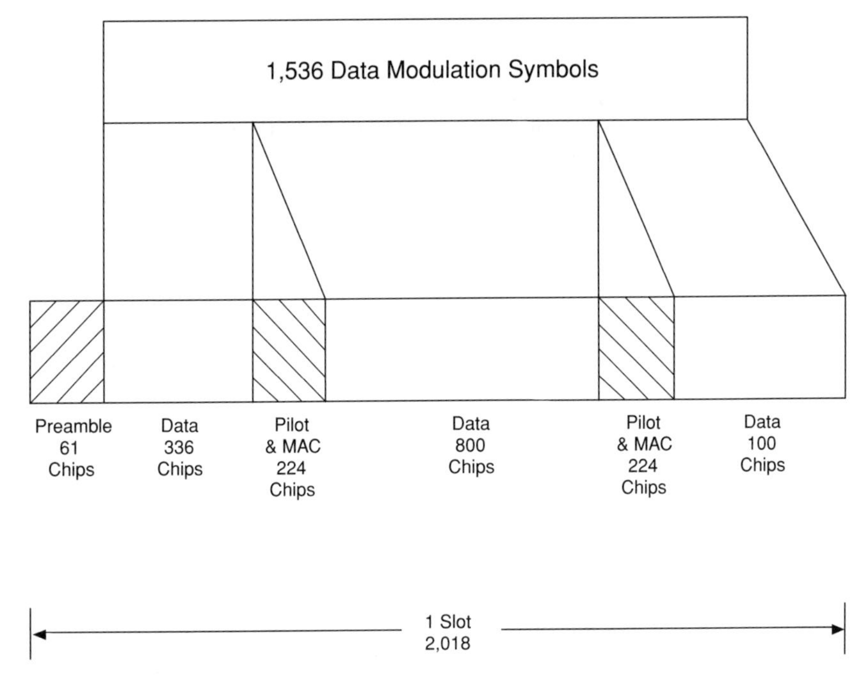

Figure 12.45 TDM for single slot transmission on rate 614.4 kbps

Such LFSRs have the maximal period of $2^{15} - 1$, and to obtain the sequences with period 2^{15}, in the only case when in these sequences there is the series of 14 subsequent zeros, the one more (15th) zero is added to the series. The initial state of these registers should be the state producing '1' and then 15 zeros.

Table 12.19 TDM parameters

	Number of Values per Physical Layer Packet					
Data Rate (kbps)	Slots	Bits	Preamble Chips	Pilot Chips	MAC Chips	Data Chips
38.4	16	1,024	2,560	1,024	4,096	24,576
76.8	8	1,024	512	1,536	2,048	12,288
153.6	4	1,024	256	768	1,024	6,144
307.2	2	1,024	128	384	512	3,072
614.4	1	1,024	64	192	256	1,536
307.2	4	2,048	128	768	1,024	6,272
614.4	2	2,048	64	384	512	3,136
1,228.8	1	2,048	64	192	256	1,536
921.6	2	3,072	64	384	512	3,136
1,843.2	1	3,072	64	192	256	1,536
1,228.8	2	4,096	64	384	512	3,136
2,457.6	1	4,096	64	192	256	1,536

References

[1] *CDMA2000 High Rate Packet Data Air Interface Specification*, 3GPP2 C.S0024-B Version 2.0, March 2007.
[2] *CDMA Development Group*, http://www.cdg.org.
[3] *Physical Layer Standard for cdma2000 Spread Spectrum System*, 3GPP2 C.S00002-0 Version 1.0, July 1999.
[4] *Telecommunications Industry Association*, http://www.tiaonline.org.
[5] Proakis, J. *Digital Communications*. McGraw Hill, 1995.
[6] Sklar, B. *Digital Communications. Fundamentals and Applications*. Prentice Hall, 2001.

Index

Modulation and Coding Techniques in Wireless Communications Edited by Evgenii Krouk and Sergei Semenov